Advanced Reservoir Management and Engineering

Advanced Reservoir Management and Engineering

Second edition

Tarek Ahmed

D. Nathan Meehan

ELSEVIER

AMSTERDAM • BOSTON • HEIDELBERG • LONDON
NEW YORK • OXFORD • PARIS • SAN DIEGO
SAN FRANCISCO • SINGAPORE • SYDNEY • TOKYO
Gulf Professional Publishing is an imprint of Elsevier

GPP

Gulf Professional Publishing is an imprint of Elsevier
225 Wyman Street, Waltham, MA 02451, USA
The Boulevard, Langford Lane, Kidlington, Oxford, OX5 1GB

First edition 2004
Second edition 2012

Library of Congress Cataloging-in-Publication Data
A catalog record for this book is available from the Library of Congress

British Library Cataloguing-in-Publication Data
A catalogue record for this book is available from the British Library

ISBN: 978-0-1238-5548-0

12 11 10 9 8 7 6 5 4 3 2 1

CONTENTS

PREFACE

The primary focus of this book is to present the basic physics of reservoir engineering using the simplest and most straightforward of mathematical techniques. It is only through having a complete understanding of physics of reservoir engineering that the engineer can hope to solve complex reservoir problems in a practical manner. The book is arranged so that it can be used as a textbook for senior and graduate students or as a reference book for practicing engineers.

Chapter 1 describes the theory and practice of well testing and pressure analysis techniques, which is probably one of the most important subjects in reservoir engineering. Chapter 2 discusses various water influx models along with detailed descriptions of the computational steps involved in applying these models. Chapter 3 presents the mathematical treatment of unconventional gas reservoirs that include abnormally pressured reservoirs, coalbed methane, tight gas, gas hydrates, and shallow gas reservoirs. Chapter 4 covers the basic principle of oil recovery mechanisms and the various forms of the material balance equation (MBE). Chapter 5 focuses on illustrating the practical application of the MBE in predicting the oil reservoir performance under different scenarios of driving mechanisms. Chapter 6, is an overview of enhanced oil recovery mechanisms and their application.

Chapter 7 covers the fundamentals of oilfield economic analysis including risk analysis, treatment of various international fiscal regimes and reserve reporting issues. Chapter 8 discusses the financial reporting and merger and acquisition topics relevant to reservoir engineers. Chapter 9 covers petroleum engineering professionalism and ethics.

Acknowledgment: D. Nathan Meehan would like to express his appreciation to Baker Hughes, Incorporated for supporting the development of the additions to this book.

CHAPTER 1

Well Testing Analysis

1.1 PRIMARY RESERVOIR CHARACTERISTICS

Flow in porous media is a complex phenomenon and cannot be described as explicitly as flow through pipes or conduits. It is easy to measure the length and diameter of a pipe and compute its flow capacity as a function of pressure; however, flow in porous media is different in that there are no clear-cut flow paths which lend themselves to measurement.

The analysis of fluid flow in porous media has evolved throughout the years along two fronts: experimental and analytical. Physicists, engineers and hydrologists have experimentally examined the behavior of various fluids as they flow through porous media ranging from sand packs to fused Pyrex glass. On the basis of their analyses they have attempted to formulate laws and correlations that can then be utilized to make analytical predictions for similar systems.

The objective of this chapter is to present the mathematical relationships designed to describe flow behavior of reservoir fluids. The mathematical forms of these relationships will vary depending upon characteristics of the reservoir. The primary reservoir characteristics that must be considered include:

- types of fluids in the reservoir;
- flow regimes;
- reservoir geometry;
- number of flowing fluids in the reservoir.

1.1.1 Types of Fluids

The isothermal compressibility coefficient is essentially the controlling factor in identifying the type of the reservoir fluid. In general, reservoir fluids are classified into three groups:

(1) incompressible fluids;
(2) slightly compressible fluids;
(3) compressible fluids.

The isothermal compressibility coefficient c is described mathematically by the following two equivalent expressions:

In terms of fluid volume:

$$c = \frac{-1}{V}\frac{\partial V}{\partial p} \tag{1.1}$$

In terms of fluid density:

$$c = \frac{1}{\rho}\frac{\partial \rho}{\partial p} \tag{1.2}$$

where

V = fluid volume
ρ = fluid density
p = pressure, psi
c = isothermal compressibility coefficient, Ψ^{-1}

Incompressible Fluids. An incompressible fluid is a fluid whose volume or density does not change with pressure. That is:

$$\frac{\partial V}{\partial p} = 0 \quad \text{and} \quad \frac{\partial \rho}{\partial p} = 0$$

Incompressible fluids do not exist; however, this behavior may be assumed in some cases to simplify the derivation and the final form of many flow equations.

Slightly Compressible Fluids. These "slightly" compressible fluids exhibit small changes in volume, or density, with changes in pressure. Knowing the volume V_{ref} of a slightly compressible liquid at a reference (initial) pressure p_{ref}, the changes in the volumetric behavior of such fluids as a function of pressure p can be mathematically described by integrating Eq. (1.1), to give:

$$-c \int_{p_{ref}}^{p} dp = \int_{V_{ref}}^{V} \frac{dV}{V}$$

$$\exp[c(p_{ref} - p)] = \frac{V}{V_{ref}} \tag{1.3}$$

$$V = V_{ref} \exp[c(p_{ref} - p)]$$

where

p = pressure, psia
V = volume at pressure p, ft^3
p_{ref} = initial (reference) pressure, psia
V_{ref} = fluid volume at initial (reference) pressure, psia

The exponential e^x may be represented by a series expansion as:

$$e^x = 1 + x + \frac{x^2}{2!} + \frac{x^2}{3!} + \cdots + \frac{x^n}{n!} \tag{1.4}$$

Because the exponent x (which represents the term $c(p_{ref} - p)$) is very small, the e^x term can be approximated by truncating Eq. (1.4) to:

$$e^x = 1 + x \tag{1.5}$$

Combining Eq. (1.5) with (1.3) gives:

$$V = V_{ref}[1 + c(p_{ref} - p)] \tag{1.6}$$

A similar derivation is applied to Eq. (1.2), to give:

$$\rho = \rho_{ref}[1 - c(p_{ref} - p)] \tag{1.7}$$

where

V = volume at pressure p
ρ = density at pressure p
V_{ref} = volume at initial (reference) pressure p_{ref}
ρ_{ref} = density at initial (reference) pressure p_{ref}

It should be pointed out that many crude oil and water systems fit into this category.

Compressible Fluids. Compressible fluids are defined as fluids that experience large changes in volume as a function of pressure. All gases and gas-liquid systems are considered compressible fluids. The truncation of the series expansion as given by Eq. (1.5) is not valid in this category and the complete expansion as given by Eq. (1.4) is used.

The isothermal compressibility of any vapor phase fluid is described by the following expression:

$$c_g = \frac{1}{p} - \frac{1}{Z}\left(\frac{\partial Z}{\partial p}\right)_T \tag{1.8}$$

Figures 1.1 and 1.2 show schematic illustrations of the volume and density changes as a function of pressure for all three types of fluids.

1.1.2 Flow Regimes

There are basically three types of flow regimes that must be recognized in order to describe the fluid flow behavior and reservoir pressure distribution as a function of time. These three flow regimes are:

(1) steady-state flow;
(2) unsteady-state flow;
(3) pseudosteady-state flow.

Steady-state Flow. The flow regime is identified as a steady-state flow if the pressure at every location in the reservoir remains constant, i.e., does not change with time. Mathematically, this condition is expressed as:

$$\left(\frac{\partial p}{\partial t}\right)_i = 0 \tag{1.9}$$

This equation states that the rate of change of pressure p with respect to time t at any location i

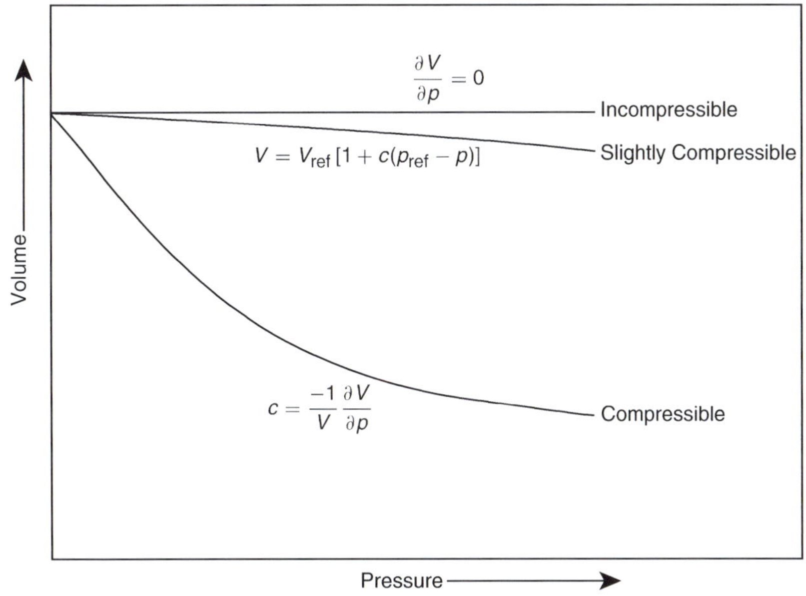

FIGURE 1.1 Pressure–volume relationship.

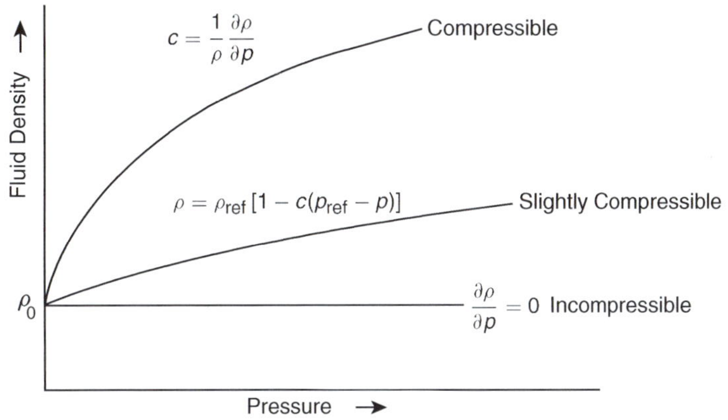

FIGURE 1.2 Fluid density vs. pressure for different fluid types.

is zero. In reservoirs, the steady-state flow condition can only occur when the reservoir is completely recharged and supported by strong aquifer or pressure maintenance operations.

Unsteady-state Flow. Unsteady-state flow (frequently called transient flow) is defined as the fluid flowing condition at which the rate of change of pressure with respect to time at any position in the reservoir is not zero or constant. This definition suggests that the pressure derivative with respect to time is essentially a function of both position i and time t, thus:

$$\left(\frac{\partial p}{\partial t}\right) = f(i, t) \tag{1.10}$$

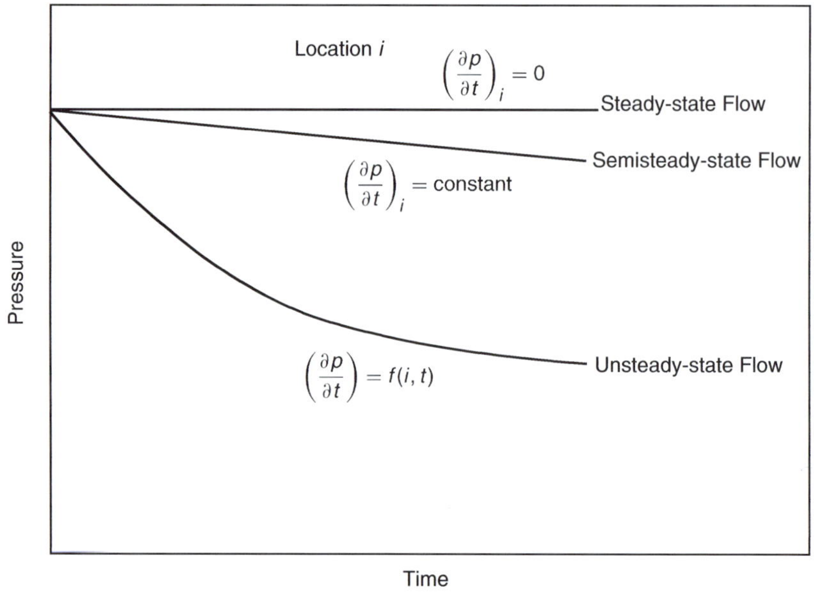

FIGURE 1.3 Flow regimes.

Pseudosteady-state Flow. When the pressure at different locations in the reservoir is declining linearly as a function of time, i.e., at a constant declining rate, the flowing condition is characterized as pseudosteady-state flow. Mathematically, this definition states that the rate of change of pressure with respect to time at every position is constant, or:

$$\left(\frac{\partial p}{\partial t}\right)_i = \text{constant} \tag{1.11}$$

It should be pointed out that pseudosteady-state flow is commonly referred to as semisteady-state flow and quasisteady-state flow and is possible for slightly compressible fluids.

Figure 1.3 shows a schematic comparison of the pressure declines as a function of time of the three flow regimes.

1.1.3 Reservoir Geometry

The shape of a reservoir has a significant effect on its flow behavior. Most reservoirs have irregular boundaries and a rigorous mathematical description of their geometry is often possible only with the use of numerical simulators. However, for many engineering purposes, the actual flow geometry may be represented by one of the following flow geometries:

- radial flow;
- linear flow;
- spherical and hemispherical flow.

Radial Flow. In the absence of severe reservoir heterogeneities, flow into or away from a wellbore will follow radial flow lines at a substantial distance from the wellbore. Because fluids move toward the well from all directions and coverage at the wellbore, the term radial flow is used to characterize the flow of fluid into the wellbore. Figure 1.4 shows idealized flow lines and isopotential lines for a radial flow system.

Linear Flow. Linear flow occurs when flow paths are parallel and the fluid flows in a single direction. In addition, the cross-sectional area to flow must be constant. Figure 1.5 shows an idealized linear flow system. A common

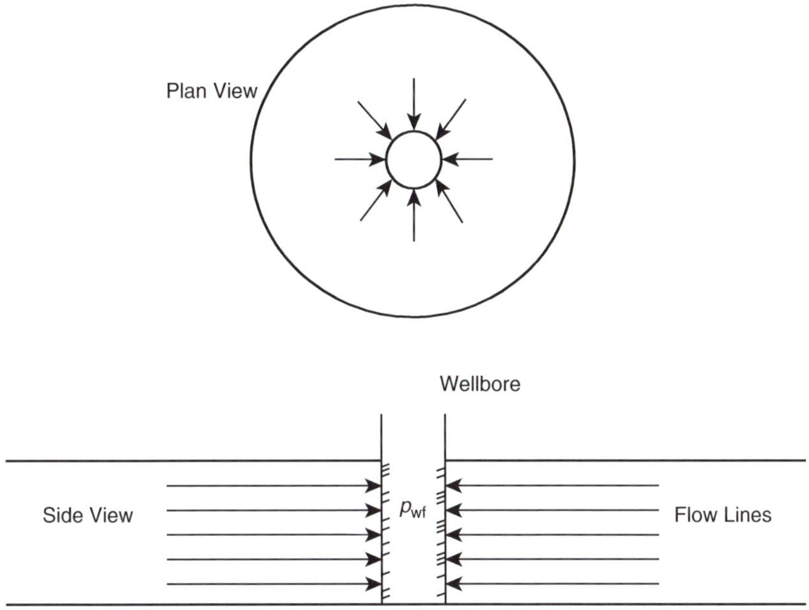

FIGURE 1.4 Ideal radial flow into a wellbore.

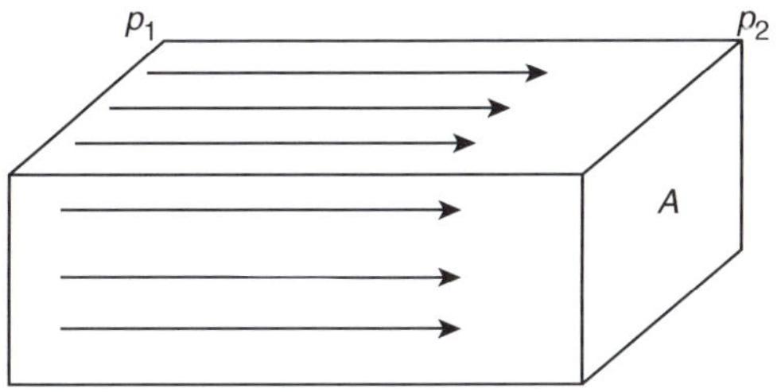

FIGURE 1.5 Linear flow.

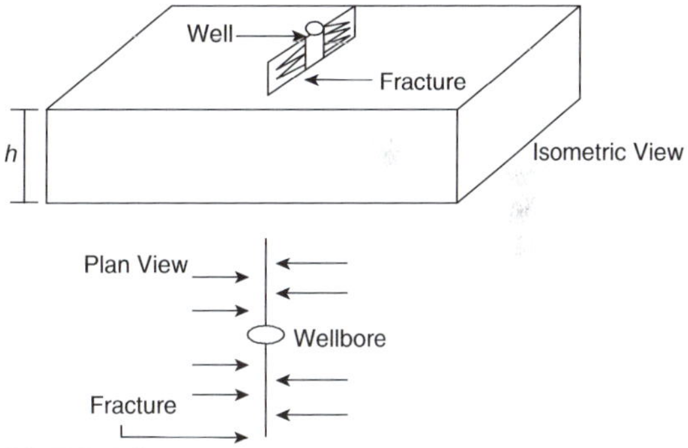

FIGURE 1.6 Ideal linear flow into vertical fracture.

application of linear flow equations is the fluid flow into vertical hydraulic fractures as illustrated in Figure 1.6.

Spherical and Hemispherical Flow. Depending upon the type of wellbore completion configuration, it is possible to have spherical or hemispherical flow near the wellbore. A well with a limited perforated interval could result in spherical flow in the vicinity of the perforations as illustrated in Figure 1.7. A well which only partially penetrates the pay zone, as shown in

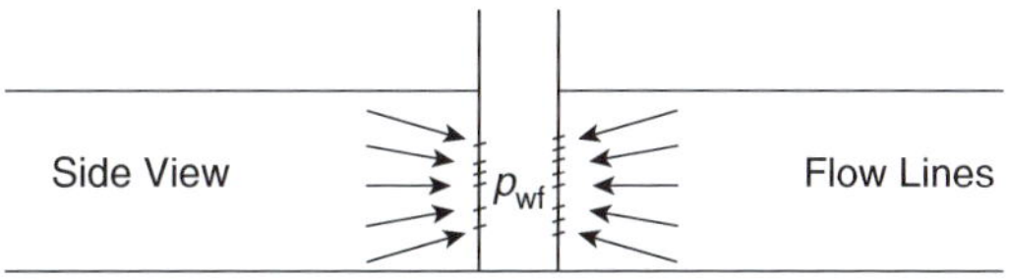

FIGURE 1.7 Spherical flow due to limited entry.

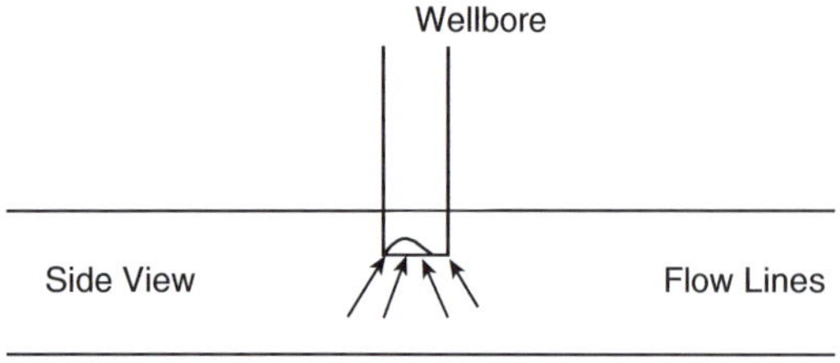

FIGURE 1.8 Hemispherical flow in a partially penetrating well.

Figure 1.8, could result in hemispherical flow. The condition could arise where coning of bottom water is important.

1.1.4 Number of Fluids Flowing in the Reservoir

The mathematical expressions that are used to predict the volumetric performance and pressure behavior of a reservoir vary in form and complexity depending upon the number of mobile fluids in the reservoir. There are generally three cases of flowing system:

(1) single-phase flow (oil, water, or gas);
(2) two-phase flow (oil–water, oil–gas, or gas–water);
(3) three-phase flow (oil, water, and gas).

The description of fluid flow and subsequent analysis of pressure data becomes more difficult as the number of mobile fluids increases.

1.2 FLUID FLOW EQUATIONS

The fluid flow equations are used to describe flow behavior in a reservoir take many forms depending upon the combination of variables presented previously (i.e., types of flow, types of fluids, etc.). By combining the conservation of mass equation with the transport equation (Darcy's equation) and various equations of state, the necessary flow equations can be developed. Since all flow equations to be considered depend on Darcy's law, it is important to consider this transport relationship first.

1.2.1 Darcy's Law

The fundamental law of fluid motion in porous media is Darcy's law. The mathematical expression developed by Darcy in 1856 states that the velocity of a homogeneous fluid in a porous medium is proportional to the pressure gradient, and inversely proportional to the fluid viscosity. For a horizontal linear system, this relationship is:

$$v = \frac{q}{A} = -\frac{k}{\mu}\frac{dp}{dx} \tag{1.12a}$$

where

v = apparent velocity, cm/s
q = volumetric flow rate, cm^3/s
A = total cross-sectional area of the rock, cm^2

In other words, A includes the area of the rock material as well as the area of the pore channels. The fluid viscosity μ is expressed in centipoise units, and the pressure gradient dp/dx is in atmospheres per centimeter, taken in the same direction as v and q. The proportionality constant k is the permeability of the rock expressed in Darcy units.

The negative sign in Eq. (1.12a) is added because the pressure gradient dp/dx is negative in the direction of flow as shown in Figure 1.9.

For a horizontal-radial system, the pressure gradient is positive (see Figure 1.10) and Darcy's equation can be expressed in the following generalized radial form:

$$v = \frac{q_r}{A_r} = \frac{k}{\mu}\left(\frac{\partial p}{\partial r}\right)_r \tag{1.12b}$$

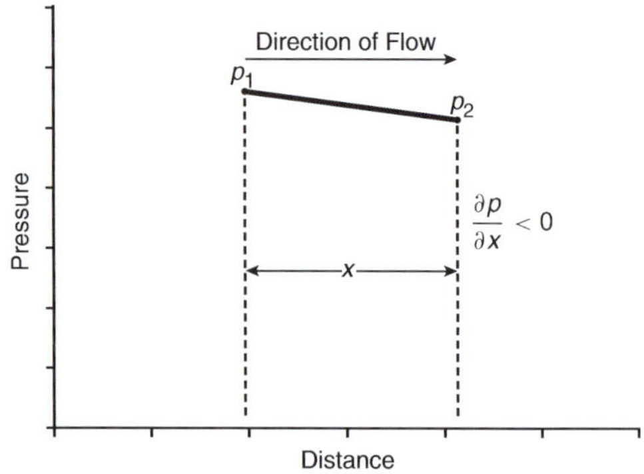

FIGURE 1.9 Pressure vs. distance in a linear flow.

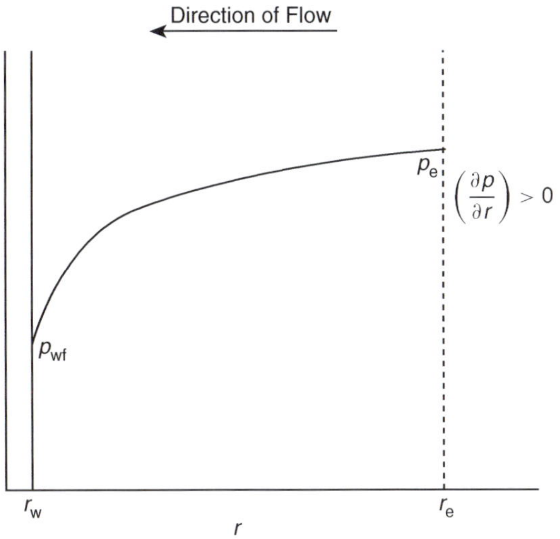

FIGURE 1.10 Pressure gradient in radial flow.

where

q_r = volumetric flow rate at radius r
A_r = cross-sectional area to flow at radius r
$(\partial p/\partial r)_r$ = pressure gradient at radius r
v = apparent velocity at radius r

The cross-sectional area at radius r is essentially the surface area of a cylinder. For a fully penetrated well with a net thickness of h, the cross-sectional area A_r is given by:

$$A_r = 2\pi rh$$

Darcy's law applies only when the following conditions exist:

- laminar (viscous) flow;
- steady-state flow;
- incompressible fluids;
- homogeneous formation.

For turbulent flow, which occurs at higher velocities, the pressure gradient increases at a greater rate than does the flow rate and a special modification of Darcy's equation is needed. When turbulent flow exists, the application of Darcy's equation can result in serious errors. Modifications for turbulent flow will be discussed later in this chapter.

1.2.2 Steady-state Flow

As defined previously, steady-state flow represents the condition that exists when the pressure throughout the reservoir does not change with time. The applications of steady-state flow to describe the flow behavior of several types of fluid in different reservoir geometries are presented below. These include:

- linear flow of incompressible fluids;
- linear flow of slightly compressible fluids;
- linear flow of compressible fluids;
- radial flow of incompressible fluids;
- radial flow of slightly compressible fluids;
- radial flow of compressible fluids;
- multiphase flow.

Linear Flow of Incompressible Fluids. In a linear system, it is assumed that the flow occurs through a constant cross-sectional area A, where both ends are entirely open to flow. It is also assumed that no flow crosses the sides, top, or bottom as shown in Figure 1.11. If an incompressible fluid is flowing across the element dx, then the fluid velocity v and the flow rate q are constant at all points. The flow behavior in this system can be expressed by the differential form of Darcy's equation, i.e., Eq. (1.12a). Separating the variables of Eq. (1.12a) and integrating over the length of the linear system:

$$\frac{q}{A}\int_0^L dx = -\frac{k}{u}\int_{p_1}^{p_2} dp$$

which results in:

$$q = \frac{kA(p_1 - p_2)}{\mu L}$$

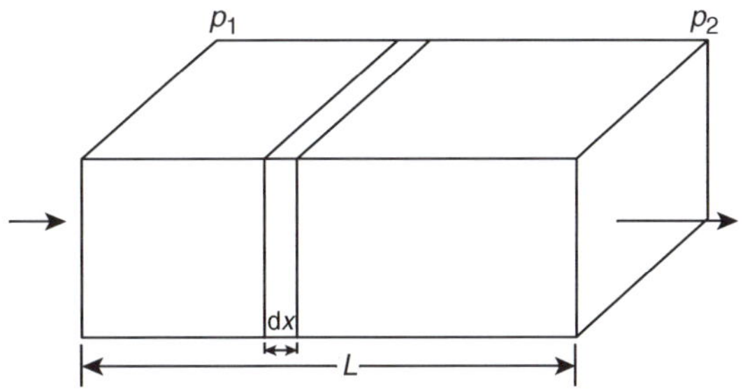

FIGURE 1.11 Linear flow model.

It is often desirable to express the above relationship in customary field units, or:

$$q = \frac{0.001127kA(p_1 - p_2)}{\mu L} \quad \textbf{(1.13)}$$

where

q = flow rate, bbl/day
k = absolute permeability, md
p = pressure, psia
μ = viscosity, cp
L = distance, ft
A = cross-sectional area, ft^2

Example 1.1

An incompressible fluid flows in a linear porous media with the following properties:

$L = 2000$ ft,	$h = 20$ ft,	width = 300 ft
$k = 100$ md,	$\phi = 15\%$,	$\mu = 2$ cp
$p_1 = 2000$ psi,	$p_2 = 1990$ psi	

Calculate:

(a) flow rate in bbl/day;
(b) apparent fluid velocity in ft/day;
(c) actual fluid velocity in ft/day.

Solution

Calculate the cross-sectional area A:

$$A = (h)(\text{width}) = (20)(100) = 6000 \text{ ft}^2$$

(a) Calculate the flow rate from Eq. (1.13):

$$q = \frac{0.001127kA(p_1 - p_2)}{\mu L} = \frac{(0.001127)(100)(6000)(2000-1990)}{(2)(2000)} = 1.6905 \text{ bbl/day}$$

(b) Calculate the apparent velocity:

$$v = \frac{q}{A} = \frac{(1.6905)(5.615)}{6000} = 0.0016 \text{ ft/day}$$

(c) Calculate the actual fluid velocity:

$$v = \frac{q}{\phi A} = \frac{(1.6905)(5.615)}{(0.15)(6000)} = 0.0105 \text{ ft/day}$$

The difference in the pressure $(p_1 - p_2)$ in Eq. (1.13) is not the only driving force in a tilted reservoir. The gravitational force is the other important driving force that must be accounted for to determine the direction and rate of flow. The fluid gradient force (gravitational force) is always directed *vertically downward* while the force that results from an applied pressure drop may be in any direction. The force causing flow would then be the *vector sum of these two*. In practice, we obtain this result by introducing a new parameter, called "fluid potential" (given by symbol Φ), which has the same dimensions as pressure, e.g., psi. The fluid potential at any point in the reservoir is defined as the pressure at that point less the pressure that would be exerted by a fluid head extending to an arbitrarily assigned datum level. Let Δz_i be the vertical distance from a point i in the reservoir to this datum level:

$$\Phi_i = p_i - \left(\frac{\rho}{144}\right)\Delta z_i \quad \textbf{(1.14)}$$

where ρ is the density in lb/ft^3.

Expressing the fluid density in g/cm^3 in Eq. (1.14) gives:

$$\Phi_i = p_i - 0.433\gamma\,\Delta z \quad \textbf{(1.15)}$$

where

Φ_i = fluid potential at point i, psi
p_i = pressure at point i, psi
Δz_i = vertical distance from point i to the selected datum level
ρ = fluid density under reservoir conditions, lb/ft^3
γ = *fluid density* under reservoir conditions, g/cm^3; this is *not* the fluid specific gravity

The datum is usually selected at the gas–oil contact, oil–water contact, or the highest point in formation. In using Eq. (1.14) or (1.15) to calculate the fluid potential Φ_i at location i, the vertical distance z_i is assigned as a positive value when the point i is below the datum level and as a negative value when it is above the datum level. That is:

If point i is above the datum level:

$$\Phi_i = p_i + \left(\frac{\rho}{144}\right)\Delta z_i$$

and equivalently:

$$\Phi_i = p_i + 0.433\gamma\,\Delta z_i$$

If point i is below the datum level:

$$\Phi_i = p_i - \left(\frac{\rho}{144}\right)\Delta z_i$$

and equivalently:

$$\Phi_i = p_i - 0.433\gamma\,\Delta z_i$$

Applying the above-generalized concept to Darcy's equation (Eq. (1.13)) gives:

$$q = \frac{0.001127kA\,(\Phi_1 - \Phi_2)}{\mu L} \quad \textbf{(1.16)}$$

It should be pointed out that the fluid potential drop $\Phi_1 - \Phi_2$ is equal to the pressure drop $(p_1 - p_2)$ only when the flow system is horizontal.

Example 1.2

Assume that the porous media with the properties as given in the previous example are tilted with a dip angle of 5°, as shown in Figure 1.12. The incompressible fluid has a density of 42 lb/ft^3. Resolve Example 1.1 using this additional information.

Solution

Step 1. For the purpose of illustrating the concept of fluid potential, select the datum level at half the vertical distance between the two points, i.e., at 87.15 ft, as shown in Figure 1.12.

Step 2. Calculate the fluid potential at points 1 and 2. Since point 1 is below the datum level, then:

$$\Phi_1 = p_1 - \left(\frac{\rho}{144}\right)\Delta z_1 = 2000 - \left(\frac{42}{144}\right)(87.15)$$
$$= 1974.58 \text{ psi}$$

Since point 2 is above the datum level, then:

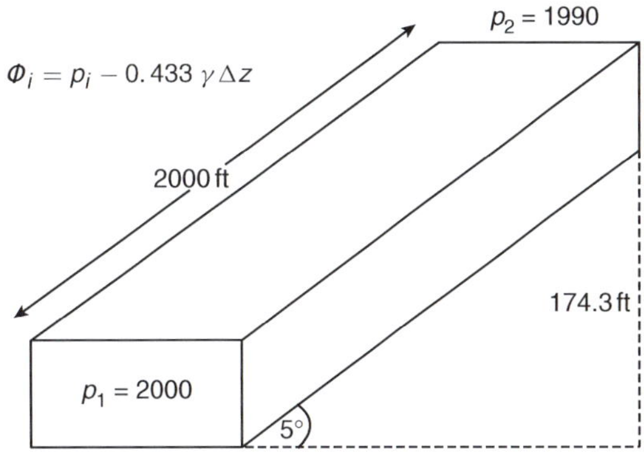

FIGURE 1.12 Example of a tilted layer.

$$\Phi_2 = p_2 + \left(\frac{\rho}{144}\right)\Delta z_2 = 1990 + \left(\frac{42}{144}\right)(87.15)$$
$$= 2015.42 \text{ psi}$$

Because $\Phi_2 > \Phi_1$, the fluid flows downward from point 2 to 1. The difference in the fluid potential is:

$$\Delta\Phi = 2015.42 - 1974.58 = 40.84 \text{ psi}$$

Notice that, if we select point 2 for the datum level, then:

$$\Phi_1 = 2000 - \left(\frac{42}{144}\right)(174.3) = 1949.16 \text{ psi}$$

$$\Phi_2 = 1990 + \left(\frac{42}{144}\right)(0) = 1990 \text{ psi}$$

The above calculations indicate that regardless of the position of the datum level, the flow is downward from point 2 to 1 with:

$$\Delta\Phi = 1990 - 1949.16 = 40.84 \text{ psi}$$

Step 3. Calculate the flow rate:

$$q = \frac{0.001127kA(\Phi_1 - \Phi_2)}{\mu L}$$
$$= \frac{(0.001127)(100)(6000)(40.84)}{(2)(2000)} = 6.9 \text{ bbl/day}$$

Step 4. Calculate the velocity:

$$\text{Apparent velocity} = \frac{(6.9)(5.615)}{6000} = 0.0065 \text{ ft/day}$$

$$\text{Actual velocity} = \frac{(6.9)(5.615)}{(0.15)(6000)} = 0.043\ \text{ft/day}$$

Linear Flow of Slightly Compressible Fluids. Eq. (1.6) describes the relationship that exists between pressure and volume for a slightly compressible fluid, or:

$$V = V_{\text{ref}}[1 + c(p_{\text{ref}} - p)]$$

This equation can be modified and written in terms of flow rate as:

$$q = q_{\text{ref}}[1 + c(p_{\text{ref}} - p)] \quad \textbf{(1.17)}$$

where q_{ref} is the flow rate at some reference pressure p_{ref}. Substituting the above relationship in Darcy's equation gives:

$$\frac{q}{A} = \frac{q_{\text{ref}}[1 + c(p_{\text{ref}} - p)]}{A} = -0.001127\frac{k}{\mu}\frac{dp}{dx}$$

Separating the variables and arranging:

$$\frac{q_{\text{ref}}}{A}\int_0^L dx = -0.001127\frac{k}{\mu}\int_{p_1}^{p_2}\left[\frac{dp}{1 + c(p_{\text{ref}} - p)}\right]$$

Integrating gives:

$$q_{\text{ref}} = \left[\frac{0.001127kA}{\mu cL}\right]\ln\left[\frac{1 + c(p_{\text{ref}} - p_2)}{1 + c(p_{\text{ref}} - p_1)}\right] \quad \textbf{(1.18)}$$

where

q_{ref} = flow rate at a reference pressure p_{ref}, bbl/day
p_1 = upstream pressure, psi
p_2 = downstream pressure, psi
k = permeability, md
μ = viscosity, cp
c = average liquid compressibility, psi^{-1}

Selecting the upstream pressure p_1 as the reference pressure p_{ref} and substituting in Eq. (1.18) gives the flow rate at point 1 as:

$$q_1 = \left[\frac{0.001127kA}{\mu cL}\right]\ln[1 + c(p_1 - p_2)] \quad \textbf{(1.19)}$$

Choosing the downstream pressure p_2 as the reference pressure and substituting in Eq. (1.18) gives:

$$q_2 = \left[\frac{0.001127kA}{\mu cL}\right]\ln\left[\frac{1}{1 + c(p_2 - p_1)}\right] \quad \textbf{(1.20)}$$

where q_1 and q_2 are the flow rates at points 1 and 2, respectively.

Example 1.3

Consider the linear system given in Example 1.1 and, assuming a slightly compressible liquid, calculate the flow rate at both ends of the linear system. The liquid has an average compressibility of 21×10^{-5} psi^{-1}.

Solution

Choosing the upstream pressure as the reference pressure gives:

$$q_1 = \left[\frac{0.001127kA}{\mu cL}\right]\ln[1 + c(p_1 - p_2)]$$
$$= \left[\frac{(0.001127)(100)(6000)}{(2)(21\times10^{-5})(2000)}\right]$$
$$\times \ln\left[1 + 21\times10^{-5}(2000 - 1990)\right] = 1.689\ \text{bbl/day}$$

Choosing the downstream pressure gives:

$$q_2 = \left[\frac{0.001127kA}{\mu cL}\right]\ln\left[\frac{1}{1 + c(p_2 - p_1)}\right]$$
$$= \left[\frac{(0.001127)(100)(6000)}{(2)(21\times10^{-5})(2000)}\right]$$
$$\times \ln\left[\frac{1}{1 + (21\times10^{-5})(1990 - 2000)}\right] = 1.689\ \text{bbl/day}$$

The above calculations show that q_1 and q_2 are not largely different, which is due to the fact that the liquid is slightly incompressible and its volume is not a strong function of pressure.

Linear Flow of Compressible Fluids (Gases). For a viscous (laminar) gas flow in a homogeneous linear system, the real-gas equation of state can be applied to calculate the number of gas moles n at the pressure p, temperature T, and volume V:

$$n = \frac{pV}{ZRT}$$

At standard conditions, the volume occupied by the above n moles is given by:

$$V_{sc} = \frac{nZ_{sc}RT_{sc}}{p_{sc}}$$

Combining the above two expressions and assuming $Z_{sc} = 1$ gives:

$$\frac{pV}{ZT} = \frac{p_{sc}V_{sc}}{T_{sc}}$$

Equivalently, the above relation can be expressed in terms of the reservoir condition flow rate q, in bbl/day, and surface condition flow rate Q_{sc}, in scf/day, as:

$$\frac{p(5.615q)}{ZT} = \frac{p_{sc}Q_{sc}}{T_{sc}}$$

Rearranging:

$$\left(\frac{p_{sc}}{T_{sc}}\right)\left(\frac{ZT}{p}\right)\left(\frac{Q_{sc}}{5.615}\right) = q \tag{1.21}$$

where

q = gas flow rate at pressure p, bbl/day
Q_{sc} = gas flow rate at standard conditions, scf/day
Z = gas compressibility factor
T_{sc}, p_{sc} = standard temperature and pressure in °R and psia, respectively.

Dividing both sides of the above equation by the cross-sectional area A and equating it with that of Darcy's law, i.e., Eq. (1.12a), gives:

$$\frac{q}{A} = \left(\frac{p_{sc}}{T_{sc}}\right)\left(\frac{ZT}{p}\right)\left(\frac{Q_{sc}}{5.615}\right)\left(\frac{1}{A}\right) = -0.001127\frac{k}{\mu}\frac{dp}{dx}$$

The constant 0.001127 is to convert Darcy's units to field units. Separating variables and arranging yields:

$$\left[\frac{Q_{sc}p_{sc}T}{0.006328kT_{sc}A}\right]\int_0^L dx = -\int_{p_1}^{p_2}\frac{p}{Z\mu_g}dp$$

Assuming that the product of $Z\mu_g$ is constant over the specified pressure range between p_1 and p_2, and integrating, gives:

$$\left[\frac{Q_{sc}p_{sc}T}{0.006328kT_{sc}A}\right]\int_0^L dx = -\frac{1}{Z\mu_g}\int_{p_1}^{p_2} p\,dp$$

or

$$Q_{sc} = \frac{0.003164T_{sc}Ak(p_1^2 - p_2^2)}{p_{sc}T(Z\mu_g)L}$$

where

Q_{sc} = gas flow rate at standard conditions, scf/day
k = permeability, md
T = temperature, °R
μ_g = gas viscosity, cp
A = cross-sectional area, ft^2
L = total length of the linear system, ft

Setting $p_{sc} = 14.7$ psi and $T_{sc} = 520$ °R in the above expression gives:

$$Q_{sc} = \frac{0.111924Ak(p_1^2 - p_2^2)}{TLZ\mu_g} \tag{1.22}$$

It is essential to note that the gas properties Z and μ_g are very strong functions of pressure, but they have been removed from the integral to simplify the final form of the gas flow equation. The above equation is valid for applications when the pressure is less than 2000 psi. The gas properties must be evaluated at the average pressure $\bar{p}$ as defined below:

$$\bar{p} = \sqrt{\frac{p_1^2 + p_2^2}{2}} \tag{1.23}$$

Example 1.4

A natural gas with a specific gravity of 0.72 is flowing in linear porous media at 140 °F. The upstream and downstream pressures are 2100 and 1894.73 psi, respectively. The cross-sectional area is constant at 4500 ft^2. The total length is 2500 ft with an absolute permeability of 60 md. Calculate the gas flow rate in scf/day ($p_{sc} = 14.7$ psia, $T_{sc} = 520$ °R).

Solution

Step 1. Calculate average pressure by using Eq. (1.23):

$$\bar{p} = \sqrt{\frac{2100^2 + 1894.73^2}{2}} = 2000 \text{ psi}$$

Step 2. Using the specific gravity of the gas, calculate its pseudo-critical properties by applying the following equations:

$$T_{pc} = 168 + 325\gamma_g - 12.5\gamma_g^2$$
$$= 168 + 325(0.72) - 12.5(0.72)^2 = 395.5\ ^\circ\text{R}$$
$$p_{pc} = 677 + 15.0\gamma_g - 37.5\gamma_g^2$$
$$= 677 + 15.0(0.72) - 37.5(0.72)^2 = 668.4 \text{ psia}$$

Step 3. Calculate the pseudo-reduced pressure and temperature:

$$p_{pr} = \frac{2000}{668.4} = 2.99$$

$$T_{pr} = \frac{600}{395.5} = 1.52$$

Step 4. Determine the *Z*-factor from a Standing–Katz chart to give:

$$Z = 0.78$$

Step 5. Solve for the viscosity of the gas by applying the Lee–Gonzales–Eakin method and using the following sequence of calculations:

$$M_a = 28.96\gamma_g$$
$$= 28.96(0.72) = 20.85$$

$$\rho_g = \frac{pM_a}{ZRT}$$
$$= \frac{(2000)(20.85)}{(0.78)(10.73)(600)} = 8.30 \text{ lb/ft}^3$$

$$K = \frac{(9.4 + 0.02M_a)T^{1.5}}{209 + 19M_a + T}$$
$$= \frac{[9.4 + 0.02(20.96)](600)^{1.5}}{209 + 19(20.96) + 600} = 119.72$$

$$X = 3.5 + \frac{986}{T} + 0.01M_a$$
$$= 3.5 + \frac{986}{600} + 0.01(20.85) = 5.35$$

$$Y = 2.4 - 0.2X$$
$$= 2.4 - (0.2)(5.35) = 1.33$$

$$\mu_g = 10^{-4}K \exp\left[X\left(\frac{\rho_g}{62.4}\right)^Y\right] = 0.0173 \text{ cp}$$
$$= 10^{-4}\left(119.72 \exp\left[5.35\left(\frac{8.3}{62.4}\right)^{1.33}\right]\right)$$
$$= 0.0173$$

Step 6. Calculate the gas flow rate by applying Eq. (1.22):

$$Q_{sc} = \frac{0.111924Ak(p_1^2 - p_2^2)}{TLZ\mu_g}$$
$$= \frac{(0.111924)(4500)(60)(2100^2 - 1894.73^2)}{(600)(2500)(0.78)(0.0173)}$$
$$= 1{,}224{,}242 \text{ scf/day}$$

Radial Flow of Incompressible Fluids. In a radial flow system, all fluids move toward the producing well from all directions. However, before flow can take place, a pressure differential must exist. Thus, if a well is to produce oil, which implies a flow of fluids through the formation to the wellbore, the pressure in the formation at the wellbore must be less than the pressure in the formation at some distance from the well.

The pressure in the formation at the wellbore of a producing well is known as the bottom-hole flowing pressure (flowing BHP, p_{wf}).

Figure 1.13 schematically illustrates the radial flow of an incompressible fluid toward a vertical well. The formation is considered to have a uniform thickness h and a constant permeability k. Because the fluid is incompressible, the flow rate q must be constant at all radii. Due to the steady-state flowing condition, the pressure profile around the wellbore is maintained constant with time.

Let p_{wf} represent the maintained bottom-hole flowing pressure at the wellbore radius r_w and p_e denote the external pressure at the external or drainage radius. Then Darcy's

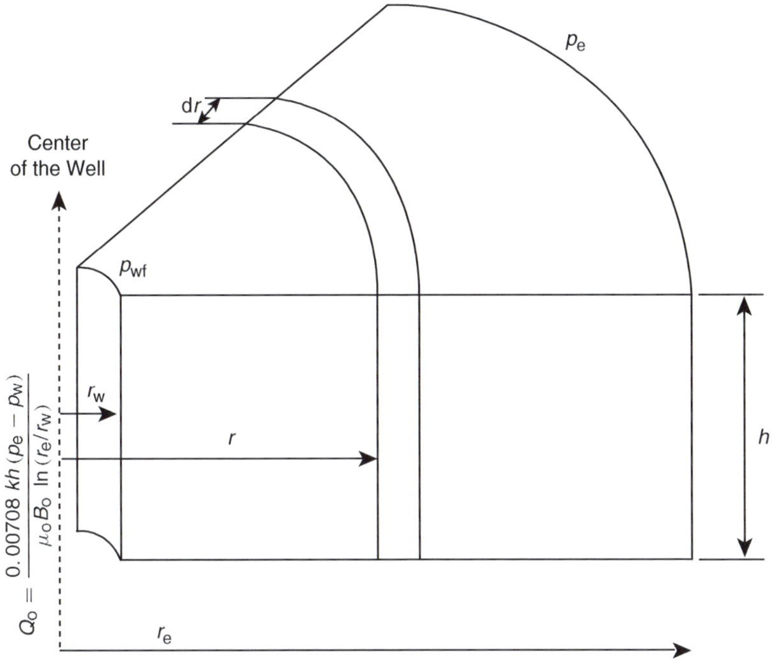

FIGURE 1.13 Radial flow model.

generalized equation as described by Eq. (1.12b) can be used to determine the flow rate at any radius r:

$$v = \frac{q}{A_r} = 0.001127 \frac{k}{\mu} \frac{dp}{dr} \quad \textbf{(1.24)}$$

where

v = apparent fluid velocity, bbl/day ft^2
q = flow rate at radius r, bbl/day
k = permeability, md
μ = viscosity, cp
0.001127 = conversion factor to express the equation in field units
A_r = cross-sectional area at radius r

The minus sign is no longer required for the radial system shown in Figure 1.13 as the radius increases in the same direction as the pressure. In other words, as the radius increases by going away from the wellbore the pressure also increases. At any point in the reservoir the cross-sectional area across which flow occurs will be the surface area of a cylinder, which is $2\pi rh$, or:

$$v = \frac{q}{A_r} = \frac{q}{2\pi rh} = 0.001127 \frac{k}{\mu} \frac{dp}{dr}$$

The flow rate for a crude oil system is customarily expressed in surface units, i.e., stock-tank barrel (STB), rather than reservoir units. Using the symbol Q_o to represent the oil flow as expressed in STB/day, then:

$$q = B_o Q_o$$

where B_o is the oil formation volume factor in bbl/STB. The flow rate in Darcy's equation can be expressed in STB/day, to give:

$$\frac{Q_o B_o}{2\pi rh} = 0.001127 \frac{k}{\mu_o} \frac{dp}{dr}$$

Integrating this equation between two radii, r_1 and r_2, when the pressures are p_1 and p_2, yields:

$$\int_{r_1}^{r_2} \left(\frac{Q_o}{2\pi h}\right) \frac{dr}{r} = 0.001127 \int_{P_1}^{P_2} \left(\frac{k}{\mu_o B_o}\right) dp \quad \textbf{(1.25)}$$

For an incompressible system in a uniform formation, Eq. (1.25) can be simplified to:

$$\frac{Q_o}{2\pi h}\int_{r_1}^{r_2}\frac{dr}{r}=\frac{0.001127k}{\mu_o B_o}\int_{P_1}^{P_2} dp$$

Performing the integration gives:

$$Q_o=\frac{0.00708kh(p_2-p_1)}{\mu_o B_o \ln(r_2/r_1)}$$

Frequently, the two radii of interest are the wellbore radius r_w and the external or drainage radius r_e. Then:

$$Q_o=\frac{0.00780kh(p_e-p_w)}{\mu_o B_o \ln(r_e/r_w)} \quad (1.26)$$

where

Q_o = oil flow rate, STB/day
p_e = external pressure, psi
p_{wf} = bottom-hole flowing pressure, psi
k = permeability, md
μ_o = oil viscosity, cp
B_o = oil formation volume factor, bbl/STB
h = thickness, ft
r_e = external or drainage radius, ft
r_w = wellbore radius, ft

The external (drainage) radius r_e is usually determined from the well spacing by equating the area of the well spacing with that of a circle. That is:

$$\pi r_e^2 = 43{,}560A$$

or

$$r_e=\sqrt{\frac{43{,}560A}{\pi}} \quad (1.27)$$

where A is the well spacing in acres.

In practice, neither the external radius nor the wellbore radius is generally known with precision. Fortunately, they enter the equation as a logarithm, so the errors in the equation will be less than the errors in the radii.

Eq. (1.26) can be arranged to solve for the pressure p at any radius r, to give:

$$p=p_{wf}+\left[\frac{Q_o B_o \mu_o}{0.00708kh}\right]\ln\left(\frac{r}{r_w}\right) \quad (1.28)$$

Example 1.5

An oil well in the Nameless Field is producing at a stabilized rate of 600 STB/day at a stabilized bottom-hole flowing pressure of 1800 psi. Analysis of the pressure buildup test data indicates that the pay zone is characterized by a permeability of 120 md and a uniform thickness of 25 ft. The well drains an area of approximately 40 acres. The following additional data is available:

$r_w = 0.25$ ft, $A = 40$ acres
$B_o = 1.25$ bbl/STB, $\mu_o = 2.5$ cp

Calculate the pressure profile (distribution) and list the pressure drop across 1 ft intervals from r_w to 1.25 ft, 4 to 5 ft, 19 to 20 ft, 99 to 100 ft, and 744 to 745 ft.

Solution

Step 1. Rearrange Eq. (1.26) and solve for the pressure p at radius r:

$$p=p_{wf}+\left[\frac{\mu_o B_o Q_o}{0.00708kh}\right]\ln\left(\frac{r}{r_w}\right)$$

$$=1800+\left[\frac{(2.5)(1.25)(600)}{(0.00708)(120)(25)}\right]\ln\left(\frac{r}{0.25}\right)$$

$$=1800+88.28\ln\left(\frac{r}{0.25}\right)$$

Step 2. Calculate the pressure at the designated radii:

r (ft)	*p* (psi)	Radius Interval	Pressure Drop
0.25	1800		
1.25	1942	0.25–1.25	1942–1800 = 142 psi
4	2045		
5	2064	4–5	2064–2045 = 19 psi
19	2182		
20	2186	19–20	2186–2182 = 4 psi
99	2328		
100	2329	99–100	2329–2328 = 1 psi
744	2506.1		
745	2506.2	744–745	2506.2–2506.1 = 0.1 psi

Figure 1.14 shows the pressure profile as a function of radius for the calculated data.

Results of Example 1.5 reveal that the pressure drop just around the wellbore (i.e., 142 psi) is 7.5 times greater than at the 4–5 ft interval, 36 times greater than at 19–20 ft, and 142 times than that at the 99–100 ft interval. The reason for this large pressure drop around the wellbore is that the fluid flows in from a large drainage area of 40 acres.

The external pressure p_e used in Eq. (1.26) cannot be measured readily, but p_e does not deviate substantially from the initial reservoir pressure if a strong and active aquifer is present.

Several authors have suggested that the average reservoir pressure p_r, which often is reported in well test results, should be used in performing material balance calculations and flow rate prediction. Craft and Hawkins (1959) showed that the average pressure is located at about 61% of the drainage radius r_e for a steady-state flow condition.

Substituting $0.61r_e$ in Eq. (1.28) gives:

$$p(\text{at } r = 0.61r_e) = p_r = p_{wf} + \left[\frac{Q_o B_o \mu_o}{0.0078kh}\right] \ln\left(\frac{0.61r_e}{r_w}\right)$$

or in terms of flow rate:

$$Q_o = \frac{0.0078kh(p_r - p_{wf})}{\mu_o B_o \ln(0.61r_e/r_w)} \tag{1.29}$$

But since $\ln(0.61r_e/r_w) = \ln(r_e/r_w) - 0.5$, then:

$$Q_o = \frac{0.00708kh(p_r - p_{wf})}{\mu_o B_o[\ln(r_e/r_w) - 0.5]} \tag{1.30}$$

Golan and Whitson (1986) suggested a method for approximating the drainage area of wells producing from a common reservoir. These authors assume that the volume drained by a single well is proportional to its rate of flow. Assuming constant reservoir properties and a uniform thickness, the approximate drainage area of a single well A_w is:

$$A_w = A_T\left(\frac{q_w}{q_T}\right) \tag{1.31}$$

where

A_w = drainage area of a well
A_T = total area of the field
q_T = total flow rate of the field
q_w = well flow rate

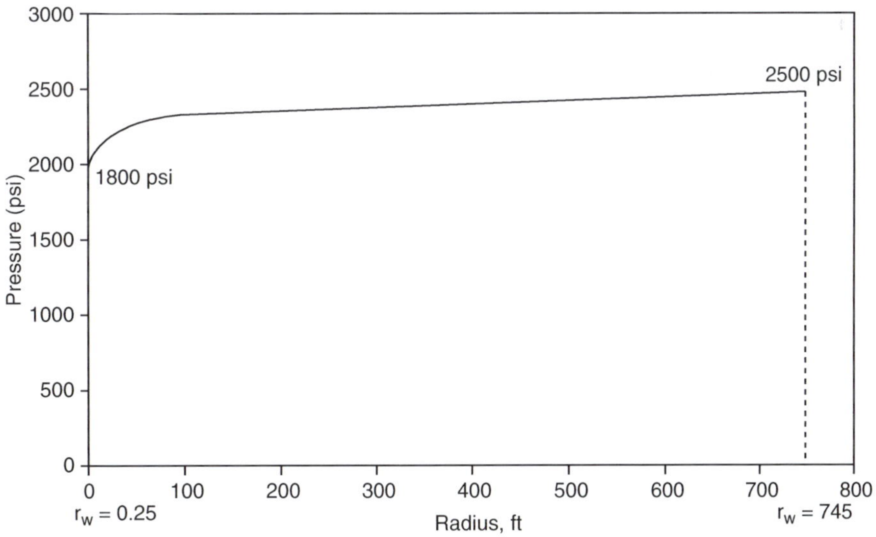

FIGURE 1.14 Pressure profile around the wellbore.

Radial Flow of Slightly Compressible Fluids. Terry et al. (1991) used Eq. (1.17) to express the dependency of the flow rate on pressure for slightly compressible fluids. If this equation is substituted into the radial form of Darcy's law, the following is obtained:

$$\frac{q}{A_r} = \frac{q_{ref}[1 + c(p_{ref} - p)]}{2\pi rh} = 0.001127\frac{k}{\mu}\frac{dp}{dr}$$

where q_{ref} is the flow rate at some reference pressure p_{ref}.

Separating the variables and assuming a constant compressibility over the entire pressure drop, and integrating over the length of the porous medium:

$$\frac{q_{ref}\mu}{2\pi kh}\int_{r_w}^{r_e}\frac{dr}{r} = 0.001127\int_{p_{wf}}^{p_e}\frac{dp}{1 + c(p_{ref} - p)}$$

gives

$$q_{ref} = \left[\frac{0.00708kh}{\mu c \ln(r_e/r_w)}\right]\ln\left[\frac{1 + c(p_e - p_{ref})}{1 + c(p_{wf} - p_{ref})}\right]$$

where q_{ref} is the oil flow rate at a reference pressure p_{ref}. Choosing the bottom-hole flow pressure p_{wf} as the reference pressure and expressing the flow rate in STB/day gives:

$$Q_o = \left[\frac{0.00708kh}{\mu_o B_o c_o \ln(r_e/r_w)}\right]\ln[1 + c_o(p_e - p_{wf})] \quad \textbf{(1.32)}$$

where

c_o = isothermal compressibility coefficient, psi^{-1}
Q_o = oil flow rate, STB/day
k = permeability, md

Example 1.6

The following data is available on a well in the Red River Field:

$p_e = 2506$ psi,	$p_{wf} = 1800$ psi
$r_e = 745$ ft,	$r_w = 0.25$ ft
$B_o = 1.25$ bbl/STB,	$\mu_o = 2.5$ cp
$k = 0.12$ darcy,	$h = 25$ ft
$c_o = 25 \times 10^{-6}\ \text{psi}^{-1}$	

Assuming a slightly compressible fluid, calculate the oil flow rate. Compare the result with that of an incompressible fluid.

Solution

For a slightly compressible fluid, the oil flow rate can be calculated by applying Eq. (1.32):

$$Q_o = \left[\frac{0.00708kh}{\mu_o B_o c_o \ln(r_e/r_w)}\right]\ln\left[1 + c_o\left(\frac{p_e}{p_{wf}}\right)\right]$$

$$= \left[\frac{(0.00708)(120)(25)}{(2.5)(1.25)(25\times10^{-6})\ln(745/0.25)}\right]$$

$$\times\ln\left[1 + (25\times10^{-6})(2506 - 1800)\right] = 595\ \text{STB/day}$$

Assuming an incompressible fluid, the flow rate can be estimated by applying Darcy's equation, i.e., Eq. (1.26):

$$Q_o = \frac{0.00708kh(p_e - p_w)}{\mu_o B_o \ln(r_e/r_w)}$$

$$= \frac{(0.00708)(120)(25)(2506 - 1800)}{(2.5)(1.25)\ln(745/0.25)} = 600\ \text{STB/day}$$

Radial Flow of Compressible Gases. The basic differential form of Darcy's law for a horizontal laminar flow is valid for describing the flow of both gas and liquid systems. For a radial gas flow, Darcy's equation takes the form:

$$q_{gr} = \frac{0.001127(2\pi rh)k}{\mu_g}\frac{dp}{dr} \quad \textbf{(1.33)}$$

where

q_{gr} = gas flow rate at radius r, bbl/day
r = radial distance, ft
h = zone thickness, ft
μ_g = gas viscosity, cp
p = pressure, psi
0.001127 = conversion constant from Darcy units to field units

The gas flow rate is traditionally expressed in scf/day. Referring to the gas flow rate at standard (surface) condition as Q_g, the gas flow rate q_{gr} under wellbore flowing condition can be converted to that of surface condition by applying the definition of the gas formation volume factor B_g to q_{gr} as:

$$Q_g = \frac{q_{gr}}{B_g}$$

where

$$B_g = \frac{p_{sc}}{5.615 T_{sc}} \frac{ZT}{p} \text{ bbl/scf}$$

or

$$\left(\frac{p_{sc}}{5.615 T_{sc}}\right)\left(\frac{ZT}{p}\right) Q_g = q_{gr} \qquad \textbf{(1.34)}$$

where

p_{sc} = standard pressure, psia
T_{sc} = standard temperature, °R
Q_g = gas flow rate, scf/day
q_{gr} = gas flow rate at radius r, bbl/day
p = pressure at radius r, psia
T = reservoir temperature, °R
Z = gas compressibility factor at p and T
Z_{sc} = gas compressibility factor at standard condition ≅ 1.0

Combining Eqs. (1.33) and (1.34) yields:

$$\left(\frac{p_{sc}}{5.615 T_{sc}}\right)\left(\frac{ZT}{p}\right) Q_g = \frac{0.001127(2\pi rh)k}{\mu_g} \frac{dp}{dr}$$

Assuming that $T_{sc} = 520$ °R and $p_{sc} = 14.7$ psia:

$$\left(\frac{TQ_g}{kh}\right)\frac{dr}{r} = 0.703\left(\frac{2p}{\mu_g Z}\right) dp \qquad \textbf{(1.35)}$$

Integrating Eq. (1.35) from the wellbore conditions (r_w and p_{wf}) to any point in the reservoir (r and p) gives:

$$\int_{r_w}^{r}\left(\frac{TQ_g}{kh}\right)\frac{dr}{r} = 0.703 \int_{p_{wf}}^{p}\left(\frac{2p}{\mu_g Z}\right) dp \qquad \textbf{(1.36)}$$

Imposing Darcy's law conditions on Eq. (1.36), i.e., steady-state flow, which requires that Q_g is constant at all radii, and homogeneous formation, which implies that k and h are constant, gives:

$$\left(\frac{TQ_g}{kh}\right) \ln\left(\frac{r}{r_w}\right) = 0.703 \int_{p_{wf}}^{p}\left(\frac{2p}{\mu_g Z}\right) dp$$

The term:

$$\int_{p_{wf}}^{p}\left(\frac{2p}{\mu_g Z}\right) dp$$

can be expanded to give:

$$\int_{p_{wf}}^{p}\left(\frac{2p}{\mu_g Z}\right) dp = \int_{0}^{p}\left(\frac{2p}{\mu_g Z}\right) dp - \int_{0}^{p_{wf}}\left(\frac{2p}{\mu_g Z}\right) dp$$

Replacing the integral in Eq. (1.35) with the above expanded form yields:

$$\left(\frac{TQ_g}{kh}\right) \ln\left(\frac{r}{r_w}\right) = 0.703\left[\int_{0}^{p}\left(\frac{2p}{\mu_g Z}\right) dp - \int_{0}^{p_{wf}}\left(\frac{2p}{\mu_g Z}\right) dp\right] \qquad \textbf{(1.37)}$$

The integral $\int_0^p 2p/(\mu_g Z)\,dp$ is called the "real-gas pseudo-potential" or "real-gas pseudopressure" and it is usually represented by $m(p)$ or ψ. Thus:

$$m(p) = \psi = \int_{0}^{p}\left(\frac{2p}{\mu_g Z}\right) dp \qquad \textbf{(1.38)}$$

Eq. (1.38) can be written in terms of the real-gas pseudopressure as:

$$\left(\frac{TQ_g}{kh}\right) \ln\left(\frac{r}{r_w}\right) = 0.703(\psi - \psi_w)$$

or

$$\psi = \psi_w + \frac{Q_g T}{0.703kh} \ln\left(\frac{r}{r_w}\right) \qquad \textbf{(1.39)}$$

Eq. (1.39) indicates that a graph of ψ vs. $\ln(r/r_w)$ yields a straight line with a slope of $Q_gT/0.703kh$ and an intercept value of ψ_w as shown in Figure 1.15. The exact flow rate is then given by:

$$Q_g = \frac{0.703kh(\psi - \psi_w)}{T \ln(r/r_w)} \qquad \textbf{(1.40)}$$

In the particular case when $r = r_e$, then:

$$Q_g = \frac{0.703kh(\psi_e - \psi_w)}{T \ln(r_e/r_w)} \qquad \textbf{(1.41)}$$

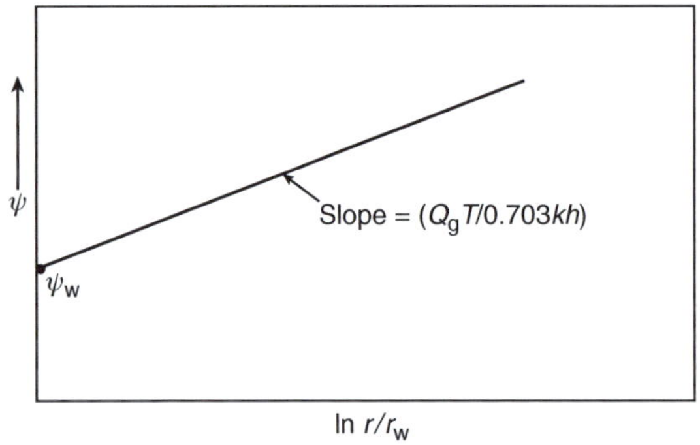

FIGURE 1.15 Graph of ψ vs. ln (r/r_w).

where

ψ_e = real-gas pseudopressure as evaluated from 0 to p_e, psi²/cp
ψ_w = real-gas pseudopressure as evaluated from 0 to p_{wf}, psi²/cp
k = permeability, md
h = thickness, ft
r_e = drainage radius, ft
r_w = wellbore radius, ft
Q_g = gas flow rate, scf/day

Because the gas flow rate is commonly expressed in Mscf/day, Eq. (1.41) can be expressed as:

$$Q_g = \frac{kh(\psi_e - \psi_w)}{1422T \ln(r_e/r_w)} \tag{1.42}$$

where

Q_g = gas flow rate, Mscf/day

Eq. (1.42) can be expressed in terms of the average reservoir pressure p_r instead of the initial reservoir pressure p_e as:

$$Q_g = \frac{kh(\psi_r - \psi_w)}{1422T[\ln(r_e/r_w) - 0.5]} \tag{1.43}$$

To calculate the integral in Eq. (1.42), the values of $2p/\mu_g Z$ are calculated for several values of pressure p. Then $2p/\mu_g Z$ vs. p is plotted on a Cartesian scale and the area under the curve is calculated either numerically or graphically, where the area under the curve from $p = 0$ to any pressure p represents the value of ψ corresponding to p. The following example will illustrate the procedure.

Example 1.7

The *PVT* data from a gas well in the Anaconda Gas Field is given below:

p (psi)	μ_g (cp)	Z
0	0.0127	1.000
400	0.01286	0.937
800	0.01390	0.882
1200	0.01530	0.832
1600	0.01680	0.794
2000	0.01840	0.770
2400	0.02010	0.763
2800	0.02170	0.775
3200	0.02340	0.797
3600	0.02500	0.827
4000	0.02660	0.860
4400	0.02831	0.896

The well is producing at a stabilized bottom-hole flowing pressure of 3600 psi. The wellbore radius is 0.3 ft. The following additional data is available:

$k = 65$ md, $h = 15$ ft, $T = 600$ °R
$p_e = 4400$ psi, $r_e = 1000$ ft

Calculate the gas flow rate in Mscf/day.

Solution

Step 1. Calculate the term $2p/\mu_g Z$ for each pressure as shown below:

p (psi)	μ_g (cp)	Z	$2p/\mu_g Z$ (psia/cp)
0	0.0127	1.000	0
400	0.01286	0.937	66,391
800	0.01390	0.882	130,508
1200	0.01530	0.832	188,537
1600	0.01680	0.794	239,894
2000	0.01840	0.770	282,326
2400	0.02010	0.763	312,983
2800	0.02170	0.775	332,986
3200	0.02340	0.797	343,167
3600	0.02500	0.827	348,247
4000	0.02660	0.860	349,711
4400	0.02831	0.896	346,924

Step 2. Plot the term $2p/\mu_g Z$ vs. pressure as shown in Figure 1.16.

Step 3. Calculate numerically the area under the curve for each value of p. These areas correspond to the real-gas pseudopressure ψ at each pressure. These ψ values are tabulated below; note that $2p/\mu_g Z$ vs. p is also plotted in Figure 1.16.

p (psi)	ψ (psi²/cp)
400	13.2×10^6
800	52.0×10^6
1200	113.1×10^6
1600	198.0×10^6
2000	304.0×10^6
2400	422.0×10^6
2800	542.4×10^6
3200	678.0×10^6
3600	816.0×10^6
4000	950.0×10^6
4400	1089.0×10^6

Step 4. Calculate the flow rate by applying Eq. (1.41):

At $p_w = 3600$ psi: gives $\psi_w = 816.0 \times 10^6$ psi²/cp

At $p_e = 4400$ psi: gives $\psi_e = 1089 \times 10^6$ psi²/cp

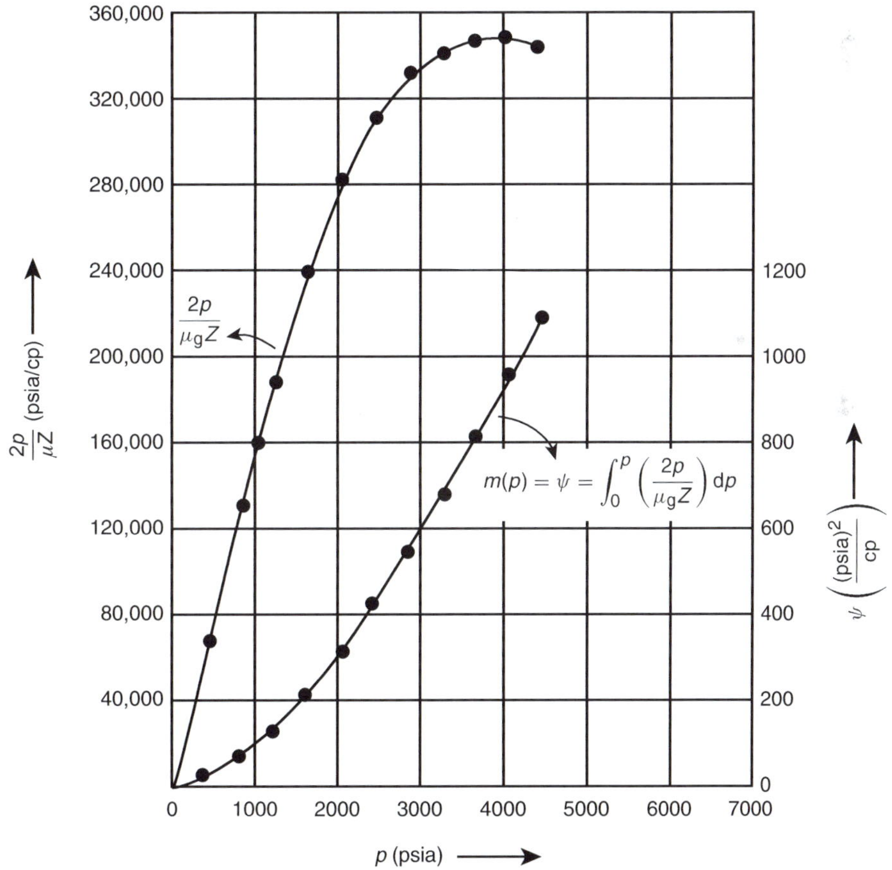

FIGURE 1.16 Real-gas pseudopressure data for Example 1.7. *(After Donohue, D., Erkekin, T., 1982. Gas Well Testing, Theory and Practice. International Human Resources Development Corporation, Boston).*

$$Q_g = \frac{0.703kh(\psi_e - \psi_w)}{T\ln(r_e/r_w)}$$
$$= \frac{(65)(15)(1089 - 816)10^6}{(1422)(600)\ln(1000/0.25)}$$
$$= 37,614 \text{ Mscf/day}$$

In the approximation of the gas flow rate, the exact gas flow rate as expressed by the different forms of Darcy's law, i.e., Eqs. (1.36)–(1.43), can be approximated by moving the term $2/\mu_g Z$ outside the integral as a constant. It should be pointed out that the product of $Z\mu_g$ is considered constant only under a pressure range of less than 2000 psi. Eq. (1.42) can be rewritten as:

$$Q_g = \left[\frac{kh}{1422T\ln(r_e/r_w)}\right]\int_{p_{wf}}^{p_e}\left(\frac{2p}{\mu_g Z}\right)dp$$

Removing the term $2/\mu_g Z$ and integrating gives:

$$Q_g = \frac{kh(p_e^2 - p_{wf}^2)}{1422T(\mu_g Z)_{avg}\ln(r_e/r_w)} \tag{1.44}$$

where

Q_g = gas flow rate, Mscf/day
k = permeability, md

The term $(\mu_g Z)_{avg}$ is evaluated at an average pressure $\bar{p}$ that is defined by the following expression:

$$\bar{p} = \sqrt{\frac{p_{wf}^2 + p_e^2}{2}}$$

The above approximation method is called the pressure-squared method and is limited to flow calculations when the reservoir pressure is less than 2000 psi. Other approximation methods are discussed in Chapter 2.

Example 1.8
Using the data given in Example 1.7, resolve the gas flow rate by using the pressure-squared method. Compare with the exact method (i.e., real-gas pseudopressure solution).

Solution

Step 1. Calculate the arithmetic average pressure:

$$\bar{p} = \sqrt{\frac{4400^2 + 3600^2}{2}} = 4020 \text{ psi}$$

Step 2. Determine the gas viscosity and gas compressibility factor at 4020 psi:

$$\mu_g = 0.0267$$
$$Z = 0.862$$

Step 3. Apply Eq. (1.44):

$$Q_g = \frac{kh(p_e^2 - p_{wf}^2)}{1422T(\mu_g Z)_{avg}\ln(r_e/r_w)}$$
$$= \frac{(65)(15)[4400^2 - 3600^2]}{(1422)(600)(0.0267)(0.862)\ln(1000/0.25)}$$
$$= 38314 \text{ Mscf/day}$$

Step 4. Results show that the pressure-squared method approximates the exact solution of 37,614 with an absolute error of 1.86%. This error is due to the limited applicability of the pressure-squared method to a pressure range of less than 2000 psi.

Horizontal Multiple-phase Flow. When several fluid phases are flowing simultaneously in a horizontal porous system, the concept of the effective permeability of each phase and the associated physical properties must be used in Darcy's equation. For a radial system, the generalized form of Darcy's equation can be applied to each reservoir as follows:

$$q_o = 0.001127\left(\frac{2\pi rh}{\mu_o}\right)k_o\frac{dp}{dr}$$
$$q_w = 0.001127\left(\frac{2\pi rh}{\mu_w}\right)k_w\frac{dp}{dr}$$
$$q_g = 0.001127\left(\frac{2\pi rh}{\mu_g}\right)k_g\frac{dp}{dr}$$

where

k_o, k_w, k_g = effective permeability to oil, water, and gas, md
μ_o, μ_w, μ_g = viscosity of oil, water, and gas, cp
q_o, q_w, q_g = flow rates for oil, water, and gas, bbl/day
k = absolute permeability, md

The effective permeability can be expressed in terms of the relative and absolute permeability as:

$$k_o = k_{ro}k$$
$$k_w = k_{rw}k$$
$$k_g = k_{rg}k$$

Using the above concept in Darcy's equation and expressing the flow rate in standard conditions yields:

$$Q_o = 0.00708(rhk)\left(\frac{k_{ro}}{\mu_o B_o}\right)\frac{dp}{dr} \tag{1.45}$$

$$Q_w = 0.00708(rhk)\left(\frac{k_{rw}}{\mu_w B_w}\right)\frac{dp}{dr} \tag{1.46}$$

$$Q_g = 0.00708(rhk)\left(\frac{k_{rg}}{\mu_g B_g}\right)\frac{dp}{dr} \tag{1.47}$$

where

Q_o, Q_w = oil and water flow rates, STB/day
B_o, B_w = oil and water formation volume factor, bbl/STB
Q_g = gas flow rate, scf/day
B_g = gas formation volume factor, bbl/scf
k = absolute permeability, md

The gas formation volume factor B_g is expressed by:

$$B_g = 0.005035\frac{ZT}{p} \text{ bbl/scf}$$

Performing the regular integration approach on Eqs. (1.45)–(1.47) yields:

- Oil phase:

$$Q_o = \frac{0.00708(kh)(k_{ro})(p_e - p_{wf})}{\mu_o B_o \ln(r_e/r_w)} \tag{1.48}$$

- Water phase:

$$Q_w = \frac{0.00708(kh)(k_{rw})(p_e - p_{wf})}{\mu_w B_w \ln(r_e/r_w)} \tag{1.49}$$

- Gas phase:

In terms of real-gas potential:

$$Q_g = \frac{(kh)k_{rg}(\psi_e - \psi_w)}{1422T \ln(r_e/r_w)} \tag{1.50}$$

In terms of pressure squared:

$$Q_g = \frac{(kh)k_{rg}(p_e^2 - p_{wf}^2)}{1422(\mu_g Z)_{avg} T \ln(r_e/r_w)} \tag{1.51}$$

where

Q_g = gas flow rate, Mscf/day
k = absolute permeability, md
T = temperature, °R

In numerous petroleum engineering calculations, it is convenient to express the flow rate of any phase as a ratio of other flowing phases. Two important flow ratios are the "instantaneous" water–oil ratio (WOR) and the "instantaneous" gas–oil ratio (GOR). The generalized form of Darcy's equation can be used to determine both flow ratios.

The water–oil ratio is defined as the ratio of the water flow rate to that of the oil. Both rates are expressed in stock-tank barrels per day, or:

$$\text{WOR} = \frac{Q_w}{Q_o}$$

Dividing Eq. (1.45) by (1.47) gives:

$$\text{WOR} = \left(\frac{k_{rw}}{k_{ro}}\right)\left(\frac{\mu_o B_o}{\mu_w B_w}\right) \tag{1.52}$$

where

WOR = water–oil ratio, STB/STB

The instantaneous GOR, as expressed in scf/STB, is defined as the *total* gas flow rate, i.e., free gas and solution gas, divided by the oil flow rate, or:

$$\text{GOR} = \frac{Q_o R_s + Q_g}{Q_o}$$

or

$$\text{GOR} = R_s + \frac{Q_g}{Q_o} \quad (1.53)$$

where

GOR = "instantaneous" gas–oil ratio, scf/STB
R_s = gas solubility, scf/STB
Q_g = free gas flow rate, scf/day
Q_o = oil flow rate, STB/day

Substituting Eqs. (1.45) and (1.47) into (1.53) yields:

$$\text{GOR} = R_s + \left(\frac{k_{rg}}{k_{ro}}\right)\left(\frac{\mu_o B_o}{\mu_g B_g}\right) \quad (1.54)$$

where

B_g = gas formation volume factor, bbl/scf

A complete discussion of the practical applications of the WOR and GOR is given in the subsequent chapters.

1.2.3 Unsteady-state Flow

Figure 1.17(a) shows a shut-in well that is centered in a homogeneous circular reservoir of radius r_e with a uniform pressure p_i throughout the reservoir. This initial reservoir condition represents the zero producing time. If the well is allowed to flow at a constant flow rate of q, a pressure disturbance will be created at the sand face. The pressure at the wellbore, i.e., p_{wf}, will drop instantaneously as the well is opened. The pressure disturbance will move away from the wellbore at a rate that is determined by:

- permeability;
- porosity;
- fluid viscosity;
- rock and fluid compressibilities.

Figure 1.17(b) shows that at time t_1, the pressure disturbance has moved a distance r_1 into the reservoir. Note that the pressure disturbance

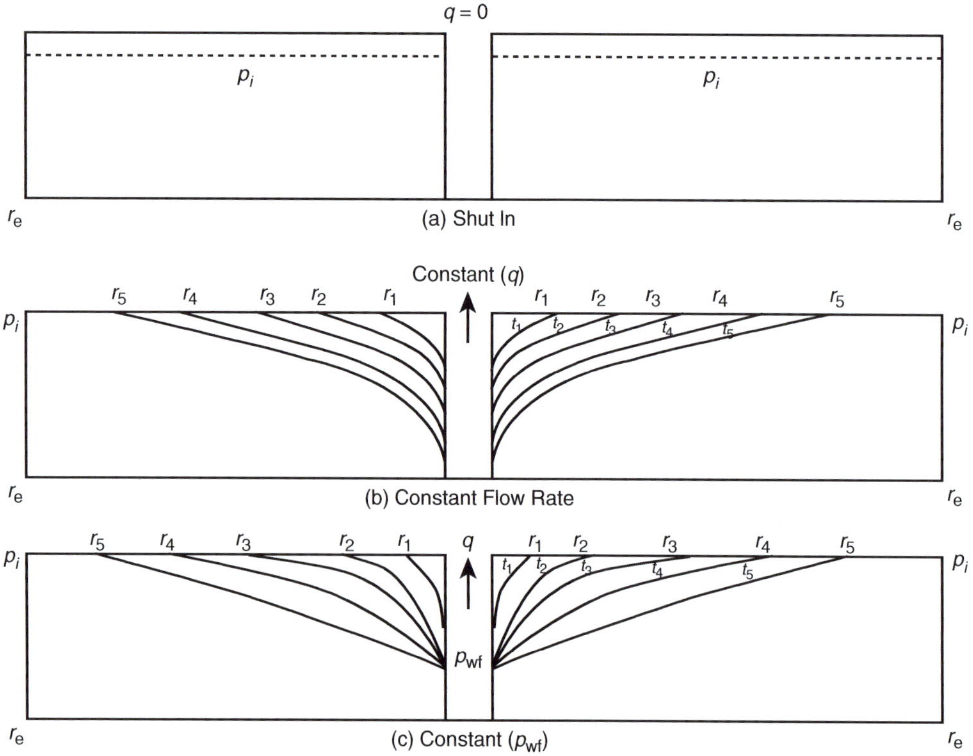

FIGURE 1.17 Pressure disturbance as a function of time.

radius is continuously increasing with time. This radius is commonly called the radius of investigation and referred to as r_{inv}. It is also important to point out that as long as the radius of investigation has not reached the reservoir boundary, i.e., r_e, the reservoir will be acting as if it is infinite in size. During this time we say that the reservoir is *infinite acting* because the outer drainage radius r_e, can be mathematically infinite, i.e., $r_e = \infty$. A similar discussion to the above can be used to describe a well that is producing at a constant bottom-hole flowing pressure. Figure 1.17(c) schematically illustrates the propagation of the radius of investigation with respect to time. At time t_4, the pressure disturbance reaches the boundary, i.e., $r_{inv} = r_e$. This causes the pressure behavior to change.

Based on the above discussion, the transient (unsteady-state) flow is defined as the time period during which the boundary has no effect on the pressure behavior in the reservoir and the reservoir will behave as if it is infinite in size. Figure 1.17(b) shows that the transient flow period occurs during the time interval $0 < t < t_t$ for the constant flow rate scenario and during the time period $0 < t < t_4$ for the constant p_{wf} scenario is depicted by Figure 1.17(c).

1.2.4 Basic Transient Flow Equation

Under the steady-state flowing condition, the same quantity of fluid enters the flow system as leaves it. In the unsteady-state flow condition, the flow rate into an element of volume of a porous medium may not be the same as the flow rate out of that element and, accordingly, the fluid content of the porous medium changes with time. The other controlling variables in unsteady-state flow *additional* to those already used for steady-state flow, therefore, become:

- time t;
- porosity ϕ;
- total compressibility c_t.

The mathematical formulation of the transient flow equation is based on combining three independent equations and a specifying set of boundary and initial conditions that constitute the unsteady-state equation. These equations and boundary conditions are briefly described below:

- *Continuity equation:* The continuity equation is essentially a material balance equation that accounts for every pound mass of fluid produced, injected, or remaining in the reservoir.
- *Transport equation:* The continuity equation is combined with the equation for fluid motion (transport equation) to describe the fluid flow rate "in" and "out" of the reservoir. Basically, the transport equation is Darcy's equation in its generalized differential form.
- *Compressibility equation:* The fluid compressibility equation (expressed in terms of density or volume) is used in formulating the unsteady-state equation with the objective of describing the changes in the fluid volume as a function of pressure.
- *Initial and boundary conditions:* There are two boundary conditions and one initial condition is required to complete the formulation and the analytical solution of the transient flow equation. The two boundary conditions are:
  (1) the formation produces at a constant rate into the wellbore;
  (2) there is no flow across the outer boundary and the reservoir behaves as if it were infinite in size, i.e., $r_e = \infty$.

  The initial condition simply states that the reservoir is at a uniform pressure when production begins, i.e., time = 0.

Consider the flow element shown in Figure 1.18. The element has a width of dr and is located at a distance of r from the center of the well. The porous element has a differential volume of dV. According to the concept of the material balance equation, the rate of mass flow into an element minus the rate of mass flow out of the element during a differential time Δt

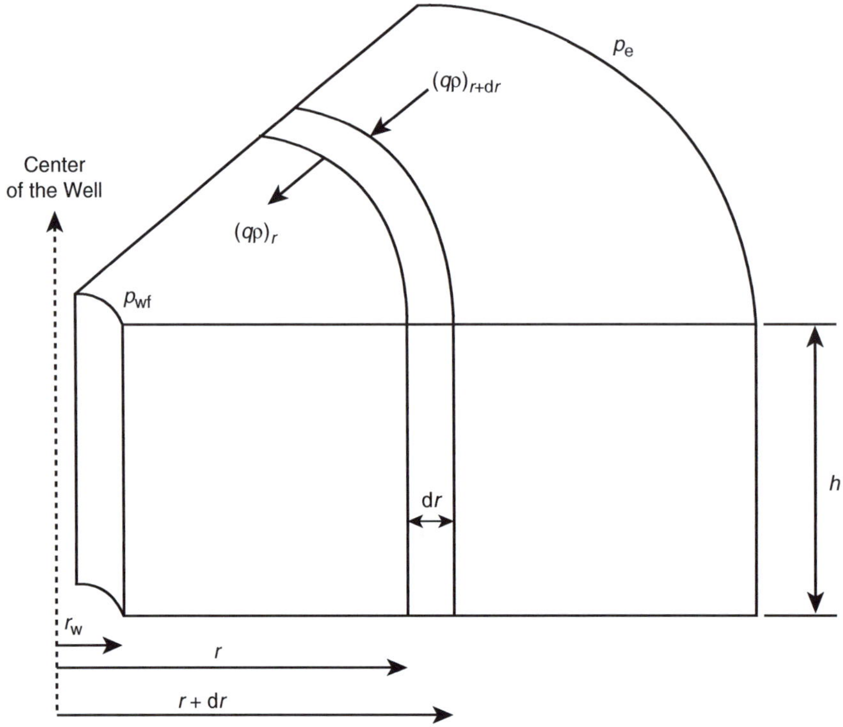

FIGURE 1.18 Illustration of radial flow.

must be equal to the mass rate of accumulation during that time interval, or:

$$\begin{bmatrix}\text{mass entering the}\\ \text{volume element}\\ \text{during interval } \Delta t\end{bmatrix} - \begin{bmatrix}\text{mass leaving the}\\ \text{volume element}\\ \text{during interval } \Delta t\end{bmatrix} = \begin{bmatrix}\text{rate of mass}\\ \text{accumulation}\\ \text{during interval } \Delta t\end{bmatrix} \quad \textbf{(1.55)}$$

The individual terms of Eq. (1.55) are described below:

- *Mass entering the volume element during time interval Δt*:

  Here

$$(\text{Mass})_{\text{in}} = \Delta t[A\nu\rho]_{r+\text{d}r} \quad \textbf{(1.56)}$$

where

ν = velocity of flowing fluid, ft/day

ρ = fluid density at $(r + \text{d}r)$, lb/ft^3

A = area at $(r + \text{d}r)$

Δt = time interval, days

The area of the element at the entering side is:

$$A_{r+\text{d}r} = 2\pi(r + \text{d}r)h \quad \textbf{(1.57)}$$

Combining Eqs. (1.57) with (1.46) gives:

$$[\text{Mass}]_{\text{in}} = 2\pi\,\Delta t(r + \text{d}r)h(\nu\rho)_{r+\text{d}r} \quad \textbf{(1.58)}$$

- *Mass leaving the volume element*: Adopting the same approach as that of the leaving mass gives:

$$[\text{Mass}]_{\text{out}} = 2\pi\,\Delta t r h(\nu\rho)_r \quad \textbf{(1.59)}$$

- *Total accumulation of mass*: The volume of the elements with radius r is given by:

$$V = \pi r^2 h$$

Differentiating the above equation with respect to r gives:

$$\frac{dV}{dr} = 2\pi rh$$

or

$$dV = (2\pi rh)dr \tag{1.60}$$

Total mass accumulation during $\Delta t = dV[(\phi\rho)_{t+\Delta t} - (\phi\rho)_t]$. Substituting for dV yields:

$$\text{Total mass accumulation} = (2\pi rh)\, dr[(\phi\rho)_{t+\Delta t} - (\phi\rho)_t] \tag{1.61}$$

Replacing the terms of Eq. (1.55) with those of the calculated relationships gives:

$$2\pi h(r + dr)\Delta t(\phi\rho)_{r+dr} - 2\pi hr\, \Delta t(\phi\rho)_r = (2\pi rh)\, dr[(\phi\rho)_{t+\Delta t} - (\phi\rho)_t]$$

Dividing the above equation by $(2\pi rh)dr$ and simplifying gives:

$$\frac{1}{(r)dr}\left[(r + dr)(v\rho)_{r+dr} - r(v\rho)_r\right] = \frac{1}{\Delta t}\left[(\phi\rho)_{t+\Delta t} - (\phi\rho)_t\right]$$

or

$$\frac{1}{r}\frac{\partial}{\partial r}[r(v\rho)] = \frac{\partial}{\partial t}(\phi\rho) \tag{1.62}$$

where

- ϕ = porosity, %
- ρ = density, lb/ft^3
- V = fluid velocity, ft/day

Eq. (1.62) is called the continuity equation and it provides the principle of conservation of mass in radial coordinates.

The transport equation must be introduced into the continuity equation to relate the fluid velocity to the pressure gradient within the control volume dV. Darcy's law is essentially the basic motion equation, which states that the velocity is proportional to the pressure gradient $\partial p/\partial r$. From Eq. (1.24):

$$v = (5.615)\,(0.001127)\frac{k}{\mu}\frac{\partial p}{\partial r} = (0.006328)\frac{k}{\mu}\frac{\partial p}{\partial r} \tag{1.63}$$

where

k = permeability, md
v = velocity, ft/day

Combining Eq. (1.63) with (1.62) results in:

$$\frac{0.006328}{r}\frac{\partial}{\partial r}\left(\frac{k}{\mu}(\rho r)\frac{\partial p}{\partial r}\right) = \frac{\partial}{\partial t}(\phi\rho) \tag{1.64}$$

Expanding the right-hand side by taking the indicated derivatives eliminates the porosity from the partial derivative term on the right-hand side:

$$\frac{\partial}{\partial t}(\phi\rho) = \phi\frac{\partial\rho}{\partial t} + \rho\frac{\partial\phi}{\partial t} \tag{1.65}$$

The porosity is related to the formation compressibility by the following:

$$c_f = \frac{1}{\phi}\frac{\partial\phi}{\partial p} \tag{1.66}$$

Applying the chain rule of differentiation to $\partial\phi/\partial t$:

$$\frac{\partial\phi}{\partial t} = \frac{\partial\phi}{\partial p}\frac{\partial p}{\partial t}$$

Substituting Eq. (1.66) into this equation:

$$\frac{\partial\phi}{\partial t} = \phi c_f\frac{\partial p}{\partial t}$$

Finally, substituting the above relation into Eq. (1.65) and the result into Eq. (1.64) gives:

$$\frac{0.006328}{r}\frac{\partial}{\partial r}\left(\frac{k}{\mu}(\rho r)\frac{\partial p}{\partial r}\right) = \rho\phi c_f\frac{\partial p}{\partial t} + \phi\frac{\partial p}{\partial t} \tag{1.67}$$

Eq. (1.67) is the general partial differential equation used to describe the flow of any fluid flowing in a radial direction in porous media. In addition to the initial assumptions, Darcy's equation has been added, which implies that the flow is laminar. Otherwise, the equation is not restricted to any type of fluid and is equally valid for gases or liquids. However, compressible and slightly compressible fluids must be treated separately in order to develop practical equations that can be used to describe the flow behavior of these two fluids. The treatments of the following systems are discussed below:

- radial flow of slightly compressible fluids;
- radial flow of compressible fluids.

1.2.5 Radial flow of Slightly Compressibility Fluids

To simplify Eq. (1.67), assume that the permeability and viscosity are constant over pressure, time, and distance ranges. This leads to:

$$\left[\frac{0.006328k}{\mu r}\right]\frac{\partial}{\partial r}\left(r\rho\frac{\partial p}{\partial r}\right)=\rho\phi c_f\frac{\partial p}{\partial t}+\phi\frac{\partial \rho}{\partial t} \tag{1.68}$$

Expanding the above equation gives:

$$0.006328\left(\frac{k}{\mu}\right)\left[\frac{\rho}{r}\frac{\partial p}{\partial r}+\rho\frac{\partial^2 p}{\partial r^2}+\frac{\partial p}{\partial r}\frac{\partial \rho}{\partial r}\right]=\rho\phi c_f\left(\frac{\partial p}{\partial t}\right)+\phi\left(\frac{\partial \rho}{\partial t}\right)$$

Using the chain rule in the above relationship yields:

$$0.006328\left(\frac{k}{\mu}\right)\left[\frac{\rho}{r}\frac{\partial p}{\partial r}+\rho\frac{\partial^2 p}{\partial r^2}+\left(\frac{\partial p}{\partial r}\right)^2\frac{\partial \rho}{\partial p}\right]=\rho\phi c_f\left(\frac{\partial p}{\partial t}\right)+\phi\left(\frac{\partial p}{\partial t}\right)\left(\frac{\partial \rho}{\partial p}\right)$$

Dividing the above expression by the fluid density ρ gives:

$$0.006328\left(\frac{k}{u}\right)\left[\frac{1}{r}\frac{\partial p}{\partial r}+\frac{\partial^2 p}{\partial r^2}+\left(\frac{\partial p}{\partial r}\right)^2\left(\frac{1}{\rho}\frac{\partial \rho}{\partial p}\right)\right]=\phi c_f\left(\frac{\partial p}{\partial t}\right)+\phi\frac{\partial p}{\partial t}\left(\frac{1}{\rho}\frac{\partial \rho}{\partial p}\right)$$

Recalling that the compressibility of any fluid is related to its density by:

$$c=\frac{1}{\rho}\frac{\partial \rho}{\partial p}$$

Combining the above two equations gives:

$$0.006328\left(\frac{k}{\mu}\right)\left[\frac{\partial^2 p}{\partial r^2}+\frac{1}{r}\frac{\partial p}{\partial r}+c\left(\frac{\partial p}{\partial r}\right)^2\right]=\phi c_f\left(\frac{\partial p}{\partial t}\right)+\phi c\left(\frac{\partial p}{\partial t}\right)$$

The term $c(\partial p/\partial r)^2$ is considered very small and may be ignored, which leads to:

$$0.006328\left(\frac{k}{\mu}\right)\left[\frac{\partial^2 p}{\partial r^2}+\frac{1}{r}\frac{\partial p}{\partial r}\right]=\phi(c_f+c)\frac{\partial p}{\partial t} \tag{1.69}$$

Defining total compressibility c_t as:

$$c_t=c+c_f \tag{1.70}$$

Combining Eq. (1.68) with (1.69) and rearranging gives:

$$\frac{\partial^2 p}{\partial r^2}+\frac{1}{r}\frac{\partial p}{\partial r}=\frac{\phi\mu c_t}{0.006328k}\frac{\partial p}{\partial t} \tag{1.71}$$

where time t is expressed in days.

Eq. (1.71) is called the diffusivity equation and is considered one of the most important and widely used mathematical expressions in petroleum engineering. The equation is particularly used in the analysis of well testing data where time t is commonly recorded in hours. The equation can be rewritten as:

$$\frac{\partial^2 p}{\partial r^2}+\frac{1}{r}\frac{\partial p}{\partial r}=\frac{\phi\mu c_t}{0.0002637k}\frac{\partial p}{\partial t} \tag{1.72}$$

where

k = permeability, md
r = radial position, ft
p = pressure, psia
c_t = total compressibility, psi^{-1}
t = time, hours
ϕ = porosity, fraction
μ = viscosity, cp

When the reservoir contains more than one fluid, total compressibility should be computed as:

$$c_t=c_oS_o+c_wS_w+c_gS_g+c_f \tag{1.73}$$

where

c_o, c_w, c_g = compressibility of oil, water, and gas, respectively
S_o, S_w, S_g = fractional saturation of oil, water, and gas, respectively

Note that the introduction of c_t into Eq. (1.71) does not make this equation applicable to multiphase flow; the use of c_t, as defined by Eq. (1.72), simply accounts for the compressibility of any immobile fluids which may be in the reservoir with the fluid that is flowing.

The term $0.000264k/\phi\mu c_t$ is called the diffusivity constant and is denoted by symbol η, or also given as:

$$\eta = \frac{0.0002637k}{\phi\mu c_t} \tag{1.74}$$

The diffusivity equation can then be written in a more convenient form as:

$$\frac{\partial^2 p}{\partial r^2} + \frac{1}{r}\frac{\partial p}{\partial r} = \frac{1}{\eta}\frac{\partial p}{\partial t} \tag{1.75}$$

The diffusivity equation as represented by relationship (1.75) is essentially designed to determine the pressure as a function of time t and position r.

Notice that for a steady-state flow condition, the pressure at any point in the reservoir is constant and does not change with time, i.e., $\partial p/\partial t = 0$, so Eq. (1.75) reduces to:

$$\frac{\partial^2 p}{\partial r^2} + \frac{1}{r}\frac{\partial p}{\partial r} = 0 \tag{1.76}$$

Eq. (1.76) is called Laplace's equation for steady-state flow.

Example 1.9

Show that the radial form of Darcy's equation is the solution to Eq. (1.76).

Solution

Step 1. Start with Darcy's law as expressed by Eq. (1.28):

$$p = p_{wf} + \left[\frac{Q_o B_o u_o}{0.00708kh}\right] \ln\left(\frac{r}{r_w}\right)$$

Step 2. For a steady-state incompressible flow, the term with the square brackets is constant and labeled as C, or:

$$p = p_{wf} + [C]\ln\left(\frac{r}{r_w}\right)$$

Step 3. Evaluate the above expression for the first and second derivative, to give:

$$\frac{\partial p}{\partial r} = [C]\left(\frac{1}{r}\right)$$

$$\frac{\partial^2 p}{\partial r_2} = [C]\left(\frac{-1}{r^2}\right)$$

Step 4. Substitute the above two derivatives in Eq. (1.76):

$$\frac{-1}{r^2}[C] + \left(\frac{1}{r}\right)[C]\left(\frac{1}{r}\right) = 0$$

Step 5. Results of step 4 indicate that Darcy's equation satisfies Eq. (1.76) and is indeed the solution to Laplace's equation.

To obtain a solution to the diffusivity equation (Eq. (1.75), it is necessary to specify an initial condition and impose two boundary conditions. The initial condition simply states that the reservoir is at a uniform pressure p_i when production begins. The two boundary conditions require that the well is producing at a constant production rate and the reservoir behaves as if it were infinite in size, i.e., $r_e = \infty$.

Based on the boundary conditions imposed on Eq. (1.75), there are two generalized solutions to the diffusivity equation. These are:

(1) the constant-terminal-pressure solution;
(2) the constant-terminal-rate solution.

The constant-terminal-pressure solution is designed to provide the cumulative flow at any particular time for a reservoir in which the pressure at one boundary of the reservoir is held constant. This technique is frequently used in water influx calculations in gas and oil reservoirs.

The constant-terminal-rate solution of the radial diffusivity equation solves for the pressure change throughout the radial system providing that the flow rate is held constant at one terminal end of the radial system, i.e., at the producing well. There are two commonly used forms of the constant-terminal-rate solution:

(1) the Ei function solution;
(2) the dimensionless pressure drop p_D solution.

Constant-terminal-pressure Solution. In the constant-rate solution to the radial diffusivity equation, the flow rate is considered to be constant at certain radius (usually wellbore radius) and the pressure profile around that radius is determined as a function of time and position. In the constant-terminal-pressure solution, the pressure is known to be constant at some particular radius and the solution is designed to provide the cumulative fluid movement across the specified radius (boundary).

The constant-pressure solution is widely used in water influx calculations. A detailed description of the solution and its practical reservoir engineering applications is appropriately discussed in Chapter 5.

Constant-terminal-rate Solution. The constant-terminal-rate solution is an integral part of most transient test analysis techniques, e.g., drawdown and pressure buildup analyses. Most of these tests involve producing the well at a constant flow rate and recording the flowing pressure as a function of time, i.e., $p(r_w, t)$. There are two commonly used forms of the constant-terminal-rate solution:

(1) the Ei function solution;
(2) the dimensionless pressure drop p_D solution.

These two popular forms of solution to the diffusivity equation are discussed below.

The Ei Function Solution. For an infinite-acting reservoir, Matthews and Russell (1967) proposed the following solution to the diffusivity equation, i.e., Eq. (1.66):

$$p(r,t) = p_i + \left[\frac{70.6 Q_o \mu B_o}{kh}\right] \text{Ei}\left[\frac{-948\phi\mu c_t r^2}{kt}\right] \quad (1.77)$$

where

$p(r, t)$ = pressure at radius r from the well after t hours
t = time, hours
k = permeability, md
Q_o = flow rate, STB/day

The mathematical function, Ei, is called the exponential integral and is defined by:

$$\text{Ei}(-x) = -\int_x^\infty \frac{e^{-u}\,du}{u} = \left[\ln x - \frac{x}{1!} + \frac{x^2}{2(2!)} - \frac{x^3}{3(3!)} + \cdots\right] \quad (1.78)$$

Craft et al. (1991) presented the values of the Ei function in tabulated and graphical forms as shown in Table 1.1 and Figure 1.19, respectively.

The Ei solution, as expressed by Eq. (1.77), is commonly referred to as the line source solution. The exponential integral "Ei" can be approximated by the following equation when its argument x is less than 0.01:

$$\text{Ei}(-x) = \ln(1.781x) \quad (1.79)$$

where the argument x in this case is given by:

$$x = \frac{948\phi\mu c_t r^2}{kt}$$

Eq. (1.79) approximates the Ei function with less than 0.25% error. Another expression that can be used to approximate the Ei function for the range of $0.01 < x < 3.0$ is given by:

$$\text{Ei}(-x) = a_1 + a_2\ln(x) + a_3[\ln(x)]^2 + a_4[\ln(x)]^3 + a_5 x + a_6 x^2 + a_7 x^3 + \frac{a_8}{x} \quad (1.80)$$

with the coefficients a_1–a_8 having the following values:

$a_1 = -0.33153973$ $a_2 = -0.81512322$
$a_3 = 5.22123384 \times 10^{-2}$ $a_4 = 5.9849819 \times 10^{-3}$
$a_5 = 0.662318450$ $a_6 = -0.12333524$
$a_7 = 1.0832566 \times 10^{-2}$ $a_8 = 8.6709776 \times 10^{-4}$

The above relationship approximated the Ei values with an average error of 0.5%.

It should be pointed out that for $x > 10.9$, Ei$(-x)$ can be considered zero for reservoir engineering calculations.

TABLE 1.1 Values of −Ei(−x) as a Function of x

x	−Ei(−x)	x	−Ei(−x)	x	−Ei(−x)
0.1	1.82292	3.5	0.00697	6.9	0.00013
0.2	1.22265	3.6	0.00616	7.0	0.00012
0.3	0.90568	3.7	0.00545	7.1	0.00010
0.4	0.70238	3.8	0.00482	7.2	0.00009
0.5	0.55977	3.9	0.00427	7.3	0.00008
0.6	0.45438	4.0	0.00378	7.4	0.00007
0.7	0.37377	4.1	0.00335	7.5	0.00007
0.8	0.31060	4.2	0.00297	7.6	0.00006
0.9	0.26018	4.3	0.00263	7.7	0.00005
1.0	0.21938	4.4	0.00234	7.8	0.00005
1.1	0.18599	4.5	0.00207	7.9	0.00004
1.2	0.15841	4.6	0.00184	8.0	0.00004
1.3	0.13545	4.7	0.00164	8.1	0.00003
1.4	0.11622	4.8	0.00145	8.2	0.00003
1.5	0.10002	4.9	0.00129	8.3	0.00003
1.6	0.08631	5.0	0.00115	8.4	0.00002
1.7	0.07465	5.1	0.00102	8.5	0.00002
1.8	0.06471	5.2	0.00091	8.6	0.00002
1.9	0.05620	5.3	0.00081	8.7	0.00002
2.0	0.04890	5.4	0.00072	8.8	0.00002
2.1	0.04261	5.5	0.00064	8.9	0.00001
2.2	0.03719	5.6	0.00057	9.0	0.00001
2.3	0.03250	5.7	0.00051	9.1	0.00001
2.4	0.02844	5.8	0.00045	9.2	0.00001
2.5	0.02491	5.9	0.00040	9.3	0.00001
2.6	0.02185	6.0	0.00036	9.4	0.00001
2.7	0.01918	6.1	0.00032	9.5	0.00001
2.8	0.01686	6.2	0.00029	9.6	0.00001
2.9	0.01482	6.3	0.00026	9.7	0.00001
3.0	0.01305	6.4	0.00023	9.8	0.00001
3.1	0.01149	6.5	0.00020	9.9	0.00000
3.2	0.01013	6.6	0.00018	10.0	0.00000
3.3	0.00894	6.7	0.00016		
3.4	0.00789	6.8	0.00014		

After Craft, B.C., Hawkins, M. (Revised by Terry, R.E.), 1991. Applied Petroleum Reservoir Engineering, second ed. Prentice Hall, Englewood Cliffs, NJ.

Example 1.10

An oil well is producing at a constant flow rate of 300 STB/day under unsteady-state flow conditions. The reservoir has the following rock and fluid properties:

$B_o = 1.25$ bbl/STB, $\mu_o = 1.5$ cp, $c_t = 12 \times 10^{-6}$ psi^{-1}
$k_o = 60$ md, $h = 15$ ft, $p_i = 4000$ psi
$\phi = 15\%$, $r_w = 0.25$ ft

(1) Calculate the pressure at radii of 0.25, 5, 10, 50, 100, 500, 1000, 1500, 2000, and 2500 ft, for 1 hour. Plot the results as:
 (a) pressure vs. the logarithm of radius;
 (b) pressure vs. radius.
(2) Repeat part 1 for $t = 12$ and 24 hours. Plot the results as pressure vs. logarithm of radius.

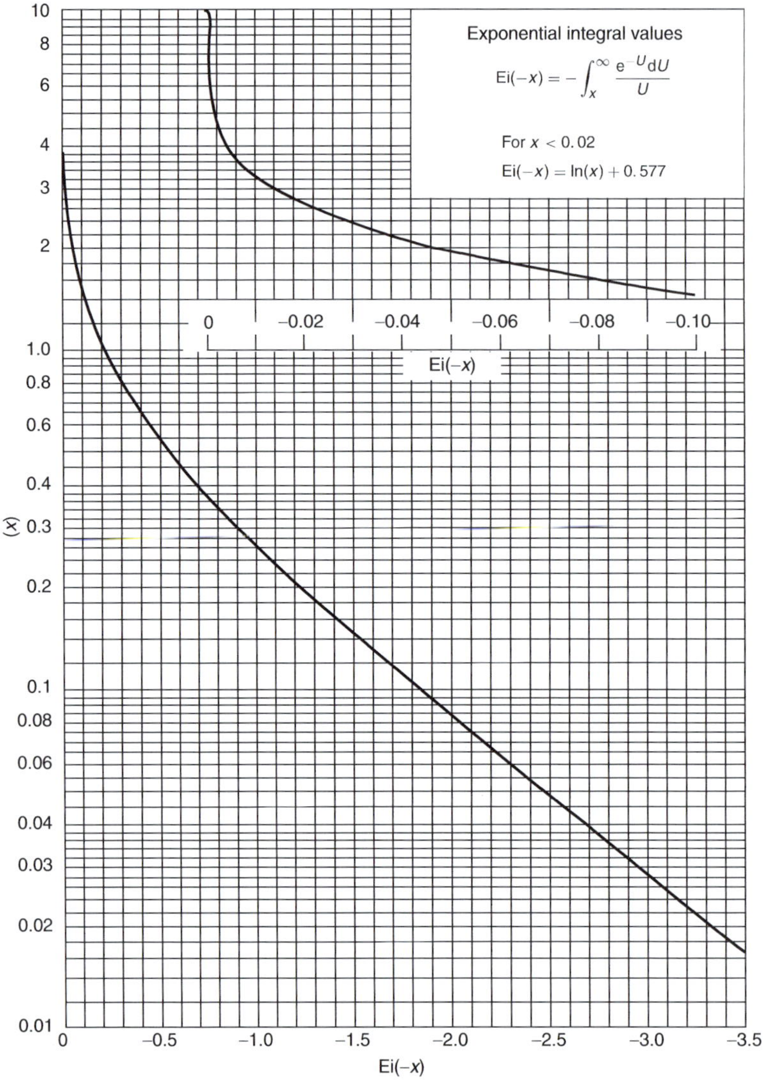

FIGURE 1.19 Ei function. *(After Craft, B.C., Hawkins, M. (Revised by Terry, R.E.), 1991. Applied Petroleum Reservoir Engineering, second ed. Prentice Hall, Englewood Cliffs, NJ).*

Solution

Step 1. From Eq. (1.77):

$$p(r,t) = 4000 + \left[\frac{70.6(300)(1.5)(1.25)}{(60)(15)}\right] \times \mathrm{Ei}\left[\frac{-948(1.5)(1.5)(12\times 10^{6})r^2}{(60)(t)}\right] = 4000 + 44.125\mathrm{Ei}\left[(-42.6\times 10^{-6})\frac{r^2}{t}\right]$$

Step 2. Perform the required calculations after 1 hour in the following tabulated form:

r (ft)	$x = (-42.6\times 10^{-6})\ r^2/1$	Ei(−x)	p(r, 12) = 4000 + 44.125 Ei(−x)
0.25	-2.6625×10^{-6}	−12.26[a]	3459
5	−0.001065	−6.27[a]	3723
10	−0.00426	−4.88[a]	3785

50	−0.1065	−1.76[b]	3922
100	−0.4260	−0.75[b]	3967
500	−10.65	0	4000
1000	−42.60	0	4000
1500	−95.85	0	4000
2000	−175.40	0	4000
2500	−266.25	0	4000

[a]As calculated from Eq. (1.29).
[b]From Figure 1.19.

Step 3. Show the results of the calculation graphically as illustrated in Figures 1.20 and 1.21.

Step 4. Repeat the calculation for $t = 12$ and 24 hours, as in the tables below:

r (ft)	$x = (42.6 \times 10^{-6})\, r^2/12$	Ei($-x$)	$p(r, 12) = 4000 + 44.125\,\text{Ei}(-x)$
0.25	0.222×10^{-6}	−14.74[a]	3350
5	88.75×10^{-6}	−8.75[a]	3614
10	355.0×10^{-6}	−7.37[a]	3675
50	0.0089	−4.14[a]	3817
100	0.0355	−2.81[b]	3876
500	0.888	−0.269	3988

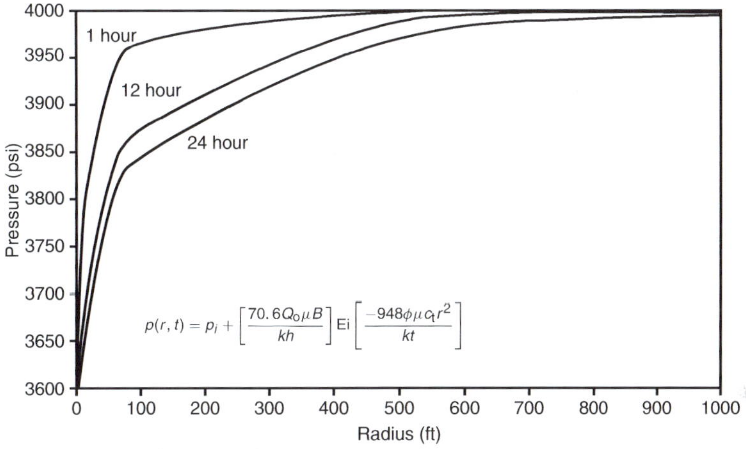

FIGURE 1.20 Pressure profiles as a function of time.

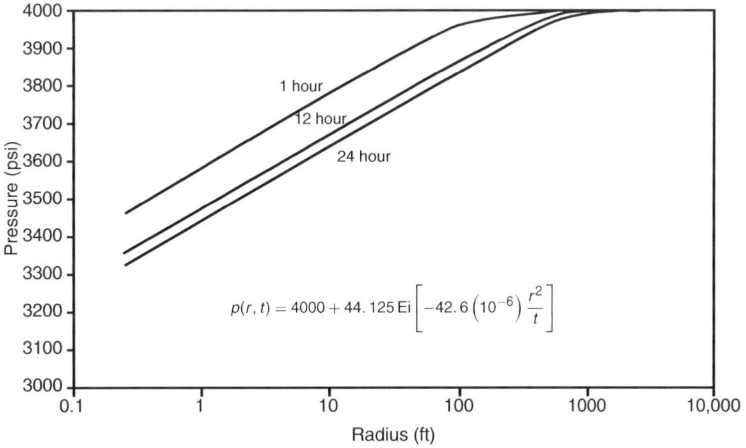

FIGURE 1.21 Pressure profiles as a function of time on a semi-log scale.

1000	3.55	−0.0069	4000
1500	7.99	-3.77×10^{-5}	4000
2000	14.62	0	4000
2500	208.3	0	4000

[a]As calculated from Eq. (1.29).
[b]From Figure 1.19.

r (ft)	$x = (-42.6\times10^{-6})\ r^2/24$	Ei(−x)	p(r, 24) = 4000 + 44.125 Ei(−x)
0.25	-0.111×10^{-6}	−15.44[a]	3319
5	-44.38×10^{-6}	−9.45[a]	3583
10	-177.5×10^{-6}	−8.06[a]	3644
50	−0.0045	−4.83[a]	3787
100	−0.0178	−8.458[b]	3847
500	−0.444	−0.640	3972
1000	−1.775	−0.067	3997
1500	−3.995	−0.0427	3998
2000	−7.310	8.24×10^{-6}	4000
2500	−104.15	0	4000

[a]As calculated from Eq. (1.29).
[b]From Figure 1.19.

Step 5. Results of step 4 are shown graphically in Figure 1.21.

Figure 1.21 indicates that as the pressure disturbance moves radially away from the wellbore, the reservoir boundary and its configuration has no effect on the pressure behavior, which leads to the definition of transient flow as: "Transient flow is that time period during which the boundary has no effect on the pressure behavior and the well acts as if it exists in an infinite size reservoir."

Example 1.10 shows that most of the pressure loss occurs close to the wellbore; accordingly, near-wellbore conditions will exert the greatest influence on flow behavior. Figure 1.21 shows that the pressure profile and the drainage radius are continuously changing with time. It is also important to notice that the production rate of the well has no effect on the velocity or the distance of the pressure disturbance, since the Ei function is independent of the flow rate.

When the Ei parameter $x<0.01$, the log approximation of the Ei function as expressed by Eq. (1.79) can be used in (1.77) to give:

$$p(r,t)=p_i-\frac{162.6Q_oB_o\mu_o}{kh}\left[\log\left(\frac{kt}{\phi\mu c_t r^2}\right)-3.23\right] \tag{1.81}$$

For most of the transient flow calculations, engineers are primarily concerned with the behavior of the bottom-hole flowing pressure at the wellbore, i.e., $r=r_w$. Eq. (1.81) can be applied at $r=r_w$ to yield:

$$p_{wf}=p_i-\frac{162.6Q_oB_o\mu_o}{kh}\left[\log\left(\frac{kt}{\phi\mu c_t r_w^2}\right)-3.23\right] \tag{1.82}$$

where

k = permeability, md
t = time, hours
c_t = total compressibility, psi^{-1}

It should be noted that Eqs. (1.81) and (1.82) cannot be used until the flow time t exceeds the limit imposed by the following constraint:

$$t>9.48\times10^4\frac{\phi\mu c_t r^2}{k} \tag{1.83}$$

where

k = permeability, md
t = time, hours

Notice that when a well is producing under unsteady-state (transient) flowing conditions at a constant flow rate, Eq. (1.82) can be expressed as the equation of a straight line by manipulating the equation to give:

$$p_{wf}=p_i-\frac{162.6Q_oB_o\mu_o}{kh}\left[\log(t)+\log\left(\frac{k}{\phi\mu c_t r_w^2}\right)-3.23\right]$$

or

$$p_{wf}=a+m\log(t)$$

The above equation indicates that a plot of p_{wf} vs. t on a semi-logarithmic scale would produce a straight line with an intercept of a and a slope of m as given by:

$$a = p_i - \frac{162.6Q_oB_o\mu_o}{kh}\left[\log\left(\frac{k}{\phi\mu c_t r_w^2}\right) - 3.23\right]$$

$$m = \frac{162.6Q_oB_o\mu_o}{kh}$$

Example 1.11

Using the data in Example 1.10, estimate the bottom-hole flowing pressure after 10 hours of production.

Solution

Step 1. Eq. (1.82) can only be used to calculate p_{wf} at any time that exceeds the time limit imposed by Eq. (1.83), or:

$$t > 9.48 \times 10^4 \frac{\phi\mu c_t r^2}{k}$$

$$t = 9.48 \times 10^4 \frac{(0.15)(1.5)(12 \times 10^{-6})(0.25)^2}{60}$$

$$= 0.000267 \text{ hours}$$

$$= 0.153 \text{ seconds}$$

For all practical purposes, Eq. (1.82) can be used anytime during the transient flow period to estimate the bottom-hole pressure.

Step 2. Since the specified time of 10 hours is greater than 0.000267 hours, the value of p_{wf} can be estimated by applying Eq. (1.82):

$$p_{wf} = p_i - \frac{162.6Q_oB_o\mu_o}{kh}\left[\log\left(\frac{kt}{\phi\mu c_t r_w^2}\right) - 3.23\right]$$

$$= 4000 - \frac{162.6(300)(1.25)(1.5)}{(60)(15)}$$

$$\times \left[\log\left(\frac{(60)(10)}{(0.15)(1.5)(12 \times 10^{-6})(0.25)^2}\right) - 3.23\right]$$

$$= 3358 \text{ psi}$$

The second form of solution to the diffusivity equation is called the dimensionless pressure drop solution and is discussed below.

The Dimensionless Pressure Drop p_D Solution. To introduce the concept of the dimensionless pressure drop solution, consider for example Darcy's equation in a radial form as given previously by Eq. (1.26):

$$Q_o = \frac{0.00708kh(p_e - p_{wf})}{\mu_o B_o \ln(r_e/r_w)} = \frac{kh(p_e - p_{wf})}{141.2\mu_o B_o \ln(r_e/r_w)}$$

Rearranging the above equation gives:

$$\frac{p_e - p_{wf}}{(141.2Q_oB_o\mu_o/kh)} = \ln\left(\frac{r_e}{r_w}\right) \quad \textbf{(1.84)}$$

It is obvious that the right-hand side of the above equation has no units (i.e., it is dimensionless) and, accordingly, the left-hand side must be dimensionless. Since the left-hand side is dimensionless, and $p_e - p_{wf}$ has the units of psi, it follows that the term $Q_oB_o\mu_o/0.00708kh$ has units of pressure. In fact, any pressure difference divided by $Q_oB_o\mu_o/0.00708kh$ is a dimensionless pressure. Therefore, Eq. (1.84) can be written in a dimensionless form as:

$$p_D = \ln(r_{eD})$$

where

$$p_D = \frac{p_e - p_{wf}}{(141.2Q_oB_o\mu_o/kh)} \quad r_{eD} = \frac{r_e}{r_w}$$

The dimensionless pressure drop concept can be extended to describe the changes in the pressure during the unsteady-state flow condition where the pressure is a function of time and radius:

$$p = p(r, t)$$

Therefore, the dimensionless pressure during the unsteady-state flowing condition is defined by:

$$p_D = \frac{p_i - p(r, t)}{(141.2Q_oB_o\mu_o/kh)} \quad \textbf{(1.85)}$$

Since the pressure $p(r, t)$, as expressed in a dimensionless form, varies with time and location, it is traditionally presented as a function of dimensionless time t_D and radius r_D as defined below:

$$t_D = \frac{0.0002637kt}{\phi \mu c_t r_w^2} \quad \textbf{(1.86a)}$$

Another common form of the dimensionless time t_D is based on the total drainage area A as given by:

$$t_{DA} = \frac{0.0002637kt}{\varphi \mu c_t A} = t_D \left(\frac{r_w^2}{A}\right) \quad \textbf{(1.86b)}$$

$$r_D = \frac{r}{r_w} \quad \textbf{(1.87)}$$

and

$$r_{eD} = \frac{r_e}{r_w} \quad \textbf{(1.88)}$$

where

p_D = dimensionless pressure drop
r_{eD} = dimensionless external radius
t_D = dimensionless time based on wellbore radius r_w
t_{DA} = dimensionless time based on well drainage area A
A = well drainage area, i.e., πr_e^2, ft^2
r_D = dimensionless radius
t = time, hours
$p(r, t)$ = pressure at radius r and time t
k = permeability, md
μ = viscosity, cp

The above dimensionless groups (i.e., p_D, t_D, and r_D) can be introduced into the diffusivity equation (Eq. 1.75) to transform the equation into the following dimensionless form:

$$\frac{\partial^2 p_D}{\partial r_D^2} + \frac{1}{r_D}\frac{\partial p_D}{\partial r_D} = \frac{\partial p_D}{\partial t_D} \quad \textbf{(1.89)}$$

van Everdingen and Hurst (1949) proposed an analytical solution to the above equation by assuming:

- a perfectly radial reservoir system;
- the producing well is in the center and producing at a constant production rate of Q;
- uniform pressure p_i throughout the reservoir before production;
- no flow across the external radius r_e.

van Everdingen and Hurst presented the solution to Eq. (1.88) in the form of an infinite series of exponential terms and Bessel functions. The authors evaluated this series for several values of r_{eD} over a wide range of values for t_D and presented the solution in terms of dimensionless pressure drop p_D as a function of dimensionless radius r_{eD} and dimensionless time t_D. Chatas (1953) and Lee (1982) conveniently tabulated these solutions for the following two cases:

(1) infinite-acting reservoir $r_{eD} = \infty$;
(2) finite-radial reservoir.

Infinite-acting Reservoir. For an infinite-acting reservoir, i.e., $r_{eD} = \infty$, the solution to Eq. (1.89) in terms of the dimensionless pressure drop p_D is strictly a function of the dimensionless time t_D, or:

$$p_D = f(t_D)$$

Chatas and Lee tabulated the p_D values for the infinite-acting reservoir as shown in Table 1.2. The following mathematical expressions can be used to approximate these tabulated values of p_D.

- For $t_D < 0.01$:

$$p_D = 2\sqrt{\frac{t_D}{\pi}} \quad \textbf{(1.90)}$$

- For $t_D > 100$:

$$p_D = 0.5[\ln(t_D) + 0.80907] \quad \textbf{(1.91)}$$

- For $0.02 < t_D \leq 1000$:

$$p_D = a_1 + a_2 \ln(t_D) + a_3[\ln(t_D)]^2 + a_4\left[\ln(t_D)^3\right] + a_5 t_D + a_6(t_D)^2 + a_7(t_D)^3 + \frac{a_8}{t_D} \quad \textbf{(1.92)}$$

where the values of the coefficients of the above equations are:

$a_1 = 0.8085064$ $\quad a_2 = 0.29302022$
$a_3 = 3.5264177 \times 10^{-2}$ $\quad a_4 = -1.4036304 \times 10^{-3}$
$a_5 = -4.7722225 \times 10^{-4}$ $\quad a_6 = 5.1240532 \times 10^{-7}$
$a_7 = -2.3033017 \times 10^{-10}$ $\quad a_8 = -2.6723117 \times 10^{-3}$

TABLE 1.2 p_D vs. t_D – Infinite Radial System, Constant Rate at the Inner Boundary

t_D	p_D	t_D	p_D	t_D	p_D
0	0	0.15	0.3750	60.0	2.4758
0.0005	0.0250	0.2	0.4241	70.0	2.5501
0.001	0.0352	0.3	0.5024	80.0	2.6147
0.002	0.0495	0.4	0.5645	90.0	2.6718
0.003	0.0603	0.5	0.6167	100.0	2.7233
0.004	0.0694	0.6	0.6622	150.0	2.9212
0.005	0.0774	0.7	0.7024	200.0	3.0636
0.006	0.0845	0.8	0.7387	250.0	3.1726
0.007	0.0911	0.9	0.7716	300.0	3.2630
0.008	0.0971	1.0	0.8019	350.0	3.3394
0.009	0.1028	1.2	0.8672	400.0	3.4057
0.01	0.1081	1.4	0.9160	450.0	3.4641
0.015	0.1312	2.0	1.0195	500.0	3.5164
0.02	0.1503	3.0	1.1665	550.0	3.5643
0.025	0.1669	4.0	1.2750	600.0	3.6076
0.03	0.1818	5.0	1.3625	650.0	3.6476
0.04	0.2077	6.0	1.4362	700.0	3.6842
0.05	0.2301	7.0	1.4997	750.0	3.7184
0.06	0.2500	8.0	1.5557	800.0	3.7505
0.07	0.2680	9.0	1.6057	850.0	3.7805
0.08	0.2845	10.0	1.6509	900.0	3.8088
0.09	0.2999	15.0	1.8294	950.0	3.8355
0.1	0.3144	20.0	1.9601	1000.0	3.8584
		30.0	2.1470		
		40.0	2.2824		
		50.0	2.3884		

Notes: For $t_D < 0.01$: $p_D \cong 2 z t_D / x$.
For $100 < t_D < 0.25 r_e^2 D$: $p_D \cong 0.5(\ln t_D + 0.80907)$.
After Lee, J., 1982. Well Testing, SPE Textbook Series, Permission to Publish by the SPE, Copyright SPE, 1982.

Finite Radial Reservoir. For a finite radial system, the solution to Eq. (1.89) is a function of both the dimensionless time t_D and dimensionless time radius r_{eD}, or:

$$p_D = f(t_D, r_{eD})$$

where

$$r_{eD} = \frac{\text{external radius}}{\text{wellbore radius}} = \frac{r_e}{r_w} \tag{1.93}$$

Table 1.3 presents p_D as a function of t_D for $1.5 < r_{eD} < 10$. It should be pointed out that van Everdingen and Hurst principally applied the p_D function solution to model the performance of water influx into oil reservoirs. Thus, the authors' wellbore radius r_w was in this case the external radius of the reservoir and r_e was essentially the external boundary radius of the aquifer. Therefore, the ranges of the r_{eD} values in Table 1.3 are practical for this application.

Consider the Ei function solution to the diffusivity equations as given by Eq. (1.77):

$$p(r, t) = p_i + \left[\frac{70.6 QB\mu}{kh}\right] \text{Ei}\left[\frac{-948 \phi \mu c_t r^2}{kt}\right]$$

TABLE 1.3 p_D vs. t_D – Finite Radial System, Constant Rate at the Inner Boundary Radial System, Constant Rate at the Inner Boundary

$r_{eD} = 1.5$		$r_{eD} = 2.0$		$r_{eD} = 2.5$		$r_{eD} = 3.0$		$r_{eD} = 3.5$		$r_{eD} = 4.0$	
t_D	p_D	t_D	p_D	t_D	p_D	t_D	p_D	t_D	p_D	t_D	p_D
0.06	0.251	0.22	0.443	0.40	0.565	0.52	0.627	1.0	0.802	1.5	0.927
0.08	0.288	0.24	0.459	0.42	0.576	0.54	0.636	1.1	0.830	1.6	0.948
0.10	0.322	0.26	0.476	0.44	0.587	0.56	0.645	1.2	0.857	1.7	0.968
0.12	0.355	0.28	0.492	0.46	0.598	0.60	0.662	1.3	0.882	1.8	0.988
0.14	0.387	0.30	0.507	0.48	0.608	0.65	0.683	1.4	0.906	1.9	1.007
0.16	0.420	0.32	0.522	0.50	0.618	0.70	0.703	1.5	0.929	2.0	1.025
0.18	0.452	0.34	0.536	0.52	0.628	0.75	0.721	1.6	0.951	2.2	1.059
0.20	0.484	0.36	0.551	0.54	0.638	0.80	0.740	1.7	0.973	2.4	1.092
0.22	0.516	0.38	0.565	0.56	0.647	0.85	0.758	1.8	0.994	2.6	1.123
0.24	0.548	0.40	0.579	0.58	0.657	0.90	0.776	1.9	1.014	2.8	1.154
0.26	0.580	0.42	0.593	0.60	0.666	0.95	0.791	2.0	1.034	3.0	1.184
0.28	0.612	0.44	0.607	0.65	0.688	1.0	0.806	2.25	1.083	3.5	1.255
0.30	0.644	0.46	0.621	0.70	0.710	1.2	0.865	2.50	1.130	4.0	1.324
0.35	0.724	0.48	0.634	0.75	0.731	1.4	0.920	2.75	1.176	4.5	1.392
0.40	0.804	0.50	0.648	0.80	0.752	1.6	0.973	3.0	1.221	5.0	1.460
0.45	0.884	0.60	0.715	0.85	0.772	2.0	1.076	4.0	1.401	5.5	1.527
0.50	0.964	0.70	0.782	0.90	0.792	3.0	1.328	5.0	1.579	6.0	1.594
0.55	1.044	0.80	0.849	0.95	0.812	4.0	1.578	6.0	1.757	6.5	1.660
0.60	1.124	0.90	0.915	1.0	0.832	5.0	1.828			7.0	1.727
0.65	1.204	1.0	0.982	2.0	1.215					8.0	1.861
0.70	1.284	2.0	1.649	3.0	1.506					9.0	1.994
0.75	1.364	3.0	2.316	4.0	1.977					10.0	2.127
0.80	1.444	5.0	3.649	5.0	2.398						

$r_{eD} = 4.5$		$r_{eD} = 5.0$		$r_{eD} = 6.0$		$r_{eD} = 7.0$		$r_{eD} = 8.5$		$r_{eD} = 9.0$		$r_{eD} = 10.0$	
t_D	p_D	t_D	p_D	t_D	p_D	t_D	p_D	t_D	p_D	t_D	p_D	t_D	p_D
2.0	1.023	3.0	1.167	4.0	1.275	6.0	1.436	8.0	1.556	10.0	1.651	12.0	1.732
2.1	1.040	3.1	1.180	4.5	1.322	6.5	1.470	8.5	1.582	10.5	1.673	12.5	1.750
2.2	1.056	3.2	1.192	5.0	1.364	7.0	1.501	9.0	1.607	11.0	1.693	13.0	1.768
2.3	1.702	3.3	1.204	5.5	1.404	7.5	1.531	9.5	1.631	11.5	1.713	13.5	1.784
2.4	1.087	3.4	1.215	6.0	1.441	8.0	1.559	10.0	1.663	12.0	1.732	14.0	1.801
2.5	1.102	3.5	1.227	6.5	1.477	8.5	1.586	10.5	1.675	12.5	1.750	14.5	1.817
2.6	1.116	3.6	1.238	7.0	1.511	9.0	1.613	11.0	1.697	13.0	1.768	15.0	1.832
2.7	1.130	3.7	1.249	7.5	1.544	9.5	1.638	11.5	1.717	13.5	1.786	15.5	1.847
2.8	1.144	3.8	1.259	8.0	1.576	10.0	1.663	12.0	1.737	14.0	1.803	16.0	1.862
2.9	1.158	3.9	1.270	8.5	1.607	11.0	1.711	12.5	1.757	14.5	1.819	17.0	1.890
3.0	1.171	4.0	1.281	9.0	1.638	12.0	1.757	13.0	1.776	15.0	1.835	18.0	1.917
3.2	1.197	4.2	1.301	9.5	1.668	13.0	1.810	13.5	1.795	15.5	1.851	19.0	1.943
3.4	1.222	4.4	1.321	10.0	1.698	14.0	1.845	14.0	1.813	16.0	1.867	20.0	1.968
3.6	1.246	4.6	1.340	11.0	1.757	15.0	1.888	14.5	1.831	17.0	1.897	22.0	2.017
3.8	1.269	4.8	1.360	12.0	1.815	16.0	1.931	15.0	1.849	18.0	1.926	24.0	2.063
4.0	1.292	5.0	1.378	13.0	1.873	17.0	1.974	17.0	1.919	19.0	1.955	26.0	2.108
4.5	1.349	5.5	1.424	14.0	1.931	18.0	2.016	19.0	1.986	20.0	1.983	28.0	2.151
5.0	1.403	6.0	1.469	15.0	1.988	19.0	2.058	21.0	2.051	22.0	2.037	30.0	2.194
5.5	1.457	6.5	1.513	16.0	2.045	20.0	2.100	23.0	2.116	24.0	2.906	32.0	2.236

(Continued)

TABLE 1.3 (Continued)

$r_{eD} = 4.5$		$r_{eD} = 5.0$		$r_{eD} = 6.0$		$r_{eD} = 7.0$		$r_{eD} = 8.5$		$r_{eD} = 9.0$		$r_{eD} = 10.0$	
t_D	p_D	t_D	p_D	t_D	p_D	t_D	p_D	t_D	p_D	t_D	p_D	t_D	p_D
6.0	1.510	7.0	1.556	17.0	2.103	22.0	2.184	25.0	2.180	26.0	2.142	34.0	2.278
7.0	1.615	7.5	1.598	18.0	2.160	24.0	2.267	30.0	2.340	28.0	2.193	36.0	2.319
8.0	1.719	8.0	1.641	19.0	2.217	26.0	2.351	35.0	2.499	30.0	2.244	38.0	2.360
9.0	1.823	9.0	1.725	20.0	2.274	28.0	2.434	40.0	2.658	34.0	2.345	40.0	2.401
10.0	1.927	10.0	1.808	25.0	2.560	30.0	2.517	45.0	2.817	38.0	2.446	50.0	2.604
11.0	2.031	11.0	1.892	30.0	2.846					40.0	2.496	60.0	2.806
12.0	2.135	12.0	1.975							45.0	2.621	70.0	3.008
13.0	2.239	13.0	2.059							50.0	2.746	80.0	3.210
14.0	2.343	14.0	2.142							60.0	2.996	90.0	3.412
15.0	2.447	15.0	2.225							70.0	3.246	100.0	3.614

Notes: For t_D smaller than values listed in this table for a given r_{eD} reservoir is infinite acting.
Find p_D in Table 1.2.
For $25 < t_D$ and t_D larger than values in table:

$$p_D \cong \frac{(1/2 + 2t_D)}{r_{eD}^2} - \frac{3r_{eD}^4 - 4r_{eD}^4 \ln r_{eD} - 2r_{eD}^2 - 1}{4(r_{eD}^2 - 1)^2}$$

For wells in rebounded reservoirs with $r_{eD}^2 \gg 1$:

$$p_D \cong \frac{2t_D}{r_{eD}^2} + \ln r_{eD} - 3/4.$$

After Lee, J., 1982. Well Testing, SPE Textbook Series, Permission to Publish by the SPE, Copyright SPE, 1982.

This relationship can be expressed in a dimensionless form by manipulating the expression to give:

$$\frac{p_i - p(r,t)}{[141.2Q_o B_o \mu_o / kh]} = -\frac{1}{2}\text{Ei}\left[\frac{-(r/r_w)^2}{4(0.0002637kt/\phi\mu c_t r_w^2)}\right]$$

From the definition of the dimensionless variables of Eqs. (1.85)–(1.88), i.e., p_D, t_D, and r_D, this relation is expressed in terms of these dimensionless variables as:

$$p_D = -\frac{1}{2}\text{Ei}\left(-\frac{r_D^2}{4t_D}\right) \tag{1.94}$$

Chatas (1953) proposed the following mathematical form for calculated p_D when $25 < t_D$ and $0.25r_{eD}^2 < t_D$:

$$p_D = \frac{0.5 + 2t_D}{r_{eD}^2 - 1} - \frac{r_{eD}^4[3 - 4\ln(r_{eD})] - 2r_{eD}^2 - 1}{4(r_{eD}^2 - 1)^2}$$

Two special cases of the above equation arise when $r_{eD}^2 \gg 1$ or when $t_D/r_{eD}^2 > 25$:

- If $r_{eD}^2 \gg 1$, then:

$$p_D = \frac{2t_D}{r_{eD}^2} + \ln(r_{eD}) - 0.75$$

- If $t_D/r_{eD}^2 > 25$, then:

$$p_D = \frac{1}{2}\left[\ln\frac{t_D}{r_D^2} + 0.80907\right] \tag{1.95}$$

The computational procedure of using the p_D function to determine the bottom-hole flowing pressure changing the transient flow period, i.e., during the infinite-acting behavior, is summarized in the following steps:

Step 1. Calculate the dimensionless time t_D by applying Eq. (1.86a):

$$t_D = \frac{0.0002637kt}{\phi\mu c_t r_w^2}$$

Step 2. Determine the dimensionless radius r_{eD}. Note that for an infinite-acting reservoir, the dimensionless radius $r_{eD} = \infty$.

Step 3. Using the calculated value of t_D, determine the corresponding pressure function p_D from the appropriate table or equations, e.g., Eq. (1.91) or (1.95):

- For an infinite-acting reservoir:

$$p_D = 0.5[\ln(t_D) + 0.80907]$$

- For a finite-acting reservoir:

$$p_D = \frac{1}{2}\left[\ln\left(\frac{t_D}{r_D^2}\right) + 0.080907\right]$$

Step 4. Solve for the pressure by applying Eq. (1.85):

$$p(r_w, t) = p_i - \left(\frac{141.2 Q_o B_o \mu_o}{kh}\right) p_D \quad \textbf{(1.96)}$$

Example 1.12

A well is producing at a constant flow rate of 300 STB/day under unsteady-state flow conditions. The reservoir has the following rock and fluid properties (see Example 1.10):

$B_o = 1.25$ bbl/STB, $\mu_o = 1.5$ cp, $c_t = 12 \times 10^{-6}$ psi^{-1}
$k = 60$ md, $h = 15$ ft, $p_i = 4000$ psi
$\phi = 15\%$, $r_w = 0.25$ ft

Assuming an infinite-acting reservoir, i.e., $r_{eD} = \infty$, calculate the bottom-hole flowing pressure after 1 hour of production by using the dimensionless pressure approach.

Solution

Step 1. Calculate the dimensionless time t_D from Eq. (1.86a):

$$t_D = \frac{0.0002637 kt}{\phi \mu c_t r_w^2} = \frac{0.000264(60)(1)}{(0.15)(1.5)(12 \times 10^{-6})(0.25)^2} = 93866.67$$

Step 2. Since $t_D > 100$, use Eq. (1.91) to calculate the dimensionless pressure drop function:

$$p_D = 0.5[\ln(t_D) + 0.80907] = 0.5[\ln(93866.67) + 0.80907] = 6.1294$$

Step 3. Calculate the bottom-hole pressure after 1 hour by applying Eq. (1.96):

$$p(r_w, t) = p_i - \left(\frac{141.2 Q_o B_o \mu_o}{kh}\right) p_D$$

$$p(0.25, 1) = 4000 - \left[\frac{141.2(300)(1.25)(1.5)}{(60)(15)}\right] \times (6.1294) = 3459 \text{ psi}$$

This example shows that the solution as given by the p_D function technique is identical to that of the Ei function approach. The main difference between the two formulations is that *the p_D function can only be used to calculate the pressure at radius r when the flow rate Q is constant and known.* In that case, the p_D function application is essentially restricted to the wellbore radius because the rate is usually known. On the other hand, the Ei function approach can be used to calculate the pressure at any radius in the reservoir by using the well flow rate Q.

It should be pointed out that, for an infinite-acting reservoir with $t_D > 100$, the p_D function is related to the Ei function by the following relation:

$$p_D = 0.5\left[-\text{Ei}\left(\frac{-1}{4t_D}\right)\right] \quad \textbf{(1.97)}$$

The previous example, i.e., Example 1.12, is not a practical problem, but it is essentially designed to show the physical significance of the p_D solution approach. In transient flow testing, we normally record the bottom-hole flowing pressure as a function of time. Therefore, the dimensionless pressure drop technique can be used to determine one or more of the reservoir properties, e.g., k or kh, as discussed later in this chapter.

1.2.6 Radial Flow of Compressible Fluids

Gas viscosity and density vary significantly with pressure and therefore the assumptions of

Eq. (1.75) are not satisfied for gas systems, i.e., compressible fluids. To develop the proper mathematical function for describing the flow of compressible fluids in the reservoir, the following two additional gas equations must be considered:

(1) Gas density equation:

$$\rho = \frac{pM}{ZRT}$$

(2) Gas compressibility equation:

$$c_g = \frac{1}{p} - \frac{1}{Z}\frac{dZ}{dp}$$

Combining the above two basic gas equations with Eq. (1.67) gives:

$$\frac{1}{r}\frac{\partial}{\partial r}\left(r\frac{p}{\mu Z}\frac{\partial p}{\partial r}\right) = \frac{\phi \mu c_t}{0.000264k}\frac{p}{\mu Z}\frac{\partial p}{\partial t} \tag{1.98}$$

where

t = time, hours
k = permeability, md
c_t = total isothermal compressibility, psi^{-1}
ϕ = porosity

Al-Hussainy et al. (1966) linearized the above basic flow equation by introducing the real-gas pseudopressure $m(p)$ into Eq. (1.98). Recalling the previously defined $m(p)$ equation:

$$m(p) = \int_0^p \frac{2p}{\mu Z}\,dp \tag{1.99}$$

and differentiating this relation with respect to p, gives:

$$\frac{\partial m(p)}{\partial p} = \frac{2p}{\mu p} \tag{1.100}$$

The following relationships are obtained by applying the chain rule:

$$\frac{\partial m(p)}{\partial r} = \frac{\partial m(p)}{\partial p}\frac{\partial p}{\partial r} \tag{1.101}$$

$$\frac{\partial m(p)}{\partial t} = \frac{\partial m(p)}{\partial p}\frac{\partial p}{\partial t} \tag{1.102}$$

Substituting Eq. (1.100) into (1.101) and (1.102), gives:

$$\frac{\partial p}{\partial r} = \frac{\mu Z}{2p}\frac{\partial m(p)}{\partial r} \tag{1.103}$$

and

$$\frac{\partial p}{\partial t} = \frac{\mu Z}{2p}\frac{\partial m(p)}{\partial t} \tag{1.104}$$

Combining Eqs. (1.103) and (1.104) with (1.98), yields:

$$\frac{\partial^2 m(p)}{\partial r^2} + \frac{1}{r}\frac{\partial m(p)}{\partial r} = \frac{\phi \mu c_t}{0.000264k}\frac{\partial m(p)}{\partial t} \tag{1.105}$$

Eq. (1.105) is the radial diffusivity equation for compressible fluids. This differential equation relates the real-gas pseudopressure (real-gas potential) to the time t and the radius r. Al-Hussainy et al. (1966) pointed out that in gas well testing analysis, the constant-rate solution has more practical applications than that provided by the constant-pressure solution. The authors provided the exact solution to Eq. (1.105) that is commonly referred to as the $m(p)$ solution method. There are also two other solutions that approximate the exact solution. These two approximation methods are called the pressure-squared method and the pressure method. In general, there are three forms of mathematical solutions to the diffusivity equation:

(1) $m(p)$ solution method (exact solution);
(2) pressure-squared method (p^2 approximation method);
(3) pressure method (p approximation method).

These three solution methods are presented below.

First Solution: *m(p)* Method (Exact Solution). Imposing the constant-rate condition as one of the boundary conditions required to solve Eq. (1.105), Al-Hussainy et al. (1966) proposed the following exact solution to the diffusivity equation:

$$m(p_{wf}) = m(p_i) - 57895.3\left(\frac{p_{sc}}{T_{sc}}\right)\left(\frac{Q_g T}{kh}\right)\left[\log\left(\frac{kt}{\phi \mu_i c_{ti} r_w^2}\right) - 3.23\right] \tag{1.106}$$

where

p_{wf} = bottom-hole flowing pressure, psi
p_e = initial reservoir pressure
Q_g = gas flow rate, Mscf/day
t = time, hours
k = permeability, md
p_{sc} = standard pressure, psi
T_{sc} = standard temperature, °R
T = reservoir temperature
r_w = wellbore radius, ft
h = thickness, ft
μ_i = gas viscosity at the initial pressure, cp
c_{ti} = total compressibility coefficient at p_i, psi^{-1}
ϕ = porosity

Setting p_{sc} = 14.7 psia and T_{sc} = 520 °R, then Eq. (1.106) reduces to:

$$m(p_{wf}) = m(p_i) - \left(\frac{1637 Q_g T}{kh}\right)\left[\log\left(\frac{kt}{\phi \mu_i c_{ti} r_w^2}\right) - 3.23\right] \quad \textbf{(1.107)}$$

The above equation can be simplified by introducing the dimensionless time (as defined previously by Eq. (1.85) into Eq. (1.107):

$$t_D = \frac{0.0002637 kt}{\phi \mu_i c_{ti} r_w^2}$$

Equivalently, Eq. (1.107) can be written in terms of the dimensionless time t_D as:

$$m(p_{wf}) = m(p_i) - \left(\frac{1637 Q_g T}{kh}\right)\left[\log\left(\frac{4t_D}{\gamma}\right)\right] \quad \textbf{(1.108)}$$

The parameter γ is called Euler's constant and is given by:

$$\gamma = e^{0.5772} = 1.781 \quad \textbf{(1.109)}$$

The solution to the diffusivity equation as given by Eqs. (1.107) and (1.108) expresses the bottom-hole real-gas pseudopressure as a function of the transient flow time t. The solution as expressed in terms of $m(p)$ is the recommended mathematical expression for performing gas well pressure analysis due to its applicability in all pressure ranges.

The radial gas diffusivity equation can be expressed in a dimensionless form in terms of the dimensionless real-gas pseudopressure drop ψ_D. The solution to the dimensionless equation is given by:

$$\psi_D = \frac{m(p_i) - m(p_{wf})}{(1422 Q_g T / kh)}$$

or

$$m(p_{wf}) = m(p_i) - \left(\frac{1422 Q_g T}{kh}\right)\psi_D \quad \textbf{(1.110)}$$

where

Q_g = gas flow rate, Mscf/day
k = permeability, md

The dimensionless pseudopressure drop ψ_D can be determined as a function of t_D by using the appropriate expression of Eqs. (1.90)–(1.95). When t_D>100, ψ_D can be calculated by applying Eq. (1.81). That is:

$$\psi_D = 0.5[\ln(t_D) + 0.80907] \quad \textbf{(1.111)}$$

Example 1.13

A gas well with a wellbore radius of 0.3 ft is producing at a constant flow rate of 2000 Mscf/day under transient flow conditions. The initial reservoir pressure (shut-in pressure) is 4400 psi at 140 °F. The formation permeability and thickness are 65 md and 15 ft, respectively. The porosity is recorded as 15%. Example 1.7 documents the properties of the gas as well as values of $m(p)$ as a function of pressures. The table is reproduced below for convenience:

P	μ_g (cp)	Z	$m(p)$ (psi^2/cp)
0	0.01270	1.000	0.000
400	0.01286	0.937	13.2×10^6
800	0.01390	0.882	52.0×10^6
1200	0.01530	0.832	113.1×10^6
1600	0.01680	0.794	198.0×10^6
2000	0.01840	0.770	304.0×10^6
2400	0.02010	0.763	422.0×10^6
2800	0.02170	0.775	542.4×10^6
3200	0.02340	0.797	678.0×10^6
3600	0.02500	0.827	816.0×10^6
4000	0.02660	0.860	950.0×10^6
4400	0.02831	0.896	1089.0×10^6

Assuming that the initial total isothermal compressibility is 3×10^{-4} psi^{-1}, calculate the bottom-hole flowing pressure after 1.5 hours.

Solution

Step 1. Calculate the dimensionless time t_D:

$$t_D = \frac{0.0002637kt}{\phi \mu_i c_{ti} r_w^2}$$
$$= \frac{(0.0002637)(65)(1.5)}{(0.15)(0.02831)(3 \times 10^{-4})(0.3^2)} = 224,498.6$$

Step 2. Solve for $m(p_{wf})$ by using Eq. (1.108):

$$m(p_{wf}) = m(p_i) - \left(\frac{1637 Q_g T}{kh}\right)\left[\log\left(\frac{4t_D}{\gamma}\right)\right]$$
$$= 1089 \times 10^6 - \frac{(1637)(2000)(600)}{(65)(15)}$$
$$\times \left[\log\left(\frac{(4)224,498.6}{e^{0.5772}}\right)\right] = 1077.5 \times 10^6$$

Step 3. From the given *PVT* data, interpolate using the value of $m(p_{wf})$ to give a corresponding p_{wf} of 4367 psi.

An identical solution can be obtained by applying the ψ_D approach as shown below:

Step 1. Calculate ψ_D from Eq. (1.111):

$$\psi_D = 0.5[\ln(t_D) + 0.80907]$$
$$= 0.5[\ln(224,498.6) + 0.8090] = 6.565$$

Step 2. Calculate $m(p_{wf})$ by using Eq. (1.110):

$$m(p_{wf}) = m(p_i) - \left(\frac{1422 Q_g T}{kh}\right)\psi_D$$
$$= 1089 \times 10^6 - \left(\frac{1422(2000)(600)}{(65)(15)}\right)(6.565)$$
$$= 1077.5 \times 10^6$$

By interpolation at $m(p_{wf}) = 1077.5 \times 10^6$, this gives a corresponding value of $p_{wf} = 4367$ psi.

Second Solution: Pressure-squared Method. The first approximation to the exact solution is to move the pressure-dependent term (μZ) outside the integral that defines $m(p_{wf})$ and $m(p_i)$, to give:

$$m(p_i) - m(p_{wf}) = \frac{2}{\overline{\mu} \overline{Z}} \int_{p_{wf}}^{p_i} p\, dp \qquad \textbf{(1.112)}$$

or

$$m(p_i) - m(p_{wf}) = \frac{p_i^2 - p_{wf}^2}{\overline{\mu} \overline{Z}} \qquad \textbf{(1.113)}$$

The bars over μ and Z represent the values of the gas viscosity and deviation factor as evaluated at the average pressure $\overline{p}$. This average pressure is given by:

$$\overline{p} = \sqrt{\frac{p_i^2 + p_{wf}^2}{2}} \qquad \textbf{(1.114)}$$

Combining Eq. (1.113) with (1.107), (1.108), or (1.110), gives:

$$p_{wf}^2 = p_i^2 - \left(\frac{1637 Q_g T \overline{\mu} \overline{Z}}{kh}\right)\left[\log\left(\frac{kt}{\phi \mu_i c_{ti} r_w^2}\right) - 3.23\right] \qquad \textbf{(1.115)}$$

or

$$p_{wf}^2 = p_i^2 - \left(\frac{1637 Q_g T \overline{\mu} \overline{Z}}{kh}\right)\left[\log\left(\frac{4t_D}{\gamma}\right)\right] \qquad \textbf{(1.116)}$$

Equivalently:

$$p_{wf}^2 = p_i^2 - \left(\frac{1422 Q_g T \overline{\mu} \overline{Z}}{kh}\right)\psi_D \qquad \textbf{(1.117)}$$

The above approximation solution forms indicate that the product (μZ) is assumed constant at an average pressure $\overline{p}$. This effectively limits the applicability of the p^2 method to reservoir pressures of less than 2000. It should be pointed out that when the p^2 method is used to determine p_{wf} it is perhaps sufficient to set $\overline{\mu} \overline{Z} = \mu_i Z$.

Example 1.14

A gas well is producing at a constant rate of 7454.2 Mscf/day under transient flow conditions. The following data is available:

$k = 50$ md, $h = 10$ ft, $\phi = 20\%$, $p_i = 1600$ psi
$T = 600\ ^\circ$R, $r_w = 0.3$ ft, $c_{ti} = 6.25 \times 10^{-4}$ psi^{-1},

The gas properties are tabulated below:

P	μ_g (cp)	Z	$m(p)$ (psi^2/cp)
0	0.01270	1.000	0.000
400	0.01286	0.937	13.2×10^6
800	0.01390	0.882	52.0×10^6
1200	0.01530	0.832	113.1×10^6
1600	0.01680	0.794	198.0×10^6

Calculate the bottom-hole flowing pressure after 4 hours by using:

(a) the $m(p)$ method;
(b) the p^2 method.

Solution

(a) The $m(p)$ method:
Step 1. Calculate t_D:

$$t_D = \frac{0.000264(50)(4)}{(0.2)(0.0168)(6.25 \times 10^{-4})(0.32^2)} = 279,365.1$$

Step 2. Calculate ψ_D:

$$\psi_D = 0.5[\ln(t_D) + 0.80907] = 0.5[\ln(279,365.1) + 0.80907] = 6.6746$$

Step 3. Solve for $m(p_{wf})$ by applying Eq. (1.110):

$$m(p_{wf}) = m(p_i) - \left(\frac{1422 Q_g T}{kh}\right)\psi_D = (198 \times 10^6) - \left[\frac{1422(7454.2)(600)}{(50)(10)}\right] 6.6746 = 113.1 \times 10^6$$

The corresponding value of $p_{wf} = 1200$ psi

(b) The p^2 method:
Step 1. Calculate ψ_D by applying Eq. (1.111):

$$\psi_D = 0.5[\ln(t_D) + 0.80907] = 0.5[\ln(279\,365.1) + 0.80907] = 6.6747$$

Step 2. Calculate p_{wf}^2 by applying Eq. (1.116):

$$p_{wf}^2 = p_i^2 - \left(\frac{1422 Q_g T \bar{\mu} \bar{Z}}{kh}\right)\psi_D = 1600^2 - \left[\frac{(1422)(7454.2)(600)(0.0168)(0.794)}{(50)(10)}\right] 6.6747 = 1,427,491$$

$$p_{wf} = 1195 \text{ psi}$$

Step 3. The absolute average error is 0.4%.

Third Solution: Pressure Approximation Method. The second method of approximation to the exact solution of the radial flow of gases is to treat the gas as a pseudo-liquid. Recall that the gas formation volume factor B_g as expressed in bbl/scf is given by:

$$B_g = \left(\frac{p_{sc}}{5.615 T_{sc}}\right)\left(\frac{ZT}{p}\right)$$

or

$$B_g = 0.00504\left(\frac{ZT}{p}\right)$$

Solving the above expression for p/Z gives:

$$\frac{p}{Z} = \left(\frac{T p_{sc}}{5.615 T_{sc}}\right)\left(\frac{1}{B_g}\right)$$

The difference in the real-gas pseudopressure is given by:

$$m(p_i) - (p_{wf}) = \int_{p_{wf}}^{p_i} \frac{2p}{\mu Z} dp$$

Combining the above two expressions gives:

$$m(p_i) - m(p_{wf}) = \frac{2 T p_{sc}}{5.615 T_{sc}} \int_{p_{wf}}^{p_i} \left(\frac{1}{\mu B_g}\right) dp \quad \textbf{(1.118)}$$

Fetkovich (1973) suggested that at high pressures above 3000 psi ($p > 3000$), $1/\mu B_g$ is nearly constant as shown schematically in Figure 1.22. Imposing Fetkovich's condition on Eq. (1.118) and integrating gives:

$$m(p_i) - m(p_{wf}) = \frac{2 T p_{sc}}{5.615 T_{sc} \bar{\mu} \bar{B}_g}(p_i - p_{wf}) \quad \textbf{(1.119)}$$

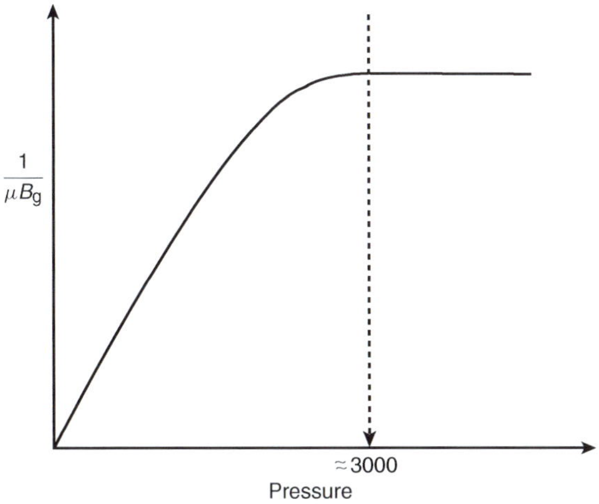

FIGURE 1.22 Plot of $1/\mu B_g$ vs. pressure.

Combining Eq. (1.119) with (1.107), (1.108), or (1.110) gives:

$$p_{wf} = p_i - \left(\frac{162.5 \times 10^3 Q_g \bar{\mu} \bar{B}_g}{kh}\right)\left[\log\left(\frac{kt}{\phi \bar{\mu} \bar{c}_t r_w^2}\right) - 3.23\right] \quad (1.120)$$

or

$$p_{wf} = p_i - \left(\frac{(162.5 \times 10^3) Q_g \bar{\mu} \bar{B}_g}{kh}\right)\left[\log\left(\frac{4t_D}{\gamma}\right)\right] \quad (1.121)$$

or, equivalently, in terms of dimensionless pressure drop:

$$p_{wf} = p_i - \left(\frac{(141.2 \times 10^3) Q_g \bar{\mu} \bar{B}_g}{kh}\right) p_D \quad (1.122)$$

where

Q_g = gas flow rate, Mscf/day
k = permeability, md
B_g = gas formation volume factor, bbl/scf
t = time, hours
p_D = dimensionless pressure drop
t_D = dimensionless

It should be noted that the gas properties, i.e., μ, B_g, and c_t, are evaluated at pressure $\bar{p}$ as defined below:

$$\bar{p} = \frac{p_i + p_{wf}}{2} \quad (1.123)$$

Again, this method is limited only to applications above 3000 psi. When solving for p_{wf}, it might be sufficient to evaluate the gas properties at p_i.

Example 1.15

The data of Example 1.13 is repeated below for convenience.

A gas well with a wellbore radius of 0.3 ft is producing at a constant flow rate of 2000 Mscf/day under transient flow conditions. The initial reservoir pressure (shut-in pressure) is 4400 psi at 140 °F. The formation permeability and thickness are 65 md and 15 ft, respectively. The porosity is recorded as 15%. The properties of the gas as well as values of $m(p)$ as a function of pressures are tabulated below:

P	μ_g (cp)	Z	$m(p)$ (psi^2/cp)
0	0.01270	1.000	0.000
400	0.01286	0.937	13.2×10^6
800	0.01390	0.882	52.0×10^6
1200	0.01530	0.832	113.1×10^6
1600	0.01680	0.794	198.0×10^6
2000	0.01840	0.770	304.0×10^6
2400	0.02010	0.763	422.0×10^6
2800	0.02170	0.775	542.4×10^6
3200	0.02340	0.797	678.0×10^6
3600	0.02500	0.827	816.0×10^6
4000	0.02660	0.860	950.0×10^6
4400	0.02831	0.896	1089.0×10^6

Assuming that the initial total isothermal compressibility is 3×10^{-4} psi^{-1}, calculate, the bottom-hole flowing pressure after 1.5 hours by using the p approximation method and compare it with the exact solution.

Solution

Step 1. Calculate the dimensionless time t_D:

$$t_D = \frac{0.0002637kt}{\phi \mu_i c_{ti} r_w^2} = \frac{(0.000264)(65)(1.5)}{(0.15)(0.02831)(3 \times 10^{-4})(0.3^2)} = 224,498.6$$

Step 2. Calculate B_g at p_i:

$$B_g = 0.00504\left(\frac{Z_i T}{p_i}\right)$$
$$= 0.00504\frac{(0.896)(600)}{4400} = 0.0006158\,\text{bbl/scf}$$

Step 3. Calculate the dimensionless pressure p_D by applying Eq. (1.91):

$$p_D = 0.5[\ln(t_D) + 0.80907]$$
$$= 0.5[\ln(224,498.6) + 0.80907] = 6.565$$

Step 4. Approximate p_{wf} from Eq. (1.122):

$$p_{wf} = p_i - \left(\frac{(141.201^3)Q_g\bar{\mu}\bar{B}_g}{kh}\right)p_D$$
$$= 4400 - \left[\frac{141.2\times10^3(2000)(0.02831)(0.0006158)}{(65)(15)}\right]6.565$$
$$= 4367\,\text{psi}$$

The solution is identical to the solution of Example 1.13.

It should be pointed out that Examples 1.10–1.15 are designed to illustrate the use of different solution methods. However, these examples are not practical because, in transient flow analysis, the bottom-hole flowing pressure is usually available as a function of time. All the previous methodologies are essentially used to characterize the reservoir by determining the permeability k or the permeability and thickness product (kh).

1.2.7 Pseudosteady State

In the unsteady-state flow cases discussed previously, it was assumed that a well is located in a very large reservoir and is producing at a constant flow rate. This rate creates a pressure disturbance in the reservoir that travels throughout this "infinite-size reservoir." During this transient flow period, reservoir boundaries have no effect on the pressure behavior of the well. Obviously, the time period when this assumption can be imposed is often very short in length. As soon as the pressure disturbance reaches all drainage boundaries, it ends the transient (unsteady-state) flow regime and the beginning of the boundary-dominated flow condition. This different type of flow regime is called pseudosteady (semisteady)-state flow. It is necessary at this point to impose different boundary conditions on the diffusivity equation and drive an appropriate solution to this flow regime.

Figure 1.23 shows a well in a radial system that is producing at a constant rate for a long enough period that eventually affects the entire drainage area. During this semisteady-state flow, the change in pressure with time becomes the same throughout the drainage area. Figure 1.23(b) shows that the pressure distributions become paralleled at successive time periods. Mathematically, this important condition can be expressed as:

$$\left(\frac{\partial p}{\partial t}\right)_r = \text{constant} \tag{1.124}$$

The "constant" referred to in the above equation can be obtained from a simple material balance using the definition of the compressibility, assuming no free gas production, thus:

$$c = \frac{-1}{V}\frac{dV}{dp}$$

Rearranging:

$$cV\,dp = -dV$$

Differentiating with respect to time t:

$$cV\frac{dp}{dt} = -\frac{dV}{dt} = q$$

or

$$\frac{dp}{dt} = -\frac{q}{cV}$$

Expressing the pressure decline rate dp/dt in the above relation in psi/hour gives:

$$\frac{dp}{dt} = -\frac{q}{24cV} = -\frac{Q_o B_o}{24cV} \tag{1.125}$$

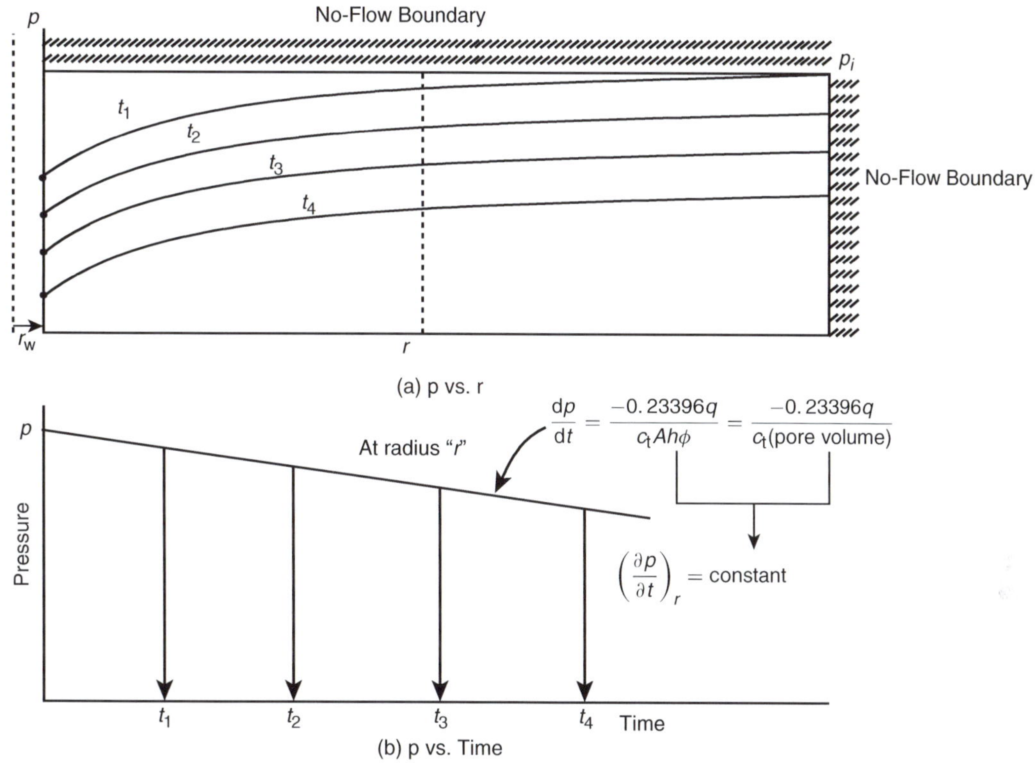

FIGURE 1.23 Semisteady-state flow regime.

where

q = flow rate, bbl/day
Q_o = flow rate, STB/day
dp/dt = pressure decline rate, psi/hour
V = pore volume, bbl

For a radial drainage system, the pore volume is given by:

$$V = \frac{\pi r_e^2 h\phi}{5.615} = \frac{Ah\phi}{5.615} \tag{1.126}$$

where

A = drainage area, ft^2

Combining Eq. (1.126) with (1.125) gives:

$$\frac{dp}{dt} = -\frac{0.23396q}{c_t(\pi r_e^2)h\phi} = -\frac{-0.23396q}{c_t A h\phi} = \frac{-0.23396q}{c_t(\text{pore volume})} \tag{1.127}$$

Eq. (1.127) reveals the following important characteristics of the behavior of the pressure decline rate dp/dt during the semisteady-state flow:

- the reservoir pressure declines at a higher rate with increasing fluid production rate;
- the reservoir pressure declines at a slower rate for reservoirs with higher total compressibility coefficients;
- the reservoir pressure declines at a lower rate for reservoirs with larger pore volumes.

In the case of water influx with an influx rate of e_w (bbl/day), the equation can be modified as:

$$\frac{dp}{dt} = \frac{-0.23396q + e_w}{c_t(\text{pore volume})}$$

Example 1.16

An oil well is producing at constant oil flow rate of 120 STB/day under a semisteady-state flow regime. Well testing data indicates that the pressure is declining at a constant rate of 0.04655 psi/hour. The following addition data is available:

$h = 72$ ft, $\phi = 25\%$
$B_o = 1.3$ bbl/STB, $c_t = 25 \times 10^{-6}$ psi^{-1}

Calculate the well drainage area.

Solution

Here

$$q = Q_o B_o = (120)(1.3) = 156 \text{ bbl/day}$$

Apply Eq. (1.127) to solve for A:

$$\frac{dp}{dt} = -\frac{0.23396q}{c_t(\pi r_e^2)h\phi} = \frac{-0.23396q}{c_t A h\phi} = -\frac{-0.23396q}{c_t(\text{pore volume})}$$

$$-0.04655 = -\frac{0.23396(156)}{(25 \times 10^{-6})(A)(72)(0.25)}$$

$$A = 1{,}742{,}400 \text{ ft}^2$$

or

$$A = \frac{1{,}742{,}400}{43{,}560} = 40 \text{ acres}$$

Matthews et al. (1954) pointed out that once the reservoir is producing *under the semisteady-state condition*, each well will drain from within its own no-flow boundary independently of the other wells. For this condition to prevail, the pressure decline rate dp/dt must be approximately constant throughout the entire reservoir, otherwise flow would occur across the boundaries causing a readjustment in their positions. Because the pressure at every point in the reservoir is changing at the same rate, it leads to the conclusion that the average reservoir pressure is changing at the same rate. This average reservoir pressure is essentially set equal to the volumetric average reservoir pressure $\bar{p}_r$. It is the pressure that is used to perform flow calculations during the semisteady-state flowing condition. The above discussion indicates that, in principle, Eq. (1.127) can be used to estimate the average pressure in the well drainage area $\bar{p}$ by replacing the pressure decline rate dp/dt with $(p_i - \bar{p})/t$, or:

$$p_i - \bar{p} = \frac{0.23396qt}{c_t(Ah\phi)}$$

or

$$\bar{p} = p_i - \left[\frac{0.23396q}{c_t(Ah\phi)}\right]t \qquad (1.128)$$

Note that the above expression is essentially an equation of a straight line, with a slope of $m^\backslash$ and intercept of p_i, as expressed by:

$$\bar{p} = a + m^\backslash t$$

$$m^\backslash = -\left[\frac{0.23396q}{c_t(Ah\phi)}\right] = -\left[\frac{0.23396q}{c_t(\text{pore volume})}\right]$$

$$a = p_i$$

Eq. (1.128) indicates that the average reservoir pressure, after producing a cumulative oil production of N_p STB, can be roughly approximated by:

$$\bar{p} = p_i - \left[\frac{0.23396 B_o N_p}{c_t(Ah\phi)}\right]$$

It should be noted that when performing material balance calculations, the volumetric average pressure of the entire reservoir is used to calculate the fluid properties. This pressure can be determined from the individual well drainage properties as follows:

$$\bar{p}_r = \frac{\Sigma_j(\bar{p}V)_j}{\Sigma_j V_j}$$

where

V_j = pore volume of the jth well drainage volume
$(\bar{p})_j$ = volumetric average pressure within the jth drainage volume

Figure 1.24 illustrates the concept of the volumetric average pressure. In practice, the V_i are difficult to determine and, therefore, it is common to use individual well flow rates q_i in

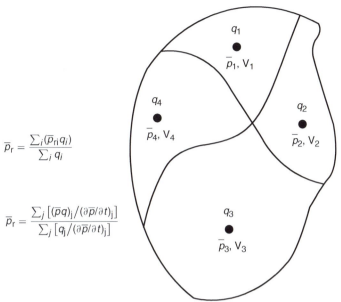

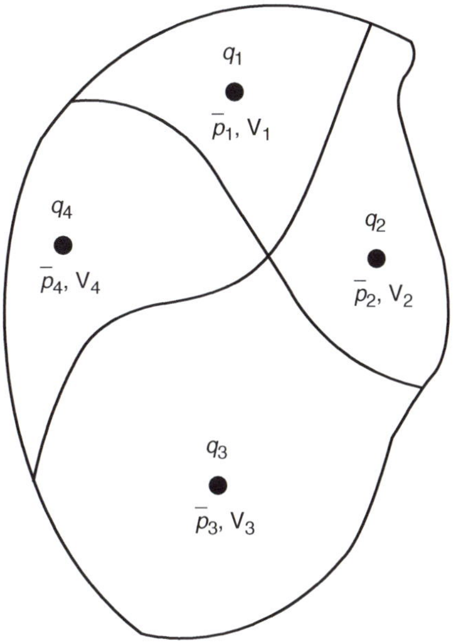

$$\bar{p}_r = \frac{\sum_i (\bar{p}_{ri} q_i)}{\sum_i q_i}$$

$$\bar{p}_r = \frac{\sum_j [(\bar{p}q)_j/(\partial \bar{p}/\partial t)_j]}{\sum_j [q_j/(\partial \bar{p}/\partial t)_j]}$$

FIGURE 1.24 Volumetric average reservoir pressure.

determining the average reservoir pressure from individual well average drainage pressure:

$$\bar{p}_r = \frac{\Sigma_j (\bar{p}q)_j}{\Sigma_j q_j}$$

The flow rates are measured on a routing basis throughout the lifetime of the field, thus facilitating the calculation of the volumetric average reservoir pressure $\bar{p}_r$. Alternatively, the average reservoir pressure can be expressed in terms of the individual well average drainage pressure decline rates and fluid flow rates by:

$$\bar{p}_r = \frac{\Sigma_j [(\bar{p}q)_j/(\partial \bar{p}/\partial t)_j]}{\Sigma_j [q_j/(\partial \bar{p}/\partial t)_j]} \tag{1.129}$$

However, since the material balance equation is usually applied at regular intervals of 3–6 months, i.e., $\Delta t = 3\text{–}6$ months, throughout the lifetime of the field, the average field pressure can be expressed in terms of the incremental net change in underground fluid withdrawal $\Delta(F)$ as:

$$\bar{p}_r = \frac{\Sigma_j \bar{p}_j\, \Delta(F)_j/\Delta \bar{p}_j}{\Sigma_j \Delta(F)_j/\Delta \bar{p}_j} \tag{1.130}$$

where the total underground fluid withdrawal at time t and $t + \Delta t$ are given by:

$$F_t = \int_0^t [Q_o B_o + Q_w B_w + (Q_g - Q_o R_s - Q_w R_{sw}) B_g]\, dt$$
$$F_{t+\Delta t} = \int_0^{t+\Delta t} [Q_o B_o + Q_w B_w + (Q_g - Q_o R_s - Q_w R_{sw}) B_g]\, dt$$

with

$$\Delta(F) = F_{t+\Delta t} - F_t$$

where

R_s = gas solubility, scf/STB
R_{sw} = gas solubility in the water, scf/STB
B_g = gas formation volume factor, bbl/scf
Q_o = oil flow rate, STB/day
q_o = oil flow rate, bbl/day
Q_w = water flow rate, STB/day
q_w = water flow rate, bbl/day
Q_g = gas flow rate, scf/day

The practical applications of using the pseudosteady-state flow condition to describe the flow behavior of the following two types of fluids are presented below:

(1) radial flow of slightly compressible fluids;
(2) radial flow of compressible fluids.

1.2.8 Radial Flow of Slightly Compressible Fluids

The diffusivity equation as expressed by Eq. (1.72) for the transient flow regime is:

$$\frac{\partial^2 p}{\partial r^2} + \frac{1}{r}\frac{\partial p}{\partial r} = \left(\frac{\phi \mu c_t}{0.000264k}\right)\frac{\partial p}{\partial t}$$

For the semisteady-state flow, the term $\partial p/\partial t$ is constant and is expressed by Eq. (1.127). Substituting Eq. (1.127) into the diffusivity equation gives:

$$\frac{\partial^2 p}{\partial r^2} + \frac{1}{r}\frac{\partial p}{\partial r} = \left(\frac{\phi \mu c_t}{0.000264k}\right)\left(\frac{-0.23396q}{c_t A h \phi}\right)$$

or

$$\frac{\partial^2 p}{\partial r^2} + \frac{1}{r}\frac{\partial p}{\partial r} = -\frac{-887.22q\mu}{Ahk}$$

This expression can be expressed as:

$$\frac{1}{r}\frac{\partial}{\partial r}\left(r\frac{\partial p}{\partial r}\right) = -\frac{887.22q\mu}{(\pi r_e^2)hk}$$

Integrating this equation gives:

$$r\frac{\partial p}{\partial r} = -\frac{887.22q\mu}{(\pi r_e^2)hk}\left(\frac{r^2}{2}\right) + c_1$$

where c_1 is the constant of integration and can be evaluated by imposing the outer no-flow boundary condition (i.e., $(\partial p/\partial r)_{re} = 0$) on the above relation, to give:

$$c_1 = \frac{141.2q\mu}{\pi hk}$$

Combining these two expressions gives:

$$\frac{\partial p}{\partial r} = \frac{141.2q\mu}{hk}\left(\frac{1}{r} - \frac{r}{r_e^2}\right)$$

Integrating again:

$$\int_{p_{wf}}^{p_i} dp = \frac{141.2q\mu}{hk}\int_{r_w}^{r_e}\left(\frac{1}{r} - \frac{r}{r_e^2}\right) dr$$

Performing the above integration and assuming r_w^2/r_e^2 is negligible gives:

$$(p_i - p_{wf}) = \frac{141.2q\mu}{kh}\left[\ln\left(\frac{r_e}{r_w}\right) - \frac{1}{2}\right]$$

A more appropriate form of the above is to solve for the flow rate as expressed in STB/day, to give:

$$Q = \frac{0.00708kh(p_i - p_{wf})}{\mu B[\ln(r_e/r_w) - 0.5]} \quad (1.131)$$

where

Q = flow rate, STB/day
B = formation volume factor, bbl/STB
k = permeability, md

The volumetric average pressure in the well drainage area $\bar{p}$ is commonly used to calculate the liquid flow rate under the semisteady-state flowing condition. Introducing $\bar{p}$ into Eq. (1.131) gives:

$$Q = \frac{0.00708kh(\bar{p} - p_{wf})}{\mu B[\ln(r_e/r_w) - 0.75]} = \frac{(\bar{p} - p_{wf})}{141.2\mu B[\ln(r_e/r_w) - 0.75]} \quad (1.132)$$

Note that:

$$\ln\left(\frac{r_e}{r_w}\right) - 0.75 = \ln\left(\frac{0.471 r_e}{r_w}\right)$$

The above observation suggests that the volumetric average pressure $\bar{p}$ occur at about 47% of the drainage radius during the semisteady-state condition. That is:

$$Q = \frac{0.00708kh(\bar{p} - p_{wf})}{\mu B[\ln(0.471 r_e/r_w)]}$$

It should be pointed out that the pseudosteady-state flow occurs regardless of the geometry of the reservoir. Irregular geometries also reach this state when they have been produced long enough for the entire drainage area to be affected.

Rather than developing a separate equation for the geometry of each drainage area, Ramey and Cobb (1971) introduced a correction factor called the shape factor C_A which is designed to account for the deviation of the drainage area from the ideal circular form. The shape factor, as listed in Table 1.4, accounts also for the location of the well within the drainage area.

TABLE 1.4 Shape Factors for Various Single-well Drainage Areas

In Bounded Reservoirs	C_A	ln C_A	$\frac{1}{2} ln\left(\frac{2.2458}{C_A}\right)$	Exact for $t_{DA} >$	Less than 1% Error for $t_{DA} >$	Use Infinite System Solution with Less than 1% Error for $t_{DA} >$
	31.62	3.4538	−1.3224	0.1	0.6	0.10
	31.6	3.4532	−1.3220	0.1	0.06	0.10
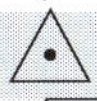	27.6	3.3178	−1.2544	0.2	0.07	0.09
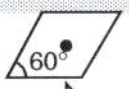	27.1	3.2995	−1.2452	0.2	0.07	0.09
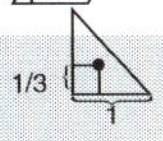	21.9	3.0865	−1.1387	0.4	0.12	0.08
	0.098	−2.3227	+ 1.5659	0.9	0.60	0.015
	30.8828	3.4302	−1.3106	0.1	0.05	0.09
	12.9851	2.5638	−0.8774	0.7	0.25	0.03
	10132	1.5070	−0.3490	0.6	0.30	0.025
	3.3351	1.2045	−0.1977	0.7	0.25	0.01
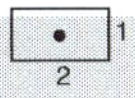	21.8369	3.0836	−1.1373	0.3	0.15	0.025
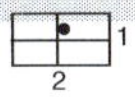	10.8374	2.3830	−0.7870	0.4	0.15	0.025
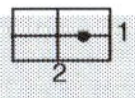	10141	1.5072	−0.3491	1.5	0.50	0.06
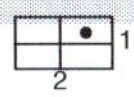	2.0769	0.7309	−0.0391	1.7	0.50	0.02
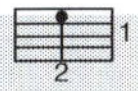	3.1573	1.1497	−0.1703	0.4	0.15	0.005
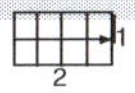	0.5813	−0.5425	+ 0.6758	2.0	0.60	0.02
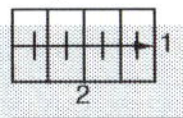	0.1109	−2.1991	+ 1.5041	3.0	0.60	0.005
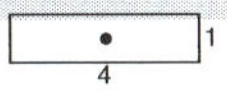	5.3790	1.6825	−0.4367	0.8	0.30	0.01
	2.6896	0.9894	−0.0902	0.8	0.30	0.01
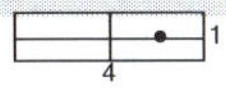	0.2318	−1.4619	+ 1.1355	4.0	2.00	0.03
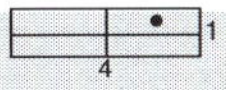	0.1155	−2.1585	+ 1.4838	4.0	2.00	0.01
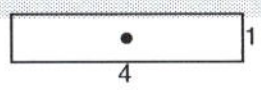	2.3606	0.8589	−0.0249	1.0	0.40	0.025
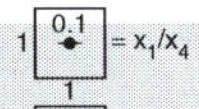	In vertically fractured reservoirs use $(x_e/x_f)^2$ in place of A/r_w^2, for fractured systems					
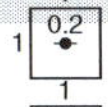	2.6541	0.9761	−0.0835	0.175	0.08	Cannot use
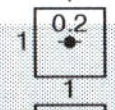	2.0348	0.7104	+ 0.0493	0.175	0.09	Cannot use
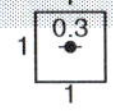	1.9986	0.6924	+ 0.0583	0.175	0.09	Cannot use

(Continued)

TABLE 1.4 (Continued)

In Bounded Reservoirs	C_A	ln C_A	$\frac{1}{2}\ln\left(\frac{2.2458}{C_A}\right)$	Exact for $t_{DA} >$	Less than 1% Error for $t_{DA} >$	Use Infinite System Solution with Less than 1% Error for $t_{DA} >$
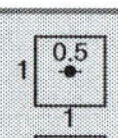	1.6620	0.5080	+0.1505	0.175	0.09	Cannot use
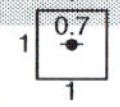	1.3127	0.2721	+0.2685	0.175	0.09	Cannot use
				In water-drive reservoirs		
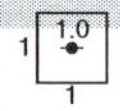	0.7887	−0.2374	+0.5232	0.175	0.09	Cannot use
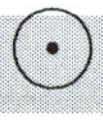	19.1	2.95	−1.07	–	–	–
			In reservoirs of unknown production character			
	25.0	3.22	−1.20	–	–	–

After Earlougher, Robert C., Jr., 1977. Advances in Well Test Analysis, Monograph, vol. 5. Society of Petroleum Engineers of AIME, Dallas, TX; Permission to publish by the SPE, Copyright SPE, 1977.

Introducing c_A into Eq. (1.132) and solving for p_{wf} gives the following two solutions:

(1) In terms of the volumetric average pressure $\bar{p}$:

$$p_{wf} = \bar{p} - \frac{162.6QB\mu}{kh}\log\left(\frac{2.2458A}{C_A r_w^2}\right) \quad \textbf{(1.133)}$$

(2) In terms of the initial reservoir pressure p_i, recall Eq. (1.128) which shows the changes of the average reservoir pressure $\bar{p}$ as a function of time and initial reservoir pressure p_i:

$$\bar{p} = p_i - \frac{0.23396qt}{c_t Ah\phi}$$

Combining this equation with Eq. (1.133) gives:

$$p_{wf} = \left(p_i - \frac{0.23396QBt}{Ah\phi c_t}\right) - \frac{162.6QB\mu}{kh}\log\left(\frac{2.2458A}{C_A r_w^2}\right) \quad \textbf{(1.134)}$$

where

k = permeability, md
A = drainage area, ft^2
C_A = shape factor
Q = flow rate, STB/day
t = time, hours
c_t = total compressibility coefficient, psi^{-1}

Eq. (1.134) can be slightly rearranged as:

$$p_{wf} = \left[p_i - \frac{162.6QB\mu}{kh}\log\left(\frac{2.2458A}{C_A r_w^2}\right)\right] - \left(\frac{0.23396QB}{Ah\phi c_t}\right)t$$

The above expression indicates that under semisteady-state flow and constant flow rate, it can be expressed as an equation of a straight line:

$$p_{wf} = a_{pss} + m_{pss}t$$

with a_{pss} and m_{pss} as defined by:

$$a_{pss} = \left[p_i - \frac{162.6QB\mu}{kh}\log\left(\frac{2.2458A}{C_A r_w^2}\right)\right]$$

$$m_{pss} = -\left(\frac{0.23396QB}{c_t(Ah\phi)}\right) = -\left(\frac{0.23396QB}{c_t(\text{pore volume})}\right)$$

It is obvious that during the pseudosteady (semisteady)-state flow condition, a plot of the bottom-hole flowing pressure p_{wf} vs. time t

would produce a straight line with a negative slope of m_{pss} and intercept of a_{pss}.

A more generalized form of Darcy's equation can be developed by rearranging Eq. (1.133) and solving for Q to give:

$$Q = \frac{kh(\overline{p} - p_{wf})}{162.6B\mu \log(2.2458A/C_A r_w^2)} \tag{1.135}$$

It should be noted that if Eq. (1.135) is applied to a circular reservoir of radius r_e, then:

$$A = \pi r_e^2$$

and the shape factor for a circular drainage area, as given in Table 1.4, as:

$$C_A = 31.62$$

Substituting in Eq. (1.135), it reduces to:

$$Q = \frac{0.00708kh(\overline{p} - p_{wf})}{B\mu[\ln(r_e/r_w) - 0.75]}$$

This equation is identical to Eq. (1.134).

Example 1.17

An oil well is developed on the center of a 40 acre square-drilling pattern. The well is producing at a constant flow rate of 100 STB/day under a semisteady-state condition. The reservoir has the following properties:

$\phi = 15\%$,	$h = 30$ ft,	$k = 20$ md
$\mu = 1.5$ cp,	$B_o = 1.2$ bbl/STB,	$c_t = 25 \times 10^{-6}$ psi^{-1}
$p_i = 4500$ psi,	$r_w = 0.25$ ft,	$A = 40$ acres

(a) Calculate and plot the bottom-hole flowing pressure as a function of time.

(b) Based on the plot, calculate the pressure decline rate. What is the decline in the average reservoir pressure from $t = 10$ to 200 hours?

Solution

(a) For the p_{wf} calculations:

Step 1. From Table 1.4, determine C_A:

$$C_A = 30.8828$$

Step 2. Convert the area A from acres to ft^2:

$$A = (40)(43{,}560) = 1{,}742{,}400 \text{ ft}^2$$

Step 3. Apply Eq. (1.134):

$$p_{wf} = \left(p_i \frac{0.23396QBt}{Ah\phi c_t}\right) - \frac{162.6QB\mu}{kh}\log\left(\frac{2.2458A}{1C_A r_w^2}\right)$$
$$= 4500 - 0.143t - 48.78\log(2027,436)$$

or

$$p_{wf} = 4192 - 0.143t$$

Step 4. Calculate p_{wf} at different assumed times, as follows:

t (hour)	**$p_{wf} = 4192 - 0.143t$**
10	4191
20	4189
50	4185
100	4178
200	4163

Step 5. Present the results of step 4 in graphical form as shown in Figure 1.25.

(b) It is obvious from Figure 1.25 and the above calculation that the bottom-hole flowing pressure is declining at a rate of 0.143 psi/hour, or:

$$\frac{dp}{dt} = -0.143 \text{ psi/hour}$$

The significance of this example is that the rate of pressure decline during the pseudosteady state is the same throughout the drainage area. This means that the *average reservoir pressure* $\overline{p_r}$ is declining at the same rate of 0.143 psi/hour, therefore the change in $\overline{p_r}$ from 10 to 200 hours is:

$$\Delta\overline{p_r} = (0.143)(200 - 10) = 27.17 \text{ psi}$$

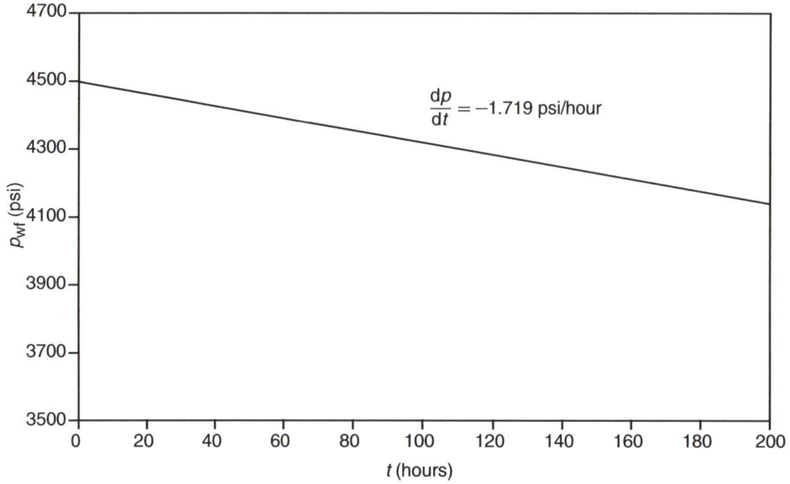

FIGURE 1.25 Bottom-hole flowing pressure as a function of time.

Example 1.18

An oil well is producing under a constant bottom-hole flowing pressure of 1500 psi. The current average reservoir pressure $\bar{p}_r$ is 3200 psi. The well is developed in the center of 40 acre square-drilling pattern. Given the following additional information:

$\phi = 16\%$,	$h = 15$ ft,	$k = 50$ md
$\mu = 26$ cp,	$B_o = 1.15$ bbl/STB	
$c_t = 10 \times 10^{-6}$ psi^{-1},	$r_w = 0.25$ ft	

Calculate the flow rate.

Solution

Because the volumetric average pressure is given, solve for the flow rate by applying Eq. (1.135):

$$Q = \frac{kh(\bar{p} - p_{wf})}{162.6 B\mu \log[2.2458A/C_A r_w^2]}$$
$$= \frac{(50)(15)(3200 - 1500)}{(162.6)(1.15)(2.6)\log[(2.2458(40)(43\,560))/((30.8828)(0.25^2))]}$$
$$= 416 \text{ STB/day}$$

It is interesting to note that Eq. (1.135) can also be presented in a dimensionless form by rearranging and introducing the dimensionless time t_D and dimensionless pressure drop p_D, to give:

$$p_D = 2\pi t_{DA} + \frac{1}{2}\ln\left(\frac{2.3458A}{C_A r_w^2}\right) + s \qquad \textbf{(1.136)}$$

with the dimensionless time based on the well drainage given by Eq. (1.86a) as:

$$t_{DA} = \frac{0.0002637kt}{\phi\mu c_t A} = t_A\left(\frac{r_w^2}{A}\right)$$

where

s = skin factor (to be introduced later in the chapter)
C_A = shape factor
t_{DA} = dimensionless time based on the well drainage area πr_e^2

Eq. (1.136) suggests that during the *boundary-dominated flow*, i.e., pseudosteady state, a plot of p_D vs. t_{DA} on a Cartesian scale would produce a straight line with a slope of 2π. That is:

$$\frac{\partial p_D}{\partial t_{DA}} = 2\pi \qquad \textbf{(1.137)}$$

For a well located in a circular drainage area with no skin, i.e., $s = 0$, and taking the logarithm of both sides of Eq. (1.136) gives:

$$\log(p_D) = \log(2\pi) + \log(t_{DA})$$

which indicates that a plot of p_D vs. t_{DA} on a log–log scale would produce a 45° straight line and an intercept of 2π.

1.2.9 Radial Flow of Compressible Fluids (Gases)

The radial diffusivity equation as expressed by Eq. (1.105) was developed to study the performance of a compressible fluid under unsteady-state conditions. The equation has the following form:

$$\frac{\partial^2 m(p)}{\partial r^2}+\frac{1}{r}\frac{\partial m(p)}{\partial r}=\frac{\phi\mu c_t}{0.000264k}\frac{\partial m(p)}{\partial t}$$

For semisteady-state flow, the rate of change of the real-gas pseudopressure with respect to time is constant. That is:

$$\frac{\partial m(p)}{\partial t}=\text{constant}$$

Using the same technique identical to that described previously for liquids gives the following exact solution to the diffusivity equation:

$$Q_g=\frac{kh[m(\overline{p}_r)-m(p_{wf})]}{1422T[\ln(r_e/r_w)-0.75]} \tag{1.138}$$

where

Q_g = gas flow rate, Mscf/day
T = temperature, °R
k = permeability, md

Two approximations to the above solution are widely used. These are:

(1) the pressure-squared approximation;
(2) the pressure approximation.

Pressure-squared Approximation Method. As outlined previously, this method provides us with results compatible to the exact solution approach when $p < 2000$ psi. The solution has the following familiar form:

$$Q_g=\frac{kh(\overline{p}_r^2-p_{wf}^2)}{1422T\overline{\mu}\overline{Z}(\ln(r_e/r_w)-0.75)} \tag{1.139}$$

The gas properties $\overline{Z}$ and $\overline{\mu}$ are evaluated at:

$$\overline{p}=\sqrt{\frac{\overline{p}_r^2+p_{wf}^2}{2}}$$

where

Q_g = gas flow rate, Mscf/day
T = temperature, °R
k = permeability, md

Pressure Approximation Method. This approximation method is applicable at $p > 3000$ psi and has the following mathematical form:

$$Q_g=\frac{kh(\overline{p}_r-p_{wf})}{1422\overline{\mu}\overline{B}_g[\ln(r_e/r_w)-0.75]} \tag{1.140}$$

with the gas properties evaluated at:

$$\overline{p}=\frac{\overline{p}_r+p_{wf}}{2}$$

where

Q_g = gas flow rate, Mscf/day
k = permeability, md
$\overline{B}_g$ = gas formation volume factor at an average pressure, bbl/scf

The gas formation volume factor is given by the following expression:

$$B_g=0.00504\overline{Z}T/\overline{p}$$

In deriving the flow equations, the following two main assumptions were made:

(1) uniform permeability throughout the drainage area;
(2) laminar (viscous) flow.

Before using any of the previous mathematical solutions to the flow equations, the solution must be modified to account for the possible deviation from the above two assumptions. Introducing the following two correction factors into the solution of the flow equation can eliminate these two assumptions:

(1) skin factor;
(2) turbulent flow factor.

1.2.10 Skin Factor

It is not unusual during drilling, completion, or workover operations for materials such as mud filtrate, cement slurry, or clay particles to enter the formation and reduce the permeability around the wellbore. This effect is commonly referred to as "wellbore damage" and the region of altered permeability is called the "skin zone." This zone can extend from a few inches to several feet from the wellbore. Many other wells are stimulated by acidizing or fracturing, which in effect increases the permeability near the wellbore. Thus, the permeability near the wellbore is always different from the permeability away from the well where the formation has not been affected by drilling or stimulation. A schematic illustration of the skin zone is shown in Figure 1.26.

The effect of the skin zone is to alter the pressure distribution around the wellbore. In case of wellbore damage, the skin zone causes an additional pressure loss in the formation. In case of wellbore improvement, the opposite to that of wellbore damage occurs. If we refer to the pressure drop in the skin zone as $i \, \Delta p_{\text{skin}}$, Figure 1.27 compares the differences in the skin zone pressure drop for three possible outcomes.

- *First outcome*: $\Delta p_{\text{skin}} > 0$, which indicates an additional pressure drop due to wellbore damage, i.e., $k_{\text{skin}} < k$.
- *Second outcome*: $\Delta p_{\text{skin}} < 0$, which indicates less pressure drop due to wellbore improvement, i.e., $k_{\text{skin}} > k$.
- *Third outcome*: $\Delta p_{\text{skin}} = 0$, which indicates no changes in the wellbore condition, i.e., $k_{\text{skin}} = k$.

Hawkins (1956) suggested that the permeability in the skin zone, i.e., k_{skin}, is uniform and the pressure drop across the zone can be

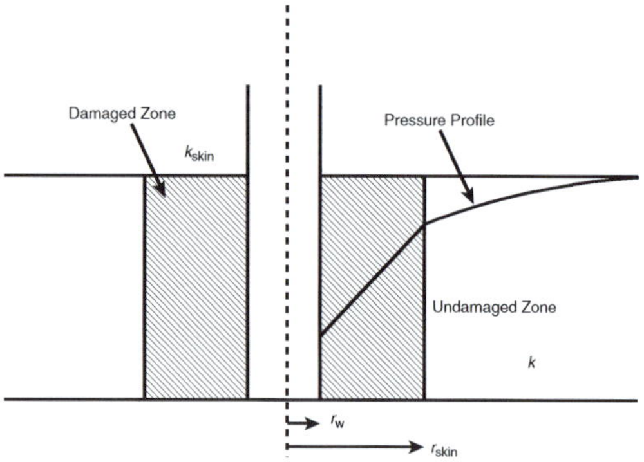

FIGURE 1.26 Near-wellbore skin effect.

$$\Delta p_{\text{skin}} = \left(\frac{Q_o B_o \mu_o}{0.00708 \, kh} \right) \left[\frac{k}{k_{\text{skin}}} - 1 \right] \ln \left(\frac{r_{\text{skin}}}{r_w} \right)$$

$$\Delta p_{\text{skin}} = \left[\frac{Q_o B_o \mu_o}{0.00708 \, kh} \right] s$$

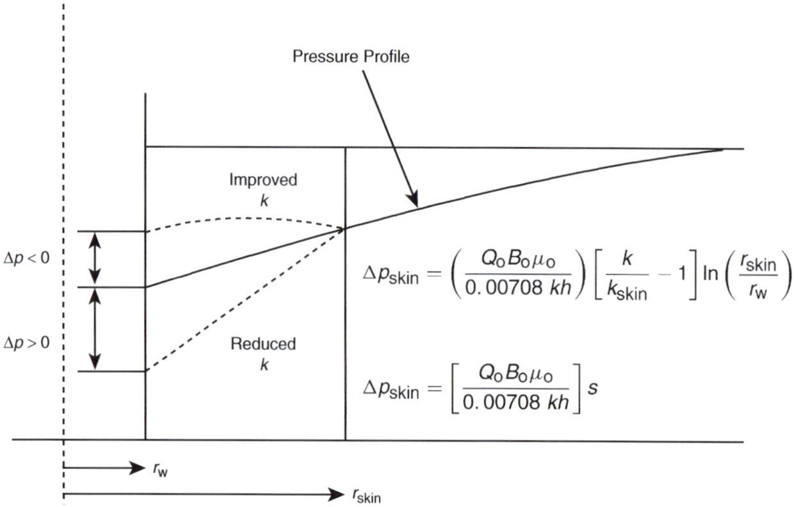

FIGURE 1.27 Representation of positive and negative skin effects.

approximated by Darcy's equation. Hawkins proposed the following approach:

$$\Delta p_{\text{skin}} = \begin{bmatrix} \Delta p \text{ in skin zone} \\ \text{due to } k_{\text{skin}} \end{bmatrix} - \begin{bmatrix} \Delta p \text{ in skin zone} \\ \text{due to } k \end{bmatrix}$$

Applying Darcy's equation gives:

$$(\Delta p)_{\text{skin}} = \left(\frac{Q_o B_o \mu_o}{0.00708 h k_{\text{skin}}}\right) \ln\left(\frac{r_{\text{skin}}}{r_w}\right) - \left(\frac{Q_o B_o \mu_o}{0.00708 h k}\right) \ln\left(\frac{r_{\text{skin}}}{r_w}\right)$$

or

$$\Delta p_{\text{skin}} = \left(\frac{Q_o B_o \mu_o}{0.00708 kh}\right) \left[\frac{k}{k_{\text{skin}}} - 1\right] \ln\left(\frac{r_{\text{skin}}}{r_w}\right)$$

where

k = permeability of the formation, md
k_{skin} = permeability of the skin zone, md

The above expression for determining the additional pressure drop in the skin zone is commonly expressed in the following form:

$$\Delta p_{\text{skin}} = \left(\frac{Q_o B_o \mu_o}{0.00708 kh}\right) s = 141.2 \left(\frac{Q_o B_o \mu_o}{kh}\right) s \quad \textbf{(1.141)}$$

where s is called the skin factor and defined as:

$$s = \left[\frac{k}{k_{\text{skin}}} - 1\right] \ln\left(\frac{r_{\text{skin}}}{r_w}\right) \quad \textbf{(1.142)}$$

Depending on the permeability ratio k/k_{skin} and if $\ln(r_{\text{skin}}/r_w)$ is always positive, there are only three possible outcomes in evaluating the skin factor s:

(1) *Positive skin factor*, $s > 0$: When the damaged zone near the wellbore exists, k_{skin} is less than k and, hence, s is a positive number. The magnitude of the skin factor increases as k_{skin} decreases and as the depth of the damage r_{skin} increases.
(2) *Negative skin factor*, $s < 0$: When the permeability around the well k_{skin} is higher than that of the formation k, a negative skin factor exists. This negative factor indicates an improved wellbore condition.
(3) *Zero skin factor*, $s = 0$: Zero skin factor occurs when no alternation in the permeability around the wellbore is observed, i.e., $k_{\text{skin}} = k$.

Eq. (1.142) indicates that a negative skin factor will result in a negative value of Δp_{skin}. This implies that a stimulated well will require less pressure drawdown to produce at rate q than an equivalent well with uniform permeability.

The proposed modification of the previous flow equation is based on the concept that the actual total pressure drawdown will increase or decrease by an amount Δp_{skin}. Assuming that $(\Delta p)_{\text{ideal}}$ represents the pressure drawdown for a drainage area with a uniform permeability k, then:

$$(\Delta p)_{\text{actual}} = (\Delta p)_{\text{ideal}} + (\Delta p)_{\text{skin}}$$

or

$$(p_i - p_{wf})_{\text{actual}} = (p_i - p_{wf})_{\text{ideal}} + \Delta p_{\text{skin}} \quad \textbf{(1.143)}$$

The above concept of modifying the flow equation to account for the change in the pressure drop due the wellbore skin effect can be applied to the previous three flow regimes:

(1) steady-state flow;
(2) unsteady-state (transient) flow;
(3) pseudosteady (semisteady)-state flow.

Basically, Eq. (1.143) can be applied as given in the following sections.

Steady-state Radial Flow (Accounting for the Skin Factor). Substituting Eqs. (1.26) and (1.141) into Eq. (1.143), gives:

$$(\Delta p)_{\text{actual}} = (\Delta p)_{\text{ideal}} + (\Delta p)_{\text{skin}}$$

$$(p_i - p_{wf})_{\text{actual}} = \left(\frac{Q_o B_o \mu_o}{0.00708 kh}\right) \ln\left(\frac{r_e}{r_w}\right) + \left(\frac{Q_o B_o \mu_o}{0.00708 kh}\right) s$$

Solving for the flow rate gives:

$$Q_o = \frac{0.00708 kh (p_i - p_{wf})}{\mu_o B_o [\ln(r_e/r_w) + s]} \quad \textbf{(1.144)}$$

where

Q_o = oil flow rate, STB/day
k = permeability, md

h = thickness, ft
s = skin factor
B_o = oil formation volume factor, bbl/STB
μ_o = oil viscosity, cp
p_i = initial reservoir pressure, psi
p_{wf} = bottom-hole flowing pressure, psi

Unsteady-State Radial Flow (Accounting for the Skin Factor) for Slightly Compressible Fluids. Combining Eqs. (1.82) and (1.141) with (1.143) yields:

$$(\Delta p)_{actual} = (\Delta p)_{ideal} + (\Delta p)_{skin}$$

$$p_i - p_{wf} = 162.6\left(\frac{Q_o B_o \mu_o}{kh}\right)\left[\log\frac{kt}{\phi\mu c_t r_w^2} - 3.23\right] + 141.2\left(\frac{Q_o B_o \mu_o}{kh}\right)s$$

or

$$p_i - p_{wf} = 162.6\left(\frac{Q_o B_o \mu_o}{kh}\right)\left[\log\frac{kt}{\phi\mu c_t r_w^2} - 3.23 + 0.87s\right] \quad (1.145)$$

For Compressible Fluids. A similar approach to that of the above gives:

$$m(p_i) - m(p_{wf}) = \frac{1637 Q_g T}{kh}\left[\log\frac{kt}{\phi\mu c_{t_i} r_w^2} - 3.23 + 0.87s\right] \quad (1.146)$$

and in terms of the pressure-squared approach, the difference $[m(p_i) - m(p_{wf})]$ can be replaced with:

$$m(p_i) - m(p_{wf}) = \int_{p_{wf}}^{p_i} \frac{2p}{\mu Z}\,dp = \frac{p_i^2 - p_{wf}^2}{\overline{\mu Z}}$$

to give

$$p_i^2 - p_{wf}^2 = \frac{1637 Q_g T\overline{Z}\overline{\mu}}{kh}\left[\log\frac{kt}{\phi\mu_i c_{ti} r_w^2} - 3.23 + 0.87s\right] \quad (1.147)$$

where

Q_g = gas flow rate, Mscf/day
T = temperature, °R
k = permeability, md
t = time, hours

Pseudosteady-State Flow (Accounting for the Skin Factor) for Slightly Compressible Fluids. Introducing the skin factor into Eq. (1.134) gives:

$$Q_o = \frac{0.00708\,kh(\overline{p}_r - p_{wf})}{\mu_o B_o[\ln(r_e/r_w) - 0.75 + s]} \quad (1.148)$$

For Compressible Fluids.

$$Q_g = \frac{kh[m(\overline{p}_r) - m(p_{wf})]}{1422\,T[\ln(r_e/r_w) - 0.75 + s]} \quad (1.149)$$

or in terms of the pressure-squared approximation:

$$Q_g = \frac{kh(\overline{p}_r^2 - p_{wf}^2)}{1422\,T\overline{\mu}\overline{Z}[\ln(r_e/r_w) - 0.75 + s]} \quad (1.150)$$

where

Q_g = gas flow rate, Mscf/day
k = permeability, md
T = temperature, °R
$\overline{\mu}_g$ = gas viscosity at average pressure $\overline{p}$, cp
$\overline{Z}_g$ = gas compressibility factor at average pressure $\overline{p}$

Example 1.19

Calculate the skin factor resulting from the invasion of the drilling fluid to a radius of 2 ft. The permeability of the skin zone is estimated at 20 md as compared with the unaffected formation permeability of 60 md. The wellbore radius is 0.25 ft.

Solution

Apply Eq. (1.142) to calculate the skin factor:

$$s = \left[\frac{60}{20} - 1\right]\ln\left(\frac{2}{0.25}\right) = 4.16$$

Matthews and Russell (1967) proposed an alternative treatment to the skin effect by introducing the "effective or apparent wellbore radius" r_{wa} that accounts for the pressure drop in the skin. They define r_{wa} by the following equation:

$$r_{wa} = r_w\,e^{-s} \quad (1.151)$$

All of the ideal radial flow equations can also be modified for the skin by simply replacing the wellbore radius r_w with that of the apparent wellbore radius r_{wa}. For example, Eq. (1.145) can be equivalently expressed as:

$$p_i - p_{wf} = 162.6\left(\frac{Q_o B_o \mu_o}{kh}\right)\left[\log\left(\frac{kt}{\phi\mu c_t r_{wa}^2}\right) - 3.23\right] \quad (1.152)$$

1.2.11 Turbulent Flow Factor

All of the mathematical formulations presented so far are based on the assumption that laminar flow conditions are observed during flow. During radial flow, the flow velocity increases as the wellbore is approached. This increase in the velocity might cause the development of turbulent flow around the wellbore. If turbulent flow does exist, it is most likely to occur with gases, and it causes an additional pressure drop similar to that caused by the skin effect. The term "non-Darcy flow" has been adopted by the industry to describe the additional pressure drop due to the turbulent (non-Darcy) flow.

Referring to the additional real-gas pseudopressure drop due to non-Darcy flow as $\Delta\psi_{\text{non-Darcy}}$, the total (actual) drop is given by:

$$(\Delta\psi)_{\text{actual}} = (\Delta\psi)_{\text{ideal}} + (\Delta\psi)_{\text{skin}} + (\Delta\psi)_{non-Darcy}$$

Wattenbarger and Ramey (1968) proposed the following expression for calculating $(\Delta\psi)_{\text{non-Darcy}}$:

$$(\Delta\psi)_{non-Darcy} = 3.161 \times 10^{-12}\left[\frac{\beta T \gamma_g}{\mu_{gw} h^2 r_w}\right] Q_g^2 \quad (1.153)$$

This equation can be simplified as:

$$(\Delta\psi)_{non-Darcy} = FQ_g^2 \quad (1.154)$$

where F is called the "non-Darcy flow coefficient" and is given by:

$$F = 3.161 \times 10^{-12}\left[\frac{\beta T \gamma_g}{\mu_{gw} h^2 r_w}\right] \quad (1.155)$$

where

Q_g = gas flow rate, Mscf/day
μ_{gw} = gas viscosity as evaluated at p_{wf}, cp
γ_g = gas specific gravity
h = thickness, ft
F = non-Darcy flow coefficient, psi^2/cp/(Mscf/day)^2
β = turbulence parameter

Jones (1987) proposed a mathematical expression for estimating the turbulence parameter β as:

$$\beta = 1.88(10^{-10})(k)^{-1.47}(\phi)^{-0.53} \quad (1.156)$$

where

k = permeability, md
ϕ = porosity, fraction

The term FQ_g^2 can be included in all the compressible gas flow equations in the same way as the skin factor. This non-Darcy term is interpreted as a *rate-dependent skin*. The modification of the gas flow equations to account for the turbulent flow condition is given below for the three flow regimes:

(1) unsteady-state (transient) flow;
(2) semisteady-state flow;
(3) steady-state flow.

Unsteady-state Radial Flow. The gas flow equation for an unsteady-state flow is given by Eq. (1.146) and can be modified to include the additional drop in the real-gas potential, as:

$$m(p_i) - m(p_{wf}) = \left(\frac{1637 Q_g T}{kh}\right)\left[\log\left(\frac{kt}{\phi\mu_i c_{ti} r_w^2}\right) - 3.23 + 0.87s\right] + FQ_g^2 \quad (1.157)$$

Eq. (1.157) is simplified as:

$$m(p_i) - m(p_{wf}) = \left(\frac{1637 Q_g T}{kh}\right)\left[\log\left(\frac{kt}{\phi\mu_i c_{ti} r_w^2}\right) - 3.23 + 0.87s + 0.87DQ_g\right] \quad (1.158)$$

where the term DQ_g is interpreted as the rate-dependent skin factor. The coefficient D is called the "inertial or turbulent flow factor" and given by:

$$D = \frac{Fkh}{1422T} \tag{1.159}$$

The true skin factor s which reflects the formation damage or stimulation is usually combined with the non-Darcy rate-dependent skin and labeled as the apparent or total skin factor $s^{\backslash}$. That is:

$$s^{\backslash} = s + DQ_g \tag{1.160}$$

or

$$m(p_i) - m(p_{wf}) = \left(\frac{1637 Q_g T}{kh}\right)\left[\log\left(\frac{kt}{\phi\mu_i c_{ti} r_w^2}\right) - 3.23 + 0.87 s^{\backslash}\right] \tag{1.161}$$

Eq. (1.61) can be expressed in the pressure-squared approximation form as:

$$p_i^2 - p_{wf}^2 \left(\frac{1637 Q_g T \bar{Z} \bar{\mu}}{kh}\right)\left[\log\frac{kt}{\phi\mu_i c_{ti} r_w^2} - 3.23 + 0.87 s^{\backslash}\right] \tag{1.162}$$

where

Q_g = gas flow rate, Mscf/day
t = time, hours
k = permeability, md
μ_i = gas viscosity as evaluated at p_i, cp

Semisteady-state Flow. Eqs. (1.149) and (1.150) can be modified to account for the non-Darcy flow as follows:

$$Q_g = \frac{kh[m(\bar{p}_r) - m(p_{wf})]}{1422T[\ln(r_e/r_w) - 0.75 + s + DQ_g]} \tag{1.163}$$

or in terms of the pressure-squared approach:

$$Q_g = \frac{kh(\bar{p}_r^2 - p_{wf}^2)}{1422T\bar{\mu}\bar{Z}[\ln(r_e/r_w) - 0.75 + s + DQ_g]} \tag{1.164}$$

where the coefficient D is defined as:

$$D = \frac{Fkh}{1422T} \tag{1.165}$$

Steady-state Flow. Similar to the above modification procedure, Eqs. (1.43) and (1.44) can be expressed as:

$$Q_g = \frac{kh[m(p_i) - m(p_{wf})]}{1422T[\ln(r_e/r_w) - 0.5 + s + DQ_g]} \tag{1.166}$$

$$Q_g = \frac{kh(p_e^2 - p_{wf}^2)}{1422T\bar{\mu}\bar{Z}[\ln(r_e/r_w) - 0.5 + s + DQ_g]} \tag{1.167}$$

Example 1.20

A gas well has an estimated wellbore damage radius of 2 ft and an estimated reduced permeability of 30 md. The formation has permeability and porosity of 55 md and 12%, respectively. The well is producing at a rate of 20 MMscf/day with a gas gravity of 0.6. The following additional data is available:

$r_w = 0.25$, $h = 20$ ft, $T = 140\,°F$, $\mu_{gw} = 0.013$ cp

Calculate the apparent skin factor.

Solution

Step 1. Calculate skin factor from Eq. (1.142):

$$s = \left[\frac{k}{k_{skin}} - 1\right]\ln\left(\frac{r_{skin}}{r_w}\right) = \left[\frac{55}{30} - 1\right]\ln\left(\frac{2}{0.25}\right) = 1.732$$

Step 2. Calculate the turbulence parameter β by applying Eq. (1.156):

$$\beta = 1.88(10^{-10})(k)^{-1.47}(\phi)^{-0.53}$$
$$= 1.88 \times 10^{10}(55)^{-1.47}(0.12)^{-0.53}$$
$$= 159.904 \times 10^6$$

Step 3. Calculate the non-Darcy flow coefficient from Eq. (1.155):

$$F = 3.161 \times 10^{-12}\left[\frac{\beta T \gamma_g}{\mu_{gw} h^2 r_w}\right]$$
$$= 3.1612 \times 10^{-12}\left[\frac{159.904 \times 10^6(600)(0.6)}{(0.013)(20)^2(0.25)}\right]$$
$$= 0.14$$

Step 4. Calculate the coefficient D from Eq. (1.159):

$$D = \frac{Fkh}{1422T} = \frac{(0.14)(55)(20)}{(1422)(600)} = 1.805 \times 10^{-4}$$

Step 5. Estimate the apparent skin factor by applying Eq. (1.160):

$$s' = s + DQ_g = 1.732 + (1.805 \times 10^{-4})(20,000) = 5.342$$

1.2.12 Principle of Superposition

The solutions to the radial diffusivity equation, as presented earlier in this chapter, appear to be applicable only for describing the pressure distribution in an infinite reservoir that was caused by constant production from a single well. Since real reservoir systems usually have several wells that are operating at varying rates, a more generalized approach is needed to study the fluid flow behavior during the unsteady-state flow period.

The principle of superposition is a powerful concept that can be applied to remove the restrictions that have been imposed on various forms of solution to the transient flow equation. Mathematically, the superposition theorem states that any sum of individual solutions to the diffusivity equation is also a solution to that equation. This concept can be applied to account for the following effects on the transient flow solution:

- effects of multiple wells;
- effects of rate change;
- effects of the boundary;
- effects of pressure change.

Slider (1976) presented an excellent review and discussion of the practical applications of the principle of superposition in solving a wide variety of unsteady-state flow problems.

Effects of Multiple Wells. Frequently, it is desired to account for the effects of more than one well on the pressure at some point in the reservoir. The superposition concept states that the total pressure drop at any point in the reservoir is the sum of the pressure changes at that point caused by the flow in each of the wells in the reservoir. In other words, we simply superimpose one effect upon another.

Figure 1.28 shows three wells that are producing at different flow rates from an infinite-acting reservoir, i.e., an unsteady-state flow reservoir. The principle of superposition states that the total pressure drop observed at any well, e.g., well 1, is:

$$(\Delta p)_{\text{total drop at well 1}} = (\Delta p)_{\text{drop due to well 1}} + (\Delta p)_{\text{drop due to well 2}} + (\Delta p)_{\text{drop due to well 3}}$$

The pressure drop at well 1 due to its own production is given by the *log approximation* to the Ei function solution presented by Eq. (1.145), or:

$$(p_i - p_{wf}) = (\Delta p)_{\text{well 1}} = \frac{162.6 Q_{o1} B_o \mu_o}{kh}\left[\log\left(\frac{kt}{\phi\mu c_t r_w^2}\right) - 3.23 + 0.87s\right]$$

where

t = time, hours
s = skin factor
k = permeability, md
Q_{o1} = oil flow rate from well 1

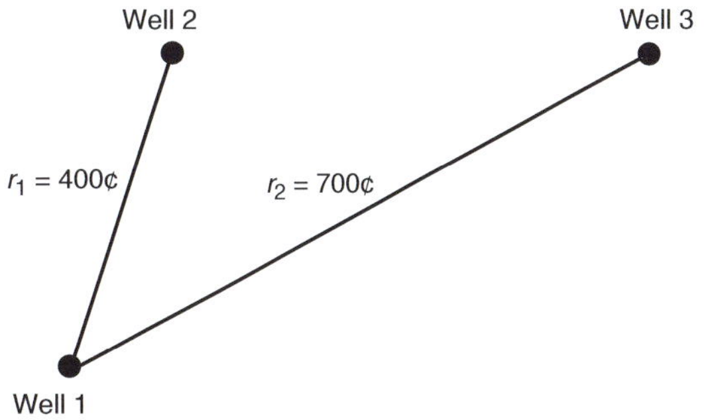

FIGURE 1.28 Well layout for Example 1.21.

The additional pressure drops at well 1 due to the production from wells 2 and 3 must be written in terms of the Ei function solution, as expressed by Eq. (1.77), since the log approximation cannot be applied in calculating the pressure at a large distance r from the well where $x > 0.1$. Therefore:

$$p(r,t) = p_i + \left[\frac{70.6Q_o\mu_o B_o}{kh}\right]\text{Ei}\left[\frac{-948\phi\mu_o c_t r^2}{kt}\right]$$

Applying the above expression to calculate the additional pressure drop due to two wells gives:

$$(\Delta p)_{\text{drop due to well 2}} = p_i - p(r_1,t) = -\left[\frac{70.6Q_{o1}\mu_o B_o}{kh}\right] \times \text{Ei}\left[\frac{-948\phi\mu_o c_t r_1^2}{kt}\right]$$

$$(\Delta p)_{\text{drop due to well 3}} = p_i - p(r_2,t) = -\left[\frac{70.6Q_{o2}\mu_o B_o}{kh}\right] \times \text{Ei}\left[\frac{-948\phi\mu_o c_t r_2^2}{kt}\right]$$

The total pressure drop is then given by:

$$(p_i - p_{wf})_{\text{total at well 1}} = \left(\frac{162.6Q_{o1}B_o\mu_o}{kh}\right)\left[\log\left(\frac{kt}{\phi\mu c_t r_w^2}\right) - 3.23 + 0.87s\right] - \left(\frac{70.6Q_{o2}B_o\mu_o}{kh}\right)\text{Ei}\left[-\frac{948\phi\mu c_t r_1^2}{kt}\right] - \left(\frac{70.6Q_{o3}B_o\mu_o}{kh}\right)\text{Ei}\left[-\frac{948\phi\mu c_t r_2^2}{kt}\right]$$

where

Q_{o1}, Q_{o2}, Q_{o3} = respective producing rates of wells 1, 2, 3

The above computational approach can be used to calculate the pressure at wells 2 and 3. Further, it can be extended to include any number of wells flowing under the unsteady-state flow condition. It should also be noted that if the point of interest is an operating well, the skin factor s must be included for that well only.

Example 1.21

Assume that the three wells, as shown in Figure 1.28, are producing under a transient flow condition for 15 hours. The following additional data is available:

$Q_{o1} = 100$ STB/day, $Q_{o2} = 160$ STB/day
$Q_{o3} = 200$ STB/day, $p_i = 4500$ psi
$B_o = 1.20$ bbl/STB, $c_t = 20 \times 10^{-6}$ psi^{-1}
$(s)_{\text{well 1}} = -0.5$, $h = 20$ ft
$\phi = 15\%$, $k = 40$ md
$r_w = 0.25$ ft, $\mu_o = 2.0$ cp
$r_1 = 400$ ft, $r_2 = 700$ ft

If the three wells are producing at a constant flow rate, calculate the sand face flowing pressure at well 1.

Solution

Step 1. Calculate the pressure drop at well 1 caused by its own production by using Eq. (1.145):

$$(p_i - p_{wf}) = (\Delta p)_{\text{well 1}} = \frac{162.6Q_{o1}B_o\mu_o}{kh} \times \left[\log\left(\frac{kt}{\phi\mu c_t r_w^2}\right) - 3.23 + 0.87s\right]$$

$$(\Delta p)_{\text{drop due to well 1}} = \frac{(162.6)(100)(1.2)(2.0)}{(40)(20)} \times [\log\left(\frac{(40)(15)}{(0.15)(2)(20\times10^{-6})(0.25)^2}\right) - 3.23 + 0.87(0)] = 270.2 \text{ psi}$$

Step 2. Calculate the pressure drop at well 1 due to the production from well 2:

$$(\Delta p)_{\text{drop due to well 2}} = p_i - p(r_1,t) = -\left[\frac{70.6Q_{o1}\mu_o B_o}{kh}\right]\text{Ei}\left[\frac{-948\phi\mu_o c_t r_1^2}{kt}\right]$$

$$(\Delta p)_{\text{drop due to well 2}} = -\frac{(70.6)(160)(1.2)(2)}{(40)(20)} \times \text{Ei}\left[-\frac{(948)(0.15)(2.0)(20\times10^{-6})(400)^2}{(40)(15)}\right]$$

$$= 33.888[-\text{Ei}(-1.5168)]$$

$$= (33.888)(0.13) = 4.41 \text{ psi}$$

Step 3. Calculate the pressure drop due to production from well 3:

$$(\Delta p)_{\text{drop due to well 3}} = p_i - p(r_2, t)$$
$$= -\left[\frac{70.6Q_{o2}\mu_o B_o}{kh}\right]\text{Ei}\left[\frac{-948\phi\mu_o c_t r_2^2}{kt}\right]$$
$$(\Delta p)_{\text{drop due to well 3}} = -\frac{(70.6)(200)(1.2)(2)}{(40)(20)}$$
$$\times\,\text{Ei}\left[-\frac{(948)(0.15)(2.0)(20\times 10^{-6})(700)^2}{(40)(15)}\right]$$
$$= (42.36)[-\text{Ei}(-4.645)]$$
$$= (42.36)(1.84\times 10^{-3}) = 0.08 \text{ psi}$$

Step 4. Calculate the total pressure drop at well 1:

$$(\Delta p)_{\text{total at well 1}} = 270.2 + 4.41 + 0.08 = 274.69 \text{ psi}$$

Step 5. Calculate p_{wf} at well 1:

$$p_{wf} = 4500 - 274.69 = 4225.31 \text{ psi}$$

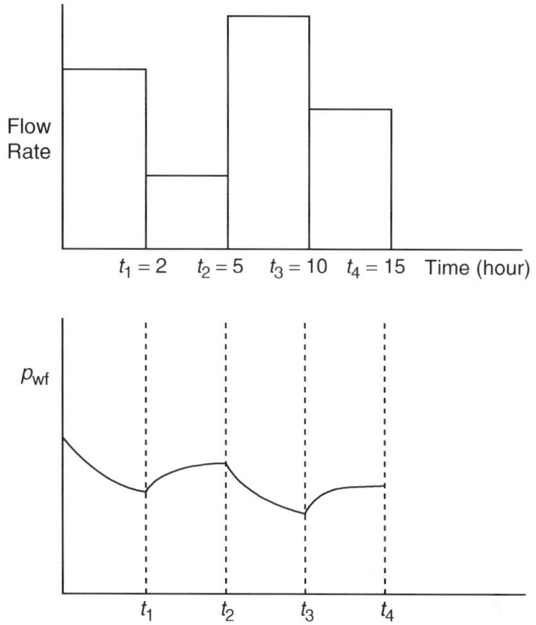

FIGURE 1.29 Production and pressure history of a well.

Effects of Variable Flow Rates. All of the mathematical expressions presented previously in this chapter require that the wells produce at a constant rate during the transient flow periods. Practically all wells produce at varying rates and, therefore, it is important that we are able to predict the pressure behavior when the rate changes. For this purpose, the concept of superposition states that "Every flow rate change in a well will result in a pressure response which is independent of the pressure responses caused by the other previous rate changes." Accordingly, the total pressure drop that has occurred at any time is the summation of pressure changes caused separately by each net flow rate change.

Consider the case of a shut-in well, i.e., $Q = 0$, that was then allowed to produce at a series of constant rates for the different time periods shown in Figure 1.29. To calculate the total pressure drop at the sand face at time t_4, the composite solution is obtained by adding the individual constant-rate solutions at the specified rate-time sequence, or:

$$(\Delta p)_{\text{total}} = (\Delta p)_{\text{due to }(Q_{o1}-0)} + (\Delta p)_{\text{due to }(Q_{o2}-Q_{o1})} + (\Delta p)_{\text{due to }(Q_{o3}-Q_{o2})} + (\Delta p)_{\text{due to }(Q_{o4}-Q_{o3})}$$

The above expression indicates that there are four contributions to the total pressure drop resulting from the four individual flow rates.

The first contribution results from increasing the rate from 0 to Q_1 and is in effect over the entire time period t_4, thus:

$$(\Delta p)_{Q_1-0} = \left[\frac{162.6(Q_1-0)B\mu}{kh}\right] \times \left[\log\left(\frac{kt_4}{\phi\mu c_t r_w^2}\right) - 3.23 + 0.87s\right]$$

It is essential to notice the *change* in the rate, i.e., (new rate − old rate), that is used in the above equation. It is the change in the rate that causes the pressure disturbance. Further, it should be noted that the "time" in the equation represents the total elapsed time since the change in the rate has been in effect.

The second contribution results from decreasing the rate from Q_1 to Q_2 at t_1, thus:

$$(\Delta p)_{Q_2-Q_1} = \left[\frac{162.6(Q_2-Q_1)B\mu}{kh}\right] \times \left[\log\left(\frac{k(t_4-t_1)}{\phi\mu c_t r_w^2}\right) - 3.23 + 0.87s\right]$$

Using the same concept, the two other contributions from Q_2 to Q_3 and from Q_3 to Q_4 can be computed as:

$$(\Delta p)_{Q_3-Q_2} = \left[\frac{162.6(Q_3-Q_2)B\mu}{kh}\right] \times \left[\log\left(\frac{k(t_4-t_2)}{\phi\mu c_t r_w^2}\right) - 3.23 + 0.87s\right]$$

$$(\Delta p)_{Q_4-Q_3} = \left[\frac{162.6(Q_4-Q_3)B\mu}{kh}\right] \times \left[\log\left(\frac{k(t_4-t_3)}{\phi\mu c_t r_w^2}\right) - 3.23 + 0.87s\right]$$

The above approach can be extended to model a well with several rate changes. Note, however, that the above approach is valid only if the well is flowing under the unsteady-state flow condition for the total time elapsed since the well began to flow at its initial rate.

Example 1.22

Figure 1.29 shows the rate history of a well that is producing under transient flow conditions for 15 hours. Given the following data:

$p_i = 5000$ psi,	$h = 20$ ft,	$B_o = 1.1$ bbl/STB
$\phi = 15\%$,	$\mu_o = 2.5$ cp,	$r_w = 0.3$ ft
$c_t = 20 \times 10^{-6}$ psi^{-1},	$s = 0$,	$k = 40$ md

Calculate the sand face pressure after 15 hours.

Solution

Step 1. Calculate the pressure drop due to the first flow rate for the entire flow period:

$$(\Delta p)_{Q_1-0} = \frac{(162.6)(100-0)(1.1)(2.5)}{(40)(20)} \times \left[\log\left(\frac{(40)(15)}{(0.15)(2.5)(20\times10^{-6})(0.3)^2}\right) - 3.23 + 0\right] = 319.6 \text{ psi}$$

Step 2. Calculate the additional pressure change due to the change of the flow rate from 100 to 70 STB/day:

$$(\Delta p)_{Q_2-Q_1} = \frac{(162.6)(70-100)(1.1)(2.5)}{(40)(20)} \times \left[\log\left[\frac{(40)(15-2)}{(0.15)(2.5)(20\times10^{-6})(0.3)^2}\right] - 3.23\right] = -94.85 \text{ psi}$$

Step 3. Calculate the additional pressure change due to the change of the flow rate from 70 to 150 STB/day:

$$(\Delta p)_{Q_3-Q_2} = \frac{(162.6)(150-70)(1.1)(2.5)}{(40)(20)} \times \left[\log\left(\frac{(40)(15-5)}{(0.15)(2.5)(20\times10^{-6})(0.3)^2}\right) - 3.23\right] = 249.18 \text{ psi}$$

Step 4. Calculate the additional pressure change due to the change of the flow rate from 150 to 85 STB/day:

$$(\Delta p)_{Q_4-Q_3} = \frac{(162.6)(85-150)(1.1)(2.5)}{(40)(20)} \times \left[\log\left[\frac{(40)(15-10)}{(0.15)(2.5)(20\times10^{-6})(0.3)^2}\right] - 3.23\right] = -190.44 \text{ psi}$$

Step 5. Calculate the total pressure drop:

$$(\Delta p)_{total} = 319.6 + (-94.85) + 249.18 + (-190.44) = 283.49 \text{ psi}$$

Step 6. Calculate the wellbore pressure after 15 hours of transient flow:

$$p_{wf} = 5000 - 283.49 = 4716.51 \text{ psi}$$

Effects of the Reservoir Boundary. The superposition theorem can also be extended to predict the pressure of a well in a bounded reservoir. Figure 1.30 shows a well that is located at a distance L from the non-flow boundary, e.g., sealing fault. The no-flow boundary can be

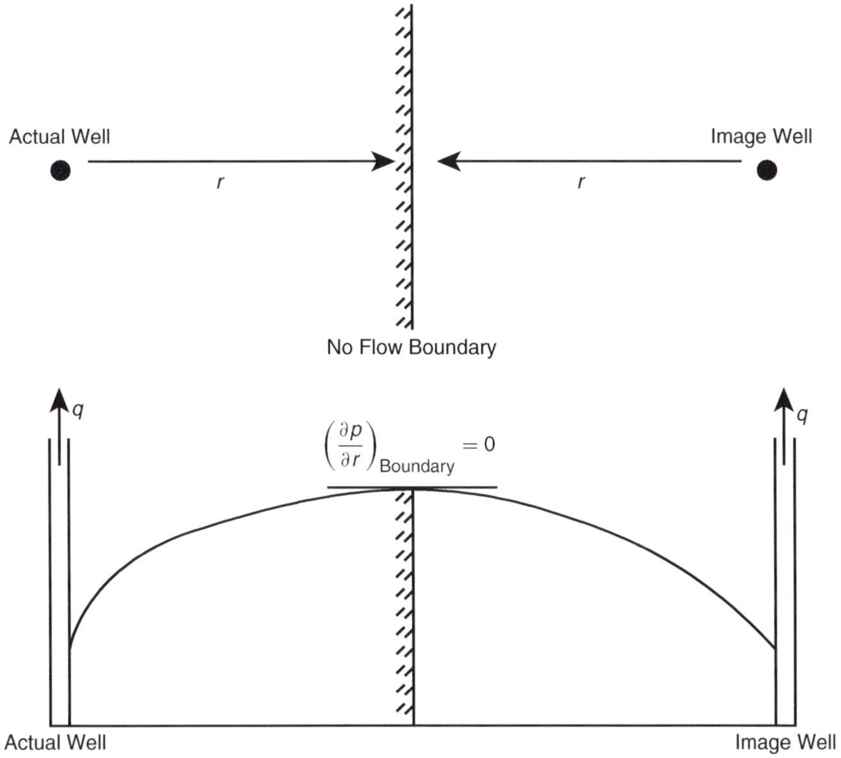

FIGURE 1.30 Method of images in solving boundary problems.

represented by the following pressure gradient expression:

$$\left(\frac{\partial p}{\partial L}\right)_{\text{Boundary}} = 0$$

Mathematically, the above boundary condition can be met by placing an *image* well, identical to that of the actual well, on the other side of the fault at exactly distance L. Consequently, the effect of the boundary on the pressure behavior of a well would be the same as the effect from an image well located at a distance $2L$ from the actual well.

The superposition method used for accounting the boundary effects is frequently called the *method of images*. Thus, for the problem of the system configuration given in Figure 1.30, the problem reduces to one of determining the effect of the image well on the actual well. The total pressure drop at the actual well will be the pressure drop due to its own production plus the additional pressure drop caused by an identical well at a distance of $2L$, or:

$$(\Delta p)_{\text{total}} = (\Delta p)_{\text{actual well}} + (\Delta p)_{\text{due to image well}}$$

or

$$(\Delta p)_{\text{total}} = \frac{162.6 Q_o B \mu}{kh}\left[\log\left(\frac{kt}{\phi \mu c_t r_w^2}\right) - 3.23 + 0.87s\right] - \left(\frac{70.6 Q_o B \mu}{kh}\right) \text{Ei}\left(-\frac{948 \phi \mu c_t (2L)^2}{kt}\right) \quad \textbf{(1.168)}$$

Eq. (1.168) assumes that the reservoir is infinite except for the indicated boundary. The effect of boundaries always causes a greater pressure drop than that calculated for infinite reservoirs.

The concept of image wells can be extended to generate the pressure behavior of a well located within a variety of boundary configurations.

Example 1.23

Figure 1.31 shows a well located between two sealing faults at 400 and 600 ft from the two faults. The well is producing under a transient flow condition at a constant flow rate of 200 STB/day given as follows:

$p_i = 500$ psi, $k = 600$ md, $B_o = 1.1$ bbl/STB
$\phi = 17\%$, $\mu_o = 2.0$ cp, $h = 25$ ft
$r_w = 0.3$ ft, $s = 0$, $c_t = 25 \times 10^{-6}$ psi^{-1}

Calculate the sand face pressure after 10 hours.

Solution

Step 1. Calculate the pressure drop due to the actual well flow rate:

$$(p_i - p_{wf}) = (\Delta p)_{actual} = \frac{162.6 Q_{o1} B_o \mu_o}{kh} \times \left[\log\left(\frac{kt}{\phi \mu c_t r_w^2}\right) - 3.23 + 0.87 s\right]$$

$$(\Delta p)_{actual} = \frac{(162.6)(200)(1.1)(2.0)}{(60)(25)} \times \left[\log\left(\frac{(60)(10)}{(0.17)(2)(25 \times 10^{-6})(0.3)^2}\right) - 3.23 + 0\right]$$

$$= 270.17 \text{ psi}$$

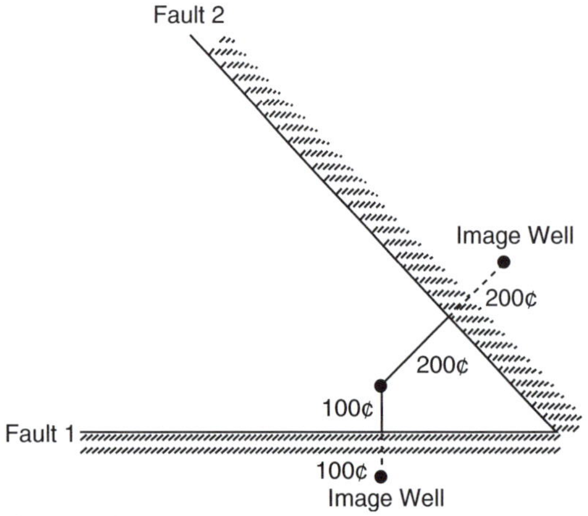

FIGURE 1.31 Well layout for Example 1.23.

Step 2. Determine the additional pressure drop due to the first fault (i.e., image well 1):

$$(\Delta p)_{\text{image well 1}} = p_i - p(2L_1, t)$$

$$= -\left[\frac{70.6 Q_{o2} \mu_o B_o}{kh}\right] \text{Ei}\left[\frac{-948 \phi \mu_o c_t (2L_1)^2}{kt}\right]$$

$$(\Delta p)_{\text{image well 1}} = -\frac{(70.6)(200)(1.1)(2.0)}{(60)(25)} \times \text{Ei}\left[-\frac{(948)(0.17)(2)(25 \times 10^{-6})(2 \times 100)^2}{(60)(10)}\right]$$

$$= 20.71[-\text{Ei}(-0.537)] = 10.64 \text{ psi}$$

Step 3. Calculate the effect of the second fault (i.e., image well 2):

$$(\Delta p)_{\text{image well 2}} = p_i - p(2L_2, t)$$

$$= -\left[\frac{70.6 Q_{o2} \mu_o B_o}{kh}\right] \text{Ei}\left[\frac{-948 \phi \mu_o c_t (2L_2)^2}{kt}\right]$$

$$(\Delta p)_{\text{image well 2}} = 20.71\left[-\text{Ei}\left(\frac{-948(0.17)(2)(25 \times 10^{-6})(2 \times 200)^2}{(60)(10)}\right)\right]$$

$$= 20.71[-\text{Ei}(-2.15)] = 1.0 \text{ psi}$$

Step 4. The total pressure drop is:

$$(\Delta p)_{total} = 270.17 + 10.64 + 1.0 = 28.18 \text{ psi}$$

Step 5.

$$p_{wf} = 5000 - 281.8 = 4718.2 \text{ psi}$$

Accounting for Pressure-change Effects. Superposition is also used in applying the constant-pressure case. Pressure changes are accounted for in this solution in much the same way that rate changes are accounted for in the constant-rate case. The description of the superposition method to account for the pressure-change effect is fully described in Chapter 2.

1.3 TRANSIENT WELL TESTING

Detailed reservoir information is essential to the petroleum engineer in order to analyze the current behavior and future performance of the reservoir. Pressure transient testing is designed to provide the engineer with a quantitative analysis of the reservoir properties. A transient test is essentially conducted by creating a pressure disturbance in the reservoir and recording the pressure response at the wellbore, i.e., bottom-hole flowing pressure p_{wf}, as a function of time. The pressure transient tests most commonly used in the petroleum industry include:

- pressure drawdown;
- pressure buildup;
- multirate;
- interference;
- pulse;
- drill stem (DST);
- falloff;
- injectivity;
- step rate.

It should be pointed out that when the flow rate is changed and the pressure response is recorded in the same well, the test is called a "single-well" test. Drawdown, buildup, injectivity, falloff, and step-rate tests are examples of a single-well test. When the flow rate is changed in one well and the pressure response is measured in another well(s), the test is called a "multiple-well" test.

Several of the above listed tests are briefly described in the following sections.

It has long been recognized that the pressure behavior of a reservoir following a rate change directly reflects the geometry and flow properties of the reservoir. Some of the information that can be obtained from a well test includes:

Drawdown tests	Pressure profile
	Reservoir behavior
	Permeability
	Skin
	Fracture length
	Reservoir limit and shape
Buildup tests	Reservoir behavior
	Permeability
	Fracture length
	Skin
	Reservoir pressure
	Boundaries
DST	Reservoir behavior
	Permeability
	Skin
	Fracture length
	Reservoir limit
	Boundaries
Falloff tests	Mobility in various banks
	Skin
	Reservoir pressure
	Fracture length
	Location of front
	Boundaries
Interference and pulse tests	Communication between wells
	Reservoir-type behavior
	Porosity
	Interwell permeability
	Vertical permeability
Layered reservoir tests	Horizontal permeability
	Vertical permeability
	Skin
	Average layer pressure
	Outer boundaries
Step-rate tests	Formation parting pressure
	Permeability
	Skin

There are several excellent technical and reference books that comprehensively and thoroughly address the subject of well testing and transient flow analysis, in particular:

- C.S. Matthews and D.G. Russell, *Pressure Buildup and Flow Test in Wells* (1967);
- Energy Resources Conservation Board (ERBC), *Theory and Practice of the Testing of Gas Wells* (1975);
- Robert Earlougher, *Advances in Well Test Analysis* (1977);
- John Lee, *Well Testing* (1982);
- M.A. Sabet, *Well Test Analysis* (1991);
- Roland Horn, *Modern Well Test Analysis* (1995).

1.3.1 Drawdown Test

A pressure drawdown test is simply a series of bottom-hole pressure measurements made during a period of flow at constant producing rate. Usually the well is shut in prior to the flow test for a period of time sufficient to allow the pressure to equalize throughout the formation, i.e., to reach static pressure. A schematic of the ideal flow rate and pressure history is shown in Figure 1.32.

The fundamental objectives of drawdown testing are to obtain the average permeability, k, of the reservoir rock within the drainage area of the well, and to assess the degree of damage of stimulation induced in the vicinity of the wellbore through drilling and completion practices. Other objectives are to determine the pore volume and to detect reservoir inhomogeneities within the drainage area of the well.

When a well is flowing at a constant rate of Q_o under the unsteady-state condition, the pressure behavior of the well will act as if it exists in an infinite-size reservoir. The pressure behavior during this period is described by Eq. (1.145) as:

$$p_{wf} = p_i - \frac{162.6 Q_o B_o \mu}{kh}\left[\log\left(\frac{kt}{\phi \mu c_t r_w^2}\right) - 3.23 + 0.87s\right]$$

where

k = permeability, md
t = time, hours
r_w = wellbore radius, ft
s = skin factor

The above expression can be written as:

$$p_{wf} = p_i - \frac{162.6 Q_o B_o \mu}{kh} \times \left[\log(t) + \log\left(\frac{k}{\phi \mu c_t r_w^2}\right) - 3.23 + 0.87s\right] \quad \textbf{(1.169)}$$

This relationship is essentially an equation of a straight line and can be expressed as:

$$p_{wf} = a + m \log(t)$$

where

$$a = p_i - \frac{162.6 Q_o B_o \mu}{kh}\left[\log\left(\frac{k}{\phi \mu c_t r_w^2}\right) - 3.23 + 0.87s\right]$$

and the slope m is given by:

$$-m = \frac{-162.6 Q_o B_o \mu_o}{kh} \quad \textbf{(1.170)}$$

Eq. (1.169) suggests that a plot of p_{wf} vs. time t on semilog graph paper would yield a straight line with a slope m in psi/cycle. This semilog straight-line portion of the drawdown data, as shown in Figure 1.33, can also be expressed in another convenient form by employing the definition of the slope:

$$m = \frac{p_{wf} - p_{1\,hour}}{\log(t) - \log(1)} = \frac{p_{wf} - p_{1\,hour}}{\log(t) - 0}$$

or

$$p_{wf} = m \log(t) + p_{1\,hour}$$

Eq. (1.170) can also be rearranged to determine the capacity kh of the drainage area of the

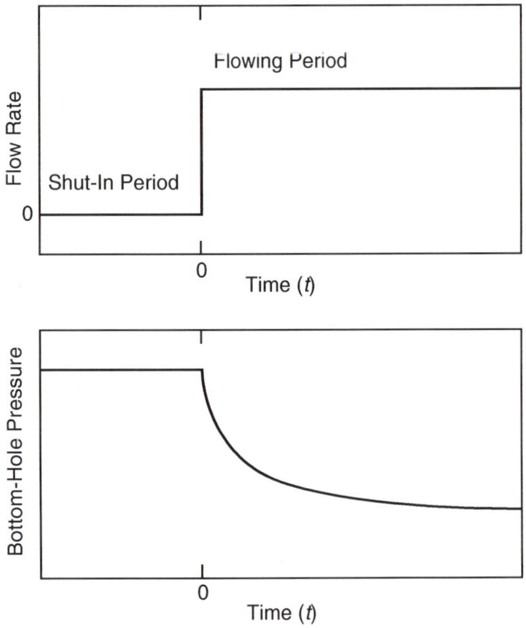

FIGURE 1.32 Idealized drawdown test.

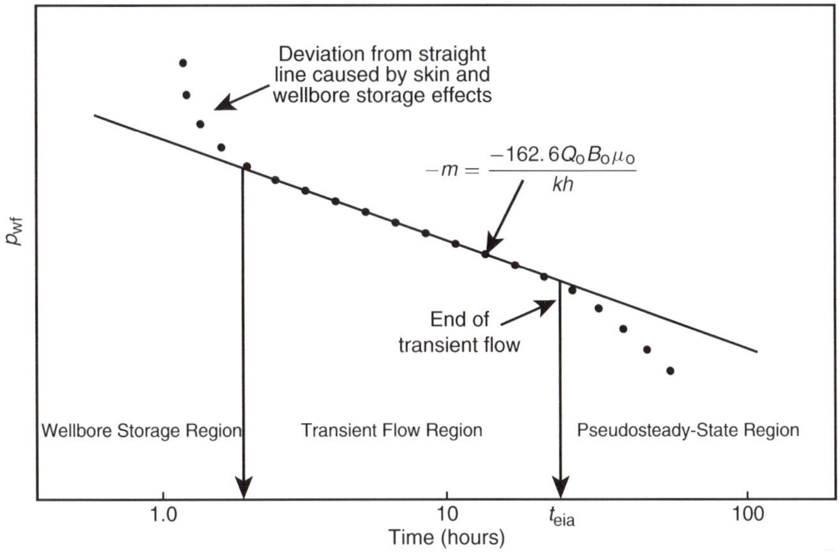

FIGURE 1.33 Semilog plot of pressure drawdown data.

well. If the thickness is known, then the average permeability is given by:

$$k = \frac{162.6Q_oB_o\mu_o}{|m|h}$$

where

k = average permeability, md
$|m|$ = absolute value of slope, psi/cycle

Clearly, kh/μ or k/μ may also be estimated.

The skin effect can be obtained by rearranging Eq. (1.169) as:

$$s = 1.151\left[\frac{p_i - p_{wf}}{|m|} - \log t - \log\left(\frac{k}{\phi\mu c_t r_w^2}\right) + 3.23\right]$$

or, more conveniently, if selecting $p_{wf} = p_{1\ \text{hour}}$ which is found on the extension of the straight line at $t = 1$ hour, then:

$$s = 1.151\left[\frac{p_i - p_{1\ \text{hour}}}{|m|} - \log\left(\frac{k}{\phi\mu c_t r_w^2}\right) + 3.23\right] \quad \textbf{(1.171)}$$

where

$|m|$ = absolute value of the slope m

In Eq. (1.14), $p_{1\ \text{hour}}$ must be obtained from the semilog straight line. If the pressure data measured at 1 hour does not fall on that line, the line must be extrapolated to 1 hour and the extrapolated value of $p_{1\ \text{hour}}$ must be used in Eq. (1.171). This procedure is necessary to avoid calculating an incorrect skin by using a wellbore-storage-influenced pressure. Figure 1.33 illustrates the extrapolation to $p_{1\ \text{hour}}$.

Note that the additional pressure drop due to the skin was expressed previously by Eq. (1.141) as:

$$\Delta p_{skin} = 141.2\left(\frac{Q_oB_o\mu_o}{kh}\right)s$$

This additional pressure drop can be equivalently written in terms of the semilog straight-line slope m by combining the above expression with that of Eq. (1.171) to give:

$$\Delta p_{skin} = 0.87|m|s$$

Another physically meaningful characterization of the skin factor is the flow coefficient E as defined by the ratio of the actual or observed productivity index J_{actual} of well and its ideal productivity index J_{ideal}. The ideal productivity index J_{ideal} is the value obtained with no alternation of permeability around the wellbore. Mathematically, the flow coefficient is given by:

$$E = \frac{J_{\text{actual}}}{J_{\text{ideal}}} = \frac{\bar{p} - p_{\text{wf}} - \Delta p_{\text{skin}}}{\bar{p} - p_{\text{wf}}}$$

where

$\bar{p}$ = average pressure in the well drainage area

If the drawdown test is long enough, the bottom-hole pressure will deviate from the semilog straight line and make the transition from infinite acting to pseudosteady state. The rate of pressure decline during the pseudosteady-state flow is defined by Eq. (1.127) as:

$$\frac{dp}{dt} = -\frac{0.23396q}{c_t(\pi r_e^2)h\phi} = \frac{-0.23396q}{c_t(A)h\phi} = \frac{-0.23396q}{c_t(\text{pore volume})}$$

Under this condition, the pressure will decline at a constant rate at any point in the reservoir including the bottom-hole flowing pressure p_{wf}. That is:

$$\frac{dp_{\text{wf}}}{dt} = m' = \frac{-0.23396q}{c_t A h\phi}$$

This expression suggests that during the semisteady-state flow, a plot of p_{wf} vs. t on a Cartesian scale would produce a straight line with a negative slope of m' that is defined by:

$$-m' = \frac{-0.23396q}{c_t A h\phi}$$

where

m' = slope of the *Cartesian straight line* during the pseudosteady state, psi/hour
q = flow rate, bbl/day
A = drainage area, ft^2

Example 1.24

Estimate the oil permeability and skin factor from the drawdown data of Figure 1.34.[1]

The following reservoir data are available:

$h = 130$ ft,	$\phi = 20\%$,	$r_w = 0.25$ ft
$p_i = 1154$ psi,	$Q_o = 348$ STB/D,	$m = -22$ psi/cycle
$B_o = 1.14$ bbl/STB/D	$\mu_o = 3.93$ cp,	$c_t = 8.74 \times 10^{-6}$ psi^{-1}

Assuming that the wellbore storage effect is not significant, calculate:

- the permeability;
- the skin factor;
- the additional pressure drop due to the skin.

Solution

Step 1. From Figure 1.34, calculate $p_{1\text{ hour}}$:

$$p_{1\text{ hour}} = 954 \text{ psi}$$

Step 2. Determine the slope of the transient flow line:

$$m = -22 \text{ psi/cycle}$$

Step 3. Calculate the permeability by applying Eq. (1.170):

$$k - \frac{-162.6Q_o B_o \mu_o}{mh} = \frac{-(162.6)(348)(1.14)(3.93)}{(-22)(130)} = 89 \text{ md}$$

Step 4. Solve for the skin factor s by using Eq. (1.171):

$$s = 1.151\left[\frac{p_i - p_{1\text{ hour}}}{|m|} - \log\left(\frac{k}{\phi\mu c_t r_w^2}\right) + 3.23\right]$$

$$= 1.151\left[\left(\frac{1154 - 954}{22}\right) - \log\left(\frac{89}{(0.2)(3.93)(8.74\times 10^{-6})(0.25)^2}\right) + 3.2275\right] = 4.6$$

Step 5. Calculate the additional pressure drop:

$$\Delta p_{\text{skin}} = 0.87|m|s = 0.87(22)(4.6) = 88 \text{ psi}$$

[1]This example problem and the solution procedure are given in Earlougher, R. *Advances in Well Test Analysis*, Monograph Series, SPE, Dallas (1997).

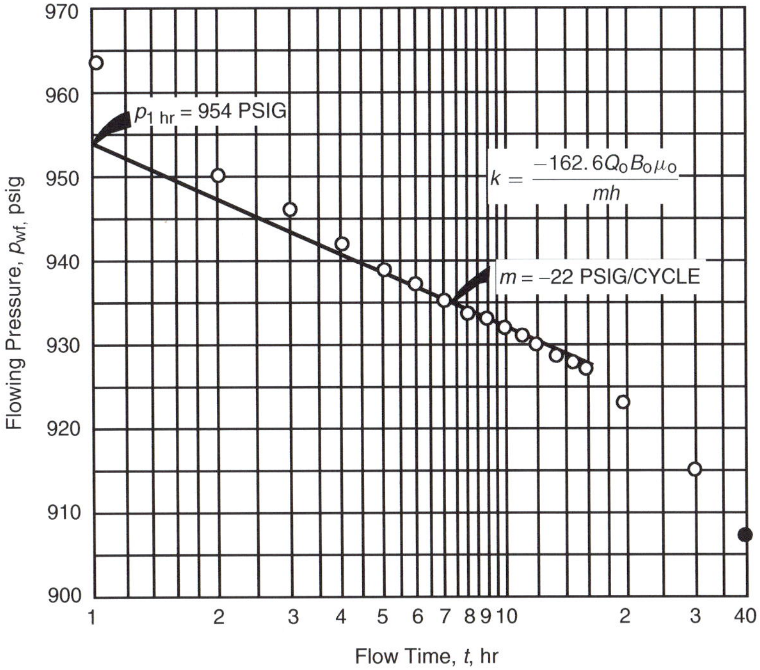

FIGURE 1.34 Earlougher's semilog data plot for the drawdown test. *(Permission to publish by the SPE, copyright SPE, 1977.)*

It should be noted that for a *multiphase flow*, Eqs. (1.169) and (1.171) become:

$$p_{wf} = p_i - \frac{162.6q_t}{\lambda_t h}\left[\log(t) + \log\left(\frac{\lambda_t}{\phi c_t r_w^2}\right) - 3.23 + 0.87s\right]$$

$$s = 1.151\left[\frac{p_i - p_{1\ hour}}{|m|} - \log\left(\frac{\lambda_t}{\phi c_t r_w^2}\right) + 3.23\right]$$

with

$$\lambda_t = \frac{k_o}{\mu_o} + \frac{k_w}{\mu_w} + \frac{k_g}{\mu_g}$$

$$q_t = Q_oB_o + Q_wB_w + (Q_g - Q_oR_s)B_g$$

or equivalently in terms of GOR as:

$$q_t = Q_oB_o + Q_wB_w + (GOR - R_s)Q_oB_g$$

where

q_t = total fluid voidage rate, bbl/day
Q_o = oil flow rate, STB/day
Q_w = water flow rate, STB/day
Q_g = total gas flow rate, scf/day
R_s = gas solubility, scf/STB
B_g = gas formation volume factor, bbl/scf
λ_t = total mobility, md/cp
k_o = effective permeability to oil, md
k_w = effective permeability to water, md
k_g = effective permeability to gas, md

The above drawdown relationships indicate that a plot of p_{wf} vs. t on a semilog scale would produce a straight line with a slope m that can be used to determine the total mobility λ_t from:

$$\lambda_t = \frac{162.6q_t}{mh}$$

Perrine (1956) showed that the effective permeability of each phase, i.e., k_o, k_w, and k_g, can be determined as:

$$k_o = \frac{162.6Q_oB_o\mu_o}{mh}$$

$$k_w = \frac{162.6Q_wB_w\mu_w}{mh}$$

$$k_g = \frac{162.6(Q_g - Q_oR_s)B_g\mu_g}{mh}$$

If the drawdown pressure data is available during both the unsteady-state flow period and the pseudosteady-state flow period, it is possible to estimate the drainage shape and the drainage area of the test well. The transient semilog plot is used to determine its slope m and $p_{1\ \text{hour}}$; the Cartesian straight-line plot of the pseudosteady-state data is used to determine its slope $m^\backslash$ and its intercept p_{int}. Earlougher (1977) proposed the following expression to determine the shape factor C_A:

$$C_A = 5.456\left(\frac{m}{m^\backslash}\right)\exp\left[\frac{2.303(p_{1\ \text{hour}} - p_{\text{int}})}{m}\right]$$

where

m = slope of transient semilog straight line, psi/log cycle
$m^\backslash$ = slope of the semisteady-state Cartesian straight line
$p_{1\ \text{hour}}$ = pressure at $t = 1$ hour from transient semilog straight line, psi
p_{int} = pressure at $t = 0$ from pseudosteady-state Cartesian straight line, psi

The calculated shape factor from applying the above relationship is compared with those values listed in Table 1.4 to select the geometry of well drainage with a shape factor closest to the calculated value. When extending the drawdown test time with the objective of reaching the drainage boundary of the test well, the test is commonly called the "reservoir limit test."

The reported data of Example 1.24 was extended by Earlougher to include the pseudosteady-state flow period and to determine the geometry of the test well drainage area as shown in Example 1.25.

Example 1.25

Use the data in Example 1.24 and the Cartesian plot of the pseudosteady-state flow period, as shown in Figure 1.35, to determine the geometry and drainage area of the test well.

Solution

Step 1. From Figure 1.35, determine the slope $m^\backslash$ and intercept p_{int}:

$$m^\backslash = -0.8 \text{ psi/hour}$$

$$p_{\text{int}} = 940 \text{ psi}$$

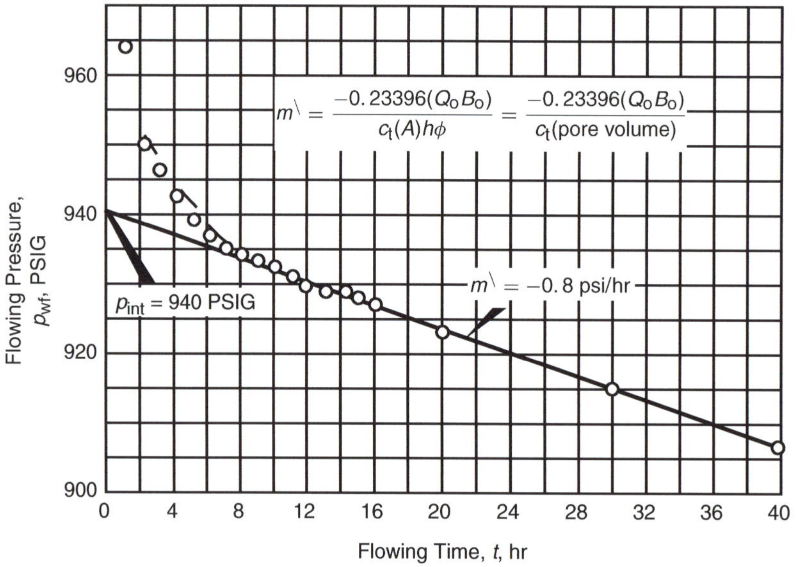

FIGURE 1.35 Cartesian plot of the drawdown test data. *(Permission to publish by the SPE, copyright SPE, 1977.)*

Step 2. From Example 1.24:

$$m = -22 \text{ psi/cycle}$$
$$p_{1\text{ hour}} = 954 \text{ psi}$$

Step 3. Calculate the shape factor C_A from Earlougher's equation:

$$C_A = 5.456\left(\frac{m}{m^{\backslash}}\right)\exp\left[\frac{2.303(p_{1\text{ hour}} - p_{\text{int}})}{m}\right]$$
$$= 5.456\left(\frac{-22}{-0.8}\right)\exp\left[\frac{2.303(954-940)}{-22}\right]$$
$$= 34.6$$

Step 4. From Table 1.4, $C_A = 34.6$ corresponds to a well in the center of a circle, square, or hexagon:
- For a circle: $C_A = 31.62$
- For a square: $C_A = 30.88$
- For a hexagon: $C_A = 31.60$

Step 5. Calculate the pore volume and drainage area from Eq. (1.127):

$$\frac{dp}{dt} = m^{\backslash} = \frac{-0.23396(Q_o B_o)}{c_t(A)h\phi} = \frac{-0.23396(Q_o B_o)}{c_t(\text{pore volume})}$$

Solving for the pore volume gives:

$$\text{Pore volume} = \frac{-0.23396q}{c_t m^{\backslash}} = \frac{-0.23396(348)(1.4)}{(8.74\times 10^{-6})(-0.8)}$$
$$= 2.37 \text{ MMbbl}$$

and the drainage area:

$$A = \frac{2.37\times 10^6(5.615)}{43,460(0.2)(130)} = 11.7 \text{ acres}$$

The above example indicates that the measured bottom-hole flowing pressures are 88 psi more than they would be in the absence of the skin. However, it should be pointed out that when the concept of positive skin factor $+s$ indicates formation damage, whereas a negative skin factor $-s$ suggests formation stimulation, this is essentially a misleading interpretation of the skin factor. The skin factor as determined from any transient well testing analysis represents the composite "total" skin factor that includes the following other skin factors:

- skin due to wellbore damage or stimulation s_d;
- skin due to partial penetration and restricted entry s_r;
- skin due to perforations s_p;
- skin due to turbulence flow s_t;
- skin due to deviated well s_{dw}.

That is:

$$s = s_d + s_r + s_p + s_t + s_{dw}$$

where s is the skin factor as calculated from transient flow analysis. Therefore, to determine if the formation is damaged or stimulated from the skin factor value s obtained from well test analysis, the individual components of the skin factor in the above relationship must be known, to give:

$$s_d = s - s_r - s_p - s_t - s_{dw}$$

There are correlations that can be used to separately estimate these individual skin quantities.

Wellbore Storage. Basically, well test analysis deals with the interpretation of the wellbore pressure response to a given change in the flow rate (from zero to a constant value for a drawdown test, or from a constant rate to zero for a buildup test). Unfortunately, the producing rate is controlled at the surface, not at the sand face. Because of the wellbore volume, a constant surface flow rate does not ensure that the entire rate is being produced from the formation. This effect is due to wellbore storage. Consider the case of a drawdown test. When the well is first opened to flow after a shut-in period, the pressure in the wellbore drops. This drop in pressure causes the following two types of wellbore storage:

(1) a wellbore storage effect caused by fluid expansion;
(2) a wellbore storage effect caused by changing fluid level in the casing–tubing annulus.

As the bottom-hole pressure drops, the wellbore fluid expands and, thus, the initial surface flow rate is not from the formation, but basically from the fluid that had been stored in the wellbore. This is defined as the *wellbore storage due to fluid expansion.*

The second type of wellbore storage is due to a change in the annulus fluid level (falling level during a drawdown test, rising level during a drawdown test, and rising fluid level during a pressure buildup test). When the well is open to flow during a drawdown test, the reduction in pressure causes the fluid level in the annulus to fall. This annulus fluid production joins that from the formation and contributes to the total flow from the well. The falling fluid level is generally able to contribute more fluid than that by expansion.

The above discussion suggests that part of the flow will be contributed by the wellbore instead of the reservoir. That is:

$$q = q_f + q_{wb}$$

where

q = surface flow rate, bbl/day
q_f = formation flow rate, bbl/day
q_{wb} = flow rate contributed by the wellbore, bbl/day

During this period when the flow is dominated by the wellbore storage, the measured drawdown pressures will not produce the ideal semilog straight-line behavior that is expected during transient flow. This indicates that the pressure data collected during the period of the wellbore storage effect cannot be analyzed by using conventional methods. As production time increases, the wellbore contribution decreases and the formation rate increases until it eventually equals the surface flow rate, i.e., $q = q_f$, which signifies the *end of the wellbore storage effect.*

The effect of fluid expansion and changing fluid level can be quantified in terms of the *wellbore storage factor* C which is defined as:

$$C = \frac{\Delta V_{wb}}{\Delta p}$$

where

C = wellbore storage coefficient, bbl/psi
ΔV_{wb} = change in the volume of fluid in the wellbore, bbl

The above relationship can be applied to mathematically represent the individual effect of wellbore fluid expansion and falling (or rising) fluid level, to give the following relationships.

Wellbore Storage Effect Caused by Fluid Expansion.

$$C_{FE} = V_{wb} c_{wb}$$

where

C_{FE} = wellbore storage coefficient due to fluid expansion, bbl/psi
V_{wb} = total wellbore fluid volume, bbl
c_{wb} = average compressibility of fluid in the wellbore, psi^{-1}

Wellbore Storage Effect Due To Changing Fluid Level.

$$C_{FL} = \frac{144 A_a}{5.615 \rho}$$

with

$$A_a = \frac{\pi[(ID_C)^2 - (OD_T)^2]}{4(144)}$$

where

C_{FL} = wellbore storage coefficient due to changing fluid level, bbl/psi
A_a = annulus cross-sectional area, ft^2
OD_T = outside diameter of the production tubing, inches
ID_C = inside diameter of the casing, inches
ρ = wellbore fluid density, lb/ft^3

This effect is essentially small if a packer is placed near the producing zone. The total storage effect is the sum of both coefficients. That is:

$$C = C_{FE} + C_{FL}$$

It should be noted during oil well testing that the fluid expansion is generally insignificant due to the small compressibility of liquids. For gas wells, the primary storage effect is due to gas expansion.

To determine the duration of the wellbore storage effect, it is convenient to express the wellbore storage factor in a dimensionless form as:

$$C_D = \frac{5.615C}{2\pi h\phi c_t r_w^2} = \frac{0.8936C}{\phi h c_t r_w^2} \tag{1.172}$$

where

C_D = dimensionless wellbore storage factor
C = wellbore storage factor, bbl/psi
c_t = total compressibility coefficient, psi^{-1}
r_w = wellbore radius, ft
h = thickness, ft

Horn (1995) and Earlougher (1977), among other authors, have indicated that the wellbore pressure is directly proportional to the time during the wellbore storage-dominated period of the test and is expressed by:

$$p_D = \frac{t_D}{C_D} \tag{1.173}$$

where

p_D = dimensionless pressure during wellbore storage domination time
t_D = dimensionless time

Taking the logarithm of both sides of this relationship gives:

$$\log(p_D) = \log(t_D) - \log(C_D)$$

This expression has a characteristic that is diagnostic of wellbore storage effects. It indicates that a plot of p_D vs. t_D on a log–log scale will yield a straight line of a *unit slope*, i.e., a straight line with a 45° angle, during the wellbore storage-dominated period. Since p_D is proportional to pressure drop Δp and t_D is proportional to time t, it is convenient to plot $\log(p_i - p_{wf})$ vs. $\log(t)$ and observe where the plot has a slope of one cycle in pressure per cycle in time. This unit slope observation is of major value in well test analysis.

The log–log plot is a valuable aid for recognizing wellbore storage effects in transient tests (e.g., drawdown or buildup tests) when early-time pressure recorded data is available. It is recommended that this plot be made a part of the transient test analysis. As wellbore storage effects become less severe, the formation begins to influence the bottom-hole pressure more and more, and the data points on the log–log plot fall below the unit-slope straight line and signify the end of the wellbore storage effect. At this point, wellbore storage is no longer important and standard semilog data-plotting analysis techniques apply here. As a rule of thumb, the time that indicates the end of the wellbore storage effect can be determined from the log–log plot by moving $1 - 1\frac{1}{2}$ cycles in time after the plot starts to deviate from the unit slope and reading the corresponding time on the x-axis. This time maybe estimated from:

$$t_D > (60 + 3.5s)C_D$$

or

$$t > \frac{(200,000 + 12,000s)C}{(kh/\mu)}$$

where

t = total time that marks the end of the wellbore storage effect and the beginning of the semilog straight line, hours
k = permeability, md
s = skin factor
μ = viscosity, cp
C = wellbore storage coefficient, bbl/psi

In practice, it is convenient to determine the wellbore storage coefficient C by selecting *a point on the log–log unit-slope straight line* and reading the coordinate of the point in terms of t and Δp, to give:

$$C = \frac{qt}{24\,\Delta p} = \frac{QBt}{24\,\Delta p}$$

where

t = time, hours
Δp = pressure difference $(p_i - p_{wf})$, psi
q = flow rate, bbl/day
Q = flow rate, STB/day
B = formation volume factor, bbl/STB

It is important to note that the volume of fluids stored in the wellbore distorts the early-time pressure response and controls the duration of wellbore storage, especially in deep wells with large wellbore volumes. If the wellbore storage effects are not minimized or if the test is not continued beyond the end of the wellbore storage-dominated period, the test data will be difficult to analyze with current conventional well testing methods. To minimize wellbore storage distortion and to keep well tests within reasonable lengths of time, it may be necessary to run tubing, packers, and bottom-hole shut-in devices.

Example 1.26

The following data is given for an oil well that is scheduled for a drawdown test:

- Volume of fluid in the wellbore = 180 bbl
- Tubing outside diameter = 2 in.
- Production oil density in the wellbore = 7.675 in.
- Average oil density in the wellbore = 45 lb/ft^3

$h = 50$ ft,	$\phi = 15\%$
$r_w = 0.25$ ft,	$\mu_o = 2$ cp
$k = 30$ md,	$s = 0$
$c_t = 20 \times 10^{-6}$ psi^{-1},	$c_o = 10 \times 10^{-6}$ psi^{-1}

If this well is placed under a constant production rate, calculate the dimensionless wellbore storage coefficient C_D. How long will it take for wellbore storage effects to end?

Solution

Step 1. Calculate the cross-sectional area of the annulus A_a:

$$A_a = \frac{\pi[(ID_C)^2 - (OD_T)^2]}{4(144)} = \frac{\pi[(7.675)^2 - (2)^2]}{(4)(144)} = 0.2995 \text{ ft}^2$$

Step 2. Calculate the wellbore storage factor caused by fluid expansion:

$$C_{FE} = V_{wb} c_{wb} = (180)(10 \times 10^{-6}) = 0.0018 \text{ bbl/psi}$$

Step 3. Determine the wellbore storage factor caused by the falling fluid level:

$$C_{FL} = \frac{144 A_a}{5.615 \rho} = \frac{144(0.2995)}{(5.615)(45)} = 0.1707 \text{ bbl/psi}$$

Step 4. Calculate the total wellbore storage coefficient:

$$C = C_{FE} + C_{FL} = 0.0018 + 0.1707 = 0.1725 \text{ bbl/psi}$$

The above calculations show that the effect of fluid expansion C_{FE} can generally be neglected in crude oil systems.

Step 5. Calculate the dimensionless wellbore storage coefficient from Eq. (1.172):

$$C_D = \frac{0.8936 C}{\phi h c_t r_w^2} = \frac{0.8936(0.1707)}{0.15(50)(20 \times 10^{-6})(0.25)^2} = 16,271$$

Step 6. Approximate the time required for wellbore storage influence to end from:

$$t = \frac{(200,000 + 12,000 s) C \mu}{kh} = \frac{(200,000 + 0)(0.1725)(2)}{(30)(50)} = 46 \text{ hours}$$

The straight-line relationship as expressed by Eq. (1.170) is only valid during the infinite-acting behavior of the well. Obviously, reservoirs are not infinite in extent, so the infinite-acting radial flow period cannot last indefinitely. Eventually, the effects of the reservoir boundaries will be felt at the well being tested. The time at which the boundary effect is felt is dependent on the following factors:

- permeability k;
- total compressibility c_t;

- porosity ϕ;
- viscosity μ;
- distance to the boundary;
- shape of the drainage area.

Earlougher (1977) suggested the following mathematical expression for estimating the duration of the infinite-acting period:

$$t_{eia} = \left[\frac{\phi \mu c_t A}{0.0002637k}\right](t_{DA})_{eia}$$

where

t_{eia} = time to the end of infinite-acting period, hours
A = well drainage area, ft^2
c_t = total compressibility, psi^{-1}
$(t_{DA})_{eia}$ = dimensionless time to the end of the infinite-acting period

This expression is designed to predict the time that marks the end of transient flow in a drainage system of any geometry by obtaining the value of t_{DA} from Table 1.4. The *last three columns* of the table provide with values of t_{DA} that allow the engineer to calculate:

- the maximum elapsed time during which a reservoir is infinite acting;
- the time required for the pseudosteady-state solution to be applied and predict pressure drawdown within 1% accuracy;
- the time required for the pseudosteady-state solution (equations) to be exact and applied.

As an example, for a well centered in a circular reservoir, the maximum time for the reservoir to remain as an infinite-acting system can be determined using the entry in the final column of Table 1.4 to give $(t_{DA})_{eia} = 0.1$, and accordingly:

$$t_{eia} = \left[\frac{\phi \mu c_t A}{0.0002637k}\right](t_{DA})_{eia} = \left[\frac{\phi \mu c_t A}{0.0002637k}\right]0.1$$

or

$$t_{eia} = \frac{380\phi \mu c_t A}{k}$$

For example, for a well that is located in the center of a 40 acre circular drainage area with the following properties:

$$k = 60 \text{ md}, \quad c_t = 6 \times 10^{-6} \text{ psi}^{-1}, \quad \mu = 1.5 \text{ cp}, \quad \phi = 0.12$$

the maximum time, in hours, for the well to remain in an infinite-acting system is:

$$t_{eia} = \frac{380\phi \mu c_t A}{k} = \frac{380(0.12)(1.4)(6 \times 10^{-6})(40 \times 43560)}{60}$$
$$= 11.1 \text{ hours}$$

Similarly, the pseudosteady-state solution can be applied any time after the semisteady-state flow begins at t_{pss} as estimated from:

$$t_{pss} = \left[\frac{\phi \mu c_t A}{0.0002637k}\right](t_{DA})_{pss}$$

where $(tDA)_{pss}$ can be found from the entry in the fifth column of the table.

Hence, the specific steps involved in a drawdown test analysis are:

Step 1. Plot $p_i - p_{wf}$ vs. t on a log–log scale.
Step 2. Determine the time at which the unit-slope line ends.
Step 3. Determine the corresponding time at $1\frac{1}{2}$ log cycle, ahead of the observed time in step 2. This is the time that marks the end of the wellbore storage effect and the start of the semilog straight line.
Step 4. Estimate the wellbore storage coefficient from:

$$C = \frac{qt}{24\,\Delta p} = \frac{QBt}{24\,\Delta p}$$

where t and Δp are values read from a point on the log–log unit-slope straight line and q is the flow rate in bbl/day.

Step 5. Plot p_{wf} vs. t on a semilog scale.
Step 6. Determine the start of the straight-line portion as suggested in step 3 and draw the best line through the points.
Step 7. Calculate the slope of the straight line and determine the permeability k and

skin factor s by applying Eqs. (1.170) and (1.171), respectively:

$$k = \frac{-162.6 Q_o B_o \mu_o}{mh}$$

$$s = 1.151\left[\frac{p_i - p_{1\text{ hour}}}{|m|} - \log\left(\frac{k}{\phi\mu c_t r_w^2}\right) + 3.23\right]$$

Step 8. Estimate the time to the end of the infinite-acting (transient flow) period, i.e., t_{eia}, which marks the beginning of the pseudosteady-state flow.

Step 9. Plot all the recorded pressure data after t_{eia} as a function of time on a regular Cartesian scale. This data should form a straight-line relationship.

Step 10. Determine the slope of the pseudosteady-state line, i.e., dp/dt (commonly referred to as $m^\backslash$) and use Eq. (1.127) to solve for the drainage area A:

$$A = \frac{-0.23396 QB}{c_t h\phi (dp/dt)} = \frac{-0.23396 QB}{c_t h\phi m^\backslash}$$

where

$m^\backslash$ = slope of the semisteady-state Cartesian straight line
Q = fluid flow rate, STB/day
B = formation volume factor, bbl/STB

Step 11. Calculate the shape factor C_A from the expression that was developed by Earlougher (1977):

$$C_A = 5.456\left(\frac{m}{m^\backslash}\right)\exp\left[\frac{2.303(p_{1\text{ hour}} - p_{\text{int}})}{m}\right]$$

where

m = slope of transient semilog straight line, psi/log cycle
$m^\backslash$ = slope of the pseudosteady-state Cartesian straight line
$p_{1\text{ hour}}$ = pressure at $t = 1$ hour from transient semilog straight line, psi
p_{int} = pressure at $t = 0$ from semisteady-state Cartesian straight line, psi

Step 12. Use Table 1.4 to determine the drainage configuration of the tested well that has a value of the shape factor C_A closest to that of the calculated one, i.e., step 11.

Radius of Investigation. The radius of investigation r_{inv} of a given test is the effective distance traveled by the pressure transients, as measured from the tested well. This radius depends on the speed with which the pressure waves propagate through the reservoir rock, which, in turn, is determined by the rock and fluid properties, such as:

- porosity;
- permeability;
- fluid viscosity;
- total compressibility.

As time t increases, more of the reservoir is influenced by the well and the radius of drainage, or investigation, increases as given by:

$$r_{\text{inv}} = 0.0325\sqrt{\frac{kt}{\phi\mu c_t}}$$

where

t = time, hours
k = permeability, md
c_t = total compressibility, psi^{-1}

It should be pointed out that the equations developed for slightly compressible liquids can be extended to describe the behavior of real gases by replacing the pressure with the real-gas pseudopressure $m(p)$, as defined by:

$$m(p) = \int_0^p \frac{2p}{\mu Z}dp$$

with the transient pressure drawdown behavior as described by Eq. (1.162), or:

$$m(p_{wf}) = m(p_i) - \left[\frac{1637 Q_g T}{kh}\right] \times \left[\log\left(\frac{kt}{\phi\mu_i c_{ti} r_w^2}\right) - 3.23 + 0.87 s^\backslash\right]$$

Under constant gas flow rate, the above relation can be expressed in a linear form as:

$$m(p_{wf}) = \left\{m(p_i) - \left[\frac{1637 Q_g T}{kh}\right] \times \left[\log\left(\frac{k}{\phi\mu_i c_{ti} r_w^2}\right) - 3.23 + 0.87 s^\backslash\right]\right\} - \left[\frac{1637 Q_g T}{kh}\right]\log(t)$$

or

$$m(p_{wf}) = a + m \log(t)$$

which indicates that a plot of $m(p_{wf})$ vs. $\log(t)$ would produce a semilog straight line with a negative slope of:

$$m = \frac{1637 Q_g T}{kh}$$

Similarly, in terms of the pressure-squared approximation form:

$$p_{wf}^2 = p_i^2 - \left[\frac{1637 Q_g T \bar{Z} \bar{\mu}}{kh}\right] \times \left[\log\left(\frac{kt}{\phi \mu_i c_{ti} r_w^2}\right) - 3.23 + 0.87 s^{\prime}\right]$$

or

$$p_{wf}^2 = \left\{p_i^2 - \left[\frac{1637 Q_g T \bar{Z} \bar{\mu}}{kh}\right] \times \left[\log\left(\frac{k}{\phi \mu_i c_{ti} r_w^2}\right) - 3.23 + 0.87 s^{\prime}\right]\right\} - \left[\frac{1637 Q_g T \bar{Z} \bar{\mu}}{kh}\right] \log(t)$$

This equation is an equation of a straight line that can be simplified to give:

$$p_{wf}^2 = a + m \log(t)$$

which indicates that a plot of p_{wf}^2 vs. $\log(t)$ would produce a semilog straight line with a negative slope of:

$$m = \frac{1637 Q_g T \bar{Z} \bar{\mu}}{kh}$$

The true skin factor s which reflects the formation damage or stimulation is usually combined with the non-Darcy rate-dependent skin and is labeled as the apparent or total skin factor:

$$s^{\prime} = s + D Q_g$$

with the term DQ_g interpreted as the rate-dependent skin factor. The coefficient D is called the inertial or turbulent flow factor and is given by Eq. (1.159):

$$D = \frac{Fkh}{1422 T}$$

where

Q_g = gas flow rate, Mscf/day
t = time, hours
k = permeability, md
μ_i = gas viscosity as evaluated at p_i, cp

The apparent skin factor $s^{\prime}$ is given by:

For pseudopressure approach:

$$s^{\prime} = 1.151\left[\frac{m(p_i) - m(p_{1\ hour})}{|m|} - \log\left(\frac{k}{\phi \mu_i c_{ti} r_w^2}\right) + 3.23\right]$$

- For pressure-squared approach:

$$s^{\prime} = 1.151\left[\frac{p_i^2 - p_{1\ hour}^2}{|m|} - \log\left(\frac{k}{\phi \bar{\mu} \bar{c}_t r_w^2}\right) + 3.23\right]$$

If the duration of the drawdown test of the gas well is long enough to reach its boundary, the pressure behavior during the boundary-dominated period (pseudosteady-state condition) is described by an equation similar to that of Eq. (1.136) as:

- For pseudopressure approach:

$$\frac{m(p_i) - m(p_{wf})}{q} = \frac{\Delta m(p)}{q} = \frac{711 T}{kh}\left(\ln \frac{4A}{1.781 C_A r_{wa}^2}\right) + \left[\frac{2.356 T}{\phi (\mu_g c_g)_i A h}\right] t$$

and as a linear equation by:

$$\frac{\Delta m(p)}{q} = b_{pss} + m^{\prime} t$$

This relationship indicates that a plot of $\Delta m(p)/q$ vs. t will form a straight line with:

Intercept: $$b_{pss} = \frac{711 T}{kh}\left(\ln \frac{4A}{1.781 C_A r_{wa}^2}\right)$$

Slope: $$m^{\prime} = \frac{2.356 T}{(\mu_g c_t)_i (\phi h A)} = \frac{2.356 T}{(\mu_g c_t)_i (\text{pore volume})}$$

- For pressure-squared approach:

$$\frac{p_i^2 - p_{wf}^2}{q} = \frac{\Delta(p^2)}{q} = \frac{711\bar{\mu}\bar{Z}T}{kh}\left(\ln\frac{4A}{1.781\,C_A r_{wa}^2}\right) + \left[\frac{2.356\bar{\mu}\bar{Z}T}{\phi(\mu_g c_g)_i Ah}\right]t$$

and in a linear form as:

$$\frac{\Delta(p^2)}{q} = b_{pss} + m^{\backslash}t$$

This relationship indicates that a plot of $\Delta(p^2)/q$ vs. t on a Cartesian scale will form a straight line with:

$$\text{Intercept:}\quad b_{pss} = \frac{711\bar{\mu}\bar{Z}T}{kh}\left(\ln\frac{4A}{1.781\,C_A r_{wa}^2}\right)$$

$$\text{Slope:}\quad m^{\backslash} = \frac{2.356\bar{\mu}\bar{Z}T}{(\mu_g c_t)_i(\phi hA)} = \frac{2.356\bar{\mu}\bar{Z}T}{(\mu_g c_t)_i(\text{pore volume})}$$

where

q = flow rate, Mscf/day
A = drainage area, ft^2
T = temperature, °R
t = flow time, hours

Meunier et al. (1987) suggested a methodology for expressing the time t and the corresponding pressure p that allows the use of liquid flow equations without special modifications for gas flow. Meunier et al. introduced the following normalized pseudopressure p_{pn} and normalized pseudotime t_{pn}:

$$p_{pn} = p_i + \left(\frac{\mu_i Z_i}{p_i}\right)\int_0^p \frac{p}{\mu Z}dp$$

$$t_{pn} = \mu_i c_{ti}\left[\int_0^t \frac{1}{\mu c_t}dp\right]$$

The subscript "i" on μ, Z, and c_t refers to the evaluation of these parameters at the initial reservoir pressure p_i. By using the Meunier et al. definition of the normalized pseudopressure and normalized pseudotime there is no need to modify any of the liquid analysis equations. However, care should be exercised when replacing the liquid flow rate with the gas flow rate. It should be noted that in all transient flow equations when applied to the oil phase, the flow rate is expressed as the product of Q_oB_o in bbl/day; that is, in reservoir barrels/day. Therefore, when applying these equations to the gas phase, the product of the gas flow rate and gas formation volume factor Q_gB_g should be given in bbl/day. For example, if the gas flow rate is expressed in scf/day, the gas formation volume factor must be expressed in bbl/scf. The recorded pressure and time are then simply replaced by the normalized pressure and normalized time to be used in all the traditional graphical techniques, including pressure buildup.

1.3.2 Pressure Buildup Test

The use of pressure buildup data has provided the reservoir engineer with one more useful tool in the determination of reservoir behavior. Pressure buildup analysis describes the buildup in wellbore pressure with time after a well has been shut in. One of the principal objectives of this analysis is to determine the static reservoir pressure without waiting weeks or months for the pressure in the entire reservoir to stabilize. Because the buildup in wellbore pressure will generally follow some definite trend, it has been possible to extend the pressure buildup analysis to determine:

- the effective reservoir permeability;
- the extent of permeability damage around the wellbore;
- the presence of faults and to some degree the distance to the faults;
- any interference between producing wells;
- the limits of the reservoir where there is no strong water drive or where the aquifer is no larger than the hydrocarbon reservoir.

Certainly all of this information will probably not be available from any given analysis, and the degree of usefulness of this information will depend on the experience and the amount of other information available for correlation purposes.

The general formulas used in analyzing pressure buildup data come from a solution of the diffusivity equation. In pressure buildup and drawdown analyses, the following assumptions, regarding the reservoir, fluid, and flow behavior, are usually made:

- *Reservoir*: homogeneous; isotropic; horizontal of uniform thickness.
- *Fluid*: single phase; slightly compressible; constant μ_o and B_o.
- *Flow*: laminar flow; no gravity effects.

Pressure buildup testing requires shutting in a producing well and recording the resulting increase in the wellbore pressure as a function of shut-in time. The most common and simplest analysis techniques require that the well produces at a constant rate for a flowing time of t_p, either from startup or long enough to establish a stabilized pressure distribution, before shut in. Traditionally, the shut-in time is denoted by the symbol Δt. Figure 1.36 schematically shows the stabilized constant flow rate before shut-in and the ideal behavior of pressure increase during the buildup period. The pressure is measured immediately before shut-in and is recorded as a function of time during the shut-in period. The resulting pressure buildup curve is then analyzed to determine reservoir properties and the wellbore condition.

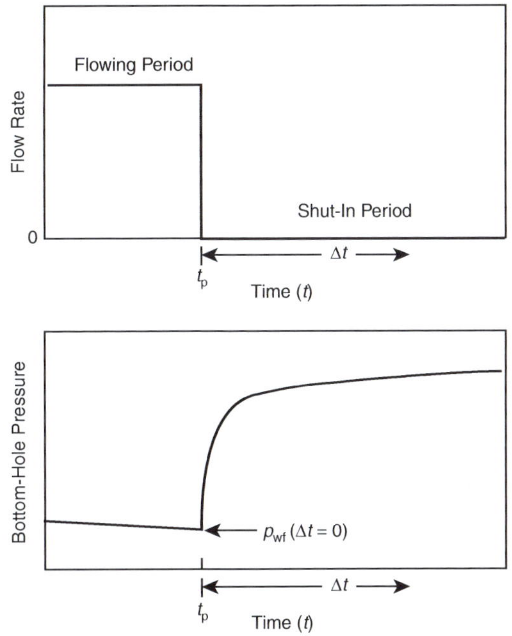

FIGURE 1.36 Idealized pressure buildup test.

Stabilizing the well at a constant rate before testing is an important part of a pressure buildup test. If stabilization is overlooked or is impossible, standard data analysis techniques may provide erroneous information about the formation.

Two widely used methods are discussed below; these are:

(1) the Horner plot;
(2) the Miller–Dyes–Hutchinson method.

1.3.3 Horner Plot

A pressure buildup test is described mathematically by using the principle of superposition. Before the shut-in, the well is allowed to flow at a constant flow rate of Q_o STB/day for t_p days. At the end of the flowing period, the well is shut in with a corresponding change in the flow rate from the "old" rate of Q_o to the "new" flow rate of $Q^{new} = 0$, i.e., $Q^{new} - Q^{old} = -Q_o$.

Calculation of the total pressure change that occurs at the sand face during the shut-in time is basically the sum of the pressure changes that are caused by:

- flowing the well at a stabilized flow rate of Q^{old}, i.e., the flow rate before shut-in Q_o, and is in effect over the entire time of $t_p + \Delta t$;
- the net change in the flow rate from Q_o to 0 and is in effect over Δt.

The composite effect is obtained by adding the individual constant-rate solutions at the specified rate–time sequence, as:

$$p_i - p_{ws} = (\Delta p)_{total} = (\Delta p)_{due\ to\ (Q_o - 0)} + (\Delta p)_{due\ to\ (0 - Q_o)}$$

where

p_i = initial reservoir pressure, psi
p_{ws} = wellbore pressure during shut in, psi

The above expression indicates that there are two contributions to the total pressure change at the wellbore resulting from the two individual flow rates.

The first contribution results from increasing the rate from 0 to Q_o and is in effect over the entire time period $t_p + \Delta t$, thus:

$$(\Delta p)_{Q_o-0} = \left[\frac{162.6(Q_o - 0)B_o\mu_o}{kh}\right] \times \left[\log\left(\frac{k(t_p + \Delta t)}{\phi\mu_o c_t r_w^2}\right) - 3.23 + 0.87s\right]$$

The second contribution results from decreasing the rate from Q_o to 0 at t_p, i.e., shut-in time, thus:

$$(\Delta p)_{0-Q_o} = \left[\frac{162.6(0 - Q_o)B_o\mu_o}{kh}\right] \times \left[\log\left(\frac{k\Delta t}{\phi\mu_o c_t r_w^2}\right) - 3.23 + 0.87s\right]$$

The pressure behavior in the well during the shut-in period is then given by:

$$p_i - p_{ws} = \frac{162.6Q_o\mu_o B_o}{kh}\left[\log\frac{k(t_p + \Delta t)}{\phi\mu_o c_t r_w^2} - 3.23\right] - \frac{162.6(-Q_o)\mu_o B_o}{kh}\left[\log\frac{k\,\Delta t}{\phi\mu_o c_t r_w^2} - 3.23\right]$$

Expanding this equation and canceling terms gives:

$$p_{ws} = p_i - \frac{162.6Q_o\mu_o B_o}{kh}\left[\log\left(\frac{t_p + \Delta t}{\Delta t}\right)\right] \quad \textbf{(1.174)}$$

where

p_i = initial reservoir pressure, psi
p_{ws} = sand face pressure during pressure buildup, psi
t_p = flowing time before shut-in, hours
Q_o = stabilized well flow rate before shut-in, STB/day
Δt = shut-in time, hours

The pressure buildup equation, i.e., Eq. (1.174) was introduced by Horner (1951) and is commonly referred to as the Horner equation.

Eq. (1.174) is basically an equation of a straight line that can be expressed as:

$$p_{ws} = p_i - m\left[\log\left(\frac{t_p + \Delta t}{\Delta t}\right)\right] \quad \textbf{(1.175)}$$

This expression suggests that a plot of p_{ws} vs. $(t_p + \Delta t)/\Delta t$ on a semilog scale would produce a straight-line relationship with intercept p_i and slope m, where:

$$m = \frac{162.6Q_o B_o\mu_o}{kh} \quad \textbf{(1.176)}$$

or

$$k = \frac{162.6Q_o B_o\mu_o}{mh}$$

and where

m = slope of straight line, psi/cycle
k = permeability, md

This plot, commonly referred to as the Horner plot, is illustrated in Figure 1.37. Note that on the Horner plot, the scale of time ratio $(t_p + \Delta t)/\Delta t$ increases from right to left. It is observed from Eq. (1.174) that $p_{ws} = p_i$ when the time ratio is unity. Graphically, this means that the initial reservoir pressure, p_i, can be obtained by extrapolating the Horner plot straight line to $(t_p + \Delta t)/\Delta t = 1$.

The time corresponding to the point of shut-in, t_p can be estimated from the following equation:

$$t_p = \frac{24N_p}{Q_o}$$

where

N_p = well cumulative oil produced before shut in, STB
Q_o = stabilized well flow rate before shut in, STB/day
t_p = total production time, hours

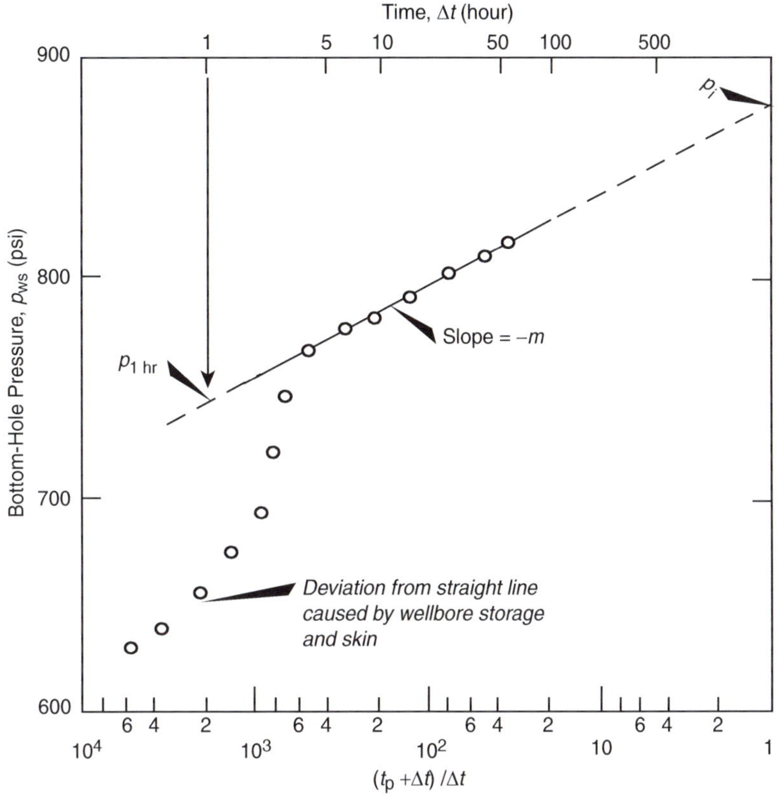

FIGURE 1.37 Horner plot. *(After Earlougher, Robert C., Jr., 1977. Advances in Well Test Analysis, Monograph, vol. 5. Society of Petroleum Engineers of AIME, Dallas, TX; Permission to publish by the SPE, copyright SPE, 1977.)*

Earlougher (1977) pointed out that a result of using the superposition principle is that the skin factor, s, does not appear in the general pressure buildup equation, Eq. (1.174). That means the Horner plot slope is not affected by the skin factor; however, the skin factor still does affect the shape of the pressure buildup data. In fact, an early-time deviation from the straight line can be caused by the skin factor as well as by wellbore storage, as illustrated in Figure 1.36. The deviation can be significant for the large negative skins that occur in hydraulically fractured wells. The skin factor does affect flowing pressure before shut-in and its value may be estimated from the buildup test data plus the flowing pressure immediately before the buildup test, as given by:

$$s = 1.151\left[\frac{p_{1\ \text{hour}} - p_{\text{wf at } \Delta t=0}}{|m|} - \log\left(\frac{k}{\phi\mu c_t r_w^2}\right) + 3.23\right] \tag{1.177}$$

with an additional pressure drop across the altered zone of:

$$\Delta p_{\text{skin}} = 0.87|m|s$$

where

$p_{\text{wf at } \Delta t=0}$ = bottom-hole flowing pressure *immediately before shut in*, psi

s = skin factor

$|m|$ = absolute value of the slope in the Horner plot, psi/cycle
r_w = wellbore radius, ft

The value of $p_{1\ \text{hour}}$ must be taken from the Horner straight line. Frequently, the pressure data does not fall on the straight line at 1 hour because of wellbore storage effects or large negative skin factors. In that case, the semilog line must be extrapolated to 1 hour and the corresponding pressure is read.

It should be noted that for a *multiphase flow*, Eqs. (1.174) and (1.177) become:

$$p_{ws} = p_i - \frac{162.36 q_t}{\lambda_t h}\left[\log\left(\frac{t_p + \Delta t}{\Delta t}\right)\right]$$

$$s = 1.151\left[\frac{p_{1\ \text{hour}} - p_{\text{wf at }\Delta t=0}}{|m|} - \log\left(\frac{\lambda_t}{\phi c_t r_w^2}\right) + 3.23\right]$$

with

$$\lambda_t = \frac{k_o}{\mu_o} + \frac{k_w}{\mu_w} + \frac{k_g}{\mu_g}$$

$$q_t = Q_o B_o + Q_w B_w + (Q_g - Q_o R_s) B_g$$

or equivalently in terms of GOR as:

$$q_t = Q_o B_o + Q_w B_w + (\text{GOR} - R_s) Q_o B_g$$

where

q_t = total fluid voidage rate, bbl/day
Q_o = oil flow rate, STB/day
Q_w = water flow rate, STB/day
Q_g = gas flow rate, scf/day
R_s = gas solubility, scf/STB
B_g = gas formation volume factor, bbl/scf
λ_t = total mobility, md /cp
k_o = effective permeability to oil, md
k_w = effective permeability to water, md
k_g = effective permeability to gas, md

The regular Horner plot would produce a semilog straight line with a slope m that can be used to determine the total mobility λ_t from:

$$\lambda_t = \frac{162.6 q_t}{mh}$$

Perrine (1956) showed that the effective permeability of each phase, i.e., k_o, k_w, and k_g, can be determined as:

$$k_o = \frac{162.6 Q_o B_o \mu_o}{mh}$$

$$k_w = \frac{162.6 Q_w B_w \mu_w}{mh}$$

$$k_g = \frac{162.6 (Q_g - Q_o R_s) B_g \mu_g}{mh}$$

For gas systems, a plot of $m(p_{ws})$ or p_{ws}^2 vs. $(t_p + \Delta t)/\Delta t$ on a semilog scale would produce a straight-line relationship with a slope of m and apparent skin factor s as defined by:

- For pseudopressure approach:

$$m = \frac{1637 Q_g T}{kh}$$

$$s' = 1.151\left[\frac{m(p_{1\ \text{hour}}) - m(p_{\text{wf at }\Delta t=0})}{|m|} - \log\left(\frac{k}{\phi \mu_i c_{ti} r_w^2}\right) + 3.23\right]$$

- For pressure-squared approach:

$$m = \frac{1637 Q_g \bar{Z} \bar{\mu}_g}{kh}$$

$$s' = 1.151\left[\frac{p_{1\ \text{hour}}^2 - p_{\text{wf at }\Delta t=0}^2}{|m|} - \log\left(\frac{k}{\phi \mu_i c_{ti} r_w^2}\right) + 3.23\right]$$

where

Q_g = the gas flow rate, Mscf/day

It should be pointed out that when a well is shut-in for a pressure buildup test, the well is usually closed at the surface rather than the sand face. Even though the well is shut in, the reservoir fluid continues to flow and accumulates in the wellbore until the well fills sufficiently to transmit the effect of shut-in to the formation. This "after-flow" behavior is caused by the wellbore storage and it has a significant influence on pressure buildup data. During the period of wellbore storage effects, the pressure data points fall below the semilog straight line. The duration of these effects may be estimated by making the log–log data plot described previously of $\log(p_{ws} - p_{wf})$ vs. $\log(\Delta t)$ with p_{wf} as the value recorded immediately before shut-in. When wellbore storage dominates, that plot will have a unit-slope straight line; as the semilog straight line is approached, the log–log plot bends over to a gently curved line with a low slope.

The wellbore storage coefficient C is calculated by selecting a point on the log–log unit-slope straight line and reading the coordinate of the point in terms of Δt and Δp:

$$C = \frac{q\,\Delta t}{24\,\Delta p} = \frac{QB\,\Delta t}{24\,\Delta p}$$

where

Δt = shut-in time, hours
Δp = pressure difference $(p_{ws} - p_{wf})$, psi
q = flow rate, bbl/day
Q = flow rate, STB/day
B = formation volume factor, bbl/STB

with a dimensionless wellbore storage coefficient as given by Eq. (1.34) as:

$$C_D = \frac{0.8936C}{\phi h c_t r_w^2}$$

In all the pressure buildup test analyses, the log–log data plot should be made before the straight line is chosen on the semilog data plot. This log–log plot is essential to avoid drawing a semilog straight line through the wellbore storage-dominated data. The beginning of the semilog line can be estimated by observing the time when the data points on the log–log plot reach the slowly curving low-slope line and adding 1 to $1\frac{1}{2}$ cycles in time after the end of the unit-slope straight line. Alternatively, the time to the beginning of the semilog straight line can be estimated from:

$$\Delta t > \frac{170{,}000C\,e^{0.14s}}{(kh/\mu)}$$

where

c = calculated wellbore storage coefficient, bbl/psi
k = permeability, md
s = skin factor
h = thickness, ft

Example 1.27

Table 1.5 shows the pressure buildup data from an oil well with an estimated drainage radius of 2640 ft. Before shut-in, the well had produced at a stabilized rate of 4900 STB/day for 310 hours. Known reservoir data is:[2]

depth = 10476 ft,	$r_w = 0.354$ ft,	$c_t = 22.3 \times 10^{-6}$ psi^{-1}
$Q_o = 4900$ STB/D,	$h = 482$ ft,	$p_{wf}(\Delta t = 0) =$ 2761 psig
$\mu_o = 0.20$ cp,	$B_o = 1.55$ bbl/STB,	$\phi = 0.09$
$t_p = 310$ hour,	$r_e = 2640$ ft	

Calculate:

- the average permeability k;
- the skin factor;
- the additional pressure drop due to skin.

Solution

Step 1. Plot p_{ws} vs. $(t_p + \Delta t)/\Delta t$ on a semilog scale as shown in Figure 1.38.

Step 2. Identify the correct straight-line portion of the curve and determine the slope m:

$$m = 40\ \text{psi/cycle}$$

Step 3. Calculate the average permeability by using Eq. (1.176):

$$k = \frac{162.6 Q_o B_o \mu_o}{mh} = \frac{(162.6)(4900)(1.55)(0.22)}{(40)(482)} = 12.8\ \text{md}$$

Step 4. Determine p_{wf} after 1 hour from the straight-line portion of the curve:

$$p_{1\,hr} = 3266\ \text{psi}$$

[2]This example problem and the solution procedure are given in Earlougher, R. *Advances in Well Test Analysis*, Monograph Series, SPE, Dallas (1997).

TABLE 1.5 Earlougher's Pressure Buildup Data

Δt (hour)	$t_p + \Delta t$ (hour)	$t_p + \Delta t \Delta t$	p_{ws} (psig)
0.0	—	—	2761
0.10	310.30	3101	3057
0.21	310.21	1477	3153
0.31	310.31	1001	3234
0.52	310.52	597	3249
0.63	310.63	493	3256
0.73	310.73	426	3260
0.84	310.84	370	3263
0.94	310.94	331	3266
1.05	311.05	296	3267
1.15	311.15	271	3268
1.36	311.36	229	3271
1.68	311.68	186	3274
1.99	311.99	157	3276
2.51	312.51	125	3280
3.04	313.04	103	3283
3.46	313.46	90.6	3286
4.08	314.08	77.0	3289
5.03	315.03	62.6	3293
5.97	315.97	52.9	3297
6.07	316.07	52.1	3297
7.01	317.01	45.2	3300
8.06	318.06	39.5	3303
9.00	319.00	35.4	3305
10.05	320.05	31.8	3306
13.09	323.09	24.7	3310
16.02	326.02	20.4	3313
20.00	330.00	16.5	3317
26.07	336.07	12.9	3320
31.03	341.03	11.0	3322
34.98	344.98	9.9	3323
37.54	347.54	9.3	3323

[a]This example problem and the solution procedure are given in Earlougher, R. *Advance Well Test Analysis*, Monograph Series, SPE, Dallas (1977).

Permission to Publish by the SPE, Copyright SPE, 1977.

Step 5. Calculate the skin factor by applying Eq. (1.177):

$$s = 1.151\left[\frac{p_{1\,\text{hour}} - p_{\text{wf}\,\Delta t=0}}{m} - \log\left(\frac{k}{\phi\mu c_t r_w^2}\right) + 3.23\right]$$

$$= 1.151\left[\frac{3266 - 2761}{40} - \log\left(\frac{(12.8)}{(0.09)(0.20)(22.6\times10^{-6})(0.354)^2}\right) + 3.23\right]$$

$$= 8.6$$

Step 6. Calculate the additional pressure drop by using:

$$\Delta p_{\text{skin}} = 0.87|m|s = 0.87(40)(8.6) = 299.3 \text{ psi}$$

It should be pointed out that Eq. (1.174) assumes the reservoir to be infinite in size, i.e., $r_e = \infty$, which implies that at some point in the reservoir the pressure would always be equal to

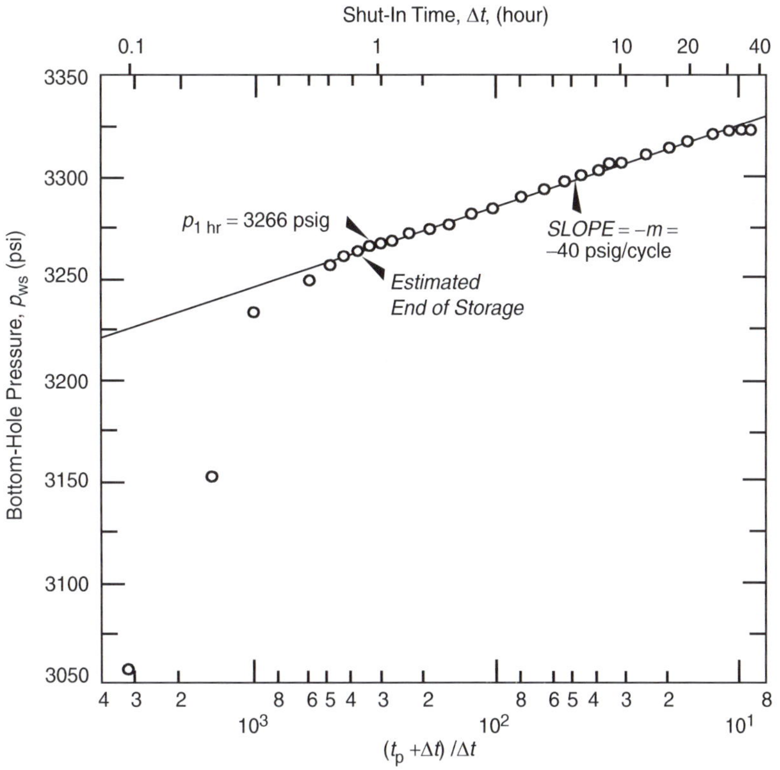

FIGURE 1.38 Earlougher's semilog data plot for the buildup test. *(Permission to publish by the SPE, copyright SPE, 1977.)*

the initial reservoir pressure p_i and the Horner straight-line plot will always extrapolate to p_i. However, reservoirs are finite and soon after production begins, fluid removal will cause a pressure decline everywhere in the reservoir system. Under these conditions, the straight line will not extrapolate to the initial reservoir pressure p_i but, instead, the pressure obtained will be a false pressure as denoted by p^*. The false pressure, as illustrated by Matthews and Russell (1967) in Figure 1.39, has no physical meaning but it is used to determine the average reservoir pressure $\bar{p}$. It is clear that p^* will *only equal* the initial (original) reservoir pressure p_i when a new well in a newly discovered field is tested. Using the concept of the false pressure p^*, Horner expressions, as given by Eqs. (1.174) and (1.175), should be expressed in terms of p^* instead of p_i as:

$$p_{ws} = p^* - \frac{162.6 Q_o \mu_o B_o}{kh}\left[\log\left(\frac{t_p + \Delta t}{\Delta t}\right)\right]$$

and

$$p_{ws} = p^* - m\left[\log\left(\frac{t_p + \Delta t}{\Delta t}\right)\right] \qquad \textbf{(1.178)}$$

Bossie-Codreanu (1989) suggested that the well drainage area can be determined from the Horner pressure buildup plot or the MDH plot, discussed next, by selecting the coordinates of any three points located on the semilog straight-line portion of the plot to determine the slope of the pseudosteady-state line m_{pss}. The coordinates of these three points are designated as:

- shut-in time Δt_1 and with a corresponding shut-in pressure p_{ws1};
- shut-in time Δt_2 and with a corresponding shut-in pressure p_{ws2};

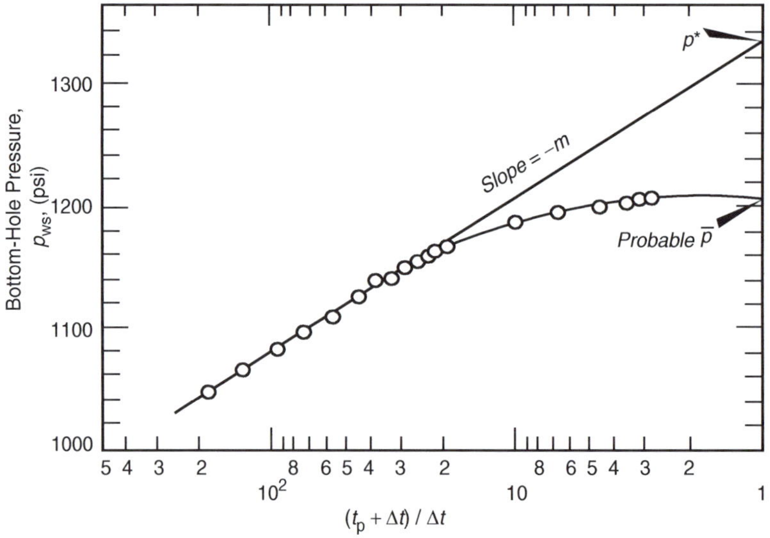

FIGURE 1.39 Typical pressure buildup curve for a well in a finite. *(After Earlougher, Robert C., Jr., 1977. Advances in Well Test Analysis, Monograph, vol. 5. Society of Petroleum Engineers of AIME, Dallas, TX; Permission to publish by the SPE, copyright SPE, 1977.)*

- shut-in time Δt_3 and with a corresponding shut-in pressure p_{ws3}.

The selected shut-in times satisfy $\Delta t_1 < \Delta t_2 < \Delta t_3$. The slope of the pseudosteady-state straight-line m_{pss} is then approximated by:

$$m_{pss} = \frac{(p_{ws2} - p_{ws1})\log(\Delta t_3/\Delta t_1) - (p_{ws3} - p_{ws1})\log[\Delta t_2/\Delta t_1]}{(\Delta t_3 - \Delta t_1)\log(\Delta t_2 \Delta t_1) - (\Delta t_2 - \Delta t_1)\log(\Delta t_3/\Delta t_1)} \quad \textbf{(1.179)}$$

The well drainage area can be calculated from Eq. (1.127):

$$m^{\backslash} = m_{pss} = \frac{0.23396 Q_o B_o}{c_t A h \phi}$$

Solving for the drainage area gives:

$$A = \frac{0.23396 Q_o B_o}{c_t m_{pss} h \phi}$$

where

m_{pss} or $m^{\backslash}$ = slope of *straight line* during the pseudosteady-state flow, psi/hour
Q_o = flow rate, bbl/day
A = well drainage area, ft^2

1.3.4 Miller–Dyes–Hutchinson Method

The Horner plot may be simplified if the well has been producing long enough to reach a pseudosteady state. Assuming that the production time t_p is much greater than the total shut-in time Δt, i.e., $t_p \gg \Delta t$, the term $t_p + \Delta t \simeq t_p$ and:

$$\log\left(\frac{t_p + \Delta t}{\Delta t}\right) \cong \log\left(\frac{t_p}{\Delta t}\right) = \log(t_p) - \log(\Delta t)$$

Applying the above mathematical assumption to Eq. (1.178), gives:

$$p_{ws} = p^* - m[\log(t_p) - \log(\Delta t)]$$

or

$$p_{ws} = [p^* - m\log(t_p)] + m\log(\Delta t)$$

This expression indicates that a plot of p_{ws} vs. $\log(\Delta t)$ would produce a semilog straight line with a positive slope of $+m$ that is identical to that obtained from the Horner plot. The

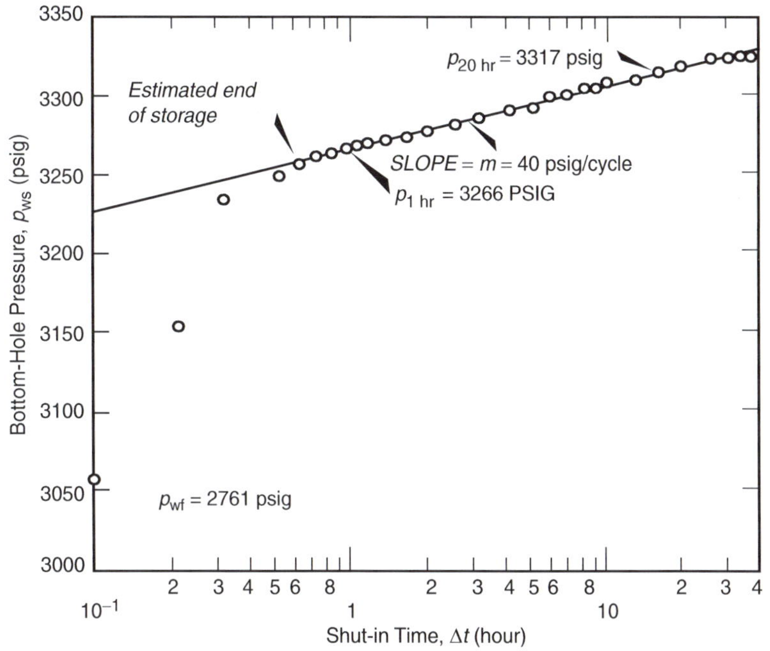

FIGURE 1.40 Miller–Dyes–Hutchinson plot for the buildup test. *(After Earlougher, Robert C., Jr., 1977. Advances in Well Test Analysis, Monograph, vol. 5. Society of Petroleum Engineers of AIME, Dallas, TX; Permission to publish by the SPE, copyright SPE, 1977.)*

slope is defined mathematically by Eq. (1.176) as:

$$m = \frac{162.6 Q_o B_o \mu_o}{kh}$$

The semilog straight-line slope m has the same value as of the Horner plot. This plot is commonly called the Miller–Dyes–Hutchinson (MDH) plot. The false pressure p^* may be estimated from the MDH plot by using:

$$p^* = p_{1\ \mathrm{hour}} + m \log(t_p + 1) \quad \textbf{(1.180)}$$

where $p_{1\ \mathrm{hour}}$ is read from the semilog straight-line plot at $\Delta t = 1$ hour. The MDH plot of the pressure buildup data given in Table 1.5 in terms of p_{ws} vs. $\log(\Delta t)$ is shown in Figure 1.40.

Figure 1.40 shows a positive slope of $m = 40$ psi/cycle that is identical to the value obtained in Example 1.26 with a $p_{1\ \mathrm{hour}} =$ 3266 psig.

As in the Horner plot, the time that marks the beginning of the MDH semilog straight line may be estimated by making the log–log plot of $(p_{ws} - p_{wf})$ vs. Δt and observing when the data points deviate from the 45° angle (unit slope). The exact time is determined by moving 1 to $1\frac{1}{2}$ cycles in time after the end of the unit-slope straight line.

The observed pressure behavior of the test well following the end of the transient flow will depend on:

- shape and geometry of the test well drainage area;
- the position of the well relative to the drainage boundaries;
- length of the producing time t_p before shut-in.

If the well is located in a reservoir with no other wells, the shut-in pressure would eventually become constant (as shown in Figure 1.38) and equal to the *volumetric average reservoir pressure* $\bar{p}_r$. This pressure is required in many reservoir engineering calculations such as:

- material balance studies;
- water influx;
- pressure maintenance projects;
- secondary recovery;
- degree of reservoir connectivity.

Finally, in making future predictions of production as a function of $\bar{p}_r$, pressure measurements throughout the reservoir's life are almost mandatory if one is to compare such a prediction to actual performance and make the necessary adjustments to the predictions. One way to obtain this pressure is to shut-in all wells producing from the reservoir for a period of time that is sufficient for pressures to equalize throughout the system to give $\bar{p}_r$. Obviously, such a procedure is not practical.

To use the MDH method to estimate average drainage region pressure $\bar{p}_r$ for a circular or square system producing at *pseudosteady state before shut-in*:

(1) Choose any convenient time on the semilog straight line Δt and read the corresponding pressure p_{ws}.

(2) Calculate the dimensionless shut-in time based on the drainage area A from:

$$\Delta t_{DA} = \frac{0.0002637 k\, \Delta t}{\phi \mu c_t A}$$

(3) Enter Figure 1.41 with the dimensionless time Δt_{DA} and determine an MDH dimensionless pressure p_{DMDH} from the upper curve of Figure 1.41.

(4) Estimate the average reservoir pressure in the closed drainage region from:

$$\bar{p}_r = p_{ws} + \frac{m p_{DMDH}}{1.1513}$$

where m is the semilog straight line of the MDH plot.

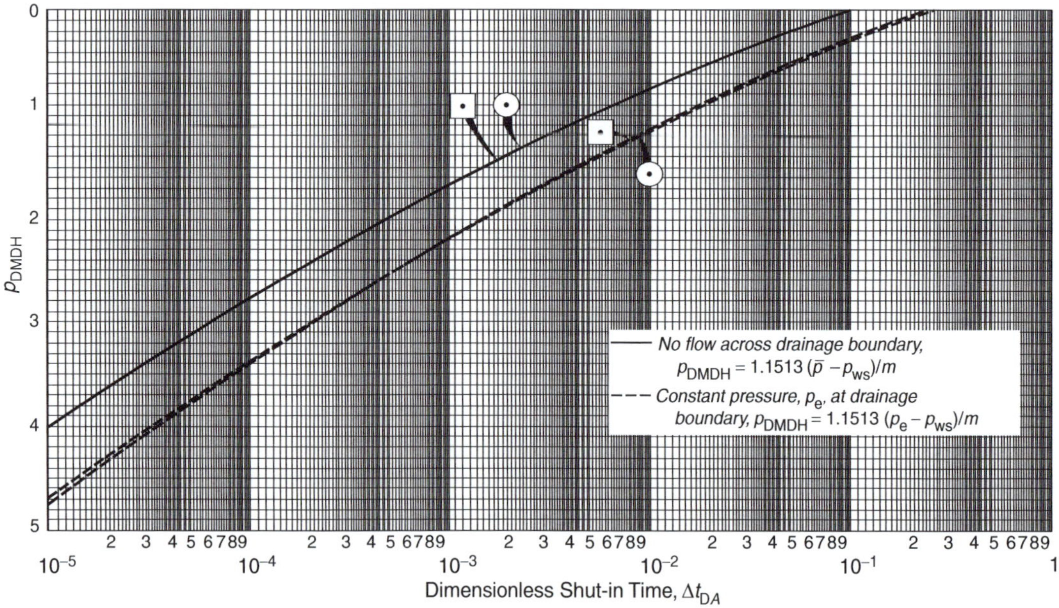

FIGURE 1.41 Miller–Dyes–Hutchinson dimensionless pressure for circular and square drainage areas. *(After Earlougher, Robert C., Jr., 1977. Advances in Well Test Analysis, Monograph, vol. 5. Society of Petroleum Engineers of AIME, Dallas, TX; Permission to publish by the SPE, copyright SPE, 1977.)*

There are several other methods for determining $\bar{p}_r$ from a buildup test. Three of these methods are briefly presented below:

(1) the Matthews–Brons–Hazebroek (MBH) method;
(2) the Ramey–Cobb method;
(3) the Dietz method.

1.3.5 MBH Method

As noted previously, the buildup test exhibits a semilog straight line which begins to bend down and become flat at the later shut-in times because of the effect of the boundaries. Matthews et al. (1954) proposed a methodology for estimating average pressure from buildup tests in bounded drainage regions. The MBH method is based on theoretical correlations between the extrapolated semilog straight line to the false pressure p^* and current average drainage area pressure $\bar{p}$. The authors point out that the average pressure in the drainage area of each well can be related to p^* if the geometry, shape, and location of the well relative to the drainage boundaries are known. They developed a set of correction charts, as shown in Figures 1.42–1.45, for various drainage geometries.

The y-axis of these figures represents the MBH dimensionless pressure p_{DMBH} that is defined by:

$$p_{DMBH} = \frac{2.303(p^* - \bar{p})}{|m|}$$

or

$$\bar{p} = p^* - \left(\frac{|m|}{2.303}\right) p_{DMBH} \tag{1.181}$$

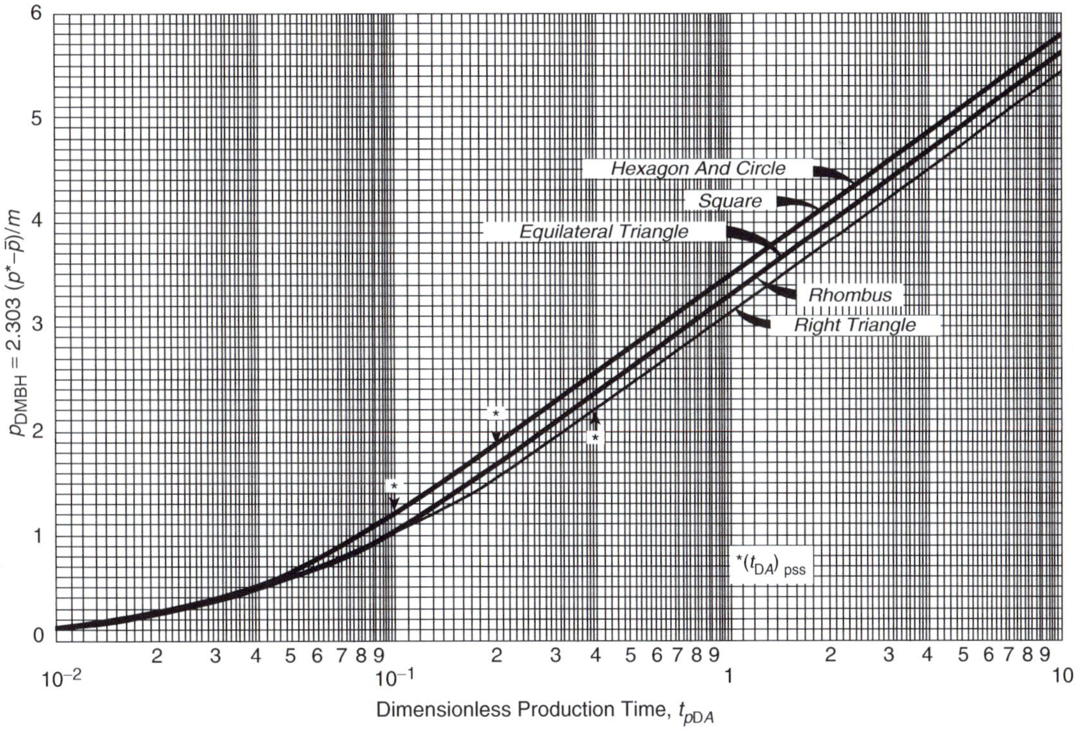

FIGURE 1.42 Matthews–Brons–Hazebroek dimensionless pressure for a well in the center of equilateral drainage areas. *(After Earlougher, Robert C., Jr., 1977. Advances in Well Test Analysis, Monograph, vol. 5. Society of Petroleum Engineers of AIME, Dallas, TX; Permission to publish by the SPE, copyright SPE, 1977.)*

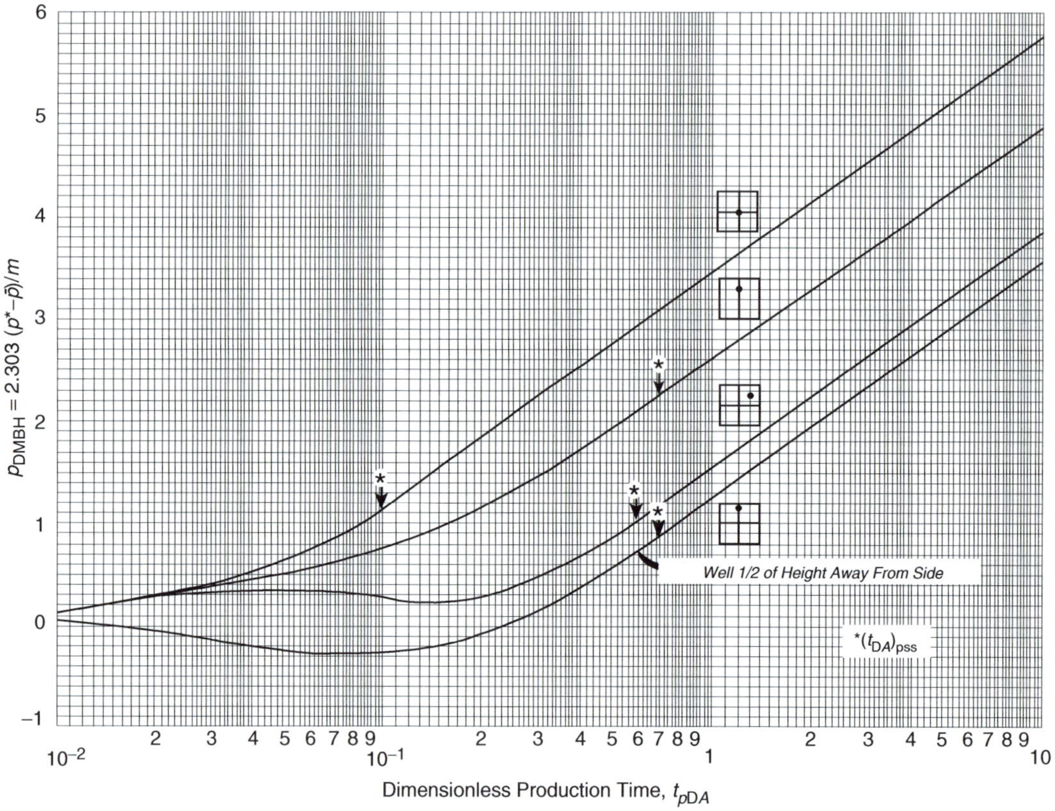

FIGURE 1.43 Matthews–Brons–Hazebroek dimensionless pressure for different well locations in a square drainage area. *(After Earlougher, Robert C., Jr., 1977. Advances in Well Test Analysis, Monograph, vol. 5. Society of Petroleum Engineers of AIME, Dallas, TX; Permission to publish by the SPE, copyright SPE, 1977.)*

where m is the *absolute* value of the slope obtained from the Horner semilog straight-line plot. The MBH dimensionless pressure is determined at the dimensionless producing time t_{pDA} that corresponds to the flowing time t_p. That is:

$$t_{pDA} = \left[\frac{0.0002637k}{\phi\mu c_t A}\right] t_p \tag{1.182}$$

where

t_p = flowing time before shut-in, hours
A = drainage area, ft^2
k = permeability, md
c_t = total compressibility, psi^{-1}

The following steps summarize the procedure for applying the MBH method:

Step 1. Make a Horner plot.
Step 2. Extrapolate the semilog straight line to the value of p^* at $(t_p + \Delta t)/\Delta t = 1.0$.
Step 3. Evaluate the slope of the semilog straight line m.
Step 4. Calculate the MBH dimensionless producing time t_{pDA} from Eq. (1.182):

$$t_{pDA} = \left[\frac{0.0002673k}{\phi\mu c_t A}\right] t_p$$

Step 5. Find the closest approximation to the shape of the well drainage area in Figures 1.41–1.44 and identify the correction curve.

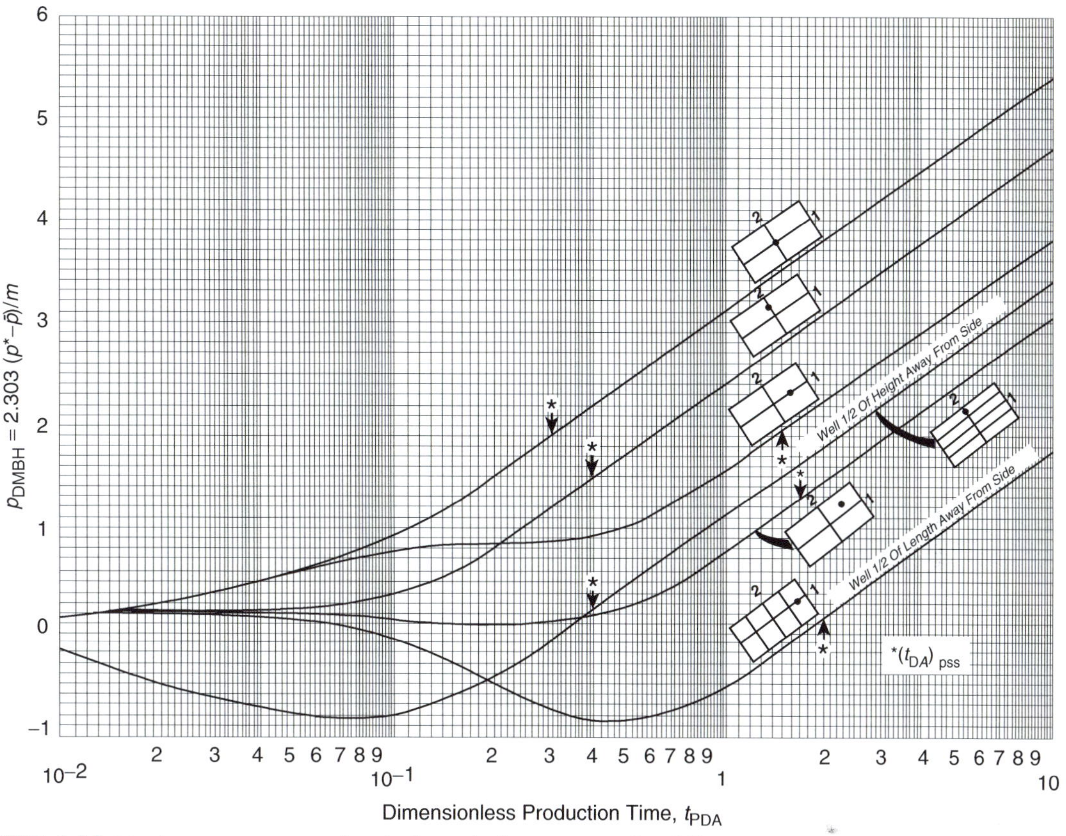

FIGURE 1.44 Matthews–Brons–Hazebroek dimensionless pressure for different well locations in a 2:1 rectangular drainage area. *(After Earlougher, Robert C., Jr., 1977. Advances in Well Test Analysis, Monograph, vol. 5. Society of Petroleum Engineers of AIME, Dallas, TX; Permission to publish by the SPE, copyright SPE, 1977.)*

Step 6. Read the value of p_{DMBH} from the correction curve at t_{PDA}

Step 7. Calculate the value of $\bar{p}$ from Eq. (1.181):

$$\bar{p} = p^* - \left(\frac{|m|}{2.303}\right) p_{DMBH}$$

As in the normal Horner analysis technique, the producing time t_p is given by:

$$t_p = \frac{24 N_p}{Q_o}$$

where N_p is the cumulative volume produced since the *last pressure buildup test* and Q_o is the constant flow rate just before shut-in. Pinson (1972) and Kazemi (1974) indicate that t_p should be compared with the time required to reach the pseudosteady state, t_{pss}:

$$t_{pss} = \left[\frac{\phi \mu c_t A}{0.0002367 k}\right] (t_{DA})_{pss} \qquad \textbf{(1.183)}$$

For a symmetric closed or circular drainage area, $(t_{DA})_{pss} = 0.1$ as listed in the fifth column of Table 1.4.

If $t_p \gg t_{pss}$, then t_{pss} should ideally replace t_p in both the Horner plot and for use with the MBH dimensionless pressure curves.

The above methodology gives the value of $\bar{p}$ in the drainage area of *one well*, e.g., well *i*. If a number of wells are producing from the

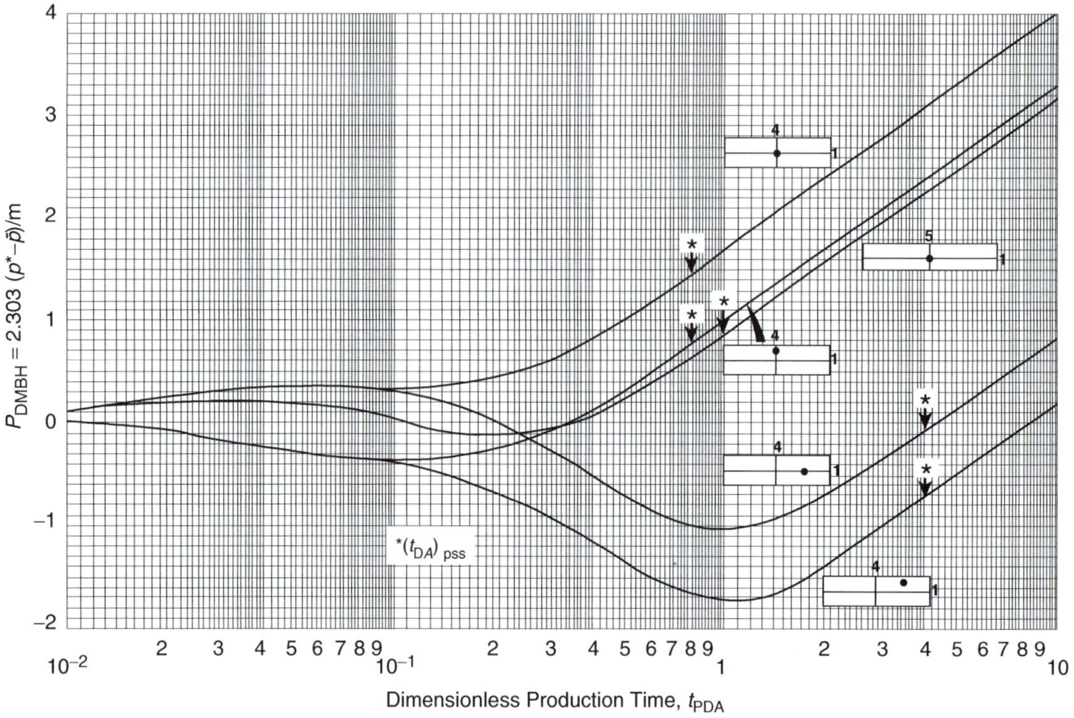

FIGURE 1.45 Matthews–Brons–Hazebroek dimensionless pressure for different well locations in 4:1 and 5:1 rectangular drainage areas. *(After Earlougher, Robert C., Jr., 1977. Advances in Well Test Analysis, Monograph, vol. 5. Society of Petroleum Engineers of AIME, Dallas, TX; Permission to publish by the SPE, copyright SPE, 1977).*

reservoir, each well can be analyzed separately to give $\overline{p}$ for its own drainage area. The reservoir average pressure $\overline{p}_r$ can be estimated from these *individual well average drainage pressures* by using one of the relationships given by Eqs. (1.129) and (1.130). That is:

$$\overline{p}_r = \frac{\sum_i (\overline{p}q)_i/(\partial\overline{p}/\partial t)_i}{\sum_i q_i/(\partial\overline{p}/\partial t)_i}$$

or

$$\overline{p}_r = \frac{\sum_i [\overline{p}\Delta(F)/\Delta\overline{p}]_i}{\sum_i [\Delta(F)/\Delta\overline{p}]_i}$$

with

$$F_t = \int_0^t [Q_o B_o + Q_w B_w + (Q_g - Q_o R_s - Q_w R_{sw}) B_g] dt$$
$$F_{t+\Delta t} = \int_0^{t+\Delta t} [Q_o B_o + Q_w B_w + (Q_g - Q_o R_s - Q_w R_{sw}) B_g] dt$$

and

$$\Delta(F) = F_{t+\Delta t} - F_t$$

Similarly, it should be noted that the MBH method and Figures 1.41–1.44 can be applied for compressible gases by defining p_{DMBH} as:

- For the pseudopressure approach:

$$p_{DMBH} = \frac{2.303[m(p^*) - m(\overline{p})]}{|m|} \quad \textbf{(1.184)}$$

- For the pressure-squared approach:

$$p_{DMBH} = \frac{2.303[(p^*)^2 - (\overline{p})^2]}{|m|} \quad \textbf{(1.185)}$$

Example 1.28

Using the information given in Example 1.27 and pressure buildup data listed in Table 1.5, calculate the average pressure in the well drainage area and the drainage area by applying Eq. (1.179). The data is listed below for convenience:

$r_e = 2640$ ft,	$r_w = 0.354$ ft,	$c_t = 22.6 \times 10^{-6}$ psi^{-1}
$Q_o = 4900$ STB/D,	$h = 482$ ft,	$p_{wf \text{ at } \Delta t = 0} =$ 2761 psig
$\mu_o = 0.20$ cp,	$B_o = 1.55$ bbl/STB,	$\phi = 0.09$
$t_p = 310$ hours,	depth = 10,476 ft,	reported average pressure = 3323 psi

Solution

Step 1. Calculate the drainage area of the well:

$$A = \pi r_e^2 = \pi(2640)^2$$

Step 2. Compare the production time t_p, i.e., 310 hours, with the time required to reach the pseudosteady state t_{pss} by applying Eq. (1.183). Estimate t_{pss} using $(t_{DA})_{pss} = 0.1$ to give:

$$t_{pss} = \left[\frac{\phi\mu c_t A}{0.0002367k}\right](t_{DA})_{pss}$$

$$= \left[\frac{(0.09)(0.2)(22.6\times10^{-6})(\pi)(2640)^2}{(0.0002637)(12.8)}\right]0.1$$

$$= 264 \text{ hours}$$

Thus, we could replace t_p by 264 hours in our analysis because $t_p > t_{pss}$. However, since t_p is only about $1.2t_{pss}$, we use the actual production time of 310 hours in the calculation.

Step 3. Figure 1.38 does not show p^* since the semilog straight line is not extended to $(t_p + \Delta t)/t = \Delta t = 1.0$. However, p^* can be calculated from p_{ws} at $(t_p + \Delta t)/\Delta t = 10.0$ by extrapolating one cycle. That is:

$$p^* = 3325 + (1 \text{ cycle})(40 \text{ psi/cycle}) = 3365 \text{ psig}$$

Step 4. Calculate t_{pDA} by applying Eq. (1.182) to give:

$$t_{pDA} = \left[\frac{0.0002637k}{\phi\mu c_t A}\right]t_p$$

$$= \left[\frac{0.0002637(12.8)}{(0.09)(0.2)(22.6\times10^{-6})(\pi)(2640)^2}\right]310$$

$$= 0.117$$

Step 5. From the curve of the circle in Figure 1.42, obtain the value of p_{DMBH} at $t_{pDA} = 0.117$, to give:

$$p_{DMBH} = 1.34$$

Step 6. Calculate the average pressure from Eq. (1.181):

$$\bar{p} = p^* - \left(\frac{|m|}{2.303}\right)p_{DMBH}$$

$$= 3365 - \left(\frac{40}{2.303}\right)(1.34) = 3342 \text{ psig}$$

This is 19 psi higher than the maximum pressure recorded that is 3323 psig.

Step 7. Select the coordinates of any three points located on the semilog straight line portion of the Horner plot, to give:

- $(\Delta t_1, p_{ws1}) = (2.52, 3280)$
- $(\Delta t_2, p_{ws2}) = (9.00, 3305)$
- $(\Delta t_3, p_{ws3}) = (20.0, 3317)$

Step 8. Calculate m_{pss} by applying Eq. (1.179):

$$m_{pss} = \frac{(p_{ws2} - p_{ws1})\log(\Delta t_3/\Delta t_1) - (p_{ws3} - p_{ws1})\log(\Delta t_2/\Delta t_1)}{(\Delta t_3 - \Delta t_1)\log(\Delta t_2/\Delta t_1) - (\Delta t_2 - \Delta t_1)\log(\Delta t_3/\Delta t_1)}$$

$$= \frac{(3305-3280)\log(20/2.51) - (3317-3280)\log(9/2.51)}{(20-2.51)\log(9/2.51) - (9-2.51)\log(20/2.51)}$$

$$= 0.52339 \text{ psi/hour}$$

Step 9. The well drainage area can then be calculated from Eq. (1.127):

$$A = \frac{0.23396 Q_o B_o}{c_t m_{pss} h\phi}$$

$$= \frac{0.23396(4900)(1.55)}{(22.6\times10^{-6})(0.52339)(482)(0.09)}$$

$$= 3,462,938 \text{ ft}^2$$

$$= \frac{3,363,938}{43,560} = 80 \text{ acres}$$

The corresponding drainage radius is 1050 ft which differs considerably from the given radius of 2640 ft. Using the calculated drainage radius of 1050 ft and repeating the MBH calculations gives:

$$t_{pss} = \left[\frac{(0.09)(0.2)(22.6\times10^{-6})(\pi)(1050)^2}{(0.0002637)(12.8)}\right]0.1$$
$$= 41.7 \text{ hours}$$
$$t_{pDA} = \left[\frac{0.0002637(12.8)}{(0.09)(0.2)(22.6\times10^{-6})(\pi)(1050)^2}\right]310 = 0.743$$
$$p_{DMBH} = 3.15$$
$$\bar{p} = 3365 - \left(\frac{40}{2.303}\right)(3.15) = 3311 \text{ psig}$$

The value is 12 psi higher than the reported value of average reservoir pressure.

1.3.6 Ramey–Cobb Method

Ramey and Cobb (1971) proposed that the average pressure in the well drainage area can be read directly from the Horner semilog straight line if the following data is available:

- shape of the well drainage area;
- location of the well within the drainage area;
- size of the drainage area.

The proposed methodology is based on calculating the dimensionless producing time t_{pDA} as defined by Eq. (1.182):

$$t_{pDA} = \left[\frac{0.0002637k}{\phi\mu c_t A}\right]t_p$$

where

t_p = producing time since the last shut-in, hours
A = drainage area, ft^2

Knowing the shape of the drainage area and well location, determine the dimensionless time to reach pseudosteady state $(t_{DA})_{pss}$, as given in Table 1.4 in the fifth column. Compare t_{pDA} with $(t_{DA})_{pss}$:

- If $t_{pDA} < (t_{DA})_{pss}$, then read the average pressure $\bar{p}$ from the Horner semilog straight line at:

$$\left(\frac{t_p + \Delta t}{\Delta t}\right) = \exp(4\pi t_{pDA}) \qquad \textbf{(1.186)}$$

or use the following expression to estimate $\bar{p}$:

$$\bar{p} = p^* - m\log[\exp(4\pi t_{pDA})] \qquad \textbf{(1.187)}$$

- If $t_{pDA} > (t_{DA})_{pss}$, then read the average pressure $\bar{p}$ from the Horner semilog straight-line plot at:

$$\left(\frac{t_p + \Delta t}{\Delta t}\right) = C_A t_{pDA} \qquad \textbf{(1.188)}$$

where

C_A = shape factor as determined from Table 1.4.

Equivalently, the average pressure can be estimated from:

$$\bar{p} = p^* - m\log(C_A t_{pDA}) \qquad \textbf{(1.189)}$$

where

m = absolute value of the semilog straight-line slope, psi/cycle
p^* = false pressure, psia
C_A = shape factor, from Table 1.4

Example 1.29

Using the data given in Example 1.27, recalculate the average pressure using the Ramey and Cobb method.

Solution

Step 1. Calculate t_{pDA} by applying Eq. (1.182):

$$t_{pDA} = \left[\frac{0.0002637k}{\phi\mu c_t A}\right]t_p$$
$$= \left[\frac{0.0002637(12.8)}{(0.09)(0.2)(22.6\times10^{-6})(\pi)(2640)^2}\right](310)$$
$$= 0.1175$$

Step 2. Determine C_A and $(t_{DA})_{pss}$ from Table 1.4 for a well located in the centre of a circle, to give:

$$C_A = 31.62$$
$$(t_{DA})_{pss} = 0.1$$

Step 3. Since $t_{pDA} > (t_{DA})_{pss}$, calculate $\bar{p}$ from Eq. (1.189):

$$\bar{p} = p^* - m \log(C_A t_{pDA})$$
$$= 3365 - 40 \log[31.62(0.1175)] = 3342 \text{ psi}$$

This value is identical to that obtained from the MBH method.

1.3.7 Dietz Method

Dietz (1965) indicated that if the test well has been producing long enough to reach the pseudosteady state before shut-in, the average pressure can be read directly from the MDH semilog straight-line plot, i.e., p_{ws} vs. $\log(\Delta t)$, at the following shut-in time:

$$(\Delta t)_{\bar{p}} = \frac{\phi \mu c_t A}{0.0002637 C_A k} \qquad \textbf{(1.190)}$$

where

Δt = shut-in time, hours
A = drainage area, ft^2
C_A = shape factor
k = permeability, md
c_t = total compressibility, psi^{-1}

Example 1.30

Using the Dietz method and the buildup data given in Example 1.27, calculate the average pressure.

Solution

Step 1. Using the buildup data given in Table 1.5, construct the MDH plot of p_{ws} vs. $\log(\Delta t)$ as shown in Figure 1.40. From the plot, read the following values:

$$m = 40 \text{ psi/cycle}$$
$$p_{1\text{ hour}} = 3266 \text{ psig}$$

Step 2. Calculate false pressure p^* from Eq. (1.180) to give:

$$p^* = p_{1\text{ hour}} + m \log(t_p + 1)$$
$$= 3266 + 40 \log(310 + 1) = 3365.7 \text{ psi}$$

Step 3. Calculate the shut-in time $(\Delta t)_{\bar{p}}$ from Eq. (1.188):

$$(\Delta t)_{\bar{p}} = \frac{(0.09)(0.2)(22.6 \times 10^{-6})(\pi)(2640)^2}{(0.0002637)(12.8)(31.62)}$$
$$= 83.5 \text{ hours}$$

Step 4. Since the MDH plot does not extend to 83.5 hours, the average pressure can be calculated from the semilog straight-line equation as given by:

$$p = p_{1\text{ hour}} + m \log(\Delta t - 1) \qquad \textbf{(1.191)}$$

or

$$\bar{p} = 3266 + 40 \log(83.5 - 1) = 3343 \text{ psi}$$

As indicated earlier, the skin factor s is used to calculate the additional pressure drop in the altered permeability area around the wellbore and to characterize the well through the calculation of the flow coefficient E. That is:

$$\Delta p_{skin} = 0.87|m|s$$

and

$$E = \frac{J_{actual}}{J_{ideal}} = \frac{\bar{p} - p_{wf} - \Delta p_{skin}}{\bar{p} - p_{wf}}$$

where $\bar{p}$ is the average pressure in the well drainage area. Lee (1982) suggested that for rapid analysis of the pressure buildup, the flow efficiency can be approximated by using the extrapolated straight-line pressure p^*, to give:

$$E = \frac{J_{actual}}{J_{ideal}} \approx \frac{p^* - p_{wf} - \Delta p_{skin}}{\bar{p} - p_{wf}}$$

Earlougher (1977) pointed out that there are a surprising number of situations where a single pressure point or "spot pressure" is the only pressure information available about a well. The average drainage region pressure $\bar{p}$ can be estimated from the spot pressure reading at shut-in time Δt using:

$$\bar{p} = p_{\text{ws at }\Delta t} + \frac{162.6Q_o\mu_o B_o}{kh}\left[\log\left(\frac{\phi\mu c_t A}{0.0002637kC_A\,\Delta t}\right)\right]$$

For a closed square drainage region $C_A = 30.8828$ and:

$$\bar{p} = p_{\text{ws at }\Delta t} + \frac{162.6Q_o\mu_o B_o}{kh}\left[\log\left(\frac{122.8\phi\mu c_t A}{k\,\Delta t}\right)\right]$$

where

$p_{\text{ws at }\Delta t}$ = spot pressure reading at shut-in time Δt
Δt = shut-in time, hours
A = drainage area, ft^2
C_A = shape factor
k = permeability, md
c_t = total compressibility, psi^{-1}

It is appropriate at this time to briefly introduce the concept of type curves and discuss their applications in well testing analysis.

1.4 TYPE CURVES

The type curve analysis approach was introduced in the petroleum industry by Agarwal et al. (1970) as a valuable tool when used in conjunction with conventional semilog plots. A type curve is a graphical representation of the theoretical solutions to flow equations. The type curve analysis consists of finding the theoretical type curve that "matches" the actual response from a test well and the reservoir when subjected to changes in production rates or pressures. The match can be found graphically by physically superposing a graph of actual test data with a similar graph of type curve(s) and searching for the type curve that provides the best match. Since type curves are plots of theoretical solutions to transient and pseudosteady-state flow equations, they are usually presented in terms of dimensionless variables (e.g., p_D, t_D, r_D, and C_D) rather than real variables (e.g., Δp, t, r, and C). The reservoir and well parameters, such as permeability and skin, can then be calculated from the dimensionless parameters defining that type curve.

Any variable can be made "dimensionless" by multiplying it by a group of constants with opposite dimensions, but the choice of this group will depend on the type of problem to be solved. For example, to create the dimensionless pressure drop p_D, the actual pressure drop Δp in psi is multiplied by the group A with units of psi^{-1}, or:

$$p_D = A\,\Delta p$$

Finding group A that makes a variable dimensionless is derived from equations that describe reservoir fluid flow. To introduce this concept, recall Darcy's equation that describes radial, incompressible, steady-state flow as expressed by:

$$Q = \left[\frac{kh}{141.2B\mu[\ln(r_e/r_{wa}) - 0.5]}\right]\Delta p \qquad (1.192)$$

where r_{wa} is the apparent (effective) wellbore radius and defined by Eq. (1.151) in terms of the skin factor s as:

$$r_{wa} = r_w\,e^{-s}$$

Group A can be defined by rearranging Darcy's equation as:

$$\ln\left(\frac{r_e}{r_{wa}}\right) - \frac{1}{2} = \left[\frac{kh}{141.2QB\mu}\right]\Delta p$$

Because the left-hand side of this equation is dimensionless, the right-hand side must be accordingly dimensionless. This suggests that the term $kh/141.\,2QB\mu$ is essentially group A with units of psi^{-1} that defines the dimensionless variable p_D, or:

$$p_D = \left[\frac{kh}{141.2QB\mu}\right]\Delta p \qquad (1.193)$$

Taking the logarithm of both sides of this equation gives:

$$\log(p_D) = \log(\Delta p) + \log\left(\frac{kh}{141.2QB\mu}\right) \qquad (1.194)$$

where

Q = flow rate, STB/day
B = formation, volume factor, bbl/STB
μ = viscosity, cp

For a constant flow rate, Eq. (1.194) indicates that the logarithm of dimensionless pressure drop, $\log(p_D)$, will differ from the logarithm of the *actual* pressure drop, $\log(\Delta p)$, by a constant amount of:

$$\log\left(\frac{kh}{141.2QB\mu}\right)$$

Similarly, the dimensionless time t_D is given by Eq. (1.86a and b) as:

$$t_D = \left[\frac{0.0002637k}{\phi\mu c_t r_w^2}\right] t$$

Taking the logarithm of both sides of this equation gives:

$$\log(t_D) = \log(t) + \log\left[\frac{0.0002637k}{\phi\mu c_t r_w^2}\right] \quad \textbf{(1.195)}$$

where

t = time, hours
c_t = total compressibility coefficient, psi^{-1}
ϕ = porosity

Hence, a graph of $\log(\Delta p)$ vs. $\log(t)$ will have an *identical shape* (i.e., parallel) to a graph of $\log(p_D)$ vs. $\log(t_D)$, although the curve will be shifted by $\log[kh/(141.2QB\mu)]$ vertically in pressure and $\log[0.0002637k/(\phi\mu c_t r_w^2)]$ horizontally in time. This concept is illustrated in Figure 1.46.

Not only do these two curves have the same shape, but if they are *moved relative to each other until they coincide or "match,"* the vertical and horizontal displacements required to achieve the match are related to the constants in Eqs. (1.194) and (1.195). Once these constants are determined from the vertical and horizontal displacements, it is possible to estimate reservoir properties such as permeability and porosity. This process of matching two curves through the vertical and horizontal displacements and

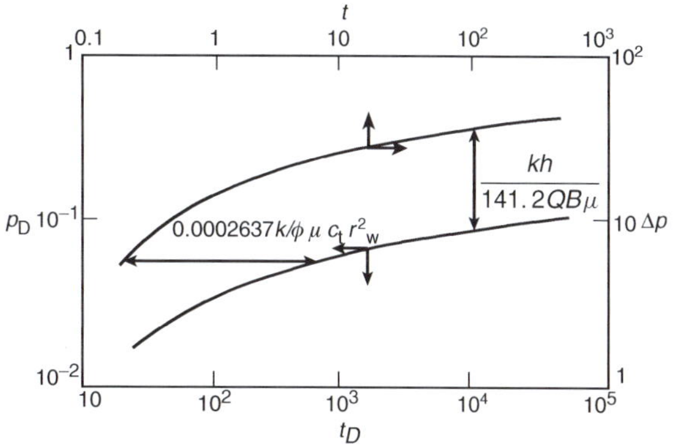

FIGURE 1.46 Concept of type curves.

determining the reservoir or well properties is called type curve matching.

Eq. (1.95) shows that the solution to the diffusivity equation can be expressed in terms of the dimensionless pressure drop as:

$$p_D = -\frac{1}{2}\text{Ei}\left(-\frac{r_D^2}{4t_D}\right)$$

Eq. (1.95) indicates that when $t_D/r_D^2 > 25, p_D$ can be approximated by:

$$p_D = \frac{1}{2}\left[\ln\left(\frac{t_D}{r_D^2}\right) + 0.080907\right]$$

Note that:

$$\frac{t_D}{r_D^2} = \left(\frac{0.0002637k}{\phi\mu c_t r^2}\right) t$$

Taking the logarithm of both sides of this equation, gives:

$$\log\left(\frac{t_D}{r_D^2}\right) = \log\left(\frac{0.0002637k}{\phi\mu c_t r^2}\right) + \log(t) \quad \textbf{(1.196)}$$

Eqs. (1.194) and (1.196) indicate that a graph of $\log(\Delta p)$ vs. $\log(t)$ will have an *identical shape* (i.e., parallel) to a graph of $\log(p_D)$ vs. $\log(t_D/r_D^2)$, although the curve will be shifted by $\log(kh141.2/QB\mu)$ vertically in pressure and $\log(0.0002637k/\phi\mu c_t r^2)$ horizontally in time. When these two curves are moved relative to each other until they coincide or "match," the vertical and horizontal movements, in mathematical terms, are given by:

$$\left(\frac{p_D}{\Delta p}\right)_{MP} = \frac{kh}{141.2QB\mu} \qquad (1.197)$$

and

$$\left(\frac{t_D/r_D^2}{t}\right)_{MP} = \frac{0.0002637k}{\phi\mu c_t r^2} \qquad (1.198)$$

The subscript "MP" denotes a match point.

A more practical solution then to the diffusivity equation is a plot of the dimensionless p_D vs. t_D/r_D^2, as shown in Figure 1.47, that can be used to determine the pressure at any time and radius from the producing well. Figure 1.47 is basically a type curve that is mostly used in interference tests when analyzing pressure response data in a shut-in observation well at a distance r from an active producer or injector well.

In general, the type curve approach employs the following procedure that will be illustrated in Figure 1.47:

Step 1. Select the proper type curve, e.g., Figure 1.47.

Step 2. Place tracing paper over Figure 1.47 and construct a log–log scale having the same dimensions as those of the type curve. This can be achieved by tracing the major and minor grid lines from the type curve to the tracing paper.

Step 3. Plot the well test data in terms of Δp vs. t on the tracing paper.

Step 4. Overlay the tracing paper on the type curve and slide the actual data plot, keeping the x and y axes of both graphs

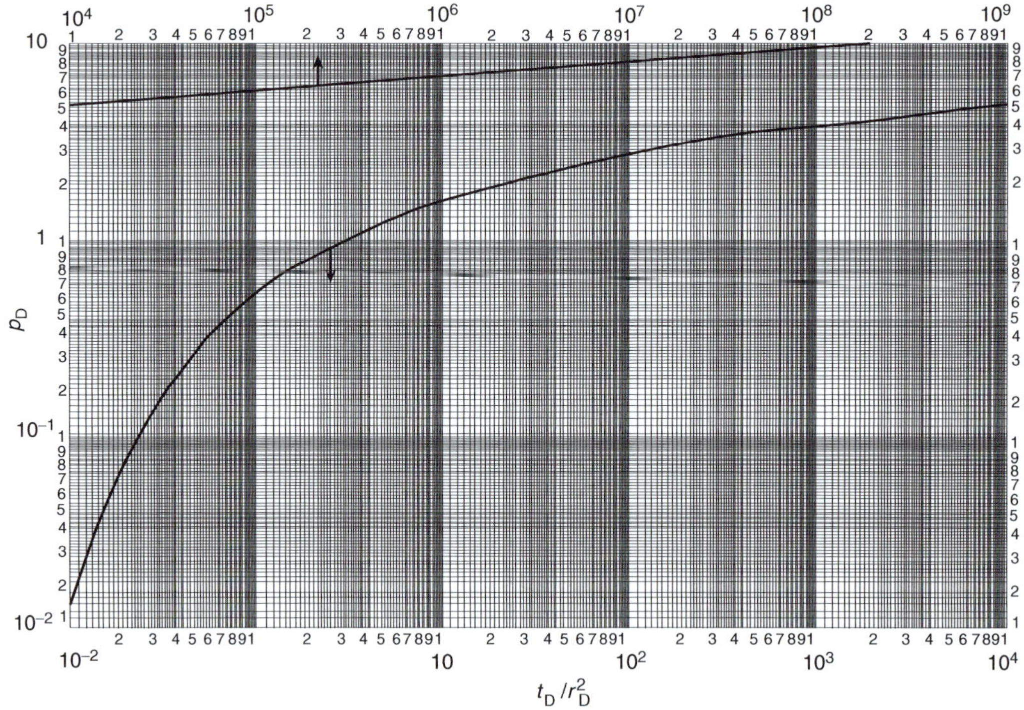

FIGURE 1.47 Dimensionless pressure for a single well in an infinite system, no wellbore storage, no skin. Exponential–integral solution. *(After Earlougher, Robert C., Jr., 1977. Advances in Well Test Analysis, Monograph, vol. 5. Society of Petroleum Engineers of AIME, Dallas, TX; Permission to publish by the SPE, copyright SPE, 1977.)*

parallel, until the actual data point curve coincides or matches the type curve.

Step 5. Select any arbitrary point match point MP, such as an intersection of major grid lines, and record $(\Delta p)_{MP}$ and $(t)_{MP}$ from the actual data plot and the corresponding values of $(P_D)_{MP}$ and $(t_D/r_D^2)_{MP}$ from the type curve.

Step 6. Using the match point, calculate the properties of the reservoir.

The following example illustrates the convenience of using the type curve approach in an interference test for 48 hours followed by a fall-off period of 100 hours.

Example 1.31

During an interference test, water was injected at a 170 bbl/day for 48 hours. The pressure response in an observation well 119 ft away from the injector is given below:[3]

t (hours)	p (psig)	$\Delta p_{ws} = p_i - p$ (psi)
0	$p_i = 0$	0
4.3	22	−22
21.6	82	−82
28.2	95	−95
45.0	119	−119
48.0		Injection ends
51.0	109	−109
69.0	55	−55
73.0	47	−47
93.0	32	−32
142.0	16	−16
148.0	15	−15

Other data includes:

$p_i = 0$ psi, $B_w = 1.00$ bbl/STB
$c_t = 9.0 \times 10^{-6}$ psi^{-1}, $h = 45$ ft
$\mu_w = 1.3$ cp, $q = -170$ bbl/day

Calculate the reservoir permeability and porosity.

Solution

Step 1. Figure 1.48 shows a plot of the well test data during the injection period, i.e., 48 hours, in terms of Δp vs. t on tracing paper with the same scale dimensions as in Figure 1.47. Using the overlay technique with the vertical and horizontal movements, find the segment of the type curve that matches the actual data.

Step 2. Select any point on the graph that will be defined as a match point MP, as shown in Figure 1.48. Record $(\Delta p)_{MP}$ and $(t)_{MP}$ from the actual data plot and the corresponding values of $(p_D)_{MP}$ and $(t_D/r_D^2)_{MP}$ from the type curve, to give:

- Type curve match values:

$$(p_D)_{MP} = 0.96, \quad \left(\frac{t_D}{r_D^2}\right)_{MP} = 0.94$$

- Actual data match values:

$$(\Delta p)_{MP} = -100 \text{ psig}, \quad (t)_{MP} = 10 \text{ hours}$$

Step 3. Using Eqs. (1.197) and (1.198), solve for the permeability and porosity:

$$k = \frac{141.2QB\mu}{h}\left(\frac{p_D}{\Delta p}\right)_{MP} = \frac{141.2(-170)(1.0)(1.0)}{45}\left(\frac{0.96}{-100}\right)_{MP} = 5.1 \text{ md}$$

and

$$\phi = \frac{0.0002637k}{\mu c_t r^2[(t_D/r_D^2)/t]_{MP}} = \frac{0.0002637(5.1)}{(1.0)(9.0\times 10^{-6})(119)^2[0.94/10]_{MP}} = 0.11$$

Eq. (1.94) shows that the dimensionless pressure is related to the dimensionless radius and time by:

$$p_D = -\frac{1}{2}\text{Ei}\left(-\frac{r_D^2}{4t_D}\right)$$

[3]This example problem and the solution procedure are given in Earlougher, R. *Advanced Well Test Analysis*, Monograph Series, SPE, Dallas (1977).

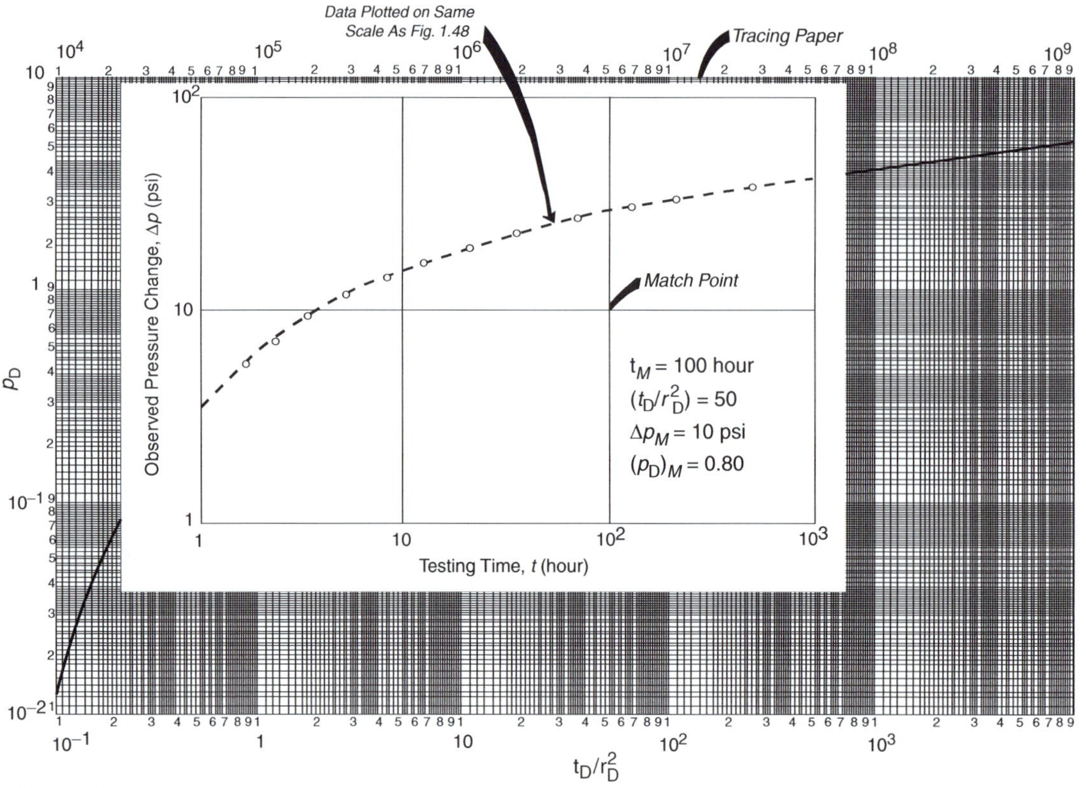

FIGURE 1.48 Illustration of type curve matching for an interference test using the type curve. *(After Earlougher, Robert C., Jr., 1977. Advances in Well Test Analysis, Monograph, vol. 5. Society of Petroleum Engineers of AIME, Dallas, TX; Permission to publish by the SPE, copyright SPE, 1977.)*

At the wellbore radius where $r = r_w$, i.e., $r_D = 1$, and $p(r, t) = p_{wf}$, the above expression is reduced to:

$$p_D = -\frac{1}{2}\mathrm{Ei}\left(\frac{-1}{4t_D}\right)$$

The log approximation as given by Eq. (1.91) can be applied to the above solution to give:

$$p_D = \frac{1}{2}[\ln(t_D) + 0.80901]$$

and, to account for the skin s, by:

$$p_D = \frac{1}{2}[\ln(t_D) + 0.80901] + s$$

or

$$p_D = \frac{1}{2}[\ln(t_D) + 0.80901 + 2s]$$

Notice that the above expressions assume zero wellbore storage, i.e., dimensionless wellbore storage $C_D = 0$. Several authors have conducted detailed studies on the effects and duration of wellbore storage on pressure drawdown and buildup data. Results of these studies were presented in the type curve format in terms of the dimensionless pressure as a function of dimensionless time, radius, and wellbore storage, i.e., $p_D = f(t_D, r_D, C_D)$. The following two methods that utilize the concept of the type curve approach are briefly introduced below:

(1) the Gringarten type curve;
(2) the pressure derivative method.

1.4.1 Gringarten Type Curve

During the *early-time period* where the flow is dominated by the wellbore storage, the wellbore pressure is described by Eq. (1.173) as:

$$p_D = \frac{t_D}{C_D}$$

or

$$\log(p_D) = \log(t_D) - \log(C_D)$$

This relationship gives the characteristic signature of wellbore storage effects on well testing data which indicates that a plot of p_D vs. t_D on a log–log scale will yield a straight line of a *unit slope*. At the end of the storage effect, which signifies the beginning of the infinite-acting period, the resulting pressure behavior produces the usual straight line on a semilog plot as described by:

$$p_D = \frac{1}{2}[\ln(t_D) + 0.80901 + 2s]$$

It is convenient when using the type curve approach in well testing to include the dimensionless wellbore storage coefficient in the above relationship. Adding and subtracting ln (C_D) inside the brackets of the above equation gives:

$$p_D = \frac{1}{2}[\ln(t_D) - \ln(C_D) + 0.80901 + \ln(C_D) + 2s]$$

or, equivalently:

$$p_D = \frac{1}{2}\left[\ln\left(\frac{t_D}{C_D}\right) + 0.80907 + \ln(C_D\, e^{2s})\right] \quad \textbf{(1.199)}$$

where

p_D = dimensionless pressure
C_D = dimensionless wellbore storage coefficient
t_D = dimensionless time
s = skin factor

Eq. (1.199) describes the pressure behavior of a well with a wellbore storage and a skin in a homogeneous reservoir during the transient (infinite-acting) flow period. Gringarten et al. (1979) expressed the above equation in the graphical type curve format shown in Figure 1.49. In this figure, the dimensionless pressure p_D is plotted

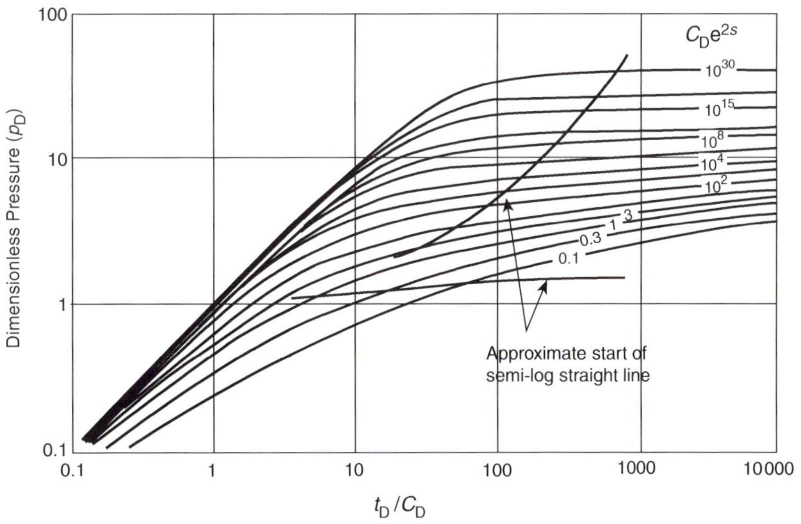

FIGURE 1.49 Type curves for a well with wellbore storage and skin in a reservoir with homogeneous behavior. *(Bourdet, D., Whittle, T.M., Douglas, A.A., Pirard, Y.M., 1983. A new set of type curves simplifies well test analysis. World Oil, May, 95–106; Copyright ©1983 World Oil.)*

on a log–log scale vs. dimensionless time group t_D/C_D. The resulting curves, characterized by the dimensionless group $C_D\,e^{2s}$, represent different well conditions ranging from damaged wells to stimulated wells.

Figure 1.49 shows that all the curves merge, in early time, into a unit-slope straight line corresponding to pure wellbore storage flow. At a later time with the end of the wellbore storage-dominated period, curves correspond to infinite-acting radial flow. The end of wellbore storage and the start of infinite-acting radial flow are marked on the type curves of Figure 1.49. There are three dimensionless groups that Gringarten et al. used when developing the type curve:

(1) dimensionless pressure p_D;
(2) dimensionless ratio t_D/C_D;
(3) dimensionless characterization group $C_D\,e^{2s}$.

The above three dimensionless parameters are defined mathematically for both the drawdown and buildup tests as follows.

For Drawdown

Dimensionless Pressure p_D.

$$p_D = \frac{kh(p_i - p_{wf})}{141.2QB\mu} = \frac{kh\,\Delta p}{141.2QB\mu} \quad (1.200)$$

where

k = permeability, md
p_{wf} = bottom-hole flowing pressure, psi
Q = flow rate, bbl/day
B = formation volume factor, bbl/STB

Taking logarithms of both sides of the above equation gives:

$$\log(p_D) = \log(p_i - p_{wf}) + \log\left(\frac{kh}{141.2QB\mu}\right)$$
$$\log(p_D) = \log(\Delta p) + \log\left(\frac{kh}{141.2QB\mu}\right) \quad (1.201)$$

Dimensionless Ratio t_D/C_D.

$$\frac{t_D}{C_D} = \left(\frac{0.0002637kt}{\phi\mu c_t r_w^2}\right)\left(\frac{\phi h c_t r_w^2}{0.8396C}\right)$$

Simplifying gives:

$$\frac{t_D}{C_D} = \left(\frac{0.0002951\,kh}{\mu C}\right)t \quad (1.202)$$

where

t = flowing time, hours
C = wellbore storage coefficient, bbl/psi

Taking logarithms gives:

$$\log\left(\frac{t_D}{C_D}\right) = \log(t) + \log\left[\frac{0.0002951\,kh}{\mu C}\right] \quad (1.203)$$

Eqs. (1.201) and (1.203) indicate that a plot of the actual drawdown data of $\log(\Delta p)$ vs. $\log(t)$ will produce a parallel curve that has an identical shape to a plot of $\log(p_D)$ vs. $\log(t_D/C_D)$. When displacing the actual plot, vertically and horizontally, to find a dimensionless curve that coincides or closely fits the actual data, these displacements are given by the constants of Eqs. (1.200) and (1.202) as:

$$\left(\frac{p_D}{\Delta p}\right)_{MP} = \frac{kh}{141.2QB\mu} \quad (1.204)$$

and

$$\left(\frac{t_D/C_D}{t}\right)_{MP} = \frac{0.0002951\,kh}{\mu C} \quad (1.205)$$

where MP denotes a match point.

Eqs. (1.204) and (1.205) can be solved for the permeability k (or the flow capacity kh) and the wellbore storage coefficient C, respectively:

$$k = \frac{141.2QB\mu}{h}\left(\frac{p_D}{\Delta p}\right)_{MP}$$

and

$$C = \frac{0.0002951\,kh}{\mu((t_D/C_D)/t)_{MP}}$$

Dimensionless Characterization Group $C_D\,e^{2s}$. The mathematical definition of the dimensionless characterization group $C_D\,e^{2s}$ as given below is valid for both the drawdown and buildup tests:

$$C_D\,e^{2s} = \left[\frac{5.615C}{2\pi\phi\mu c_t r_w^2}\right]e^{2s} \quad (1.206)$$

where

ϕ = porosity
c_t = total isothermal compressibility, psi^{-1}
r_w = wellbore radius, ft

When the match is achieved, the dimensionless group $C_D\,e^{2s}$ describing the matched curve is recorded.

For Buildup. It should be noted that *all type curve solutions* are obtained for the drawdown solution. Therefore, these type curves cannot be used for buildup tests without restriction or modification. The only restriction is that the flow period, i.e., t_p, before shut-in must be somewhat large. However, Agarwal (1980) empirically found that by plotting the buildup data $p_{ws} - p_{wf\ at\ \Delta t=0}$ vs. "equivalent time" Δt_e instead of the shut-in time Δt, on a log–log scale, the type curve analysis can be made without the requirement of a long drawdown flowing period before shut-in. Agarwal introduced the equivalent time Δt_e as defined by:

$$\Delta t_e = \frac{\Delta t}{1 + (\Delta t/t_p)} = \left[\frac{\Delta t}{t_p} + \Delta t\right] t_p \tag{1.207}$$

where

Δt = shut-in time, hours
t_p = total flowing time since the last shut-in, hours
Δt_e = Agarwal equivalent time, hours

Agarwal's equivalent time Δt_e is simply designed to account for the effects of producing time t_p on the pressure buildup test. The concept of Δt_e is that the pressure change $\Delta p = p_{ws} - p_{wf}$ at time Δt during a buildup test is the same as the pressure change $\Delta p = p_i - p_{wf}$ at Δt_e during a drawdown test. Thus, a graph of buildup test in terms of $p_{ws} - p_{wf}$ vs. Δt_e will overlay a graph of pressure change vs. flow time for a drawdown test. Therefore, when applying the type curve approach in analyzing pressure buildup data, the actual shut-in time Δt is replaced by the equivalent time Δt_e.

In addition to the characterization group $C_D e^{2s}$ as defined by Eq. (1.206), the following two dimensionless parameters are used when applying the Gringarten type curve in analyzing pressure buildup test data.

Dimensionless Pressure p_D.

$$p_D = \frac{kh(p_{ws} - p_{wf})}{141.2QB\mu} = \frac{kh\,\Delta p}{141.2QB\mu} \tag{1.208}$$

where

p_{ws} = shut-in pressure, psi
p_{wf} = flow pressure just before shut-in, i.e., at $\Delta t = 0$, psi

Taking the logarithms of both sides of the above equation gives:

$$\log(p_D) = \log(\Delta p) + \log\left(\frac{kh}{141.2QB\mu}\right) \tag{1.209}$$

Dimensionless Ratio t_D/C_D.

$$\frac{t_D}{C_D} = \left[\frac{0.0002951\,kh}{\mu C}\right]\Delta t_e \tag{1.210}$$

Taking the logarithm of each side of Eq. (1.200) gives:

$$\log\left(\frac{t_D}{C_D}\right) = \log(\Delta t_e) + \log\left(\frac{0.0002951\,kh}{\mu C}\right) \tag{1.211}$$

Similarly, a plot of actual pressure buildup data of $\log(\Delta p)$ vs. $\log(\Delta t_e)$ would have a shape identical to that of $\log(p_D)$ vs. $\log(t_D/C_D)$. When the actual plot is matched to one of the curves of Figure 1.49, then:

$$\left(\frac{p_D}{\Delta p}\right)_{MP} = \frac{kh}{141.2QB\mu}$$

which can be solved for the flow capacity kh or the permeability k. That is:

$$k = \left[\frac{141.2QB\mu}{h}\right]\left(\frac{p_D}{\Delta p}\right)_{MP} \tag{1.212}$$

and

$$\left(\frac{t_D/C_D}{\Delta t_e}\right)_{MP} = \frac{0.0002951\,kh}{\mu C} \tag{1.213}$$

Solving for C gives:

$$C = \left[\frac{0.0002951\,kh}{\mu}\right]\frac{(\Delta t_e)_{MP}}{(t_D/C_D)_{MP}} \quad \textbf{(1.214)}$$

The recommended procedure for using the Gringarten type curve is given by the following steps:

Step 1. Using the test data, perform *conventional* test analysis and determine:
- wellbore storage coefficient C and C_D;
- permeability k;
- false pressure p^*;
- average pressure $\bar{p}$;
- skin factor s;
- shape factor C_A;
- drainage area A.

Step 2. Plot $p_i - p_{wf}$ vs. flowing time t for a drawdown test or $(p_{ws} - p_{wp})$ vs. equivalent time Δt_e for a buildup test on log–log paper (tracing paper) with the same size log cycles as the Gringarten type curve.

Step 3. Check the early-time points on the actual data plot for the unit-slope (45° angle) straight line to verify the presence of the wellbore storage effect. If a unit-slope straight line presents, calculate the wellbore storage coefficient C and the dimensionless C_D from any point on the unit-slope straight line with coordinates of $(\Delta p, t)$ or $(\Delta p, \Delta t_e)$, to give:

$$\text{For drawdown:} \quad C = \frac{QBt}{24(p_i - p_{wf})} = \frac{QB}{24}\left(\frac{t}{\Delta p}\right) \quad \textbf{(1.215)}$$

$$\text{For buildup:} \quad C = \frac{QB\,\Delta t_e}{24(p_{ws} - p_{wf})} = \frac{QB}{24}\left(\frac{\Delta t_e}{\Delta p}\right) \quad \textbf{(1.216)}$$

Estimate the dimensionless wellbore storage coefficient from:

$$C_D = \left[\frac{0.8936}{\phi h c_t r_w^2}\right]C \quad \textbf{(1.217)}$$

Step 4. Overlay the graph of the test data on the type curves and find the type curve that nearly fits most of the actual plotted data. Record the type curve dimensionless group $(C_D\,e^{2s})_{MP}$.

Step 5. Select a match point MP and record the corresponding values of $(p_D, \Delta p)_{MP}$ from the y-axis and $(t_D/C_D, t)_{MP}$ or $(t_D/C_D, \Delta t_e)_{MP}$ from the x-axis.

Step 6. From the match, calculate:

$$k = \left[\frac{141.2\,QB\mu}{h}\right]\left(\frac{p_D}{\Delta p}\right)_{MP}$$

and

$$\text{For drawdown:} \quad C = \left[\frac{0.0002951\,kh}{\mu}\right]\left(\frac{t}{(t_D/C_D)}\right)_{MP}$$

or

$$\text{For buildup:} \quad C = \left[\frac{0.0002951\,kh}{\mu}\right]\left(\frac{\Delta t_e}{(t_D/C_D)}\right)_{MP}$$

and

$$C_D = \left[\frac{0.8936}{\phi h c_t r_w^2}\right]C$$

$$s = \frac{1}{2}\ln\left[\frac{(C_D\,e^{2s})_{MP}}{C_D}\right] \quad \textbf{(1.218)}$$

Sabet (1991) used the buildup data presented by Bourdet et al. (1983) to illustrate the use of Gringarten type curves. The data is used in Example 1.32.

Example 1.32

Table 1.6 summarizes the pressure buildup data for an oil well that has been producing at a constant flow rate of 174 STB/day before shut-in. Additional pertinent data is given below:

$\phi = 25\%$, $c_t = 4.2 \times 10^{-6}\ \text{psi}^{-1}$
$Q = 174$ STB/day, $t_p = 15$ hour
$B = 1.06$ bbl/STB, $r_w = 0.29$ ft
$\mu = 2.5$ cp, $h = 107$ ft

Perform the conventional pressure buildup analysis by using the Horner plot approach and

TABLE 1.6 Pressure Buildup Test with Afterflow

Δt (hour)	p_{ws} (psi)	Δp (psi)	$\frac{t_p+\Delta t}{\Delta t}$	Δt_e
0.00000	3086.33	0.00	–	0.00000
0.00417	3090.57	4.24	3600.71	0.00417
0.00833	3093.81	7.48	1801.07	0.00833
0.01250	3096.55	10.22	1201.00	0.01249
0.01667	3100.03	13.70	900.82	0.01666
0.02083	3103.27	16.94	721.12	0.02080
0.02500	3106.77	20.44	601.00	0.02496
0.02917	3110.01	23.68	515.23	0.02911
0.03333	3113.25	26.92	451.05	0.03326
0.03750	3116.49	30.16	401.00	0.03741
0.04583	3119.48	33.15	328.30	0.04569
0.05000	3122.48	36.15	301.00	0.04983
0.05830	3128.96	42.63	258.29	0.05807
0.06667	3135.92	49.59	225.99	0.06637
0.07500	3141.17	54.84	201.00	0.07463
0.08333	3147.64	61.31	181.01	0.08287
0.09583	3161.95	75.62	157.53	0.09522
0.10833	3170.68	84.35	139.47	0.10755
0.12083	3178.39	92.06	125.14	0.11986
0.13333	3187.12	100.79	113.50	0.13216
0.14583	3194.24	107.91	103.86	0.14443
0.16250	3205.96	119.63	93.31	0.16076
0.17917	3216.68	130.35	84.72	0.17706
0.19583	3227.89	141.56	77.60	0.19331
0.21250	3238.37	152.04	71.59	0.20953
0.22917	3249.07	162.74	66.45	0.22572
0.25000	3261.79	175.46	61.00	0.24590
0.29167	3287.21	200.88	52.43	0.28611
0.33333	3310.15	223.82	46.00	0.32608
0.37500	3334.34	248.01	41.00	0.36585
0.41667	3356.27	269.94	37.00	0.40541
0.45833	3374.98	288.65	33.73	0.44474
0.50000	3394.44	308.11	31.00	0.48387
0.54167	3413.90	327.57	28.69	0.52279
0.58333	3433.83	347.50	26.71	0.56149
0.62500	3448.05	361.72	25.00	0.60000
0.66667	3466.26	379.93	23.50	0.63830
0.70833	3481.97	395.64	22.18	0.67639
0.75000	3493.69	407.36	21.00	0.71429
0.81250	3518.63	432.30	19.46	0.77075
0.87500	3537.34	451.01	18.14	0.82677
0.93750	3553.55	467.22	17.00	0.88235
1.00000	3571.75	485.42	16.00	0.93750
1.06250	3586.23	499.90	15.12	0.99222
1.12500	3602.95	516.62	14.33	1.04651
1.18750	3617.41	531.08	13.63	1.10039
1.25000	3631.15	544.82	13.00	1.15385
1.31250	3640.86	554.53	12.43	1.20690

(*Continued*)

TABLE 1.6 (Continued)

Δt (hour)	p_{ws} (psi)	Δp (psi)	$\frac{t_p + \Delta t}{\Delta t}$	Δt_e
1.37500	3652.85	566.52	11.91	1.25954
1.43750	3664.32	577.99	11.43	1.31179
1.50000	3673.81	587.48	11.00	1.36364
1.62500	3692.27	605.94	10.23	1.46617
1.75000	3705.52	619.19	9.57	1.56716
1.87500	3719.26	632.93	9.00	1.66667
2.00000	3732.23	645.90	8.50	1.76471
2.25000	3749.71	663.38	7.67	1.95652
2.37500	3757.19	670.86	7.32	2.05036
2.50000	3763.44	677.11	7.00	2.14286
2.75000	3774.65	688.32	6.45	2.32394
3.00000	3785.11	698.78	6.00	2.50000
3.25000	3794.06	707.73	5.62	2.67123
3.50000	3799.80	713.47	5.29	2.83784
3.75000	3809.50	723.17	5.00	3.00000
4.00000	3815.97	729.64	4.75	3.15789
4.25000	3820.20	733.87	4.53	3.31169
4.50000	3821.95	735.62	4.33	3.46154
4.75000	3823.70	737.37	4.16	3.60759
5.00000	3826.45	740.12	4.00	3.75000
5.25000	3829.69	743.36	3.86	3.88889
5.50000	3832.64	746.31	3.73	4.02439
5.75000	3834.70	748.37	3.61	4.15663
6.00000	3837.19	750.86	3.50	4.28571
6.25000	3838.94	752.61	3.40	4.41176
6.75000	3838.02	751.69	3.22	4.65517
7.25000	3840.78	754.45	3.07	4.88764
7.75000	3843.01	756.68	2.94	5.10989
8.25000	3844.52	758.19	2.82	5.32258
8.75000	3846.27	759.94	2.71	5.52632
9.25000	3847.51	761.18	2.62	5.72165
9.75000	3848.52	762.19	2.54	5.90909
10.25000	3850.01	763.68	2.46	6.08911
10.75000	3850.75	764.42	2.40	6.26214
11.25000	3851.76	765.43	2.33	6.42857
11.75000	3852.50	766.17	2.28	6.58879
12.25000	3853.51	767.18	2.22	6.74312
12.75000	3854.25	767.92	2.18	6.89189
13.25000	3855.07	768.74	2.13	7.03540
13.75000	3855.50	769.17	2.09	7.17391
14.50000	3856.50	770.17	2.03	7.37288
15.25000	3857.25	770.92	1.98	7.56198
16.00000	3857.99	771.66	1.94	7.74194
16.75000	3858.74	772.41	1.90	7.91339
17.50000	3859.48	773.15	1.86	8.07692
18.25000	3859.99	773.66	1.82	8.23308
19.00000	3860.73	774.40	1.79	8.38235
19.75000	3860.99	774.66	1.76	8.52518

(*Continued*)

TABLE 1.6 (Continued)

Δt (hour)	p_{ws} (psi)	Δp (psi)	$\frac{t_p+\Delta t}{\Delta t}$	Δt_e
20.50000	3861.49	775.16	1.73	8.66197
21.25000	3862.24	775.91	1.71	8.79310
22.25000	3862.74	776.41	1.67	8.95973
23.25000	3863.22	776.89	1.65	9.11765
24.25000	3863.48	777.15	1.62	9.26752
25.25000	3863.99	777.66	1.59	9.40994
26.25000	3864.49	778.16	1.57	9.54545
27.25000	3864.73	778.40	1.55	9.67456
28.50000	3865.23	778.90	1.53	9.82759
30.00000	3865.74	779.41	1.50	10.0000

Adapted from Bourdet, D., Whittle, T.M., Douglas, A.A., Pirard, Y.M., 1983. A new set of type curves simplifies well test analysis. World Oil, May, 95–106.
After Sabet, M., 1991. Well Test Analysis. Gulf Publishing, Dallas, TX.

compare the results with those obtained by using the Gringarten type curve approach.

Solution

Step 1. Plot Δp vs. Δt_e on a log–log scale, as shown in Figure 1.50. The plot shows that the early data form a straight line with a 45° angle, which indicates the wellbore storage effect. Determine the coordinates of a point on the straight line, e.g., $\Delta p = 50$ and $\Delta t_e = 0.06$, and calculate C and C_D:

$$C = \frac{QB\,\Delta t_e}{24\,\Delta p} = \frac{(174)(1.06)(0.06)}{(24)(50)} = 0.0092\,\text{bbl/psi}$$

$$C_D = \frac{0.8936C}{\phi h c_t r_w^2} = \frac{0.8936(0.0092)}{(0.25)(107)(4.2\times10^{-6})(0.29)^2} = 872$$

Step 2. Make a Horner plot of p_{ws} vs. $(t_p + \Delta t)/\Delta t$ on semilog paper, as shown in Figure 1.51, and perform the conventional well test analysis, to give:

$$m = 65.62\ \text{psi/cycle}$$

$$k = \frac{162.6QB\mu}{mh} \frac{(162.6)(174)(2.5)}{(65.62)(107)} = 10.1\ \text{md}$$

$$p_{1\,\text{hour}} = 3797\ \text{psi}$$

$$s = 1.151\left[\frac{p_{1\,\text{hour}} - p_{wf}}{(m)} - \log\left(\frac{k}{\phi\mu c_t r_w^2}\right) + 3.23\right]$$

$$= 1.151\left[\frac{3797 - 3086.33}{65.62} - \log\left(\frac{10.1}{(0.25)(2.5)(4.2\times10^{-6})(0.29)^2}\right) + 3.23\right]$$

$$= 7.37$$

$$\Delta p_{skin} = (0.87)(65.62)(7.37) = 421\ \text{psi}$$

$$p^* = 3878\ \text{psi}$$

Step 3. Plot Δp vs. Δt_e, on log–log graph paper with the same size log cycles as the Gringarten type curve. Overlay the actual test data plot on the type curve and find the type curve that matches the test data. As shown in Figure 1.52, the data matched the curve with the dimensionless group of $C_D\,e^{2s} = 10^{10}$ and a match point of:

$$(p_D)_{MP} = 1.79$$

$$(\Delta p)_{MP} = 100$$

$$\left(\frac{t_D}{C_D}\right) = 14.8$$

$$(\Delta t_e) = 1.0$$

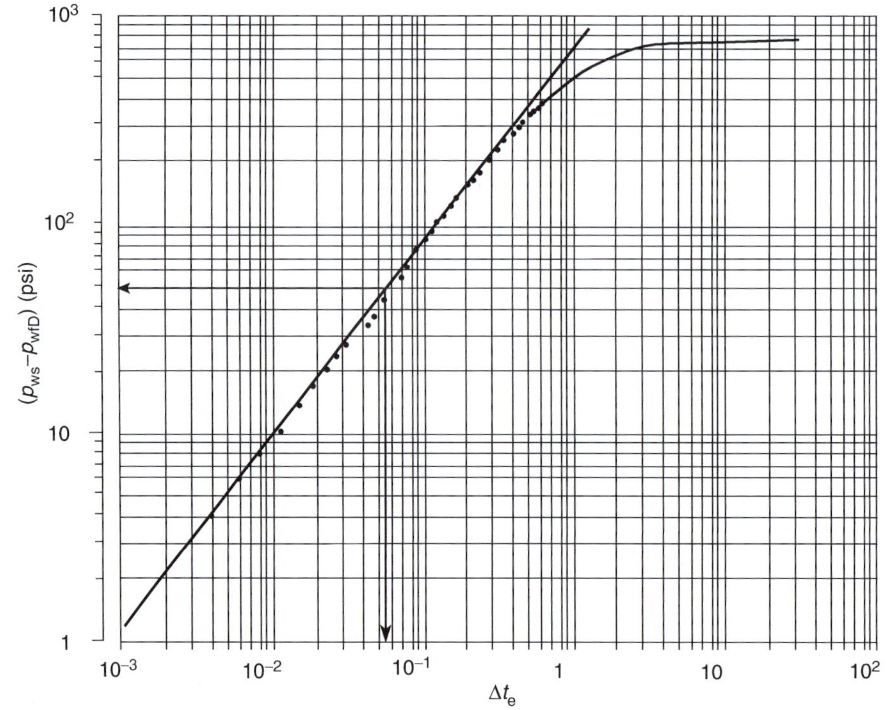

FIGURE 1.50 Log–log plot. Data from Table 1.6. *(After Sabet, M.A., 1991. Well Test Analysis, Gulf Publishing Company.)*

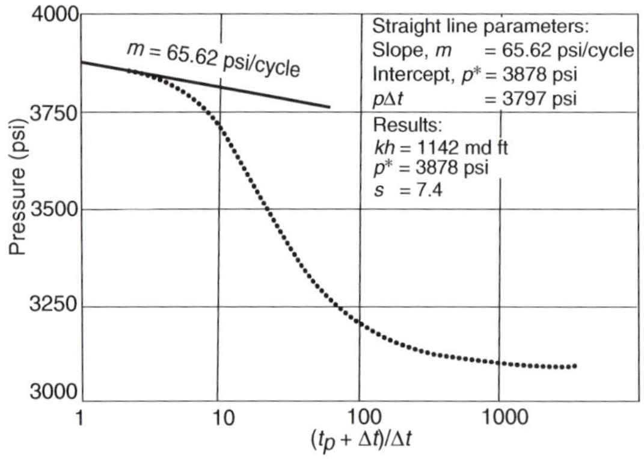

FIGURE 1.51 The Horner plot: data from Table 1.6. *(Bourdet, D., Whittle, T.M., Douglas, A.A., Pirard, Y.M., 1983. A new set of type curves simplifies well test analysis. World Oil, May, 95–106; Copyright ©1983 World Oil.)*

Step 4. From the match, calculate the following properties:

$$k = \left[\frac{141.2 QB\mu}{h}\right]\left(\frac{p_D}{\Delta p}\right)_{MP}$$

$$= \frac{141.2(174)(1.06)(2.5)}{(107)}\left(\frac{1.79}{100}\right) = 10.9 \text{ md}$$

$$C = \left[\frac{0.0002951\,kh}{\mu}\right]\left[\frac{\Delta t_e}{(t_D/C_D)}\right]_{MP}$$

$$= \left[\frac{0.002951(10.9)(107)}{2.5}\right]\left[\frac{1.0}{14.8}\right] = 0.0093$$

$$C_D = \left[\frac{0.8936}{\phi h c_t r_w^2}\right] C$$

$$= \left[\frac{0.8936}{(0.25)(107)(4.2\times 10^{-6})(0.29)^2}\right](0.0093)$$

$$= 879$$

$$s = \frac{1}{2}\ln\left[\frac{(C_D e^{2s})_{MP}}{C_D}\right] = \frac{1}{2}\ln\left[\frac{10^{10}}{879}\right] = 8.12$$

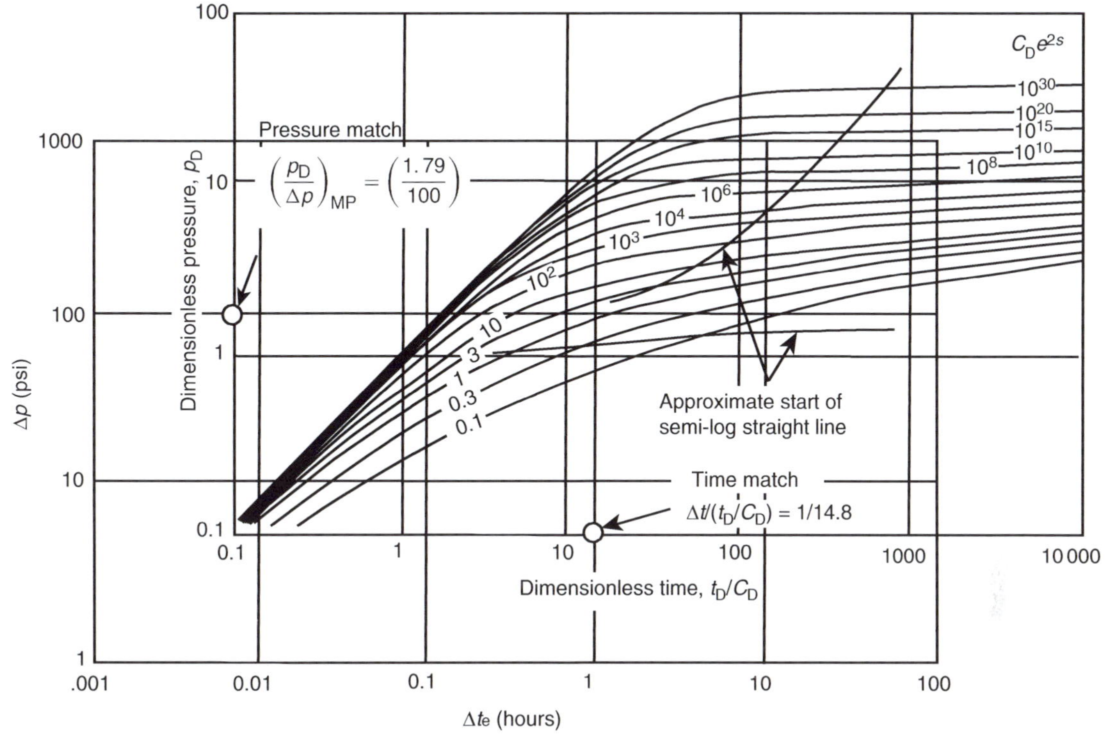

FIGURE 1.52 Buildup data plotted on log–log graph paper and matched to type curve by Gringarten et al. *(Bourdet, D., Whittle, T.M., Douglas, A.A., Pirard, Y.M., 1983. A new set of type curves simplifies well test analysis. World Oil, May, 95–106; Copyright ©1983 World Oil.)*

Results of the example show a good agreement between the conventional well testing analysis and that of the Gringarten type curve approach.

Similarly, the Gringarten type curve can also be used for gas systems by redefining the dimensionless pressure drop and time as:

- For the gas pseudopressure approach:

$$p_D = \frac{kh\,\Delta[m(p)]}{1422\,Q_g T}$$

- For the pressure-squared approach:

$$p_D = \frac{kh\,\Delta[p^2]}{1422\,Q_g \mu_i Z_i T}$$

with the dimensionless time as:

$$t_D = \left[\frac{0.0002637k}{\phi \mu c_t r_w^2}\right] t$$

where

Q_g = gas flow rate, Mscf/day
T = temperature, °R

$$\Delta[m(p)] = m(p_{ws}) - m(p_{wf\ at\ \Delta t=0}) \quad \text{for the buildup test}$$
$$= m(p_i) - m(p_{wf}) \quad \text{for the drawdown test}$$
$$\Delta[p^2] = (p_{ws})^2 - (p_{wf\ at\ \Delta t=0})^2 \quad \text{for the buildup test}$$
$$= (p_i)^2 - (p_{wf})^2 \quad \text{for the drawdown test}$$

1.5 PRESSURE DERIVATIVE METHOD

The type curve approach for the analysis of well testing data was developed to allow for the identification of flow regimes during the wellbore storage-dominated period and the infinite-acting radial flow. Example 1.31 illustrates that it can be used to estimate the reservoir properties and wellbore condition. However, because of the similarity of curves shapes, it is difficult to obtain a unique solution. As shown in Figure 1.49, all type curves have very similar shapes for high values of $C_D\,e^{2s}$ which leads to the problem of finding a unique match by a simple comparison of shapes and determining the correct values of k, s, and C.

Tiab and Kumar (1980) and Bourdet et al. (1983) addressed the problem of identifying the correct flow regime and selecting the proper interpretation model. Bourdet et al. proposed that flow regimes can have clear characteristic shapes if the "pressure derivative" rather than pressure is plotted vs. time on the log–log coordinates. Since the introduction of the pressure derivative type curve, well testing analysis has been greatly enhanced by its use. The use of this pressure derivative type curve offers the following advantages:

- Heterogeneities hardly visible on the conventional plot of well testing data are amplified on the derivative plot.
- Flow regimes have clear characteristic shapes on the derivative plot.
- The derivative plot is able to display in a single graph many separate characteristics that would otherwise require different plots.
- The derivative approach improves the definition of the analysis plots and therefore the quality of the interpretation.

Bourdet et al. (1983) defined the pressure derivative as the derivative of p_D with respect to t_D/C_D as:

$$p_D^{\backslash} = \frac{d(p_D)}{d(t_D/C_D)} \quad \textbf{(1.219)}$$

It has been shown that during the wellbore storage-dominated period the pressure behavior is described by:

$$p_D = \frac{t_D}{C_D}$$

Taking the derivative of p_D with respect to t_D/C_D gives:

$$\frac{d(p_D)}{d(t_D/C_D)} = p_D^{\backslash} = 1.0$$

Since $p_D^{\backslash} = 1$, this implies that multiplying $p_D^{\backslash}$ by t_D/C_D gives t_D/C_D, or:

$$p_D^{\backslash}\left(\frac{t_D}{C_D}\right) = \frac{t_D}{C_D} \quad \textbf{(1.220)}$$

Eq. (1.220) indicates that a plot of $p_D^{\backslash}(t_D/C_D)$ vs. t_D/C_D in log–log coordinates will produce a unit-slope straight line during the wellbore storage-dominated flow period.

Similarly, during the radial infinite-acting flow period, the pressure behavior is given by Eq. (1.219) as:

$$p_D = \frac{1}{2}\left[\ln\left(\frac{t_D}{C_D}\right) + 0.80907 + \ln(C_D\,e^{2s})\right]$$

Differentiating with respect to t_D/C_D, gives:

$$\frac{d(p_D)}{d(t_D/C_D)} = p_D^{\backslash} = \frac{1}{2}\left[\frac{1}{(t_D/C_D)}\right]$$

Simplifying gives:

$$p_D^{\backslash}\left(\frac{t_D}{C_D}\right) = \frac{1}{2} \quad \textbf{(1.221)}$$

This indicates that a plot of $p_D^{\backslash}(t_D/C_D)$ vs. t_D/C_D on a log–log scale will produce a *horizontal line* at $p_D^{\backslash}(t_D/C_D) = 0.5$ during the transient flow (radial infinite-acting) period. As shown by Eqs. (1.220) and (1.221) the derivative plot of $p_D^{\backslash}(t_D/C_D)$ vs. t_D/C_D for the entire well test data will produce *two straight lines* that are characterized by:

- a unit-slope straight line during the wellbore storage-dominated flow;
- a horizontal line $p_D^{\backslash}(t_D/C_D) = 0.5$ during the transient flow period.

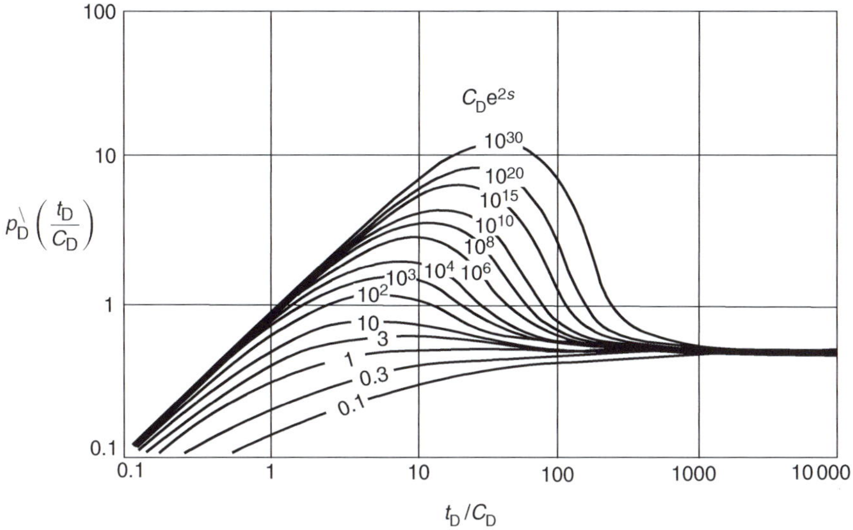

FIGURE 1.53 Pressure derivative type curve in terms of $P'_D(t_D/C_D)$. *(Bourdet, D., Whittle, T.M., Douglas, A.A., Pirard, Y. M., 1983. A new set of type curves simplifies well test analysis. World Oil, May, 95–106; Copyright ©1983 World Oil.)*

The fundamental basis for the pressure derivative approach is essentially based on identifying these two straight lines that can be used as reference lines when selecting the proper well test data interpreting model.

Bourdet et al. replotted the Gringarten type curve in terms of $p'_D(t_D/C_D)$ vs. t_D/C_D on a log–log scale as shown in Figure 1.53. It shows that at the early time during the wellbore storage-dominated flow, the curves follow a unit-slope log–log straight line. When infinite-acting radial flow is reached, the curves become horizontal at a value of $p'_D(t_D/C_D) = 0.5$ as indicated by Eq. (1.221). In addition, notice that the transition from pure wellbore storage to infinite-acting behavior gives a "hump" with a height that characterizes the value of the skin factor/s.

Figure 1.53 illustrates that the effect of skin is only manifested in the curvature between the straight line due to wellbore storage flow and the *horizontal straight line* due to the infinite-acting radial flow. Bourdet et al. indicated that data in this curvature portion of the curve is not always well defined. For this reason, the authors found it useful to combine their derivative type curves with that of the Gringarten type curve by superimposing the two types of curves, i.e., Figures 1.49 and 1.53, on the same scale. The result of superimposing the two sets of type curves on the same graph is shown in Figure 1.54. The use of the new type curve allows the *simultaneous* matching of pressure-change data and derivative data, since both are plotted on the same scale. The derivative pressure data provides, without ambiguity, the pressure match and the time match, while the $C_D\, e^{2s}$ value is obtained by comparing the label of the match curves for the derivative pressure data and pressure drop data.

The procedure for analyzing well test data using the derivative type curve is summarized in the following steps:

Step 1. Using the actual well test data, calculate the pressure difference Δp and the pressure derivative plotting functions as defined below for drawdown and buildup tests.

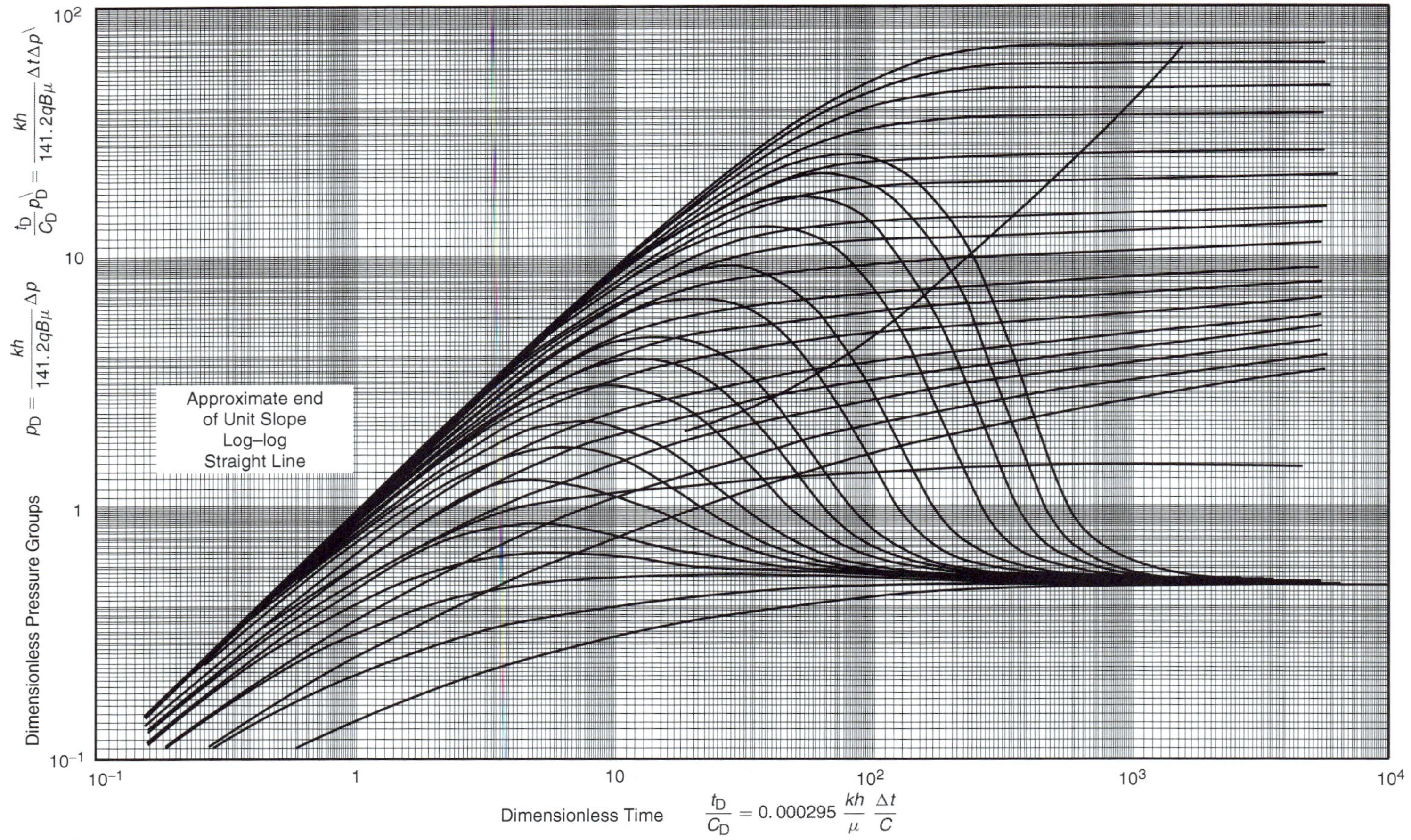

FIGURE 1.54 Pressure derivative type curves. *(Bourdet, D., Whittle, T.M., Douglas, A.A., Pirard, Y.M., 1983. A new set of type curves simplifies well test analysis. World Oil, May, 95–106; Copyright ©1983 World Oil.)*

For the drawdown tests, for every recorded drawdown pressure point, i.e., flowing time t and a corresponding bottom-hole flowing pressure p_{wf}, calculate:

The pressure difference: $\Delta p = p_i - p_{wf}$

The derivative function: $t\,\Delta/p^{\prime} = -t\left(\frac{d(\Delta p)}{d(t)}\right)$ **(1.222)**

For the buildup tests, for every recorded buildup pressure point, i.e., shut-in time Δt and corresponding shut-in pressure p_{ws}, calculate:

The pressure difference: $\Delta p = p_{ws} - p_{wf\ at\ \Delta t=0}$

The derivative function:

$$\Delta t_e\,\Delta p^{\prime} = \Delta t\left(\frac{t_p + \Delta t}{\Delta t}\right)\left[\frac{d(\Delta p)}{d(\Delta t)}\right] \quad \textbf{(1.223)}$$

The derivatives included in Eqs. (1.222) and (1.223), i.e., $[dp_{wf}/dt]$ and $[d(\Delta p_{ws})/d(\Delta t)]$, can be determined numerically at any data point i by using the central difference formula for *evenly spaced time* or the three-point weighted average approximation as shown graphically in Figure 1.55 and mathematically by the following expressions:

- Central differences:

$$\left(\frac{dp}{dx}\right)_i = \frac{p_{i+1} - p_{i-1}}{x_{i+1} - x_{i-1}} \quad \textbf{(1.224)}$$

- Three-point weighted average:

$$\left(\frac{dp}{dx}\right)_i = \frac{(\Delta p_1/\Delta x_1)\Delta x_2 + (\Delta p_2/\Delta x_2)\Delta x_1}{\Delta x_1 + \Delta x_2} \quad \textbf{(1.225)}$$

It should be pointed out that selection of the method of numerical differentiation is a problem that must be considered and examined when applying the pressure derivative method. There are many differentiation methods that use only two points, e.g., backward difference, forward difference, and central difference formulas, and very complex algorithms that utilize several pressure points. It is important to try several different methods in order to find one which best smoothes the data.

Step 2. On tracing paper with the same size log cycles as the Bourdet–Gringarten

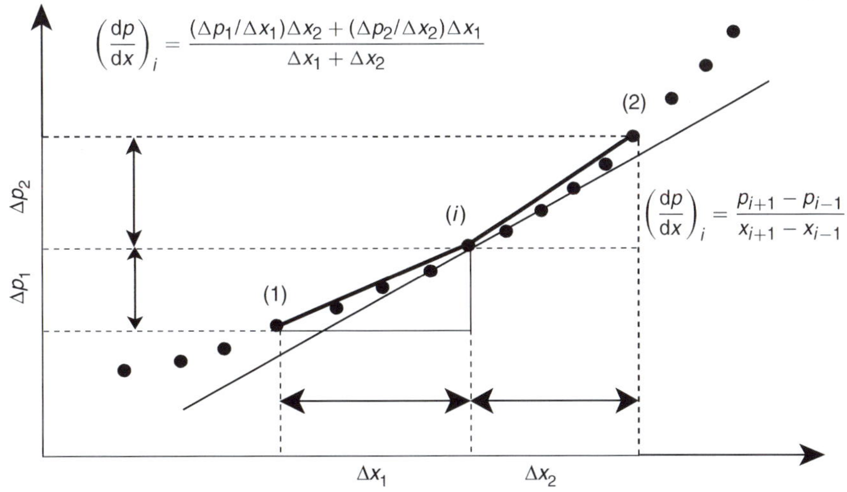

FIGURE 1.55 Differentiation algorithm using three points.

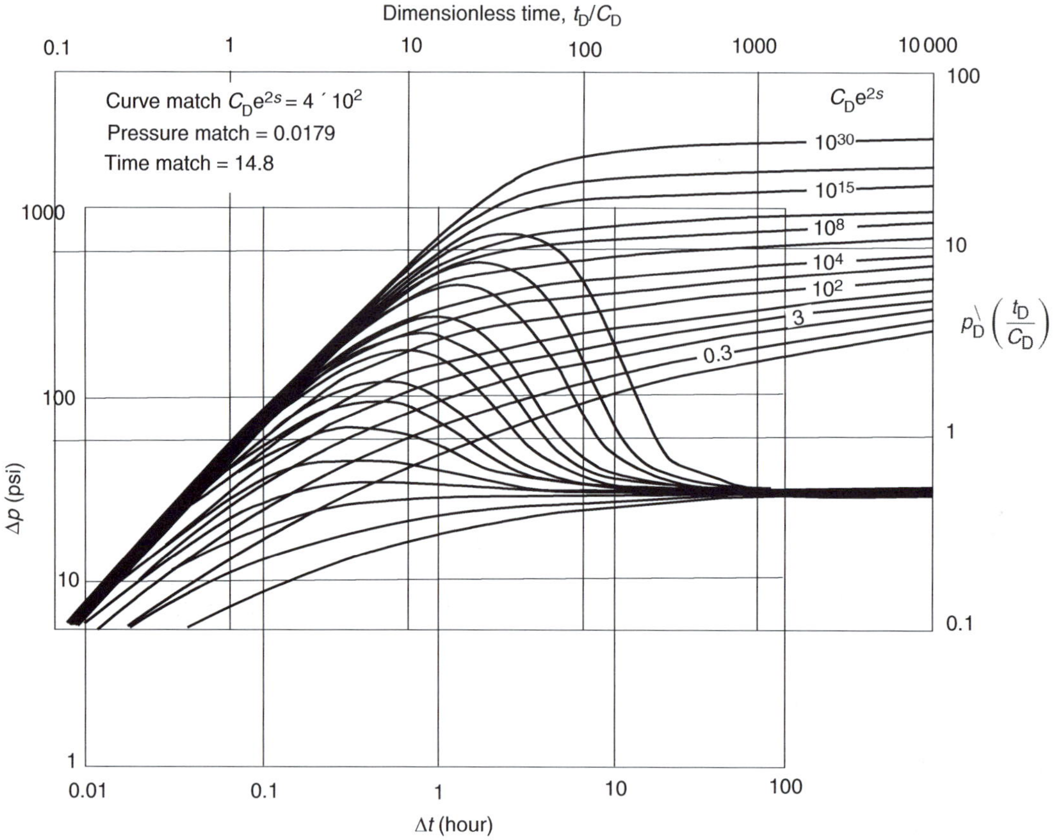

FIGURE 1.56 Type curve matching. Data from Table 1.6. *(Bourdet, D., Whittle, T.M., Douglas, A.A., Pirard, Y.M., 1983. A new set of type curves simplifies well test analysis. World Oil, May, 95–106; Copyright ©1983 World Oil.)*

type curve graph, i.e., Figure 1.54, plot:

- (Δp) and ($t\,\Delta p^{\backslash}$) as a function of the flowing time t when analyzing drawdown test data. Note that there are two sets of data on the same log–log graph as illustrated in Figure 1.56; the first is the analytical solution and the second is the actual drawdown test data.
- The pressure difference Δp vs. the equivalent time Δt_e and the derivative function ($\Delta t_e\,\Delta p^{\backslash}$) vs. the *actual shut-in time* Δt. Again, there are two sets of data on the same graph as shown in Figure 1.56.

Step 3. Check the actual early-time pressure points, i.e., pressure difference vs. time on a log–log scale, for the unit-slope line. If it exists, draw a line through the points and calculate the wellbore storage coefficient C by selecting a point on the unit-slope line as identified with coordinates of (t, Δp) or (Δt_e, Δp) and applying Eq. (1.215) or Eq. (1.216), as follows:

$$\text{For drawdown:}\quad C = \frac{QB}{24}\left(\frac{t}{\Delta p}\right)$$

$$\text{For buildup:}\quad C = \frac{QB}{24}\left(\frac{\Delta t_e}{\Delta p}\right)$$

Step 4. Calculate the dimensionless wellbore storage coefficient C_D by applying Eq. (1.217) and using the value of C as calculated in Step 3. That is:

$$C_D = \left[\frac{0.8936}{\phi h c_t r_w^2}\right] C$$

Step 5. Check the late-time data points on the *actual pressure derivative* plot to see if they form a horizontal line which indicates the occurrence of transient (unsteady-state) flow. If it exists, draw a horizontal line through these derivative plot points.

Step 6. Place the actual two sets of plots, i.e., the pressure difference plot and derivative function plot, on the Gringarten–Bourdet type curve in Figure 1.54, and force a simultaneous match of the two plots to Gringarten–Bourdet type curves. The unit-slope line should overlay the unit slope on the type curve and the late-time horizontal line should overlay the horizontal line on the type curve which corresponds to a value of 0.5. Note that it is convenient to match both pressure and pressure derivative curves, even though it is redundant. With the double match, a high degree of confidence in the results is obtained.

Step 7. From the match of the best fit, select a match point MP and record the corresponding values of the following:

- From the Gringarten type curve, determine $(p_D, \Delta p)_{MP}$ and the corresponding $(t_D/C_D, t)_{MP}$ or $(t_D/C_D, \Delta t_e)_{MP}$.
- Record the value of the type curve dimensionless group $(C_D e^{2s})_{MP}$ from the Bourdet type curves.

Step 8. Calculate the permeability by applying Eq. (1.212):

$$k = \left[\frac{141.2 QB\mu}{h}\right]\left[\frac{p_D}{\Delta p}\right]_{MP}$$

Step 9. Recalculate the wellbore storage coefficient C and C_D by applying Eqs. (1.214) and (1.217), or:

For drawdown: $$C = \left[\frac{0.0002951\,kh}{\mu}\right]\frac{(t)_{MP}}{(t_D/C_D)_{MP}}$$

For buildup: $$C = \left[\frac{0.0002951\,kh}{\mu}\right]\frac{(\Delta t_e)_{MP}}{(t_D/C_D)_{MP}}$$

with

$$C_D = \left[\frac{0.8936}{\phi h c_t r_w^2}\right] C$$

Compare the calculated values of C and C_D with those calculated in steps 3 and 4.

Step 10. Calculate the skin factor s by applying Eq. (1.218) and using the value of C_D in step 9 and the value of $(C_D e^{2s})_{MP}$ in step 7, to give:

$$s = \frac{1}{2}\ln\left[\frac{(C_D\, e^{2s})_{MP}}{C_D}\right]$$

Example 1.33

Using the data from Example 1.31, analyze the given well test data using the pressure derivative approach.

Solution

Step 1. Calculate the derivative function for every recorded data point by applying Eq. (1.223) or the approximation method of Eq. (1.224) as tabulated Table 1.7 and shown graphically in Figure 1.57.

Step 2. Draw a straight line with a 45° angle that fits the early-time test points, as shown in Figure 1.57, and select the coordinates of a point on the straight line, to give (0.1, 70). Calculate C and C_D:

$$C = \frac{QB\,\Delta t}{24\,\Delta p} = \frac{1740(1.06)(0.1)}{(24)(70)} = 0.00976$$

$$C_D = \left[\frac{0.8936}{\phi h c_t r_w^2}\right] = \frac{0.8936(0.00976)}{(0.25)(107)(4.2\times 10^{-6})(0.29)^2}$$

$$= 923$$

TABLE 1.7 Pressure Derivative Method Using Data of Table 6.6

Δt (hour)	Δp (psi)	Slope (psi/hour)	$\Delta p^{\backslash}$ (psi/hour)	$\Delta t \Delta t^{\backslash}\ (t_p + \Delta t) t_p$
0.00000	0.00	1017.52	–	–
0.00417	4.24	777.72	897.62	3.74
0.00833	7.48	657.55	717.64	5.98
0.01250	10.22	834.53	746.04	9.33
0.01667	13.70	778.85	806.69	13.46
0.02083	16.94	839.33	809.09	16.88
0.02500	20.44	776.98	808.15	20.24
0.02917	23.68	778.85	777.91	22.74
0.03333	26.92	776.98	777.91	25.99
0.03750	30.16	358.94	567.96	21.35
0.04583	33.15	719.42	539.18	24.79
0.05000	36.15	780.72	750.07	37.63
0.05830	42.63	831.54	806.13	47.18
0.06667	49.59	630.25	730.90	48.95
0.07500	54.84	776.71	703.48	53.02
0.08333	61.31	1144.80	960.76	80.50
0.09583	75.62	698.40	921.60	88.88
0.10833	84.35	616.80	657.60	71.75
0.12083	92.06	698.40	657.60	80.10
0.13333	100.79	569.60	634.00	85.28
0.14583	107.91	703.06	636.33	93.70
0.16250	119.63	643.07	673.07	110.56
0.17917	130.35	672.87	657.97	119.30
0.19583	141.56	628.67	650.77	129.10
0.21250	152.04	641.87	635.27	136.91
0.22917	162.74	610.66	626.26	145.71
0.25000	175.46	610.03	610.34	155.13
0.29167	200.88	550.65	580.34	172.56
0.33333	223.82	580.51	565.58	192.71
0.37500	248.01	526.28	553.40	212.71
0.41667	269.94	449.11	487.69	208.85
0.45833	288.65	467.00	458.08	216.36
0.50000	308.11	467.00	467.00	241.28
0.54167	327.57	478.40	472.70	265.29
0.58333	347.50	341.25	409.82	248.36
0.62500	361.72	437.01	389.13	253.34
0.66667	379.93	377.10	407.05	283.43
0.70833	395.64	281.26	329.18	244.18
0.75000	407.36	399.04	340.15	267.87
0.81250	432.30	299.36	349.20	299.09
0.87500	451.01	259.36	279.36	258.70
0.93750	467.22	291.20	275.28	274.20
1.00000	485.42	231.68	261.44	278.87
1.06250	499.90	267.52	249.60	283.98
1.12500	516.62	231.36	249.44	301.67
1.18750	531.08	219.84	225.60	289.11
1.25000	544.82	155.36	187.60	254.04
1.31250	554.53	191.84	173.60	247.79

(*Continued*)

TABLE 1.7 (Continued)

Δt (hour)	Δp (psi)	Slope (psi/hour)	$\Delta p^{\backslash}$ (psi/hour)	$\Delta t \Delta t^{\backslash} (t_p + \Delta t) t_p$
1.37500	566.52	183.52	187.68	281.72
1.43750	577.99	151.84	167.68	264.14
1.50000	587.48	147.68	149.76	247.10
1.62500	605.94	106.00	126.84	228.44
1.75000	619.19	109.92	107.96	210.97
1.87500	632.93	103.76	106.84	225.37
2.00000	645.90	69.92	86.84	196.84
2.25000	663.38	59.84	64.88	167.88
2.37500	670.66	50.00	54.92	151.09
2.50000	677.11	44.84	47.42	138.31
2.75000	688.32	41.84	43.34	141.04
3.00000	698.78	35.80	38.82	139.75
3.25000	707.73	22.96	29.38	118.17
3.50000	713.47	38.80	30.88	133.30
3.75000	723.17	25.88	32.34	151.59
4.00000	729.64	16.92	21.40	108.43
4.25000	733.87	7.00	11.96	65.23
4.50000	735.62	7.00	7.00	40.95
4.75000	737.37	11.00	9.00	56.29
5.00000	740.12	12.96	11.98	79.87
5.25000	743.36	11.80	12.38	87.74
5.50000	746.31	8.24	10.02	75.32
5.75000	748.37	9.96	9.10	72.38
6.00000	750.86	7.00	8.48	71.23
6.25000	752.51	−1.84	2.58	22.84
6.75000	751.69	5.52	1.84	18.01
7.25000	754.45	4.46	4.99	53.66
7.75000	756.68	3.02	3.74	43.96
8.25000	758.19	3.50	3.26	41.69
8.75000	759.94	2.48	2.99	41.42
9.25000	761.18	2.02	2.25	33.65
9.75000	762.19	2.98	2.50	40.22
10.25000	763.68	1.48	2.23	38.48
10.75000	764.42	2.02	1.75	32.29
11.25000	765.43	1.48	1.75	34.45
11.75000	766.17	2.02	1.75	36.67
12.25000	767.18	1.48	1.75	38.94
12.75000	767.92	1.64	1.56	36.80
13.25000	768.74	0.86	1.25	31.19
13.75000	769.17	1.33	1.10	28.90
14.50000	770.17	1.00	1.17	33.27
15.25000	770.92	0.99	0.99	30.55
16.00000	771.66	1.00	0.99	32.85
16.75000	772.41	0.99	0.99	35.22
17.50000	773.15	0.68	0.83	31.60
18.25000	773.66	0.99	0.83	33.71

(*Continued*)

TABLE 1.7 (Continued)

Δt (hour)	Δp (psi)	Slope (psi/hour)	$\Delta p^{\backslash}$ (psi/hour)	$\Delta t \Delta t^{\backslash}$ $(t_p + \Delta t) t_p$
19.00000	774.40	0.35	0.67	28.71
19.75000	774.66	0.67	0.51	23.18
20.50000	775.16	1.00	0.83	40.43
21.25000	775.91	0.50	0.75	38.52
22.25000	776.41	0.48	0.49	27.07
23.25000	776.89	0.26	0.37	21.94
24.25000	777.15	0.51	0.38	24.43
25.25000	777.66	0.50	0.50	34.22
26.25000	778.16	0.24	0.37	26.71
27.25000	778.40	0.40[a]	0.32[b]	24.56[c]
28.50000	778.90	0.34	0.37	30.58
30.00000	779.41	25.98	13.16	1184.41

[a](778.9 −778.4)/(28.5 −27.25) = 0.40.
[b](0.40 + 0.24)/2 = 0.32.
[c]27.25 −0.32 − (15 + 27.25)/15 = 24.56.
After Sabet, M., 1991. Well Test Analysis. Gulf Publishing, Dallas, TX.

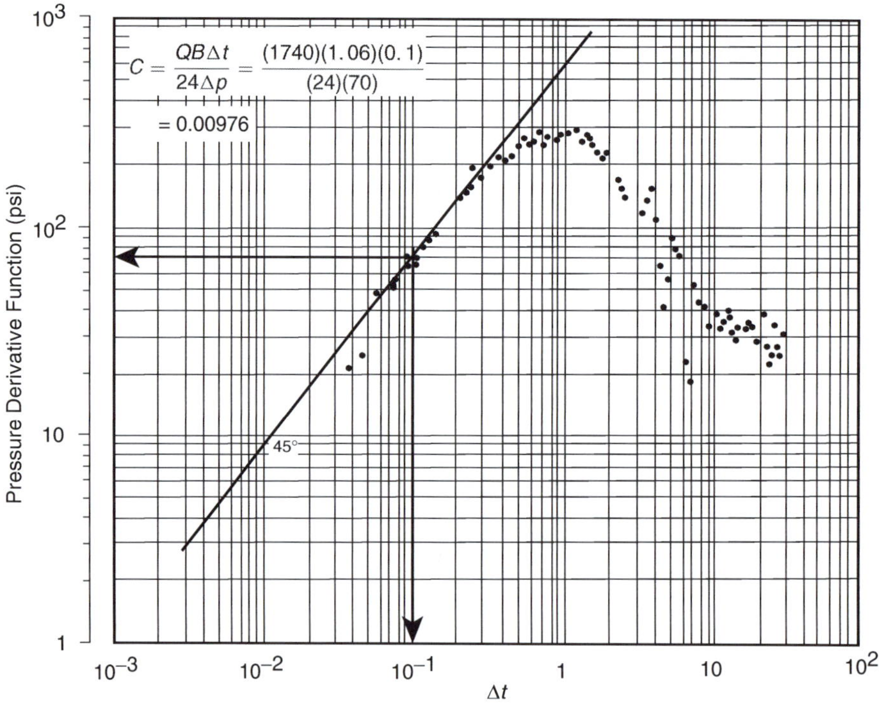

FIGURE 1.57 Log–log plot. Data from Table 1.7.

Step 3. Overlay the pressure difference data and pressure derivative data over the Gringarten–Bourdet type curve to match the type curve, as shown in Figure 1.57, with the following match points:

$$(C_D\, e^{2s})_{MP} = 4 \times 10^9$$

$$\left(\frac{p_D}{\Delta p}\right)_{MP} = 0.0179$$

$$\left[\left(\frac{(t_D/C_D)}{\Delta t}\right)\right]_{MP} = 14.8$$

Step 4. Calculate the permeability k:

$$k = \left[\frac{141.2QB\mu}{h}\right]\left(\frac{p_D}{\Delta p}\right)_{MP} = \left[\frac{141.2(174)(1.06)(2.5)}{107}\right](0.0179) = 10.9 \text{ md}$$

Step 5. Calculate C and C_D:

$$C = \left[\frac{0.0002951\,kh}{\mu}\right]\frac{(\Delta t_e)_{MP}}{(t_D/C_D)_{MP}} = \left[\frac{0.0002951(10.9)(107)}{2.5}\right]\left(\frac{1}{14.8}\right) = 0.0093 \text{ bbl/psi}$$

$$C_D = \frac{0.8936C}{\phi h c_t r_w^2} = \frac{0.8936(0.093)}{(0.25)(107)(4.2\times10^{-6})(0.29)^2} = 879$$

Step 6. Calculate the skin factor s:

$$s = \frac{1}{2}\ln\left[\frac{(C_D\, e^{2s})_{MP}}{C_D}\right] = \frac{1}{2}\ln\left[\frac{4\times10^9}{879}\right] = 7.7$$

Note that the derivative function, as plotted in Figure 1.57, shows an appreciable amount of scatter points, and the horizontal line that signifies the radial infinite-acting state is not clear. A practical limitation associated with the use of the pressure derivative approach is the ability to measure pressure transient data with sufficient frequency and accuracy so that it can be differentiated. Generally, the derivative function will show severe oscillations unless the data is smoothed before taking the derivative.

Smoothing of any time series, such as pressure–time data, is not an easy task, and unless it is done with care and know-how, a part of the data which is representative of the reservoir (signal) could be lost. Signal filtering, smoothing, and interpolation are considered topics of science and engineering, and unless the proper smoothing techniques are applied to the field data, the results could be utterly misleading.

In addition to the reservoir heterogeneity, there are many inner and outer reservoir boundary conditions that will cause the transient state plot to deviate from the expected semilog straight-line behavior during the infinite-acting behavior of the test well, such as:

- faults and other impermeable flow barriers;
- partial penetration;
- phase separation and packer failures;
- interference;
- stratified layers;
- naturally and hydraulically fractured reservoirs;
- boundary;
- lateral increase in mobility.

The theory which describes the unsteady-state flow data is based on the ideal radial flow of fluids in a homogeneous reservoir system of uniform thickness, porosity, and permeability. Any deviation from this ideal concept can cause the predicted pressure to behave differently from the actual measured pressure. In addition, a well test response may have different behavior at different times during the test. In general, the following four different time periods can be identified on a log–log plot of Δp vs. Δt as shown in Figure 1.58:

(1) The *wellbore storage effect* is always the first flow regime to appear.
(2) Evidence of the well and reservoir *heterogeneities effect* will then appear in the pressure behavior response. This behavior maybe a result of multilayered formation, skin, hydraulic fracture, or fissured formation.

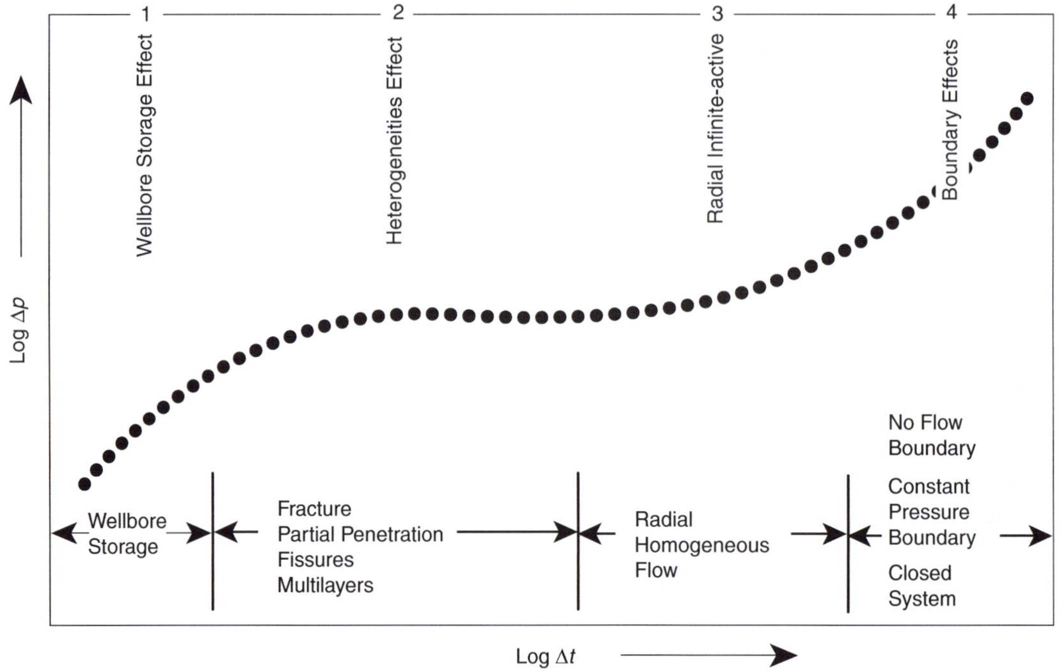

FIGURE 1.58 Log–log plot of a typical drawdown.

(3) The pressure response exhibits the *radial infinite-active* behavior and represents an equivalent homogeneous system.
(4) The last period represents the *boundary effects* that may occur at late time.

Thus, many types of flow regimes can appear before and after the actual semilog straight line develops, and they follow a very strict chronology in the pressure response. Only global diagnosis, with identification of all successive regimes present, will indicate exactly when conventional analysis, e.g., the semilog plot technique, is justified. Recognition of the above four different sequences of responses is perhaps the most important element in well test analysis. The difficulty arises from the fact that some of these responses could be missing, overlapping, or undetectable through the traditional graphical semilog straight-line approach. Selection of the correct *reservoir interpretation model* is a prerequisite and an important step before analyzing well test data and interpreting the test results. With proper well test design and sufficient test length for the response to be detected, most pressure transient data can provide an unambiguous indicator of the type and the associated characteristics of the reservoir. However, many well tests cannot or are not run for sufficient test duration to eliminate ambiguity in selecting the proper model to analyze test data. With a sufficient length of well testing time, the reservoir response during well testing is then used to identify a well test interpretation model from which well and reservoir parameters, such as permeability and skin, can be determined. This *model identification* requirement holds for both traditional graphical analyses as well as for computer-aided techniques.

It should be pointed out that both the semilog and log–log plots of pressure vs. time data are often insensitive to pressure changes and cannot be solely used as diagnostic plots to find the interpretation model that best represents the dynamic behavior of the well and reservoir during the test. The pressure derivative type curve, however, is the most definitive of the type curves for identifying the proper interpretation model. The pressure derivative approach has been applied with tremendous success as a diagnostic tool for the following reasons:

- It magnifies small pressure changes.
- Flow regimes have *clear characteristic shapes* on the pressure derivative plot.
- It clearly differentiates between responses of various reservoir models, such as:
  - dual-porosity behaviors;
  - naturally and hydraulically fractured reservoirs;
  - closed boundary systems;
  - constant pressure boundaries;
  - faults and impermeable boundaries;
  - Infinite-acting systems.
- It identifies various reservoir behaviors and conditions that are not apparent in the traditional well analysis approach.
- It defines a clear recognizable pattern of various flow periods.
- It improves the overall accuracy of test interpretation.
- It provides an accurate estimation of relevant reservoir parameters.

Al-Ghamdi and Issaka (2001) pointed out that there are three major difficulties during the process of identifying the proper interpretation model:

(1) The limited number of available interpretation models that are restricted to prespecified setting and idealized conditions.
(2) The limitation of the majority of existing heterogeneous reservoir models to one type of heterogeneities and the ability to accommodate multiple heterogeneities within the same model.
(3) The non-uniqueness problem where identical responses are generated by completely different reservoir models of totally different geological configuration.

Lee (1982) suggested that the best approach of identifying the correct interpretation model incorporates the following three plotting techniques:

(1) The traditional log–log type curve plot of pressure difference Δp vs. time.
(2) The derivative type curve.
(3) The "specialized graph" such as the Horner plot for a homogeneous system among other plots.

Based on the knowledge of the shapes of different flow regimes, the double plot of pressure and its derivative is used to diagnose the system and choose a well/reservoir model to match the well test data. The specialized plots can then be used to confirm the results of the pressure-derivative type curve match. Therefore, after reviewing and checking the quality of the test raw data, the analysis of well tests can be divided into the following two steps:

(1) The reservoir model identification and various flow regimes encountered during the tests are determined.
(2) The values of various reservoir and well parameters are calculated.

1.5.1 Model Identification

The validity of the well test interpretation is totally dependent on two important factors, the accuracy of the measured field data and the applicability of the selected interpretation model. Identifying the correct model for analyzing the well test data can be recognized by plotting the data in several formats to eliminate the ambiguity in model selection. Gringarten (1984) pointed out that the interoperation model consists of three main components that are independent of each other and dominate at different times during the test and they follow the chronology of the pressure response. These are:

(I) *Inner boundaries*. Identification of the inner boundaries is performed on the early-time test data. There are only five possible inner boundaries and flow conditions in and around the wellbore:

(1) wellbore storage;
(2) skin;
(3) phase separation;
(4) partial penetration;
(5) fracture.

(II) *Reservoir behavior*. Identification of the reservoir is performed on the middle-time data during the infinite-acting behavior and includes two main types:

(1) homogeneous;
(2) heterogeneous.

(III) *Outer boundaries*. Identification of the outer boundaries is performed on the late-time data. There are two outer boundaries:

(1) no-flow boundary;
(2) constant-pressure boundary.

Each of the above three components exhibits a distinctly different characteristic that can be not only identified separately, but also described in different mathematical forms.

1.5.2 Analysis of Early-time Test Data

Early-time data is meaningful and can be used to obtain unparalleled information on the reservoir around the wellbore. During this early-time period, wellbore storage, fractures, and other inner boundary flow regimes are the dominant flowing conditions and exhibit a distinct different behavior. These inner boundary conditions and their associated flow regimes are briefly discussed below.

Wellbore Storage and Skin. The most effective procedure for analyzing and understanding the entire recorded transient well test data is by employing the log–log plot of the pressure difference $\Delta p^{\backslash}$ and its derivative $\Delta p^{\backslash}$ vs. elapsed time. Identification of the inner boundaries is performed on early-time test data and starts with the wellbore storage. During this time when the wellbore storage dominates, Δp and its derivative $\Delta p^{\backslash}$ are proportional to the elapsed time and produce a 45° straight line on the log–log plot, as shown in Figure 1.59. On the derivative plot, the transition from the wellbore storage to the infinite-acting radial flow gives a "hump" with a maximum that indicates wellbore damage (positive skin). Conversely, the absence of a maximum indicates a non-damaged or stimulated well.

Phase Separation in Tubing. Stegemeier and Matthews (1958), in a study of anomalous pressure buildup behavior, graphically illustrated and discussed the effects of several reservoir conditions on the Horner straight-line plot, as shown in Figure 1.60. The problem occurs when gas and oil are segregated in the tubing and annulus during shut-in, which can cause the wellbore pressure to increase. This increase in the pressure could exceed the reservoir pressure and force the liquid to flow back into the formation with a resulting decrease in the wellbore pressure. Stegemeier and Matthews investigated this "humping" effect, as shown in Figure 1.60, which means that bottom-hole pressure builds up to a maximum and then decreases. They attributed this behavior to the rise of bubbles of gas and the redistribution of fluids within the wellbore. Wells which show the humping behavior have the following characteristics:

- They are completed in moderately permeable formations with a considerable skin effect or restriction to flow near the wellbore.
- The annulus is packed off.

The phenomenon does not occur in tighter formations because the production rate is small and, thus, there is ample space for the segregated gas to move into and expand. Similarly, if there is no restriction to flow near the wellbore, fluid can easily flow back into the formation to equalize the pressure and prevent humping. If the annulus is not packed off, bubble rise in the tubing will simply unload liquid into the casing–tubing annulus rather than displace the fluid back into the formation.

Stegemeier and Matthews also showed how *leakage through the wellbore* between dually completed zones at different pressure can cause an anomalous hump in measured pressures. When this leakage occurs, the pressure differential between zones becomes small, allowing

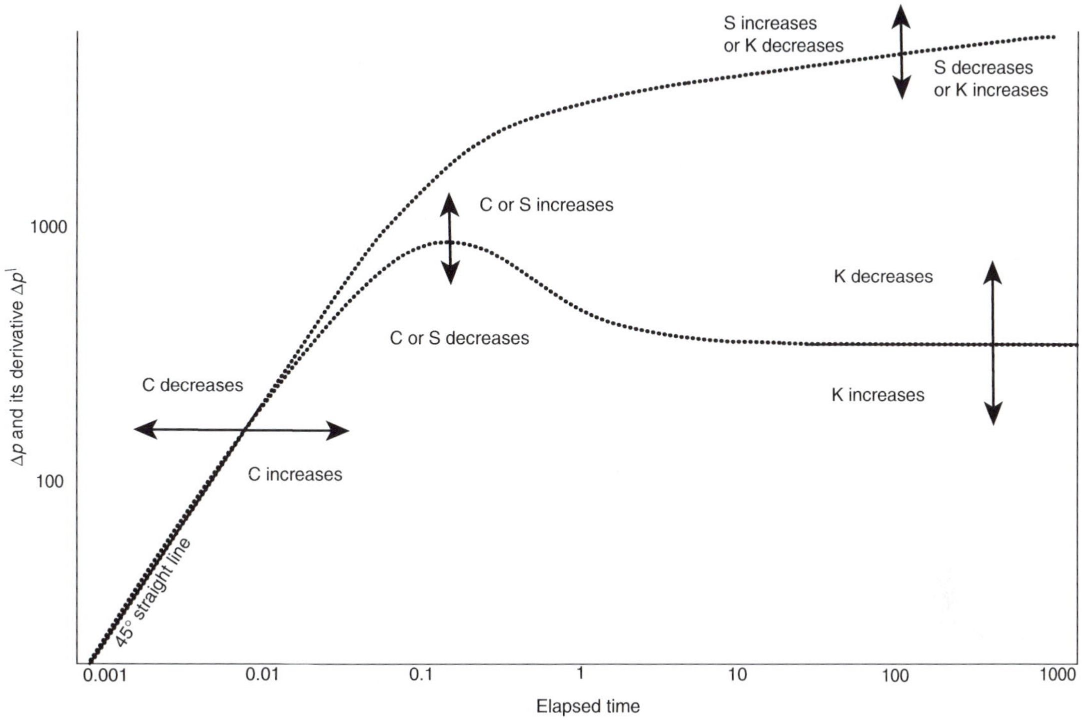

FIGURE 1.59 Δp and its derivative vs. elapsed time.

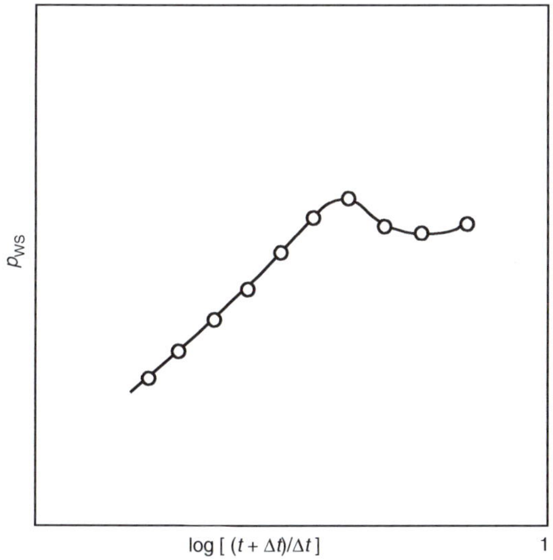

FIGURE 1.60 Phase separation in tubing. *(After Stegemeier, G., Matthews, C., 1958. A study of anomalous pressure buildup behavior. Trans. AIME 213, 44–50.)*

fluid to flow, and causes a hump in the pressure observed in the other zone.

Effect of Partial Penetration. Depending on the type of wellbore completion configuration, it is possible to have spherical or hemispherical flow near the wellbore. If the well penetrates the reservoir for a short distance below the cap rock, the flow will be hemispherical. When the well is cased through a thick pay zone and only a small part of the casing is perforated, the flow in the immediate vicinity of the wellbore will be spherical. Away from the wellbore, the flow is essentially radial. However, for a short duration of transient test, the flow will remain spherical during the test.

In the case of a pressure buildup test of a partially depleted well, Culham (1974) described the flow by the following expression:

$$p_i - p_{ws} = \frac{2453 QB\mu}{k^{2/3}} \left[\frac{1}{\sqrt{\Delta t}} - \frac{1}{\sqrt{t_p + \Delta t}} \right]$$

This relationship suggests that a plot of $(p_i - p_{ws})$ vs. $[(1/\sqrt{\Delta t}) - (1/\sqrt{t_p + \Delta t})]$ on a Cartesian scale would be a straight line that passes through the origin with a slope of m as given by:

For spherical flow: $m = \dfrac{2453 QB\mu}{k^{2/3}}$

For hemispherical flow: $m = \dfrac{1226 QB\mu}{k^{2/3}}$

with the *total* skin factor s defined by:

$$s = 34.7 r_{ew} \sqrt{\frac{\phi \mu c_t}{k}} \left[\frac{(p_{ws})_{\Delta t} - p_{wf \text{ at } \Delta t = 0}}{m} + \frac{1}{\sqrt{\Delta t}} \right] - 1$$

The dimensionless parameter r_{ew} is given by:

For spherical flow: $r_{ew} = \dfrac{h_p}{2\ln(h_p / r_w)}$

For hemispherical flow: $r_{ew} = \dfrac{h_p}{\ln(2h_p / r_w)}$

where

$(p_{ws})_{\Delta t}$ = the shut-in pressure at any shut-in time Δt, hours
h_p = perforated length, ft
r_w = wellbore radius, ft

An important factor in determining the partial penetration skin factor is the ratio of the horizontal permeability k_h to the vertical permeability k_v, i.e., k_h/k_v. If the vertical permeability is small, the well will tend to behave as if the formation thickness h is equal to the completion thickness h_p. When the vertical permeability is high, the effect of the partial penetration is to introduce an extra pressure drop near the wellbore. This extra pressure drop will cause a large positive skin factor or smaller apparent wellbore radius when analyzing well test data. Similarly, opening only a few holes in the casing can also cause additional skin damage. Saidikowski (1979) indicated that the total skin factor s as calculated from a pressure transient test is related to the true skin factor caused by formation damage s_d and skin factor due to partial penetration s_P by the following relationship:

$$s = \left(\frac{h}{h_p}\right) s_d + s_p$$

Saidikowski estimated the skin factor due to partial penetration from the following expression:

$$s_p = \left(\frac{h}{h_p} - 1\right) \left[\ln\left(\frac{h}{r_w} \sqrt{\frac{k_h}{k_v}}\right) - 2 \right]$$

where

r_w = wellbore radius, ft
h_p = perforated interval, ft
h = total thickness, ft
k_h = horizontal permeability, md
k_v = vertical permeability, md

1.5.3 Analysis of Middle-time Test Data

Identification of the basic reservoir characteristics is performed during the reservoir infinite-acting period and by using the middle-time test data. Infinite-acting flow occurs after the inner boundary effects have disappeared (e.g., wellbore storage, skin) and before the outer boundary effects have been felt. Gringarten et al. (1979) suggested that all reservoir behaviors can be classified as homogeneous or heterogeneous systems. The homogeneous system is described by only one porous medium that can be characterized by average rock properties through the conventional well testing approach. Heterogeneous systems are subclassified into the following two categories:

(1) double porosity reservoirs;
(2) multilayered or double-permeability reservoirs.

A brief discussion of the above two categories is given below.

Naturally Fractured (Double-porosity) Reservoirs. Naturally fractured reservoirs are typically characterized by a double-porosity behavior; a primary porosity ϕ_m that represents the matrix and a secondary porosity ϕ_f that represents the fissure system. Basically, "fractures" are created hydraulically for well stimulation, whereas "fissures" are considered natural fractures. The double- or dual-porosity model assumes two porous regions of distinctly different porosities and permeabilities within the formation. Only one, the "fissure system," has a permeability k_f high enough to produce to the well. The matrix system does not produce directly to the well but acts as a source of fluid to the fissure system. A very important characteristic of the double-porosity system is the nature of the fluid exchange between the two distinct porous systems. Gringarten (1984) presented a comprehensive treatment and an excellent review of the behavior of fissured reservoirs and the appropriate methodologies of analyzing well test data.

Warren and Root (1963) presented extensive theoretical work on the behavior of naturally fractured reservoirs. They assumed that the formation fluid flows from the matrix system into the fractures under pseudosteady-state conditions with the fractures acting like conduits to the wellbore. Kazemi (1969) proposed a similar model with the main assumption that the interporosity flow occurs under transient flow. Warren and Root indicated that two characteristic parameters, in addition to permeability and skin, control the behavior of double-porosity systems. These are:

(1) The dimensionless parameter ω that defines the storativity of the fractures as a ratio to that of the total reservoir. Mathematically, it is given by:

$$\omega = \frac{(\phi h c_t)_f}{(\phi h c_t)_{f+m}} = \frac{(\phi h c_t)_f}{(\phi h c_t)_f + (\phi h c_t)_m} \tag{1.226}$$

where
ω = storativity ratio
h = thickness
c_t = total compressibility, psi^{-1}
ϕ = porosity

The subscripts "f" and "m" refer to the fissure and matrix, respectively. A typical range of ω is 0.1–0.001.

(2) The second parameter λ is the interporosity flow coefficient which describes the ability of the fluid to flow from the matrix into the fissures and is defined by the following relationship:

$$\lambda = \alpha\left(\frac{k_m}{k_f}\right) r_w^2 \tag{1.227}$$

where
λ = interporosity flow coefficient
k = permeability
r_w = wellbore radius

The factor α is the block-shaped parameter that depends on the geometry and the characteristic shape of the matrix–fissure system and has the dimension of a reciprocal of the area defined by the following expression:

$$\alpha = \frac{A}{Vx}$$

where

A = surface area of the matrix block, ft^2
V = volume of the matrix block
x = characteristic length of the matrix block, ft

Most of the proposed models assume that the matrix–fissure system can be represented by one the following four geometries:

(a) *Cubic* matrix blocks separated by fractures with λ as given by:

$$\lambda = \frac{60}{l_m^2}\left(\frac{k_m}{k_f}\right) r_w^2$$

where
l_m = length of a block side

(b) *Spherical* matrix blocks separated by fractures with λ as given by:

$$\lambda = \frac{15}{r_m^2}\left(\frac{k_m}{k_f}\right) r_w^2$$

where
r_m = radius of the sphere

(c) *Horizontal strata* (rectangular slab) matrix blocks separated by fractures with λ as given by:

$$\lambda = \frac{12}{h_f^2}\left(\frac{k_m}{k_f}\right) r_w^2$$

where
h_f = thickness of an individual fracture or high-permeability layer

(d) *Vertical cylinder* matrix blocks separated by fractures with λ as given by:

$$\lambda = \frac{8}{r_m^2}\left(\frac{k_m}{k_f}\right)r_w^2$$

where
r_m = radius of each cylinder

In general, the value of the interporosity flow parameter ranges between 10^{-3} and 10^{-9}. Cinco and Samaniego (1981) identified the following extreme interporosity flow conditions:

- Restricted interporosity flow which corresponds to a high skin between the least permeable media (matrix) and the highest permeable media (fissures) and is mathematically equivalent to the pseudosteady-state solution, i.e., the Warren and Root model.
- Unrestricted interporosity flow that corresponds to zero skin between the most and highest permeable media and is described by the unsteady-state (transient) solution.

Warren and Root proposed the first identification method of the double-porosity system, as shown by the drawdown semilog plot of Figure 1.61. The curve is characterized by *two parallel straight lines* due to the two separate porosities in the reservoir. Because the secondary porosity (fissures) has the greater transmissivity and is connected to the wellbore, it responds first as described by the first semilog straight line. The primary porosity (matrix), having a much lower transmissivity, responds much later. The combined effect of the two porosities gives rise to the second semilog straight line. The two straight

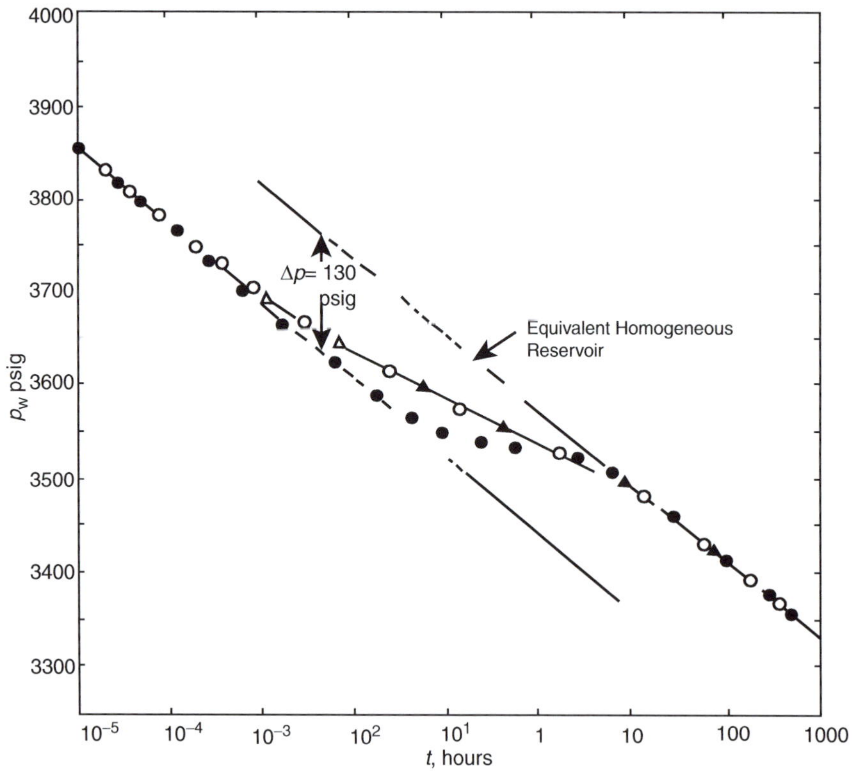

FIGURE 1.61 Pressure drawdown according to the model by Warren and Root. *(Kazemi, H., 1969. Pressure transient analysis of naturally fractured reservoirs with uniform fracture distribution. SPE J. 9 (4), 451–462; Copyright © 1969 SPE.)*

lines are separated by a transition period during which the pressure tends to stabilize.

The first straight line reflects the transient radial flow through the fractures and, thus, *its slope is used to determine the system permeability–thickness product*. However, because the fracture storage is small, the fluid in the fractures is quickly depleted with a combined rapid pressure decline in the fractures. This pressure drop in the fracture allows more fluid to flow from the matrix into the fractures, which causes a slowdown in the pressure decline rate (as shown in Figure 1.61 by the transition period). As the matrix pressure approaches the pressure of the fractures, the pressure is stabilized in two systems and yields the *second semilog straight line*. It should be pointed out that the first semilog straight line may be shadowed by wellbore storage effects and might not be recognized. Therefore, in practice, only parameters characterizing the homogeneous behavior of the *total* system $k_f h$ can be obtained.

Figure 1.62 shows the pressure buildup data for a naturally fractured reservoir. As for the drawdown, wellbore storage effects may obscure the first semilog straight line. If both semilog straight lines develop, analysis of the total permeability–thickness product is estimated from the slope m of either straight line and the use of Eq. (1.176), or:

$$(k_f h) = \frac{162.6QB\mu}{m}$$

The skin factor s and the false pressure p^* are calculated as described by using the *second*

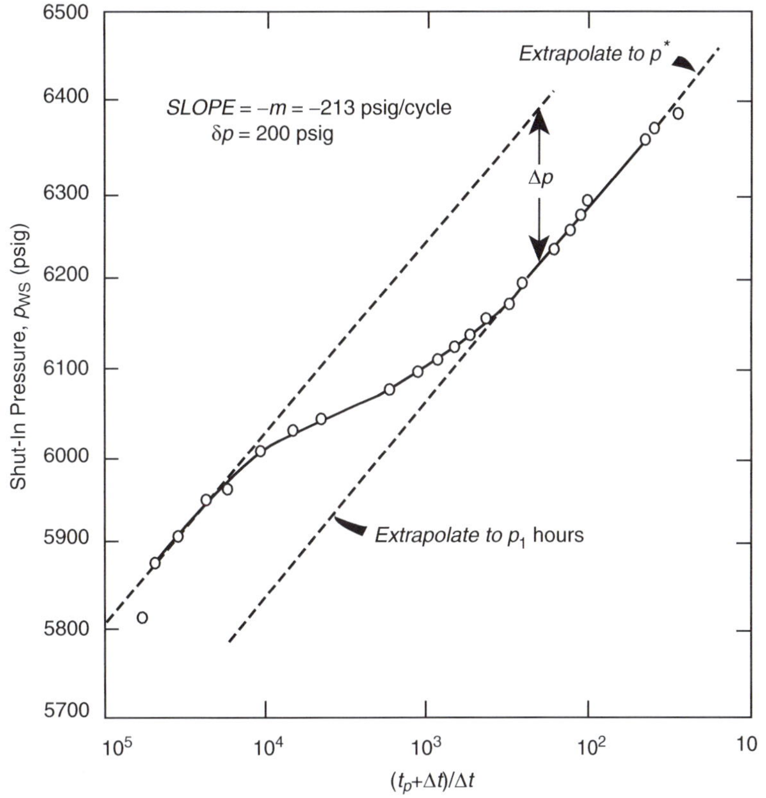

FIGURE 1.62 Buildup curve from a fractured reservoir. *(After Warren, J.E., Root, P.J., 1963. The behavior of naturally fractured reservoirs. SPE J. 3 (3), 245–255.)*

straight line. Warren and Root indicated that the storativity ratio ω can be determined from the vertical displacement between the two straight lines, identified as Δp in Figures 1.61 and 1.62, by the following expression:

$$\omega = 10^{(-\Delta p/m)} \quad (1.228)$$

Bourdet and Gringarten (1980) indicated that by drawing a horizontal line through the *middle* of the transition curve to intersect with both semilog straight lines, as shown in Figures 1.61 and 1.62, the interporosity flow coefficient λ can be determined by reading the corresponding time at the *intersection* of either of the two straight lines, e.g., t_1 or t_2, and applying the following relationships:

- In drawdown tests:

$$\lambda = \left[\frac{\omega}{1-\omega}\right]\left[\frac{(\phi h c_t)_m \mu r_w^2}{1.781 k_f t_1}\right] = \left[\frac{1}{1-\omega}\right]\left[\frac{(\phi h c_t)_m \mu r_w^2}{1.781 k_f t_2}\right] \quad (1.229)$$

- In buildup tests:

$$\lambda = \left[\frac{\omega}{1-\omega}\right]\left[\frac{(\phi h c_t)_m \mu r_w^2}{1.781 k_f t_p}\right]\left(\frac{t_p+\Delta t}{\Delta t}\right)_1$$

or

$$\lambda = \left[\frac{1}{1-\omega}\right]\left[\frac{(\phi h c_t)_m \mu r_w^2}{1.781 k_f t_p}\right]\left(\frac{t_p+\Delta t}{\Delta t}\right)_2 \quad (1.230)$$

where

k_f = permeability of the fracture, md
t_p = producing time before shut-in, hours
r_w = wellbore radius, ft
μ = viscosity, cp

The subscripts 1 and 2 (e.g., t_1) refer to the first and second line time intersection with the horizontal line drawn through the middle of the transition region pressure response during drawdown or buildup tests.

The above relationships indicate that the value of λ is dependent on the value of ω. Since ω is the ratio of fracture to matrix storage, as defined in terms of the *total* isothermal compressibility coefficients of the matrix and fissures by Eq. (1.226), thus:

$$\omega = \frac{1}{1 + \left[((\phi h)_m/(\phi h)_f)((c_t)_m/(c_t)_f)\right]}$$

it suggests that ω is also dependent on the *PVT* properties of the fluid. It is quite possible for the oil contained in the fracture to be below the bubble point, while the oil contained in the matrix is above the bubble point. Thus, ω is pressure dependent and, therefore, λ is greater than 10, so the level of heterogeneity is insufficient for dual porosity effects to be of importance and the reservoir can be treated with a single porosity.

Example 1.34

The pressure buildup data as presented by Najurieta (1980) and Sabet (1991) for a double-porosity system is tabulated below:

Δt (hour)	p_{ws} (psi)	$\frac{t_p+\Delta t}{\Delta t}$
0.003	6617	31,000,000
0.017	6632	516,668
0.033	6644	358,334
0.067	6650	129,168
0.133	6654	64,544
0.267	6661	32,293
0.533	6666	16,147
1.067	6669	8074
2.133	6678	4038
4.267	6685	2019
8.533	6697	1010
17.067	6704	506
34.133	6712	253

The following additional reservoir and fluid properties are available:

$p_i = 6789.5$ psi,	$p_{wf\ at\ \Delta t=0} = 6352$ psi,
$Q_o = 2554$ STB/day,	$B_o = 2.3$ bbl/STB,
$\mu_o = 1$ cp,	$t_p = 8611$ hour
$r_w = 0.375$ ft,	$c_t = 8.17 \times 10^{-6}$ psi^{-1},
	$\phi_m = 0.21$
$k_m = 0.1$ md,	$h_m = 17$ ft

Estimate ω and λ.

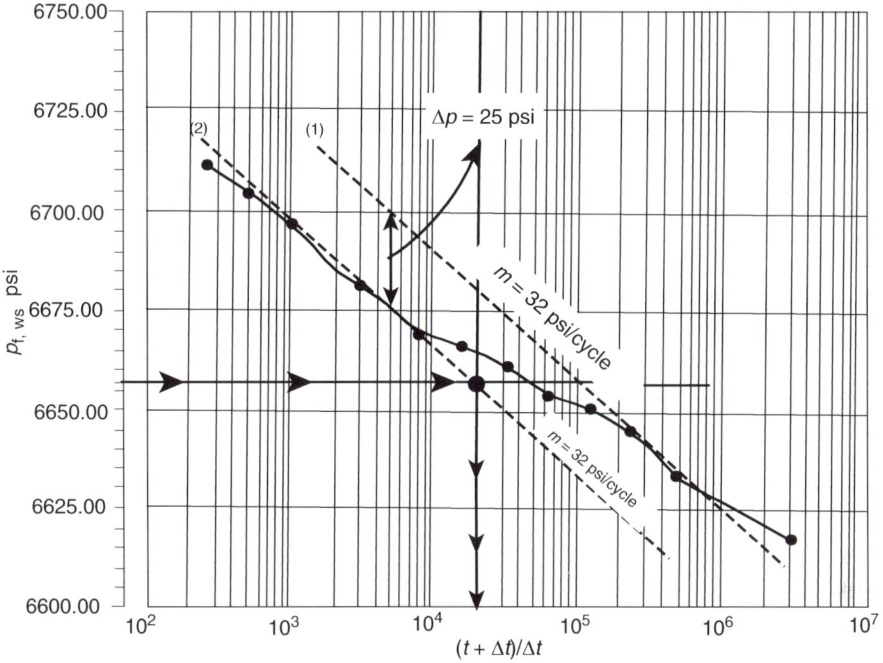

FIGURE 1.63 Semilog plot of the buildup test data. *(After Sabet, M., 1991. Well Test Analysis. Gulf Publishing, Dallas, TX.)*

Solution

Step 1. Plot p_{ws} vs. $(t_p + \Delta t)/\Delta t$ on a semilog scale as shown in Figure 1.63.

Step 2. Figure 1.63 shows two parallel semilog straight lines with a slope of $m = 32$ psi/cycle.

Step 3. Calculate $(k_f h)$ from the slope m:

$$(k_f h) = \frac{162.6 Q_o B_o \mu_o}{m} = \frac{162.6(2556)(2.3)(1.0)}{32}$$
$$= 29{,}848.3 \text{ md ft}$$

and

$$k_f = \frac{29{,}848.3}{17} = 1756 \text{ md}$$

Step 4. Determine the vertical distance Δp between the two straight lines:

$$\Delta p = 25 \text{ psi}$$

Step 5. Calculate the storativity ratio ω from Eq. (1.228):

$$\omega = 10^{-(\Delta p/m)} = 10^{-(25/32)} = 0.165$$

Step 6. Draw a horizontal line through the middle of the transition region to intersect with the two semilog straight lines. Read the corresponding time at the second intersection, to give:

$$\left(\frac{t_p + \Delta t}{\Delta t}\right)_2 = 20{,}000$$

Step 7. Calculate from Eq. (1.230):

$$\lambda = \left[\frac{1}{1-\omega}\right]\left[\frac{(\phi h c_t)_m \mu r_w^2}{1.781 k_f t_p}\right]\left(\frac{t_p + \Delta t}{\Delta t}\right)_2$$
$$= \left[\frac{1}{1 - 0.165}\right]$$
$$\times \left[\frac{(0.21)(17)(8.17 \times 10^{-6})(1)(0.375)^2}{1.781(1756)(8611)}\right](20{,}000)$$
$$= 3.64 \times 10^{-9}$$

It should be noted that pressure behavior in a naturally fractured reservoir is similar to that obtained in a *layered reservoir with no cross-flow*. In fact, in any reservoir system with two predominant rock types, the pressure buildup behavior is similar to that of Figure 1.62.

Gringarten (1987) pointed out that the two straight lines on the semilog plot may or may not be present depending on the condition of the well and duration of the test. He concluded that the semilog plot is not an efficient or sufficient tool for identifying double-porosity behavior. In the log–log plot, as shown in Figure 1.62, the double-porosity behavior yields an S-shaped curve. The *initial portion* of the curve represents the homogeneous behavior resulting from depletion in the most permeable medium, e.g., fissures. A *transition period* follows and corresponds to the interporosity flow. Finally, the *last portion* represents the homogeneous behavior of both media when recharge from the least permeable medium (matrix) is fully established and pressure is equalized. The log–log analysis represents a significant improvement over conventional semilog analysis for identifying double-porosity behavior. However, S-shape behavior is difficult to recognize in highly damaged wells and well behavior can then be erroneously diagnosed as homogeneous. Furthermore, a similar S-shape behavior may be found in irregularly bounded well drainage systems.

Perhaps the most efficient means for identifying double-porosity systems is the use of the pressure derivative plot. It allows unambiguous identification of the system, provided the quality of the pressure data is adequate and, more importantly, an accurate methodology is used in calculating pressure derivatives. As discussed earlier, the pressure derivative analysis involves a log–log plot of the derivative of the pressure with respect to time vs. elapsed time. Figure 1.64 shows the combined log–log plot of pressure and derivative vs. time for a dual-porosity system. The derivative plot shows a "minimum" or a "dip" on the pressure derivative curve caused by the interporosity flow during the transition period. The "minimum" is between two horizontal lines; the first represents the radial flow controlled by the fissures and the second describes the combined behavior of the double-porosity system. Figure 1.64 shows, at early time, the typical behavior of wellbore storage effects with the deviation from the 45° straight line to a maximum representing a wellbore damage. Gringarten (1987) suggested that the shape of the minimum depends on the double-porosity behavior. For a restricted interporosity flow, the minimum takes a V-shape, whereas unrestricted interporosity yields an open U-shaped minimum.

Based on Warren and Root's double-porosity theory and the work of Mavor and Cinco (1979), Bourdet and Gringarten (1980) developed specialized pressure type curves that can be used for analyzing well test data in dual-porosity systems. They showed that double-porosity behavior is controlled by the following independent variables:

- p_D
- t_D/C_D
- $C_D\,e^{2s}$
- ω
- $\lambda\,e^{-2s}$

with the dimensionless pressure p_D and time t_D as defined below:

$$p_D = \left[\frac{k_f h}{141.2QB\mu}\right]\Delta p$$

$$t_D = \frac{0.0002637 k_f t}{[(\phi\mu c_t)_f + (\phi\mu c_t)_m]\mu r_w^2} = \frac{0.0002637 k_f t}{(\phi\mu c_t)_{f+m}\mu r_w^2}$$

where

k = permeability, md
t = time, hours
μ = viscosity, cp
r_w = wellbore radius, ft

and subscripts:

f = fissure
m = matrix
f + m = total system
D = dimensionless

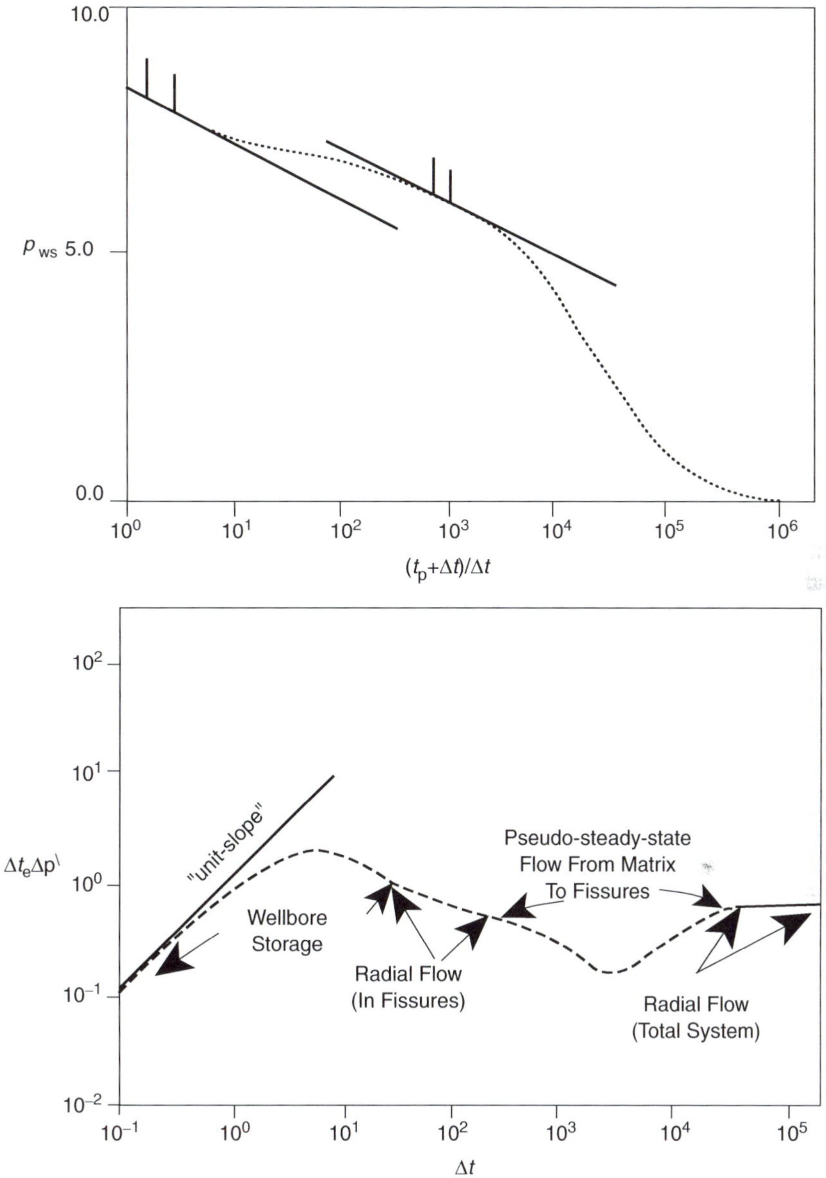

FIGURE 1.64 Dual-porosity behavior shows as two parallel semilog straight lines on a semilog plot, as a minimum on a derivative plot.

Bourdet et al. (1984) extended the practical applications of these curves and enhanced their use by introducing the pressure derivative type curves to the solution. They developed two sets of pressure derivative type curves as shown in Figures 1.65 and 1.66. The first set, i.e., Figure 1.65, is based on the assumption that the interporosity flow obeys the pseudosteady-state flowing condition and the other set (Figure 1.66) assumes transient interporosity flow. The use of either set involves plotting the

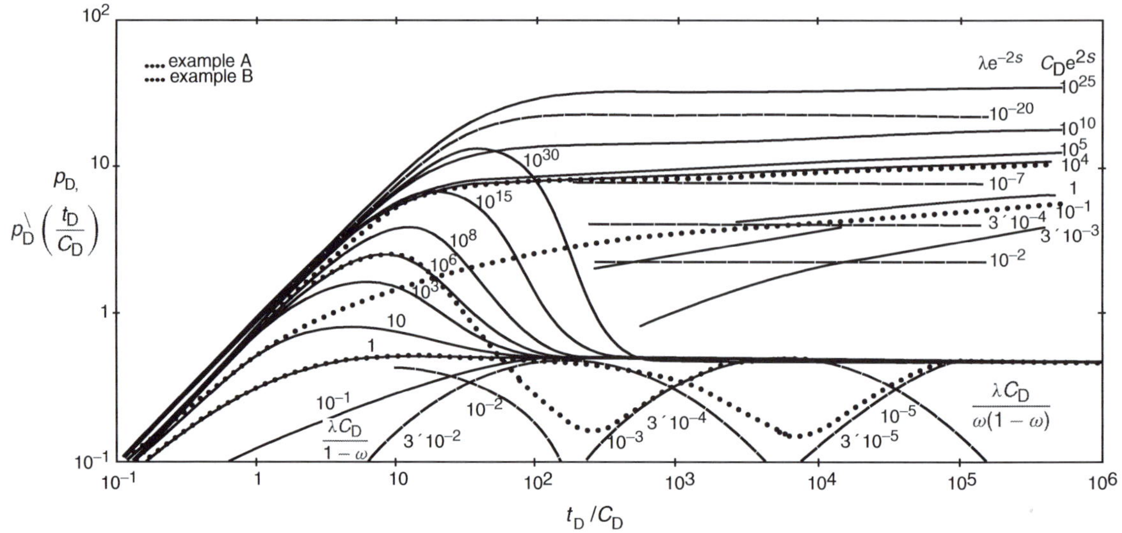

FIGURE 1.65 Type curve matching. *(Bourdet, D., Alagoa, A., Ayoub, J.A., Pirard, Y.M., 1984. New type curves aid analysis of fissured zone well tests. World Oil, April, pp. 111–124; Copyright ©1984 World Oil.)*

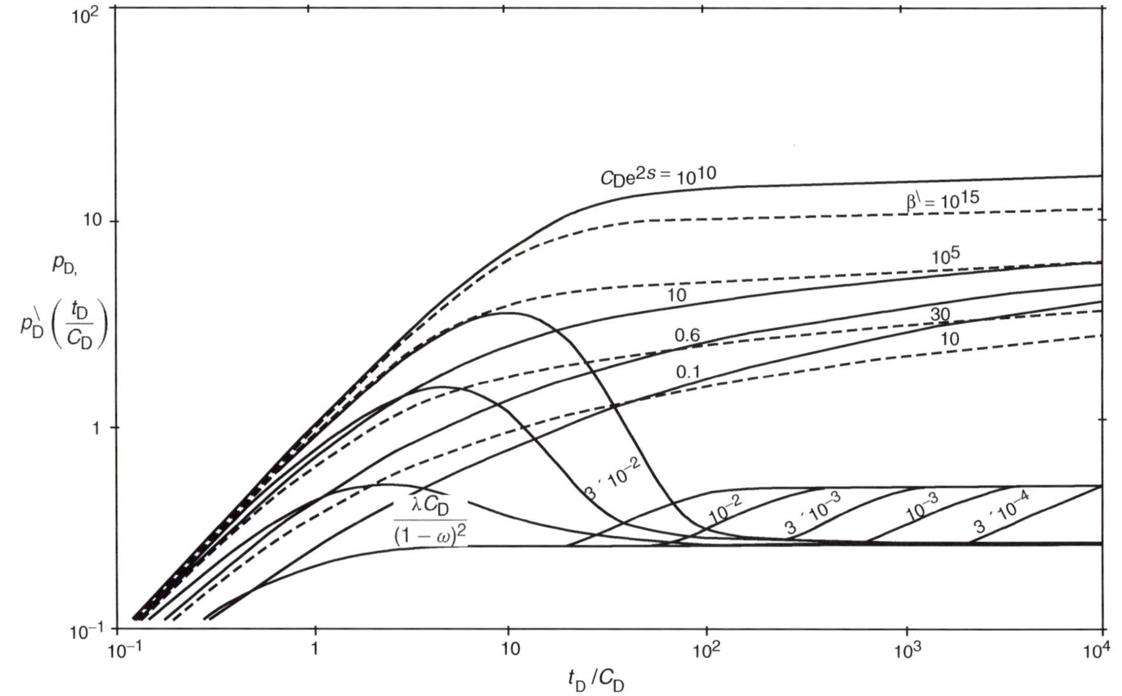

FIGURE 1.66 Type curve matching. *(Bourdet, D., Alagoa, A., Ayoub, J.A., Pirard, Y.M., 1984. New type curves aid analysis of fissured zone well tests. World Oil, April, pp. 111–124; Copyright ©1984 World Oil.)*

pressure difference Δp and the derivative function, as defined by Eq. (1.222) for drawdown tests or Eq. (1.223) for buildup tests, vs. time with same size log cycles as the type curve. The controlling variables in each of the two type curve sets are given below.

First Type Curve Set

Pseudo Steady-state Interporosity Flow. The actual *pressure response*, i.e., pressure difference Δp, is described by the following three component curves:

(1) At early times, the flow comes from the fissures (most permeable medium) and the actual pressure difference plot, i.e., Δp curve, matches one of the homogeneous curves that is labeled ($C_D\, e^{2s}$) with a corresponding value of $(C_D\, e^{2s})_f$ that describes the *fissure flow*. This value is designated as $[(C_D\, e^{2s})_f]_M$.

(2) As the pressure difference response reaches the transition regime, Δp deviates from the $C_D\, e^{2s}$ curve and follows one of the transition curves that describes this flow regime by $\lambda\, e^{-2s}$, designated as $[\lambda\, e^{-2s}]_M$.

(3) Finally, the pressure difference response leaves the transition curve and matches a new $C_D\, e^{2s}$ curve below the first one with a corresponding value of $(C_D\, e^{2s})_{f+m}$ that describes the *total system* behavior, i.e., matrix and fissures. This value is recorded as $[(C_D\, e^{2s})_{f+m}]_M$.

On the pressure derivative response, the storativity ratio ω defines the shape of the derivative curve during the transition regime that is described by a "depression" or a "minimum." The duration and depth of the depression are linked by the value of ω; a small ω produces a long and, therefore, deep transition. The interporosity coefficient λ is the second parameter defining the position of the time axis of the transition regime. A decrease of λ value moves the depression to the right side of the plot.

As shown in Figure 1.65, the pressure derivative plots match on four component curves:

(1) The derivative curve follows the fissure flow curve $[(C_D\, e^{2s})_f]_M$.

(2) The derivative curve reaches an early transition period, expressed by a depression and described by an early transition curve $[\lambda(C_D)_{f+m}/\omega(1-\omega)]_M$.

(3) The derivative pressure curve then matches a late transition curve labeled $[\lambda(C_D)_{f+m}/(1-\omega)]_M$.

(4) The total system behavior is reached on the 0.5 line.

Second Type Curve Set

Transient Interporosity Flow. As developed by Bourdet and Gringarten (1980) and expanded by Bourdet et al. (1984) to include the pressure derivative approach, this type curve is built in the same way as for the pseudosteady-state interporosity flow. As shown in Figure 1.66, the pressure behavior is defined by three component curves, $(C_D\, e^{2s})_f$, $\beta^\backslash$, and $(C_D\, e^{2s})_{f+m}$. The authors defined $\beta^\backslash$ as the interporosity dimensionless group and is given by:

$$\beta^\backslash = \delta\left[\frac{(C_D\, e^{2s})_{f+m}}{\lambda\, e^{-2s}}\right]$$

where the parameter δ is the shape coefficient with assigned values as given below:

- $\delta = 1.0508$ for spherical blocks
- $\delta = 1.8914$ for slab matrix blocks

As the first fissure flow is short-lived with transient interporosity flow models, the $(C_D\, e^{2s})_f$ curves are not seen in practice and therefore have not been included in the derivative curves. The dual-porosity derivative response starts on the derivative of a $\beta^\backslash$ transition curve, then follows a late transition curve labeled $\lambda(C_D)_{f+m}/(1-\omega)^2$ until it reaches the total system regime on the 0.5 line.

Bourdet (1985) points out that the pressure derivative responses during the transition flow regime are very different between the two types of double-porosity model. With the transient interporosity flow solutions, the transition starts from early time and does not drop to a very low level. With pseudosteady-state interporosity flow, the transition starts later and the shape of the depression is much more pronounced. There is *no lower limit* for the depth of the depression when the flow from the matrix to the fissures follows the pseudosteady-state model, whereas for the interporosity transient flow the depth of the *depression does not exceed* 0.25.

In general, the matching procedure and reservoir parameters estimation as applied to the type-curve of Figure 1.66 can be summarized in the following steps:

Step 1. Using the actual well test data, calculate the pressure difference Δp and the pressure derivative plotting functions as defined by Eq. (1.222) for drawdown or Eq. (1.223) for buildup tests, i.e.:

- For drawdown tests:

The pressure difference: $\Delta p = p_i - p_{wf}$

The derivative function: $t\Delta p' = -t\left(\frac{d(\Delta p)}{d(t)}\right)$

- For buildup tests:

The pressure difference: $\Delta p = p_{ws} - p_{wf\ at\ \Delta t=0}$

The derivative function: $\Delta t_e \Delta p' = \Delta t\left(\frac{t_p + \Delta t}{\Delta t}\right)\left[\frac{d(\Delta p)}{d(\Delta t)}\right]$

Step 2. On tracing paper with the same size log cycles as in Figure 1.66, plot the data of step 1 as a function of flowing time t for drawdown tests or equivalent time Δt_e for buildup tests.

Step 3. Place the actual two sets of plots, i.e., Δp and derivative plots, on Figure 1.65 or Figure 1.66 and force a simultaneous match of the two plots to Gringarten–Bourdet type curves. Read the matched derivative curve $[\lambda(C_D)_{f+m}/(1-\omega)^2]_M$.

Step 4. Choose any point and read its coordinates on both figures to give:

$(\Delta p, p_D)_{MP}$ and $(t \text{ or } \Delta t_e, t_D/C_D)_{MP}$

Step 5. With the match still maintained, read the values of the curves labeled $(C_D e^{2s})$ which match the initial segment of the curve $[(C_D e^{2s})_f]_M$ and the final segment $[(C_D e^{2s})_{f+m}]_M$ of the data curve.

Step 6. Calculate the well and reservoir parameters from the following relationships:

$$\omega = \frac{[(C_D e^{2s})_{f+m}]_M}{[(C_D e^{2s})_f]_M} \quad (1.231)$$

$$k_f h = 141.2 QB\mu \left(\frac{p_D}{\Delta p}\right)_{MP} \text{ md ft} \quad (1.232)$$

$$C = \left[\frac{0.000295 k_f h}{\mu}\right] \frac{(\Delta t)_{MP}}{(C_D/C_D)_{MP}} \quad (1.233)$$

$$(C_D)_{f+m} = \frac{0.8926 C}{\phi c_t h r_w^2} \quad (1.234)$$

$$s = 0.5 \ln\left[\frac{[(C_D e^{2s})_{f+m}]_M}{(C_D)_{f+m}}\right] \quad (1.235)$$

$$\lambda = \left[\frac{\lambda(C_D)_{f+m}}{(1-\omega)^2}\right]_M \frac{(1-\omega)^2}{(C_D)_{f+m}} \quad (1.236)$$

The selection of the best solution between the pseudosteady-state and the transient interporosity flow is generally straightforward; with the pseudosteady-state model, the drop of the derivative during transition is a function of the transition duration. Long transition regimes, corresponding to small ω values, produce derivative levels much smaller than the practical 0.25 limit of the transient solution.

The following pressure buildup data as given by Bourdet et al. and reported conveniently by Sabet (1991) is used below as an example to illustrate the use of pressure derivative type curves.

Example 1.35

Table 1.8 shows the pressure buildup and pressure derivative data for a naturally fractured reservoir. The following flow and reservoir data is also given:

$Q = 960$ STB/day, $B_o = 1.28$ bbl/STB
$c_t = 1 \times 10^{-5}$ psi^{-1}, $\phi = 0.007$
$\mu = 1$ cp, $r_w = 0.29$ ft, $h = 36$ ft

TABLE 1.8 Pressure Buildup Test, Naturally Fractured Reservoir

Δt (hour)	Δp_{ws} (psi)	$\frac{t_p+\Delta t}{\Delta t}$	Slope (psi/hour)	$\Delta p \backslash \frac{t_p+\Delta t}{t_p}$
0.00000E + 00	0.000		3180.10	
3.48888E −03	11.095	14 547.22	1727.63	8.56
9.04446E −03	20.693	5 612.17	847.26	11.65
1.46000E −02	25.400	3 477.03	486.90	9.74
2.01555E −02	28.105	2 518.92	337.14	8.31
2.57111E −02	29.978	1 974.86	257.22	7.64
3.12666E −02	31.407	1 624.14	196.56	7.10
3.68222E −02	32.499	1 379.24	159.66	6.56
4.23777E −02	33.386	1 198.56	127.80	6.10
4.79333E −02	34.096	1 059.76	107.28	5.64
5.90444E −02	35.288	860.52	83.25	5.63
7.01555E −02	36.213	724.39	69.48	5.36
8.12666E −02	36.985	625.49	65.97	5.51
9.23777E −02	37.718	550.38	55.07	5.60
0.10349	38.330	491.39	48.83	5.39
0.12571	39.415	404.71	43.65	5.83
0.14793	40.385	344.07	37.16	5.99
0.17016	41.211	299.25	34.38	6.11
0.19238	41.975	264.80	29.93	6.21
0.21460	42.640	237.49	28.85	6.33
0.23682	43.281	215.30	30.96	7.12
0.25904	43.969	196.92	25.78	7.39
0.28127	44.542	181.43	24.44	7.10
0.30349	45.085	168.22	25.79	7.67
0.32571	45.658	156.81	20.63	7.61
0.38127	46.804	134.11	18.58	7.53
0.43682	47.836	117.18	17.19	7.88
0.49238	48.791	104.07	16.36	8.34
0.54793	49.700	93.62	15.14	8.72
0.60349	50.541	85.09	12.50	8.44
0.66460	51.305	77.36	12.68	8.48
0.71460	51.939	72.02	11.70	8.83
0.77015	52.589	66.90	11.14	8.93
0.82571	53.208	62.46	10.58	9.11
0.88127	53.796	58.59	10.87	9.62
0.93682	54.400	55.17	8.53	9.26
0.99238	54.874	52.14	10.32	9.54
1.04790	55.447	49.43	7.70	9.64
1.10350	55.875	46.99	8.73	9.26
1.21460	56.845	42.78	7.57	10.14
1.32570	57.686	39.28	5.91	9.17
1.43680	58.343	36.32	6.40	9.10
1.54790	59.054	33.79	6.05	9.93
1.65900	59.726	31.59	5.57	9.95
1.77020	60.345	29.67	5.44	10.08
1.88130	60.949	27.98	4.74	9.93
1.99240	61.476	26.47	4.67	9.75
2.10350	61.995	25.13	4.34	9.87

(*Continued*)

TABLE 1.8 (Continued)

Δt (hour)	Δp_{ws} (psi)	$\frac{t_p+\Delta t}{\Delta t}$	Slope (psi/hour)	$\Delta p \backslash \frac{t_p+\Delta t}{t_p}$
2.21460	62.477	23.92	3.99	9.62
2.43680	63.363	21.83	3.68	9.79
2.69240	64.303	19.85	3.06[a]	9.55[b]
2.91460	64.983	18.41	3.16	9.59
3.13680	65.686	17.18	2.44	9.34
3.35900	66.229	16.11	19.72	39.68

[a](64.983 −64.303)/(2.9146 −2.69240) = 3.08.
[b][(3.68 + 3.06)/2] × 19.85 × 2.69240²/50.75 = 9.55.
Adapted from Bourdet, D., Alagoa, A., Ayoub, J.A., Pirard, Y.M., 1984. New type curves aid analysis of fissured zone well tests. World Oil, April, 111–124.
After Sabet, M., 1991. Well Test Analysis. Gulf Publishing, Dallas, TX.

It is reported that the well was opened to flow at a rate of 2952 STB/day for 1.33 hours, shut-in for 0.31 hours, opened again at the same rate for 5.05 hours, closed for 0.39 hours, opened for 31.13 hours at the rate of 960 STB/day, and then shut-in for the pressure buildup test.

Analyze the buildup data and determine the well and reservoir parameters assuming transient interporosity flow.

Solution

Step 1. Calculate the flowing time t_p as follows:

Total oil produced:

$$N_p = \frac{2952}{4}[1.33 + 5.05] + \frac{960}{24} 31.13 \simeq 2030 \text{ STB}$$

$$t_p = \frac{(24)(2030)}{960} = 50.75 \text{ hours}$$

Step 2. Confirm the double-porosity behavior by constructing the Horner plot as shown in Figure 1.67. The graph shows the two parallel straight lines confirming the dual-porosity system.

Step 3. Using the same grid system of Figure 1.66, plot the *actual pressure derivative* vs. shut-in time as shown in Figure 1.68(a) and Δp_{ws} vs. time (as shown in Figure 1.68(b)). The 45° line shows that the test was slightly affected by the wellbore storage.

Step 4. Overlay the pressure difference and pressure derivative plots over the transient interporosity type curve, as shown in Figure 1.69, to give the following matching parameters:

$$\left[\frac{p_D}{\Delta p}\right]_{MP} = 0.053$$

$$\left[\frac{t_D/C_D}{\Delta t}\right]_{MP} = 270$$

$$\left[\frac{\lambda (C_D)_{f+m}}{(1-\omega)^2}\right]_M = 0.03$$

$$[(C_D\, e^{2s})_f]_M = 33.4$$

$$[(C_D\, e^{2s})_{f+m}]_M = 0.6$$

Step 5. Calculate the well and reservoir parameters by applying Eqs. (1.231)–(1.236) to give:

$$\omega = \frac{[(C_D\, e^{2s})_{f+m}]_M}{[(C_D\, e^{2s})_f]_M} = \frac{0.6}{33.4} = 0.018$$

Kazemi (1969) pointed out that if the vertical separation between the two parallel slopes Δp is less than 100 psi, the calculation of ω by Eq. (1.228) will produce a significant error in its values. Figure 1.67 shows that Δp is about 11 psi and Eq. (1.228) gives an *erroneous value* of:

$$\omega = 10^{-(\Delta p/m)} = 10^{-(11/22)} = 0.316$$

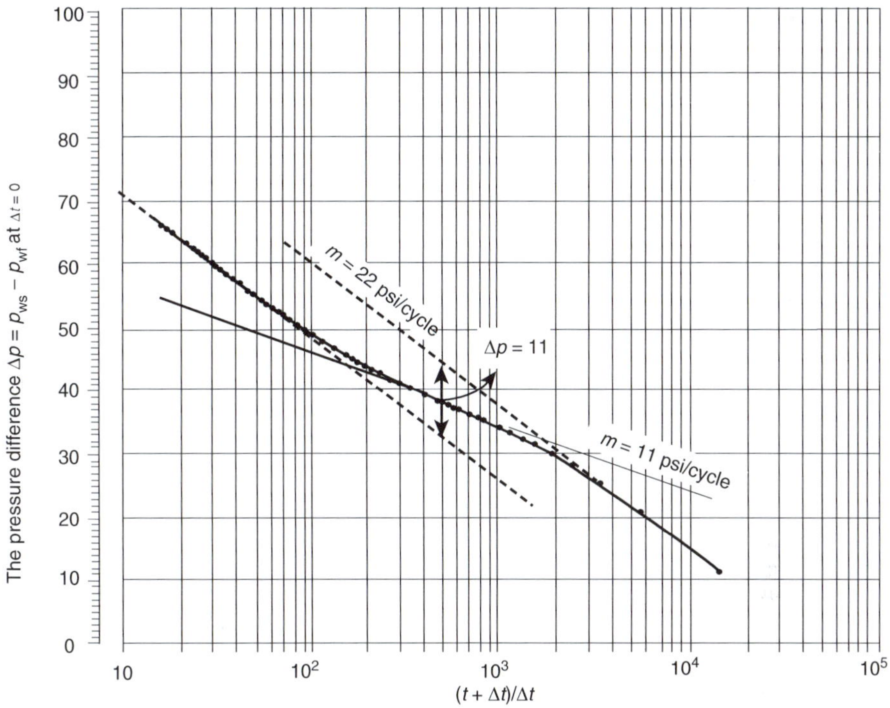

FIGURE 1.67 The Horner plot; data from Table 1.8. *(After Sabet, M., 1991. Well Test Analysis. Gulf Publishing, Dallas, TX.)*

Also

$$k_f h = 141.2 QB\mu \left(\frac{p_D}{\Delta p}\right)_{MP}$$

$$= 141.2(960)(1)(1.28)(0.053) = 9196 \text{ md ft}$$

$$C = \left[\frac{0.000295 k_f h}{\mu}\right] \frac{(\Delta t)_{MP}}{(C_D/C_D)_{MP}}$$

$$= \frac{(0.000295)(9196)}{(1.0)(270)} = 0.01 \text{bbl/psi}$$

$$(C_D)_{f+m} = \frac{0.8926C}{\phi c_t h r_w^2}$$

$$= \frac{(0.8936)(0.01)}{(0.07)(1 \times 10^{-5})(36)90.29)^2} = 4216$$

$$s = 0.5 \ln \left[\frac{[(C_D\, e^{2s})_{f+m}]_M}{(C_D)_{f+m}}\right]$$

$$= 0.5 \ln \left[\frac{0.6}{4216}\right] = -4.4$$

$$\lambda = \left[\frac{\lambda(C_D)_{f+m}}{(1-\omega)^2}\right]_M \frac{(1-\omega)^2}{(C_D)_{f+m}}$$

$$= (0.03)\left[\frac{(1-0.018)^2}{4216}\right] = 6.86 \times 10^{-6}$$

Layered Reservoirs. The pressure behavior of a no-crossflow multilayered reservoir with communication only at the wellbore will behave significantly different from a single-layer reservoir. Layered reservoirs can be classified into the following three categories:

(1) *Crossflow layered reservoirs* are those which communicate both in the wellbore and in the reservoir.
(2) *Commingled layered reservoirs* are those which communicate only in the wellbore. A complete permeability barrier exists between the various layers.
(3) *Composite reservoirs* are made up of commingled zones and some of the zones consist of crossflow layers. Each crossflow layer behaves on tests as if it were a homogeneous and isotropic layer; however, the composite reservoir should behave exactly as a commingled reservoir.

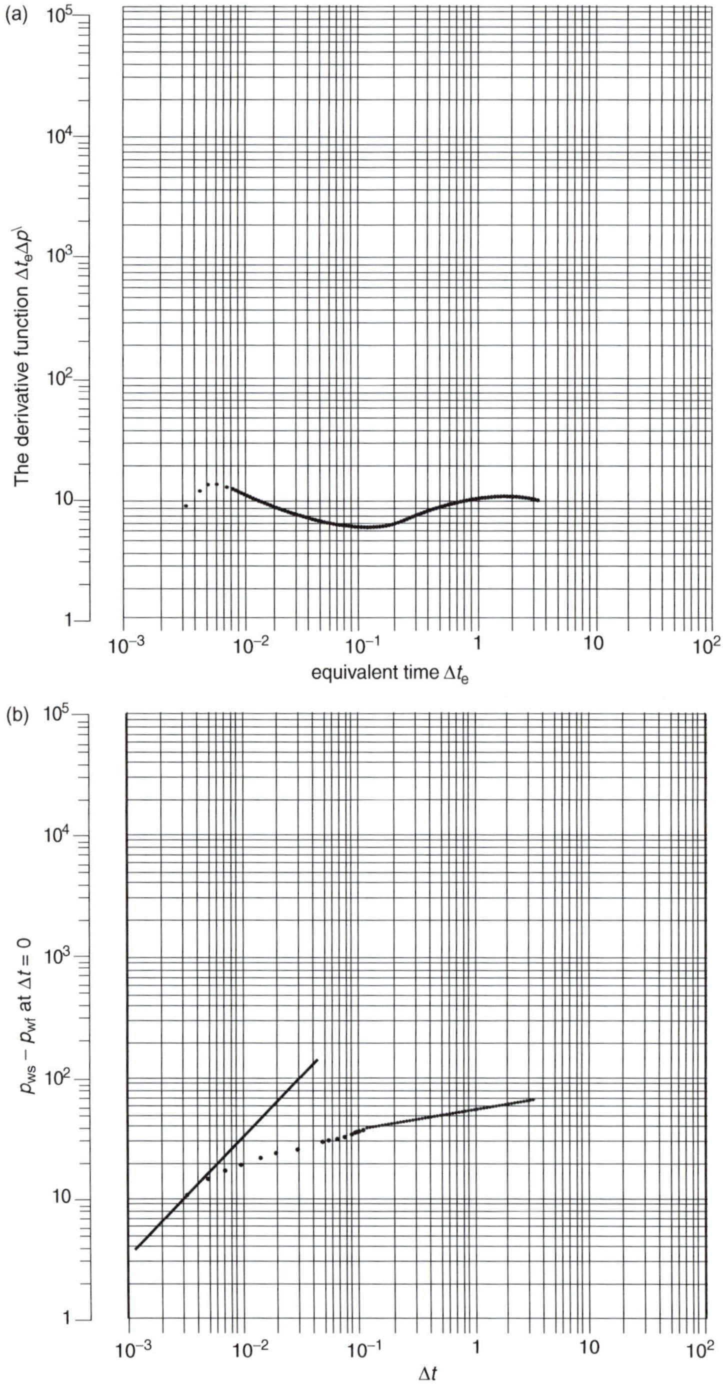

FIGURE 1.68 (a) Derivative function. (b) Log–log plot of Δp vs. Δt_e. *(After Sabet, M., 1991. Well Test Analysis. Gulf Publishing, Dallas, TX.)*

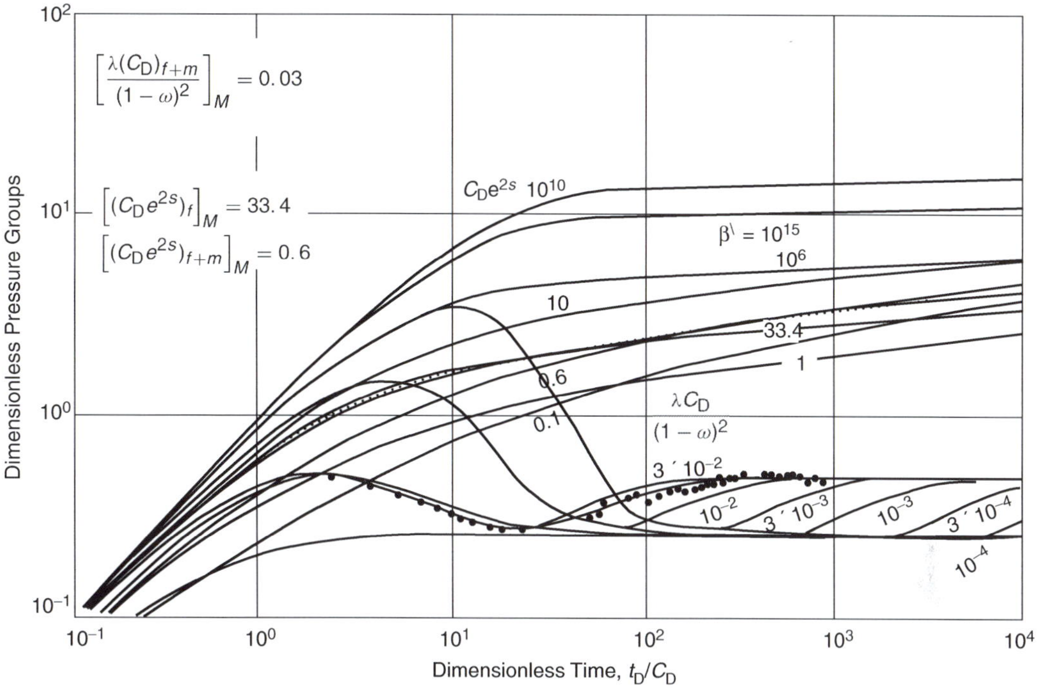

FIGURE 1.69 Type curve matching. *(Bourdet, D., Alagoa, A., Ayoub, J.A., Pirard, Y.M., 1984. New type curves aid analysis of fissured zone well tests. World Oil, April, pp. 111–124; Copyright ©1984 World Oil.)*

Some layered reservoirs behave as double-porosity reservoirs when in fact they are not. When reservoirs are characterized by layers of very low permeabilities interbedded with relatively thin high-permeability layers, they could behave on well tests exactly as if they were naturally fractured systems and could be treated with the interpretation models designed for double-porosity systems. Whether the well produces from a commingled, crossflow, or composite system, the test objectives are to determine skin factor, permeability, and average pressure.

The pressure response of crossflow layered systems during well testing is similar to that of homogeneous systems and can be analyzed with the appropriate conventional semilog and log–log plotting techniques. Results of the well test should be interpreted in terms of the arithmetic total permeability–thickness and porosity–compressibility–thickness products as given by:

$$(kh)_t = \sum_{i=1}^{n \text{ layers}} (kh)_i$$

$$(\phi c_t h)_t = \sum_{i=1}^{n \text{ layers}} (\phi c_t h)_i$$

Kazemi and Seth (1969) proposed that if the total permeability–thickness product $(kh)_t$ is known from a well test, the individual layer permeability k_i may be approximated from the layer flow rate q_i and the total flow rate q_t by applying the following relationship:

$$k_i = \frac{q_i}{q_t}\left[\frac{(kh)_t}{h_i}\right]$$

The pressure buildup behavior of a commingled two-layer system without crossflow is shown schematically in Figure 1.70. The straight line

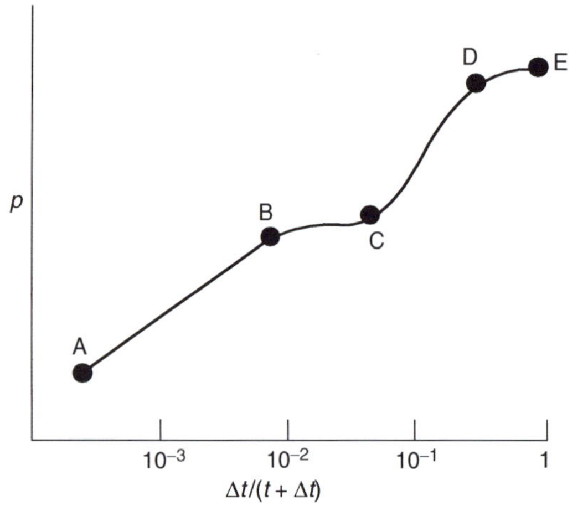

FIGURE 1.70 Theoretical pressure buildup curve for two-layer reservoir. *(Lefkovits, H., Hazebroek, P., Allen, E., Matthews, C., 1961. A study of the behavior of bounded reservoirs. SPE. J. 1 (1), 43–58; Copyright ©1961 SPE.)*

AB that follows the early-time data gives the proper value of the average flow capacity $(kh)_t$ of the reservoir system. The flattening portion BC analogous to a single-layer system attaining statistic pressure indicates that the pressure in the more permeable zone has almost reached its average value. The portion CD represents a repressurization of the more permeable layer by the less depleted, less permeable layer with a final rise DE at the stabilized average pressure. Notice that the buildup is somewhat similar to the buildup in naturally fractured reservoirs.

Sabet (1991) points out that when a commingled system is producing under the pseudo-steady-state flow condition, the flow rate from any layer q_i can be approximated from total flow rate and the layer storage capacity $\phi c_t h$ from:

$$q_i = q_t \left[\frac{(\phi c_t h)_i}{\sum_{j=1} (\phi c_t h_i) j} \right]$$

1.5.4 Hydraulically Fractured Reservoirs

A fracture is defined as a single crack initiated from the wellbore by hydraulic fracturing. It should be noted that fractures are different from "fissures," which are the formation of natural fractures. Hydraulically induced fractures are usually vertical, but can be horizontal if the formation is less than approximately 3000 ft deep. Vertical fractures are characterized by the following properties:

- fracture half-length x_f, ft;
- dimensionless radius $r_e D$, where $r_e D = r_e/x_f$;
- fracture height h_f, which is often assumed equal to the formation thickness, ft;
- fracture permeability k_f, md;
- fracture width w_f, ft;
- fracture conductivity F_C, where $F_C = k_f\, w_f$.

The analysis of fractured well tests deals with the identification of well and reservoir variables that would have an impact on future well performance. However, fractured wells are substantially more complicated. The well-penetrating fracture has unknown geometric features, i.e., x_f, w_f, and h_f, and unknown conductivity properties.

Gringarten et al. (1974) and Cinco and Samaniego (1981), among others, propose three transient flow models to consider when analyzing transient pressure data from vertically fractured wells. These are:

(1) infinite conductivity vertical fractures;
(2) finite conductivity vertical fractures;
(3) uniform flux fractures.

Descriptions of the above three types of fractures are given below.

Infinite Conductivity Vertical Fractures. These fractures are created by conventional hydraulic fracturing and are characterized by a very high conductivity, which for all practical purposes can be considered as infinite. In this case, the fracture acts similar to a large-diameter pipe with *infinite permeability* and, therefore, there is essentially no pressure drop from the tip of the fracture to the wellbore, i.e., no pressure loss in the fracture. This model assumes that the flow into the wellbore is only through the fracture and exhibits three flow periods:

(1) fracture linear flow period;
(2) formation linear flow period;
(3) infinite-acting pseudoradial flow period.

Several specialized plots are used to identify the start and end of each flow period. For example, an early-time log–log plot of Δp vs. Δt will exhibit a straight line of half-unit slope. These flow periods associated with infinite conductivity fractures and the diagnostic specialized plots will be discussed later in this section.

Finite Conductivity Fractures. These are very long fractures created by massive hydraulic fracture (MHF). These types of fractures need large quantities of propping agent to keep them open and, as a result, the fracture permeability k_f is reduced as compared to that of the infinite conductivity fractures. These finite conductivity vertical fractures are characterized by measurable pressure drops in the fracture and, therefore, exhibit unique pressure responses when testing hydraulically fractured wells. The transient pressure behavior for this system can include the following four sequence flow periods (to be discussed later):

(1) initially "linear flow within the fracture";
(2) followed by "bilinear flow";
(3) then "linear flow in the formation";
(4) eventually, "infinite acting pseudoradial flow."

Uniform Flux Fractures. A uniform flux fracture is the one in which the reservoir fluid flow rate from the formation into the fracture is uniform along the entire fracture length. This model is similar to the infinite conductivity vertical fracture in several aspects. The difference between these two systems occurs at the boundary of the fracture. The system is characterized by a variable pressure along the fracture and exhibits essentially two flow periods:

(1) linear flow;
(2) infinite-acting pseudoradial flow.

Except for highly propped and conductive fractures, it is thought that the uniform-influx fracture theory better represents reality than the infinite conductivity fracture; however, the difference between the two is rather small.

The fracture has a much greater permeability than the formation it penetrates; hence, it influences the pressure response of a well test significantly. The general solution for the pressure behavior in a reservoir is expressed in terms of dimensionless variables. The following dimensionless groups are used when analyzing pressure transient data in a hydraulically fractured well:

Diffusivity group:
$$\eta_{fD} = \frac{k_f \phi c_t}{k \phi_f c_{ft}} \tag{1.237}$$

Time group:
$$t_{Dx_f} = \left[\frac{0.0002637k}{\phi \mu c_t x_f^2}\right] t = t_D \left(\frac{r_w^2}{x_f^2}\right) \tag{1.238}$$

Conductivity group:
$$F_{CD} = \frac{k_f}{k}\frac{w_f}{x_f} = \frac{F_C}{k x_f} \tag{1.239}$$

Storage group:
$$C_{Df} = \frac{0.8937C}{\phi c_t h x_f^2} \tag{1.240}$$

Pressure group:
$$p_D = \frac{kh\Delta p}{141.2QB\mu} \quad \text{for oil} \tag{1.241}$$

$$p_D = \frac{kh\,\Delta m(p)}{1424QT} \quad \text{for gas} \tag{1.242}$$

Fracture group:
$$r_{eD} = \frac{r_e}{x_f}$$

where

x_f = fracture half-length, ft
w_f = fracture width, ft
k_f = fracture permeability, md
k = pre-frac formation permeability, md
t_{Dx_f} = dimensionless time based on the fracture half-length x_f
t = flowing time in drawdown, Δt or Δt_e in buildup, hours
T = temperature, °R
F_C = fracture conductivity, md ft
F_{CD} = dimensionless fracture conductivity
η = hydraulic diffusivity
c_{ft} = total compressibility of the fracture, psi^{-1}

Notice that the above equations are written in terms of the pressure drawdown tests. These equations should be modified for buildup tests by replacing the pressure and time with the appropriate values as shown below:

Test	Pressure	Time
Drawdown	$\Delta p = p_i - p_{wf}$	t
Buildup	$\Delta p = p_{ws} - p_{wf}$ at $\Delta t = 0$	Δt or Δt_e

In general, a fracture could be classified as an infinite conductivity fracture when the dimensionless fracture conductivity is greater than 300, i.e., $F_{CD} > 300$.

There are four flow regimes, as shown conceptually in Figure 1.71, associated with the three types of vertical fractures. These are:

(1) fracture linear flow;
(2) bilinear flow;
(3) formation linear flow;
(4) infinite-acting pseudoradial flow.

These flow periods can be identified by expressing the pressure transient data in different types of graphs. Some of these graphs are excellent tools for diagnosis and identification of regimes, since test data may correspond to different flow periods.

The specialized graphs of analysis for each flow period include:

- a graph of Δp vs. $\sqrt{\text{time}}$ for linear flow;
- a graph of Δp vs. $\sqrt[4]{\text{time}}$ for bilinear flow;
- a graph of Δp vs. log(time) for infinite-acting pseudoradial flow.

These types of flow regimes and the diagnostic plots are discussed below.

Fracture Linear Flow. This is the first flow period which occurs in a fractured system. Most of the fluid enters the wellbore during this period of time as a result of expansion within

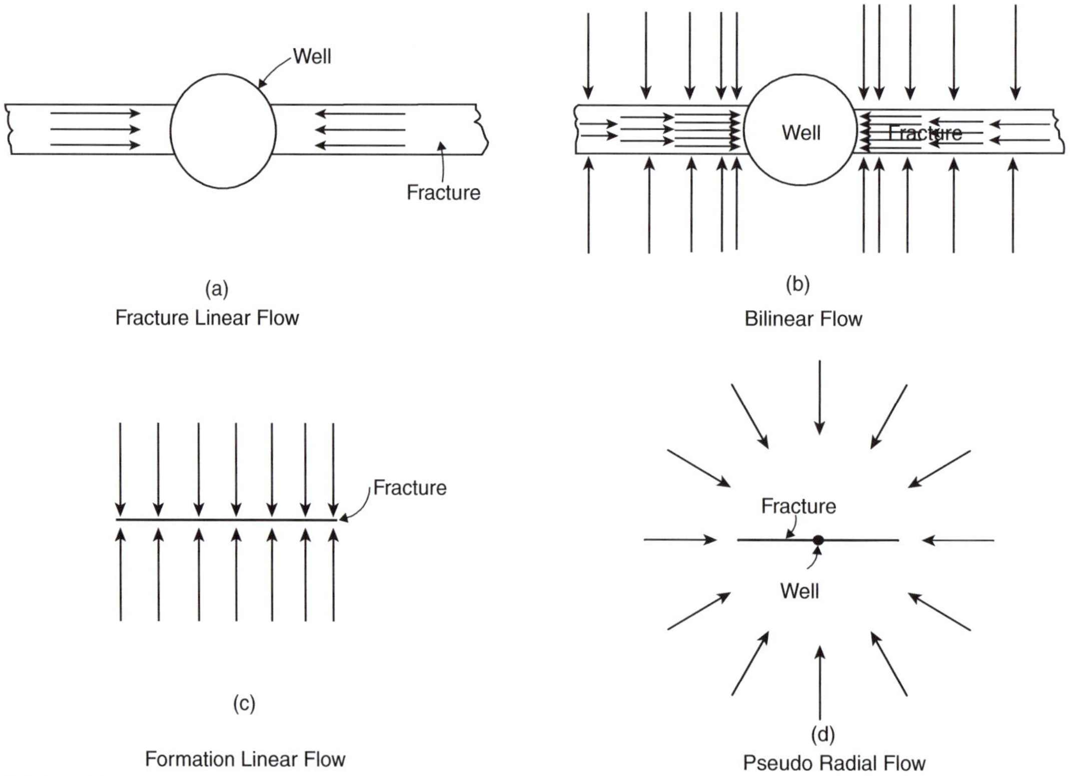

FIGURE 1.71 Flow periods for a vertically fractured well. *(After Cinco and Samaniego, JPT, 1981.)*

the fracture, i.e., there is negligible fluid coming from the formation. Flow within the fracture and from the fracture to the wellbore during this time period is linear and can be described by the diffusivity equation as expressed in a linear form and is applied to both the fracture linear flow and formation linear flow periods. The pressure transient test data during the linear flow period can be analyzed with a graph of Δp vs. $\sqrt{\text{time}}$. Unfortunately, the fracture linear flow occurs at very early time to be of practical use in well test analysis. However, if the fracture linear flow exists (for fractures with $F_{CD} > 300$), the formation linear flow relationships as given by Eqs. (1.237)–(1.242) can be used in an exact manner to analyze the pressure data during the formation linear flow period.

If fracture linear flow occurs, the duration of the flow period is short, as it often is in finite conductivity fractures with $F_{CD} < 300$, and care must be taken not to misinterpret the early pressure data. It is common in this situation for skin effects or wellbore storage effects to alter pressures to the extent that the linear flow straight line does not occur or is very difficult to recognize. If the early-time slope is used in determining the fracture length, the slope m_{vf} will be erroneously high, the computed fracture length will be unrealistically small, and no quantitative information will be obtained regarding flow capacity in the fracture.

Cinco and Samaniego (1981) observed that the fracture linear flow ends when:

$$t_{Dx_f} \approx \frac{0.01(F_{CD})^2}{(\eta_{fD})^2}$$

Bilinear Flow. This flow period is called bilinear flow because two types of linear flow occur simultaneously. As originally proposed by Cinco (1981), one flow is a linear incompressible flow within the fracture and the other is a linear compressible flow in the formation. Most of the fluid which enters the wellbore during this flow period comes from the formation. Fracture tip effects do not affect well behavior during bilinear flow and, accordingly, it will not be possible to determine the fracture length from the well bilinear flow period data. However, the actual value of the fracture conductivity F_C can be determined during this flow period. The pressure drop through the fracture is significant for the finite conductivity case and the bilinear flow behavior is observed; however, the *infinite conductivity case does not exhibit bilinear flow behavior* because the pressure drop in the fracture is negligible. Thus, identification of the bilinear flow period is very important for two reasons:

(1) It will not be possible to determine a unique fracture length from the well bilinear flow period data. If this data is used to determine the length of the fracture, it will produce a much smaller fracture length than the actual.
(2) The actual fracture conductivity $k_f w_f$ can be determined from the bilinear flow pressure data.

Cinco and Samaniego suggested that during this flow period, the change in the wellbore pressure can be described by the following expressions.

For fractured oil wells in terms of dimensionless pressure:

$$p_D = \left[\frac{2.451}{\sqrt{F_{CD}}}\right](t_{Dx_f})^{1/4} \tag{1.243}$$

Taking the logarithm of both sides of Eq. (1.243) gives:

$$\log(p_D) = \log\left[\frac{2.451}{\sqrt{F_{CD}}}\right] + \frac{1}{4}\log(t_{Dx_f}) \tag{1.244}$$

In terms of pressure:

$$\Delta p = \left[\frac{44.1\,QB\mu}{h\sqrt{F_C}(\phi\mu c_t k)^{1/4}}\right]t^{1/4} \tag{1.245}$$

or equivalently:

$$\Delta p = m_{bf} t^{1/4}$$

Taking the logarithm of both sides of the above expression gives:

$$\log(\Delta p) = \log(m_{bf}) + \frac{1}{4}\log(t) \tag{1.246}$$

with the bilinear slope m_{bf} as given by:

$$m_{bf} = \left[\frac{44.1QB\mu}{h\sqrt{F_C}(\phi\mu c_t k)^{1/4}}\right]$$

where F_C is the fracture conductivity as defined by:

$$F_C = k_f w_f \tag{1.247}$$

For fractured gas wells in a dimensionless form:

$$m_D = \left[\frac{2.451}{\sqrt{F_{CD}}}\right](t_{Dx_f})^{1/4}$$

or

$$\log(m_D) = \log\left[\frac{2.451}{\sqrt{F_{CD}}}\right] + \frac{1}{4}\log(t_{Dx_f}) \tag{1.248}$$

In terms of $m(p)$:

$$\Delta m(p) = \left[\frac{444.6QT}{h\sqrt{F_C}(\phi\mu c_t k)^{1/4}}\right]t^{1/4} \tag{1.249}$$

or equivalently:

$$\Delta m(p) = m_{bf}t^{1/4} \tag{1.250}$$

Taking the logarithm of both sides gives:

$$\log[\Delta m(p)] = \log(m_{bf}) + \frac{1}{4}\log(t)$$

Eqs. (1.245) and (1.249) indicate that a plot of Δp or $\Delta m(p)$ vs. $(\text{time})^{1/4}$ on a *Cartesian scale* would produce a straight line *passing through the origin* with a slope of "m_{bf} (bilinear flow slope)" as given by:

For oil:

$$m_{bf} = \frac{44.1QB\mu}{h\sqrt{F_C}(\phi\mu c_t k)^{1/4}} \tag{1.251}$$

The slope can then be used to solve for fracture conductivity F_C:

$$F_C = \left[\frac{44.1QB\mu}{m_{bf}h(\phi\mu c_t k)^{1/4}}\right]^2$$

For gas:

$$m_{bf} = \frac{444.6QT}{h\sqrt{F_C}(\phi\mu c_t k)^{1/4}} \tag{1.252}$$

with

$$F_C = \left[\frac{444.6QT}{m_{bf}h(\phi\mu c_t k)^{1/4}}\right]^2$$

It should be noted that *if the straight-line plot does not pass through the origin*, it indicates an additional pressure drop "Δp_s" caused by flow restriction within the fracture in the vicinity of the wellbore (chocked fracture, where the fracture permeability just away from the wellbore is reduced). Examples of restrictions that cause a loss of resulting production include:

- inadequate perforations;
- turbulent flow which can be reduced by increasing the proppant size or concentration;
- overdisplacement of proppant;
- kill fluid was dumped into the fracture.

Similarly, Eqs. (1.246) and (1.250) suggest that a plot of Δp or $\Delta m(p)$ vs. (time) on a log–log *scale* would produce a straight line with a slope of $m_{bf} = \frac{1}{4}$ and which can be used as a diagnostic tool for bilinear flow detection.

When the bilinear flow ends, the plot will exhibit curvature which could concave upwards or downwards depending upon the value of the dimensionless fracture conductivity F_{CD}, as shown in Figure 1.72. When the value of F_{CD} is less than 1.6, the curve will concave downward, and will concave upward if the value of F_{CD} is greater than 1.6. The upward trend indicates that the fracture tip begins to affect wellbore behavior. If the test is not run sufficiently long for bilinear flow to end when $F_{CD} > 1.6$, it is not possible to determine the length of the fracture. When the dimensionless fracture conductivity F_{CD} is less than 1.6, it indicates that the fluid flow *in the reservoir has* changed from a predominantly one-dimensional linear flow to a two-dimensional flow regime. In this particular

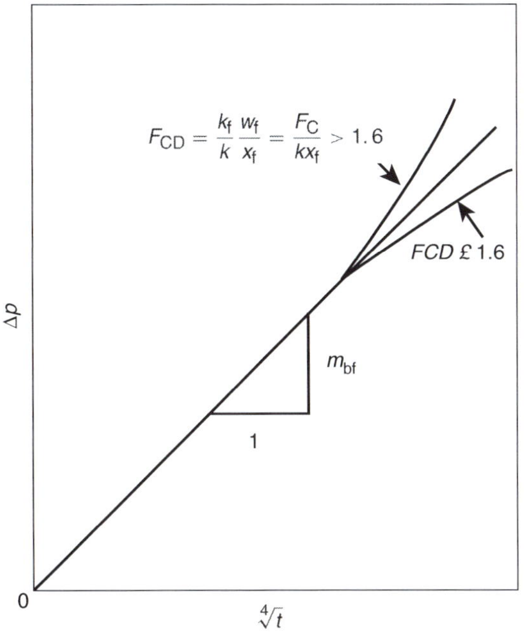

FIGURE 1.72 Graph for analysis of pressure data of bilinear flows. *(After Cinco-Ley, H., Samaniego, F., 1981. Transient pressure analysis for finite conductivity fracture case versus damage fracture case. SPE Paper 10179.)*

case, it is not possible to uniquely determine fracture length even if bilinear flow does end during the test.

Cinco and Samaniego pointed out that the dimensionless fracture conductivity F_{CD} can be estimated from the bilinear flow straight line, i.e., Δp vs. $(\text{time})^{1/4}$, by reading the value of the pressure difference Δp at which the line ends Δp_{ebf} and applying the following approximation:

For oil: $$F_{CD} = \frac{194.9QB\mu}{kh\,\Delta p_{ebf}} \quad (1.253)$$

For gas: $$F_{CD} = \frac{1965.1QT}{kh\,\Delta m(p)_{ebf}} \quad (1.254)$$

where

Q = flow rate, STB/day or Mscf/day
T = temperature, °R

The end of the bilinear flow, "ebf," straight line depends on the fracture conductivity and can be estimated from the following relationships:

For $F_{CD} > 3$: $$t_{Debf} \simeq \frac{0.1}{(F_{CD})^2}$$

For $1.6 \le F_{CD} \le 3$: $$t_{Debf} \simeq 0.0205[F_{CD} - 1.5]^{-1.53}$$

For $F_{CD} \le 1.6$: $$t_{Debf} \simeq \left[\frac{4.55}{\sqrt{F_{CD}}} - 2.5\right]^{-4}$$

The procedure for analyzing the bilinear flow data is summarized by the following steps:

Step 1. Make a plot of Δp vs. time on a log–log scale.
Step 2. Determine if any data fall on a straight line with a $\frac{1}{4}$ slope.
Step 3. If data points do fall on the straight line with a $\frac{1}{4}$ slope, replot the data in terms of Δp vs. $(\text{time})^{1/4}$ on a Cartesian scale and identify the data which forms the bilinear straight line.
Step 4. Determine the slope of the bilinear straight line m_{bf} formed in step 3.
Step 5. Calculate the fracture conductivity $F_C = k_f w_f$ from Eq. (1.251) or (1.252):

For oil: $$F_C = (k_f w_f) = \left[\frac{44.1QB\mu}{m_{bf}h(\phi\mu c_t k)^{1/4}}\right]^2$$

For gas: $$F_C = (k_f w_f) = \left[\frac{444.6QT}{m_{bf}h(\phi\mu c_t k)^{1/4}}\right]^2$$

Step 6. Read the value of the pressure difference at which the line ends, Δp_{ebf} or $\Delta m(p)_{ebf}$.
Step 7. Approximate the dimensionless facture conductivity from:

For oil: $$F_{CD} = \frac{194.9QB\mu}{kh\,\Delta p_{ebf}}$$

For gas: $$F_{CD} = \frac{1956.1QT}{kh\,\Delta m(p)_{ebf}}$$

Step 8. Estimate the fracture length from the mathematical definition of F_{CD} as expressed by Eq. (1.239) and the value of F_C of step 5:

$$x_f = \frac{F_C}{F_{CD}k}$$

Example 1.36

A buildup test was conducted on a fractured well producing from a tight gas reservoir. The following reservoir and well parameters are available:

$Q = 7350$ Mscf/day, $t_p = 2640$ hours
$h = 118$ ft, $\phi = 0.10$
$k = 0.025$ md, $\mu = 0.0252$
$T = 690\,°R$, $c_t = 0.129 \times 10^{-3}$ psi^{-1}
$p_{wf \text{ at } \Delta t=0} = 1320$, $r_w = 0.28$ ft

The graphical presentation of the buildup data is given in terms of the log–log plot of $\Delta m(p)$ vs. $(\Delta t)^{1/4}$, as shown in Figure 1.73.

Calculate the fracture and reservoir parameters by performing conventional well testing analysis.

Solution

Step 1. From the plot of $\Delta m(p)$ vs. $(\Delta t)^{1/4}$, in Figure 1.73, determine:

$m_{bf} = 1.6 \times 10^8$ psi^2/cp hour$^{1/4}$
$t_{sbf} \approx 0.35$ hours(start of bilinear flow)
$t_{ebf} \approx 2.5$ hours(end of bilinear flow)
$\Delta m(p)_{ebf} \approx 2.05 \times 10^8$ psi^2/cp

Step 2. Perform the bilinear flow analysis, as follows:

- Using Eq. (1.252), calculate fracture conductivity F_C:

$$F_C = \left[\frac{444.6QT}{m_{bf}h(\phi\mu c_t k)^{1/4}}\right]^2$$

$$= \left[\frac{444.6(7350)(690)}{(1.62\times 10^8)(118)[(0.1)(0.0252)(0.129\times 10^{-3})(0.025)]^{1/4}}\right]^2$$

$$= 154 \text{ md ft}$$

- Calculate the dimensionless conductivity F_{CD} by using Eq. (1.254):

$$F_{CD} = \frac{1965.1QT}{kh\,\Delta m(p)_{ebf}}$$

$$= \frac{1965.1(7350)(690)}{(0.025)(118)(2.02\times 10^8)} = 16.7$$

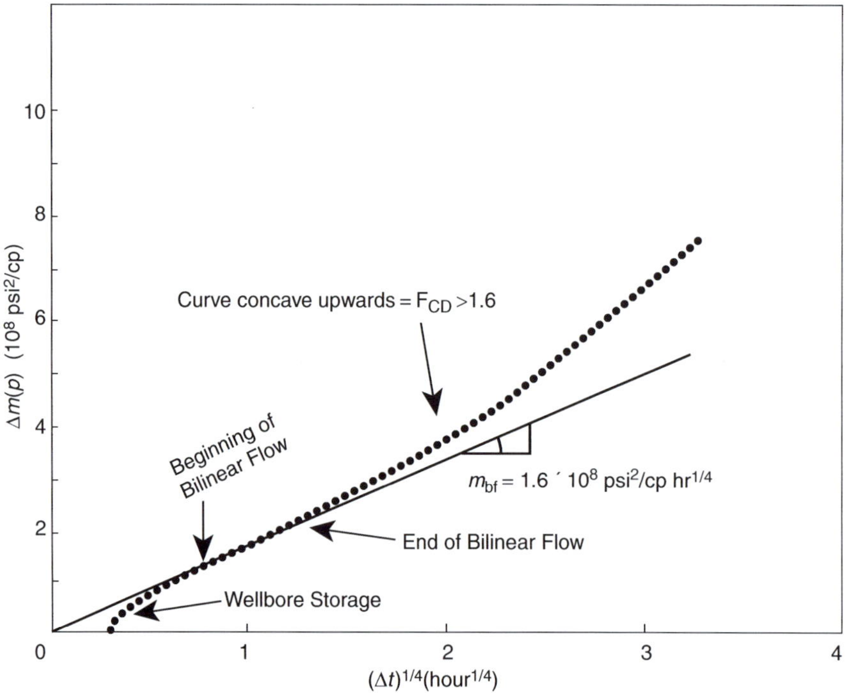

FIGURE 1.73 Bilinear flow graph for data of Example 1.36. *(After Sabet, M., 1991. Well Test Analysis. Gulf Publishing, Dallas, TX.)*

- Estimate the fracture half-length from Eq. (1.239):

$$x_f = \frac{F_C}{F_{CD}k} = \frac{154}{(16.7)(0.025)} = 368 \text{ ft}$$

Formation Linear Flow. At the end of the bilinear flow, there is a transition period after which the fracture tips begin to affect the pressure behavior at the wellbore and a linear flow period might develop. This linear flow period is exhibited by vertical fractures whose dimensionless conductivity is greater that 300, i.e., $F_{CD} > 300$. As in the case of fracture linear flow, the formation linear flow pressure data collected during this period is a function of the fracture length x_f and fracture conductivity F_C. The pressure behavior during this linear flow period can be described by the diffusivity equation as expressed in linear form:

$$\frac{\partial^2 p}{\partial x^2} = \frac{\phi \mu c_t}{0.002637k} \frac{\partial p}{\partial t}$$

The solution to the above linear diffusivity equation can be applied to both fracture linear flow and the formation linear flow, with the solution given in a dimensionless form by:

$$p_D = (\pi t_{Dx_f})^{1/2}$$

or in terms of real pressure and time, as:

For oil fractured wells: $\Delta p = \left[\frac{4.064QB}{hx_f}\sqrt{\frac{\mu}{k\phi c_t}}\right] t^{1/2}$

or in simplified form as: $\Delta p = m_{vf}\sqrt{t}$

For gas fractured wells: $\Delta m(p) = \left[\frac{40.925QT}{hx_f}\sqrt{\frac{1}{k\phi\mu c_t}}\right] t^{1/2}$

or equivalently as: $\Delta m(p) = m_{vf}\sqrt{t}$

The linear flow period may be recognized by pressure data that exhibits a straight line of a $\frac{1}{2}$ slope on a log–log plot of Δp vs. time, as illustrated in Figure 1.74. Another diagnostic presentation of pressure data points is the plot of

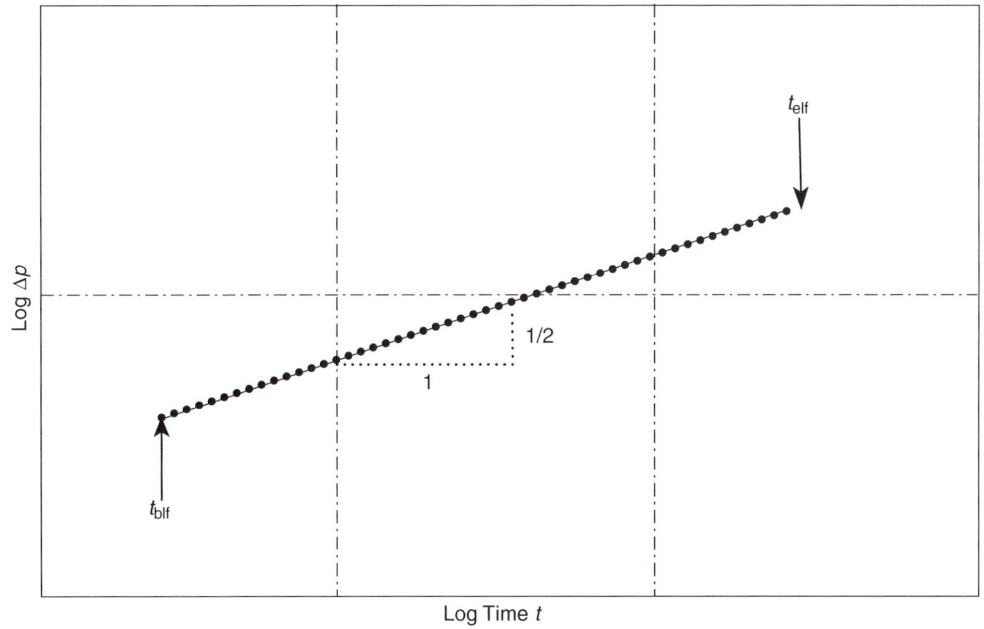

FIGURE 1.74 Pressure data for a $\frac{1}{2}$-slope straight line in a log–log graph. *(After Cinco-Ley, H., Samaniego, F., 1981. Transient pressure analysis for finite conductivity fracture case versus damage fracture case. SPE Paper 10179.)*

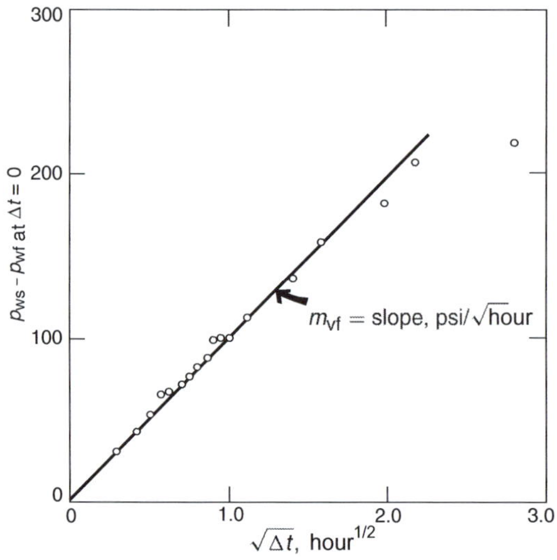

FIGURE 1.75 Square-root data plot for buildup test.

$\Delta(p)$ or $\Delta m(p)$ vs. $\sqrt{\text{time}}$ on a Cartesian scale (as shown in Figure 1.75) that would *produce a straight line* with a slope of m_{vf} related to the fracture length by the following equations:

Oil fractured well: $$x_f = \left[\frac{4.064 QB}{m_{vf} h}\right]\sqrt{\frac{\mu}{k\phi c_t}} \quad \textbf{(1.255)}$$

Gas fractured well: $$x_f = \left[\frac{40.925 QT}{m_{vf} h}\right]\sqrt{\frac{1}{k\phi\mu c_t}} \quad \textbf{(1.256)}$$

where

Q = flow rate, STB/day or Mscf/day
T = temperature, °R
m_{vf} = slope, psi/$\sqrt{\text{hour}}$ or psi^2/cp$\sqrt{\text{hour}}$
k = permeability, md
c_t = total compressibility, psi^{-1}

The straight-line relationships as illustrated by Figures 1.74 and 1.75 provide distinctive and easily recognizable evidence of a fracture. When properly applied, these plots are the best diagnostic tools available for the purpose of detecting a fracture. In practice, the $\frac{1}{2}$ slope is rarely seen except in fractures with high conductivity. Finite conductivity fracture responses generally enter a transition period after the bilinear flow (the $\frac{1}{4}$ slope) and reach the infinite-acting pseudoradial flow regime before ever achieving a $\frac{1}{2}$ slope (linear flow). For a long duration of wellbore storage effect, the bilinear flow pressure behavior may be masked and data analysis becomes difficult with current interpretation methods.

Agarwal et al. (1979) pointed out that the pressure data during the transition period displays a curved portion before straightening to a line of proper slope that represents the fracture linear flow. The duration of the curved portion that represents the transition flow depends on the fracture flow capacity. The lower the fracture flow capacity, the longer the duration of the curved portion. The beginning of formation linear flow, "blf," depends on F_{CD} and can be approximated from the following relationship:

$$t_{Dblf} \approx \frac{100}{(F_{CD})^2}$$

and the end of this linear flow period, "elf," occurs at approximately:

$$t_{Dblf} \approx 0.016$$

Identifying the coordinates of these two points (i.e., beginning and end of the straight line) in terms of time can be used to estimate F_{CD} from:

$$F_{CD} \approx 0.0125\sqrt{\frac{t_{elf}}{t_{blf}}}$$

where t_{elf} and t_{blf} are given in hours.

Infinite-acting Pseudoradial Flow. During this period, the flow behavior is similar to the radial reservoir flow with a negative skin effect caused by the fracture. The traditional semilog and log–log plots of the transient pressure data can be used during this period; for example, the drawdown pressure data can be analyzed by using Eqs. (1.169)–(1.171). That is:

$$p_{wf} = p_i - \frac{162.6 Q_o B_o \mu}{kh} \times \left[\log(t) + \log\left(\frac{k}{\phi\mu c_t r_w^2}\right) - 3.23 + 0.87s\right]$$

or in a linear form as:

$$p_i - p_{wf} = \Delta p = a + m \log(t)$$

with the slope m of:

$$m = \frac{162.6 Q_o B_o \mu_o}{kh}$$

Solving for the formation capacity gives:

$$kh = \frac{162.6 Q_o B_o \mu_o}{|m|}$$

The skin factor s can be calculated by Eq. (1.171):

$$s = 1.151 \left[\frac{p_i - p_{1\,\text{hour}}}{|m|} - \log\left(\frac{k}{\phi \mu c_t r_w^2}\right) + 3.23 \right]$$

If the semilog plot is made in terms of Δp vs. t, notice that the slope m is the same when making the semilog plot in terms of p_{wf} vs. t. Then:

$$s = 1.151 \left[\frac{\Delta p_{1\,\text{hour}}}{|m|} - \log\left(\frac{k}{\phi \mu c_t r_w^2}\right) + 3.23 \right]$$

$\Delta p_{1\,\text{hour}}$ can then be calculated from the mathematical definition of the slope m, i.e., rise/run, by using two points on the semilog straight line (conveniently, one point could be Δp at log (10)) to give:

$$m = \frac{\Delta p_{\text{at}\,\log(10)} - \Delta p_{1\,\text{hour}}}{\log(10) - \log(1)}$$

Solving this expression for $\Delta p_{1\,\text{hour}}$ gives:

$$\Delta p_{1\,\text{hour}} = \Delta p_{\text{at}\,\log(10)} - m \quad \textbf{(1.257)}$$

Again, $\Delta p_{\text{at}\,\log(10)}$ must be read at the corresponding point on the straight line at log(10).

Wattenbarger and Ramey (1968) have shown that an approximate relationship exists between the pressure change Δp at the end of the linear flow, i.e., Δp_{elf}, and the beginning of the infinite-acting pseudoradial flow, Δp_{bsf}, as given by:

$$\Delta p_{bsf} \geq 2\Delta p_{elf} \quad \textbf{(1.258)}$$

The above rule is commonly referred to as the "double-Δp rule" and can be obtained from the log–log plot when the $\frac{1}{2}$ slope ends and by reading the value of Δp, i.e., Δp_{elf}, at this point. For fractured wells, doubling the value of Δp_{elf} will mark the beginning of the infinite-acting pseudoradial flow period. Equivalently, a time rule as referred to as the "10Δt rule" can be applied to mark the beginning of pseudoradial flow by:

$$\text{For drawdown:} \quad t_{bsf} \geq 10 t_{elf} \quad \textbf{(1.259)}$$

$$\text{For buildup:} \quad \Delta t_{bsf} \geq 10 \Delta t_{elf} \quad \textbf{(1.260)}$$

which indicates that correct infinite-acting pseudoradial flow occurs one log cycle beyond the end of the linear flow. The concept of the above two rules is illustrated graphically in Figure 1.76.

Another approximation that can be used to mark the start of the infinite-acting radial flow period for a finite conductivity fracture is given by:

$$t_{Dbs} \approx 5 \exp[-0.5 (F_{CD})^{-0.6}] \quad \text{for} \quad F_{CD} > 0.1$$

Sabet (1991) used the following drawdown test data, as originally given by Gringarten et al. (1975), to illustrate the process of analyzing a hydraulically fractured well test data.

Example 1.37

The drawdown test data for an infinite conductivity fractured well is tabulated below:

t (hour)	p_{wf} (psi)	Δp (psi)	$\sqrt{t}$ (hour$^{1/2}$)
0.0833	3759.0	11.0	0.289
0.1670	3755.0	15.0	0.409
0.2500	3752.0	18.0	0.500
0.5000	3744.5	25.5	0.707
0.7500	3741.0	29.0	0.866
1.0000	3738.0	32.0	1.000
2.0000	3727.0	43.0	1.414
3.0000	3719.0	51.0	1.732
4.0000	3713.0	57.0	2.000
5.0000	3708.0	62.0	2.236
6.0000	3704.0	66.0	2.449
7.0000	3700.0	70.0	2.646
8.0000	3695.0	75.0	2.828
9.0000	3692.0	78.0	3.000
10.0000	3690.0	80.0	3.162
12.0000	3684.0	86.0	3.464
24.0000	3662.0	108.0	4.899
48.0000	3635.0	135.0	6.928
96.0000	3608.0	162.0	9.798
240.0000	3570.0	200.0	14.142

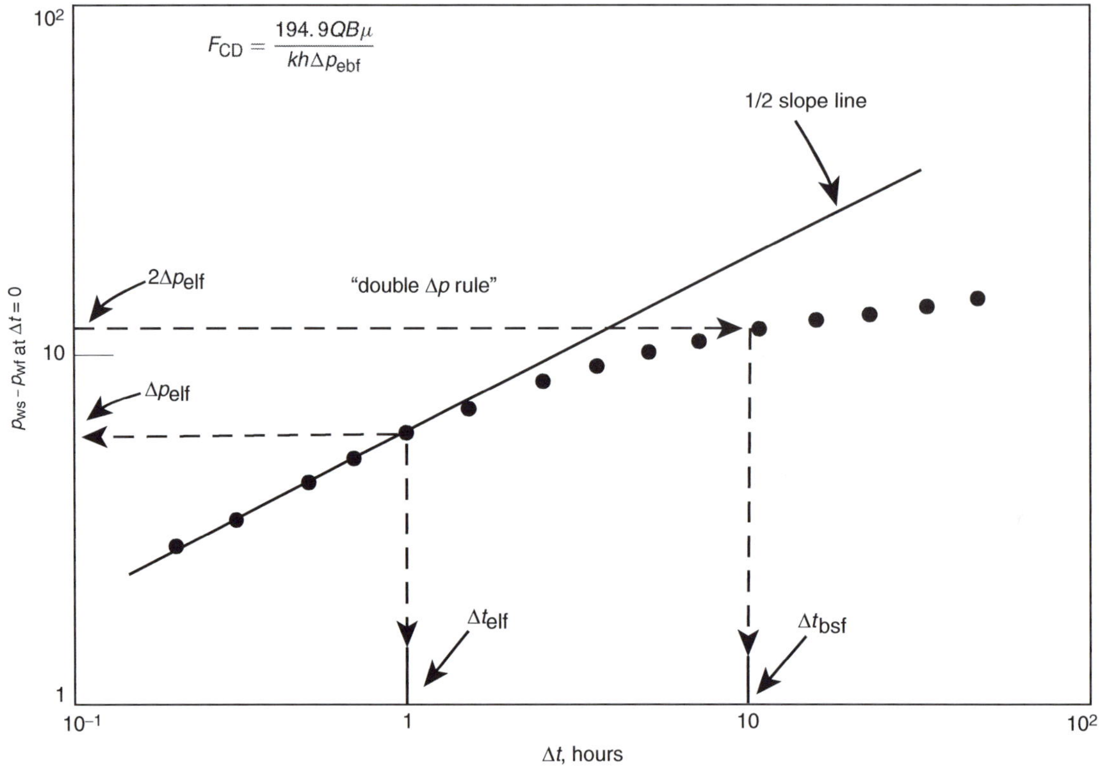

FIGURE 1.76 Use of the log–log plot to approximate the beginning of pseudoradial flow.

Additional reservoir parameters are:

$h = 82$ ft,	$\phi = 0.12$
$c_t = 21 \times 10^{-6}$ psi^{-1},	$\mu = 0.65$ cp
$B_o = 1.26$ bbl/STB,	$r_w = 0.28$ ft
$Q = 419$ STB/day,	$p_i = 3770$ psi

Estimate

- permeability, k;
- fracture half-length, x_f;
- skin factor, s.

Solution

Step 1. Plot:
- Δp vs. t on a log–log scale, as shown in Figure 1.77;
- Δp vs. $\sqrt{t}$ on a Cartesian scale, as shown in Figure 1.78;
- Δp vs. t on a semilog scale, as shown in Figure 1.79.

Step 2. Draw a straight line through the early points representing log(Δp) vs. log(t), as shown in Figure 1.77, and determine the slope of the line. Figure 1.77 shows a slope of $\frac{1}{2}$ (not 45° angle) indicating linear flow with no wellbore storage effects. This linear flow lasted for approximately 0.6 hours. That is:

$$t_{elf} = 0.6 \text{ hours} \Delta p_{elf} = 30 \text{ psi}$$

and therefore the beginning of the infinite-acting pseudoradial flow can be approximated by the "double Δp rule" or "one log cycle rule," i.e., Eqs. (1.258) and (1.259), to give:

$$t_{bsf} \geq 10 t_{elf} \geq 6 \text{ hours}$$
$$\Delta p_{bsf} \geq 2\Delta p_{elf} \geq 60 \text{ psi}$$

Step 3. From the Cartesian scale plot of Δp vs. $\sqrt{t}$, draw a straight line through the

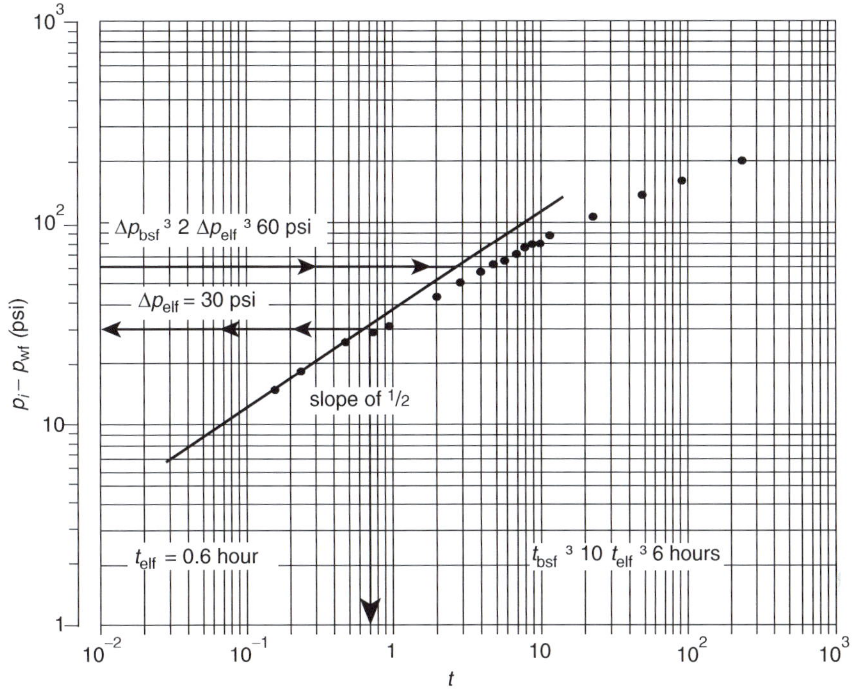

FIGURE 1.77 Log–log plot, drawdown test data of Example 1.37. *(After Sabet, M., 1991. Well Test Analysis. Gulf Publishing, Dallas, TX).*

early pressure data points representing the first 0.3 hours of the test (as shown in Figure 1.79) and determine the slope of the line, to give:

$$m_{vf} = 36 \text{ psi/hour}^{1/2}$$

Step 4. Determine the slope of the semilog straight line representing the unsteady-state radial flow in Figure 1.79, to give:

$$m = 94.1 \text{ psi/cycle}$$

Step 5. Calculate the permeability k from the slope:

$$k = \frac{162.6 Q_o B_o \mu_o}{mh} = \frac{162.6(419)(1.26)(0.65)}{(94.1)(82)}$$
$$= 7.23 \text{ md}$$

Step 6. Estimate the length of the fracture half-length from Eq. (1.255), to give:

$$x_f = \left[\frac{4.064 QB}{m_{vf} h}\right] \sqrt{\frac{\mu}{k \phi c_t}}$$
$$= \left[\frac{4.064(419)(1.26)}{(36)(82)}\right] \sqrt{\frac{0.65}{(7.23)(0.12)(21 \times 10^{-6})}}$$
$$= 137.3 \text{ ft}$$

Step 7. From the semilog straight line of Figure 1.78, determine Δp at $t = 10$ hours, to give:

$$\Delta p_{\text{at } \Delta t = 10} = 71.7 \text{ psi}$$

Step 8. Calculate $\Delta p_{1 \text{ hour}}$ by applying Eq. (1.257):

$$\Delta p_{1 \text{ hour}} = \Delta p_{\text{at } \Delta t = 10} - m$$
$$= 71.7 - 94.1 = -22.4 \text{ psi}$$

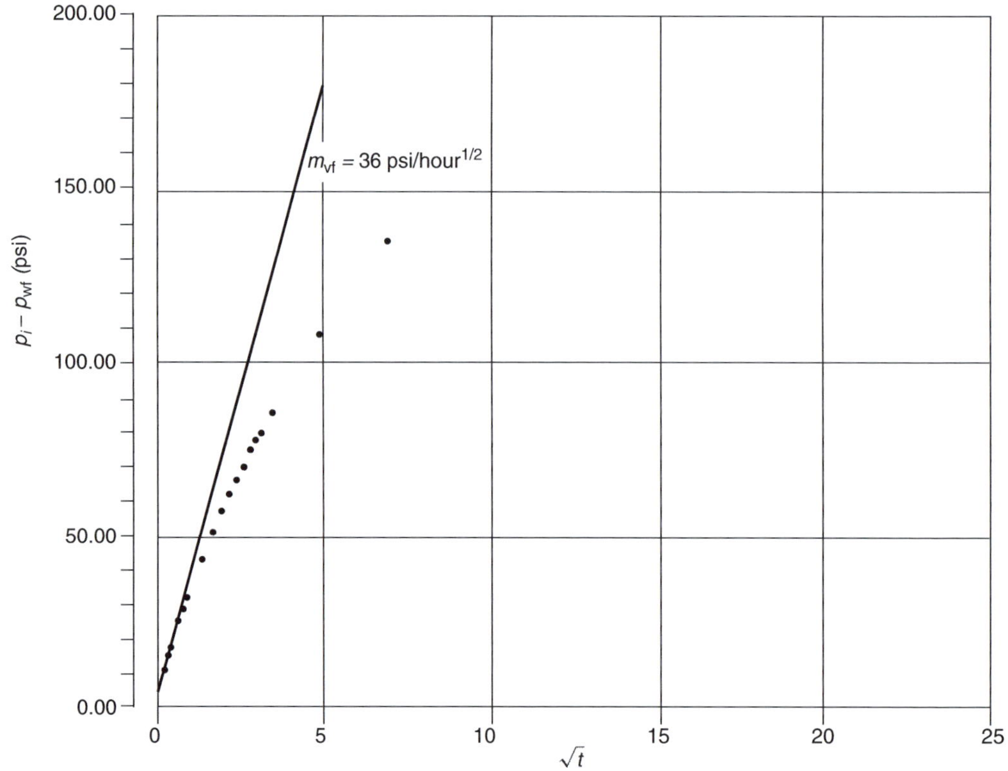

FIGURE 1.78 Linear plot, drawdown test data of Example 1.37. *(After Sabet, M., 1991. Well Test Analysis. Gulf Publishing, Dallas, TX.)*

Step 9. Solve for the "total" skin factor s, to give

$$s = 1.151\left[\frac{\Delta p_{1\ \text{hour}}}{|m|} - \log\left(\frac{k}{\phi \mu c_t r_w^2}\right) + 3.23\right]$$

$$= 1.151\left[\frac{-22.4}{94.1} - \log\left(\frac{7.23}{0.12(0.65)(21 \times 10^{-6})(0.28)^2} + 3.23\right)\right.$$

$$= -5.5$$

with an apparent wellbore ratio of:

$$r'_w = r_w\, e^{-s} = 0.28\, e^{5.5} = 68.5\ \text{ft}$$

Notice that the "total" skin factor is a composite of effects that include:

$$s = s_d + s_f + s_t + s_p + s_{sw} + s_r$$

where

s_d = skin due to formation and fracture damage
s_f = skin due to the fracture, large negative value $s_f \ll 0$
s_t = skin due to turbulence flow
s_p = skin due to perforations
s_{sw} = skin due to slanted well
s_r = skin due to restricted flow

For fractured oil well systems, several of the skin components are negligible or cannot be applied, mainly s_t, s_p, s_{sw}, and s_r; therefore:

$$s = s_d + s_f$$

or

$$s_d = s - s_f$$

Smith and Cobb (1979) suggested that the best approach to evaluate damage in a fractured

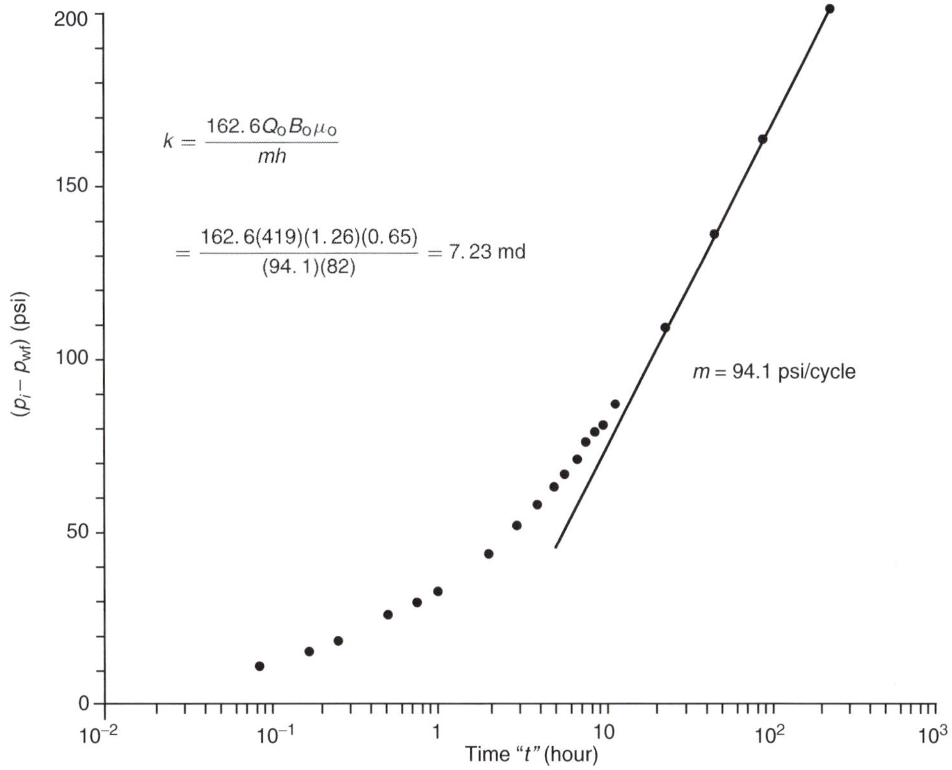

FIGURE 1.79 Semilog plot, drawdown test data from Example 1.37.

well is to use the square root plot. In an ideal well without damage, the square root straight line will extrapolate to p_{wf} at $\Delta t = 0$, i.e., p_{wf} at $\Delta t = 0$; however, when a well is damaged the intercept pressure p_{int} will be greater than $p_{wf \text{ at } \Delta t = 0}$, as illustrated in Figure 1.80. Note that the well shut-in pressure is described by Eq. (1.253) as:

$$p_{ws} = p_{wf \text{ at } \Delta t = 0} + m_{vf}\sqrt{t}$$

Smith and Cobb pointed out that the total skin factor exclusive of s_f, i.e., $s - s_f$, can be determined from the square root plot by extrapolating the straight line to $\Delta t = 0$ and an intercept pressure p_{int} to give the pressure loss due to skin damage, $(\Delta p_s)_d$, as:

$$(\Delta p_s)_d = p_{int} - p_{wf \text{ at } \Delta t = 0} = \left[\frac{141.2 QB\mu}{kh}\right] s_d$$

Eq. (1.253) indicates that if $p_{int} = p_{wf \text{ at } \Delta t = 0}$, then the skin due to fracture s_f is equal to the total skin.

It should be pointed out that the external boundary can distort the semilog straight line if the fracture half-length is greater than one-third of the drainage radius. The pressure behavior during this infinite-acting period is dependent on the fracture length. For relatively short fractures, the flow is radial but becomes linear as the fracture length increases as it reaches the drainage radius. As noted by Russell and Truitt (1964), the slope obtained from the traditional well test analysis of a fractured well is erroneously too small and the calculated value of the slope progressively decreases with increasing fracture length. This dependency of the pressure response behavior on the fracture length is illustrated by the theoretical Horner buildup curves given by Russell and Truitt and shown in Figure 1.81. If the fracture penetration ratio x_f/x_e is defined as the ratio of the fracture half-length x_f to the half-length x_e of a closed square-drainage area, then Figure 1.81

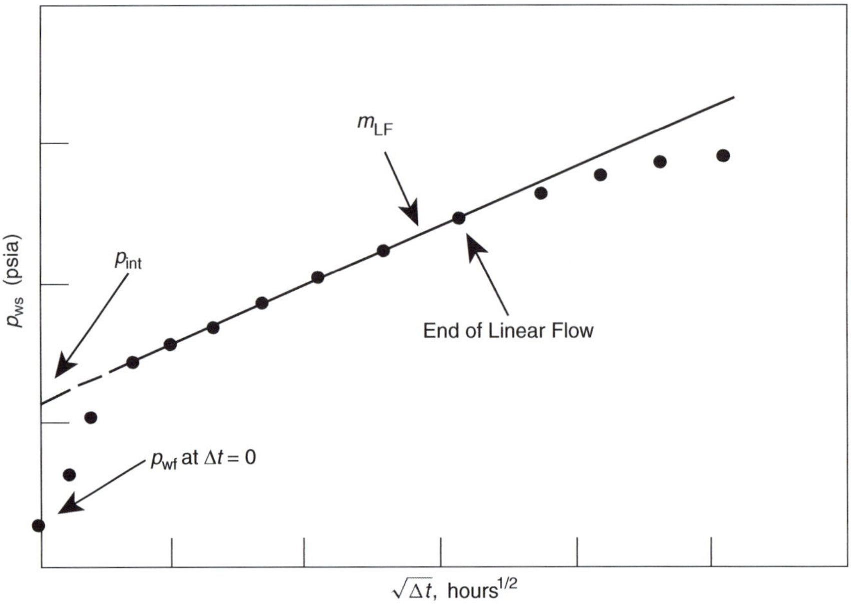

FIGURE 1.80 Effect of skin on the square root plot.

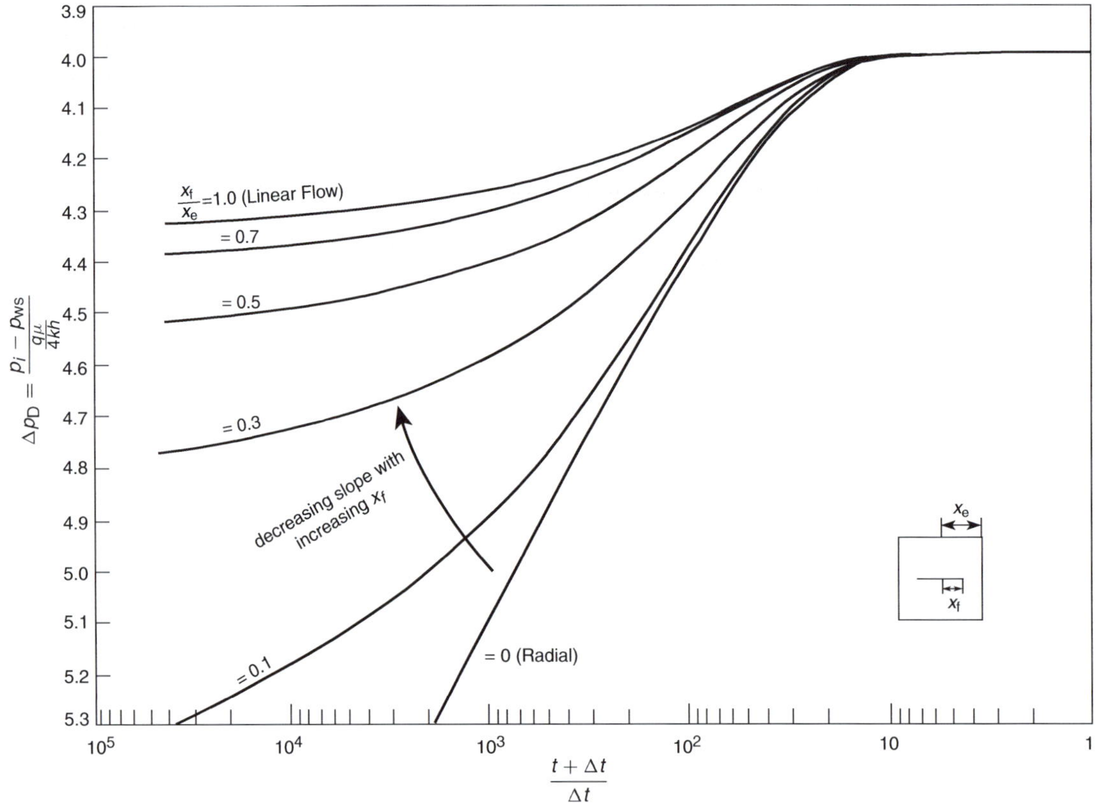

FIGURE 1.81 Vertically fractured reservoir, calculated pressure buildup curves. *(After Russell, D., Truitt, N., 1964. Transient pressure behaviour in vertically fractured reservoirs. J. Pet. Technol. 16 (10), 1159–1170.)*

shows the effects of fracture penetration on the slope of the buildup curve. For fractures of small penetration, the slope of the buildup curve is only slightly less than that for the unfractured "radial flow" case. However, the slope of the buildup curve becomes progressively smaller with increasing fracture penetrations. This will result in a calculated flow capacity *kh* which is too large, an erroneous average pressure, and a skin factor which is too small. Obviously a modified method for analyzing and interpreting the data must be employed to account for the effect of length of the fracture on the pressure response during the infinite-acting flow period. Most of the published correction techniques require the use of iterative procedures. The type curve matching approach and other specialized plotting techniques have been accepted by the oil industry as accurate and convenient approaches for analyzing pressure data from fractured wells, as briefly discussed below.

An alternative and convenient approach to analyzing fractured well transient test data is type curve matching. The type curve matching approach is based on plotting the pressure difference Δp vs. time on the same scale as the selected type curve and matching one of the type curves. Gringarten et al. (1974) presented the type curves shown in Figures 1.82 and 1.83 for infinite conductivity vertical fracture and uniform flux vertical fracture, respectively, in a square well drainage area. Both figures present log–log plots of the dimensionless pressure drop p_d (equivalently referred to as dimensionless wellbore pressure p_{wd}) vs. dimensionless time t_{Dx_f}. The fracture solutions show an initial period controlled by linear flow where the pressure is a function of the square root of

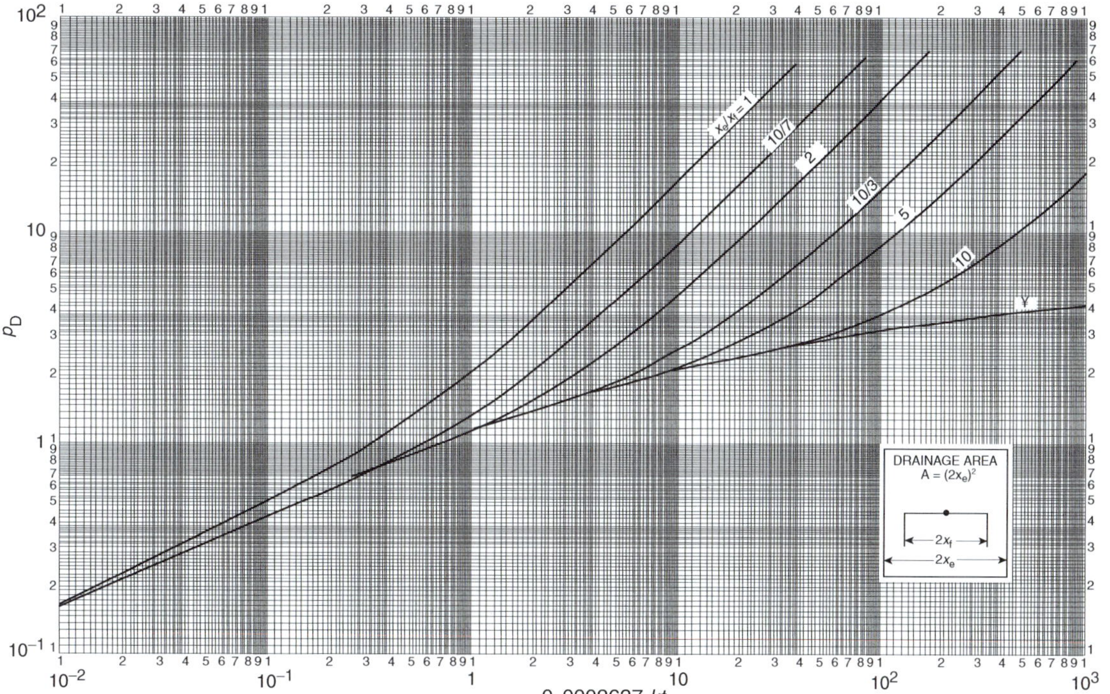

$$t_{Dxf} = \frac{0.0002637\,kt}{\phi \mu c_t x_f^2}$$

FIGURE 1.82 Dimensionless pressure for vertically fractured well in the center of a closed square, no wellbore storage, infinite conductivity fracture. *(After Gringarten, A.C., Ramey, H.J., Jr., Raghavan, R., 1974. Unsteady-state pressure distributions created by a well with a single infinite-conductivity vertical fracture. SPE J. 14 (4), 347–360.)*

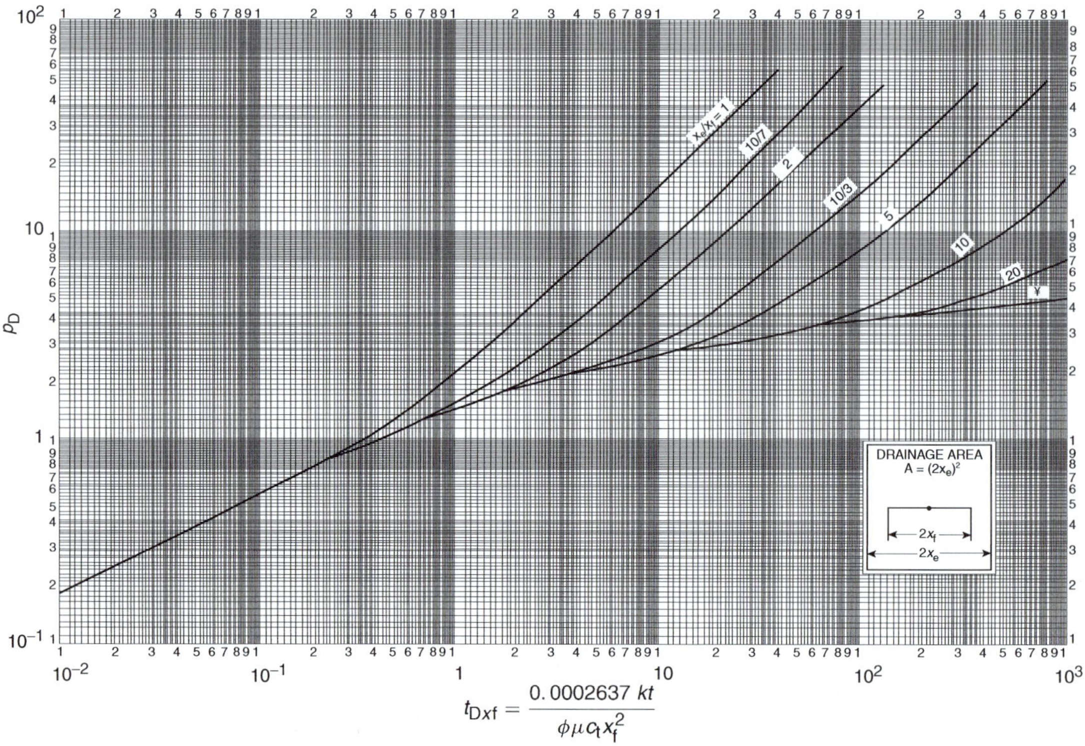

FIGURE 1.83 Dimensionless pressure for vertically fractured well in the center of a closed square, no wellbore storage, uniform-flux fracture. *(After Gringarten, A.C., Ramey, H.J., Jr., Raghavan, R., 1974. Unsteady-state pressure distributions created by a well with a single infinite-conductivity vertical fracture. SPE J. 14 (4), 347–360.)*

time. In log–log coordinates, as indicated before, this flow period is characterized by a straight line with $\frac{1}{2}$ slope. The infinite-acting pseudoradial flow occurs at a t_{Dx_f} between 1 and 3. Finally, all solutions reach pseudosteady state.

During the matching process a match point is chosen; the dimensionless parameters on the axis of the type curve are used to estimate the formation permeability and fracture length from:

$$k = \frac{141.2QB\mu}{h}\left[\frac{p_D}{\Delta p}\right]_{MP} \tag{1.261}$$

$$x_f = \sqrt{\frac{0.0002637k}{\phi \mu C_t}\left(\frac{\Delta t}{t_{Dx_f}}\right)_{MP}} \tag{1.262}$$

For large ratios of x_e/x_f, Gringarten et al. suggested that the apparent wellbore radius r'_w can be approximated from:

$$r'_w \approx \frac{x_f}{2} = r_w\, e^{-s}$$

Thus, the skin factor can be approximated from:

$$s = \ln\left(\frac{2r_w}{x_f}\right) \tag{1.263}$$

Earlougher (1977) points out that if all the test data falls on the $\frac{1}{2}$-slope line on the log(Δp) vs. log (time) plot, i.e., the test is not long enough to reach the infinite-acting pseudoradial flow period, then the *formation permeability k* cannot be estimated by either type curve matching or semilog plot. This situation often occurs in tight gas wells. However, the last point on the $\frac{1}{2}$ slope line, i.e., $(\Delta p)_{last}$ and $(t)_{last}$, may be used to estimate an upper limit of the permeability and a minimum fracture length from:

$$k \le \frac{30.358QB\mu}{h(\Delta p)_{last}} \quad (1.264)$$

$$x_f \ge \sqrt{\frac{0.01648k(t)_{last}}{\phi\mu c_t}} \quad (1.265)$$

The above two approximations are only valid for $x_e/x_f \gg 1$ and for infinite conductivity fractures. For uniform-flux fracture, the constants 30.358 and 0.01648 become 107.312 and 0.001648.

To illustrate the use of the Gringarten type curves in analyzing well test data, the authors presented Example 1.38.

Example 1.38

Tabulated below is the pressure buildup data for an infinite conductivity fractured well:

Δt (hour)	p_{ws} (psi)	$p_{ws} - p_{wf\ at\ \Delta t=0}$ (psi)	$\frac{(t_p+\Delta t)}{\Delta t}$
0.000	3420.0	0.0	0.0
0.083	3431.0	11.0	93,600.0
0.167	3435.0	15.0	46,700.0
0.250	3438.0	18.0	31,200.0
0.500	3444.5	24.5	15,600.0
0.750	3449.0	29.0	10,400.0
1.000	3542.0	32.0	7800.0
2.000	3463.0	43.0	3900.0
3.000	3471.0	51.0	2600.0
4.000	3477.0	57.0	1950.0
5.000	3482.0	62.0	1560.0
6.000	3486.0	66.0	1300.0
7.000	3490.0	70.0	1120.0
8.000	3495.0	75.0	976.0
9.000	3498.0	78.0	868.0
10.000	3500.0	80.0	781.0
12.000	3506.0	86.0	651.0
24.000	3528.0	108.0	326.0
36.000	3544.0	124.0	218.0
48.000	3555.0	135.0	164.0
60.000	3563.0	143.0	131.0
72.000	3570.0	150.0	109.0
96.000	3582.0	162.0	82.3
120.000	3590.0	170.0	66.0
144.000	3600.0	180.0	55.2
192.000	3610.0	190.0	41.6
240.000	3620.0	200.0	33.5

Other available data:

$p_i = 3700$,	$r_w = 0.28$ ft
$\phi = 12\%$,	$h = 82$ ft
$c_t = 21 \times 10^{-6}\ psi^{-1}$,	$\mu = 0.65$ cp
$B = 1.26$ bbl/STB,	$Q = 419$ STB/day
$t_p = 7800$ hours,	drainage area = 1600 acres (not fully developed)

Calculate:

- permeability;
- fracture half-length, x_f;
- skin factor.

Solution

Step 1. Plot Δp vs. Δt on tracing paper with the same scale as the Gringarten type curve of Figure 1.82. Superimpose the tracing paper on the type curve, as shown in Figure 1.84, with the following match points:

$$(\Delta p)_{MP} = 100 \text{ psi}$$
$$(\Delta t)_{MP} = 10 \text{ hours}$$
$$(p_D)_{MP} = 1.22$$
$$(t_D)_{MP} = 0.68$$

Step 2. Calculate k and x_f by using Eqs. (1.261) and (1.262):

$$k = \frac{141.2QB\mu}{h}\left[\frac{p_D}{\Delta p}\right]_{MP}$$
$$= \frac{(141.2)(419)(1.26)(0.65)}{(82)}\left[\frac{1.22}{100}\right] = 7.21 \text{ md}$$

$$x_f = \sqrt{\frac{0.0002637k}{\phi\mu C_t}\left(\frac{\Delta t}{t_{Dx_f}}\right)_{MP}}$$
$$= \sqrt{\frac{0.0002637(7.21)}{(0.12)(0.65)(21\times 10^{-6})}\left(\frac{10}{0.68}\right)} = 131 \text{ ft}$$

Step 3. Calculate the skin factor by applying Eq. (1.263):

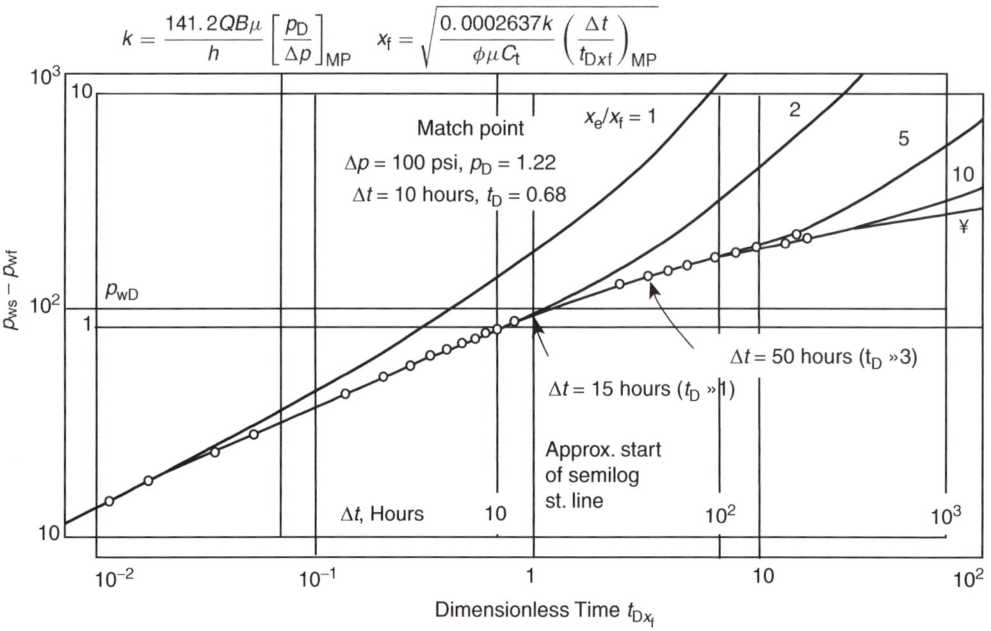

FIGURE 1.84 Type curve matching. Data from Example 1.38. *(After Gringarten, A.C., Ramey, H.J., Jr., Raghavan, R., 1974. Unsteady-state pressure distributions created by a well with a single infinite-conductivity vertical fracture. SPE J. 14 (4), 347–360; Copyright ©1974 SPE.)*

$$s = \ln\left(\frac{2r_w}{x_f}\right)$$

$$\approx \ln\left[\frac{(2)(0.28)}{131}\right] = 5.46$$

Step 4. Approximate the time that marks the start of the semilog straight line based on the Gringarten et al. criterion. That is:

$$t_{Dx_f} = \left[\frac{0.0002637k}{\phi\mu c_t x_f^2}\right] t \geq 3$$

or

$$t \geq \frac{(3)(0.12)(0.68)(21\times10^{-6})(131)^2}{(0.0002637)(7.21)} \geq 50 \text{ hours}$$

All the data beyond 50 hours can be used in the conventional Horner plot approach to estimate permeability and skin factor. Figure 1.85 shows a Horner graph with the following results:

$m = 95$ psi/cycle
$p^* = 3764$ psi
$p_{1\ hour} = 3395$ psi
$k = 7.16$ md
$s = -5.5$
$x_f = 137$ ft

Cinco and Samaniego (1981) developed the type curves shown in Figure 1.86 for finite conductivity vertical fracture. The proposed type curve is based on the bilinear flow theory and presented in terms of $(p_D F_{CD})$ vs. $(t_{Dx_f} F_{CD}^2)$ on a log–log scale for various values of F_{CD} ranging from 0.1π to 1000π. The main feature of this graph is that for all values of F_{CD} the behavior of the bilinear flow ($\frac{1}{4}$ *slope*) and the formation linear flow ($\frac{1}{2}$ *slope*) is given by a single curve. Note that there is a transition period between the bilinear and linear flows. The dashed line in this figure indicates the approximate start of the infinite-acting pseudoradial flow.

The pressure data is plotted in terms of log (Δp) vs. log (t) and the resulting graph is

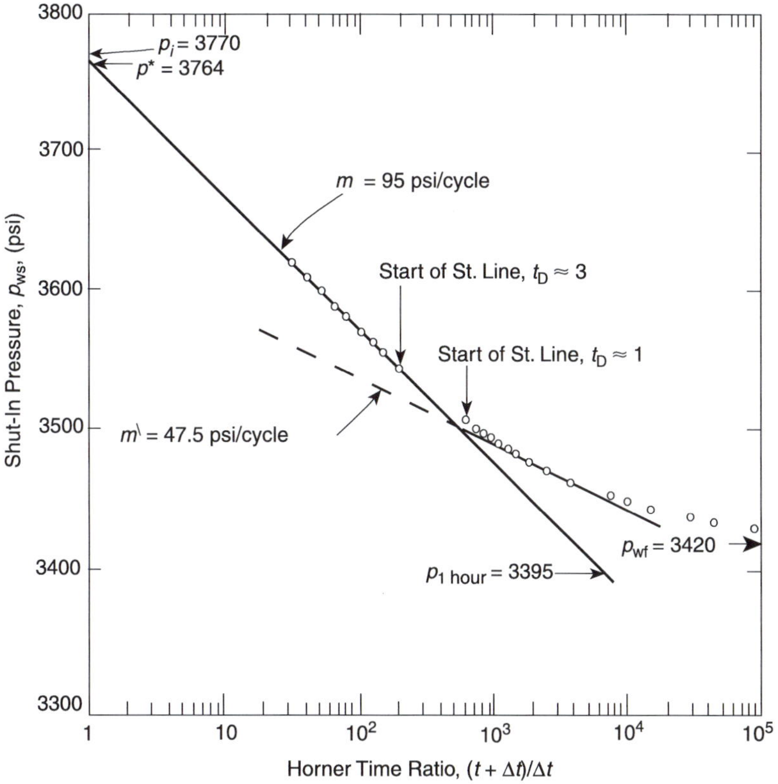

FIGURE 1.85 Horner graph for a vertical fracture (infinite conductivity).

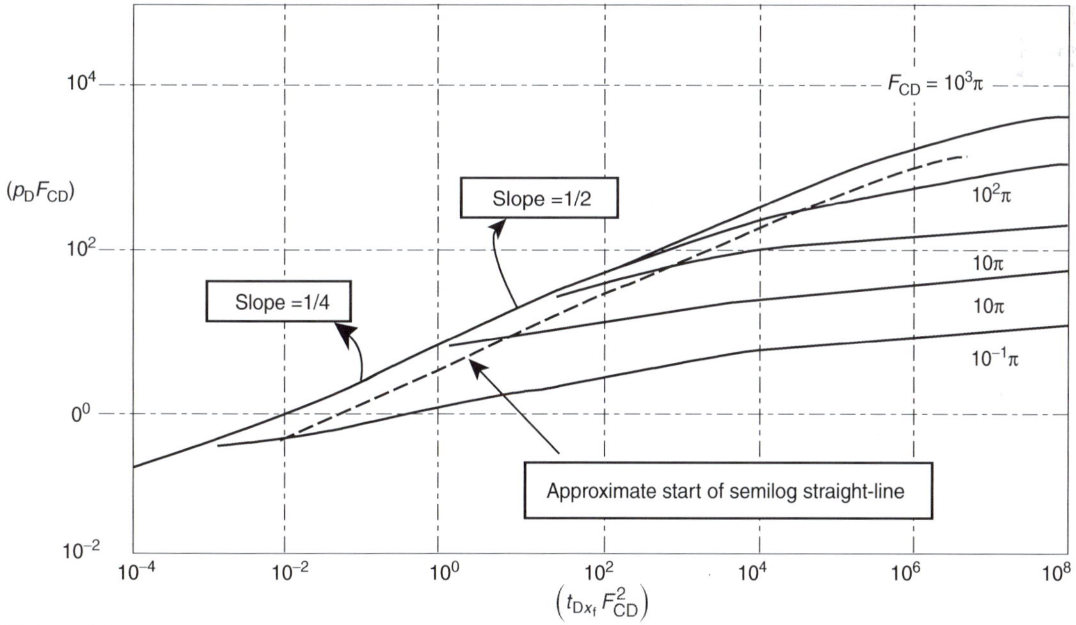

FIGURE 1.86 Type curve for vertically fractured gas wells graph. *(After Cinco-Ley, H., Samaniego, F., 1981. Transient pressure analysis for finite conductivity fracture case versus damage fracture case. SPE Paper 10179.)*

matched to a type curve that is characterized by a dimensionless finite conductivity, $(F_{CD})_M$, with match points of:

- $(\Delta p)_{MP}$, $(p_D F_{CD})_{MP}$;
- $(t)_{MP}$, $(t_{Dx_f} F_{CD}^2)_{MP}$;
- end of bilinear flow $(t_{ebf})_{MP}$;
- beginning of formation linear flow $(t_{blf})_{MP}$;
- beginning of semilog straight line $(t_{bssl})_{MP}$.

From the above match F_{CD} and x_f can be calculated:

For oil: $$F_{CD} = \left[\frac{141.2 QB\mu}{hk}\right] \frac{(p_D F_{CD})_{MP}}{(\Delta p)_{MP}} \quad (1.266)$$

For gas: $$F_{CD} = \left[\frac{1424 QT}{hk}\right] \frac{(p_D F_{CD})_{MP}}{(\Delta m(p))_{MP}} \quad (1.267)$$

The fracture half-length is given by:

$$x_f = \left[\frac{0.0002637k}{\phi \mu c_t}\right] \frac{(t)_{MP} (F_{CD})_M^2}{(t_{Dx_f} F_{CD}^2)_{MP}}$$

Defining the dimensionless effective wellbore radius r'_{wD} as the ratio of the apparent wellbore radius r'_{wD} to the fracture half-length x_f, i.e., $r'_{wD} = r'_w/x_f$, Cinco and Samaniego correlated r'_{wD} with the dimensionless fracture conductivity F_{CD} and presented the resulting correlation in graphical form, as shown in Figure 1.87.

Figure 1.87 indicates that when the dimensionless fracture conductivity is greater than 100, the dimensionless effective wellbore radius r'_{wD} is independent of the fracture conductivity with a fixed value of 0.5, i.e., $r'_{wD} = 0.5$ for $F_{CD} > 100$. The apparent wellbore radius is expressed in terms of the fracture skin factor s_f by:

$$r'_w = r_w \, e^{-s_f}$$

Introducing r'_{wD} into the above expression and solving for s_f gives:

$$s_f = \ln\left[\left(\frac{x_f}{r_w}\right) r'_{wD}\right]$$

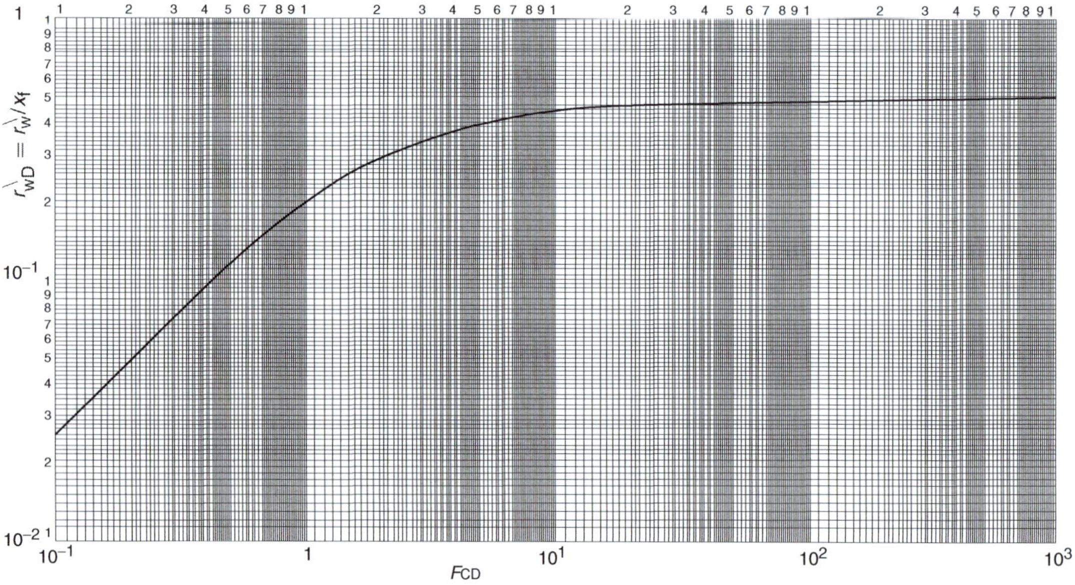

FIGURE 1.87 Effective wellbore radius vs. dimensionless fracture conductivity for a vertical fracture graph. *(After Cinco-Ley, H., Samaniego, F., 1981. Transient pressure analysis for finite conductivity fracture case versus damage fracture case. SPE Paper 10179.)*

For $F_{CD} > 100$, this gives:

$$s_f = -\ln\left(\frac{x_f}{2r_w}\right)$$

where

s_f = skin due to fracture
r_w = wellbore radius, ft

It should be kept in mind that specific analysis graphs must be used for different flow regimes to obtain a better estimate of both fracture and reservoir parameters. Cinco and Samaniego used the pressure buildup data given in Example 1.39 to illustrate the use of their type curve to determine the fracture and reservoir parameters.

Example 1.39

The buildup test data as given in Example 1.36 is given below for convenience:

$Q = 7350$ Mscf/day,	$t_p = 2640$ hours
$h = 118$ ft,	$\phi = 0.10$
$k = 0.025$ md,	$\mu = 0.0252$
$T = 690$ °R,	$c_t = 0.129 \times 10^{-3}$ psi^{-1}
$p_{wf\ \text{at}\ \Delta t=0} = 1320$ psia,	$r_w = 0.28$ ft

The graphical presentation of the buildup data is given in the following two forms:

(1) The log–log plot of $\Delta m(p)$ vs. $(\Delta t)^{1/4}$, as shown earlier in Figure 1.73.
(2) The log–log plot of $\Delta m(p)$ vs. (Δt), on the type curve of Figure 1.86 with the resulting match as shown in Figure 1.88.

Calculate the fracture and reservoir parameters by performing conventional and type curve analysis. Compare the results.

Solution

Step 1. From the plot of $\Delta m(p)$ vs. $(\Delta t)^{1/4}$, in Figure 1.73, determine:

- $m_{bf} = 1.6 \times 10^8$ psi^2/cp hour$^{1/4}$
- $t_{sbf} \approx 0.35$ hours (start of bilinear flow)
- $t_{ebf} \approx 2.5$ hours (end of bilinear flow)
- $\Delta m(p)_{ebf} \approx 2.05 \times 10^8$ psi^2/cp

Step 2. Perform the bilinear flow analysis, as follows:

- Using Eq. (1.252), calculate fracture conductivity F_C:

$$F_C = \left[\frac{444.6QT}{m_{bf}h(\phi\mu c_t k)^{1/4}}\right]^2$$

$$= \left[\frac{444.6(7350)(690)}{(1.62\times10^8)(118)[(0.1)(0.0252)(0.129\times10^{-3})(0.025)]^{1/4}}\right]^2$$

$$= 154 \text{ md ft}$$

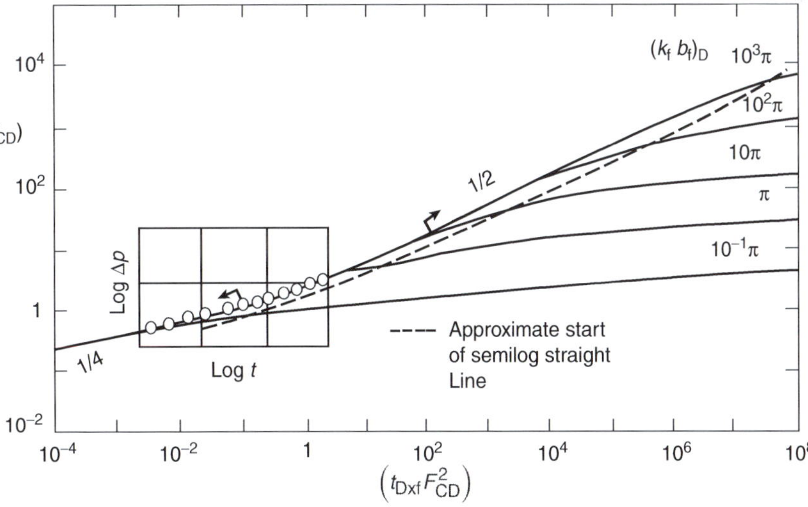

FIGURE 1.88 Type curve matching for data in bilinear and transitional flow graph. *(After Cinco-Ley, H., Samaniego, F., 1981. Transient pressure analysis for finite conductivity fracture case versus damage fracture case. SPE Paper 10179.)*

- Calculate the dimensionless conductivity F_{CD} by using Eq. (1.254):

$$F_{CD} = \frac{1965.1QT}{kh\,\Delta m(p)_{ebf}} = \frac{1965.1(7350)(690)}{(0.025)(118)(2.02\times 10^8)} = 16.7$$

- Estimate the fracture half-length from Eq. (1.239):

$$x_f = \frac{F_C}{F_{CD}k} = \frac{154}{(16.7)(0.025)} = 368 \text{ ft}$$

- Estimate the dimensionless ratio r'_w/x_f from Figure 1.86:

$$\frac{r'_w}{x_f} \approx 0.46$$

- Calculate the apparent wellbore radius r'_w:

$$r'_w = (0.46)(368) = 169 \text{ ft}$$

- Calculate the apparent skin factor

$$s = \ln\left(\frac{r_w}{r'_w}\right) = \ln\left(\frac{0.28}{169}\right) = -6.4$$

Step 3. Perform the type curve analysis as follows:

- Determine the match points from Figure 1.88, to give:

$$\Delta m(p)_{MP} = 10^9 \text{ psi}^2/\text{cp}$$
$$(p_D F_{CD})_{MP} = 6.5$$
$$(\Delta t)_{mp} = 1 \text{ hour}$$
$$[t_{Dx_f}(F_{CD})^2]_{MP} = 3.69\times 10^{-2}$$
$$t_{sbf} \simeq 0.35 \text{ hour}$$
$$t_{ebf} = 2.5 \text{ hour}$$

- Calculate F_{CD} from Eq. 1.267

$$F_{CD} = \left[\frac{1424(7350)(690)}{(118)(0.025)}\right]\frac{6.5}{(10^9)} = 15.9$$

- Calculate the fracture half-length from Eq. (1.267):

$$x_f = \left[\frac{0.0002637(0.025)}{(0.1)(0.02525)(0.129\times 10^{-3})}\,\frac{(1)(15.9)^2}{3.69\times 10^{-2}}\right]^{1/2} = 373 \text{ ft}$$

- Calculate F_C from Eq. (1.239):

$$F_C = F_{CD}x_f k = (15.9)(373)(0.025) = 148 \text{ md ft}$$

- From Figure 1.86:

$$\frac{r'_w}{x_f} = 0.46$$
$$r'_w = (373)\,(0.46) = 172 \text{ ft}$$

Test Results	Type Curve Analysis	Bilinear Flow Analysis
F_C	148.0	154.0
x_f	373.0	368.0
F_{CD}	15.9	16.7
r'_w	172.0	169.0

The concept of the pressure derivative can be effectively employed to identify different flow regime periods associated with hydraulically fractured wells. As shown in Figure 1.89, a finite conductivity fracture shows a $\frac{1}{4}$ straight-

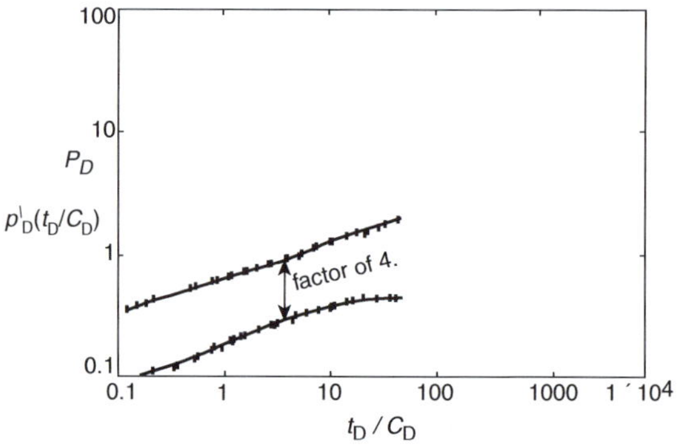

FIGURE 1.89 Finite conductivity fracture shows as a $\frac{1}{4}$ slope line on a log–log plot, same on a derivative plot. Separation between pressure and derivative is a factor of 4.

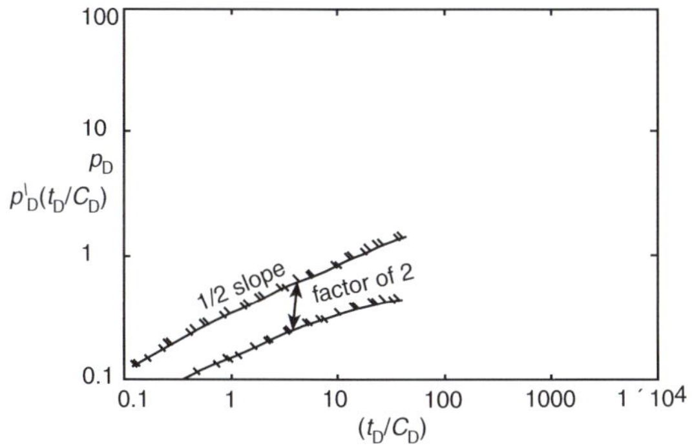

FIGURE 1.90 Infinite conductivity fracture shows as a $\frac{1}{2}$ slope line on a log–log plot, same on a derivative plot. Separation between pressure and derivative is a factor of 2.

line slope for both the pressure difference Δp and its derivative; however, the two parallel lines are separated by a factor of 4. Similarly, for an infinite conductivity fracture, two straight parallel lines represent Δp and its derivative with a $\frac{1}{2}$ slope and separation between the lines of a factor of 2 (as shown in Figure 1.90).

In tight reservoirs where the productivity of wells is enhanced by massive hydraulic fracturing (MHF), the resulting fractures are characterized as long vertical fractures with finite conductivities. These wells tend to produce at a constant and low bottom-hole flowing pressure, rather than constant flow rate. The diagnostic plots and the conventional analysis of bilinear flow data can be used when analyzing well test data under constant flowing pressure. Eqs. (1.245)–(1.249) can be rearranged and expressed in the following forms.

For fractured oil wells:

$$\frac{1}{Q} = \left[\frac{44.1 B\mu}{h\sqrt{F_C}(\phi\mu c_t k)^{1/4}\,\Delta p}\right] t^{1/4}$$

or equivalently:

$$\frac{1}{Q} = m_{bf} t^{1/4}$$

and

$$\log\left(\frac{1}{Q}\right) = \log(m_{bf}) + \frac{1}{4}\log(t)$$

where

$$m_{bf} = \frac{44.1 B\mu}{h\sqrt{F_C}(\phi\mu c_t k)^{1/4}\,\Delta p}$$

$$F_C = k_f w_f = \left[\frac{44.1 B\mu}{h m_{bf}(\phi\mu c_t k)^{1/2}\,\Delta p}\right]^2 \qquad \textbf{(1.268)}$$

For fractured gas wells:

$$\frac{1}{Q} = m_{bf} t^{1/4}$$

or

$$\log\left(\frac{1}{Q}\right) = \log(m)$$

where

$$m_{bf} = \frac{444.6 T}{h\sqrt{F_C}(\phi\mu c_t k)^{1/4}\,\Delta m(p)}$$

Solving for F_C:

$$F_C = \left[\frac{444.6 T}{h m_{bf}(\phi\mu c_t k)^{1/4}\,\Delta m(p)}\right]^2 \qquad \textbf{(1.269)}$$

The following procedure can be used to analyze bilinear flow data under constant flow pressure:

Step 1. Plot $1/Q$ vs. t on a log–log scale and determine if any data falls on a straight line of a $\frac{1}{4}$ slope.

Step 2. If any data forms a $\frac{1}{4}$ slope in step 1, plot $1/Q$ vs. $t^{1/4}$ on a Cartesian role and determine the slope m_{bf}.

Step 3. Calculate the fracture conductivity F_C from Eq. (1.268) or (1.269):

For oil: $$F_C = \left[\frac{44.1 B\mu}{h m_{bf}(\phi\mu c_t k)^{1/4}(p_i - p_{wf})}\right]^2$$

For gas: $$F_C = \left[\frac{444.6 T}{h m_{bf}(\phi\mu c_t k)^{1/4}[m(p_i) - m(p_{wf})]}\right]^2$$

Step 4. Determine the value of Q when the bilinear straight line ends and designate it as Q_{ebf}.

Step 5. Calculate F_{CD} from Eq. (1.253) or (1.254):

For oil: $$F_{CD} = \frac{194.9 Q_{ebf} B \mu}{kh(p_i - p_{wf})}$$

For gas: $$F_{CD} = \frac{1965.1 Q_{ebf} T}{kh[m(p_i) - m(p_{wf})]}$$

Step 6. Estimate the fracture half-length from:

$$x_f = \frac{F_C}{F_{CD} k}$$

Agarwal et al. (1979) presented constant-pressure type curves for finite conductivity fractures, as shown in Figure 1.91. The reciprocal of the dimensionless rate $1/Q_D$ is expressed as a function of dimensionless time t_{Dx_f}, on log–log paper, with the dimensionless fracture conductivity F_{CD} as a correlating parameter. The reciprocal dimensionless rate $1/Q_D$ is given by:

For oil wells: $$\frac{1}{Q_D} = \frac{kh(p_i - p_{wf})}{141.2 Q \mu B} \quad (1.270)$$

For gas wells: $$\frac{1}{Q_D} = \frac{kh[m(p_i) - m(p_{wf})]}{1424 QT} \quad (1.271)$$

with

$$t_{Dx_f} = \frac{0.0002637 kt}{\phi (\mu c_t)_i x_f^2} \quad (1.272)$$

where

p_{wf} = wellbore pressure, psi
Q = flow rate, STB/day or Mscf/day
T = temperature, °R
t = time, hours

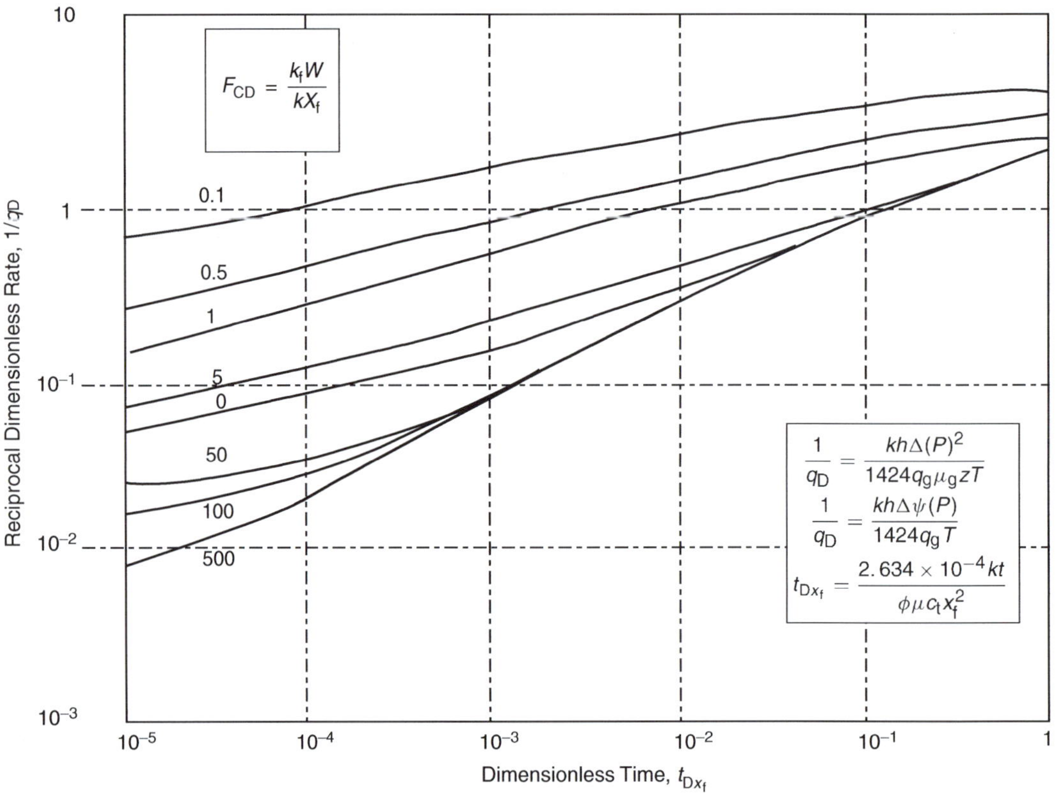

FIGURE 1.91 Log–log type curves for finite capacity vertical fractures; constant wellbore pressure. *(After Agarwal, R.G., Carter, R.D., Pollock, C.B., 1979. Evaluation and performance prediction of low-permeability gas wells stimulated by massive hydraulic fracturing. J. Pet. Technol. 31 (3), 362–372; also in SPE Reprint Series No. 9.)*

subscripts denote

i = initial
D = dimensionless

Example 1.40, as adopted from Agarwal et al. (1979), illustrates the use of these type curves.

Example 1.40

A pre-frac buildup test was performed on a well producing from a tight gas reservoir, to give a formation permeability of 0.0081 md. Following an MHF treatment, the well produced at a constant pressure with recorded rate-time data as given below:

t (days)	*Q* (Mscf/day)	1/*Q* (day/Mscf)
20	625	0.00160
35	476	0.00210
50	408	0.00245
100	308	0.00325
150	250	0.00400
250	208	0.00481
300	192	0.00521

The following additional data is available:

$p_i = 2394$ psi, $\Delta m(p) = 396 \times 10^6$ psi²/cp
$h = 32$ ft, $\phi = 0.107$
$T = 720$ °R, $c_{ti} = 2.34 \times 10^{-4}$ psi^{-1}
$\mu_i = 0.0176$ cp, $k = 0.0081$ md

Calculate:

- fracture half-length, x_f;
- fracture conductivity, F_C.

Solution

Step 1. Plot $1/Q$ vs. t on tracing paper, as shown in Figure 1.92, using the log–log scale of the type curves.

Step 2. We must make use of the available values of k, h, and $\Delta m(p)$ by arbitrarily choosing a convenient value of the flow rate and calculating the corresponding $1/Q_D$. Selecting $Q = 1000$ Mscf/day, calculate the corresponding value of $1/Q_D$ by applying Eq. (1.271):

$$\frac{1}{Q_D} = \frac{kh\,\Delta m(p)}{1424QT} = \frac{(0.0081)(32)(396 \times 10^6)}{1424(1000)(720)} = 0.1$$

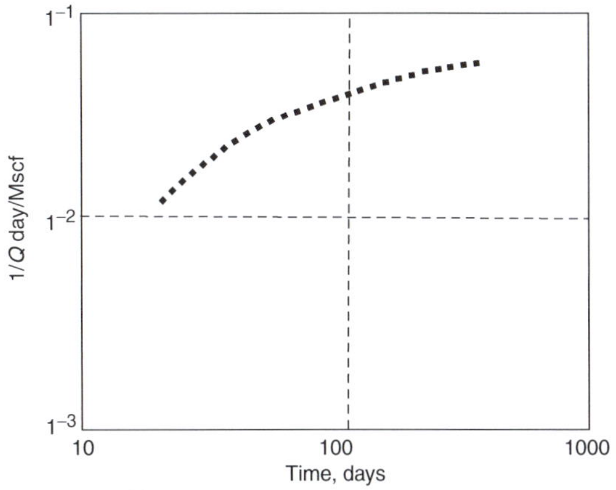

FIGURE 1.92 Reciprocal smooth rate vs. time for MHF, Example 1.42.

Step 3. Thus, the position of $1/Q = 10^{-3}$ on the y-axis of the tracing paper is fixed in relation to $1/Q_D = 0.1$ on the y-axis of the type curve graph paper; as shown in Figure 1.93.

Step 4. Move the tracing paper horizontally along the x-axis until a match is obtained, to give:

$$t = 100 \text{ days} = 2400 \text{ hours}$$
$$t_{Dx_f} = 2.2 \times 10^{-2}$$
$$F_{CD} = 50$$

Step 5. Calculate the fracture half-length from Eq. (1.272):

$$x_f^2 = \left[\frac{0.0002637k}{\phi(\mu c_t)_i}\right]\left(\frac{t}{t_{Dxf}}\right)_{MP} = \left[\frac{0.0002637(0.0081)}{(0.107)(0.0176)(2.34 \times 10^{-4})}\right]\left(\frac{2400}{2.2 \times 10^{-2}}\right) = 528{,}174 \text{ ft}$$

$$x_f \approx 727 \text{ ft}$$

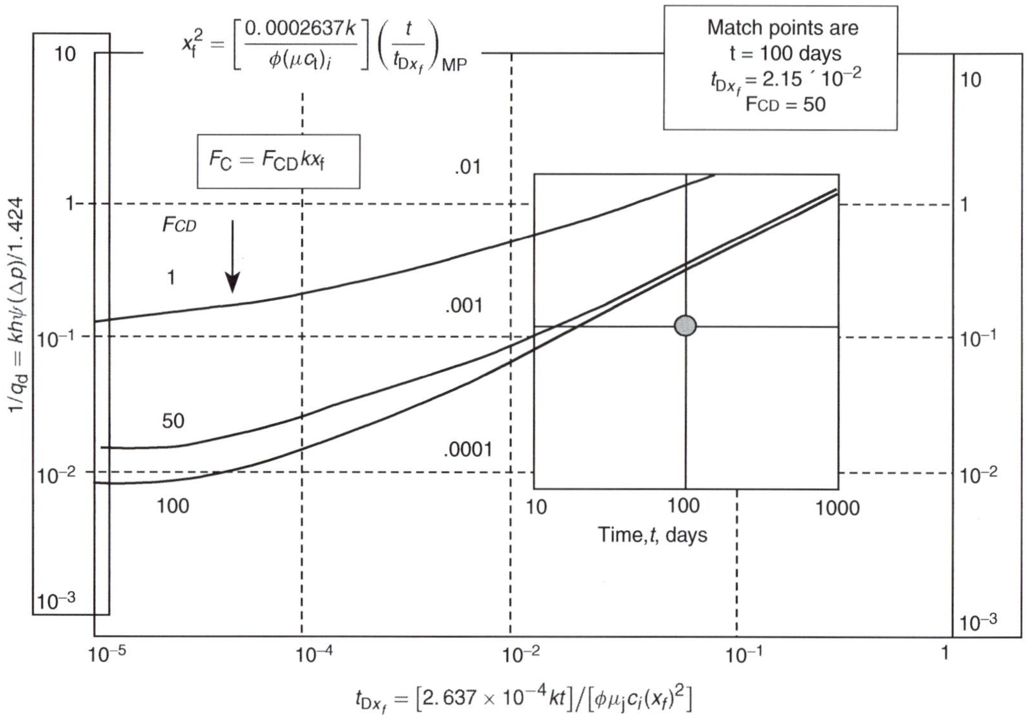

FIGURE 1.93 Type curve matching for MHF gas well, Example 1.42.

Thus, the total fracture length is:

$$2x_f = 1454 \text{ ft}$$

Step 6. Calculate the fracture conductivity F_C from Eq. (1.220):

$$F_C = F_{CD}kx_f = (50)(0.0081)(727) = 294 \text{ md ft}$$

It should be pointed out that if the pre-fracturing buildup test were not available, matching would require shifting the tracing paper along both the x and y-axes to obtain the proper match. This emphasizes the need for determining kh from a pre-fracturing test.

Faults or Impermeable Barriers. One of the important applications of a pressure buildup test is analyzing the test data to detect or confirm the existence of faults and other flow barriers. When a sealing fault is located near a test well, it significantly affects the recorded well pressure behavior during the buildup test. This pressure behavior can be described mathematically by applying the principle of superposition as given by the method of images. Figure 1.94 shows a test well that is located at a distance L from a sealing fault. Applying method images, as given Eq. (1.168), the total pressure drop as a function of time t is:

$$(\Delta p)_{\text{total}} = \frac{162.6Q_o B\mu}{kh}\left[\log\left(\frac{kt}{\phi\mu c_t r_w^2}\right) - 3.23 + 0.87s\right] - \left(\frac{70.6Q_o B\mu}{kh}\right)\text{Ei}\left(-\frac{948\phi\mu c_t(2L)^2}{kt}\right)$$

When both the test well and image well are shut-in for a buildup test, the principle of

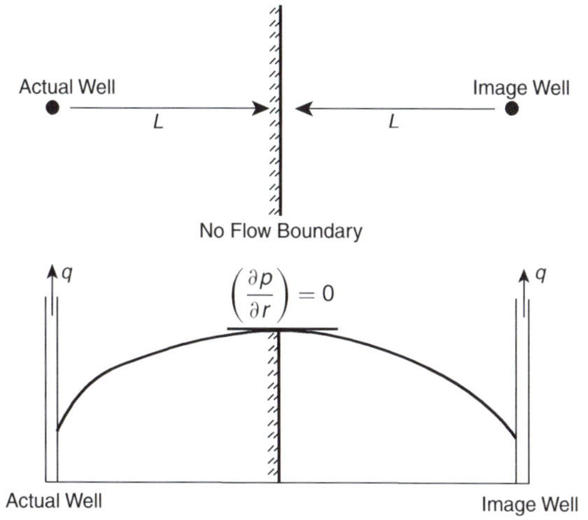

FIGURE 1.94 Method of images in solving boundary problems.

superposition can be applied to Eq. (1.68) to predict the buildup pressure at Δt as:

$$p_{ws} = p_i - \frac{162.6Q_o B_o \mu_o}{kh}\left[\log\left(\frac{t_p + \Delta t}{\Delta t}\right)\right] - \left(\frac{70.6Q_o B_o \mu_o}{kh}\right)\text{Ei}\left[\frac{-948\phi\mu c_t(2L)^2}{k(t_p + \Delta t)}\right] - \left(\frac{70.6(-Q_o)B_o \mu_o}{kh}\right)\text{Ei}\left[\frac{-948\phi\mu c_t(2L)^2}{k\,\Delta t}\right] \quad \textbf{(1.273)}$$

Recalling that the exponential integral $\text{Ei}(-x)$ can be approximated by Eq. (1.79) when $x < 0.01$ as:

$$\text{Ei}(-x) = \ln(1.781x)$$

the value of the $\text{Ei}(-x)$ can be set equal to zero when x is greater than 10.9, i.e., $\text{Ei}(-x) = 0$ for $x > 10.9$. Note that the value of $(2L)^2$ is large and for early buildup times, when Δt is small, the last two terms can be set equal to zero, or:

$$p_{ws} = p_i - \frac{162.6Q_o B_o \mu_o}{kh}\left[\log\left(\frac{t_p + \Delta t}{\Delta t}\right)\right] \quad \textbf{(1.274)}$$

which is essentially the regular Horner equation with a semilog straight-line slope of:

$$m = \frac{162.6Q_o B_o \mu_o}{kh}$$

For a shut-in time sufficiently large that the *logarithmic approximation is accurate for the Ei functions*, Eq. (1.273) becomes:

$$p_{ws} = p_i - \frac{162.6Q_o B_o \mu_o}{kh}\left[\log\left(\frac{t_p + \Delta t}{\Delta t}\right)\right] - \frac{162.6Q_o B_o \mu_o}{kh}\left[\log\left(\frac{t_p + \Delta t}{\Delta t}\right)\right]$$

Rearranging this equation by recombining terms gives:

$$p_{ws} = p_i - 2\left(\frac{162.6Q_o B_o \mu_o}{kh}\right)\left[\log\left(\frac{t_p + \Delta t}{\Delta t}\right)\right]$$

Simplifying

$$p_{ws} = p_i - 2m\left[\log\left(\frac{t_p + \Delta t}{\Delta t}\right)\right] \quad \textbf{(1.275)}$$

Three observations can be made by examining Eqs. (1.274) and (1.275):

(1) For early shut-in time buildup data, Eq. (1.274) indicates that the data from the early shut-in times will form a straight line on the Horner plot with a slope that is identical to a reservoir without sealing fault.
(2) At longer shut-in times, the data will form a *second straight line* on the Horner plot with a slope that is twice that of the first line, i.e., second slope $= 2m$. The presence of the second straight line with a double slope of the first straight line provides a means of recognizing the presence of a fault from pressure buildup data.
(3) The shut-in time required for the slope to double can be approximated from the following expression:

$$\frac{948\phi\mu c_t(2L)^2}{k\,\Delta t} < 0.01$$

Solving for Δt gives:

$$\Delta t > \frac{380,000\phi\mu c_t L^2}{k}$$

where

Δt = minimum shut-in time, hours
k = permeability, md
L = distance between well and the sealing fault, ft

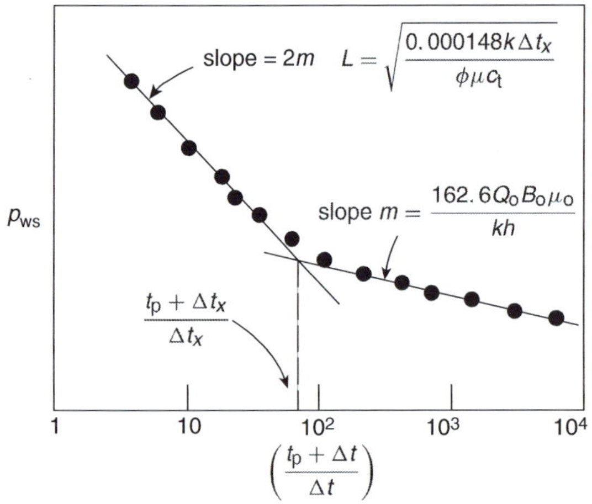

FIGURE 1.95 Theoretical Horner plot for a faulted system.

Notice that the value of p^* for use in calculating the average drainage region pressure $\bar{p}$ is obtained by extrapolating the *second straight line* to a unit-time ratio, i.e., to $(t_p + \Delta t)/\Delta t = 1.0$. The permeability and skin factor are calculated in the normal manner described before using the slope of the *first straight line*.

Gray (1965) suggested that for the case in which the slope of the buildup test has the time to double, as shown schematically in Figure 1.95, the distance L from the well to the fault can be calculated by finding the time Δt_x at which the two semilog straight lines intersect. That is:

$$L = \sqrt{\frac{0.000148k\,\Delta t_x}{\phi\mu c_t}} \tag{1.276}$$

Lee (1982) illustrated Gray's method through Example 1.41.

Example 1.41

A pressure buildup test was conducted to confirm the existence of a sealing fault near a newly drilled well. Data from the test is given below:

Δt (hour)	p_{ws} (psi)	$(t_p + \Delta t)/\Delta t$
6	3996	47.5
8	4085	35.9
10	4172	28.9
12	4240	24.3
14	4298	20.9
16	4353	18.5
20	4435	15.0
24	4520	12.6
30	4614	10.3
36	4700	8.76
42	4770	7.65
48	4827	6.82
54	4882	6.17
60	4931	5.65
66	4975	5.23

Other data include the following:

$\phi = 0.15$, $\mu_o = 0.6$ cp
$c_t = 17 \times 10^{-6}$ psi^{-1}, $r_w = 0.5$ ft
$Q_o = 1221$ STB/day, $h = 8$ ft
$B_o = 1.31$ bbl/STB

A total of 14,206 STB of oil had been produced before shut-in. Determine whether the sealing fault exists and the distance from the well to the fault.

Solution

Step 1. Calculate total production time t_p:

$$t_p = \frac{24N_p}{Q_o} = \frac{(24)(14,206)}{1221} = 279.2 \text{ hours}$$

Step 2. Plot p_{ws} vs. $(t_p + \Delta t)/\Delta t$ as shown in Figure 1.96. The plot clearly shows two straight lines with the first slope of 650 psi/cycle and the second with 1300 psi/cycle. Notice that the second slope is twice that of the first slope indicating the existence of the sealing fault.

Step 3. Using the value of the *first slope*, calculate the permeability k:

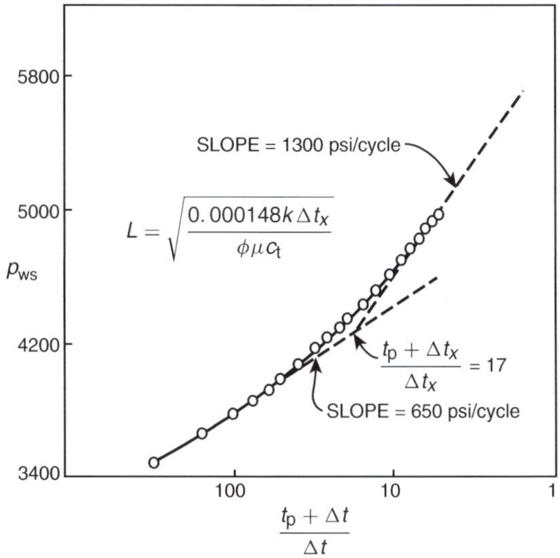

FIGURE 1.96 Estimating distance to a no-flow boundary.

$$k = \frac{162.6 Q_o B_o \mu_o}{mh} = \frac{162.6(1221)(1.31)(0.6)}{(650)(8)}$$
$$= 30 \text{ md}$$

Step 4. Determine the value of Horner's time ratio at the intersection of the two semilog straight lines shown in Figure 1.96, to give:

$$\frac{t_p + \Delta t_x}{\Delta t_x} = 17$$

or

$$\frac{279.2 + \Delta t_x}{\Delta t_x} = 17$$

from which:

$$\Delta t_x = 17.45 \text{ hours}$$

Step 5. Calculate the distance L from the well to the fault by applying Eq. (1.276):

$$L = \sqrt{\frac{0.000148 k \Delta t_x}{\phi \mu c_t}}$$
$$= \sqrt{\frac{0.000148(30)(17.45)}{(0.15)(0.6)(17 \times 10^{-6})}} = 225 \text{ ft}$$

Qualitative Interpretation of Buildup Curves. The Horner plot has been the most widely accepted means for analyzing pressure buildup data since its introduction in 1951. Another widely used aid in pressure transient analysis is the plot of change in pressure Δp vs. time on a log–log scale. Economides (1988) pointed out that this log–log plot serves the following two purposes:

(1) the data can be matched to type curves;
(2) the type curves can illustrate the expected trends in pressure transient data for a large variety of well and reservoir systems.

The visual impression afforded by the log–log presentation has been greatly enhanced by the introduction of the pressure derivative which represents the changes of the slope of buildup data with respect to time. When the data produces a straight line on a semilog plot, the pressure derivative plot will, therefore, be constant. This means the pressure derivative plot will be flat for that portion of the data that can be correctly analyzed as a straight line on the Horner plot.

Many engineers rely on the log–log plot of Δp and its derivative vs. time to diagnose and select the proper interpretation model for a given set of pressure transient data. Patterns visible in the log–log diagnostic and Horner plots for five frequently encountered reservoir systems are illustrated graphically by Economides as shown in Figure 1.97. The curves on the right represent buildup responses for five different patterns, a–e, with the curves on the left representing the corresponding responses when the data is plotted in the log–log format of Δp and $(\Delta t \, \Delta p^{\backslash})$ vs. time.

The five different buildup examples shown in Figure 1.97 were presented by Economides (1988) and are briefly discussed below:

Example a illustrates the most common response—that of a homogeneous reservoir with wellbore storage and skin. Wellbore storage derivative transients are

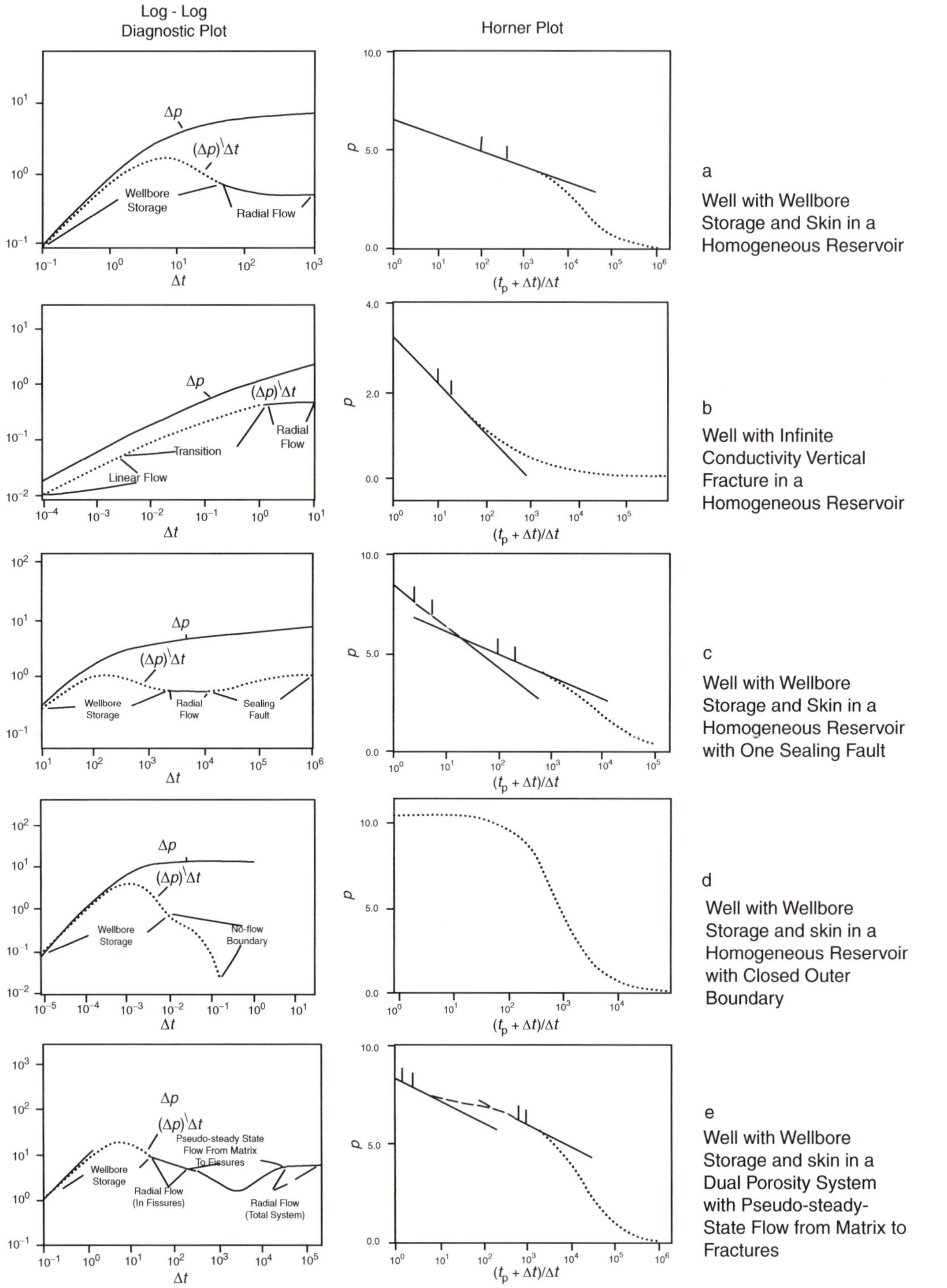

FIGURE 1.97 Qualitative interpretation of buildup curves. *(After Economides, C., 1988. Use of the pressure derivative for diagnosing pressure-transient behavior. J. Pet. Technol. 40 (10), 1280–1282.)*

recognized as a "hump" in early time. The flat derivative portion in late time is easily analyzed as the Horner semilog straight line.

Example b shows the behavior of an infinite conductivity, which is characteristic of a well that penetrates a natural fracture. The $\frac{1}{2}$ slopes in both the pressure change and its derivative result in two parallel lines during the flow regime, representing linear flow to the fracture.

Example c shows the homogeneous reservoir with a single vertical planar barrier to flow or a fault. The level of the second-derivative plateau is twice the value of the level of the first-derivative plateau, and the Horner plot shows the familiar slope-doubling effect.

Example d illustrates the effect of a closed drainage volume. Unlike the drawdown pressure transient, this has a unit-slope line in late time that is indicative of pseudosteady-state flow; the buildup pressure derivative drops to zero. The permeability and skin cannot be determined from the Horner plot because no portion of the data exhibits a flat derivative for this example. When transient data resembles example d, the only way to determine the reservoir parameters is with a type curve match.

Example e exhibits a valley in the pressure derivative that is indicative of reservoir heterogeneity. In this case, the feature results from dual-porosity behavior of the pseudosteady flow from matrix to fractures.

Figure 1.97 clearly shows the value of the pressure/pressure derivative presentation. An important advantage of the log–log presentation is that the transient patterns have a standard appearance as long as the data is plotted with square log cycles. The visual patterns in semilog plots are amplified by adjusting the range of the vertical axis. Without adjustment, many or all of the data may appear to lie on one line and subtle changes can be overlooked.

Some of the pressure derivative patterns shown are similar to the characteristics of other models. For example, the pressure derivative doubling associated with a fault (example c) can also indicate transient interporosity flow in a dual-porosity system. The sudden drop in the pressure derivative in buildup data can indicate either a closed outer boundary or constant-pressure outer boundary resulting from a gas cap, an aquifer, or pattern injection wells. The valley in the pressure derivative (example e) could indicate a layered system instead of dual porosity. For these cases and others, the analyst should consult geological, seismic, or core analysis data to decide which model to use in an interpretation. With additional data, a more conclusive interpretation for a given transient data set may be found.

An important place to use the pressure/pressure derivative diagnosis is on the well site. If the objective of the test is to determine permeability and skin, the test can be terminated once the derivative plateau is identified. If heterogeneities or boundary effects are detected in the transient, the test can be run longer to record the entire pressure/pressure derivative response pattern needed for the analysis.

1.6 INTERFERENCE AND PULSE TESTS

When the flow rate is changed and the pressure response is recorded in the same well, the test is called a "single-well" test. Examples of single-well tests are drawdown, buildup, injectivity, falloff, and step-rate tests. When the flow rate is changed in one well and the pressure response is recorded in another well, the test is called a "multiple-well" test. Examples of multiple-well tests are interference and pulse tests.

Single-well tests provide valuable reservoir and well characteristics that include flow capacity kh, wellbore conditions, and fracture length as examples of these important properties.

However, these tests do not provide the directional nature of reservoir properties (such as permeability in the x, y, and z direction) and have inabilities to indicate the degree of communication between the test wells and adjacent wells. Multiple-well tests are run to determine:

- the presence or lack of communication between the test well and surrounding wells;
- the mobility–thickness product kh/μ;
- the porosity–compressibility–thickness product $\phi c_t h$;
- the fracture orientation if intersecting one of the test wells;
- the permeability in the direction of the major and minor axes.

The multiple-well test requires at least one active (producing or injecting) well and at least one pressure observation well, as shown schematically in Figure 1.98. In an interference test, all the test wells are shut-in until their wellbore pressures stabilize. The active well is then allowed to produce or inject at constant rate and the pressure response in the observation well(s) is observed. Figure 1.98 indicates this concept with one active well and one observation well. As the figure indicates, when the active well starts to produce, the pressure in the shut-in observation well begins to respond after some "time lag" that depends on the reservoir rock and fluid properties.

Pulse testing is a form of interference testing. The producer or injector is referred to as "the pulser or the active well" and the observation well is called "the responder." The tests are conducted by sending a series of short-rate pulses from the active well (producer or injector) to a shut-in observation well(s). Pulses

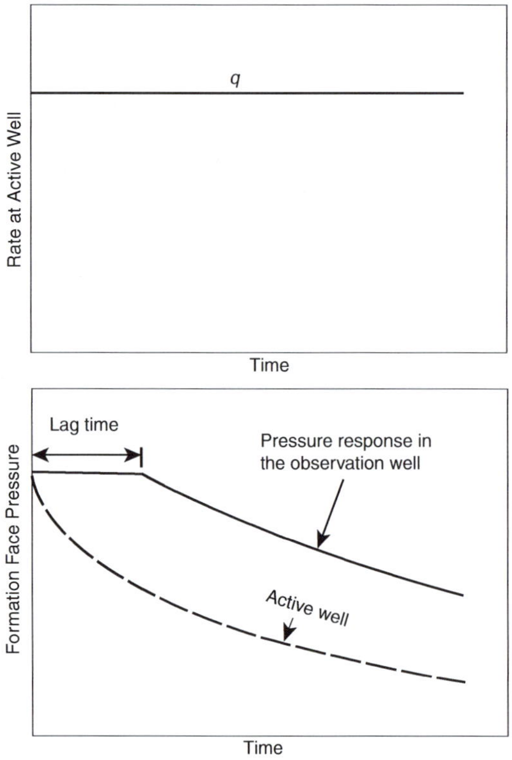

FIGURE 1.98 Rate history and pressure response of a two-well interference test conducted by placing the active well on production at constant rate.

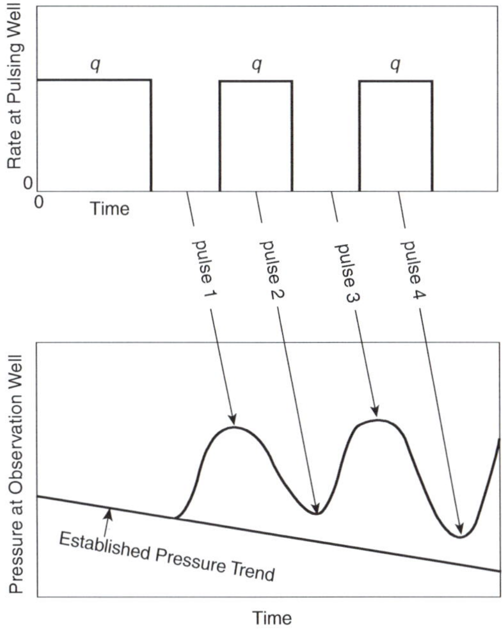

FIGURE 1.99 Illustration of rate history and pressure response for a pulse test. *(After Earlougher, Robert C., Jr., 1977. Advances in Well Test Analysis, Monograph, vol. 5. Society of Petroleum Engineers of AIME, Dallas, TX; Permission to publish by the SPE, copyright SPE, 1977.)*

generally are alternating periods of production (or injection) and shut-in, with the same rate during each production (injection) period, as illustrated in Figure 1.99 for a two-well system.

Kamal (1983) provided an excellent review of interference and pulse testing and summarized various methods that are used to analyze test data. These methods for analyzing interference and pulse tests are presented below.

1.6.1 Interference Testing in Homogeneous Isotropic Reservoirs

A reservoir is classified as "homogeneous" when the porosity and thickness do not change significantly with location. An "isotropic" reservoir indicates that the permeability is the same throughout the system. In these types of reservoirs, the type curve matching approach is perhaps the most convenient to use when analyzing interference test data in a homogeneous reservoir system. As given previously by Eq. (1.77), the pressure drop at any distance r from an active well (i.e., distance between an active well and a shut-in observation well) is expressed as:

$$p_i - p(r,t) = \Delta p = \left[\frac{-70.6QB\mu}{kh}\right] \mathrm{Ei}\left[\frac{-948\phi c_t r^2}{kt}\right]$$

Earlougher (1977) expressed the above expression in a dimensionless form as:

$$\frac{(p_i - p(r,t))/141.2QB\mu}{kh} = -\frac{1}{2}\mathrm{Ei}\left[\left(\frac{-1}{4}\right)\left(\frac{\phi\mu c_t r_w^2}{0.0002637kt}\right)\left(\frac{r}{r_w}\right)^2\right]$$

From the definitions of the dimensionless parameters p_D, t_D, and r_D, the above equations can be expressed in a dimensionless form as:

$$p_D = -\frac{1}{2}\mathrm{Ei}\left[\frac{-r_D^2}{4t_D}\right] \tag{1.277}$$

with the dimensionless parameters as defined by:

$$p_D = \frac{[p_i - p(r,t)]kh}{141.2QB\mu}$$

$$r_D = \frac{r}{r_w}$$

$$t_D = \frac{0.0002637kt}{\phi\mu c_t r_w^2}$$

where

$p(r, t)$ = pressure at distance r and time t, psi
r = distance between the active well and a shut-in observation well
t = time, hours
p_i = reservoir pressure
k = permeability, md

Earlougher expressed in Eq. (1.277) a type curve form as shown previously in Figure 1.47 and reproduced for convenience as Figure 1.100.

To analyze an interference test by type curve matching, plot the observation well(s) pressure change Δp vs. time on tracing paper laid over Figure 1.100 using the matching procedure described previously. When the data is matched to the curve, any convenient match point is selected and match point values from the tracing paper and the underlying type curve grid are read. The following expressions can then be applied to estimate the average reservoir properties:

$$k = \left[\frac{141.2QB\mu}{h}\right]\left[\frac{p_D}{\Delta p}\right]_{MP} \quad (1.278)$$

$$\phi = \frac{0.0002637}{c_t r^2}\left[\frac{k}{\mu}\right]\left[\frac{t}{t_D/r_D^2}\right]_{MP} \quad (1.279)$$

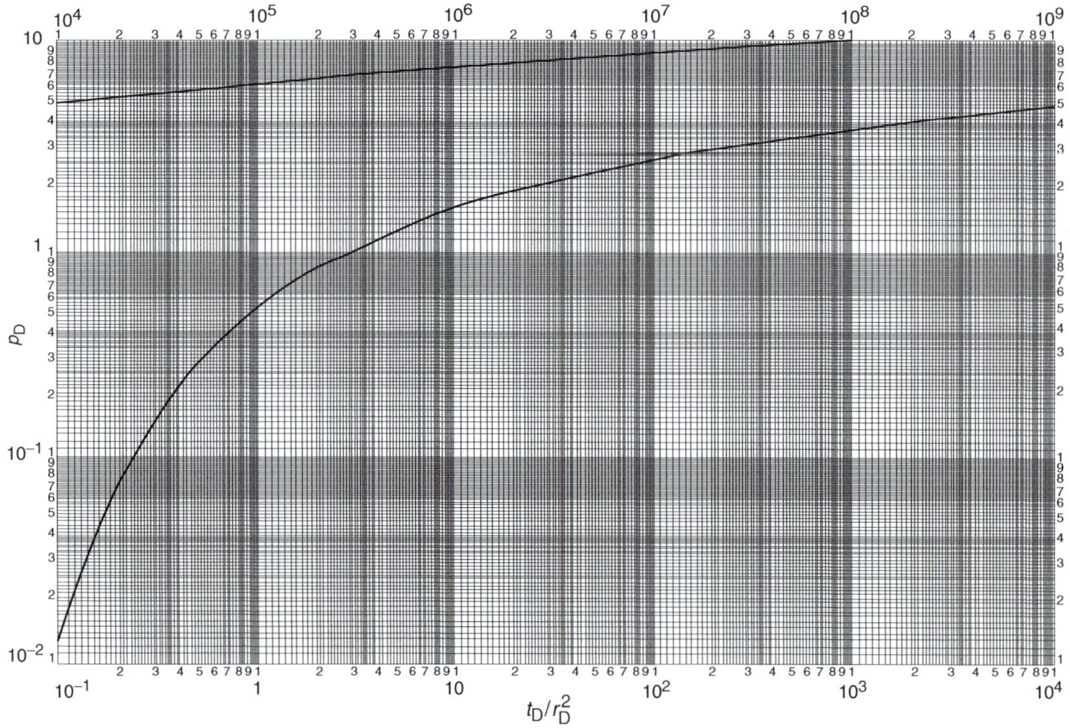

FIGURE 1.100 Dimensionless pressure for a single well in an infinite system, no wellbore storage, no skin. Exponential–integral solution. *(After Earlougher, Robert C., Jr., 1977. Advances in Well Test Analysis, Monograph, vol. 5. Society of Petroleum Engineers of AIME, Dallas, TX; Permission to publish by the SPE, copyright SPE, 1977.)*

where

r = distance between the active and observation wells, ft
k = permeability, md

Sabet (1991) presented an excellent discussion on the use of the type curve approach in analyzing interference test data by making use of test data given by Strobel et al. (1976). The data, as given by Sabet, is used in Example 1.42 to illustrate the type curve matching procedure.

Example 1.42

An interference test was conducted in a dry gas reservoir using two observation wells, designated as Well 1 and Well 3, and an active well, designated as Well 2. The interference test data is listed below:

- Well 2 is the producer, $Q_g = 12.4$ MMscf/day;
- Well 1 is located 8 miles east of Well 2, i.e., $r_{12} = 8$ miles;
- Well 3 is located 2 miles west of Well 2, i.e., $r_{23} = 2$ miles.

Flow Rate	Time	Observed Pressure (psia)			
Q (MMscf/day)	T (hour)	Well 1		Well 3	
		p_1	Δp_1	p_3	Δp_3
0.0	24	2912.045	0.000	2908.51	0.00
12.4	0	2912.045	0.000	2908.51	0.00
12.4	24	2912.035	0.010	2907.66	0.85
12.4	48	2912.032	0.013	2905.80	2.71
12.4	72	2912.015	0.030	2903.79	4.72
12.4	96	2911.997	0.048	2901.85	6.66
12.4	120	2911.969	0.076	2899.98	8.53
12.4	144	2911.918	0.127	2898.25	10.26
12.4	169	2911.864	0.181	2896.58	11.93
12.4	216	2911.755	0.290	2893.71	14.80
12.4	240	2911.685	0.360	2892.36	16.15
12.4	264	2911.612	0.433	2891.06	17.45
12.4	288	2911.533	0.512	2889.79	18.72
12.4	312	2911.456	0.589	2888.54	19.97
12.4	336	2911.362	0.683	2887.33	21.18
12.4	360	2911.282	0.763	2886.16	22.35
12.4	384	2911.176	0.869	2885.01	23.50
12.4	408	2911.108	0.937	2883.85	24.66
12.4	432	2911.030	1.015	2882.69	25.82
12.4	444	2910.999	1.046	2882.11	26.40
0.0	450	Well 2 shut-in			
0.0	480	2910.833	1.212	2881.45	27.06
0.0	504	2910.714	1.331	2882.39	26.12
0.0	528	2910.616	1.429	2883.52	24.99
0.0	552	2910.520	1.525	2884.64	23.87
0.0	576	2910.418	1.627	2885.67	22.84
0.0	600	2910.316	1.729	2886.61	21.90
0.0	624	2910.229	1.816	2887.46	21.05
0.0	648	2910.146	1.899	2888.24	20.27
0.0	672	2910.076	1.969	2888.96	19.55
0.0	696	2910.012	2.033	2889.60	18.91

The following additional reservoir data is available:

$T = 671.6\ °R$, $h = 75$ ft, $c_{ti} = 2.74 \times 10^{-4}\ \text{psi}^{-1}$
$B_{gi} = 920.9$ bbl/MMscf, $r_w = 0.25$ ft, $Z_i = 0.868$
$S_w = 0.21$, $\gamma_g = 0.62$, $\mu_{gi} = 0.0186$ cp

Using the type curve approach, characterize the reservoir in terms of permeability and porosity.

Solution

Step 1. Plot Δp vs. t on a log–log tracing paper with the same dimensions as those of Figure 1.100, as shown in Figures 1.101 and 1.102 for Wells 1 and 3, respectively.

Step 2. Figure 1.103 shows the match of interference data for Well 3, with the following matching points:

$$(p_D)_{MP} = 0.1 \quad \text{and} \quad (\Delta p)_{MP} = 2 \text{ psi}$$

$$\left(\frac{t_D}{r_D^2}\right)_{MP} = 1 \quad \text{and} \quad (t)_{MP} = 159 \text{ hours}$$

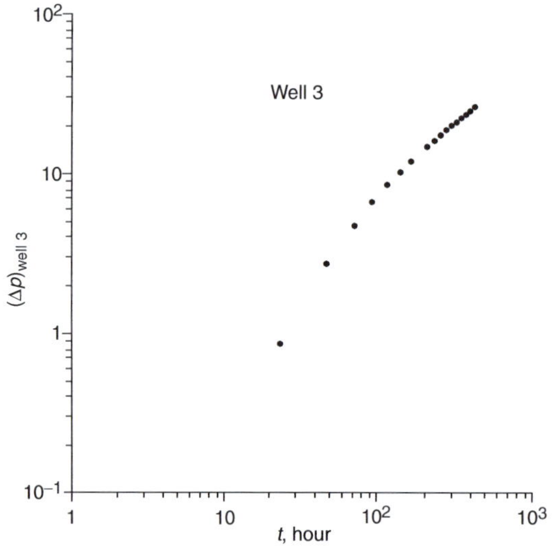

FIGURE 1.101 Interference data of Well 3. *(After Sabet, M., 1991. Well Test Analysis. Gulf Publishing, Dallas, TX.)*

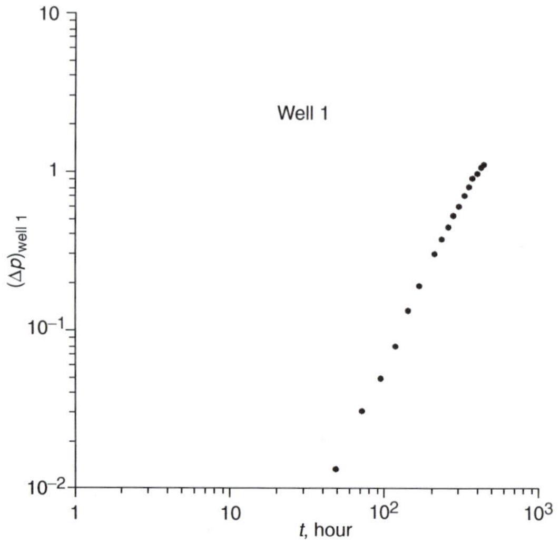

FIGURE 1.102 Interference data of Well 1. *(After Sabet, M., 1991. Well Test Analysis. Gulf Publishing, Dallas, TX.)*

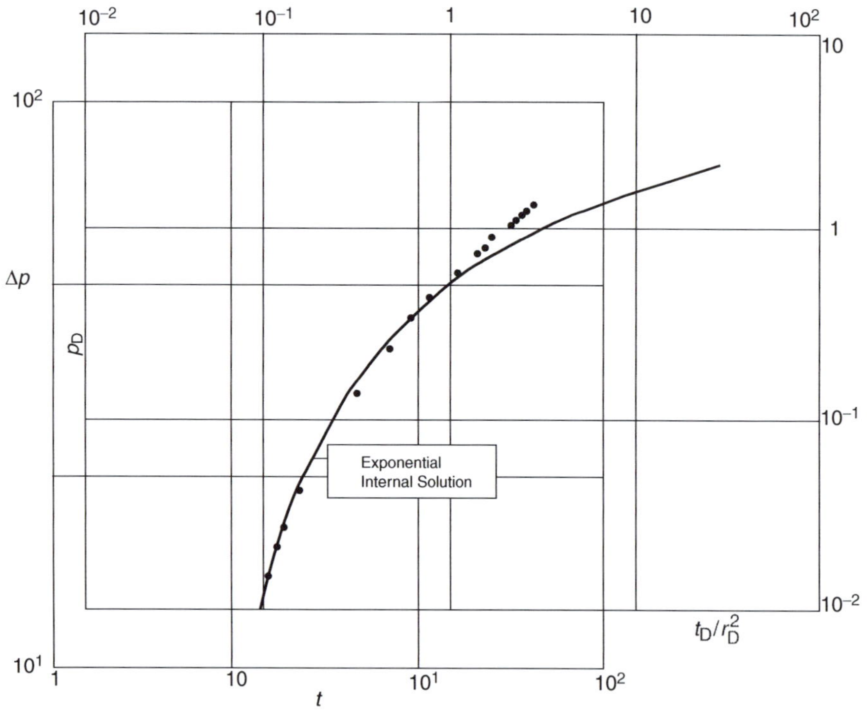

FIGURE 1.103 Match of interference data of Well 3. *(After Sabet, M., 1991. Well Test Analysis. Gulf Publishing, Dallas, TX.)*

Step 3. Solve for k and ϕ between Wells 2 and 3 by applying Eqs. (1.278) and (1.279)

$$k = \left[\frac{141.2QB\mu}{h}\right]\left[\frac{p_D}{\Delta p}\right]_{MP}$$

$$= \left[\frac{141.2(12.4)(920.9)(0.0186)}{75}\right]\left(\frac{0.1}{2}\right) = 19.7 \text{ md}$$

$$\phi = \frac{0.002637}{c_t r^2}\left[\frac{k}{\mu}\right]\left[\frac{t}{t_D/r_D^2}\right]_{MP}$$

$$= \frac{0.0002637}{(2.74\times 10^{-4})(2\times 5280)^2}\left(\frac{19.7}{0.0186}\right)\left(\frac{159}{1}\right)$$

$$= 0.00144$$

Step 4. Figure 1.104 shows the match of the test data for Well 1 with the following matching points:

$$(p_D)_{MP} = 1 \quad \text{and} \quad (\Delta p)_{MP} = 5.6 \text{ psi}$$

$$\left(\frac{t_D}{r_D^2}\right)_{MP} = 0.1 \quad \text{and} \quad (t)_{MP} = 125 \text{ hours}$$

Step 5. Calculate k and ϕ:

$$k = \left[\frac{141.2(12.4)(920.9)(0.0186)}{75}\right]\left(\frac{1}{5.6}\right)$$

$$= 71.8 \text{ md}$$

$$\phi = \frac{0.0002637}{(2.74\times 10^{-4})(8\times 5280)^2}\left(\frac{71.8}{0.0180}\right)\left(\frac{125}{0.1}\right)$$

$$= 0.0026$$

In a homogeneous and isotropic reservoir, i.e., permeability is the same throughout the reservoir, the minimum area of the reservoir

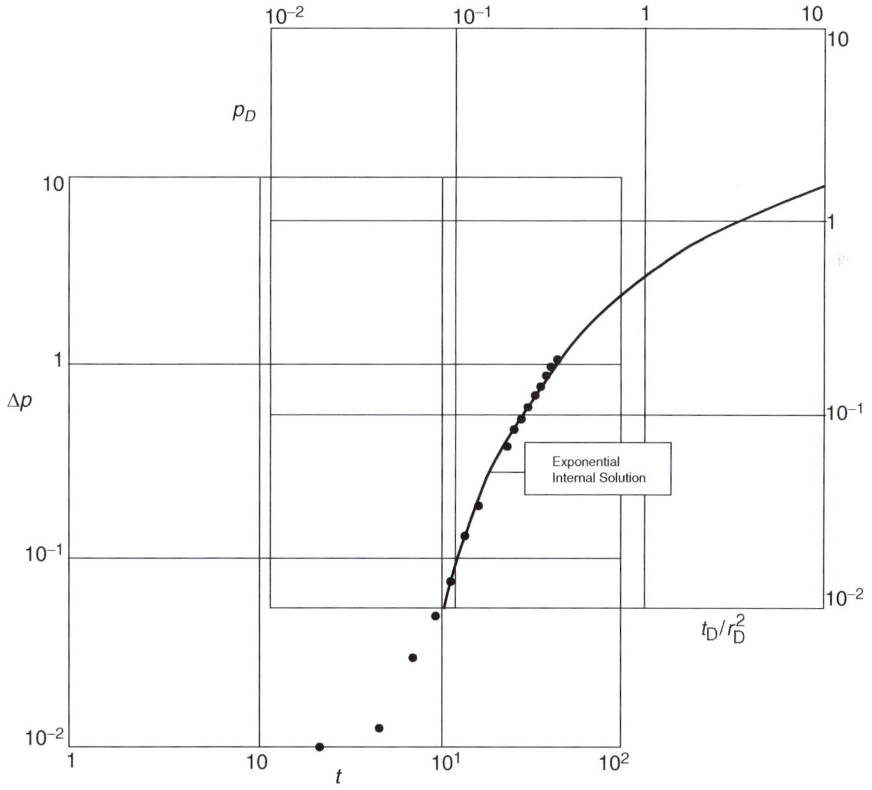

FIGURE 1.104 Match of interference data of Well 1.

investigated during an interference test between two wells located at a distance r apart is obtained by drawing two circles of radius r centered at each well.

1.6.2 Interference Testing in Homogeneous Anisotropic Reservoirs

A homogeneous anisotropic reservoir is one in which the porosity ϕ and thickness h are the same throughout the system, but the permeability varies with direction. Using multiple observation wells when conducting an interference test in a homogeneous anisotropic reservoir, it is possible to determine the maximum and minimum permeabilities, i.e., k_{max} and k_{min}, and their directions relative to well locations. Based on the work of Papadopulos (1965), Ramey (1975) adopted the Papadopulos solution for estimating anisotropic reservoir properties from an interference test that requires at least three observation wells for analysis. Figure 1.105 defines the necessary nomenclature used in the analysis of interference data in a homogeneous anisotropic reservoir.

Figure 1.105 shows an active well, with its coordinates at the *origin*, and several observation wells are each located at coordinates defined by (x, y). Assuming that all the wells in the testing area have been shut in for a sufficient time to equalize the pressure to p_i, placing the active well on production (or injection) will cause a change in pressure of Δp, i.e., $\Delta p = p_i - p(x, y, t)$, at all observation wells. This change in the pressure will occur after a lag period with a length that depends, among other parameters, on:

- the distance between the active well and observation well;
- permeability;
- wellbore storage in the active well;
- the skin factor following a lag period.

Ramey (1975) showed that the change in pressure at an observation well with coordinates of (x, y) at any time t is given by the Ei function as:

$$p_D = -\frac{1}{2}\mathrm{Ei}\left[\frac{-r_D^2}{4t_D}\right]$$

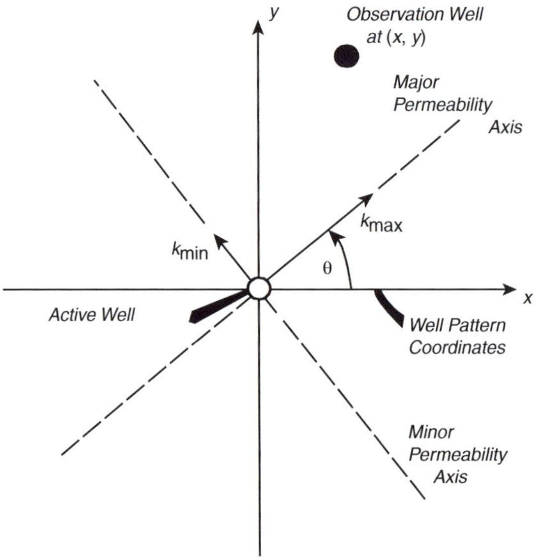

FIGURE 1.105 Nomenclature for anisotropic permeability system. *(After Ramey Jr., H.J., 1975. Interference analysis for anisotropic formations. J. Pet. Technol., 27 (10) 1290–1298).*

The dimensionless variables are defined by:

$$p_D = \frac{\bar{k}h[p_i - p(x, y, t)]}{141.2QB\mu} \quad (1.280)$$

$$\frac{t_D}{r_D^2} = \left[\frac{(\bar{k})^2}{y^2 k_x + x^2 k_y - 2xyk_{xy}}\right]\left(\frac{0.0002637t}{\phi\mu c_t}\right) \quad (1.281)$$

with

$$\bar{k} = \sqrt{k_{max}k_{min}} = \sqrt{k_x k_y - k_{xy}^2} \quad (1.282)$$

Ramey also developed the following relationships:

$$k_{max} = \frac{1}{2}\left[(k_x + k_y) + \sqrt{(k_x k_y)^2 + 4k_{xy}^2}\right] \quad (1.283)$$

$$k_{min} = \frac{1}{2}\left[(k_x + k_y)^2 - \sqrt{(k_x k_y)^2 + 4k_{xy}^2}\right] \quad (1.284)$$

$$\theta_{max} = \arctan\left(\frac{k_{max} - k_x}{k_{xy}}\right) \quad (1.285)$$

$$\theta_{min} = \arctan\left(\frac{k_{min} - k_y}{k_{xy}}\right) \quad (1.286)$$

where

k_x = permeability in x direction, md
k_y = permeability in y direction, md
k_{xy} = permeability in xy direction, md
k_{min} = minimum permeability, md
k_{max} = maximum permeability, md
$\bar{k}$ = average system permeability, md
θ_{max} = direction (angle) of k_{max} as measured from the $+x$-axis
θ_{min} = direction (angle) of k_{min} as measured from the $+y$-axis
x, y = coordinates, ft
t = time, hours

Ramey pointed out that if $\phi\mu c_t$ is not known, solution of the above equations will require that a minimum of three observation wells is used in the test, otherwise the required information can be obtained with only two observation wells. Type curve matching is the first step of the analysis technique. Observed pressure changes at each observation well, i.e., $\Delta p = p_i - p(x, y, t)$ are plotted on log–log paper and matched with the exponential–integral type curve shown in Figure 1.100. The associated specific steps of the methodology of using the type curve in determining the properties of a homogeneous anisotropic reservoir are summarized below:

Step 1. From at least three observation wells, plot the observed pressure change Δp vs. time t for each well on the same size scale as the type curve given in Figure 1.100.

Step 2. Match each of the observation well data set to the type curve of Figure 1.100. Select a convenient match point for each data set so that the pressure match point $(\Delta p, p_D)_{MP}$ is the same for all observation well responses, while the time match points $(t, t_D/r_D^2)_{MP}$ vary.

Step 3. From the pressure match point $(\Delta p, p_D)_{MP}$, calculate the average system permeability from:

$$\bar{k} = \sqrt{k_{min}k_{max}} = \left[\frac{141.2QB\mu}{h}\right]\left(\frac{p_D}{\Delta p}\right)_{MP} \quad (1.287)$$

Notice from Eq. (1.282) that:

$$(\bar{k})^2 = k_{min}k_{max} = k_x k_y - k_{xy}^2 \quad (1.288)$$

Step 4. Assuming *three observation wells*, use the time match $[(t, (t_D/r_D^2)]_{MP}$ for each observation well to write:

Well 1:

$$\left[\frac{(t_D/r_D^2)}{t}\right]_{MP} = \left(\frac{0.0002637}{\phi\mu c_t}\right) \times \left(\frac{(\bar{k})^2}{y_1^2 k_x + x_1^2 k_y - 2x_1 y_1 k_{xy}}\right)$$

Rearranging gives:

$$y_1^2 k_x + x_1^2 k_y - 2x_1 y_1 k_{xy} = \left(\frac{0.0002637}{\phi\mu c_t}\right) \times \left(\frac{(\bar{k})^2}{[(t_D/r_D^2)/t]_{MP}}\right) \quad (1.289)$$

Well 2:

$$\left[\frac{(t_D/r_D^2)}{t}\right]_{MP} = \left(\frac{0.0002637}{\phi\mu c_t}\right) \times \left(\frac{(\bar{k})^2}{y_2^2 k_x + x_2^2 k_y - 2x_2 y_2 k_{xy}}\right)$$

$$y_2^2 k_x + x_2^2 k_y - 2x_2 y_2 k_{xy} = \left(\frac{0.0002637}{\phi\mu c_t}\right) \times \left(\frac{(\bar{k})^2}{[(t_D/r_D^2)/t]_{MP}}\right) \tag{1.290}$$

Well 3:

$$\left[\frac{(t_D/r_D^2)}{t}\right]_{MP} = \left(\frac{0.0002637}{\phi\mu c_t}\right) \times \left(\frac{(\bar{k})^2}{y_3^2 k_x + x_3^2 k_y - 2x_3 y_3 k_{xy}}\right)$$

$$y_3^2 k_x + x_3^2 k_y - 2x_3 y_3 k_{xy} = \left(\frac{0.0002637}{\phi\mu c_t}\right) \times \left(\frac{(\bar{k})^2}{[(t_D/r_D^2)/t]_{MP}}\right) \tag{1.291}$$

Eqs. (1.288)–(1.291) contain the following four unknowns:

- k_x = permeability in x direction
- k_y = permeability in y direction
- k_{xy} = permeability in xy direction
- $\phi\mu c_t$ = porosity group

These four equations can be solved simultaneously for the above four unknowns. Example 1.43 as given by Ramey (1975) and later by Earlougher (1977) is used to clarify the use of the proposed methodology for determining the properties of an anisotropic reservoir.

Example 1.43

The following data is for an interference test in a nine-spot pattern with one active well and eight observation wells. Before testing, all wells were shut in. The test was conducted by injecting at −115 STB/day and observing the fluid levels in the remaining eight shut-in wells. Figure 1.106 shows the well locations. For simplicity, only the recorded pressure data for three observation wells, as tabulated below, is used to illustrate the methodology. These selected wells are labeled Well 5-E, Well 1-D, and Well 1-E.

Well 1-D		Well 5-E		Well 1-E	
t (hour)	**Δp (psi)**	**t (hour)**	**Δp (psi)**	**t (hour)**	**Δp (psi)**
23.5	−6.7	21.0	−4.0	27.5	−3.0
28.5	−7.2	47.0	−11.0	47.0	−5.0
51.0	−15.0	72.0	−16.3	72.0	−11.0
77.0	−20.0	94.0	−21.2	95.0	−13.0
95.0	−25.0	115.0	−22.0	115.0	−16.0
			−25.0		

The well coordinates (x, y) are as follows:

	Well	x (ft)	y (ft)
1	1-D	0	475
2	5-E	475	0
3	1-E	475	514

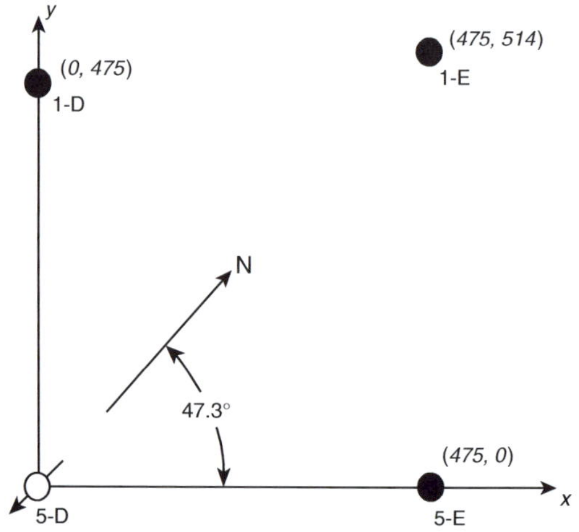

FIGURE 1.106 Well locations for Example 1.43. *(After Earlougher, Robert C., Jr., 1977. Advances in Well Test Analysis, Monograph, vol. 5. Society of Petroleum Engineers of AIME, Dallas, TX; Permission to publish by the SPE, copyright SPE, 1977.)*

$i_w = -115$ STB/day,	$B_w = 1.0$ bbl/STB,	$\mu_w = 1.0$ cp
$\phi = 20\%$,	$T = 75\ ^\circ$F,	$h = 25$ ft
$c_o = 7.5 \times 10^{-6}\ \text{psi}^{-1}$,	$c_w = 3.3 \times 10^{-6}\ \text{psi}^{-1}$	
$c_f = 3.7 \times 10^{-6}\ \text{psi}^{-1}$,	$r_w = 0.563$ ft,	$p_i = 240$ psi

Calculate k_{max}, k_{min}, and their directions relative to the x-axis.

Solution

Step 1. Plot Δp vs. time t for each of the three observation wells on a log–log plot of the same scale as that of Figure 1.100. The resulting plots with the associated match on the type curve are shown in Figure 1.107.

Step 2. Select the same pressure match point on the pressure scale for all the observation wells; however, the match point on the time scale is different for all wells.

Match Point	Well 1-D	Well 5-E	Well 1-E
$(p_D)_{MP}$	0.26	0.26	0.26
$(t_D/r_D^2)_{MP}$	1.00	1.00	1.00
$(\Delta p)_{MP}$	−10.00	−10.00	−10.00
$(t)_{MP}$	72.00	92.00	150.00

Step 3. From the pressure match point, use Eq. (1.287) to solve for $\overline{k}$:

$$\overline{k} = \sqrt{k_{min} k_{max}} = \left[\frac{141.2 QB\mu}{h}\right]\left(\frac{p_D}{\Delta p}\right)_{MP}$$

$$= \sqrt{k_{min} k_{max}} = \left[\frac{141.2(-115)(1.0)(1.0)}{25}\right]\left(\frac{0.26}{-10}\right)$$

$$= 16.89\,\text{md}$$

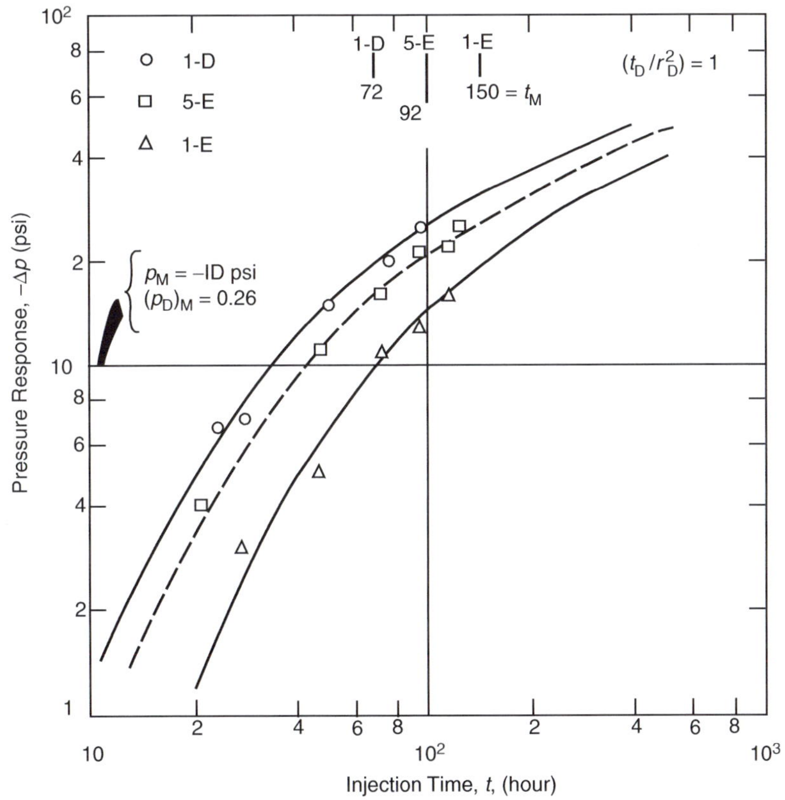

FIGURE 1.107 Interference data of Example 1.6 matched to Figure 1.100. Pressure match is same for all curves. *(After Earlougher, Robert C., Jr., 1977. Advances in Well Test Analysis, Monograph, vol. 5. Society of Petroleum Engineers of AIME, Dallas, TX; Permission to publish by the SPE, copyright SPE, 1977.)*

or

$$k_{min} k_{max} = (16.89)^2 = 285.3$$

Step 4. Using the time match point $(t, t_D/r_D^2)_{MP}$ for each observation well, apply Eqs. (1.289)–(1.291) to give:

For Well 1-D with $(x_1, y_1) = (0, 475)$:

$$y_1^2 k_x + x_1^2 k_y - 2x_1 y_1 k_{xy} = \left(\frac{0.0002637}{\phi\mu c_t}\right) \times \left(\frac{(\bar{k})^2}{[(t_D/r_D^2)/t]_{MP}}\right)(475)^2 k_x$$

$$+ (0)^2 k_y - 2(0)(475) = \frac{0.0002637(285.3)}{\phi\mu c_t}\left(\frac{72}{1.0}\right)$$

Simplifying gives:

$$k_x = \frac{2.401 \times 10^{-5}}{\phi\mu c_t} \quad \text{(A)}$$

For Well 5-E with $(x_2, y_2) = (475, 0)$:

$$(0)^2 k_x + (475)^2 k_y - 2(475)(0)k_{xy} = \frac{0.0002637(285.3)}{\phi\mu C_t}\left(\frac{92}{1.0}\right)$$

or

$$k_y = \frac{3.068 \times 10^{-5}}{\phi\mu c_t} \quad \text{(B)}$$

For Well 1-E with $(x_3, y_3) = (475, 514)$:

$$(514)^2 k_x + (475)^2 k_y - 2(475)(514)k_{xy} = \frac{0.0002637(285.3)}{\phi\mu c_t}\left(\frac{150}{1.0}\right)$$

or

$$0.5411 k_x + 0.4621 k_y - k_{xy} = \frac{2.311 \times 10^{-5}}{\phi\mu c_t} \quad \text{(C)}$$

Step 5. Combine Eqs. A–C to give:

$$k_{xy} = \frac{4.059 \times 10^{-6}}{\phi\mu c_t} \quad \text{(D)}$$

Step 6. Using Eqs. A, B, and D in Eq. (1.288) gives:

$$[k_x y_y] - k_{xy}^2 = (\bar{k})^2$$

$$\left[\frac{(2.401 \times 10^{-5})}{(\phi\mu c_t)}\frac{(3.068 \times 10^{-5})}{(\phi\mu c_t)}\right] - \frac{(4.059) \times 10^{-6})^2}{(\phi\mu c_t)} = (16.89)^2 = 285.3$$

or

$$\phi\mu c_t = \sqrt{\frac{(2.401\times10^{-5})(3.068\times10^{-5}) - (4.059\times10^{-6})^2}{285.3}}$$

$$= 1.589\times10^{-6}\ \text{cp/psi}$$

Step 7. Solve for c_t:

$$c_t = \frac{1.589 \times 10^{-6}}{(0.20)(1.0)} = 7.95 \times 10^{-6}\ \text{psi}^{-1}$$

Step 8. Using the calculated value of $\phi\mu c_t$ from step 6, i.e., $\phi\mu c_t = 1.589 \times 10^{-6}$, in Eqs. A, B, and D, solve for k_x, k_y, and k_{xy}:

$$k_x = \frac{2.401 \times 10^{-5}}{1.589 \times 10^{-6}} = 15.11\ \text{md}$$

$$k_y = \frac{3.068 \times 10^{-5}}{1.589 \times 10^{-6}} = 19.31\ \text{md}$$

$$k_{xy} = \frac{4.059 \times 10^{-6}}{1.589 \times 10^{-6}} = 2.55\ \text{md}$$

Step 9. Estimate the maximum permeability value by applying Eq. (1.283), to give:

$$k_{max} = \frac{1}{2}\left[(k_x + k_y) + \sqrt{(k_x k_y)^2 + 4k_{xy}^2}\right]$$

$$= \frac{1}{2}\left[(15.11 + 19.31) + \sqrt{(15.11 - 19.31)^2 + 4(2.55)^2}\right]$$

$$= 20.5\ \text{md}$$

Step 10. Estimate the minimum permeability value by applying Eq. (1.284):

$$k_{min} = \frac{1}{2}\left[(k_x + k_y)^2 - \sqrt{(k_x k_y)^2 + 4k_{xy}^2}\right]$$

$$= \frac{1}{2}\left[(15.11 + 19.31) - \sqrt{(15.11 - 19.31)^2 + 4(2.55)^2}\right]$$

$$= 13.9\ \text{md}$$

Step 11. Estimate the direction of k_{max} from Eq. (1.285):

$$\theta_{max} = \arctan\left(\frac{k_{max} - k_x}{k_{xy}}\right)$$
$$= \arctan\left(\frac{20.5 - 15.11}{2.55}\right)$$
$$= 64.7^\circ \text{ as measured from the } +x\text{–axis}$$

1.6.3 Pulse Testing in Homogeneous Isotropic Reservoirs

Pulse tests have the same objectives as conventional interference tests, which include:

- estimation of permeability k;
- estimation of porosity–compressibility product ϕc_t;
- existence of pressure communication between wells.

The tests are conducted by sending a sequence of flow disturbances, "pulses," into the reservoir from an active well and monitoring the pressure responses to these signals at shut-in observation wells. The pulse sequence is created while producing from (or injecting into) the active well, then shutting it in, and repeating that sequence in a regular pattern, as depicted by Figure 1.108. The figure is for an active producing well that is pulsed by shutting in, continuing production, and repeating the cycle.

The production (or injection) rate should be the same during each period. The lengths of all production periods and all shut-in periods should be equal; however, production periods do not have to equal shut-in periods. These pulses create a very distinctive pressure response at the observation well which can be easily distinguished from any pre-existing trend in reservoir pressure, or random pressure perturbations, "noise," which could otherwise be misinterpreted.

It should be noted that pulse testing offers several advantages over conventional interference tests:

- Because the pulse length used in a pulse test is short, ranging from a few hours to a few days, boundaries seldom affect the test data.
- Because of the distinctive pressure response, there are fewer interpretation problems caused by random "noise" and by trends in reservoir pressure at the observation well.
- Because of shorter test times, pulse tests cause less disruption of normal field operations than interference test.

For each pulse, the pressure response at the observation well is recorded (as illustrated in Figure 1.109) with a very sensitive pressure gauge. In pulse tests, pulse 1 and pulse 2 have characteristics that differ from those of all subsequent pulses. Following these pulses, all odd pulses have similar characteristics and all even pulses also have similar characteristics. Any one of the pulses can be analyzed for k and ϕc_t. Usually, several pulses are analyzed and compared.

Figure 1.109, which depicts the rate history of the active well and the pressure response at an observation well, illustrates the following five parameters which are required for the analysis of a pulse test:

(1) The "pulse period" Δt_p represents the length of the shut-in time.
(2) The "cycle period" Δt_C represents the total time length of a cycle, i.e., the shut-in period plus the flow or injection period.
(3) The "flowing or injection period" Δt_f represents the length of the flow or injection time.
(4) The "time lag" t_L represents the elapsed time between the end of a pulse and the pressure peak caused by the pulse. This time lag t_L is associated with each pulse and essentially describes the time required for a pulse created when the rate is changed to move from the active well to the observation well. It should be pointed out that a flowing (or injecting) period is a "pulse" and a shut-in period is another pulse; the combined two pulses constitute a "cycle."
(5) The "pressure response amplitude" Δp is the vertical distance between two adjacent

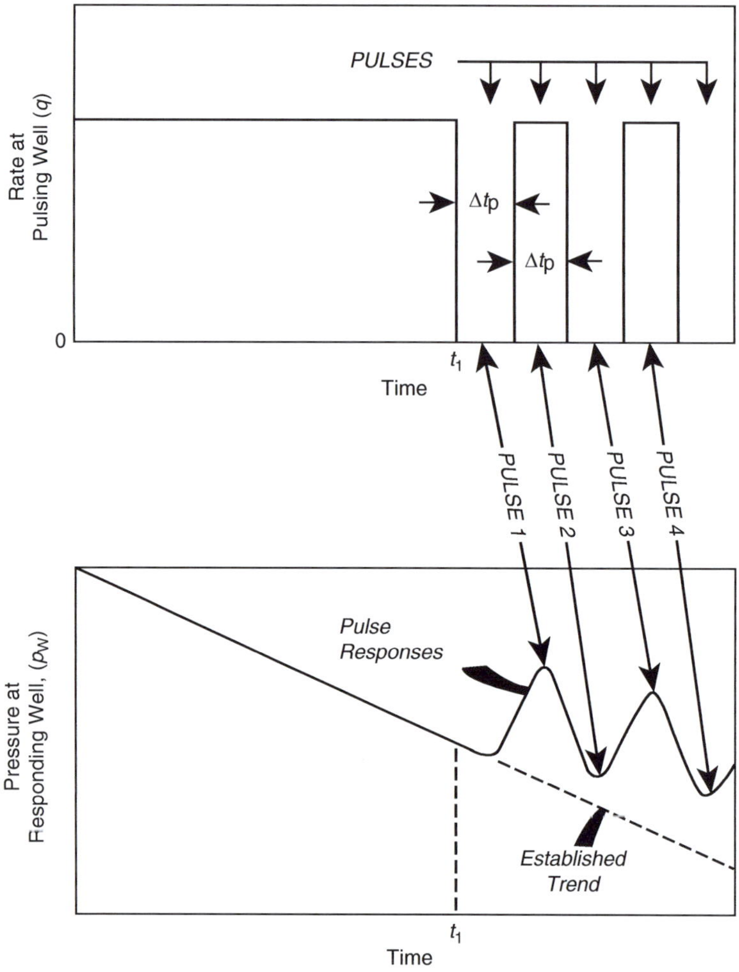

FIGURE 1.108 Schematic illustration of rate (pulse) history and pressure response for a pulse test. *(After Earlougher, Robert C., Jr., 1977. Advances in Well Test Analysis, Monograph, vol. 5. Society of Petroleum Engineers of AIME, Dallas, TX; Permission to publish by the SPE, copyright SPE, 1977.)*

peaks (or valleys) and a line parallel to this through the valley (or peak), as illustrated in Figure 1.109. Analysis of simulated pulse tests show that pulse 1, i.e., the "first odd pulse," and pulse 2, i.e., the "first even pulse," have characteristics that differ from all subsequent pulses. Beyond these *initial pulses*, all odd pulses have similar characteristics, and all even pulses exhibit similar behavior.

Kamal and Bigham (1975) proposed a pulse test analysis technique that uses the following four dimensionless groups:

(1) Pulse ratio $F^{\backslash}$, as defined by:

$$F^{\backslash} = \frac{\text{Pulse period}}{\text{Cycle period}} = \frac{\Delta t_p}{\Delta t_p + \Delta t_f} = \frac{\Delta t_p}{\Delta t_C} \quad \textbf{(1.292)}$$

where the time is expressed in hours.

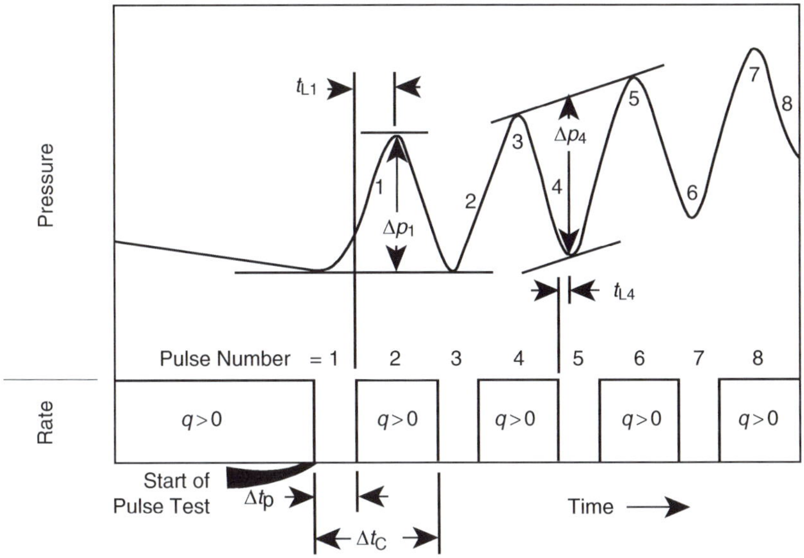

FIGURE 1.109 Schematic pulse test rate and pressure history showing definition of time lag (t_L) and pulse response amplitude (Δp) curves. *(After Earlougher, Robert C., Jr., 1977. Advances in Well Test Analysis, Monograph, vol. 5. Society of Petroleum Engineers of AIME, Dallas, TX; Permission to publish by the SPE, copyright SPE, 1977.)*

(2) Dimensionless time lag $(t_L)_D$, as given by:

$$(t_L)_D = \frac{t_L}{\Delta t_C} \tag{1.293}$$

(3) Dimensionless distance (r_D) between the active and observation wells, as given by:

$$r_D = \frac{r}{r_w} \tag{1.294}$$

where

r = distance between the active well and the observation well, ft

(4) Dimensionless pressure response amplitude Δp_D, as given by:

$$\Delta p_D = \left[\frac{\bar{k}h}{141.2B\mu}\frac{\Delta p}{Q}\right] \tag{1.295}$$

where

Q = rate at the active well while it is active, with the sign convention that $\Delta p/Q$ is always positive, i.e., the absolute value of $|\Delta p/Q|$.

Kamal and Bigham developed a family of curves, as shown in Figures 1.110–1.117, that correlates the pulse ratio $F^\backslash$ and the dimensionless time lag $(t_L)_D$ to the dimensionless pressure Δp_D. These curves are specifically designated to analyze the pulse test data for the following conditions:

- *First odd pulse*: Figures 1.110 and 1.114.
- *First even pulse*: Figures 1.111 and 1.115.
- All the *remaining odd pulses* except the first: Figures 1.112 and 1.116.
- All the *remaining even pulses* except the first: Figures 1.113 and 1.117.

The time lag t_L and pressure response amplitude Δp from one or more pulse responses are used to estimate the average reservoir permeability from:

$$\bar{k} = \left[\frac{141.2QB\mu}{h\,\Delta p[(t_L)_D]^2}\right]\left[\Delta p_D\left(\frac{t_L}{\Delta t_C}\right)^2\right]_{\text{Fig}} \tag{1.296}$$

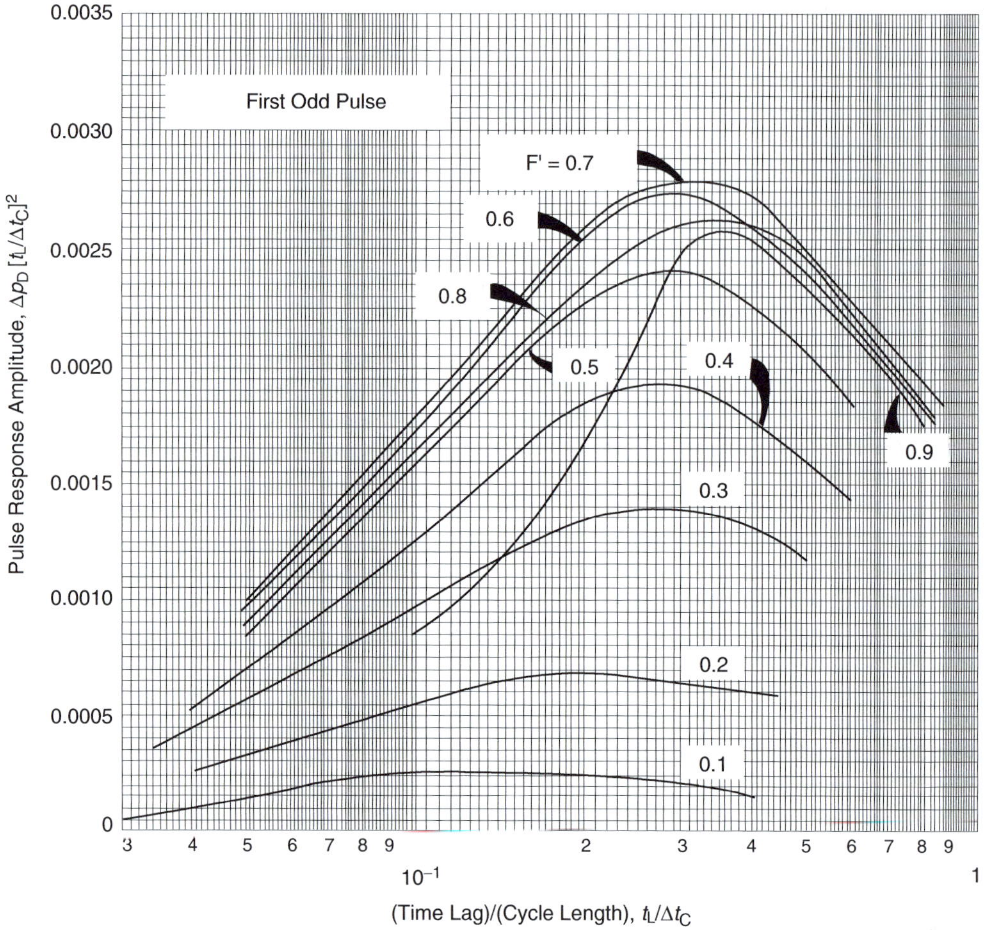

FIGURE 1.110 Pulse testing: relation between time lag and response amplitude for first odd pulse. *(After Kamal and Bigham, 1975.)*

The term $[\Delta p_D(t_L/\Delta t_C)^2]_{Fig}$ is determined from Figures 1.110, 1.111, 1.112, or 1.113 for the appropriate values of $t_L/\Delta t_C$ and $F^\backslash$. The other parameters of Eq. (1.296) are defined below:

- Δp = amplitude of the pressure response from the observation well for the *pulse being analyzed*, psi
- Δt_C = cycle length, hours
- Q = production (injection) rate during active period, STB/day
- $\overline{k}$ = average permeability, md

Once the permeability is estimated from Eq. (1.296), the porosity–compressibility product can be estimated from:

$$\phi c_t = \left[\frac{0.002637\overline{k}(t_L)}{\mu r^2}\right]\frac{1}{[(t_L)_D/r_D^2]_{Fig}} \tag{1.297}$$

where

t_L = time lag, hours
r = distance between the active well and observation well, ft

The term $[(t_L)_D/r_D^2]_{Fig}$ is determined from Figures 1.114, 1.115, 1.116, or 1.117. Again,

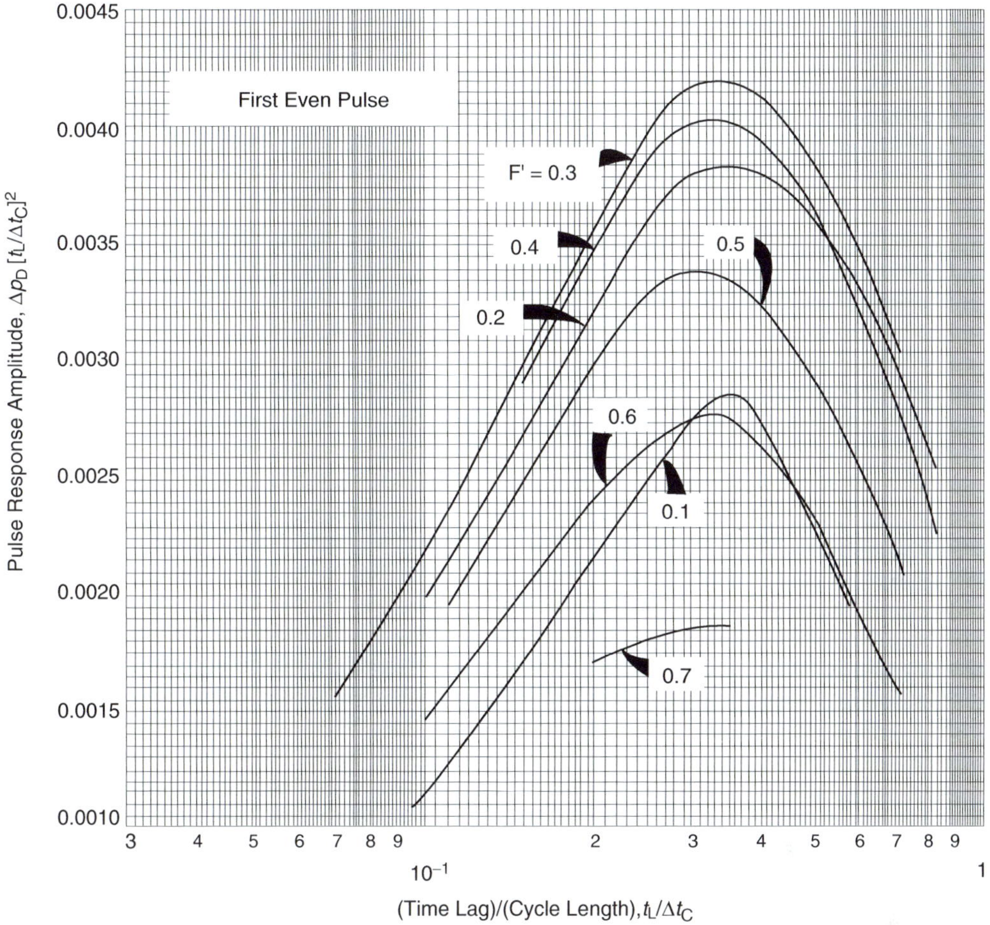

FIGURE 1.111 Pulse testing: relation between time lag and response amplitude for first even pulse. *(After Kamal and Bigham, 1975.)*

the appropriate figure to be used in analyzing the pressure response data depends on whether the first odd or first even pulse or one of the remaining pulses is being analyzed.

Example 1.44

In a pulse test following rate stabilization, the active well was shut in for 2 hours, then produced for 2 hours, and the sequence was repeated several times.[4]

An observation well at 933 ft from the active well recorded an amplitude pressure response of 0.639 psi during the *fourth* pulse and a time lag of 0.4 hours. The following additional data is also available:

$Q = 425$ STB/day, $B = 1.26$ bbl/STB
$r = 933$ ft, $h = 26$ ft
$\mu = 0.8$ cp, $\phi = 0.08$

Estimate $\overline{k}$ and ϕc_t.

Solution

Step 1. Calculate the pulse ratio $F^{\prime}$ from Eq. (1.292), to give:

[4]After John Lee, *Well Testing* (1982).

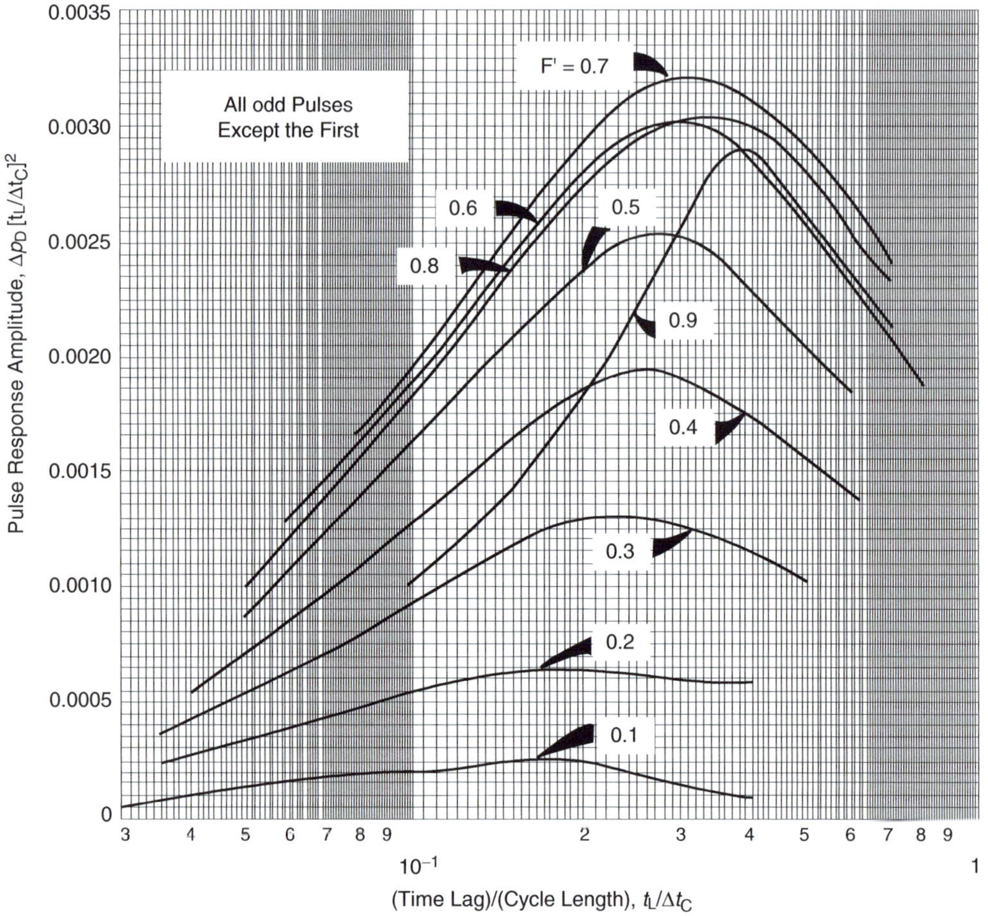

FIGURE 1.112 Pulse testing: relation between time lag and response amplitude for all odd pulses after the first. *(After Kamal and Bigham, 1975.)*

$$F' = \frac{\Delta t_p}{\Delta t_C} = \frac{\Delta t_p}{\Delta t_p + \Delta t_f} = \frac{2}{2+2} = 0.5$$

Step 2. Calculate the dimensionless time lag $(t_L)_D$ by applying Eq. (1.293):

$$(t_L)_D = \frac{t_L}{\Delta t_C} = \frac{0.4}{4} = 0.1$$

Step 3. Using the values of $(t_L)_D = 0.1$ and $F' = 0.5$, use Figure 1.113 to get:

$$\left[\Delta p_D \left(\frac{t_L}{\Delta t_C}\right)^2\right]_{\text{Fig}} = 0.00221$$

Step 4. Estimate the average permeability from Eq. (1.296), to give:

$$\bar{k} = \left[\frac{141.2 QB\mu}{h\,\Delta p[(t_L)_D]^2}\right]\left[\Delta p_D \left(\frac{t_L}{\Delta t_C}\right)^2\right]_{\text{Fig}}$$

$$= \left[\frac{(141.2)(425)(1.26)(0.8)}{(26)(0.269)[0.1]^2}\right](0.00221) = 817 \text{ md}$$

Step 5. Using $(t_L)_D = 0.1$ and $F' = 0.5$, use Figure 1.117 to get:

$$\left[\frac{(t_L)_D}{r_D^2}\right]_{\text{Fig}} = 0.091$$

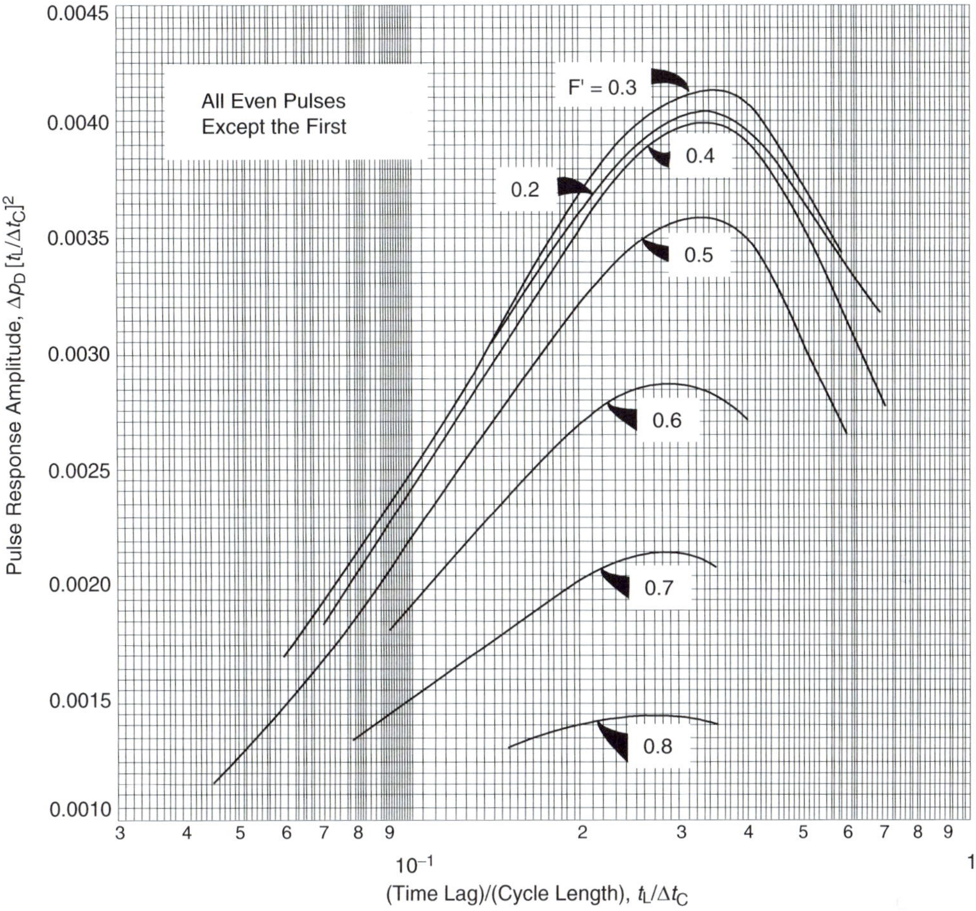

FIGURE 1.113 Pulse testing: relation between time lag and response amplitude for all even pulses after the first. *(After Kamal and Bigham, 1975.)*

Step 6. Estimate the product ϕc_t by applying Eq. (1.297):

$$\phi c_t = \left[\frac{0.00026387\bar{k}(t_L)}{\mu r^2}\right]\frac{1}{[(t_L)_D/r_D^2]_{\text{Fig}}}$$

$$= \left[\frac{0.0002637(817)(0.4)}{(0.8)(933)^2}\right]\frac{1}{(0.091)} = 1.36 \times 10^{-6}$$

Step 7. Estimate c_t as:

$$c_t = \frac{1.36 \times 10^{-6}}{0.08} = 17 \times 10^{-6}\ \text{psi}^{-1}$$

Example 1.45

A pulse test was conducted using an injection well as the pulsing well in a five-spot pattern with the four offsetting production wells as the responding wells. The reservoir was at its static pressure conditions when the first injection pulse was initiated at 9:40 a.m., with an injection rate of 700 bbl/day. The injection rate was maintained for 3 hours followed by a shut-in period for 3 hours. The injection shut-in periods were repeated several times and the results of pressure observation are given in

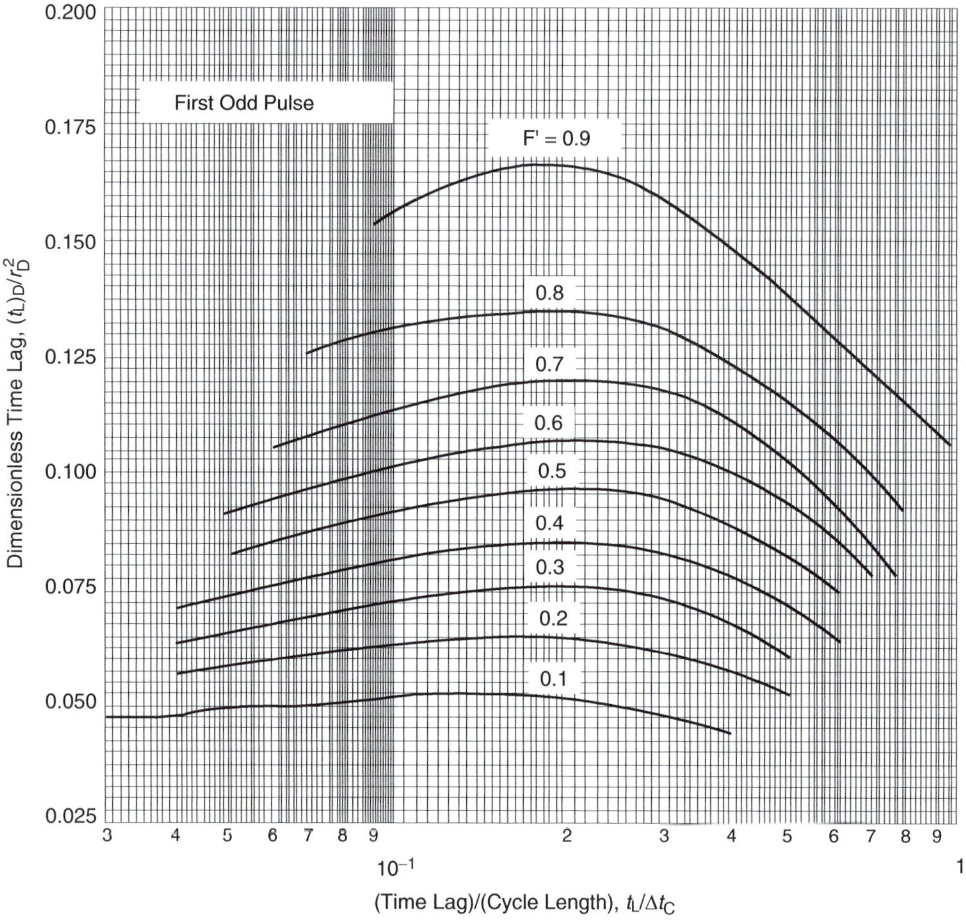

FIGURE 1.114 Pulse testing: relation between time lag and cycle length for first odd pulse. *(After Kamal and Bigham, 1975.)*

Table 1.9. The following additional data is available:[5]

$\mu = 0.87$ cp, $c_t = 9.6 \times 10^{-6}$ psi^{-1}
$\phi = 16\%$, $r = 330$ ft

Calculate the permeability and average thickness.

Solution

Step 1. Plot the pressure response from one of the observation wells as a function of time, as shown in Figure 1.118.

Analyzing First Odd-pulse Pressure Data

Step 1. From Figure 1.118 determine the amplitude pressure response and time lag during the first pulse, to give:

$$\Delta p = 6.8 \text{ psi}$$
$$t_L = 0.9 \text{ hour}$$

[5]Data reported by H. C. Slider, *Worldwide Practical Petroleum Reservoir Engineering Methods*, Penn Well Books, 1976.

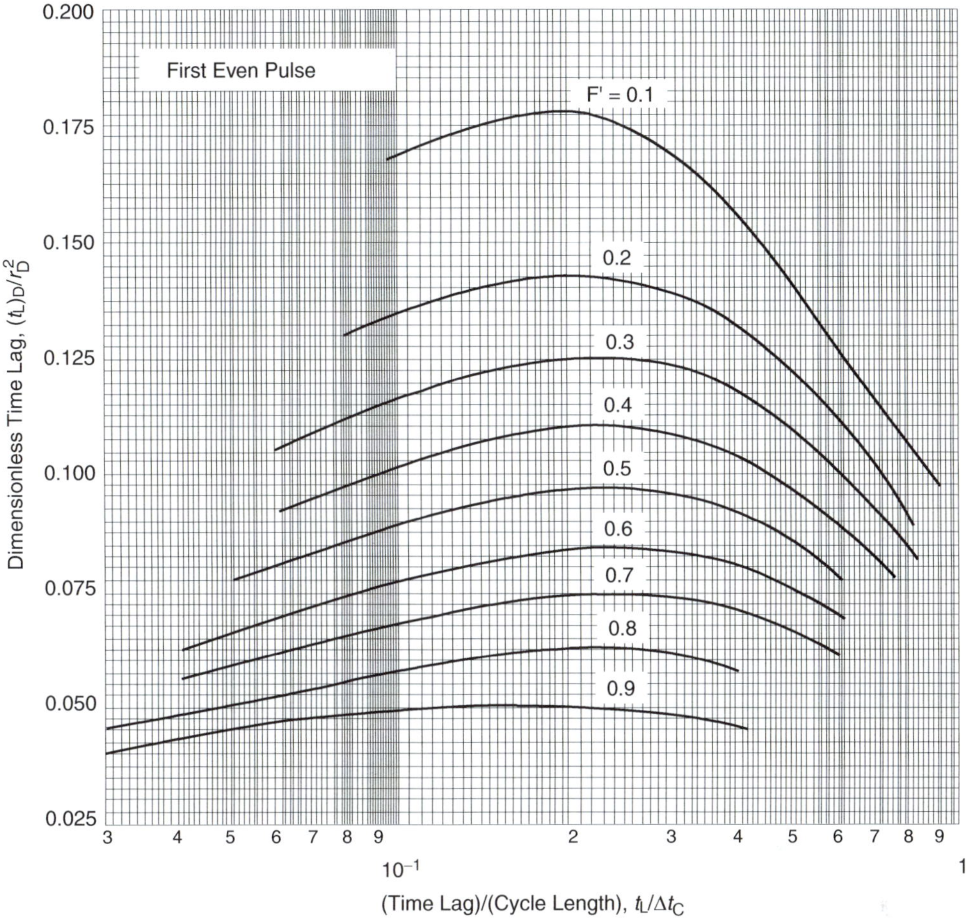

FIGURE 1.115 Pulse testing: relation between time lag and cycle length for first even pulse. *(After Kamal and Bigham, 1975.)*

Step 2. Calculate the pulse ratio $F^\backslash$ from Eq. (1.292) to give:

$$F^\backslash = \frac{\Delta t_p}{\Delta t_C} = \frac{3}{3+3} = 0.5$$

Step 3. Calculate the dimensionless time lag $(t_L)_D$ by applying Eq. (1.293):

$$(t_L)_D = \frac{t_L}{\Delta t_C} = \frac{0.9}{6} = 0.15$$

Step 4. Using the values of $(t_L)_D = 0.15$ and $F^\backslash = 0.5$, use Figure 1.110 to get:

$$\left[\Delta p_D \left(\frac{t_L}{\Delta t_C}\right)^2\right]_{\text{Fig}} = 0.0025$$

Step 5. Estimate average hk from Eq. (1.296), to give:

$$h\bar{k} = \left[\frac{141.2 QB\mu}{\Delta p[(t_L)_D]^2}\right]\left[\Delta p_D\left(\frac{t_L}{\Delta t_C}\right)^2\right]_{\text{Fig}}$$

$$= \left[\frac{(141.2)(700)(1.0)(0.86)}{(6.8)[0.15]^2}\right](0.0025) = 1387.9 \text{ md ft}$$

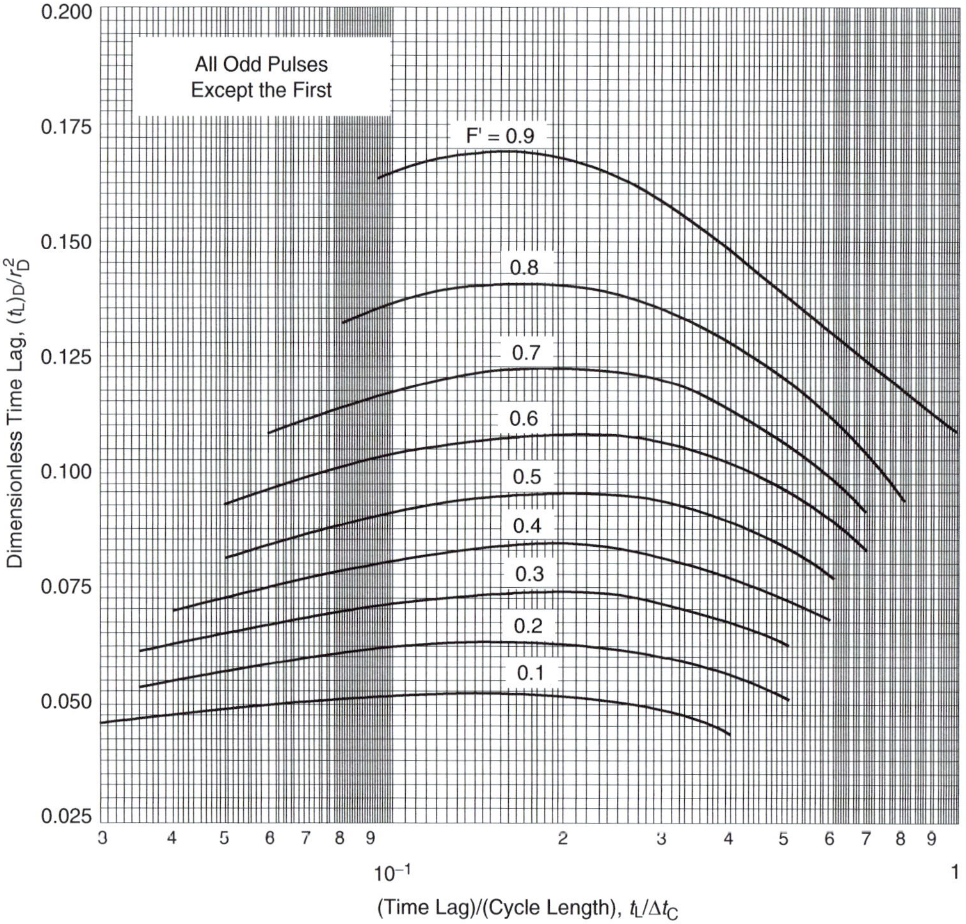

FIGURE 1.116 Pulse testing: relation between time lag and cycle length for all odd pulses after the first. *(After Kamal and Bigham, 1975.)*

Step 6. Using $(t_L)_D = 0.15$ and $F^{\backslash} = 0.5$, use Figure 1.114 to get:

$$\left[\frac{(t_L)_D}{r_D^2}\right]_{\text{Fig}} = 0.095$$

Step 7. Estimate the average permeability by rearranging Eq. (1.297) as:

$$\bar{k} = \left[\frac{\phi c_t \mu r^2}{0.0002637(t_L)}\right]\left[\frac{(t_L)_D}{r_D^2}\right]_{\text{Fig}}$$

$$= \left[\frac{(0.16)(9.6\times 10^{-6})(0.86)(330)^2}{0.0002637(0.9)}\right](0.095) = 57.6\text{ md}$$

Estimate the thickness h from the value of the product hk as calculated in step 5 and the above average permeability. That is:

$$h = \left[h\bar{k}/\bar{k}\right] = \left[\frac{1387.9}{57.6}\right] = 24.1\text{ ft}$$

Analyzing the Fifth Pulse Pressure Data

Step 1. From Figure 1.110 determine the amplitude pressure response and time lag during the fifth pulse, to give:

- $\Delta p = 9.2$ psi
- $t_L = 0.7$ hour

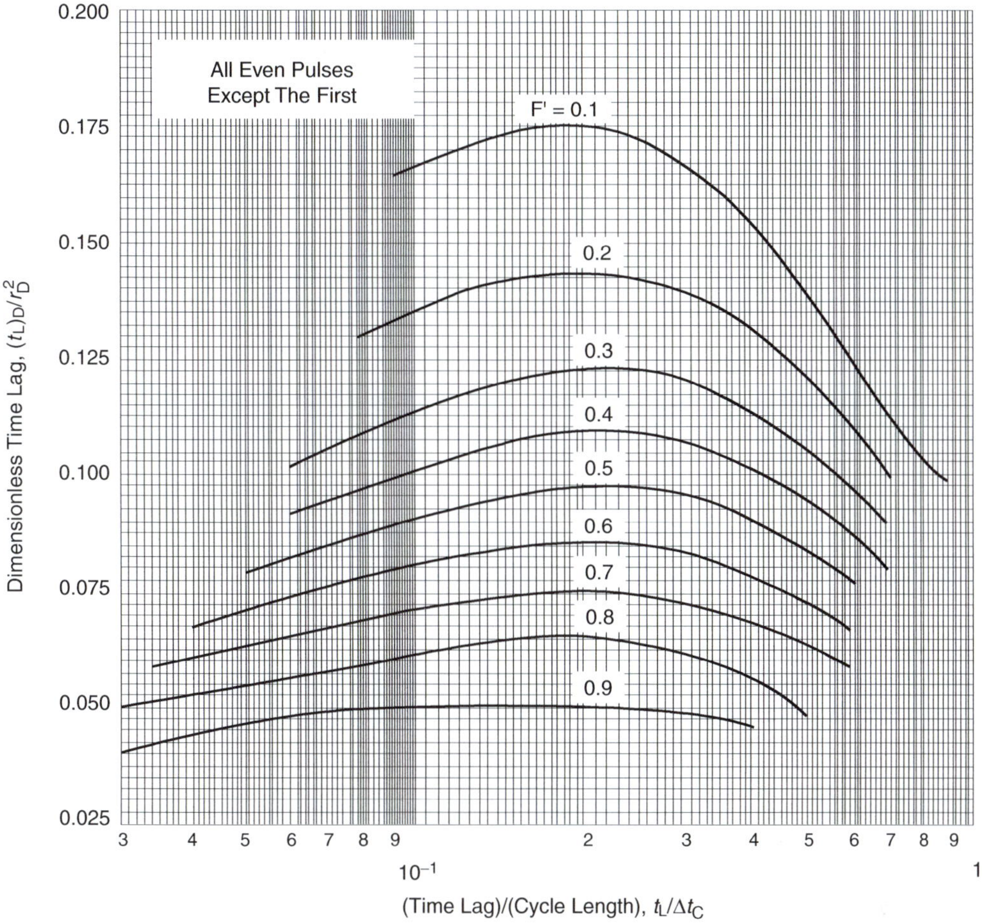

FIGURE 1.117 Pulse testing: relation between time lag and cycle length for all even pulses after the first. *(After Kamal and Bigham, 1975.)*

Step 2. Calculate the pulse ratio $F^\backslash$ from Eq. (1.292) to give:

$$F^\backslash = \frac{\Delta t_p}{\Delta t_C} = \frac{\Delta t_p}{\Delta t_p + \Delta t_f} = \frac{3}{3+3} = 0.5$$

Step 3. Calculate the dimensionless time lag $(t_L)_D$ by applying Eq. (1.293):

$$(t_L)_D = \frac{t_L}{\Delta t_C} = \frac{0.7}{6} = 0.117$$

Step 4. Using the values of $(t_L)_D = 0.117$ and $F^\backslash = 0.5$, use Figure 1.111 to get:

$$\left[\Delta p_D \left(\frac{t_L}{\Delta t_C}\right)^2\right]_{Fig} = 0.0018$$

Step 5. Estimate average hk from Eq. (1.296), to give:

$$h\bar{k} = \left[\frac{141.2 Q B \mu}{\Delta p[(t_L)_D]^2}\right] \left[\Delta p_D \left(\frac{t_L}{\Delta t_C}\right)^2\right]_{Fig}$$

$$= \left[\frac{(141.2)(700)(1.0)(0.86)}{(9.2)[0.117]^2}\right](0.0018) = 1213 \text{ md ft}$$

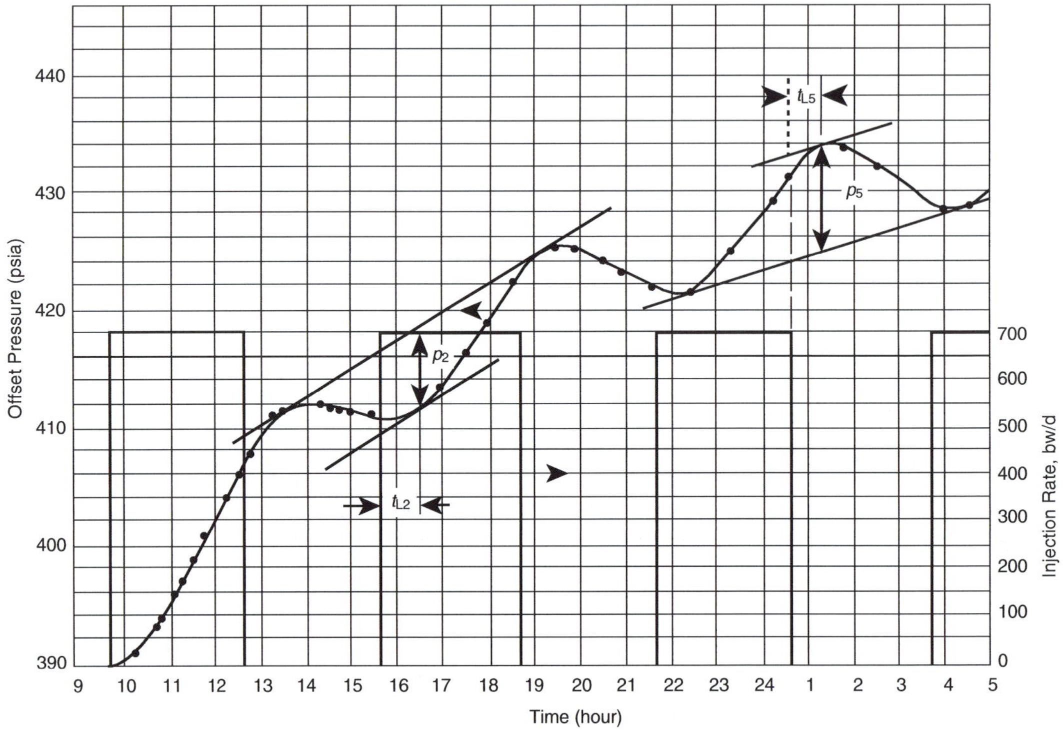

FIGURE 1.118 Pulse pressure response for Example 1.45.

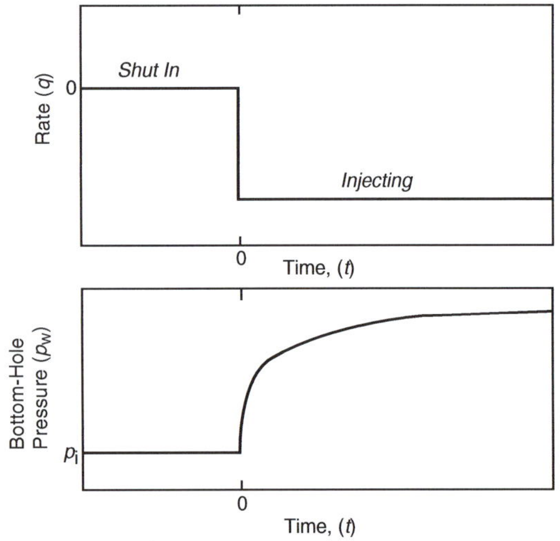

FIGURE 1.119 Idealized rate schedule and pressure response for injectivity testing.

Step 6. Using $(t_L)_D = 0.117$ and $F^{\backslash} = 0.5$, use Figure 1.115 to get:

$$\left[\frac{(t_L)_D}{r_D^2}\right]_{Fig} = 0.093$$

Step 7. Estimate the average permeability by rearranging Eq. (1.297) as:

$$\overline{k} = \left[\frac{\phi c_t \mu r^2}{0.0002637(t_L)}\right]\left[\frac{(t_L)_D}{r_D^2}\right]_{Fig}$$

$$= \left[\frac{(0.16)(9.6\times10^{-6})(0.86)(330)^2}{0.0002637(0.7)}\right](0.095) = 72.5 \text{ md}$$

Estimate the thickness h from the value of the product hk as calculated in step 5 and the above average permeability. That is:

$$h = \left[\overline{hk}/\overline{k}\right] = \left[\frac{1213}{72.5}\right] = 16.7 \text{ ft}$$

The above calculations should be repeated for all other pulses and the results should be compared with core and conventional well testing analysis to determine the best values that describe these properties.

1.6.4 Pulse Testing in Homogeneous Anisotropic Reservoirs

The analysis for the pulse test case is the same as that for the homogeneous isotropic case, except the average permeability $\bar{k}$ as defined by Eq. (1.282) is introduced into (1.296) and (1.297), to give:

$$\bar{k} = \sqrt{k_x k_y - k_{xy}^2} = \left[\frac{141.2QB\mu}{h\,\Delta p[(t_L)_D]^2}\right]\left[\Delta p_D\left(\frac{t_L}{\Delta t_C}\right)^2\right]_{\text{Fig}} \quad (1.298)$$

and

$$\phi c_t = \left[\frac{0.0002637(t_L)}{\mu r^2}\right]\left[\frac{(\bar{k})^2}{y^2 k_x + x^2 k_y - 2xyk_{xy}}\right] \times \frac{1}{[(t_L)_D/r_D^2]_{\text{Fig}}} \quad (1.299)$$

The solution methodology outlined in analyzing interference test data in homogeneous anisotropic reservoirs can be employed when estimating various permeability parameters from pulse testing.

1.6.5 Pulse Test Design Procedure

Prior knowledge of the expected pressure response is important so that the range and sensitivity of the pressure gauge and length of time needed for the test can be predetermined. To design a pulse test, Kamal and Bigham (1975) recommend the following procedure:

Step 1. The first step in designing a pulse test is to select the appropriate pulse ratio $F^\backslash$ as defined by Eq. (1.292), i.e., pulse ratio = pulse period/cycle period. A pulse ratio near 0.7 is recommended if analyzing the odd pulses; and near 0.3 if analyzing the even pulses. It should be noted the $F^\backslash$ should not exceed 0.8 or drop below 0.2.

Step 2. Calculate the dimensionless time lag from one of the following approximations:

$$\text{For odd pulses:} \quad (t_L)_D = 0.09 + 0.3F^\backslash \quad (1.300)$$

$$\text{For even pulses:} \quad (t_L)_D = 0.027 - 0.027F^\backslash \quad (1.301)$$

Step 3. Using the values of $F^\backslash$ and $(t_L)_D$ from step 1 and step 2, respectively, determine the dimensionless parameter $[(t_L)_D/r_D^2]$ from Figure 1.114 or Figure 1.115.

Step 4. Using the values of $F^\backslash$ and $(t_L)_D$, determine the dimensionless response amplitude $[\Delta p_D(t_L/\Delta t_C)^2]_{\text{Fig}}$ from the appropriate curve in Figure 1.110 or Figure 1.111.

Step 5. Using the following parameters:

- estimates of k, h, ϕ, μ, and c_t,
- values of $\lfloor (t_L)_D/r_D^2 \rfloor_{\text{Fig}}$ and $[\Delta p_D(t_L/\Delta t_C)^2]_{\text{Fig}}$ from steps 3 and 4, and
- Eqs. (1.277) and (1.278)

calculate the cycle period (Δt_C) and the response amplitude Δp from:

$$t_L = \left[\frac{\phi\mu c_t r^2}{0.0002637\bar{k}}\right]\left[\frac{(t_L)_D}{r_D^2}\right]_{\text{Fig}} \quad (1.302)$$

$$\Delta t_C = \frac{t_L}{(t_L)_D} \quad (1.303)$$

$$\Delta p = \left[\frac{141.2QB\mu}{h\bar{k}[(t_L)_D]^2}\right]\left[\Delta p_D\left(\frac{t_L}{\Delta t_C}\right)^2\right]_{\text{Fig}} \quad (1.304)$$

Step 6. Using the pulse ratio $F^\backslash$ and cycle period Δt_C, calculate the pulsing (shut-in) period and flow period from:

Pulse (shut-in) period: $\Delta t_p = F^\backslash \Delta t_C$
Flow period: $\Delta t_f = \Delta t_C - \Delta t_p$

Example 1.46

Design a pulse test using the following approximate properties:

$\mu = 3$ cp, $\phi = 0.18$, $k = 200$ md
$h = 25$ ft, $r = 600$ ft, $c_t = 10 \times 10^{-6}$ psi^{-1}
$B = 1$ bbl/STB, $Q = 100$ bbl/day, $F^{\prime} = 0.6$

Solution

Step 1. Calculate $(t_L)_D$ from Eq. (1.300) or (1.301). Since $F^{\prime}$ is 0.6, the odd pulses should be used and therefore from Eq. (1.300):

$$(t_L)_D = 0.09 + 0.3(0.6) = 0.27$$

Step 2. Selecting the first odd pulse, determine the dimensionless cycle period from Figure 1.114 to get:

$$\left[\frac{(t_L)_D}{r_D^2}\right]_{Fig} = 0.106$$

Step 3. Determine the dimensionless response amplitude from Figure 1.110 to get:

$$\left[\Delta p_D\left(\frac{t_L}{\Delta t_C}\right)\right]^2_{Fig} = 0.00275$$

Step 4. Solve for t_L, Δt_C, and Δp by applying Eqs. (1.302) through (1.304), to give:

Time lag:

$$t_L = \left[\frac{\phi\mu C_t r^2}{0.0002637\bar{k}}\right]\left[\frac{(t_L)_D}{r_D^2}\right]_{Fig}$$
$$= \left[\frac{(0.18)(3)(10 \times 10^{-6})(660)^2}{(0.0002637)(200)}\right](0.106) = 4.7 \text{ hours}$$

Cycle time:

$$\Delta t_C = \frac{t_L}{(t_L)_D} = \frac{4.7}{0.27} = 17.5 \text{ hours}$$

Pulse length (shut-in):

$$\Delta t_P = \Delta t_C F^{\prime} = (17.5)(0.27) \approx 5 \text{ hours}$$

Flow period:

$$\Delta t_f = \Delta t_C - \Delta t_p = 17.5 - 4.7 \approx 13 \text{ hours}$$

Step 5. Estimate the pressure response from Eq. (1.304):

$$\Delta p = \left[\frac{141.2QB\mu}{h\bar{k}[(t_L)_D]^2}\right]\left[\Delta p_D\left(\frac{t_L}{\Delta t_C}\right)^2\right]_{Fig}$$
$$= \left[\frac{(141.2)(100)(1)(3)}{(25)(200)(0.27)^2}\right](0.00275) = 0.32 \text{ psi}$$

This is the expected response amplitude for *odd-pulse* analysis. We shut-in the well for 5 hours and produced for 13 hours and repeated each cycle with a period of 18 hours.

The above calculations can be repeated if we desire to analyze the first even-pulse response.

1.7 FORMATION TESTING

Reservoir evaluation during the drilling and completion process involves many tools, including mud logs, Logging While Drilling (LWD), open hole logs, coring, etc. Estimates of reservoir permeability and productivity along with pore fluid contents usually require some sort of flow test to achieve a desired level of accuracy. Most of the pressure transient analysis methods discussed so far are applied to completed wells or to a conventional drill-stem test (DST).

One of the most powerful well testing tools in newly drilled wells is the formation tester (FT), which allows operators enormous flexibility during the drilling of a well (Figure 1.120). It can be conveyed on wireline or on the drillstring using advanced LWD technology. The latter is important when it is difficult to convey the tool to the desired depth using wireline, such as in steep directional wells and horizontal laterals or in difficult hole environments. FT devices offer the operator a chance to measure pressures at many locations in a well accurately and efficiently. They can be verified by repeat measurements and are valid over a wide range of mobilities. With advanced pressure transient techniques, directional permeabilities can be measured quantitatively, offering improved reservoir characterization and the ability to correlate petrophysical properties with permeability and productivity measures.

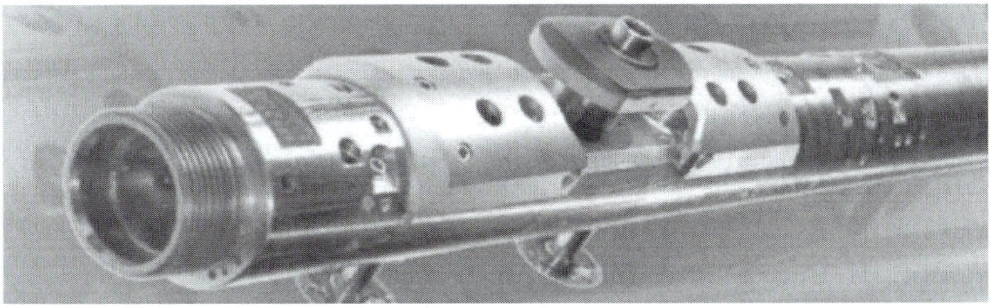

FIGURE 1.120 Example formation tester. *(RCI, courtesy Baker Hughes.)*

FT devices offer the operator the chance to recover representative formation fluids captured and maintained above saturation pressures, preserving concentrations of nonhydrocarbon diluents[6] such as H_2S and CO_2 with minimal contamination. Advances in downhole measurements allow rapid estimation of *in situ* PVT properties including density, viscosity, GOR, FVF, bubble point pressure, sulfur content, and compressibility along with fluid typing and identifying drilling fluid contamination. Other advanced applications include mini-DSTs and determination of parameters vital to hydraulic fracturing using Micro-Frac.

FT devices use one or more snorkels that are pushed flush into the formation face at a desired depth by backup arms on the opposite side of the tool. They may or may not have straddle packers adjacent to them to provide isolation. The snorkel can allow a small amount of fluids into the tool for pressure identification or larger volumes of fluids that can allow clean samples of reservoir fluids to be analyzed and/or recovered to surface. In either case, high-resolution quartz crystal pressure gauges are used for accurate transient testing and pressure measurements.

Figure 1.121 shows the conceptual design of such a tool along with the measured pressures as a function of time.

[6]FT devices constructed of nonreactive metals such as titanium are required to maintain H_2S concentrations.

1.7.1 Supercharging

Most of the pressure transient analysis techniques discussed so far have been applicable for production or injection wells that have been completed and "cleaned up." DSTs and FTs require the reservoir engineer to ensure that the correct reservoir conditions are being identified. Invasion of mud filtrate during the drilling process may cause an excessively high pressure in the near wellbore region called "supercharging." It occurs in environments where the mud cake does not adequately isolate the wellbore pressure from the formation. The supercharging is a bigger problem in LWD environments, where active mud circulation limits filtrate cake growth. As a result, the leak-off rates are higher in dynamic mud conditions in LWD than in wireline where mud condition is static. Supercharging is larger if mud cake permeability K_m is high and formation permeability K_f is low. It is commonly observed both with wireline and LWD measurements if the permeability of the formation is less than 1 mD.

1.7.2 Flow Analysis

Flow toward a single point in a reservoir typically results in spherical flow. In a layered reservoir, this may give way over time to cylindrical flow. In an FT, flow is restricted due to the probe and results in semi-hemispherical flow (Figure 1.122).

Various analysis techniques have been borrowed from the conventional well testing

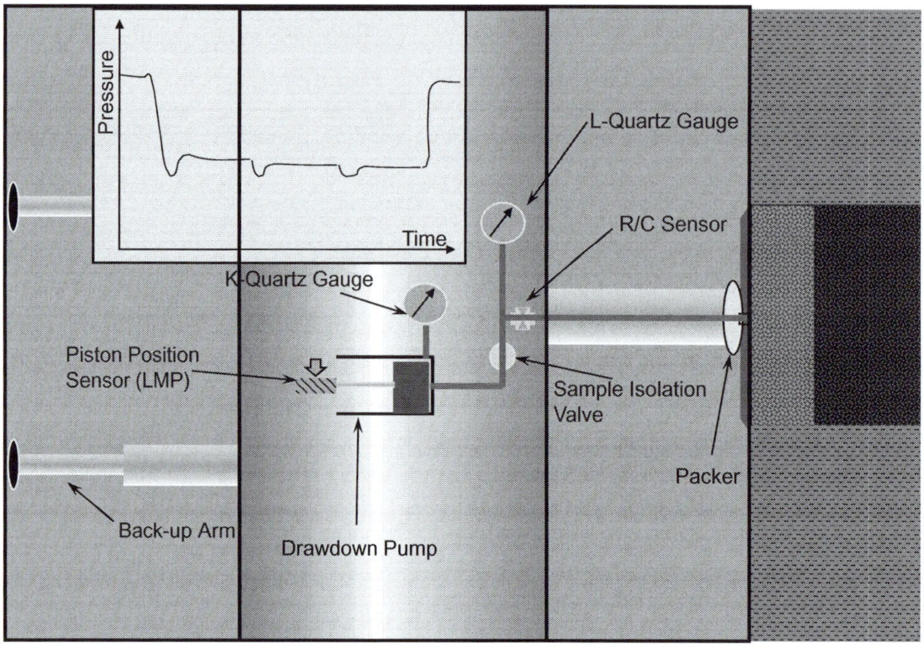

FIGURE 1.121 Illustrative pressure schematic. *(courtesy Baker Hughes).*

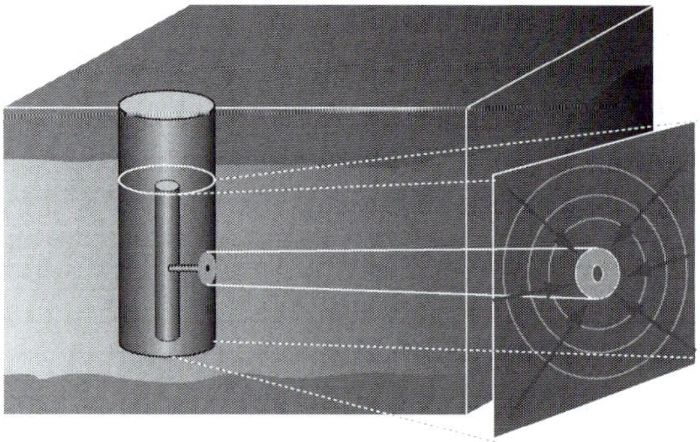

FIGURE 1.122 Flow restricted by a probe on a wellbore is created in semi-hemispherical domain.

studies to analyze the FT-derived data. The "drawdown mobility" calculation incorporates the pressure drawdown corresponding to the piston drawdown rate in Darcy's equation to calculate the near wellbore mobility.

$$\frac{k_{dd}}{\mu} = \frac{q_{dd}}{4\pi\Delta p}\left(\frac{\left[1 + S_{geom}\left(\frac{r_p}{r_w}\right)\right]}{r_p}\right)$$

where

S_{geom} = flow geometry effect due to hemispherical flow
k_{dd} = drawdown permeability
q_{dd} = drawdown piston rate
r_p = = probe radius
r_w = wellbore radius
μ = viscosity

There is a drawback in this analysis, particularly in very low permeability formations where transitional behavior of pressure with respect to the piston rate becomes more significant due to tool storage effect. In low permeable formation, the flow rate from the formation can be different from the piston drawdown rate.

Mobility calculation by Formation Rate Analysis (FRA) accounts for the tool storage effect by calculating the system compressibility during the drawdown within the fixed tool volume. In FRA, the formation flow rate is calculated from the piston drawdown rate using

Darcy's equation and the material balance in the tool.

$$q_{ac} = q_f - q_{dd}$$

The subscripts for flow rate q represent accumulation, formation flow, and piston drawdown and implicitly assume a small density variation. Darcy's law for this system becomes

$$q_f = \frac{kG_o r_p}{\mu}(P^* - P(t))$$

where

G_o = geometric factor (4.67)
r_p = probe radius (in cm)
k = permeability, md
μ = viscosity, cp

Adding the mass balance with respect to time and liquid accumulation rate:

$$C_t = \frac{1}{V_{sys}}\frac{\partial V_t}{\partial P(t)}$$

$$C_t V_{sys} = \frac{\partial V_t/\partial t}{\partial P_t/\partial t}$$

$$q_{ac} = C_t V_{sys}\frac{\partial P_t}{\partial t}$$

We can then solve for pressure as a function of time as follows:

$$q_{ac} = q_f - q_{dd}$$

$$C_t V_{sys}\frac{\partial P_t}{\partial t} = (P^* - P(t))\left(\frac{kG_o r_i}{\mu}\right) - q_{dd}$$

$$P(t) = P^* - \left(\frac{\mu}{kG_o r_i}\right)\left(C_t V_{sys}\frac{\partial P_t}{\partial t} + q_{dd}\right)$$

Which then simplifies to

$$P(t) = P^* - \left(\frac{\mu}{kG_o r_i}\right)(q_f)$$

A plot of $P(t)$ vs. formation rate should approach a straight line with negative slope and intercept P^* at the $P(t)$ axis, and the mobility is calculated from the slope (Figure 1.123). The FRA plot should yield identical slopes for both build up and drawdown if there was constant

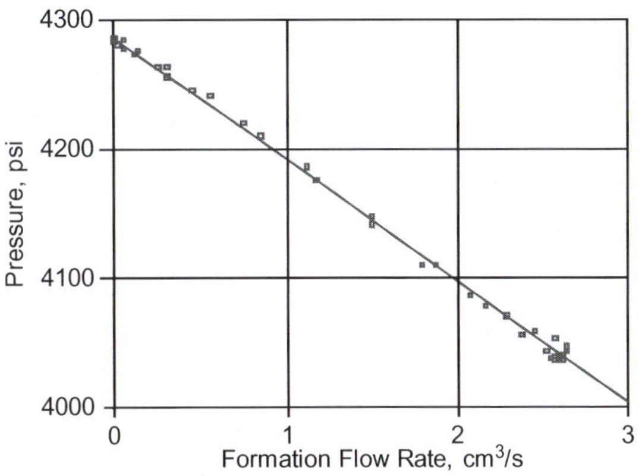

FIGURE 1.123 Example flow rate analysis interpretation.

compressibility. Compressibility effects can be resolved using multilinear regression techniques.

1.7.3 Example use of gradients

In a reservoir that is hydrostatic the pressures in the continuous phases vary by depth based on the density of the fluid in the continuous phase. In an oil reservoir, oil density is primarily a function of the gravity of the oil, the amount of dissolved gas, and the pressure. Although compositional variations in oil density and dissolved gas are not uncommon over large areas and in very thick reservoirs, it is often common to have an approximately constant pressure gradient over the thickness of an oil reservoir among all wells that are hydraulically continuous. Similarly, water gradients are usually even more constant. Except for very heavy oils (whose density approaches that of water), it is usually possible to distinguish oil- and water-bearing formations by obtaining multiple formation pressures at different depths in the reservoir (Figure 1.124). If there is no clear oil/water contact in a wellbore, the use of these gradients can often identify an oil/water contact depth (Figure 1.125). Natural gas gradients are typically much less than liquid gradients and can serve a similar purpose. In thick reservoirs, the density of oil usually varies with changing depth. This

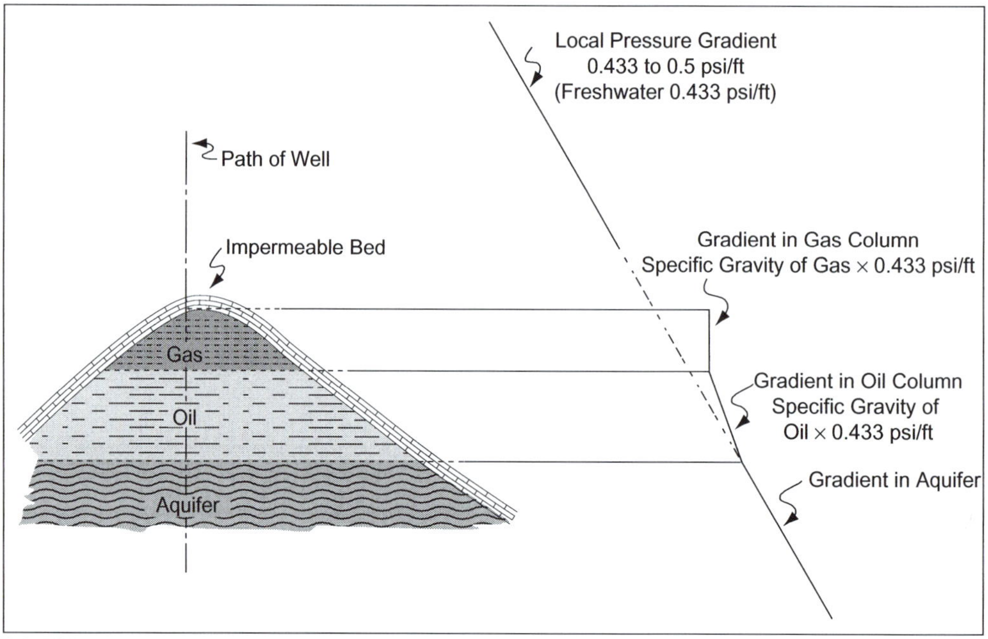

FIGURE 1.124 Illustrative pressure gradients reflecting reservoir fluid contents.

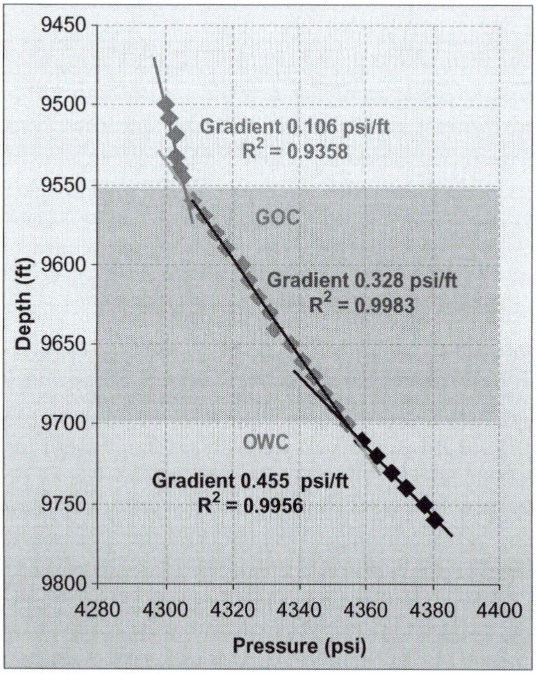

FIGURE 1.125 Interpretation of fluid gradients to detect contacts.

is quite common, and it is possible to detect varying density (compositional grading) from a depth vs. pressure plot using FT data.

The geology of the reservoir is very important when considering gradient analysis; prior knowledge of its structure can serve as lead indicators to what the gradient trends will look like and is demonstrated by the diagram of the single well producing multiple zones from the same field but not necessarily the same HC deposit or communicating reservoirs (Fig 1.126). A discontinuity in the pressures obtained from formation tests can probably be used to identify hydraulic isolation among reservoirs due to layering, faulting, or other geological heterogeneity.

Similarly, varying free water levels will result in different gradient interpretations. In Figure 1.127, the first illustration shows Well A encountering an oil sand and a water sand. Because the top layer has no water contact, the reservoir engineer could conclude that if the reservoirs were in hydraulic communication

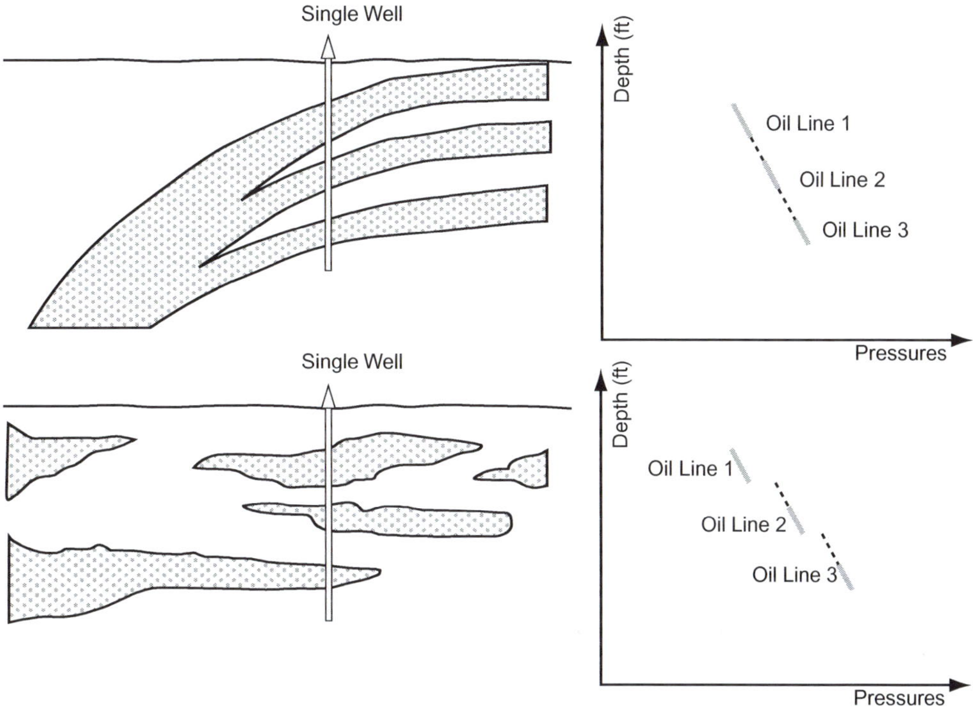

FIGURE 1.126 Interpretation of gradients to detect reservoir compartmentalization.

then the FWL in the top layer could be at the subsea depth associated with the water encountered in the lower, water-bearing layer. Because the gradient slopes intersect at a subsea depth greater than the known oil levels, the layers are hydraulically discontinuous and such a conclusion cannot be reached.

The two oil layers in the second Well A example are shown to be hydraulically separated and will behave independently during production.

Similarly, pressure gradients and magnitudes shift during production, in transition zones, and due to injection. One of the most powerful indicators of bypassed oil in a mature waterflood is a low-pressure layer in an infill well. While pressure reductions can be communicated over time in permeable reservoirs, the increase in pressure due to fluid injection requires good hydraulic communication and sufficient volumetric sweep. Low-pressure layers identify poorly swept zones that can often result in significant increases in production and recovery when flooded. Pressure barriers with layers can also be identified using FTs.

In the following example (Table 1.9), a series of formation tests were obtained for three adjacent wells in a sand that was indicated from open hole logs to be oil-bearing in Well 1 and water-bearing in Wells 2 and 3. Geological mapping suggests that they are in a continuous reservoir and that there should be an oil–water contact Wells 2 and 3 below the bottom of Well 1. Unfortunately, no fluid samples were obtained with any of the formation tests.

Use the following data to answer (if possible) the following questions:

- Which wells are in hydraulic communication?
- What contacts are present in any of the wells?
- What is the *in situ* density of the oil, and of the water?

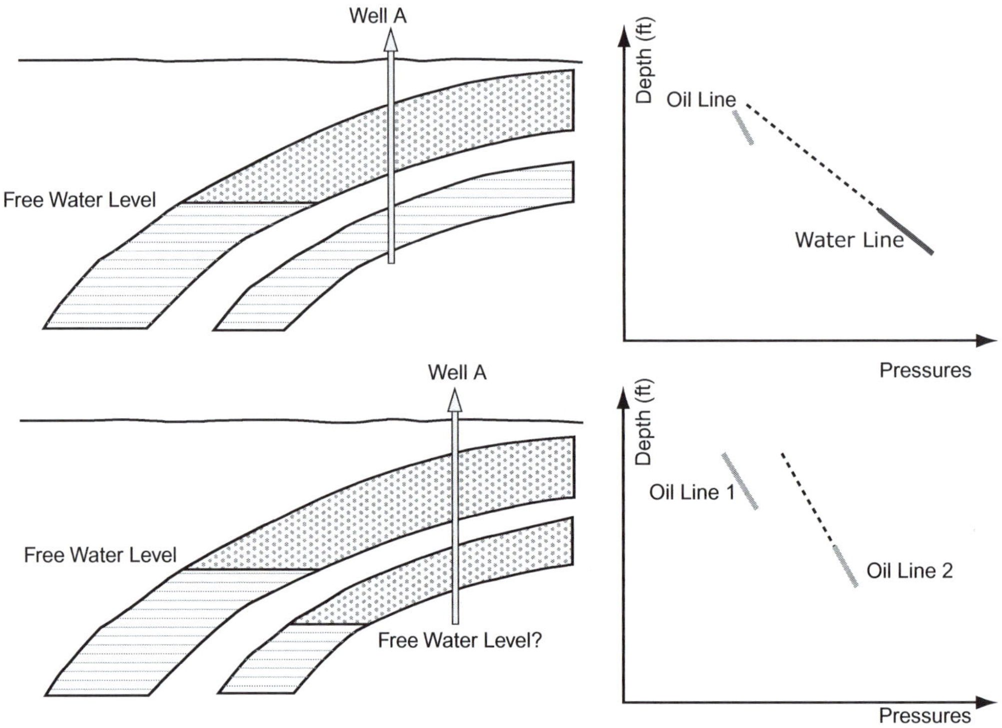

FIGURE 1.127 Interpretation of gradients to detect hydraulic isolation.

TABLE 1.9

	RCI Pressure		
Subsea Depth, ft	**Well 1**	**Well 2**	**Well 3**
−6123.0	3431.6		
−6125.0	3431.7		
−6130.9	3433.9		
−6137.0	3436.2		
−6142.0	3438.1		
−6158.0	3444.1		
−6182.0		3453.7	
−6190.0		3457.3	
−6201.0		3462.3	
−6256.0			3505.1
−6275.0			3513.7
−6278.0			3515.1

1.7.4 Solution

Graphing the pressure points as a function of depth yields Figure 1.128.

The gradient between the Well 1 top data point at −6123 is much less than that at −6125 and below. This suggests that this top point is potentially in a gas cap. With only one data point, graphical methods cannot be used to determine the gas gradient. However, if gas density is estimated (perhaps from nearby wells), with the known values for pressure and temperature, an approximate gas–oil contact of −6124 ft can be obtained. The pressure gradient of the water-bearing sands is approximately 0.4546 psi/ft. The density of fresh water is 0.433 psi/ft, so the *in situ* density of the salt water is 0.4546/0.433 or 1.05. The oil gradient suggests an *in situ* density of 0.87 g/cm^3. Specific and API gravity are normally quoted at

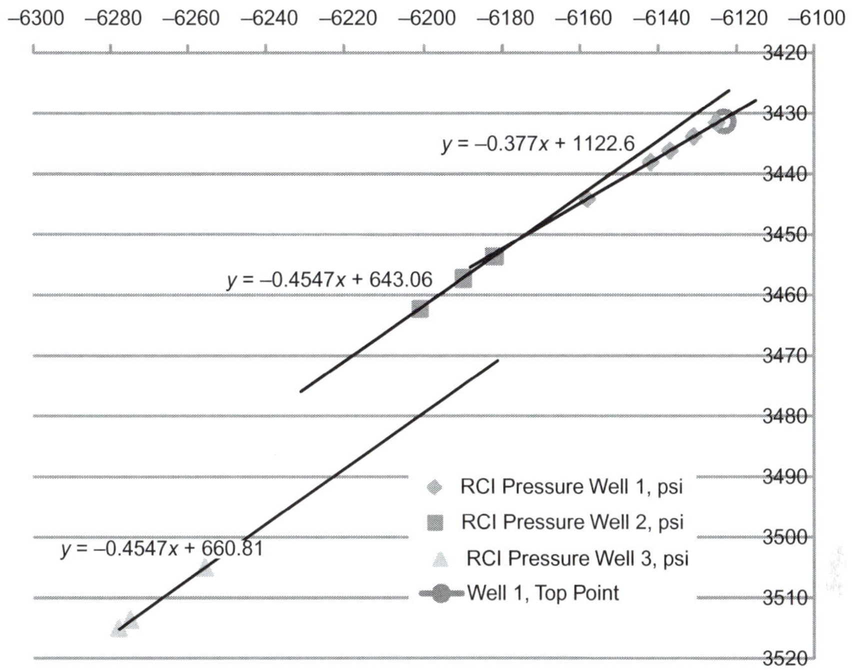

FIGURE 1.128 Example gradient interpretation.

standard conditions, so both of these densities would need to be corrected for temperature, pressure, and gas saturation to indicate surface values.

Examining the gradients, it appears that Well 3 is hydraulically isolated from Wells 1 and 2. This has a relatively high degree of certainty. Wells 1 and 2 appear to be hydraulically connected, and simultaneously solving the gradients measured in the oil and water zones suggests an oil–water contact of −6179 ft. The closer the gradients are in the oil and water zones, the more difficult it is to be accurate about the value for the OWC. Additionally, the further apart the data points are (vertically or areally), the greater the chance that other errors are possible in the OWC determination. While caution is warranted, the use of formation test data to interpolate a best technical estimate for OWC is normal practice. Since the capillary pressure effect may be important in pressure measurements, the calculated OWC may be shifted up or down based on the wettability conditions. Therefore, it can be misleading to rely on the pressure data alone to determine the OWC. It needs to be integrated with other log data to make a reliable determination of OWC.

1.7.5 Fluid Identification

The produced fluid recovered by formation testing can be analyzed *in situ* by the most advanced FTs and/or recovered to the surface for additional analysis. Advanced FTs like the In-Situ Fluids eXplorer™ (IFX) available in the Reservoir Characterization Instrument™ (RCI)[7] have multiple-channel visible and near-infrared spectrometers, methane detection devices, fluorescence spectrometers, refractometers, and tools to calculate *in situ* density, viscosity, sound speed, GOR, and compressibility. Figure 1.129 illustrates typical responses of gas, oil, and water under near-infrared spectrometry. The 17 wavelength channels at x-axis indicate

[7]Both are trademarks of Baker Hughes Incorporated.

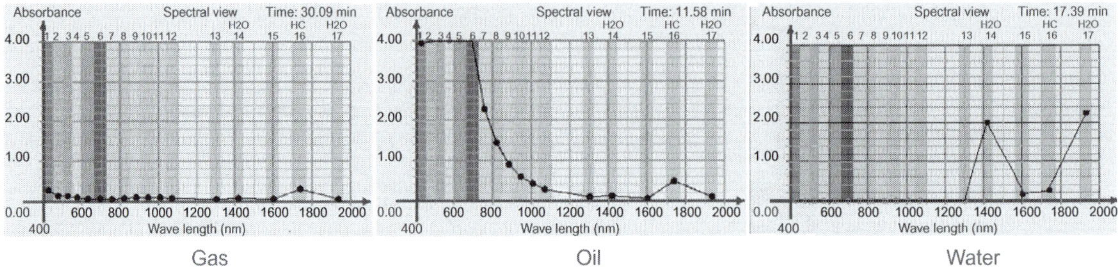

FIGURE 1.129 Responses of oil, gas, and water under near-infrared spectroscopy.

the color darkness of the flowing fluid at a particular wavelength.

Once the mud filtrate contamination is minimized, the fluid flow is directed to the sample chambers for fluid collection. Typical fluid volumes that can be recovered to the surface range from about 500 to 20,000 cc with pressure ratings up to 25,000 psi. Samples recovered at near-reservoir condition may prove invaluable in early reservoir characterization, particularly in layered reservoirs that are likely to be commingled during the completion or stimulation process. The authors have experienced situations with significant compositional variations obtained along a horizontal lateral in a reservoir previously believed to have been a common compartment.

1.7.6 Advanced applications

The state-of-the-art FTs allow the operator to perform a mini-DST as well as tests to optimize hydraulic fracturing processes. In a standard DST, drillers isolate an interval of the borehole and induce formation fluids to flow to surface, where they measure flow volumes before burning or sending the fluids to a disposal tank. The RCI tool, in particular the straddle packer module, provides similar functions to DST but on a wireline and at a smaller scale.

While the mini-DST is less expensive than a conventional DST, it also provides a significant safety advantage in that fluids are not produced to the surface. Cost benefits come from cheaper downhole equipment, shorter operating time, and the avoidance of any surface handling equipment. There are no problems of fluid disposal, no safety issues, and no problems with environmental regulations. The mini-DST investigates a smaller volume of formation due to smaller packed-off interval and withdraws a smaller amount of fluid at a lower flow rate. Mini-DST can be applied to individual hydraulic flow unit to characterize its flow properties, which could be a significant input in understanding and quantifying the reservoir heterogeneity. This is in contrast with conventional well testing, which provides average properties of all units and is insensitive of the reservoir heterogeneity.

The pressure transients measured during the drawdown or buildup periods are interpreted to obtain the reservoir parameters. The pressure transients obey the same laws of physics as that measured during a conventional DST or well testing. Therefore, they can be analyzed and interpreted the same way. As a result, mini-DST conducted by wireline tools such as RCI can provide reservoir permeability, assessment of formation damage (skin factor), and formation flow capacity as well as single-phase fluid sampling at *in situ* conditions.

In a homogeneous layer, three flow regimes are observed in a min-DST: early radial flow around the packed-off interval, pseudo-spherical flow until the pressure pulse reaches a boundary, and total radial flow between upper and lower no-flow boundaries (Figure 1.130). Rarely are all three seen because tool storage effects can mask the early radial flow, whereas the distance to the nearest barrier determines whether or not the other regimes are developed

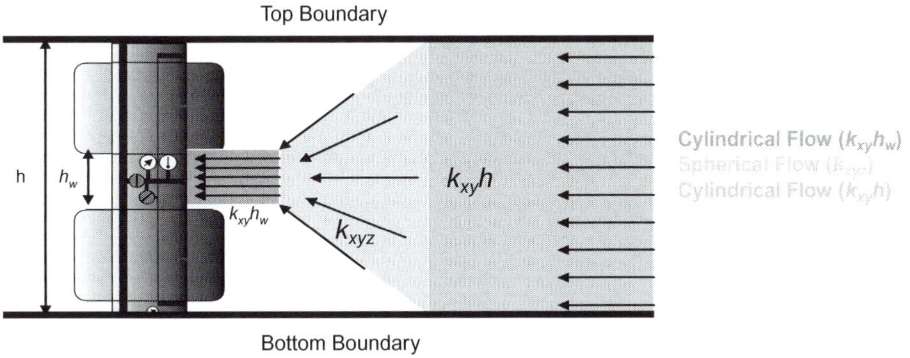

FIGURE 1.130 Flow regimes for mini-DSTs.

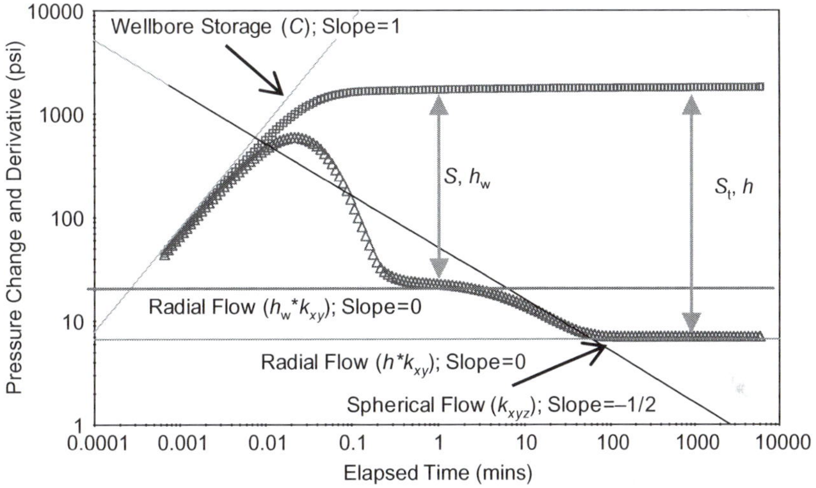

FIGURE 1.131 Interpretation approach for mini-DSTs.

during the test period. It has been common to observe a pseudo-spherical flow regime and, occasionally, total radial flow in buildup tests (Figure 1.130).

To analyze pressure and rate response during pressure testing, it is necessary to have knowledge on the nature of the formation and the fluids therein. For the packer configuration, an analytical model for a partially completed well with storage and skin is used. The partially penetrating well model with homogenous reservoir behavior assumes uniform reservoir thickness, h, and porosity, ϕ, with the well completed over a limited section, h_w. The distance from the center of the isolated section to the bottom of the reservoir is designated as Z_w.

On a log–log plot of the pressure derivative vs. a particular function of time (Figure 1.131), spherical flow is identified by a negative half slope and radial flow by a stabilized horizontal line. Tool storage includes the compressibility of the fluid between the packers. A common model is to relate the sandface flow rate, q_{sf} to the measured flow rate, q and the rate of change of pressure by a constant, C. The very early part of a buildup is dominated by wellbore storage, also called afterflow. C can be estimated from the rate of change of pressure

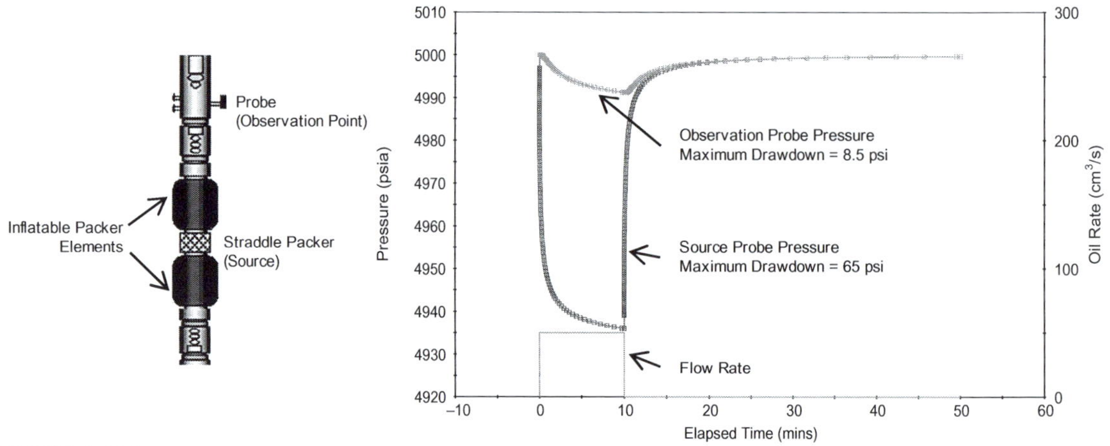

FIGURE 1.132 VIT interpretation approach.

at this time. Skin due to partial completion and formation damage can be calculated separately using the pressure and derivative data together.

For a very short time during the early fluid flow, it is expected to have radial flow due to thickness of the isolated zone ($k_{xy} \times h_w$). It is unlikely to observe this flow regime in most tests because it might be masked by the tool storage effect in early time. The spherical flow regime, which represents the geometric mean of three-directional permeability (k_{xyz}), is the dominant flow regime in early time with straddle packer configuration. Flow regime becomes radial once the flow is restricted by top and bottom no-flow boundaries. The radial flow represents the product of horizontal permeability and reservoir thickness ($k_{xy} \times h$). The observation of both spherical and radial flow regimes in a mini-DST provides an opportunity to calculate both spherical permeability (k_{xyz}) and horizontal permeability (k_x) and consequently the vertical permeability (k_z) from the integration of both parameters.

Permeability anisotropy (k_z/k_x) is an important parameter for coning studies and completion/perforation policy determination in vertical wells. It is also one of the key factors influencing the production performance in horizontal wells. It is usually impossible to determine the vertical permeability in conventional well testing. Vertical interference test (VIT) performed by the RCI provides unique information on horizontal and vertical permeability of the tested formation in reservoirs. In a VIT, the combination of the straddle packer and conventional probe is used to conduct the test. A single probe or a combination of probes can be run directly above or below the straddle packer to collect pressure responses generated by the fluid flow within the straddle packer section (Figure 1.132).

The time required the pressure transients to propagate from the source to the observation probe is a function of the storativity and vertical permeability. For a given reservoir fluids and formation rock conditions, the storativity can be calculated and considered constant. The simultaneous analysis of pressure transients from the source and observation points provides the unique calculation of k_z values for the distance between two packers. Horizontal permeability (k_x), skin factor, and productivity index are also calculated in a VIT (Figure 1.133).

Micro-Frac Testing

Micro-Frac testing (MFT) is performed in vertical wells with the RCI straddle packer module to determine stress and fracture gradient in

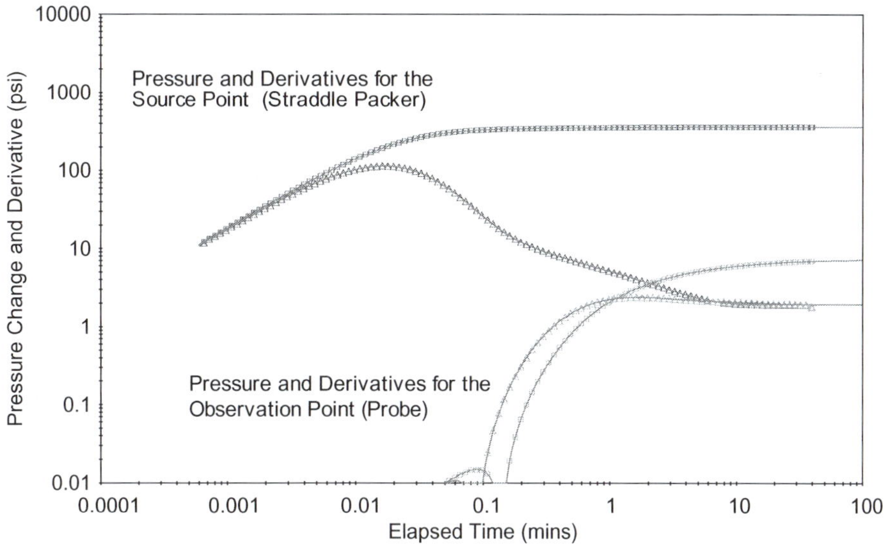

FIGURE 1.133 VIT interpretation for resolving anisotropy.

reservoir and/or caprock formations for well injection plans, gas storage, or geomechanics-related issues and fracture closure pressure for *in situ* stress determination. Instead of producing reservoir fluids, fluids are injected between straddle packers to create a small hydraulic fracture in the formation. After a brief pumping period that extends the created fracture, fluid injection is ceased and pressures are observed during the closure of the created fracture. Further pumping and shut-in periods are used to characterize certain rock properties and the minimum compressive stress. These properties are used in the design and analysis of subsequent stimulation treatments or geomechanical interpretations. Petroleum geomechanical models are used to optimize wellbore stability and pore pressure prediction, manage subsidence, etc.

1.8 INJECTION WELL TESTING

Injectivity testing is a pressure transient test during injection into a well. Injection well testing and the associated analysis are essentially simple, as long as the mobility ratio between the injected fluid and the reservoir fluid is unity. Earlougher (1977) pointed out that the unit-mobility ratio is a reasonable approximation for many reservoirs under water floods. The objectives of injection tests are similar to those of production tests, namely the determination of:

- permeability;
- skin;
- average pressure;
- reservoir heterogeneity;
- front tracking.

Injection well testing involves the application of one or more of the following approaches:

- injectivity test;
- pressure falloff test;
- step-rate injectivity test.

The above three analyses of injection well testing are briefly presented in the following sections.

1.8.1 Injectivity Test Analysis

In an injectivity test, the well is shut-in until the pressure is stabilized at initial reservoir pressure

p_i. At this time, the injection begins at a constant rate q_{inj}, as schematically illustrated in Figure 1.119, while recording the bottom-hole pressure p_{wf}. For a unit-mobility ratio system, the injectivity test would be identical to a pressure drawdown test except that the constant rate is negative with a value of q_{inj}. However, in all the preceding relationships, the injection rate will be treated as a positive value, i.e., $q_{inj} > 0$.

For a constant injection rate, the bottom-hole pressure is given by the linear form of Eq. (1.169) as:

$$p_{wf} = p_{1\ hour} + m \log(t) \tag{1.305}$$

The above relationship indicates that a plot of bottom-hole injection pressure vs. the logarithm of injection time would produce a straight-line section as shown in Figure 1.136, with an intercept of $p_{1\ hour}$ and a slope m as defined by:

$$m = \frac{162.6 q_{inj} B\mu}{kh}$$

where

q_{inj} = absolute value of injection rate, STB/day
m = slope, psi/cycle
k = permeability, md
h = thickness, ft

Sabet (1991) pointed out that, depending on whether the density of the injected fluid is higher or lower than the reservoir fluid, the injected fluid will tend to override or underride the reservoir fluid and, therefore the net pay h which should be used in interpreting injectivity tests would not be the same as the net pay which is used in interpreting drawdown tests.

Earlougher (1977) pointed out that, as in drawdown testing, the wellbore storage has great effects on the recorded injectivity test data due to the expected large value of the wellbore storage coefficient. Earlougher recommended that all injectivity test analyses must include the log–log plot of $(p_{wf} - p_i)$ vs. injection time with the objective of determining the duration of the wellbore storage effects. As defined previously, the beginning of the semilog straight line, i.e., the end of the wellbore storage effects, can be estimated from the following expression:

$$t > \frac{(200{,}000 + 12{,}000s)C}{kh/\mu} \tag{1.306}$$

where

t = time that marks the end of wellbore storage effects, hours
k = permeability, md
s = skin factor
C = wellbore storage coefficient, bbl/psi
μ = viscosity, cp

Once the semilog straight line is identified, the permeability and skin can be determined as outlined previously by:

$$k = \frac{162.6 q_{inj} B\mu}{mh} \tag{1.307}$$

$$s = 1.1513\left[\frac{p_{1\ hour} - p_i}{m} - \log\left(\frac{k}{\phi\mu c_t r_w^2}\right) + 3.2275\right] \tag{1.308}$$

The above relationships are valid as long as the mobility ratio is approximately equal to 1. If the reservoir is under water flood and a water injection well is used for the injectivity test, the following steps summarize the procedure of analyzing the test data assuming a unit-mobility ratio:

Step 1. Plot $(p_{wf} - p_i)$ vs. injection time on a log–log scale.
Step 2. Determine the time at which the unit-slope line, i.e., 45° line, ends.
Step 3. Move $1\frac{1}{2}$ log cycles ahead of the observed time in step 2 and read the corresponding time which marks the start of the semilog straight line.
Step 4. Estimate the wellbore storage coefficient C by selecting any point on the unit-slope line and reading its coordinates, i.e., Δp and t, and applying the following expression:

$$C = \frac{q_{inj}Bt}{24\Delta p} \quad (1.309)$$

Step 5. Plot p_{wf} vs. t on a semilog scale and determine the slope m of the straight line that represents the transient flow condition.

Step 6. Calculate the permeability k and skin factor from Eqs. (1.307) and (1.308), respectively.

Step 7. Calculate the radius of investigation r_{inv} at the end of injection time. That is:

$$r_{inv} = 0.0359\sqrt{\frac{kt}{\phi\mu c_t}} \quad (1.310)$$

Step 8. Estimate the radius to the leading edge of the water bank r_{wb} before the initiation of the injectivity test from:

$$r_{wb} = \sqrt{\frac{5.615W_{inj}}{\pi h\phi(\bar{S}_w - S_{wi})}} = \sqrt{\frac{5.615W_{inj}}{\pi h\phi(\Delta S_w)}} \quad (1.311)$$

where

r_{wb} = radius to the water bank, ft

W_{inj} = cumulative water injected at the start of the test, bbl

$\bar{S}_w$ = average water saturation at the start of the test

S_{wi} = initial water saturation

Step 9. Compare r_{wb} with r_{inv}: if $r_{inv} < r_{wb}$, the unit-mobility ratio assumption is justified.

Example 1.47

Figures 1.134 and 1.135 show pressure response data for a 7-hour injectivity test in a water-flooded reservoir in terms of $\log(p_{wf} - p_i)$ vs. $\log(t)$ and $\log(p_{wf})$ vs. $\log(t)$, respectively. Before the test, the reservoir had been under water flood for 2 years with a constant injection rate of 100 STB/day. The injectivity test was initiated after shutting in all

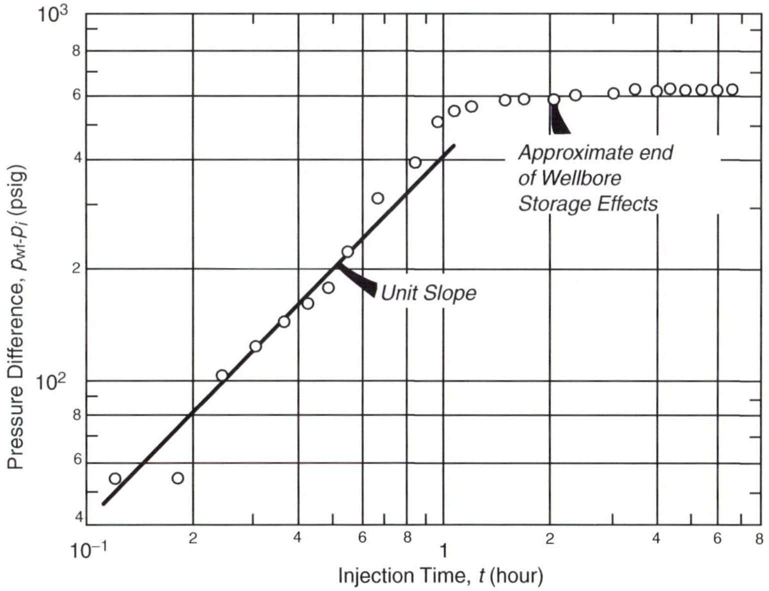

FIGURE 1.134 Log–log data plot for the injectivity test of Example 1.47. Water injection into a reservoir at static conditions. *(After Earlougher, Robert C., Jr., 1977. Advances in Well Test Analysis, Monograph, vol. 5. Society of Petroleum Engineers of AIME, Dallas, TX; Permission to publish by the SPE, copyright SPE, 1977.)*

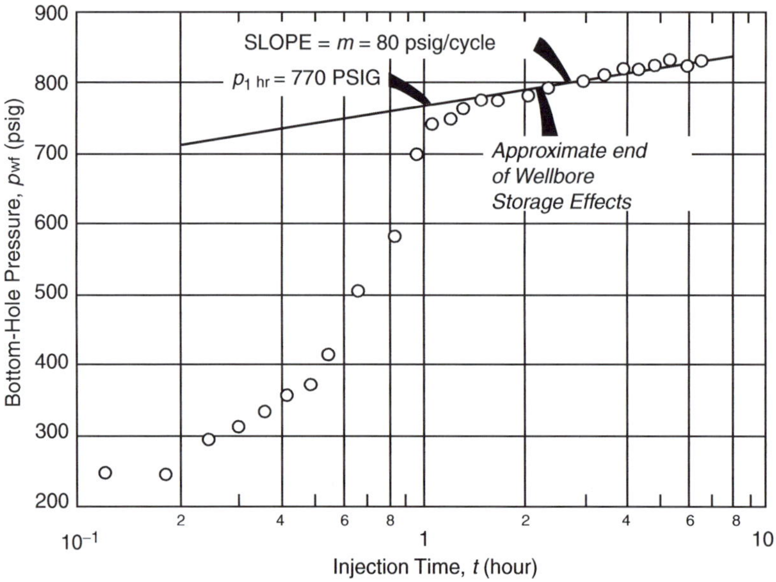

FIGURE 1.135 Semilog plot for the injectivity test of Example 1.47. Water injection into a reservoir at static conditions. *(After Earlougher, Robert C., Jr., 1977. Advances in Well Test Analysis, Monograph, vol. 5. Society of Petroleum Engineers of AIME, Dallas, TX; Permission to publish by the SPE, copyright SPE, 1977).*

wells for several weeks to stabilize the pressure at p_i. The following data is available:[8]

$c_t = 6.67 \times 10^{-6}\ \text{psi}^{-1}$		
$B = 1.0$ bbl/STB,	$\mu = 1.0$ cp,	
$S_w = 62.4\ \text{lb/ft}^3$,	$\phi = 0.15$,	$q_{inj} = 100$ STB/day
$h = 16$ ft,	$r_w = 0.25$ ft,	$p_i = 194$ psig
$\Delta S_w = 0.4$,	depth = 1002 ft,	total test time = 7 hours

The well is completed with 2-inch tubing set on a packer. Estimate the reservoir permeability and skin factor.

Solution

Step 1. The log–log data plot of Figure 1.134 indicates that the data begins to deviate from the unit-slope line at about 0.55 hours. Using the rule of thumb of moving $1-1\frac{1}{2}$ cycles in time after the data starts deviating from the unit-slope line, suggests that the start of the semilog straight line begins after 5–10 hours of testing. However, Figures 1.134 and 1.135 clearly show that the wellbore storage effects have ended after 2–3 hours.

Step 2. From the unit-slope portion of Figure 1.134, select the coordinates of a point (i.e., Δp and t) and calculate the wellbore storage coefficient C by applying Eq. (1.309):

$$\Delta p = 408 \text{ psig}$$

$$t = 1 \text{ hour}$$

$$C = \frac{q_{inj} B t}{24 \Delta p} = \frac{(100)(1.0)(1)}{(24)(408)} = 0.0102 \text{ bbl/psi}$$

Step 3. From the semilog plot in Figure 1.135, determine the slope of the straight line m to give:

$$m = 770 \text{ psig/cycle}$$

[8]After Robert Earlougher, *Advances in Well Test Analysis*, 1977.

Step 4. Calculate the permeability and skin factor by using Eqs. (1.307) and (1.308):

$$k = \frac{162.6 q_{inj} B\mu}{mh} = \frac{(162.6)(100)(1.0)(1.0)}{(80)(16)} - 12.7 \text{ md}$$

$$s = 1.1513\left[\frac{p_{1\,hr} - p_i}{m} - \log\left(\frac{k}{\phi\mu c_t r_w^2}\right) + 3.2275\right]$$

$$= 1.1513\left[\frac{770 - 194}{80} - \log\left(\frac{12.7}{(0.15)(1.0)(6.67\times10^{-6})(0.25)^2}\right) + 3.2275\right] = 2.4$$

Step 5. Calculate the radius of investigation after 7 hours by applying Eq. (1.310):

$$r_{inv} = 0.0359\sqrt{\frac{kt}{\phi\mu c_t}} = 0.0359\sqrt{\frac{(12.7)(7)}{(0.15)(1.0)(6.67\times10^{-6})}} \simeq 338 \text{ ft}$$

Step 6. Estimate the distance of the leading edge of the water bank before the start of the test from Eq. (1.311):

$$W_{inj} \cong (2)(365)(100)(1.0) = 73{,}000 \text{ bbl}$$

$$r_{wb} = \sqrt{\frac{5.615 W_{inj}}{\pi h\phi(\Delta S_w)}} = \sqrt{\frac{(5.165)(73\,000)}{\pi(16)(0.15)(0.4)}} \cong 369 \text{ ft}$$

Since $r_{inv} < r_{wb}$, the use of the unit–mobility ratio analysis is justified.

1.8.2 Pressure Falloff Test

A pressure falloff test is usually preceded by an injectivity test of a long duration. As illustrated schematically in Figure 1.136, falloff testing is analogous to pressure buildup testing in a production well. After the injectivity test that lasted for a total injection time of t_p at a constant injection rate of q_{inj}, the well is then shut in. The pressure data taken immediately before and during the shut in period is analyzed by the Horner plot method.

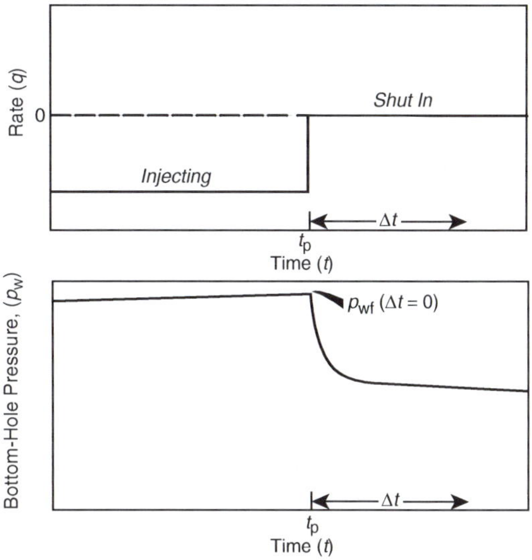

FIGURE 1.136 Idealized rate schedule and pressure response for falloff testing.

The recorded pressure falloff data can be represented by Eq. (1.179), as:

$$p_{ws} = p^* + m\left[\log\left(\frac{t_p + \Delta t}{\Delta t}\right)\right]$$

with

$$m = \left|\frac{162.6 q_{inj} B\mu}{kh}\right|$$

where p^* is the false pressure that is only equal to the initial (original) reservoir pressure in a newly discovered field. As shown in Figure 1.137, a plot of p_{ws} vs. $\log[(t_p + \Delta t)/\Delta t]$ would form a straight-line portion with an intercept of p^* at $(t_p + \Delta t)/\Delta t = 1$ and a negative slope of m.

It should be pointed out that the log–log data plot should be constructed to identify the end of the wellbore storage effects and

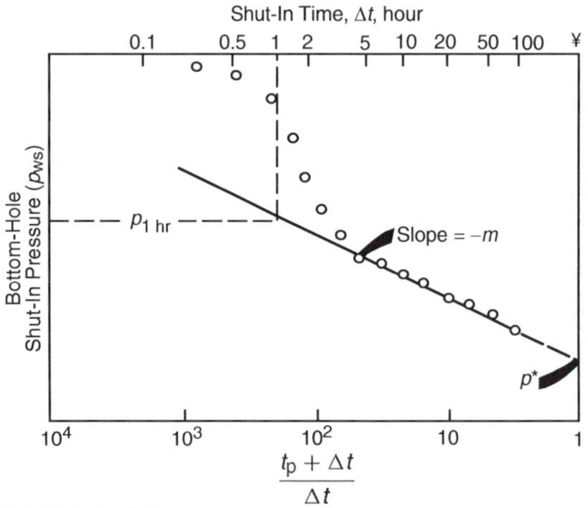

FIGURE 1.137 Horner plot of a typical falloff test.

beginning of the proper semilog straight line. The permeability and skin factor can be estimated as outlined previously by the expressions:

$$k = \frac{162.6 q_{inj} B \mu}{|m| h}$$

$$s = 1.513 \left[\frac{p_{wf\ at\ \Delta t=0} - p_{1\ hour}}{|m|} - \log\left(\frac{k}{\phi \mu c_t r_w^2} \right) + 3.2275 \right]$$

Earlougher (1977) indicated that if the injection rate varies before the falloff test, the equivalent injection time may be approximated by:

$$t_p = \frac{24 W_{inj}}{q_{inj}}$$

where

W_{inj} = cumulative volume injected since the last pressure equalization, i.e., last shut-in
q_{inj} = injection rate just before shut-in

It is not uncommon for a falloff test to experience a change in wellbore storage after the test begins at the end of the injectivity test. This will occur in any well which goes on vacuum during the test. An injection well will go on vacuum when the bottom-hole pressure decreases to a value which is insufficient to support a column of water to the surface. Prior to going on vacuum, an injection well will experience storage due to water expansion; after going on vacuum, the storage will be due to a falling fluid level. This change in storage will generally exhibit itself as a decrease in the rate of pressure decline.

The falloff data can also be expressed in graphical form by plotting p_{ws} vs. $\log(\Delta t)$ as proposed by MDH (Miller–Dyes–Hutchinson). The mathematical expression for estimating the false pressure p^* from the MDH analysis is given by Eq. (1.180) as:

$$p^* = p_{1\ hour} - |m| \log(t_p + 1) \quad \textbf{(1.312)}$$

Earlougher pointed out that the MDH plot is more practical to use unless t_p is less than about twice the shut-in time.

The following example, as adopted from the work of McLeod and Coulter (1969) and Earlougher (1977), is used to illustrate the methodology of analyzing the falloff pressure data.

Example 1.48

During a stimulation treatment, brine was injected into a well and the falloff data, as reported by McLeod and Coulter (1969), is shown graphically in Figures 1.138–1.140. Other available data includes:[9]

Total injection $t_p = 6.82$ hours,
Total falloff time = 0.67 hours
$q_{inj} = 807$ STB/day, $B_w = 1.0$ bbl/STB,
$c_w = 3.0 \times 10^{-6}\ psi^{-1}$, $\phi = 0.25$
$h = 28$ ft, $\mu_w = 1.0$ cp
$c_t = 1.0 \times 10^{-5}\ psi^{-1}$, $r_w = 0.4$ ft
$S_w = 67.46\ lb/ft^3$, depth = 4819 ft
Hydrostatic fluid gradient = 0.4685 psi/ft

The recorded shut-in pressures are expressed in terms of *wellhead pressures* p_{ts} with $p_{tf\ at\ \Delta t=0}$ = 1310 psig. Calculate:

- the wellbore storage coefficient;
- the permeability;
- the skin factor;
- the average pressure.

---

[9]Robert Earlougher, *Advances in Well Test Analysis*, 1977.

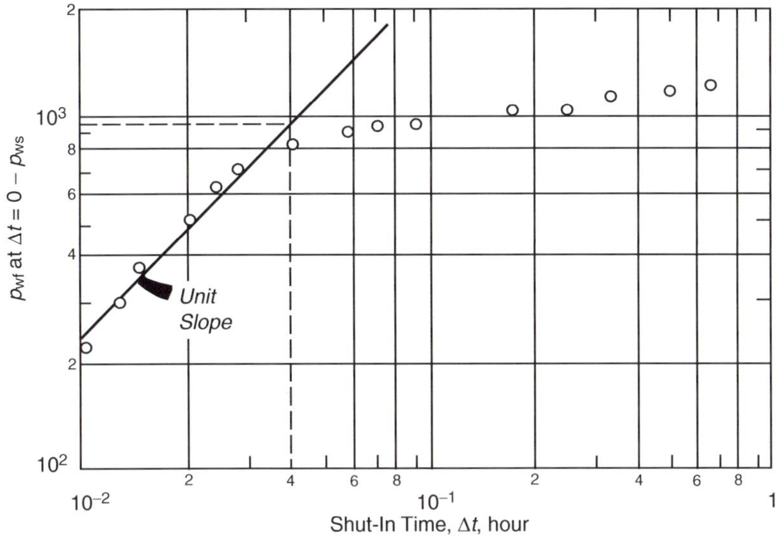

FIGURE 1.138 Log–log data plot for a falloff test after brine injection, Example 1.48. *(After Earlougher, Robert C., Jr., 1977. Advances in Well Test Analysis, Monograph, vol. 5. Society of Petroleum Engineers of AIME, Dallas, TX; Permission to publish by the SPE, copyright SPE, 1977.)*

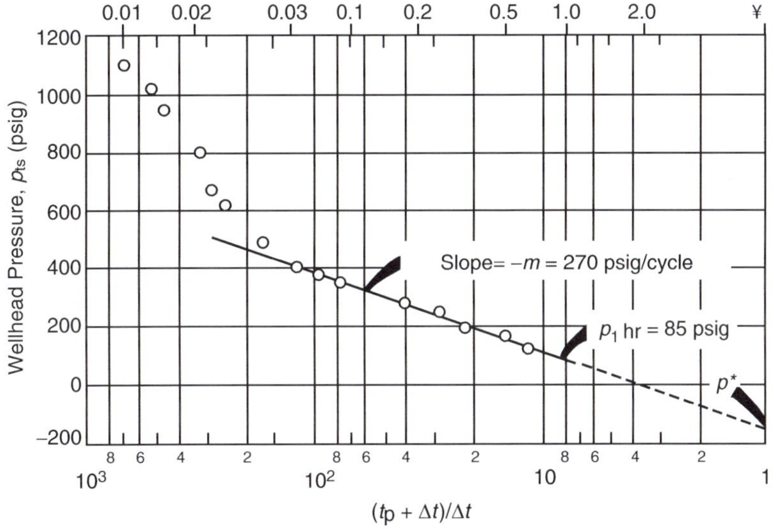

FIGURE 1.139 Horner plot of pressure falloff after brine injection, Example 1.48.

Solution

Step 1. From the log–log plot of Figure 1.138, the semilog straight line begins around 0.1–0.2 hours after shut-in. Using $\Delta p = 238$ psi at $\Delta t = 0.01$ hours as the selected coordinates of a point on the unit slope straight line, calculate the wellbore storage coefficient from Eq. (1.309), to give:

$$C = \frac{q_{inj} B t}{24 \Delta p} = \frac{(807)(1.0)(0.01)}{(24)(238)} = 0.0014 \text{ bbl/psi}$$

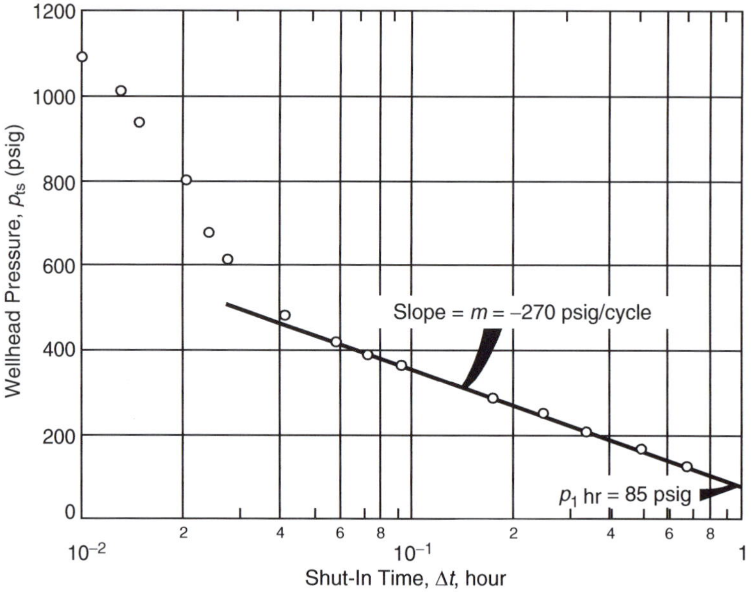

FIGURE 1.140 Miller–Dyes–Hutchinson plot of pressure falloff after brine injection, Example 1.48.

Step 2. Figures 1.139 and 1.140 show the Horner plot, i.e., "wellhead pressures vs. log $[(t_p + \Delta t)/\Delta t]$," and the MDH plot, i.e., "wellhead pressures vs. $\log(\Delta t)$," respectively, with both plots giving:

$$m = 270 \text{ psig/cycle}$$

$$p_{1\ \text{hour}} = 85 \text{ psig}$$

Using these two values, calculate k and s:

$$k = \frac{162.6 q_{\text{inj}} B\mu}{|m|h}$$

$$= \frac{(162.6)(807)(1.0)(1.0)}{(270)(28)}$$

$$= 17.4 \text{md}$$

$$s = 1.513\left[\frac{p_{\text{wf at } \Delta t=0} - p_{1\text{ hour}}}{|m|} - \log\left(\frac{k}{\phi\mu c_t r_w^2}\right) + 3.2275\right]$$

$$= 1.513\left[\frac{1310-85}{270} - \log\left(\frac{17.4}{(0.25)(1.0)(1.0\times10^{-5})(0.4)^2}\right)\right]$$

$$+3.3375$$

$$= 0.15$$

Step 3. Determine p^* from the extrapolation of the Horner plot of Figure 1.139 to $(t_p + \Delta t)/\Delta t = 1$, to give:

$$p^*_{ts} = -151 \text{ psig}$$

Eq. (1.312) can be used to approximate p^*:

$$p^* = p_{1\text{ hour}} - |m| \log(t_p + 1)$$

$$p^*_{ts} = 85 - (270)\log(6.82 + 1) = -156 \text{ psig}$$

This is the false pressure at the wellhead, i.e., the surface. Using the hydrostatic gradient of 0.4685 psi/ft and the depth of 4819 ft, the reservoir false pressure is:

$$p^* = (4819)(0.4685) - 151 = 2107 \text{ psig}$$

and since injection time t_p is short compared with the shut-in time, we can assume that:

$$\bar{p} = p^* = 2107 \text{ psig}$$

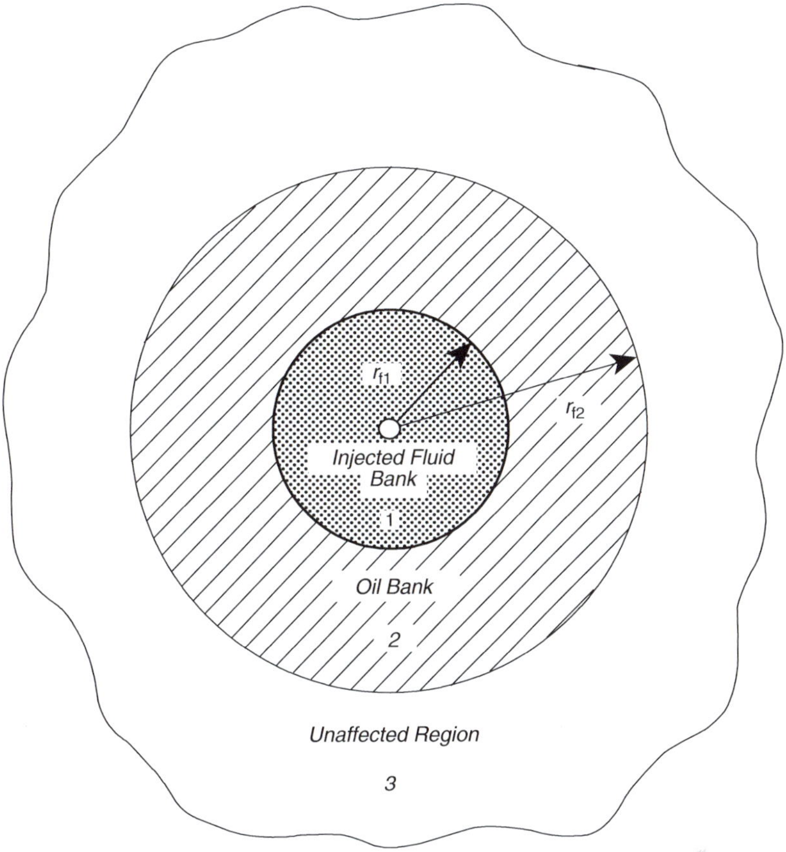

FIGURE 1.141 Schematic diagram of fluid distribution around an injection well (composite reservoir).

Pressure Falloff Analysis in Non-unit–Mobility Ratio Systems. Figure 1.141 shows a plan view of the saturation distribution in the vicinity of an injection well. This figure shows two distinct zones.

Zone 1. It represents the water bank with its leading edge at a distance of r_{f1} from the injection well. The mobility λ of the injected fluid in this zone, i.e., zone 1, is defined as the ratio of effective permeability of the injected fluid at its average saturation to its viscosity, or:

$$\lambda_1 = \left(\frac{k}{\mu}\right)_1$$

Zone 2. It represents the oil bank with the leading edge at a distance of r_{f2} from the injection well. The mobility λ of the oil bank in this zone, i.e., zone 2, is defined as the ratio of effective oil permeability as evaluated at initial or connate water saturation to its viscosity, or:

$$\lambda_2 = \left(\frac{k}{\mu}\right)_2$$

The assumption of a two-bank system is applicable if the reservoir is filled with liquid or if the maximum shut-in time of the falloff test is such that the radius of investigation of the test does not exceed the outer radius of the oil bank. The ideal behavior of the falloff test in a two-bank system as expressed in terms of the Horner plot is illustrated in Figure 1.142.

Figure 1.142 shows two distinct straight lines with slopes of m_1 and m_2, which intersect at Δt_{fx}. The slope m_1 of the first line is used to

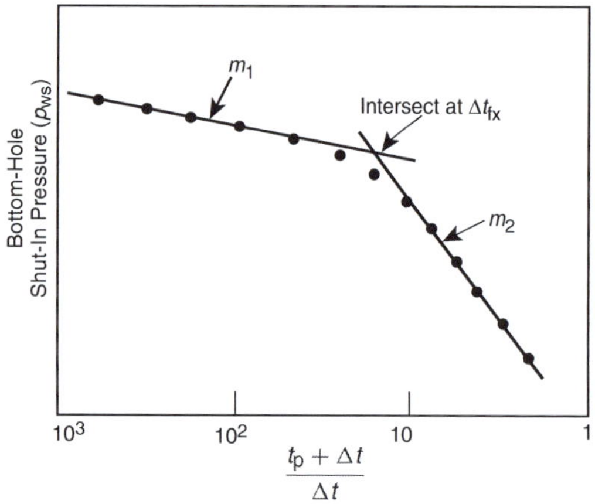

FIGURE 1.142 Pressure falloff behavior in a two-bank system.

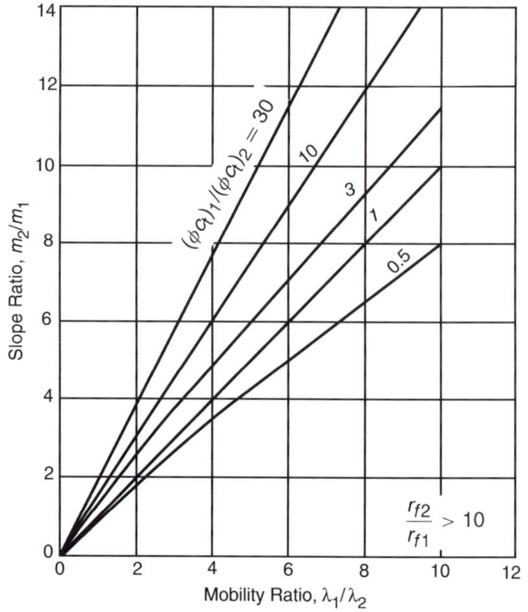

FIGURE 1.143 Relationship between mobility ratio, slope ratio, and storage ratio. *(After Merrill, L.S., Kazemi, H., Cogarty, W.B., 1974. Pressure falloff analysis in reservoirs with fluid banks. J. Pet. Technol. 26 (7), 809–818.)*

estimate the effective permeability to water k_w in the flooded zone and the skin factor s. It is commonly believed that the slope of the second line m_2 will yield the mobility of the oil bank λ_o. However, Merrill et al. (1974) pointed out that the slope m_2 can be used only to determine the oil zone mobility if $r_{f2} > 10r_{f1}$ and $(\phi c_t)_1 = (\phi c_t)_2$, and developed a technique that can be used to determine the distance r_{f1} and mobility of each bank. The technique requires knowing the values of (ϕc_t) in the first and second zone, i.e., $(\phi c_t)_1$ and $(\phi c_t)_2$. The authors proposed the following expression:

$$\lambda = \frac{k}{\mu} = \frac{162.6QB}{m_2 h}$$

The authors also proposed two graphical correlations, as shown in Figures 1.143 and 1.144, that can be used with the Horner plot to analyze the pressure falloff data.

The proposed technique is summarized by the following:

Step 1. Plot Δp vs. Δt on a log–log scale and determine the end of the wellbore storage effect.

Step 2. Construct the Horner plot or the MDH plot and determine m_1, m_2, and Δt_{fx}.

Step 3. Estimate the effective permeability in the first zone, i.e., injected fluid invaded zone, "zone 1," and the skin factor from:

$$k_1 = \frac{162.6 q_{inj} B \mu}{|m_1| h}$$

$$s = 1.513\left[\frac{p_{wf\ at\ \Delta t=0} - p_{1\ hour}}{|m_1|} - \log\left(\frac{k_1}{\phi \mu_1 (c_t)_1 r_w^2}\right) + 3.2275\right] \quad \textbf{(1.313)}$$

where the subscript "1" denotes zone 1, the injected fluid zone.

Step 4. Calculate the following dimensionless ratios:

$$\frac{m_2}{m_1} \quad \text{and} \quad \frac{(\phi c_t)_1}{(\phi c_t)_2}$$

with the subscripts "1" and "2" denoting zone 1 and zone 2, respectively.

Step 5. Use Figure 1.143 with the two dimensionless ratios of step 4 and read the mobility ratio λ_1/λ_2.

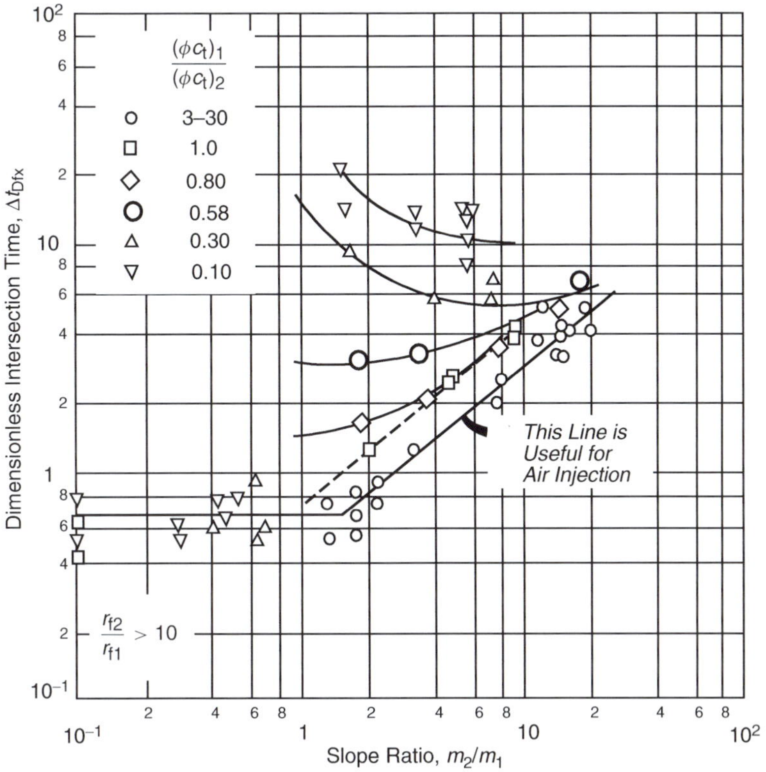

FIGURE 1.144 Correlation of dimensionless intersection time, Δt_{Dfx}, for falloff data from a two-zone reservoir. *(After Merrill, L.S., Kazemi, H., Cogarty, W.B., 1974. Pressure falloff analysis in reservoirs with fluid banks. J. Pet. Technol. 26 (7), 809–818.)*

Step 6. Estimate the effective permeability in the second zone from the following expression:

$$k_2 = \left(\frac{\mu_2}{\mu_1}\right)\frac{k_1}{\lambda_1/\lambda_2} \tag{1.314}$$

Step 7. Obtain the dimensionless time Δt_{Dfx} from Figure 1.144.

Step 8. Calculate the distance to the leading edge of the injected fluid bank r_{f1} from:

$$r_{f1} = \sqrt{\left[\frac{0.0002637(k/\mu)_1}{(\phi c_t)_1}\right]\left(\frac{\Delta t_{fx}}{\Delta t_{Dfx}}\right)} \tag{1.315}$$

To illustrate the technique, Merrill et al. (1974) presented Example 1.49.

Example 1.49

Figure 1.145 shows the MDH semilog plot of simulated falloff data for a two-zone water flood with no apparent wellbore storage effects. Data used in the simulation is given below:

$r_w = 0.25$ ft, $h = 20$ ft, $r_{f1} = 30$ ft

$r_{f2} = r_e = 3600$ ft, $\left(\frac{k}{\mu}\right)_1 = \eta_1 = 100$ md/cp

$\left(\frac{k}{\mu}\right)_2 = \eta_2 = 50$ md/cp, $(\phi c_t)_1 = 8.95 \times 10^{-7}$ psi^{-1}

$(\phi c_t)_2 = 1.54 \times 10^{-6}$ psi^{-1}, $q_{inj} = 400$ STB/day

$B_w = 1.0$ bbl/STB

Calculate λ_1, λ_2, and r_{f1} and compare with the simulation data.

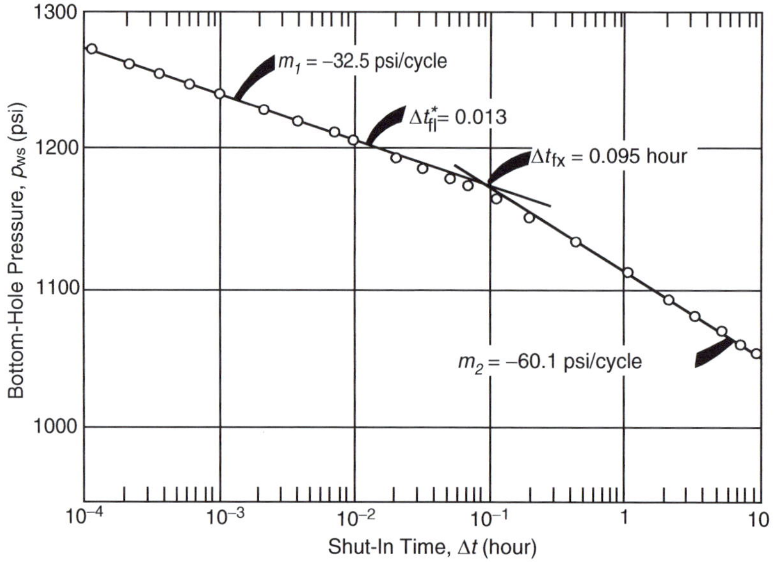

FIGURE 1.145 Falloff test data for Example 1.49. *(After Merrill, L.S., Kazemi, H., Cogarty, W.B., 1974. Pressure falloff analysis in reservoirs with fluid banks. J. Pet. Technol. 26 (7), 809–818.)*

Solution

Step 1. From Figure 1.145, determine m_1, m_2, and Δt_{fx} to give:

$$m_1 = 32.5 \text{ psi/cycle}$$

$$m_2 = 60.1 \text{ psi/cycle}$$

$$\Delta t_{fx} = 0.095 \text{ hour}$$

Step 2. Estimate $(k/\mu)_1$, i.e., mobility of water bank, from Eq. (1.313):

$$\left(\frac{k}{\mu}\right)_1 = \frac{162.6 q_{inj} B}{|m_1| h} = \frac{162.6(400)(1.0)}{(32.5)(20)} = 100 \text{ md/cp}$$

The obtained value matches the value used in the simulation.

Step 3. Calculate the following dimensionless ratios:

$$\frac{m_2}{m_1} = \frac{-60.1}{-32.5} = 1.85$$

$$\frac{(\phi c_t)_1}{(\phi c_t)_2} = \frac{8.95 \times 10^{-7}}{1.54 \times 10^{-6}} = 0.581$$

Step 4. Using the two dimensionless ratios as calculated in step 4, determine the ratio λ_1/λ_2 from Figure 1.143:

$$\frac{\lambda_1}{\lambda_2} = 2.0$$

Step 5. Calculate the mobility in the second zone, i.e., oil bank mobility $\lambda_2 = (k/\mu)_2$, from Eq. (1.314):

$$\left(\frac{k}{\mu}\right)_2 = \frac{(k/\mu)_1}{(\lambda_1/\lambda_2)} = \frac{100}{2.0} = 50 \text{ md/cp}$$

with the exact match of the input data.

Step 6. Determine Δt_{Dfx} from Figure 1.130:

$$\Delta t_{Dfx} = 3.05$$

Step 7. Calculate r_{f1} from Eq. (1.315):

$$r_{f1} = \sqrt{\frac{(0.0002637)(100)(0.095)}{(8.95 \times 10^{-7})(3.05)}} = 30 \text{ ft}$$

Yeh and Agarwal (1989) presented a different approach of analyzing the recorded data from the injectivity and falloff tests. Their methodology uses the pressure derivate Δp and Agarwal equivalent time Δt_e (see Eq. (1.207))

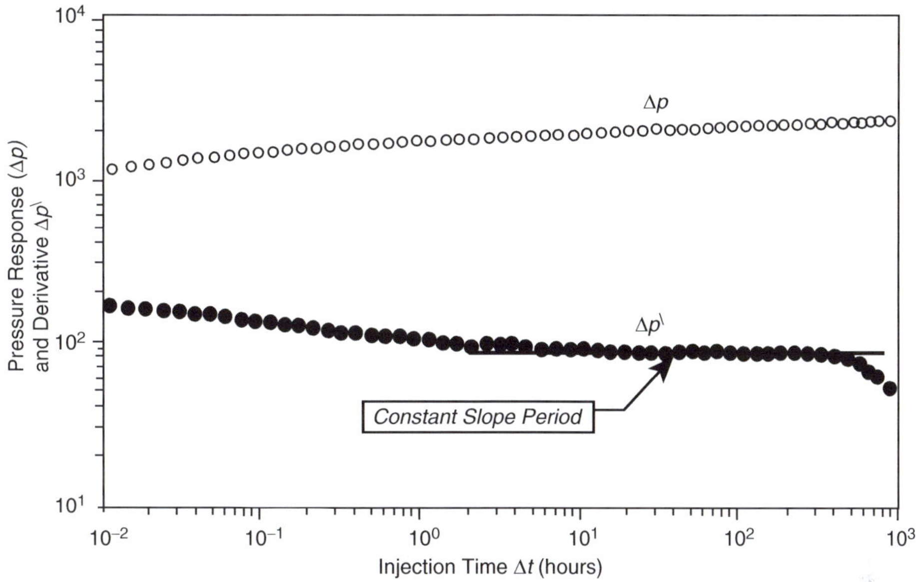

FIGURE 1.146 Injection pressure response and derivative (base case).

in performing the analysis. The authors defined the following nomenclature:

- During the injectivity test period:

$$\Delta p_{wf} = p_{wf} - p_i$$

$$\Delta p'_{wf} = \frac{d(\Delta p_{wf})}{d(\ln t)}$$

where
p_{wf} = bottom-hole pressure at time t during injection, psi
t = injection time, hours
$\ln t$ = natural logarithm of t

- During the falloff test period:

$$\Delta p_{ws} = p_{wf\ at\ \Delta t=0} - p_{ws}$$

$$\Delta p'_{ws} = \frac{d(\Delta p_{ws})}{d(\ln \Delta t_e)}$$

with

$$\Delta t_e = \frac{t_p\,\Delta t}{t_p + \Delta t}$$

where
Δt = shut-in time, hours
t_p = injection time, hours

Through the use of a numerical simulator, Yeh and Agarwal simulated a large number of injectivity and falloff tests and made the following observations for both tests.

Pressure Behavior During Injectivity Tests

(1) A log–log plot of the injection pressure difference Δp_{wf} and its derivative $\Delta p'_{wf}$ vs. injection time will exhibit a constant-slope period, as shown in Figure 1.146, and designated as $(\Delta p'_{wf})_{const}$. The water mobility λ_1 in the floodout zone, i.e., water bank, can be estimated from:

$$\lambda_1 = \left(\frac{k}{\mu}\right)_1 = \frac{70.62\, q_{inj} B}{h(\Delta p'_{wf})_{const}}$$

Notice that the constant 70.62 is used instead of 162.6 because the pressure derivative is calculated with respect to the natural logarithm of time.

(2) The skin factor as calculated from the semi-log analysis method is usually in excess of its true value because of the contrast between injected and reservoir fluid properties.

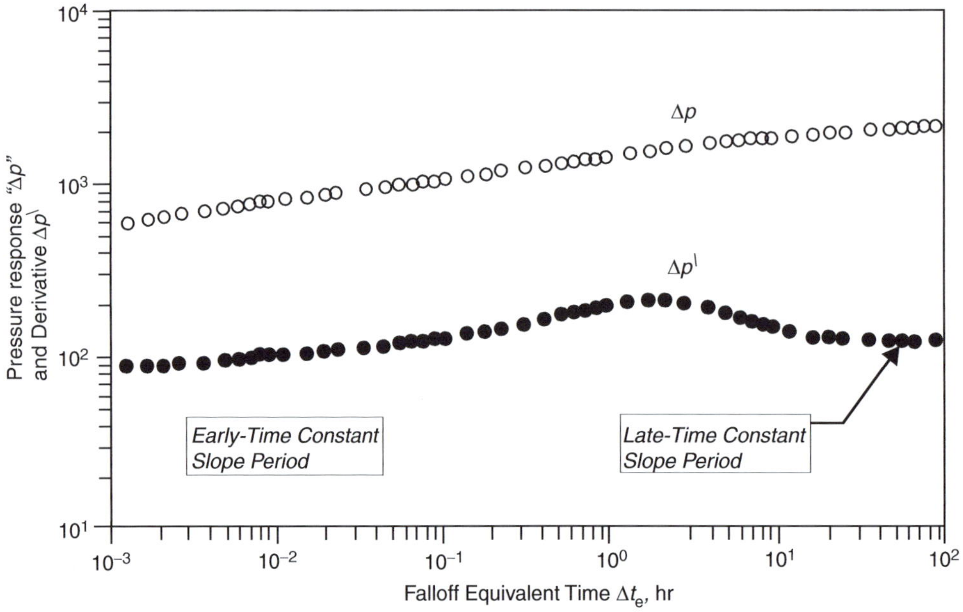

FIGURE 1.147 Falloff pressure response and derivative (base case).

Pressure Behavior During Falloff Tests

(1) The log–log plot of the pressure falloff response in terms of Δp and its derivative as a function of the falloff equivalent time Δt_e is shown in Figure 1.147. The resulting derivative curve shows two constant-slope periods, $(\Delta p^{\backslash}_{ws})_1$ and $(\Delta p^{\backslash}_{ws})_2$, which reflect the radial flow in the floodout zone, i.e., water bank, and, the radial flow in the unflooded zone, i.e., oil bank.

These two derivative constants can be used to estimate the mobility of the water bank λ_1 and the oil bank λ_2 from:

$$\lambda_1 = \frac{70.62 q_{inj} B}{h(\Delta p^{\backslash}_{ws})_1}$$

$$\lambda_2 = \frac{70.62 q_{inj} B}{h(\Delta p^{\backslash}_{ws})_2}$$

(2) The skin factor can be estimated from the first semilog straight line and closely represents the actual mechanical skin on the wellbore.

1.8.3 Step-rate Test

Step-rate injectivity tests are specifically designed to determine the pressure at which fracturing could be induced in the reservoir rock. In this test, water is injected at a constant rate for about 30 minutes before the rate is increased and maintained for successive periods, each of which also lasts for 30 minutes. The pressure observed at the end of each injection rate is plotted vs. the rate. This plot usually shows two straight lines which intersect at the fracture pressure of the formation, as shown schematically in Figure 1.148. The suggested procedure is summarized below:

Step 1. Shut-in the well and allow the bottom-hole pressure to stabilize (if shutting in the well is not possible, or not practical, stabilize the well at a low flow rate). Measure the stabilized pressure.

Step 2. Open the well at a low injection rate and maintain this rate for a preset time.

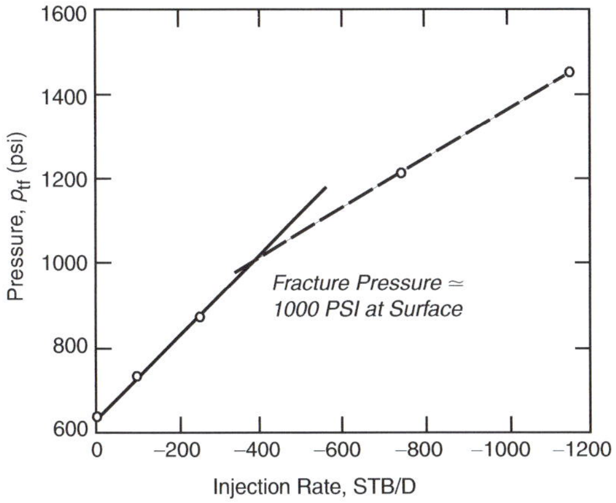

FIGURE 1.148 Step-rate injectivity data plot.

Record the pressure at the end of the flow period.

Step 3. Increase the rate, and at the end of an interval of time equal to that used in step 2, again record the pressure.

Step 4. Repeat step 3 for a number of increasing rates until the parting pressure is noted on the step-rate plot depicted by Figure 1.148.

As pointed out by Horn (1995), data presented in graphical form is much easier to understand than a single table of numbers. Horn proposed the following "toolbox" of graphing functions that is considered an essential part of computer-aided well test interpretation system:

Flow Period	Characteristic	Plot Used
Infinite-acting radial flow (drawdown)	Semilog straight line	p vs. $\log(\Delta t)$ (semilog plot, sometimes called MDH plot)
Infinite-acting radial flow (buildup)	Horner straight line	p vs. $\log(t_p + \Delta t)/\Delta t$ (Horner plot)
Wellbore storage	Straight line p vs. t, or unit-slope $\log \Delta p$ vs. $\log(\Delta t)$	$\log \Delta p$ vs. $\log(\Delta t)$ (log–log plot, type curve)
Finite conductivity fracture	Straight-line slope $\frac{1}{4}$, $\log(\Delta p)$ Δp vs. $\log(\Delta t)$ plot	$\log \Delta p$ vs. $\log(\Delta t)$, or Δp vs. $\Delta t^{1/4}$
Infinite conductivity fracture	Straight-line slope $\frac{1}{2}$, $\log(\Delta p)$ vs. $\log(\Delta t)$ plot	$\log \Delta p$ vs. $\log(\Delta t)$, or Δp vs. $\Delta t^{1/2}$
Dual-porosity behavior	S-shaped transition between parallel semilog straight lines	p vs. $\log(\Delta t)$ (semilog plot)
Closed boundary	Pseudosteady state, pressure linear with time	p vs. Δt (Cartesian plot)
Impermeable fault	Doubling of slope on semilog straight line	p vs. $\log(\Delta t)$ (semilog plot)
Constant-pressure boundary	Constant pressure, flat line on all p, t plots	Any

Chaudhry (2003) presented another useful "toolbox" that summarizes the pressure derivative trends for common flow regimes that have been presented in this chapter, as shown in Table 1.10.

Kamal et al. (1995) conveniently summarized; in tabulated form, various plots and flow regimes most commonly used in transient tests and the information obtained from each test as shown in Tables 1.11 and 1.12.

1.9 PROBLEMS

1 An incompressible fluid flows in a linear porous media with the following properties.

$L = 2500$ ft, $h = 30$ ft, width $= 500$ ft, $k = 50$ md
$\phi = 17\%$, $\mu = 2$ cp, inlet pressure $= 2100$ psi
$Q = 4$ bbl/day, $\rho = 45$ lb/ft^3

Calculate and plot the pressure profile throughout the linear system.

2 Assume the reservoir linear system as described in Problem 1 is tilted with a dip angle of 7°. Calculate the fluid potential through the linear system.

3 A gas with specific gravity 0.7 is flowing in a linear reservoir system at 150 °F. The upstream and downstream pressures are

TABLE 1.10 Pressure Derivative Trends for Common Flow Regimes

Wellbore storage dual-porosity matrix to fissure flow	Semilog straight lines with slope 1.151
	Parallel straight-line responses are characteristics of naturally fractured reservoirs
Dual porosity with pseudosteady-state interporosity flow	Pressure change slope → increasing, leveling off, increasing
	Pressure derivative slope = 0, valley = 0
	Additional distinguishing characteristic is middle-time valley trend during more than 1 log cycle
Dual porosity with transient inter-porosity flow	Pressure change slope → steepening
	Pressure derivative slope = 0, upward trend = 0
	Additional distinguishing characteristic → middle-time slope doubles
Pseudosteady state	Pressure change slope → for drawdown and zero for buildup
	Pressure derivative slope → for drawdown and steeply descending for buildup
	Additional distinguishing characteristic → late time drawdown pressure change and derivative are overlain; slope of 1 occurs much earlier in the derivative
Constant-pressure boundary (steady state)	Pressure change slope → 0
	Pressure derivative slope → steeply descending
	Additional distinguishing characteristic → cannot be distinguished from psuedosteady state in pressure buildup test
Single sealing fault (pseudoradial flow)	Pressure change slope → steeping
	Pressure derivative slope → 0, upward trend → 0
	Additional distinguishing characteristic → late-time slope doubles
Elongated reservoir linear flow	Pressure change slope → 0.5
	Pressure derivative slope → 0.5
	Additional distinguishing characteristic → late-time pressure change and derivative are offset by factor of 2; slope of 0.5 occurs much earlier in the derivative
Wellbore storage infinite-acting radial flow	Pressure change slope = 1, pressure derivative slope = 1
	Additional distinguishing characteristics are: early time pressure change, and derivative are overlain
Wellbore storage, partial penetration, infinite-acting radial flow	Pressure change increases and pressure derivative slope = 0
	Additional distinguishing characteristic is: middle-time flat derivative
Linear flow in an infinite conductivity vertical fracture	$K(x_f)^2$ → calculate from specialized plot
	Pressure slope = 0.5 and pressure derivative slope = 0.5
	Additional distinguishing characteristics are: early-time pressure change and the derivative are offset by a factor of 2
Bilinear flow to an infinite conductivity vertical fracture	$K_f\, w$ → calculate from specialized plot
	Pressure slope = 0.25 and pressure derivative slope = 0.25
	Additional distinguishing characteristic are: early-time pressure change and derivative are offset by factor of 4
Wellbore storage infinite-acting radial flow	Sealing fault
Wellbore storage	No flow boundary
Wellbore storage linear flow	Kb^2 → calculate from specialized plot

TABLE 1.11 Reservoir Properties Obtainable from Various Transient Tests

Drill item tests	Reservoir behavior	Step-rate tests	Formation parting pressure
	Permeability		Permeability
	Skin		Skin
	Fracture length	Falloff tests	Mobility in various banks
	Reservoir pressure		Skin
	Reservoir limit		Reservoir pressure
	Boundaries		Fracture length
Repeat/multiple-formation test	Pressure profile		Location of front
			Boundaries
Drawdown tests	Reservoir behavior	Interference and pulse test	Communication between wells
	Permeability		
	Skin		Reservoir type behavior
	Fracture length		Porosity
	Reservoir limit		Interwell permeability
	Boundaries		Vertical permeability
Buildup tests	Reservoir behavior	Layered reservoir tests	Properties of individual layers
	Permeability		Horizontal permeability
	Skin		Vertical permeability
	Fracture length		Skin
	Reservoir pressure		Average layer pressure
	Boundaries		Outer Boundaries

After Kamal, M., Freyder, D., Murray, M., 1995. Use of transient testing in reservoir management. J. Pet. Technol. 47 (11), 992–999.

2000 and 1800 psi, respectively. The system has the following properties:

$L = 2000$ ft, $W = 300$ ft, $h = 15$ ft
$k = 40$ md, $\phi = 15\%$

Calculate the gas flow rate.

4. An oil well is producing a crude oil system at 1000 STB/day and 2000 psi of bottom-hole flowing pressure. The pay zone and the producing well have the following characteristics.

$h = 35$ ft, $r_w = 0.25$ ft, drainage area $= 40$ acres
API $= 45°$, $\gamma_g = 0.72$, $R_s = 700$ scf/STB
$k = 80$ md

Assuming steady-state flowing conditions, calculate and plot the pressure profile around the wellbore.

5. Assuming steady-state flow and an incompressible fluid, calculate the oil flow rate under the following conditions:

$p_e = 2500$ psi, $p_{wf} = 2000$ psi, $r_e = 745$ ft
$r_w = 0.3$ ft, $\mu_o = 2$ cp, $B_o = 1.4$ bbl/STB
$h = 30$ ft, $k = 60$ md

6. A gas well is flowing under a bottom-hole flowing pressure of 900 psi. The current reservoir pressure is 1300 psi. The following additional data is available:

$T = 140$ °F, $\gamma_g = 0.65$, $r_w = 0.3$ ft
$k = 60$ md, $h = 40$ ft, $r_e = 1000$ ft

Calculate the gas flow rate by using:
(a) the real-gas pseudopressure approach;
(b) the pressure-squared method.

7. After a period of shut-in of an oil well, the reservoir pressure has stabilized at 3200 psi. The well is allowed to flow at a constant flow rate of 500 STB/day under a transient flow condition. Given:

$B_o = 1.1$ bbl/STB, $\mu_o = 2$ cp, $c_t = 15 \times 10^{-6}$ psi^{-1}
$k = 50$ md, $h = 20$ ft, $\phi = 20\%$
$r_w = 0.3$ ft, $p_i = 3200$ psi

TABLE 1.12 Plots and Flow Regimes of Transient Tests

Flow regime	Plot: Cartesian	$\sqrt{\Delta t}$	$\sqrt{4\Delta t}$	Log–log	Semilog
Wellbore storage	Straight line			Unit slope on Δp and p'	Positive s
	Slope → C			Δp and p' coincide	Negative s
	Intercepts → Δt_c				
	Δp_c				
Linear flow		Straight line		Slope = $\frac{1}{2}$ on p' and on	
		Slope = $m_f \rightarrow l_f$		Δp if $s = 0$	
		Intercept = fracture damage		Slope < $\frac{1}{2}$ on Δp if $s \neq 0$	
				p' at half the level of Δp	
Bilinear flow			Straight line	Slope = $\frac{1}{4}$	
			Slope = $m_{bf} \rightarrow C_{fd}$	p' at $\frac{1}{4}$ level of Δp	
First IARF[a] (high-k layer, fractures)	Decreasing slope			p' horizontal at $p'_D = 0.5$	Straight line
					Slope = $m \rightarrow kh$
					$\Delta p_{1\ hour} \rightarrow s$
Transition	More decreasing slope			$\Delta p = \lambda\, e^{-2s}$ or B'	Straight line
				$p'_D = 0.25$ (transition) = <0.25 (pseudo-steady state)	Slope = $m/2$ (transition) = 0 (pseudo-steady state)
Second IARF (total system)	Similar slope to first IARF			p' horizontal at $p'_D = 0.5$ $p' = 0.5$	Straight line
					Slope = $m \rightarrow kh, p^*$
					$\Delta p_{1\ hour} \rightarrow s$
Single no-flow boundary				p' horizontal at $p'_D = 1.0$	Straight line
					Slope = $2m$
					Intersection with IARF → distance to boundary
Outer no-flow boundaries (drawdown test only)	Straight line			Unit slope for Δp and p'	Increasing slope
	Slope = $m^* \rightarrow \phi Ah$			Δp and p' coincide	

[a]IARF = Infinite-Acting Radial Flow.

After Kamal, M., Freyder, D., Murray, M., 1995. Use of transient testing in reservoir management. J. Pet. Technol. 47 (11), 992–999.

Calculate and plot the pressure profile after 1, 5, 10, 15, and 20 hours.

8. An oil well is producing at a constant flow rate of 800 STB/day under a transient flow condition. The following data is available:

$B_o = 1.2$ bbl/STB, $\mu_o = 3$ cp, $c_t = 15 \times 10^{-6}$ psi^{-1}
$k = 100$ md, $h = 25$ ft, $\phi = 15\%$
$r_w = 0.5$, $p_i = 4000$ psi,

Using the Ei function approach and the p_D method, calculate the bottom-hole flowing pressure after 1, 2, 3, 5, and 10 hours. Plot the results on a semilog scale and Cartesian scale.

9. A well is flowing under a drawdown pressure of 350 psi and produces at a constant flow rate of 300 STB/day The net thickness is 25 ft. Given:

$r_e = 660$ ft, $r_w = 0.25$ ft
$\mu_o = 1.2$ cp, $B_o = 1.25$ bbl/STB

Calculate:
(a) the average permeability;
(b) the capacity of the formation.

10. An oil well is producing from the center of a 40 acre square drilling pattern. Given:

$\phi = 20\%$, $h = 15$ ft, $k = 60$ md
$\mu_o = 1.5$ cp, $B_o = 1.4$ bbl/STB, $r_w = 0.25$ ft
$p_i = 2000$ psi, $p_{wf} = 1500$ psi

Calculate the oil flow rate.

11. A shut-in well is located at a distance of 700 ft from one well and 1100 ft from a second well. The first well flows for 5 days at 180 STB/day, at which time the second well begins to flow at 280 STB/day Calculate the pressure drop in the shut-in well when the second well has been flowing for 7 days. The following additional data is given:

$p_i = 3000$ psi, $B_o = 1.3$ bbl/STB, $\mu_o = 1.2$ cp
$h = 60$ ft, $c_t = 15 \times 10^{-6}$ psi^{-1}, $\phi = 15\%$, $k = 45$ md

12. A well is opened to flow at 150 STB/day for 24 hours. The flow rate is then increased to 360 STB/day and lasts for another 24 hours. The well flow rate is then reduced to 310 STB/day for 16 hours. Calculate the pressure drop in a shut-in well 700 ft away from the well, given:

$\phi = 15\%$, $h = 20$ ft, $k = 100$ md
$\mu_o = 2$ cp, $B_o = 1.2$ bbl/STB, $r_w = 0.25$ ft
$p_i = 3000$ psi, $c_t = 12 \times 10^{-6}$ psi^{-1}

13. A well is flowing under unsteady-state flowing conditions for 5 days at 300 STB/day The well is located at a distance of 350 and 420 ft from two sealing faults. Given:

$\phi = 17\%$, $c_t = 16 \times 10^{-6}$ psi^{-1}, $k = 80$ md
$p_i = 3000$ psi, $B_o = 1.3$ bbl/STB, $\mu_o = 1.1$ cp
$r_w = 0.25$ ft, $h = 25$ ft

Calculate the pressure in the well after 5 days.

14. A drawdown test was conducted on a new well with results as given below:

t (hour)	p_{wf} (psi)
1.50	2978
3.75	2949
7.50	2927
15.00	2904
37.50	2876
56.25	2863
75.00	2848
112.50	2810
150.00	2790
225.00	2763

Given:

$p_i = 3400$ psi, $h = 25$ ft, $Q = 300$ STB/day
$c_t = 18 \times 10^{-6}$ psi^{-1}, $\mu_o = 1.8$ cp
$B_o = 1.1$ bbl/STB, $r_w = 0.25$ ft, $\phi = 12\%$

and assuming no wellbore storage, calculate:
(a) the average permeability;
(b) the skin factor.

15. A drawdown test was conducted on a discovery well. The well was allowed to flow at a constant flow rate of 175 STB/day. The fluid and reservoir data is given below:

$S_{wi} = 25\%$, $\phi = 15\%$, $h = 30$ ft, $c_t = 18 \times 10^{-6}$ psi^{-1}
$r_w = 0.25$ ft, $p_i = 4680$ psi, $\mu_o = 1.5$ cp
$B_o = 1.25$ bbl/STB

The drawdown test data is given below:

t (hour)	p_{wf} (psi)
0.6	4388
1.2	4367
1.8	4355
2.4	4344
3.6	4334
6.0	4318
8.4	4309
12.0	4300
24.0	4278
36.0	4261
48.0	4258
60.0	4253
72.0	4249
84.0	4244
96.0	4240
108.0	4235
120.0	4230
144.0	4222
180.0	4206

Calculate:

(a) the drainage area;
(b) the skin factor;
(c) the oil flow rate at a bottom-hole flowing pressure of 4300 psi, assuming semisteady-state flowing conditions.

16. A pressure buildup test was conducted on a well that had been producing at 146 STB/day for 53 hours.

The reservoir and fluid data is given below.

$B_o = 1.29$ bbl/STB, $\mu_o = 0.85$ cp
$c_t = 12 \times 10^{-6}$ psi^{-1}, $\phi = 10\%$, $p_{wf} = 1426.9$ psig
$A = 20$ acres

The buildup data is as follows:

Time	p_{ws} (psig)
0.167	1451.5
0.333	1476.0
0.500	1498.6
0.667	1520.1
0.833	1541.5
1.000	1561.3
1.167	1581.9
1.333	1599.7
1.500	1617.9
1.667	1635.3
2.000	1665.7
C	1691.8
2.667	1715.3
3.000	1736.3
3.333	1754.7
3.667	1770.1
4.000	1783.5
4.500	1800.7
5.000	1812.8
5.500	1822.4
6.000	1830.7
6.500	1837.2
7.000	1841.1
7.500	1844.5
8.000	1846.7
8.500	1849.6
9.000	1850.4
10.000	1852.7
11.000	1853.5
12.000	1854.0
12.667	1854.0
14.620	1855.0

Calculate:

(a) the average reservoir pressure;
(b) the skin factor;
(c) the formation capacity;
(d) an estimate of the drainage area and compare with the given value.

CHAPTER 2

Water Influx

Water-bearing rocks called aquifers surround nearly all hydrocarbon reservoirs. These aquifers may be substantially larger than the oil or gas reservoirs they adjoin as to appear infinite in size, and they may be so small in size as to be negligible in their effect on reservoir performance.

As reservoir fluids are produced and reservoir pressure declines, a pressure differential develops from the surrounding aquifer into the reservoir. Following the basic law of fluid flow in porous media, the aquifer reacts by encroaching across the original hydrocarbon–water contact. In some cases, water encroachment occurs due to hydrodynamic conditions and recharge of the formation by surface waters at an outcrop. In many cases, the pore volume of the aquifer is not significantly larger than the pore volume of the reservoir itself. Thus, the expansion of the water in the aquifer is negligible relative to the overall energy system, and the reservoir behaves volumetrically. In this case, the effects of water influx can be ignored. In other cases, the aquifer permeability may be sufficiently low such that a very large pressure differential is required before an appreciable amount of water can encroach into the reservoir. In this instance, the effects of water influx can be ignored as well.

The objective of this chapter, however, concerns those reservoir–aquifer systems in which the size of the aquifer is large enough and the permeability of the rock is high enough that water influx occurs as the reservoir is depleted. This chapter is designed to provide the various water influx calculation models and a detailed description of the computational steps involved in applying these models.

2.1 CLASSIFICATION OF AQUIFERS

Many gas and oil reservoirs are produced by a mechanism termed "water drive." Often this is called natural water drive to distinguish it from artificial water drive that involves the injection of water into the formation. Hydrocarbon production from the reservoir and the subsequent pressure drop prompt a response from the aquifer to offset the pressure decline. This response comes in the form of a water influx, commonly called water encroachment, which is attributed to:

- expansion of the water in the aquifer;
- compressibility of the aquifer rock;
- artesian flow where the water-bearing formation outcrop is located structurally higher than the pay zone.

Reservoir–aquifer systems are commonly classified on the basis described in the following subsections.

2.1.1 Degree of Pressure Maintenance

Based on the degree of reservoir pressure maintenance provided by the aquifer, the natural water drive is often qualitatively described as:

- the active water drive;
- the partial water drive;
- the limited water drive.

The term “active” water drive refers to the water encroachment mechanism in which the rate of water influx equals the reservoir *total* production rate. Active water drive reservoirs are typically characterized by a gradual and slow reservoir pressure decline. If during any long period the production rate and reservoir pressure remain reasonably constant, the reservoir voidage rate must be equal to the water influx rate:

$$\begin{bmatrix}\text{Water influx}\\ \text{rate}\end{bmatrix} = \begin{bmatrix}\text{Oil flow}\\ \text{rate}\end{bmatrix} + \begin{bmatrix}\text{Free gas}\\ \text{flow rate}\end{bmatrix} + \begin{bmatrix}\text{Water}\\ \text{production}\\ \text{rate}\end{bmatrix}$$

or

$$e_w = Q_o B_o + Q_g B_g + Q_w B_w \tag{2.1}$$

where

e_w = water influx rate, bbl/day
Q_o = oil flow rate, STB/day
B_o = oil formation volume factor, bbl/STB
Q_g = free gas flow rate, scf/day
B_g = gas formation volume factor, bbl/scf
Q_w = water flow rate, STB/day
B_w = water formation volume factor, bbl/STB

Eq. (2.1) can be equivalently expressed in terms of cumulative production by introducing the following derivative terms:

$$e_w = \frac{dW_e}{dt} = B_o \frac{dN_p}{dt} + (\text{GOR} - R_s)\frac{dN_p}{dt} B_g + \frac{dW_p}{dt} B_w \tag{2.2}$$

where

W_e = cumulative water influx, bbl
t = time, days
N_p = cumulative oil production, STB
GOR = current gas–oil ratio, scf/STB
R_s = current gas solubility, scf/STB
B_g = gas formation volume factor, bbl/scf
W_p = cumulative water production, STB
dN_p/dt = daily oil flow rate Q_o, STB/day
dW_p/dt = daily water flow rate Q_w, STB/day
dW_e/dt = daily water influx rate e_w, bbl/day
$(\text{GOR} - R_s)dN_p/dt$ = daily free gas rate, scf/day

Example 2.1

Calculate the water influx rate e_w in a reservoir whose pressure is stabilized at 3000 psi. Given:

initial reservoir pressure = 3500 psi
dN_p/dt = 32,000 STB/day
B_o = 1.4 bbl/STB, GOR = 900 scf/STB, R_s = 700 scf/STB
B_g = 0.00082 bbl/scf, dW_p/dt = 0, B_w = 1.0 bbl/STB

Solution

Applying Eq. (2.1) or (2.2) gives:

$$\begin{aligned} e_w &= \frac{dW_e}{dt} = B_o \frac{dN_p}{dt} + (\text{GOR} - R_s)\frac{dN_p}{dt} B_g + \frac{dW_p}{dt} B_w \\ &= (1.4)(32{,}000) + (900 - 700)(32{,}000)(0.00082) + 0 \\ &= 50{,}048 \text{ bbl/day} \end{aligned}$$

2.1.2 Outer Boundary Conditions

The aquifer can be classified as infinite or finite (bounded). Geologically all formations are finite but may act as infinite if the changes in the pressure at the oil–water contact are not “felt” at the aquifer boundary. Some aquifers outcrop and are infinite acting because of surface replenishment. In general, the outer boundary governs the behavior of the aquifer and can be classified as follows:

- Infinite system indicates that the effect of the pressure changes at the oil–aquifer boundary can never be felt at the outer boundary. This boundary is for all intents and purposes at a constant pressure equal to initial reservoir pressure.
- Finite system indicates that the aquifer outer limit is affected by the influx into the oil zone and that the pressure at this outer limit changes with time.

2.1.3 Flow Regimes

There are basically three flow regimes that influence the rate of water influx into the reservoir. As previously described in Chapter 1, these flow regimes are:

(1) steady state;
(2) semi(pseudo)steady state;
(3) unsteady state.

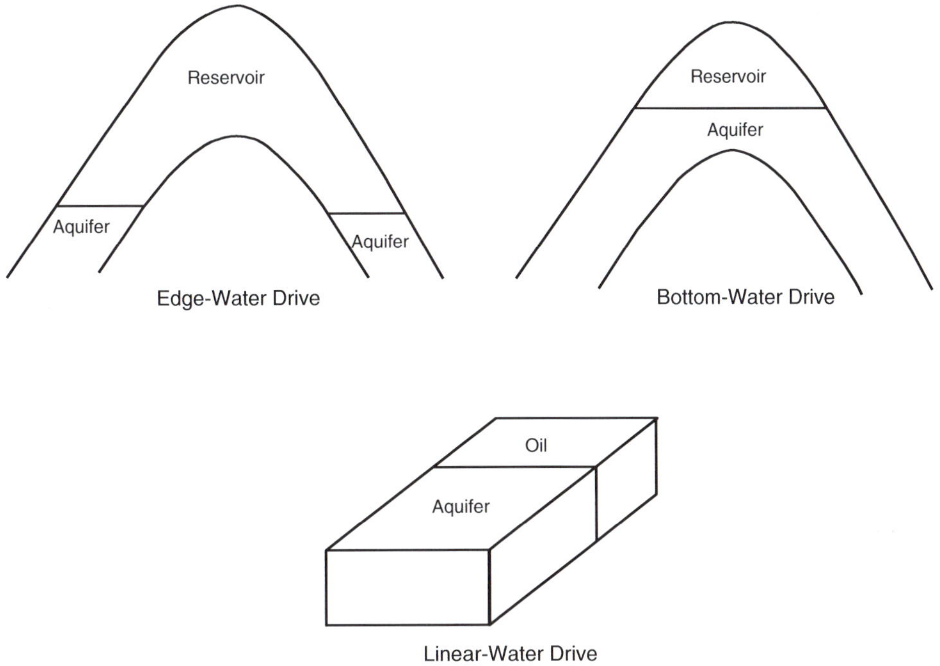

FIGURE 2.1 Flow geometries.

2.1.4 Flow Geometries

Reservoir–aquifer systems can be classified on the basis of flow geometry as:

- edge-water drive;
- bottom-water drive;
- linear-water drive.

In edge-water drive, as shown in Figure 2.1, water moves into the flanks of the reservoir as a result of hydrocarbon production and pressure drop at the reservoir–aquifer boundary. The flow is essentially radial with negligible flow in the vertical direction.

Bottom-water drive occurs in reservoirs with large areal extent and a gentle dip where the reservoir–water contact completely underlies the reservoir. The flow is essentially radial and, in contrast to the edge-water drive, the bottom-water drive has significant vertical flow.

In linear-water drive, the influx is from one flank of the reservoir. The flow is strictly linear with a constant cross-sectional area.

2.2 RECOGNITION OF NATURAL WATER INFLUX

Normally very little information is obtained during the exploration and development period of a reservoir concerning the presence or characteristics of an aquifer that could provide a source of water influx during the depletion period. Natural water drive may be assumed by analogy with nearby producing reservoirs, but early reservoir performance trends can provide clues. A comparatively low, and decreasing, rate of reservoir pressure decline with increasing cumulative withdrawals is indicative of fluid influx. Successive calculations of barrels withdrawn per psi change in reservoir pressure can supplement performance graphs. However, if the reservoir limits have not been delineated by the developmental dry holes, the influx could be from an undeveloped area of the reservoir not accounted for in averaging reservoir pressure. If the reservoir pressure is below the oil saturation pressure, a low rate of increase in produced GOR is also indicative of fluid influx.

Early water production from edge wells is indicative of water encroachment. Such observations must be tempered by the possibility that the early water production is due to formation fractures, thin high-permeability streaks, or coning in connection with a limited aquifer. The water production may be due to casing leaks.

Calculation of increasing original oil-in-place from successive reservoir pressure surveys by using the material balance and assuming no water influx is also indicative of fluid influx.

2.3 WATER INFLUX MODELS

It should be appreciated that there are more uncertainties attached to this part of reservoir engineering than to any other. This is simply because one seldom drills wells into an aquifer to gain the necessary information about the porosity, permeability, thickness, and fluid properties. Instead, these properties have frequently to be inferred from what has been observed in the reservoir. Even more uncertain, however, is the geometry and areal continuity of the aquifer itself.

Several models have been developed for estimating water influx that is based on assumptions that describe the characteristics of the aquifer. Because of the inherent uncertainties in the aquifer characteristics, all of the proposed models require historical reservoir performance data to evaluate constants representing aquifer property parameters. However, these parameters are rarely known from exploration and development drilling with sufficient accuracy for direct application in various aquifer models. The material balance equation can be used to determine historical water influx provided original oil-in-place is known from pore volume estimates. This permits evaluation of the constants in the influx equations so that future water influx rate can be forecast.

The mathematical water influx models that are commonly used in the petroleum industry include:

- pot aquifer;
- Schilthuis steady state;
- Hurst modified steady state;
- van Everdingen and Hurst unsteady state:
  - edge-water drive;
  - bottom-water drive;
- Carter–Tracy unsteady state;
- Fetkovich method:
  - radial aquifer;
  - linear aquifer.

The following sections describe the above models and their practical applications in water influx calculations.

2.3.1 The Pot Aquifer Model

The simplest model that can be used to estimate the water influx into a gas or oil reservoir is based on the basic definition of compressibility. A drop in the reservoir pressure, due to the production of fluids, causes the aquifer water to expand and flow into the reservoir. The compressibility is defined mathematically as:

$$c = \frac{1}{V}\frac{\partial V}{\partial p} = \frac{1}{V}\frac{\Delta V}{\Delta p}$$

or

$$\Delta V = cV\,\Delta p$$

Applying the above basic compressibility definition to the aquifer gives:

Water influx = (Aquifer compressibility) × (Initial volume of water) (Pressure drop)

or

$$W_e = c_t W_i (p_i - p), \quad c_t = c_w + c_f \tag{2.3}$$

where

W_e = cumulative water influx, bbl
c_t = aquifer total compressibility, psi^{-1}
c_w = aquifer water compressibility, psi^{-1}
c_f = aquifer rock compressibility, psi^{-1}
W_i = initial volume of water in the aquifer, bbl
p_i = initial reservoir pressure, psi
p = current reservoir pressure (pressure at oil–water contact), psi

Calculating the initial volume of water in the aquifer requires knowledge of aquifer dimensions and properties. These, however, are seldom measured since wells are not deliberately drilled into the aquifer to obtain such information. For instance, if the aquifer shape is radial, then:

$$W_i = \left[\frac{\pi(r_a^2 - r_e^2)h\phi}{5.615}\right] \quad \textbf{(2.4)}$$

where

r_a = radius of the aquifer, ft
r_e = radius of the reservoir, ft
h = thickness of the aquifer, ft
ϕ = porosity of the aquifer

Eq. (2.4) suggests that water is encroaching in a radial form from all directions. Quite often, water does not encroach on all sides of the reservoir, or the reservoir is not circular in nature. To account for these cases, a modification to Eq. (2.4) must be made in order to properly describe the flow mechanism. One of the simplest modifications is to include the fractional encroachment angle f in the equation, as illustrated in Figure 2.2, to give:

$$W_e = (c_w + c_f)W_i f(p_i - p) \quad \textbf{(2.5)}$$

where the fractional encroachment angle f is defined by:

$$f = \frac{(\text{Encroachment angle})^\circ}{360^\circ} = \frac{\theta}{360^\circ} \quad \textbf{(2.6)}$$

The above model is only applicable to a small aquifer, i.e., pot aquifer, whose dimensions are of the same order of magnitude as the reservoir itself. Dake (1978) pointed out that because the aquifer is considered relatively small, a pressure drop in the reservoir is instantaneously transmitted throughout the entire reservoir–aquifer system. Dake suggested that for large aquifers, a mathematical model is required that includes time dependence to account for the fact that it takes a finite time for the aquifer to respond to a pressure change in the reservoir.

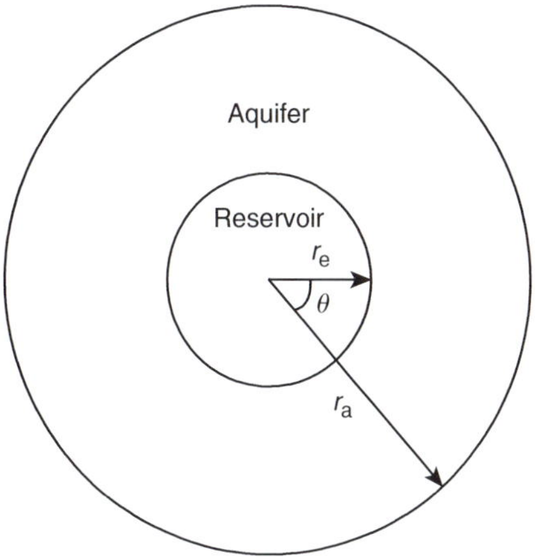

FIGURE 2.2 Radial aquifer geometries.

Example 2.2

Calculate the cumulative water influx that results from a pressure drop of 200 psi at the oil–water contact with an encroachment angle of 80°. The reservoir–aquifer system is characterized by the following properties:

	Reservoir	Aquifer
Radius, ft	2600	10,000
Porosity	0.18	0.12
c_f, psi^{-1}	4×10^{-6}	3×10^{-6}
c_w, psi^{-1}	5×10^{-6}	4×10^{-6}
h, ft	20	25

Solution

Step 1. Calculate the initial volume of water in the aquifer from Eq. (2.4):

$$W_i = \left[\frac{\pi(r_a^2 - r_e^2)h\phi}{5.615}\right] = \left[\frac{\pi(10,000^2 - 2600^2)(25)(0.12)}{5.615}\right] = 156.5 \text{ MMbbl}$$

Step 2. Determine the cumulative water influx by applying Eq. (2.5):

$$W_e = (c_w + c_f)W_i f(p_i - p) = (4.0 + 3.0)10^{-6}(156.5\times10^6)\left(\frac{80}{360}\right)(200) = 48,689 \text{ bbl}$$

2.3.2 The Schilthuis Steady-State Model

Schilthuis (1936) proposed that for an aquifer that is flowing under the steady-state flow regime, the flow behavior could be described by Darcy's equation. The rate of water influx e_w can then be determined by applying Darcy's equation:

$$\frac{dW_e}{dt} = e_w = \left[\frac{0.00708\,kh}{\mu_w \ln(r_a/r_e)}\right](p_i - p) \quad \textbf{(2.7)}$$

This relationship can be more conveniently expressed as:

$$\frac{dW_e}{dt} = e_w = C(p_i - p) \quad \textbf{(2.8)}$$

where

e_w = rate of water influx, bbl/day
k = permeability of the aquifer, md
h = thickness of the aquifer, ft
r_a = radius of the aquifer, ft
r_e = radius of the reservoir, ft
t = time, days

The parameter C is called the "water influx constant" and expressed in bbl/day/psi. This water influx constant C may be calculated from the reservoir historical production data over a number of selected time intervals, provided the rate of water influx e_w has been determined independently from a different expression. For instance, the parameter C may be estimated by combining Eq. (2.1) with Eq. (2.8). Although the influx constant can only be obtained in this manner when the reservoir pressure stabilizes, once it has been found it may be applied to both stabilized and changing reservoir pressures.

Example 2.3

The data given in Example 2.1 is used in this example:

$p_i = 3500$ psi, $p = 3000$ psi, $Q_o = 32{,}000$ STB/day
$B_o = 1.4$ bbl/STB, GOR = 900 scf/STB, $R_s = 700$ scf/STB
$B_g = 0.00082$ bbl/scf, $Q_w = 0$, $B_w = 1.0$ bbl/STB

Calculate the Schilthuis water influx constant.

Solution

Step 1. Solve for the rate of water influx e_w by using Eq. (2.1):

$$\begin{aligned} e_w &= Q_o B_o + Q_g B_g + Q_w B_w \\ &= (1.4)(32{,}000) + (900-700)(32{,}000)(0.0082) + 0 \\ &= 50{,}048 \text{ bbl/day} \end{aligned}$$

Step 2. Solve for the water influx constant from Eq. (2.8):

$$\frac{dW_e}{dt} = e_w = C(p_i - p)$$

or

$$C = \frac{e_w}{p_i - p} = \frac{50{,}048}{3500-3000} = 100 \text{ bbl/day/psi}$$

If the steady-state approximation is considered to adequately describe the aquifer flow regime, the values of the calculated water influx constant C will be constant over the historical period.

Note that the pressure drops contributing to the influx are the cumulative pressure drops from the initial pressure.

In terms of the cumulative water influx W_e, Eq. (2.8) is integrated to give the common Schilthuis expression for water influx as:

$$\int_0^{W_e} dW_e = \int_0^t C(p_i - p)\,dt$$

or

$$W_e = C\int_0^t (p_i - p)dt \quad \textbf{(2.9)}$$

where

W_e = cumulative water influx, bbl
C = water influx constant, bbl/day/psi
t = time, days
p_i = initial reservoir pressure, psi
p = pressure at the oil–water contact at time t, psi

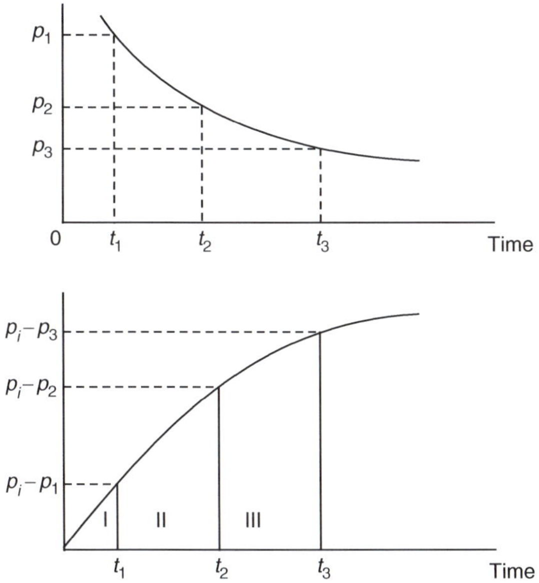

FIGURE 2.3 Calculating the area under the curve.

When the pressure drop $(p_i - p)$ is plotted versus the time t, as shown in Figure 2.3, the area under the curve represents the integral $\int_0^t (p_i - p)\,dt$. This area at time t can be determined numerically by using the trapezoidal rule (or any other numerical integration method) as:

$$\int_0^t (p_i - p)\,dt = \text{area}_{\text{I}} + \text{area}_{\text{II}} + \text{area}_{\text{III}} + \cdots$$
$$= \left(\frac{p_i - p_1}{2}\right)(t_1 - 0)$$
$$+ \frac{(p_i - p_1) + (p_i - p_2)}{2}(t_2 - t_1)$$
$$+ \frac{(p_i - p_2) + (p_i - p_3)}{2}(t_3 - t_2) + \cdots$$

Equation (2.9) can then be written as:

$$W_e = C \sum_0^t (\Delta p)\,\Delta t \tag{2.10}$$

Example 2.4

The pressure history of a water drive oil reservoir is given below:

t (days)	**p (psi)**
0	
100	3450
200	3410
300	3380
400	3340

The aquifer is under a steady-state flowing condition with an estimated water influx constant of 130 bbl/day/psi. Given the initial reservoir pressure is 3500 psi, calculate the cumulative water influx after 100, 200, 300, and 400 days using the steady-state model.

Solution

Step 1. Calculate the total pressure drop at each time t:

t (days)	**p**	**$p_i - p$**
0	3500	0
100	3450	50
200	3410	90
300	3380	120
400	3340	160

Step 2. Calculate the cumulative water influx after 100 days:

$$W_e = C\left[\left(\frac{p_i - p_1}{2}\right)(t_1 - 0)\right]$$
$$= 130\left(\frac{50}{2}\right)(100 - 0) = 325{,}000 \text{ bbl}$$

Step 3. Determine W_e after 200 days:

$$W_e = C\left\{\left(\frac{p_i - p_1}{2}\right)(t_1 - 0) + \left[\frac{(p_i - p_1) + (p_i - p_2)}{2}\right](t_2 - t_1)\right\}$$
$$= 130\left[\left(\frac{50}{2}\right)(100 - 0) + \left(\frac{50 + 90}{2}\right)(200 - 100)\right]$$
$$= 1{,}235{,}000 \text{ bbl}$$

Step 4. Determine W_e after 300 days:

$$W_e = C\left\{\left(\frac{p_i - p_1}{2}\right)(t_1 - 0)\right.$$
$$+ \left[\frac{(p_i - p_1) + (p_i - p_2)}{2}\right](t_2 - t_1)$$
$$\left. + \frac{(p_i - p_2) + (p_i - p_3)}{2}(t_3 - t_2)\right\}$$
$$= 130\left[\left(\frac{50}{2}\right)(100) + \left(\frac{50 + 90}{2}\right)(200 - 100)\right.$$

$$+\left(\frac{120+90}{2}\right)(300-200)\bigg] = 2,600,000 \text{ bbl}$$

Step 5. Similarly, calculate W_e after 400 days:

$$W_e = 130\bigg[2500 + 7000 + 10,500$$

$$+\left(\frac{160+120}{2}\right)(400-300)\bigg] = 4,420,000 \text{ bbl}$$

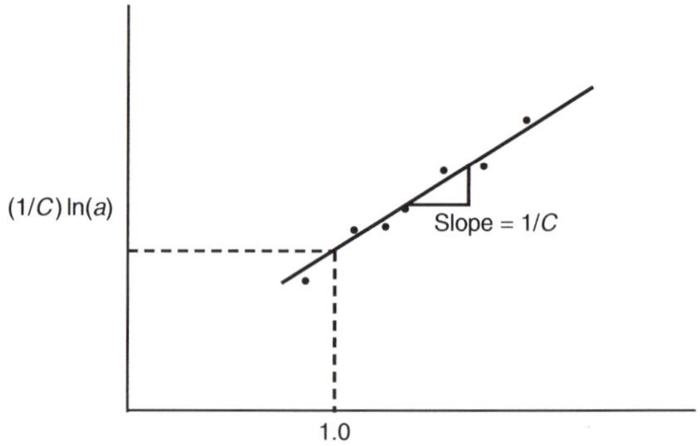

FIGURE 2.4 Graphical determination of C and a.

2.3.3 The Hurst Modified Steady-State Equation

One of the problems associated with the Schilthuis steady-state model is that as the water is drained from the aquifer, the aquifer drainage radius r_a will increase as the time increases. Hurst (1943) proposed that the "apparent" aquifer radius r_a would increase with time and, therefore, the dimensionless radius r_a/r_e may be replaced with a *time-dependent function* as given below:

$$\frac{r_a}{r_e} = at \tag{2.11}$$

Substituting Eq. (2.11) into Eq. (2.7) gives:

$$e_w = \frac{dW_e}{dt} = \frac{0.00708kh(p_i - p)}{\mu_w \ln(at)} \tag{2.12}$$

The Hurst modified steady-state equation can be written in a more simplified form as:

$$e_w = \frac{dW_e}{dt} = \frac{C(p_i - p)}{\ln(at)} \tag{2.13}$$

and in terms of the cumulative water influx:

$$W_e = C\int_0^t \left[\frac{p_i - p}{\ln(at)}\right] dt \tag{2.14}$$

Approximating the integral with a summation gives:

$$W_e = C\sum_0^t \left[\frac{\Delta p}{\ln(at)}\right]\Delta t \tag{2.15}$$

The Hurst modified steady-state equation contains two unknown constants, i.e., a and C, that must be determined from the reservoir–aquifer pressure and water influx historical data. The procedure for determining the constants a and C is based on expressing Eq. (2.13) as a linear relationship:

$$\left(\frac{p_i - p}{e_w}\right) = \frac{1}{C}\ln(at)$$

or

$$\frac{p_i - p}{e_w} = \left(\frac{1}{C}\right)\ln(a) + \left(\frac{1}{C}\right)\ln(t) \tag{2.16}$$

Eq. (2.16) indicates that a plot of the term $(p_i - p)/e_w$ vs. $\ln(t)$ would produce a straight line with a slope of $1/C$ and intercept of $(1/C)\ln(a)$, as shown schematically in Figure 2.4.

Example 2.5

The following data, as presented by Craft and Hawkins (1959), documents the reservoir pressure as a function of time for a water drive reservoir. Using the reservoir historical data, Craft and Hawkins calculated the water influx by applying the material balance equation (see Chapter 4). The

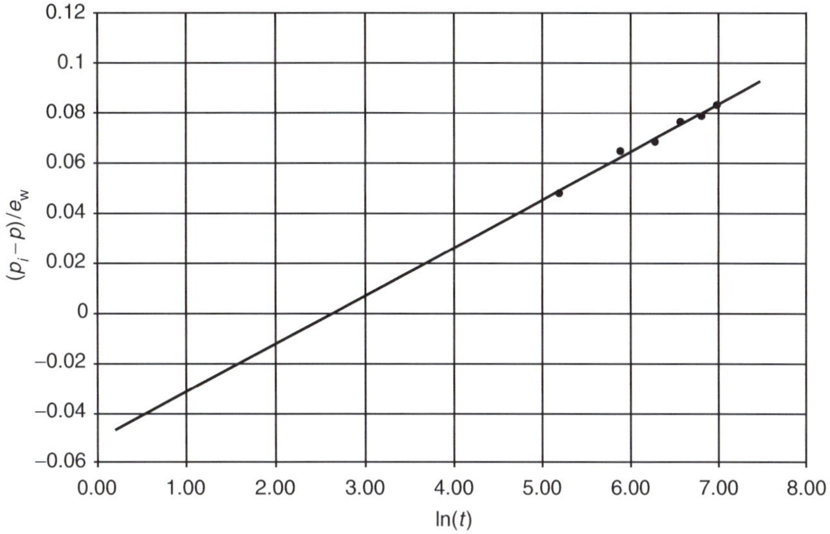

FIGURE 2.5 Determination of *C* and *n* for Example 2.5.

rate of water influx was also calculated numerically at each time period:

Time (days)	Pressure (psi)	W_e (M bbl)	e_w (bbl/day)	$p_i - p$ (psi)
0	3793	0	0	0
182.5	3774	24.8	389	19
365.0	3709	172.0	1279	84
547.5	3643	480.0	2158	150
730.0	3547	978.0	3187	246
912.5	3485	1616.0	3844	308
1095.0	3416	2388.0	4458	377

It is predicted that the boundary pressure would drop to 3379 psi after 1186.25 days of production. Calculate the cumulative water influx at that time.

Solution

Step 1. Construct the following table:

t (days)	ln(t)	$p_i - p$	e_w (bbl/day)	$(p_i - p)/e_w$
0	–	0	0	–
182.5	5.207	19	389	0.049
365.0	5.900	84	1279	0.066
547.5	6.305	150	2158	0.070
730.0	6.593	246	3187	0.077
912.5	6.816	308	3844	0.081
1095.0	6.999	377	4458	0.085

Step 2. Plot the term $(p_i - p)/e_w$ vs. ln(t) and draw the best straight line through the points as shown in Figure 2.5, and determine the slope of the line:

$$\text{Slope} = \frac{1}{C} = 0.020$$

Step 3. Determine the coefficient C of the Hurst equation from the slope:

$$C = \frac{1}{\text{Slope}} = \frac{1}{0.02} = 50$$

Step 4. Use any point on the straight line and solve for the parameter a by applying Eq. (2.13):

$$a = 0.064$$

Step 5. The Hurst equation is represented by:

$$W_e = 50 \int_0^t \left[\frac{p_i - p}{\ln(0.064t)}\right] dt$$

Step 6. Calculate the cumulative water influx after 1186.25 days from:

$$W_e = 2388 \times 10^3 + \int_{1095}^{1186.25} 50\left[\frac{p_i - p}{\ln(0.064t)}\right] dt$$

$$= 2388 \times 10^3 + 50\left[((3793 - 3379)/\ln(0.064 \times 1186.25)\right.$$

$$\left. + (3793 - 3416)/\ln(0.064 \times 1095))/2\right](1186.25 - 1095)$$

$$= 2388 \times 10^3 + 420.508 \times 10^3 = 2809 \text{ Mbbl}$$

2.3.4 The van Everdingen and Hurst Unsteady-State Model

The mathematical formulations that describe the flow of a crude oil system into a wellbore are identical in form to those equations that describe the flow of water from an aquifer into a cylindrical reservoir, as shown schematically in Figure 2.6. When an oil well is brought on production at a constant flow rate after a shut-in period, the pressure behavior is essentially controlled by the transient (unsteady-state) flowing condition. This flowing condition is defined as the time period during which the boundary has no effect on the pressure behavior.

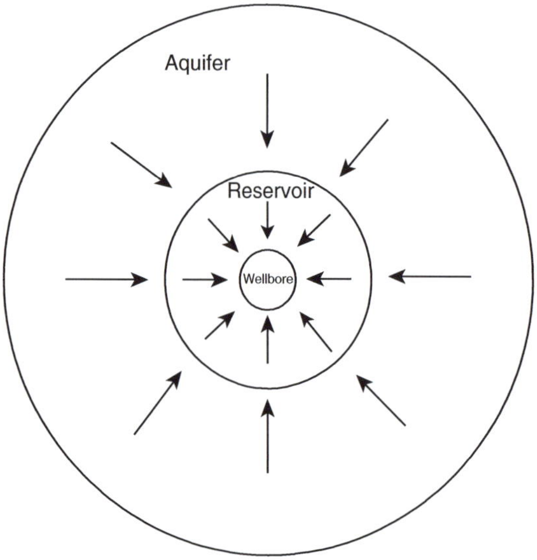

FIGURE 2.6 Water influx into a cylindrical reservoir.

The dimensionless form of the diffusivity equation, as presented in Chapter 1 by Eq. (1.89), is basically the general mathematical equation that is designed to model the transient flow behavior in reservoirs or aquifers. In a dimensionless form, the diffusivity equation is:

$$\frac{\partial^2 P_D}{\partial r_D^2} + \frac{1}{r_D}\frac{\partial P_D}{\partial r_D} = \frac{\partial P_D}{\partial t_D}$$

van Everdingen and Hurst (1949) proposed solutions to the dimensionless diffusivity equation for the following two reservoir–aquifer boundary conditions:

(1) constant terminal rate;
(2) constant terminal pressure.

For the constant-terminal-rate boundary condition, the rate of water influx is assumed constant for a given period, and the pressure drop at the reservoir–aquifer boundary is calculated.

For the constant-terminal-pressure boundary condition, a boundary pressure drop is assumed constant over some finite time period, and the water influx rate is determined.

In the description of water influx from an aquifer into a reservoir, there is greater interest in calculating the influx rate rather than the pressure. This leads to the determination of the water influx as a function of a given pressure drop at the inner boundary of the reservoir–aquifer system.

van Everdingen and Hurst (1949) solved the diffusivity equation for the aquifer–reservoir system by applying the Laplace transformation to the equation. The authors' solution can be used to determine the water influx in the following systems:

- edge-water drive system (radial system);
- bottom-water drive system;
- linear-water drive system.

Edge-Water Drive. Figure 2.7 shows an idealized radial flow system that represents an edge-water drive reservoir. The inner boundary is

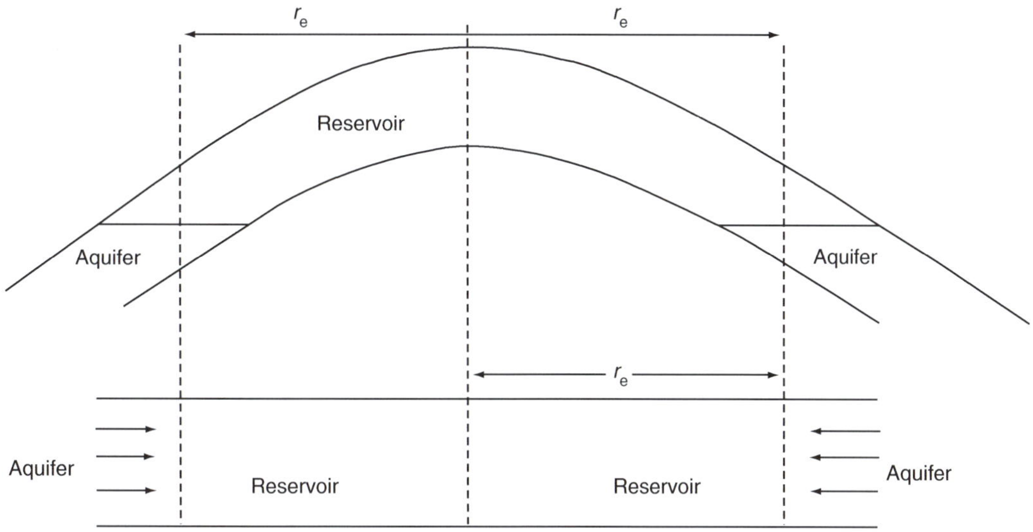

FIGURE 2.7 Idealized radial flow model.

defined as the interface between the reservoir and the aquifer. The flow across this inner boundary is considered horizontal and encroachment occurs across a cylindrical plane encircling the reservoir. With the interface as the inner boundary, it is possible to impose a constant terminal pressure at the inner boundary and determine the rate of water influx across the interface.

van Everdingen and Hurst proposed a solution to the dimensionless diffusivity equation that utilizes the constant-terminal-pressure condition in addition to the following initial and outer boundary conditions:

Initial conditions:

$$p = p_i \quad \text{for all values of radius } r$$

Outer boundary conditions:

- For an infinite aquifer:

$$p = p_i \quad \text{at } r = \infty$$

- For a bounded aquifer:

$$\frac{\partial p}{\partial r} = 0 \quad \text{at } r = r_a$$

van Everdingen and Hurst assumed that the aquifer is characterized by:

- uniform thickness;
- constant permeability;
- uniform porosity;
- constant rock compressibility;
- constant water compressibility.

The authors expressed their mathematical relationship for calculating the water influx in the form of a dimensionless parameter called dimensionless water influx W_{eD}. They also expressed the dimensionless water influx as a function of the dimensionless time t_D and dimensionless radius r_D; thus, they made the solution to the diffusivity equation generalized and it can be applied to any aquifer where the flow of water into the reservoir is essentially radial. The solutions were derived for the cases of bounded aquifers and aquifers of infinite extent. The authors presented their solution in tabulated and graphical forms as reproduced here in Figures 2.8–2.11 and Tables 2.1 and 2.2. The two dimensionless parameters t_D and r_D are given by:

$$t_D = 6.328 \times 10^{-3} \frac{kt}{\phi \mu_w c_t r_e^2} \tag{2.17}$$

$$r_D = \frac{r_a}{r_e} \tag{2.18}$$

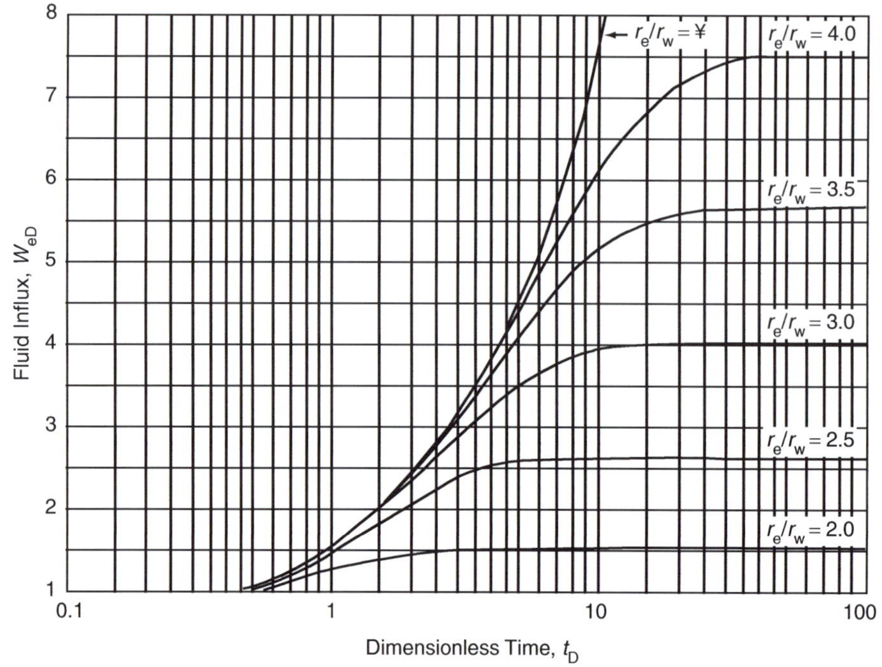

FIGURE 2.8 Dimensionless water influx W_{eD} for several values of r_e/r_R, i.e., r_a/r_e. (*van Everdingen and Hurst W_{eD}. Permission to publish by the SPE*).

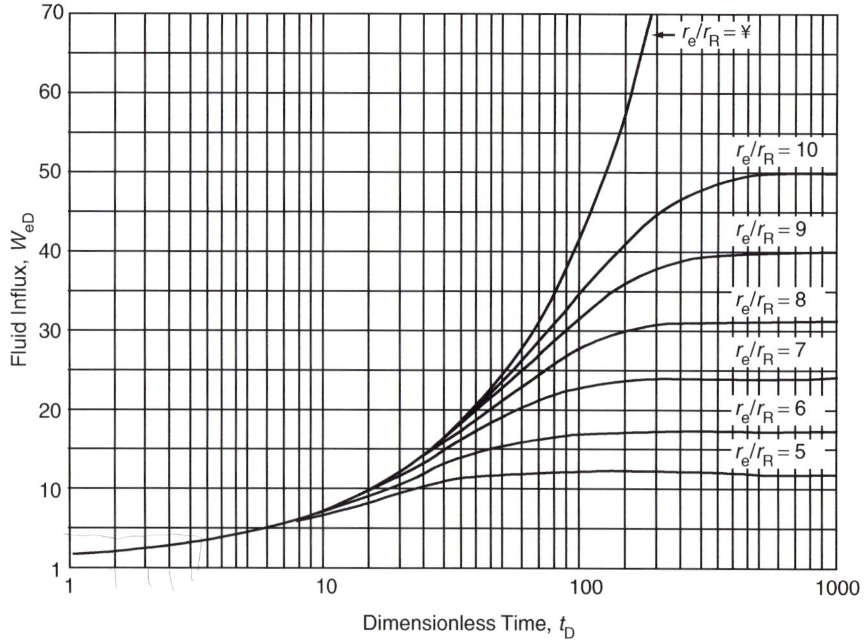

FIGURE 2.9 Dimensionless water influx W_{eD} for several values of r_e/r_R, i.e., r_a/r_e. (*van Everdingen and Hurst W_{eD} values. Permission to publish by the SPE*).

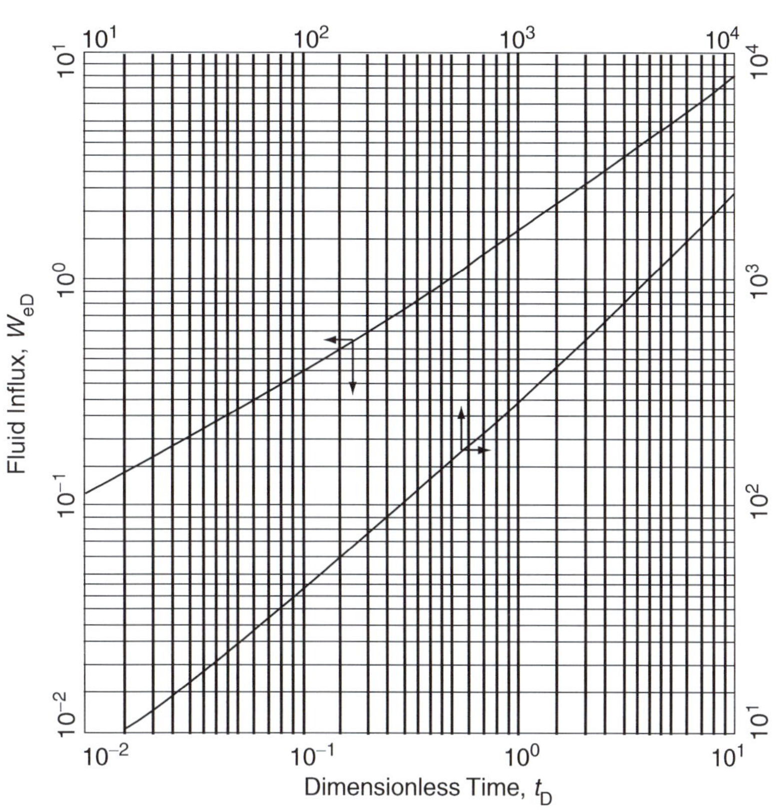

FIGURE 2.10 Dimensionless water influx W_{eD} for infinite aquifer. (*van Everdingen and Hurst W_{eD} values. Permission to publish by the SPE*).

$$c_t = c_w + c_f \tag{2.19}$$

where

t = time, days
k = permeability of the aquifer, md
ϕ = porosity of the aquifer
μ_w = viscosity of water in the aquifer, cp
r_a = radius of the aquifer, ft
r_e = radius of the reservoir, ft
c_w = compressibility of the water, psi^{-1}
c_f = compressibility of the aquifer formation, psi^{-1}
c_t = total compressibility coefficient, psi^{-1}

The water influx is then given by:

$$W_e = B\,\Delta p W_{eD} \tag{2.20}$$

with

$$B = 1.119\phi c_t r_e^2 h \tag{2.21}$$

where

W_e = cumulative water influx, bbl
B = water influx constant, bbl/psi
Δp = pressure drop at the boundary, psi
W_{eD} = dimensionless water influx

Eq. (2.21) assumes that the water is encroaching in a radial form. Quite often water does not encroach on all sides of the reservoir, or the reservoir is not circular in nature. In these cases, some modifications must be made in Eq. (2.21) to properly describe the flow mechanism. One of the simplest modifications is to introduce the encroachment angle, as a

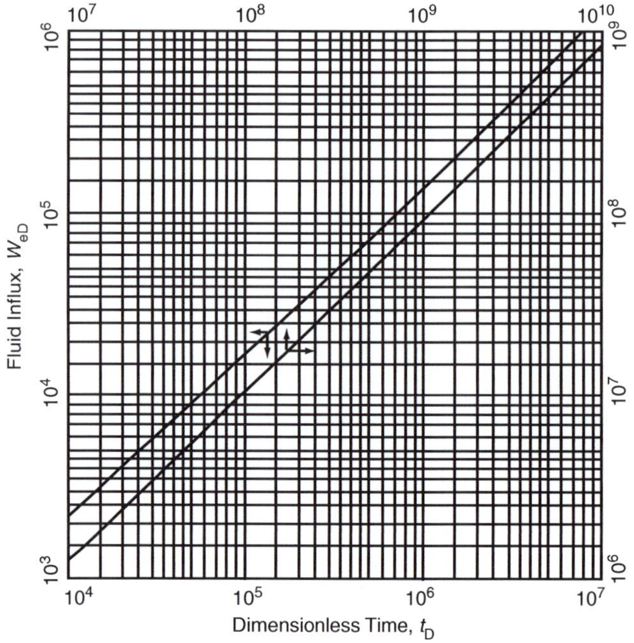

FIGURE 2.11 Dimensionless water influx W_{eD} for infinite aquifer. (*van Everdingen and Hurst W_{eD} values. Permission to publish by the SPE*).

dimensionless parameter f, to the water influx constant B, as follows:

$$f = \frac{\theta}{360} \tag{2.22}$$

$$B = 1.119\phi c_t r_e^2 hf \tag{2.23}$$

θ is the angle subtended by the reservoir circumference, i.e., for a full circle $\theta = 360°$ and for a semicircular reservoir against a fault $\theta = 180°$, as shown in Figure 2.12.

Example 2.6

Calculate the water influx at the end of 1, 2, and 5 years into a circular reservoir with an aquifer of infinite extent, i.e., $r_{eD} = \infty$. The initial and current reservoir pressures are 2500 and 2490 psi, respectively. The reservoir–aquifer system has the following properties[1]:

	Reservoir	Aquifer
Radius, ft	2000	∞
h, ft	20	22.7
k, md	50	100
ϕ, %	15	20
μ_w, cp	0.5	0.8
c_w, psi^{-1}	1×10^{-6}	0.7×10^{-6}
c_f, psi^{-1}	2×10^{-6}	0.3×10^{-6}

[1]Data for this example was reported by Cole, F.W., 1969. Reservoir Engineering Manual, Gulf Publishing, Houston, TX.

Solution

Step 1. Calculate the aquifer total compressibility coefficient c_t from Eq. (2.19):

$$c_t = c_w + c_f = 0.7(10^{-6}) + 0.3(10^{-6}) = 1 \times 10^{-6}\ \text{psi}^{-1}$$

Step 2. Determine the water influx constant from Eq. (2.23):

$$B = 1.119\phi c_t r_e^2 hf$$
$$= 1.119(0.2)(1 \times 10^{-6})(2000)^2(22.7)(360/360) = 20.4$$

Step 3. Calculate the corresponding dimensionless time after 1, 2, and 5 years:

$$t_D = 6.328 \times 10^{-3} \frac{kt}{\phi \mu_w c_t r_e^2}$$

TABLE 2.1 Dimensionless Water Influx W_{eD} for Infinite Aquifer

Dimensionless Time t_D	Fluid Influx W_{eD}	Dimensionless Time t_D	Fluid Influx W_{eD}	Dimensionless Time t_D	Fluid Influx W_{eD}	Dimensionless Time t_D	Fluid Influx W_{eD}	Dimensionless Time t_D	Fluid Influx W_{eD}	Dimensionless Time t_D	Fluid Influx W_{eD}
0.00	0.000	79	35.697	455	150.249	1190	340.843	3250	816.090	35.000	6780.247
0.01	0.112	80	36.058	460	151.640	1200	343.308	3300	827.088	40.000	7650.096
0.05	0.278	81	36.418	465	153.029	1210	345.770	3350	838.067	50.000	9363.099
0.10	0.404	82	36.777	470	154.416	1220	348.230	3400	849.028	60.000	11,047.299
0.15	0.520	83	37.136	475	155.801	1225	349.460	3450	859.974	70.000	12,708.358
0.20	0.606	84	37.494	480	157.184	1230	350.688	3500	870.903	75.000	13,531.457
0.25	0.689	85	37.851	485	158.565	1240	353.144	3550	881.816	80.000	14,350.121
0.30	0.758	86	38.207	490	159.945	1250	355.597	3600	892.712	90.000	15,975.389
0.40	0.898	87	38.563	495	161.322	1260	358.048	3650	903.594	100.000	17,586.284
0.50	1.020	88	38.919	500	162.698	1270	360.496	3700	914.459	125.000	21,560.732
0.60	1.140	89	39.272	510	165.444	1275	361.720	3750	925.309	$1.5(10)^5$	$2.538(10)^4$
0.70	1.251	90	39.626	520	168.183	1280	362.942	3800	936.144	2.0″	3.308″
0.80	1.359	91	39.979	525	169.549	1290	365.386	3850	946.966	2.5″	4.066″
0.90	1.469	92	40.331	530	170.914	1300	367.828	3900	957.773	3.0″	4.817″
1	1.569	93	40.684	540	173.639	1310	370.267	3950	968.566	4.0″	6.267″
2	2.447	94	41.034	550	176.357	1320	372.704	4000	979.344	5.0″	7.699″
3	3.202	95	41.385	560	179.069	1325	373.922	4050	990.108	6.0″	9.113″
4	3.893	96	41.735	570	181.774	1330	375.139	4100	1000.858	7.0″	$1.051\ (10)^5$
5	4.539	97	42.084	575	183.124	1340	377.572	4150	1011.595	8.0″	1.189″
6	5.153	98	42.433	580	184.473	1350	380.003	4200	1022.318	9.0″	1.326″
7	5.743	99	42.781	590	187.166	1360	382.432	4250	1033.028	1.0(10)6′	1.462″
8	6.314	100	43.129	600	189.852	1370	384.859	4300	1043.724	1.5″	2.126″
9	6.869	105	44.858	610	192.533	1375	386.070	4350	1054.409	2.0″	2.781″
10	7.411	110	46.574	620	195.208	1380	387.283	4400	1065.082	2.5″	3.427″
11	7.940	115	48.277	625	196.544	1390	389.705	4450	1075.743	3.0″	4.064″
12	8.457	120	49.968	630	197.878	1400	392.125	4500	1086.390	4.0″	5.313″
13	8.964	125	51.648	640	200.542	1410	394.543	4550	1097.024	5.0″	6.544″
14	9.461	130	53.317	650	203.201	1420	396.959	4600	1107.646	6.0″	7.761″
15	9.949	135	54.976	660	205.854	1425	398.167	4650	1118.257	7.0″	8.965″
16	10.434	140	56.625	670	208.502	1430	399.373	4700	1128.854	8.0″	$1.016(10)^6$
17	10.913	145	58.265	675	209.825	1440	401.786	4750	1139.439	9.0″	1.134″
18	11.386	150	59.895	680	211.145	1450	404.197	4800	1150.012	$1.0(10)^7$	1.252″
19	11.855	155	61.517	690	213.784	1460	406.606	4850	1160.574	1.5″	1.828″
20	12.319	160	63.131	700	216.417	1470	409.013	4900	1171.125	2.0″	2.398″

(Continued)

TABLE 2.1 (Continued)

Dimensionless Time t_D	Fluid Influx W_{eD}	Dimensionless Time t_D	Fluid Influx W_{eD}	Dimensionless Time t_D	Fluid Influx W_{eD}	Dimensionless Time t_D	Fluid Influx W_{eD}	Dimensionless Time t_D	Fluid Influx W_{eD}	Dimensionless Time t_D	Fluid Influx W_{eD}
21	12.778	165	64.737	710	219.046	1475	410.214	4950	1181.666	2.5″	2.961″
22	13.233	170	66.336	720	221.670	1480	411.418	5000	1192.198	3.0″	3.517″
23	13.684	175	67.928	725	222.980	1490	413.820	5100	1213.222	4.0″	4.610″
24	14.131	180	69.512	730	224.289	1500	416.220	5200	1234.203	5.0″	5.689″
25	14.573	185	71.090	740	226.904	1525	422.214	5300	1255.141	6.0″	6.758″
26	15.013	190	72.661	750	229.514	1550	428.196	5400	1276.037	7.0″	7.816″
27	15.450	195	74.226	760	232.120	1575	434.168	5500	1296.893	8.0″	8.866″
28	15.883	200	75.785	770	234.721	1600	440.128	5600	1317.709	9.0″	9.911″
29	16.313	205	77.338	775	236.020	1625	446.077	5700	1338.486	$1.0(10)^8$	$1.095(10)^7$
30	16.742	210	78.886	780	237.318	1650	452.016	5800	1359.225	1.5″	1.604″
31	17.167	215	80.428	790	239.912	1675	457.945	5900	1379.927	2.0″	2.108″
32	17.590	220	81.965	800	242.501	1700	463.863	6000	1400.593	2.5″	2.607″
33	18.011	225	83.497	810	245.086	1725	469.771	6100	1421.224	3.0″	3.100″
34	18.429	230	85.023	820	247.668	1750	475.669	6200	1441.820	4.0″	4.071″
35	18.845	235	86.545	825	248.957	1775	481.558	6300	1462.383	5.0″	5.032″
36	19.259	240	88.062	830	250.245	1800	487.437	6400	1482.912	6.0″	5.984″
37	19.671	245	89.575	840	252.819	1825	493.307	6500	1503.408	7.0″	6.928″
38	20.080	250	91.084	850	255.388	1850	499.167	6600	1523.872	8.0″	7.865″
39	20.488	255	92.589	860	257.953	1875	505.019	6700	1544.305	9.0″	8.797″
40	20.894	260	94.090	870	260.515	1900	510.861	6800	1564.706	$1.0(10)^9$	9.725″
41	21.298	265	95.588	875	261.795	1925	516.695	6900	1585.077	1.5″	$1.429(10)^8$
42	21.701	270	97.081	880	263.073	1950	522.520	7000	1605.418	2.0″	1.880″
43	22.101	275	98.571	890	265.629	1975	528.337	7100	1625.729	2.5″	2.328″
44	22.500	280	100.057	900	268.181	2000	534.145	7200	1646.011	3.0″	2.771″
45	22.897	285	101.540	910	270.729	2025	539.945	7300	1666.265	4.0″	3.645″
46	23.291	290	103.019	920	273.274	2050	545.737	7400	1686.490	5.0″	4.510″
47	23.684	295	104.495	925	274.545	2075	551.522	7500	1706.688	6.0″	5.368″
48	24.076	300	105.968	930	275.815	2100	557.299	7600	1726.859	7.0″	6.220″
49	24.466	305	107.437	940	278.353	2125	563.068	7700	1747.002	8.0″	7.066″
50	24.855	310	108.904	950	280.888	2150	568.830	7800	1767.120	9.0″	7.909″
51	25.244	315	110.367	960	283.420	2175	574.585	7900	1787.212	$1.0(10)^{10}$	8.747″
52	25.633	320	111.827	970	285.948	2200	580.332	8000	1807.278	1.5″	$1.288''(10)^9$
53	26.020	325	113.284	975	287.211	2225	586.072	8100	1827.319	2.0″	1.697″
54	26.406	330	114.738	980	288.473	2250	591.806	8200	1847.336	2.5″	2.103″
55	26.791	335	116.189	990	290.995	2275	597.532	8300	1867.329	3.0″	2.505″
56	27.174	340	117.638	1000	293.514	2300	603.252	8400	1887.298	4.0″	3.299″

57	27.555	345	119.083	1010	296.030	2325	608.965	8500	1907.243	5.0″	4.087″
58	27.935	350	120.526	1020	298.543	2350	614.672	8600	1927.166	6.0″	4.868″
59	28.314	355	121.966	1025	299.799	2375	620.372	8700	1947.065	7.0″	5.643″
60	28.691	360	123.403	1030	301.053	2400	626.066	8800	1966.942	8.0″	6.414″
61	29.068	365	124.838	1040	303.560	2425	631.755	8900	1986.796	9.0″	7.183″
62	29.443	370	126.720	1050	306.065	2450	637.437	9000	2006.628	$1.0(10)^{11}$	7.948″
63	29.818	375	127.699	1060	308.567	2475	643.113	9100	2026.438	1.5″	$1.17(10)^{10}$
64	30.192	380	129.126	1070	311.066	2500	648.781	9200	2046.227	2.0″	1.55″
65	30.565	385	130.550	1075	312.314	2550	660.093	9300	2065.996	2.5″	1.92″
66	30.937	390	131.972	1080	313.562	2600	671.379	9400	2085.744	3.0″	2.29″
67	31.308	395	133.391	1090	316.055	2650	682.640	9500	2105.473	4.0″	3.02″
68	31.679	400	134.808	1100	318.545	2700	693.877	9600	2125.184	5.0″	3.75″
69	32.048	405	136.223	1110	321.032	2750	705.090	9700	2144.878	6.0″	4.47″
70	32.417	410	137.635	1120	323.517	2800	716.280	9800	2164.555	7.0″	5.19″
71	32.785	415	139.045	1125	324.760	2850	727.449	9900	2184.216	8.0″	5.89″
72	33.151	420	140.453	1130	326.000	2900	738.598	10,000	2203.861	9. 0″	6.58″
73	33.517	425	141.859	1140	328.480	2950	749.725	12,500	2688.967	$1.0(10)^{12}$	7.28″
74	33.883	430	143.262	1150	330.958	3000	760.833	15,000	3164.780	1.5″	$1.08(10)^{11}$
75	34.247	435	144.664	1160	333.433	3050	771.922	17,500	3633.368	2.0″	1.42″
76	34.611	440	146.064	1170	335.906	3100	782.992	20,000	4095.800		
77	34.974	445	147.461	1175	337.142	3150	794.042	25,000	5005.726		
78	35.336	450	148.856	1180	338.376	3200	805.075	30,000	5899.508		

van Everdingen and Hurst W_{eD}. Permission to publish by the SPE.

TABLE 2.2 Dimensionless Water Influx W_{eD} for Several Values of r_e/r_R, i.e., r_a/r_e

$r_e/r_R = 1.5$		$r_e/r_R = 2.0$		$r_e/r_R = 2.5$		$r_e/r_R = 3.0$		$r_e/r_R = 3.5$		$r_e/r_R = 4.0$		$r_e/r_R = 4.5$	
Dimensionless Time t_D	Fluid Influx W_{eD}	Dimensionless Time t_D	Fluid Influx W_{eD}	Dimensionless Time t_D	Fluid Influx W_{eD}	Dimensionless Time t_D	Fluid Influx W_{eD}	Dimensionless Time t_D	Fluid Influx W_{eD}	Dimensionless Time t_D	Fluid Influx W_{eD}	Dimensionless Time t_D	Fluid Influx W_{eD}
$5.0(10)^{-2}$	0.276	$5.0(10)^{-2}$	0.278	$1.0(10)^{-1}$	0.408	$3.0(10)^{-1}$	0.755	1.00	1.571	2.00	2.442	2.5	2.835
6.0″	0.304	7.5″	0.345	1.5″	0.509	4.0″	0.895	1.20	1.761	2.20	2.598	3.0	3.196
7.0″	0.330	$1.0(10)^{-1}$	0.404	2.0″	0.599	5.0″	1.023	1.40	1.940	2.40	2.748	3.5	3.537
8.0″	0.354	1.25″	0.458	2.5″	0.681	6.0″	1.143	1.60	2.111	2.60	2.893	4.0	3.859
9.0″	0.375	1.50″	0.507	3.0″	0.758	7.0″	1.256	1.80	2.273	2.80	3.034	4.5	4.165
$1.0(10)^{-1}$	0.395	1.75″	0.553	3.5″	0.829	8.0″	1.363	2.00	2.427	3.00	3.170	5.0	4.454
1.1″	0.414	2.00″	0.597	4.0″	0.897	9.0″	1.465	2.20	2.574	3.25	3.334	5.5	4.727
1.2″	0.431	2.25″	0.638	4.5″	0.962	1.00	1.563	2.40	2.715	3.50	3.493	6.0	4.986
1.3″	0.446	2.50″	0.678	5.0″	1.024	1.25	1.791	2.60	2.849	3.75	3.645	6.5	5.231
1.4″	0.461	2.75″	0.715	5.5″	1.083	1.50	1.997	2.80	2.976	4.00	3.792	7.0	5.464
1.5″	0.474	3.00″	0.751	6.0″	1.140	1.75	2.184	3.00	3.098	4.25	3.932	7.5	5.684
1.6″	0.486	3.25″	0.785	6.5″	1.195	2.00	2.353	3.25	3.242	4.50	4.068	8.0	5.892
1.7″	0.497	3.50″	0.817	7.0″	1.248	2.25	2.507	3.50	3.379	4.75	4.198	8.5	6.089
1.8″	0.507	3.75″	0.848	7.5″	1.299	2.50	2.646	3.75	3.507	5.00	4.323	9.0	6.276
1.9″	0.517	4.00″	0.877	8.0″	1.348	2.75	2.772	4.00	3.628	5.50	4.560	9.5	6.453
2.0″	0.525	4.25″	0.905	8.5″	1.395	3.00	2.886	4.25	3.742	6.00	4.779	10	6.621
2.1″	0.533	4.50″	0.932	9.0″	2.440	3.25	2.990	4.50	3.850	6.50	4.982	11	6.930
2.2″	0.541	4.75″	0.958	9.5″	1.484	3.50	3.084	4.75	3.951	7.00	5.169	12	7.208
2.3″	0.548	5.00″	0.993	1.0	1.526	3.75	3.170	5.00	4.047	7.50	5.343	13	7.457
2.4″	0.554	5.50″	1.028	1.1	1.605	4.00	3.247	5.50	4.222	8.00	5.504	14	7.680
2.5″	0.559	6.00″	1.070	1.2	1.679	4.25	3.317	6.00	4.378	8.50	5.653	15	7.880
2.6″	0.565	6.50″	1.108	1.3	1.747	4.50	3.381	6.50	4.516	9.00	5.790	16	8.060
2.8″	0.574	7.00″	1.143	1.4	1.811	4.75	3.439	7.00	4.639	9.50	5.917	18	8.365
3.0″	0.582	7.50″	1.174	1.5	1.870	5.00	3.491	7.50	4.749	10	6.035	20	8.611
3.2″	0.588	8.00″	1.203	1.6	1.924	5.50	3.581	8.00	4.846	11	6.246	22	8.809
3.4″	0.594	9.00″	1.253	1.7	1.975	6.00	3.656	8.50	4.932	12	6.425	24	8.968
3.6″	0.599	1.00″	1.295	1.8	2.022	6.50	3.717	9.00	5.009	13	6.580	26	9.097
3.8″	0.603	1.1	1.330	2.0	2.106	7.00	3.767	9.50	5.078	14	6.712	28	9.200
4.0″	0.606	1.2	1.358	2.2	2.178	7.50	3.809	10.00	5.138	15	6.825	30	9.283
4.5″	0.613	1.3	1.382	2.4	2.241	8.00	3.843	11	5.241	16	6.922	34	9.404
5.0″	0.617	1.4	1.402	2.6	2.234	9.00	3.894	12	5.321	17	7.004	38	9.481
6.0″	0.621	1.6	1.432	2.8	2.340	10.00	3.928	13	5.385	18	7.076	42	9.532
7.0″	0.623	1.7	1.444	3.0	2.380	11.00	3.951	14	5.435	20	7.189	46	9.565
8.0″	0.624	1.8	1.453	3.4	2.444	12.00	3.967	15	5.476	22	7.272	50	9.586

2.0	1.468	3.8	2.491	14.00	3.985	16	5.506	24	7.332	60	9.612
2.5	1.487	4.2	2.525	16.00	3.993	17	5.531	26	7.377	70	9.621
3.0	1.495	4.6	2.551	18.00	3.997	18	5.551	30	7.434	80	9.623
4.0	1.499	5.0	2.570	20.00	3.999	20	5.579	34	7.464	90	9.624
5.0	1.500	6.0	2.599	22.00	3.999	25	5.611	38	7.481	100	9.625
		7.0	2.613	24.00	4.000	30	5.621	42	7.490		
		8.0	2.619			35	5.624	46	7.494		
		9.0	2.622			40	5.625	50	7.499		
		10.0	2.624								

$r_e/r_R = 5.0$		$r_e/r_R = 6.0$		$r_e/r_R = 7.0$		$r_e/r_R = 8.0$		$r_e/r_R = 9.0$		$r_e/r_R = 10.0$	
Dimensionless Time t_D	Fluid Influx W_{eD}	Dimensionless Time t_D	Fluid Influx W_{eD}	Dimensionless Time t_D	Fluid Influx W_{eD}	Dimensionless Time t_D	Fluid Influx W_{eD}	Dimensionless Time t_D	Fluid Influx W_{eD}	Dimensionless Time t_D	Fluid Influx W_{eD}
3.0	3.195	6.0	5.148	9.0	6.861	9	6.861	10	7.417	15	9.96
3.5	3.542	6.5	5.440	9.50	7.127	10	7.398	15	9.945	20	12.32
4.0	3.875	7.0	5.724	10	7.389	11	7.920	20	12.26	22	13.22
4.5	4.193	7.5	6.002	11	7.902	12	8.431	22	13.13	24	14.95
5.0	4.499	8.0	6.273	12	8.397	13	8.930	24	13.98	26	14.95
5.5	4.792	8.5	6.537	13	8.876	14	9.418	26	14.79	28	15.78
6.0	5.074	9.0	6.795	14	9.341	15	9.895	26	15.59	30	16.59
6.5	5.345	9.5	7.047	15	9.791	16	10.361	30	16.35	32	17.38
7.0	5.605	10.0	7.293	16	10.23	17	10.82	32	17.10	34	18.16
7.5	5.854	10.5	7.533	17	10.65	18	11.26	34	17.82	36	18.91
8.0	6.094	11	7.767	18	11.06	19	11.70	36	18.52	38	19.65
8.5	6.325	12	8.220	19	11.46	20	12.13	38	19.19	40	20.37
9.0	6.547	13	8.651	20	11.85	22	12.95	40	19.85	42	21.07
9.5	6.760	14	9.063	22	12.58	24	13.74	42	20.48	44	21.76
10	6.965	15	9.456	24	13.27	26	14.50	44	21.09	46	22.42
11	7.350	16	9.829	26	13.92	28	15.23	46	21.69	48	23.07
12	7.706	17	10.19	28	14.53	30	15.92	48	22.26	50	23.71
13	8.035	18	10.53	30	15.11	34	17.22	50	22.82	52	24.33
14	8.339	19	10.85	35	16.39	38	18.41	52	23.36	54	24.94
15	8.620	20	11.16	40	17.49	40	18.97	54	23.89	56	25.53
16	8.879	22	11.74	45	18.43	45	20.26	56	24.39	58	26.11
18	9.338	24	12.26	50	19.24	50	21.42	58	24.88	60	26.67
20	9.731	25	12.50	60	20.51	55	22.46	60	25.36	65	28.02
22	10.07	31	13.74	70	21.45	60	23.40	65	26.48	70	29.29
24	10.35	35	14.40	80	22.13	70	24.98	70	27.52	75	30.49
26	10.59	39	14.93	90	22.63	80	26.26	75	28.48	80	31.61
28	10.80	51	16.05	100	23.00	90	27.28	80	29.36	85	32.67

(Continued)

TABLE 2.2 (Continued)

$r_e/r_R = 5.0$		$r_e/r_R = 6.0$		$r_e/r_R = 7.0$		$r_e/r_R = 8.0$		$r_e/r_R = 9.0$		$r_e/r_R = 10.0$	
Dimensionless Time t_D	Fluid Influx W_{eD}	Dimensionless Time t_D	Fluid Influx W_{eD}	Dimensionless Time t_D	Fluid Influx W_{eD}	Dimensionless Time t_D	Fluid Influx W_{eD}	Dimensionless Time t_D	Fluid Influx W_{eD}	Dimensionless Time t_D	Fluid Influx W_{eD}
30	10.98	60	16.56	120	23.47	100	28.11	85	30.18	90	33.66
34	11.26	70	16.91	140	23.71	120	29.31	90	30.93	95	34.60
38	11.46	80	17.14	160	23.85	140	30.08	95	31.63	100	35.48
42	11.61	90	17.27	180	23.92	160	30.58	100	32.27	120	38.51
46	11.71	100	17.36	200	23.96	180	30.91	120	34.39	140	40.89
50	11.79	110	17.41	500	24.00	200	31.12	140	35.92	160	42.75
60	11.91	120	17.45			240	31.34	160	37.04	180	44.21
70	11.96	130	17.46			280	31.43	180	37.85	200	45.36
80	11.98	140	17.48			320	31.47	200	38.44	240	46.95
90	11.99	150	17.49			360	31.49	240	39.17	280	47.94
100	12.00	160	17.49			400	31.50	280	39.56	320	48.54
120	12.00	180	17.50			500	31.50	320	39.77	360	48.91
		200	17.50					360	39.88	400	49.14
		220	17.50					400	39.94	440	49.28
								440	39.97	480	49.36
								480	39.98		

van Everdingen and Hurst W_{eD}. Permission to publish by the SPE.

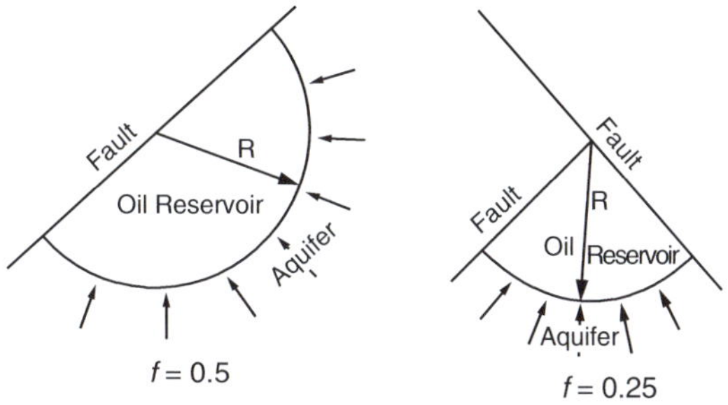

FIGURE 2.12 Gas cap drive reservoir. (*After Cole, F.W., 1969.* Reservoir Engineering Manual. Gulf Publishing Company, Houston, TX).

$$= 6.328 \times 10^{-3} \frac{100t}{(0.8)(0.2)(1 \times 10^{-6})(2000)^2} = 0.9888t$$

Thus in tabular form:

t (days)	$t_D = 0.9888t$
365	361
730	722
1825	1805

Step 4. Using Table 2.1 determine the dimensionless water influx W_{eD}:

t (days)	t_D	W_{eD}
365	361	123.5
730	722	221.8
1825	1805	484.6

Step 5. Calculate the cumulative water influx by applying Eq. (2.20):

$$W_e = B\,\Delta p W_{eD}$$

t (days)	W_{eD}	$W_e = (20.4)(2500 \times 2490)\,W_{eD}$
365	123.5	25,194 bbl
730	221.8	45,247 bbl
1825	484.6	98,858 bbl

Example 2.6 shows that, for a given pressure drop, doubling the time interval will not double the water influx. This example also illustrates how to calculate water influx as a result of a single pressure drop. As there will usually be many of these pressure drops occurring throughout the prediction period, it is necessary to analyze the procedure to be used where these multiple pressure drops are present.

Consider Figure 2.13 that illustrates the decline in the boundary pressure as a function of time for a radial reservoir–aquifer system. If the boundary pressure in the reservoir shown in Figure 2.13 is suddenly reduced at time t, from p_i to p_1, a pressure drop of $(p_i - p_1)$ will be imposed across the aquifer. Water will continue to expand and the new reduced pressure will continue to move outward into the aquifer. Given a sufficient length of time, the pressure at the outer edge of the aquifer will finally be reduced to p_1.

If some time after the boundary pressure has been reduced to p_1, a second pressure p_2 is suddenly imposed at the boundary, a new pressure wave will begin moving outward into the aquifer. This new pressure wave will also cause water expansion and therefore encroachment into the reservoir. However, *this new pressure drop will not* be $p_i - p_2$ but will be $p_1 - p_2$. This second pressure wave will be moving behind the first pressure wave. Just ahead of the second pressure wave will be the pressure at the end of the first pressure drop p_1.

Since these pressure waves are assumed to occur at different times, they are entirely independent of each other. Thus, water expansion will continue to take place as a result of the first pressure drop, even though additional water influx is also taking place as a result of one or more later pressure drops. This is essentially an application of the principle of superposition. To determine the total water influx into a reservoir at any given time, it is necessary to determine the water influx as a result of each successive pressure drop that has been imposed on the reservoir and aquifer.

In calculating the cumulative water influx into a reservoir at successive intervals, it is necessary to calculate the total water influx from the beginning. This is required because of the different times during which the various pressure drops have been effective.

The van Everdingen and Hurst computational procedure for determining the water influx as a function of time and pressure is summarized by the following steps and described conceptually in Figure 2.14:

Step 1. Assume that the boundary pressure has declined from its initial value of p_i to p_1 after t_1 days. To determine the cumulative water influx in response to this first pressure drop $\Delta p_1 = p_i - p_1$ can be simply calculated from Eq. (2.20), or:

$$W_e = B\,\Delta p_1 (W_{eD})_{t_1}$$

where W_e is the cumulative water influx due to the first pressure drop Δp_1. The dimensionless water influx $(W_{eD})_{t_1}$ is

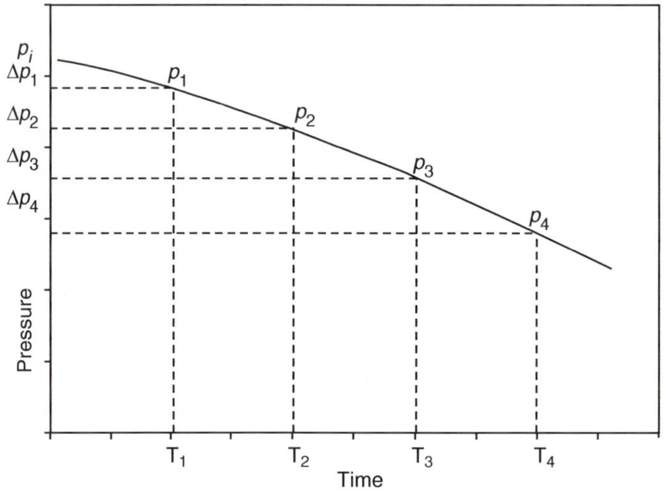

FIGURE 2.13 Boundary pressure versus time.

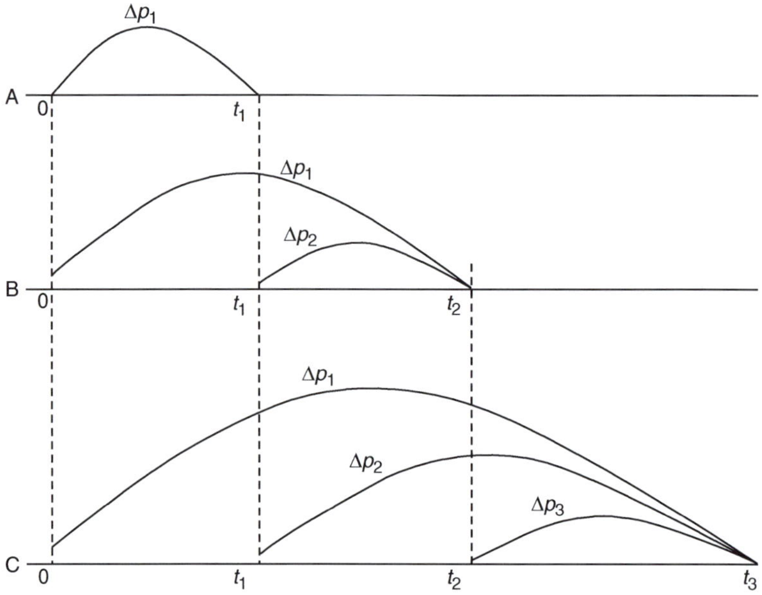

FIGURE 2.14 Illustration of the superposition concept.

evaluated by calculating the dimensionless time at t_1 days. This simple calculation step is shown by A in Figure 2.14.

Step 2. Let the boundary pressure declines again to p_2 after t_2 days with a pressure drop of $\Delta p_2 = p_1 - p_2$. The total cumulative water influx after t_2 days will result from the first pressure drop Δp_1 and the second pressure drop Δp_2, or:

$$W_e = \text{Water influx due to } \Delta p_1 + \text{Water influx due to } \Delta p_2$$

$$W_e = (W_e)_{\Delta p_1} + (W_e)_{\Delta p_2}$$

where

$$(W_e)_{\Delta p_1} = B\,\Delta p_1 (W_{eD})_{t_2}$$
$$(W_e)_{\Delta p_2} = B\,\Delta p_2 (W_{eD})_{t_2 - t_1}$$

The above relationships indicate that the effect of the first pressure drop Δp_1 will continue for the entire time t_2, while the effect of the second pressure drop will continue only for $(t_2 - t_1)$ days as shown by B in Figure 2.14.

Step 3. A third pressure drop of $\Delta p_3 = p_2 - p_3$ would cause an additional water influx as illustrated by C in Figure 2.14. The total cumulative water influx can then be calculated from:

$$W_e = (W_e)_{\Delta p_1} + (W_e)_{\Delta p_2} + (W_e)_{\Delta p_3}$$

where

$$(W_e)_{\Delta p_1} = B\,\Delta p_1 (W_{eD})_{t_3}$$
$$(W_e)_{\Delta p_2} = B\,\Delta p_2 (W_{eD})_{t_3 - t_1}$$
$$(W_e)_{\Delta p_3} = B\,\Delta p_3 (W_{eD})_{t_3 - t_2}$$

The van Everdingen and Hurst water influx relationship can then be expressed in a more generalized form as:

$$W_e = B \sum \Delta p W_{eD} \quad \textbf{(2.24)}$$

The authors also suggested that instead of using the entire pressure drop for the first period, a better approximation is to consider that one-half of the pressure drop, $\frac{1}{2}(p_i - p_1)$, is effective during the entire first period. For the second period, the effective pressure drop then is one-half of the pressure drop during the first period, $\frac{1}{2}(p_i - p_2)$, which simplifies to:

$$\frac{1}{2}(p_i - p_1) + \frac{1}{2}(p_1 - p_2) = \frac{1}{2}(p_i - p_2)$$

Similarly, the effective pressure drop for use in the calculations for the third period would be one-half of the pressure drop during the second period, $\frac{1}{2}(p_1 - p_2)$, plus one-half of the pressure drop during the third period, $\frac{1}{2}(p_2 - p_3)$, which simplifies to $\frac{1}{2}(p_1 - p_3)$. The time intervals must all be equal in order to preserve the accuracy of these modifications.

Example 2.7

Using the data given in Example 2.6, calculate the cumulative water influx at the end of 6, 12, 18, and 24 months. The predicted boundary pressure at the end of each specified time period is given below:

Time (days)	Time (months)	Boundary Pressure (psi)
0	0	2500
182.5	6	2490
365.0	12	2472
547.5	18	2444
730.0	24	2408

Data from Example 2.6 is listed below:

$$B = 20.4$$
$$t_D = 0.9888\,t$$

Solution

Water influx after 6 months:

Step 1. Determine water influx constant B. Example 2.6 gives a value of:

$$B = 20.4 \text{ bbl/psi}$$

Step 2. Calculate the dimensionless time t_D at $t = 182.5$ days:

$$t_D = 0.9888t = 0.9888(182.5) = 180.5$$

Step 3. Calculate the first pressure drop Δp_1. This pressure is taken as one-half of the actual pressured drop, or:

$$\Delta p_1 = \frac{p_i - p_1}{2} = \frac{2500 - 2490}{2} = 5 \text{ psi}$$

Step 4. Determine the dimensionless water influx W_{eD} from Table 2.1 at $t_D = 180.5$, to give:

$$(W_{eD})_{t_1} = 69.46$$

Step 5. Calculate the cumulative water influx at the end of 182.5 days *due to the first pressure drop* of 5 psi, i.e., $(W_e)_{\Delta p_1 = 5}$, by using the van Everdingen and Hurst equation, or:

$$(W_e)_{\Delta p_1 = 5 \text{ psi}} = B\,\Delta p_1 (W_{eD})_{t_1} = (20.4)(5)(69.46)$$
$$= 7085 \text{ bbl}$$

Cumulative water influx after 12 months:

Step 1. After an additional 6 months, the pressure has declined from 2490 to 2472 psi. This second pressure Δp_2 is taken as one-half the actual pressure drop during the first period, plus one-half the actual pressure drop during the second period, or:

$$\Delta p_2 = \frac{p_i - p_2}{2} = \frac{2500 - 2472}{2} = 14 \text{ psi}$$

Step 2. The total cumulative water influx at the end of 12 months would result from the first pressure drop Δp_1 and the second pressure drop Δp_2.

The first pressure drop Δp_1 has been effective for a year, but the second pressure drop Δp_2 has been effective for only 6 months, as shown in Figure 2.15. Separate calculations must be made for the two pressure drops because of this time difference, and the results added in order to determine the total water influx. That is:

$$W_e = (W_e)_{\Delta p_1} + (W_e)_{\Delta p_2}$$

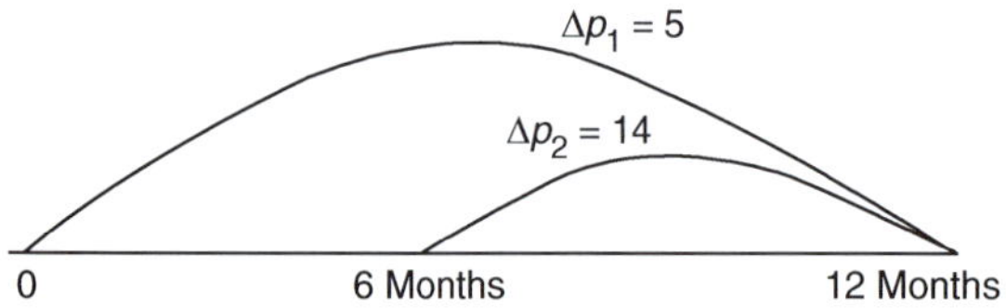

FIGURE 2.15 Duration of the pressure drop in Example 2.7.

Step 3. Calculate the dimensionless time at 365 days, as:

$$t_D = 0.9888t = 0.9888(365) = 361$$

Step 4. Determine the dimensionless water influx at $t_D = 361$ from Table 2.1, to give:

$$W_{eD} = 123.5$$

Step 5. Calculate the water influx due to the first and second pressure drop, i.e., $(W_e)_{\Delta p_1}$ and $(W_e)_{\Delta p_2}$, or:

$$(W_e)_{\Delta p_1 = 5} = (20.4)(5)(123.5) = 12,597 \text{ bbl}$$
$$(W_e)_{\Delta p_2 = 14} = (20.4)(14)(69.46) = 19,838 \text{ bbl}$$

Step 6. Calculate the total cumulative water influx after 12 months:

$$W_e = (W_e)_{\Delta p_1} + (W_e)_{\Delta p_2}$$
$$= 12,597 + 19{,}938 = 32{,}435 \text{ bbl}$$

Water influx after 18 months:

Step 1. Calculate the third pressure drop Δp_3, which is taken as one-half of the actual pressure drop during the second period plus one-half of the *actual* pressure drop during the third period, or:

$$\Delta p_3 = \frac{p_1 - p_3}{2} = \frac{2490 - 2444}{2} = 23 \text{ psi}$$

Step 2. Calculate the dimensionless time after 6 months:

$$t_D = 0.9888t = 0.9888(547.5) = 541.5$$

Step 3. Determine the dimensionless water influx from Table 2.1 at $t_D = 541.5$:

$$W_{eD} = 173.7$$

Step 4. The first pressure drop will have been effective for the entire 18 months, the second pressure drop will have been effective for 12 months, and the last pressure drop will have been effective for only 6 months, as shown in Figure 2.16. Therefore, the cumulative water influx is as calculated below:

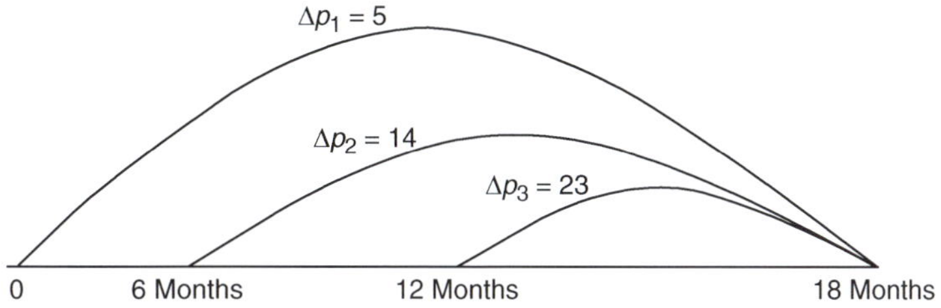

FIGURE 2.16 Pressure drop data for Example 2.7.

Time (days)	t_D	Δp	W_{eD}	$B\,\Delta p W_{eD}$
547.5	541.5	5	173.7	17,714
365	361	14	123.5	35,272
182.5	180.5	23	69.40	32,291
				$W_e = 85{,}277$ bbl

Water influx after 24 months:

The first pressure drop has now been effective for the entire 24 months, the second pressure drop has been effective for 18 months, the third pressure drop has been effective for 12 months, and the fourth pressure drop has been effective for only 6 months. A summary of the calculations is given below:

Time (days)	t_D	Δp	W_{eD}	$B\,\Delta p W_{eD}$
730	722	5	221.8	22,624
547.5	541.5	14	173.7	49,609
365	361	23	123.5	57,946
182.5	180.5	32	69.40	45,343
				$W_e = 175{,}522$ bbl

Edwardson et al. (1962) developed three sets of simple polynomial expressions for calculating the dimensionless water influx W_{eD} for infinite-acting aquifers. The proposed three expressions essentially approximate the W_{eD} data in three dimensionless time regions.

(1) For $t_D<0.01$:

$$W_{eD} = \sqrt{\frac{t_D}{\pi}} \tag{2.25}$$

(2) For $0.01<t_D<200$:

$$W_{eD} = \frac{1.2838\sqrt{t_D} + 1.19328 t_D + 0.269872 (t_D)^{3/2} + 0.00855294 (t_D)^2}{1 + 0.616599\sqrt{t_D} + 0.413008 t_D} \tag{2.26}$$

(3) For $t_D>200$:

$$W_{eD} = \frac{-4.29881 + 2.02566 t_D}{\ln(t_D)} \tag{2.27}$$

Bottom-Water Drive. The van Everdingen and Hurst solution to the *radial* diffusivity equation is considered the most rigorous aquifer influx model to date. However, the proposed solution technique is not adequate to describe the vertical water encroachment in bottom-water drive systems. Coats (1962) presented a mathematical model that takes into account the vertical flow effects from bottom-water aquifers. He correctly noted that in many cases reservoirs are situated on top of an aquifer with a continuous horizontal interface between the reservoir fluid and the aquifer water and with a significant aquifer thickness. He stated that in such situations significant bottom-water drive would occur. He modified the diffusivity equation to account for the vertical flow by including an additional term in the equation, to give:

$$\frac{\partial^2 p}{\partial r^2} + \frac{1}{r}\frac{\partial p}{\partial r} + F_k \frac{\partial^2 p}{\partial z^2} = \frac{\mu \phi c}{k}\frac{\partial p}{\partial t} \tag{2.28}$$

where F_k is the ratio of vertical to horizontal permeability, or:

$$F_k = \frac{k_v}{k_h} \tag{2.29}$$

where

k_v = vertical permeability
k_h = horizontal permeability

Allard and Chen (1988) pointed out that there are an infinite number of solutions to Eq. (2.28), representing all possible reservoir–aquifer configurations. They suggested that it is possible to derive a general solution that is applicable to a variety of systems by the solution to Eq. (2.28) in terms of the dimensionless time t_D, dimensionless radius r_D, and a newly introduced dimensionless variable z_D.

$$z_D = \frac{h}{r_e\sqrt{F_k}} \tag{2.30}$$

where

z_D = dimensionless vertical distance
h = aquifer thickness, ft

Allen and Chen used a numerical model to solve Eq. (2.28). The authors developed a solution to the bottom-water influx that is comparable in form with that of van Everdingen and Hurst:

$$W_e = B\sum \Delta p W_{eD} \tag{2.31}$$

They defined the water influx constant B as identical to that of Eq. (2.21), or

$$B = 1.119\phi c_t r_e^2 h \tag{2.32}$$

Note that the water influx constant B in bottom-water drive reservoirs *does not include* the encroachment angle θ.

The actual values of W_{eD} are different from those of the van Everdingen and Hurst model because W_{eD} for the bottom-water drive is also a function of the vertical permeability. Allard and Chen tabulated the values of W_{eD} as a function of r_D, t_D, and z_D. These values are presented in Tables 2.3–2.7.

The solution procedure of a bottom-water influx problem is identical to the edge-water influx problem outlined in Example 2.7. Allard and Chen illustrated results of their method in the following example.

Example 2.8

An infinite-acting bottom-water aquifer is characterized by the following properties:

$r_a = \infty$,	$k_h = 50$ md, $F_k = 0.04$,	$\phi = 0.1$
$\mu_w = 0.395$ cp,	$c_t = 8\times 10^{-6}$ psi^{-1},	$h = 200$ ft
$r_e = 2000$ ft,	$\theta = 360°$	

The boundary pressure history is given below:

Time (days)	p (psi)
0	3000
30	2956
60	2917
90	2877
120	2844
150	2811
180	2791
210	2773
240	2755

Calculate the cumulative water influx as a function of time by using the bottom-water drive solution and compare with the edge-water drive approach.

Solution

Step 1. Calculate the dimensionless radius for an infinite-acting aquifer:

$$r_D = \infty$$

Step 2. Calculate z_D from Eq. (2.30):

$$z_D = \frac{h}{r_e\sqrt{F_k}} = \frac{200}{2000\sqrt{0.04}} = 0.5$$

TABLE 2.3 Dimensionless Water Influx W_{eD} for Infinite Aquifer

	Z'_D						
t_D	0.05	0.1	0.3	0.5	0.7	0.9	1.0
0.1	0.700	0.677	0.508	0.349	0.251	0.195	0.176
0.2	0.793	0.786	0.696	0.547	0.416	0.328	0.295
0.3	0.936	0.926	0.834	0.692	0.548	0.440	0.396
0.4	1.051	1.041	0.952	0.812	0.662	0.540	0.486
0.5	1.158	1.155	1.059	0.918	0.764	0.631	0.569
0.6	1.270	1.268	1.167	1.021	0.862	0.721	0.651
0.7	1.384	1.380	1.270	1.116	0.953	0.806	0.729
0.8	1.503	1.499	1.373	1.205	1.039	0.886	0.803
0.9	1.621	1.612	1.477	1.286	1.117	0.959	0.872
1	1.743	1.726	1.581	1.347	1.181	1.020	0.932
2	2.402	2.393	2.288	2.034	1.827	1.622	1.509
3	3.031	3.018	2.895	2.650	2.408	2.164	2.026
4	3.629	3.615	3.477	3.223	2.949	2.669	2.510
5	4.217	4.201	4.048	3.766	3.462	3.150	2.971
6	4.784	4.766	4.601	4.288	3.956	3.614	3.416
7	5.323	5.303	5.128	4.792	4.434	4.063	3.847
8	5.829	5.808	5.625	5.283	4.900	4.501	4.268
9	6.306	6.283	6.094	5.762	5.355	4.929	4.680
10	6.837	6.816	6.583	6.214	5.792	5.344	5.080
11	7.263	7.242	7.040	6.664	6.217	5.745	5.468
12	7.742	7.718	7.495	7.104	6.638	6.143	5.852
13	8.196	8.172	7.943	7.539	7.052	6.536	6.231
14	8.648	8.623	8.385	7.967	7.461	6.923	6.604
15	9.094	9.068	8.821	8.389	7.864	7.305	6.973
16	9.534	9.507	9.253	8.806	8.262	7.682	7.338
17	9.969	9.942	9.679	9.218	8.656	8.056	7.699
18	10.399	10.371	10.100	9.626	9.046	8.426	8.057
19	10.823	10.794	10.516	10.029	9.432	8.793	8.411
20	11.241	11.211	10.929	10.430	9.815	9.156	8.763
21	11.664	11.633	11.339	10.826	10.194	9.516	9.111
22	12.075	12.045	11.744	11.219	10.571	9.874	9.457
23	12.486	12.454	12.147	11.609	10.944	10.229	9.801
24	12.893	12.861	12.546	11.996	11.315	10.581	10.142
25	13.297	13.264	12.942	12.380	11.683	10.931	10.481
26	13.698	13.665	13.336	12.761	12.048	11.279	10.817
27	14.097	14.062	13.726	13.140	12.411	11.625	11.152

(Continued)

TABLE 2.3 (Continued)

	Z'_D						
t_D	0.05	0.1	0.3	0.5	0.7	0.9	1.0
28	14.493	14.458	14.115	13.517	12.772	11.968	11.485
29	14.886	14.850	14.501	13.891	13.131	12.310	11.816
30	15.277	15.241	14.884	14.263	13.488	12.650	12.145
31	15.666	15.628	15.266	14.634	13.843	12.990	12.473
32	16.053	16.015	15.645	15.002	14.196	13.324	12.799
33	16.437	16.398	16.023	15.368	14.548	13.659	13.123
34	16.819	16.780	16.398	15.732	14.897	13.992	13.446
35	17.200	17.160	16.772	16.095	15.245	14.324	13.767
36	17.579	17.538	17.143	16.456	15.592	14.654	14.088
37	17.956	17.915	17.513	16.815	15.937	14.983	14.406
38	18.331	18.289	17.882	17.173	16.280	15.311	14.724
39	18.704	18.662	18.249	17.529	16.622	15.637	15.040
40	19.088	19.045	18.620	17.886	16.964	15.963	15.356
41	19.450	19.407	18.982	18.240	17.305	16.288	15.671
42	19.821	19.777	19.344	18.592	17.644	16.611	15.985
43	20.188	20.144	19.706	18.943	17.981	16.933	16.297
44	20.555	20.510	20.065	19.293	18.317	17.253	16.608
45	20.920	20.874	20.424	19.641	18.651	17.573	16.918
46	21.283	21.237	20.781	19.988	18.985	17.891	17.227
47	21.645	21.598	21.137	20.333	19.317	18.208	17.535
48	22.006	21.958	21.491	20.678	19.648	18.524	17.841
49	22.365	22.317	21.844	21.021	19.978	18.840	18.147
50	22.722	22.674	22.196	21.363	20.307	19.154	18.452
51	23.081	23.032	22.547	21.704	20.635	19.467	18.757
52	23.436	23.387	22.897	22.044	20.962	19.779	19.060
53	23.791	23.741	23.245	22.383	21.288	20.091	19.362
54	24.145	24.094	23.593	22.721	21.613	20.401	19.664
55	24.498	24.446	23.939	23.058	21.937	20.711	19.965
56	24.849	24.797	24.285	23.393	22.260	21.020	20.265
57	25.200	25.147	24.629	23.728	22.583	21.328	20.564
58	25.549	25.496	24.973	24.062	22.904	21.636	20.862
59	25.898	25.844	25.315	24.395	23.225	21.942	21.160
60	26.246	26.191	25.657	24.728	23.545	22.248	21.457
61	26.592	26.537	25.998	25.059	23.864	22.553	21.754
62	26.938	26.883	26.337	25.390	24.182	22.857	22.049
63	27.283	27.227	26.676	25.719	24.499	23.161	22.344
64	27.627	27.570	27.015	26.048	24.616	23.464	22.639

65	27.970	27.913	27.352	26.376	25.132	23.766	22.932
66	28.312	28.255	27.688	26.704	25.447	24.088	23.225
67	28.653	28.596	28.024	27.030	25.762	24.369	23.518
68	28.994	28.936	28.359	27.356	26.075	24.669	23.810
69	29.334	29.275	28.693	27.681	26.389	24.969	24.101
70	29.673	29.614	29.026	28.008	26.701	25.268	24.391
71	30.011	29.951	29.359	28.329	27.013	25.566	24.881
72	30.349	30.288	29.691	28.652	27.324	25.864	24.971
73	30.686	30.625	30.022	28.974	27.634	26.161	25.260
74	31.022	30.960	30.353	29.296	27.944	26.458	25.548
75	31.357	31.295	30.682	29.617	28.254	26.754	25.836
76	31.692	31.629	31.012	29.937	28.562	27.049	26.124
77	32.026	31.963	31.340	30.257	28.870	27.344	26.410
78	32.359	32.296	31.668	30.576	29.178	27.639	26.697
79	32.692	32.628	31.995	30.895	29.485	27.933	25.983
80	33.024	32.959	32.322	31.212	29.791	28.226	27.268
81	33.355	33.290	32.647	31.530	30.097	28.519	27.553
82	33.686	33.621	32.973	31.846	30.402	28.812	27.837
83	34.016	33.950	33.297	32.163	30.707	29.104	28.121
84	34.345	34.279	33.622	32.478	31.011	29.395	28.404
85	34.674	34.608	33.945	32.793	31.315	29.686	28.687
86	35.003	34.935	34.268	33.107	31.618	29.976	28.970
87	35.330	35.263	34.590	33.421	31.921	30.266	29.252
88	35.657	35.589	34.912	33.735	32.223	30.556	29.534
89	35.984	35.915	35.233	34.048	32.525	30.845	29.815
90	36.310	36.241	35.554	34.360	32.826	31.134	30.096
91	36.636	36.566	35.874	34.672	33.127	31.422	30.376
92	36.960	36.890	36.194	34.983	33.427	31.710	30.656
93	37.285	37.214	36.513	35.294	33.727	31.997	30.935
94	37.609	37.538	36.832	35.604	34.026	32.284	31.215
95	37.932	37.861	37.150	35.914	34.325	32.570	31.493
96	38.255	38.183	37.467	36.223	34.623	32.857	31.772
97	38.577	38.505	37.785	36.532	34.921	33.142	32.050
98	38.899	38.826	38.101	36.841	35.219	33.427	32.327
99	39.220	39.147	38.417	37.149	35.516	33.712	32.605
100	39.541	39.467	38.733	37.456	35.813	33.997	32.881
105	41.138	41.062	40.305	38.987	37.290	35.414	34.260
110	42.724	42.645	41.865	40.508	38.758	36.821	35.630
115	44.299	44.218	43.415	42.018	40.216	38.221	36.993

(*Continued*)

TABLE 2.3 (Continued)

t_D	Z'_D 0.05	0.1	0.3	0.5	0.7	0.9	1.0
120	45.864	45.781	44.956	43.520	41.666	39.612	38.347
125	47.420	47.334	46.487	45.012	43.107	40.995	39.694
130	48.966	48.879	48.009	46.497	44.541	42.372	41.035
135	50.504	50.414	49.523	47.973	45.967	43.741	42.368
140	52.033	51.942	51.029	49.441	47.386	45.104	43.696
145	53.555	53.462	52.528	50.903	48.798	46.460	45.017
150	55.070	54.974	54.019	52.357	50.204	47.810	46.333
155	56.577	56.479	55.503	53.805	51.603	49.155	47.643
160	58.077	57.977	56.981	55.246	52.996	50.494	48.947
165	59.570	59.469	58.452	56.681	54.384	51.827	50.247
170	61.058	60.954	59.916	58.110	55.766	53.156	51.542
175	62.539	62.433	61.375	59.534	57.143	54.479	52.832
180	64.014	63.906	62.829	60.952	58.514	55.798	54.118
185	65.484	65.374	64.276	62.365	59.881	57.112	55.399
190	66.948	66.836	65.718	63.773	61.243	58.422	56.676
195	68.406	68.293	67.156	65.175	62.600	59.727	57.949
200	69.860	69.744	68.588	66.573	63.952	61.028	59.217
205	71.309	71.191	70.015	67.967	65.301	62.326	60.482
210	72.752	72.633	71.437	69.355	66.645	63.619	61.744
215	74.191	74.070	72.855	70.740	67.985	64.908	63.001
220	75.626	75.503	74.269	72.120	69.321	66.194	64.255
225	77.056	76.931	75.678	73.496	70.653	67.476	65.506
230	78.482	78.355	77.083	74.868	71.981	68.755	66.753
235	79.903	79.774	78.484	76.236	73.306	70.030	67.997
240	81.321	81.190	79.881	77.601	74.627	71.302	69.238
245	82.734	82.602	81.275	78.962	75.945	72.570	70.476
250	84.144	84.010	82.664	80.319	77.259	73.736	71.711
255	85.550	85.414	84.050	81.672	78.570	75.098	72.943
260	86.952	86.814	85.432	83.023	79.878	76.358	74.172
265	88.351	88.211	86.811	84.369	81.182	77.614	75.398
270	89.746	89.604	88.186	85.713	82.484	78.868	76.621
275	91.138	90.994	89.558	87.053	83.782	80.119	77.842
280	92.526	92.381	90.926	88.391	85.078	81.367	79.060
285	93.911	93.764	92.292	89.725	86.371	82.612	80.276
290	95.293	95.144	93.654	91.056	87.660	83.855	81.489
295	96.672	96.521	95.014	92.385	88.948	85.095	82.700

300	98.048	97.895	96.370	93.710	90.232	86.333	83.908
305	99.420	99.266	97.724	95.033	91.514	87.568	85.114
310	100.79	100.64	99.07	96.35	92.79	88.80	86.32
315	102.16	102.00	100.42	97.67	94.07	90.03	87.52
320	103.52	103.36	101.77	98.99	95.34	91.26	88.72
325	104.88	104.72	103.11	100.30	96.62	92.49	89.92
330	106.24	106.08	104.45	101.61	97.89	93.71	91.11
335	107.60	107.43	105.79	102.91	99.15	94.93	92.30
340	108.95	108.79	107.12	104.22	100.42	96.15	93.49
345	110.30	110.13	108.45	105.52	101.68	97.37	94.68
350	111.65	111.48	109.78	106.82	102.94	98.58	95.87
355	113.00	112.82	111.11	108.12	104.20	99.80	97.06
360	114.34	114.17	112.43	109.41	105.45	101.01	98.24
365	115.68	115.51	113.76	110.71	106.71	102.22	99.42
370	117.02	116.84	115.08	112.00	107.96	103.42	100.60
375	118.36	118.18	116.40	113.29	109.21	104.63	101.78
380	119.69	119.51	117.71	114.57	110.46	105.83	102.95
385	121.02	120.84	119.02	115.86	111.70	107.04	104.13
390	122.35	122.17	120.34	117.14	112.95	108.24	105.30
395	123.68	123.49	121.65	118.42	114.19	109.43	106.47
400	125.00	124.82	122.94	119.70	115.43	110.63	107.64
405	126.33	126.14	124.26	120.97	116.67	111.82	108.80
410	127.65	127.46	125.56	122.25	117.90	113.02	109.97
415	128.97	128.78	126.86	123.52	119.14	114.21	111.13
420	130.28	130.09	128.16	124.79	120.37	115.40	112.30
425	131.60	131.40	129.46	126.06	121.60	116.59	113.46
430	132.91	132.72	130.75	127.33	122.83	117.77	114.62
435	134.22	134.03	132.05	128.59	124.06	118.96	115.77
440	135.53	135.33	133.34	129.86	125.29	120.14	116.93
445	136.84	136.64	134.63	131.12	126.51	121.32	118.08
450	138.15	137.94	135.92	132.38	127.73	122.50	119.24
455	139.45	139.25	137.20	133.64	128.96	123.68	120.39
460	140.75	140.55	138.49	134.90	130.18	124.86	121.54
465	142.05	141.85	139.77	136.15	131.39	126.04	122.69
470	143.35	143.14	141.05	137.40	132.61	127.21	123.84
475	144.65	144.44	142.33	138.66	133.82	128.38	124.98
480	145.94	145.73	143.61	139.91	135.04	129.55	126.13
485	147.24	147.02	144.89	141.15	136.25	130.72	127.27
490	148.53	148.31	146.16	142.40	137.46	131.89	128.41

(Continued)

TABLE 2.3 (Continued)

t_D	Z'_D 0.05	0.1	0.3	0.5	0.7	0.9	1.0
495	149.82	149.60	147.43	143.65	138.67	133.06	129.56
500	151.11	150.89	148.71	144.89	139.88	134.23	130.70
510	153.68	153.46	151.24	147.38	142.29	136.56	132.97
520	156.25	156.02	153.78	149.85	144.70	138.88	135.24
530	158.81	158.58	156.30	152.33	147.10	141.20	137.51
540	161.36	161.13	158.82	154.79	149.49	143.51	139.77
550	163.91	163.68	161.34	157.25	151.88	145.82	142.03
560	166.45	166.22	163.85	159.71	154.27	148.12	144.28
570	168.99	168.75	166.35	162.16	156.65	150.42	146.53
580	171.52	171.28	168.85	164.61	159.02	152.72	148.77
590	174.05	173.80	171.34	167.05	161.39	155.01	151.01
600	176.57	176.32	173.83	169.48	163.76	157.29	153.25
610	179.09	178.83	176.32	171.92	166.12	159.58	155.48
620	181.60	181.34	178.80	174.34	168.48	161.85	157.71
630	184.10	183.85	181.27	176.76	170.83	164.13	159.93
640	186.60	186.35	183.74	179.18	173.18	166.40	162.15
650	189.10	188.84	186.20	181.60	175.52	168.66	164.37
660	191.59	191.33	188.66	184.00	177.86	170.92	166.58
670	194.08	193.81	191.12	186.41	180.20	173.18	168.79
680	196.57	196.29	193.57	188.81	182.53	175.44	170.99
690	199.04	198.77	196.02	191.21	184.86	177.69	173.20
700	201.52	201.24	198.46	193.60	187.19	179.94	175.39
710	203.99	203.71	200.90	195.99	189.51	182.18	177.59
720	206.46	206.17	203.34	198.37	191.83	184.42	179.78
730	208.92	208.63	205.77	200.75	194.14	186.66	181.97
740	211.38	211.09	208.19	203.13	196.45	188.89	184.15
750	213.83	213.54	210.62	205.50	198.76	191.12	186.34
760	216.28	215.99	213.04	207.87	201.06	193.35	188.52
770	218.73	218.43	215.45	210.24	203.36	195.57	190.69
780	221.17	220.87	217.86	212.60	205.66	197.80	192.87
790	223.61	223.31	220.27	214.96	207.95	200.01	195.04
800	226.05	225.74	222.68	217.32	210.24	202.23	197.20
810	228.48	228.17	225.08	219.67	212.53	204.44	199.37
820	230.91	230.60	227.48	222.02	214.81	206.65	201.53
830	233.33	233.02	229.87	224.36	217.09	208.86	203.69
840	235.76	235.44	232.26	226.71	219.37	211.06	205.85

850	238.18	237.86	234.65	229.05	221.64	213.26	208.00
860	240.59	240.27	237.04	231.38	223.92	215.46	210.15
870	243.00	242.68	239.42	233.72	226.19	217.65	212.30
880	245.41	245.08	241.80	236.05	228.45	219.85	214.44
890	247.82	247.49	244.17	238.37	230.72	222.04	216.59
900	250.22	249.89	246.55	240.70	232.98	224.22	218.73
910	252.62	252.28	248.92	243.02	235.23	226.41	220.87
920	255.01	254.68	251.28	245.34	237.49	228.59	223.00
930	257.41	257.07	253.65	247.66	239.74	230.77	225.14
940	259.80	259.46	256.01	249.97	241.99	232.95	227.27
950	262.19	261.84	258.36	252.28	244.24	235.12	229.39
960	264.57	264.22	260.72	254.59	246.48	237.29	231.52
970	266.95	266.60	263.07	256.89	248.72	239.46	233.65
980	269.33	268.98	265.42	259.19	250.96	241.63	235.77
990	271.71	271.35	267.77	261.49	253.20	243.80	237.89
1000	274.08	273.72	270.11	263.79	255.44	245.96	240.00
1010	276.35	275.99	272.35	265.99	257.58	248.04	242.04
1020	278.72	278.35	274.69	268.29	259.81	250.19	244.15
1030	281.08	280.72	277.03	270.57	262.04	252.35	246.26
1040	283.44	283.08	279.36	272.86	264.26	254.50	248.37
1050	285.81	285.43	281.69	275.15	266.49	256.66	250.48
1060	288.16	287.79	284.02	277.43	268.71	258.81	252.58
1070	290.52	290.14	286.35	279.71	270.92	260.95	254.69
1080	292.87	292.49	288.67	281.99	273.14	263.10	256.79
1090	295.22	294.84	290.99	284.26	275.35	265.24	258.89
1100	297.57	297.18	293.31	286.54	277.57	267.38	260.98
1110	299.91	299.53	295.63	288.81	279.78	269.52	263.08
1120	302.28	301.87	297.94	291.07	281.98	271.66	265.17
1130	304.60	304.20	300.25	293.34	284.19	273.80	267.26
1140	306.93	308.54	302.56	295.61	286.39	275.93	269.35
1150	309.27	308.87	304.87	297.87	288.59	278.06	271.44
1160	311.60	311.20	307.18	300.13	290.79	280.19	273.52
1170	313.94	313.53	309.48	302.38	292.99	282.32	275.61
1180	316.26	315.86	311.78	304.64	295.19	284.44	277.69
1190	318.59	318.18	314.08	306.89	297.38	286.57	279.77
1200	320.92	320.51	316.38	309.15	299.57	288.69	281.85
1210	323.24	322.83	318.67	311.39	301.76	290.81	283.92
1220	325.56	325.14	320.96	313.64	303.95	292.93	286.00
1230	327.88	327.46	323.25	315.89	306.13	295.05	288.07

(Continued)

TABLE 2.3 (Continued)

	Z'_D						
t_D	0.05	0.1	0.3	0.5	0.7	0.9	1.0
1240	330.19	329.77	325.54	318.13	308.32	297.16	290.14
1250	332.51	332.08	327.83	320.37	310.50	299.27	292.21
1260	334.82	334.39	330.11	322.61	312.68	301.38	294.28
1270	337.13	336.70	332.39	324.85	314.85	303.49	296.35
1280	339.44	339.01	334.67	327.08	317.03	305.60	298.41
1290	341.74	341.31	336.95	329.32	319.21	307.71	300.47
1300	344.05	343.61	339.23	331.55	321.38	309.81	302.54
1310	346.35	345.91	341.50	333.78	323.55	311.92	304.60
1320	348.65	348.21	343.77	336.01	325.72	314.02	306.65
1330	350.95	350.50	346.04	338.23	327.89	316.12	308.71
1340	353.24	352.80	348.31	340.46	330.05	318.22	310.77
1350	355.54	355.09	350.58	342.68	332.21	320.31	312.82
1360	357.83	357.38	352.84	344.90	334.38	322.41	314.87
1370	360.12	359.67	355.11	347.12	336.54	324.50	316.92
1380	362.41	361.95	357.37	349.34	338.70	326.59	318.97
1390	364.69	364.24	359.63	351.56	340.85	328.68	321.02
1400	366.98	366.52	361.88	353.77	343.01	330.77	323.06
1410	369.26	368.80	364.14	355.98	345.16	332.86	325.11
1420	371.54	371.08	366.40	358.19	347.32	334.94	327.15
1430	373.82	373.35	368.65	360.40	349.47	337.03	329.19
1440	376.10	375.63	370.90	362.61	351.62	339.11	331.23
1450	378.38	377.90	373.15	364.81	353.76	341.19	333.27
1460	380.65	380.17	375.39	367.02	355.91	343.27	335.31
1470	382.92	382.44	377.64	369.22	358.06	345.35	337.35
1480	385.19	384.71	379.88	371.42	360.20	347.43	339.38
1490	387.46	386.98	382.13	373.62	362.34	349.50	341.42
1500	389.73	389.25	384.37	375.82	364.48	351.58	343.45
1525	395.39	394.90	389.96	381.31	369.82	356.76	348.52
1550	401.04	400.55	395.55	386.78	375.16	361.93	353.59
1575	406.68	406.18	401.12	392.25	380.49	367.09	358.65
1600	412.32	411.81	406.69	397.71	385.80	372.24	363.70
1625	417.94	417.42	412.24	403.16	391.11	377.39	368.74
1650	423.55	423.03	417.79	408.60	396.41	382.53	373.77
1675	429.15	428.63	423.33	414.04	401.70	387.66	378.80
1700	434.75	434.22	428.85	419.46	406.99	392.78	383.82
1725	440.33	439.79	434.37	424.87	412.26	397.89	388.83

1750	445.91	445.37	439.89	430.28	417.53	403.00	393.84
1775	451.48	450.93	445.39	435.68	422.79	408.10	398.84
1880	457.04	456.48	450.88	441.07	428.04	413.20	403.83
1825	462.59	462.03	456.37	446.46	433.29	418.28	408.82
1850	468.13	467.56	461.85	451.83	438.53	423.36	413.80
1875	473.67	473.09	467.32	457.20	443.76	428.43	418.77
1900	479.19	478.61	472.78	462.56	448.98	433.50	423.73
1925	484.71	484.13	478.24	467.92	454.20	438.56	428.69
1950	490.22	489.63	483.69	473.26	459.41	443.61	433.64
1975	495.73	495.13	489.13	478.60	464.61	448.66	438.59
2000	501.22	500.62	494.56	483.93	469.81	453.70	443.53
2025	506.71	506.11	499.99	489.26	475.00	458.73	448.47
2050	512.20	511.58	505.41	494.58	480.18	463.76	453.40
2075	517.67	517.05	510.82	499.89	485.36	468.78	458.32
2100	523.14	522.52	516.22	505.19	490.53	473.80	463.24
2125	528.60	527.97	521.62	510.49	495.69	478.81	468.15
2150	534.05	533.42	527.02	515.78	500.85	483.81	473.06
2175	539.50	538.86	532.40	521.07	506.01	488.81	477.96
2200	544.94	544.30	537.78	526.35	511.15	493.81	482.85
2225	550.38	549.73	543.15	531.62	516.29	498.79	487.74
2250	555.81	555.15	548.52	536.89	521.43	503.78	492.63
2275	561.23	560.56	553.88	542.15	526.56	508.75	497.51
2300	566.64	565.97	559.23	547.41	531.68	513.72	502.38
2325	572.05	571.38	564.58	552.66	536.80	518.69	507.25
2350	577.46	576.78	569.92	557.90	541.91	523.65	512.12
2375	582.85	582.17	575.26	563.14	547.02	528.61	516.98
2400	588.24	587.55	580.59	568.37	552.12	533.56	521.83
2425	593.63	592.93	585.91	573.60	557.22	538.50	526.68
2450	599.01	598.31	591.23	578.82	562.31	543.45	531.53
2475	604.38	603.68	596.55	584.04	567.39	548.38	536.37
2500	609.75	609.04	601.85	589.25	572.47	553.31	541.20
2550	620.47	619.75	612.45	599.65	582.62	563.16	550.86
2600	631.17	630.43	623.03	610.04	592.75	572.99	560.50
2650	641.84	641.10	633.59	620.40	602.86	582.80	570.13
2700	652.50	651.74	644.12	630.75	612.95	592.60	579.73
2750	663.13	662.37	654.64	641.07	623.02	602.37	589.32
2800	673.75	672.97	665.14	651.38	633.07	612.13	598.90
2850	684.34	683.56	675.61	661.67	643.11	621.88	608.45
2900	694.92	694.12	686.07	671.94	653.12	631.60	617.99

(Continued)

TABLE 2.3 (Continued)

	Z'_D						
t_D	0.05	0.1	0.3	0.5	0.7	0.9	1.0
2950	705.48	704.67	696.51	682.19	663.13	641.32	627.52
3000	716.02	715.20	706.94	692.43	673.11	651.01	637.03
3050	726.54	725.71	717.34	702.65	683.08	660.69	646.53
3100	737.04	736.20	727.73	712.85	693.03	670.36	656.01
3150	747.53	746.68	738.10	723.04	702.97	680.01	665.48
3200	758.00	757.14	748.45	733.21	712.89	689.64	674.93
3250	768.45	767.58	758.79	743.36	722.80	699.27	684.37
3300	778.89	778.01	769.11	753.50	732.69	708.87	693.80
3350	789.31	788.42	779.42	763.62	742.57	718.47	703.21
3400	799.71	798.81	789.71	773.73	752.43	728.05	712.62
3450	810.10	809.19	799.99	783.82	762.28	737.62	722.00
3500	820.48	819.55	810.25	793.90	772.12	747.17	731.38
3550	830.83	829.90	820.49	803.97	781.94	756.72	740.74
3600	841.18	840.24	830.73	814.02	791.75	766.24	750.09
3650	851.51	850.56	840.94	824.06	801.55	775.76	759.43
3700	861.83	860.86	851.15	834.08	811.33	785.27	768.76
3750	872.13	871.15	861.34	844.09	821.10	794.76	778.08
3800	882.41	881.43	871.51	854.09	830.86	804.24	787.38
3850	892.69	891.70	881.68	864.08	840.61	813.71	796.68
3900	902.95	901.95	891.83	874.05	850.34	823.17	805.96
3950	913.20	912.19	901.96	884.01	860.06	832.62	815.23
4000	923.43	922.41	912.09	893.96	869.77	842.06	824.49
4050	933.65	932.62	922.20	903.89	879.47	851.48	833.74
4100	943.86	942.82	932.30	913.82	889.16	860.90	842.99
4150	954.06	953.01	942.39	923.73	898.84	870.30	852.22
4200	964.25	963.19	952.47	933.63	908.50	879.69	861.44
4250	974.42	973.35	962.53	943.52	918.16	889.08	870.65
4300	984.58	983.50	972.58	953.40	927.60	898.45	879.85
4350	994.73	993.64	982.62	963.27	937.43	907.81	889.04
4400	1004.9	1003.8	992.7	973.1	947.1	917.2	898.2
4450	1015.0	1013.9	1002.7	983.0	956.7	926.5	907.4
4500	1025.1	1024.0	1012.7	992.8	966.3	935.9	916.6
4550	1035.2	1034.1	1022.7	1002.6	975.9	945.2	925.7
4600	1045.3	1044.2	1032.7	1012.4	985.5	954.5	934.9
4650	1055.4	1054.2	1042.6	1022.2	995.0	963.8	944.0
4700	1065.5	1064.3	1052.6	1032.0	1004.6	973.1	953.1

4750	1075.5	1074.4	1062.6	1041.8	1014.1	982.4	962.2
4800	1085.6	1084.4	1072.5	1051.6	1023.7	991.7	971.4
4850	1095.6	1094.4	1082.4	1061.4	1033.2	1000.9	980.5
4900	1105.6	1104.5	1092.4	1071.1	1042.8	1010.2	989.5
4950	1115.7	1114.5	1102.3	1080.9	1052.3	1019.4	998.6
5000	1125.7	1124.5	1112.2	1090.6	1061.8	1028.7	1007.7
5100	1145.7	1144.4	1132.0	1110.0	1080.8	1047.2	1025.8
5200	1165.6	1164.4	1151.7	1129.4	1099.7	1065.6	1043.9
5300	1185.5	1184.3	1171.4	1148.8	1118.6	1084.0	1062.0
5400	1205.4	1204.1	1191.1	1168.2	1137.5	1102.4	1080.0
5500	1225.3	1224.0	1210.7	1187.5	1156.4	1120.7	1098.0
5600	1245.1	1243.7	1230.3	1206.7	1175.2	1139.0	1116.0
5700	1264.9	1263.5	1249.9	1226.0	1194.0	1157.3	1134.0
5800	1284.6	1283.2	1269.4	1245.2	1212.8	1175.5	1151.9
5900	1304.3	1302.9	1288.9	1264.4	1231.5	1193.8	1169.8
6000	1324.0	1322.6	1308.4	1283.5	1250.2	1211.9	1187.7
6100	1343.6	1342.2	1327.9	1302.6	1268.9	1230.1	1205.5
6200	1363.2	1361.8	1347.3	1321.7	1287.5	1248.3	1223.3
6300	1382.8	1381.4	1366.7	1340.8	1306.2	1266.4	1241.1
6400	1402.4	1400.9	1386.0	1359.8	1324.7	1284.5	1258.9
6500	1421.9	1420.4	1405.3	1378.8	1343.3	1302.5	1276.6
6600	1441.4	1439.9	1424.6	1397.8	1361.9	1320.6	1294.3
6700	1460.9	1459.4	1443.9	1416.7	1380.4	1338.6	1312.0
6800	1480.3	1478.8	1463.1	1435.6	1398.9	1356.6	1329.7
6900	1499.7	1498.2	1482.4	1454.5	1417.3	1374.5	1347.4
7000	1519.1	1517.5	1501.5	1473.4	1435.8	1392.5	1365.0
7100	1538.5	1536.9	1520.7	1492.3	1454.2	1410.4	1382.6
7200	1557.8	1556.2	1539.8	1511.1	1472.6	1428.3	1400.2
7300	1577.1	1575.5	1559.0	1529.9	1491.0	1446.2	1417.8
7400	1596.4	1594.8	1578.1	1548.6	1509.3	1464.1	1435.3
7500	1615.7	1614.0	1597.1	1567.4	1527.6	1481.9	1452.8
7600	1634.9	1633.2	1616.2	1586.1	1545.9	1499.7	1470.3
7700	1654.1	1652.4	1635.2	1604.8	1564.2	1517.5	1487.8
7800	1673.3	1671.6	1654.2	1623.5	1582.5	1535.3	1505.3
7900	1692.5	1690.7	1673.1	1642.2	1600.7	1553.0	1522.7
8000	1711.6	1709.9	1692.1	1660.8	1619.0	1570.8	1540.1
8100	1730.8	1729.0	1711.0	1679.4	1637.2	1588.5	1557.6
8200	1749.9	1748.1	1729.9	1698.0	1655.3	1606.2	1574.9
8300	1768.9	1767.1	1748.8	1716.6	1673.5	1623.9	1592.3

(Continued)

TABLE 2.3 (Continued)

t_D	Z'_D						
	0.05	0.1	0.3	0.5	0.7	0.9	1.0
8400	1788.0	1786.2	1767.7	1735.2	1691.6	1641.5	1609.7
8500	1807.0	1805.2	1786.5	1753.7	1709.8	1659.2	1627.0
8600	1826.0	1824.2	1805.4	1772.2	1727.9	1676.8	1644.3
8700	1845.0	1843.2	1824.2	1790.7	1746.0	1694.4	1661.6
8800	1864.0	1862.1	1842.9	1809.2	1764.0	1712.0	1678.9
8900	1883.0	1881.1	1861.7	1827.7	1782.1	1729.6	1696.2
9000	1901.9	1900.0	1880.5	1846.1	1800.1	1747.1	1713.4
9100	1920.8	1918.9	1899.2	1864.5	1818.1	1764.7	1730.7
9200	1939.7	1937.4	1917.9	1882.9	1836.1	1782.2	1747.9
9300	1958.6	1956.6	1936.6	1901.3	1854.1	1799.7	1765.1
9400	1977.4	1975.4	1955.2	1919.7	1872.0	1817.2	1782.3
9500	1996.3	1994.3	1973.9	1938.0	1890.0	1834.7	1799.4
9600	2015.1	2013.1	1992.5	1956.4	1907.9	1852.1	1816.6
9700	2033.9	2031.9	2011.1	1974.7	1925.8	1869.6	1833.7
9800	2052.7	2050.6	2029.7	1993.0	1943.7	1887.0	1850.9
9900	2071.5	2069.4	2048.3	2011.3	1961.6	1904.4	1868.0
1.00×10^4	2.090×10^3	2.088×10^3	2.067×10^3	2.029×10^3	1.979×10^3	1.922×10^3	1.885×10^3
1.25×10^4	2.553×10^3	2.551×10^3	2.526×10^3	2.481×10^3	2.421×10^3	2.352×10^3	2.308×10^3
1.50×10^4	3.009×10^3	3.006×10^3	2.977×10^3	2.925×10^3	2.855×10^3	2.775×10^3	2.724×10^3
1.75×10^4	3.457×10^3	3.454×10^3	3.421×10^3	3.362×10^3	3.284×10^3	3.193×10^3	3.135×10^3
2.00×10^4	3.900×10^3	3.897×10^3	3.860×10^3	3.794×10^3	3.707×10^3	3.605×10^3	3.541×10^3
2.50×10^4	4.773×10^3	4.768×10^3	4.724×10^3	4.646×10^3	4.541×10^3	4.419×10^3	4.341×10^3
3.00×10^4	5.630×10^3	5.625×10^3	5.574×10^3	5.483×10^3	5.361×10^3	5.219×10^3	5.129×10^3
3.50×10^4	6.476×10^3	6.470×10^3	6.412×10^3	6.309×10^3	6.170×10^3	6.009×10^3	5.906×10^3
4.00×10^4	7.312×10^3	7.305×10^3	7.240×10^3	7.125×10^3	6.970×10^3	6.790×10^3	6.675×10^3
4.50×10^4	8.139×10^3	8.132×10^3	8.060×10^3	7.933×10^3	7.762×10^3	7.564×10^9	7.437×10^3
5.00×10^4	8.959×10^3	8.951×10^3	8.872×10^3	8.734×10^3	8.548×10^3	8.331×10^3	8.193×10^3
6.00×10^4	1.057×10^4	1.057×10^4	1.047×10^4	1.031×10^4	1.010×10^4	9.846×10^3	9.684×10^3
7.00×10^4	1.217×10^4	1.217×10^4	1.206×10^4	1.188×10^4	1.163×10^4	1.134×10^4	1.116×10^4
8.00×10^4	1.375×10^4	1.375×10^4	1.363×10^4	1.342×10^4	1.315×10^4	1.283×10^4	1.262×10^4
9.00×10^4	1.532×10^4	1.531×10^4	1.518×10^4	1.496×10^4	1.465×10^4	1.430×10^4	1.407×10^4
1.00×10^5	1.687×10^4	1.686×10^4	1.672×10^4	1.647×10^4	1.614×10^4	1.576×10^4	1.551×10^4
1.25×10^5	2.071×10^4	2.069×10^4	2.052×10^4	2.023×10^4	1.982×10^4	1.936×10^4	1.906×10^4
1.50×10^5	2.448×10^4	2.446×10^4	2.427×10^4	2.392×10^4	2.345×10^4	2.291×10^4	2.256×10^4
2.00×10^5	3.190×10^4	3.188×10^4	3.163×10^4	3.119×10^4	3.059×10^4	2.989×10^4	2.945×10^4
2.50×10^5	3.918×10^4	3.916×10^4	3.885×10^4	3.832×10^4	3.760×10^4	3.676×10^4	3.622×10^4

3.00×10^5	4.636×10^4	4.633×10^4	4.598×10^4	4.536×10^4	4.452×10^4	4.353×10^4	4.290×10^4
4.00×10^5	6.048×10^4	6.044×10^4	5.999×10^4	5.920×10^4	5.812×10^4	5.687×10^4	5.606×10^4
5.00×10^5	7.438×10^4	7.431×10^4	7.376×10^4	7.280×10^4	7.150×10^4	6.998×10^4	6.900×10^4
6.00×10^5	8.805×10^4	8.798×10^4	8.735×10^4	8.623×10^4	8.471×10^4	8.293×10^4	8.178×10^4
7.00×10^5	1.016×10^5	1.015×10^5	1.008×10^5	9.951×10^4	9.777×10^4	9.573×10^4	9.442×10^4
8.00×10^5	1.150×10^5	1.149×10^5	1.141×10^5	1.127×10^5	1.107×10^5	1.084×10^5	1.070×10^5
9.00×10^5	1.283×10^5	1.282×10^5	1.273×10^5	1.257×10^5	1.235×10^5	1.210×10^5	1.194×10^5
1.00×10^6	1.415×10^5	1.412×10^5	1.404×10^5	1.387×10^5	1.363×10^5	1.335×10^5	1.317×10^5
1.50×10^6	2.059×10^5	2.060×10^5	2.041×10^5	2.016×10^5	1.982×10^5	1.943×10^5	1.918×10^5
2.00×10^6	2.695×10^5	2.695×10^5	2.676×10^5	2.644×10^5	2.601×10^5	2.551×10^5	2.518×10^5
2.50×10^6	3.320×10^5	3.319×10^5	3.296×10^5	3.254×10^5	3.202×10^5	3.141×10^5	3.101×10^5
3.00×10^6	3.937×10^5	3.936×10^5	3.909×10^5	3.864×10^5	3.803×10^5	3.731×10^5	3.684×10^5
4.00×10^6	5.154×10^5	5.152×10^5	5.118×10^5	5.060×10^5	4.981×10^5	4.888×10^5	4.828×10^5
5.00×10^6	6.352×10^5	6.349×10^5	6.308×10^5	6.238×10^5	6.142×10^5	6.029×10^5	5.956×10^5
6.00×10^6	7.536×10^5	7.533×10^5	7.485×10^5	7.402×10^5	7.290×10^5	7.157×10^5	7.072×10^5
7.00×10^6	8.709×10^5	8.705×10^5	8.650×10^5	8.556×10^5	8.427×10^5	8.275×10^5	8.177×10^5
8.00×10^6	9.972×10^5	9.867×10^5	9.806×10^5	9.699×10^5	9.555×10^5	9.384×10^5	9.273×10^5
9.00×10^6	1.103×10^6	1.102×10^6	1.095×10^6	1.084×10^6	1.067×10^6	1.049×10^6	1.036×10^6
1.00×10^7	1.217×10^6	1.217×10^6	1.209×10^6	1.196×10^6	1.179×10^6	1.158×10^6	1.144×10^6
1.50×10^7	1.782×10^6	1.781×10^6	1.771×10^6	1.752×10^6	1.727×10^6	1.697×10^6	1.678×10^6
2.00×10^7	2.337×10^6	2.336×10^6	2.322×10^6	2.298×10^6	2.266×10^6	2.227×10^6	2.202×10^6
2.50×10^7	2.884×10^6	2.882×10^6	2.866×10^6	2.837×10^6	2.797×10^6	2.750×10^6	2.720×10^6
3.00×10^7	3.425×10^6	3.423×10^6	3.404×10^6	3.369×10^6	3.323×10^6	3.268×10^6	3.232×10^6
4.00×10^7	4.493×10^6	4.491×10^6	4.466×10^6	4.422×10^6	4.361×10^6	4.290×10^6	4.244×10^6
5.00×10^7	5.547×10^6	5.544×10^6	5.514×10^6	5.460×10^6	5.386×10^6	5.299×10^6	5.243×10^6
6.00×10^7	6.590×10^6	6.587×10^6	6.551×10^6	6.488×10^6	6.401×10^6	6.299×10^6	6.232×10^6
7.00×10^7	7.624×10^6	7.620×10^6	7.579×10^6	7.507×10^6	7.407×10^6	7.290×10^6	7.213×10^6
8.00×10^7	8.651×10^6	8.647×10^6	8.600×10^6	8.519×10^6	8.407×10^6	8.274×10^6	8.188×10^6
9.00×10^7	9.671×10^6	9.666×10^6	9.615×10^6	9.524×10^6	9.400×10^6	9.252×10^6	9.156×10^6
1.00×10^8	1.069×10^7	1.067×10^7	1.062×10^7	1.052×10^7	1.039×10^7	1.023×10^7	1.012×10^7
1.50×10^8	1.567×10^7	1.567×10^7	1.555×10^7	1.541×10^7	1.522×10^7	1.499×10^7	1.483×10^7
2.00×10^8	2.059×10^7	2.059×10^7	2.048×10^7	2.029×10^7	2.004×10^7	1.974×10^7	1.954×10^7
2.50×10^8	2.546×10^7	2.545×10^7	2.531×10^7	2.507×10^7	2.476×10^7	2.439×10^7	2.415×10^7
3.00×10^8	3.027×10^7	3.026×10^7	3.010×10^7	2.984×10^7	2.947×10^7	2.904×10^7	2.875×10^7
4.00×10^8	3.979×10^7	3.978×10^7	3.958×10^7	3.923×10^7	3.875×10^7	3.819×10^7	3.782×10^7
5.00×10^8	4.920×10^7	4.918×10^7	4.894×10^7	4.851×10^7	4.793×10^7	4.724×10^7	4.679×10^7
6.00×10^8	5.852×10^7	5.850×10^7	5.821×10^7	5.771×10^7	5.702×10^7	5.621×10^7	5.568×10^7
7.00×10^8	6.777×10^7	6.774×10^7	6.741×10^7	6.684×10^7	6.605×10^7	6.511×10^7	6.450×10^7
8.00×10^8	7.700×10^7	7.693×10^7	7.655×10^7	7.590×10^7	7.501×10^7	7.396×10^7	7.327×10^7

(*Continued*)

TABLE 2.3 (Continued)

t_D	Z'_D 0.05	0.1	0.3	0.5	0.7	0.9	1.0
9.00×10^{8}	8.609×10^{7}	8.606×10^{7}	8.564×10^{7}	8.492×10^{7}	8.393×10^{7}	8.275×10^{7}	8.199×10^{7}
1.00×10^{9}	9.518×10^{7}	9.515×10^{7}	9.469×10^{7}	9.390×10^{7}	9.281×10^{7}	9.151×10^{7}	9.066×10^{7}
1.50×10^{9}	1.401×10^{8}	1.400×10^{8}	1.394×10^{8}	1.382×10^{8}	1.367×10^{8}	1.348×10^{8}	1.336×10^{8}
2.00×10^{9}	1.843×10^{8}	1.843×10^{8}	1.834×10^{8}	1.819×10^{8}	1.799×10^{8}	1.774×10^{8}	1.758×10^{8}
2.50×10^{9}	2.281×10^{8}	2.280×10^{8}	2.269×10^{8}	2.251×10^{8}	2.226×10^{8}	2.196×10^{8}	2.177×10^{8}
3.00×10^{9}	2.714×10^{8}	2.713×10^{8}	2.701×10^{8}	2.680×10^{8}	2.650×10^{8}	2.615×10^{8}	2.592×10^{8}
4.00×10^{9}	3.573×10^{8}	3.572×10^{8}	3.558×10^{8}	3.528×10^{8}	3.489×10^{8}	3.443×10^{8}	3.413×10^{8}
5.00×10^{9}	4.422×10^{8}	4.421×10^{8}	4.401×10^{8}	4.367×10^{8}	4.320×10^{8}	4.263×10^{8}	4.227×10^{8}
6.00×10^{9}	5.265×10^{8}	5.262×10^{8}	5.240×10^{8}	5.199×10^{8}	5.143×10^{8}	5.077×10^{8}	5.033×10^{8}
7.00×10^{9}	6.101×10^{8}	6.098×10^{8}	6.072×10^{8}	6.025×10^{8}	5.961×10^{8}	5.885×10^{8}	5.835×10^{8}
8.00×10^{9}	6.932×10^{8}	6.930×10^{8}	6.900×10^{8}	6.847×10^{8}	6.775×10^{8}	6.688×10^{8}	6.632×10^{8}
9.00×10^{9}	7.760×10^{8}	7.756×10^{8}	7.723×10^{8}	7.664×10^{8}	7.584×10^{8}	7.487×10^{8}	7.424×10^{8}
1.00×10^{10}	8.583×10^{8}	8.574×10^{8}	8.543×10^{8}	8.478×10^{8}	8.389×10^{8}	8.283×10^{8}	8.214×10^{8}
1.50×10^{10}	1.263×10^{9}	1.264×10^{9}	1.257×10^{9}	1.247×10^{9}	1.235×10^{9}	1.219×10^{9}	1.209×10^{9}
2.00×10^{10}	1.666×10^{9}	1.666×10^{9}	1.659×10^{9}	1.646×10^{9}	1.630×10^{9}	1.610×10^{9}	1.596×10^{9}
2.50×10^{10}	2.065×10^{9}	2.063×10^{9}	2.055×10^{9}	2.038×10^{9}	2.018×10^{9}	1.993×10^{9}	1.977×10^{9}
3.00×10^{10}	2.458×10^{9}	2.458×10^{9}	2.447×10^{9}	2.430×10^{9}	2.405×10^{9}	2.376×10^{9}	2.357×10^{9}
4.00×10^{10}	3.240×10^{9}	3.239×10^{9}	3.226×10^{9}	3.203×10^{9}	3.171×10^{9}	3.133×10^{9}	3.108×10^{9}
5.00×10^{10}	4.014×10^{9}	4.013×10^{9}	3.997×10^{9}	3.968×10^{9}	3.929×10^{9}	3.883×10^{9}	3.852×10^{9}
6.00×10^{10}	4.782×10^{9}	4.781×10^{9}	4.762×10^{9}	4.728×10^{9}	4.682×10^{9}	4.627×10^{9}	4.591×10^{9}
7.00×10^{10}	5.546×10^{9}	5.544×10^{9}	5.522×10^{9}	5.483×10^{9}	5.430×10^{9}	5.366×10^{9}	5.325×10^{9}
8.00×10^{10}	6.305×10^{9}	6.303×10^{9}	6.278×10^{9}	6.234×10^{9}	6.174×10^{9}	6.102×10^{9}	6.055×10^{9}
9.00×10^{10}	7.060×10^{9}	7.058×10^{9}	7.030×10^{9}	6.982×10^{9}	6.914×10^{9}	6.834×10^{9}	6.782×10^{9}
1.00×10^{11}	7.813×10^{9}	7.810×10^{9}	7.780×10^{9}	7.726×10^{9}	7.652×10^{9}	7.564×10^{9}	7.506×10^{9}
1.50×10^{11}	1.154×10^{10}	1.153×10^{10}	1.149×10^{10}	1.141×10^{10}	1.130×10^{10}	1.118×10^{10}	1.109×10^{10}
2.00×10^{11}	1.522×10^{10}	1.521×10^{10}	1.515×10^{10}	1.505×10^{10}	1.491×10^{10}	1.474×10^{10}	1.463×10^{10}
2.50×10^{11}	1.886×10^{10}	1.885×10^{10}	1.878×10^{10}	1.866×10^{10}	1.849×10^{10}	1.828×10^{10}	1.814×10^{10}
3.00×10^{11}	2.248×10^{10}	2.247×10^{10}	2.239×10^{10}	2.224×10^{10}	2.204×10^{10}	2.179×10^{10}	2.163×10^{10}
4.00×10^{11}	2.965×10^{10}	2.964×10^{10}	2.953×10^{10}	2.934×10^{10}	2.907×10^{10}	2.876×10^{10}	2.855×10^{10}
5.00×10^{11}	3.677×10^{10}	3.675×10^{10}	3.662×10^{10}	3.638×10^{10}	3.605×10^{10}	3.566×10^{10}	3.540×10^{10}
6.00×10^{11}	4.383×10^{10}	4.381×10^{10}	4.365×10^{10}	4.337×10^{10}	4.298×10^{10}	4.252×10^{10}	4.221×10^{10}
7.00×10^{11}	5.085×10^{10}	5.082×10^{10}	5.064×10^{10}	5.032×10^{10}	4.987×10^{10}	4.933×10^{10}	4.898×10^{10}
8.00×10^{11}	5.783×10^{10}	5.781×10^{10}	5.706×10^{10}	5.723×10^{10}	5.673×10^{10}	5.612×10^{10}	5.572×10^{10}
9.00×10^{11}	6.478×10^{10}	6.746×10^{10}	6.453×10^{10}	6.412×10^{10}	6.355×10^{10}	6.288×10^{10}	6.243×10^{10}
1.00×10^{12}	7.171×10^{10}	7.168×10^{10}	7.143×10^{10}	7.098×10^{10}	7.035×10^{10}	6.961×10^{10}	6.912×10^{10}
1.50×10^{12}	1.060×10^{11}	1.060×10^{11}	1.056×10^{11}	1.050×10^{11}	1.041×10^{11}	1.030×10^{11}	1.022×10^{11}
2.00×10^{12}	1.400×10^{11}	1.399×10^{11}	1.394×10^{11}	1.386×10^{11}	1.374×10^{11}	1.359×10^{11}	1.350×10^{11}

Permission to publish by the SPE.

TABLE 2.4 Dimensionless Water Influx W_{eD} for $r'_D = 4$

	Z'_D						
t_D	0.05	0.1	0.3	0.5	0.7	0.9	1.0
2	2.398	2.389	2.284	2.031	1.824	1.620	1.507
3	3.006	2.993	2.874	2.629	2.390	2.149	2.012
4	3.552	3.528	3.404	3.158	2.893	2.620	2.466
5	4.053	4.017	3.893	3.627	3.341	3.045	2.876
6	4.490	4.452	4.332	4.047	3.744	3.430	3.249
7	4.867	4.829	4.715	4.420	4.107	3.778	3.587
8	5.191	5.157	5.043	4.757	4.437	4.096	3.898
9	5.464	5.434	5.322	5.060	4.735	4.385	4.184
10	5.767	5.739	5.598	5.319	5.000	4.647	4.443
11	5.964	5.935	5.829	5.561	5.240	4.884	4.681
12	6.188	6.158	6.044	5.780	5.463	5.107	4.903
13	6.380	6.350	6.240	5.983	5.670	5.316	5.113
14	6.559	6.529	6.421	6.171	5.863	5.511	5.309
15	6.725	6.694	6.589	6.345	6.044	5.695	5.495
16	6.876	6.844	6.743	6.506	6.213	5.867	5.671
17	7.014	6.983	6.885	6.656	6.371	6.030	5.838
18	7.140	7.113	7.019	6.792	6.523	6.187	5.999
19	7.261	7.240	7.140	6.913	6.663	6.334	6.153
20	7.376	7.344	7.261	7.028	6.785	6.479	6.302
22	7.518	7.507	7.451	7.227	6.982	6.691	6.524
24	7.618	7.607	7.518	7.361	7.149	6.870	6.714
26	7.697	7.685	7.607	7.473	7.283	7.026	6.881
28	7.752	7.752	7.674	7.563	7.395	7.160	7.026
30	7.808	7.797	7.741	7.641	7.484	7.283	7.160
34	7.864	7.864	7.819	7.741	7.618	7.451	7.350
38	7.909	7.909	7.875	7.808	7.719	7.585	7.496
42	7.931	7.931	7.909	7.864	7.797	7.685	7.618
46	7.942	7.942	7.920	7.898	7.842	7.752	7.697
50	7.954	7.954	7.942	7.920	7.875	7.808	7.764
60	7.968	7.968	7.965	7.954	7.931	7.898	7.864
70	7.976	7.976	7.976	7.968	7.965	7.942	7.920
80	7.982	7.982	7.987	7.976	7.976	7.965	7.954
90	7.987	7.987	7.987	7.984	7.983	7.976	7.965
100	7.987	7.987	7.987	7.987	7.987	7.983	7.976
120	7.987	7.987	7.987	7.987	7.987	7.987	7.987

Permission to publish by the SPE.

Step 3. Calculate the water influx constant B:

$$B = 1.119\phi c_t r_e^2 h$$
$$= 1.119(0.1)(8 \times 10^{-6})(2000)^2(200)$$
$$= 716 \text{ bbl/psi}$$

Step 4. Calculate the dimensionless time t_D:

$$t_D = 6.328 \times 10^{-3} \frac{kt}{\phi \mu_w c_t r_e^2}$$

$$= 6.328 \times 10^{-3} \left[\frac{50}{(0.1)(0.395)(8 \times 10^{-6})(2000)^2}\right] t$$

$$= 0.2503t$$

Step 5. Calculate the water influx by using the bottom-water model and edge-water model. Note that the difference between the two models lies in the approach used

TABLE 2.5 Dimensionless Water Influx W_{eD} for $r'_D = 6$

	Z'_D						
t_D	0.05	0.1	0.3	0.5	0.7	0.9	1.0
6	4.780	4.762	4.597	4.285	3.953	3.611	3.414
7	5.309	5.289	5.114	4.779	4.422	4.053	3.837
8	5.799	5.778	5.595	5.256	4.875	4.478	4.247
9	6.252	6.229	6.041	5.712	5.310	4.888	4.642
10	6.750	6.729	6.498	6.135	5.719	5.278	5.019
11	7.137	7.116	6.916	6.548	6.110	5.648	5.378
12	7.569	7.545	7.325	6.945	6.491	6.009	5.728
13	7.967	7.916	7.719	7.329	6.858	6.359	6.067
14	8.357	8.334	8.099	7.699	7.214	6.697	6.395
15	8.734	8.709	8.467	8.057	7.557	7.024	6.713
16	9.093	9.067	8.819	8.398	7.884	7.336	7.017
17	9.442	9.416	9.160	8.730	8.204	7.641	7.315
18	9.775	9.749	9.485	9.047	8.510	7.934	7.601
19	10.09	10.06	9.794	9.443	8.802	8.214	7.874
20	10.40	10.37	10.10	9.646	9.087	8.487	8.142
22	10.99	10.96	10.67	10.21	9.631	9.009	8.653
24	11.53	11.50	11.20	10.73	10.13	9.493	9.130
26	12.06	12.03	11.72	11.23	10.62	9.964	9.594
28	12.52	12.49	12.17	11.68	11.06	10.39	10.01
30	12.95	12.92	12.59	12.09	11.46	10.78	10.40
35	13.96	13.93	13.57	13.06	12.41	11.70	11.32
40	14.69	14.66	14.33	13.84	13.23	12.53	12.15
45	15.27	15.24	14.94	14.48	13.90	13.23	12.87
50	15.74	15.71	15.44	15.01	14.47	13.84	13.49
60	16.40	16.38	16.15	15.81	15.34	14.78	14.47
70	16.87	16.85	16.67	16.38	15.99	15.50	15.24
80	17.20	17.18	17.04	16.80	16.48	16.06	15.83
90	17.43	17.42	17.30	17.10	16.85	16.50	16.29
100	17.58	17.58	17.49	17.34	17.12	16.83	16.66
110	17.71	17.69	17.63	17.50	17.34	17.09	16.93
120	17.78	17.78	17.73	17.63	17.49	17.29	17.17
130	17.84	17.84	17.79	17.73	17.62	17.45	17.34
140	17.88	17.88	17.85	17.79	17.71	17.57	17.48
150	17.92	17.91	17.88	17.84	17.77	17.66	17.58
175	17.95	17.95	17.94	17.92	17.87	17.81	17.76
200	17.97	17.97	17.96	17.95	17.93	17.88	17.86
225	17.97	17.97	17.97	17.96	17.95	17.93	17.91
250	17.98	17.98	17.98	17.97	17.96	17.95	17.95
300	17.98	17.98	17.98	17.98	17.98	17.97	17.97
350	17.98	17.98	17.98	17.98	17.98	17.98	17.98
400	17.98	17.98	17.98	17.98	17.98	17.98	17.98
450	17.98	17.98	17.98	17.98	17.98	17.98	17.98
500	17.98	17.98	17.98	17.98	17.98	17.98	17.98

Permission to publish by the SPE.

TABLE 2.6 Dimensionless Water Influx W_{eD} for $r'_D = 8$

	Z'_D						
t_D	0.05	0.1	0.3	0.5	0.7	0.9	1.0
9	6.301	6.278	6.088	5.756	5.350	4.924	4.675
10	6.828	6.807	6.574	6.205	5.783	5.336	5.072
11	7.250	7.229	7.026	6.650	6.204	5.732	5.456
12	7.725	7.700	7.477	7.086	6.621	6.126	5.836
13	8.173	8.149	7.919	7.515	7.029	6.514	6.210
14	8.619	8.594	8.355	7.937	7.432	6.895	6.578
15	9.058	9.032	8.783	8.351	7.828	7.270	6.940
16	9.485	9.458	9.202	8.755	8.213	7.634	7.293
17	9.907	9.879	9.613	9.153	8.594	7.997	7.642
18	10.32	10.29	10.01	9.537	8.961	8.343	7.979
19	10.72	10.69	10.41	9.920	9.328	8.691	8.315
20	11.12	11.08	10.80	10.30	9.687	9.031	8.645
22	11.89	11.86	11.55	11.02	10.38	9.686	9.280
24	12.63	12.60	12.27	11.72	11.05	10.32	9.896
26	13.36	13.32	12.97	12.40	11.70	10.94	10.49
28	14.06	14.02	13.65	13.06	12.33	11.53	11.07
30	14.73	14.69	14.30	13.68	12.93	12.10	11.62
34	16.01	15.97	15.54	14.88	14.07	13.18	12.67
38	17.21	17.17	16.70	15.99	15.13	14.18	13.65
40	17.80	17.75	17.26	16.52	15.64	14.66	14.12
45	19.15	19.10	18.56	17.76	16.83	15.77	15.21
50	20.42	20.36	19.76	18.91	17.93	16.80	16.24
55	21.46	21.39	20.80	19.96	18.97	17.83	17.24
60	22.40	22.34	21.75	20.91	19.93	18.78	18.19
70	23.97	23.92	23.36	22.55	21.58	20.44	19.86
80	25.29	25.23	24.71	23.94	23.01	21.91	21.32
90	26.39	26.33	25.85	25.12	24.24	23.18	22.61
100	27.30	27.25	26.81	26.13	25.29	24.29	23.74
120	28.61	28.57	28.19	27.63	26.90	26.01	25.51
140	29.55	29.51	29.21	28.74	28.12	27.33	26.90
160	30.23	30.21	29.96	29.57	29.04	28.37	27.99
180	30.73	30.71	30.51	30.18	29.75	29.18	28.84
200	31.07	31.04	30.90	30.63	30.26	29.79	29.51
240	31.50	31.49	31.39	31.22	30.98	30.65	30.45
280	31.72	31.71	31.66	31.56	31.39	31.17	31.03
320	31.85	31.84	31.80	31.74	31.64	31.49	31.39
360	31.90	31.90	31.88	31.85	31.78	31.68	31.61
400	31.94	31.94	31.93	31.90	31.86	31.79	31.75
450	31.96	31.96	31.95	31.94	31.91	31.88	31.85
500	31.97	31.97	31.96	31.96	31.95	31.93	31.90
550	31.97	31.97	31.97	31.96	31.96	31.95	31.94
600	31.97	31.97	31.97	31.97	31.97	31.96	31.95
700	31.97	31.97	31.97	31.97	31.97	31.97	31.97
800	31.97	31.97	31.97	31.97	31.97	31.97	31.97

TABLE 2.7 Dimensionless Water Influx W_{eD} for $r'_D = 10$

	Z'_D						
t_D	0.05	0.1	0.3	0.5	0.7	0.9	1.0
22	12.07	12.04	11.74	11.21	10.56	9.865	9.449
24	12.86	12.83	12.52	11.97	11.29	10.55	10.12
26	13.65	13.62	13.29	12.72	12.01	11.24	10.78
28	14.42	14.39	14.04	13.44	12.70	11.90	11.42
30	15.17	15.13	14.77	14.15	13.38	12.55	12.05
32	15.91	15.87	15.49	14.85	14.05	13.18	12.67
34	16.63	16.59	16.20	15.54	14.71	13.81	13.28
36	17.33	17.29	16.89	16.21	15.35	14.42	13.87
38	18.03	17.99	17.57	16.86	15.98	15.02	14.45
40	18.72	18.68	18.24	17.51	16.60	15.61	15.02
42	19.38	19.33	18.89	18.14	17.21	16.19	15.58
44	20.03	19.99	19.53	18.76	17.80	16.75	16.14
46	20.67	20.62	20.15	19.36	18.38	17.30	16.67
48	21.30	21.25	20.76	19.95	18.95	17.84	17.20
50	21.92	21.87	21.36	20.53	19.51	18.38	17.72
52	22.52	22.47	21.95	21.10	20.05	18.89	18.22
54	23.11	23.06	22.53	21.66	20.59	19.40	18.72
56	23.70	23.64	23.09	22.20	21.11	19.89	19.21
58	24.26	24.21	23.65	22.74	21.63	20.39	19.68
60	24.82	24.77	24.19	23.26	22.13	20.87	20.15
65	26.18	26.12	25.50	24.53	23.34	22.02	21.28
70	27.47	27.41	26.75	25.73	24.50	23.12	22.36
75	28.71	28.55	27.94	26.88	25.60	24.17	23.39
80	29.89	29.82	29.08	27.97	26.65	25.16	24.36
85	31.02	30.95	30.17	29.01	27.65	26.10	25.31
90	32.10	32.03	31.20	30.00	28.60	27.03	26.25
95	33.04	32.96	32.14	30.95	29.54	27.93	27.10
100	33.94	33.85	33.03	31.85	30.44	28.82	27.98
110	35.55	35.46	34.65	33.49	32.08	30.47	29.62
120	36.97	36.90	36.11	34.98	33.58	31.98	31.14
130	38.28	38.19	37.44	36.33	34.96	33.38	32.55
140	39.44	39.37	38.64	37.56	36.23	34.67	33.85
150	40.49	40.42	39.71	38.67	37.38	35.86	35.04
170	42.21	42.15	41.51	40.54	39.33	37.89	37.11
190	43.62	43.55	42.98	42.10	40.97	39.62	38.90
210	44.77	44.72	44.19	43.40	42.36	41.11	40.42
230	45.71	45.67	45.20	44.48	43.54	42.38	41.74
250	46.48	46.44	46.01	45.38	44.53	43.47	42.87
270	47.11	47.06	46.70	46.13	45.36	44.40	43.84
290	47.61	47.58	47.25	46.75	46.07	45.19	44.68
310	48.03	48.00	47.72	47.26	46.66	45.87	45.41
330	48.38	48.35	48.10	47.71	47.16	46.45	46.03
350	48.66	48.64	48.42	48.08	47.59	46.95	46.57
400	49.15	49.14	48.99	48.74	48.38	47.89	47.60
450	49.46	49.45	49.35	49.17	48.91	48.55	48.31
500	49.65	49.64	49.58	49.45	49.26	48.98	48.82
600	49.84	49.84	49.81	49.74	49.65	49.50	49.41
700	49.91	49.91	49.90	49.87	49.82	49.74	49.69

(Continued)

TABLE 2.7 (Continued)

t_D	Z'_D 0.05	0.1	0.3	0.5	0.7	0.9	1.0
800	49.94	49.94	49.93	49.92	49.90	49.85	49.83
900	49.96	49.96	49.94	49.94	49.93	49.91	49.90
1000	49.96	49.96	49.96	49.96	49.94	49.93	49.93
1200	49.96	49.96	49.96	49.96	49.96	49.96	49.96

Data for this example was reported by Cole, Frank *Reservoir Engineering Manual*, Gulf Publishing Company, 1969. Permission to publish by the SPE.

in calculating the dimensionless water influx W_{eD}:

$$W_e = B\sum \Delta p W_{eD}$$

t (days)	t_D	Δp (psi)	Bottom-Water Model W_{eD}	Bottom-Water Model W_e (Mbbl)	Edge-Water Model W_{eD}	Edge-Water Model W_e (Mbbl)
0	0	0	–	–	–	–
30	7.5	22	5.038	79	6.029	95
60	15.0	41.5	8.389	282	9.949	336
90	22.5	39.5	11.414	572	13.459	678
120	30.0	36.5	14.994	933	16.472	1103
150	37.5	33.0	16.994	1353	19.876	1594
180	45.0	26.5	19.641	1810	22.897	2126
210	52.5	19.0	22.214	2284	25.827	2676
240	60.0	18.0	24.728	2782	28.691	3250

Linear-Water Drive. As shown by van Everdingen and Hurst, the water influx from a linear aquifer is proportional to the square root of time. The van Everdingen and Hurst dimensionless water influx is replaced by the square root of time, as given by:

$$W_e = B_L \sum \left[\Delta p_n \sqrt{t - t_n}\right]$$

where

B_L = linear-aquifer water influx constant, bbl/psi/$\sqrt{\text{time}}$
t = time (any convenient time units, e.g., months, years)
Δp = pressure drop as defined previously for the radial edge-water drive

The linear-aquifer water influx constant B_L is determined for the material balance equation as described in Chapter 4.

2.3.5 The Carter and Tracy Water Influx Model

The van Everdingen and Hurst methodology provides the exact solution to the radial diffusivity equation and therefore is considered the correct technique for calculating water influx. However, because superposition of solutions is required, their method involves tedious calculations. To reduce the complexity of water influx calculations, Carter and Tracy (1960) proposed a calculation technique that does not require superposition and allows direct calculation of water influx.

The primary difference between the Carter–Tracy technique and the van Everdingen and Hurst technique is that Carter–Tracy technique assumes constant water influx rates over each finite time interval. Using the Carter–Tracy technique, the cumulative water influx at any time, t_n, can be calculated directly from the previous value obtained at t_{n-1}, or:

$$(W_e)_n = (W_e)_{n-1} + \left[(t_D)_n - (t_D)_{n-1}\right] \times \left[\frac{B\,\Delta p_n - (W_e)_{n-1}(p'_D)_n}{(p_D)_n - (t_D)_{n-1}(p'_D)_n}\right] \quad (2.33)$$

where

B = the van Everdingen and Hurst water influx constant as defined by Eq. (2.23)

t_D = the dimensionless time as defined by Eq. (2.17)
n = the *current* time step
$n-1$ = the *previous* time step
Δp_n = total pressure drop, $p_i - p_n$, psi
p_D = dimensionless pressure
p'_D = dimensionless pressure derivative

Values of the dimensionless pressure p_D as a function of t_D and r_D are tabulated in Chapter 1, Table 1.2. In addition to the curve-fit equations given in Chapter 1 (Eqs. (1.90) through (1.95), Edwardson et al. (1962) developed the following approximation of p_D for an infinite-acting aquifer:

$$p_D = \frac{370.529\sqrt{t_D} + 137.582 t_D + 5.69549(t_D)^{1.5}}{328.834 + 265.488\sqrt{t_D} + 45.2157 t_D + (t_D)^{1.5}} \quad (2.34)$$

The dimensionless pressure derivative can then be approximated by:

$$p'_D = \frac{E}{F} \quad (2.35)$$

where

$$E = 716.441 + 46.7984 (t_D)^{0.5} + 270.038 t_D + 71.0098 (t_D)^{1.5}$$
$$F = 1296.86(t_D)^{0.5} + 1204.73 t_D + 618.618 (t_D)^{1.5} + 538.072(t_D)^2 + 142.41(t_D)^{2.5}$$

When the dimensionless time $t_D > 100$, the following approximation can be used for p_D:

$$p_D = \frac{1}{2}[\ln(t_D) + 0.80907]$$

with the derivative given by:

$$p'_D = \frac{1}{2t_D}$$

Fanchi (1985) matched the van Everdingen and Hurst tabulated values of the dimensionless pressure p_D as a function of t_D and r_D in Table 1.2 by using a regression model and proposed the following expression:

$$p_D = a_0 + a_1 t_D + a_2 \ln(t_D) + a_2[\ln(t_D)]^2$$

in which the regression coefficients are given below:

r_{eD}	a_0	a_1	a_2	a_3
1.5	0.10371	1.6665700	−0.04579	−0.01023
2.0	0.30210	0.6817800	−0.01599	−0.01356
3.0	0.51243	0.2931700	0.015340	−0.06732
4.0	0.63656	0.1610100	0.158120	−0.09104
5.0	0.65106	0.1041400	0.309530	−0.11258
6.0	0.63367	0.0694000	0.41750	−0.11137
8.0	0.40132	0.0410400	0.695920	−0.14350
10.0	0.14386	0.0264900	0.896460	−0.15502
∞	0.82092	−0.000368	0.289080	0.028820

It should be noted that the Carter and Tracy method is not an exact solution to the diffusivity equation and should be considered as an approximation.

Example 2.9

Rework Example 2.7 by using the Carter and Tracy method.

Solution Example 2.7 shows the following preliminary results:

- Water influx constant $B = 20.4$ bbl/psi;
- $t_D = 0.9888t$.

Step 1. For each time step n, calculate the total pressure drop $\Delta p_n = p_i - p_n$ and the corresponding t_D:

n	t_1 (days)	p_n	Δp_n	t_D
0	0	2500	0	0
1	182.5	2490	10	180.5
2	365.0	2472	28	361.0
3	547.5	2444	56	541.5
4	730.0	2408	92	722.0

Step 2. Since the values of t_D are greater than 100, use Eq. (1.91) to calculate p_D and its derivative p'_D. That is:

$$p_D = \frac{1}{2}[\ln(t_D) + 0.80907]$$

$$p'_D = \frac{1}{2t_D}$$

n	t	t_D	p_D	p'_D
0	0	0	–	–
1	182.5	180.5	3.002	2.770×10^{-3}
2	365.0	361.0	3.349	1.385×10^{-3}
3	547.5	541.5	3.552	0.923×10^{-3}
4	730.0	722.0	3.696	0.693×10^{-3}

Step 3. Calculate cumulative water influx by applying Eq. (2.33)

W_e after 182.5 days:

$$(W_e)_n = (W_e)_{n-1} + \left[(t_D)_n - (t_D)_{n-1}\right] \times \left[\frac{B\,\Delta p_n - (W_e)_{n-1}(p'_D)_n}{(p_D)_n - (t_D)_{n-1}(p'_D)_n}\right] = 0 + [180.5 - 0]$$

$$\times \left[\frac{(20.4)(10) - (0)(2.77 \times 10^{-3})}{3.002 - (0)(2.77 \times 10^{-3})}\right]$$

$$= 12{,}266 \text{ bbl}$$

W_e after 365 days:

$$W_e = 12{,}266 + [361 - 180.5]$$

$$\times \left[\frac{(20.4)(28) - (12{,}266)(1.385 \times 10^{-3})}{3.349 - (180.5)(1.385 \times 10^{-3})}\right]$$

$$= 42{,}545 \text{ bbl}$$

W_e after 547.5 days:

$$W_e = 42{,}546 + [541.5 - 361]$$

$$\times \left[\frac{(20.4)(56) - (42{,}546)(0.923 \times 10^{-3})}{3.552 - (361)(0.923 \times 10^{-3})}\right]$$

$$= 104{,}406 \text{ bbl}$$

W_e after 720 days:

$$W_e = 104{,}406 + [722 - 541.5]$$

$$\times \left[\frac{(20.4)(92) - (104{,}406)(0.693 \times 10^{-3})}{3.696 - (541.5)(0.693 \times 10^{-3})}\right]$$

$$= 202{,}477 \text{ bbl}$$

The following table compares the results of the Carter and Tracy water influx calculations with those of the van Everdingen and Hurst method:

Time (months)	Carter and Tracy, W_e (bbl)	van Everdingen and Hurst, W_e (bbl)
0	0	0
6	12,266	7085
12	42,546	32,435
18	104,400	85,277
24	202,477	175,522

The above comparison indicates that the Carter and Tracy method considerably overestimates the water influx. However, this is due to

Time (months)	Time (days)	p (psi)	Δp (psi)	t_D	p_D	p'_D	Carter–Tracy W_e (bbl)	van Everdingen and Hurst W_e (bbl)
0	0	2500.0	0.00	0	0.00	0	0.0	0
1	30	2498.9	1.06	30.0892	2.11	0.01661	308.8	
2	61	2497.7	2.31	60.1784	2.45	0.00831	918.3	
3	91	2496.2	3.81	90.2676	2.66	0.00554	1860.3	
4	122	2494.4	5.56	120.357	2.80	0.00415	3171.7	
5	152	2492.4	7.55	150.446	2.91	0.00332	4891.2	
6	183	2490.2	9.79	180.535	3.00	0.00277	7057.3	7088.9
7	213	2487.7	12.27	210.624	3.08	0.00237	9709.0	
8	243	2485.0	15.00	240.713	3.15	0.00208	12,884.7	
9	274	2482.0	17.98	270.802	3.21	0.00185	16,622.8	
10	304	2478.8	21.20	300.891	3.26	0.00166	20,961.5	
11	335	2475.3	24.67	330.981	3.31	0.00151	25,938.5	
12	365	2471.6	28.38	361.070	3.35	0.00139	31,591.5	32,435.0
13	396	2467.7	32.34	391.159	3.39	0.00128	37,957.8	
14	426	2463.5	36.55	421.248	3.43	0.00119	45,074.5	
15	456	2459.0	41.00	451.337	3.46	0.00111	52,978.6	
16	487	2454.3	45.70	481.426	3.49	0.00104	61,706.7	

17	517	2449.4	50.64	511.516	3.52	0.00098	71,295.3	
18	547	2444.3	55.74	541.071	3.55	0.00092	81,578.8	85,277.0
19	578	2438.8	61.16	571.130	3.58	0.00088	92,968.2	
20	608	2433.2	66.84	601.190	3.60	0.00083	105,323.0	
21	638	2427.2	72.75	631.249	3.63	0.00079	118,681.0	
22	669	2421.1	78.92	661.309	3.65	0.00076	133,076.0	
23	699	2414.7	85.32	691.369	3.67	0.00072	148,544.0	
24	730	2408.0	91.98	721.428	3.70	0.00069	165,119.0	175,522.0

the fact that a large time step of 6 months was used in the Carter and Tracy method to determine the water influx. The accuracy of this method can be increased substantially by restricting the time step to 1 month. Recalculating the water influx on a monthly basis produces an excellent match with the van Everdingen and Hurst method as shown above.

2.3.6 The Fetkovich Method

Fetkovich (1971) developed a method of describing the approximate water influx behavior of a finite aquifer for radial and linear geometries. In many cases, the results of this model closely match those determined using the van Everdingen and Hurst approach. The Fetkovich theory is much simpler, and, like the Carter–Tracy technique, this method does not require the use of superposition. Hence, the application is much easier, and this method is also often utilized in numerical simulation models.

The Fetkovich model is based on the premise that the productivity index concept will adequately describe water influx from a finite aquifer into a hydrocarbon reservoir. That is, the water influx rate is directly proportional to the pressure drop between the average aquifer pressure and the pressure at the reservoir–aquifer boundary. The method neglects the effects of any transient period. Thus, in cases where pressures are changing rapidly at the aquifer–reservoir interface, predicted results may differ somewhat from the more rigorous van Everdingen and Hurst or Carter–Tracy approaches. However, in many cases pressure changes at the waterfront are gradual and this method offers an excellent approximation to the two methods discussed above.

This approach begins with two simple equations. The first is the productivity index (PI) equation for the aquifer, which is analogous to the PI equation used to describe an oil or gas well:

$$e_w = \frac{dW_e}{dt} = J(\overline{p}_a - p_r) \qquad (2.36)$$

where

e_w = water influx rate from aquifer, bbl/day
J = productivity index for the aquifer, bbl/day/psi
$\overline{p}_a$ = average aquifer pressure, psi
p_r = inner aquifer boundary pressure, psi

The second equation is an aquifer material balance equation for a constant compressibility, which states that the amount of pressure depletion in the aquifer is directly proportional to the amount of water influx from the aquifer, or:

$$W_e = c_t W_i(p_i - \overline{p}_a)f \qquad (2.37)$$

where

W_i = initial volume of water in the aquifer, bbl
c_t = total aquifer compressibility, $c_w + c_f$, psi^{-1}
p_i = initial pressure of the aquifer, psi
$f = \theta/360$

Eq. (2.27) suggests that the *maximum* possible water influx occurs if $\overline{p}_a = 0$, or:

$$W_{ei} = c_t W_i p_i f \qquad (2.38)$$

where

W_{ei} = maximum water influx, bbl

Combining Eq. (2.38) with (2.37) gives:

$$\overline{p}_a = p_i\left(1 - \frac{W_e}{c_t W_i p_i}\right) = p_i\left(1 - \frac{W_e}{W_{ei}}\right) \tag{2.39}$$

Eq. (2.37) provides a simple expression to determine the average aquifer pressure $\overline{p}_a$ after removing W_e bbl of water from the aquifer to the reservoir, i.e., cumulative water influx.

Differentiating Eq. (2.39) with respect to time gives:

$$\frac{dW_e}{dt} = -\frac{W_{ei}}{p_i}\frac{d\overline{p}_a}{dt} \tag{2.40}$$

Fetkovich combined Eq. (2.40) with (2.36) and integrated to give the following form:

$$W_e = \frac{W_{ei}}{p_i}(p_i - p_r)\exp\left(\frac{-Jp_i t}{W_{ei}}\right) \tag{2.41}$$

where

W_e = cumulative water influx, bbl
p_r = reservoir pressure, i.e., pressure at the oil or gas–water contact
t = time, days

Eq. (2.41) has no practical applications since it was derived for a constant inner boundary pressure. To use this solution in the case in which the boundary pressure is varying continuously as a function of time, the superposition technique must be applied. Rather than using superposition, Fetkovich suggested that, if the reservoir–aquifer boundary pressure history is divided into a finite number of time intervals, the incremental water influx during the nth interval is:

$$(\Delta W_e)_n = \frac{W_{ei}}{p_i}\left[(\overline{p}_a)_{n-1} - (\overline{p}_r)_n\right]\left[1 - \exp\left(-\frac{Jp_i\,\Delta t_n}{W_{ei}}\right)\right] \tag{2.42}$$

where $(\overline{p}_a)_{n-1}$ is the average aquifer pressure at the end of the previous time step. This average pressure is calculated from Eq. (2.39) as:

$$(\overline{p}_a)_{n-1} = p_i\left(1 - \frac{(W_e)_{n-1}}{W_{ei}}\right) \tag{2.43}$$

The average reservoir boundary pressure $(\overline{p}_r)_n$ is estimated from:

$$(\overline{p}_r)_n = \frac{(p_r)_n + (p_r)_{n-1}}{2} \tag{2.44}$$

The productivity index J used in the calculation is a function of the geometry of the aquifer. Fetkovich calculated the productivity index from Darcy's equation for bounded aquifers. Lee and Wattenbarger (1996) pointed out that the Fetkovich method can be extended to infinite-acting aquifers by requiring that the ratio of water influx rate to pressure drop is approximately constant throughout the productive life of the reservoir. The productivity index J of the aquifer is given by the following expressions:

Type of Outer Aquifer Boundary	J for Radial Flow (bbl/day/psi)	J for Linear Flow (bbl/day/psi)	
Finite, no flow	$J = \frac{0.00708khf}{\mu_w[\ln(r_D) - 0.75]}$	$J = \frac{0.003381kwh}{\mu_w L}$	(2.45)
Finite, constant pressure	$J = \frac{0.00708khf}{\mu_w[\ln(r_D)]}$	$J = \frac{0.001127kwh}{\mu_w L}$	(2.46)
Infinite	$J = \frac{0.00708khf}{\mu_w \ln(a/r_e)}$ $a = \sqrt{0.0142kt/(\phi\mu_w c_t)}$	$J = \frac{0.001kwh}{\mu_w\sqrt{0.0633kt/(\phi\mu_w c_t)}}$	(2.47)

where

w = width of the linear aquifer, ft
L = length of the linear aquifer, ft
r_D = dimensionless radius, r_a/r_e
k = permeability of the aquifer, md
t = time, days
θ = encroachment angle
h = thickness of the aquifer
$f = \theta/360$

The following steps describe the methodology of using the Fetkovich model in predicting the cumulative water influx:

Step 1. Calculate the initial volume of water in the aquifer from:

$$W_i = \frac{\pi}{5.615}(r_a^2 - r_e^2)h\phi$$

Step 2. Calculate the maximum possible water influx W_{ei} by applying Eq. (2.38), or:

$$W_{ei} = c_t W_i p_i f$$

Step 3. Calculate the productivity index J based on the boundary conditions and aquifer geometry.

Step 4. Calculate the incremental water influx $(\Delta W_e)_n$ from the aquifer during the nth time interval by using Eq. (2.42). For example, during the first time step Δt_1:

$$(\Delta W_e)_1 = \frac{W_{ei}}{p_i}\left[p_i - (\overline{p}_r)_1\right]\left[1 - \exp\left(\frac{-Jp_i\,\Delta t_1}{W_{ei}}\right)\right]$$

with

$$(\overline{p}_r)_1 = \frac{p_i + (p_r)_1}{2}$$

For the second time interval Δt_2:

$$(\Delta W_e)_2 = \frac{W_{ei}}{p_i}\left[(\overline{p}_a) - (\overline{p}_r)_2\right]\left[1 - \exp\left(\frac{-Jp_i\,\Delta t_2}{W_{ei}}\right)\right]$$

where $(\overline{p}_a)_1$ is the average aquifer pressure at the end of the first period and removing $(\Delta W_e)_1$ barrels of water from the aquifer to the reservoir. From Eq. (2.43):

$$(\overline{p}_a)_1 = p_i\left(1 - \frac{(\Delta W_e)_1}{W_{ei}}\right)$$

Step 5. Calculate the cumulative (total) water influx at the end of any time period from:

$$W_e = \sum_{i=1}^{n}(\Delta W_e)_i$$

Example 2.10

Using the Fetkovich method, calculate the water influx as a function of time for the following reservoir–aquifer and boundary pressure data[2]:

$p_i = 2740$ psi, $h = 100$ ft, $c_t = 7\times10^{-6}$ psi^{-1}
$\mu_w = 0.55$ cp, $k = 200$ md, $\theta = 140°$
Reservoir area = 40,363 acres
Aquifer area = 1,000,000 acres

Time (days)	p_r (psi)
0	2740
365	2500
730	2290
1095	2109
1460	1949

Figure 2.17[3] shows the wedge reservoir–aquifer system with an encroachment angle of 140°.

Solution

Step 1. Calculate the reservoir radius r_e:

$$r_e = \left(\frac{\theta}{360}\right)\sqrt{\frac{43{,}560A}{\pi}} = 9200 \text{ ft}$$

$$= \left(\frac{140}{360}\right)\sqrt{\frac{(43{,}560)(2374)}{\pi}} = 9200 \text{ ft}$$

Step 2. Calculate the equivalent aquifer radius r_a:

$$r_a = \left(\frac{140}{360}\right)\sqrt{\frac{(43{,}560)(1{,}000{,}000)}{\pi}} = 46{,}000 \text{ ft}$$

[2]Data for this example is given by Dake, L.P., 1978. *Fundamentals of Reservoir Engineering*, Elsevier, Amsterdam.
[3]Data for this example is given by Dake, L.P., 1978. *Fundamentals of Reservoir Engineering*, Elsevier, Amsterdam.

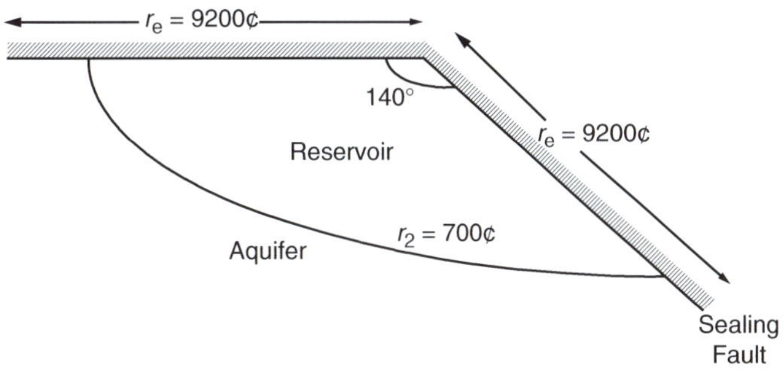

FIGURE 2.17 Aquifer–reservoir geometry for Example 2.10.

Step 3. Calculate the dimensionless radius r_D:

$$r_D = \frac{r_a}{r_e} = \frac{46,000}{9200} = 5$$

Step 4. Calculate initial water-in-place W_i:

$$W_i = \frac{\pi(r_a^2 - r_e^2)h\theta}{5.615}$$
$$= \frac{\pi(46,000^2 - 9200^2)(100)(0.25)}{5.615}$$
$$= 28.41 \text{ MMMbbl}$$

Step 5. Calculate W_{ei} from Eq. (2.38):

$$W_{ei} = c_t W_i p_i f$$
$$= 7 \times 10^{-6}(28.41 \times 10^9)(2740)\left(\frac{140}{360}\right)$$
$$= 211.9 \text{ MMMbbl}$$

Step 6. Calculate the productivity index J of the radial aquifer from Eq. (2.45)

$$J = \frac{0.00708(200)(100)(140/360)}{0.55\ln(5)} = 116.bbl/day/psi$$

and therefore:

$$\frac{Jp_i}{W_{ei}} = \frac{(116.5)(2740)}{211.9 \times 10^6} = 1.506 \times 10^{-3}$$

Since the time step Δt is fixed at 365 days, then:

$$1 - \exp^{(-Jp_i\,\Delta t/W_{ei})} = 1 - \exp^{(-1.506 \times 10^{-3} \times 365)} = 0.4229$$

Substituting in Eq. (2.43) gives:

$$(\Delta W_e)_n = \frac{W_{ei}}{p_i}\left[(\overline{p}_a)_{n-1} - (\overline{p}_r)_n\right] \times \left[1 - \exp\left(-\frac{Jp_i\,\Delta t_n}{W_{ei}}\right)\right]$$
$$= \frac{211.9 \times 10^6}{2740}\left[(\overline{p}_a)_{n-1} - (\overline{p}_r)_n\right](0.4229)$$
$$= 32,705\left[(\overline{p}_a)_{n-1} - (\overline{p}_r)_n\right]$$

Step 7. Calculate the cumulative water influx as shown in the following table:

n	t (days)	p_r	$(\overline{p}_r)_n$	$(\overline{p}_a)_{n-1}$	$(\overline{p}_a)_{n-1} - (\overline{p}_r)_n$	$(\Delta W_e)_n$ (MMbbl)	W_e (MMbbl)
0	0	2740	2740	–	–	–	–
1	365	2500	2620	2740	120	3.925	3.925
2	730	2290	2395	2689	294	9.615	13.540
3	1095	2109	2199	2565	366	11.970	25.510
4	1460	1949	2029	2409	381	12.461	37.971

Problems

(1) Calculate the cumulative water influx that results from a pressure drop of 200 psi at the oil–water contact with an encroachment angle of 50°. The reservoir–aquifer system is characterized by the following properties:

	Reservoir	Aquifer
Radius, ft	6000	20,000
Porosity	0.18	0.15
c_f, psi^{-1}	4×10^{-6}	3×10^{-6}
c_w, psi^{-1}	5×10^{-6}	4×10^{-6}
h, ft	25	20

(2) An active water drive oil reservoir is producing under steady-state flowing conditions. The following data is available:

$p_i = 4000$ *psi*, $p = 3000$ psi
$Q_o = 40,000$ *STB/day*, $B_o = 1.3$ *bbl/STB*
GOR $= 700$ *scf/STB*, $R_s = 500$ *scf/STB*
$Z = 0.82$, $T = 140$ °F
$Q_w = 0$, $B_w = 1.0$ *bbl/STB*

Calculate the Schilthuis water influx constant.

(3) The pressure history of a water drive oil reservoir is given below:

t (days)	*p* (psi)
0	4000
120	3950
220	3910
320	3880
420	3840

The aquifer is under a steady-state flowing condition with an estimated water influx constant of 80 bbl/day/psi. Using the steady-state model, calculate and plot the cumulative water influx as a function of time.

(4) A water drive reservoir has the following boundary pressure history:

Time (months)	Boundary Pressure (psi)
0	2610
6	2600
12	2580
18	2552
24	2515

The aquifer–reservoir system is characterized by the following data:

	Reservoir	Aquifer
Radius, ft	2000	∞
h, ft	25	30
k, md	60	80
ϕ, %	17	18
μ_w, cp	0.55	0.85
c_w, psi^{-1}	0.7×0^{-6}	0.8×10^{-6}
c_f, psi^{-1}	0.2×0^{-6}	0.3×10^{-6}

If the encroachment angle is 360°, calculate the water influx as a function of time by using:

(a) the van Everdingen and Hurst method;
(b) the Carter and Tracy Method.

(5) The following table summarizes the original data available on the West Texas water drive reservoir:

	Oil Zone	Aquifer
Geometry	Circular	Semicircular
Area, acres	640	Infinite
Initial reservoir pressure, psia	4000	4000
Initial oil saturation	0.80	0
Porosity, %	22	–
B_{oi}, bbl/STB	1.36	–
B_{wi}, bbl/STB	1.00	1.05
c_o, psi	6×10^{-6}	–
c_w, psi^{-1}	3×10^{-6}	7×10^{-6}

The geological data of the aquifer estimates the water influx constant at 551 bbl/psi. After 1120 days of production, the reservoir average pressure has dropped to

3800 psi and the field has produced 860,000 STB of oil.

The field condition after 1120 days of production is given below:

$p = 3800\ psi$, $N_p = 860,000\ STB$,
$B_o = 1.34\ bbl/STB$, $B_w = 1.05\ bbl/STB$,
$W_e = 991,000$ bbl
$t_D = 32.99$ (*dimensionless* time after 1120 *days*),
$W_p = 0$ bbl

It is expected that the average reservoir pressure will drop to 3400 psi after 1520 days (i.e., from the start of production). Calculate the cumulative water influx after 1520 days.

(6) A wedge reservoir–aquifer system with an encroachment angle of 60° has the following boundary pressure history:

Time (days)	Boundary Pressure (psi)
0	2850
365	2610
730	2400
1095	2220
1460	2060

Given the following aquifer data:

$h = 120$ ft, $c_f = 5 \times 10^{-6}\ psi^{-1}$
$c_w = 4 \times 10^{-6}\ psi^{-1}$, $\mu_w = 0.7$ cp
$k = 60$ md, $\phi = 12\%$
Reservoir area = 40,000 acres
Aquifer area = 980,000 acres, $T = 140\ °F$

calculate the cumulative influx as a function of time by using:

(a) the van Everdingen and Hurst method;
(b) the Carter and Tracy method;
the Fetkovich method.

CHAPTER

3

Unconventional Gas Reservoirs

Efficient development and operation of a natural gas reservoir depend on understanding the reservoir characteristics and the well performance. Predicting the future recovery of the reservoir and the producing wells is the most important part in the economic analysis of the field for further development and expenditures. To forecast the performance of a gas field and its existing production wells, sources of energy for producing the hydrocarbon system must be identified and their contributions to reservoir behavior must be evaluated.

The objective of this chapter is to document the methods that can be used to evaluate and predict:

- vertical and horizontal gas well performance;
- conventional and nonconventional gas field performance.

3.1 VERTICAL GAS WELL PERFORMANCE

Determination of the flow capacity of a gas well requires a relationship between the inflow gas rate and the sand face pressure or flowing bottom-hole pressure. This inflow performance relationship (IPR) may be established by the proper solution of Darcy's equation. Solution of Darcy's law depends on the conditions of the flow existing in the reservoir or the flow regime.

When a gas well is first produced after being shut in for a period of time, the gas flow in the reservoir follows an unsteady-state behavior until the pressure drops at the drainage boundary of the well. Then the flow behavior passes through a short transition period, after which it attains a steady-state or semisteady (pseudosteady)-state condition. The objective of this chapter is to describe the empirical as well as analytical expressions that can be used to establish the IPRs under the pseudosteady-state flow condition.

3.1.1 Gas Flow under Laminar (Viscous) Flowing Conditions

The exact solution to the differential form of Darcy's equation for compressible fluids under the pseudosteady-state flow condition was given previously by Eq. (1.149), as:

$$Q_g = \frac{kh[\overline{\psi}_r - \psi_{wf}]}{1422T[\ln(r_e/r_w) - 0.75 + s]} \tag{3.1}$$

with

$$\overline{\psi}_r = m(\overline{p}_r) = 2\int_0^{\overline{p}_r} \frac{p}{\mu Z}\,dp$$

$$\psi_{wf} = m(p_{wf}) = 2\int_0^{p_{wf}} \frac{p}{\mu Z}\,dp$$

where

Q_g = gas flow rate, Mscf/day
k = permeability, md

$m(\overline{p}_r) = \overline{\psi}_r$ = average reservoir real-gas pseudo-pressure, psi^2/cp
T = temperature, °R
s = skin factor
h = thickness
r_e = drainage radius
r_w = wellbore radius

Note that the shape factor C_A, which is designed to account for the deviation of the drainage area from the ideal circular form as introduced in Chapter 1 and given in Table 1.4, can be included in Darcy's equation to give:

$$Q_g = \frac{kh[\overline{\psi}_r - \psi_{wf}]}{1422T\left[\frac{1}{2}\ln(4A/1.781C_A r_w^2) + s\right]}$$

with

$$A = \pi r_e^2$$

where

A = drainage area, ft^2
C_A = shape factor with values as given in Table 1.4

For example, a circular drainage area has a shape factor of 31.62, i.e., $C_A = 31.62$, as shown in Table 1.4, and reduces the above equation into Eq. (3.1).

The productivity index J for a gas well can be written analogously to that for oil wells with the definition as the production rate per unit pressure drop. That is:

$$J = \frac{Q_g}{[\overline{\psi}_r - \psi_{wf}]} = \frac{kh}{1422T\left[\frac{1}{2}\ln(4A/1.781C_A r_w^2) + s\right]}$$

For the most commonly used flow geometry, i.e., a circular drainage area, the above equation is reduced to:

$$J = \frac{Q_g}{\overline{\psi}_r - \psi_{wf}} = \frac{kh}{1422T[\ln(r_e/r_w) - 0.75 + s]} \tag{3.2}$$

or

$$Q_g = J(\overline{\psi}_r - \psi_{wf}) \tag{3.3}$$

With the absolute open flow potential (AOF), i.e., maximum gas flow rate $(Q_g)_{max}$, as calculated by setting $\psi_{wf} = 0$, then:

$$\text{AOF} = (Q_g)_{max} = J(\overline{\psi}_r - 0)$$

or

$$\text{AOF} = (Q_g)_{max} = J\overline{\psi}_r \tag{3.4}$$

where

J = productivity index, Mscf/day/psi^2/cp
$(Q_g)_{max}$ = maximum gas flow rate, Mscf/day
AOF = absolute open flow potential, Mscf/day

Eq. (3.3) can be expressed in a linear relationship as:

$$\psi_{wf} = \overline{\psi}_r - \left(\frac{1}{J}\right)Q_g \tag{3.5}$$

Eq. (3.5) indicates that a plot of ψ_{wf} vs. Q_g would produce a straight line with a slope of 1/J and intercept of $\overline{\psi}_r$, as shown in Figure 3.1. If two different stabilized flow rates are available, the line can be extrapolated and the slope is determined to estimate AOF, J, and $\overline{\psi}_r$.

Eq. (3.1) can be written alternatively in the following integral form:

$$Q_g = \frac{kh}{1422T[\ln(r_e/r_w) - 0.75 + s]}\int_{p_{wf}}^{\overline{p}_r}\left(\frac{2p}{\mu_g Z}\right)dp \tag{3.6}$$

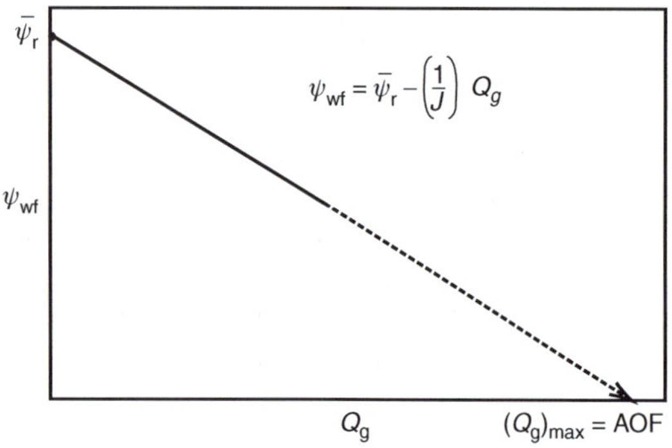

FIGURE 3.1 Steady-state gas well flow.

Note that $(p/\mu_g Z)$ is directly proportional to $(1/\mu_g B_g)$, where B_g is the gas formation volume factor and defined as:

$$B_g = 0.00504 \frac{ZT}{p} \tag{3.7}$$

where

B_g = gas formation volume factor, bbl/scf
Z = gas compressibility factor
T = temperature, °R

Eq. (3.6) can then be written in terms of B_g of Eq. (3.7), as follows. Arrange Eq. (3.6) to give:

$$\frac{p}{ZT} = \frac{0.00504}{B_g}$$

Arrange Eq. (3.7) in the following form:

$$Q_g = \frac{kh}{1422[\ln(r_e/r_w) - 0.75 + s]} \int_{p_{wr}}^{\bar{p}_r} \left(\frac{2}{\mu_g} \frac{p}{TZ}\right) dp$$

Combining the above two expressions:

$$Q_g = \left[\frac{7.08(10^{-6})kh}{\ln(r_e/r_w) - 0.75 + s}\right] \int_{p_{wf}}^{\bar{p}_r} \left(\frac{1}{\mu_g B_g}\right) dp \tag{3.8}$$

where

Q_g = gas flow rate, Mscf/day
μ_g = gas viscosity, cp
k = permeability, md

Figure 3.2 shows a typical plot of the gas pressure functions $(2p/\mu Z)$ and $(1/\mu_g B_g)$ vs. pressure. The integral in Eqs. (3.6) and (3.8) represents the area under the curve between $\bar{p}_r$ and p_{wf}. As illustrated in Figure 3.2, the pressure function exhibits the following three distinct pressure application regions.

High-Pressure Region. When the bottom-hole flowing pressure p_{wf} and average reservoir pressure $\bar{p}_r$ are both higher than 3000 psi, the pressure functions $(2p/\mu_g Z)$ and $(1/\mu_g B_g)$ are nearly constant, as shown by Region III in Figure 3.2. This observation suggests that the pressure term $(1/\mu_g B_g)$ in Eq. (3.8) can be treated as a constant and can be removed outside the integral, to give:

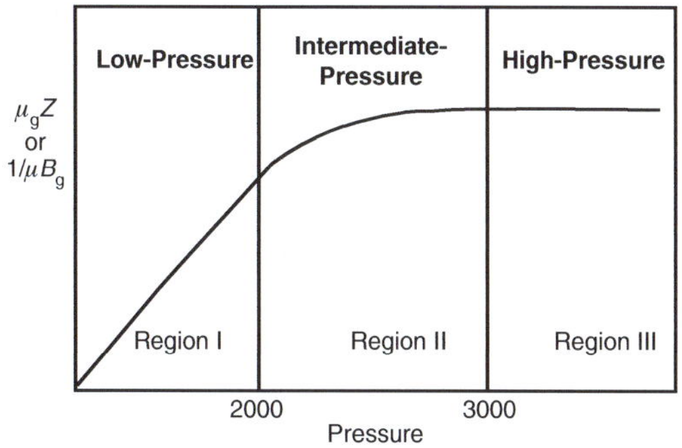

FIGURE 3.2 Gas PVT data.

$$Q_g = \left[\frac{7.08(10^{-6})kh}{\ln(r_e/r_w) - 0.75 + s}\right] \left(\frac{1}{\mu_g B_g}\right) \int_{p_{wf}}^{\bar{p}_r} dp$$

or

$$Q_g = \frac{7.08(10^{-6})kh(\bar{p}_r - p_{wf})}{(\mu_g B_g)_{avg}[\ln(r_e/r_w) - 0.75 + s]} \tag{3.9}$$

where

Q_g = gas flow rate, Mscf/day
B_g = gas formation volume factor, bbl/scf
k = permeability, md

The gas viscosity μ_g and formation volume factor B_g should be evaluated at the average pressure p_{avg} as given by:

$$p_{avg} = \frac{\bar{p}_r + p_{wf}}{2} \tag{3.10}$$

The method of determining the gas flow rate by using Eq. (3.9) is commonly called the "pressure approximation method."

It should be pointed out that the concept of the productivity index J cannot be introduced into Eq. (3.9) since this equation is only valid for applications when both p_{wf} and $\bar{p}_r$ are more than 3000 psi.

Note that deviation from the circular drainage area can be treated as an additional skin by including the shape factor C_A in Eq. (3.9), to give:

$$Q_g = \frac{7.08(10^{-6})kh(\overline{p}_r - p_{wf})}{(\mu_g B_g)_{avg}\left[\frac{1}{2}\ln(4A/1.781\,C_A r_w^2)\right] + s]}$$

Intermediate-Pressure Region. Between 2000 and 3000 psi, the pressure function shows distinct curvature. When the bottom-hole flowing pressure and average reservoir pressure are both between 2000 and 3000 psi, the pseudopressure gas pressure approach (i.e., Eq. (3.1)) should be used to calculate the gas flow rate:

$$Q_g = \frac{kh[\overline{\psi}_r - \psi_{wf}]}{1422T[\ln(r_e/r_w) - 0.75 + s]}$$

and for a noncircular drainage area, the above flow should be modified to include the shape factor C_A and the drainage area, to give:

$$Q_g = \frac{kh[\overline{\psi}_r - \psi_{wf}]}{1422T\left[\frac{1}{2}\ln(4A/1.781\,C_A r_w^2) + s\right]}$$

Low-Pressure Region. At low pressures, usually less than 2000 psi, the pressure functions $(2p/\mu Z)$ and $(1/\mu_g B_g)$ exhibit a linear relationship with pressure as shown in Figure 3.2 and is identified as Region I. Golan and Whitson (1986) indicated that the product $(\mu_g Z)$ is essentially constant when evaluating any pressure below 2000 psi. Implementing this observation in Eq. (3.6) and integrating gives:

$$Q_g = \frac{kh}{1422T[\ln(r_e/r_w) - 0.75 + s]}\left(\frac{2}{\mu_g Z}\right)\int_{p_{wf}}^{\overline{p}_r} p\,dp$$

or

$$Q_g = \frac{kh(\overline{p}_r^2 - p_{wf}^2)}{1422T(\mu_g Z)_{avg}[\ln(r_e/r_w) - 0.75 + s]} \quad (3.11)$$

and for a noncircular drainage area:

$$Q_g = \frac{kh(\overline{p}_r^2 - p_{wf}^2)}{1422T(\mu_g Z)_{avg}\left[\frac{1}{2}\ln(4A/1.781\,C_A r_w^2) + s\right]}$$

where

Q_g = gas flow rate, Mscf/day
k = permeability, md
T = temperature, °R
Z = gas compressibility factor
μ_g = gas viscosity, cp

It is recommended that the Z factor and gas viscosity be evaluated at the average pressure p_{avg} as defined by:

$$p_{avg} = \sqrt{\frac{\overline{p}_r^2 + p_{wf}^2}{2}}$$

It should be pointed out that, for the remainder of this chapter, it will be assumed that the well is draining a circular area with a shape factor of 31.16.

The method of calculating the gas flow rate by Eq. (3.11) is called the "pressure-squared approximation method."

If both $\overline{p}_r$ and p_{wf} are lower than 2000 psi, Eq. (3.11) can be expressed in terms of the productivity index J as:

$$Q_g = J(\overline{p}_r^2 - p_{wf}^2) \quad (3.12)$$

with

$$(Q_g)_{max} = AOF = J\overline{p}_r^2 \quad (3.13)$$

where

$$J = \frac{kh}{1422T(\mu_g Z)_{avg}[\ln(r_e/r_w) - 0.75 + s]} \quad (3.14)$$

Example 3.1

The *PVT* properties of a gas sample taken from a dry gas reservoir are given below:

p (psi)	μ_g (cp)	Z	ψ (psi²/cp)	B_g (bbl/scf)
0	0.01270	1.000	0	–
400	0.01286	0.937	13.2×10^6	0.007080
1200	0.01530	0.832	113.1×10^6	0.002100
1600	0.01680	0.794	198.0×10^6	0.001500
2000	0.01840	0.770	304.0×10^6	0.001160
3200	0.02340	0.797	678.0×10^6	0.000750
3600	0.02500	0.827	816.0×10^6	0.000695
4000	0.02660	0.860	950.0×10^6	0.000650

The reservoir is producing under the pseudosteady-state condition. The following additional data is available:

$k = 65$ md, $h = 15$ ft, $T = 600\ ^\circ R$
$r_e = 1000$ ft, $r_w = 0.25$ ft, $s = -0.4$

Calculate the gas flow rate under the following conditions:

(a) $\overline{p}_r = 4000$ psi, $p_{wf} = 3200$ psi;
(b) $\overline{p}_r = 2000$ psi, $p_{wf} = 1200$ psi.

Use the appropriate approximation methods and compare results with the exact solution.

Solution

(a) Calculation of Q_g at $\overline{p}_r = 4000$ psi and $p_{wf} = 3200$ psi:

Step 1. Select the approximation method. Because $\overline{p}_r$ and p_{wf} are both greater than 3000, the pressure approximation method is used, i.e., Eq. (3.9).

Step 2. Calculate average pressure and determine the corresponding gas properties.

$$\overline{p} = \frac{4000 + 3200}{2} = 3600 \text{ psi}$$

$$\mu_g = 0.025, \quad B_g = 0.000695$$

Step 3. Calculate the gas flow rate by applying Eq. (3.9):

$$Q_g = \frac{7.08(10^{-6})kh(\overline{p}_r - p_{wf})}{(\mu_g B_g)_{avg}[\ln(r_e/r_w) - 0.75 + s]}$$
$$= \frac{7.08(10^{-6})(65)(15)(4000 - 3200)}{(0.025)(0.000695)[\ln(1000/0.25) - 0.75 - 0.4]}$$
$$= 44{,}490 \text{ Mscf/day}$$

Step 4. Recalculate Q_g by using the pseudopressure equation, i.e., Eq. (3.1), to give:

$$Q_g = \frac{kh[\overline{\psi}_r - \psi_{wf}]}{1422T[\ln(r_e/r_w) - 0.75 + s]}$$
$$= \frac{(65)(15)(950.0 - 678.0)10^6}{(1422)(600)[\ln(1000/0.25) - 0.75 - 0.4]}$$
$$= 43{,}509 \text{ Mscf/day}$$

Comparing results of the pressure approximation method with the pseudopressure approach indicates that the gas flow rate can be approximated using the "pressure method" with an absolute percentage error of 2.25%.

(b) Calculation of Q_g at $\overline{p}_r = 2000$ and $p_{wf} = 1200$:

Step 1. Select the appropriate approximation method. Because $\overline{p}_r$ and $p_{wf} \leq 2000$, use the pressure-squared approximation.

Step 2. Calculate average pressure and the corresponding μ_g and Z:

$$\overline{p} = \sqrt{\frac{2000^2 + 1200^2}{2}} = 1649 \text{ psi}$$

$$\mu_g = 0.017, \quad Z = 0.791$$

Step 3. Calculate Q_g by using the pressure-squared equation, i.e., Eq. (3.11):

$$Q_g = \frac{kh(\overline{p}_r^{\,2} - p_{wf}^2)}{1422T(\mu_g Z)_{avg}[\ln(r_e/r_w) - 0.75 + s]}$$
$$= \frac{(65)(15)(2000^2 - 1200^2)}{1422(600)(0.017)(0.791)[\ln(1000/0.25) - 0.75 - 0.4]}$$
$$= 30{,}453 \text{ Mscf/day}$$

Step 4. Using the tabulated values of real-gas pseudopressure, calculate the exact Q_g by applying Eq. (3.1):

$$Q_g = \frac{kh[\overline{\psi}_r - \psi_{wf}]}{1422T[\ln(r_e/r_w) - 0.75 + s]}$$
$$= \frac{(65)(15)(304.0 - 113.1)10^6}{(1422)(600)[\ln(1000/0.25) - 0.75 - 0.4]}$$
$$= 30{,}536 \text{ Mscf/day}$$

Comparing results of the two methods, the pressure-squared approximation predicted the gas flow rate with an average absolute error of 0.27%.

3.1.2 Gas Flow under Turbulent Flow Conditions

All of the mathematical formulations presented thus far in this chapter are based on the

assumption that laminar (viscous) flow conditions are observed during the gas flow. During radial flow, the flow velocity increases as the wellbore is approached. This increase of the gas velocity might cause the development of a turbulent flow around the wellbore. If turbulent flow does exist, it causes an additional pressure drop similar to that caused by the mechanical skin effect.

As presented in Chapter 1 by Eqs. (1.163)–(1.165), the semisteady-state flow equation for compressible fluids can be modified to account for the additional pressure drop due to the turbulent flow by *including the rate-dependent skin factor* DQ_g, where the term D is called the turbulent flow factor. The resulting pseudosteady-state equations are given in the following three forms:

(1) Pressure-squared approximation form:

$$Q_g = \frac{kh(\bar{p}_r^2 - p_{wf}^2)}{1422T(\mu_g Z)_{avg}[\ln(r_e/r_w) - 0.75 + s + DQ_g]} \quad (3.15)$$

where D is the inertial or turbulent flow factor and is given by Eq. (1.159) as:

$$D = \frac{Fkh}{1422T} \quad (3.16)$$

and where the non-Darcy flow coefficient F is defined by Eq. (1.155) as:

$$F = 3.161(10^{-12})\left[\frac{\beta T \gamma_g}{\mu_g h^2 r_w}\right] \quad (3.17)$$

where

F = non-Darcy flow coefficient
k = permeability, md
T = temperature, °R
γ_g = gas gravity
r_w = wellbore radius, ft
h = thickness, ft
β = turbulence parameter as given by Eq. (1.156): $\beta = 1.88(10^{-10})k^{-1.47}\phi^{-0.53}$
ϕ = porosity

(2) Pressure approximation form:

$$Q_g = \frac{7.08(10^{-6})kh(\bar{p}_r - p_{wf})}{(\mu_g B_g)_{avg} T[\ln(r_e/r_w) - 0.75 + s + DQ_g]} \quad (3.18)$$

(3) Real-gas pseudopressure form:

$$Q_g = \frac{kh(\bar{\psi}_r - \psi_{wf})}{1422T[\ln(r_e/r_w) - 0.75 + s + DQ_g]} \quad (3.19)$$

Eqs. (3.15), (3.18), and (3.19) are essentially quadratic relationships in Q_g and, thus, they do not represent explicit expressions for calculating the gas flow rate. There are two separate empirical treatments that can be used to represent the turbulent flow problem in gas wells. Both treatments, with varying degrees of approximation, are directly derived and formulated from the three forms of the pseudosteady-state equations, i.e., Eqs. (3.15)–(3.17). These two treatments are called:

(1) the simplified treatment approach;
(2) the laminar–inertial–turbulent (LIT) treatment.

These two empirical treatments of the gas flow equation are presented below.

Simplified Treatment Approach. Based on the analysis for flow data obtained from a large number of gas wells, Rawlins and Schellhardt (1936) postulated that the relationship between the gas flow rate and pressure can be expressed in the pressure-squared form, i.e., Eq. (3.11), by including an exponent n to account for the additional pressure drop due to the turbulent flow as:

$$Q_g = \frac{kh}{1422T(\mu_g Z)_{avg}[\ln(r_e/r_w) - 0.75 + s]}[\bar{p}_r^2 - p_{wf}^2]^n$$

Introducing the performance coefficient C into the above equation, as defined by:

$$C = \frac{kh}{1422T(\mu_g Z)_{avg}[\ln(r_e/r_w) - 0.75 + s]}$$

gives:

$$Q_g = C[\bar{p}_r^2 - p_{wf}^2]^n \quad (3.20)$$

where

Q_g = gas flow rate, Mscf/day
$\overline{p}_r$ = average reservoir pressure, psi
n = exponent
C = performance coefficient, Mscf/day/psi^2

The exponent n is intended to account for the additional pressure drop caused by the high-velocity gas flow, i.e., turbulence. Depending on the flowing conditions, the exponent n may vary from 1.0 for completely laminar flow to 0.5 for fully turbulent flow, i.e., $0.5 \le n \le 1.0$.

The performance coefficient C in Eq. (3.20) is included to account for:

- reservoir rock properties;
- fluid properties;
- reservoir flow geometry.

It should be pointed out that Eq. (3.20) is based on the assumption that the gas flow obeys the pseudosteady-state or the steady-state flowing condition as required by Darcy's equation. This condition implies that the well has established a constant drainage radius r_e, and therefore, the performance coefficient C should remain constant. On the other hand, during the unsteady-state (transient) flow condition, the well drainage radius is continuously changing.

Eq. (3.20) is commonly called the deliverability or back-pressure equation. If the coefficients of the equation (i.e., n and C) can be determined, the gas flow rate Q_g at any bottom-hole flow pressure p_{wf} can be calculated and the IPR curve constructed. Taking the logarithm of both sides of Eq. (3.20) gives:

$$\log(Q_g) = \log(C) + n \log(\overline{p}_r^2 - p_{wf}^2) \tag{3.21}$$

Eq. (3.21) suggests that a plot of Q_g vs. $(\overline{p}_r^2 - p_{wf}^2)$ on a log–log scale should yield a straight line having a slope of n. In the natural gas industry, the plot is traditionally reversed by plotting $(\overline{p}_r^2 - p_{wf}^2)$ vs. Q_g on a logarithmic scale to produce a straight line with a slope of $1/n$. This plot as shown schematically in Figure 3.3 is commonly referred to as the deliverability graph or the back-pressure plot.

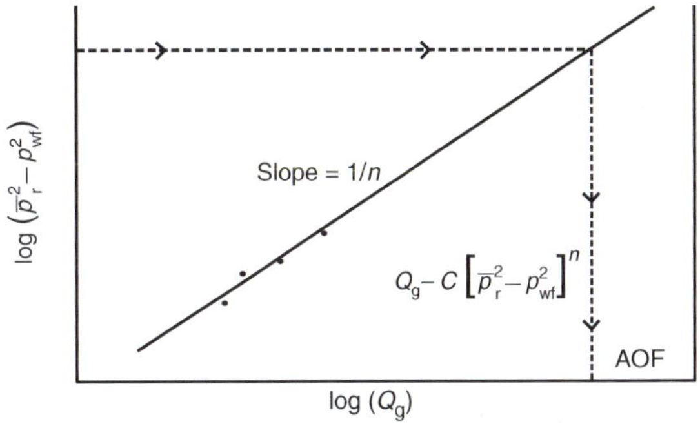

FIGURE 3.3 Well deliverability graph.

The deliverability exponent n can be determined from any two points on the *straight line*, i.e., $(Q_{g1}, \Delta p_1^2)$ and $(Q_{g2}, \Delta p_2^2)$, according to the following expression:

$$n = \frac{\log(Q_{g1}) - \log(Q_{g2})}{\log(\Delta p_1^2) - \log(\Delta p_2^2)} \tag{3.22}$$

Given n, any point on the straight line can be used to compute the performance coefficient C from:

$$C = \frac{Q_g}{(\overline{p}_r^2 - p_{wf}^2)^n} \tag{3.23}$$

The coefficients of the back-pressure equation or any of the other empirical equations are traditionally determined from analyzing gas well testing data. Deliverability testing has been used for more than 60 years by the petroleum industry to characterize and determine the flow potential of gas wells. There are essentially three types of deliverability tests:

(1) conventional deliverability (back-pressure) test;
(2) isochronal test;
(3) modified isochronal test.

These tests basically allow the wells to flow at multiple rates and measure the bottom-hole flowing pressure as a function of time. When the recorded data is properly analyzed, it is possible to determine the flow potential and establish the IPRs of the gas well. The deliverability test is discussed later in this chapter for the purpose of introducing basic techniques used in analyzing the test data.

Laminar–Inertial–Turbulent (LIT) Approach. Essentially, this approach is based on expressing the total pressure drop in terms of the pressure drop due to Darcy's (laminar) flow and the additional pressure drop due to the turbulent flow. That is:

$$(\Delta p)_{\text{Total}} = (\Delta p)_{\text{Laminar flow}} + (\Delta p)_{\text{Turbulent flow}}$$

The three forms of the semisteady-state equation as presented by Eqs. (3.15), (3.18), and (3.19), i.e., the pseudopressure, pressure-squared, and pressure approach, can be rearranged in quadratic forms for the purpose of separating the "laminar" and "inertial–turbulent" terms and composing these equations as follows.

Pressure-Squared Quadratic Form. Eq. (3.15) can be written in a more simplified form as:

$$Q_g = \frac{kh(\bar{p}_r^2 - p_{wf}^2)}{1422T(\mu_g Z)_{avg}[\ln(r_e/r_w) - 0.75 + s + DQ_g]}$$

Rearranging this equation gives:

$$\bar{p}_r^2 - p_{wf}^2 = aQ_g + bQ_g^2 \qquad (3.24)$$

with

$$a = \left(\frac{1422T\mu_g Z}{kh}\right)\left[\ln\left(\frac{r_e}{r_w}\right) - 0.75 + s\right] \qquad (3.25)$$

$$b = \left(\frac{1422T\mu_g Z}{kh}\right)D \qquad (3.26)$$

where

a = laminar flow coefficient
b = inertial–turbulent flow coefficient
Q_g = gas flow rate, Mscf/day
Z = gas deviation factor
k = permeability, md
μ_g = gas viscosity, cp

Eq. (3.24) indicates that the first term on the right-hand side of the equation (i.e., aQ_g) represents the pressure drop due to laminar (Darcy) flow, while the second term represents aQ_g^2, the pressure drop due to the turbulent flow.

The term aQ_g in Eq. (3.26) represents the pressure-squared drop due to laminar flow, while the term bQ_g^2 accounts for the pressure-squared drop due to inertial–turbulent flow effects.

Eq. (3.24) can be liberalized by dividing both sides of the equation by Q_g, to yield:

$$\frac{\bar{p}_r^2 - p_{wf}^2}{Q_g} = a + bQ_g \qquad (3.27)$$

The coefficients a and b can be determined by plotting $(\bar{p}_r^2 - p_{wf}^2/2)$ vs. Q_g on a Cartesian scale and should yield a straight line with a slope of b and intercept of a. As presented later in this chapter, data from deliverability tests can be used to construct the linear relationship as shown schematically in Figure 3.4.

Given the values of a and b, the quadratic flow equation, i.e., Eq. (3.24), can be solved for Q_g at any p_{wf} from:

$$Q_g = \frac{-a + \sqrt{a^2 + 4b(\bar{p}_r^2 - p_{wf}^2)}}{2b} \qquad (3.28)$$

Furthermore, by assuming various values of p_{wf} and calculating the corresponding Q_g from

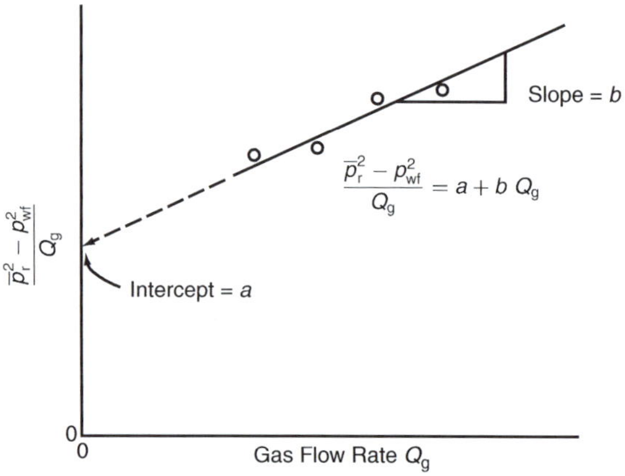

FIGURE 3.4 Graph of the pressure-squared data.

Eq. (3.28), the current IPR of the gas well at the current reservoir pressure $\overline{p}_r$ can be generated.

It should be pointed out that the following assumptions were made in developing Eq. (3.24):

(1) The flow is single phase.
(2) The reservoir is homogeneous and isotropic.
(3) The permeability is independent of pressure.
(4) The product of the gas viscosity and compressibility factor, i.e., $(\mu_g Z)$, is constant.

This method is recommended for applications at pressures less than 2000 psi.

Pressure Quadratic Form. The pressure approximation equation, i.e., Eq. (3.18), can be rearranged and expressed in the following quadratic form:

$$Q_g = \frac{7.08(10^{-6})kh(\overline{p}_r - p_{wf})}{(\mu_g B_g)_{avg} T[\ln(r_e/r_w) - 0.75 + s + DQ_g]}$$

Rearranging gives:

$$\overline{p}_r - p_{wf} = a_1 Q_g + b_1 Q_g^2 \quad (3.29)$$

where

$$a_1 = \frac{141.2(10^{-3})(\mu_g B_g)}{kh}\left[\ln\left(\frac{r_e}{r_w}\right) - 0.75 + s\right] \quad (3.30)$$

$$b_1 = \left[\frac{141.2(10^{-3})(\mu_g B_g)}{kh}\right] D \quad (3.31)$$

The term $a_1 Q_g$ represents the pressure drop due to laminar flow, while the term $b_1 Q_g^2$ accounts for the additional pressure drop due to the turbulent flow condition. In a linear form, Eq. (3.17) can be expressed as:

$$\frac{\overline{p}_r - p_{wf}}{Q_g} = a_1 + b_1 Q_g \quad (3.32)$$

The laminar flow coefficient a_1 and inertial–turbulent flow coefficient b_1 can be determined from the linear plot of the above equation as shown in Figure 3.5.

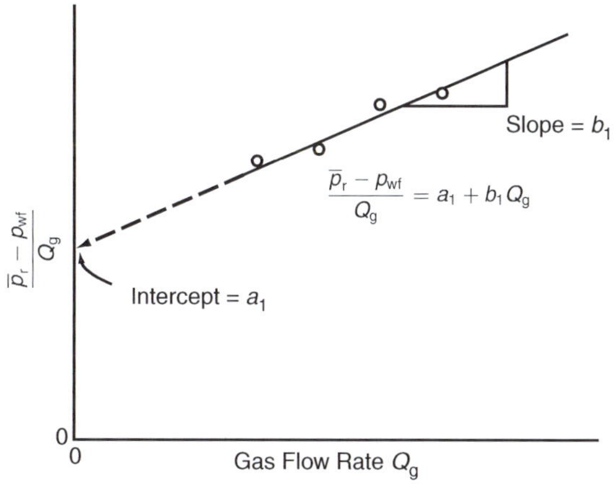

FIGURE 3.5 Graph of the pressure method data.

Once the coefficients a_1 and b_1 are determined, the gas flow rate can be determined at any pressure from:

$$Q_g = \frac{-a_1 + \sqrt{a_1^2 + 4b_1(\overline{p}_r - p_{wf})}}{2b_1} \quad (3.33)$$

The application of Eq. (3.29) is also restricted by the assumptions listed for the pressure-squared approach. However, the pressure method is applicable at pressures higher than 3000 psi.

Pseudopressure Quadratic Approach. The pseudopressure equation has the form:

$$Q_g = \frac{kh(\overline{\psi}_r - \psi_{wf})}{1422T[\ln(r_e/r_w) - 0.75 + s + DQ_g]}$$

This expression can be written in a more simplified form as:

$$\overline{\psi}_r - \psi_{wf} = a_2 Q_g + b_2 Q_g^2 \quad (3.34)$$

where

$$a_2 = \left(\frac{1422}{kh}\right)\left[\ln\left(\frac{r_e}{r_w}\right) - 0.75 + s\right] \quad (3.35)$$

$$b_2 = \left(\frac{1422}{kh}\right) D \quad (3.36)$$

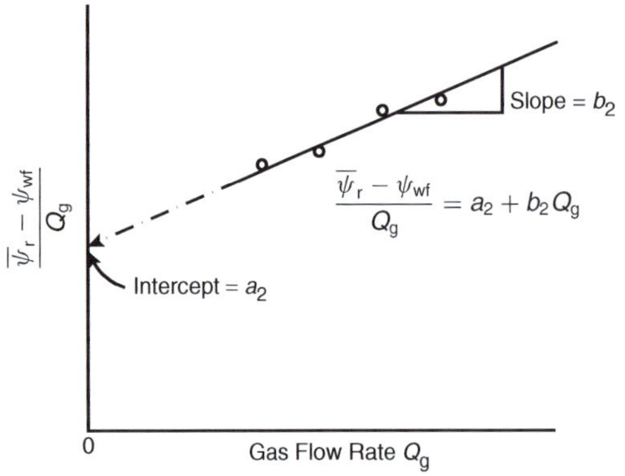

FIGURE 3.6 Graph of real-gas pseudopressure data.

The term $a_2 Q_g$ in Eq. (3.34) represents the pseudopressure drop due to laminar flow, while the term $b_2 Q_g^2$ accounts for the pseudopressure drop due to inertial–turbulent flow effects.

Eq. (3.34) can be liberalized by dividing both sides of the equation by Q_g, to yield:

$$\frac{\overline{\psi}_r - \psi_{wf}}{Q_g} = a_2 + b_2 Q_g \tag{3.37}$$

The above expression suggests that a plot of $(\overline{\psi}_r - \psi_{wf}/Q_g)$ vs. Q_g on a Cartesian scale should yield a straight line with a slope of b_2 and intercept of a_2 as shown in Figure 3.6.

Given the values of a_2 and b_2, the gas flow rate at any p_{wf} is calculated from:

$$Q_g = \frac{-a_2 + \sqrt{a_2^2 + 4b_2(\overline{\psi}_r - \psi_{wf})}}{2b_2} \tag{3.38}$$

It should be pointed out that the pseudopressure approach is more rigorous than either the pressure-squared or pressure method and is applicable to all ranges of pressure.

In the next subsection, the back-pressure test is introduced. However, the material is intended only to be an introduction. There are several excellent books by the following authors that address transient flow and well testing in great detail:

- Earlougher (1977);
- Matthews and Russell (1967);
- Lee (1982);
- Canadian Energy Resources Conservation Board (ERCB) (1975).

3.1.3 Back-Pressure Test

Rawlins and Schellhardt (1936) proposed a method for testing gas wells by gauging the ability of the well to flow against particular pipeline back pressures greater than atmospheric pressure. This type of flow test is commonly referred to as the "conventional deliverability test." The required procedure for conducting this back-pressure test consists of the following steps:

Step 1. Shut in the gas well sufficiently long for the formation pressure to equalize at the volumetric average pressure $\overline{p}_r$.

Step 2. Place the well on production at a constant flow rate Q_{g1} for a sufficient time to allow the bottom-hole flowing pressure to stabilize at p_{wf1}, i.e., to reach the pseudosteady state.

Step 3. Repeat step 2 for several rates and record the stabilized bottom-hole flow pressure at each corresponding flow rate. If three or four rates are used, the test may be referred to as a three-point or four-point flow test.

The rate and pressure history of a typical four-point test is shown in Figure 3.7. The figure illustrates a normal sequence of rate changes where the rate is increased during the test. Tests may also be run, however, using a reverse sequence. Experience indicates that a normal rate sequence gives better data in most wells. The most important factor to be considered in performing the conventional deliverability test is the length of the flow periods. It is required that each rate be maintained sufficiently long for the well to stabilize, i.e., to reach the pseudosteady state. The pseudosteady-state time is defined as the time when the rate of change of pressure with respect to time, i.e., dp/dt, is constant through the reservoir at a

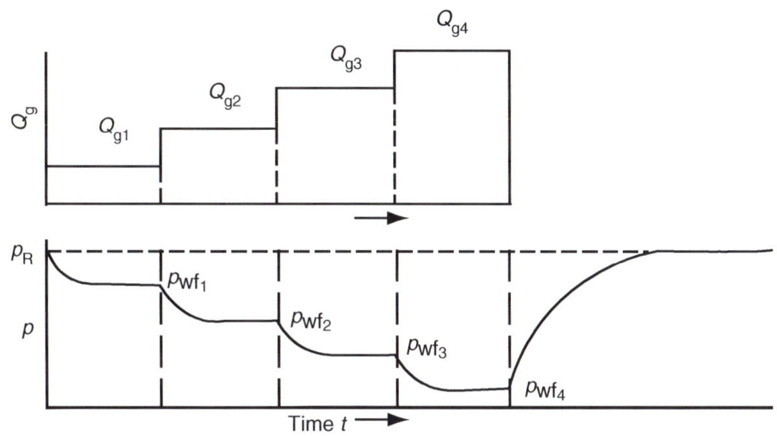

FIGURE 3.7 Conventional back-pressure test.

constant flow rate. This stabilization time for a well in the center of a circular or square drainage area may be estimated from:

$$t_{pss} = \frac{15.8\phi\mu_{gi}c_{ti}A}{k} \tag{3.39}$$

with

$$c_{ti} = S_w c_{wi} + (1 - S_w)c_{gi} + c_f$$

where

t_{pss} = stabilization (pseudosteady-state) time, days
c_{ti} = total compressibility coefficient at initial pressure, psi^{-1}
c_{wi} = water compressibility coefficient at initial pressure, psi^{-1}
c_f = formation compressibility coefficient, psi^{-1}
c_{gi} = gas compressibility coefficient at initial pressure, psi^{-1}
ϕ = porosity, fraction
μ_g = gas viscosity, cp
k = effective gas permeability, md
A = drainage area, ft^2

To properly apply Eq. (3.39), the fluid properties and system compressibility must be determined at the average reservoir pressure. However, evaluating these parameters at initial reservoir pressure has been found to provide a good first-order approximation of the time required to reach the pseudosteady-state condition and establish a constant drainage area. The recorded bottom-hole flowing pressure p_{wf} vs. flow rate Q_g can be analyzed in several graphical forms to determine the coefficients of the selected flow gas flow equation. That is:

Back-pressure equation: $\log(Q_g) = \log(C) + n\log(\overline{p}_r^2 - p_{wf}^2)$

Pressure-squared equation: $\overline{p}_r^2 - p_{wf}^2 = aQ_g + bQ_g^2$

Pressure equation: $\dfrac{\overline{p}_r - p_{wf}}{Q_g} = a_1 + b_1 Q_g$

Pseudopressure equation: $\overline{\psi}_r - \psi_{wf} = a_2 Q_g + b_2 Q_g^2$

The application of the back-pressure test data to determine the coefficients of any of the empirical flow equations is illustrated in the following example.

Example 3.2

A gas well was tested using a three-point conventional deliverability test with an initial average reservoir pressure of 1952 psi. The recorded data during the test is given below:

p_{wf} (psia)	$m(p_{wf}) = \psi_{wf}$ (psi^2/cp)	Q_g (Mscf/day)
1952	316×10^6	0
1700	245×10^6	2624.6
1500	191×10^6	4154.7
1300	141×10^6	5425.1

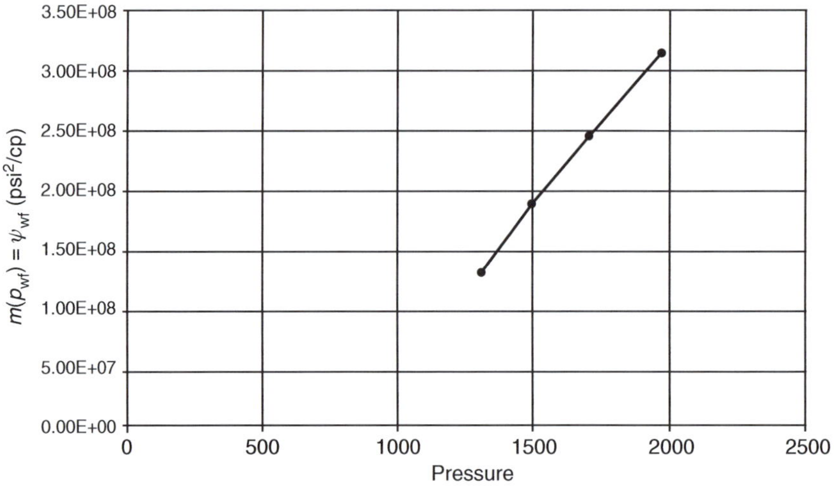

FIGURE 3.8 Real-gas potential vs. pressure.

Figure 3.8 shows the gas pseudopressure ψ as a function of pressure. Generate the current IPR by using the following methods:

(a) simplified back-pressure equation;
(b) LIT methods:
  (i) pressure-squared approach, Eq. (3.29);
  (ii) pressure approach, Eq. (3.33);
  (iii) pseudopressure approach, Eq. (3.26);
(c) compare results of the calculation.

Solution

(a) Back-pressure equation:
  Step 1. Prepare the following table:

p_{wf}	p_{wf}^2 (psi² × 10³)	$(\bar{p}_r^2 - p_{wf}^2)$ (psi² × 10³)	Q_g (Mscf/day)
$\bar{p}_r = 1952$	3810	0	0
1700	2890	920	2624.6
1500	2250	1560	4154.7
1300	1690	2120	5425.1

  Step 2. Plot $(\bar{p}_r^2 - p_{wf}^2)$ vs. Q_g on a log–log scale as shown in Figure 3.9. Draw the best straight line through the points.

  Step 3. Using any two points on the straight line, calculate the exponent n from Eq. (3.22), as:

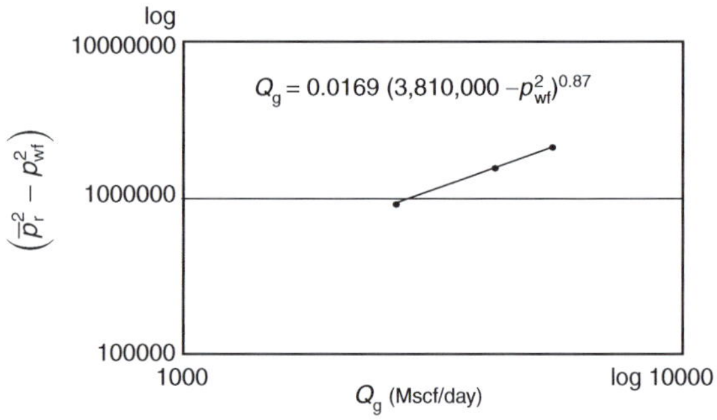

FIGURE 3.9 Back-pressure curve.

$$n = \frac{\log(Q_{g1}) - \log(Q_{g2})}{\log(\Delta p_2^2) - \log(\Delta p_2^2)}$$
$$= \frac{\log(4000) - \log(1800)}{\log(1500) - \log(600)} = 0.87$$

  Step 4. Determine the performance coefficient C from Eq. (3.23) by using the coordinate of any point on the straight line, or:

$$C = \frac{Q_g}{(\bar{p}_r^2 - p_{wf}^2)^n}$$
$$= \frac{1800}{(600,000)^{0.87}} = 0.0169\ \text{Mscf/psi}^2$$

Step 5. The back-pressure equation is then expressed as:

$$Q_g = 0.0169(3,810,000 - p_{wf}^2)^{0.87}$$

Step 6. Generate the IPR data by assuming various values of p_{wf} and calculate the corresponding Q_g:

p_{wf}	Q_g (Mscf/day)
1952	0
1800	1720
1600	3406
1000	6891
500	8465
0	8980

where the AOF $= (Q_g)_{max} = 8980$ Mscf/day.

(b) LIT method:

(i) Pressure-squared method:

Step 1. Construct the following table:

p_{wf}	$(\bar{p}_r^2 - p_{wf}^2)$ (psi$^2 \times 10^3$)	Q_g (Mscf/day)	$(\bar{p}_r^2 - p_{wf}^2)/Q_g$
$\bar{p}_r = 1952$	0	0	–
1700	920	2624.6	351
1500	1560	4154.7	375
1300	2120	5425.1	391

Step 2. Plot $(\bar{p}_r^2 - p_{wf}^2)/Q_g$ vs. Q_g on a Cartesian scale and draw the best straight line as shown in Figure 3.10.

Step 3. Determine the intercept and the slope of the straight line, to give:

Intercept $a = 318$

Slope $b = 0.01333$

Step 4. The quadratic form of the pressure-squared approach is given by Eq. (3.24) as:

$$\bar{p}_r^2 - p_{wf}^2 = aQ_g + bQ_g^2$$

$$(3,810,000 - p_{wf}^2) = 318Q_g + 0.01333Q_g^2$$

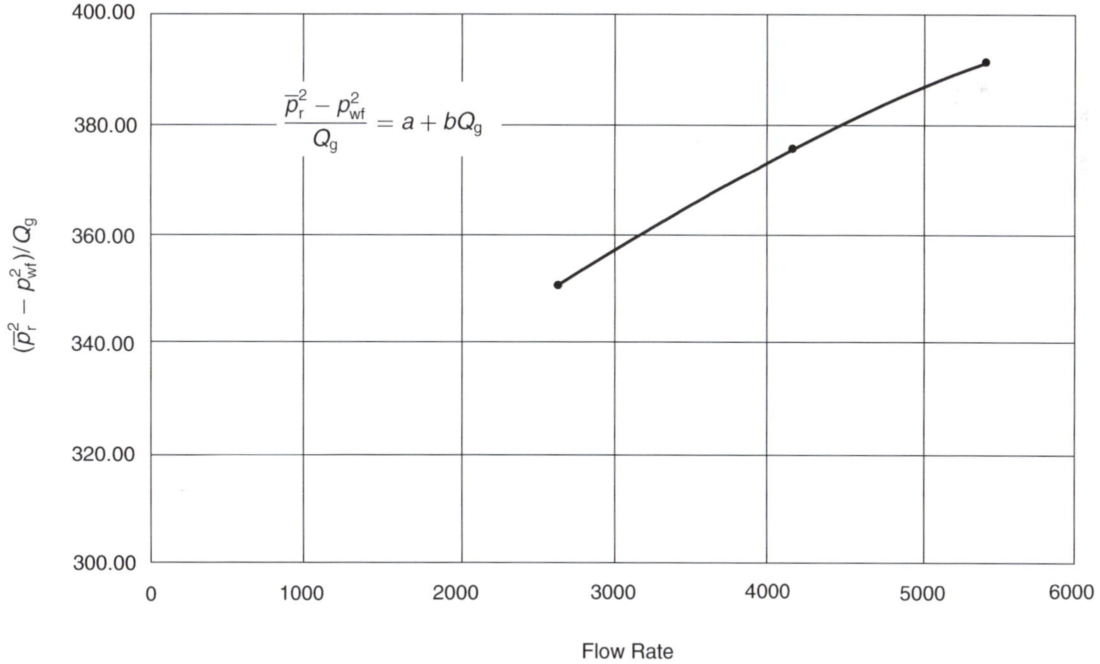

FIGURE 3.10 Pressure-squared method.

Step 5. Construct the IPR data by assuming various values of p_{wf} and solving for Q_g by using Eq. (3.28):

p_{wf}	$(\bar{p}_r^2 - p_{wf}^2)$ psi$^2 \times 10^3$	Q_g (Mscf/day)
1952	0	0
1800	570	1675
1600	1250	3436
1000	2810	6862
500	3560	8304
0	3810	8763 = AOF = $(Q_g)_{max}$

(ii) Pressure method:

Step 1. Construct the following table:

p_{wf}	$(\bar{p}_r - p_{wf})$	Q_g (Mscf/day)	$(\bar{p}_r - p_{wf})/Q_g$
$\bar{p}_r = 1952$	0	0	–
1700	252	262.6	0.090
1500	452	4154.7	0.109
1300	652	5425.1	0.120

Step 2. Plot $(\bar{p}_r - p_{wf})/Q_g$ vs. Q_g on a Cartesian scale as shown in Figure 3.11. Draw the best straight line and determine the intercept and slope as:

Intercept $a_1 = 0.06$

Slope $b_1 = 1.11 \times 10^{-5}$

Step 3. The quadratic form of the pressure method is then given by:

$$\bar{p}_r - p_{wf} = a_1 Q_g + b_1 Q_g^2$$

or

$$(1952 - p_{wf}) = 0.06 Q_g + (1.111 \times 10^{-5}) Q_g^2$$

Step 4. Generate the IPR data by applying Eq. (3.33):

p_{wf}	$(\bar{p}_r - p_{wf})$	Q_g (Mscf/day)
1952	0	0
1800	152	1879
1600	352	3543
1000	952	6942
500	1452	9046
0	1952	10,827

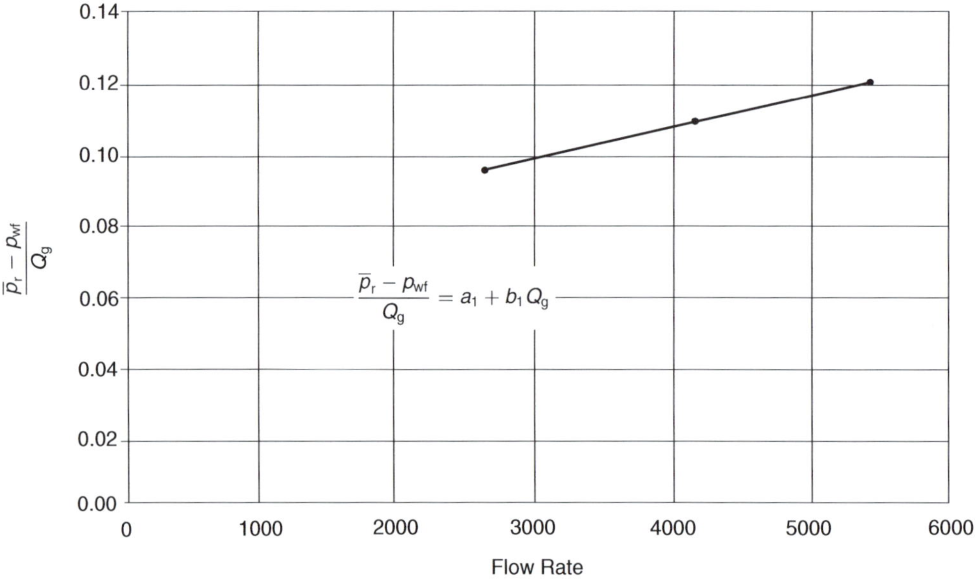

FIGURE 3.11 Pressure approximation method.

(iii) Pseudopressure approach:

Step 1. Construct the following table:

p_{wf}	ψ (psi²/cp)	$(\overline{\psi}_r - \psi_{wf})$	Q_g (Mscf/day)	$(\overline{\psi}_r - \psi_{wf})/Q_g$
$\overline{p}_r = 1952$	316×10^6	0	0	–
1700	245×10^6	71×10^6	262.6	27.05×10^3
1500	191×10^6	125×10^6	4154.7	30.09×10^3
1300	141×10^6	175×10^6	5425.1	32.26×10^3

Step 2. Plot $(\overline{\psi}_r - \psi_{wf})/Q_g$ on a Cartesian scale as shown in Figure 3.12 and determine the intercept a_2 and slope b_2 as:

$$a_2 = 22.28 \times 10^3$$
$$b_2 = 1.727$$

Step 3. The quadratic form of the gas pseudopressure method is given by Eq. (3.34):

$$\overline{\psi}_r - \psi_{wf} = a_2 Q_g + b_2 Q_g^2$$

$$316 \times 10^6 - \psi_{wf} = 22.28 \times 10^3 Q_g + 1.727 Q_g^2$$

Step 4. Generate the IPR data by assuming various values of p_{wf}, i.e., ψ_{wf}, and calculating the corresponding Q_g from Eq. (3.38):

p_{wf}	$m(p)$ or ψ	$\overline{\psi}_r - \psi_{wf}$	Q_g (Mscf/day)
1952	316×10^6	0	0
1800	270×10^6	46×10^6	1794
1600	215×10^6	101×10^6	3503
1000	100×10^6	216×10^6	6331
500	40×10^6	276×10^6	7574
0	0	316×10^6	8342 = AOF $(Q_g)_{max}$

(c) Compare the gas flow rates as calculated by the four different methods. Results of the IPR calculation are documented below:

Gas Flow Rate (Mscf/day)

Pressure	Back Pressure	p^2 Approach	p Approach	ψ Approach
1952	0	0	0	0
1800	1720	1675	1879	1811
1600	3406	3436	3543	3554
1000	6891	6862	6942	6460
500	8465	8304	9046	7742
0	8980	8763	10827	8536
	6.0%	5.4%	11%	–

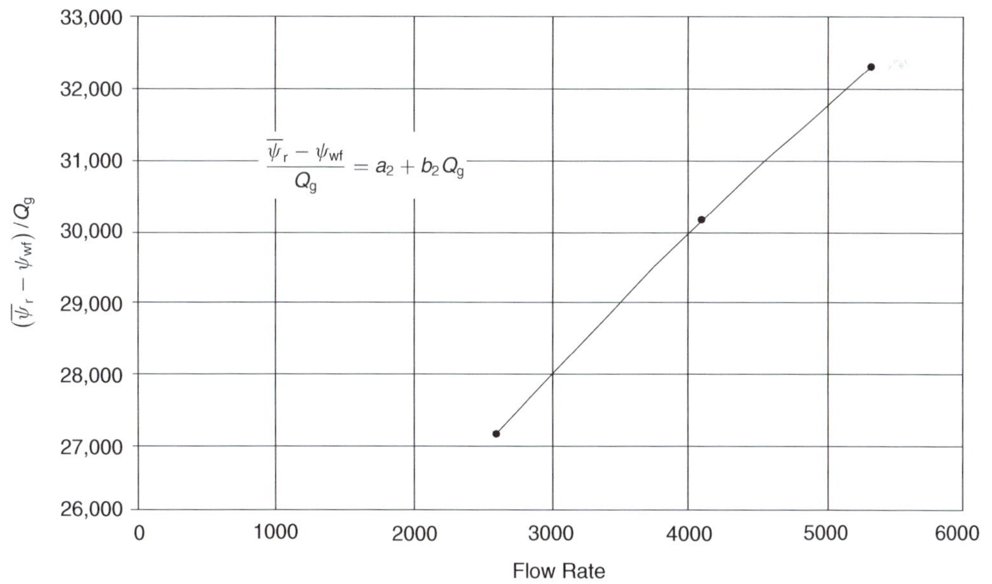

FIGURE 3.12 Pseudopressure method.

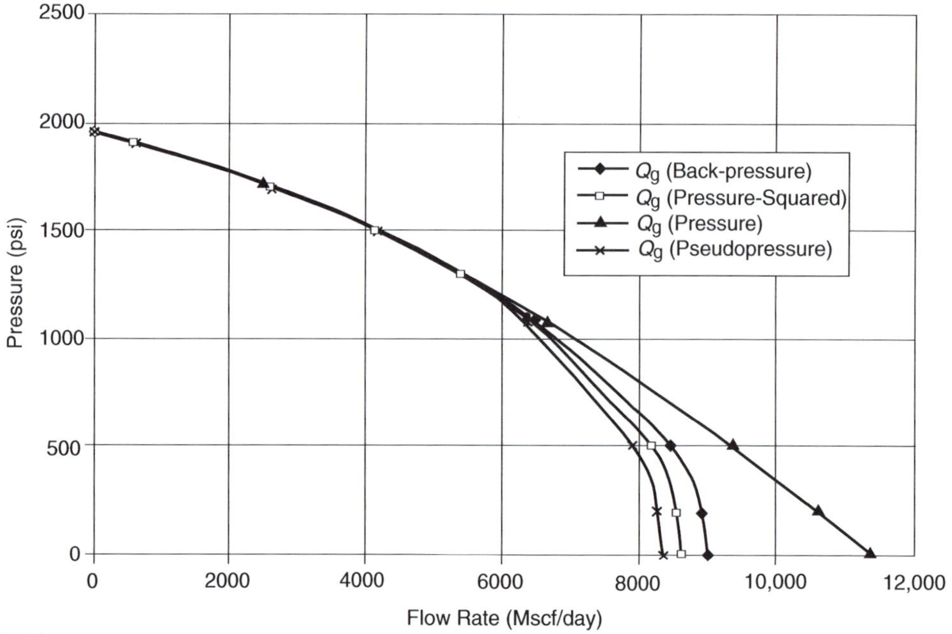

FIGURE 3.13 IPR for all methods.

Since the pseudopressure analysis is considered more accurate and rigorous than the other three methods, the accuracy of each of the methods in predicting the IPR data is compared with that of the ψ approach. Figure 3.13 compares graphically the performance of each method with that of the ψ approach. Results indicate that the pressure-squared equation generated the IPR data with an absolute average error of 5.4% as compared with 6% and 11% for the back-pressure equation and the pressure approximation method, respectively.

It should be noted that the pressure approximation method is limited to applications for pressures greater than 3000 psi.

3.1.4 Future Inflow Performance Relationships

Once a well has been tested and the appropriate deliverability or inflow performance equation established, it is essential to predict the IPR data as a function of average reservoir pressure. The gas viscosity μ_g and gas compressibility factor Z are considered the parameters that are subject to the greatest change as reservoir pressure $\overline{p}_r$ changes.

Assume that the current average reservoir pressure is $\overline{p}_{r1}$ with gas viscosity of μ_{g1} and compressibility factor of Z_1. At a selected future average reservoir pressure $\overline{p}_{r2}$, μ_{g2} and Z_2 represent the corresponding gas properties. To approximate the effect of reservoir pressure changes, i.e., from $\overline{p}_{r1}$ to $\overline{p}_{r2}$, on the coefficients of the deliverability equation, the following methodology is recommended.

Back-Pressure Equation. Recall the back-pressure equation:

$$Q_g = C[\overline{p}_r^2 - p_{wf}^2]^n$$

where the coefficient C describes the gas and reservoir properties by:

$$C = \frac{kh}{1422T(\mu_g Z)_{avg}[\ln(r_e/r_w) - 0.75 + s]}$$

The performance coefficient C is considered a pressure-dependent parameter and should be adjusted with each change of the reservoir pressure. Assuming that the reservoir pressure has declined from p_{r1} to p_{r2}, the performance coefficient at p_1 can be adjusted to reflect the pressure drop by applying the following simple approximation:

$$C_2 = C_1 \left[\frac{\mu_{g1} Z_1}{\mu_{g2} Z_2}\right] \tag{3.40}$$

The value of n is considered essentially constant. Subscripts 1 and 2 refer to the properties at p_{r1} and p_{r2}.

LIT Methods. The laminar flow coefficients a and the inertial–turbulent flow coefficient b of any of the previous LIT methods, i.e., Eqs. (3.24), (3.29), and (3.34), are modified according to the following simple relationships:

Pressure-Squared Method. The pressure-squared equation is written as:

$$\bar{p}_r^2 - p_{wf}^2 = aQ_g + bQ_g^2$$

The coefficients of the above expression are given by:

$$a = \left(\frac{1422 T \mu_g Z}{kh}\right)\left[\ln\left(\frac{r_e}{r_w}\right) - 0.75 + s\right]$$

$$b = \left(\frac{1422 T \mu_g Z}{kh}\right) D$$

Obviously, the coefficients a and b are pressure dependent and should be modified to account for the change of the reservoir pressure from $\bar{p}_{r1}$ to $\bar{p}_{r2}$. The proposed relationships for adjusting the coefficients are as follows:

$$a_2 = a_1 \left[\frac{\mu_{g2} Z_2}{\mu_{g1} Z_1}\right] \tag{3.41}$$

$$b_2 = b_1 \left[\frac{\mu_{g2} Z_2}{\mu_{g1} Z_1}\right] \tag{3.42}$$

where the subscripts 1 and 2 represent conditions at reservoir pressures $\bar{p}_{r1}$ to $\bar{p}_{r2}$, respectively.

Pressure Approximation Method. The pressure approximation equation for calculating the gas rate is given by:

$$\bar{p}_r - p_{wf} = a_1 Q_g + b_1 Q_g^2$$

with

$$a_1 = \frac{141.2(10^{-3})(\mu_g B_g)}{kh}\left[\ln\left(\frac{r_e}{r_w}\right) - 0.75 + s\right]$$

$$b_1 = \left[\frac{141.2(10^{-3})(\mu_g B_g)}{kh}\right] D$$

The recommended methodology for adjusting the coefficients a and b is based on applying the following two simple expressions:

$$a_2 = a_1 \frac{\mu_{g2} B_{g2}}{\mu_{g1} B_{g1}} \tag{3.43}$$

$$b_2 = b_1 \left[\frac{\mu_{g2} B_{g2}}{\mu_{g1} B_{g1}}\right] \tag{3.44}$$

where B_g is the gas formation volume factor in bbl/scf.

Pseudopressure Approach. Recall the pseudopressure equation:

$$\bar{\psi}_r - \psi_{wf} = a_2 Q_g + b_2 Q_g^2$$

The coefficients are described by:

$$a_2 = \left(\frac{1422}{kh}\right)\left[\ln\left(\frac{r_e}{r_w}\right) - 0.75 + s\right]$$

$$b_2 = \left(\frac{1422}{kh}\right) D$$

Note that the coefficients a and b of the pseudopressure approach are essentially independent of the reservoir pressure and can be treated as constants.

Example 3.3

In addition to the data given in Example 3.2, the following information is available:

- $(\mu_g Z) = 0.01206$ at 1952 psi;
- $(\mu_g Z) = 0.01180$ at 1700 psi.

Using the following methods:

(a) back-pressure equation,
(b) pressure-squared equation, and
(c) pseudopressure equation

generate the IPR data for the well when the reservoir pressure drops from 1952 to 1700 psi.

Solution

Step 1. Adjust the coefficients a and b of each equation.

- For the back-pressure equation: Adjust C by using Eq. (3.40):

$$C_2 = C_1\left[\frac{\mu_{g_1} Z_1}{\mu_{g2} Z_2}\right]$$

$$C = 0.0169\left(\frac{0.01206}{0.01180}\right) = 0.01727$$

and therefore the future gas flow rate is expressed by:

$$Q_g = 0.01727(1700^2 - p_{wf}^2)^{0.87}$$

- Pressure-squared method: Adjust a and b by applying Eqs. (3.41) and (3.42):

$$a_2 = a_1\left[\frac{\mu_{g2} B_{g2}}{\mu_{g1} B_{g1}}\right]$$

$$a = 318\left(\frac{0.01180}{0.01206}\right) = 311.14$$

$$b_2 = b_1\left[\frac{\mu_{g2} B_{g2}}{\mu_{g1} B_{g1}}\right]$$

$$b = 0.01333\left(\frac{0.01180}{0.01206}\right) = 0.01304$$

$$(1700^2 - p_{wf}^2) = 311.14 Q_g + 0.01304 Q_g^2$$

- Pseudopressure method: No adjustments are needed because the coefficients are independent of the pressure:

$$(245 \times 10^6 - \psi_{wf}) = 22.28 \times 10^3 Q_g + 1.727 Q_g^2$$

Step 2. Generate the IPR data:

Gas Flow Rate Q_g (Mscf/day)

p_{wf}	Back Pressure	p^2 Method	ψ Method
$\bar{p}_r = 1700$	0	0	0
1600	1092	1017	1229
1000	4987	5019	4755
500	6669	6638	6211
0	7216	7147	7095

Figure 3.14 compares graphically the IPR data as predicted by the above three methods.

It should be pointed out that all the various well tests and IPRs previously discussed are intended to evaluate the formation capacity to deliver gas to the wellbore for a specified average reservoir pressure $\bar{p}_r$ and a bottom-hole flowing pressure p_{wf}. The volume of gas that can actually be delivered to the surface will also depend on the surface tubing head pressure p_t and the pressure drop from the wellbore to the surface due to the weight of the gas column and friction loss through the tubing. Cullender and Smith (1956) described the pressure loss by the following expression:

$$p_{wf}^2 = e^S p_t^2 + \frac{L}{H}(F_r Q_g \overline{TZ})^2(e^S - 1)$$

with

$$S = \frac{0.0375\gamma_g H}{\overline{TZ}}$$

$$F_r = \frac{0.004362}{d^{0.224}}, \quad \text{where } d \le 4.277 \text{ in.}$$

$$F_r = \frac{0.004007}{d^{0.164}}, \quad \text{where } d > 4.277 \text{ in.}$$

where

p_{wf} = bottom-hole flowing pressure, psi
p_t = tubing head (wellhead) pressure, psi

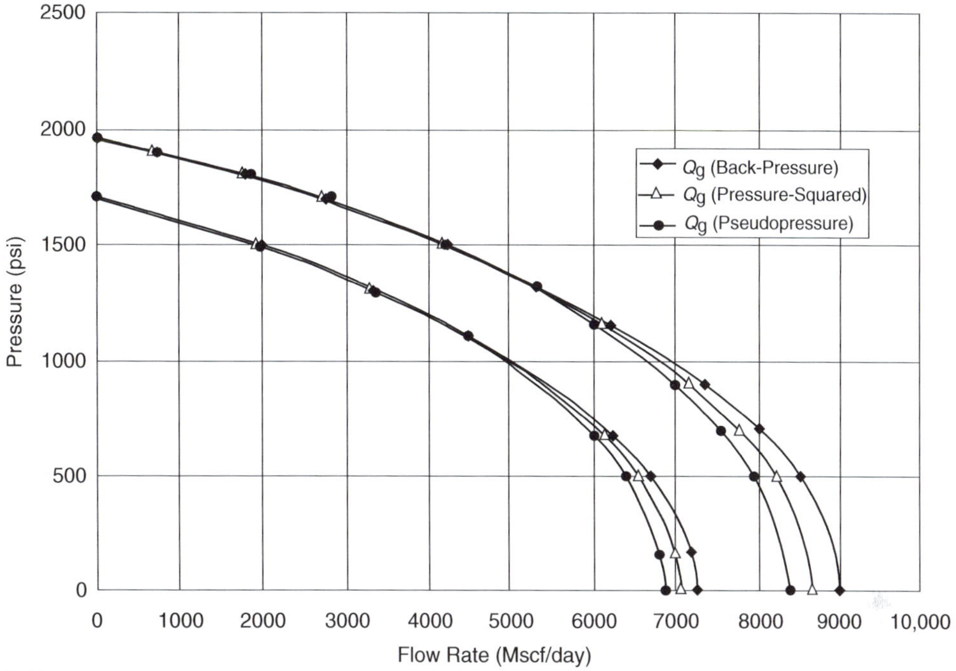

FIGURE 3.14 IPR comparison.

Q_g = gas flow rate, Mscf/day
L = actual tubing flow length, ft
H = vertical depth of the well to midpoint of perforation, ft
$\overline{T}$ = arithmetic average temperature, i.e., $(T_t + T_b)/2$, °R
T_t = tubing head temperature, °R
T_b = wellbore temperature, °R
$\overline{Z}$ = gas deviation factor at arithmetic average pressure, i.e., $(p_t + p_{wf})/2$
F_r = friction factor for tubing ID
d = inside tubing diameter, in.
γ_g = specific gravity of the gas

The Cullender and Smith equation can be combined with the back-pressure equation by the gas flow rate Q_g to give:

$$\frac{p_{wf}^2 - e^S p_t^2}{(L/H)(F_r\overline{T}\overline{Z})^2(e^S - 1)} = C(p_r^2 - p_{wf}^2)^{2n}$$

It should be pointed out that the above nonlinear equation can be solved for pwf using any of the numerical iterative techniques. The correct value of p_{wf} can then be used to establish the gas deliverability of the well.

3.2 HORIZONTAL GAS WELL PERFORMANCE

Many low-permeability gas reservoirs are historically considered to be noncommercial due to low production rates. Most vertical wells drilled in tight gas reservoirs are stimulated using hydraulic fracturing and/or acidizing treatments to attain economical flow rates. In addition, to deplete a tight gas reservoir, vertical wells must be drilled at close spacing to efficiently drain the reservoir. This would require a large number of vertical wells. In such reservoirs, horizontal wells provide an attractive alternative to effectively deplete tight gas reservoirs and attain high flow rates. Joshi (1991) pointed out that horizontal wells are applicable in both low- and high-permeability reservoirs. The excellent reference textbook by Joshi (1991) gives a comprehensive treatment of horizontal well performance in oil and gas reservoirs.

In calculating the gas flow rate from a horizontal well, Joshi (1991) introduced the concept of the effective wellbore radius $r_w^\backslash$ into the gas flow equation. The effective wellbore radius is given by:

$$r_w^\backslash = \frac{r_{eh}(L/2)}{a\left[1+\sqrt{1-(L/2a)^2}\right][h/(2r_w)]^{h/L}} \quad \textbf{(3.45)}$$

with

$$a = \left(\frac{L}{2}\right)\left[0.5+\sqrt{0.25+(2r_{eh}/L)^4}\right]^{0.5} \quad \textbf{(3.46)}$$

and

$$r_{eh} = \sqrt{\frac{43,560A}{\pi}} \quad \textbf{(3.47)}$$

where

L = length of the horizontal well, ft
h = thickness, ft
r_w = wellbore radius, ft
r_{eh} = horizontal well drainage radius, ft
a = half the major axis of the drainage ellipse, ft
A = drainage area of the horizontal well, acres

For a pseudosteady-state flow, Joshi (1991) expressed Darcy's equation of a laminar flow in the following two familiar forms:

(1) Pressure-squared form:

$$Q_g = \frac{kh(\bar{p}_r^2 - p_{wf}^2)}{1422T(\mu_g Z)_{avg}\left[\ln(r_{eh}/r_w^\backslash) - 0.75 + s\right]} \quad \textbf{(3.48)}$$

where

Q_g = gas flow rate, Mscf/day
s = skin factor
k = permeability, md
T = temperature, °R

(2) Pseudopressure form:

$$Q_g = \frac{kh(\bar{\psi}_r - \psi_{wf})}{1422T\left[\ln(r_{eh}/r_w^\backslash) - 0.75 + s\right]} \quad \textbf{(3.49)}$$

Example 3.4

A horizontal gas well, 2000 foot long, is draining an area of approximately 120 acres. The following data is available:

$\bar{p}_r = 2000$ psi, $\bar{\psi}_r = 340 \times 10^6$ psi²/cp
$p_{wf} = 1200$ psi, $\psi_{wf} = 128 \times 10^6$ psi²/cp
$(\mu_g Z)_{avg} = 0.011826$, $r_w = 0.3$ *ft*, $s = 0.5$
$h = 20$ ft, $T = 180$ °F, $k = 1.5$ md

Assuming a pseudosteady-state flow, calculate the gas flow rate by using the pressure-squared and pseudopressure methods.

Solution

Step 1. Calculate the drainage radius of the horizontal well:

$$r_{eh} = \sqrt{\frac{(43,560)(120)}{\pi}} = 1290 \text{ ft}$$

Step 2. Calculate half the major axis of the drainage ellipse by using Eq. (3.46):

$$a = \left[\frac{2000}{2}\right]\left[0.5+\sqrt{0.25+\left[\frac{(2)(1290)}{2000}\right]^4}\right]^{0.5} = 1495.8$$

Step 3. Calculate the effective wellbore radius $r_w^\backslash$ from Eq. (3.45):

$$\left(\frac{h}{2r_w}\right)^{h/L} = \left[\frac{20}{(2)(0.3)}\right]^{20/2000} = 1.0357$$

$$1+\sqrt{1-\left(\frac{L}{2a}\right)^2} = 1+\sqrt{1-\left(\frac{2000}{2(1495.8)}\right)^2} = 1.7437$$

Applying Eq. (3.45) gives:

$$r_w^\backslash = \frac{1290(2000/2)}{1495.8(1.7437)(1.0357)} = 477.54 \text{ ft}$$

Step 4. Calculate the flow rate by using the pressure-squared approximation approach by using Eq. (3.48):

$$Q_g = \frac{(1.5)(20)(2000^2 - 1200^2)}{(1422)(640)(0.011826)\left[\ln(1290/477.54) - 0.75 + 0.5\right]}$$
$$= 9594 \text{ Mscf/day}$$

Step 5. Calculate the flow rate by using the ψ approach as described by Eq. (3.49):

$$Q_g = \frac{(1.5)(20)(340-128)(10^6)}{(1422)(640)\left[\ln(1290/477.54)-0.75+0.5\right]} = 9396 \text{ Mscf/day}$$

For turbulent flow, Darcy's equation must be modified to account for the additional pressure caused by the non-Darcy flow by including the rate-dependent skin factor DQ_g. In practice, the back-pressure equation and the LIT approach are used to calculate the flow rate and construct the IPR curve for the horizontal well. Multirate tests, i.e., deliverability tests, must be performed on the horizontal well to determine the coefficients of the selected flow equation.

3.3 MATERIAL BALANCE EQUATION FOR CONVENTIONAL AND UNCONVENTIONAL GAS RESERVOIRS

Reservoirs that initially contain free gas as the only hydrocarbon system are termed as gas reservoirs. Such a reservoir contains a mixture of hydrocarbon components that exists wholly in the gaseous state. The mixture may be a "dry," "wet," or "condensate" gas, depending on the composition of the gas and the pressure and temperature at which the accumulation exists.

Gas reservoirs may have water influx from a contiguous water-bearing portion of the formation or may be volumetric (i.e., have no water influx).

Most gas engineering calculations involve the use of gas formation volume factor B_g and gas expansion factor E_g. The equations for both these factors are summarized below for convenience.

- Gas formation volume factor B_g is defined as the volume occupied by n moles of gas at certain pressure p and temperature T to that occupied at standard conditions. Applying the real-gas equation of state to both conditions gives:

$$B_g = \frac{p_{sc}}{T_{sc}}\frac{ZT}{p} = 0.02827\frac{ZT}{p} \text{ ft}^3/\text{scf} \quad \textbf{(3.50)}$$

Expressing B_g in bb/scf gives:

$$B_g = \frac{p_{sc}}{5.616 T_{sc}}\frac{ZT}{p} = 0.00504\frac{ZT}{p} \text{ bbl/scf}$$

- The gas expansion factor is simply the reciprocal of B_g, or:

$$E_g = \frac{1}{B_g} = \frac{T_{sc}}{p_{sc}}\frac{p}{ZT} = 35.37\frac{p}{ZT} \text{ scf/ft}^3 \quad \textbf{(3.51)}$$

Expressing E_g in scf/bbl gives

$$E_g = \frac{5.615 T_{sc}}{p_{sc}}\frac{p}{ZT} = 198.6\frac{p}{ZT} \text{ scf/bbl}$$

One of the primary concerns when conducting a reservoir study on a gas field is the determination of the initial gas-in-place G. There are commonly two approaches that are extensively used in natural gas engineering:

(1) the volumetric method;
(2) the material balance approach.

3.3.1 The Volumetric Method

Data used to estimate the gas-bearing reservoir pore volume (PV) include, but are not limited to, well logs, core analyses, bottom-hole pressure (BHP) and fluid sample information, and well tests. This data typically is used to develop various subsurface maps. Of these, structural and stratigraphic cross-sectional maps help to establish the reservoir's areal extent and to identify reservoir discontinuities such as pinch-outs, faults, or gas–water contacts. Subsurface contour maps, usually drawn relative to a known or marker formation, are constructed with lines connecting points of equal elevation and therefore portray the geologic structure. Subsurface isopachous maps are constructed with lines of equal net gas-bearing formation thickness. With these maps, the reservoir PV can then be estimated by planimetering the areas between the isopachous lines and using an approximate volume calculation technique, such as pyramidal or trapezoidal methods.

The volumetric equation is useful in reserve work for estimating gas-in-place at any stage of depletion. During the development period before reservoir limits have been accurately defined, it is convenient to calculate gas-in-place per acre-foot of bulk reservoir rock. Multiplication of this unit figure by the best available estimate of bulk reservoir volume then gives gas-in-place for the lease, tract, or reservoir under consideration. Later in the life of the reservoir, when the reservoir volume is defined and performance data is available, volumetric calculations provide valuable checks on gas-in-place estimates obtained from material balance methods.

The equation for calculating gas-in-place is:

$$G = \frac{43,560Ah\phi(1 - S_{wi})}{B_{gi}} \tag{3.52}$$

with

$$B_{gi} = 0.02827 \frac{Z_i T}{p_i} \ \text{ft}^3/\text{scf}$$

where

G = gas-in-place, scf
A = area of reservoir, acres
h = average reservoir thickness, ft
ϕ = porosity
S_{wi} = water saturation
B_{gi} = gas formation volume factor at initial pressure p_i, ft^3/scf

This equation can be applied at the initial pressure p_i and at a depletion pressure p in order to calculate the cumulative gas production G_p:

$$\text{Gas produced} = \text{Initial gas in place} - \text{Remaining gas}$$

$$G_p = \frac{43,560Ah\phi(1 - S_{wi})}{B_{gi}} - \frac{43,560Ah\phi(1 - S_{wi})}{B_g}$$

or

$$G_p = 43,560Ah\phi(1 - S_{wi})\left(\frac{1}{B_{gi}} - \frac{1}{B_g}\right)$$

Rearranging gives:

$$\frac{1}{B_g} = \frac{1}{B_{gi}} - \left[\frac{1}{43,560Ah\phi(1 - S_{wi})}\right] G_p$$

From the definition of the gas expansion factor E_g, i.e., $E_g = 1/B_g$, the above form of the material balance equation can be expressed as:

$$E_g = E_{gi} - \left[\frac{1}{43,560Ah\phi(1 - S_{wi})}\right] G_p$$

or

$$E_g = E_{gi} - \left[\frac{1}{(\text{PV})(1 - S_{wi})}\right] G_p$$

This relationship indicates that a plot of E_g vs. G_p will produce a straight line with an intercept on the x-axis with a value of E_{gi} and on the y-axis with a value that represents the initial gas-in-place. Note that when $p = 0$, the gas expansion factor is also zero, $E_g = 0$, and that will reduce the above equation to:

$$G_p = (\text{Pore volume})(1 - S_{wi})E_{gi} = G$$

The same approach can be applied at both initial and *abandonment conditions* in order to calculate the recoverable gas.

Applying Eq. (3.52) to the above expression gives:

$$G_p = \frac{43,560Ah\phi(1 - S_{wi})}{B_{gi}} - \frac{43,560Ah\phi(1 - S_{wi})}{B_{ga}}$$

or

$$G_p = 43,560Ah\phi(1 - S_{wi})\left(\frac{1}{B_{gi}} - \frac{1}{B_{ga}}\right) \tag{3.53}$$

where B_{ga} is evaluated at abandonment pressure. Application of the volumetric method assumes that the pore volume occupied by gas is constant. If water influx is occurring, A, h, and S_w will change.

Example 3.5

A gas reservoir has the following characteristics:

$A = 3000$ acres, $h = 30$ ft, $\phi = 0.15$, $S_{wi} = 20\%$
$T = 150\,°F$, $p_i = 2600$ psi, $Z_i = 0.82$

p	Z
2600	0.82
1000	0.88
400	0.92

Calculate the cumulative gas production and recovery factor at 1000 and 400 psi.

Solution

Step 1. Calculate the reservoir PV:

$$PV = 43,560Ah\phi = 43,560(3000)(30)(0.15)$$
$$= 588.06 \ \text{MMft}^3$$

Step 2. Calculate B_g at every given pressure by using Eq. (3.50):

$$B_g = 0.02827 \frac{ZT}{p} \ \text{ft}^3/\text{scf}$$

p	Z	B_g (ft³/scf)
2600	0.82	0.0054
1000	0.88	0.0152
400	0.92	0.0397

Step 3. Calculate initial gas-in-place at 2600 psi:

$$G = \frac{43,560Ah\phi(1 - S_{wi})}{B_{gi}} = \frac{(PV)(1 - S_{wi})}{B_{gi}}$$
$$= 588.06(10^6)(1 - 0.2)/0.0054 = 87.12 \ \text{MMMscf}$$

Step 4. Since the reservoir is assumed volumetric, calculate the remaining gas at 1000 and 400 psi.

Remaining gas at 1000 psi:

$$G_{1000\ \text{psi}} = \frac{(PV)(1 - S_{wi})}{(B_g)_{1000\ \text{psi}}}$$
$$= 588.06(10^6)^n(1 - 0.2)/0.0152$$
$$= 30.95 \ \text{MMMscf}$$

Remaining gas at 400 psi:

$$G_{400\ \text{psi}} = \frac{(PV)(1 - S_{wi})}{(B_g)_{400\ \text{psi}}}$$
$$= 588.06(10^6)(1 - 0.2)/0.0397$$
$$= 11.95 \ \text{MMMscf}$$

Step 5. Calculate cumulative gas production G_p and the recovery factor RF at 1000 and 400 psi.

At 1000 psi:

$$G_p = (G - G_{1000\ \text{psi}}) = (87.12 - 30.95) \times 10^9$$
$$= 56.17 \ \text{MMMscf}$$
$$RF = \frac{56.17 \times 10^9}{87.12 \times 10^9} = 64.5\%$$

At 400 psi:

$$G_p = (G - G_{400\ \text{psi}}) = (87.12 - 11.95) \times 10^9$$
$$= 75.17 \ \text{MMMscf}$$
$$RF = \frac{75.17 \times 10^9}{87.12 \times 10^9} = 86.3\%$$

The recovery factors for volumetric gas reservoirs will range from 80% to 90%. If a strong water drive is present, trapping of residual gas at higher pressures can reduce the recovery factor substantially, to the range of 50–80%.

3.3.2 The Material Balance Method

Material balance is one of the fundamental tools of reservoir engineering. Pletcher (2000) presented excellent documentation of the material balance equation in its various forms and discussed some procedures of improving their performances in predicting gas reserves. If enough production–pressure history is available for a gas reservoir in terms of:

- cumulative gas production G_p as a function of pressure,
- gas properties as a function of pressure at reservoir temperature, and
- the initial reservoir pressure, p_i,

then the gas reserves can be calculated without knowing the areal extent of the reservoir or the drainage area of the well A, thickness h, porosity ϕ, or water saturation S_w. This can be accomplished by forming a mass or mole balance on the gas, as:

$$n_p = n_i - n_f \tag{3.54}$$

where

n_p = moles of gas produced
n_i = moles of gas initially in the reservoir
n_f = moles of gas remaining in the reservoir

Representing the gas reservoir by an idealized gas container, as shown schematically in Figure 3.15, the gas moles in Eq. (3.54) can be replaced by their equivalents using the real-gas law, to give:

$$n_p = \frac{p_{sc} G_p}{Z_{sc} R T_{sc}}$$

$$n_i = \frac{p_i V}{ZRT}$$

$$n_f = \frac{p[V - (W_e - B_w W_p)]}{ZRT}$$

Substituting the above three relationships into Eq. (3.54) and knowing $Z_{sc} = 1$ gives:

$$\frac{p_{sc} G_p}{RT_{sc}} = \frac{p_i V}{ZRT} - \frac{p[V - (W_e - B_w W_p]}{ZRT} \tag{3.55}$$

where

p_i = initial reservoir pressure
G_p = cumulative gas production, scf

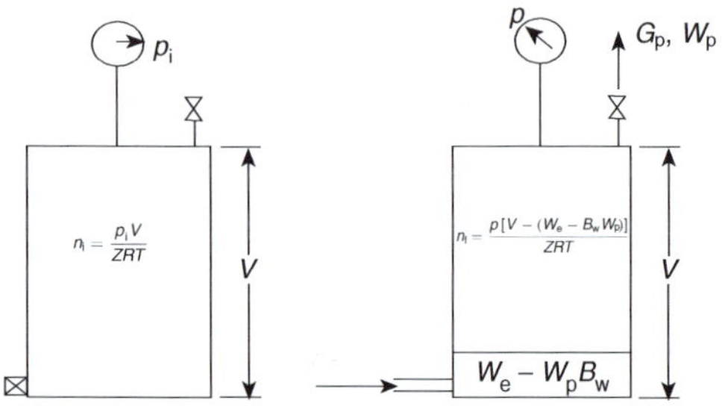

FIGURE 3.15 Idealized water-drive gas reservoir.

p = current reservoir pressure
V = original gas volume, ft^3
Z_i = gas deviation factor at p_i
Z = gas deviation factor at p
T = temperature, °R
W_e = cumulative water influx, ft^3
W_p = cumulative water production, stock-tank ft^3

Eq. (3.55) is essentially the general material balance equation (MBE). It can be expressed in numerous forms depending on the type of the application and the driving mechanism. In general, dry gas reservoirs can be classified into two categories:

(1) volumetric gas reservoirs;
(2) water drive gas reservoirs.

These two types of gas reservoirs are presented next.

3.3.3 Volumetric Gas Reservoirs

For a volumetric reservoir and assuming no water production, Eq. (3.55) is reduced to:

$$\frac{p_{sc} G_p}{T_{sc}} = \left(\frac{p_i}{Z_i T}\right) V - \left(\frac{p}{ZT}\right) V \tag{3.56}$$

Eq. (3.56) is commonly expressed in the following two forms:

(1) in terms of p/Z;
(2) in terms of B_g.

The above two forms of the MBE for volumetric gas reservoirs are discussed below.

Form 1: MBE as Expressed in Terms of *p/Z*. Rearranging Eq. (3.7) and solving for p/Z gives:

$$\frac{p}{Z} = \frac{p_i}{Z_i} - \left(\frac{p_{sc} T}{T_{sc} V}\right) G_p \tag{3.57}$$

or equivalently:

$$\frac{p}{Z} = \frac{p_i}{Z_i} - (m) G_p$$

Eq. (3.57) is the equation of a straight line with a negative slope m, when p/Z is plotted vs.

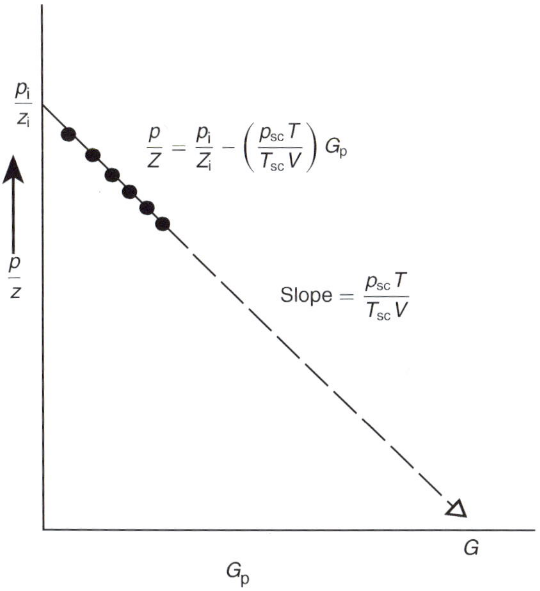

FIGURE 3.16 Gas material balance equation.

the cumulative gas production G_p as shown in Figure 3.16. This straight-line relationship is perhaps one of the most widely used relationships in gas-reserve determination. Eq. (3.57) reveals the straight-line relationship and provides the engineer with the following four characteristics of plot:

(1) Slope of the straight line is equal to:

$$-m = -\frac{p_{sc}T}{T_{sc}V}$$

or

$$V = \frac{p_{sc}T}{T_{sc}m} \tag{3.58}$$

The calculated reservoir gas volume V can be used to determine the areal extent of the reservoir from:

$$V = 43,560Ah\phi(1 - S_{wi})$$

That is:

$$A = \frac{V}{43,560h\phi(1 - S_{wi})}$$

If reserve calculations are performed on a well-by-well basis, the drainage radius of the well can then estimated from:

$$r_e = \sqrt{\frac{43,560A}{\pi}}$$

where A is the area of the reservoir in acres.

(2) Intercept at $G_p = 0$ gives p_i/Z_i.

(3) Intercept at $p/Z = 0$ gives the gas initially in place G in scf. Notice that when $p/Z = 0$, Eq. (3.57) is reduced to:

$$0 = \frac{p_i}{Z_i} - \left(\frac{p_{sc}T}{T_{sc}V}\right)G_p$$

Rearranging:

$$\frac{T_{sc}}{p_{sc}}\frac{p_i}{TZ_i}V = G_p$$

This equation is essentially $E_{gi}V$ and therefore:

$$E_{gi}V = G$$

(4) Cumulative gas production or gas recovery at any pressure.

Example 3.6

A volumetric gas reservoir has the following production history[1] :

Time, *t* (years)	Reservior Pressure, *p* (psia)	*Z*	Cumulative Production, G_p (MMMscf)
0.0	1798	0.869	0.00
0.5	1680	0.870	0.96
1.0	1540	0.880	2.12
1.5	1428	0.890	3.21
2.0	1335	0.900	3.92

The following data is also available:

$\phi = 13\%$, $S_{wi} = 0.52$, $A = 1060$ acres, $h = 54$ ft
$T = 164\ ^\circ F$

Calculate the gas initially in place volumetrically and from the MBE.

[1]After Ikoku, C., 1984. *Natural Gas Reservoir Engineering*. John Wiley & Sons, New York, NY.

Solution

Step 1. Calculate B_{gi} from Eq. (3.50):

$$B_{gi} = 0.02827\frac{(0.869)(164+460)}{1798} = 0.00853\ \text{ft}^3/\text{scf}$$

Step 2. Calculate the gas initially in place volumetrically by applying Eq. (3.52):

$$G = \frac{43,560Ah\phi(1-S_{wi})}{B_{gi}}$$
$$= 43,560(1060)(54)(0.13)(1-0.52)/0.00853$$
$$= 18.2\ \text{MMMscf}$$

Step 3. Plot p/Z vs. G_p as shown in Figure 3.17 and determine G as:

$$G = 14.2\ \text{MMMscf}$$

The value of the gas initially in place as calculated from the MBE compares reasonably with the volumetric value.

The reservoir gas volume V can be expressed in terms of the volume of gas at standard conditions by:

$$V = B_{gi}G = \left(\frac{p_{sc}Z_iT}{T_{sc}\ p_i}\right)G$$

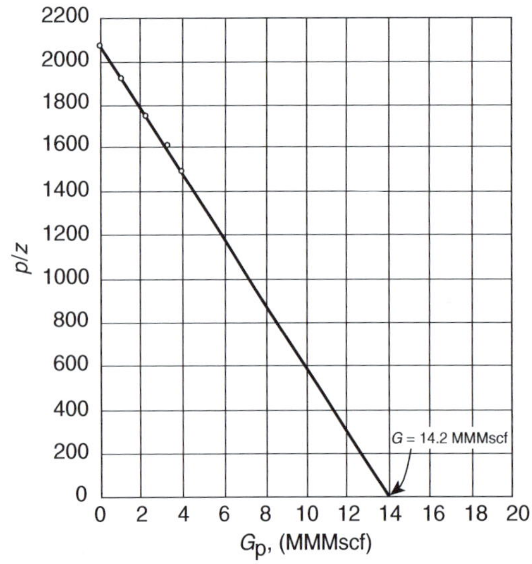

FIGURE 3.17 Relationship of p/z vs. G_p for Example 3.6.

Combining the above relationship with that of Eq. (3.57):

$$\frac{p}{Z} = \frac{p_i}{Z_i} = \left(\frac{p_{sc}T}{T_{sc}V}\right)G_p$$

gives:

$$\frac{p}{Z} = \frac{p_i}{Z_i} - \left[\left(\frac{p_i}{Z_i}\right)\frac{1}{G}\right]G_p \tag{3.59}$$

or

$$\frac{p}{Z} = \frac{p_i}{Z_i} - [m]G_p$$

The above equation indicates that a plot of p/Z vs. G_p would produce a straight line with a slope of m and intercept of p_i/Z_i, with the slope m defined by:

$$m = \left(\frac{p_i}{Z_i}\right)\frac{1}{G}$$

Eq. (3.59) can be rearranged to give:

$$\frac{p}{Z} = \frac{p_i}{Z_i}\left[1 - \frac{G_p}{G}\right] \tag{3.60}$$

Again, Eq. (3.59) shows that for a volumetric reservoir, the relationship between p/Z and G_p is essentially linear. This popular equation indicates that extrapolation of the straight line to the abscissa, i.e., at $p/Z = 0$, will give the value of the gas initially in place as $G = G_p$. Note that when $p/Z = 0$, Eqs. (3.59) and (3.60) give:

$$G = G_p$$

The graphical representation of Eq. (3.59) can be used to detect the presence of water influx, as shown in Figure 3.18. When the plot of p/Z vs. G_p deviates from the linear relationship, it indicates the presence of water encroachment.

Field Average p/Z. From the individual well performance in terms of p/Z vs. G_p, the recovery performance of the entire field can be estimated from the following relationship:

$$\left(\frac{p}{N}\right)_{\text{Field}} = \frac{p_i}{Z_i} - \frac{\sum_{j=1}^{n}(G_p)_j}{\sum_{j=1}^{n}\left[G_p/\frac{p_i}{Z_i} - \frac{p}{Z}\right]_j}$$

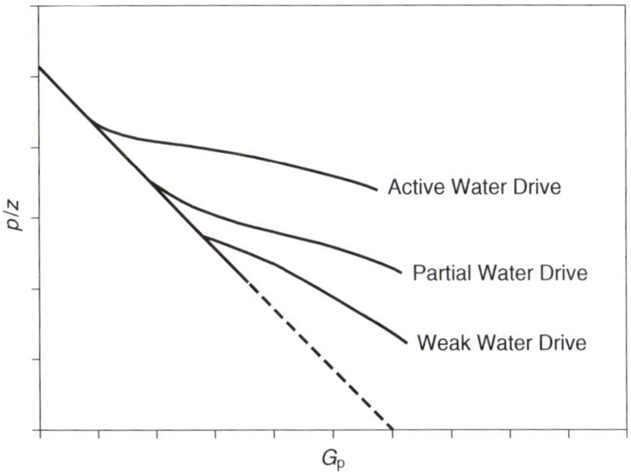

FIGURE 3.18 Effect of water drive on p/z vs. G_p relationship.

The summation $\sum$ is taken over the total number of the field gas wells n, i.e., $j = 1, 2, \ldots, n$. The total field performance in terms of $(p/Z)_{\text{Field}}$ vs. $(G_p)_{\text{Field}}$ can then be constructed from the estimated values of the field p/Z and actual total field production, i.e., $(p/Z)_{\text{Field}}$ vs. $\sum G_p$. The above equation is applicable as long as all wells are producing with defined static boundaries, i.e., under pseudosteady-state conditions.

When using the MBE for reserve analysis for the entire reservoir that is characterized by a distinct lack of pressure equilibrium throughout the reservoir, the following average pressure decline $(p/Z)_{\text{Field}}$ can be used:

$$\left(\frac{p}{Z}\right)_{\text{Field}} = \frac{\sum_{j=1}^{n}\left(\frac{p\,\Delta G_p}{\Delta p}\right)_j}{\sum_{j=1}^{n}\left(\frac{\Delta G_p}{\Delta p/Z}\right)_j}$$

where Δp and ΔG_p are the incremental pressure difference and cumulative production, respectively.

Form 2: MBE as Expressed in Terms of B_g. From the definition of the initial gas formation volume factor, it can be expressed as:

$$B_{gi} = \frac{V}{G}$$

Replacing B_{gi} in the relation with Eq. (3.50) gives:

$$\frac{p_{sc}}{T_{sc}}\frac{Z_i T}{p_i} = \frac{V}{G} \tag{3.61}$$

where

V = volume of gas originally in place, ft^3
G = volume of gas originally in place, scf
p_i = original reservoir pressure
Z_i = gas compressibility factor at p_i

Recalling Eq. (3.57):

$$\frac{p}{Z} = \frac{p_i}{Z_i} - \left(\frac{p_{sc}T}{T_{sc}V}\right)G_p$$

Eq. (3.61) can be combined with Eq. (3.57) to give:

$$G = \frac{G_p B_g}{B_g - B_{gi}} \tag{3.62}$$

Eq. (3.62) suggests that to calculate the initial gas volume, the information required is production data, pressure data, gas specific gravity for obtaining Z factors, and reservoir temperature. However, early in the producing life of a reservoir, *the denominator of the right-hand side of the MBE is very small*, while the numerator is relatively large. A small change in the denominator will result in a large discrepancy in the calculated value of initial gas-in-place. Therefore, the MBE should not be relied on early in the producing life of the reservoir.

Material balances on volumetric gas reservoirs are simple. Initial gas-in-place may be computed from Eq. (3.62) by substituting cumulative gas produced and appropriate gas formation volume factors at corresponding reservoir pressures during the history period. If successive calculations at various times during the history give consistent and constant values for initial gas-in-place, the reservoir is operating *under volumetric control and the computed G is reliable*, as shown in Figure 3.19. Once G has been determined and the absence of water influx established in this fashion, the same equation can be used to make future predictions of cumulative gas production as a function of reservoir pressure.

It should be pointed out that the successive application of Eq. (3.62) can yield increasing or

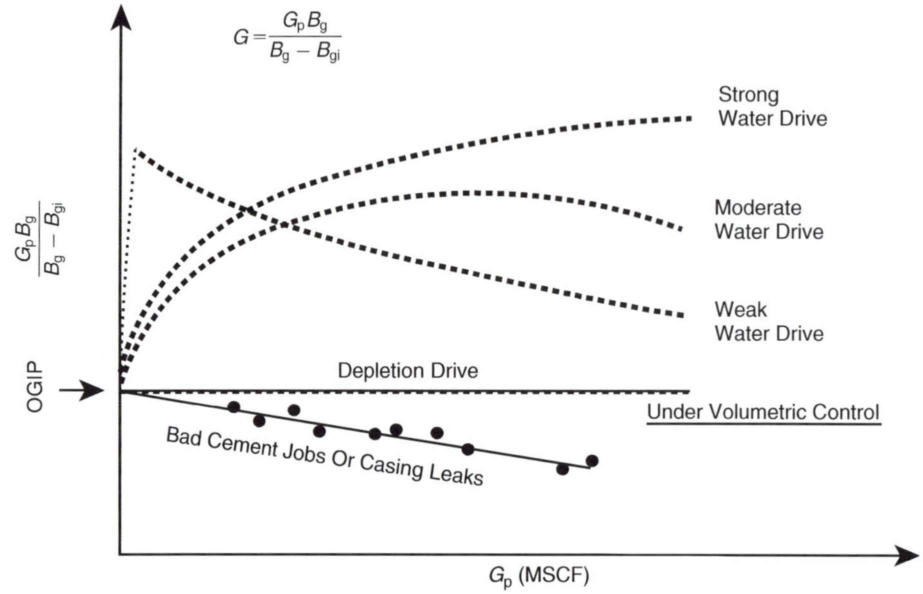

FIGURE 3.19 Gas-in-place in a depletion driver reservoir.

decreasing values of the gas initially in place G. Two different situations therefore exist:

(1) When the calculated value of the gas initially in place G appears to increase with time, the reservoir might be under drive. The invasion of water reduces the pressure drop for a given amount of production, making the reservoir appear larger as time progresses. The reservoir should in this case be classified as a water drive gas reservoir. Another possibility, if no known aquifer exists in the region, is that gas from a different reservoir or zone might migrate through fractures or leaky faults.

(2) If the calculated value of G decreases with time, the pressure drops more rapidly than would be the case in a volumetric reservoir. This implies loss of gas to other zones, leaky cementing job or casing leaks, among other possibilities.

Example 3.7

After producing 360 MMscf of gas from a volumetric gas reservoir, the pressure has declined from 3200 to 3000 psi.

(a) Calculate the gas initially in place, given:

$$B_{gi} = 0.005278 \text{ ft}^3/\text{scf}, \text{ at } p_i = 3200 \text{ psi}$$
$$B_g = 0.005390 \text{ ft}^3/\text{scf}, \text{ at } p = 3000 \text{ psi}$$

(b) Recalculate the gas initially in place assuming that the pressure measurements were incorrect and the true average pressure is 2900 psi, instead of 2900 psi. The gas formation volume factor at this pressure is 0.00558 ft^3/scf.

Solution

(a) Using Eq. (3.14), calculate G:

$$G = \frac{G_p B_g}{B_g - B_{gi}}$$
$$= \frac{360 \times 10^6 (0.00539)}{0.00539 - 0.005278} = 17.325 \text{ MMMscf}$$

(b) Recalculate G by using the correct value of B_g:

$$G = \frac{360 \times 10^6 (0.00558)}{0.00558 - 0.005278} = 6.652 \text{ MMMscf}$$

Thus, an error of 100 psia, which is only 3.5% of the total reservoir pressure, resulted in an increase in calculated gas-in-place of approximately 160%. Note that a similar error in reservoir pressure later in the producing life of the reservoir will not result in an error as large as that calculated early in the producing life of the reservoir.

Gas Recovery Factor. The gas recovery factor (RF) at any depletion pressure is defined as the cumulative gas produced G_p at this pressure divided by the gas initially in place G:

$$\mathrm{RF} = \frac{G_p}{G}$$

Introducing the gas RF into Eq. (3.60) gives:

$$\frac{p}{Z} = \frac{p_i}{Z_i}\left[1 - \frac{G_p}{G}\right]$$

or

$$\frac{p}{Z} = \frac{p_i}{Z_i}[1 - RF]$$

Solving for the RF at any depletion pressure gives:

$$\mathrm{RF} = 1 - \left[\frac{p}{Z}\frac{Z_i}{p_i}\right]$$

3.3.4 Water Drive Gas Reservoirs

The plot of p/Z vs. cumulative gas production G_p is a widely accepted method for solving gas material balance under depletion drive conditions. The extrapolation of the plot to atmospheric pressure provides a reliable estimate of the original gas-in-place. If a water drive is present, the plot often appears to be linear, but the extrapolation will give an erroneously high value for gas-in-place. If the gas reservoir has a water drive, then there will be two unknowns in the MBE, even though production data, pressure, temperature, and gas gravity are known. These two unknowns are initial gas-in-place and cumulative water influx. To use the MBE to calculate initial gas-in-place, some independent method of estimating W_e, the cumulative water influx, must be developed.

Eq. (3.13) can be modified to include the cumulative water influx and water production, to give:

$$G = \frac{G_p B_g - (W_e - W_p B_w)}{B_g - B_{gi}} \tag{3.63}$$

The above equation can be arranged and expressed as:

$$G + \frac{W_e}{B_g - B_{gi}} = \frac{G_p B_g + W_p B_w}{B_g - B_{gi}} \tag{3.64}$$

where

B_g = gas formation volume factor, bbl/scf
W_e = cumulative water influx, bbl

Eq. (3.64) reveals that for a volumetric reservoir, i.e., $W_e = 0$, the right-hand side of the equation will be constant and equal to the initial gas-in-place "G" regardless of the amount of gas G_p that has been produced. That is:

$$G + 0 = \frac{G_p B_g + W_p B_w}{B_g - B_{gi}}$$

For a water drive reservoir, the values of the right-hand side of Eq. (3.64) will continue to increase because of the $W_e/(B_g - B_{gi})$ term. A plot of several of these values at successive time intervals is illustrated in Figure 3.20. Extrapolation of the line formed by these points back to the point where $G_p = 0$ shows the true value of G, because when $G_p = 0$, then $W_e/(B_g - B_{gi})$ is also zero.

This graphical technique can be used to estimate the value of W_e, because at any time the difference between the horizontal line (i.e., true value of G) and the straight line $G + [W_e/(B_g - B_{gi})]$ will give the value of $W_e/(B_g - B_{gi})$.

Because gas often is bypassed and trapped by the encroaching water, recovery factors for gas reservoirs with water drive can be significantly lower than for volumetric reservoirs produced by simple gas expansion. In addition, the presence of reservoir heterogeneities, such as low-permeability stringers or layering, may reduce gas recovery further. As noted previously,

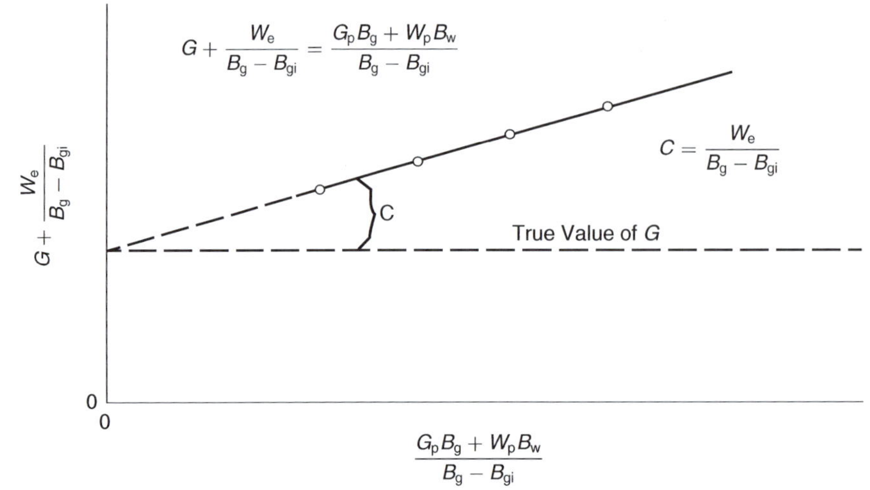

FIGURE 3.20 Effect of water influx on calculating the gas initially in place.

ultimate recoveries of 80–90% are common in volumetric gas reservoirs, while typical recovery factors in water drive gas reservoirs can range from 50% to 70%. The amount of gas that is trapped in the region that has been flooded by water encroachment can be estimated by defining the following characteristic reservoir parameters and the steps as outlined below:

(PV) = reservoir pore volume, ft^3
$(PV)_{water}$ = pore volume of the water-invaded zone, ft^3
S_{grw} = residual gas saturation to water displacement
S_{wi} = initial water saturation
G = gas initially in place, scf
G_p = cumulative gas production at depletion pressure p, scf
B_{gi} = initial gas formation volume factor, ft^3/scf
B_g = gas formation volume factor at depletion pressure p, ft/scf
Z = gas deviation factor at depletion pressure p

Step 1. Express the reservoir pore volume (PV) in terms of the initial gas-in-place G as follows:

$$GB_{gi} = (PV)(1 - S_{wi})$$

Solving for the reservoir pore volume gives:

$$PV = \frac{GB_{gi}}{1 - S_{wi}}$$

Step 2. Calculate the pore volume in the water-invaded zone, as:

$$W_e - W_p B_w - (PV)_{water}(1 - S_{wi} - S_{grw})$$

Solving for the pore volume of the water-invaded zone, $(PV)_{water}$, gives:

$$(PV)_{water} = \frac{W_e - W_p B_w}{1 - S_{wi} - S_{grw}}$$

Step 3. Calculate trapped gas volume in the water-invaded zone, or:

$$\text{Trapped gas volume} = (PV)_{water} S_{grw} = \left(\frac{W_e - W_p B_w}{1 - S_{wi} - S_{grw}}\right) S_{grw}$$

Step 4. Calculate the number of moles of gas n trapped in the water-invaded zone by using the equation of state, or:

$$p\,(\text{trapped gas volume}) = ZnRT$$

Solving for n gives:

$$n = \frac{p((W_e - W_p B_w)/(1 - S_{wi} - S_{grw}))S_{grw}}{ZRT}$$

This indicates that the higher the pressure, the greater the quantity of trapped gas. Dake (1994) pointed out that if the pressure is reduced by rapid gas withdrawal, the volume of gas trapped in each individual pore space, i.e., S_{grw}, will remain unaltered but its quantity n will be reduced.

Step 5. The gas saturation at any pressure can be adjusted to account for the trapped gas as follows:

$$S_g = \frac{\text{Remaining gas volume} - \text{Trapped gas volume}}{\text{Reservior pore volume} - \text{Pore volume of water–invaded zone}}$$

$$S_g = \frac{(G - G_p)B_g - ((W_e - W_p B_w)/(1 - S_{wi} S_{grw}))S_{grw}}{(GB_{gi}/(1 - S_{wi})) - ((W_e - W_p B_w)/(1 - S_{wi} - S_{grw}))}$$

There are several methods of expressing the MBE in a convenient graphical form that can be used to describe the recovery performance of a volumetric or water drive gas reservoir including:

- energy plot;
- MBE as a straight line;
- Cole plot;
- modified Cole plot;
- Roach plot;
- modified Roach plot;
- Fetkovich et al. plot;
- Paston et al. plot;
- Hammerlindl method.

These methods are presented below.

The Energy Plot. Many graphical methods have been proposed for solving the gas MBE that are useful in detecting the presence of water influx. One such graphical technique is called the energy plot, which is based on arranging Eq. (3.60):

$$\frac{p}{Z} = \frac{p_i}{Z_i}\left[1 - \frac{G_p}{G}\right]$$

to give:

$$1 - \left[\frac{p\,Z_i}{Z\,p_i}\right] = \frac{G_p}{G}$$

Taking the logarithm of both sides of this equation:

$$\log\left[1 - \frac{Z_i p}{p_i Z}\right] = \log G_p - \log G \qquad (3.65)$$

Figure 3.21 shows a schematic illustration of the plot.

From Eq. (3.65), it is obvious that a plot of $[1 - (Z_i p)/(p_i Z)]$ vs. G_p on log–log coordinates will yield a straight line with a slope of 1 (45° angle). An extrapolation to 1 on the vertical axis ($p = 0$) yields a value for initial gas-in-place, G. The graphs obtained from this type of analysis have been referred to as energy plots. They have been found to be useful in detecting water influx early in the life of a reservoir. If W_e is not 0, the slope of the plot will be less than 1, and will also decrease with time, since W_e increases with time. *An increasing slope can only occur as a result of either gas leaking from the reservoir or bad data*, since the increasing slope would imply that the gas occupied PV was increasing with time.

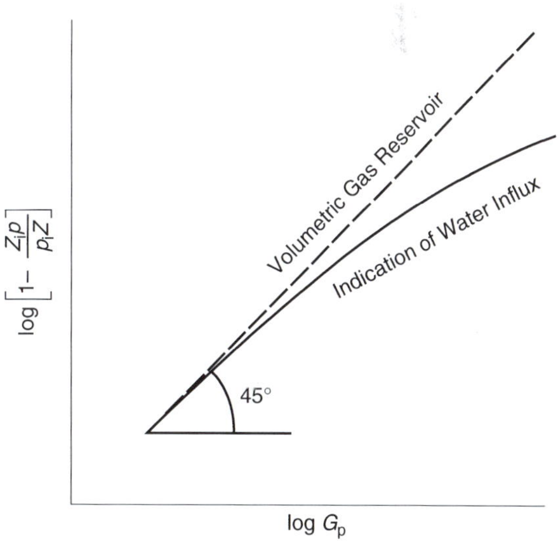

FIGURE 3.21 Energy plot.

Generalized MBE as a Straight Line. Havlena and Odeh (1963, 1964) expressed the material balance in terms of gas production, fluid expansion, and water influx as:

$$\begin{bmatrix}\text{Underground}\\ \text{withdrawal}\end{bmatrix} = [\text{Gas expansion}] + \begin{bmatrix}\text{Water expansion and}\\ \text{pore compaction}\end{bmatrix} + \begin{bmatrix}\text{Water}\\ \text{influx}\end{bmatrix} + [\text{Fluid injection}]$$

and mathematically as:

$$G_pB_g + W_pB_w = G(B_g - B_{gi}) + GB_{gi}\frac{c_wS_{wi}+c_f}{1-S_{wi}}\Delta p + W_e + (W_{inj}B_w + G_{inj}B_{ginj})$$

Assuming no water or gas injection, i.e., W_{inj} and $G_{inj} = 0$, the above generalized MBE reduces to:

$$G_pB_g + W_pB_w = G(B_g - B_{gi}) + GB_{gi}\frac{c_wS_{wi}+c_f}{1-S_{wi}}\Delta p + W_e \tag{3.66}$$

where

$\Delta p = p_i - p$
B_g = gas formation volume factor, bbl/scf

Using the nomenclature of Havlena and Odeh, Eq. (3.66) can be written in the following form:

$$F = G(E_G + E_{f,w}) + W_e \tag{3.67}$$

with the terms F, E_G, and $E_{f,w}$ as defined by:
Underground fluid withdrawal F:

$$F = G_pB_g + W_pB_w \tag{3.68}$$

Gas expansion term E_G:

$$E_G = B_g - B_{gi} \tag{3.69}$$

Water and rock expansion $E_{f,w}$:

$$E_{f,w} = B_{gi}\frac{c_wS_{wi}+c_f}{1-S_{wi}}\Delta p \tag{3.70}$$

Eq. (3.67) can be further simplified by introducing the total system expansion term E_t that combined both compressibilities E_G and $E_{f,w}$ as defined by:

$$E_t = E_G + E_{f,w}$$

to give:

$$F = GE_t + W_e$$

Note that for a volumetric gas reservoir with no water influx or production, Eq. (3.66) is expressed in an expanded form as:

$$G_pB_g = G(B_g - B_{gi}) + GB_{gi}\frac{c_wS_{wi}+c_f}{1-S_{wi}}\Delta p$$

Dividing both sides of the above equation by G and rearranging gives:

$$\frac{G_p}{G} = 1 - \left[1 - \frac{(c_wS_{wi}+c_f)\,\Delta p}{1-S_{wi}}\right]\frac{B_{gi}}{B_g}$$

Inserting the typical values of $c_w = 3\times10^{-6}\ \text{psi}^{-1}$, $c_f = 10\times10^{-6}\ \text{psi}^{-1}$, and $S_{wi} = 0.25$ in the above relationship and considering a large pressure drop of $\Delta p = 1000$ psi, the term in the square brackets becomes:

$$\left[1 - \frac{(c_wS_{wi}+c_f)\,\Delta p}{1-S_{wi}}\right] = 1 - \frac{[3\times0.25+10]10^{-6}(1000)}{1-0.25} = 1-0.014$$

The above value of 0.014 suggests that the inclusion of the term accounting for the reduction in the hydrocarbon PV due to connate water expansion and shrinkage of the PV only alters the material balance by 1.4%, and therefore the term is frequently neglected. The main reason for the omission is that the water and formation compressibilities are usually, although not always, insignificant in comparison with the gas compressibility.

Assuming that the rock and water expansion term $E_{f,w}$ is negligible in comparison with the gas expansion term E_G, Eq. (3.57) is reduced to:

$$F = GE_G + W_e \tag{3.71}$$

Finding the proper model that can be used to determine the cumulative water influx W_e is perhaps the biggest unknown when applying the MBE. The water influx is usually replaced with the analytical aquifer model that must be known or determined from the MBE. The MBE can be expressed as the equation of a straight line by dividing both sides of the above equation by the gas expansion E_G to give:

$$\frac{F}{E_G} = G + \frac{W_e}{E_G} \tag{3.72}$$

The graphical presentation of Eq. (3.72) is given in Figure 3.22. Assuming that the water influx can be adequately described by the van Everdingen and Hurst (1949) unsteady-state model, the selected water influx model can be integrated into Eq. (3.72), to give:

$$\frac{F}{E_G} = G + B\frac{\sum[\Delta p W_{eD}]}{E_G}$$

This expression suggests that a graph of F/E_G vs. $\sum \Delta p W_{eD}/E_G$ will yield a straight line, provided the unsteady-state influx summation, $\sum \Delta p W_{eD}$, is accurately assumed. The resulting straight line intersects the y-axis at the initial gas-in-place G and has a slope equal to the water influx constant B, as illustrated in Figure 3.23.

Nonlinear plots will result if the aquifer is improperly characterized. A systematic upward or downward curvature suggests that the summation term is too small or too large, respectively, while an S-shaped curve indicates that a linear (instead of a radial) aquifer should be assumed. The points should plot sequentially from left to right. A reversal of this plotting sequence indicates that an unaccounted aquifer

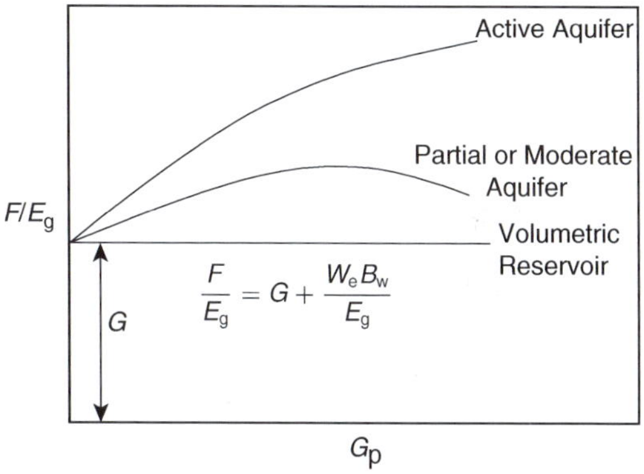

FIGURE 3.22 Defining the reservoir driving mechanism.

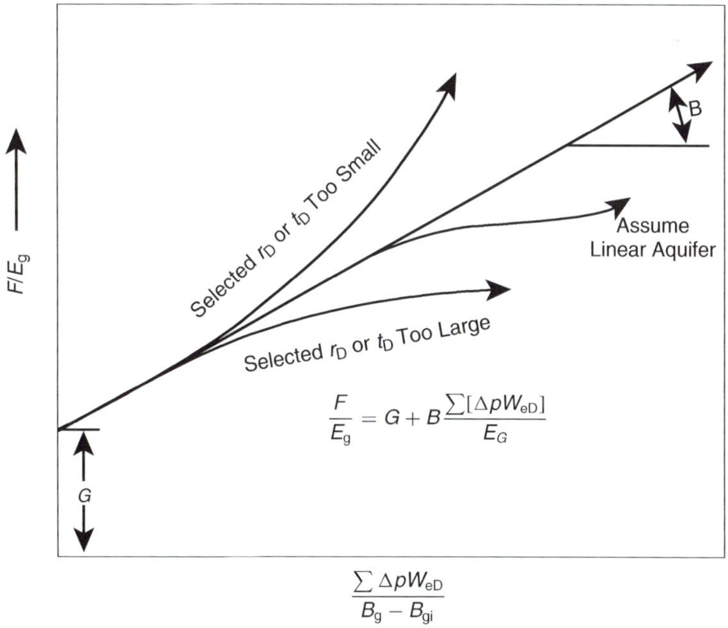

FIGURE 3.23 Havlena–Odeh MBE plot for a gas reservoir.

boundary has been reached and that a smaller aquifer should be assumed in computing the water influx term.

A linear infinite system rather than a radial system might better represent some reservoirs, such as reservoirs formed as fault blocks in salt domes. The van Everdingen and Hurst dimensionless water influx W_{eD} is replaced by the square root of time as:

$$W_e = C \sum \Delta p_n \sqrt{t - t_n} \quad \textbf{(3.73)}$$

where

C = water influx constant, ft³/psi
t = time (any convenient units, i.e., days, years, etc.)

The water influx constant C must be determined by using the past production and pressure of the field in conjunction with the Havlena and Odeh methodology. For the linear system, the underground withdrawal F is plotted vs. $[\sum \Delta p_n \sqrt{t - t_n}/(B_g - B_{gi})]$ on a Cartesian coordinate graph. The plot should result in a straight line with G being the intercept and the water influx constant C being the slope of the straight line.

To illustrate the use of the linear aquifer model in the gas MBE expressed as the equation of a straight line, Havlena and Odeh proposed the following problem.

Example 3.8

The volumetric estimate of the gas initially in place for a dry gas reservoir ranges from 1.3 to 1.65×10^{12} scf. Production, pressures, and the pertinent gas expansion term, i.e., $E_g = B_g - B_{gi}$, are presented in Table 3.1. Calculate the original gas-in-place G.

Solution

Step 1. Assume a volumetric gas reservoir.
Step 2. Plot p/Z vs. G_p or $G_p B_g/(B_g - B_{gi})$ vs. G_p.
Step 3. A plot of $G_p B_g/(B_g - B_{gi})$ vs. $G_p B_g$ shows upward curvature, as shown in Figure 3.24 indicating water influx.
Step 4. Assuming a linear water influx, plot $G_p B_g/(B_g - B_{gi})$ vs. $(\sum \Delta p_n \sqrt{t - t_n})/(B_g - B_{gi})$ as shown in Figure 3.25.
Step 5. As evident from Figure 3.25, the necessary straight-line relationship is regarded as satisfactory evidence for the presence of the linear aquifer.
Step 6. From Figure 3.25, determine the original gas-in-place G and the linear water influx constant C, to give:

$$G = 1.325 \times 10^{12} \text{ scf}$$

$$C = 212.7 \times 10^3 \text{ ft}^3/\text{psi}$$

Drive Indices for Gas Reservoirs. Drive indices have been defined for oil reservoirs (see Chapter 4) to indicate the relative magnitude of the various energy forces contributing to the driving mechanism of the reservoir. Similarly, drive indices can be defined for gas reservoirs by dividing Eq. (3.66) by $G_p B_g + W_p B_w$, to give:

$$\frac{G}{G_p}\left(1 - \frac{B_{gi}}{B_g}\right) + \frac{G}{G_p}\frac{E_{f,w}}{B_g} + \frac{W_e - W_p B_w}{G_p B_g} = 1$$

Define the following three drive indices:

(1) Gas drive index (GDI) as:

$$\text{GDI} = \frac{G}{G_p}\left(1 - \frac{B_{gi}}{B_g}\right)$$

(2) Compressibility drive index (CDI) as:

$$\text{CDI} = \frac{G}{G_p}\frac{E_{f,w}}{B_g}$$

(3) Water drive index (WDI) as:

$$\text{WDI} = \frac{W_e - W_p B_w}{G_p B_g}$$

Substituting the above three indices into the MBE gives:

$$\text{GDI} + \text{CDI} + \text{WDI} = 1$$

TABLE 3.1 Havlena–Odeh Dry Gas Reservoir Data for Example 8.8

Time (months)	Average Reservoir Pressure (psi)	$E_g = (B_g - B_{gi}) \times 10^{-6}$ (ft^3/scf)	$E_g = (G_g - B_g) \times 10^6$ (ft^3)	$\frac{\sum \Delta p_n Z \sqrt{t - t_n}}{B_g - B_{gi}} (10^6)$	$\frac{F}{E_g} = \frac{G_p B_g}{B_g - B_{gi}} (10^{12})$
0	2883	0.0	—	—	—
2	2881	4.0	5.5340	0.3536	1.3835
4	2874	18.0	24.5967	0.4647	1.3665
6	2866	34.0	51.1776	0.6487	1.5052
8	2857	52.0	76.9246	0.7860	1.4793
10	2849	68.0	103.3184	0.9306	1.5194
12	2841	85.0	131.5371	1.0358	1.5475
14	2826	116.5	180.0178	1.0315	1.5452
16	2808	154.5	240.7764	1.0594	1.5584
18	2794	185.5	291.3014	1.1485	1.5703
20	2782	212.0	336.6281	1.2426	1.5879
22	2767	246.0	392.8592	1.2905	1.5970
24	2755	273.5	441.3134	1.3702	1.6136
26	2741	305.5	497.2907	1.4219	1.6278
28	2726	340.0	556.1110	1.4672	1.6356
30	2712	373.5	613.6513	1.5714	1.6430
32	2699	405.0	672.5969	1.5714	1.6607
34	2688	432.5	723.0868	1.6332	1.6719
36	2667	455.5	771.4902	1.7016	1.6937

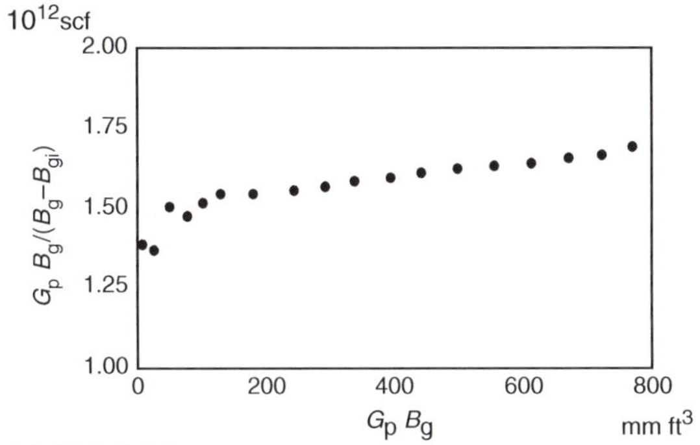

FIGURE 3.24 Indication of the water influx.

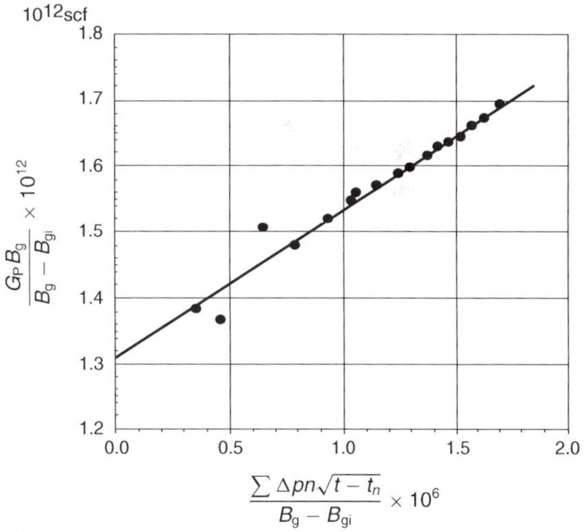

FIGURE 3.25 Havlena–Odeh MBE plot for Example 3.8.

Pletcher (2000) pointed out that if the drive indices do not sum to 1.0, this indicates that the solution to the MBE has not been obtained or is simply incorrect. In practice, however, drive indices calculated from actual field data rarely sum exactly to 1.0 unless accurate recording of production data is achieved. The summed drive indices typically fluctuate above or below 1 depending on the quality of the collected production data with time.

The Cole Plot. The Cole plot is a useful tool for distinguishing between water drive and depletion drive reservoirs. The plot is derived from the generalized MBE as given in an expanded form by Eq. (3.64) as:

$$\frac{G_p B_g + W_p B_w}{B_g - B_{gi}} = G + \frac{W_e}{B_g - B_{gi}}$$

or in a compact form by Eq. (3.72) as:

$$\frac{F}{E_G} = G + \frac{W_e}{E_G}$$

Cole (1969) proposed ignoring the water influx term W_e/E_G and simply plotting the left-hand side of the above expression as a function of the cumulative gas production, G_p. This is simply for display purposes to inspect its variation during depletion. Plotting F/E_G vs. production time or pressure decline, Δp, can be equally illustrative.

Dake (1994) presented an excellent discussion of the strengths and weaknesses of the MBE as a straight line. He pointed out that the plot will have one of the three shapes depicted previously in Figure 3.19. If the reservoir is of the volumetric depletion type, $W_e = 0$, then the values of F/E_G evaluated, say, at six monthly intervals, should plot as a straight line parallel to the abscissa, whose ordinate value is the gas initially in place. Alternatively, if the reservoir is affected by natural water influx, then the plot of F/E_G will usually produce a concave-downward-shaped arc whose exact form is dependent upon the aquifer size and strength and the gas off-take rate. Backward extrapolation of the F/E_G trend to the ordinate should nevertheless provide an estimate of the gas initially in place ($W_e \sim 0$); however, the plot can be highly nonlinear in this region yielding a rather uncertain result. The main advantage in the F/E_G vs. G_p plot, however, is that it is much more sensitive than other methods in establishing whether the reservoir is being influenced by natural water influx or not.

However, in the presence of a weak water drive, the far right-hand term in the above expression, i.e., $[W_e/(B_g - B_{gi})]$, would decrease with time because the denominator would increase faster than the numerator. Therefore, the plotted points will exhibit a negative slope as shown in Figure 3.19. As reservoir depletion progresses in a weak water drive reservoir, the points migrate vertically down and to the right toward the time value of G. Therefore, under a weak water drive, the apparent initial gas-in-place decreases with time, contrary to that for a strong or moderate water drive. Pletcher (2000) pointed out that the weak water drive curve begins with a positive slope in the very early stages of reservoir depletion (as shown in Figure 3.19) prior to developing the signature negative slope. The very early points are difficult to use for determining G because they frequently exhibit a wide scatter behavior that is introduced by even small errors in pressure measurements early in the reservoir life. Therefore, the curve is a "hump-shaped" curve similar to the moderate water drive with the exception that the positive-slope portion of the hump is very short and in practice will not appear if early data is *not* obtained.

Modified Cole Plot. Pore compressibility can be very large in shallow unconsolidated reservoirs with values in excess of $100 \times 10^{-6}\ \text{psi}^{-1}$. Such large values have been measured, for instance, in the Bolivar Coast Fields in Venezuela, and therefore it would be inadmissible to omit c_f from the gas MBE. In such cases, the term $E_{f,w}$ should be included when constructing the Cole plot and the equation should be written as:

$$\frac{F}{E_t} = G + \frac{W_e}{E_t}$$

As pointed out by Pletcher, the left-hand term F/E_t now incorporates energy contributions from the formation (and water) compressibility as well as the gas expansion. The modified Cole plot shows F/E_t on the y-axis vs. G_p on the x-axis. Vertically, the points will lie closer to the true value of G than the original Cole plot. In reservoirs where formation compressibility is a significant contributor to reservoir energy, such as abnormally pressured reservoirs, the original Cole plot will exhibit a negative slope even if no water drive is present.

The modified plot, however, will be a horizontal line assuming the correct value of c_f is used in calculating the term F/E_t. Thus, constructing *both* the original and modified Cole plots will distinguish between the following two possibilities:

(1) Reservoirs that are subject to *weak aquifer and significant* c_f. In this case, both plots, i.e., the original and modified Cole plots, will have a negative slope.
(2) Reservoirs where c_f *is significant but there is no aquifer attached.* In this particular case, the original Cole plot will have a negative slope, while the modified plot will be horizontal.

It should be pointed out that negative slopes in the original and modified Cole plots could result from any unaccounted-for source of energy that is decreasing with time relative to gas expansion. This could include, for example, communication with other depleting reservoirs.

An "abnormally pressured" gas reservoir (sometimes called an "overpressured" or "geopressured" gas reservoir) is defined as a reservoir with pressures greater than a normal pressure gradient, i.e., over 0.5 psi/ft. A typical p/Z vs. G_p plot for an abnormally pressured gas reservoir will exhibit two straight lines as shown in Figure 3.26:

(1) The first straight line corresponds to the "apparent" gas reservoir behavior with an extrapolation that gives the "apparent gas-in-place G_{app}."
(2) The second straight line corresponds to the "normal pressure behavior" with an extrapolation that gives the "actual initial gas-in-place G."

Hammerlindl (1971) pointed out that in abnormally high-pressure volumetric gas reservoirs, two distinct slopes are evident when the plot of p/Z vs. G_p is used to predict reserves because of the formation and fluid compressibility effects as shown in Figure 3.26. The final slope of the p/Z plot is steeper than the initial slope; consequently, reserve estimates based on the early life portion of the curve are

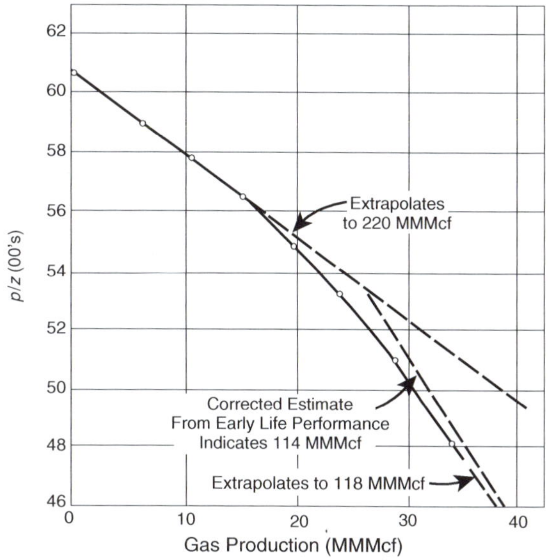

FIGURE 3.26 p/Z vs. cumulative production—North Ossum Field, Lafayette Parish, Louisiana NS2B reservoir. *(After Hammerlindl, D.J., 1971. Predicting gas reserves in abnormally pressure reservoirs. Paper SPE 3479, presented at the 46th Annual Fall Meeting of SPE–AIME. New Orleans, LA, October, 1971).*

erroneously high. The initial slope is due to gas expansion and significant pressure maintenance brought about by formation compaction, crystal expansion, and water expansion. At approximately normal pressure gradient, the formation compaction is essentially complete and the reservoir assumes the characteristics of a normal gas expansion reservoir. This accounts for the second slope. Most early decisions are made based on the early life extrapolation of the p/Z plot; therefore, the effects of hydrocarbon PV change on reserve estimates, productivity, and abandonment pressure must be understood.

All gas reservoir performance is related to effective compressibility, not gas compressibility. When the pressure is abnormal and high, the effective compressibility may equal two or more times the gas compressibility. If the effective compressibility is equal to twice the gas compressibility, then the first cubic foot of gas produced is due to 50% gas expansion and 50% formation compressibility and water expansion. As the pressure is lowered in the reservoir, the contribution due to gas expansion becomes greater because the gas compressibility is approaching the effective compressibility. Using formation compressibility, gas production, and shut-in bottom-hole pressures, two methods are presented for correcting the reserve estimates from the early life data (assuming no water influx).

Gunawan Gan and Blasingame (2001) provided a comprehensive literature review of the methods and theories that have been proposed to explain the nonlinear behavior of p/Z vs. G_p. There are essentially two theories for such behavior:

(1) rock collapse theory;
(2) shale water influx theory.

These theories are briefly addressed below.

Rock Collapse Theory. Harville and Hawkins (1969) suggested that the nonlinear behavior that is characterized with two straight-line plots can be attributed to "pore collapse" and formation compaction. They concluded from a study on the North Ossum Field (Louisiana) that the initial slope is a result of the continuous increase in the *net* overburden pressure as the pore pressure declines with production. This increase in the net overburden pressure causes rock failure, i.e., rock collapse, which subsequently causes a continuous decrease in the rock compressibility c_f. This process continues until c_f eventually reaches a "normal value" that marks the beginning of the second slope. At this point, the reservoir performance becomes similar to that for a constant-volume, normally pressured, gas reservoir system.

Shale Water Influx Theory. Several investigators have attributed the nonlinear behavior of p/Z vs. G_p to shale water influx or peripheral water influx from a limited aquifer and the treatment of PV compressibility as a constant. Bourgoyne (1990) demonstrated that reasonable values of shale permeability and compressibility, treated as a function of pressure, can be used to match abnormal gas reservoir performance behavior to yield the first straight line. The second straight line is a result of a decrease in pressure support from the surrounding shales as the gas reservoir is depleted.

Fetkovich et al. (1998) differentiated between two different PV compressibilities: the "total" and the "instantaneous." The total PV compressibility is defined mathematically by the following expression:

$$\bar{c}_f = \frac{1}{(PV)_i}\left[\frac{(PV)_i - (PV)_p}{p_i - p}\right]$$

The term in the square brackets is the slope of the chord from initial condition (P_i, $(PV)_i$) to any lower pressure (P, $(PV)_p$), where

$\bar{c}_f$ = cumulative pore volume (formation or rock) compressibility, psi^{-1}
p_i = initial pressure, psi
p = pressure, psi
$(PV)_i$ = pore volume at initial reservoir pressure
$(PV)_p$ = pore volume at pressure p

The instantaneous pore volume (rock or formation) compressibility is defined as:

$$c_f = \frac{1}{(PV)_p}\frac{\partial(PV)}{\partial p}$$

The instantaneous compressibility c_f should be used in reservoir simulation, while the cumulative compressibility $\bar{c}_f$ must be used with forms of the material balance that apply cumulative pressure drop $(p_i - p)$.

Both types of compressibilities are pressure dependent and best determined by special core analysis. An example of this analysis is shown below for a Gulf Coast sandstone as given by Fetkovich et al.:

p (psia)	**$p_i - p$ (psi)**	**$(PV)_i - (PV)_p$ (cm³)**	**$\bar{c}_f$ (10^{-6} psi^{-1})**	**c_f (10^{-6} psi^{-1})**
p_i = 9800	0	0.000	16.50	16.50
9000	800	0.041	14.99	13.70
8000	1800	0.083	13.48	11.40
7000	2800	0.117	12.22	9.10
6000	3800	0.144	11.08	6.90
5000	4800	0.163	9.93	5.00
4000	5800	0.177	8.92	3.80
3000	6800	0.190	8.17	4.10
2000	7800	0.207	7.76	7.30
1000	8800	0.243	8.07	16.80
500	9300	0.276	8.68	25.80

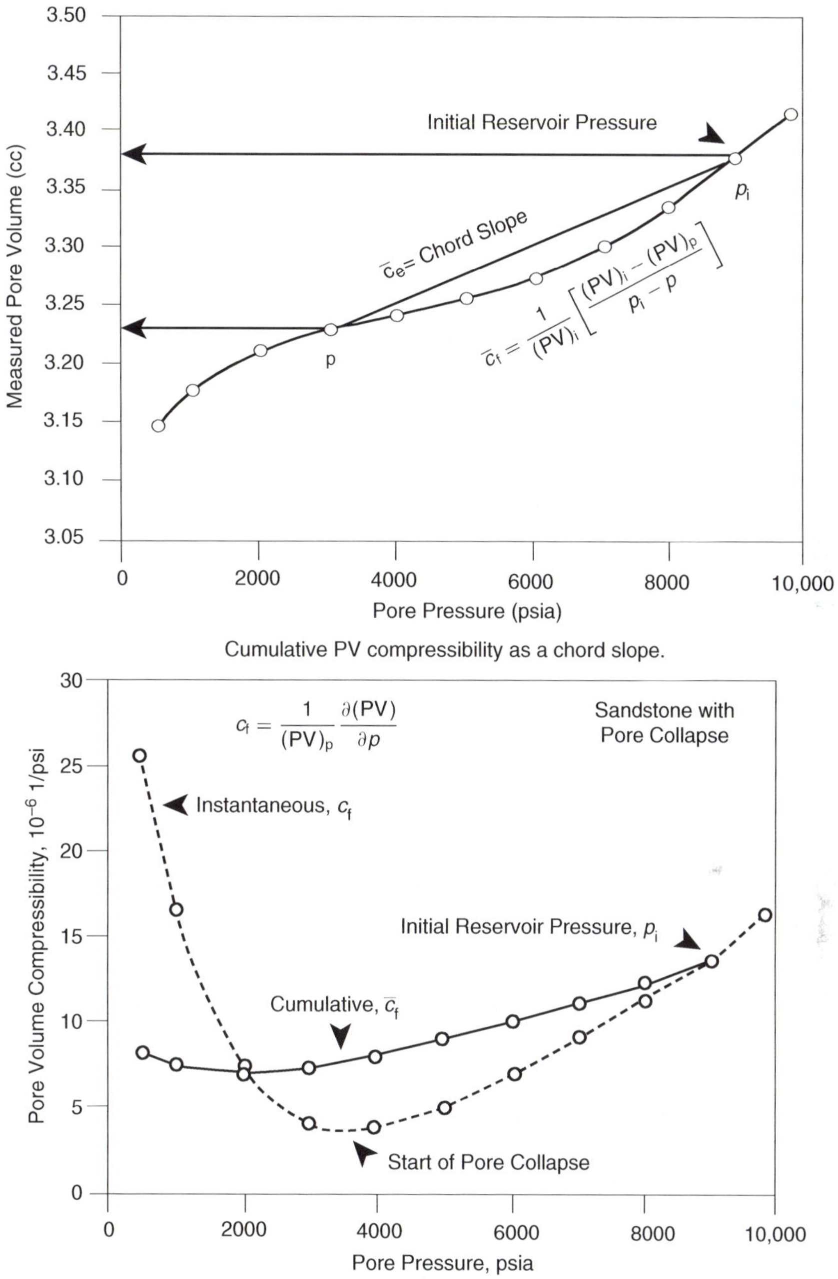

FIGURE 3.27 Cumulative and instantaneous c_f.

Figure 3.27 shows how c_f and $\bar{c}_f$ vary as a function of pressure for this overpressured Gulf Coast sandstone reservoir. Figure 3.27 gives the proper definition of the "pore collapse," which is the condition when the instantaneous PV compressibility begins to increase at decreasing reservoir pressure.

Roach Plot for Abnormally Pressured Gas Reservoirs. Roach (1981) proposed a graphical technique for analyzing abnormally pressured gas reservoirs. The MBE as expressed by Eq. (3.66) may be written in the following form for a volumetric gas reservoir:

$$\left(\frac{p}{Z}\right)c_t = \left(\frac{p_i}{Z_i}\right)\left[1 - \frac{G_p}{G}\right] \quad (3.74)$$

where

$$c_t = 1 - \frac{(c_f + c_w S_{wi})(p_i - p)}{1 - S_{wi}} \quad (3.75)$$

Defining the rock expansion term E_R as:

$$E_R = \frac{c_f + c_w S_{wi}}{1 - S_{wi}} \quad (3.76)$$

Eq. (3.75) can be expressed as:

$$c_t = 1 - E_R(p_i - p) \quad (3.77)$$

Eq. (3.74) indicates that plotting the term $(p/Z)c_t$ vs. cumulative gas production G_p on Cartesian coordinates results in a straight line with an x intercept at the original gas-in-place and a y intercept at the original $(p/Z)_i$. Since c_t is unknown and must be found by choosing the compressibility values resulting in the best straight-line fit, this method is a trial-and-error procedure.

Roach used the data published by Duggan (1972) for the Mobil–David Anderson Gas Field to illustrate the application of Eqs. (3.74) and (3.77) to determine graphically the gas initially in place. Duggan reported that the reservoir had an initial pressure of 9507 psig at 11,300 ft. Volumetric estimates of original gas-in-place indicated that the reservoir contains 69.5 MMMscf. The historical p/Z vs. G_p plot produced an initial gas-in-place of 87 MMMscf, as shown in Figure 3.28.

Using the trial-and-error approach, Roach showed that a value of the rock expansion term E_r of 1805×10^{-6} would result in a straight line with an initial gas-in-place of 75 MMMscf, as shown in Figure 3.28.

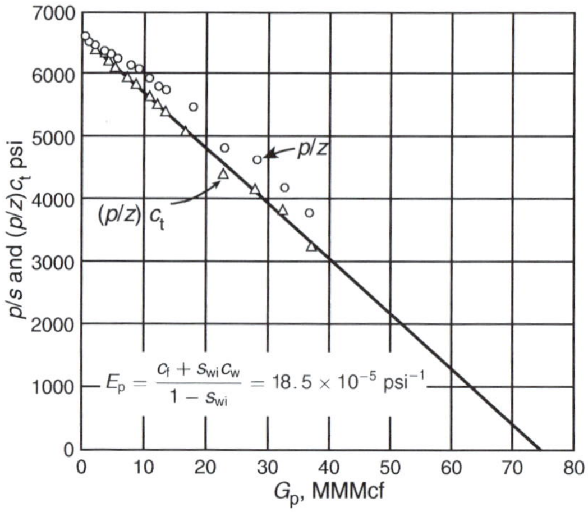

FIGURE 3.28 Mobil–David Anderson "L" p/Z vs. cumulative production. *(After Roach, R.H., 1981. Analyzing geopressured reservoirs—a material balance technique. SPE Paper 9968, Society of Petroleum Engineers of AIME, Dallas, TX, December, 1981).*

To avoid the trial-and-error procedure, Roach proposed that Eqs. (3.74) and (3.77) can be combined and expressed in a linear form by:

$$\frac{(p/Z)_i/(p/Z) - 1}{p_i - p} = \frac{1}{G}\left[\frac{(p/Z)_i/(p/Z)}{p_i - p}\right]G_p - \frac{S_{wi} c_w + c_f}{1 - S_{wi}} \quad (3.78)$$

or equivalently as:

$$\alpha = \left(\frac{1}{G}\right)\beta - E_R \quad (3.79)$$

with

$$\alpha = \frac{[(p_i/Z_i)/(p/Z)] - 1}{p_i - p} \quad (3.80)$$

$$\beta = \left[\frac{(p_i/Z_i)/(p/Z)}{p_i - p}\right]G_p \quad (3.81)$$

$$E_R = \frac{S_{wi} c_w + c_f}{1 - S_{wi}}$$

where

G = initial gas-in-place, scf
E_R = rock and water expansion term, psi^{-1}
S_{wi} = initial water saturation

Eq. (3.79) shows that a plot of α vs. β will yield a straight line with

$$\text{Slope} = \frac{1}{G}$$

$$y \text{ intercept} = -E_R$$

To illustrate the proposed methodology, Roach applied Eq. (3.79) to the Mobil–David Gas Field with the results as shown graphically in Figure 3.29. The slope of the straight line gives $G = 75.2$ MMMscf and the intercept gives $E_R = 1805 \times 10^{-6}$.

Begland and Whitehead (1989) proposed a method to predict the percentage recovery of volumetric, high-pressured gas reservoirs from the initial pressure to the abandonment pressure when only initial reservoir data is available. The proposed technique allows the PV and water compressibilities to be pressure dependent. The authors derived the following form of the MBE for a volumetric gas reservoir:

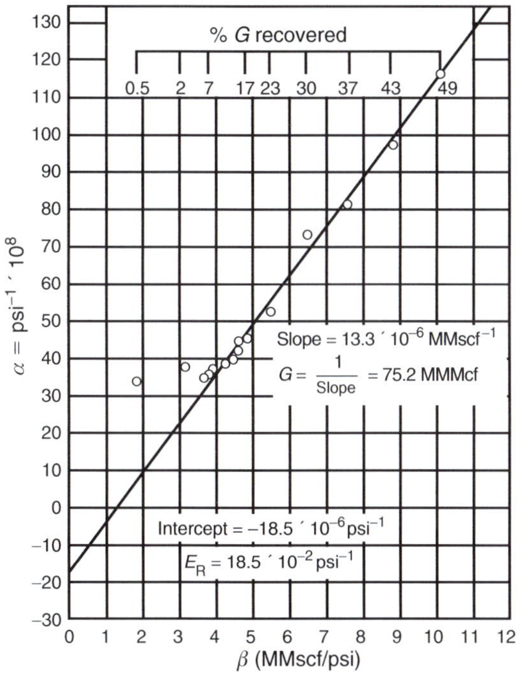

FIGURE 3.29 Mobil–David Anderson "L" *p/Z* gas material balance. *(After Roach, R.H., 1981. Analyzing geopressured reservoirs—a material balance technique. SPE Paper 9968, Society of Petroleum Engineers of AIME, Dallas, TX, December, 1981).*

$$r = \frac{G_p}{G} = \frac{B_g - B_{gi}}{B_g} + \frac{(B_{gi} S_{wi}/(1 - S_{wi}))[((B_{tw}/B_{twi}) - 1) + ((c_f(p_i - p))/S_{wi})]}{B_g} \quad \textbf{(3.82)}$$

where

r = recovery factor
B_g = gas formation volume factor, bbl/scf
c_f = formation compressibility, psi^{-1}
B_{tw} = two-phase water formation volume factor, bbl/STB
B_{twi} = initial two-phase water formation volume factor, bbl/STB

The water two-phase formation volume factor (FVF) is determined from:

$$B_{tw} = B_w + B_g(R_{swi} - R_{sw})$$

where

R_{sw} = gas solubility in the water phase, scf/STB
B_w = water FVF, bbl/STB
B_g = gas FVF, bbl/scf

The following three assumptions are inherent in Eq. (3.82):

(1) a volumetric, single-phase gas reservoir;
(2) no water production;
(3) the formation compressibility c_f remains constant over the pressure drop $(p_i - p)$.

The authors point out that the change in water compressibility c_w is implicit in the change of B_{tw} with pressure as determined above.

Begland and Whitehead suggested that because c_f is pressure dependent, Eq. (3.82) is not correct as reservoir pressure declines from the initial pressure to some value several hundred psi lower. The pressure dependence of c_f can be accounted for in Eq. (3.82) and is solved in an incremental manner.

Modified Roach Plot for Pot Aquifer Gas Reservoirs. Assuming that the aquifer can be described adequately by a pot aquifer model with a total water volume of W_{aq}, the MBE can be arranged to give:

$$\frac{(p/Z)_i/(p/Z)-1}{p_i-p} = \frac{1}{G}\left[\frac{(p/Z)_i/(p/Z)G_p+(W_pB_w/B_{gi})}{p_i-p}\right] - \left[\frac{S_{wi}c_w+c_f}{1-S_{wi}}+\frac{(c_w+c_f)W_{aq}}{GB_{gi}}\right]$$

or equivalently as the equation of a straight line:

$$\alpha = \left(\frac{1}{G}\right)\beta - E_R$$

with

$$\alpha = \frac{[(p_i/Z_i)/(p/Z)]-1}{p_i-p}$$

$$\beta = \left[\frac{(p_i/Z_i)/(p/Z)G_p+(W_pB_w/B_{gi})}{p_i-p}\right]$$

$$E_R = \frac{S_{wi}c_w+c_f}{1-S_{wi}}+\frac{(c_w+c_f)W_{aq}}{GB_{gi}}$$

Plotting α vs. β will produce a straight line with a correct slope of $1/G$ and constant intercept of E_R.

Fetkovich et al. Plot for Abnormal Pressure Gas Reservoirs. Fetkovich et al. (1998) adopted the *shale water influx theory* and developed a general gas MBE that accounts for the total *cumulative* effects of the various reservoir compressibilities as well as the total water associated with the reservoir. The "associated" water includes:

- connate water;
- water within interbedded shales and nonpay reservoir rock;
- volume of water in the attached aquifer.

The authors expressed the associated water as a ratio of the total volume of the associated water to that of the reservoir pore volume, or total associated water volume

$$M = \frac{\text{Total associated water volume}}{\text{Reservoir pore volume}}$$

where M is a dimensionless volume ratio.

In the development of the general MBE, the authors also introduced the cumulative effective compressibility term $\bar{c}_e$ as defined by:

$$\bar{c}_e = \frac{S_{wi}\bar{c}_w + M(\bar{c}_f+\bar{c}_w)+\bar{c}_f}{1-S_{wi}} \quad (3.83)$$

where

$\bar{c}_e$ = cumulative effective compressibility, psi^{-1}
$\bar{c}_f$ = total PV (formation) compressibility, psi^{-1}
$\bar{c}_w$ = cumulative total water compressibility, psi^{-1}
S_{wi} = initial water saturation

The gas MBE can then be expressed as:

$$\frac{p}{Z}[1-\bar{c}_e(p_i-p)] = \frac{p_i}{Z_i} - \left[\frac{p_i/Z_i}{G}\right]G_p \quad (3.84)$$

The $\bar{c}_e$ function represents the cumulative change in hydrocarbon PV caused by compressibility effects and water influx from interbedded shales and nonpay reservoir rock, and water influx from a small limited aquifer. The effect of the compressibility function $\bar{c}_e$ on the MBE depends strongly on the magnitude of $\bar{c}_w, \bar{c}_f$, and the dimensionless parameter M. The nonlinear behavior of the p/Z vs. G_p plot is basically attributed to changes in the magnitude of $\bar{c}_e$ with declining reservoir pressure, as follows:

- The first straight line in the "early-time" trend is developed in the abnormal pressure period where the effect of $\bar{c}_w$ and $\bar{c}_f$ (as described by the $\bar{c}_e$ function) is significant.
- The second straight line in the "late-time" trend is a result of increasing the magnitude of the gas compressibility significantly to dominate the reservoir driving mechanism.

The procedure for estimating the initial gas-in-place G from Eq. (3.84) is summarized in the following steps:

Step 1. Using the available rock and water compressibilities ($\bar{c}_f$ and $\bar{c}_w$ as a function of pressure) in Eq. (3.83), generate a family of $\bar{c}_e$ curves for several assumed values of the dimensionless volume rates M:

$$\overline{c}_e = \frac{S_{wi}\overline{c}_w + M(\overline{c}_f + \overline{c}_w) + \overline{c}_f}{1 - S_{wi}}$$

Step 2. Assume a range of values for G with the largest value based on extrapolation of the early depletion data, and the lowest value being somewhat larger than the current G_p. For an *assumed value* of G, calculate $\overline{c}_e$ from Eq. (3.84) for each measured p/Z and G_p data point, or:

$$\overline{c}_e = \left[1 - \frac{(p/Z)_i}{(p/Z)}\left(1 - \frac{G_p}{G}\right)\right]\frac{1}{p_i - p}$$

Step 3. For a given assumed value of G, plot the calculated values of $\overline{c}_e$ from step 2 as a function of pressure and repeat for all other values of G. This family of $\overline{c}_e$ curves is essentially generated independently from the MBE to *match* the $\overline{c}_e$ values as calculated in step 1.

Step 4. The match gives G, the M value, and the $\overline{c}_e$ function that can be used to predict the p/Z vs. G_p plot by rearranging Eq. (3.84) and assuming several values of p/Z and calculating the corresponding G_p, to give:

$$G_p = G\left\{1 - \left(\frac{Z_i\,p}{p_i\,Z}\right)[1 - \overline{c}_e(p_i - p)]\right\}$$

Paston et al. Plot for Abnormal Pressure Gas Reservoirs. Harville and Hawkins (1969) attributed the concave-downward shape of the p/Z vs. G_p curve for overpressured gas reservoirs to pore collapse and formation compaction. Hammerlindl (1971) calculated the changes in the PV and indicated that the system isothermal compressibility changed from 28×10^{-6} psi^{-1} at initial conditions to 6×10^{-6} psi^{-1} at final condition. Poston and Berg (1997) suggested that the gas MBE can be arranged to solve for the original gas-in-place, formation compressibility, and water influx values simultaneously. The MBE as presented by Eq. (3.66) can be rearranged to give:

$$\frac{1}{\Delta p}\left[\left(\frac{p_i Z}{p Z_i}\right) - 1\right] = \left(\frac{1}{G}\right)\left[\left(\frac{Z p_i}{Z_i p}\right)\left(\frac{G_p}{\Delta p}\right)\right] - (c_e + W_{en})$$

where the energy term for the net water influx W_{en} and effective compressibility c_e are given by:

$$W_{en} = \frac{(W_e - W_p)B_w}{\Delta p G B_{gi}}$$

$$c_e = \frac{c_w S_{wi} + c_f}{1 - S_{wi}}$$

where

G = gas initially in place, scf
B_{gi} = initial gas FVF, bbl/scf
c_w = water compressibility coefficient, psi^{-1}
$\Delta p = p_i - p$

The above form of the MBE indicates that for a volumetric gas reservoir (i.e., $W_e = 0$) with a constant effective compressibility, a plot of the left-hand side of the equation vs. $(Zp_i/Z_ip)(G_p/\Delta p)$ would produce a straight line with a slope of $1/G$ and a negative intercept of $-c_e$ that can be used to solve the above equation for the formation compressibility c_f, to give:

$$c_f = -\,c_e(1 - S_{wi}) - c_w S_{wi}$$

Experience has shown that c_f values should range over $6\times10^{-6}<c_f<25\times10^{-6}$ psi^{-1}, a value over 25×10^{-6} as calculated from the above expression; that is, c_e might indicate water influx.

Hammerlindl Method for Abnormal Pressure Gas Reservoirs. Hammerlindl (1971) proposed two methods to correct apparent gas-in-place G_{app} obtained by extrapolation of the early straight line of the p/Z vs. G_p graph. Both methods use the initial reservoir pressure p_i and another average reservoir pressure p_1 at some time while the reservoir is still behaving as an abnormally pressured reservoir. The proposed mathematical expressions for both methods are given below.

Method I. Hammerlindl suggested that the actual gas-in-place G can be estimated by correcting the apparent gas-in-place G_{app} by incorporating the ratio R of the effective total system compressibility to the gas compressibility, to give:

$$G = \frac{G_{app}}{R}$$

with

$$R = \frac{1}{2}\left(\frac{c_{eff,i}}{c_{gi}} + \frac{c_{eff,1}}{c_{g1}}\right)$$

where the effective total system compressibility $c_{eff,i}$ at the initial reservoir pressure and the effective system compressibility $c_{eff,1}$ at reservoir pressure p_1 are given by:

$$c_{eff,i} = \frac{S_{gi}c_{gi} + S_{wi}c_{wi} + c_f}{S_{gi}}$$

$$c_{eff,1} = \frac{S_{gi}c_{g1} + S_{wi}c_{w1} + c_f}{S_{gi}}$$

where

p_i = initial reservoir pressure, psi
p_1 = average reservoir pressure during the abnormally pressured behavior, psi
c_{gi} = gas compressibility at p_i, psi^{-1}
c_{g1} = gas compressibility at p_1, psi^{-1}
c_{wi} = water compressibility at p_i, psi^{-1}
c_{w1} = water compressibility at p_1, psi^{-1}
S_{wi} = initial water saturation

Method II. Hammerlindl's second method also uses two pressures p_i and p_1 to compute actual gas-in-place from the following relationship:

$$G = \text{Corr}\, G_{app}$$

where the correction factor "Corr" is given by:

$$\text{Corr} = \frac{(B_{g1} - B_{gi})S_{gi}}{(B_{g1} - B_{gi})S_{gi} + B_{gi}(p_i - p_1)(c_f + c_w S_{wi})}$$

and B_g is the gas formation volume factor at p_i and p_1 as expressed in ft^3/scf by:

$$B_g = 0.02827\frac{ZT}{p}$$

Effect of Gas Production Rate on Ultimate Recovery. Volumetric gas reservoirs are essentially depleted by expansion and, therefore, the ultimate gas recovery is independent of the field production rate. The gas saturation in this type of reservoirs is never reduced, only the number of pounds of gas occupying the pore spaces is reduced. Therefore, it is important to reduce the abandonment pressure to the lowest possible level. In closed gas reservoirs, it is not uncommon to recover as much as 90% of the initial gas-in-place.

Cole (1969) pointed out that for water drive gas reservoirs, recovery may be rate dependent. There are two possible influences that the producing rate may have on ultimate recovery. First, in an active water drive reservoir, the abandonment pressure may be quite high, sometimes only a few psi below the initial pressure. In such a case, the gas remaining in the pore spaces at abandonment will be relatively great. However, the encroaching water reduces the initial gas saturation. Therefore, the high abandonment pressure is somewhat offset by the reduction in initial gas saturation. If the reservoir can be produced at a rate greater than the water influx rate, without water coning, then a high producing rate could result in maximum recovery by taking advantage of a combination of reduced abandonment pressure and reduction in initial gas saturation. Second, the water-coning problems may be very severe in gas reservoirs, in which case it will be necessary to restrict withdrawal rates to reduce the magnitude of this problem.

Cole suggested that recovery from water drive gas reservoirs is substantially less than that from closed gas reservoirs. As a rule of thumb, recovery from a water drive reservoir will be approximately 50% to 0% of the initial gas-in-place. The structural location of producing wells and the degree of water coning are important considerations in determining ultimate recovery. A set of circumstances could exist—such as the location of wells very high on the structure with very little coning tendencies—where water drive recovery would be greater than depletion drive recovery. Abandonment pressure is a major factor in determining recovery efficiency, and permeability is usually the most important factor in determining the magnitude of the abandonment pressure. Reservoirs with low permeability will have higher abandonment pressures than reservoirs with high permeability. A certain minimum flow rate must be sustained, and a higher permeability will permit this minimum flow rate at a lower pressure.

3.4 COALBED METHANE

The term "coal" refers to sedimentary rocks that contain more than 50% by weight and more than 70% by volume of organic materials consisting mainly of carbon, hydrogen, and oxygen in addition to inherent moisture. Coals generate an extensive suite of hydrocarbons and nonhydrocarbon components. Although the term "methane" is used frequently in the industry, in reality the produced gas is typically a mixture of C_1, C_2, traces of C_3, and heavier N_2 and CO_2. Methane, as one such hydrocarbon constituent of coal, is of special interest for two reasons:

(1) Methane is usually present in high concentration, in coal, depending on composition, temperature, pressure, and other factors.
(2) Of the many molecular species trapped within coal, methane can be easily liberated by simply reducing the pressure in the bed. Other hydrocarbon components are tightly held and generally can be liberated only through different extraction methods.

Levine (1991) suggested that the materials comprising a coalbed fall broadly into the following two categories:

(1) "volatile" low-molecular-weight materials (components) that can be liberated from the coal by pressure reduction, mild heating, or solvent extraction;
(2) materials that will remain in the solid state after the separation of volatile components.

Most of the key data needed for estimating gas-in-place and performing other performance calculations is obtained mainly from the following core tests:

- *Canister desorption tests*: These tests are conducted on coal samples to determine:
  - the total adsorbed gas content G_c of the coal sample as measured in scf/ton of coal;
  - desorption time t that is defined by the time required to disrobe 63% of the total adsorbed gas.
- *Proximate tests*: These tests are designed to determine coal composition in terms of:
  - percentage of ash;
  - fixed carbon;
  - moisture content;
  - volatile matter.

Remner et al. (1986) presented a comprehensive study on the effects of coal seam properties on the coalbed methane drainage process. The authors pointed out that reservoir characteristics of coalbeds are complex because they are naturally fractured reservoirs that are characterized by two distinct porosity systems, i.e., dual-porosity systems. These are:

(1) *Primary porosity system*: The matrix primary porosity system in these reservoirs is composed of very fine pores, "micropores," with extremely low permeability. These micropores contain a large internal surface area on which substantial quantities of gas may be adsorbed. With such low permeability, the primary porosity is both impermeable to gas and inaccessible to water. However, the desorbed gas can flow (transport) through the primary porosity system by the diffusion process, as discussed later in this section. The micropores are essentially responsible for most of the porosity in coal.
(2) *Secondary porosity system*: The secondary porosity system (macropores) of coal seams consists of the natural fracture network of cracks and fissures inherent in all coals. The macropores, known as cleats, act as a sink to the primary porosity system and provide

the permeability for fluid flow. They act as conduits to the production wells as shown in Figure 3.30. The cleats are mainly composed of the following two major components:

(a) *The face cleat*: The face cleat, as shown conceptually in Figure 3.30 by Remner et al., is continuous throughout the reservoir and is capable of draining large areas.
(b) *The butt cleat*: Butt cleats contact a much smaller area of the reservoir and thus are limited in their drainage capacities.

In addition to the cleat system, a fracture system caused by tectonic activity may also be present in coals. Water and gas flow to coalbed methane wells occurs in the cleat and fracture systems. These cleats and fractures combine to make up the bulk permeability measured from well tests conducted on coalbed methane wells.

The bulk of the methane, i.e., gas-in-place, is stored in an adsorbed state on internal coal surfaces and is considered a near liquid-like state as opposed to a free gas phase. The coal cleats are considered initially saturated with water and must be removed (produced) from the natural fractures, i.e., cleats, to lower the reservoir pressure. When the pressure is reduced, the gas is released (desorbed) from the coal matrix into the fractures. The gas production is then controlled by a four-step process that includes:

Step 1. Removal of water from the coal cleats and lowering the reservoir pressure to

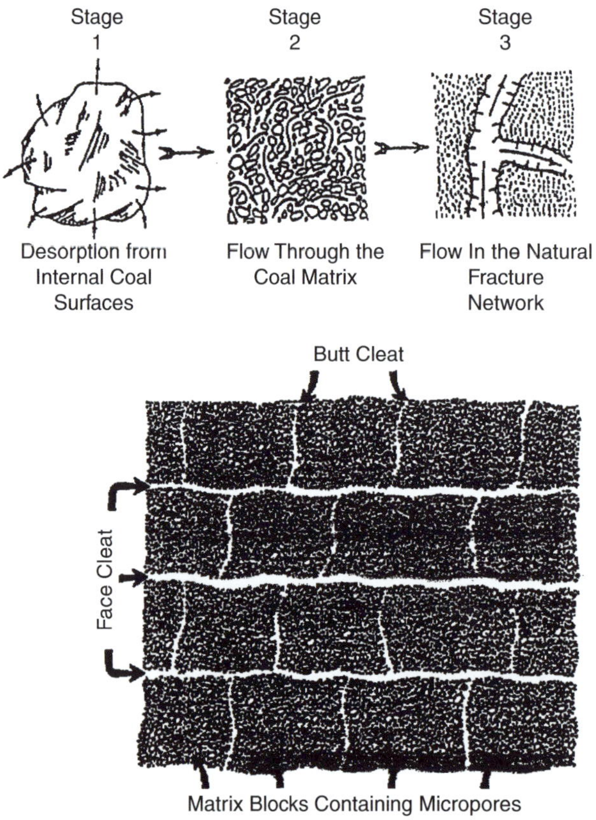

FIGURE 3.30 Schematic of methane flow dynamics in a coal seam system. *(After King, G., Ertekin, T., Schwerer, F., 1986. Numerical simulation of the transient behavior of coal seam wells. SPE Form. Eval. 1 (2), 165–183).*

that of the gas desorption pressure. This process is called dewatering the reservoir.

Step 2. Desorption of gas from the coal internal surface area.

Step 3. Diffusion of the desorbed gas to the coal cleat system.

Step 4. Flow of the gas through fractures to the wellbore.

The economical development of coalbed methane (CBM) reservoirs depends on the following four coal seam characteristics:

(1) gas content G_c;
(2) density of the coal ρ_B;
(3) deliverability and drainage efficiency;
(4) permeability and porosity.

Hughes and Logan (1990) pointed out that an economic reservoir must first contain a sufficient amount of adsorbed gas (gas content), must have adequate permeability to produce that gas, must have enough pressure for adequate gas storage capacity, and, finally, the desorption time must be such that it is economical to produce that gas. These four characteristic coal seam parameters that are required to economically develop the reservoir are discussed below.

3.4.1 Gas Content

The gas present in the coal is molecularly adsorbed on the coal's extensive surface area. Gas content estimation methods involve placing freshly cut reservoir coal samples in airtight gas desorption canisters and measuring the volume of gas that desorbs as a function of time at ambient temperature and pressure conditions. A disadvantage of this analysis procedure is that the measured desorbed gas volume is not equal to the total gas content since a large amount of gas is commonly lost by desorption during sample recovery. The volume of gas lost during this core recovery time is referred to as "lost gas." The volume of the lost gas can be estimated by using the USBM direct method, as illustrated in Figure 3.31. The method simply involves plotting the desorbed gas volume vs. the square root of time, $\sqrt{t}$, on a Cartesian scale and extrapolating the early-time desorption data back to time zero. Experience has shown that

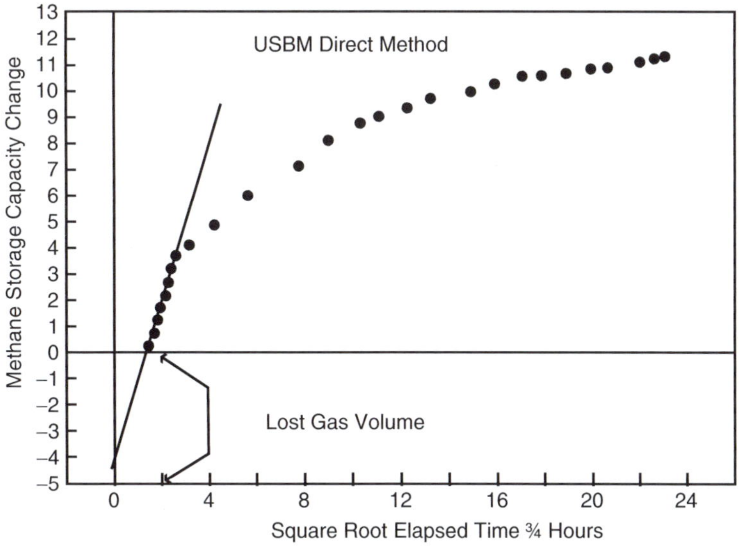

FIGURE 3.31 Plot of test data used to determine lost gas volume.

this technique works adequately in shallow, low-pressure, low-temperature coals with a lost gas volume in the range of 5–10% of the total adsorbed gas content of the coal. However, in higher-pressure coal seams, the lost gas volume may exceed 50% of the total adsorbed gas content.

It should be pointed out that some of the gas may not desorb from coal by the end of desorption measurements and remains absorbed in the core sample. The term "residual gas" is commonly referred to the gas that remains at the end of the desorption test. McLennan and Schafer (1995) and Nelson (1999) pointed out that the rate of gas desorption from coals is so very slow that impracticably long time intervals would be required for complete gas desorption to occur. This residual gas content remaining at the end of desorption measurements is determined by crushing the sample and measuring the released gas volume. The chief limitation of this direct method analysis procedure is that it yields different gas content values depending upon the coal sample type, gas desorption testing conditions, and lost gas estimation method. Nelson (1999) pointed out that the failure to quantify and account for any residual gas volume that may remain in the coal sample at the end of gas desorption measurements would result in significant underestimation error in coalbed gas-in-place evaluations. This residual gas volume can be a significant fraction, ranging between 5% and 50%, of the total adsorbed gas content.

Another important laboratory measurement is known as the "sorption isotherm" and is required to relate the gas storage capacity of a coal sample to pressure. This information is required to predict the volume of gas that will be released from the coal as the reservoir pressure declines. *Note that the gas content G_c is a measurement of the actual (total) gas contained in a given coal reservoir, while the sorption isotherm defines the relationship of pressure to the capacity of a given coal to hold gas at a constant temperature.* Accurate determinations of both gas content and the sorption isotherm are required to estimate recoverable reserve and production profiles. An example of a typical sorption isotherm relationship is shown in Figure 3.32 as given by Mavor et al. (1990). This sorption isotherm was measured

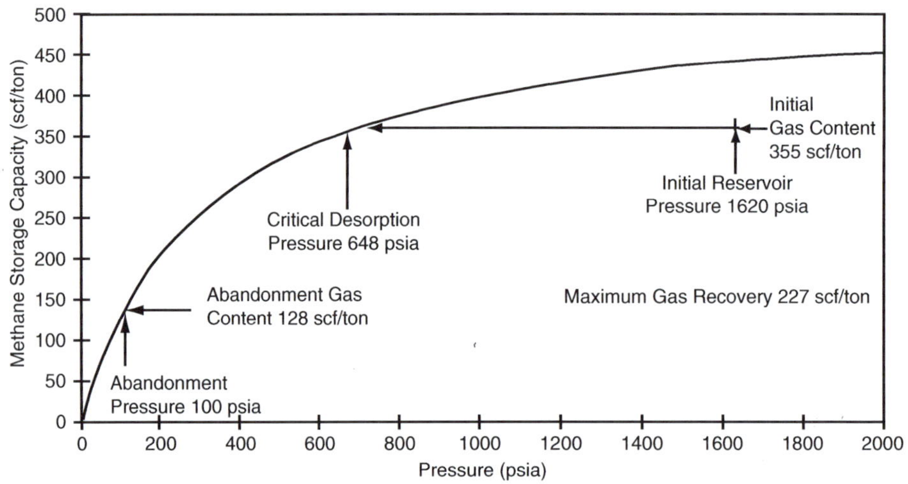

FIGURE 3.32 Sorption isotherm curve. *(After Mavor, M., Close, J., Mcbane, R., 1990. Formation evaluation of coalbed methane wells. Pet. Soc. CIM, CIM/SPE Paper 90-101).*

on a sample collected from a well in the Fruitland Formation Coal Seam of the San Juan Basin, New Mexico. The authors pointed out that the total gas content G_c of the coal was determined to be 355 scf/ton by desorption canister tests performed on whole core samples at the well location. The gas content is less than the sorption isotherm gas storage capacity of 440 scf/ton at the initial reservoir pressure of 1620 psia. This implies that the pressure must be reduced to 648 psia, which corresponds to 355 scf/ton on the sorption isotherm curve. This pressure is known as the critical or desorption pressure p_d. This value will determine whether a coal seam is saturated or undersaturated. A saturated coal seam holds as much adsorbed gas as it possibly can for the given reservoir pressure and temperature. An analogy would be an oil reservoir having a bubble point equal to the initial reservoir pressure. If the initial reservoir pressure is greater than the critical desorption pressure, the coalbed is considered an undersaturated one as in the case of Fruitland Formation Coal. An undersaturated coal seam is undesirable since more water will have to be produced (dewatering process) before gas begins to flow.

For an undersaturated reservoir, i.e., $p_i > p_d$, the total volume of water that must be removed to drop from the initial reservoir pressure p_i to the desorption pressure p_d can be estimated from the total isothermal compressibility coefficient:

$$c_t = \frac{1}{W_i} \frac{W_p}{p_i - p_d} \tag{3.85}$$

where

W_p = total volume of water removed, bbl
W_i = total volume of water in the reservoir (area), bbl
p_i = initial reservoir pressure, psi
p_d = desorption pressure, psi
c_t = total system compressibility coefficient in psi^{-1} as given by:
$c_t = c_w + c_f$

with

c_w = water compressibility
c_f = formation compressibility

Solving Eq. (3.85) for water removed gives:

$$W_p = c_t W_i (p_i - p_d) \tag{3.86}$$

Example 3.9

An undersaturated coal system has the following reservoir parameters:

drainage area = 160 acres, thickness = 15 ft,
porosity = 3%
initial pressure = 650 psia,
desorption pressure = 450 psia
total compressibility = 16×10^{-5} psi^{-1}

Estimate the total volume of water that must be produced for the reservoir pressure to decline from initial pressure to desorption pressure.

Solution

Step 1. Calculate the total volume of water initially in the drainage area:

$$W_i = 7758 A h \phi S_{wi}$$
$$W_i = 7758(160)(15)(0.03)(1.0) = 558,576 \text{ bbl}$$

Step 2. Estimate the total water volume to be produced to reach the desorption pressure from Eq. (3.86):

$$W_p = 16(10^{-5})(558,576)(650 - 450) = 17,874 \text{ bbl}$$

Step 3. Assuming the area is being drained with only one well that is discharging at 300 bbl/day, the total time to reach the desorption pressure is:

$$t = \frac{17,874}{300} = 60 \text{ days}$$

For most coal seams, the quantity of gas held in the coal is primarily a function of coal rank, ash content, and the initial reservoir pressure. The adsorbed capacity of the coal seam varies nonlinearly with pressure. A common method of utilizing sorption isotherm data is to assume that the relationship between gas storage capacity and pressure can be described by a relationship that was originally proposed by Langmuir (1918). The sorption isotherm data that fits this relationship is known as a "Langmuir isotherm" and is given by:

$$V = V_L \frac{p}{p + p_L} \tag{3.87}$$

where

V = volume of gas currently adsorbed at p, scf/ft^3 of coal
V_L = Langmuir's volume, scf/ft^3
p_L = Langmuir's pressure, psi
p = reservoir pressure, psi

Because the amount of gas adsorbed depends on mass of coal, not volume, a more useful form of the Langmuir equation that expresses the adsorbed volume in scf/ton is:

$$V = V_m \frac{bp}{1 + bp} \tag{3.88}$$

where

V = volume of gas currently adsorbed at p, scf/ton
V_m = Langmuir's isotherm constant, scf/ton
b = Langmuir's pressure constant, psi^{-1}
p = pressure, psi

The two sets of Langmuir's constants are related by:

$$V_L = 0.031214 V_m \rho_B$$

and

$$p_L = \frac{1}{b}$$

where ρ_B is the bulk density of the coal deposit in g/cm^3.

The Langmuir pressure b and volume V_m can be estimated by fitting the sorption isotherm data to Eq. (3.88). The equation can be linearized as follows:

$$V = V_m - \left(\frac{1}{b}\right)\frac{V}{p} \tag{3.89}$$

The above relationship suggests that a plot of the desorbed gas volume V vs. the ratio V/p on a Cartesian scale would produce a straight line with a slope of $-1/b$ and intercept of V_m.

Similarly, when expressing the adsorbed gas volume in scf/ft^3, Eq. (3.87) can be expressed as the equation of a straight line to give:

$$V = V_L - p_L\left(\frac{V}{p}\right)$$

A plot of V in scf/ft^3 as a function of V/P would produce a straight line with an intercept of V_L and a negative slope of $-p_L$.

Example 3.10

The following sorption isotherm data is given by Mavor et al. (1990) for a coal sample from the San Juan Basin:

p (psi)	76.0	122.0	205.0	221.0	305.0	504.0	507.0	756.0	1001.0	1008.0
V (scf/ton)	77.0	113.2	159.8	175.0	206.4	265.3	267.2	311.9	339.5	340.5

Calculate the Langmuir isotherm constant V_m and the Langmuir pressure constant b for the San Juan Basin coal sample.

Solution

Step 1. Calculate V/p for each of the measured data points and construct the following table:

p	V	V/p
76.0	77.0	1.013158
122.0	113.2	0.927869
205.0	159.8	0.779512
221.0	175.0	0.791855
305.0	206.4	0.676721
504.0	265.3	0.526389
507.0	267.2	0.527022
756.0	311.9	0.412566
1001.0	339.5	0.339161
1108.0	340.5	0.307310

Step 2. Plot V vs. V/p on a Cartesian scale, as shown in Figure 3.33, and draw the best straight line through the points.

Step 3. Determine the coefficient of the straight line, i.e., slope and intercept, to give:

$$\text{Intercept} = V_m = 465.2 \text{ scf/ton}$$
$$\text{Slope} = -1/b = -380.26, \text{ or } b = 0.00263 \text{ psi}^{-1}$$

Step 4. The Langmuir equation, i.e., Eq. (3.88), can be written as:

$$V = 465.2\frac{0.00263p}{1 + 0.00263p}$$

Seidle and Arrl (1990) proposed that the desorbed gas will begin to flow through the cleats at the time that is required for a well to reach the semisteady state. For a gas well centered in a circular or square drainage area, the semisteady-state flow begins when the dimension time t_{DA} is 0.1, or:

$$t_{DA} = 0.1 = \frac{2.637(10^{-4})k_g t}{\phi(\mu_g c_t)_i A}$$

Solving for the time t gives:

$$t = \frac{379.2\varphi(\mu_g c_t)_i A}{k_g}$$

where

t = time, hours
A = drainage area, ft^2
k_g = gas effective compressibility, md

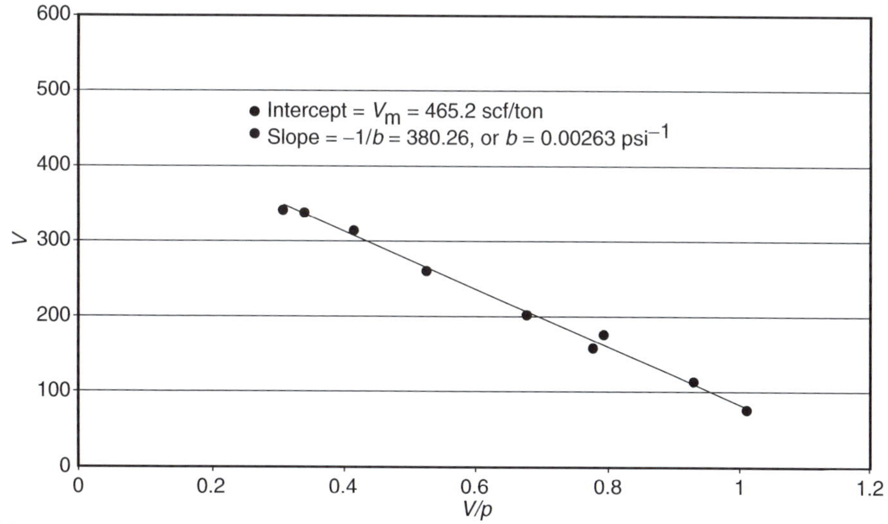

FIGURE 3.33 Volume V vs. the ratio V/p of Example 3.10.

ϕ = cleat porosity, fraction
μ_g = gas viscosity, cp
c_t = total system compressibility, psi^{-1}

Both gas viscosity and system compressibility are calculated at the desorption pressure. The total system compressibility is given by:

$$c_t = c_p + S_w c_w + S_g c_g + c_s$$

where

c_p = cleat volume compressibility, psi^{-1}
S_w = water saturation
S_g = gas saturation
c_w = water compressibility, psi^{-1}
c_g = gas compressibility, psi^{-1}
c_s = apparent sorption compressibility, psi^{-1}

The authors pointed out that the adsorption of the gas on the coal surface increases the total system compressibility by c_s, i.e., apparent sorption compressibility, that is given by:

$$c_s = \frac{0.17525 B_g V_m \rho_B b}{\phi(1 + bp)^2} \quad \textbf{(3.90)}$$

whcre

B_g = gas formation volume factor, bbl/scf
ρ_B = bulk density of the coal deposit, g/cm^3
V_m, b = Langmuir's constants

Example 3.11

In addition to the data given in Example 3.10 for the San Juan coal, the following properties are also available:

$\rho_B = 1.3\ g/cm^3$, $\phi = 2\%$, $T = 575\ °R$
$p_d = 600$ psi, $S_w = 0.9$, $S_g = 0.1$
$c_f = 15 \times 10^{-6}\ psi^{-1}$,
$c_w = 10 \times 10^{-6}\ psi^{-1}$,
$c_g = 2.3 \times 10^{-3}\ psi^{-1}$
$A = 40$ acres, $k_g = 5$ md,
$\mu_g = 0.012$ cp
Z = factor at 600 psi = 0.86

Calculate the time required to achieve the semisteady state.

Solution

Step 1. From Example 3.10:

$$V_m = 465.2\ \text{scf/ton}$$
$$b = 0.00263\ psi^{-1}$$

Step 2. Calculate B_g in bbl/scf from Eq. (3.7), or:

$$B_g = 0.00504 \frac{ZT}{P} = 0.00504 \frac{(0.86)(575)}{600} = 0.00415\ \text{bbl/scf}$$

Step 3. Apply Eq. (3.90) to calculate c_s to give:

$$c_s = \frac{0.17525(0.00415)(465.2)(1.3)(0.00263)}{0.02[1 + (0.00263)(600)]^2} = 8.71 \times 10^{-3}\ psi^{-1}$$

Step 4. Calculate c_t:

$$c_t = 15(10^{-6}) + 0.9(10)(10^{-6}) + 0.1(2.3)(10^{-3}) + 8.71(10^{-3}) = 0.011\ psi^{-1}$$

Step 5. Calculate the time to reach semisteady state:

$$t = \frac{(379.2)(0.03)(0.012)(0.011)(40)(43{,}560)}{5} = 523\ \text{hours}$$

Seidle and Arrl (1990) proposed the use of conventional black-oil simulators to model the production behavior of coalbed methane. The authors pointed out that the amount of gas held by coal at a given pressure is analogous to the amount of gas dissolved in a crude oil system at a given pressure. The Langmuir isotherm of coalbeds is comparable to the solution gas–oil ratio R_s of conventional oil reservoirs. A conventional reservoir simulator can be used to describe coalbed methane by treating the gas adsorbed to the surface of the coal as a dissolved gas in immobile oil.

Seidle and Arrl suggested that the introduction of the oil phase requires increasing the porosity and altering the initial saturations. The gas–water relative permeability curves must be modified and fluid properties of the immobile oil must also be adjusted. The required adjustments for use in a conventional black-oil simulator are summarized below:

Step 1. Select any arbitrary initial oil saturation S_{om} for the model, with the subscript m denoting a model value. The initial value may be set as the residual oil saturation and must remain constant throughout the simulation.

Step 2. Adjust the actual coalbed cleat porosity ϕ_m by the following expression:

$$\phi_m = \frac{\phi}{1 - S_{om}} \tag{3.91}$$

Step 3. Adjust the actual water and gas saturations, i.e., S_w and S_g, to equivalent model saturations S_{wm} and S_{gm} from:

$$S_{wm} = (1 - S_{om})S_w \tag{3.92}$$

$$S_{gm} = (1 - S_{om})S_g \tag{3.93}$$

These two equations are used to adjust gas–water relative permeability data for input into the simulator. The relative permeability corresponding to the actual S_g or S_w is assigned to the equivalent model saturation S_{gm} or S_{wm}.

Step 4. To ensure that the oil phase will remain immobile, assign a zero oil relative permeability $K_{ro} = 0$ for all saturations and/or specifying a very large oil viscosity, i.e., $\mu_o = 10^6$ cp.

Step 5. To link the gas dissolved in the immobile oil, i.e., R_s in immobile oil, convert the sorption isotherm data to gas solubility data using the following expression:

$$R_s = \left(\frac{0.17525\rho_B}{\phi_m S_{om}}\right) V \tag{3.94}$$

where

R_s = equivalent gas solubility, scf/STB

V = gas content, scf/STB

ρ_B = bulk coal seam density, g/cm^3

Eq. (3.94) can be expressed equally in terms of Langmuir's constants by replacing the gas content V with Eq. (3.88) to give:

$$R_s = \left(\frac{0.17525\rho_B}{\phi_m S_{om}}\right)(V_m)\left(\frac{bp}{1 + bp}\right) \tag{3.95}$$

Step 6. To conserve mass over the course of simulation, the oil formation volume factor must be constant with a value of 1.0 bbl/STB.

Using the relative permeability and coal seam properties as given by Ancell et al. (1980) and Seidle and Arrl (1990), the following example illustrates the use of the above methodology.

Example 3.12

The following coal seam properties and relative permeability are available:

$S_{gi} = 0.0$, $V_m = 660$ scf/ton, $b = 0.00200$ psi^{-1}
$\rho_B = 1.3$ g/cm^3, $\phi = 3\%$

S_g	$S_w = 1 - S_g$	k_{rg}	k_{rw}
0.000	1.000	0.000	1.000
0.100	0.900	0.000	0.570
0.200	0.800	0.000	0.300
0.225	0.775	0.024	0.256
0.250	0.750	0.080	0.210
0.300	0.700	0.230	0.140
0.350	0.650	0.470	0.090
0.400	0.600	0.750	0.050
0.450	0.550	0.940	0.020
0.475	0.525	0.980	0.014
0.500	0.500	1.000	0.010
0.600	0.400	1.000	0.000
1.000	0.000	1.000	0.000

Adjust the above relative permeability data and convert the sorption isotherm data into gas solubility for use in a black-oil model.

Solution

Step 1. Select any arbitrary initial oil saturation, to get:

$$S_{om} = 0.1$$

Step 2. Adjust the actual cleat porosity by using Eq. (3.91):

$$\phi_m = \frac{0.03}{1-0.1} = 0.0333$$

Step 3. Re-tabulate the relative permeability data by only readjusting the saturation values using Eqs. (3.92) and (3.93), to give:

S_g	S_w	$S_{gm} = 0.9S_g$	$S_{wm} = 0.9S_w$	k_{rg}	k_{rw}
0.0000	1.0000	0.0000	0.9000	0.0000	1.0000
0.1000	0.9000	0.9000	0.8100	0.0000	0.5700
0.2000	0.8000	0.1800	0.7200	0.0000	0.3000
0.2250	0.7750	0.2025	0.6975	0.0240	0.2560
0.2500	0.7500	0.2250	0.6750	0.0800	0.2100
0.3000	0.7000	0.2700	0.6300	0.2300	0.1400
0.3500	0.6500	0.3150	0.5850	0.4700	0.0900
0.4000	0.6000	0.3600	0.5400	0.7500	0.0500
0.4500	0.5500	0.4045	0.4950	0.9400	0.0200
0.4750	0.5250	0.4275	0.4275	0.9800	0.0140
0.5000	0.5000	0.4500	0.4500	1.0000	0.0100
0.6000	0.4000	0.5400	0.3600	1.0000	0.0000
1.0000	0.0000	0.9000	0.0000	1.0000	0.0000

Step 4. Calculate R_s from either Eq. (3.92) or (3.93) at different assumed pressures:

$$R_s = \left[\frac{(0.17525)(1.30)}{(0.0333)(0.1)}\right] V = 68{,}354V$$

with

$$V = (660)\frac{0.0002p}{1+0.002p}$$

to give:

p (psia)	V (scf/ton)	R_s (scf/STB)
0.0	0.0	0.0
50.0	60.0	4101.0
100.0	110.0	7518.0
150.0	152.3	10,520.0
200.0	188.6	12,890.0
250.0	220.0	15,040.0
300.0	247.5	16,920.0
350.0	271.8	18,570.0
400.0	293.3	20,050.0
450.0	312.6	21,370.0
500.0	330.0	22,550.0

For pressures below the critical desorption pressure, the fractional gas recovery could be roughly estimated from the following relationship:

$$RF = 1 - \left[\left(\frac{V_m}{G_c}\right)\left(\frac{bp}{1+bp}\right)\right]^a \quad \textbf{(3.96)}$$

where

RF = gas recovery factor
V_m, b = Langmuir's constants
V = gas content at pressure p, scf/ton
G_c = gas content at critical desorption pressure, scf/ton
p = reservoir pressure, psi
a = recovery exponent

The recovery exponent a is included to account for the deliverability, heterogeneity, well spacing, among other factors that affect the gas recovery. The recovery exponent a is usually <0.5 and can be roughly estimated from the following relationship that was generated from evaluating several CBM case studies:

$$a = -2371.9\left(\frac{b\,p}{V}\right)^2 - 16.336\left(\frac{b\,p}{V}\right) + 0.5352$$

A detailed discussion of the MBE calculations and predicting the recovery performance of coal seems are presented later in this chapter.

Example 3.13

Using the data in Example 3.10 and assuming $G_c = 330$ scf/ton at 500 psia, estimate the gas recovery factor as a function of pressure to an abandonment pressure of 100 psia.

Solution

Step 1. Substitute Langmuir's constants, i.e., V_m and b, and the recovery exponent into Eq. (3.96), to give:

$$\mathrm{RF} = 1 - \left[\left(\frac{660}{330}\right)\left(\frac{0.002p}{1+0.002p}\right)\right]^a$$

$$\mathrm{RF} = 1 - \left[\frac{0.004p}{1+0.002p}\right]^a$$

where

$$a = -2371.9\left(\frac{bp}{V}\right)^2 - 16.336\left(\frac{bp}{V}\right) + 0.5352$$

Step 2. Assume several reservoir pressures and calculate the recovery factor in the following tabulated form:

p, psi	V	bp/V	(bp/V)^2	a	RF
450	312.6	0.002879	8.28909E−06	0.468506	0.025013
400	293.3	0.002728	7.43971E−06	0.472996	0.054187
350	271.8	0.002575	6.6328E−06	0.477396	0.088523
300	247.5	0.002424	5.87695E−06	0.481658	0.129393
250	220	0.002273	5.16529E−06	0.485821	0.178796
200	188.6	0.002121	4.49818E−06	0.489884	0.239780
150	152.3	0.00197	3.8801E−06	0.493818	0.317379
100	110	0.001818	3.30579E−06	0.497657	0.421162

Many factors influence the measured gas content G_c and sorption isotherm and, consequently, affect the determination of the initial gas-in-place. Among these factors are:

- moisture content of the coal;
- temperature;
- rank of the coal.

These parameters are briefly discussed below.

- *Moisture content*: One of the major difficulties in measuring the gas content and sorption isotherm is the reproduction of the coal content at reservoir conditions. The moisture content of coal is the weight of the water in the coal matrix, not the water contained as free water in the fracture system. The gas storage capacity of coal is significantly affected by moisture content as shown in Figures 3.34 and 3.35. Figure 3.34 illustrates Langmuir isotherms as the moisture increases from 0.37% to 7.41% with apparent reduction of the methane storage capacity. Figure 3.35 shows that the quantity of methane adsorbed in coal is inversely proportional to the inherent moisture content. As evidenced by these two figures, an increase in the moisture content decreases the ability of coal to store gas.
- *Temperature*: This affects both the volume of gas retained by the coal and the rate at which it is desorbed. Numerous laboratory studies confirmed the following two observations:
  (a) the rate of gas desorption from the coal is exponentially dependent upon temperature (i.e., the higher the temperature, the faster the desorption);
  (b) the gas sorption capacity of the coal is inversely proportional to temperature (i.e., the storage capacity of the coal decreases with increasing temperature as shown in Figure 3.34).

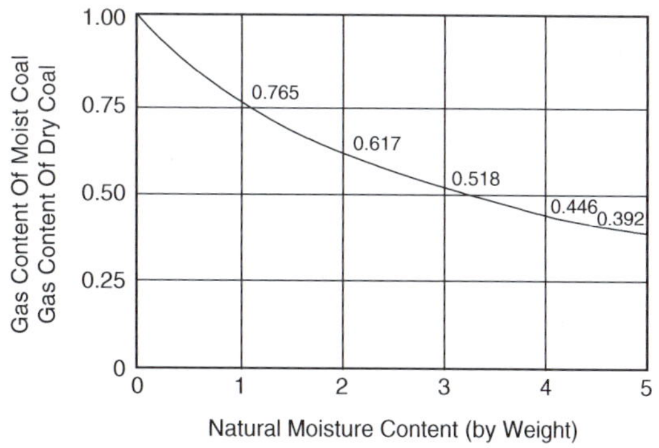

FIGURE 3.34 Effect of moisture content on gas storage capacity.

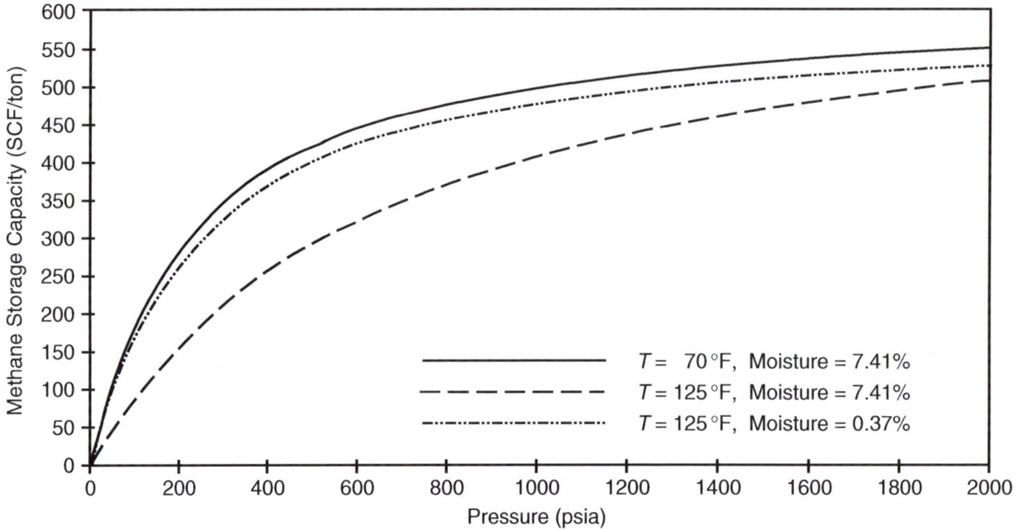

FIGURE 3.35 Sorption isotherm temperature and moisture content sensitivity.

- *Rank of the coal*: According to the American Society for Testing and Materials (ASTM), coal rank is the assignment of a distinct maturation level to a coal derived through the measurement of chemical and physical properties of the coal. The properties most commonly used for rank classification include the fixed carbon content, volatile matter content, and calorific value, among older properties. Coal rank determination is important as the capability of the coal to have generated gas is related to the rank of the coal. Figure 3.36 shows that the gas content and the storage capacity of the coal increase with higher coal ranks. Coals with higher ranks have more capacities to both store and generate gas.

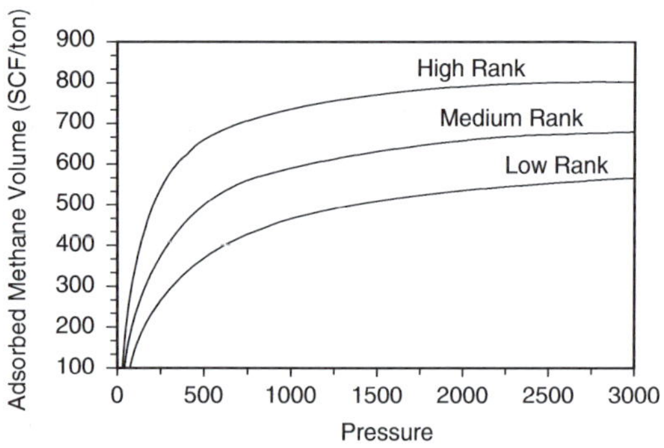

FIGURE 3.36 Relationship between rank and sorptive capacity.

3.4.2 Density of the Coal

Gas-in-place volume G is the total amount of gas stored within a specific reservoir rock volume. The basic equation used to calculate G is:

$$G = 1359.7 A h \rho_B G_c \tag{3.97}$$

where

G = initial gas-in-place, scf
A = drainage area, acres
h = thickness, ft
ρ_B = average coal bulk density, g/cm^3
G_c = average gas content, scf/ton

Mavor and Nelson (1997) pointed out that the use of Eq. (3.97) requires accurate determination of the four parameters in the equation, i.e., A, h, G_c, and ρ_B. The accuracy of G

estimates is limited by uncertainties or errors in the parameters. Nelson (1999) pointed out the density of the coal is a strong function of its composition. Since the mineral matter component of coal has a significantly higher density than the bulk organic matter, coal density will be directly correlated to the mineral matter content. Coal density and compositional properties are not uniform throughout the coal seam but vary vertically and laterally as a function of coal rank, moisture content, and mineral matter content, among other depositional environment geological variables. To illustrate the significant vertical and lateral changes in coal density, Mavor and Nelson (1997) used the basal Fruitland Formation coalbed reservoirs at three well locations in the San Juan Basin as examples for this density variation. As shown below, these examples list the variations in ash content, gas content, and average density.

Well	Interval	Average Ash Content (%)	Average Density (g/cm^3)	Average Gas Content (scf/ton)
1	Intermediate	27.2	1.49	370
	Basal	20.4	1.44	402
2	Intermediate	36.4	1.56	425
	Basal	31.7	1.52	460
3	Intermediate	61.3	1.83	343
	Basal	43.3	1.63	512

It is commonly assumed that interbedded rocks having densities greater than 1.75 g/cm^3 have negligible gas storage capacity.

Because of its organic richness, coal has a much lower bulk density than, for example, shale or sandstone, and, as a result, the gross thickness of coal-bearing intervals can be readily quantified using geophysical log data. Nelson (1999) pointed out that the commonly used analysis practice for coalbed reservoir thickness is to use 1.75 g/cm^3 as the maximum log density value for the gas-bearing seams. The author stated that the density of ash in San Juan Basin coal is typically 2.4–2.5 g/cm^3. The amount of gas stored in coalbed reservoir rocks between the density values of 1.75 and 2.5 g/cm^3 can be significant. This suggests that if the reservoir thickness analysis is based upon a maximum log density value of 1.75 g/cm^3, the calculated gas-in-place volume as expressed by Eq. (3.97) can greatly underestimate the gas-in-place. It should be pointed out that the moisture content, which varies inversely as a function of coal rank, substantially affects the coal density. As shown by Eq. (3.97), the gas initially in place G is a function of coal density ρ_c. Neavel et al. (1999), Unsworth et al. (1989), Pratt et al. (1999), and Nelson (1989) observed that high-rank coals (bituminous coals) have a low moisture content of less than 10%, whereas low-rank coals (sub-bituminous coals) have a very high moisture content (>25%). The authors pointed out that at 5% ash content, Powder River Basin sub-bituminous coal has a dry-basis density of 1.4 g/cm^3; however, with a moisture content of 27% and ash content of 5%, the density is only 1.33 g/cm^3. This density value difference indicates how crucial the accurate moisture content is for a reliable estimate of gas-in-place.

3.4.3 Deliverability and Drainage Efficiency

Interest has grown recently in utilizing the vast resources of coalbed methane reservoirs. As indicated earlier, methane is held in an adsorbed state on the surface of the coal pores by reservoir pressure; this pressure must be reduced to allow desorption of methane from coal surfaces and subsequent methane production. The reservoir pressure is caused by an existing static pressure due to groundwater. Hence, unlike a conventional gas reservoir, gas production is obtained from coal seams by first dewatering and depressurizing the coal seam. Typically, coal seams are naturally fractured and contain laterally extensive, closed, spaced vertical fractures (i.e., cleats). Because the intrinsic permeability of the coal matrix is usually very small, these cleats must be well developed with the minimum required permeability (usually >1 md) to economically develop the reservoir. Holditch et al. (1988) proposed that to produce gas at economic rates from a coal seam, the following three criteria must be met:

(1) An extensive cleat system must exist to provide the needed permeability.
(2) The gas content must be large enough to provide a source that is worth developing.
(3) The cleat system must be connected to the wellbore.

Therefore, large-scale coalbed methane field development requires significant initial investment before any gas production can occur. Most coalbed methane reservoirs require:

- hydraulic fracture stimulation to supplement the coal cleats and to interconnect the cleat system to the wellbore;
- artificial lift of the reservoir water;
- water disposal facilities;
- complete well pattern development.

In general, proper well spacing and stimulation govern the economic attractiveness of the gas production from coalbeds.

Construction of a complete theory of coal well deliverability is difficult as it is neccssary to consider the two-phase flow of gas and water in the coalbed. However, coal wells produce substantial amounts of water before the reservoir pressure declines to the desorption pressure. Once the drainage area of a coal well has been dewatered and the gas rate peaks, water production often declines to negligible rates. This peak in gas rate is essentially a function of:

- the ability of the primary porosity, i.e., porosity of the coal matrix, to supply gas to the secondary porosity system (cleat system);
- the conductivity of the cleat system to water.

Unlike conventional gas and oil reservoirs where minimal well interference is desired, the design of efficient dewatering and depressurizing systems requires maximum well interference for *maximum drawdown*. Well performance in coalbed reservoirs is strongly dependent on this amount of pressure interference between wells, which allows the reservoir pressure to be lowered rapidly and consequently allows gas to be released from the coal matrix. This objective can be accomplished by optimizing the following two decision variables:

(1) optimal well spacing;
(2) optimal drilling pattern shape.

Wick et al. (1986) used a numerical simulator to examine the effect of well spacing on single-well production. The investigation examined the recovery factors from a 160-acre coalbed that contains 1676 MMscf of gas as a function of well spacing for a total of simulation time of 15 years. Results of the study for 20-, 40-, 80-, and 160-acre well spacing are given below:

Well Spacing (acres)	Wells on 160 acres	Gas-in-Place Per Well (MMscf)	Cumulative Gas Production Per Well (MMscf)		Recovery Factor (%)	Total Gas Production from 160 acres (MMscf)
			5 Years	15 Years		
160	1	1676	190	417	25	417
80	2	838	208	388	46	776
40	4	419	197	292	70	1170
20	8	209.5	150	178	85	1429

These results suggest that gas recovery over 15 years from an *individual well increases* with larger well spacing, while gas recoveries from the first 5 years are very similar for the 40-, 80-, and 160-acre cases. This is largely a result of the need to dewater the drainage area for a particular well before gas production becomes efficient. Percentage gas recovery ranges from 25% on 160-acre spacing to 85% on 20-acre spacing. Drilling on 20-acre spacing produces the most gas from a 160-acre area in 15 years. At this time, 85% of the gas-in-place has been produced but only 25% gas recovery with one well on the 160-acre spacing. In determining optimal well spacing, an *economic evaluation that includes current and predicted future gas price* must be made by the operator to maximize both gas recovery and profit.

Selecting the optimum pattern depends heavily on the following variables:

- the coal characteristics, i.e., isotropic or anisotropic permeability behavior;
- reservoir configuration;
- locations of existing wells and total number of wells;
- initial water pressure and desorption pressure;
- volume of water to be removed and the required drawdown.

3.4.4 Permeability and Porosity

Permeability in coals is essentially controlled by the magnitude of the net stress in the reservoir. The variations in the net stress throughout the coal seam can cause local variations in permeability. It has been also shown by several investigators that the coal permeability can increase as gas is desorbed from the coal matrix. Numerous laboratory measurements have shown the dependence of permeability and porosity on the stress conditions in coal seams with relationships that are unique for each coal seam. With the production, cleat properties experience changes due to the following two distinct and opposing mechanisms:

(1) Cleat porosity and permeability *decline* due to compaction and the reduction of net stress $\Delta\sigma$.
(2) Cleat porosity and permeability *increase* due to coal matrix shrinkage as a result of gas desorption.

Walsh (1981) suggested that the change in the net stress $\Delta\sigma$ can be expressed in terms of reservoir pressure by:

$$\sigma = \sigma - \sigma_o = s(p_o - p) = s\,\Delta p \qquad (3.98)$$

where

Δp = pressure drop from p_o to p, psi
p_o = original pressure, psia
p = current pressure, psia
σ_o = original effective stress, psia
σ = effective stress, psia
s = constant relating change in psia pressure to change in effective stress

The effective stress is defined as the total stress minus the seam fluid pressure. The effective stress tends to close the cleats and to reduce permeability within the coal. If the effective stress σ is not known, it can be approximated at any given depth D by:

$$\sigma = 0.572D$$

Eq. (3.98) can be simplified by setting the constant s equal to 0.572, to give:

$$\Delta\sigma = 0.572\,\Delta p$$

Defining the average pore compressibility by the following expression:

$$\overline{c} = \frac{1}{p_o - p}\int_{p}^{p_o} c_p\,dp$$

where

$\overline{c}_p$ = average pore compressibility, psi^{-1}
c_p = pore volume compressibility, psi^{-1}

the desired relationships for expressing the changes in porosity and permeability as a function of the reservoir pressure are given by:

$$\phi = \frac{A}{1+A} \tag{3.99}$$

with

$$A = \frac{\phi_o}{1+\phi_o} \exp^{-s\bar{c}_p(\Delta p)} \tag{3.100}$$

and

$$k = k_o \left(\frac{\phi}{\phi_o}\right)^3$$

where ϕ is the porosity and the subscript o represents the value at initial conditions.

Somerton et al. (1975) proposed a correlation that allows the formation permeability to vary with the changes in the net stress $\Delta\sigma$ as follows:

$$k = k_o \left[\exp\left(\frac{-0.003\ \Delta\sigma}{(k_o)^{0.1}}\right) + 0.0002(\Delta\sigma)^{1/3}(k_o)^{1/3}\right]$$

where

k_o = original permeability at zero net stress, md
k = permeability at net stress $\Delta\sigma$, md
$\Delta\sigma$ = net stress, psia

3.4.5 Material Balance Equation for Coalbed Methane

The MBE is a fundamental tool for estimating the original gas-in-place G and predicting the recovery performance of conventional gas reservoirs. The MBE as expressed by Eq. (3.57) is:

$$\frac{p}{Z} = \frac{p_i}{Z_i} - \left(\frac{p_{sc}T}{T_{sc}V}\right)G_p$$

The great utility of the p/Z plots and the ease of their constructions for conventional gas reservoirs have led to many efforts, in particular the work of King (1992) and Seidle (1999), to extend this approach to unconventional gas resources such as coalbed methane (CBM).

The MBE for CBM can be expressed in the following generalized form:

Gas produced G_P = Gas originally adsorbed G + Original free gas G_F − Gas currently adsorbed at this pressure G_A − Remaining free G_R

or

$$G_p = G + G_F - G_A - G_R \tag{3.101}$$

For a saturated reservoir (i.e., initial reservoir pressure p_i = desorption pressure p_d) with no water influx, the four main components of the right-hand side of the above equality can be determined individually as follows.

Gas Originally Adsorbed *G*. As defined previously by Eq. (3.97), the gas-in-place G is given by:

$$G = 1359.7 A h_{\rho B} G_c$$

where

G = gas initially in place, scf
ρ_B = bulk density of coal, g/cm^3
G_c = gas content, scf/ton
A = drainage area, acres
h = average thickness, ft

Original Free Gas G_F. For this:

$$G_F = 7758 A h \phi (1 - S_{wi}) E_{gi} \tag{3.102}$$

where

G_F = original free gas-in-place, scf
S_{wi} = initial water saturation
ϕ = porosity
E_{gi} = gas expansion factor at p_i in scf/bbl and given by:

$$E_{gi} = \frac{5.615 Z_{sc} T_{sc}}{p_{sc_i}} \frac{p_i}{T Z_i} = 198.6 \frac{p_i}{T Z_i}$$

Gas Currently Adsorbed at *p*, G_A. The gas stored by adsorption at any pressure p is typically expressed with the adsorption isotherm or mathematically by Langmuir's equation, i.e., Eq. (3.88), as:

$$V = V_m \frac{bp}{1 + bp}$$

where

V = volume of gas currently adsorbed at p, scf/ton
V_m = Langmuir's isotherm constant, scf/ton
b = Langmuir's pressure constant, psi^{-1}

The volume of the adsorbed gas V as expressed in scf/ton at reservoir pressure p can be converted into scf by the following relationship:

$$G_A = 1359.7 A h P_B V \quad \textbf{(3.103)}$$

where

G_A = adsorbed gas at p, scf
P_B = average bulk density of the coal, g/cm^3
V = adsorbed gas at p, scf/ton

Remaining Free Gas G_R. During the dewatering phase of the reservoir, formation compaction (matrix shrinkage) and water expansion will significantly affect water production. Some of the desorbed gas remains in the coal–cleat system and occupies a PV that will be available with water production. King (1992) derived the following expression for calculating the average water saturation remaining in the coal cleats during the dewatering phase:

$$S_w = \frac{S_{wi}[1 + c_w(p_i - p)] - (B_w W_p / 7758 A h \phi)}{1 - (p_i - p) c_f} \quad \textbf{(3.104)}$$

where

p_i = initial pressure, psi
p = current reservoir pressure, psi
W_p = cumulative water produced, bbl
B_w = water formation volume factor, bbl/STB
A = drainage area, acres
c_w = isothermal compressibility of the water, psi^{-1}
c_f = isothermal compressibility of the formation, psi^{-1}
S_{wi} = initial water saturation

Using the above estimated average water saturation, the following relationship for the remaining gas in cleats is developed:

$$G_R = 7758 A h \phi \times \left[\frac{(B_w W_p / 7758 A h \phi) + (1 - S_{wi}) - (p_i - p)(c_f + c_w S_{wi})}{1 - (p_i - p) c_f}\right] E_g \quad \textbf{(3.105)}$$

where

G_R = remaining gas at pressure p, scf
W_p = cumulative water produced, bbl
A = drainage area, acres

and with the gas expansion factor given by:

$$E_g = 198.6 \frac{p}{TZ} \ \text{scf/bbl}$$

Substituting the above derived four terms into Eq. (3.101) and rearranging gives:

$$G_p = G + G_F - G_A - G_R$$

or

$$G_p + \frac{B_w W_p E_g}{1 - (c_f \Delta P)} = Ah\left[1359.7 P_B \left(G_c - \frac{V_m bp}{1 + bp}\right) + \frac{7758\phi[\Delta P(c_f + S_{wi} c_{wi}) - (1 - S_{wi})] E_g}{1 - (c_f \Delta P)}\right] + 7758 A h \phi (1 - S_{wi}) E_{gi} \quad \textbf{(3.106)}$$

In terms of the volume of gas adsorbed V, this equation can be expressed as:

$$G_p + \frac{B_w W_p E_g}{1 - (c_f \Delta P)} = Ah\left[1359.7 P_B (G_c - V) + \frac{7758\phi[\Delta P(c_f + S_{wi} c_{wi}) - (1 - S_{wi})] E_g}{1 - (c_f \Delta P)}\right] + 7758 A h \phi (1 - S_{wi}) E_{gi} \quad \textbf{(3.107)}$$

Each of the above two forms of the generalized MBE is the equation of a straight line and can be written as:

$$y = mx + a$$

with

$$y = G_p + \frac{B_w W_p E_g}{1 - (c_f \, \Delta P)}$$

$$x = 1359.7 P_B \left(G_c - \frac{V_m bp}{1 + bp} \right) + \frac{7758\phi[\Delta P(c_f + S_{wi} c_{wi}) - (1 - S_{wi})] E_g}{1 - (c_f \, \Delta P)}$$

or equivalently:

$$x = 1359.7 P_B (G_c - V) + \frac{7758\phi[\Delta P(c_f + S_{wi} c_{wi}) - (1 - S_{wi})] E_g}{1 - (c_f \, \Delta P)}$$

with a slope of:

$$m = Ah$$

and intercept as:

$$a = 7758 Ah\phi(1 - S_{wi}) E_{g_i}$$

A plot of y as defined above and using the production and pressure drop data vs. the term x would produce a straight line with a slope m of Ah and intercept of a. The drainage area A as calculated from the slope m and the intercept a must be the same. That is:

$$A = \frac{m}{h} = \frac{a}{7758 h\phi(1 - S_{wi}) E_{gi}}$$

For scattered points, the correct straight line must satisfy the above equality.

Neglecting the rock and fluid compressibility, Eq. (3.107) is reduced to:

$$G_p + B_w W_p E_g = Ah \left[1359.7 P_B \left(G_c - V_m \frac{bp}{1 + bp} \right) - 7758\phi(1 - S_{wi}) E_g \right] + 7758 Ah\phi(1 - S_{wi}) E_{gi} \tag{3.108}$$

This expression is again the equation of a straight line, i.e., $y = mx + a$, where

$$y = G_p + B_w W_p E_g$$

$$x = 1359.7 P_B \left(G_c - V_m \frac{bp}{1 + bp} \right) - 7758\phi(1 - S_{wi}) E_g$$

Slope $m = Ah$

Intercept $a = 7758 Ah\phi(1 - S_{wi}) E_{gi}$

In terms of the adsorbed gas volume V, Eq. (3.108) is expressed as:

$$G_p + B_w W_p E_g = Ah[1359.7 P_B (G_c - V) - 7758\phi(1 - S_{wi}) E_g] + 7758 Ah\phi(1 - S_{wi}) E_{gi} \tag{3.109}$$

With the calculation of the bulk volume Ah, the original gas-in-place G can then be calculated from:

$$G = 1359.7 (Ah) P_B G_c$$

Example 3.14

A coal well is draining a homogeneous 320-acre coal deposit.

The actual well production and pertinent coal data is given below:

Time (days)	G_p (MMscf)	W_p (MSTB)	p (psia)	p/Z (psia)
0	0	0	1500	1704.5
730	265.086	157,490	1315	1498.7
1460	968.41	290,238	1021	1135.1
2190	1704.033	368,292	814.4	887.8
2920	2423.4	425,473	664.9	714.1
3650	2992.901	464,361	571.1	607.4

Langmuir's pressure constant	$b = 0.00276 \text{ psi}^{-1}$
Langmuir's volume constant	$V_m = 428.5$ scf/ton
Average bulk density	$P_B = 1.70 \text{ g/cm}^3$
Average thickness	$h = 50$ ft
Initial water saturation	$S_{wi} = 0.95$
Drainage area	$A = 320$ acres
Initial pressure	$p_i = 1500$ psia
Critical (desorption) pressure	$p_d = 1500$ psia
Temperature	$T = 105\ ^\circ\text{F}$
Initial gas content	$G_c = 345.1$ scf/ton
Formation volume factor	$B_w = 1.00$ bbl/STB
Porosity	$\phi = 0.01$
Water compressibility	$c_w = 3 \times 10^{-6} \text{ psi}^{-1}$
Formation compressibility	$c_f = 6 \times 10^{-6} \text{ psi}^{-1}$

(a) Neglecting formation and water compressibility coefficients, calculate the well drainage area and original gas-in-place.

(b) Repeat the above calculations by including water and formation compressibilities.

Solution

Step 1. Calculate E_g and V as a function of pressure by applying the following expressions:

$$E_g = 198.6\frac{p}{Tz} = 0.3515\frac{p}{z} \text{ scf/bbl}$$

$$V = V_m\frac{bp}{1+bp} = 0.18266\frac{p}{1+0.00276p} \text{ scf/ton}$$

p (psi)	**p/Z (psi)**	**E_g (scf/bbl)**	**V (scf/ton)**
1500	1704.5	599.21728	345.0968
1315	1498.7	526.86825	335.903
1021	1135.1	399.04461	316.233
814.4	887.8	312.10625	296.5301
664.9	714.1	251.04198	277.3301
571.1	607.5	213.56673	262.1436

Step 2. Neglecting c_w and c_f, the MBE is given by Eq. (3.109), or:

$$G_p + B_w W_p E_g = Ah[1359.7 P_B(G_c - V) - 7758\phi(1 - S_{wi})E_g] + 7758Ah\phi(1 - S_{wi})E_{gi}$$

or

$$G_p + B_w W_p E_g = Ah[2322.66(345.1 - V) - 3.879E_g] + 2324.64(Ah)$$

Use the given data in the MBE to construct the following table:

Step 3. Plot $G_p + B_w W_p E_g$ vs. 2322.66 $(345.1 - V) - 3.879E_g$ on a Cartesian scale, as shown in Figure 3.37.

Step 4. Draw the best straight line through the points and determine the slope, to give:

$$\text{Slope} = Ah = 15,900 \text{ acre-ft}$$

or

$$\text{Area } A = \frac{15,900}{50} = 318 \text{ acres}$$

Step 5. Calculate the initial gas-in-place:

$$G = 1359.7Ah_{\rho B}G_c$$
$$= 1359.7(318)(50)(1.7)(345.1)$$
$$= 12.68 \text{ Bscf}$$
$$G_F = 77.58Ah\phi(1 - S_{wi})E_{gi}$$
$$= 7758(318)(50)(0.01)(0.05)(599.2)$$
$$= 0.0369 \text{ Bscf}$$
$$\text{Total gas-in-place} = G + G_F = 12.68 + 0.0369$$
$$= 12.72 \text{ Bscf}$$

Step 1. Using the given values of c_w and c_f in Eq. (3.107), calculate the Y and X terms and tabulate the results as a function of pressure as follows:

$$y = G_p + \frac{W_p E_g}{1 - [6(10^{-6})(1500 - p)]}$$

$$x = 1359.7(1.7)(345.1 - V) + \frac{7758(0.01)(1500 - p)(6(10^{-6}) + 0.95c_{wi}) - (1 - 0.95)]E_g}{1 - [6.(10^{-6})(1500 - p)]}$$

p (psi)	**V (scf/ton)**	**G_p (MMscf)**	**W_p (MMETB)**	**E_g (scf/bbl)**	**$y = G_p + W_pE_g$ (MMscf)**	**$x = 2322.66\ (345.1 - V)\ -3.879E_g$ (scf/acre-ft)**
1500	345.097	0	0	599.21	0	0
1315	335.90	265.086	0.15749	526.87	348.06	19,310
1021	316.23	968.41	0.290238	399.04	1084.23	65,494
814.4	296.53	1704.033	0.368292	312.11	1818.98	111,593
664.9	277.33	2423.4	0.425473	251.04	2530.21	156,425
571.1	262.14	2992.901	0.464361	213.57	3092.07	191,844

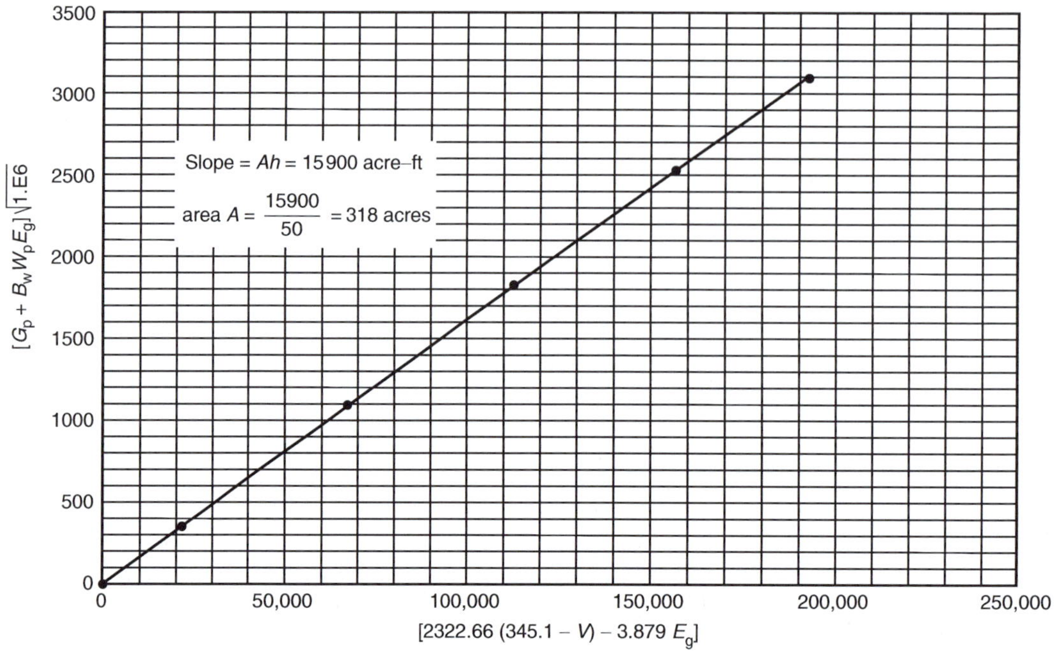

FIGURE 3.37 Graphical determination of drainage area.

p (psi)	**V (scf/ton)**	**X**	**Y**
1315	335.903	1.90E + 04	3.48E + 08
1021	316.233	6.48E + 04	1.08E + 09
814.4	296.5301	1.11E + 05	1.82E + 09
664.9	277.3301	1.50E + 05	2.53E + 09
571.1	262.1436	1.91E + 05	3.09E + 09

Step 2. Plot the x and y values on a Cartesian scale, as shown in Figure 3.38, and draw the best straight line through the points.

Step 3. Calculate the slope and intercept of the line, to give:

$$\text{Slope} = Ah = 15{,}957 \text{ acre-ft}$$

or

$$A = \frac{15{,}957}{50} = 319 \text{ acres}$$

To confirm the above calculated drainage area of the well, it can be also determined from the intercept of the straight line, to give:

$$\text{Intercept} = 3.77(10^7) = 7758 Ah\phi(1 - S_{wi})E_{gi}$$

or

$$A = \frac{3.708(10^7)}{7758(50)(0.01)(0.05)(599.2)} - 324 \text{ acres}$$

Step 4. Calculate the initial gas-in-place to give:

$$\text{Total} = G + G_F = 12.72 + 0.037 = 12.76 \text{ Bscf}$$

Under the conditions imposed on Eq. (3.108) and assuming 100% initial water saturation, the usefulness of the equation can be extended to estimate the average reservoir pressure p from the historical production data, i.e., G_p and W_p. Eq. (3.108) is given as:

$$G_p + W_p E_g = (1359.7 P_B Ah)\left[\left(G_c - V_m \frac{bp}{1 + bp}\right)\right]$$

Or in terms of G:

$$G_p + W_p E_g = G - (1359.7 P_B Ah) V_m \frac{bp}{1 + bp} \qquad \textbf{(3.110)}$$

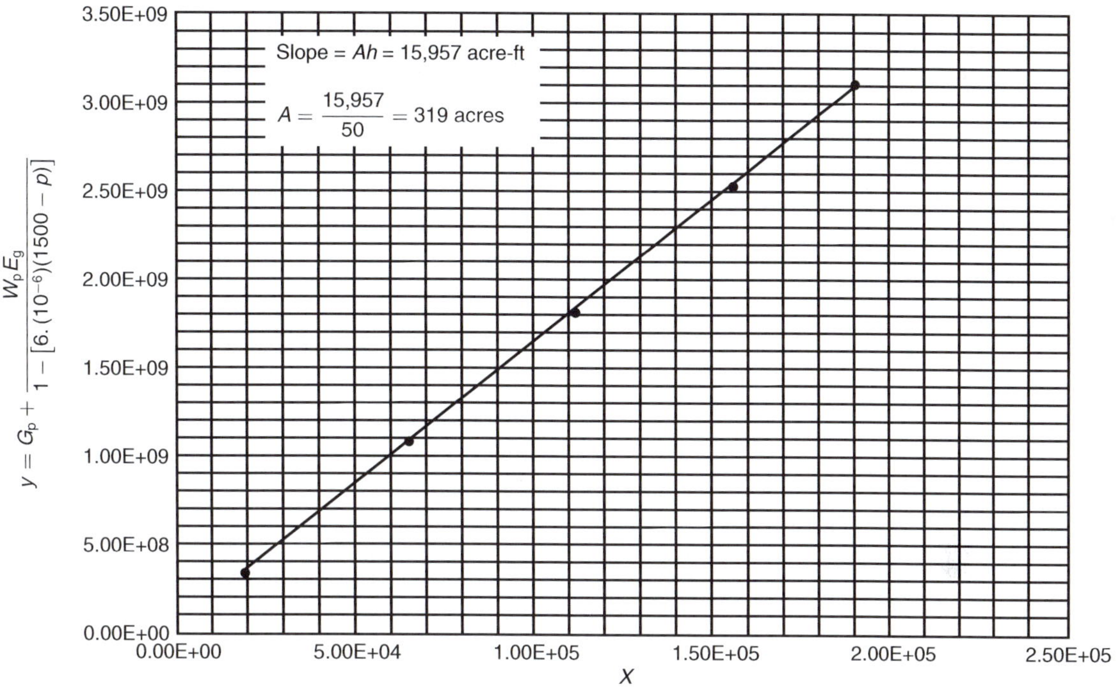

FIGURE 3.38 Straight-line relationship of y as a function of x.

At the initial reservoir pressure p_i, initial gas-in-place G is given by:

$$G = [1359.7 P_B A h] G_c = [1359.7 P_B A h]\left(V_m \frac{b p_i}{1 + b p_i}\right) \quad \textbf{(3.111)}$$

Combining Eq. (3.111) with (3.110) and rearranging gives:

$$\left[\left(\frac{p}{p_i}\right)\left(\frac{1 + b p_i}{1 + b p}\right)\right] = 1 - \left[\frac{1}{G}(G_p + B_w W_p E_g)\right]$$

or

$$\left[\left(\frac{p}{p_i}\right)\left(\frac{1 + b p_i}{1 + b p}\right)\right] = 1 - \frac{1}{G}\left(G_p + 198.6 \frac{p}{ZT} B_w W_p\right) \quad \textbf{(3.112)}$$

where

G = initial gas-in-place, scf
G_p = cumulative gas produced, scf
W_p = cumulative water produced, STB
E_g = gas formation volume factor, scf/bbl
p_i = initial pressure
T = temperature, °R
Z = z factor at pressure p

Eq. (3.112) is the equation of a straight line with a slope of $-1/G$ and intercept of 1.0. In a more convenient form, Eq. (3.112) is written as:

$$y = 1 + mx$$

where

$$y = \left[\left(\frac{p}{p_i}\right)\left(\frac{1 + b p_i}{1 + b p}\right)\right] \quad \textbf{(3.113)}$$

$$x = G_p + 198.6 \frac{p}{ZT} B_w W_p \quad \textbf{(3.114)}$$

$$m = \frac{1}{G}$$

Figure 3.39 shows the graphical linear relationship of Eq. (3.112). Solving this linear relationship for the average reservoir pressure p requires an

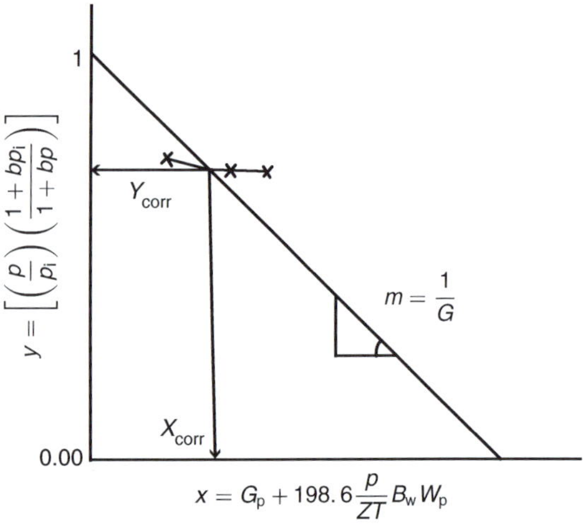

FIGURE 3.39 Graphical determination of reservoir pressure.

iterative procedure as summarized in the following steps:

Step 1. On a Cartesian scale, draw a straight line that originates from 1 on the *y*-axis and with a negative slope of $1/G$, as shown in Figure 3.39.

Step 2. At a given G_p and W_p, guess the reservoir pressure p and calculate the y and x terms as given by Eqs. (3.113) and (3.114), respectively.

Step 3. Plot the coordinate of the calculated point, i.e., (x, y), in Figure 3.39. If the coordinate of the point falls on the straight line, it indicates that the assumed reservoir pressure is correct, otherwise the process is repeated at a different pressure. The process can be rapidly converged by assuming three different pressure values and connecting the coordinate plotted points with a smooth curve that intersects with the straight line at (x_{corr}, y_{corr}). The reservoir pressure at the given W_p and G_p is calculated from:

$$p = \frac{p_i y_{corr}}{1 + bp_i(1 - y_{corr})}$$

3.4.6 Prediction of CBM Reservoir Performance

The MBE as given by its various mathematical forms, i.e., Eqs. (3.106)–(3.109), can be used to predict future performance of CBM reservoirs as a function of reservoir pressure. Assuming, for simplicity, that the water and formation compressibility coefficients are negligible, Eq. (3.106) can be expressed as:

$$G_p + B_w W_p E_g = G - (1359.7AhP_B V_m b)\frac{p}{1+bp} - 7758\phi Ah(1 - S_{wi})E_g + 7758Ah\phi(1 - S_{wi})E_{gi}$$

where

G = initial gas-in-place, scf
A = drainage area, acres
h = average thickness, ft
S_{wi} = initial water saturation
E_g = gas formation volume factor, scf/bbl
b = Langmuir's pressure constant, psi^{-1}
V_m = Langmuir's volume constant, scf/ton

In a more convenient form, the above expression is written as:

$$G_p + B_w W_p E_g = G - \frac{a_1 p}{1+bp} + a_2(E_{gi} - E_g) \quad \textbf{(3.115)}$$

where the coefficients a_1 and a_2 are given by:

$$a_1 = 1359.7AhbV_m$$
$$a_2 = 7758Ah\phi(1 - S_{wi})$$

Differentiating with respect to pressure gives:

$$\frac{\partial(G_p + B_w W_p E_g)}{\partial p} = -\frac{a_1}{(1+bp)^2} - a_2\frac{\partial E_g}{\partial p}$$

Expressing the above derivative in finite difference form gives:

$$G_p^{n+1} + B_w^{n+1}W_p^{n+1}E_g^{n+1} = G_p^n + B_w^n W_p^n E_g^n + \frac{a_1(p^n - p^{n+1})}{(1+bp^{n+1})} + a_2(E_g^n - E_g^{n+1}) \quad \textbf{(3.116)}$$

where the superscripts n and $n+1$ indicate the current and future time levels, respectively, and:

p^n, p^{n+1} = current and future reservoir pressures, psia

G_p^n, G_p^{n+1} = current and future cumulative gas production, scf
W_p^n, W_p^{n+1} = current and future cumulative water production, STB
E_g^n, E_g^{n+1} = current and future gas expansion factor, scf/bbl

Eq. (3.116) contains two unknowns, G_p^{n+1} and W_p^{n+1}, and requires two additional relations:

(1) the producing gas–water ratio (GWR) equation;
(2) the gas saturation equation.

The gas–water ratio relationship is given by:

$$\frac{Q_g}{Q_w} = \mathrm{GWR} = \frac{k_{rg}}{k_{rw}}\frac{\mu_w B_w}{\mu_g B_g} \tag{3.117}$$

where

GWR = gas–water ratio, scf/STB
k_{rg} = relative permeability to gas
k_{rw} = relative permeability to water
μ_w = water viscosity, cp
μ_g = gas viscosity, cp
B_w = water formation volume factor, bbl/STB
B_g = gas formation volume factor, bbl/STB

The cumulative gas produced G_p is related to the EWR by the following expression:

$$G_p = \int_0^{W_p} (\mathrm{GWR})\, dW_p \tag{3.118}$$

This expression suggests that the cumulative gas production at any time is essentially the area under the curve of the GWR vs. the W_p relationship, as shown in Figure 3.40.

Also, the incremental cumulative gas produced ΔG_p between W_p^n and W_p^{n+1} is given by:

$$G_p^{n+1} - G_p^n = \Delta G_p = \int_{W_p^n}^{W_p^{n+1}} (\mathrm{GWR})\, dW_p \tag{3.119}$$

This expression can be approximated by using the trapezoidal rule, to give:

$$G_p^{n+1} - G_p^n = \Delta G_p = \left[\frac{(\mathrm{GWR})^{n+1} + (\mathrm{GWR})^n}{2}\right](W_p^{n+1} - W_p^n) \tag{3.120}$$

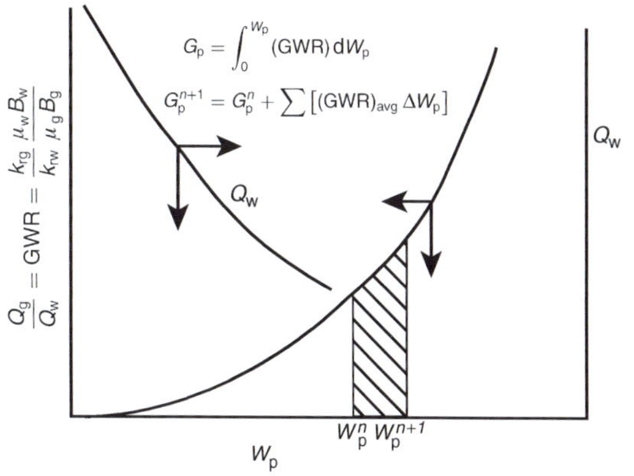

FIGURE 3.40 Relationships between GWR, Q_w, and W_p.

or

$$G_p^{n+1} = G_p^n + \sum[(\mathrm{GWR})_{avg}\,\Delta W_p] \tag{3.121}$$

The other auxiliary mathematical expression needed to predict the recovery performance of a coalbed gas reservoir is the gas saturation equation. Neglecting the water and formation compressibilities, the gas saturation is given by:

$$S_g^{n+1} = \frac{(1 - S_{wi}) - (p_i - p^{n+1})(c_f + c_w S_{wi}) + \frac{B_w^{n+1} W_p^{n+1}}{7758Ah\phi}}{1 - [(p_i - p^{n+1})c_f]} \tag{3.122}$$

The required computations are performed in a series of pressure drops that proceed from a known reservoir condition at pressure p^n to the new lower pressure p^{n+1}. It is accordingly assumed that the cumulative gas and water production has increased from G_p^n and W_p^n to G_p^{n+1} and W_p^{n+1}, while flow rates have changed from Q_g^n and Q_w^n to Q_g^{n+1} and Q_w^{n+1}. The proposed methodology for predicting the reservoir performance consists of the following steps:

Step 1. Using the gas–water relative permeability data, prepare a plot of the relative permeability ratio k_{rg}/k_{rw} vs. gas saturation S_g on a semilog scale.

Step 2. Knowing the reservoir temperature T and specific gravity of the gas γ_g, calculate and prepare a plot of E_g, B_g, and gas viscosity μ_g as a function of pressure, where

$$E_g = 198.6\frac{p}{ZT}, \quad \text{scf/bbl}$$

$$B_g = \frac{1}{E_g} = 0.00504\frac{ZT}{p}, \quad \text{bbl/scf}$$

Step 3. Select a future reservoir pressure p^{n+1} below the current reservoir pressure p^n. If the current reservoir pressure p^n is the initial reservoir pressure, set W_p^n and G_p^n equal to zero.

Step 4. Calculate B_w^{n+1}, E_g^{n+1}, and B_g^{n+1} at the selected pressure p^{n+1}.

Step 5. Estimate or guess the cumulative water production W_p^{n+1} and solve Eq. (3.116) for G_p^{n+1}, to give:

$$G_p^{n+1} = G_p^n + (B_w^n W_p^n E_g^n - B_w^{n+1} W_p^{n+1} E_g^{n+1}) + \frac{a_1(p^n - p^{n+1})}{1 + bp^{n+1}} + a_2(E_g^n - E_g^{n+1})$$

Step 6. Calculate the gas saturation at p^{n+1} and W_p^{n+1} by applying Eq. (3.122):

$$S_g^{n+1} = \frac{(1 - S_{wi}) - (p_i - p^{n+1})(c_f + c_w S_{wi}) + \frac{B_w^{n+1} W_p^{n+1}}{7758Ah\phi}}{1 - [(p_i - p^{n+1})c_f]}$$

Step 7. Determine the relative permeability ratio k_{rg}/k_{rw} at S_g^{n+1} and estimate the GWR from Eq. (3.117), or:

$$(\text{GWR})^{n+1} = \frac{k_{rg}}{k_{rw}}\left(\frac{\mu_w B_w}{\mu_g B_g}\right)^{n+1}$$

Step 8. Recalculate the cumulative gas production G_p^{n+1} by applying Eq. (3.120):

$$G_p^{n+1} = G_p^n + \frac{(\text{GWR})^{n+1} + (\text{GWR})^n}{2}(W_p^{n+1} - W_p^n)$$

Step 9. The total gas produced G_p^{n+1} as calculated from the MBE in step 5 and that of the GWR in step 8 provide two independent methods for determining the cumulative gas production. If the two values agree, the assumed value of W_p^{n+1} and the calculated G_p^{n+1} are correct. Otherwise, assume a new value for W_p^{n+1} and repeat steps 5–9. To simplify this iterative process, three values of W_p^{n+1} can be assumed that yield three different solutions of G_p^{n+1} for each of the equations (i.e., MBE and GWR equations). When the computed values of G_p^{n+1} are plotted vs. the assumed values of W_p^{n+1}, the resulting two curves (one representing results of step 5 and the other representing results of step 8) will intersect. The coordinates of the intersection give the correct G_p^{n+1} and W_p^{n+1}.

Step 10. Calculate the incremental cumulative gas production ΔG_p from:

$$\Delta G_p = G_p^{n+1} - G_p^n$$

Step 11. Calculate the gas and water flow rates from Eqs. (3.11) and (3.117), to give:

$$Q_g^{n+1} = \frac{0.703hk(k_{rg})^{n+1}(p^{n+1} - p_{wf})}{T(\mu_g Z)_{avg}[\ln(r_e/r_w) - 0.75 + s]}$$

$$Q_w^{n+1} = \left(\frac{k_{rw}}{k_{rg}}\right)^{n+1}\left(\frac{\mu_g B_g}{\mu_w B_w}\right)^{n+1} Q_g^{n+1}$$

where

Q_g = gas flow rate, scf/day
Q_w = water flow rate, STB/day
k = absolute permeability, md
T = temperature, °R
r_e = drainage radius, ft
r_w = wellbore radius, ft
s = skin factor

Step 12. Calculate the average gas flow rate as the reservoir pressure declines from p^n to p^{n+1}, as:

$$(Q_g)_{avg} = \frac{Q_g^n + Q_g^{n+1}}{2}$$

Step 13. Calculate the incremental time Δt required for the incremental gas production ΔG_p during the pressure drop from p^n to p^{n+1}, as:

$$\Delta t = \frac{\Delta G_P}{(Q_g)_{avg}} = \frac{G_p^{n+1} - G_p^n}{(Q_g)_{avg}}$$

where
Δt = incremental time, days
Step 14. Calculate the total time t:

$$t=\sum \Delta t$$

Step 15. Get:

$$W_p^n = W_p^{n+1}$$
$$G_p^n = G_p^{n+1}$$
$$Q_g^n = Q_g^{n+1}$$
$$Q_w^n = Q_w^{n+1}$$

and repeat steps 3–15.

3.4.7 Flow of Desorbed Gas in Cleats and Fractures

Flow in conventional gas reservoirs obeys, depending on the flow regime, Darcy's equation in response to a pressure gradient. In coal seams, the gas is physically adsorbed on the internal surfaces of the coal matrix. As discussed in previous sections, coal seam reservoirs are characterized by a dual-porosity system: primary (matrix) porosity and secondary (cleats) porosity. The secondary porosity system ϕ_2 of coal seams consists of the natural fracture (cleats) system inherent in these reservoirs. These cleats act as a sink to the primary porosity (porosity of the coal matrix) and as a conduit to production wells. The porosity ϕ_2 in this system ranges between 2% and 4%. Therefore, methane production from coal seams occurs by a three-stage process in which the methane:

(1) diffuses through the coal matrix to the cleat and obeys Fick's law;
(2) desorbs at the matrix–cleat interface; then
(3) flows through the cleat system to the wellbore as described by Darcy's equation.

The primary porosity system in these seams is composed of very fine pores that contain a large internal surface area on which large quantities of gas are stored. The permeability of the coal matrix system is extremely low, and, in effect, the primary porosity system (coal matrix) is both impermeable to gas and inaccessible to water. In the absence of gas flow in the matrix, the gas is transported according to gradients in concentration, i.e., diffusion process. Diffusion is a process where flow occurs via random motion of molecules from a high concentration area to an area with lower concentration, in which the flow obeys Fick's law as given by:

$$Q_g = -379.4DA\frac{dC_m}{ds} \quad (3.123)$$

where

Q_g = matrix-fracture gas flow rate, scf/day
s = fracture spacing, ft
D = diffusion coefficient, ft^2/day
C_m = molar concentration, lb$_m$-mole/ft^3
A = surface area of the coal matrix, ft^2

The volume of the adsorbed gas can be converted into molar concentration C_m from the following expression:

$$C_m = 0.5547(10^{-6})\gamma_g \rho_B V \quad (3.124)$$

where

C_m = molar concentration lb$_m$-mole/ft^3
ρ_B = coal bulk density, g/cm^3
V = adsorbed gas volume, scf/ton
γ_g = specific gravity of the gas

Zuber et al. (1987) pointed out that the diffusion coefficient D can be determined indirectly from the canister desorption test. The authors correlated the diffusion coefficient with the coal cleat spacing s and desorption time t. The average cleat spacing can be determined by visual observation of coal cores. The proposed expression is given by:

$$D=\frac{s^2}{8\pi t} \quad (3.125)$$

where

D = diffusion coefficient, ft^2/day
t = desorption time from the canister test, days
s = coal cleat spacing, ft

The desorption time t is determined from canister tests on a core sample as defined by the time required to disrobe 63% of the total adsorbed gas.

3.5 TIGHT GAS RESERVOIRS

Gas reservoirs with permeability less than 0.1 md are considered "tight gas" reservoirs. They present unique problems to reservoir engineers when applying the MBE to predict the gas-in-place and recovery performance.

The use of the conventional material balance in terms of p/Z plot is commonly utilized as a powerful tool for evaluating the performance of gas reservoirs. For a volumetric gas reservoir, the MBE is expressed in different forms that will produce a linear relationship between p/Z and the cumulative gas production G_p. Two such forms are given by Eqs. (3.59) and (3.60) as:

$$\frac{p}{Z} = \frac{p_i}{Z_i} - \left[\left(\frac{p_i}{Z_i}\right)\frac{1}{G}\right]G_p$$

$$\frac{p}{Z} = \frac{p_i}{Z_i}\left[1 - \frac{G_p}{G}\right]$$

The MBE as expressed by any of the above equations is very simple to apply because it is not dependent on flow rates, reservoir configuration, rock properties, or well details. However, there are fundamental assumptions that must be satisfied when applying the equation, including:

- uniform saturation throughout the reservoir at any time;
- there is small or no pressure variation within the reservoir;
- the reservoir can be represented by a single weighted average pressure at any time;
- the reservoir is represented by a tank, i.e., constant drainage area, of homogeneous properties.

Payne (1996) pointed out that the assumption of uniform pressure distributions is required to ensure that pressure measurements taken at different well locations represent true average reservoir pressures. This assumption implies that the average reservoir pressure to be used in the MBE can be described with one pressure value. In high-permeability reservoirs, small pressure gradients exist away from the wellbore and the average reservoir pressure estimates can be readily made with short-term shut-in buildups or static pressure surveys.

Unfortunately, the concept of the straight-line p/Z plot as described by the conventional MBE fails to produce this linear behavior when applied to tight gas reservoirs that had not established a constant drainage area. Payne (1996) suggested that the essence of the errors associated with the use of p/Z plots in tight gas reservoirs is that substantial pressure gradients exist within the formation, resulting in a violation of the basic tank assumption. These gradients manifest themselves in terms of scattered, generally curved, and rate-dependent p/Z plot behavior. This nonlinear behavior of p/Z plots, as shown in Figure 3.41, may significantly underestimate gas initially in place (GIIP) when interpreting by the

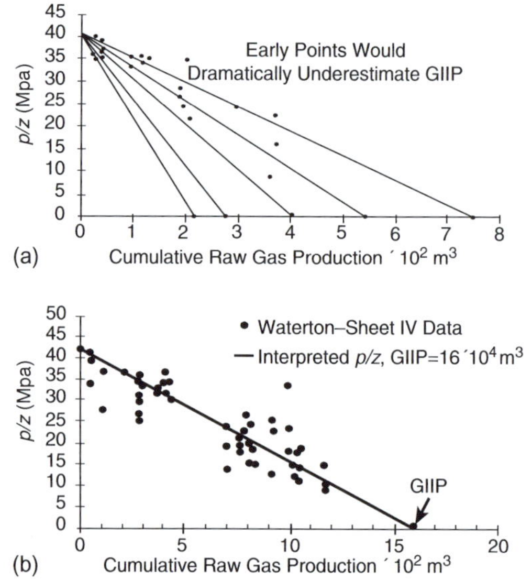

FIGURE 3.41 (a) Real-life example of p/Z plot from Sheet IVc in the Waterton Gas Field. (b) Real-life example of p/Z plot from Sheet IV in the Waterton Gas Field.

conventional straight-line method. Figure 3.41(a) reveals that the reservoir pressure declines very rapidly as the area surrounding the well cannot be recharged as fast as it is depleted by the well. This early, rapid pressure decline is seen often in tight gas reservoirs and is an indication that the use of p/Z plot analysis may be inappropriate. It is clearly apparent that the use of early points would dramatically underestimate GIIP, as shown in Figure 3.41(a) for the Waterton Gas Field with an apparent GIIP of 7.5 Bm^3. However, late-time production and pressure data shows a nearly double GIIP of 16.5 Bm^3, as shown in Figure 3.41(b).

The main problem with tight gas reservoirs is the difficulty of accurately estimating the average reservoir pressure required for p/Z plots as a function of G_p or time. If the pressures obtained during shut-in do not reflect the average reservoir pressure, the resulting analysis will be inaccurate. In tight gas reservoirs, excessive shut-in times of months or years may be required to obtain accurate estimates of average reservoir pressure. The minimum shut-in time that is required to obtain a reservoir pressure that represents the average reservoir pressure must be at least equal to time to reach the pseudosteady state t_{pss}. This time is given by Eq. (3.39) for a well in the centre of a circular or square drainage area, as:

$$t_{pss} = \frac{15.8\phi\mu_{gi} c_{ti} A}{k}$$

with

$$c_{ti} = S_{wi} c_{wi} + S_g c_{gi} + c_f$$

where

t_{pss} = stabilization (pseudosteady-state) time, days
c_{ti} = total compressibility coefficient at initial pressure, psi^{-1}
c_{wi} = water compressibility coefficient at initial pressure, psi^{-1}
c_f = formation compressibility coefficient, psi^{-1}
c_{gi} = gas compressibility coefficient at initial pressure, psi^{-1}
ϕ = porosity, fraction

With most tight gas reservoirs being hydraulically fractured, Earlougher (1977) proposed the following expression for estimating the minimum shut-in time to reach the semisteady state:

$$t_{pss} = \frac{474\phi\mu_g c_t x_f^2}{k} \quad (3.126)$$

where

x_f = fracture half-length, ft
k = permeability, md

Example 3.15

Estimate the time required for a shut-in gas well to reach its 40-acre drainage area. The well is located in the centre of a square drainage boundary with the following properties:

$\phi = 14\%$, $\mu_{gi} = 0.016$ cp, $c_{ti} = 0.0008$ psi
$A = 40$ acres, $k = 0.1$ md

Solution

Calculate the stabilization time by applying Eq. (3.39) to give:

$$t_{pss} = \frac{15.8(0.14)(0.016)(0.0008)(40)(43,560)}{0.1} = 493 \text{ days}$$

The above example indicates that an excessive shut-in time of approximately 16 months is required to obtain a reliable average reservoir pressure.

Unlike curvature in the p/Z plot, which can be caused by:

- an aquifer,
- an oil leg,
- formation compressibility, or
- liquid condensation,

scatter in the p/Z plot is diagnostic of substantial reservoir pressure gradients. Hence, if substantial scatter is seen in a p/Z plot, the tank assumption is being violated and the plot should not be used to determine GIIP. One obvious solution to the material balance problem in tight gas reservoirs is the use of a numerical simulator. Two other relatively new approaches for solving the material balance problem that can be used if reservoir simulation software is not available are:

(1) the compartmental reservoir approach;
(2) the combined decline curve and type curve approach.

These two methodologies are discussed below.

3.5.1 Compartmental Reservoir Approach

A compartmental reservoir is defined as a reservoir that consists of two or more distinct regions that are allowed to communicate. Each compartment or "tank" is described by its own material balance, which is coupled to the material balance of the neighboring compartments through influx or efflux gas across the common boundaries. Payne (1996) and Hagoort and Hoogstra (1999) proposed two different robust and rigorous schemes for the numerical solution of the MBEs, of compartmented gas reservoirs. The main difference between the two approaches is that Payne solves for the pressure in each compartment explicitly and Hagoort and Hoogstra implicitly. However, *both schemes* employ the following basic approach:

- Divide the reservoir into a number of compartments with each compartment containing one or more production wells that are proximate and that measure consistent reservoir pressures. The initial division should be made with as few tanks as possible with each compartment having different dimensions in terms of length L, width W, and height h.
- Each compartment must be characterized by a historical production and pressure decline data as a function of time.
- If the initial division is not capable of matching the observed pressure decline, additional compartments can be added either by subdividing the previously defined tanks or by adding tanks that do not contain drainage points, i.e., production wells.

The practical application of the compartmental reservoir approach is illustrated by the following two methods:

(1) the Payne method;
(2) the Hagoort and Hoogstra method.

Payne Method. Rather than using the conventional single-tank MBE in describing the performance of tight gas reservoirs, Payne (1996) suggested a different approach that is based on subdividing the reservoir into a number of tanks, i.e., compartments, which are allowed to communicate. Such compartments can be depleted either directly by wells or indirectly through other tanks. The flow rate between tanks is set proportionally to either the difference in the square of tank pressure or the difference in pseudopressures, i.e., $m(p)$. To illustrate the concept, consider a reservoir that consists of two compartments, 1 and 2, as shown schematically in Figure 3.42.

Initially, i.e., before the start of production, both compartments are in equilibrium with the same initial reservoir pressure. Gas can be produced from either one or both compartments. With gas production, the pressures in the reservoir compartments will decline at a different rate depending on the production rate from

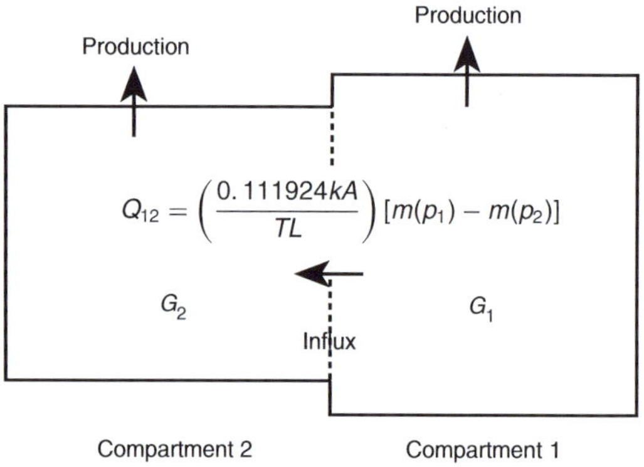

FIGURE 3.42 Schematic representation of compartmented reservoir consisting of two reservoir compartments separated by a permeable boundary.

each compartment and the crossflow rate between the two compartments. Adopting the convention that influx is positive if gas flows from compartment 1 into compartment 2, the linear gas flow rate between the two compartments in terms of gas pseudopressure is given by Eq. (1.22) of Chapter 1 as:

$$Q_{12} = \left(\frac{0.111924kA}{TL}\right)[m(p_1) - m(p_2)]$$

where

Q_{12} = flow rate between the two compartments, scf/day
$m(p_1)$ = gas pseudopressure in compartment (tank) 1, psi^2/cp
$m(p_2)$ = gas pseudopressure in compartment (tank) 2, psi^2/cp
k = permeability, md
L = distance between the center of the two compartments, ft
A = cross-sectional area, i.e., width height, ft^2
T = temperature, °R

The above equation can be expressed in a more compact form by including a "communication factor" C_{12} between the two compartments, as:

$$Q_{12} = C_{12}[m(p_1) - m(p_2)] \tag{3.127}$$

The communication factor C_{12} between the two compartments is computed by calculating the individual communication factor for each compartment and employing an average technique. The communication factor for each of the two compartments is given by:

$$\text{For compartment 1:} \quad C_1 = \frac{0.111924k_1A_1}{TL_1}$$

$$\text{For compartment 2:} \quad C_2 = \frac{0.11192k_2A_2}{TL_2}$$

And the communication factor between two compartments, C_{12}, is given by the following average technique:

$$C_{12} = \frac{2C_1C_2}{C_1 + C_2}$$

where

C_{12} = communication factor between two compartments, scf/day/psi^2/cp
C_1 = communication factor for compartment 1, scf/day/psi^2/cp
C_2 = communication factor for compartment 2, scf/day/psi^2/cp
L_1 = length of compartment 1, ft
L_2 = length of compartment 2, ft
A_1 = cross-sectional area of compartment 1, ft^2
A_2 = cross-sectional area of compartment 2, ft^2

The cumulative gas in flux G_{p12} from compartment 1 to compartment 2 is given by the integration of flow rate over time t as:

$$G_{p12} = \int_0^t Q_{12}\, dt = \sum_0^t (\Delta Q_{12})\, \Delta t \tag{3.128}$$

Payne proposed that *individual* compartment pressures are determined by assuming a straight-line relationship of p/Z vs. G_{pt} with the total gas production G_{pt} from an individual compartment as defined by the following expression:

$$G_{pt} = G_p + G_{p12}$$

where G_p is the cumulative gas produced from wells in the compartment and G_{p12} is the cumulative gas efflux/influx between the connected compartments. Solving Eq. (3.59) for the pressure in each compartment and assuming a positive flow from compartment 1 to 2 gives:

$$p_1 = \left(\frac{p_i}{Z_i}\right)Z_1\left(1 - \frac{G_{p1} + G_{p12}}{G_1}\right) \tag{3.129}$$

$$p_2 = \left(\frac{p_i}{Z_i}\right)Z_2\left(1 - \frac{G_{p2} + G_{p12}}{G_2}\right) \tag{3.130}$$

with

$$G_1 = \frac{43,560A_1h_1\phi_1(1S_{wi})}{B_{gi}} \tag{3.131}$$

$$G_2 = \frac{43,560 A_2 h_2 \phi_2 (1 S_{wi})}{B_{gi}} \tag{3.132}$$

where

G_1 = initial gas-in-place in compartment 1, scf
G_2 = initial gas-in-place in compartment 2, scf
G_{p1} = actual cumulative gas production from compartment 1, scf
G_{p2} = actual cumulative gas production from compartment 2, scf
A_1 = areal extent of compartment 1, acres
A_2 = areal extent of compartment 2, acres
h_1 = average thickness of compartment 1, ft
h_2 = average thickness of compartment 2, ft
B_{gi} = initial gas formation volume factor, ft^3/scf
ϕ_1 = average porosity in compartment 1
ϕ_2 = average porosity in compartment 2

The subscripts 1 and 2 denote the two compartments 1 and 2, while the subscript i refers to initial condition. The required input data for the Payne method consists of:

- amount of gas contained in each tank, i.e., tank dimensions, porosity, and saturation;
- intercompartment communication factors C_{12};
- initial pressure in each compartment;
- production data profiles from the individual tanks.

Payne's technique is performed fully explicit in time. At each time step, the pressures in various tanks are calculated, yielding a pressure profile that can be matched to the actual pressure decline. The specific steps of this iterative method are summarized below:

Step 1. Prepare the available gas properties data in tabulated and graphical forms that include:

Z vs. p
μ_g vs. p
$\dfrac{2p}{\mu_g Z}$ vs. p
$m(p)$ vs. p

Step 2. Divide the reservoir into compartments and determine the dimensions of each compartment in terms of:

length L
height h
width W
cross-sectional area A

Step 3. For each compartment, determine the initial gas-in-place G. Assuming two compartments for example, calculate G_1 and G_2 from Eqs. (3.131) and (3.132):

$$G_1 = \frac{43,560 A_1 h_1 \phi_1 (1 S_{wi})}{B_{gi}}$$

$$G_2 = \frac{43,560 A_2 h_2 \phi_2 (1 S_{wi})}{B_{gi}}$$

Step 4. For each compartment, make a plot of p/Z vs. G_p that can be constructed by simply drawing a straight line between p_i/Z_i with initial gas-in-place in both compartments, i.e., G_1 and G_2.

Step 5. Calculate the communication factors for each compartment and between compartments. For two compartments:

$$C_1 = \frac{0.111924 k_1 A_1}{T L_1}$$

$$C_2 = \frac{0.111924 k_2 A_2}{T L_2}$$

$$C_{12} = \frac{2 C_1 C_2}{C_1 + C_2}$$

Step 6. Select a small time step Δt and determine the corresponding *actual* cumulative gas production G_p from each compartment. Assign $G_p = 0$ if the compartment does not include a well.

Step 7. Assume (guess) the pressure distributions throughout the selected compartmental system and determine the gas deviation factor Z at each pressure. For a two-compartment system, let the initial values be denoted by p_1^k and p_2^k.

Step 8. Using the assumed values of the pressure p_1^k and p_2^k, determine the corresponding $m(p_1)$ and $m(p_2)$ from the data of step 1.

Step 9. Calculate the gas influx rate Q_{12} and cumulative gas influx G_{p12} by applying Eqs. (3.127) and (3.128), respectively.

$$Q_{12} = C_{12}[m(p_1) - m(p_2)]$$

$$G_{p12} = \int_0^t Q_{12} dt = \sum_0^t (\Delta Q_{12})\, \Delta t$$

Step 10. Substitute the values of G_{p12}, the Z factor, and actual values of G_{p1} and G_{p2} in Eqs. (3.129) and (3.130) to calculate the pressure in each compartment as denoted by p_1^{k+1} and p_2^{k+1}:

$$p_1^{k+1} = \left(\frac{p_i}{Z_i}\right) Z_1 \left(1 - \frac{G_{p1} + G_{p12}}{G_1}\right)$$

$$p_2^{k+1} = \left(\frac{p_i}{Z_i}\right) Z_2 \left(1 - \frac{G_{p2} + G_{p12}}{G_2}\right)$$

Step 11. Compare the assumed and calculated values, i.e., $|p_1^k - p_1^{k+1}|$ and $|p_2^k - p_2^{k+1}|$. If a satisfactory match is achieved within a tolerance of 5–10 psi for all the pressure values, then steps 3–7 are repeated at the new time level with the corresponding historical gas production data. If the match is not satisfactory, repeat the iterative cycle of steps 4–7 and set $p_1^k = p_1^{k+1}$ and $p_2^k = p_2^{k+1}$.

Step 12. Repeat steps 6–11 to produce a pressure decline profile for each compartment that can be compared with the actual pressure profile for each compartment or that from step 4.

Performing a material balance history match consists of varying the number of compartments required, the dimension of the compartments, and the communication factors until an acceptable match of the pressure decline is obtained. The improved accuracy in estimating the original gas-in-place, resulting from determining the optimum number and size of compartments, stems from the ability of the proposed method to incorporate reservoir pressure gradients, which are completely neglected in the single-tank conventional p/Z plot method.

Hagoort and Hoogstra Method. Based on the Payne method, Hagoort and Hoogstra (1999) developed a numerical method to solve the MBE of compartmental gas reservoirs that employs an implicit, iterative procedure and that recognizes the pressure dependency of the gas properties. The iterative technique relies on adjusting the size of compartments and the transmissibility values to match the historical pressure data for each compartment as a function of time. Referring to Figure 3.42, the authors assume a thin permeable layer with a transmissibility of Γ_{12} separating the two compartments. Hagoort and Hoogstra expressed the instantaneous gas influx through the thin permeable layer by Darcy's equation as given by (in Darcy's units):

$$Q_{12} = \frac{\Gamma_{12}(p_1^2 - p_2^2)}{2p_1(\mu_g B_g)_{avg}}$$

where

Γ_{12} = the transmissibility between compartments

Here, we suggest a slightly different approach for estimating the gas influx between compartments by modifying Eq. (1.22) in Chapter 1 to give:

$$Q_{12} = \frac{0.111924\Gamma_{12}(p_1^2 - p_2^2)}{TL} \tag{3.133}$$

with

$$\Gamma_{12} = \frac{\Gamma_1 \Gamma_2 (L_1 + L_2)}{L_1 \Gamma_2 + L_2 \Gamma_1} \tag{3.134}$$

$$\Gamma_1 = \left[\frac{kA}{Z\mu_g}\right]_1 \tag{3.135}$$

$$\Gamma_2 = \left[\frac{kA}{Z\mu_g}\right]_2 \tag{3.136}$$

where

Q_{12} = influx gas rate, scf/day
L = distance between the centers of compartments 1 and 2, ft

A = cross-sectional area, ft^2
μ_g = gas viscosity, cp
Z = gas deviation factor
k = permeability, md
p = pressure, psia
T = temperature, °R
L_1 = length of compartment 1, ft
L_2 = length of compartment 2, ft

The subscripts 1 and 2 refer to compartments 1 and 2, respectively.

Using Eq. (3.59), the material balance for the two reservoir compartments can be modified to include the gas influx from compartment 1 to compartment 2 as:

$$\frac{p_1}{Z_1} = \frac{p_1}{Z_1}\left(1 - \frac{G_{p1} + G_{p12}}{G_1}\right) \quad \textbf{(3.137)}$$

$$\frac{p_2}{Z_2} = \frac{p_1}{Z_1}\left(1 - \frac{G_{p2} - G_{p12}}{G_2}\right) \quad \textbf{(3.138)}$$

where

p_1 = initial reservoir pressure, psi
Z_1 = initial gas deviation factor
G_p = actual (historical) cumulative gas production, scf
G_1, G_2 = initial gas-in-place in compartments 1 and 2, scf
G_{p12} = cumulative gas influx from compartment 1 to 2 in scf, as given in Eq. (3.138)

Again, subscripts 1 and 2 represent compartments 1 and 2, respectively.

To solve the MBEs as represented by the relationships (3.132) and (3.135) for the two unknowns p_1 and p_2, the two expressions can be arranged and equated to zero, to give:

$$F_1(p_1, p_2) = p_1 - \left(\frac{p_i}{Z_i}\right) Z_1 \left(1 - \frac{G_{p1} + G_{p12}}{G_1}\right) = 0 \quad \textbf{(3.139)}$$

$$F_2(p_1, p_2) = p_2 - \left(\frac{p_i}{Z_i}\right) Z_2 \left(1 - \frac{G_{p2} + G_{p12}}{G_2}\right) = 0 \quad \textbf{(3.140)}$$

The general methodology of applying the method is very similar to that of Payne method and involves the following specific steps:

Step 1. Prepare the available gas properties data in tabulated and graphical forms that include, Z vs. p and μ_g vs. p.

Step 2. Divide the reservoir into compartments and determine the dimensions of each compartments in terms of:

length L
height h
width W
cross-sectional area A

Step 3. For each compartment, determine the initial gas-in-place G. For reasons of clarity, assume two gas compartments and calculate G_1 and G_2 from Eqs. (3.131) and (3.132):

$$G_1 = \frac{43,560 A_1 h_1 \phi_1 (1 S_{wi})}{B_{gi}}$$

$$G_2 = \frac{43,560 A_2 h_2 \phi_2 (1 S_{wi})}{B_{gi}}$$

Step 4. For each compartment, make a plot of p/Z vs. G_p that can be constructed by simply drawing a straight line between p_i/Z_i with initial gas-in-place in both compartments, i.e., G_1 and G_2.

Step 5. Calculate the transmissibility by applying Eq. (3.134):

Step 6. Select a time step Δt and determine the corresponding actual cumulative gas production G_{p1} and G_{p2}.

Step 7. Calculate the gas influx rate Q_{12} and cumulative gas influx G_{p12} by applying Eqs. (3.133) and (3.128), respectively:

$$Q_{12} = \frac{0.11194 \Gamma_{12}(p_1^2 - p_2^2)}{TL}$$

$$G_{p12} = \int_0^t Q_{12}\, dt = \sum_0^t (\Delta Q_{12})\, \Delta t$$

Step 8. Start the iterative solution by assuming initial estimates of the pressure for compartments 1 and 2 (i.e., p_1^k and

p_2^k). Using the Newton–Raphson iterative scheme, calculate new improved values of the pressure p_1^{k+1} and p_2^{k+1} by solving the following linear equations as expressed in a matrix form:

$$\begin{bmatrix} p_1^{k+1} \\ p_2^{k+1} \end{bmatrix} = \begin{bmatrix} p_1^{k} \\ p_2^{k} \end{bmatrix} - \begin{bmatrix} \dfrac{\partial F_1(p_1^k, p_2^k)}{\partial p_1} & \dfrac{\partial F_1(p_1^k, p_2^k)}{\partial p_2} \\ \dfrac{\partial F_2(p_1^k, p_2^k)}{\partial p_1} & \dfrac{\partial F_2(p_1^k, p_2^k)}{\partial p_2} \end{bmatrix}^{-1} \times \begin{bmatrix} -F_1(p_1^k, p_2^k) \\ -F_2(p_1^k, p_2^k) \end{bmatrix}$$

where the superscript 1 denotes the inverse of the matrix. The partial derivatives in the above system of equations can be expressed in analytical form by differentiating Eqs. (3.139) and (3.140) with respect to p_1 and p_2. During an iterative cycle, the derivatives are evaluated at the updated new pressures, i.e., p_1^{k+1} and p_2^{k+1}. The iteration is stopped when $|p_1^{k+1} - p_1^k|$ and $|p_2^{k+1} - p_2^k|$ are less than a certain pressure tolerance, i.e., 5–10 psi.

Step 9. Generate the pressure profile as a function of time for each compartment by repeating steps 2 and 3.

Step 10. Repeat steps 6–11 to produce a pressure decline profile for each compartment that can be compared with the actual pressure profile for each compartment or that from step 4.

Compare the calculated pressure profiles with those of the observed pressures. If a match has not been achieved, adjust the size and number of compartments (i.e., initial gas-in-place) and repeat steps 2–10.

3.5.2 Combined Decline Curve and Type Curve Analysis Approach

Production decline analysis is the analysis of past trends of declining production performance, i.e., rate vs. time and rate vs. cumulative production plots, for wells and reservoirs. During the past 30 years, various methods have been developed f
gas reservoirs. T
basic MBE to de
techniques. There
analysis technique

(1) the classical c
tion data;
(2) the type curve

Some graphical
of decline curves
limitations. General principles of both types and methods of combining both approaches to determine gas reserves are briefly presented below.

Decline Curve Analysis. Decline curves are one of the most extensively used forms of data analysis employed in evaluating gas reserves and predicting future production. The decline curve analysis technique is based on the assumption that the past production trend with its controlling factors will continue in the future and, therefore, can be extrapolated and described by a mathematical expression.

The method of extrapolating a "trend" for the purpose of estimating future performance must satisfy the condition that the factors that caused changes in the past performance, i.e., decline in the flow rate, will operate in the same way in the future. These decline curves are characterized by three factors:

(1) initial production rate, or the rate at some particular time;
(2) curvature of the decline;
(3) rate of decline.

These factors are a complex function of numerous parameters within the reservoir, wellbore, and surface-handling facilities.

Ikoku (1984) presented a comprehensive and rigorous treatment of production decline curve analysis. He pointed out that the following three conditions must be considered when performing production decline curve analysis:

(1) Certain conditions must prevail before we can analyze a production decline curve with any degree of reliability. The production

n stable over the period being hat is, a flowing well must have oduced with constant choke size or ant wellhead pressure and a pumping ll must have been pumped off or produced with constant fluid level. These indicate that the well must have been produced at capacity under a given set of conditions. The production decline observed should truly reflect reservoir productivity and not be the result of external causes, such as a change in production conditions, well damage, production controls, and equipment failure.

(2) Stable reservoir conditions must also prevail in order to extrapolate decline curves with any degree of reliability. This condition will normally be met as long as the producing mechanism is not altered. However, when action is taken to improve the recovery of gas, such as infill drilling, fluid injection, fracturing, and acidizing, decline curve analysis can be used to estimate the performance of the well or reservoir in the absence of the change and compare it to the actual performance with the change. This comparison will enable us to determine the technical and economic success of our efforts.

(3) Production decline curve analysis is used in the evaluation of new investments and the audit of previous expenditures. Associated with this is the sizing of equipment and facilities such as pipelines, plants, and treating facilities. Also associated with the economic analysis is the determination of reserves for a well, lease, or field. This is an independent method of reserve estimation, the result of which can be compared with volumetric or material balance estimates.

Arps (1945) proposed that the "curvature" in the production rate vs. time curve can be expressed mathematically by one of the hyperbolic family of equations. Arps recognized the following three types of rate decline behavior:

(1) exponential decline;
(2) harmonic decline;
(3) hyperbolic decline.

Each type of decline curve has a different curvature as shown in Figure 3.43. This figure depicts the characteristic shape of each type of decline when the flow rate is plotted vs. time or vs. cumulative production on Cartesian, semilog, and log–log scales. The main characteristics of these decline curves are discussed below and can be used to select the flow rate decline model that is appropriate for describing the rate–time relationship of the hydrocarbon system:

- *For exponential decline*: A straight-line relationship will result when flow rate is plotted vs. time on a *semilog scale* and also when the flow rate vs. cumulative production is plotted on a *Cartesian scale.*
- *For harmonic decline*: Rate vs. cumulative production is a straight line on a *semilog scale* with all other types of decline curves having some curvature. There are several shifting techniques that are designed to straighten out the resulting curve of plotting flow rate vs. time when plotted on a log–log scale.
- *For hyperbolic decline*: None of the above plotting scales, i.e., Cartesian, semilog, or log–log, will produce a straight-line relationship for a hyperbolic decline. However, if the flow rate is plotted vs. time on log–log paper, the resulting curve can be straightened out by using shifting techniques.

Nearly all conventional decline curve analysis is based on empirical relationships of production rate vs. time given by Arps (1945) as:

$$q_t = \frac{q_i}{(1 + bD_i t)^{1/b}} \tag{3.141}$$

where

q_t = gas flow rate at time t, MMscf/day
q_i = initial gas flow rate, MMscf/day
t = time, days
D_i = initial decline rate, day^{-1}
b = Arps's decline curve exponent

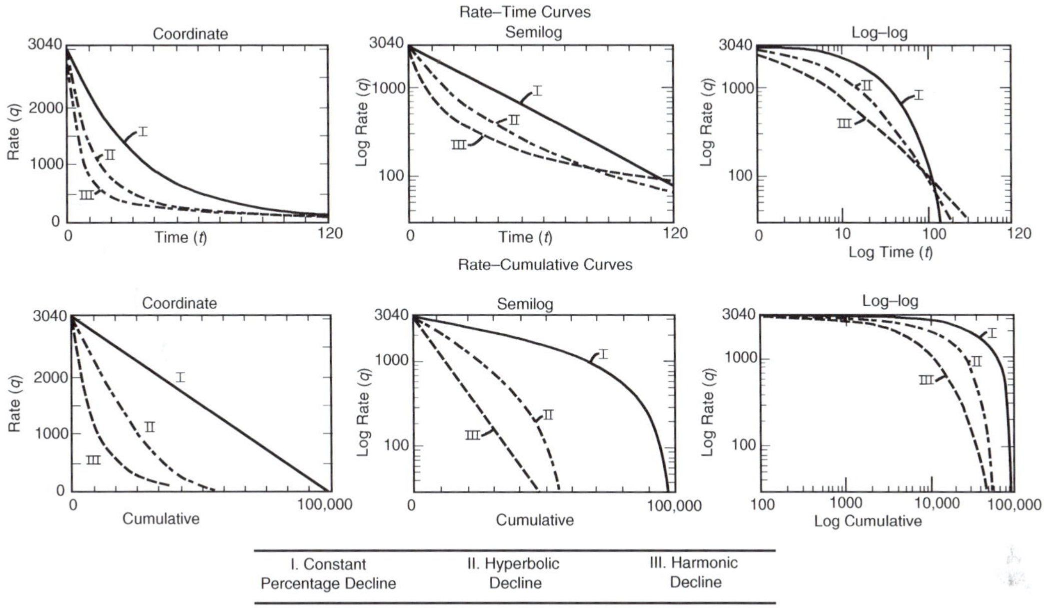

FIGURE 3.43 Classification of production decline curves. *(After Arps, J.J. Estimation of Primary Oil Reserves, Courtesy of Trans. AIME, vol. 207, 1956).*

The mathematical description of these production decline curves is greatly simplified with the use of the instantaneous (nominal) decline rate D. This decline rate is defined as the rate of change of the natural logarithm of the production rate, i.e., $\ln(q)$, with respect to time t, or:

$$D = -\frac{d(\ln q)}{dt} = -\frac{1}{q}\frac{dq}{dt} \quad (3.142)$$

The minus sign has been added since dq and dt have opposite signs and it is convenient to have D always positive. Note that the decline rate equation, i.e., Eq. (3.142), describes the instantaneous changes in the slope of the curvature dq/dt with changing of the flow rate q with time.

The parameters determined from the classical fit of the historical data, namely the decline rate D and the exponent b, can be used to predict future production. This type of decline curve analysis can be applied to individual wells or the entire reservoir. The accuracy of the entire reservoir application is sometimes better than for individual wells due to smoothing of the rate data. Based on the type of rate decline behavior of the hydrocarbon system, the value of b ranges from 0 to 1 and, accordingly, Arps's equation can be conveniently expressed in the following three forms:

Case	b	Rate–Time Relationship
Exponential	$b=0$	$q_t = q_i \exp(-D_i t)$ (3.143)
Hyperbolic	$0<b<1$	$q_t = \dfrac{q_i}{(1+bD_i t)^{1/b}}$ (3.144)
Harmonic	$b=1$	$q_t = \dfrac{q_i}{1+D_i t}$ (3.145)

Figure 3.44 illustrates the general shape of the three curves at different possible values of b.

It should be pointed out that the above forms of decline curve equations are strictly applicable *only* when the well/reservoir is under pseudosteady (semisteady)-state flow conditions, i.e., boundary-dominated flow conditions. Arps's equation has been often misused and applied to

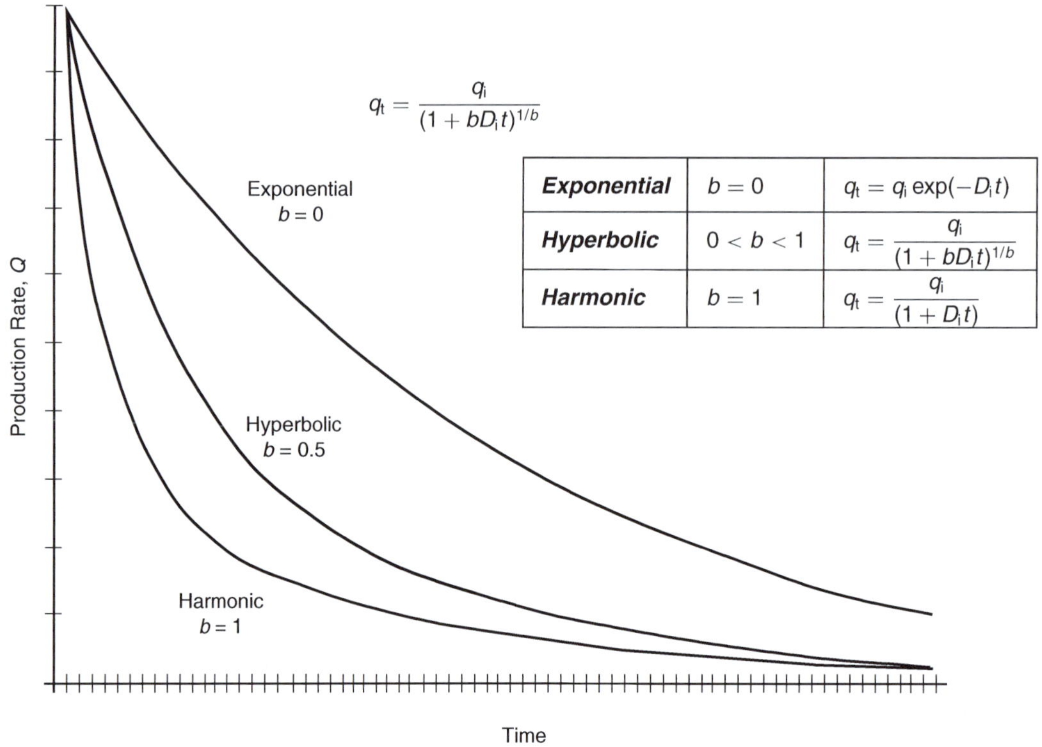

FIGURE 3.44 Decline curve—rate/time (exponential, harmonic, hyperbolic).

model the performance of oil and gas wells whose flow regimes are in a transient flow. As presented in Chapter 1, when a well is first open to flow, it is under a transient (unsteady-state) condition. It remains under this condition until the production from the well affects the total reservoir system by reaching its drainage boundary, then the well is said to be flowing under pseudosteady-state or boundary-dominated flow conditions. The following is a list of inherent assumptions that must be satisfied before performing rate–time decline curve analysis:

- The well is draining a constant drainage area, i.e., the well is under boundary-dominated flow conditions.
- The well is produced at or near capacity.
- The well is produced at a *constant bottom-hole pressure.*

Again, the above conditions must be satisfied before applying any of the decline curve analysis methods to describe the production performance of a reservoir. In most cases, tight gas wells are producing at capacity and approach a constant bottom-hole pressure, if produced at a constant line pressure. However, it can be extremely difficult to determine when a tight gas well has defined its drainage area and the start of the pseudosteady-state flowing condition.

The area under the decline curve of q vs. time between times t_1 and t_2 is a measure of the cumulative gas production G_p during this period as expressed mathematically by:

$$G_p = \int_{t_1}^{t_2} q_t \, dt \tag{3.146}$$

Replacing the flow rate q_t in the above equation with the three individual expressions that describe types of decline curves, i.e., Eqs. (3.143)–(3.145), and integrating gives:

Exponential $b = 0$

$$G_{p(t)} = \frac{1}{D_i}(q_i - q_t) \tag{3.147}$$

Hyperbolic $0<b<1$

$$G_{p(t)} = \left[\frac{(q_i)}{D_i(1-b)}\right]\left[1 - \left(\frac{q_t}{q_i}\right)^{1-b}\right] \tag{3.148}$$

Harmonic $b = 1$

$$G_{p(t)} = \left(\frac{q_i}{D_i}\right)\ln\left(\frac{q_i}{q_t}\right) \tag{3.149}$$

where

$G_{p(t)}$ = cumulative gas production at time t, MMscf
q_i = initial gas flow rate at time $t = 0$, MMscf/unit time
t = time, unit time
q_t = gas flow rate at time t, MMscf/unit time
D_i = nominal (initial) decline rate, 1/unit time

All the above expressions as given by Eqs. (3.143)–(3.149) require consistent units. Any convenient unit time can be used but, again, care should be taken to make certain that the time base of rates, i.e., q_i and q_t, matches the time unit of the decline rate D_i, e.g., for flow rate q in scf/month with D_i in month^{-1}.

Note that the *traditional* Arps decline curve analysis, as given by Eqs. (3.147)–(3.149), gives a reasonable estimation of reserves, but it has its failings, the most important one being that it *completely ignores the flowing pressure data.* As a result, it can underestimate or overestimate the reserves. The practical applications of these three commonly used decline curves are documented below.

Exponential Decline,* b = *0. The graphical presentation of this type of decline curve indicates that a plot of q_t vs. t on a semilog scale or a plot of q_t vs. $G_{p(t)}$ on a Cartesian scale will produce linear relationships that can be described mathematically by:

$$q_t = q_i \exp(-D_i t)$$

or linearly as:

$$\ln(q_t) = \ln(q_i) - D_i t$$

And similarly:

$$G_{p(t)} = \frac{q_i - q_t}{D_i}$$

or linearly as:

$$q_t = q_i - D_i G_{p(t)}$$

This type of decline curve is perhaps the simplest to use and perhaps the most conservative. It is widely used in the industry for the following reasons:

- Many wells follow a constant decline rate over a great portion of their productive life and will deviate significantly from this trend toward the end of this period.
- The mathematics involved, as described by the above line expressions, is easier to apply than the other line types.

Assuming that the historical production from a well or field is recognized by its exponential production decline behavior, the following steps summarize the procedure to predict the behavior of the well or the field as a function of time:

Step 1. Plot q_t vs. G_p on a Cartesian scale and q_t vs. t on semilog paper.
Step 2. For both plots, draw the best straight line through the points.
Step 3. Extrapolate the straight line on q_t vs. G_p to $G_p = 0$ that intercepts the y-axis with a flow rate value that is identified as q_i.

Step 4. Calculate the initial decline rate D_i by selecting a point on the Cartesian straight line with coordinates of (q_t, G_{pt}) or on a semilog line with coordinates of (q_t, t) and solve for D_i by applying Eq. (3.145) or (3.147):

$$D_i = \frac{\ln(q_i/q_t)}{t} \quad (3.150)$$

or equivalently as:

$$D_i = \frac{q_i - q_t}{G_{p(t)}} \quad (3.151)$$

If the method of least squares is used to determine the decline rate by analyzing the entire production data, then

$$D_i = \frac{\sum_t [t \ln(q_i/q_t)]}{\sum_t t^2} \quad (3.152)$$

or equivalently as:

$$D_i = \frac{q_i \sum_t G_{p(t)} - \sum_t q_t G_{p(t)}}{\sum_t [G_{p(t)}]^2} \quad (3.153)$$

Step 5. Calculate the time to reach the economic flow rate q_a (or any rate) and the corresponding cumulative gas production from Eqs. (3.143) and (3.147):

$$t_a = \frac{\ln(q_i/q_a)}{D_i}$$

$$G_{pa} = \frac{q_i - q_a}{t_a}$$

where

G_{pa} = cumulative gas production when reaching the economic flow rate or at abandonment, MMscf

q_i = initial gas flow rate at time $t = 0$, MMscf/unit time

t = abandonment time, unit time

q_a = economic (abandonment) gas flow rate, MMscf/unit time

D_i = nominal (initial) decline rate, 1/time unit

Example 3.16

The following production data is available from a dry gas field:

q_t (MMscf/day)	G_p (MMscf)	q_t (MMscf/day)	G_p (MMscf)
320	16,000	208	304,000
336	32,000	197	352,000
304	48,000	184	368,000
309	96,000	176	384,000
272	160,000	184	400,000
248	240,000		

Estimate:

(a) the future cumulative gas production when gas flow rate reaches 80 MMscf/day;
(b) the *extra* time to reach 80 MMscf/day.

Solution

(a) Use the following steps:

Step 1. A plot of G_p vs. q_t on a Cartesian scale as shown in Figure 3.45 produces a straight line indicating an exponential decline.

Step 2. From the graph, cumulative gas production is 633,600 MMscf at $q_t = 80$ MMscf/day indicating an extra production of 633.6 − 400.0 = 233.6 MMMscf.

Step 3. The intercept of the straight line with the *y*-axis gives a value of $q_i = 344$ MMscf/day.

Step 4. Calculate the initial (nominal) decline rate D_i by selecting a point *on* the straight line and solving for D_i by applying Eq. (3.150). At $G_{p(t)}$ of 352 MMscf, q_t is 197 MMscf/day, or:

$$D_i = \frac{q_i - q_t}{G_{p(t)}} = \frac{344 - 197}{352,000} = 0.000418 \ \text{day}^{-1}$$

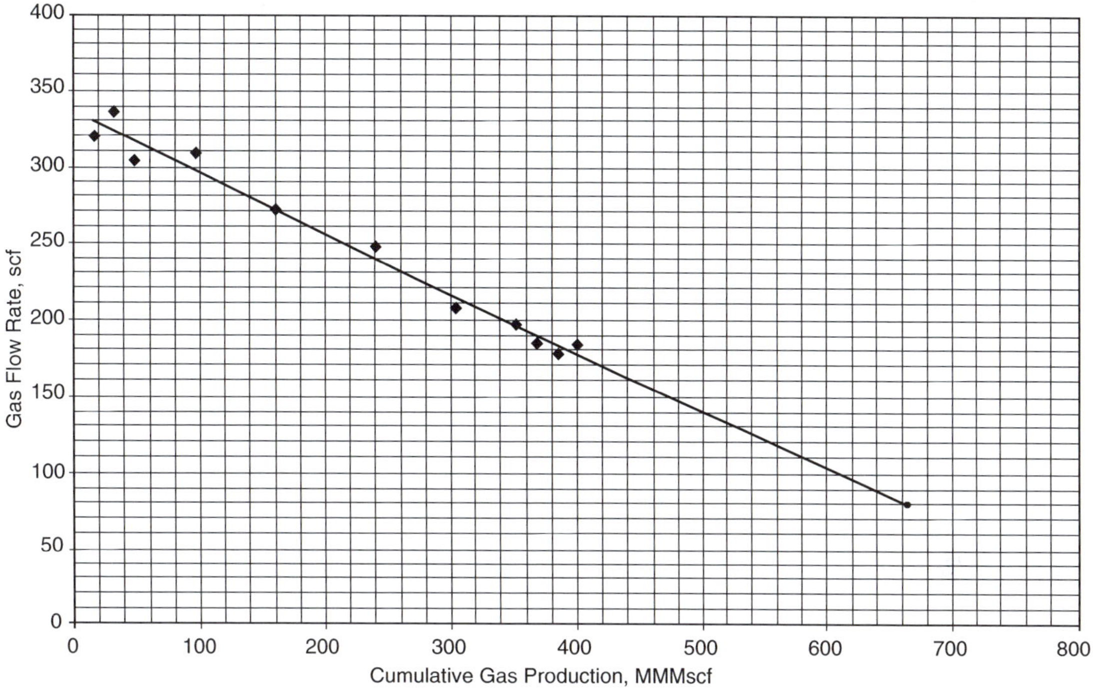

FIGURE 3.45 Decline curve data for Example 3.16.

It should be pointed out that the monthly and yearly nominal decline can be determined as:

$$D_{\mathrm{im}} = (0.000418)(30.4) = 0.0126 \ \text{month}^{-1}$$
$$D_{\mathrm{iy}} = (0.0126)(12) = 0.152 \ \text{year}^{-1}$$

Using the least-squares approach, i.e., Eq. (3.153), gives:

$$D_{\mathrm{i}} = \frac{0.3255(10^9) - 0.19709(10^9)}{0.295(10^{12})} = 0.000425 \ \text{day}^{-1}$$

(b) To calculate the *extra* time to reach 80 MMscf/day, use the following steps:

Step 1. Calculate the time to reach the last recorded flow rate of 184 MMscf from Eq. (3.150).

$$t = \frac{\ln(344/184)}{0.000425} = 1472 \ \text{days} = 4.03 \ \text{years}$$

Step 2. Calculate total time to reach a gas flow rate of 80 MMscf/day:

$$t = \frac{\ln(344/80)}{0.000425} = 3432 \ \text{days} = 9.4 \ \text{years}$$

Step 3. Extra time = 9.4 − 4.03 = 5.37 years.

Example 3.17

A gas well has the following production history:

Date	Time (months)	q_t (MMscf/month)
1/1/02	0	1240
2/1/02	1	1193
3/1/02	2	1148
4/1/02	3	1104
5/1/02	4	1066
6/1/02	5	1023
7/1/02	6	986
8/1/02	7	949
9/1/02	8	911
10/1/02	9	880
11/1/02	10	843
12/1/02	11	813
1/1/03	12	782

(a) Use the first 6 months of the production history data to determine the coefficient of the decline curve equation.
(b) Predict flow rates and cumulative gas production from August 1, 2002 through January 1, 2003.
(c) Assuming that the economic limit is 30 MMscf/month, estimate the time to reach the economic limit and the corresponding cumulative gas production.

Solution

(a) Use the following steps:

Step 1. A plot of q_t vs. t on a semilog scale as shown in Figure 3.46 indicates an exponential decline.

Step 2. Determine the initial decline rate D_i by selecting a point on the straight line and substituting the coordinates of the point in Eq. (3.150) or using the least-squares method, to give, from Eq. (3.150):

$$D_i = \frac{\ln(q_i/q_t)}{t} = \frac{\ln(1240/986)}{6} = 0.0382 \text{ month}^{-1}$$

Similarly, from Eq. (3.152):

$$D_i = \frac{\sum_t[t \ln(q_i/q_t)}{\sum_t t^2} = \frac{3.48325}{91} = 0.0383 \text{ month}^{-1}$$

(b) Use Eqs. (3.143) and (3.147) to calculate q_t and $G_{p(t)}$ in the following tabulated form:

$$q_t = 1240 \exp(-0.0383t)$$

$$G_{pt} = \frac{q_i - q_t}{0.0383}$$

Date	Time (months)	Actual q_t (MMscf/month)	Calculated q_t (MMscf/month)	$G_{p(t)}$ (MMscf/month)
2/1/02	1	1193	1193	1217
3/1/02	2	1148	1149	2387
4/1/02	3	1104	1105	3514
5/1/02	4	1066	1064	4599
6/1/02	5	1023	1026	4643
7/1/02	6	986	986	6647
8/1/02	7	949	949	7614
9/1/02	8	911	913	8545
10/1/02	9	880	879	9441
11/1/02	10	843	846	10303
12/1/02	11	813	814	11132
1/1/03	12	782	783	11931

(c) Use Eqs. (3.150) and (3.151) to calculate the time to reach an economic flow rate of 30 MMscf/month and the corresponding reserves:

$$t = \frac{\ln(1240/30)}{0.0383} = 97 \text{ months} = 8 \text{ years}$$

$$G_{pt} = \frac{(1240-30)10^6}{0.0383} = 31.6 \text{ MMMscf}$$

Harmonic Decline, b = 1. The production recovery performance of a hydrocarbon system that follows a harmonic decline, i.e., $b = 1$ in Eq. (3.141), is described by Eqs. (3.145) and (3.149):

$$q_t = \frac{q_i}{1 + D_i t}$$

$$G_{p(t)} = \left(\frac{q_i}{D_i}\right) \ln\left(\frac{q_i}{q_t}\right)$$

The above two expressions can be rearranged and expressed respectively as:

$$\frac{1}{q_t} = \frac{1}{q_i} + \left(\frac{D_i}{q_i}\right) t \tag{3.154}$$

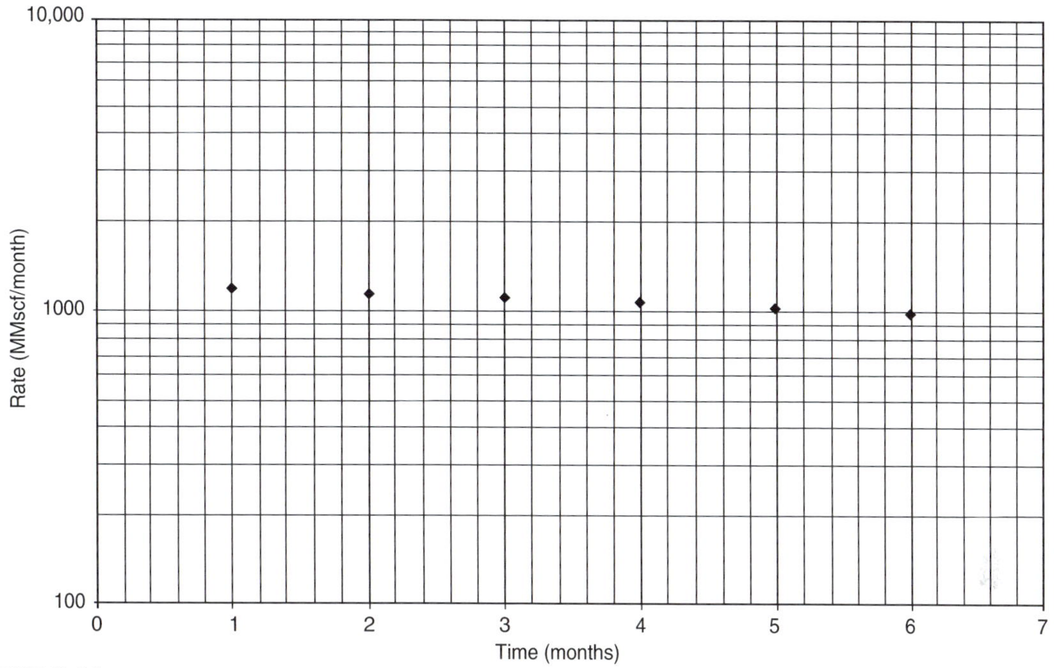

FIGURE 3.46 Decline curve data for Example 3.17.

$$\ln(q_t) = \ln(q_i) - \left(\frac{D_i}{q_i}\right) G_{p(t)} \tag{3.155}$$

The basic two plots for harmonic decline curve analysis are based on the above two relationships. Eq. (3.154) indicates that a plot of $1/q_t$ vs. t on a Cartesian scale will yield a straight line with a slope of (D_i/q_t) and intercept of $1/q_i$. Eq. (3.155) suggests that a plot of q_t vs. $G_{p(t)}$ on a semilog scale will yield a straight line with a negative slope of (D_i/q_i) and an intercept of q_i. The method of least squares can also be used to calculate the decline rate D_i, to give:

$$D_i = \frac{\sum_t (tq_i/q_t) - \sum_t t}{\sum_t t^2}$$

Other relationships that can be derived from Eqs. (3.154) and (3.155) include the time to reach the economic flow rate q_a (or any flow rate) and the corresponding cumulative gas production $G_{p(a)}$:

$$t_a = \frac{q_i - q_a}{q_a D_i}$$

$$G_{p(a)} = \left(\frac{q_i}{D_i}\right) \ln\left(\frac{q_a}{q_t}\right) \tag{3.156}$$

Hyperbolic Decline, 0<b<1. The two governing relationships for a reservoir or a well when its production follows the hyperbolic decline behavior are given by Eqs. (3.144) and (3.148), or:

$$q_t = \frac{q_i}{(1 + bD_i t)^{1/b}}$$

$$G_{p(t)} = \left[\frac{q_i}{(D_i(1-b)}\right]\left[1 - \left(\frac{q_t}{q_i}\right)^{1-b}\right]$$

The following simplified iterative method is designed to determine D_i and b from the historical production data:

Step 1. Plot q_t vs. t on a semilog scale and draw a *smooth curve* through the points.
Step 2. Extend the curve to intercept the y-axis at $t=0$ and read q_i.
Step 3. Select the other end point of the smooth curve and record the coordinates of the point and refer to it as (t_2, q_2).
Step 4. Determine the coordinates of the middle point on the smooth curve that corresponds to (t_1, q_1) with the value of q_1 as obtained from the following expression:

$$q_1 = \sqrt{q_i q_2} \tag{3.157}$$

The corresponding value of t_1 is read from the smooth curve at q_1.

Step 5. Solve the following equation iteratively for b:

$$f(b) = t_2\left(\frac{q_i}{q_1}\right)^b - t_1\left(\frac{q_i}{q_2}\right)^b - (t_2 - t_1) = 0 \tag{3.158}$$

The Newton–Raphson iterative method can be employed to solve the above nonlinear function by using the following recursion technique:

$$b^{k+1} = b^k - \frac{f(b^k)}{f'(b^k)} \tag{3.159}$$

where the derivative $f^\backslash(b^k)$ is given by:

$$f^{\backslash}(b^k) = t_2\left(\frac{q_i}{q_1}\right)^{b^k} \ln\left(\frac{q_i}{q_1}\right) - t_1\left(\frac{q_i}{q_2}\right)^{b_k} \ln\left(\frac{q_i}{q_2}\right) \tag{3.160}$$

Starting with an initial value of $b=0.5$, i.e., $b^k=0.5$, the method will usually converge after four to five iterations when setting the convergence criterion at $b^{k+1} - b^k \le 10^{-6}$.

Step 6. Solve for D_i by solving Eq. (3.144) for D_i and using the calculated value of b from step 5 and the coordinates of a point on the smooth graph, i.e., (t_2, q_2), to give:

$$D_i = \frac{(q_i/q_2)^b - 1}{bt_2} \tag{3.161}$$

The following example illustrates the proposed methodology for determining b and D_i.

Example 3.18

The following production data is reported by Ikoku for a gas well[2]:

Date	Time (years)	q_t (MMscf/day)	$G_{p(t)}$ (MMscf)
1/1/79	0.0	10.00	0.00
7/1/79	0.5	8.40	1.67
1/1/80	1.0	7.12	3.08
7/1/80	1.5	6.16	4.30
1/1/81	2.0	5.36	5.35
7/1/81	2.5	4.72	6.27
1/1/82	3.0	4.18	7.08
7/1/82	3.5	3.72	7.78
1/1/83	4.0	3.36	8.44

Estimate the future production performance for the next 16 years.

Solution

Step 1. Determine the type of decline that adequately represents the historical data. This can be done by constructing the following two plots:
(1) Plot q_t vs. t on a semilog scale as shown in Figure 3.47. The plot does not yield a straight line and, thus, the decline is *not exponential.*
(2) Plot q_t vs. $G_{p(t)}$ on semilog paper as shown in Figure 3.48. The plot again does not produce a straight line and, therefore, the decline is *not harmonic.*

---

[2]Ikoku, C., 1984. *Natural Gas Reservoir Engineering.* John Wiley & Sons, New York, NY.

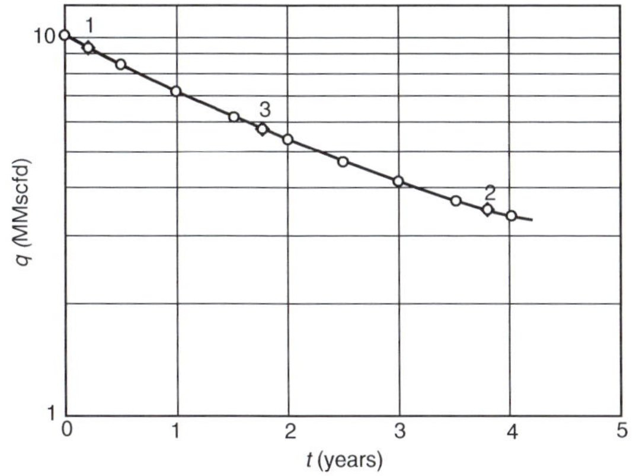

FIGURE 3.47 Rate–time plot for Example 3.18.

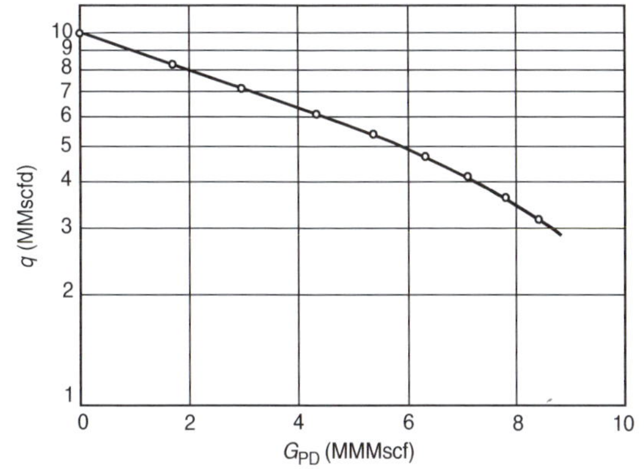

FIGURE 3.48 Rate—cumulative plot for Example 3.18.

The generated two plots indicate that the decline must be hyperbolic.

Step 2. From Figure 3.47, determine the initial flow rate q_i by extending the smooth curve to intercept with the y-axis, i.e., at $t = 0$, to give:

$$q_i = 10 \text{ MMscf/day}$$

Step 3. Select the coordinates of the other end point on the smooth curve as (t_2, q_2), to give:

$$t_2 = 4 \text{ years}$$

$$q_2 = 3.36 \text{ MMscf/day}$$

Step 4. Calculate q_1 from Eq. (3.157) and determine the corresponding time:

$$q_1 = \sqrt{q_i q_2} = \sqrt{(10)(3.36)} = 5.8 \text{ MMscf/day}$$

The corresponding time $t_1 = 1.719$ years.

Step 5. Assume $b = 0.5$, and solve Eq. (3.158) iteratively for b:

$$f(b) = t_2\left(\frac{q_i}{q_1}\right)^b - t_1\left(\frac{q_i}{q_2}\right)^b - (t_2 - t_1)$$

$$f(b) = 4(1.725)^b - 1.719(2.976)^b - 2.26$$

and

$$f^{\backslash}(b^k) = t_2\left(\frac{q_i}{q_1}\right)^{b^k} \ln\left(\frac{q_i}{q_1}\right) - t_1\left(\frac{q_i}{q_2}\right)^{b^k} \ln\left(\frac{q_i}{q_2}\right)$$

$$f^{\backslash}(b) = 2.18(1.725)^b - 1.875(2.976)^b$$

with

$$b^{k+1} = b^k - \frac{f(b^k)}{f^{\backslash}(b^k)}$$

The iterative method can be conveniently performed by constructing the following table:

k	b^k	$f(b)$	$f^{\backslash}(b)$	b^{k+1}
0	0.500000	7.57×10^{-3}	−0.36850	0.520540
1	0.520540	-4.19×10^{-4}	−0.40950	0.519517
2	0.519517	-1.05×10^{-6}	−0.40746	0.519514
3	0.519514	-6.87×10^{-9}	−0.40745	0.519514

The method converges after three iterations with a value of $b = 0.5195$.

Step 6. Solve for D_i by using Eq. (3.161):

$$D_i = \frac{(q_i/q_2)^b - 1}{bt_2}$$

$$= \frac{(10/3.36)^{0.5195} - 1}{(0.5195)(4)} = 0.3668 \text{ year}^{-1}$$

or on a monthly basis $D_i = 0.3668/12 = 0.0306$ month^{-1}or on a daily basis $D_i = 0.3668/365 = 0.001$ day^{-1}

Step 7. Use Eqs. (3.144) and (3.148) to predict the future production performance of the gas well. Note in Eq. (3.144) that the denominator contains $D_i t$ and, therefore, the product must be dimensionless, or:

$$q_t = \frac{10(10^6)}{[1+0.5195 D_i t]^{(1/0.5195)}} = \frac{(10)(10^6)}{[1+0.5195(0.3668)(t)]^{(1/0.5195)}}$$

$$G_{p(t)} = \left[\frac{q_i}{D_i(1-b)}\right]\left[1-\left(\frac{q_t}{q_i}\right)^{1-b}\right] = \left[\frac{(10)(10^6)}{(0.001)(1-0.5195)}\right] \times \left[1-\left(\frac{q_t}{(10)(10^6)}\right)^{1-0.5195}\right]$$

Results of step 7 are tabulated below and shown graphically in Figure 3.49:

Time (years)	Actual q (MMscf/day)	Calculated q (MMscf/day)	Actual Cumulative Gas (MMMscf)	Calculated Cumulative Gas (MMMscf)
0	10	10	0	0
0.5	8.4	8.392971	1.67	1.671857
1	7.12	7.147962	3.08	3.08535
1.5	6.16	6.163401	4.3	4.296641
2	5.36	5.37108	5.35	5.346644
2.5	4.72	4.723797	6.27	6.265881
3	4.18	4.188031	7.08	7.077596
3.5	3.72	3.739441	7.78	7.799804
4	3.36	3.36	8.44	8.44669
5		2.757413		9.557617
6		2.304959		10.477755
7		1.956406		11.252814
8		1.68208		11.914924
9		1.462215		12.487334
10		1.283229		12.987298
11		1.135536		13.427888
12		1.012209		13.819197
13		0.908144		14.169139
14		0.819508		14.484015
15		0.743381		14.768899
16		0.677503		15.027928
17		0.620105		15.264506
18		0.569783		15.481464
19		0.525414		15.681171
20		0.486091		15.86563

where

q_t = flow rate, MMscf/day
t = time, years
D_i = decline rate, year^{-1}

In Eq. (3.148), the time basis in q_i is expressed in days and, therefore, D_i must be expressed in day^{-1}, or:

Gentry (1972) developed a graphical method for the coefficients b and D_i as shown in Figures 3.50 and 3.51. Arps's decline curve exponent b is expressed in Figure 3.50 in terms of the ratios q_i/q and $G_p/(tq_i)$ with an upper limit for q_i/q at 100. To determine the exponent

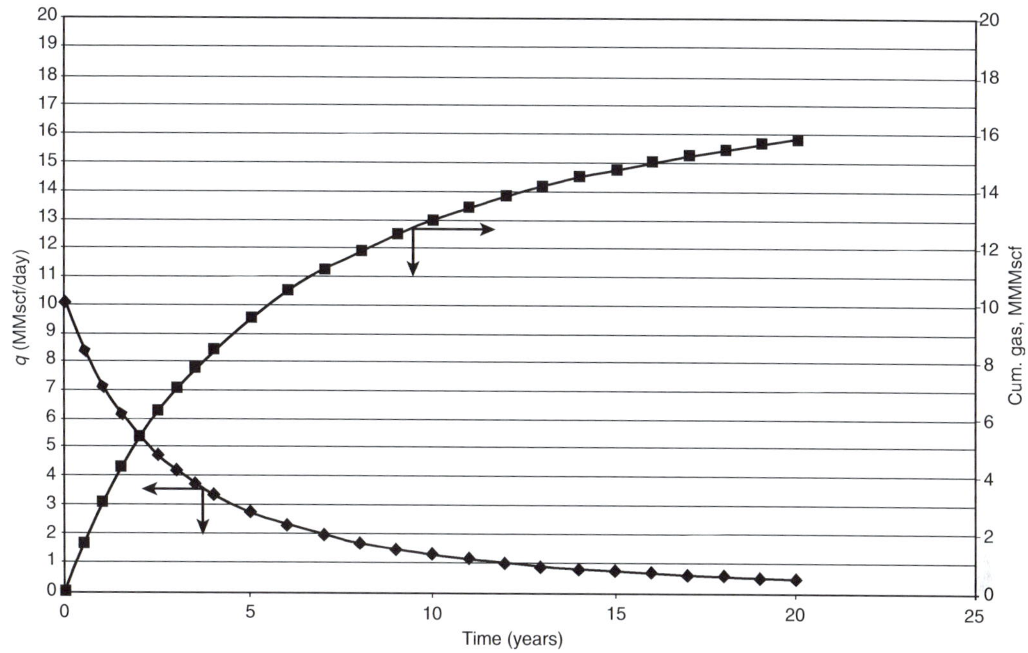

FIGURE 3.49 Decline curve data for Example 3.18.

b, enter the graph with the abscissa with a value of $G_p/(tq_i)$ that corresponds to the last data point on the decline curve and enter the coordinates with the value of the ratio of initial production rate to that of the last rate on the decline curve q_i/q. The exponent b is read by the intersection of these two values. The initial decline rate D_i can be determined from Figure 3.51 by entering the ordinate with the value of q_i/q and moving to the right to the curve that corresponds to the value of b. The initial decline rate D_i can be found by reading the value on the abscissa divided by the time t from q_i to q.

Example 3.19

Using the data given in Example 3.18, recalculate the coefficients b and D_i by using Gentry's graphs.

Solution

Step 1. Calculate the ratios q_i/q and $G_p/(tq_i)$ as:

$$\frac{q_i}{q} = \frac{10}{3.36} = 2.98$$

$$\frac{G_p}{tq_i} = \frac{8440}{(4 \times 365)(10)} = 0.58$$

Step 2. From Figure 3.50, draw a horizontal line from the y-axis at 2.98 and a vertical line from the x-axis at 0.58 and read the value of b at the intersection of the two lines, to give:

$$b = 0.5$$

Step 3. Enter Figure 3.51 with the values of 2.98 and 0.5 to give:

$$D_i t = 1.5 \quad \text{or} \quad D_i = \frac{1.5}{4} = 0.38 \text{ year}^{-1}$$

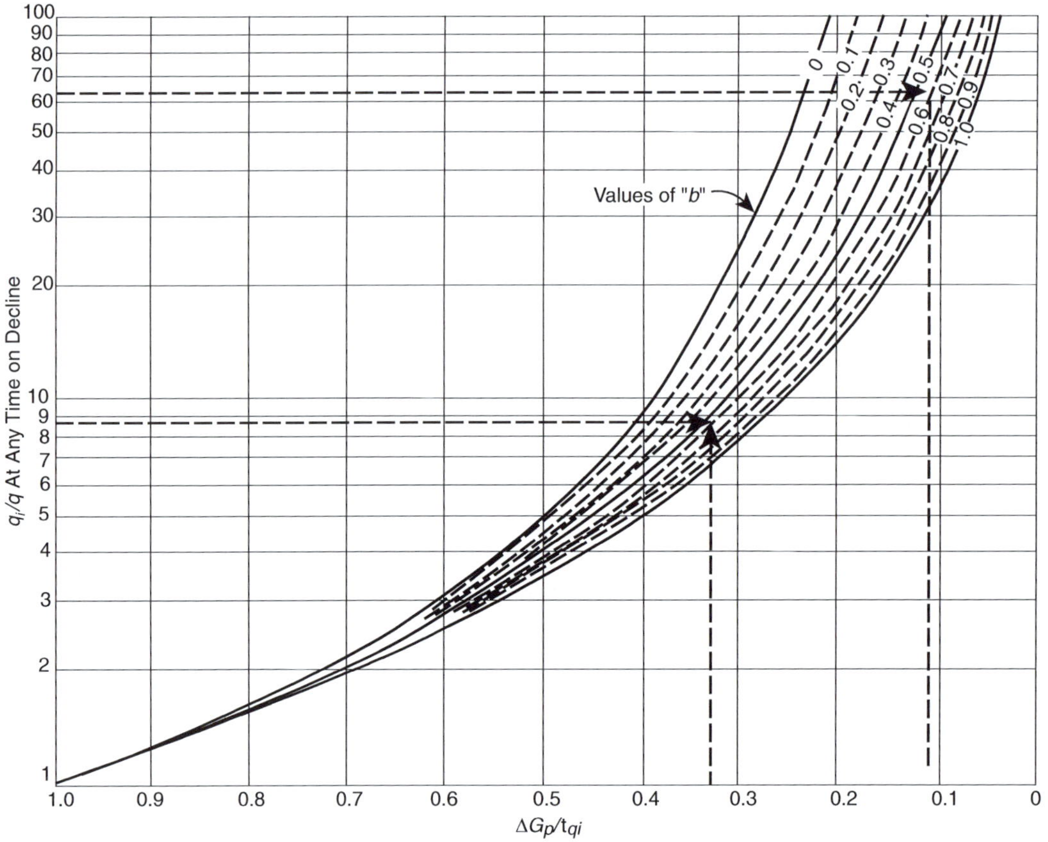

FIGURE 3.50 Relationship between production rate and cumulative production. *(After Gentry, R.W., 1972. Decline curve analysis. J. Pet. Technol. 24 (1), 38–41).*

In many cases, gas wells are not produced at their full capacity during their early life for various reasons, such as limited capacity of flow lines, transportation, low demand, or other types of restrictions. Figure 3.52 illustrates a model for estimating the time pattern of production where the rate is restricted.

Figure 3.52 shows that the well produces at a restricted flow rate of q_r for a total time of t_r with a cumulative production of G_{pr}. The proposed methodology for estimating the restricted time t_r is to set the total cumulative production $G_{p(tr)}$ that would have occurred under normal decline from the initial well capacity q_i down to q_r equal to G_{pr}. Eventually, the well will reach the time t_r where it begins to decline with a behavior that is similar to other wells in the area. The proposed method for predicting the decline rate behavior for a well under restricted flow is based on the assumption that the following data is available and applicable to the well:

- coefficients of Arps's equation, i.e., D_i and b by analogy with other wells;
- abandonment (economic) gas flow rate q_a;
- ultimate recoverable reserves G_{pa};
- allowable (restricted) flow rate q_r.

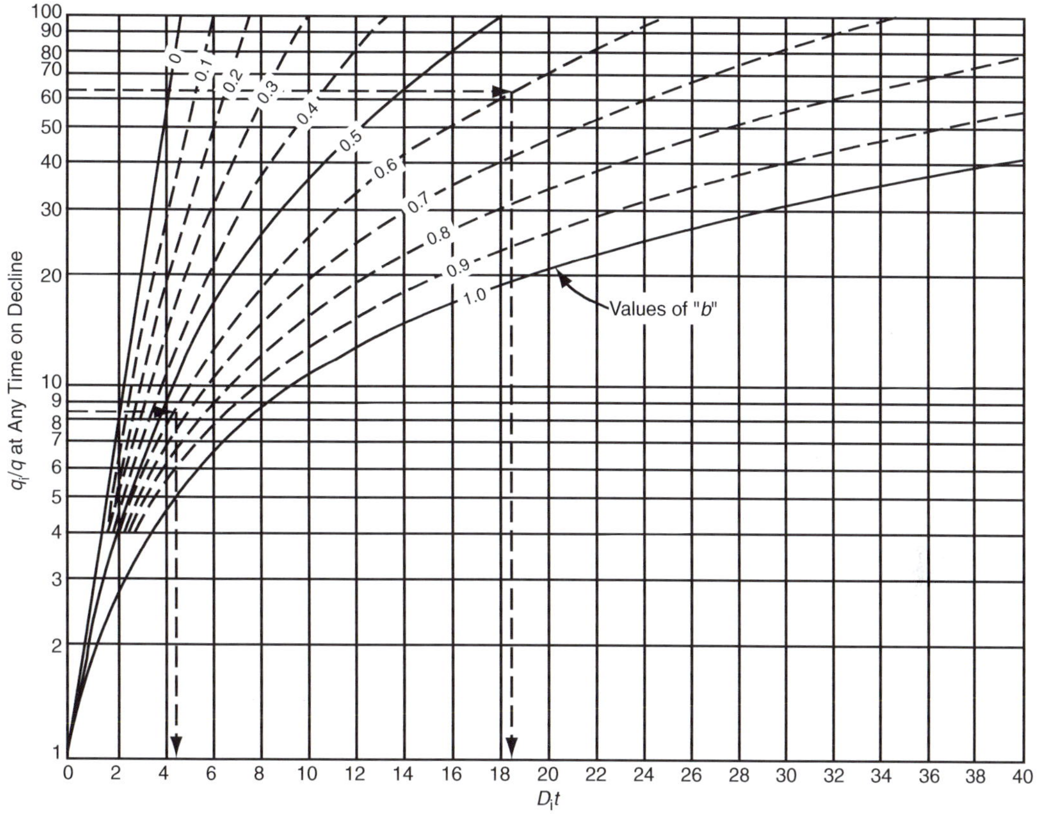

FIGURE 3.51 Relationship between production rate and time. *(After Gentry, R.W., 1972. Decline curve analysis. J. Pet. Technol. 24 (1), 38–41).*

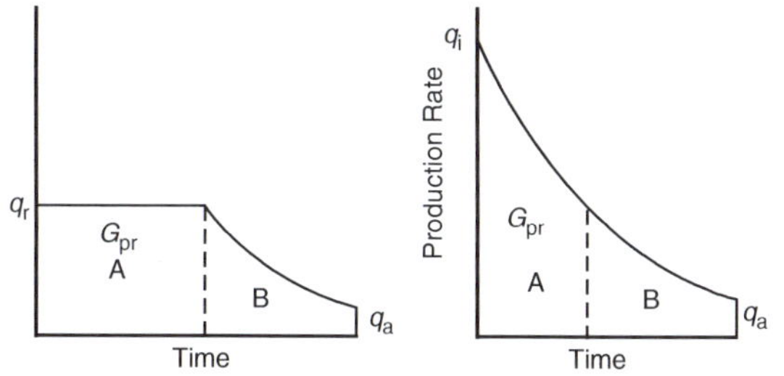

FIGURE 3.52 Estimation of the effect of restricting maximum production rate.

The methodology is summarized in the following steps:

Step 1. Calculate the initial well flow capacity q_i that would have occurred with no restrictions, as follows:

For exponential:
$$q_i = G_{pa}D_i + q_a \tag{3.162}$$

For harmonic:
$$q_i = q_r\left[1 + \frac{D_i G_{pa}}{q_r} - \ln\left(\frac{q_r}{q_a}\right)\right] \tag{3.163}$$

For hyperbolic:
$$q_i = \left\{(q_r)^b + \frac{D_i b G_{pa}}{(q_r)^{1-b}} - \frac{b(q_r)^b}{1-b} \times \left[1 - \left(\frac{q_a}{q_r}\right)^{1-b}\right]\right\}^{1/b} \tag{3.164}$$

Step 2. Calculate the cumulative gas production during the restricted flow rate period:

For exponential:
$$G_{pr} = \frac{q_i - q_r}{D_i} \tag{3.165}$$

For harmonic: $$G_{pr} = \left(\frac{q_i}{D_i}\right) \ln\left(\frac{q_i}{q_r}\right) \quad (3.166)$$

For hyperbolic: $$G_{pr} = \left[\frac{q_i}{D_i(1-b)}\right]\left[1-\left(\frac{q_r}{q_i}\right)^{1-b}\right] \quad (3.167)$$

Step 3. Regardless of the type of decline, calculate the total time of the restricted flow rate from:

$$t_r = \frac{G_{pr}}{q_r} \quad (3.168)$$

Step 4. Generate the well production performance as a function of time by applying the appropriate decline relationships as given by Eqs. (3.143)–(3.154).

Example 3.20

The volumetric calculations on a gas well show that the ultimate recoverable reserves G_{pa} are 25 MMMscf of gas. By analogy with other wells in the area, the following data is assigned to the well:

exponential decline allowable (restricted) production rate = 425 MMscf/month;
economic limit = 30 MMscf/month;
nominal decline rate = 0.044 month^{-1}.

Calculate the yearly production performance of the well.

Solution

Step 1. Estimate the initial flow rate q_i from Eq. (3.162):

$$q_i = G_{pa} D_i + q_a$$
$$= (0.044)(25,000) + 30 = 1130 \text{ MMscf/month}$$

Step 2. Calculate the cumulative gas production during the restricted flow period by using Eq. (3.165):

$$G_{pr} = \frac{q_i - q_r}{D_i}$$
$$= \frac{1130 - 425}{0.044} = 16.023 \text{ MMscf}$$

Step 3. Calculate the total time of the restricted flow from Eq. (3.168):

$$t_r = \frac{G_{pr}}{q_r}$$
$$= \frac{16,023}{425} = 37.7 \text{ months} = 3.14 \text{ years}$$

Step 4. The yearly production during the first 3 years is:

$$q = (425)(12) = 5100 \text{ MMscf/year}$$

The fourth year is divided into 1.68 months, i.e., 0.14 years, of constant production plus 10.32 months of declining production, or:

For the first 1.68 months: $(1.68)(425) = 714$ MMscf

At the end of the fourth year:

$$q = 425 \exp[-0.044(10.32)] = 270 \text{ MMscf/month}$$

Cumulative gas production for the last 10.32 months is:

$$\frac{425 - 270}{0.044} = 3523 \text{ MMscf}$$

Total production for the fourth year is:

$$714 + 3523 = 4237 \text{ MMscf}$$

Year	Production (MMscf/year)
1	5100
2	5100
3	5100
4	4237

The flow rate at the end of the fourth year, i.e., 270 MMscf/month, is set equal to the *initial flow rate at the beginning of the fifth year*. The flow rate at the end of the fifth year, q_{end}, is calculated from Eq. (3.165) as:

$$q_{end} = q_i \exp[-D_i(12)]$$
$$= 270 \exp[-0.044(12)] = 159 \text{ MMscf/month}$$

with a cumulative gas production of:

$$G_p = \frac{q_i - q_{end}}{D_i} = \frac{270 - 159}{0.044} = 2523 \text{ MMscf}$$

And for the sixth year:

$$q_{end} = 159 \exp[-0.044(12)] = 94 \text{ MMscf/month}$$

as:

$$G_p = \frac{159 - 94}{0.044} = 1482 \text{ MMscf}$$

Results of the above repeated procedure are tabulated below:

t (years)	**q_i (MMscf/month)**	**q_{end} (MMSCF/month)**	**Yearly Production (MMscf/year)**	**Cumulative Production (MMMscf)**
1	425	425	5100	5.100
2	425	425	5100	10.200
3	425	425	5100	15.300
4	425	270	4237	19.537
5	270	159	2523	22.060
6	159	94	1482	23.542
7	94	55	886	24.428
8	55	33	500	24.928

Reinitialization of Data. Fetkovich (1980) pointed out that there are several obvious situations where rate–time data must be reinitialized for reasons that include among others:

- the drive or production mechanism has changed;
- an abrupt change in the number of wells on a lease or a field due to infill drilling;
- changing the size of tubing would change q_i and also the decline exponent b.

Provided a well is not limited by tubing size or equipment capacity, the effects of stimulation will result in a change in deliverability q_i and possibly the remaining recoverable gas. However, the decline exponent b normally can be assumed constant. Fetkovich et al. (1996) suggested a "rule-of-thumb" equation to approximate an increase in rate due to stimulation as:

$$(q_i)_{new} = \left[\frac{7 + s_{old}}{7 + s_{new}}\right](q_t)_{old}$$

where

$(q_t)_{old}$ = producing rate just prior to stimulation
s = skin factor

Arps's equation, i.e., Eq. (3.141), can be expressed as:

$$q_t = \frac{(q_i)_{new}}{(1 + bt(D_i)_{new})^{1/b}}$$

with

$$(D_i)_{new} = \frac{(q_i)_{new}}{(1 - b)G}$$

where

G = gas-in-place, scf

Type Curve Analysis. As presented in Chapter 1, type curve analysis of production data is a technique where actual production rate and time are history matched to a theoretical model. The production data and theoretical model are generally expressed graphically in dimensionless forms. Any variable can be made "dimensionless" by multiplying it by a group of constants with opposite dimensions, but the choice of this group will depend on the type of problem to be solved. For example, to create the dimensionless pressure drop p_D, the actual pressure drop in psi is multiplied by the group A with units of psi^{-1}, or:

$$p_D = A\,\Delta p$$

Finding group A that makes a variable dimensionless is derived from equations that describe reservoir fluid flow. To introduce this concept, recall Darcy's equation that describes the radial, incompressible, steady-state flow as expressed by:

$$Q = \left[\frac{0.00708\,kh}{B\mu[\ln(r_e/r_{wa}) - 0.5}\right]\Delta p$$

where r_{wa} is the apparent (effective) wellbore radius and is defined by Eq. (1.151) in terms of the skin factor s as:

$$r_{wa} = r_w\,e^{-s}$$

Group A can be defined by rearranging Darcy's equation as:

$$\ln\left(\frac{r_e}{r_{wa}}\right) - \frac{1}{2} = \left[\frac{0.00708\,kh}{QB\mu}\right]\Delta p$$

Because the left-hand slide of the above equation is dimensionless, the right-hand side must be accordingly dimensionless. This suggests that the term $0.00708kh/QB\mu$ is essentially group A with units of psi^{-1} that defines the dimensionless variable p_D, or:

$$p_D = \left[\frac{0.00708\, kh}{QB\mu}\right]\Delta p$$

Or the ratio of p_D to Δp as:

$$\frac{p_D}{\Delta p} = \left[\frac{kh}{141.2QB\mu}\right]$$

Taking the logarithm of both sides of this equation gives:

$$\log(p_D) = \log(\Delta p) + \log\left(\frac{0.00708\, kh}{QB\mu}\right) \tag{3.169}$$

where

Q = flow rate, STB/day
B = formation volume factor, bbl/STB
μ = viscosity, cp

For a constant flow rate, Eq. (3.169) indicates that the logarithm of the dimensionless pressure drop, $\log(p_D)$, will differ from the logarithm of the *actual* pressure drop, $\log(\Delta p)$, by a constant amount of:

$$\log\left(\frac{0.00708\, kh}{QB\mu}\right)$$

Similarly, the dimensionless time t_D is given in Chapter 1 by Eq. (1.86a) and (1.86b), with time t given in days, as:

$$t_D = \left[\frac{0.006328\, k}{\phi\mu c_t r_w^2}\right]t$$

Taking the logarithm of both sides of this equation gives:

$$\log(t_D) = \log(t) + \log\left[\frac{0.006328\, k}{\phi\mu c_t r_w^2}\right] \tag{3.170}$$

where

t = time, days
c_t = total compressibility coefficient, psi^{-1}
ϕ = porosity

Hence, a graph of $\log(\Delta p)$ vs. $\log(t)$ will have an *identical shape* (i.e., parallel) to a graph of $\log(p_D)$ vs. $\log(t_D)$, although the curve will be shifted by $\log(0.00708kh/QB\mu)$ vertically in pressure and $\log(0.000264k/\phi\mu c_t r_w^2)$ horizontally in time. This concept is illustrated in Chapter 1 by Figure 1.46 and reproduced in this chapter for convenience.

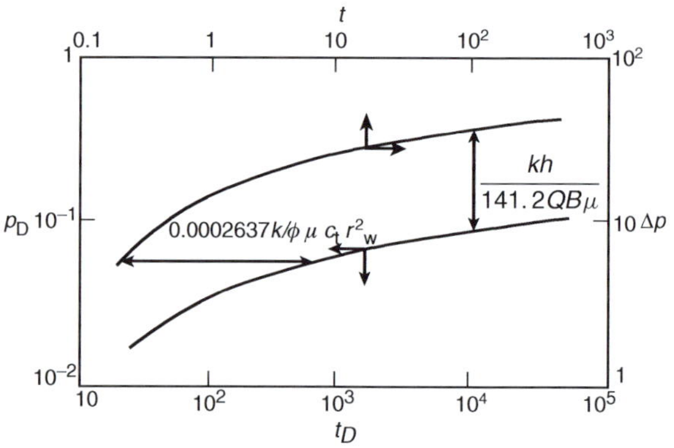

FIGURE 1.46 Concept of type curves.

Not only do these two curves have the same shape, but if they are *moved relative to each other until they coincide or "match,"* the vertical and horizontal displacements required to achieve the match are related to these constants in Eqs. (3.169) and (3.170). Once these constants are determined from the vertical and horizontal displacements, it is possible to estimate reservoir properties such as permeability and porosity. This process of matching two curves through the vertical and horizontal displacements and determining the reservoir or well properties is called type curve matching.

To fully understand the power and convenience of using the dimensionless concept approach in solving engineering problems, this concept is illustrated through the following example.

Example 3.21

A well is producing under transient (unsteady-state) flow conditions. The following properties are given:

$p_i = 3500$ psi,	$B = 1.44$ bbl/STB
$c_t = 17.6 \times 10^{-6}$ psi^{-1},	$\phi = 15\%$
$\mu = 1.3$ cp,	$h = 20$ ft
$Q = 360$ STB/day,	$k = 22.9$ md
$s = 0$	

(a) Calculate the pressure at radii 10 and 100 ft for the flowing times 0.1, 0.5, 1.0, 2.0, 5.0, 10, 20, 50, and 100 hours. Plot $p_i - p(r, t)$ vs. t on a log–log scale.

(b) Present the data from part (a) in terms of $p_i - p(r, t)$ vs. (t/r^2) on a log–log scale.

Solution

(a) During transient flow, Eq. (1.77) is designed to describe the pressure at any radius r and any time t as given by:

$$p(r,t) = p_i + \left[\frac{70.6 Q B \mu}{kh}\right] \text{Ei}\left[\frac{-948 \phi \mu c_t r^2}{kt}\right]$$

or

$$p_i - p(r,t) = \left[\frac{-70.6(360)(1.444)(1.3)}{(22.9)(20)}\right] \times \text{Ei}\left[\frac{-948(0.15)(1.3)(17.6 \times 10^{-6} r^2)}{(22.9)t}\right]$$

$$p_i - p(r,t) = -104\text{Ei}\left[-0.0001418\frac{r^2}{t}\right]$$

Values of $p_i - p(r, t)$ are presented as a function of time and radius (i.e., at $r = 10$ and 100 ft) in the following table and graphically in Figure 3.53:

Assumed t (hours)	**r = 10 ft**		
	t/r^2	$\text{Ei}[-0.0001418r^2/t]$	$p_i - p$ (r, t)
0.1	0.001	−1.51	157
0.5	0.005	−3.02	314
1.0	0.010	−3.69	384
2.0	0.020	−4.38	455
5.0	0.050	−5.29	550
10.0	0.100	−5.98	622
20.0	0.200	−6.67	694
50.0	0.500	−7.60	790
100.0	1.000	−8.29	862

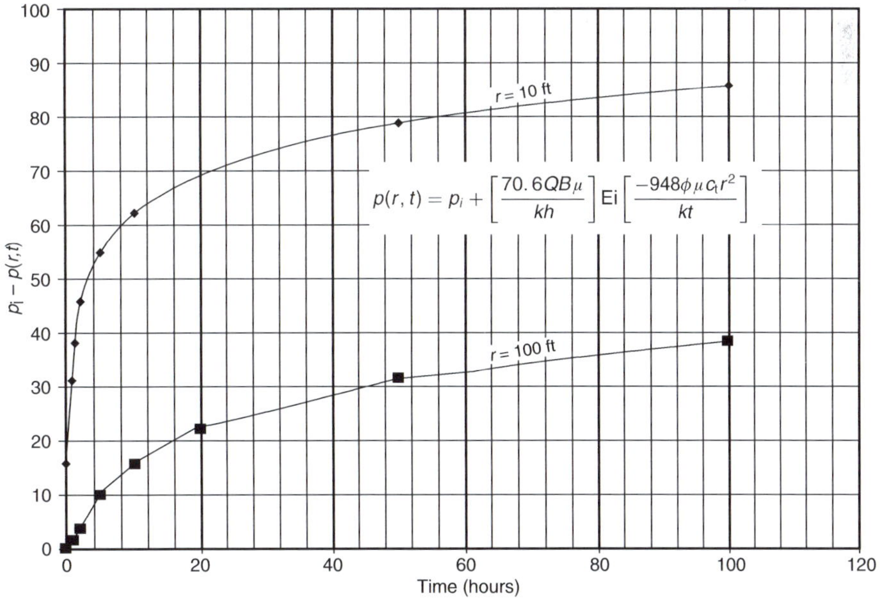

FIGURE 3.53 Pressure profile at 10 and 100 ft as a function of time.

Assumed t (hours)	$r = 100$ ft		
	t/r^2	$Ei[-0.0001418r^2/t]$	$p_i - p$ (r, t)
0.1	0.00001	0.00	0
0.5	0.00005	−0.19	2
1.0	0.00010	−0.12	12
2.0	0.00020	−0.37	38
5.0	0.00050	−0.95	99
10.0	0.00100	−1.51	157
20.0	0.00200	−2.14	223
50.0	0.00500	−3.02	314
100.0	0.00100	−3.69	386

(b) Figure 3.53 shows two different curves for the 10 and 100 ft radii. Obviously, the same calculations can be repeated for any number of radii and, consequently, the same number of curves will be generated. However, the solution can be greatly simplified by examining Figure 3.54. This plot shows that when the pressure difference $p_i - p(r, t)$ is plotted vs. t/r^2, the data for both radii forms a common curve. In fact, the pressure difference for any reservoir radius will plot on this exact same curve.

For example, in the same reservoir if we have to calculate the pressure p at 150 ft after 200 hours of transient flow, then:

$$\frac{t}{r^2} = \frac{200}{150^2} = 0.0089$$

From Figure 3.54:

$$p_i - p(r, t) = 370 \text{ psi}$$

Thus:

$$p(r, t) = p_i - 370 = 5000 - 370 = 4630 \text{ psi}$$

Several investigators have employed the dimensionless variables approach to determine reserves and to describe the recovery performance of hydrocarbon systems with time, notably:

- Fetkovich;
- Carter;
- Palacio and Blasingame;
- flowing material balance;
- Anash et al.;
- decline curve analysis for fractured wells.

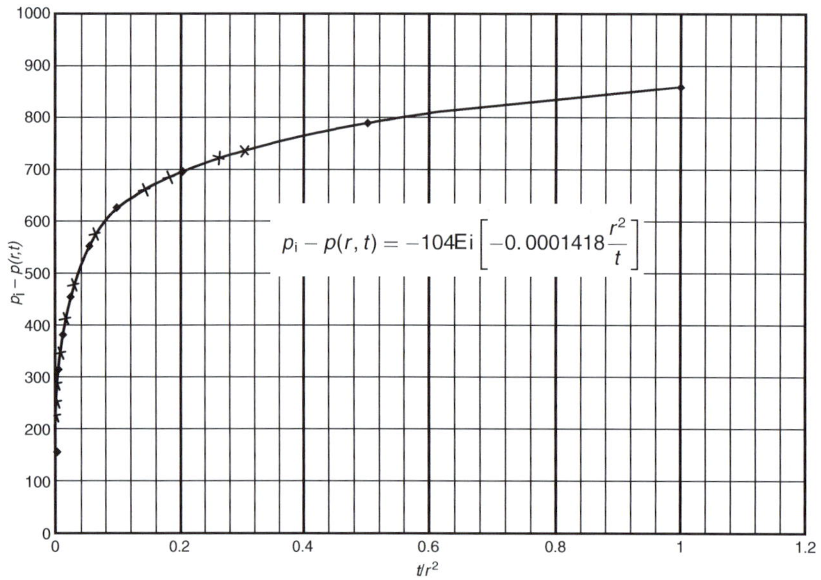

FIGURE 3.54 Pressure profile at 10 and 100 ft as a function of t/r^2.

All the methods are based on defining a set of "Decline curve dimensionless variables" that includes:

- decline curve dimensionless rate q_{Dd};
- decline curve dimensionless cumulative production Q_{Dd};
- decline curve dimensionless time t_{Dd}.

These methods were developed with the objective of providing the engineer with an additional convenient tool for estimating reserves and determining other reservoir properties for oil and gas wells using the available performance data. A review of these methods and their practical applications are documented below.

Fetkovich Type Curve. Type curve matching is an advanced form of decline analysis proposed by Fetkovich (1980). The author proposed that the concept of the dimensionless variables approach can be extended for use in decline curve analysis to simplify the calculations. He introduced the decline curve dimensionless flow rate variable q_{dD} and decline curve dimensionless time t_{dD} that are used in all decline curve and type curve analysis techniques. Arps's relationships can be expressed in the following dimensionless forms:

$$\text{Hyperbolic:}\quad \frac{q_t}{q_i} = \frac{1}{[1 + bD_i t]^{1/b}}$$

In a dimensionless form:

$$q_{Dd} = \frac{1}{[1 + bt_{Dd}]^{1/b}} \tag{3.171}$$

where the decline curve dimensionless variables q_{Dd} and t_{Dd} are defined by:

$$q_{Dd} = \frac{q_t}{q_i} \tag{3.172}$$

$$t_{Dd} = D_i t \tag{3.173}$$

$$\text{Exponential:}\quad \frac{q_t}{q_i} = \frac{1}{\exp[D_i t]}$$

Similarly:

$$q_{Dd} = \frac{1}{\exp[t_{Dd}]} \tag{3.174}$$

$$\text{Harmonic:}\quad \frac{q_t}{q_i} = \frac{1}{1 + D_i t}$$

or

$$q_{Dd} = \frac{1}{1 + t_{Dd}} \tag{3.175}$$

where q_{Dd} and t_{Dd} are the decline curve dimensionless variables as defined by Eqs. (3.172) and (3.173), respectively. During the boundary-dominated flow period, i.e., steady-state or semisteady-state flow conditions, Darcy's equation can be used to describe the initial flow rate q_i as:

$$q_i = \frac{0.00708kh\,\Delta p}{B\mu\left[\ln(r_e/r_{wa}) - \frac{1}{2}\right]} = \frac{kh(p_i - p_{wf})}{142.2B\mu\left[\ln(r_e/r_{wa}) - \frac{1}{2}\right]}$$

where

q = flow rate, STB/day
B = formation, volume factor, bbl/STB
μ = viscosity, cp
k = permeability, md
h = thickness, ft
r_e = drainage radius, ft
r_{wa} = apparent (effective) wellbore radius, ft

The ratio r_e/r_{wa} is commonly referred to as the dimensionless drainage radius r_D. That is:

$$r_D = \frac{r_e}{r_{wa}} \tag{3.176}$$

with

$$r_{wa} = r_w\, e^{-s}$$

The ratio r_e/r_{wa} in Darcy's equation can be replaced with r_D, to give:

$$q_i = \frac{kh(p_i - p_{wf})}{141.2B\mu\left[\ln(r_D) - \frac{1}{2}\right]}$$

Rearranging Darcy's equation gives:

$$\left[\frac{141.2B\mu}{kh\,\Delta p}\right]q_i = \frac{1}{\ln(r_D) - \frac{1}{2}}$$

It is obvious that the right-hand side of this equation is dimensionless, which indicates that the left-hand side of the equation is also dimensionless. The above relationship then defines the dimensionless rate q_D as:

$$q_D = \left[\frac{141.2B\mu}{kh\,\Delta p}q_i\right] = \frac{1}{\ln(r_D) - \frac{1}{2}} \quad \textbf{(3.177)}$$

Recalling the dimensionless form of the diffusivity equation, i.e., Eq. (1.89), as:

$$\frac{\partial^2 p_D}{\partial r_D^2} + \frac{1}{r_D}\frac{\partial p_D}{\partial r_D} = \frac{\partial p_D}{\partial r_D}$$

Fetkovich demonstrated that the analytical solutions to the above transient flow diffusivity equation and the pseudosteady-state decline curve equations could be combined and presented in a family of "log–log" dimensionless curves. To develop this link between the two flow regimes, Fetkovich expressed the decline curve dimensionless variables q_{Dd} and t_{Dd} in terms of the transient dimensionless rate q_D and time t_D. Combining Eq. (3.172) with Eq. (3.177) gives:

$$q_{Dd} = \frac{q_t}{q_i} = \frac{q_t/(kh(p_i - p))}{141.2B\mu\left[\ln(r_D) - \frac{1}{2}\right]}$$

or

$$q_{Dd} = q_D\left[\ln(r_D) - \tfrac{1}{2}\right] \quad \textbf{(3.178)}$$

Fetkovich expressed the decline curve dimensionless time t_{Dd} in terms of the transient dimensionless time t_D by:

$$t_{Dd} = \frac{t_D}{\frac{1}{2}\left[r_D^2 - 1\right]\left[\ln(r_D) - \frac{1}{2}\right]} \quad \textbf{(3.179)}$$

Replacing the dimensionless time t_D by Eq. (1.86a) and (1.86b) gives:

$$t_{Dd} = \frac{1}{\frac{1}{2}\left[r_D^2 - 1\right]\left[\ln(r_D) - \frac{1}{2}\right]}\left[\frac{0.006328t}{\phi(\mu c_t)r_{wa}^2}\right] \quad \textbf{(3.180)}$$

Although Arps's exponential and hyperbolic equations were developed empirically on the basis of production data, Fetkovich was able to place a physical basis to Arps's coefficients. Eqs. (3.173) and (3.180) indicate that the initial decline rate D_i can be defined mathematically by the following expression:

$$D_i = \frac{1}{\frac{1}{2}\left[r_D^2 - 1\right]\left[\ln(r_D) - \frac{1}{2}\right]}\left[\frac{0.006328}{\phi(\mu c_t)r_{wa}^2}\right] \quad \textbf{(3.181)}$$

Fetkovich arrived at his unified type curve, as shown in Figure 3.55, by solving the dimensionless form of the diffusivity equation using the constant-terminal solution approach for several assumed values of r_D and t_{Dd} and the solution to Eq. (3.171) as a function of t_{Dd} for several values of b ranging from 0 to 1.

Notice in Figure 3.55 that all curves coincide and become indistinguishable at $t_{Dt}\approx 0.3$. Any data existing before a t_{Dt} of 0.3 will appear to be exponentially declining regardless of the true value of b and, thus, will plot as a straight line on semilog paper.

With regard to the initial rate q_i, it is not simply a producing rate at early time; it is very specifically a pseudosteady-state rate at the surface. It can be substantially less than the actual early-time transient flow rates as would be produced from low-permeability wells with large negative skins.

The basic steps used in Fetkovich type curve matching of declining rate–time data are given below:

Step 1. Plot the historical flow rate q_t vs. time t in any convenient units on log–log paper or tracing paper with the same logarithmic cycles as the Fetkovich type curve.

Step 2. Place the tracing paper data plot over the type curve and slide the tracing paper with the plotted data, keeping the coordinate axes parallel, until the actual data points match one of the type curves with a specific value of b.

Because decline type curve analysis is based on *boundary-dominated flow conditions*, there is no basis for choosing the proper b values for future

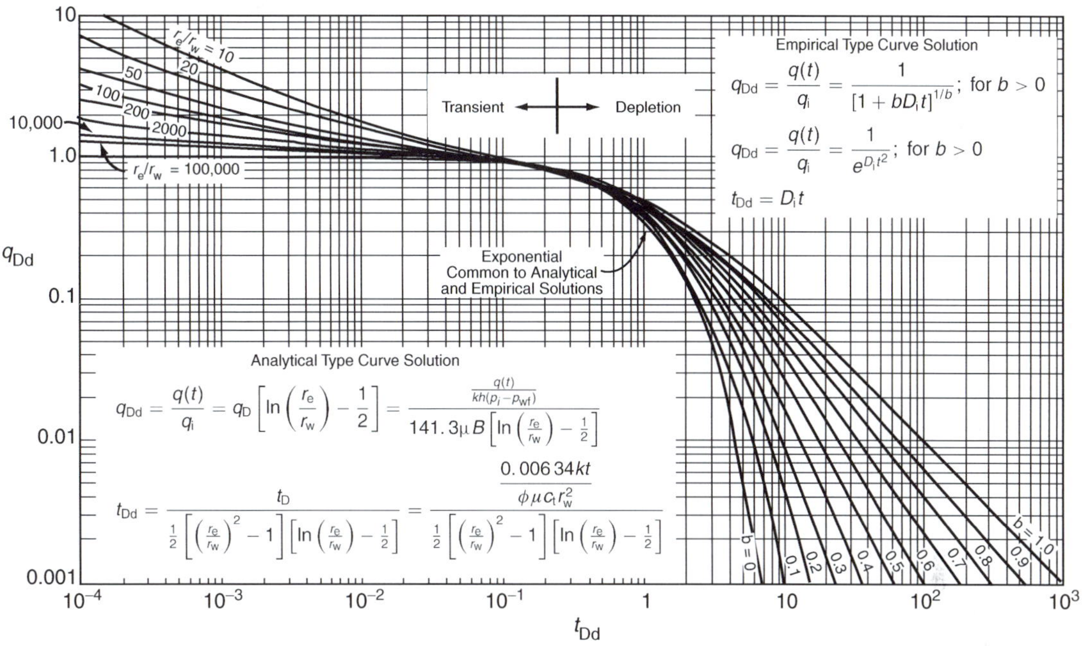

FIGURE 3.55 Fetkovich type curves. *(After Fetkovich, M.J,. 1980. Decline curve analysis using type curves. SPE 4629, SPE J., June, copyright SPE 1980).*

boundary-dominated production if only transient data is available. In addition, because of the similarity of curve shapes, unique type curve matches are difficult to obtain with transient data only. If it is apparent that boundary-dominated (i.e., pseudosteady-state) data is present and can be matched on a curve for a particular value of b, the actual curve can simply be extrapolated following the trend of the type curve into the future.

Step 3. From the match of that particular type curve of step 2, record values of the reservoir dimensionless radius r_e/r_{wa} and the parameter b.

Step 4. Select any convenient match point "MP" on the actual data plot (q_t and $t)_{MP}$ and the corresponding values lying beneath that point on the type curve grid $(q_{Dd}, t_{Dd})_{MP}$.

Step 5. Calculate the initial surface gas flow rate q_i at $t = 0$ from the rate match point:

$$q_i = \left[\frac{q_t}{q_{D_i}}\right]_{MP} \tag{3.182}$$

Step 6. Calculate the initial decline rate D_i from the time match point:

$$D_i = \left[\frac{t_{Dd}}{t}\right]_{MP} \tag{3.183}$$

Step 7. Using the value of r_e/r_{wa} from step 3 and the calculated value of q_i, calculate the formation permeability k by applying Darcy's equation in one of the following three forms:

- Pseudopressure form:

$$k = \frac{1422[\ln(r_e/r_{wa}) - 0.5]q_i}{h[m(p_i) - m(p_{wf})]} \tag{3.184}$$

- Pressure-squared form:

$$k = \frac{1422T(\mu_g Z)_{avg}[\ln(r_e/r_{wa}) - 0.5]q_i}{h(p_i^2 - p_{wf}^2)} \quad (3.185)$$

- Pressure approximation form:

$$k = \frac{141.2(10^3)T(\mu_g B_g)[\ln(r_e/r_{wa}) - 0.5]q_i}{h(p_i - p_{wf})} \quad (3.186)$$

where

k = permeability, md
p_i = initial pressure, psia
p_{wf} = bottom-hole flowing pressure, psia
$m(P)$ = pseudopressure, psi^2/cp
q_i = initial gas flow rate, Mscf/day
T = temperature, °R
h = thickness, ft
μ_g = gas viscosity, cp
Z = gas deviation factor
B_g = gas formation volume factor, bbl/scf

Step 8. Determine the reservoir pore volume (PV) of the well drainage area at the beginning of the boundary-dominated flow from the following expression:

$$PV = \frac{56.54T}{(\mu_g c_t)_i[m(p_i) - m(p_{wf})]}\left(\frac{q_i}{D_i}\right) \quad (3.187)$$

or in terms of pressure squared:

$$PV = \frac{28.27T(\mu_g Z)_{avg}}{(\mu_g c_t)_i[p_i^2 - p_{wf}^2]}\left(\frac{q_i}{D_i}\right) \quad (3.188)$$

with

$$r_e = \sqrt{\frac{PV}{\pi h \phi}} \quad (3.189)$$

$$A = \frac{\pi r e^2}{43,560} \quad (3.190)$$

where

PV = pore volume, ft^3
ϕ = porosity, fraction
μ_g = gas viscosity, cp
c_t = total compressibility coefficient, psi^{-1}
q_i = initial gas rate, Mscf/day
D_i = decline rate, day^{-1}
r_e = drainage radius of the well, ft
A = drainage area, acres

subscripts:
i = initial
avg = average

Step 9. Calculate the skin factor s from the r_e/r_{wa} matching parameter and the calculated values of A and r_e from step 8:

$$s = \ln\left[\left(\frac{r_e}{r_{wa}}\right)_{MP}\left(\frac{r_w}{r_e}\right)\right] \quad (3.191)$$

Step 10. Calculate the initial gas-in-place G from:

$$G = \frac{(PV)[1 - S_w]}{5.615B_{gi}} \quad (3.192)$$

The initial gas-in-place can also estimated from the following relationship:

$$G = \frac{q_i}{D_i(1-b)} \quad (3.193)$$

where

G = initial gas-in-place, scf
S_w = initial water saturation
B_{gi} = gas formation volume factor at p_i, bbl/scf
PV = pore volume, ft^3

An inherent problem when applying decline curve analysis is having sufficient rate–time data to determine a unique value for b as shown in the Fetkovich type curve. It illustrates that for a shorter producing time, the b value curves approach one another, which leads to difficulty in obtaining a unique match. Arguably, applying the type curve approach with only 3 years of production history may be too short for some pools. Unfortunately, since time is plotted on a log scale, the production history becomes compressed so that even when incremental history is added, it may still be difficult to differentiate and clearly identify the appropriate decline exponent b.

The following example illustrates the use of the type curve approach to determine reserves and other reservoir properties.

Example 3.22

Well A is a low-permeability gas well located in West Virginia. It produces from the Onondaga chert that has been hydraulically fractured with 50,000 gal of 3% gelled acid and 30,000 lb of sand. A conventional Horner analysis of pressure buildup data on the well indicated the following:

$p_i = 3268$ psia, $m(p_i) = 794.8 \times 10^6$ psi^2/cp
$k = 0.082$ md, $s = -5.4$

Fetkovich et al. (1987) provided the following additional data on the gas well:

$p_{wf} = 500$ psia, $m(p_{wf}) = 20.8 \times 10^6$ psi^2/cp
$\mu_{gi} = 0.0172$ cp, $c_{ti} = 177 \times (10^{-6})$ psi^{-1}
$T = 620$ °R, $h = 70$ ft
$\phi = 0.06$, $B_{gi} = 0.000853$ bbl/scf
$S_w = 0.35$, $r_w = 0.35$ ft

The historical rate time data for 8 years was plotted and matched to r_e/r_{wa} stem of 20 and $b = 0.5$, as shown in Figure 3.56, with the following match point:

$q_t = 1000$ Mscf/day, $t = 100$ days
$q_{Dd} = 0.58$, $t_{Dd} = 0.126$

Using the above data, calculate:

- permeability k;
- drainage area A;
- skin factor s;
- gas-in-place G.

Solution

Step 1. Using the match point, calculate q_i and D_i by applying Eqs. (3.182) and (3.183), respectively:

$$q_i = \left[\frac{q_t}{q_{D_t}}\right]_{MP} = \frac{1000}{0.58} = 1724 \text{ Mscf/day}$$

and:

$$D_i = \left[\frac{t_{Dd}}{t}\right]_{MP} = \frac{0.126}{100} = 0.00126 \text{ day}^{-1}$$

Step 2. Calculate the permeability k from Eq. (3.184)

$$k = \frac{1442\,T[\ln(r_e/r_{wa}) - 0.5]q_i}{h[m(p_i) - m(p_{wf})]} = \frac{1422(620)[\ln(20) - 0.5](1724.1)}{(70)[794.8 - 20.8](10^6)} = 0.07 \text{ md}$$

Step 3. Calculate the reservoir PV of the well drainage area by using Eq. (3.187):

$$PV = \frac{56.54T}{(\mu_g c_t)_i[m(p_i) - m(p_{wf})]}\left(\frac{q_i}{D_i}\right) = \frac{56.54(620)}{(0.0172)(177)(10^{-6})[794.8 - 20.8](10^6)} \times \frac{1724.1}{0.00126} = 20.36 \times 10^6 \text{ ft}^3$$

Step 4. Calculate the drainage radius and area by applying Eqs. (3.189) and (3.190):

$$r_e = \sqrt{\frac{PV}{\pi h \phi}} = \sqrt{\frac{(20.36)10^6}{\pi(70)(0.06)}} = 1242 \text{ ft}$$

and

$$A = \frac{\pi re^2}{43,560} = \frac{\pi(1242)^2}{43,560} = 111 \text{ acres}$$

Step 5. Determine the skin factor from Eq. (3.191):

$$s = \ln\left[\left(\frac{r_e}{r_{wa}}\right)_{MP}\left(\frac{r_w}{r_e}\right)\right] = \ln\left[(20)\left(\frac{0.35}{1242}\right)\right] = -5.18$$

$$PV = \frac{56.54T}{(\mu_g c_t)_i\,[m(p_i) - m(p_{wf})]}\left(\frac{q_i}{D_i}\right)$$

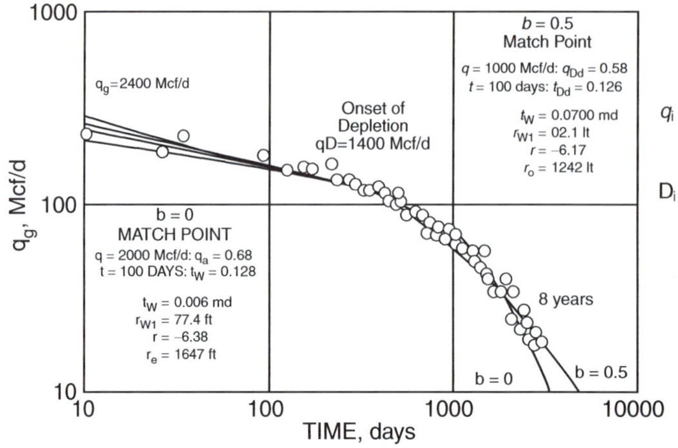

$$q_i = \left[\frac{q}{q_{D_i}}\right]_{MP}$$

$$D_i = \left[\frac{t_{D_i}}{t}\right]_{MP}$$

FIGURE 3.56 West Virginia gas well A type curve fit. *(Copyright SPE 1987).*

Step 6. Calculate the initial gas-in-place by using Eq. (3.192):

$$G = \frac{(PV)[1 - S_w]}{5.615 B_{gi}}$$

$$= \frac{(20.36)(10^6)[1 - 0.35]}{(5.615)(0.000853)} = 2.763 \text{ Bscf}$$

The initial gas G can also be estimated from Eq. (3.193), to give:

$$G = \frac{q_i}{D_i(1 - b)}$$

$$= \frac{1.7241(10^6)}{0.00126(1 - 0.5)} = 2.737 \text{ Bscf}$$

Limits of Exponent* b *and Decline Analysis of Stratified No-Crossflow Reservoirs. Most reservoirs consist of several layers with varying reservoir properties. Because of the fact that no-crossflow reservoirs are perhaps the most prevalent and important, reservoir heterogeneity is of considerable significance in long-term prediction and reserve estimates. In layered reservoirs with crossflow, adjacent layers can simply be combined into a single equivalent layer that can be described as a homogeneous layer with averaging reservoir properties of the crossflowing layers. As shown later in this section, the decline curve exponent b for a single homogeneous layer ranges between 0 and a maximum value of 0.5. For layered no-crossflow systems, values of the decline curve exponent b range between 0.5 and 1 and therefore can be used to identify the stratification. These separated layers might have the greatest potential for increasing current production and recoverable reserves.

Recalling the back-pressure equation, i.e., Eq. (3.20):

$$q_g = C(p_r^2 - p_{wf}^2)^n$$

where

n = back-pressure curve exponent
c = performance coefficient
p_r = reservoir pressure

Fetkovich et al. (1996) suggested that the Arps decline exponent b and the decline rate can be expressed in terms of the exponent n by:

$$b = \frac{1}{2n}\left[(2n - 1) - \left(\frac{p_{wf}}{p_i}\right)^2\right] \qquad (3.194)$$

$$D_i = 2n\left(\frac{q_i}{G}\right) \quad (3.195)$$

where G is the initial gas-in-place. Eq. (3.194) indicates that as the reservoir pressure p_i approaches p_{wf} with depletion, all the nonexponential decline ($b \neq 0$) will shift toward exponential decline ($b = 0$) as depletion proceeds. Eq. (3.194) also suggests that if the well is producing at a very low bottom-hole flowing pressure ($p_{wf} = 0$) or $p_{wf} \ll p_i$, it can be reduced to the following expression:

$$b = 1 - \frac{1}{2n} \quad (3.196)$$

The exponent n from a gas well back-pressure performance curve can therefore be used to calculate or estimate b and D_i. Eq. (3.195) provides the physical limits of b, which is between 0 and 0.5, over the accepted theoretical range of n, which is between 0.5 and 1.0 for a single-layer homogeneous system, as follows:

n	b
(High k) 0.50	0.0
0.56	0.1
0.62	0.2
0.71	0.3
0.83	0.4
(Low k) 1.00	0.5

However, the harmonic decline exponent, $b = 1$, cannot be obtained from the back-pressure exponent. The b value of 0.4 should be considered as a good limiting value for gas wells when not clearly defined by actual production data.

The following is a tabulation of the values of b that should be expected for single-layer homogeneous or layered crossflow systems.

b	System Characterization and Identification
0.0	Gas wells undergoing liquid loading
	Wells with high back-pressure
	High-pressure gas
	Low-pressure gas with back-pressure curve exponent of $n \approx 0.5$
	Poor water-flood performance (oil wells)
	Gravity drainage with no solution gas (oil wells)
	Solution gas drive with unfavorable k_g/k_o (oil wells)
0.3	Typical for solution gas drive reservoirs
0.4–0.5	Typical for gas wells, $b = 0$, for $p_{wf} \approx 0$; $b = 0$, for $p_{wf} \approx 0.1 p_i$
0.5	Gravity drainage for solution gas and for water-drive oil reservoirs
Undeterminable	Constant-rate or increasing-rate production period
	Flow rates are all in transient or infinite-acting period
$0.5 < b < 0.9$	Layered or composite reservoir

The significance of the decline curve exponent b value is that for a single-layer reservoir, the value of b will lie between 0 and 0.5. With layered no-crossflow performance, however, the b value can be between 0.5 and 1.0. As pointed out by Fetkovich et al. (1996), the further the b value is driven toward *a* value of 1.0, the more the unrecovered reserves remain in the tight low-permeability layer and the greater the potential to increase production and recoverable reserves through stimulation of the low-permeability layer. This suggests that decline curve analysis can be used to recognize and identify layered′ no-crossflow performance using only readily available historical production data. Recognition of the layers that are not being adequately drained compared to other layers, i.e., differential depletion, is where the opportunity lies. Stimulation of the less productive layers can allow both increased production and reserves. Figure 3.57 presents the standard Arps depletion decline curves, as presented by Fetkovich et al. (1996). Eleven curves are shown with each being described by a b value that ranges between 0 and 1 in increments of 0.1. All of the values have meaning and should be understood in order to apply decline curve analysis properly. *When decline curve analysis yields a b value greater than 0.5 (layered*

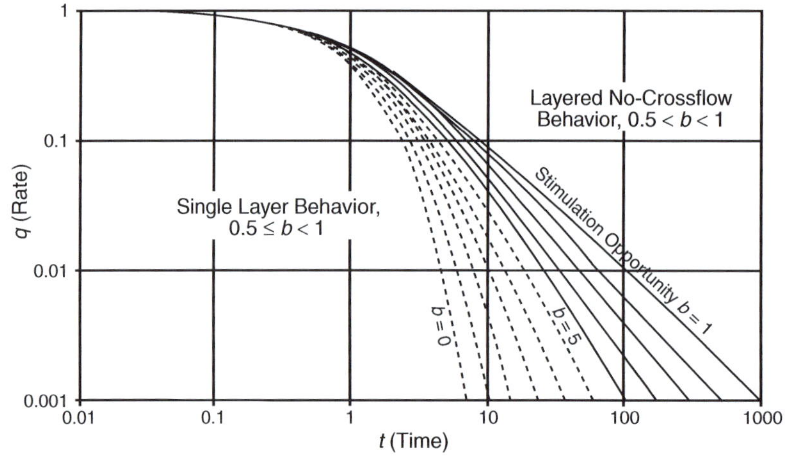

FIGURE 3.57 Depletion decline curves. *(After Fetkovich, 1997, copyright SPE 1997).*

no-crossflow production), it is inaccurate simply to make a prediction from the match point values. This is because the match point represents a best fit of the surface production data that includes production from all layers. Multiple combinations of layer production values can give the same composite curve and, therefore, unrealistic forecasts in late time may be generated.

To demonstrate the effect of the layered no-crossflow reservoir system on the exponent b, Fetkovich et al. (1996) evaluated the production depletion performance of a two-layered gas reservoir producing from two noncommunicated layers. The field produces from 10 wells and contains an estimated 1.5 Bscf of gas-in-place at an initial reservoir pressure of 428 psia. The reservoir has a gross thickness of 350 ft with a shale barrier averaging 50 ft thick that is clearly identified across the field and separates the two layers. Core data indicates a bimodal distribution with a permeability ratio between 10:1 and 20:1.

A type curve analysis and regression fit of the total field composite $\log(q_i)$ vs. $\log(t)$ yielded $b = 0.89$ that is identical to all values obtained from individual well analysis. To provide a quantitative analysis and an early recognition of a no-crossflow layered reservoir, Fetkovich (1980) expressed the rate–time equation for a gas well in terms of the back-pressure exponent n with a constant p_{wf} of zero. The derivation is based on combining Arps's hyperbolic equation with the MBE (i.e., p/Z vs. G_p) and back-pressure equation to give:

For $0.5<n<1$, $0<b<0.5$:

$$q_t = \frac{q_i}{\left[1 + (2n-1)(q_i/G)t\right]^{(2n/(2n-1))}} \tag{3.197}$$

$$G_{p(t)} = G\left\{1 - \left[1 + (2n-1)\left(\frac{q_i}{G}\right)t\right]^{(1/(2n-1))}\right\} \tag{3.198}$$

For $n = 0.5$, $b = 0$:

$$q_t = q_i \exp\left[-\left(\frac{q_i}{G}\right)t\right] \tag{3.199}$$

$$G_{p(t)} = G\left\{1 - \exp\left[-\left(\frac{q_i}{G}\right)t\right]\right\} \tag{3.200}$$

For $n = 1$, $b = 0.5$:

$$q_t = \frac{q_i}{\left[1 + (q_i/G)t\right]^2} \tag{3.201}$$

$$G_{p(t)} = G - \frac{G}{1 + (q_i t/G)} \quad (3.202)$$

The above relationships are based on $p_{wf} = 0$, which implies that $q_i = q_{max}$ as given by:

$$q_i = q_{i\,max} = \frac{khp_i^2}{1422T(\mu_g Z)_{avg}[\ln(r_e/r_w) - 0.75 + s]} \quad (3.203)$$

where

$q_{i\,max}$ = stabilized absolute open flow potential, i.e., at $P_{wf} = 0$, Mscf/day
G = initial gas-in-place, Mscf
q_t = gas flow rate at time t, Mscf/day
t = time
$G_{p(t)}$ = cumulative gas production at time t, Mscf

For a commingled well producing from two layers at a constant p_{wf}, the total flow rate $(q_t)_{total}$ is essentially the sum of the flow rate from each layer, or:

$$(q_t)_{total} = (q_t)_1 + (q_t)_2$$

where the subscripts 1 and 2 represent the more permeable layer and less permeable layer, respectively. For a hyperbolic exponent of $b = 0.5$, Eq. (3.201) can be substituted into the above expression to give:

$$\frac{(q_{max})_{total}}{[1 + t(q_{max}/G)_{total}]^2} = \frac{(q_{max})_1}{[1 + t(q_{max}/G)_1]^2} + \frac{(q_{max})_2}{[1 + t(q_{max}/G)_2]^2} \quad (3.204)$$

Eq. (3.204) indicates that only if $(q_{max}/G)_1 = (q_{max}/G)_2$, the value of $b = 0.5$ for each layer will yield a composite rate–time value of $b = 0.5$.

Mattar and Anderson (2003) presented an excellent review of methods that are available for analyzing production data using traditional and modern type curves. Basically, modern type curve analysis methods incorporate the flowing pressure data along with production rates and they use the analytical solutions to calculate hydrocarbon-in-place. Two important features of modern decline analysis that improve upon the traditional techniques are:

(1) *Normalizing of rates using flowing pressure drop*: Plotting a normalized rate $(q/\Delta p)$ enables the effects of back-pressure changes to be accommodated in the reservoir analysis.
(2) *Handling the changes in gas compressibility with pressure*: Using pseudotime as the time function, instead of real time, enables the gas material balance to be handled rigorously as the reservoir pressure declines with time.

Carter Type Curve. Fetkovich originally developed his type curves for gas and oil wells that are producing at constant pressures. Carter (1985) presented a new set of type curves developed exclusively for the analysis of gas rate data. Carter noted that the changes in fluid properties with pressure significantly affect reservoir performance predictions. Of utmost importance is the variation in the gas viscosity–compressibility product $\mu_g c_g$, which was ignored by Fetkovich. Carter developed another set of decline curves for boundary-dominated flow that uses a new correlating parameter λ to represent the changes in $\mu_g c_g$ during depletion. The λ parameter, called the "dimensionless drawdown correlating parameter," is designated to reflect the magnitude of pressure drawdown on $\mu_g c_g$ and defined by:

$$\lambda = \frac{(\mu_g c_g)_i}{(\mu_g c_g)_{avg}} \quad (3.205)$$

or equivalently:

$$\lambda = \frac{(\mu_g c_g)_i}{2}\left[\frac{m(p_i) - m(p_{wf})}{(p_i/Z_i) - (p_{wf}/Z_{wf})}\right] \quad (3.206)$$

where

c_g = gas compressibility coefficient, psi^{-1}
$m(p)$ = real-gas pseudopressure, psi^2/cp
p_{wf} = bottom-hole flowing pressure, psi
p_i = initial pressure, psi

μ_g = gas viscosity, cp
Z = gas deviation factor

For $\lambda = 1$, this indicates a negligible drawdown effect and corresponds to $b = 0$ on the Fetkovich exponential decline curve. Values of λ range between 0.55 and 1.0. The type curves presented by Carter are based on specially defined dimensionless parameters:

(1) dimensionless time t_D;
(2) dimensionless rate q_D;
(3) dimensionless geometry parameter (η) that characterizes the dimensionless radius r_{eD} and flow geometry;
(4) dimensionless drawdown correlating parameter λ.

Carter used a finite difference radial gas model to generate the data for constructing the type curves shown in Figure 3.58.

The following steps summarize the type curve matching procedure:

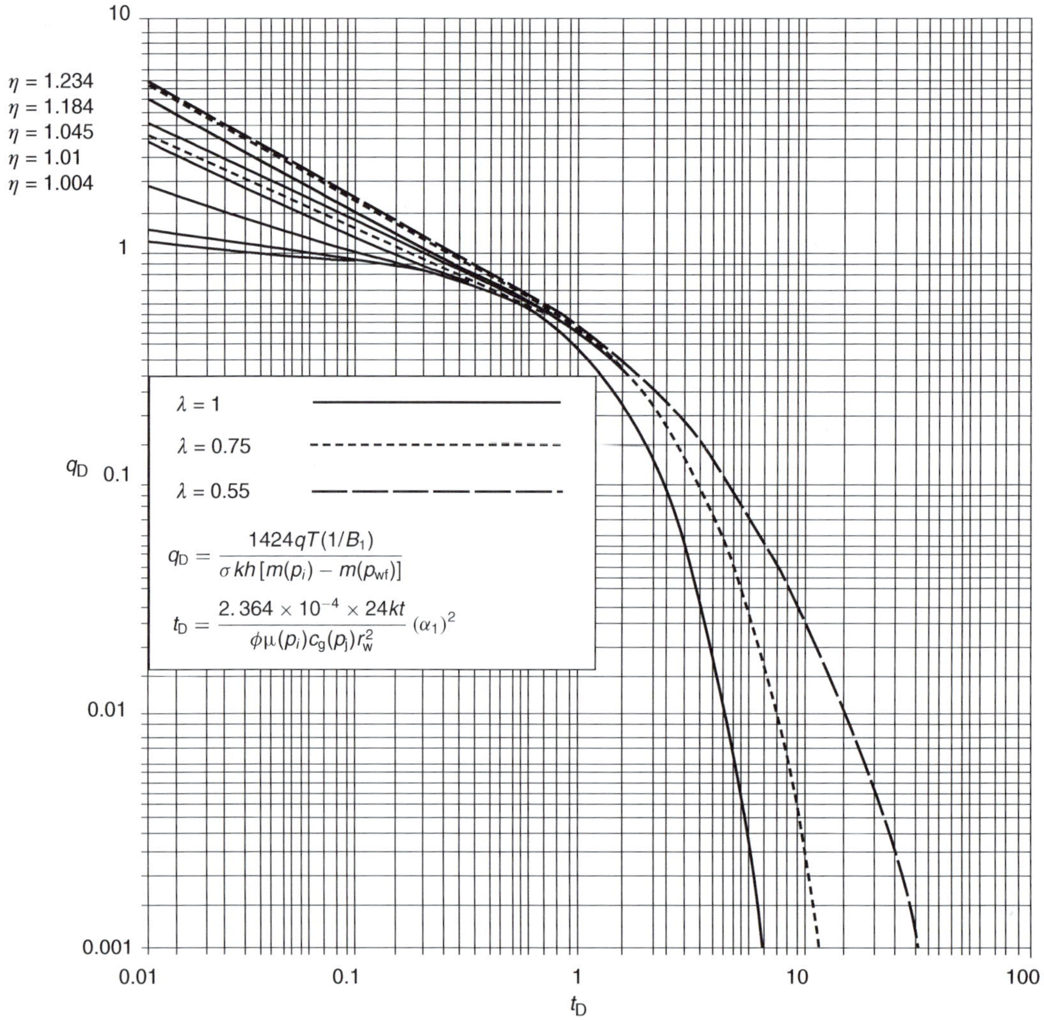

FIGURE 3.58 Radial–linear gas reservoir type curves. *(After Carter, R., 1985. Type curves for finite radial and linear gas-flow systems. SPE J. 25 (5), 719–728, copyright SPE 1985).*

Step 1. Using Eq. (3.205) or (3.206), calculate the parameter λ:

$$\lambda = \frac{(\mu_g c_g)_i}{(\mu_g c_g)_{avg}}$$

or

$$\lambda = \frac{(\mu_g c_g)_i}{2}\left[\frac{m(p_i) - m(p_{wf})}{(p_i/Z_i) - (p_{wf}/Z_{wf})}\right]$$

Step 2. Plot gas rate q in Mscf/day or MMscf/day as a function of time (t) in days using the same log–log scale as the type curves. If the actual rate values are erratic or fluctuate, it may be best to obtain averaged values of rate by determining the slope of straight lines drawn through adjacent points spaced at regular intervals on the plot of cumulative production G_p vs. time, i.e., slope $= dG_p/dt = q_g$. The resulting plot of q_g vs. t should be made on tracing paper or on a transparency so that it can be laid over the type curves for matching.

Step 3. Match the rate data to a type curve corresponding to the computed value of λ in step 1. If the computed value of λ is not as one of the values for which a type curve is shown, the needed curve can be obtained by interpolation and graphical construction.

Step 4. From the match, values of $(q_D)_{MP}$ and $(t_D)_{MP}$ corresponding to specific values for $(q)_{MP}$ and $(t)_{MP}$ are recorded. Also, a value for the dimensionless geometry parameter η is also obtained from the match. It is strongly emphasized that late-time data points (boundary-dominated pseudosteady-state flow condition) are to be matched in preference to early-time data points (unsteady-state flow condition) because matching some early rate data often will be impossible.

Step 5. Estimate the gas that would be recoverable by reducing the average reservoir pressure from its initial value to p_{wf} from the following expression:

$$\Delta G = G_i - G_{pwf} = \frac{(qt)_{MP}}{(q_D t_D)_{MP}}\frac{\eta}{\lambda} \quad (3.207)$$

Step 6. Calculate the initial gas-in-place G_i from:

$$G_i = \left[\frac{p_i/Z_i}{(p_i/Z_i) - (p_{wf}/Z_{wf})}\right]\Delta G \quad (3.208)$$

Step 7. Estimate the drainage area of the gas well from:

$$A = \frac{B_{gi} G_i}{43{,}560\phi h(1 - S_{wi})} \quad (3.209)$$

where

B_{gi} = gas formation volume factor at p_i, ft^3/scf
A = drainage area, acres
h = thickness, ft
ϕ = porosity
S_{wi} = initial water saturation

Example 3.23

The following production and reservoir data was used by Carter to illustrate the proposed calculation procedure.

p (psia)	μ_g (cp)	Z
1	0.0143	1.0000
601	0.0149	0.9641
1201	0.0157	0.9378
1801	0.0170	0.9231
2401	0.0188	0.9207
3001	0.0208	0.9298
3601	0.0230	0.9486
4201	0.0252	0.9747
4801	0.0275	1.0063
5401	0.0298	1.0418

$p_i = 5400$ psia, $p_{wf} = 500$ psi
$T = 726\ °R$, $h = 50$ ft
$\phi = 0.070$, $S_{wi} = 0.50$
$\lambda = 0.55$

Time (days)	q_t (MMscf/day)
1.27	8.300
10.20	3.400
20.50	2.630
40.90	2.090
81.90	1.700
163.80	1.410
400.00	1.070
800.00	0.791
1600.00	0.493
2000.00	0.402
3000.00	0.258
5000.00	0.127
10,000.00	0.036

Calculate the initial gas-in-place and the drainage area.

Solution

Step 1. The calculated value of λ is given as 0.55 and, therefore, the type curve for a λ value of 0.55 can be used directly from Figure 3.58.

Step 2. Plot the production data, as shown in Figure 3.59, on the same log–log scale as Figure 3.55 and determine the match points of:

$$(q)_{MP} = 1.0 \text{ MMscf/day}$$
$$(t)_{MP} = 1000 \text{ days}$$
$$(q_D)_{MP} = 0.605$$
$$(t_D)_{MP} = 1.1$$
$$\eta = 1.045$$

Step 3. Calculate ΔG from Eq. (3.207):

$$\Delta G = G_i - G_{pwf} = \frac{(qt)_{MP}}{(q_D t_D)_{MP}} \frac{\eta}{\lambda} = \frac{(1)(1000)}{(0.605)(1.1)} \frac{1.045}{0.55} = 2860 \text{ MMscf}$$

Step 4. Estimate the initial gas-in-place by applying Eq. (3.208).

$$G_i = \left[\frac{p_i/Z_i}{(p_i/Z_i) - (p_{wf}/Z_{wf})}\right]\Delta G = \left[\frac{5400/1.0418}{(5400/1.0418) - (500/0.970)}\right]2860 = 3176 \text{ MMscf}$$

Step 5. Calculate the gas formation volume factor B_{gi} at p_i.

$$B_{gi} = 0.0287\frac{Z_i T}{p_i} = 0.02827\frac{(1.0418)(726)}{5400} = 0.00396 \text{ ft}^3/\text{scf}$$

Step 6. Determine the drainage area from Eq. (3.209):

$$A = \frac{B_{gi} G_i}{43{,}560\phi h(1 - S_{wi})} = \frac{0.00396(3176)(10^6)}{43,560(0.070)(50)(1 - 0.50)} = 105 \text{ acres}$$

Palacio–Blasingame Type Curves. Palacio and Blasingame (1993) presented an innovative technique for converting gas well production data with variable rates and bottom-hole flowing pressures into "equivalent constant-rate liquid data" that allows the liquid solutions to be used to model gas flow. The reasoning for this approach is that the constant-rate type curve solutions for liquid flow problems are well established from the traditional well test analysis approach. The new solution for the gas problem is based on a material balance like time function and an algorithm that allows:

- the use of decline curves that are specifically developed for liquids;
- modeling of actual variable rate–variable pressure drop production conditions;
- explicit computation of gas-in-place.

Under pseudosteady-state flow conditions, Eq. (1.134) in Chapter 1 describes the radial flow of slightly compressible liquids as:

$$p_{wf} = \left[p_i - \frac{0.23396 QBt}{Ah\phi c_t}\right] - \frac{162.6 QB\mu}{kh} \log\left[\frac{4A}{1.781 C_A r_w^2}\right]$$

where

k = permeability, md
A = drainage area, ft^2
C_A = shape factor
Q = flow rate, STB/day

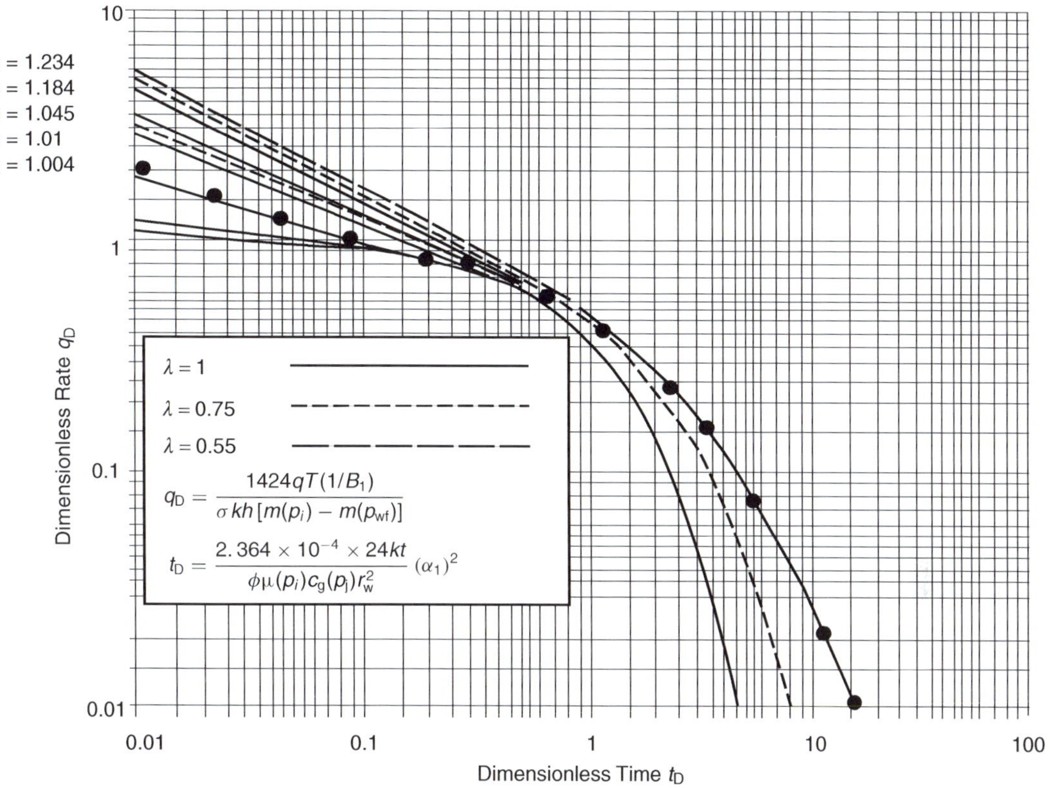

FIGURE 3.59 Carter type curves for Example 3.23.

t = time, hours
c_t = total compressibility coefficient, psi^{-1}

Expressing the time t in days and converting from "log" to natural logarithm "ln," the above relation can be written as:

$$\frac{p_i - p_{wf}}{q} = \frac{\Delta p}{q} = 70.6\frac{B\mu}{kh}\ln\left[\frac{4A}{1.781\,C_A r_{wa}^2}\right] + \left[\frac{5.615B}{Ah\phi C_t}\right]t \quad (3.210)$$

or more conveniently as:

$$\frac{\Delta p}{q} = b_{pss} + mt \quad (3.211)$$

This expression suggests that under a pseudosteady-state flowing condition, a plot of $\Delta p/q$ vs. t on a Cartesian scale would yield a straight line with an intercept of b_{pss} and slope of m, with

$$\text{Intercept } b_{pss} = 70.6\frac{B\mu}{kh}\ln\left[\frac{4A}{1.781\,C_A r_{wa}^2}\right] \quad (3.212)$$

$$\text{Slope } m = \frac{5.615B}{Ah\phi c_t} \quad (3.213)$$

where

b_{pss} = constant in the pseudosteady-state "pss" equation
t = time, days
k = permeability, md
A = drainage area, ft^2
q = flow rate, STB/day
B = formation volume factor, bbl/STB
C_A = shape factor
c_t = total compressibility, psi^{-1}
r_{wa} = apparent (effective) wellbore radius, ft

For a gas system flowing under pseudosteady-state conditions, an equation similar to Eq. (3.210) can be expressed as:

$$\frac{m(p_i)-m(p_{wf})}{q}=\frac{\Delta m(p)}{q}=\frac{711T}{kh}\left(\ln\frac{4A}{1.781C_A r_{wa}^2}\right)+\left[\frac{56.54T}{\phi(\mu_g c_g)_i Ah}\right]t \tag{3.214}$$

And in a linear form as:

$$\frac{\Delta m(p)}{q}=b_{pss}+mt \tag{3.215}$$

Similar to the liquid system, Eq. (3.215) indicates that a plot of $\Delta m(p)/q$ vs. t will form a straight line with

$$\text{Intercept } b_{pss}=\frac{711T}{kh}\left(\ln\frac{4A}{1.781C_A r_{wa}^2}\right)$$

$$\text{Slope } m=\frac{56.54T}{(\mu_g c_t)_i(\phi hA)}=\frac{56.54T}{(\mu_g c_t)_i(\text{PV})}$$

where

q = flow rate, Mscf/day
A = drainage area, ft^2
T = temperature, °R
t = flow time, days

The linkage that allows for the transformation of converting gas production data into equivalent constant-rate liquid data is based on the use of a new time function called "pseudoequivalent time or normalized material balance pseudotime," as defined by:

$$t_a=\frac{(\mu_g c_g)_i}{q_t}\int_0^t\left[\frac{q_t}{\overline{\mu}_g\overline{c}_g}\right]dt=\frac{(\mu_g c_g)_i}{q_t}\frac{Z_i G}{2p_i}\left[\overline{m}(\overline{p}_i)-\overline{m}(p)\right] \tag{3.216}$$

where

t_a = pseudoequivalent (normalized material balance) time, days
t = time, days
G = original gas-in-place, Mscf
q_t = gas flow rate at time t, Mscf/day
$\overline{p}$ = average pressure, psi
$\overline{\mu}_g$ = gas viscosity at $\overline{p}$, cp
$\overline{c}_g$ = gas compressibility at $\overline{p}$, psi^{-1}
$\overline{m}(p)$ = normalized gas pseudopressure, psi^2/cp

To perform decline curve analysis under variable rates and pressures, the authors derived a theoretical expression for decline curve analysis that combines:

- the material balance relation,
- the pseudosteady-state equation, and
- the normalized material balance time function t_a

to give the following relationship:

$$\left[\frac{q_g}{\overline{m}(p_i)-\overline{m}(p_{wf})}\right]b_{pss}=\frac{1}{1+(m/b_{pss})t_a} \tag{3.217}$$

where $\overline{m}(p)$ is the normalized pseudopressure as defined by:

$$\overline{m}(p_i)=\frac{\mu_{gi}Z_i}{p_i}\int_0^{p_i}\left[\frac{p}{\mu_g Z}\right]dp \tag{3.218}$$

$$\overline{m}(p)=\frac{\mu_{gi}Z_i}{p_i}\int_0^{p}\left[\frac{p}{\mu_g Z}\right]dp \tag{3.219}$$

and

$$m=\frac{1}{Gc_{ti}} \tag{3.220}$$

$$b_{pss}=\frac{70.6\mu_{gi}B_{gi}}{k_g h}\left[\ln\left(\frac{4A}{1.781C_A r_{wa}^2}\right)\right] \tag{3.221}$$

where

G = original gas-in-place, Mscf
c_{gi} = gas compressibility at p_i, psi^{-1}
c_{ti} = total system compressibility at p_i, psi^{-1}
q_g = gas flow rate, Mscf/day
k_g = effective permeability to gas, md

$\overline{m}(p)$ = normalized pseudopressure, psia
p_i = initial pressure
r_{wa} = effective (apparent) wellbore radius, ft
B_{gi} = gas formation volume factor at p_i, bbl/Mscf

Notice that Eq. (3.217) is essentially expressed in the same dimensionless form as the Fetkovich equation, i.e., Eq. (3.171), or:

$$q_{Dd} = \frac{1}{1+(t_a)_{Dd}} \tag{3.222}$$

with

$$q_{Dd} = \left[\frac{q_g}{\overline{m}(p_i)-\overline{m}(p_{wf})}\right] b_{pss} \tag{3.223}$$

$$(t_a)_{Dd} = \left(\frac{m}{b_{pss}}\right) t_a \tag{3.224}$$

It must be noted that the q_{Dd} definition is now in terms of normalized pseudopressures and the modified dimensionless decline time function $(t_a)_{Dt}$ is not in terms of real time but in terms of the material balance pseudotime. *Also note that Eq. (3.223) traces the path of a harmonic decline on the Fetkovich type curve with a hyperbolic exponent of $b = 1$.*

However, there is a computational problem when applying Eq. (3.216) because it requires the value of G or the average pressure $\overline{p}$, which is itself a function of G. The method is iterative in nature and requires rearranging of Eq. (3.217) in the following familiar form of linear relationship:

$$\frac{\overline{m}(p_i)-\overline{m}(p)}{q_g} = b_{pss} + m t_a \tag{3.225}$$

The iterative procedure for determining G and $\overline{p}$ is described in the following steps:

Step 1. Using the available gas properties, set up a table of Z, μ, p/Z, $(p/Z\mu)$ vs. p for the gas system:

Time	p	Z	μ	p/Z	$p/(Z\mu)$
0	p_i	Z_i	μ_i	p_i/Z_i	$p_i/(Z\mu)_i$
.	.	.	.	.	.
.	.	.	.	.	.
.	.	.	.	.	.

Step 2. Plot $(p/Z\mu)$ vs. p on a Cartesian scale and numerically determine the area under the curve for several values of p. Multiply each area by $(Z_i\mu_i/pi)$ to give the normalized pseudopressure as:

$$\overline{m}(p) = \frac{\mu_{gi}Z_i}{p_i}\int_0^p \left[\frac{p}{\mu_g Z}\right] dp$$

The required calculations of this step can be performed in the following tabulated form:

p	Area $= \int_0^p \left[\frac{p}{\mu_g Z}\right] dp$	$\overline{m}(p) = (\text{area})\frac{\mu_{gi}Z_i}{p_i}$
0	0	0
.	.	.
p_i	.	.

Step 3. Draw plots of $\overline{m}(p)$ and p/Z vs. p on a Cartesian scale.

Step 4. Assume a value for the initial gas-in-place G.

Step 5. For *each* production data point of G_p and t, calculate $\overline{p}/\overline{Z}$ from the gas MBE, i.e., Eq. (3.60):

$$\frac{\overline{p}}{\overline{Z}} = \frac{p_i}{Z_i}\left(1-\frac{G_p}{G}\right)$$

Step 6. From the plot generated in step 3, use the graph of p vs. p/Z for *each* value of the ratio $\overline{p}/\overline{Z}$ and determine the value of the corresponding average reservoir pressure $\overline{p}$. For each value of the average reservoir pressure $\overline{p}$, determine the values $\overline{m}/(\overline{p})$ for each $\overline{p}$.

Step 7. For *each* production data point, calculate t_a by applying Eq. (3.216):

$$t_a = \frac{(\mu_g c_g)_i}{q_t}\frac{Z_i G}{2p_i}\left[\overline{m}(p_i)-\overline{m}(\overline{p})\right]$$

The calculation of t_a can be conveniently performed in the following tabulated form:

t	q_t	G_p	$\bar{p}$	$\overline{m}(\bar{p})$	$t_a = \frac{(\mu_g c_g)_i}{q_i}\frac{Z_i G}{2p_i}[\overline{m}p_i - \overline{m}(\bar{p})]$
.	.	.	.	.	.
.	.	.	.	.	.
.	.	.	.	.	.

Step 8. Based on the linear relationship given by Eq. (3.225), plot $[\overline{m}(p_i) - \overline{m}(\bar{p})]/q_g$ vs. t_a on a Cartesian scale and determine the slope m.

Step 9. Recalculate the initial gas-in-place G by using the value m from step 8 and applying Eq. (3.220) to give:

$$G = \frac{1}{c_{ti} m}$$

Step 10. The new value of G from step 8 is used for the next iteration, i.e., step 4, and this process could continue until some convergence tolerance for G is met.

Palacio and Blasingame developed a modified Fetkovich–Carter type curve, as shown in Figure 3.60, to give the performance of constant-rate and constant-pressure gas flow solutions, the traditional Arps curve stems. To obtain a more accurate match to decline type curves than using flow rate data alone, the authors introduced the following two complementary plotting functions:

Integral function $(q_{Dd})_i$:

$$(q_{Dd})_i = \frac{1}{t_a}\int_0^{t_a}\left(\frac{q_g}{\overline{m}(p_i) - \overline{m}(p_{wf})}\right)dt_a \quad (3.226)$$

Derivative of the integral function $(q_{Dd})_{id}$:

$$(q_{Dd})_{id} = \left(\frac{-1}{t_a}\right)\frac{d}{dt_a}\left[\frac{1}{t_a}\int_0^{t_a}\left(\frac{q_g}{\overline{m}(p_i) - \overline{m}(p_{wf})}\right)dt_a\right] \quad (3.227)$$

Both functions can be easily generated by using simple numerical integration and differentiation methods.

To analyze gas production data, the proposed method involves the following basic steps:

Step 1. Calculate the initial gas-in-place G as outlined previously.

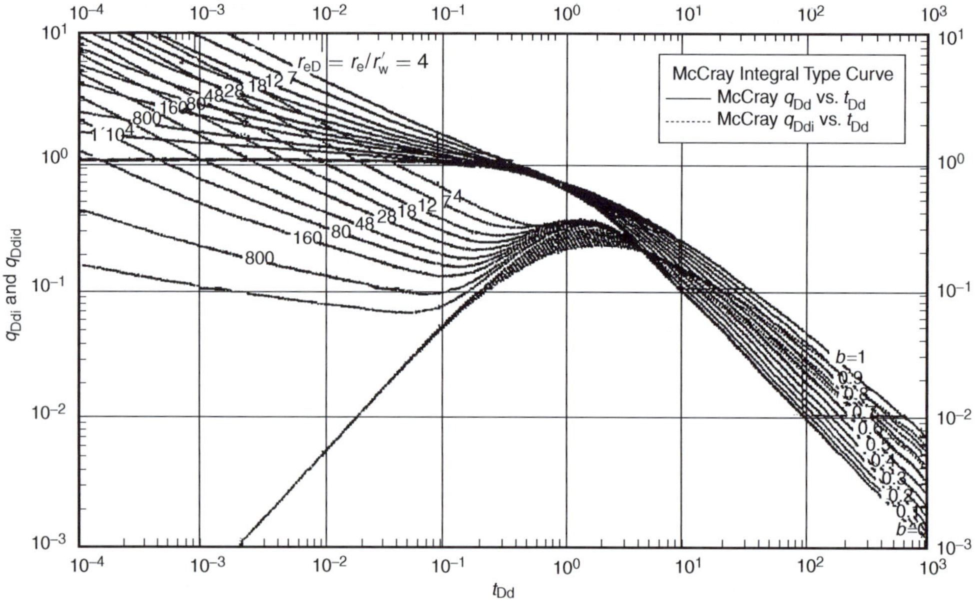

FIGURE 3.60 Palacio–Blasingame type curve.

Step 2. Construct the following table:

t	q_g	t_a	p_{wf}	$\overline{m}(p_{wf})$	$\frac{q_g}{[\overline{m}(p_i)-\overline{m}(p_{wf})]}$
.	.	.	.	.	.
.	.	.	.	.	.
.	.	.	.	.	.

Plot $q_g/[\overline{m}(p_i)-\overline{m}(\overline{p})]$ vs. t_a on a Cartesian scale.

Step 3. Using the well production data as tabulated and plotted in step 2, compute the two complementary plotting functions as given by Eqs. (3.226) and (3.227) as a function of t_a:

$$(q_{Dd})_i = \frac{1}{t_a}\int_0^{t_a}\left(\frac{q_g}{\overline{m}(p_i)-\overline{m}(p_{wf})}\right)dt_a$$

$$(q_{Dd})_{id} = \left(\frac{-1}{t_a}\right)\frac{d}{dt_a}\left[\frac{1}{t_a}\int_0^{t_a}\left(\frac{q_g}{\overline{m}(p_i)-\overline{m}(p_{wf})}\right)dt_a\right]$$

Step 4. Plot both functions, i.e., $(q_{Dd})_i$ and $(q_{Dd})_{id}$, vs. t_a on tracing paper so it can be laid over the type curve of Figure 3.60 for matching.

Step 5. Establish a match point MP and the corresponding dimensionless radius r_{eD} value to confirm the final value of G and to determine other properties:

$$G = \frac{1}{c_{ti}}\left[\frac{t_a}{t_{Dd}}\right]_{MP}\left[\frac{(q_{Dd})_i}{q_{Dd}}\right]_{MP} \tag{3.228}$$

$$A = \frac{5.615 G B_{gi}}{h\phi(1-S_{wi})}$$

$$r_e = \sqrt{\frac{A}{\pi}}$$

$$r_{wa} = \frac{r_e}{r_{eD}} \tag{3.229}$$

$$s = -\ln\left(\frac{r_{wa}}{r_w}\right)$$

$$k = \frac{141.2 B_{gi}\mu_{gi}}{h}\left[\ln\left(\frac{r_e}{r_w}\right)-\frac{1}{2}\right]\left[\frac{(q_{Dd})_i}{q_{Dd}}\right]_{MP}$$

where

G = gas-in-place, Mscf

B_{gi} = gas formation volume factor at p_i, bbl/Mscf

A = drainage area, ft^2

s = skin factor

r_{eD} = dimensionless drainage radius

S_{wi} = connate water saturation

The authors used the West Virginia gas well "A," as given by Fetkovich in Example 3.22, to demonstrate the use of the proposed type curve. The resulting fit of the data given in Example 3.22 to Placio and Blasingame is shown in Figure 3.61.

Flowing Material Balance. The flowing material balance method is a new technique that can be used to estimate the original gas-in-place (OGIP). The method as introduced by Mattar and Anderson (2003) uses the concept of the normalized rate and material balance pseudotime to create a simple linear plot, which extrapolates to fluids-in-place. The method uses the available production data in a manner similar to that of Palacio and Blasingame's approach. The authors showed that for a depletion drive gas reservoir flowing under pseudosteady-state conditions, the flow system can be described by:

$$\frac{q}{m(p_i)-m(p_{wf})} = \frac{q}{\Delta m(p)} = \left(\frac{-1}{G b^{\backslash}_{pss}}\right)Q_N + \frac{1}{b^{\backslash}_{pss}}$$

where Q_N is the normalized cumulative production as given by:

$$Q_N = \frac{2 q_t p_i t_a}{(c_t \mu_i Z_i)\,\Delta m(p)}$$

and t_a is the Palacio and Blasingame normalized material balance pseudotime as given by:

$$t_a = \frac{(\mu_g c_g)_i}{q_t}\frac{Z_i G}{2p_i}\left[\overline{m}(p_i)-\overline{m}(\overline{p})\right]$$

The authors defined $b^{\backslash}_{pss}$ as the inverse productivity index, in psi^2/cp-MMscf, as:

$$b^{\backslash}_{pss} = \frac{1.417\times 10^6 T}{kh}\left[\ln\left(\frac{r_e}{r_{wa}}\right)-\frac{3}{4}\right]$$

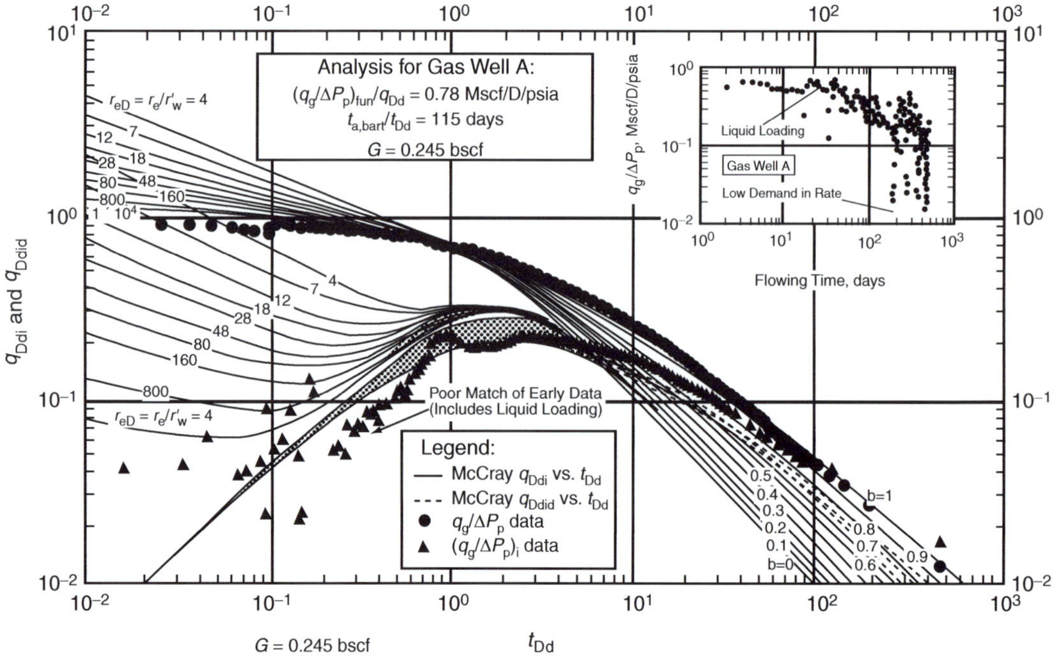

FIGURE 3.61 Palacio–Blasingame West Virginia gas well example.

where

p_i = initial pressure, psi
G = original gas-in-place
r_e = drainage radius, ft
r_{wa} = apparent wellbore radius, ft

Thus, the above expression suggests that a plot of $q/\Delta m(p)$ vs. $2qp_it_a/(c_{ti}\mu_iZ_i\,\Delta m(p))$ on a Cartesian scale would produce a straight line with following characteristics:

- x-axis intercept gives gas-in-place G;
- y-axis intercept gives $b^{\backslash}_{pss}$;
- slope gives $(-1/Gb^{\backslash}_{pss})$.

Specific steps in estimating G are summarized below:

Step 1. Using the available gas properties, set up a table of Z, μ, p/Z, $(p/Z\mu)$ vs. p for the gas system.

Step 2. Plot $(p/Z\mu)$ vs. p on a Cartesian scale and numerically determine the area under the curve for several values of p to give $m(p)$ at each pressure.

Step 3. Assume a value for the initial gas-in-place G.

Step 4. Using the assumed value of G and for *each* production data point of G_p at time t, calculate p/Z from the gas MBE, i.e., Eq. (3.60):

$$\frac{\overline{p}}{\overline{Z}} = \frac{p_i}{Z_i}\left(1 - \frac{G_p}{G}\right)$$

Step 5. For each production data point of q_t and t, calculate t_a and the normalized cumulative production Q_N:

$$t_a = \frac{(\mu_g c_g)_i}{q_t}\frac{Z_i G}{2p_i}[\overline{m}(p_i) - \overline{m}(\overline{p})]$$

$$Q_N = \frac{2q_t p_i t_a}{(c_t\mu_i Z_i)\,\Delta m(p)}$$

Step 6. Plot $q/\Delta p$ vs. Q_N on a Cartesian scale and obtain the best line through the data points. Extrapolate the line to the x-axis and read the original-gas-in-place G.

Step 7. The new value of G from step 5 is used for the next iteration, i.e., step 3, and this process could continue until some convergence tolerance for G is met.

Anash et al. Type Curves. The changes in gas properties can significantly affect reservoir performance during depletion; of utmost importance is the variation in the gas viscosity–compressibility product $\mu_g c_g$, which was ignored by Fetkovich in developing his type curves. Anash et al. (2000) proposed three functional forms to describe the product $\mu_g c_t$ as a function of pressure. They conveniently expressed the pressure in a dimensionless form as generated from the gas MBE, to give:

$$\frac{p}{Z} = \frac{p_i}{Z_i}\left(1 - \frac{G_p}{G}\right)$$

In a dimensionless form, the above MBE is expressed as:

$$p_D = (1 - G_{pD})$$

where

$$p_D = \frac{p/Z}{p_i/Z_i}, \quad G_{pD} = \frac{G_p}{G} \tag{3.230}$$

Anash and his coauthors indicated that the product ($\mu_g c_t$) can be expressed in a "dimensionless ratio" of ($\mu_g c_{ti}/\mu_g c_t$) as a function of the dimensionless pressure p_D by one of the following three forms:

(1) *First-order polynomial*: The first form is a first-degree polynomial that is adequate in describing the product $\mu_g c_t$ as a function of pressure at low gas reservoir pressure below 5000 psi, i.e., $p_i < 5000$. The polynomial is expressed in a dimensionless form as:

$$\frac{\mu_i c_{ti}}{\mu c_t} = p_D \tag{3.231}$$

where

c_{ti} = total system compressibility at p_i, psi^{-1}
μ_i = gas viscosity at p_i, cp

(2) *Exponential model*: The second form is adequate in describing the product $\mu_g c_t$ for high-pressure gas reservoirs, i.e., $p_i > 8000$ psi:

$$\frac{\mu_i c_{ti}}{\mu c_t} = \beta_o \exp(\beta_1 p_D) \tag{3.232}$$

(3) *General polynomial model*: A third- or fourth-degree polynomial is considered by the authors as a general model that is applicable to all gas reservoir systems with any range of pressures, as given by:

$$\frac{\mu_i c_{ti}}{\mu c_t} = a_0 + a_1 p_D + a_2 p_D^2 + a_3 p_D^3 + a_4 p_D^4 \tag{3.233}$$

The coefficients in Eqs. (3.232) and (3.233), i.e., β_0, β_1, a_0, a_1, etc., can be determined by plotting the dimensionless ratio $\mu_i c_{ti}/\mu c_t$ vs. p_D on a Cartesian scale, as shown in Figure 3.62, and using the least-squares type regression model to determine the coefficients.

The authors also developed the following fundamental form of the stabilized gas flow equation as:

$$\frac{dG_p}{dt} = q_g = \frac{J_g}{c_{ti}} \int_{p_{wD}}^{p_D} \left[\frac{\mu_i c_{ti}}{\mu c_t}\right] dp_D$$

with the dimensionless bottom-hole flowing pressure as defined by:

$$p_{wD} = \frac{p_{wf}/Z_{wf}}{p_i/Z_i}$$

where

q_g = gas flow rate, scf/day
p_{wf} = flowing pressure, psia
Z_{wf} = gas deviation factor at p_{wf}
J_g = productivity index, scf/day, psia

Anash et al. presented their solutions in a "type curve" format in terms of a set of the familiar dimensionless variables q_{Dd}, t_{Dd}, r_{eD}, and a newly introduced correlating parameter β that is a function of the dimensionless pressure. They presented three type curve sets, as shown

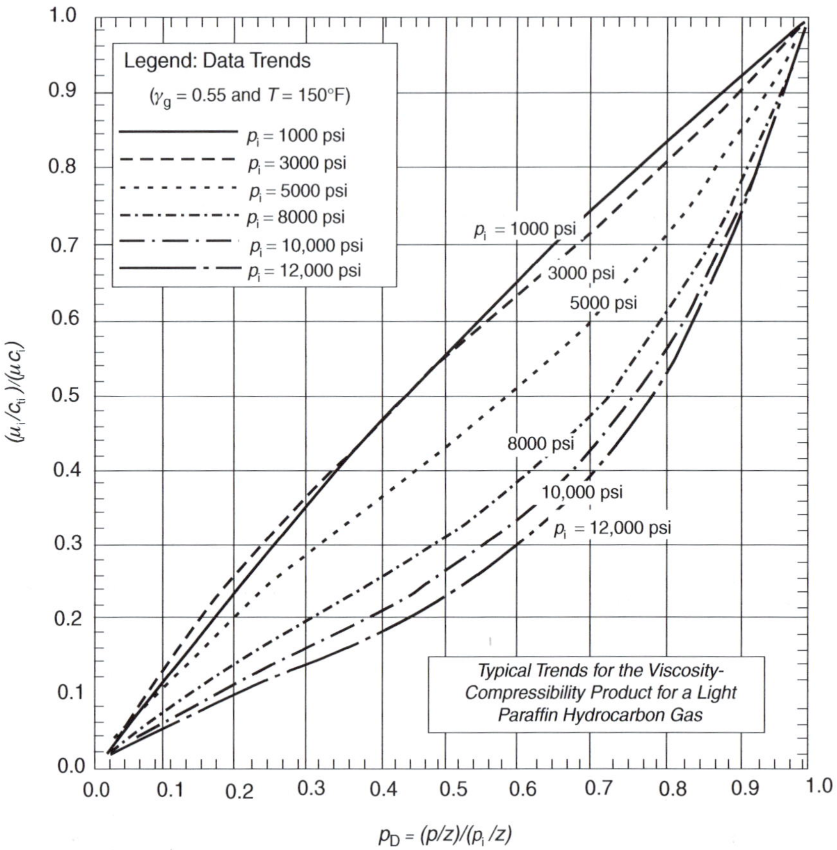

FIGURE 3.62 Typical distribution of the viscosity–compressibility function. *(After Anash, J., Blasingame, T.A., Knowles, R.S., 2000. A semianalytic (p/Z) rate-time relation for the analysis and prediction of gas well performance. SPE Reservoir Eval. Eng. 3, 525–533).*

in Figures 3.63–3.65, one for each of the functional forms selected to describe the product μc_t (i.e., first-order polynomial, exponential model, or general polynomial).

The methodology of employing the Anash et al. type curve is summarized by the following steps:

Step 1. Using the available gas properties, prepare a plot of $(\mu_i c_{ti}/\mu c_t)$ vs. p_D, where

$$p_D = \frac{p/Z}{p_i/Z_i}$$

Step 2. From the generated plot, select the appropriate functional form that describes the resulting curve. That is:

First-order polynomial:

$$\frac{\mu_i c_{ti}}{\mu c_t} = p_D$$

Exponential model:

$$\frac{\mu_i c_{ti}}{\mu c_t} = \beta_0 \exp(\beta_1 p_D)$$

General polynomial model:

$$\frac{\mu_i c_{ti}}{\mu c_t} = a_0 + a_1 p_D + a_2 p_D^2 + a_3 p_D^3 + a_4 p_D^4$$

Using a regression model, i.e., least squares, determine the coefficient of the selected functional form that adequately describes $(\mu_i c_{ti}/\mu c_t)$ vs. p_D.

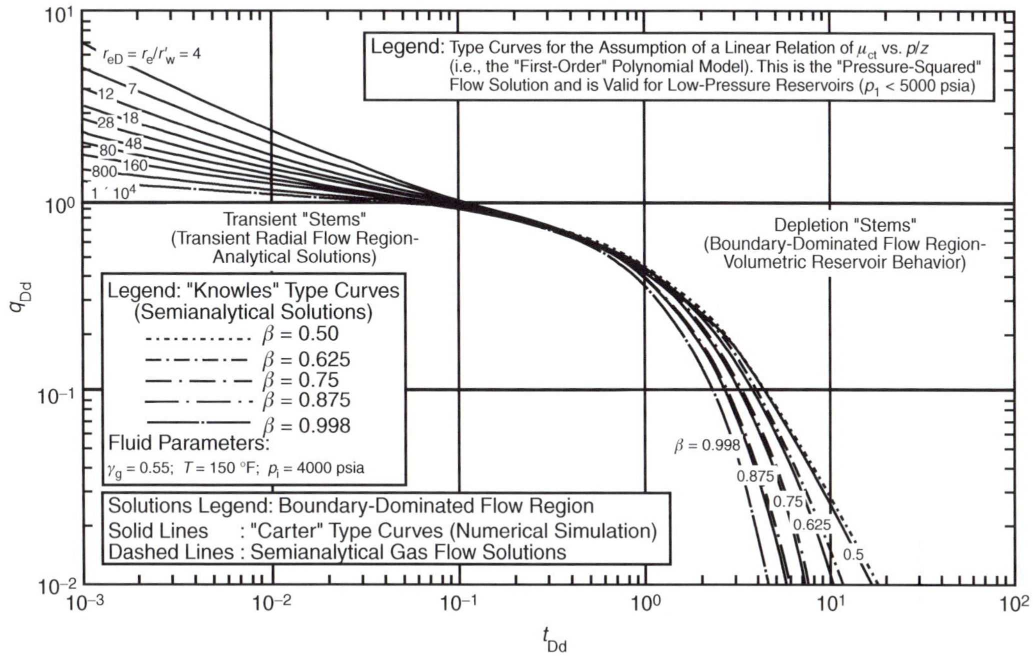

FIGURE 3.63 "First-order" polynomial solution for real-gas flow under boundary-dominated flow conditions. Solution assumes a μ c_t profile that is linear with p_D. *(Permission to copy by the SPE, 2000).*

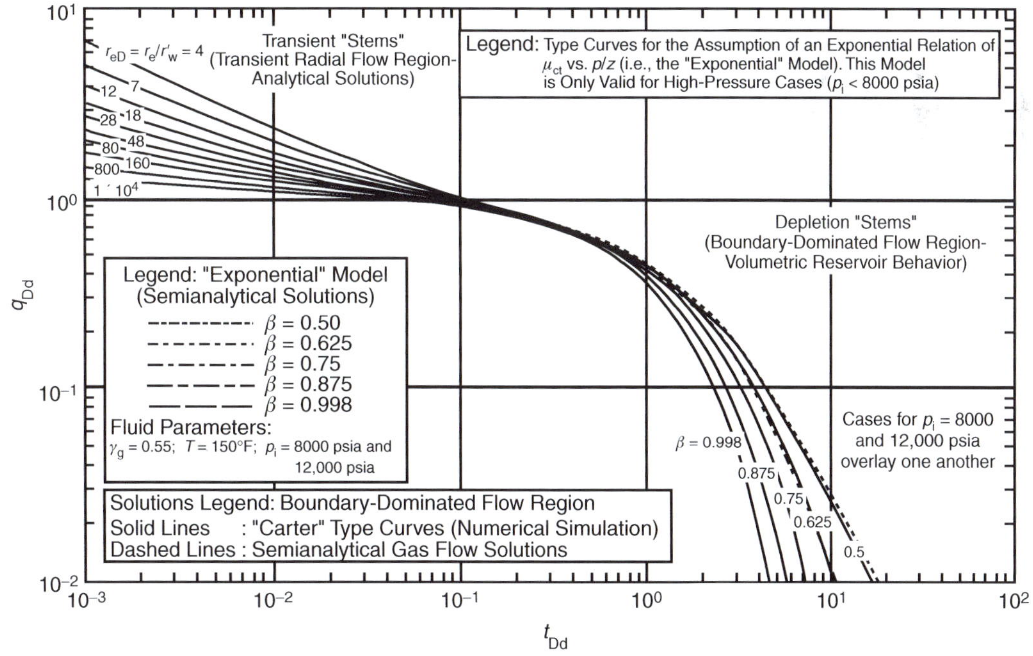

FIGURE 3.64 "Exponential" solutions for real-gas flow under boundary-dominated flow conditions. *(Permission to copy by the SPE, 2000).*

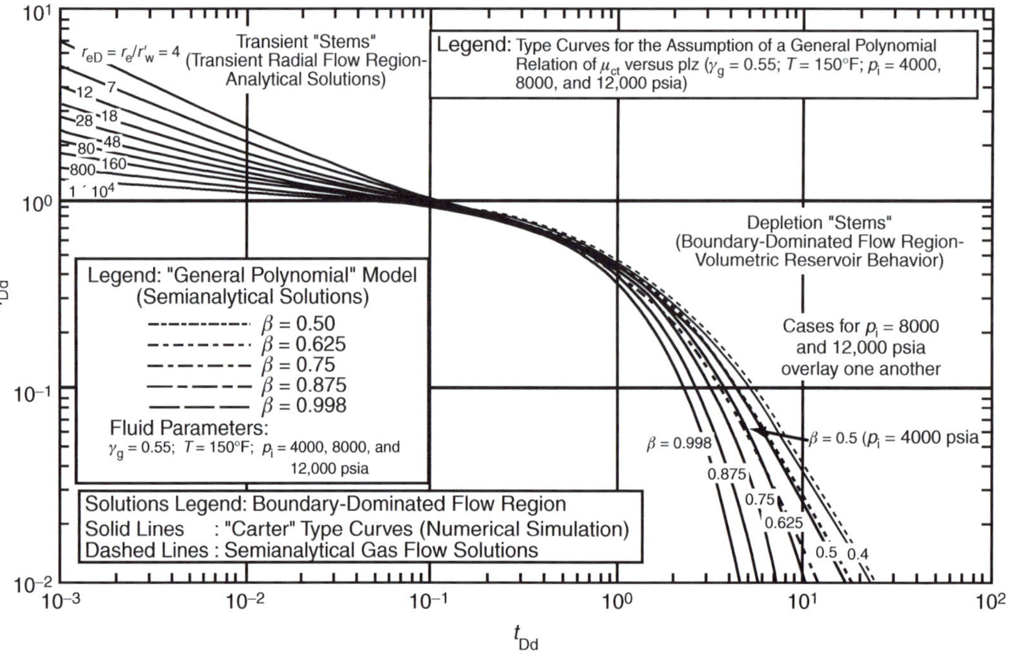

FIGURE 3.65 "General polynomial" solution for real-gas flow under boundary-dominated flow conditions. *(Permission to copy by the SPE, 2000).*

Step 3. Plot the historical flow rate q_g vs. time t on log–log scale with the same logarithmic cycles as the one given by the selected type curves (i.e., Figures 3.63–3.65).

Step 4. Using the type curve matching technique described previously, select a match point and record:

$$(q_g)_{MP} \text{ and } (q_{Dd})_{MP}$$
$$(t)_{MP} \text{ and } (t_{Dd})_{MP}$$
$$(r_{eD})_{MP}$$

Step 5. Calculate the dimensionless pressure p_{wD} using the bottom-hole flowing pressure:

$$p_{wD} = \frac{p_{wf}/Z_{wf}}{p_i/Z_i}$$

Step 6. Depending on the selected functional form in step 2, calculate the constant α for the selected functional model:

For the first-order polynomial:

$$\alpha = \tfrac{1}{2}(1 - p_{wD}^2) \quad \textbf{(3.234)}$$

For the exponential model:

$$\alpha = \frac{\beta_0}{\beta_1}\left[\exp(\beta_1) - \exp(\beta_1 p_{wD})\right] \quad \textbf{(3.235)}$$

where β_0 and β_1 are the coefficients of the exponential model.

For the polynomial function (assuming a fourth-degree polynomial):

$$\alpha = A_0 + A_1 + A_2 + A_3 + A_4 \quad \textbf{(3.236)}$$

where

$$A_0 = -(A_1 p_{wD} + A_2 p_{wD}^2 + A_3 p_{wD}^3 + A_4 p_{wD}^4) \quad \textbf{(3.237)}$$

where

$$A_1 = a_0, \quad A_2 = \frac{a_1}{2}, \quad A_3 = \frac{a_2}{3}, \quad A_4 = \frac{a_3}{4}$$

Step 7. Calculate the well productivity index J_g, in scf/day-psia, by using the flow rate match point and the constant α of step 6 in the following relation:

$$J_g = \frac{C_{ti}}{\alpha}\left(\frac{q_g}{q_{Dd}}\right)_{MP} \tag{3.238}$$

Step 8. Estimate the original gas-in-place G, in scf, from the time match point:

$$G = \frac{J_g}{C_{ti}}\left(\frac{t}{t_{Dd}}\right)_{MP} \tag{3.239}$$

Step 9. Calculate the reservoir drainage area A, in ft^2, from the following expression:

$$A = \frac{5.615 B_{gi} G}{\phi h(1 - S_{wi})} \tag{3.240}$$

where

A = drainage area, ft^2
B_{gi} = gas formation volume factor at p_i, bbl/scf
S_{wi} = connate water saturation

Step 10. Calculate the permeability k, in md, from the match curve of the dimensionless drainage radius r_{eD}:

$$k = \frac{141.2\mu_i B_{gi} J_g}{h}\left(\ln[r_{eD}]_{MP} - \frac{1}{2}\right) \tag{3.241}$$

Step 11. Calculate the skin factor from the following relationships:

$$\text{Drainage radius } r_e = \sqrt{\frac{A}{\pi}} \tag{3.242}$$

$$\text{Apparent wellbore radius } r_{wa} = \frac{r_e}{(r_{eD})_{MP}} \tag{3.243}$$

$$\text{Skin factor } s = -\ln\left(\frac{r_{wa}}{r_w}\right) \tag{3.244}$$

Example 3.24

The West Virginia gas well "A" is a vertical gas well that has been hydraulically fractured and is undergoing depletion. The production data was presented by Fetkovich and used in Example 3.22. A summary of the reservoir and fluid properties is given below:

$r_w = 0.354$ ft,	$h = 70$ ft
$\phi = 0.06$,	$T = 160$ °F
$s = 5.17$,	$k = 0.07$ md
$\gamma_g = 0.57$,	$B_{gi} = 0.00071$ bbl/scf
$\mu_{gi} = 0.0225$ cp,	$c_{ti} = 0.000184$ psi^{-1}
$p_i = 4{,}175$ psia,	$p_{wf} = 710$ psia
$\alpha = 0.4855$ (first-order polynomial)	
$S_{wi} = 0.35$	

Solution

Step 1. Figure 3.66 shows the type curve match of the production data with that of Figure 3.63 to give:

$$(q_{Dd})_{MP} = 1.0$$
$$(q_g)_{MP} = 1.98 \times 10^6 \text{ scf/day}$$
$$(t_{Dd})_{MP} = 1.0$$
$$(t)_{MP} = 695 \text{ days}$$
$$(r_{eD})_{MP} = 28$$

Step 2. Calculate the productivity index from Eq. (3.238):

$$J_g = \frac{C_{ti}}{\alpha}\left(\frac{q_g}{q_{Dd}}\right)_{MP} = \frac{0.000184}{0.4855}\left(\frac{1.98 \times 10^6}{1.0}\right) = 743.758 \text{ scf/day-psi}$$

Step 3. Solve for G by applying Eq. (3.239):

$$G = \frac{J_g}{C_{ti}}\left(\frac{t}{t_{Dd}}\right)_{MP} = \frac{743.758}{0.0001824}\left(\frac{695}{1.0}\right) = 2.834 \text{ Bscf}$$

Step 4. Calculate the drainage area from Eq. (3.240):

$$A = \frac{5.615 B_{gi} G}{\phi h(1 - S_{wi})} = \frac{5.615(0.00071)(2.834 \times 10^9)}{(0.06)(70)(1-0.35)} = 4.1398 \times 10^6 \text{ ft}^2 = 95 \text{ acres}$$

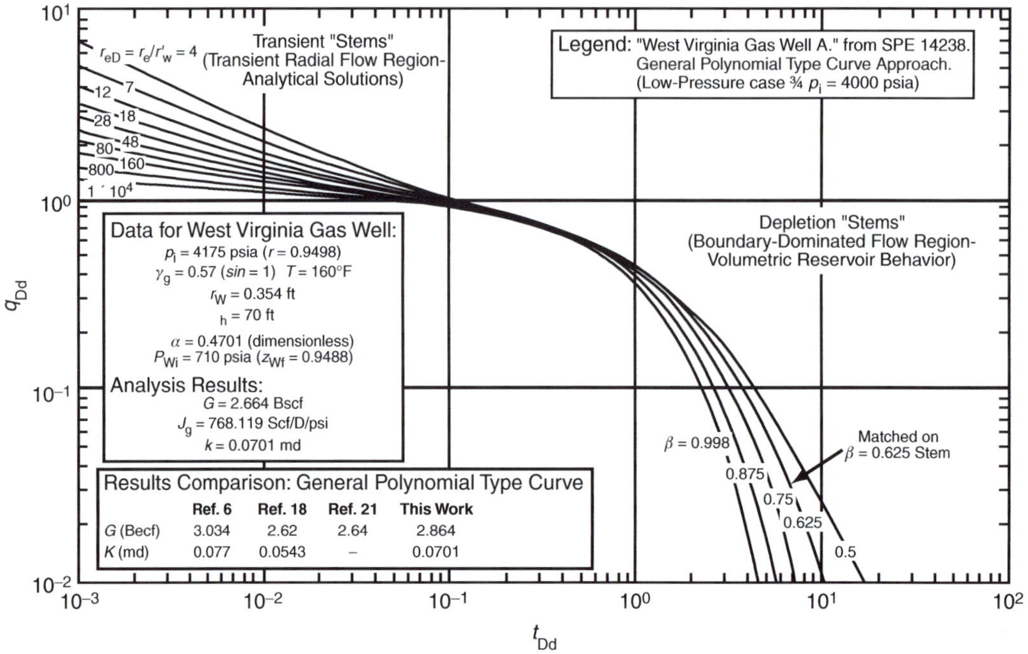

FIGURE 3.66 Type curve analysis of West Virginia gas well "A" (SPE 14238). "General polynomial" type curve analysis approach. *(Permission to copy by the SPE, 2000).*

Step 5. Compute the permeability from the match on the $r_{eD} = 28$ transient stem by using Eq. (3.241):

$$k = \frac{(141.2)(0.0225)(0.00071)(743.76)}{70}\left(\ln(28) - \frac{1}{2}\right)$$
$$= 0.0679 \text{ md}$$

Step 6. Calculate the skin factor by applying Eqs. (3.242) and (3.243):

$$r_e = \sqrt{\frac{A}{\pi}} = \sqrt{\frac{4.1398 \times 10^6}{\pi}} = 1147.9 \text{ ft}$$

$$r_{wa} = \frac{r_e}{(r_{eD})_{MP}} = \frac{1147.9}{28} = 40.997 \text{ ft}$$

$$s = -\ln\left(\frac{r_{wa}}{r_w}\right) = -\ln\left(\frac{40.997}{0.354}\right) = -4.752$$

Decline Curve Analysis for Fractured Wells. Pratikno et al. (2003) developed a new set of type curves specifically for finite conductivity, vertically fractured wells centered in bounded circular reservoirs. The authors used analytical solutions to develop these type curves and to establish a relation for the decline variables.

Recall that the general dimensionless pressure equation for a bounded reservoir during pseudosteady-state flow is given by Eq. (1.136):

$$p_D = 2\pi t_{DA} + \frac{1}{2}\left[\ln\left(\frac{A}{r_w^2}\right)\right] + \frac{1}{2}\left[\ln\left(\frac{2.2458}{C_A}\right)\right] + s$$

with the dimensionless time based on the wellbore radius t_D or drainage area t_{DA} as given by Eqs. (1.86a) and (1.86b) as:

$$t_D = \frac{0.0002637kt}{\phi\mu c_t r_w^2}$$

$$t_{DA} = \frac{0.0002637kt}{\phi\mu c_t A} = t_A\left(\frac{r_w^2}{A}\right)$$

The authors adopted the above form and suggested that for a well producing under pseudosteady state (pss) at a constant rate with a finite conductivity fracture in a circular reservoir, the dimensionless pressure drop can be expressed as:

$$p_D = 2\pi t_{DA} + b_{Dpss}$$

or

$$b_{Dpss} = p_D - 2\pi t_{DA}$$

where the term b_{Dpss} is the dimensionless pseudosteady-state constant that is independent of time; however, b_{Dpss} is a function of:

- the dimensionless radius r_{eD} and
- the dimensionless fracture conductivity F_{CD}.

The above two dimensionless parameters were defined in Chapter 1 by:

$$F_{CD} = \frac{k_f}{k}\frac{w_f}{x_f} = \frac{F_C}{kx_f}\, r_{eD} = \frac{r_e}{x_f}$$

The authors noted that during pseudosteady-state flow, the equation describing the flow during this period yields a constant value for given values of r_{eD} and F_{CD}, given closely by the following relationship:

$$b_{Dpss} = \ln(r_{eD}) - 0.049298 + \frac{0.43464}{r_{eD}^2} + \frac{a_1 + a_2 u + a_3 u^2 + a_4 u^3 + a_5 u^4}{1 + b_1 u + b_2 u^2 + b_3 u^3 + b_4 u^4}$$

with

$$u = \ln(F_{CD})$$

where

$a_1 = 0.93626800$ $\quad b_1 = -0.38553900$
$a_2 = -1.0048900$ $\quad b_2 = -0.06988650$
$a_3 = 0.31973300$ $\quad b_3 = -0.04846530$
$a_4 = -0.0423532$ $\quad b_4 = -0.\ 00813558$
$a_5 = 0.00221799$

Based on the above equations, Pratikno et al. used Palacio and Blasingame's previously defined functions (i.e., t_a, $(q_{Dd})_i$, and $(q_{Dd})_{id}$) and the parameters r_{eD} and F_{CD} to generate a set of decline curves for a sequence of 13 values for F_{CD} with a sampling of $r_{eD} = 2$, 3, 4, 5, 10, 20, 30, 40, 50, 100, 200, 300, 400, 500, and 1000. Type curves for F_{CD} of 0.1, 1, 10, 100, and 1000 are shown in Figures 3.67–3.71.

The authors recommended the following type curve matching procedure that is similar to the methodology used in applying Palacio and Blasingame's type curve:

Solution

Step 1. Analyze the available well testing data using the Gringarten or Cinco–Samaniego method, as presented in Chapter 1, to calculate the dimensionless fracture conductivity F_{CD} and the fracture half-length x_f.

Step 2. Assemble the available well data in terms of bottom-hole pressure and the flow rate q_t (in STB/day for oil or Mscf/day for gas) as a function of time. Calculate the material balance pseudotime t_a for *each given data point* by using:

For oil: $$t_a = \frac{N_p}{q_t}$$

For gas: $$t_a = \frac{(\mu_g c_g)_i}{q_t}\frac{Z_i G}{2p_i}[\overline{m}(p_i) - \overline{m}(\overline{p})]$$

where $\overline{m}(p_i)$ and $\overline{m}(p)$ are the normalized pseudopressures as defined by Eqs. (3.218) and (3.219):

$$\overline{m}(p_i) = \frac{\mu_{gi} Z_i}{p_i}\int_0^{p_i}\left[\frac{p}{\mu_g Z}\right]dp$$

$$\overline{m}(p) = \frac{\mu_{gi} Z_i}{p_i}\int_0^{p}\left[\frac{p}{\mu_g Z}\right]dp$$

Note that the initial gas-in-place G must be calculated iteratively, as illustrated previously by Palacio and Blasingame.

Step 3. Using the well production data as tabulated and plotted in step 2, compute the following three complementary plotting functions:

(1) pressure drop normalized rate q_{Dd};
(2) pressure drop normalized rate integral function $(q_{Dd})_i$;

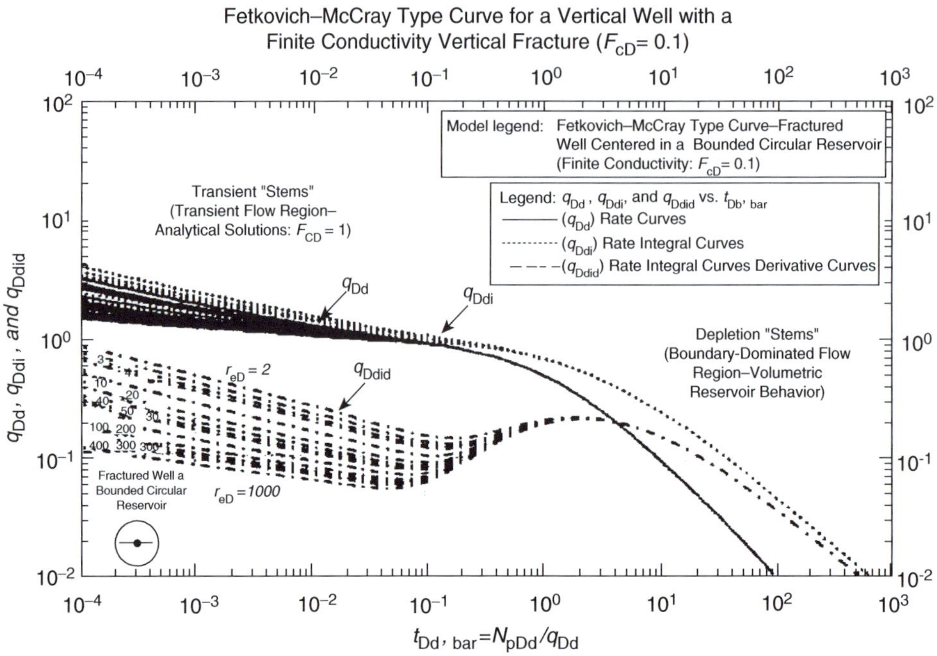

FIGURE 3.67 Fetkovich–McCray decline type curve-rate vs. material balance time format for a well with a finite conductivity vertical fracture (F_{cD} = 0. 1). *(Permission to copy by the SPE, 2003).*

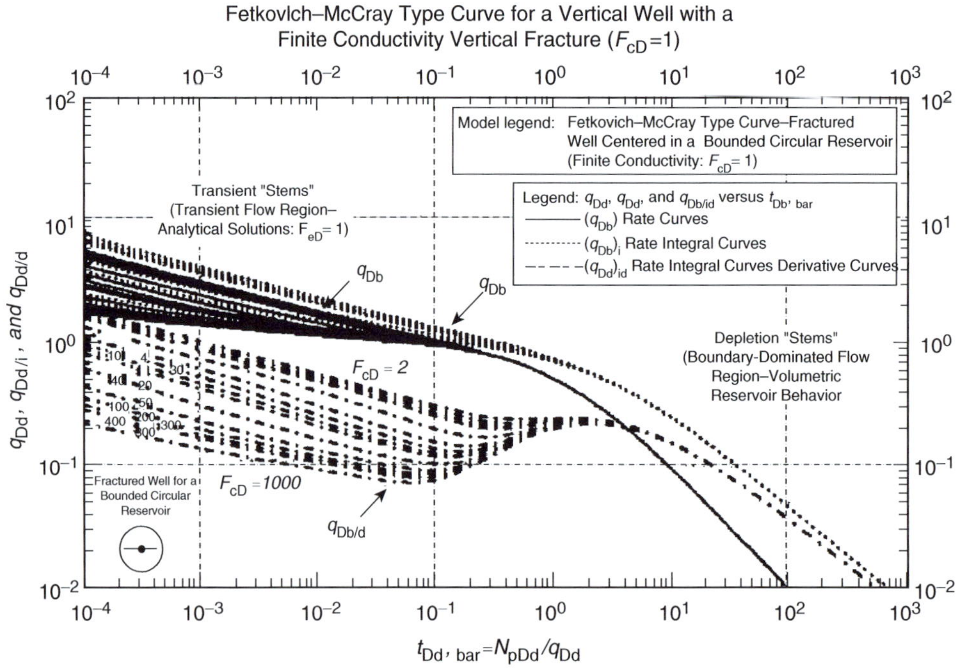

FIGURE 3.68 Fetkovich–McCray decline type curve-rate vs. material balance time format for a well with a finite conductivity vertical fracture (F_{cD} = 1). *(Permission to copy by the SPE, 2003).*

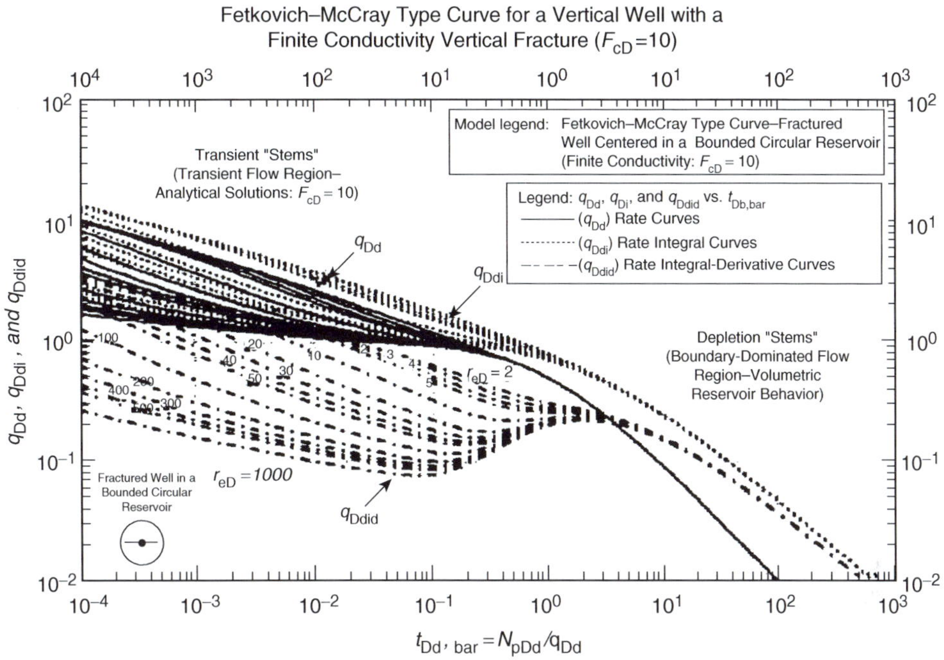

FIGURE 3.69 Fetkovich–McCray decline type curve-rate vs. material balance time format for a well with a finite conductivity vertical fracture ($F_{cD} = 10$). *(Permission to copy by the SPE, 2003).*

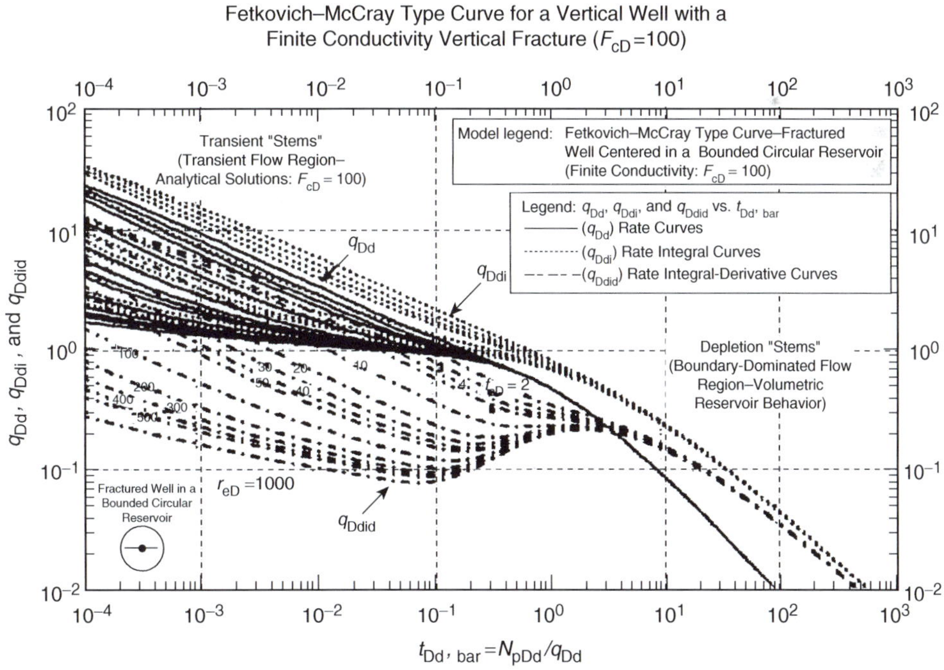

FIGURE 3.70 Fetkovich–McCray decline type curve-rate vs. material balance time format for a well with a finite conductivity vertical fracture ($F_{cD} = 100$). *(Permission to copy by the SPE, 2003).*

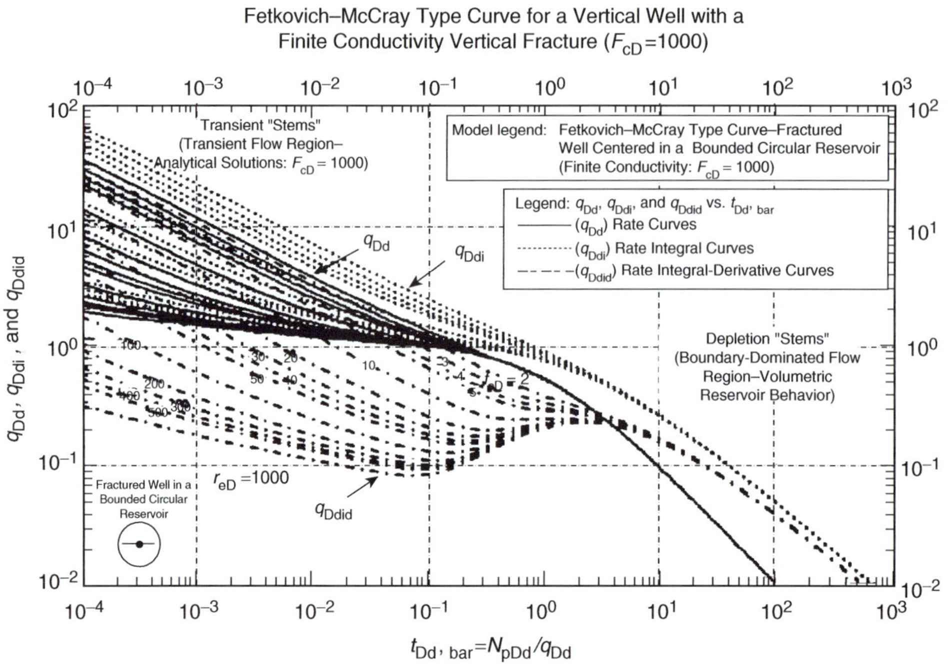

FIGURE 3.71 Fetkovich–McCray decline type curve-rate vs. material balance time format for a well with a finite conductivity vertical fracture (F_{cD} = 1000). *(Permission to copy by the SPE, 2003).*

(3) pressure drop normalized rate integral–derivative function $(q_{Dd})_{id}$.

For gas:

$$q_{Dd} = \frac{q_g}{\overline{m}(p_i) - \overline{m}(p_{wf})}$$

$$(q_{Dd})_i = \frac{1}{t_a}\int_0^{t_a}\left(\frac{q_g}{\overline{m}(p_i) - \overline{m}(p_{wf})}\right)\mathrm{d}t_a$$

$$(q_{Dd})_{id} = \left(\frac{-1}{t_a}\right)\frac{\mathrm{d}}{\mathrm{d}t_a}\left[\frac{1}{t_a}\int_0^{t_a}\left(\frac{q_g}{\overline{m}(p_i) - \overline{m}(p_{wf})}\right)\mathrm{d}t_a\right]$$

For oil:

$$q_{Dd} = \frac{q_o}{p_i - p_{wf}}$$

$$(q_{Dd})_i = \frac{1}{t_a}\int_0^{t_a}\left(\frac{q_o}{p_i - p_{wf}}\right)\mathrm{d}t_a$$

$$(q_{Dd})_{id} = \left(\frac{-1}{t_a}\right)\frac{\mathrm{d}}{\mathrm{d}t_a}\left[\frac{1}{t_a}\int_0^{t_a}\left(\frac{q_o}{p_i - p_{wf}}\right)\mathrm{d}t_a\right]$$

Step 4. Plot the three gas or oil functions, i.e., q_{Dd}, $(q_{Dd})_i$, and $(q_{Dd})_{id}$, vs. t_a on tracing paper so it can be laid over the type curve with the appropriate value F_{CD}.

Step 5. Establish a match point "MP" for each of the three functions $(q_{Dd}$, $(q_{Dd})_i$, and $(q_{Dd})_{id})$. Once a "match" is obtained, record the "time" and "rate" match points as well as the dimensionless radius r_{eD} value:

(1) Rate-axis "match point": Any $(q/\Delta p)_{MP} - (q_{Dd})_{MP}$ pair.

(2) Time-axis "match point": Any $(\bar{t})_{MP} - (t_{Dd})_{MP}$ pair.

(3) Transient flow stem: Select $(q/\Delta p)$, $(q/\Delta p)_i$, and $(q/\Delta p)_{id}$ functions that best match the transient data stem and record r_{eD}.

Step 6. Solve for b_{Dpss} by using the values of F_{CD} and r_{eD}:

$$u = \ln(F_{CD})$$

$$b_{Dpss} = \ln(r_{eD}) - 0.049298 + \frac{0.43464}{r_{eD}^2} + \frac{a_1 + a_2u + a_3u^2 + a_4u^3 + a_5u^4}{1 + b_1u + b_2u^2 + b_3u^3 + b_4u^4}$$

Step 7. Using the results of the match point, estimate the following reservoir properties:

For gas:

$$G = \frac{1}{c_{ti}}\left[\frac{t_a}{t_{Dd}}\right]_{MP}\left[\frac{(q_g/\Delta m(\overline{p}))}{q_{Dd}}\right]_{MP}$$

$$k_g = \frac{141.2B_{gi}\mu_{gi}}{h}\left[\frac{(q_g/\Delta m(\overline{p})_{MP})}{(q_{Dd})_{MP}}\right]b_{Dpss}$$

$$A = \frac{5.615GB_{gi}}{h\phi(1-S_{wi})}$$

$$r_e = \sqrt{\frac{A}{\pi}}$$

For oil:

$$N = \frac{1}{c_t}\left[\frac{t_a}{t_{Dd}}\right]_{MP}\left[\frac{(q_o/\Delta p)_i}{q_{Dd}}\right]_{MP}$$

$$k_o = \frac{141.2B_{oi}\mu_{goi}}{h}\left[\frac{(q_o/\Delta p)_{MP}}{(q_{Dd})_{MP}}\right]b_{Dpss}$$

$$A = \frac{5.615NB_{oi}}{h\phi(1-S_{wi})}$$

$$r_e = \sqrt{\frac{A}{\pi}}$$

where

G = gas-in-place, Mscf
N = oil-in-place, STB
B_{gi} = gas formation volume factor at p_i, bbl/Mscf
A = drainage area, ft^2
r_e = drainage radius, ft
S_{wi} = connate water saturation

Step 8. Calculate the fracture half-length x_f and compare with step 1:

$$x_f = \frac{r_e}{r_{eD}}$$

Example 3.25

The Texas Field vertical gas well has been hydraulically fractured and is undergoing depletion. A summary of the reservoir and fluid properties is given below:

$r_w = 0.333$ ft,	$h = 170$ ft
$\phi = 0.088$,	$T = 300$ °F
$\gamma_g = 0.70$,	$B_{gi} = 0.5498$ bblM/scf
$\mu_{gi} = 0.0361$ cp,	$c_{ti} = 5.1032 \times 10^{-5}$ psi^{-1}
$p_i = 9330$ psia,	$p_{wf} = 710$ psia
$S_{wi} = 0.131$,	$F_{CD} = 5.0$

Figure 3.72 shows the type curve match for $F_{CD} = 5$, with the matching points as given below:

$(q_{Dd})_{MP} = 1.0$
$[(q_g/\Delta m(\overline{p}))]_{MP} = 0.89$ Mscf/psi
$(t_{Dd})_{MP} = 1.0$
$(t_a)_{MP} = 58$ days
$(r_{eD})_{MP} = 2.0$

Perform type curve analysis on this gas well.

Solution:

Step 1. Solve for b_{Dpss} by using the values of F_{CD} and r_{eD}:

$$u = \ln(F_{CD}) = \ln(5) = 1.60944$$

$$b_{Dpss} = \ln(r_{eD}) - 0.049298 + \frac{0.43464}{r_{eD}^2} + \frac{a_1 + a_2u + a_3u^2 + a_4u^3 + a_5u^4}{1 + b_1u + b_2u^2 + b_3u^3 + b_4u^4}$$

$$= \ln(2) - 0.049298 + \frac{0.43464}{2^2} + \frac{a_1 + a_2u + a_3u^2 + a_4u^3 + a_5u^4}{1 + b_1u + b_2u^2 + b_3u^3 + b_4u^4} = 1.00222$$

Step 2. Using the results of the match point, estimate the following reservoir properties:

$$G = \frac{1}{c_{ti}}\left[\frac{t_a}{t_{Dd}}\right]_{MP}\left[\frac{(q_g/\Delta m(\overline{p}))}{q_{Dd}}\right]_{MP} = \frac{1}{5.1032\times10^{-5}}\left[\frac{58}{1.0}\right]_{MP}\left[\frac{0.89}{1.0}\right] = 1.012\times10^6 \text{ MMscf}$$

$$k_g = \frac{141.2B_{gi}\mu_{gi}}{h}\left[\frac{(q_g/\Delta m(\overline{p})_{MP})}{(q_{Dd})_{MP})}\right]b_{Dpss} = \frac{141.2(0.5498)(0.0361)}{170}\left[\frac{0.89}{1.0}\right]1.00222 = 0.015 \text{ md}$$

$$A = \frac{5.615GB_{gi}}{h\phi(1-S_{wi})} = \frac{5.615(1.012\times10^6)(0.5498)}{(170)(0.088)(1-0.131)} = 240{,}195 \text{ ft}^2 = 5.51 \text{ acres}$$

$$r_e = \sqrt{\frac{A}{\pi}} = \sqrt{\frac{240{,}195}{\pi}} = 277 \text{ ft}$$

Step 3. Calculate the fracture half-length x_f and compare with step 1:

$$x_f = \frac{r_e}{r_{eD}} = \frac{277}{2} = 138 \text{ ft}$$

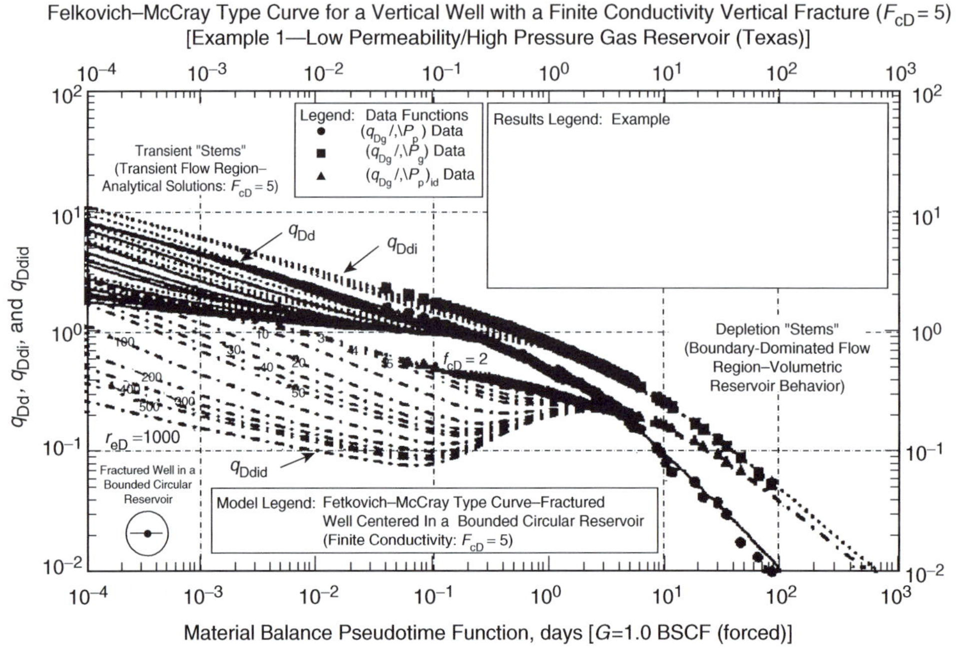

FIGURE 3.72 Match of production data for Example 1 on the Fetkovich–McCray decline type curve (pseudopressure drop normalized rate vs. material balance time format) for a well with a finite conductivity vertical fracture (F_{cD} = 5). *(Permission to copy by the SPE, 2003).*

3.6 GAS HYDRATES

Gas hydrates are solid crystalline compounds formed by the physical combination of gas and water under pressure and temperatures considerably above the freezing point of water. In the presence of free water, hydrate will form when the temperature is below a certain degree; this temperature is called "hydrate temperature T_h." Gas hydrate crystals resemble ice or wet snow in appearance but do not have the solid structure of ice. The main framework of the hydrate crystal is formed with water molecules. The gas molecules occupy void spaces (cages) in the water crystal lattice; however, enough cages must be filled with hydrocarbon molecules to stabilize the crystal lattice. When the hydrate "snow" is tossed on the ground, it causes a distinct cracking sound resulting from the escaping of gas molecules as they rupture the crystal lattice of the hydrate molecules.

Two types of hydrate crystal lattices are known, with each containing void spaces of two different sizes:

(1) Structure I of the lattice has voids of the size to accept small molecules such as methane and ethane. These "guest" gas molecules are called "hydrate formers." In general, light components such as C_1, C_2, and CO_2 form structure I hydrates.
(2) Structure II of the lattice has larger voids (i.e., "cages or cavities") that allow the entrapment of the heavier alkanes with medium-sized molecules, such as C_3, i-C_4, and n-C_4, in addition to methane and ethane, to form structure II hydrates. Several studies have shown that a stable hydrate structure is hydrate structure II. However, the gases are very lean; structure I is expected to be the hydrate stable structure.

All components heavier than C_4, i.e., C_{5+}, do not contribute to the formation of hydrates and are therefore identified as "nonhydrate components."

Gas hydrates generate considerable operational and safety concerns in subsea pipelines and process equipment. The current practice in the petroleum industry for avoiding gas hydrate is to operate outside the hydrate stability zone. During the flow of natural gas, it becomes necessary to define, and thereby avoid, conditions that promote the formation of hydrates. This is essential since hydrates can cause numerous problems such as:

- choking the flow string, surface lines, and other equipment;
- completely blocking flow lines and surface equipment;
- hydrate formation in the flow string resulting in a lower value of measured wellhead pressures.

Sloan (2000) listed several conditions that tend to promote the formation of gas hydrates. These are:

- the presence of free water and gas molecules that range in size from methane to butane;
- the presence of H_2S or CO_2 as a substantial factor contributing to the formation of hydrate since these acid gases are more soluble in water than hydrocarbons;
- temperatures below the "hydrate formation temperature" for the pressure and gas composition considered;
- high operating pressures that increase the "hydrate formation temperature";
- high velocity or agitation through piping or equipment;
- the presence of small "seed" crystal of hydrate;
- natural gas at or below its water dewpoint with liquid water present.

The above conditions necessary for hydrate formation lead to the following four classic, thermodynamic prevention methods:

(1) Water removal provides the best protection.
(2) Maintaining a high temperature throughout the flow system, i.e., insulation, pipe bundling, or electrical heating.
(3) Hydrate prevention is achieved most frequently by injecting an inhibitor, such as methanol or monoethylene glycol, which acts as antifreezes.
(4) Kinetic inhibitors are low-molecular-weight polymers dissolved in a carrier solvent and injected into the water phase in the pipeline. These inhibitors bond to the hydrate surface and prevent significant crystal growth for a period longer than the free water residence time in a pipeline.

3.6.1 Phase Diagrams for Hydrates

The temperature and pressure conditions for hydrate formation in surface gas processing facilities are generally much lower than those considered in production and reservoir engineering. The conditions of initial hydrate formation are often given by simple $p-T$ phase diagrams for water–hydrocarbon systems. A schematic illustration of the phase diagram for a typical mixture of water and light hydrocarbon is shown in Figure 3.73. This graphical illustration of the diagram shows a lower quadruple point "Q_1" and upper quadruple point "Q_2." The quadruple point defines the condition at which four phases are in equilibrium.

Each quadruple point is at the intersection of four three-phase lines. The lower quadruple

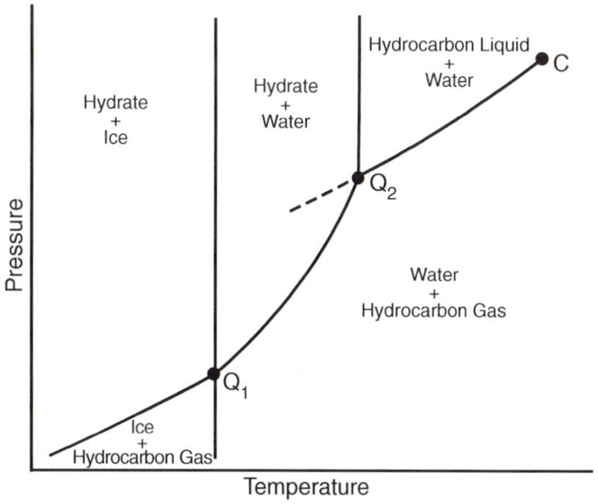

FIGURE 3.73 Phase diagram for a typical mixture of water and light hydrocarbon.

point Q_1 represents the point at which ice, hydrate, water, and hydrocarbon gas exist in equilibrium. At temperatures below the temperature that corresponds to point Q_1, hydrates form from vapor and ice. The upper quadruple point Q_2 represents the point at which water, liquid hydrocarbon, hydrocarbon gas, and hydrate exist in equilibrium, and marks the upper temperature limit for hydrate formation for that particular gas–water system. Some of the lighter natural gas components, such as methane and nitrogen, do not have an upper quadruple point, so no upper temperature limit exists for hydrate formation. This is the reason that hydrates can still form at high temperatures (up to 120 °F) in the surface facilities of high-pressure wells.

The line $Q_1 - Q_2$ separates the area in which water and gas combine to form hydrates. The vertical line extending from point Q_2 separates the area of water and hydrocarbon liquid from the area of hydrate and water.

It is convenient to divide hydrate formation into the following two categories:

Category I: Hydrate formation due to a decrease in temperature with no sudden pressure drop, such as in the flow string or surface line.

Category II: Hydrate formation where sudden expansion occurs, such as in orifices, back-pressure regulators, or chokes.

Figure 3.74 presents a graphical method for approximating hydrate formation conditions and for estimating the permissible expansion condition of natural gases without the formation of hydrates. This figure shows the hydrate-forming conditions as described by a family of "hydrate formation lines" representing natural gases with various specific gravities. Hydrates will form whenever the coordinates of the point representing the pressure and temperature are located to the *left of the hydrate formation line for the gas in question.* This graphical correlation can be used to approximate the hydrate-forming temperature as the temperature decreases along flow string and flow lines, i.e., category I.

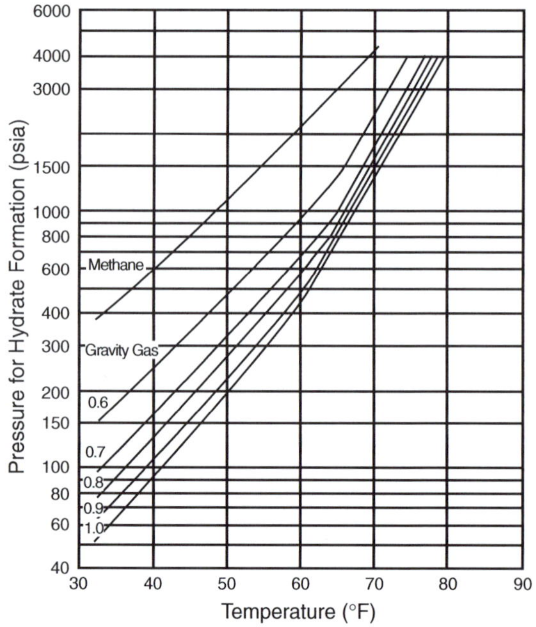

FIGURE 3.74 Pressure–temperature curves for predicting hydrate. *(Courtesy Gas Processors Suppliers Association).*

Example 3.26

A gas of 0.8 specific gravity is at a pressure of 1000 psia. To what extent can the temperature be lowered without hydrate formation in the presence of free water?

Solution

From Figure 3.74, at a specific gravity of 0.8 and a pressure of 1000 psia, hydrate temperature is 66 °F. Thus, hydrates may form at or below 66 °F.

Example 3.27

A gas has a specific gravity of 0.7 and exists at 60 °F. What would be the pressure above which hydrates could be expected to form?

Solution

From Figure 3.74, hydrate will form above 680 psia.

It should be pointed out that the graphical correlation presented in Figure 3.74 was developed for pure water–gas systems; however, the presence of dissolved solids in the water will reduce the temperatures at which natural gases will form hydrates.

When a water–wet gas expands rapidly through a valve, orifice, or other restrictions, hydrates may form because of rapid gas cooling caused by Joule–Thomson expansion. That is:

$$\frac{\partial T}{\partial p} = \frac{RT^2}{pC_P}\left(\frac{\partial Z}{\partial T}\right)_P$$

where

T = temperature
p = pressure
Z = gas compressibility factor
C_P = specific heat at constant pressure

This reduction in temperature due to the sudden reduction in pressure, i.e., $\partial T/\partial p$, could cause the condensation of water vapor from the gas and bring the mixture to the conditions necessary for hydrate formation. Figures 3.75–3.79 can be

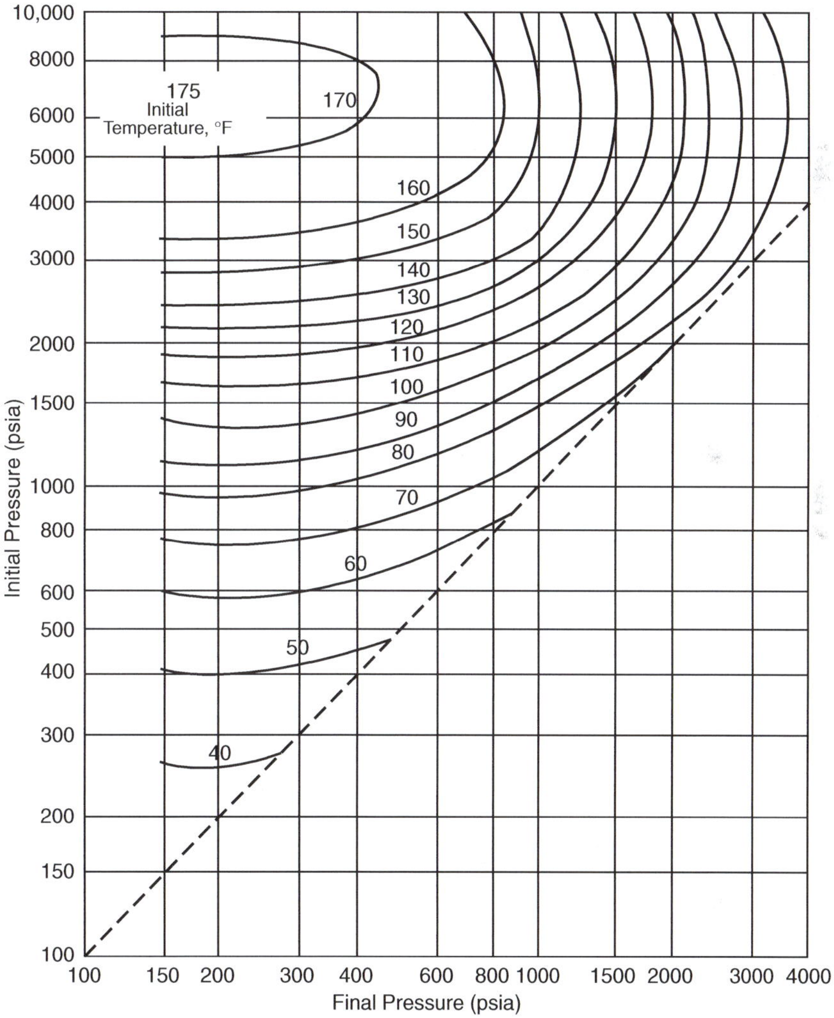

FIGURE 3.75 Permissible expansion of a 0.6 gravity natural gas without hydrate formation. *(Courtesy Gas Processors Suppliers Association).*

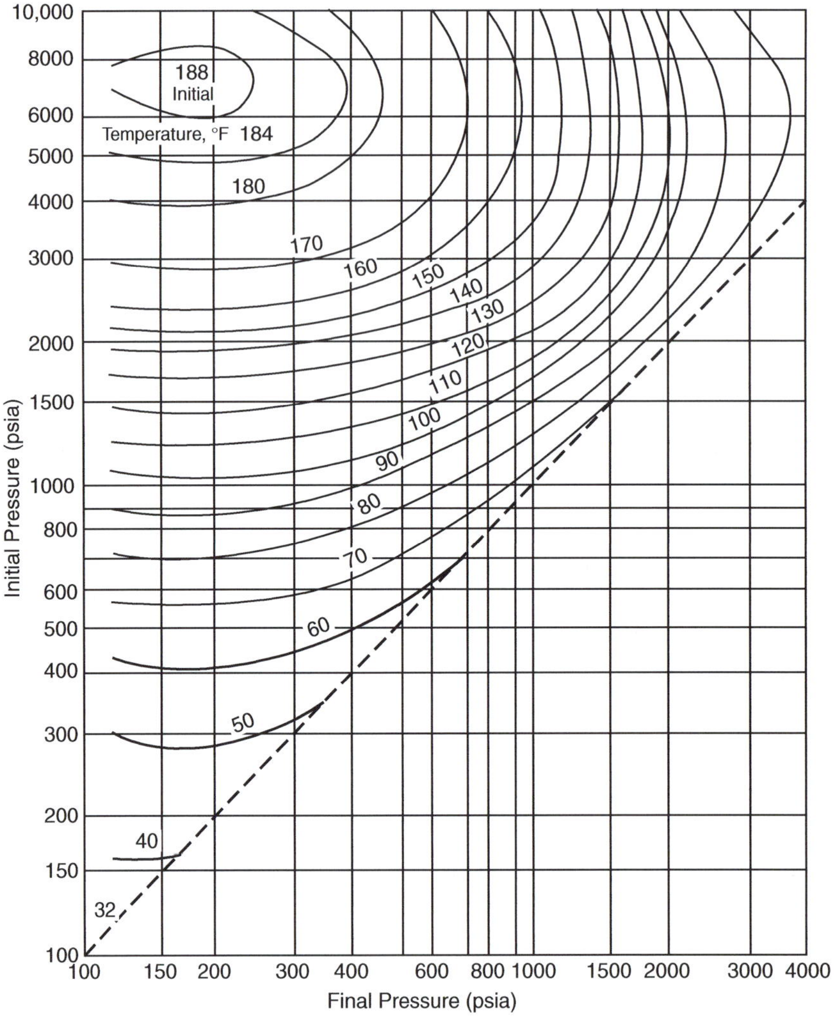

FIGURE 3.76 Permissible expansion of a 0.7 gravity natural gas without hydrate formation. *(Courtesy Gas Processors Suppliers Association).*

used to estimate the maximum reduction in pressure without causing the formation of hydrates.

The chart is entered at the intersection of the initial pressure and initial temperature isotherm; and the lowest pressure to which the gas can be expanded without forming hydrate is read directly from the x-axis below the intersection.

Example 3.28

How far can a gas of 0.7 specific gravity at 1500 psia and 120 °F be expanded without hydrate formation?

Solution:

From Figure 3.76, select the graph on the y-axis with the initial pressure of 1500 psia and move horizontally to the right to intersect with

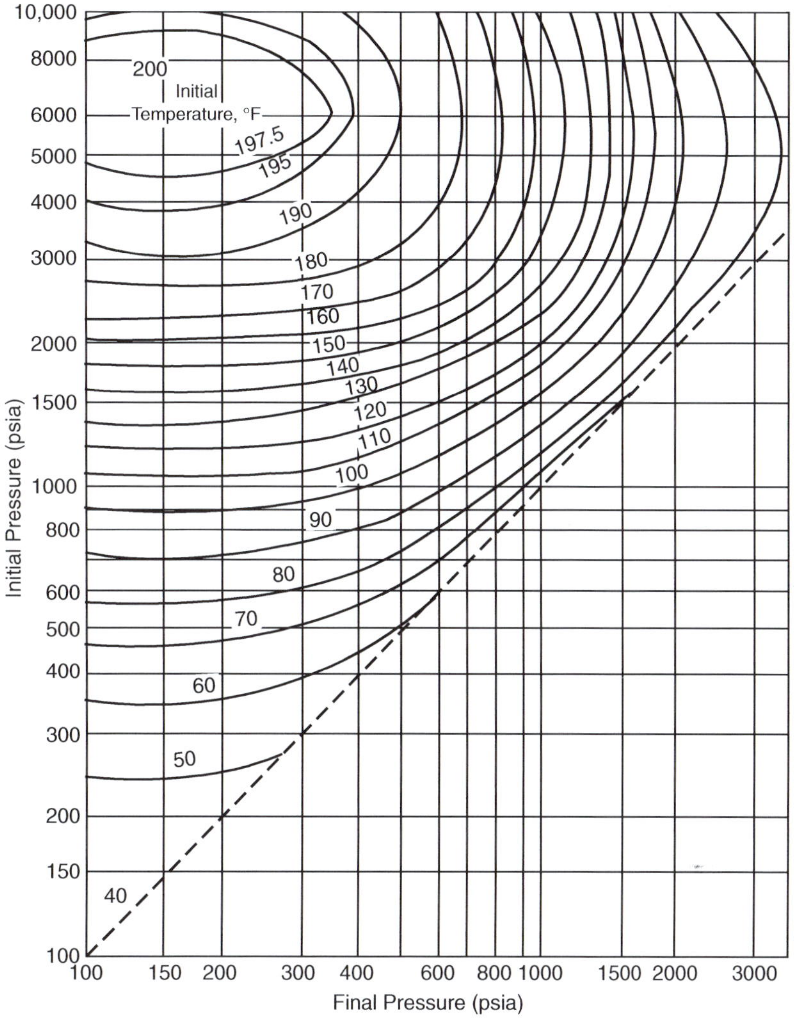

FIGURE 3.77 Permissible expansion of a 0.8 gravity natural gas without hydrate formation. *(Courtesy Gas Processors Suppliers Association).*

the 120 °F temperature isotherm. Read the "final" pressure on the *x*-axis, to give 300 psia. Hence, this gas may be expanded to a final pressure of 300 psia without the possibility of hydrate formation.

Ostergaard et al. (2000) proposed a new correlation to predict the hydrate-free zone of reservoir fluids that range in composition from black oil to lean natural gas systems. The authors separated the components of the hydrocarbon system into the following two groups:

(1) hydrate-forming hydrocarbons "h" that include methane, ethane, propane, and butanes;

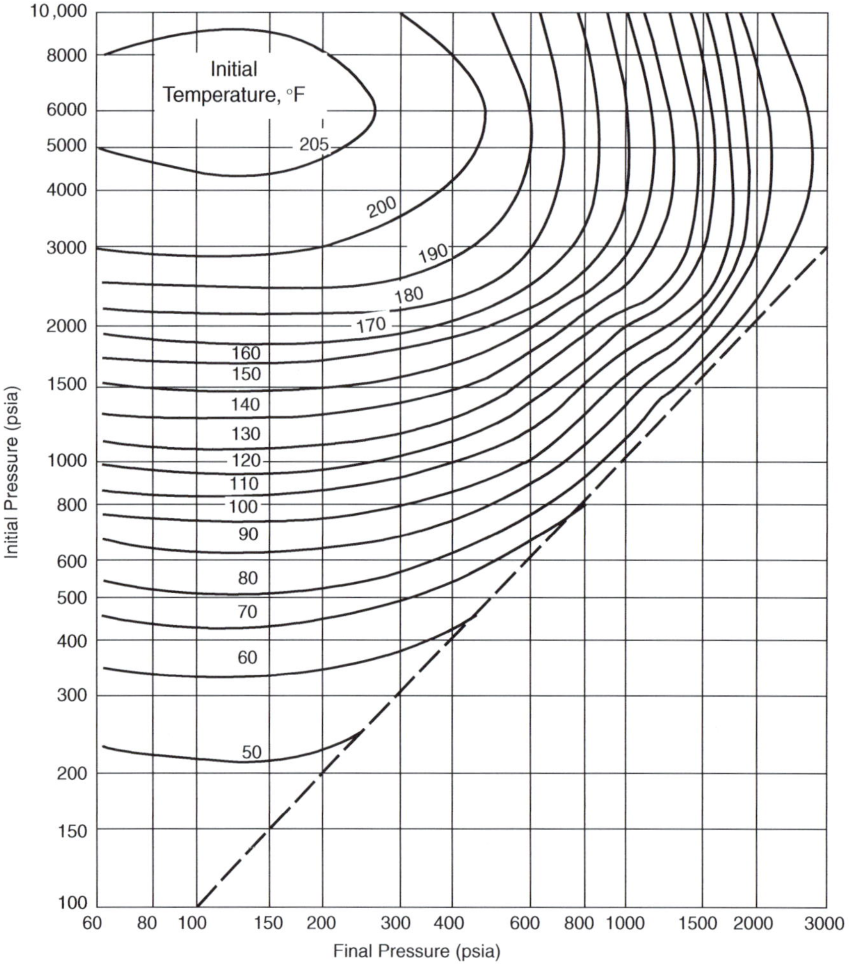

FIGURE 3.78 Permissible expansion of a 0.9 gravity natural gas without hydrate formation. *(Courtesy Gas Processors Suppliers Association).*

(2) nonhydrate-forming hydrocarbons "nh" that include pentanes and heavier components.

Define the following correlating parameters:

$$f_h = y_{C_1} + y_{C_2} + y_{C_3} + y_{i-C_4} + y_{n-C_4} \quad (3.245)$$

$$f_{nh} = y_{C_{5+}} \quad (3.246)$$

$$F_m = \frac{f_{nh}}{f_h} \quad (3.247)$$

$$\gamma_h = \frac{M_h}{28.96} = \frac{\sum_{i-C_1}^{n-C_4} y_i M_i}{28.96 \sum_{i-C_1}^{n-C_4} y_i} \quad (3.248)$$

where

h = hydrate-forming components C_1 through C_4
M_h = hydrate molecular weight

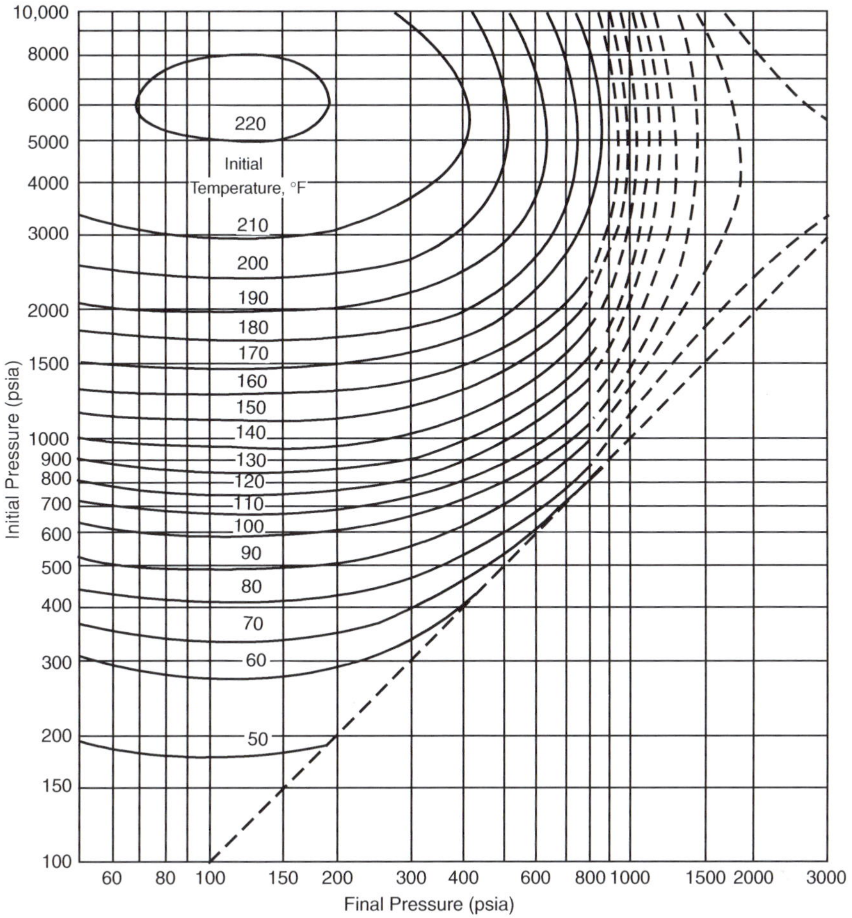

FIGURE 3.79 Permissible expansion of a 1.0 gravity natural gas without hydrate formation. *(Courtesy Gas Processors Suppliers Association).*

M_i = molecular weight of the hydrate-forming components, C_1 through C_4

nh = nonhydrate-forming components, C_5 and heavier

F_m = molar ratio between the nonhydrate-forming and hydrate-forming components

γ_h = specific gravity of hydrate-forming components

The authors correlated the hydrate dissociation pressure p_h of fluids containing only hydrocarbons as a function of the above-defined parameters by the following expression:

$$p_h = 0.1450377\exp\left\{\left[\frac{a_1}{(\gamma_h + a_2)^3} + a_3F_m + a_4F_m^2 + a_5\right]T + \frac{a_6}{(\gamma_h + a_7)^3} + a_8F_m + a_9F_m^2 + a_{10}\right\} \quad (3.249)$$

where

p_h = hydrate dissociation pressure, psi

T = temperature, °R

a_i = constants as given below

a_i	Value
a_1	2.5074400×10^{-3}
a_2	0.4685200
a_3	1.2146440×10^{-2}
a_4	$-4.6761110 \times 10^{-4}$
a_5	0.0720122
a_6	3.6625000×10^{-4}
a_7	−0.4850540
a_8	−5.4437600
a_9	3.8900000×10^{-3}
a_{10}	−29.9351000

Eq. (3.249) was developed using data on black oil, volatile oil, gas condensate, and natural gas systems in the range of 32–68 °F, which covers the practical range of hydrate formation for reservoir fluids transportation.

Eq. (3.249) can also be arranged and solved for the temperature, to give:

$$T = \frac{\ln(6.89476 p_h) - (a_6/(\gamma_h + a_7)^3) + a_8 F_m + a_9 F_m^2 + a_{10}}{[(a_1/(\gamma_h + a_2)^3) + a_3 F_m + a_4 F_m^2 + a_5]}$$

The authors pointed out that N_2 and CO_2 do not obey the general trend given for hydrocarbons in Eq. (3.249). Therefore, to account for the pressure of N_2 and CO_2 in the hydrocarbon system, they treated each of these two nonhydrocarbon fractions separately and developed the following correction factors:

$$E_{CO_2} = 1.0 + \left[(b_1 F_m + b_2)\frac{y_{CO_2}}{1 - y_{N_2}}\right] \quad (3.250)$$

$$E_{N_2} = 1.0 + \left[(b_3 F_m + b_4)\frac{y_{N_2}}{1 - \gamma_{CO_2}}\right] \quad (3.251)$$

with

$$b_1 = -2.0943 \times 10^{-4}\left(\frac{T}{1.8} - 273.15\right)^3 + 3.809 \times 10^{-3} \times \left(\frac{T}{1.8} - 273.15\right)^2 - 2.42 \times 10^{-2}\left(\frac{T}{1.8} - 273.15\right) + 0.423 \quad (3.252)$$

$$b_2 = 2.3498 \times 10^{-4}\left(\frac{T}{1.8} - 273.15\right)^2 - 2.086 \times 10^{-3}\left(\frac{T}{1.8} - 273.15\right)^2 + 1.63 \times 10^{-2}\left(\frac{T}{1.8} - 273.15\right) + 0.650 \quad (3.253)$$

$$b_3 = 1.1374 \times 10^{-4}\left(\frac{T}{1.8} - 273.15\right)^3 + 2.61 \times 10^{-4}\left(\frac{T}{1.8} - 273.15\right)^2 + 1.26 \times 10^{-2}\left(\frac{T}{1.8} - 273.15\right) + 1.123 \quad (3.254)$$

$$b_4 = 4.335 \times 10^{-5}\left(\frac{T}{1.8} - 273.15\right)^3 - 7.7 \times 10^{-5}\left(\frac{T}{1.8} - 273.15\right)^2 + 4.0 \times 10^{-3}\left(\frac{T}{1.8} - 273.15\right) + 1.048 \quad (3.255)$$

where

y_{N_2} = mole fraction of N_2
y_{CO_2} = mole fraction of CO_2
T = temperature, °R
F_m = molar ratio as defined by Eq. (3.247)

The total, i.e., corrected, hydrate dissociation pressure p_{corr} is given by:

$$p_{corr} = p_h E_{N_2} E_{CO_2} \quad (3.256)$$

To demonstrate these correlations, Ostergaard et al. presented the following example:

Example 3.29

A gas condensate system has the following composition:

Component	y_{i_1}(%)	M_i
CO_2	2.38	44.01
N_2	0.58	28.01
C_1	73.95	16.04
C_2	7.51	30.07
C_3	4.08	44.10
i-C_4	0.61	58.12
n-C_4	1.58	58.12
i-C_5	0.50	72.15
n-C_5	0.74	72.15
C_6	0.89	84.00
C_{7+}	7.18	–

Calculate the hydrate dissociation pressure at 45 °F, i.e., 505 °R.

Solution

Step 1. Calculate f_h and f_{nh} from Eqs. (3.245) and (3.246):

$$f_h = y_{C_1} + y_{C_2} + y_{C_3} + y_{i-C_4} + y_{n-C_4}$$
$$= 73.95 + 7.51 + 4.08 + 0.61 + 1.58 = 87.73\%$$
$$f_{nh} = y_{C_{5+}} = y_{i-C_5} + y_{n-C_5} + y_{C_6} + y_{C_{7+}}$$
$$= 0.5 + 0.74 + 0.89 + 7.18 = 9.31\%$$

Step 2. Calculate F_m by applying Eq. (3.247):

$$F_m = \frac{f_{nh}}{f_h} = \frac{9.31}{87.73} = 0.1061$$

Step 3. Determine the specific gravity of the hydrate-forming components by normalizing their mole fractions as shown below:

Component	y_i	**Normalized** y_i^*	M_i	$M_i y_i^*$
C_1	0.7395	0.8429	16.04	13.520
C_2	0.0751	0.0856	30.07	2.574
C_3	0.0408	0.0465	44.10	2.051
i-C_4	0.0061	0.0070	58.12	0.407
n-C_4	0.0158	0.0180	58.12	1.046
	$\Sigma = 0.8773$	$\Sigma = 1.0000$		$\Sigma = 19.5980$

$$\gamma_h = \frac{19.598}{28.96} = 0.6766$$

Step 4. Using the temperature T and the calculated values of F_m and γ_h in Eq. (3.249) gives:

$$p_h = 236.4 \text{ psia}$$

Step 5. Calculate the constants b_1 and b_2 for CO_2 by applying Eqs. (3.252) and (3.253) to give:

$$b_1 = -2.0943 \times 10^{-4}\left(\frac{505}{1.8} - 273.15\right)^3 + 3.809 \times 10^{-3}\left(\frac{505}{1.8} - 273.15\right)^2 - 2.42 \times 10^{-2} \times \left(\frac{505}{1.8} - 273.15\right) + 0.423 = 0.368$$

$$b_2 = 2.3498 \times 10^{-4}\left(\frac{505}{1.8} - 273.15\right)^2 - 2.086 \times 10^{-3}\left(\frac{505}{1.8} - 273.15\right)^2 + 1.63 \times 10^{-2} \times \left(\frac{505}{1.8} - 273.15\right) + 0.650 = 0.752$$

Step 6. Calculate the CO_2 correction factor E_{CO_2} by using Eq. (3.250):

$$E_{CO_2} = 1.0 + \left[(b_1 F_m + b_2)\frac{y_{CO_2}}{1 - y_{N_2}}\right]$$
$$= 1.0 + \left[(0.368 \times 0.1061 + 0.752)\frac{0.0238}{1 - 0.0058}\right]$$
$$= 1.019$$

Step 7. Correct for the presence of N_2, to give:

$$b_3 = 1.1374 \times 10^{-4}\left(\frac{505}{1.8} - 273.15\right)^3 + 2.61 \times 10^{-4}\left(\frac{505}{1.8} - 273.15\right)^2 + 1.26 \times 10^{-2} \times \left(\frac{505}{1.8} - 273.15\right) + 1.123 = 1.277$$

$$b_4 = 4.335 \times 10^{-5}\left(\frac{505}{1.8} - 273.15\right)^3 - 7.7 \times 10^{-5}\left(\frac{505}{1.8} - 273.15\right)^2 + 4.0 \times 10^{-3} \times \left(\frac{505}{1.8} - 273.15\right) + 1.048 = 1.091$$

$$E_{N_2} = 1.0 + \left[(b_3 F_m + b_4)\frac{y_{N_2}}{1 - \gamma_{CO_2}}\right]$$
$$= 1.0 + \left[(1.277 \times 0.1061 + 1.091)\frac{0.0058}{1 - 0.00238}\right]$$
$$= 1.007$$

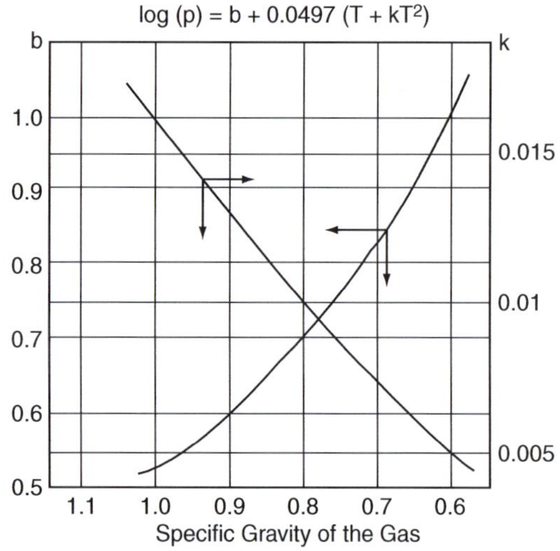

FIGURE 3.80 Coefficients b and k of Eq. (3.258).

Step 8. Estimate the total (corrected) hydrate dissociation pressure by using Eq. (3.256), to give:

$$p_{corr} = p_h E_{N_2} E_{CO_2}$$
$$= (236.4)(1.019)(1.007) = 243 \text{ psia}$$

Makogon (1981) developed an analytical relationship between hydrate and conditions in terms of pressure and temperature as a function of specific gravity of the gas. The expression is given by:

$$\log(p) = b + 0.0497(T + kT^2) \quad \textbf{(3.257)}$$

where

T = temperature, °C
p = pressure, atm

The coefficients b and k are expressed graphically as a function of the specific gravity of the gas, as shown in Figure 3.80.

Example 3.30

Find the pressure at which hydrate forms at $T = 40$ °F for a natural gas with a specific gravity of 0.631, using Eq. (3.257).

Solution

Step 1. Convert the given temperature from °F to °C:

$$T = \frac{40 - 32}{1.8} = 4.4 \text{ °C}$$

Step 2. Determine values of the coefficients b and k from Figure 3.80, to give:

$$b = 0.91 \quad k = 0.006$$

Step 3. Solve for p by applying Eq. (3.257)

$$\log(p) = b + 0.0497(T + kT^2)$$
$$= 0.91 + 0.0497[4.4 + 0.006(4.4)^2]$$
$$= 1.1368$$
$$p = 10^{1.1368} = 13.70 \text{ atm} = 201 \text{ psia}$$

Figure 3.76 gives a value of 224 psia as compared with the above value of 201.

Carson and Katz (1942) adopted the concept of the equilibrium ratios, i.e., K values, for estimating hydrate-forming conditions. They proposed that hydrates are the equivalent of solid solutions and not mixed crystals, and therefore postulated that hydrate-forming conditions could be estimated from empirically determined vapor–solid equilibrium ratios as defined by:

$$K_{i(v-s)} = \frac{y_i}{x_{i(s)}} \quad \textbf{(3.258)}$$

where

$K_{i(v-s)}$ = equilibrium ratio of component i between vapor and solid
y_i = mole fraction of component i in the vapor (gas) phase
$x_{i(s)}$ = mole fraction of component i in the solid phase on a water-free basis

The calculation of the hydrate-forming conditions in terms of pressure or temperature is analogous to the dewpoint calculation of gas mixtures. In general, a gas in the presence of free water phase will form a hydrate when:

$$\sum_{i=1}^{n} \frac{y_i}{K_{i(v-s)}} = 1 \qquad (3.259)$$

Whitson and Brule (2000) pointed out that the vapor–solid equilibrium ratio cannot be used to perform flash calculations and determine hydrate-phase splits or equilibrium-phase compositions, since $K_{i(s)}$ is based on the mole fraction of a "guest" component in the solid-phase hydrate mixture on a water-free basis.

Carson and Katz developed K value charts for the hydrate-forming molecules that include methane through butanes, CO_2, and H_2S, as shown in Figures 3.81–3.87. It should be noted that $K_{i(s)}$ for nonhydrate formers are assumed to be infinity, i.e., $K_{i(s)} = \infty$.

The solution of Eq. (3.259) for the hydrate-forming pressure or temperature is an iterative process. The process involves assuming several values of p or T and calculating the equilibrium ratios at each assumed value until the constraint represented by Eq. (3.259) is met, i.e., summation is equal to 1.

Example 3.31

Using the equilibrium ratio approach, calculate the hydrate formation pressure p_h at 50 °F for the following gas mixture:

Component	y_i
CO_2	0.002
N_2	0.094
C_1	0.784
C_2	0.060
C_3	0.036
$i\text{-}C_4$	0.005
$n\text{-}C_4$	0.019

The experimentally observed hydrate formation pressure is 325 psia at 50 °F.

Solution

Step 1. For simplicity, assume two different pressures, 300 and 350 psia, and calculate the equilibrium ratios at these pressures, to give:

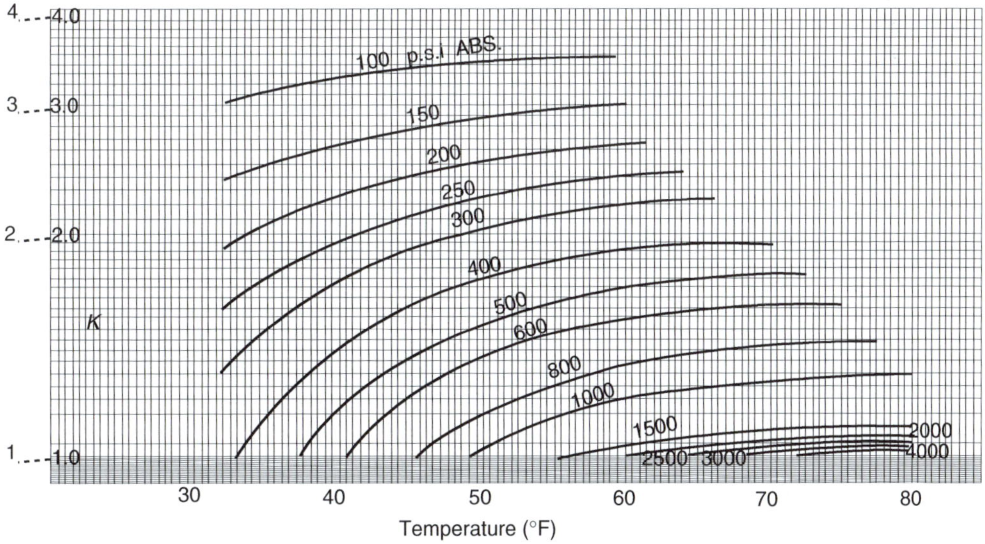

FIGURE 3.81 Vapor–solid equilibrium constant for methane. *(Carson, D., Katz, D., 1942. Natural gas hydrates. Trans. AIME 146, 150–159, courtesy SPE-AIME).*

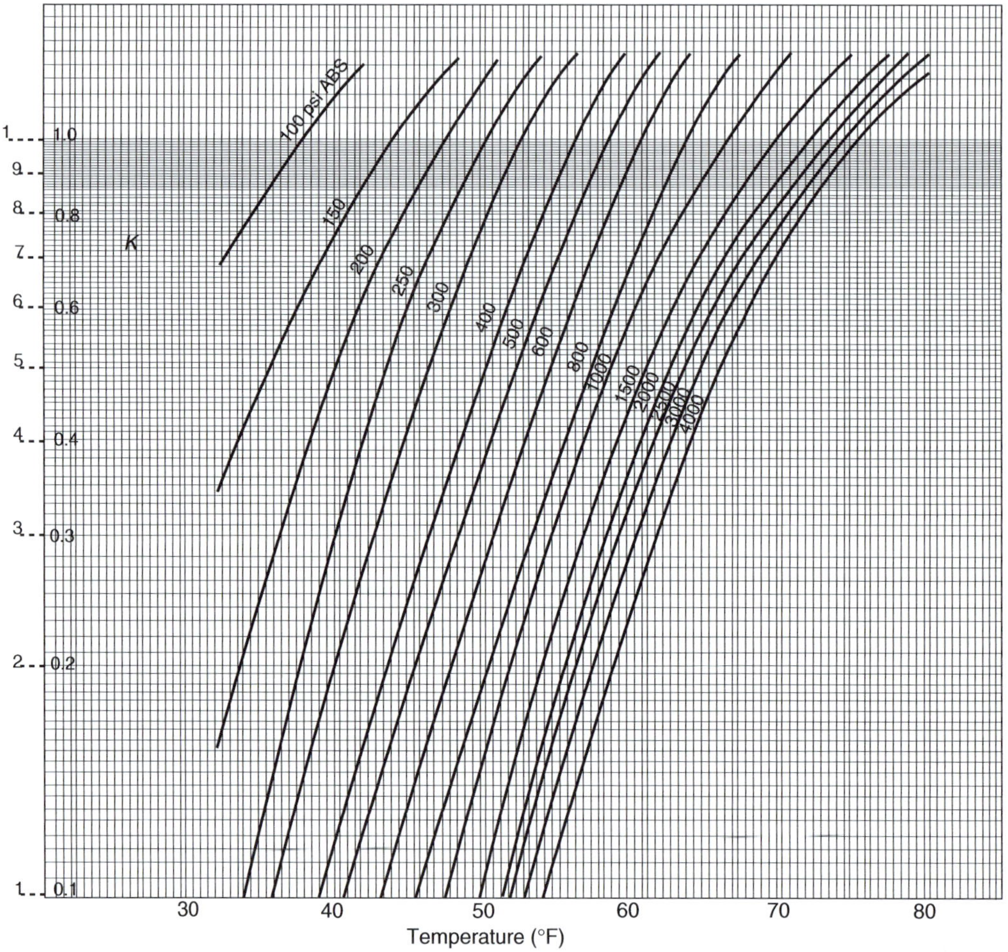

FIGURE 3.82 Vapor–solid equilibrium constant for ethane. *(Carson, D., Katz, D., 1942. Natural gas hydrates. Trans. AIME 146, 150–159, courtesy SPE-AIME).*

Component	y_i	At 300 psia		At 350 psia	
		$K_{i(v-s)}$	$y_i/K_{i(v-s)}$	$K_{i(v-s)}$	$y_i/K_{i(v-s)}$
CO_2	0.002	3.0	0.0007	2.300	0.0008
N_2	0.094	∞	0	∞	0
C_1	0.784	2.04	0.3841	1.900	0.4126
C_2	0.060	0.79	0.0759	0.630	0.0952
C_3	0.036	0.113	0.3185	0.086	0.4186
i-C_4	0.005	0.0725	0.0689	0.058	0.0862
n-C_4	0.019	0.21	0.0900	0.210	0.0900
Σ	1.000		0.9381		1.1034

Step 2. Interpolating linearly at $\Sigma y_i/K_{i(v-s)} = 1$ gives:

$$\frac{350 - 300}{1.1035 - 0.9381} = \frac{p_h - 300}{1.0 - 0.9381}$$

Hydrate-forming pressure $p_h = 319$ psia, which compares favorably with the observed value of 325 psia.

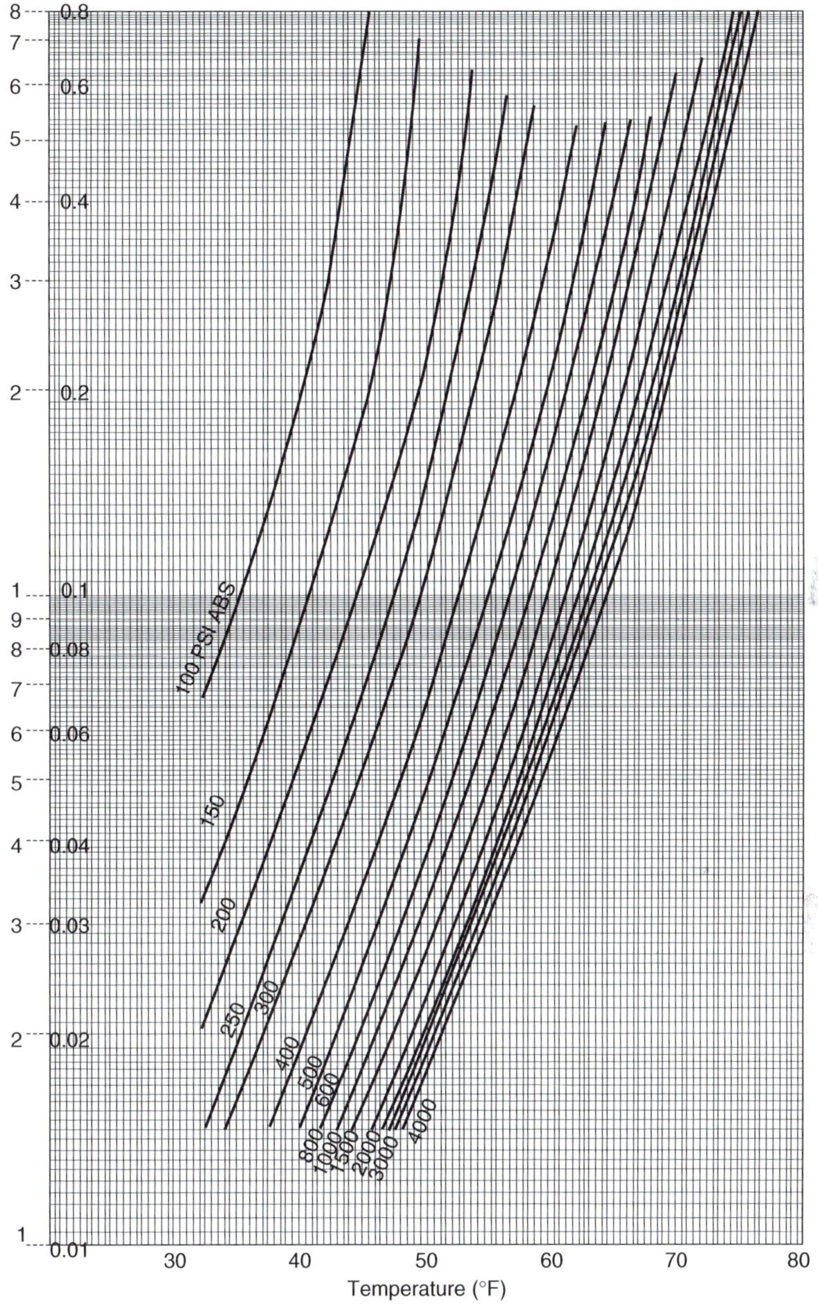

FIGURE 3.83 Vapor–solid equilibrium constant for propane. *(Carson, D., Katz, D., 1942. Natural gas hydrates. Trans. AIME 146, 150–159, courtesy SPE-AIME).*

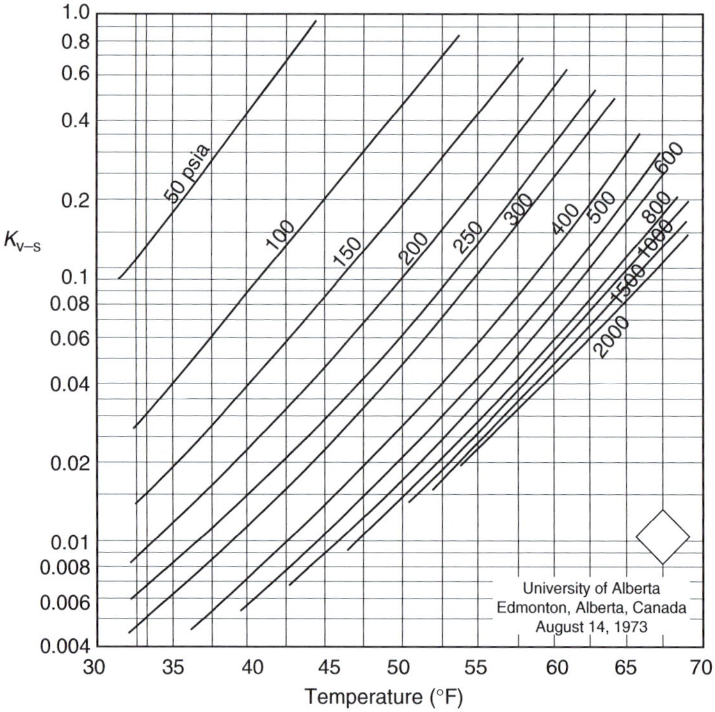

FIGURE 3.84 Vapor–solid equilibrium constant for *i*-butane. *(Carson, D., Katz, D., 1942. Natural gas hydrates. Trans. AIME 146, 150–159, courtesy SPE-AIME).*

Example 3.32

Calculate the temperature for hydrate formation at 435 psi for a gas with a 0.728 specific gravity with the following composition:

Component	y_i
CO_2	0.04
N_2	0.06
C_1	0.78
C_2	0.06
C_3	0.03
$i\text{-}C_4$	0.01
C_{5+}	0.02

Solution

The iterative procedure for estimating the hydrate-forming temperature is given in the following tabulated form:

Component	y_i	$T = 59\ °F$		$T = 50\ °F$		$T = 54\ °F$	
		K_i (v–s)	y_i/K_i (v–s)	$K_{i(v-s)}$	y_i/K_i (v–s)	$K_{i(v-s)}$	y_i/K_i (v–s)
CO_2	0.04	5.00	0.0008	1.700	0.0200	3.000	0.011
N_2	0.06	∞	0	∞	0	∞	0
C_1	0.78	1.80	0.4330	1.650	0.4730	1.740	0.448
C_2	0.06	1.30	0.0460	0.475	0.1260	0.740	0.081
C_3	0.03	0.27	0.1100	0.066	0.4540	0.120	0.250
$i\text{-}C_4$	0.01	0.08	0.1250	0.026	0.3840	0.047	0.213
C_{5+}	0.02	∞	0	∞	0	∞	0
Total	1.00				1.457		1.003

The temperature at which hydrate will form is approximately 54 °F.

Sloan (1984) curve-fitted the Katz–Carson charts by the following expression:

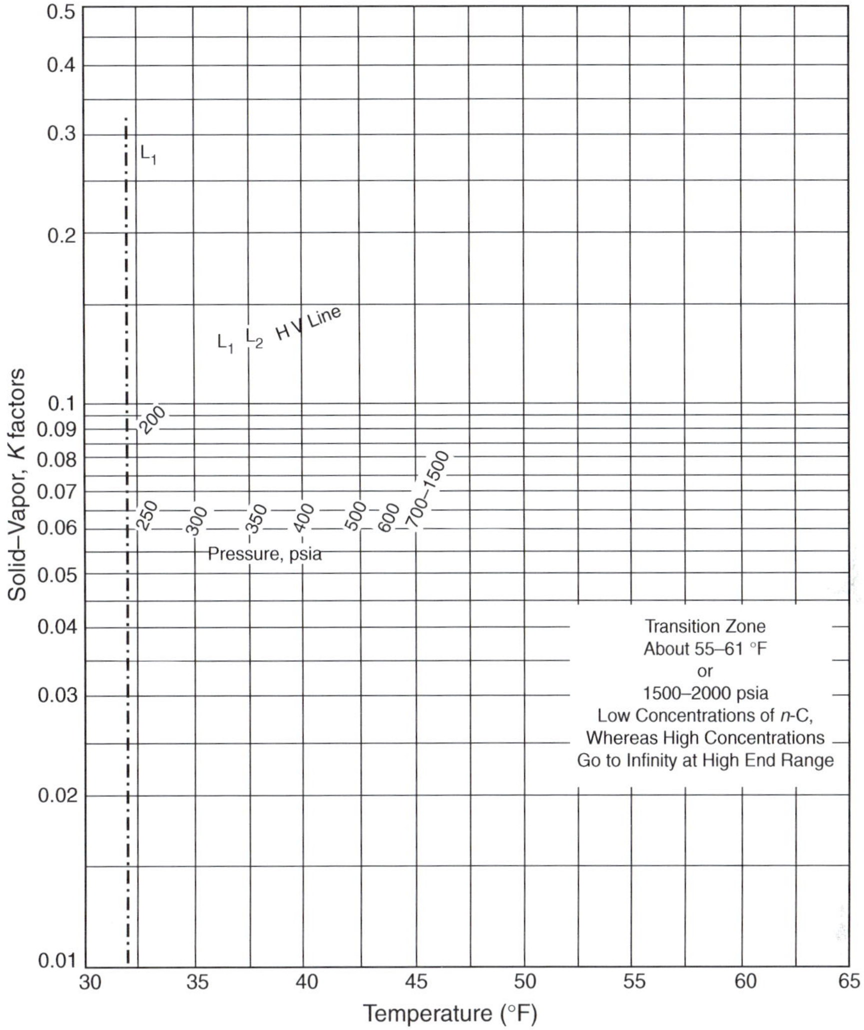

FIGURE 3.85 Vapor–solid equilibrium constant for *n*-butane. *(Carson, D., Katz, D., 1942. Natural gas hydrates. Trans. AIME 146, 150–159, courtesy SPE-AIME).*

$$
\begin{aligned}
\ln(K_{i(v-s)}) = {} & A_0 + A_1 T + A_2 p + \frac{A_3}{T} + \frac{A_4}{p} + A_5 pT \\
& + A_6 T^2 + A_7 p^2 + A_8\left(\frac{p}{T}\right) + A_9 \ln\left(\frac{p}{T}\right) \\
& + \frac{A_{10}}{p^2} + A_{11}\left(\frac{T}{p}\right) + A_{12}\left(\frac{T^2}{p}\right) + A_{13}\left(\frac{p}{T^2}\right) \\
& + A_{14}\left(\frac{T}{p^3}\right) + A_{15} T^3 + A_{16}\left(\frac{p^3}{T^2}\right) + A_{17} T^4
\end{aligned}
$$

where

T = temperature, °F
p = pressure, psia

The coefficients A_0 through A_{17} are given in Table 3.2.

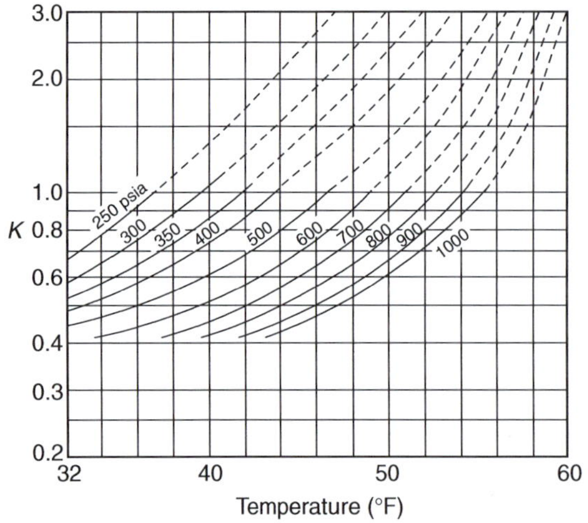

FIGURE 3.86 Vapor–solid equilibrium constant for CO_2. *(Carson, D., Katz, D., 1942. Natural gas hydrates. Trans. AIME 146, 150–159, courtesy SPE-AIME).*

Example 3.33

Resolve Example 3.32 by using Eq. (3.257).

Solution

Step 1. Convert the given pressure from psia to atm:

$$p = \frac{435}{14.7} = 29.6$$

Step 2. Determine the coefficients b and k from Figure 3.82 at the specific gravity of the gas, i.e., 0.728, to give:

$$b = 0.8 \quad k = 0.0077$$

Step 3. Apply Eq. (3.257), to give:

$$\log(p) = b + 0.0497(T + kT^2)$$
$$\log(29.6) = 0.8 + 0.0497(T + 0.0077T^2)$$
$$0.000383T^2 + 0.0497T - 0.6713 = 0$$

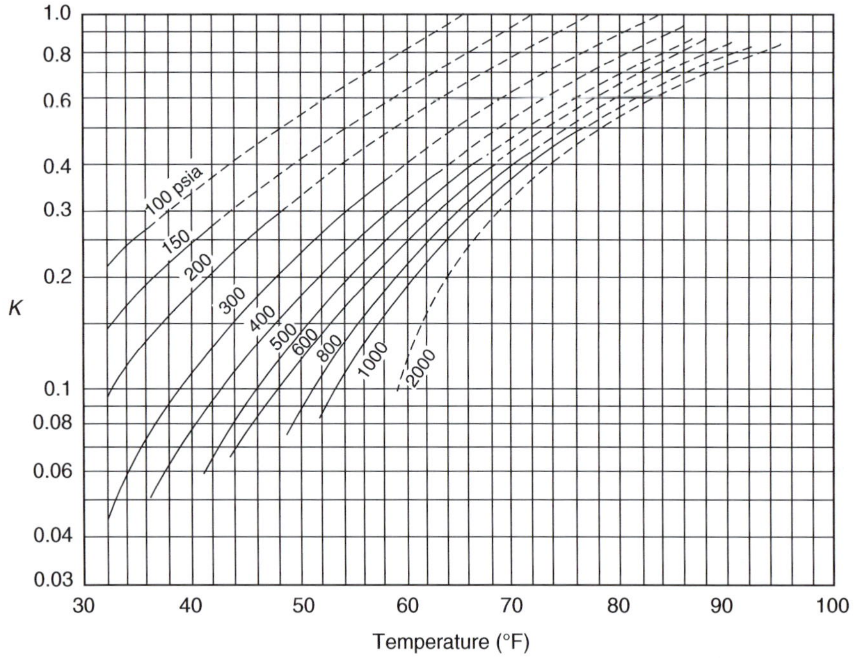

FIGURE 3.87 Vapor–solid equilibrium constant for H_2S. *(Carson, D., Katz, D., 1942. Natural gas hydrates. Trans. AIME 146, 150–159, courtesy SPE-AIME).*

TABLE 3.2 Values of Coefficients A_0 through A_{17} in Sloans Equation

Component	A_0	A_1	A_2	A_3	A_4	A_5
CH_4	1.63636	0.0	0.0	31.6621	−49.3534	5.31×10^{-6}
C_2H_6	6.41934	0.0	0.0	−290.283	2629.10	0.0
C_3H_8	−7.8499	0.0	0.0	47.056	0.0	-1.17×10^{-6}
i-C_4H_{10}	−2.17137	0.0	0.0	0.0	0.0	0.0
n-C_4H_{10}	−37.211	0.86564	0.0	732.20	0.0	0.0
N_2	1.78857	0.0	−0.001356	−6.187	0.0	0.0
CO_2	9.0242	0.0	0.0	−207.033	0.0	4.66×10^{-5}
H_2S	−4.7071	0.06192	0.0	82.627	0.0	-7.39×10^{-6}
	A_6	A_7	A_8	A_9	A_{10}	A_{11}
CH_4	0.0	0.0	0.128525	−0.78338	0.0	0.0
C_2H_6	0.0	9.0×10^{-8}	0.129759	−1.19703	-8.46×10^4	−71.0352
C_3H_8	7.145×10^{-4}	0.0	0.0	0.12348	1.669×10^4	0.0
i-C_4H_{10}	1.251×10^{-3}	1.0×10^{-8}	0.166097	−2.75945	0.0	0.0
n-C_4H_{10}	0.0	9.37×10^{-6}	−1.07657	0.0	0.0	−66.221
N_2	0.0	2.5×10^{-7}	0.0	0.0	0.0	0.0
CO_2	-6.992×10^{-3}	2.89×10^{-6}	-6.223×10^{-3}	0.0	0.0	0.0
H_2S	0.0	0.0	0.240869	−0.64405	0.0	0.0
	A_{12}	A_{13}	A_{14}	A_{15}	A_{16}	A_{17}
CH_4	0.0	−5.3569	0.0	-2.3×10^{-7}	-2.0×10^{-8}	0.0
C_2H_6	0.596404	−4.7437	7.82×10^4	0.0	0.0	0.0
C_3H_8	0.23319	0.0	-4.48×10^4	5.5×10^{-6}	0.0	0.0
i-C_4H_{10}	0.0	0.0	-8.84×10^2	0.0	-5.7×10^{-7}	-1.0×10^{-8}
n-C_4H_{10}	0.0	0.0	9.17×10^5	0.0	4.98×10^{-6}	-1.26×10^{-6}
N_2	0.0	0.0	5.87×10^5	0.0	1.0×10^{-8}	1.1×10^{-7}
CO_2	0.27098	0.0	0.0	8.82×10^{-5}	2.55×10^{-6}	0.0
H_2S	0.0	−12.704	0.0	-1.3×10^{-6}	0.0	0.0

Using the quadratic formula gives:

$$T = \frac{-0.497 + \sqrt{(0.0497)^2 - (4)(0.000383)(-0.6713)}}{(2)(0.000383)}$$

$$= 12.33\ ^\circ\text{C}$$

or

$$T = (1.8)(12.33) + 32 = 54.2\ ^\circ\text{F}$$

3.6.2 Hydrates in Subsurface

One explanation for hydrate formation is that the entrance of the gaseous molecules into vacant lattice cavities in the liquid water structure causes the water to solidify at temperatures above the freezing point of water. In general, ethane, propane, and butane raise the hydrate formation temperature for methane. For example, 1% of propane raises the hydrate-forming temperature from 41 to 49 °F at 600 psia. Hydrogen sulfide and carbon dioxide are also relatively significant contributors in causing hydrates, whereas N_2 and C_{5+} have no noticeable effect. These solid ice-like mixtures of natural gas and water have been found in formations under deep water along the continental margins of America and beneath the permafrost (i.e., permanently frozen ground) in Arctic basins. The permafrost occurs where the mean atmospheric temperature is just under 32 °F.

Muller (1947) suggested that lowering of the earth's temperature took place in early Pleistocene times, "perhaps a million years ago." If formation natural gases were cooled under pressure in the presence of free water, hydrates would form in the cooling process before ice temperatures were reached. If further lowering of temperature brought the layer into a permafrost condition, then the hydrates would remain as such. In colder climates (such as Alaska, northern Canada, and Siberia) and beneath the oceans, conditions are appropriate for gas hydrate formation.

The essential condition for gas hydrate stability at a given depth is that the actual earth temperature at that depth is lower than the hydrate-forming temperature corresponding to the pressure and gas composition conditions. The thickness of a potential hydrate zone can be an important variable in drilling operations where drilling through hydrates requires special precautions. It can also be of significance in determining regions where hydrate occurrences might be sufficiently thick to justify gas recovery. The existence of a gas hydrate stability condition, however, *does not ensure that hydrates exist in that region*, but only that they can exist. In addition, if gas and water coexist within the hydrate stability zone, then they must exist in gas hydrate form.

Consider the earth temperature curve for the Cape Simpson area of Alaska, as shown in Figure 3.88. Pressure data from a drill stem test (DST) and a repeated formation test (RFT) indicates a pressure gradient of 0.435 psi/ft. Assuming a 0.6 gas gravity with its hydrate-forming pressure and temperature as given in Figure 3.74, this hydrate $p-T$ curve can be converted into a depth vs. temperature plot by dividing the pressures by 0.435, as shown by Katz (1971) in Figure 3.88. These two curves intersect at 2100 ft in depth. Katz pointed out that at Cape Simpson, we would expect to find water in the form of ice down to 900 ft and hydrates between 900 and 2100 ft of 0.6 gas gravity.

Using the temperature profile as a function of depth for the Prudhoe Bay Field as shown in Figure 3.89, Katz (1971) estimated that the hydrate zone thickness at Prudhoe Bay for a 0.6 gravity gas might occur between 2000 and 4000 ft. Godbole et al. (1988) pointed out that the first confirmed evidence of the presence of gas hydrates in Alaska was obtained on March 15, 1972, when Arco and Exxon recovered gas hydrate core samples in pressurized core barrels at several depths between 1893 and 2546 ft from the Northwest Eileen well 2 in the Prudhoe Bay Field.

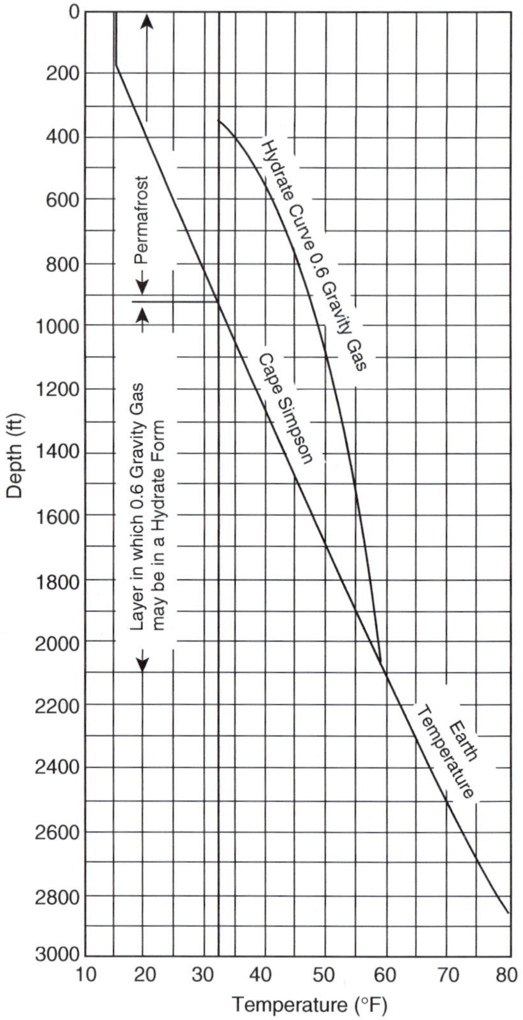

FIGURE 3.88 Method for locating the thickness of hydrate layer. *(Permission to copy SPE, copyright SPE 1971).*

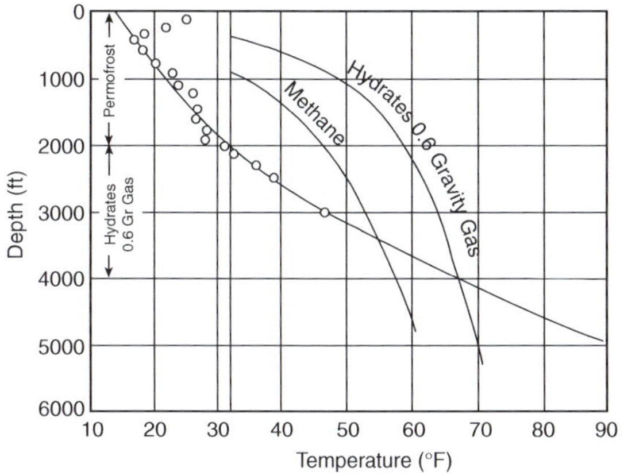

FIGURE 3.89 Hydrate zone thickness for temperature gradient at Prudhoe Bay. *(Permission to copy SPE, copyright SPE 1971).*

Studies by Holder et al. (1987) and Godbole et al. (1988) on the occurrence of in situ natural gas hydrates in the Arctic North Slope of Alaska and beneath the ocean floor suggest that the factors controlling the depth and thickness of natural gas–hydrate zones in these regions and affecting their stabilities include:

- geothermal gradient;
- pressure gradient;
- gas composition;
- permafrost thickness;
- ocean-bottom temperature;
- mean average annual surface temperature;
- water salinity.

Various methods have been proposed for harvesting the gas in hydrate form that essentially require heat to melt the hydrate or lowering the pressure on the hydrate to release the gas. Specifically:

- steam injection;
- hot brine injection;
- fire flood;
- chemicals injection;
- depressurizing.

Holder and Anger (1982) suggested that in the depressurizing scheme, pressure reduction causes destabilization of hydrates. As hydrates dissociate, they absorb heat from the surrounding formation. The hydrates continue to dissociate until they generate enough gas to raise the reservoir pressure to the equilibrium pressure of hydrates at a new temperature, which is lower than the original value. A temperature gradient is thus generated between the hydrates (sink) and surrounding media (source), and heat flows to the hydrates. The rate of dissociation of hydrates, however, is controlled by the rate of heat influx from the surrounding media or by the thermal conductivity of the surrounding rock matrix.

Many questions need to be answered if gas is to be produced from hydrates. For example:

- The form in which hydrates exist in a reservoir should be known. Hydrates may exist in different types (all hydrates, excess water, and excess ice, in conjunction with free gas or oil) and in different forms (massive, laminated, dispersed, or nodular). Each case will have a different effect on the method of production and on the economics.
- The saturation of hydrates in the reservoir.
- There could be several problems associated with gas production, such as pore blockage by ice and blockage of the wellbore resulting from re-formation of hydrates during flow of gas through the production well.
- Economics of the project is perhaps the most important impacting factor for the success of gas recovery from subsurface hydrate accumulations.

Despite the above concerns, subsurface hydrates exhibit several characteristics, especially compared with other unconventional gas resources, that increase their importance as potential energy resources and make their future recovery likely. These include a higher concentration of gas in hydrated form, enormously large deposits of hydrates, and their wide occurrence in the world.

3.7 SHALLOW GAS RESERVOIRS

Tight, shallow gas reservoirs present a number of unique challenges in determining reserves accurately. Traditional methods such as decline analysis and material balance are inaccurate due to the formation's low permeability and the usually poor-quality pressure data. The low permeabilities cause long transient periods that are not separated early from production decline with conventional decline analysis, resulting in lower confidence in selecting the appropriate decline characteristics that affect recovery factors and remaining reserves significantly. In an excellent paper, West and Cochrane (1994) used the Medicine Hat Field in western Canada as an example of these types of reservoirs and developed a methodology, called the extended material balance technique, to evaluate gas reserves and potential infill drilling.

The Medicine Hat Field is a tight, shallow gas reservoir producing from multiple highly interbedded, silty sand formations with poor permeabilities of less than 0.1 md. This poor permeability is the main characteristic of these reservoirs that affects conventional decline analysis. Because of these low permeabilities, and in part because of commingled multilayer production effects, wells experience long transient periods before they begin experiencing pseudosteady-state flow that represents the decline portion of their lives. One of the principal assumptions often neglected when conducting decline analysis is that the pseudosteady state must have been achieved. The initial transient production trend of a well or group of wells is not indicative of the long-term decline of the well. Distinguishing the transient production of a well from its pseudosteady-state production is often difficult, and this can lead to errors in determining the decline characteristic (exponential, hyperbolic, or harmonic) of the well. Figure 3.90 shows the production history from a tight, shallow gas well and illustrates the difficulty in selecting the correct decline. Another characteristic of tight, shallow gas reservoirs that affects conventional decline analysis is that constant reservoir conditions, an assumption required for conventional decline analysis, do not exist because of increasing drawdown, changing operating strategies, erratic development, and deregulation.

Material balance is affected by tight, shallow gas reservoirs because the pressure data is limited, of poor quality, and nonrepresentative of a majority of the wells. Because the risk of drilling dry holes is low and DSTs are not cost-effective in the development of shallow gas, DST data is very limited. Reservoir pressures are recorded only for government-designated

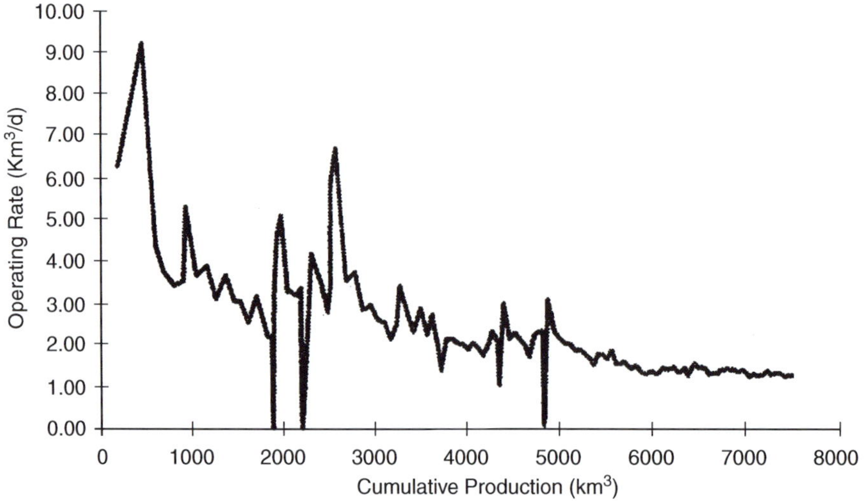

FIGURE 3.90 Production history for a typical Medicine Hat property. *(Permission to copy SPE, copyright SPE 1995).*

"control" wells that account for only 5% of all wells. Shallow gas produces from multiple formations, and production from these formations is typically commingled, exhibiting some degree of pressure equalization. Unfortunately, the control wells are segregated by tubing/packers, and consequently the control-well pressure data is not representative of most commingled wells. In addition, pressure monitoring has been very inconsistent. Varied measurement points (downhole or wellhead), inconsistent shut-in times, and different analysis types (e.g., buildup and static gradient) make quantitative pressure tracking difficult. As Figure 3.91 shows, both these problems result in a scatter of data, which makes material balance extremely difficult.

Wells in the Medicine Hat shallow gas area are generally cased, perforated, and fractured in one, two, or all three formations, as ownerships vary not only areally but also between formations. The Milk River and Medicine Hat formations are usually produced commingled. Historically, the Second White Specks formation has been segregated from the other two; recently, however, commingled production from all three formations has been approved. Spacing for shallow gas is usually two to four wells per section.

As a result of the poor reservoir quality and low pressure, well productivity is very low. Initial rates rarely exceed 700 Mscf/day. Current average production per well is approximately 50 Mscf/day for a three-formation completion. There are approximately 24,000 wells producing from the Milk River formation in southern Alberta and Saskatchewan with total estimated gas reserves of 5.3 Tscf. West and Cochrane (1994) developed an iterative methodology, called extended material balance "EMB," to determine gas reserves in 2300 wells in the Medicine Hat Field.

The EMB technique is essentially an iterative process for obtaining a suitable p/Z vs. G_p line for a reservoir where pressure data is inadequate. It combines the principles of volumetric gas depletion with the gas deliverability equation. The deliverability equation for radial flow of gas describes the relationship between the pressure differential in the wellbore and the gas flow rate from the well:

$$Q_g = C[p_r^2 - p_{wf}^2]^n \tag{3.260}$$

Because of the very low production rates from the wells in Medicine Hat shallow gas, a laminar flow regime exists that can be described

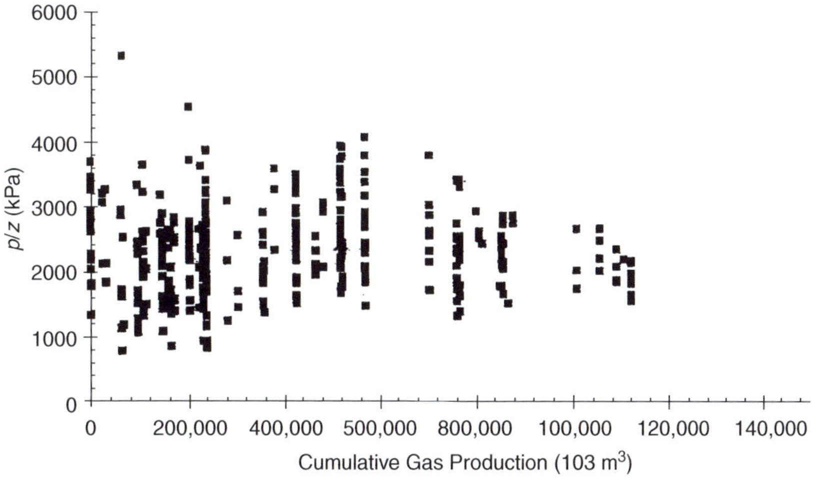

FIGURE 3.91 Scatter pressure data for a typical Medicine Hat property. *(Permission to copy SPE, copyright SPE 1995).*

with an exponent $n = 1$. The terms making up the coefficient C in Eq. (3.260) are either fixed reservoir parameters (kh, r_e, r_w, and T) that do not vary with time or terms that fluctuate with pressure, temperature, and gas composition, i.e., μ_g and Z. The performance coefficient C is given by:

$$C = \frac{kh}{1422T\mu_g Z[\ln(r_e/r_w) - 0.5]} \quad \textbf{(3.261)}$$

Because the original reservoir pressure in these shallow formations is low, the differences between initial and abandonment pressures are not significant and the variation in the pressure-dependent terms over time can be assumed negligible. C may be considered constant for a given Medicine Hat shallow gas reservoir over its life. With these simplifications for shallow gas, the deliverability equation becomes:

$$Q_g = C[p_r^2 - p_{wf}^2] \quad \textbf{(3.262)}$$

The sum of the instantaneous production rates with time will yield the relationship between G_p and reservoir pressure, similar to the MBE. By use of this common relationship, with the unknowns being reservoir pressure p and the performance coefficient C, the EMB method involves iterating to find the correct p/Z vs. G_p relationship to give a constant C with time. The proposed iterative method is applied as outlined in the following steps:

Step 1. To avoid calculating individual reserves for each of the 2300 wells, West and Cochrane (1995) grouped wells by formation and by date on production. The authors verified this simplification on a test group by ensuring that the reserves from the group of wells yielded the same results as the sum of the individual well reserves. These groupings were used for each of the 10 properties, and the results of the groupings combined to give a property production forecast. Also, to estimate the reservoir decline characteristics more accurately, the rates were normalized to reflect changes in the bottom-hole flowing pressure (BHFP).

Step 2. Using the gas specific gravity and reservoir temperature, calculate the gas deviation factor Z as a function of pressure and plot p/Z vs. p on a Cartesian scale.

Step 3. An initial estimate for the p/Z variation with G_p is made by guessing an initial pressure p_i, and a linear slope m of Eq. (3.59):

$$\frac{p}{Z} = \frac{p_i}{Z_i} - [m]G_p$$

with the slope m as defined by:

$$m = \left(\frac{p_i}{Z_i}\right)\frac{1}{G}$$

Step 4. Starting at the initial production date for the property, the p/Z vs. time relationship is established by simply substituting the actual cumulative production G_p into the MBE with estimated slope m and p_i because actual cumulative production G_p vs. time is known. The *reservoir pressure* p can then be constructed as a function of time from the plot of p/Z as a function of p, i.e., step 2.

Step 5. Knowing the actual production rates, Q_g, and BHFPs p_{wf} for each monthly time interval, and having estimated reservoir pressures p from step 3, C is calculated for each time interval with Eq. (3.262):

$$C = \frac{Q_g}{p^2 - p_{wf}^2}$$

Step 6. C is plotted vs. time. If C is not constant (i.e., the plot is not a horizontal line), a new p/Z vs. G_p is guessed and the process repeated from step 3 through step 5.

Step 7. Once a constant C solution is obtained, the representative p/Z relationship has been defined for reserves determination.

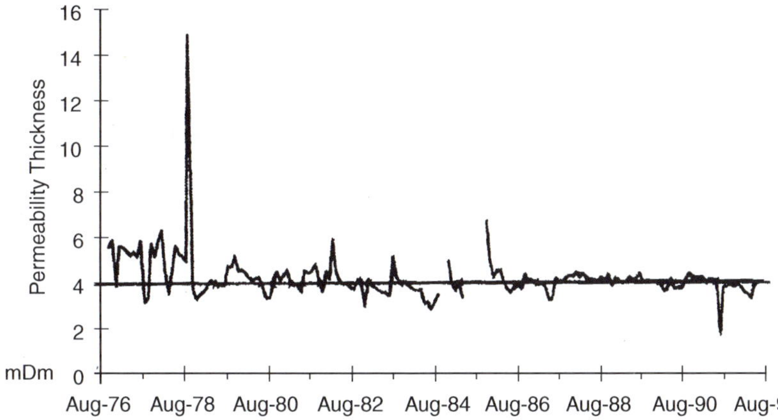

FIGURE 3.92 Example of a successful EMB solution—flat kh profile. *(Permission to copy SPE, copyright SPE 1995).*

The use of the EMB method in the Medicine Hat shallow gas makes the fundamental assumptions (1) that the gas pool depletes volumetrically (i.e., no water influx) and (2) that all wells behave like an average well with the same deliverability constant, turbulence constant, and BHFP, which is a reasonable assumption given the number of wells in the area, the homogeneity of the rocks, and the observed well production trends.

In the EMB evaluation, West and Cochrane pointed out that wells for each property were grouped according to producing interval so that the actual production from the wells could be related to a particular reservoir pressure trend. When calculating the coefficient C as outlined above, a total C based on grouped production was calculated and then divided by the number of wells producing in a given time interval to give an average C value. This average C value was used to calculate an average permeability/thickness, *kh*, for comparison with actual *kh* data obtained through buildup analysis for the reservoir from:

$$kh = 1422 T \mu_g Z \left[\ln\left(\frac{r_e}{r_w}\right) - 0.5 \right] C$$

For that reason *kh* vs. time was plotted instead of C vs. time in the method. Figure 3.92 shows a flat *kh* vs. time profile indicating a valid p/Z vs. G_p relationship.

Problems

1. The following information is available on a volumetric gas reservoir:

   Initial reservoir temperature, $T_i = 155\ ^\circ F$
   Initial reservoir pressure, $p_i = 3500$ psia
   Specific gravity of gas, $\gamma_g = 0.65$ (air $= 1$)
   Thickness of the reservoir, $h = 20$ ft
   Porosity of the reservoir, $\phi = 10\%$
   Initial water saturation, $S_{wi} = 25\%$

   After producing 300 MMscf, the reservoir pressure declined to 2500 psia. Estimate the areal extent of this reservoir.

2. The following pressures and cumulative production data[3] is available for a natural gas reservoir:

p (psia)	*Z*	G_p (MMMscf)
2080	0.759	0
1885	0.767	6.873
1620	0.787	14.002
1205	0.828	23.687
888	0.866	31.009
645	0.900	36.207

   (a) Estimate the initial gas-in-place.
   (b) Estimate the recoverable reserves at an abandonment pressure of 500 psia. Assume $z_a = 1.00$.

[3] Ikoku, C., 1984. *Natural Gas Reservoir Engineering.* John Wiley & Sons, New York, NY.

(c) What is the recovery factor at the abandonment pressure of 500 psia?

3. A gas field with an active water drive showed a pressure decline from 3000 to 2000 psia over a 10-month period. From the following production data, match the past history and calculate the original hydrocarbon gas in the reservoir. Assume $z = 0.8$ in the range of reservoir pressures and $T = 140$ °F.

Data[a]

t, month	0	2.5	5.0	7.5	10.0
p, psia	3000	2750	2500	2250	2000
G_p, MMscf	0	97.6	218.9	355.4	500.0

4. A volumetric gas reservoir produced 600 MMscf of gas of 0.62 specific gravity when the reservoir pressure declined from 3600 to 2600 psi. The reservoir temperature is reported at 140 °F. Calculate:
   (a) gas initially in place;
   (b) remaining reserves to an abandonment pressure of 500 psi;
   (c) ultimate gas recovery at abandonment.

5. The following information on a water drive gas reservoir is given:

   Bulk volume = 100, 000 acre-ft
   Gas gravity = 0.6
   Porosity = 15%
   $S_{wi} = 25\%$
   $T = 140$ °F
   $p_i = 3500$ psi

   Reservoir pressure has declined to 3000 psi while producing 30 MMMscf of gas and no water production. Calculate the cumulative water influx.

6. The pertinent data for the Mobil–David Field is given below:

   $G = 70$ MMMscf
   $p_i = 9507$ psi
   $f = 24\% fS_{wi} = 35\%$
   $c_w = 401 \times 10^{-6}$ psi^{-1}
   $c_f = 3.4 \times 10^{-6}$ psi^{-1}
   $\gamma_g = 0.74$
   $T = 266$ °F

   For this volumetric abnormally pressured reservoir, calculate and plot cumulative gas production as a function of pressure.

7. A gas well is producing under a constant bottom-hole flowing pressure of 1000 psi. The specific gravity of the produced gas is 0.65. Given:

   $p_i = 1500$ psi
   $r_w = 0.33$ ft
   $r_e = 1000$ ft
   $k = 20$ md
   $h = 29$ ft
   $T = 140$ °F
   $s = 0.40$

   calculate the gas flow rate by using:
   (a) the real-gas pseudopressure approach;
   (b) the pressure-squared approximation.

8. The following data was obtained from a back-pressure test on a gas well[4]:

Q_g (Mscf/day)	p_{wf} (psi)
0	481
4928	456
6479	444
8062	430
9640	415

   (a) Calculate values of C and n.
   (b) Determine the AOF.
   (c) Generate the IPR curves at reservoir pressures of 481 and 300 psi.

9. The following back-pressure test data is available:

Q_g (Mscf/day)	p_{wf} (psi)
0	5240
1000	4500
1350	4191
2000	3530
2500	2821

[4]Ikoku, C., 1984. *Natural Gas Reservoir Engineering*. John Wiley & Sons, New York, NY.

Given:

Gas gravity = 0.78
Porosity = 12%
S_{wi} = 15%
T = 281 °F

(a) generate the current IPR curve by using:
  (i) the simplified back-pressure equation;
  (ii) the laminar–inertial–turbulent (LIT) methods:
   - pressure-squared approach;
   - pressure approach;
   - pseudopressure approach;

(b) repeat part (a) for a future reservoir pressure of 4000 psi.

10. A 3000-foot horizontal gas well is draining an area of approximately 180 acres, given:

p_i = 2500 psi,	p_{wf} = 1500 psi,	k = 25 md
T = 120 °F,	r_w = 0.25,	h = 20 ft
γ_g = 0.65		

Calculate the gas flow rate.

11. Given the sorption isotherm data below for a coal sample from the CBM field, calculate Langmuir's isotherm constant V_m and Langmuir's pressure constant b:

p (psi)	*V* (scf/ton)
87.4	92.4
140.3	135.84
235.75	191.76
254.15	210
350.75	247.68
579.6	318.36
583.05	320.64
869.4	374.28
1151.15	407.4
1159.2	408.6

12. The following production data is available from a dry gas field:

q_t (MMscf/day)	G_p (MMscf)	q_t (MMscf/day)	G_p (MMscf)
384	19,200	249.6	364,800
403.2	38,400	236.4	422,400
364.8	57,600	220.8	441,600
370.8	115,200	211.2	460,800
326.4	192,000	220.8	480,000
297.6	288,000		

Estimate:

(a) the future cumulative gas production when gas flow rate reaches 100 MMscf/day;

(b) extra time to reach 100 MMscf/day.

13. A gas well has the following production history:

Date	Time (months)	q_t (MMscf/month)
1/1/2000	0	1017
2/1/2000	1	978
3/1/2000	2	941
4/1/2000	3	905
5/1/2000	4	874
6/1/2000	5	839
7/1/2000	6	809
8/1/2000	7	778
9/1/2000	8	747
10/1/2000	9	722
11/1/2000	10	691
12/1/2000	11	667
1/1/2001	12	641

(a) Use the first 6 months of the production history data to determine the coefficient of the decline curve equation.

(b) Predict flow rates and cumulative gas production from August 1, 2000 through January 1, 2001.

(c) Assuming that the economic limit is 20 MMscf/month, estimate the time to reach the economic limit and the corresponding cumulative gas production.

14. The volumetric calculations on a gas well show that the ultimate recoverable reserves G_{pa} are 18 MMMscf of gas. By analogy

with other wells in the area, the following data is assigned to the well:

Exponential decline;
Allowable (restricted) production rate = 425 MMscf/month;
Economic limit = 20 MMscf/month;
Nominal decline rate = 0.034 month^{-1}.

Calculate the yearly production performance of the well.

15. The following data is available on a gas well production:

$p_i = 4100$ psia, $p_{wf} = 400$ psi, $T = 600$ °R
$h = 40$ ft, $\phi = 0.10$, $S_{wi} = 0.30$
$\gamma_g = 0.65$,

Time (days)	q_t (MMscf/day)
0.7874	5.146
6.324	2.108
12.71	1.6306
25.358	1.2958
50.778	1.054
101.556	0.8742
248	0.6634
496	0.49042
992	0.30566
1240	0.24924
1860	0.15996
3100	0.07874
6200	0.02232

Calculate the initial gas-in-place and the drainage area.

16. A gas of 0.7 specific gravity is at 800 psia. To what extent can the temperature be lowered without hydrate formation in the presence of free water?
17. A gas has a specific gravity of 0.75 and exists at 70 °F. What would be the pressure above which hydrates could be expected to form?
18. How far can a 0.76 gravity gas at 1400 psia and 110 °F be expanded without hydrate formation?

CHAPTER

4

Performance of Oil Reservoirs

Each reservoir is composed of a unique combination of geometric form, geological rock properties, fluid characteristics, and primary drive mechanism. Although no two reservoirs are identical in all aspects, they can be grouped according to the primary recovery mechanism by which they produce. It has been observed that each drive mechanism has certain typical performance characteristics in terms of:

- ultimate recovery factor;
- pressure decline rate;
- gas–oil ratio;
- water production.

The recovery of oil by any of the natural drive mechanisms is called "primary recovery." The term refers to the production of hydrocarbons from a reservoir without the use of any process (such as fluid injection) to supplement the natural energy of the reservoir.

The two main objectives of this chapter are:

(1) To introduce and give a detailed discussion of the various primary recovery mechanisms and their effects on the overall performance of oil reservoirs.
(2) To provide the basic principles of the material balance equation and other governing relationships that can be used to predict the volumetric performance of oil reservoirs.

4.1 PRIMARY RECOVERY MECHANISMS

For a proper understanding of reservoir behavior and predicting future performance, it is necessary to have knowledge of the driving mechanisms that control the behavior of fluids within reservoirs.

The overall performance of oil reservoirs is largely determined by the nature of the energy, i.e., driving mechanism, available for moving the oil to the wellbore. There are basically six driving mechanisms that provide the natural energy necessary for oil recovery:

(1) rock and liquid expansion drive;
(2) depletion drive;
(3) gas cap drive;
(4) water drive;
(5) gravity drainage drive;
(6) combination drive.

These six driving mechanisms are presented below.

4.1.1 Rock and Liquid Expansion

When an oil reservoir initially exists at a pressure higher than its bubble point pressure, the reservoir is called an "undersaturated oil reservoir." At pressures above the bubble point pressure, crude oil, connate water, and rock are the only materials present. As the reservoir pressure declines, the rock and fluids expand due to their individual compressibilities. The reservoir rock compressibility is the result of two factors:

(1) expansion of the individual rock grains;
(2) formation compaction.

Both of these factors are the results of a decrease of fluid pressure within the pore spaces, and both tend to reduce the pore volume through the reduction of the porosity.

As the expansion of the fluids and reduction in the pore volume occur with the decreasing reservoir pressure, the crude oil and water will be forced out of the pore space to the wellbore. Because liquids and rocks are only slightly compressible, the reservoir will experience a rapid pressure decline. The oil reservoir under this driving mechanism is characterized by a constant gas–oil ratio that is equal to the gas solubility at the bubble point pressure.

This driving mechanism is considered the least efficient driving force and usually results in the recovery of only a small percentage of the total oil-in-place.

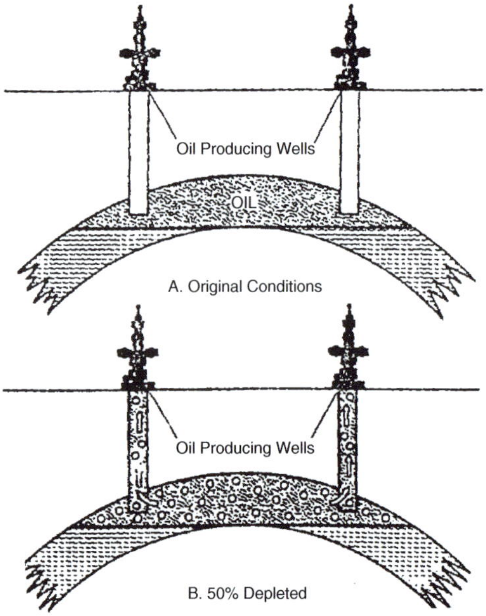

FIGURE 4.1 Solution gas drive reservoir. (*After Clark, N., 1969. Elements of Petroleum Reservoirs. Society of Petroleum Engineers, Dallas, TX*).

4.1.2 Depletion Drive Mechanism

This driving form may also be referred to by the following various terms:

- solution gas drive;
- dissolved gas drive;
- internal gas drive.

In this type of reservoir, the principal source of energy is a result of gas liberation from the crude oil and the subsequent expansion of the solution gas as the reservoir pressure is reduced. As pressure falls below the bubble point pressure, gas bubbles are liberated within the microscopic pore spaces. These bubbles expand and force the crude oil out of the pore space as shown conceptually in Figure 4.1.

Cole (1969) suggests that a depletion drive reservoir can be identified by the following characteristics:

Pressure behavior: The reservoir pressure declines rapidly and continuously. This reservoir pressure behavior is attributed to the fact that no extraneous fluids or gas caps are available to provide a replacement of the gas and oil withdrawals.

Water production: The absence of a water drive means there will be little or no water production with the oil during the entire producing life of the reservoir.

A depletion drive reservoir is characterized by a rapidly increasing gas–oil ratio from all wells, regardless of their structural position. After the reservoir pressure has been reduced below the bubble point pressure, gas evolves from solution throughout the reservoir. Once the gas saturation exceeds the critical gas saturation, free gas begins to flow toward the wellbore and the gas–oil ratio increases. The gas will also begin a vertical movement due to gravitational forces, which may result in the formation of a secondary gas cap. Vertical permeability is an important factor in the formation of a secondary gas cap.

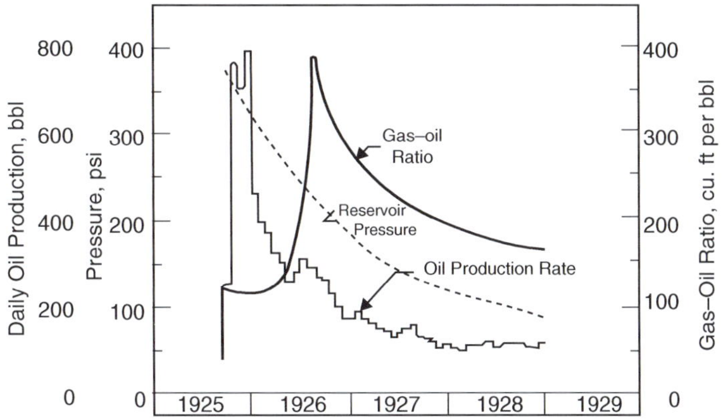

FIGURE 4.2 Production data for a solution gas drive reservoir. (*After Clark, N., 1969. Elements of Petroleum Reservoirs. Society of Petroleum Engineers, Dallas, TX*).

Unique oil recovery: Oil production by depletion drive is usually the least efficient recovery method. This is a direct result of the formation of gas saturation throughout the reservoir. Ultimate oil recovery from depletion drive reservoirs may vary from less than 5% to about 30%. The low recovery from this type of reservoir suggests that large quantities of oil remain in the reservoir, and therefore, depletion drive reservoirs are considered the best candidates for secondary recovery applications.

The above characteristic trends occurring during the production life of depletion drive reservoirs are shown in Figure 4.2 and summarized below:

Characteristics	Trend
Reservoir pressure	Declines rapidly and continuously
Gas–oil ratio	Increases to maximum and then declines
Water production	None
Well behavior	Requires pumping at early stage
Oil recovery	5–30%

4.1.3 Gas Cap Drive

Gas cap drive reservoirs can be identified by the presence of a gas cap with little or no water drive as shown in Figure 4.3. Due to the ability of the gas cap to expand, these reservoirs are characterized by a slow decline in the reservoir pressure. The natural energy available to produce the crude oil comes from the following two sources:

(1) expansion of the gas cap gas;
(2) expansion of the solution gas as it is liberated.

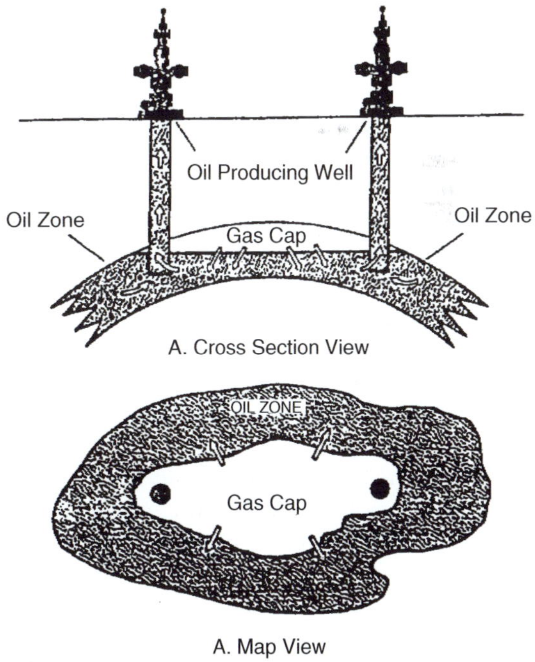

FIGURE 4.3 Gas cap drive reservoir. (*After Clark, N., 1969. Elements of Petroleum Reservoirs. Society of Petroleum Engineers, Dallas, TX*).

Cole (1969) and Clark (1969) presented a comprehensive review of the characteristic trends associated with gas cap drive reservoirs. These characteristic trends are summarized below:

Reservoir pressure: The reservoir pressure falls slowly and continuously. Pressure tends to be maintained at a higher level than in a depletion drive reservoir. The degree of pressure maintenance depends upon the volume of gas in the gas cap compared to the oil volume.

Water production: Absent or negligible water production.

Gas–oil ratio: The gas–oil ratio rises continuously in upstructure wells. As the expanding gas cap reaches the producing intervals of upstructure wells, the gas–oil ratio from the affected wells will increase to high values.

Ultimate oil recovery: Oil recovery by gas cap expansion is actually a frontal drive displacing mechanism, which, therefore, yields considerably larger recovery efficiency than that of depletion drive reservoirs. This larger recovery efficiency is also attributed to the fact that no gas saturation is being formed throughout the reservoir at the same time. Figure 4.4 shows the relative positions of the gas–oil contact at different times in the producing life of the reservoir. The expected oil recovery ranges from 20% to 40%.

The ultimate oil recovery from a gas cap drive reservoir will vary depending largely on the following six important parameters:

(1) *Size of the original gas cap*: As shown graphically in Figure 4.5, the ultimate oil recovery increases with increasing size of the gas cap.

(2) *Vertical permeability*: Good vertical permeability will permit the oil to move downward with less bypassing of gas.

(3) *Oil viscosity*: As the oil viscosity increases, the amount of gas bypassing will also increase, which leads to a lower oil recovery.

(4) *Degree of conservation of the gas*: In order to conserve gas, and thereby increase ultimate oil recovery, it is necessary to shut in the wells that produce excessive gas.

(5) *Oil production rate*: As the reservoir pressure declines with production, solution gas evolves from the crude oil and

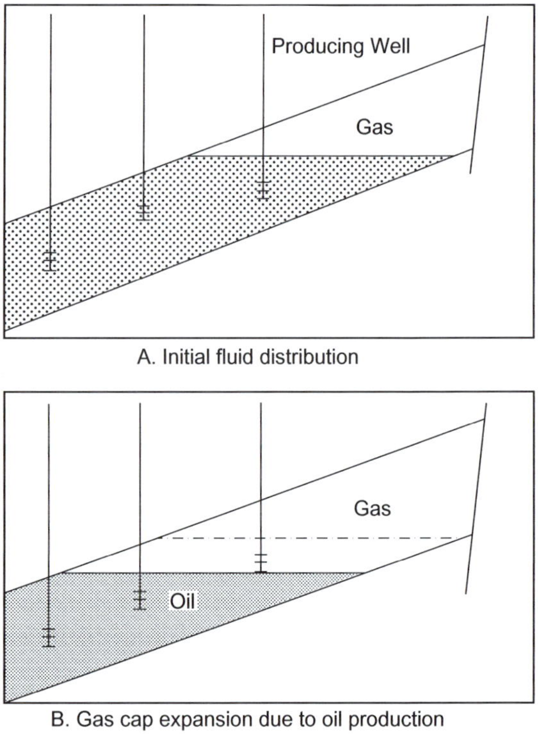

FIGURE 4.4 Gas cap drive reservoir. (*After Cole, F.W., 1969. Reservoir Engineering Manual. Gulf Publishing, Houston, TX*).

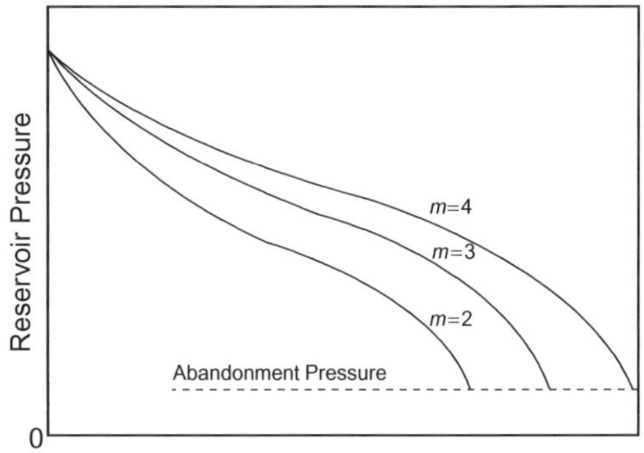

FIGURE 4.5 Effect of gas cap size on ultimate oil recovery. (*After Cole, F.W., 1969. Reservoir Engineering Manual. Gulf Publishing, Houston, TX*).

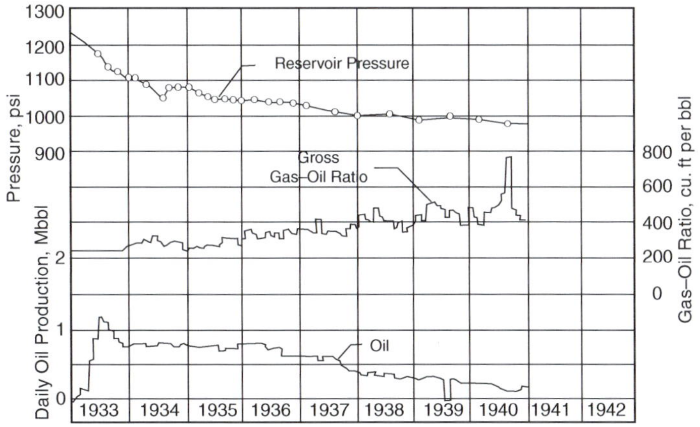

FIGURE 4.6 Production data for a gas cap drive reservoir. (*After Clark, N., 1969. Elements of Petroleum Reservoirs. Society of Petroleum Engineers, Dallas, TX. Courtesy of API*).

the gas saturation increases continuously. If the gas saturation exceeds the critical gas saturation, the evolved gas begins to flow in the oil zone. As a result of creating a mobile gas phase in the oil zone, the following two events will occur: (1) the effective permeability to oil will be decreased as a result of the increased gas saturation; (2) the effective permeability to gas will be increased, thereby increasing the flow of gas.

The formation of the free gas saturation in the oil zone cannot be prevented without resorting to pressure maintenance operations. Therefore, in order to achieve maximum benefit from a gas cap drive-producing mechanism, gas saturation in the oil zone must be kept to an absolute minimum. This can be accomplished by taking advantage of gravitational segregation of the fluids. In fact, an efficiently operated gas cap drive reservoir must also have an efficient gravity segregation drive. As the gas saturation is formed in the oil zone, it must be allowed to migrate upstructure to the gas cap. Thus, a gas cap drive reservoir is in reality a combination drive reservoir, although it is not usually considered as such.

Lower producing rates will permit the maximum amount of free gas in the oil zone to migrate to the gas cap. Therefore, gas cap drive reservoirs are rate sensitive, as lower producing rates will usually result in increased recovery.

(6) *Dip angle*: The size of the gas cap determines the overall field oil recovery. When the gas cap is considered the main driving mechanism, its size is a measure of the reservoir energy available to produce the crude oil system. Such recovery normally will be 20–40% of the original oil-in-place, but if some other features are present to assist, such as steep angle of dip, which allows good oil drainage to the bottom of the structure, considerably higher recoveries (up to 60% or greater) may be obtained. Conversely, extremely thin oil columns (where early breakthrough of the advancing gas cap occurs in producing wells) may limit oil recovery to lower figures regardless of the size of the gas cap. Figure 4.6 shows typical production and pressure data for a gas cap drive reservoir.

Well behavior: Because of the effects of gas cap expansion on maintaining reservoir pressure and the effect of decreased liquid column weight as it is produced out of the well, gas

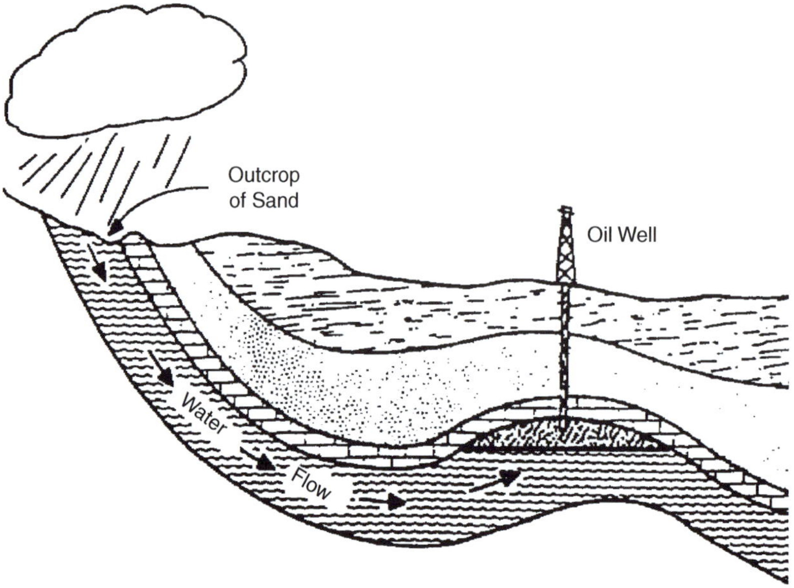

FIGURE 4.7 Reservoir having artesian water drive. (*After Clark, N., 1969. Elements of Petroleum Reservoirs. Society of Petroleum Engineers, Dallas, TX*).

cap drive reservoirs tend to flow longer than depletion drive reservoirs.

4.1.4 Water Drive Mechanism

Many reservoirs are bounded on a portion or all of their peripheries by water-bearing rocks called aquifers. The aquifers may be so large compared to the reservoir they adjoin as to appear infinite for all practical purposes, and they may range down to those so small as to be negligible in their effects on the reservoir performance.

The aquifer itself may be entirely bounded by impermeable rock so that the reservoir and aquifer together form a closed (volumetric) unit. On the other hand, the reservoir may outcrop at one or more places where it may be replenished by surface water as shown schematically in Figure 4.7.

It is common to speak of edge water or bottom water in discussing water influx into a reservoir. Bottom water occurs directly beneath the oil and edge water occurs off the flanks of the structure at the edge of the oil as illustrated in Figure 4.8. Regardless of the source of water, the water drive is the result of water moving into the pore spaces originally occupied by oil, replacing the oil and displacing it to the producing wells.

Cole (1969) presented the following discussion on the characteristics that can be used for identification of the water-driving mechanism.

Reservoir Pressure. The decline in the reservoir pressure is usually very gradual. Figure 4.9 shows the pressure–production history of a typical water drive reservoir. It is not uncommon for many thousands of barrels of oil to be produced for each pound per square inch drop in reservoir pressure. The reason for the small decline in reservoir pressure is that oil and gas withdrawals from the reservoir are replaced almost volume for volume by water encroaching into the oil zone. Several large oil reservoirs in the Gulf Coast areas of the United States have such active water drives that the reservoir pressure has declined by only about 1 psi per million barrels of oil produced. Although pressure history is normally plotted vs. cumulative oil

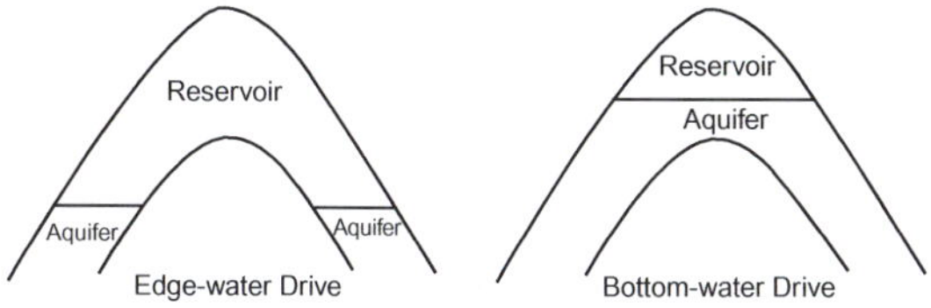

FIGURE 4.8 Aquifer Geometries.

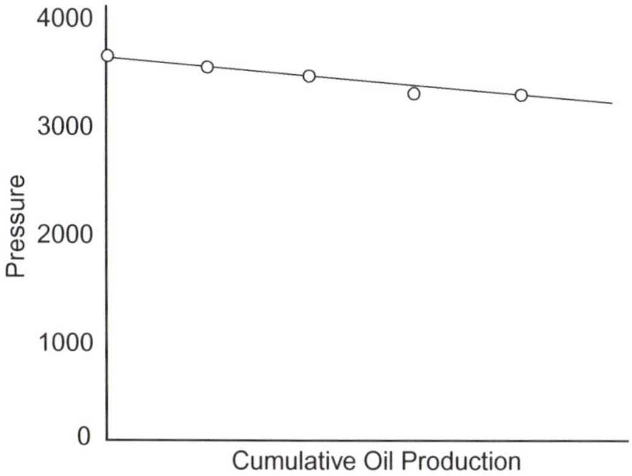

FIGURE 4.9 Pressure–production history for a water drive reservoir.

production, it should be understood that total reservoir fluid withdrawals are the really important criteria in the maintenance of reservoir pressure. In a water drive reservoir, only a certain number of barrels of water can move into the reservoir as a result of a unit pressure drop within the reservoir. Since the principal income generation is from oil, if the withdrawals of water and gas can be minimized, then the withdrawal of oil from the reservoir can be maximized with minimum pressure decline. Therefore, it is extremely important to reduce water and gas production to an absolute minimum. This can usually be accomplished by shutting in wells that are producing large quantities of these fluids and where possible transferring their allowable oil production to other wells producing with lower water–oil or gas–oil ratios.

Water Production. Early excess water production occurs in structurally low wells. This is characteristic of a water drive reservoir, and provided the water is encroaching in a uniform manner, nothing can or should be done to restrict this encroachment, as the water will probably provide the most efficient displacing mechanism possible. If the reservoir has one or more lenses of very high permeability, then the water may be moving through this more permeable zone. In this case, it may be economically feasible to perform remedial operations to shut off this permeable zone producing water. It should be realized that in most cases the oil that is being recovered from a structurally low well will be recovered from wells located higher on the structure, and any expenses involved in remedial work to reduce the water–oil ratio of structurally low wells may be needless expenditure.

Gas–Oil Ratio. There is normally a little change in the producing gas–oil ratio during the life of the reservoir. This is especially true if the reservoir does not have an initial free gas cap. Pressure will be maintained as a result of water encroachment, and therefore, there will be relatively little gas released from the solution.

Ultimate Oil Recovery. Ultimate oil recovery from water drive reservoirs is usually much larger than recovery under any other producing mechanism. Recovery is dependent upon the efficiency of the flushing action of the water as it displaces the oil. In general, as the reservoir heterogeneity increases, the recovery will decrease due to the uneven advance of the displacing water. The rate of water advance is normally faster in zones of high permeability. This results in earlier high water–oil ratios and consequent earlier economic limits. Where the reservoir is more or less homogeneous, the advancing waterfront will be more uniform, and when the economic limit, due primarily to high water–oil ratios, has been reached, a greater portion of the reservoir will have been contacted by the advancing water.

Ultimate oil recovery is also affected by the degree of activity of the water drive. In a very active water drive where the degree of pressure maintenance is good, the role of solution gas in the recovery process is reduced to almost zero,

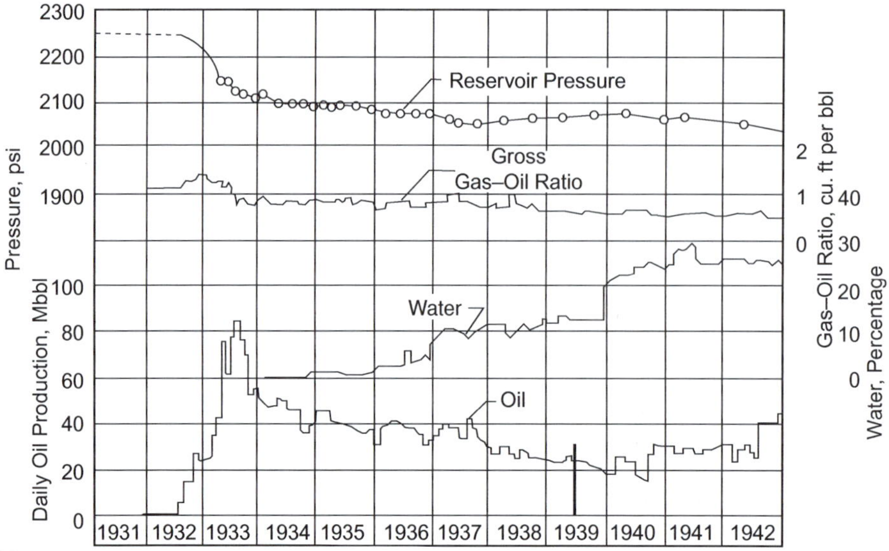

FIGURE 4.10 Production data for a water drive reservoir. (*After Clark, N., 1969. Elements of Petroleum Reservoirs. Society of Petroleum Engineers, Dallas, TX. Courtesy of API*).

with maximum advantage being taken of the water as a displacing force. This should result in maximum oil recovery from the reservoir. The ultimate oil recovery normally ranges from 35% to 75% of the original oil-in-place. The characteristic trends of a water drive reservoir are shown graphically in Figure 4.10 and summarized below:

Characteristics	Trend
Reservoir pressure	Remains high
Surface gas–oil ratio	Remains low
Water production	Starts early and increases to appreciable amounts
Well behavior	Flow until water production gets excessive
Expected oil recovery	35–75%

4.1.5 Gravity Drainage Drive

The mechanism of gravity drainage occurs in petroleum reservoirs as a result of differences in densities of the reservoir fluids. The effects of gravitational forces can be simply illustrated by placing a quantity of crude oil and a quantity of water in a jar and agitating the contents. After agitation, the jar is placed at rest, and the more dense fluid (normally water) will settle to the bottom of the jar, while the less dense fluid (normally oil) will rest on top of the denser fluid. The fluids have separated as a result of the gravitational forces acting on them.

The fluids in petroleum reservoirs have all been subjected to the forces of gravity, as evidenced by the relative positions of the fluids, i.e., gas on top, oil underlying the gas, and water underlying oil. The relative positions of the reservoir fluids are shown in Figure 4.11. Due to the long periods of time involved in the petroleum accumulation and migration process, it is generally assumed that the reservoir fluids are in equilibrium. If the reservoir fluids are in equilibrium, then the gas–oil and oil–water contacts should be essentially horizontal. Although it is difficult to determine precisely the reservoir fluid contacts, the best available data indicates that, in most reservoirs, the fluid contacts actually are essentially horizontal.

Gravity segregation of fluids is probably present to some degree in all petroleum reservoirs,

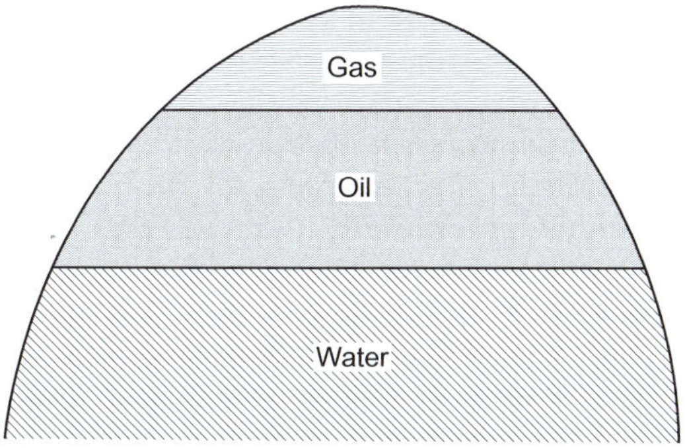

FIGURE 4.11 Initial fluids' distribution in an oil reservoir.

but it may contribute substantially to oil production in some reservoirs.

Cole (1969) stated that reservoirs operating largely under a gravity drainage producing mechanism are characterized by the following factors:

Reservoir pressure: Variable rates of pressure decline depend principally upon the amount of gas conservation. Strictly speaking, where the gas is conserved and the reservoir pressure is maintained, the reservoir would be operating under combined gas cap drive and gravity drainage mechanisms. Therefore, for the reservoir to be operating solely as a result of gravity drainage, the reservoir would show a rapid pressure decline. This would require the upstructure migration of the evolved gas where it later was produced from structurally high wells, resulting in rapid loss of pressure.

Gas–oil ratio: These types of reservoirs typically show low gas–oil ratios from structurally located low wells. This is caused by migration of the evolved gas upstructure due to gravitational segregation of the fluids. On the other hand, the structurally high wells will experience an increasing gas–oil ratio as a result of the upstructure migration of the gas released from the crude oil.

Secondary gas cap: A secondary gas cap can be found in reservoirs that initially were undersaturated. Obviously the gravity drainage mechanism does not become operative until the reservoir pressure has declined below the saturation pressure, since above the saturation pressure there will be no free gas in the reservoir.

Water production: Gravity drainage reservoirs have little or no water production. Water production is essentially indicative of a water drive reservoir.

Ultimate oil recovery: Ultimate oil recovery from gravity drainage reservoirs will vary widely, due primarily to the extent of depletion by gravity drainage alone. Where gravity drainage is good, or where producing rates are restricted to take maximum advantage of the gravitational forces, recovery will be high. There are reported cases where recovery from gravity drainage reservoirs has exceeded 80% of the initial oil-in-place. In other reservoirs where depletion drive also plays an important role in the oil recovery process, the ultimate recovery will be less.

In operating gravity drainage reservoirs, it is essential that the oil saturation in the vicinity of the wellbore must be maintained as high as possible. There are two obvious reasons for this requirement:

(1) high oil saturation means a higher oil flow rate;
(2) high oil saturation means a lower gas flow rate.

If the liberated solution gas is allowed to flow upstructure instead of toward the wellbore, then high oil saturation in the vicinity of the wellbore can be maintained.

In order to take maximum advantage of the gravity drainage producing mechanism, wells should be located as low as structurally possible. This will result in maximum conservation of the reservoir gas. A typical gravity drainage reservoir is shown in Figure 4.12.

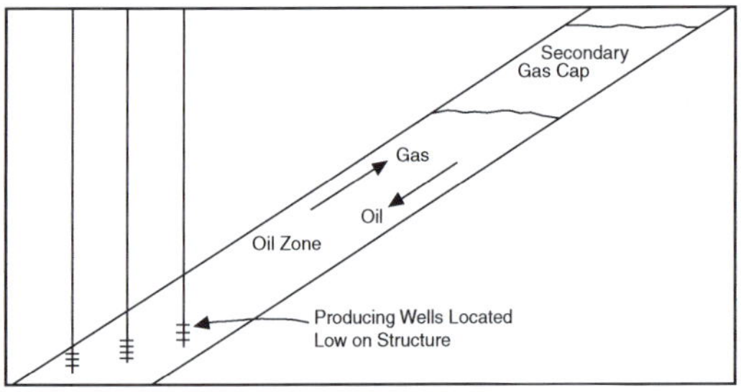

FIGURE 4.12 Gravity drainage reservoir. (*After Cole, F.W., 1969. Reservoir Engineering Manual. Gulf Publishing, Houston, TX*).

As discussed by Cole (1969), there are five factors that affect ultimate recovery from gravity drainage reservoirs:

(1) *Permeability in the direction of dip*: Good permeability, particularly in the vertical direction and in the direction of migration of the oil, is a prerequisite for efficient gravity drainage. For example, a reservoir with little structural relief, which also contained many more or less continuous shale "breaks," could probably not be operated under gravity drainage because the oil could not flow to the base of the structure.
(2) *Dip of the reservoir*: In most reservoirs, the permeability in the direction of dip is considerably larger than the permeability transverse to the direction of dip. Therefore, as the dip of the reservoir increases, the oil and gas can flow along the direction of dip (which is also the direction of greatest permeability) and still achieve their desired structural positions.
(3) *Reservoir producing rates*: Since the gravity drainage rate is limited, the reservoir producing rates should be limited to the gravity drainage rate, and then maximum recovery will result. If the reservoir producing rate exceeds the gravity drainage rate, the depletion drive-producing mechanism will become more significant with a consequent reduction in ultimate oil recovery.
(4) *Oil viscosity*: Oil viscosity is important because the gravity drainage rate is dependent upon the viscosity of the oil. In the fluid flow equations, as the viscosity decreases the flow rate increases. Therefore, the gravity drainage rate will increase as the reservoir oil viscosity decreases.
(5) *Relative permeability characteristics*: For an efficient gravity drive mechanism to be operative, the gas must flow upstructure while the oil flows downstructure. Although this situation involves counterflow of the oil and gas, both fluids are flowing and therefore relative permeability characteristics of the formation are very important.

4.1.6 Combination Drive Mechanism

The driving mechanism most commonly encountered is one in which both water and free gas are available in some degree to displace the oil toward the producing wells. The most common type of drive encountered, therefore, is a combination drive mechanism as illustrated in Figure 4.13.

Two combinations of driving forces are usually present in combination drive reservoirs:

(1) depletion drive and a weak water drive;
(2) depletion drive with a small gas cap and a weak water drive.

In addition, gravity segregation can also play an important role in any of these two drives. In general, combination drive reservoirs can be recognized by the occurrence of a combination of some of the following factors:

Reservoir pressure: These types of reservoirs usually experience a relatively rapid pressure decline. Water encroachment and/or external gas cap expansion are insufficient to maintain reservoir pressures.

Water production: The producing wells that are structurally located near the initial oil–water contact will slowly exhibit increasing water producing rates due to the increase in the water encroachment from the associated aquifer.

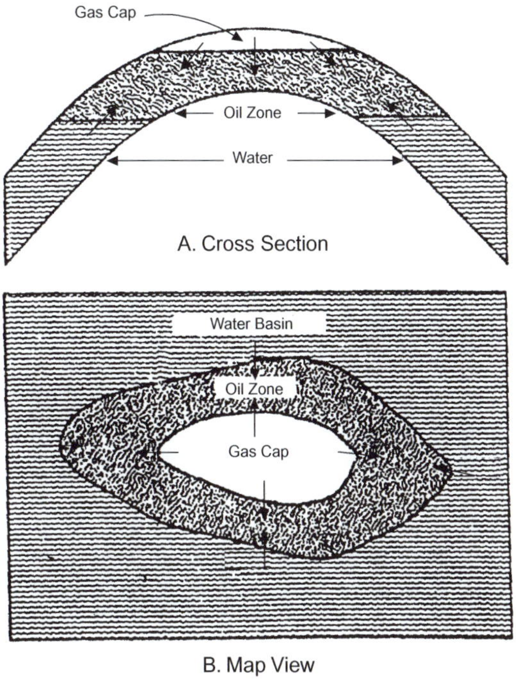

FIGURE 4.13 Combination drive reservoir. (*After Clark, N., 1969. Elements of Petroleum Reservoirs. Society of Petroleum Engineers, Dallas, TX*).

Gas–oil ratio: If a small gas cap is present, the structurally high wells will exhibit continually increasing gas–oil ratios, provided the gas cap is expanding. It is possible that the gas cap will shrink due to the production of excess free gas, in which case the structurally high wells will exhibit a decreasing gas–oil ratio. This condition should be avoided whenever possible, as large volumes of oil can be lost as a result of a shrinking gas cap.

Ultimate oil recovery: As a substantial percentage of the total oil recovery may be due to the depletion drive mechanism, the gas–oil ratio of structurally low wells will also continue to increase, due to the evolution of solution gas from the crude oil throughout the reservoir as pressure is reduced. Ultimate recovery from combination drive reservoirs is usually greater than recovery from depletion drive reservoirs but less than recovery from water drive or gas cap drive reservoirs. Actual recovery will depend upon the degree to which it is possible to reduce the magnitude of recovery by depletion drive. In most combination drive reservoirs it will be economically feasible to institute some type of pressure maintenance operation, either gas injection or water injection, or both gas and water injection, depending upon the availability of the fluids.

4.2 THE MATERIAL BALANCE EQUATION

The material balance equation (MBE) has long been recognized as one of the basic tools of reservoir engineers for interpreting and predicting reservoir performance. The MBE, when properly applied, can be used to:

- estimate initial hydrocarbon volumes in place;
- predict reservoir pressure;
- calculate water influx;
- predict future reservoir performance;
- predict ultimate hydrocarbon recovery under various types of primary drive mechanisms.

Although in some cases it is possible to solve the MBE simultaneously for the initial hydrocarbon volumes, i.e., oil and gas volumes, and the water influx, generally one or the other must be known from other data or methods that do not depend on the material balance calculations. The accuracy of the calculated values depends on the reliability of the available data, and whether the reservoir characteristics meet the assumptions that are associated with the development of the MBE. The equation is structured to simply keep inventory of all materials entering, leaving, and accumulating in the reservoir.

The concept of the MBE was presented by Schilthuis in 1936 and is simply based on the principle of the volumetric balance. It states that the cumulative withdrawal of reservoir fluids is equal to the combined effects of fluid expansion, pore volume compaction, and water influx. In its simplest form, the equation can be written on a volumetric basis as:

$$\text{Initial volume} = \text{Volume remaining} + \text{Volume removed}$$

Since oil, gas, and water are present in petroleum reservoirs, the MBE can be expressed for the total fluids or for any one of the fluids present. Three different forms of the MBE are presented below in details. These are:

(1) generalized MBE;
(2) MBE as an equation of a straight line;
(3) Tracy's form of the MBE.

4.3 GENERALIZED MBE

The MBE is designed to treat the reservoir as a *single tank* or a region that is characterized by homogeneous rock properties and described by an average pressure, i.e., no pressure variation throughout the reservoir at any particular time or stage of production. Therefore, the MBE is commonly referred to as a tank model or a zero-dimensional (0-D) model. These assumptions are of course unrealistic since reservoirs are generally considered heterogeneous with considerable variation in pressures throughout the reservoir. However, it is shown that the tank-type model accurately predicts the behavior of the reservoir in most cases if accurate average pressures and production data are available.

4.3.1 Basic Assumptions in the MBE

The MBE keeps an inventory on all material entering, leaving, or accumulating within a region over discrete periods of time during the production history. The calculation is most vulnerable to many of its underlying assumptions early in the depletion sequence when fluid movements are limited and pressure changes are small. Uneven depletion and partial reservoir development compound the accuracy problem.

The basic assumptions in the MBE are discussed in the following subsections.

Constant Temperature. Pressure–volume changes in the reservoir are assumed to occur without any temperature changes. If any temperature changes occur, they are usually sufficiently small to be ignored without significant error.

Reservoir Characteristics. The reservoir has uniform porosity, permeability, and thickness characteristics. In addition, the shifting in the gas–oil contact or oil–water contact is uniform throughout the reservoir.

Fluid Recovery. The fluid recovery is considered independent of the rate, number of wells, or location of the wells. The *time* element is not explicitly expressed in the material balance when applied to predict future reservoir performance.

Pressure Equilibrium. All parts of the reservoir have the same pressure, and fluid properties are therefore constant throughout. Minor variations in the vicinity of the wellbores may usually be ignored. Substantial pressure variation across the reservoir may cause excessive calculation error.

It is assumed that the *PVT* samples or datasets represent the actual fluid compositions and that reliable and representative laboratory procedures have been used. Notably, the vast majority of material balances assume that differential depletion data represents reservoir flow and that separator flash data may be used to correct for the wellbore transition to surface conditions. Such "black-oil" *PVT* treatments relate volume changes to temperature and pressure only. They lose validity in cases of volatile oil or gas condensate reservoirs where compositions are also important. Special laboratory procedures may be used to improve *PVT* data for volatile fluid situations.

Constant Reservoir Volume. Reservoir volume is assumed to be constant except for those conditions of rock and water expansion or water influx that are specifically considered in the equation. The formation is considered to be sufficiently competent that no significant volume change will occur through movement or reworking of the formation due to overburden pressure as the internal reservoir pressure is reduced. The constant-volume assumption also relates to an area of interest to which the equation is applied.

Reliable Production Data. All production data should be recorded with respect to the same time period. If possible, gas cap and solution gas production records should be maintained separately.

Gas and oil gravity measurements should be recorded in conjunction with the fluid volume data. Some reservoirs require a more detailed analysis and the material balance to be solved for volumetric segments. The produced fluid gravities will aid in the selection of the volumetric segments and also in the averaging of fluid properties. There are essentially three types of production data that must be recorded in order to use the MBE in performing reliable reservoir calculations. These are:

(1) Oil production data, even for properties not of interest, can usually be obtained from various sources and is usually fairly reliable.
(2) Gas production data is becoming more available and reliable as the market value of this commodity increases; unfortunately, this data will often be more questionable where gas is flared.
(3) The water production data represent only the net withdrawals of water; therefore, where subsurface disposal of produced brine is to the same source formation, most of the error due to poor data will be eliminated.

Developing the MBE. Before deriving the material balance, it is convenient to denote certain terms by symbols for brevity. The symbols used conform, where possible, to the standard nomenclature adopted by the Society of Petroleum Engineers.

p_i	Initial reservoir pressure, psi
p	Volumetric average reservoir pressure
Δp	Change in reservoir pressure $= p_i - p$, psi
p_b	Bubble point pressure, psi
N	Initial (original) oil-in-place, STB
N_p	Cumulative oil produced, STB
G_p	Cumulative gas produced, scf
W_p	Cumulative water produced
R_p	Cumulative gas–oil ratio, scf/STB
GOR	Instantaneous gas–oil ratio, scf/STB
R_{si}	Initial gas solubility, scf/STB
R_s	Gas solubility, scf/STB
B_{oi}	Initial oil formation volume factor, bbl/STB
B_o	Oil formation volume factor, bbl/STB
B_{gi}	Initial gas formation volume factor, bbl/scf
B_g	Gas formation volume factor, bbl/scf
W_{inj}	Cumulative water injected, STB
G_{inj}	Cumulative gas injected, scf
W_e	Cumulative water influx, bbl
m	Ratio of initial gas cap gas reservoir volume to initial reservoir oil volume, bbl/bbl
G	Initial gas cap gas, scf
PV	Pore volume, bbl
c_w	Water compressibility, psi^{-1}
c_f	Formation (rock) compressibility, psi^{-1}

Several of the material balance calculations require the total pore volume (PV) as expressed in terms of the initial oil volume N and the volume of the gas cap. The expression for the total PV can be derived by conveniently introducing the parameter m into the relationship as follows.

Define the ratio m as:

$$m = \frac{\text{Initial volume of gas cap in bbl}}{\text{Volume of oil initially in place in bbl}} = \frac{GB_{gi}}{NB_{oi}}$$

Solving for the volume of the gas cap gives:

$$\text{Initial volume of the gas cap,}\quad GB_{gi} = mNB_{oi},\ \text{bbl}$$

The total initial volume of the hydrocarbon system is then given by:

$$\text{Initial oil volume} + \text{Initial gas cap volume} = (PV)(1 - S_{wi})$$
$$NB_{oi} + mNB_{oi} = (\text{PV})(1 - S_{wi})$$

Solving for PV gives:

$$\text{PV} = \frac{NB_{oi}(1 + m)}{1 - S_{wi}} \qquad \textbf{(4.1)}$$

where

S_{wi} = initial water saturation
N = initial oil-in-place, STB
PV = total pore volume, bbl
m = ratio of initial gas cap gas reservoir volume to initial reservoir oil volume, bbl/bbl

Treating the reservoir PV as an idealized container as illustrated in Figure 4.14, volumetric balance expressions can be derived to account

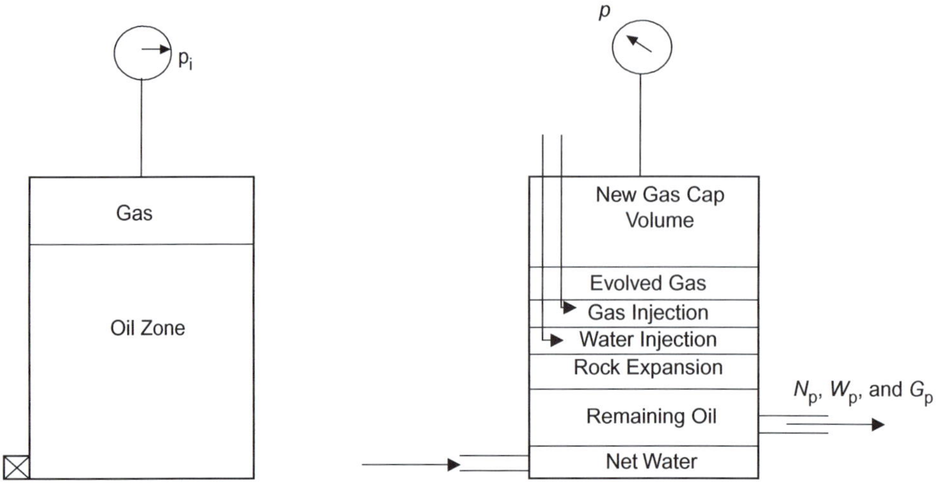

FIGURE 4.14 Tank-model concept.

for all volumetric changes that occur during the natural productive life of the reservoir. The MBE can be written in a generalized form as follows:

PV occupied by the oil initially in place at p_i
+ PV occupied by the gas in the gas cap at p_i
= PV occupied by the remaining oil at p
+ PV occupied by the gas in the gas cap at p
+ PV occupied by the evolved solution gas at p
+ PV occupied by the net water influx at p
+ change in PV due to connate water expansion and
+ pore volume reduction due to rock expansion
+ PV occupied by the injected gas at p
+ PV occupied by the injected water at p

(4.2)

The above eight terms composing the MBE can be determined separately from the hydrocarbon *PVT* and rock properties as follows.

Hydrocarbon PV occupied by the oil initially in place:

$$\text{Volume occupied by initial } \textit{oil-in-place} = NB_{oi}, \text{ bbl} \quad \textbf{(4.3)}$$

where

N = oil initially in place, STB
B_{oi} = oil formation volume factor at initial reservoir pressure p_i, bbl/STB

Hydrocarbon PV occupied by the gas in the gas cap:

$$\text{Volume of gas cap} = mNB_{oi}, \text{ bbl} \quad \textbf{(4.4)}$$

where m is a dimensionless parameter and defined as the ratio of gas cap volume to the oil zone volume.

Hydrocarbon PV occupied by the remaining oil:

$$\text{Volume of the remaining oil} = (N - N_p)B_o, \text{ bbl} \quad \textbf{(4.5)}$$

where

N_p = cumulative oil production, STB
B_o = oil formation volume factor at reservoir pressure p, bbl/STB

Hydrocarbon PV occupied by the gas cap at reservoir pressure p: As the reservoir pressure drops to a new level p, the gas in the gas cap expands and occupies a larger volume. Assuming no gas is produced from the gas cap during the pressure declines, the new volume of the gas cap can be determined as:

$$\text{Volume of the gas cap at } p = \left[\frac{mNB_{oi}}{B_{gi}}\right]B_g, \text{ bbl} \quad \textbf{(4.6)}$$

where

B_{gi} = gas formation volume factor at initial reservoir pressure, bbl/scf
B_g = current gas formation volume factor, bbl/scf

Hydrocarbon PV occupied by the evolved solution gas: Some of the solution gas that has been evolved from the oil will remain in the pore space and occupies a certain volume that can be determined by applying the following material balance on the solution gas:

$$\begin{bmatrix}\text{Volume of the evolved gas}\\ \text{that remains in the PV}\end{bmatrix} = \begin{bmatrix}\text{Volume of gas initially}\\ \text{in solution}\end{bmatrix} - [\text{Volume of gas produced}] - \begin{bmatrix}\text{Volume of gas remaining}\\ \text{in solution}\end{bmatrix} \quad (4.7)$$

where

N_p = cumulative oil produced, STB
R_p = net cumulative produced gas–oil ratio, scf/STB
R_s = current gas solubility factor, scf/STB
B_g = current gas formation volume factor, bbl/scf
R_{si} = gas solubility at initial reservoir pressure, scf/STB

PV occupied by the net water influx:

$$\text{Net water influx} = W_e - W_p B_w \quad (4.8)$$

where

W_e = cumulative water influx, bbl
W_p = cumulative water produced, STB
B_w = water formation volume factor, bbl/STB

Change in PV due to initial water and rock expansion: The component describing the reduction in the hydrocarbon PV due to the expansion of initial (connate) water and the reservoir rock cannot be neglected for an undersaturated oil reservoir. The water compressibility c_w and rock compressibility c_f are generally of the same order of magnitude as the compressibility of the oil. However, the effect of these two components can generally be neglected for gas cap drive reservoirs or when the reservoir pressure drops below the bubble point pressure.

The compressibility coefficient c, which describes the changes in the volume (expansion) of the fluid or material with changing pressure is given by:

$$c = \frac{-1}{V}\frac{\partial V}{\partial p}$$

or

$$\Delta V = Vc\,\Delta p$$

where ΔV represents the net changes or expansion of the material as a result of changes in the pressure. Therefore, the reduction in the PV due to the expansion of the connate water in the oil zone and the gas cap is given by:

$$\text{Connate water expansion} = [(PV)S_{wi}]\,c_w\,\Delta p$$

Substituting for PV with Eq. (4.1) gives:

$$\text{Expansion of connate } \textit{water} = \left[\frac{NB_{oi}(1+m)}{1-S_{wi}}S_{wi}\right]c_w\,\Delta p \quad (4.9)$$

where

Δp = change in reservoir pressure, $p_i - p$
c_w = water compressibility coefficient, psi^{-1}
m = ratio of the volume of the gas cap gas to the reservoir oil volume, bbl/bbl

Similarly, as fluids are produced and pressure declines, the entire reservoir PV is reduced (compaction), and this negative change in PV expels an equal volume of fluid as production. The reduction in the PV due to the expansion of the reservoir rock is given by:

$$\text{Change in PV} = \frac{NB_{oi}(1+m)}{1-S_{wi}}c_f\,\Delta p \quad (4.10)$$

Combining the expansions of the connate water and formation as represented by Eqs. (4.9) and (4.10) gives:

$$\text{Total changes in the PV} = NB_{oi}(1+m)\left(\frac{S_{wi}c_w + c_f}{1-S_{wi}}\right)\Delta p \quad (4.11)$$

The connate water and formation compressibilities are generally small in comparison to the compressibility of oil and gas. However, values of c_w and c_f are significant for undersaturated oil reservoirs, and they account for an appreciable fraction of the production above the bubble point. Ranges of compressibilities are given below:

Undersaturated oil	$5–50 \times 10^{-6}\ \text{psi}^{-1}$
Water	$2–4 \times 10^{-6}\ \text{psi}^{-1}$
Formation	$3–10 \times 10^{-6}\ \text{psi}^{-1}$
Gas at 1000 psi	$500–1000 \times 10^{-6}\ \text{psi}^{-1}$
Gas at 5000 psi	$50–200 \times 10^{-6}\ \text{psi}^{-1}$

PV occupied by the injection gas and water: Assuming that G_{inj} (volumes of gas) and W_{inj} volumes of water have been injected for pressure maintenance, the total PV occupied by the two injected fluids is given by:

$$\text{Total volume} = G_{inj} B_{ginj} + W_{inj} B_w \quad (4.12)$$

where

G_{inj} = cumulative gas injected, scf
B_{ginj} = injected gas formation volume factor, bbl/scf
W_{inj} = cumulative water injected, STB
B_w = water formation volume factor, bbl/STB

Combining Eqs. (4.3)–(4.12) with Eq. (4.2) and rearranging gives:

$$N = \frac{N_p B_o + (G_p - N_p R_s) B_g - (W_e - W_p B_w) - G_{inj} B_{ginj} - W_{inj} B_w}{(B_o - B_{oi}) + (R_{si} - R_s) B_g + m B_{oi}[(B_g/B_{gi}) - 1] + B_{oi}(1 + m)[(S_{wi} c_w + c_f)/(1 - S_{wi})]\Delta p} \quad (4.13)$$

where

N = initial oil-in-place, STB
G_p = cumulative gas produced, scf
N_p = cumulative oil produced, STB
R_{si} = gas solubility at initial pressure, scf/STB
m = ratio of gas cap gas volume to oil volume, bbl/bbl
B_{gi} = gas formation volume factor at p_i, bbl/scf
B_{ginj} = gas formation volume factor of the injected gas, bbl/scf

Recognizing that the cumulative gas produced G_p can be expressed in terms of the cumulative gas–oil ratio R_p and cumulative oil produced, then gives:

$$G_p = R_p N_p \quad (4.14)$$

Combining Eq. (4.14) with Eq. (4.13) gives:

$$N = \frac{N_p[B_o + (R_p - R_s)B_g] - (W_e - W_p B_w) - G_{inj} B_{ginj} - W_{inj} B_{wi}}{(B_o - B_{oi}) + (R_{si} - R_s)B_g + m B_{oi}[(B_g/B_{gi}) - 1] + B_{oi}(1 + m) \times [(S_{wi} c_w + c_f)/(1 - S_{wi})]\,\Delta p} \quad (4.15)$$

This relationship is referred to as the generalized MBE. A more convenient form of the MBE can be arrived at, by introducing the concept of the total (two-phase) formation volume factor B_t into the equation. This oil *PVT* property is defined as:

$$B_t = B_o + (R_{si} - R_s)B_g \quad (4.16)$$

Introducing B_t into Eq. (4.15) and assuming, for the sake of simplicity, that there is no water or gas injection gives:

$$N = \frac{N_p[B_t + (R_p - R_{si})B_g] - (W_e - W_p B_w)}{(B_t - B_{ti}) + m B_{ti}[B_g/B_{gi}/ - 1] + B_{ti}(1 + m)[(S_{wi} c_w + c_f)/(1 - S_{wi})]\,\Delta p} \quad (4.17)$$

(note that $B_{ti} = B_{oi}$) where

S_{wi} = initial water saturation
R_p = cumulative produced gas–oil ratio, scf/STB
Δp = change in the volumetric average reservoir pressure, psi
B_g = gas formation volume factor, bbl/scf

Example 4.1

The Anadarko Field is a combination drive reservoir. The current reservoir pressure is estimated at 2500 psi. The reservoir production data and *PVT* information are given below:

	Initial Reservoir Condition	Current Reservoir Condition
p, psi	3000	2500
B_o, bbl/STB	1.35	1.33
R_s, scf/STB	600	500
N_p, MMSTB	0	5
G_p, MMMscf	0	5.5
B_w, bbl/STB	1.00	1.00
W_e, MMbbl	0	3
W_p, MMbbl	0	0.2
B_g, bbl/scf	0.0011	0.0015
c_f, c_w	0	0

The following additional information is available:

Volume of bulk oil zone = 100,000 acres-ft

Volume of bulk gas zone = 20,000 acres-ft

Calculate the initial oil-in-place.

Solution

Step 1. Assuming the same porosity and connate water for the oil and gas zones, calculate m:

$$m = \frac{7758\phi(1 - S_{wi})(Ah)_{\text{gas cap}}}{7758\phi(1 - S_{wi})(Ah)_{\text{oil zone}}} = \frac{7758\phi(1 - S_{wi})20,000}{7758\phi(1 - S_{wi})100,000} = \frac{20,000}{100,000} = 0.2$$

Step 2. Calculate the cumulative gas–oil ratio R_p:

$$R_p = \frac{G_p}{N_p} = \frac{5.5 \times 10^9}{5 \times 10^6} = 1100 \text{ scf/STB}$$

Step 3. Solve for the initial oil-in-place by applying Eq. (4.15):

$$N = \frac{N_p[B_o + (R_p - R_s)B_g] - (W_e - W_p B_w)}{(B_o - B_{oi}) + (R_{si} - R_s)B_g + mB_{oi}[(B_g/B_{gi}) - 1] + B_{oi}(1 + m)[(S_{wi}c_w + c_f)(1 - S_{wi})]\,\Delta p}$$

$$= \frac{5 \times 10^6[1.33 + (1100 - 500)0.0015] - (3 \times 10^6 - 0.2 \times 10^6)}{(1.35 - 1.33) + (600 - 500)0.0015 + (0.2)(1.35) \times [(0.0015/0.0011) - 1]}$$

$$= 31.14 \text{ MMSTB}$$

4.3.2 Increasing Primary Recovery

It should be obvious that many steps can be taken to increase the ultimate primary recovery from a reservoir. Some of these steps can be surmised from the previous discussions, and others have been specifically noted when various subjects have been discussed. At this point we get involved with the problem of semantics when we attempt to define primary recovery. Strictly speaking, we can define secondary recovery as any production obtained using artificial energy in the reservoir. This automatically places pressure maintenance through gas or water injection in the secondary recovery category. Traditionally, most engineers in the oil patch prefer to think of pressure maintenance as an aid to primary recovery. It appears that we can logically classify the measures available for improving oil recovery during primary production as:

- well control procedures;
- reservoir control procedures, e.g., pressure maintenance.

Well Control. It should be stated that any steps taken to increase the oil or gas producing rate from an oil or gas reservoir generally increase the ultimate recovery from that reservoir by placing the economic limit further along the cumulative production scale. It is recognized that there is a particular rate of production at which the producing costs equal the operating expenses. Producing from an oil or gas well below this particular rate results in a net loss. If the productive capacity of a well can be increased, it is clear that additional oil will be produced before the economic rate is reached. Consequently, acidizing, paraffin control, sand control, clean-out, and other means actually increase ultimate production from that well.

It is clear that production of gas and water decreases the natural reservoir energy. If the production of gas and water from an oil reservoir can be minimized, a larger ultimate production may be obtained. The same concept can be similarly applied for minimizing the production of water from a gas reservoir.

Proper control of the individual well rate is a big factor in the control of gas and water coning or fingering. This general problem is not restricted to water drive and gas cap drive reservoirs. In a solution gas drive reservoir it may be possible to produce a well at a too high rate from an ultimate recovery standpoint because excessive drawdown of the producing well pressure results in an excessive gas–oil ratio and corresponding waste of the solution gas. The engineer should be aware of this possibility and test wells in a solution gas drive reservoir to see if the gas–oil ratio is sensitive.

It should be observed that excessive drawdown in a solution gas drive reservoir through excessive producing rates often causes excessive deposition of paraffin in the tubing and occasionally in the reservoir itself. Keeping gas-in-solution in the oil by keeping the well pressure as high as possible minimizes the paraffin deposition. Of course, deposition of paraffin in the tubing is not serious when compared to the deposition of paraffin in the reservoir. Given enough time and money, the paraffin can be cleaned from the tubing and flow lines. However, it is problematic whether paraffin deposited in the pores of the formation around the wellbore can be cleaned from these pores. Consequently, the operator should be very careful to avoid such deposition in the formation.

Another adverse effect that may be caused by an excess producing rate is the production of sand. Many unconsolidated formations tend to flow sand through perforations and into the producing system when flow rates are excessive. It may be possible to improve this situation with screens, gravel packing, or consolidating materials.

The proper positioning of wells in a reservoir also plays a big part in the control of gas and water production. It is obvious that wells should be positioned as far as possible from the original gas–oil, water–oil, and gas–water contacts in order to minimize the production of unwanted gas and water. The positioning of the producing wells must, of course, be consistent with the needs for reservoir drainage, the total reservoir producing capacity, and the cost of development.

In determining the proper well spacing to use in a particular reservoir, the engineer should make certain that full recognition is given to the pressure distribution that will prevail in the drainage area of a well when the economic limit is reached. In a continuous reservoir, there is no limit on the amount of reservoir that can be affected by one well. However, the engineer should be concerned with the additional oil that can be recovered prior to reaching the economic limit rate by increasing the drainage volume, or radius, of a well. In very tight reservoirs, we may be able to accomplish only a small reduction in the reservoir pressure in the additional reservoir volume. This effect may be nearly offset by the reduction of the well rate caused by the increase in the drainage radius. Thus, carc should be exercised to ensure that the greatest well spacing possible is also the most economical.

Total Reservoir Control. The effect of water and gas production on the recovery in an oil reservoir can be shown by solving Eq. (4.15) for the produced oil:

$$N_p = \frac{N[B_o - B_{oi} + (R_{si} - R_s)B_g + (c_f + c_w S_{wc})\Delta p B_{oi}/(1 - S_{wc})]}{B_o - R_s B_g} - \frac{B_g G_p - mNB_{oi}((B_g/B_{gi}) - 1) - W_e + W_p B_w}{B_o - R_s B_g}$$

It should be noted that the oil production obtainable at a particular reservoir pressure is almost directly reduced by the reservoir volume of gas ($G_p B_g$) and water produced ($W_p B_w$). Furthermore, the derivation of the MBE shows that the cumulative gas production G_p is the net produced gas defined as the produced gas less the injected gas. Similarly, if the water encroachment W_e is defined as the natural water encroachment, the produced water W_p must represent the net water produced, defined as the water produced less the water injected. Therefore, if produced water or produced gas can be injected without adversely affecting the amount of water or gas produced, the amount of oil produced at a particular reservoir pressure can be increased.

It is well known that the most efficient natural reservoir drive is water encroachment. The next most efficient is gas cap expansion, and the least efficient is solution gas drive. Consequently, it is important for the reservoir engineer to control production from a reservoir so that as little oil as possible is produced by solution gas drive and as much oil as possible is produced by water drive. However, when two or more drives operate in a reservoir, it is not always clear how much production results from each drive. One convenient method of estimating the amount of production resulting from each drive is to use material balance drive indices.

4.3.3 Reservoir Driving Indices

In a combination drive reservoir where all the driving mechanisms are simultaneously present, it is of a practical interest to determine the relative magnitude of each of the driving mechanisms and its contribution to the production. This objective can be achieved by rearranging Eq. (4.15) in the following generalized form:

$$\frac{N(B_t - B_{ti})}{A} + \frac{NmB_{ti}(B_g - B_{gi})/B_{gi}}{A} + \frac{W_e - W_p B_w}{A} + \frac{NB_{oi}(1+m)[(c_w S_{wi} + c_f)/(1 - S_{wi})](p_i - p)}{A} + \frac{W_{inj} B_{winj}}{A} + \frac{G_{inj} B_{ginj}}{A} = 1 \tag{4.18}$$

with the parameter A as defined by:

$$A = N_p[B_t + (R_p - R_{si})B_g] \tag{4.19}$$

Eq. (4.18) can be abbreviated and expressed as:

$$\text{DDI} + \text{SDI} + \text{WDI} + \text{EDI} + \text{WII} + \text{GII} = 1.0 \tag{4.20}$$

where

DDI = depletion drive index
SDI = segregation (gas cap) drive index
WDI = water drive index
EDI = expansion (rock and liquid) depletion index
WII = injected water index
GII = injected gas index

The numerators of the six terms in Eq. (4.18) represent the total net change in the volume due to gas cap and fluid expansions, net water influx, and fluid injection, while the denominator represents the cumulate reservoir voidage of produced oil and gas. Since the total volume increase must be equal to the total voidage, the sum of the four indices must therefore be necessarily equal to 1. Furthermore, the value of each index must be less than or equal to unity, but cannot be negative. The six terms on the left-hand side of Eq. (4.20) represent the six major primary driving mechanisms by which oil may be recovered from oil reservoirs. As presented earlier in this chapter, these driving forces are as follows:

Depletion drive: Depletion drive is the oil recovery mechanism wherein the production of the oil from its reservoir rock is achieved by the expansion of the original oil volume with all its original dissolved gas. This driving mechanism is represented mathematically by the first term of Eq. (4.18), or:

$$\text{DDI} = \frac{N(B_t - B_{ti})}{A} \tag{4.21}$$

where DDI is termed the depletion drive index.

Segregation drive: Segregation drive (gas cap drive) is the mechanism wherein the displacement of oil from the formation is accomplished by the expansion of the original free gas cap. This driving force is described by the second term of Eq. (4.18), or:

$$\mathrm{SDI} = \frac{NmB_{ti}(B_g - B_{gi})/B_{gi}}{A} \tag{4.22}$$

where SDI is termed the segregation drive index. It should be pointed out that it is usually impossible to eliminate the production of the gas cap gas and, thus, causes gas cap shrinkage. This distinct possibility of the shrinkage of the gas cap, and, therefore, reducing SDI, could be a result of the random location of producing wells. It will be necessary to eliminate gas cap shrinkage by either shutting in wells that produce gas from the gas cap or returning fluid to the gas cap to replace the gas that has been produced. It is a common practice to return some of the produced gas to the reservoir in order to maintain the size of the gas cap. In some cases, it has been more economical to return water instead of gas to the gas cap. This may be feasible when there are no facilities readily available for compressing the gas. Cole (1969) pointed out that this particular technique has been successfully applied in several cases, although the possibility of gravity segregation has to be considered.

Water drive: Water drive is the mechanism wherein the displacement of the oil is accomplished by the net encroachment of water into the oil zone. This mechanism is represented by the third term of Eq. (4.18), or:

$$\mathrm{WDI} = \frac{W_e - W_p B_w}{A} \tag{4.23}$$

where WDI is referred to as the water drive index.

Expansion drive index: For undersaturated oil reservoirs with no water influx, the principal source of energy is a result of the rock and fluid expansion as represented by the fourth term in Eq. (4.18) as:

$$\mathrm{EDI} = \frac{NB_{oi}(1+m)[(c_w S_{wi} + c_f)/(1 - S_{wi})](p_i - p)}{A}$$

When all the other three driving mechanisms are contributing to the production of oil and gas from the reservoir, the contribution of the rock and fluid expansion to the oil recovery is usually too small and essentially negligible and can be ignored.

Injected water drive index: The relative efficiency of the water injection pressure maintenance operations is expressed by:

$$\mathrm{WII} = \frac{W_{inj} B_{winj}}{A}$$

The magnitude of WII indicates the importance of the injected water as an improved recovery agent.

Injected gas drive index: Similar to the injected water drive index, the magnitude of its value indicates the relative importance of this drive index as compared to the other indices, as given by:

$$\mathrm{GII} = \frac{G_{inj} B_{ginj}}{A}$$

Note that for a depletion drive reservoir under pressure maintenance operations by gas injection, Eq. (4.20) is reduced to:

$$\mathrm{DDI} + \mathrm{EDI} + \mathrm{GII} = 1.0$$

Since the recovery by depletion drive and the expansion of the fluid and rock are usually poor, it is essential to maintain a high injected gas drive index. If the reservoir pressure can be maintained constant or declining at a slow rate, the values of DDI and EDI will be minimized because the changes in the numerators of both terms will essentially approach zeros. Theoretically, the highest recovery would occur at constant reservoir pressure; however, economic factors and feasibility of operation may dictate some pressure reduction.

In the absence of gas or water injection, Cole (1969) pointed out that since the sum of the remaining four driving indices is equal to 1, it follows that if the magnitude of one of the index terms is reduced, then one or both of

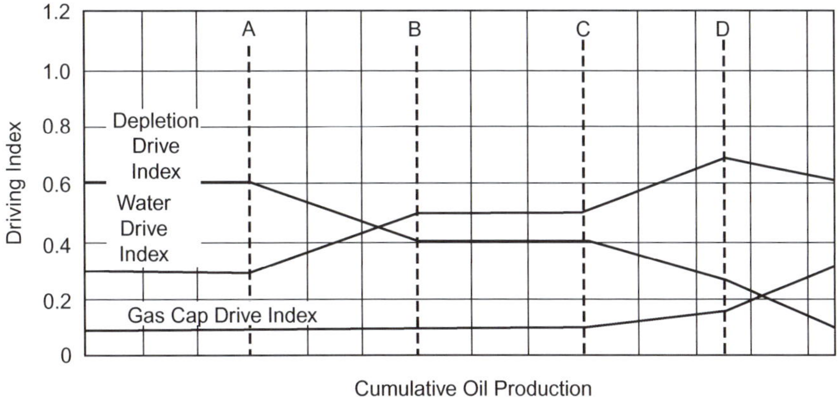

FIGURE 4.15 Driving indices in a combination drive reservoir. (*After Clark N.J., 1969. Elements of Petroleum Reservoirs, SPE*).

the remaining terms must be correspondingly increased. An effective water drive will usually result in maximum recovery from the reservoir. Therefore, if possible, the reservoir should be operated to yield a maximum water drive index and minimum values for the depletion drive index and the gas cap drive index. Maximum advantage should be taken of the most efficient drive available, and where the water drive is too weak to provide an effective displacing force, it may be possible to utilize the displacing energy of the gas cap. In any event, the depletion drive index should be maintained as low as possible at all times, as this is normally the most inefficient driving force available.

Eq. (4.20) can be solved at any time to determine the magnitude of the various driving indices. The forces displacing the oil and gas from the reservoir are subject to change from time to time, and for this reason, Eq. (4.20) should be solved periodically to determine whether there has been any change in the driving indices. Changes in fluid withdrawal rates are primarily responsible for changes in the driving indices. For example, reducing the oil producing rate could result in an increased water drive index and a correspondingly reduced depletion drive index in a reservoir containing a weak water drive. Also, by shutting in wells producing large quantities of water, the water drive index could be increased, as the net water influx (gross water influx minus water production) is the important factor.

When the reservoir has a very weak water drive but has a fairly large gas cap, the most efficient reservoir producing mechanism may be the gas cap drive, in which case a large gas cap drive index is desirable. Theoretically, recovery by gas cap drive is independent of producing rate, as the gas is readily expansible. Low vertical permeability could limit the rate of expansion of the gas cap, in which case the gas cap drive index would be rate sensitive. Also, gas coning into producing wells will reduce the effectiveness of the gas cap expansion due to the production of free gas. Gas coning is usually a rate-sensitive phenomenon: the higher the producing rates, the greater the amount of coning.

An important factor in determining the effectiveness of a gas cap drive is the degree of conservation of the gas cap gas. As a practical matter, it will often be impossible, because of royalty owners or lease agreements, to completely eliminate gas cap gas production. Where free gas is being produced, the gas cap drive index can often be markedly increased by shutting in high gas–oil-ratio wells and, if possible, transferring their allowables to other low gas–oil-ratio wells.

Figure 4.15 shows a set of plots that represent various driving indices for a combination drive reservoir. At point A some of the structurally low wells are reworked to reduce water production. This results in an effective increase in the water drive index. At point B workover operations are complete; water, gas, and oil producing rates are relatively stable; and the driving indices show no change. At point C some of the wells that have been producing relatively large, but constant, volumes of water are shut in, which results in an increase in the water drive index. At the same time some of the upstructure, high gas–oil-ratio wells have been shut in and their allowables transferred to wells lower on the structure producing with normal gas–oil ratios. At point D gas is being returned to the reservoir, and the gas cap drive index is exhibiting a decided increase. The water drive index is relatively constant, although it is decreasing somewhat, and the depletion drive index is showing a marked decline. This is indicative of a more efficient reservoir operation, and if the depletion drive index can be reduced to zero, relatively good recovery can be expected from the reservoir. Of course, to achieve a zero depletion drive index would require the complete maintenance of reservoir pressure, which is often difficult to accomplish. It can be noted from Figure 4.15 that the sum of the various drive indices is always equal to 1.

Example 4.2

A combination drive reservoir contains 10 MMSTB of oil initially in place. The ratio of the original gas cap volume to the original oil volume, i.e., m is estimated at 0.25. The initial reservoir pressure is 3000 psia at 150°F. The reservoir produced 1 MMSTB of oil, 1100 MMscf of gas of 0.8 specific gravity, and 50,000 STB of water by the time the reservoir pressure dropped to 2800 psi. The following *PVT* data is available:

	3000 psi	2800 psi
B_o, bbl/STB	1.58	1.48
R_s, scf/STB	1040	850
B_g, bbl/scf	0.00080	0.00092
B_t, bbl/STB	1.58	1.655
B_w, bbl/STB	1.000	1.000

The following data is also available:

$$S_{wi} = 0.20, \quad c_w = 1.5 \times 10^{-6}\ \text{psi}^{-1},$$
$$c_f = 1 \times 10^{-6}\ \text{psi}^{-1}$$

Calculate:

(a) the cumulative water influx;
(b) the net water influx;
(c) the primary driving indices at 2800 psi.

Solution

Because the reservoir contains a gas cap, the rock and fluid expansion can be neglected, i.e., set c_f and $c_w = 0$. However, for illustration purposes, the rock and fluid expansion term will be included in the calculations.

(a) The cumulative water influx:

Step 1. Calculate the cumulative gas–oil ratio R_p:

$$R_p = \frac{G_p}{N_p} = \frac{1100 \times 10^6}{1 \times 10^6} = 1100\ \text{scf/STB}$$

Step 2. Arrange Eq. (4.17) to solve for W_e:

$$\begin{aligned} W_e &= N_p\left[B_t + (R_p - R_{si})B_g\right] \\ &\quad - N\left[(B_t - B_{ti}) + mB_{ti}\left(\frac{B_g}{B_{gi}} - 1\right)\right. \\ &\quad \left. + B_{ti}(1+m)\left(\frac{S_{wi}c_w + c_f}{1 - S_{wi}}\right)\Delta p\right] + W_p B_{wp} \\ &= 10^6[1.655 + (1100 - 1040)0.00092] - 10^7 \\ &\quad \times \left[(1.655 - 1.58) + 0.25(1.58)\left(\frac{0.00092}{0.00080} - 1\right)\right. \\ &\quad + 1.58(1 + 0.25)\left(\frac{0.2(1.5 \times 10^{-6})}{1 - 0.2}\right) \\ &\quad \left. \times (3000 - 2800)\right] + 50,000 = 411,281\ \text{bbl} \end{aligned}$$

Neglecting the rock and fluid expansion term, the cumulative water influx is 417,700 bbl.

(b) The net water influx:

$$\begin{aligned} \text{Net water influx} &= W_e - W_p B_w = 411,281 - 50,000 \\ &= 361,281\ \text{bbl} \end{aligned}$$

(c) The primary recovery indices:

Step 1. Calculate the parameter A by using Eq. (4.19):

$$A = N_p[B_t + (R_p - R_{si})B_g]$$
$$= (1.0 \times 10^6)[1.655 + (1100 - 1040)0.00092]$$
$$= 1,710,000$$

Step 2. Calculate DDI, SDI, and WDI by applying Eqs. (4.21)–(4.23), respectively:

$$DDI = \frac{N(B_t - B_{ti})}{A} = \frac{10 \times 10^6(1.655 - 1.58)}{1,710,000} = 0.4385$$

$$SDI = \frac{NmB_{ti}(B_g - B_{gi})/B_{gi}}{A}$$
$$= \frac{10 \times 10^6(0.25)(1.58)(0.00092 - 0.0008)/0.0008}{1,710,000}$$
$$= 0.3465$$

$$WDI = \frac{W_e - W_p B_w}{A}$$
$$= \frac{411,281 - 50,000}{1,710,000} = 0.2112$$

Since

$$DDI + SDI + WDI + EDI = 1.0$$

then

$$EDI = 1 - 0.4385 - 0.3465 - 0.2112 = 0.0038$$

The above calculations show that 43.85% of the recovery was obtained by depletion drive, 34.65% by gas cap drive, 21.12% by water drive, and only 0.38% by connate water and rock expansion. The results suggest that the expansion drive index term can be neglected in the presence of a gas cap or when the reservoir pressure drops below the bubble point pressure. However, in high-PV compressibility reservoirs, such as chalks and unconsolidated sands, the energy contribution of the rock and water expansion cannot be ignored even at high gas saturations.

A source of error is often introduced in the MBE calculations when determining the average reservoir pressure and the associated problem of correctly weighting or averaging the individual well pressures. An example of such a problem is seen when the producing formations are composed of two or more zones of different permeabilities. In this case, the pressures are generally higher in the zone of low permeability and because the measured pressures are nearer to those in high-permeability zones, the measured static pressures tend to be lower and the reservoir behaves as if it contained less oil. Schilthuis explained this phenomenon by referring to the oil in the more permeable zones as active oil and by observing that the *calculated active oil usually increases with time* because the oil and gas in low-permeability zones slowly expand to offset the pressure decline. This is also true for fields that are not fully developed, because the average pressure can be that of the developed portion only, whereas the pressure is higher in the undeveloped portions. Craft and Hawkins (1991) pointed out that the effect of pressure errors on the calculated values of initial oil and water influx depends on the size of the errors in relation to the reservoir pressure decline. Notice that the pressure enters the MBE mainly when determining the *PVT* differences in terms of:

$$(B_o - B_{oi})$$
$$(B_g - B_{gi})$$
$$(R_{si} - R_s)$$

Because water influx and gas cap expansion tend to offset pressure decline, the pressure errors are more serious than for the undersaturated reservoirs. In the case of very active water drives or gas caps that are large compared to the oil zone, the MBE usually produces considerable errors when determining the initial oil-in-place because of the very small pressure decline.

Dake (1994) pointed out that there are two "necessary" conditions that must be satisfied for a meaningful application of the MBE to a reservoir:

(1) There should be adequate data collection in terms of production pressure, and *PVT*, in both frequency and quality for proper use of the MBE.

(2) It must be possible to define an average reservoir pressure trend as a function of time or production for the field.

Establishing an average pressure decline trend can be possible even if there are large pressure differentials across the field under normal conditions. Averaging *individual well pressure* declines can possibly be used to determine a uniform trend in the entire reservoir. The concept of average well pressure and its use in determining the reservoir volumetric average pressure was introduced in Chapter 1 as illustrated by Figure 1.24. This figure shows that if $(\overline{p})_j$ and V_j represent the pressure and volume drained by the jth well, respectively, the volumetric average pressure of the entire reservoir can be estimated from:

$$\overline{p}_r = \frac{\sum_j (\overline{p}V)_j}{\sum_j V_j}$$

where

V_j = the PV of the jth well drainage volume;
$(\overline{p})_j$ = volumetric average pressure *within the jth drainage volume.*

In practice, the V_j are difficult to determine, and therefore, it is common to use individual well flow rates q_i in determining the average reservoir pressure from individual well average drainage pressure. From the definition of the isothermal compressibility coefficient:

$$c = \frac{1}{V}\frac{\partial V}{\partial P}$$

differentiating with time gives:

$$\frac{\partial p}{\partial t} = \frac{1}{cV}\frac{\partial V}{\partial t}$$

or

$$\frac{\partial p}{\partial t} = \frac{1}{cV}(q)$$

This expression suggests that for a reasonably constant c at the time of measurement:

$$V \propto \frac{q}{\partial p/\partial t}$$

Since the flow rates are measured on a routine basis throughout the lifetime of the field, the average reservoir pressure can be alternatively expressed in terms of the individual well average drainage pressure decline rates and fluid flow rates by:

$$\overline{p}_r = \frac{\sum_j [(\overline{p}q)_j/(\partial\overline{p}/\partial t)_j]}{\sum_j [q_j/(\partial\overline{p}/\partial t)_j]}$$

However, since the MBE is usually applied at regular intervals of 3–6 months, i.e., $\Delta t = 3–6$ months, throughout the lifetime of the field, the average field pressure can be expressed in terms of the incremental net change in underground fluid withdrawal $\Delta(F)$ as:

$$\overline{p}_r = \frac{\sum_j \overline{p}_j \Delta(F)_j/\Delta\overline{p}_j}{\sum_j \Delta(F)_j/\Delta\overline{p}_j}$$

where the total underground fluid withdrawal at time t and $t + \Delta t$ are given by:

$$F_t = \int_0^t [Q_o B_o + Q_w B_w + (Q_g - Q_o R_s - Q_w R_{sw})B_g]\, dt$$

$$F_{t+\Delta t} = \int_0^{t+\Delta t} [Q_o B_o + Q_w B_w + (Q_g - Q_o R_s - Q_w R_{sw})B_g]\, dt$$

with

$$\Delta(F) = F_{t+\Delta t} - F_t$$

where

R_s = gas solubility, scf/STB
R_{sw} = gas solubility in the water, scf/STB
B_g = gas formation volume factor, bbl/scf
Q_o = oil flow rate, STB/day
Q_w = water flow rate, STB/day
Q_g = gas flow rate, scf/day

For a volumetric reservoir with total fluid production and initial reservoir pressure as the only available data, the average pressure can be *roughly approximated* using the following expression:

$$\overline{p}_r = p_i - \left[\frac{5.371 \times 10^{-6} F_t}{c_t(Ah\phi)}\right]$$

with the total fluid production F_t as defined above by:

$$F_t = \int_0^t [Q_o B_o + Q_w B_w + (Q_g - Q_o R_s - Q_w R_{sw})B_g]\, dt$$

where

A = well or reservoir drainage area, acres
h = thickness, ft
c_t = total compressibility coefficient, psi^{-1}
ϕ = porosity
p_i = initial reservoir pressure, psi

The above expression can be employed in an incremental manner, i.e., from time t to $t + \Delta t$, by:

$$(\bar{p}_r)_{t+\Delta t} = (\bar{p}_r)_t - \left[\frac{5.371 \times 10^{-6}\,\Delta F}{c_t(Ah\phi)}\right]$$

with

$$\Delta(F) = F_{t+\Delta t} - F_t$$

4.4 THE MBE AS AN EQUATION OF A STRAIGHT LINE

An insight into the general MBE, i.e., Eq. (4.15), may be gained by considering the physical significance of the following groups of terms from which it is comprised:

- $N_p[B_o + (R_p - R_s)B_g]$ represents the reservoir volume of cumulative oil and gas produced;
- $[W_e - W_pB_w]$ refers to the net water influx that is retained in the reservoir;
- $[G_{inj}B_{ginj} + W_{inj}B_w]$, the pressure maintenance term, represents cumulative fluid injection in the reservoir;
- $[mB_{oi}(B_g/B_{gi} - 1)]$ represents the net expansion of the gas cap that occurs with the production of N_p stock-tank barrels of oil (as expressed in bbl/STB of original oil-in-place).

  There are essentially three unknowns in Eq. (4.15):
  (1) the original oil-in-place N,
  (2) the cumulative water influx W_e,
  (3) the original size of the gas cap as compared to the oil zone size m.

In developing a methodology for determining the above three unknowns, Havlena and Odeh (1963, 1964) expressed Eq. (4.15) in the following form:

$$\begin{aligned} N_p\left[B_o + (R_p - R_s)B_g\right] + W_pB_w &= N\left[(B_o - B_{oi}) + (R_{si} - R_s)B_g\right] + mNB_{oi}\left(\frac{B_g}{B_{gi}} - 1\right) \\ &+ N(1+m)B_{oi} \times \left(\frac{c_wS_{wi} + c_f}{1 - S_{wi}}\right)\Delta p \\ &+ W_e + W_{inj}B_w + G_{inj}B_{ginj} \end{aligned} \quad (4.24)$$

Havlena and Odeh further expressed Eq. (4.24) in a more condensed form as:

$$F = N[E_o + mE_g + E_{f,w}] + (W_e + W_{inj}B_w + G_{inj}B_{ginj})$$

Assuming, for the purpose of simplicity, that no pressure maintenance by gas or water injection is being considered, the above relationship can be further simplified and written as:

$$F = N[E_o + mE_g + E_{f,w}] + W_e \quad (4.25)$$

in which the terms F, E_o, E_g, and $E_{f,w}$ are defined by the following relationships:

- F represents the underground withdrawal and is given by:

$$F = N_p[B_o + (R_p - R_s)B_g] + W_pB_w \quad (4.26)$$

  In terms of the two-phase formation volume factor B_t, the underground withdrawal F can be written as:

$$F = N_p[B_t + (R_p - R_{si})B_g] + W_pB_w \quad (4.27)$$

- E_o describes the expansion of oil and its originally dissolved gas and is expressed in terms of the oil formation volume factor as:

$$E_o = (B_o - B_{oi}) + (R_{si} - R_s)B_g \quad (4.28)$$

or, equivalently, in terms of B_t:

$$E_o = B_t - B_{ti} \quad (4.29)$$

- E_g is the term describing the expansion of the gas cap gas and is defined by the following expression:

$$E_g = B_{oi}\left(\frac{B_g}{B_{gi}} - 1\right) \quad \textbf{(4.30)}$$

In terms of the two-phase formation volume factor B_t, essentially $B_{ti} = B_{oi}$, or:

$$E_g = B_{ti}\left(\frac{B_g}{B_{gi}} - 1\right)$$

- $E_{f,w}$ represents the expansion of the initial water and the reduction in the PV and is given by:

$$E_{f,w} = (1+m)B_{oi}\left[\frac{c_w S_{wi} + c_f}{1 + S_{wi}}\right]\Delta p \quad \textbf{(4.31)}$$

Havlena and Odeh examined several cases of varying reservoir types with Eq. (4.25) and pointed out that the relationship can be rearranged in the form of a straight line. For example, in the case of a reservoir that has no initial gas cap (i.e., $m = 0$) or water influx (i.e., $W_e = 0$), and negligible formation and water compressibilities (i.e., c_f and $c_w = 0$), Eq. (4.25) reduces to:

$$F = NE_o$$

This expression suggests that a plot of the parameter F as a function of the oil expansion parameter E_o would yield a straight line with slope N and intercept equal to 0.

The straight-line method requires the plotting of a variable group vs. another variable group, with the variable group selection depending on the mechanism of production under which the reservoir is producing. The most important aspect of this method of solution is that it attaches significance to the sequence of the plotted points, to the direction in which they plot, and to the shape of the resulting plot.

The significance of the straight-line approach is that the sequence of plotting is important, and if the plotted data deviates from this straight line there is some reason for it. This significant observation will provide the engineer with valuable information that can be used in determining the following unknowns:

- initial oil-in-place N;
- size of the gas cap m;
- water influx W_e;
- driving mechanism;
- average reservoir pressure.

The applications of the straight-line form of the MBE in solving reservoir engineering problems are presented next to illustrate the usefulness of this particular form. Six cases of applications are presented and include:

Case 1: Determination of N in volumetric undersaturated reservoirs.
Case 2: Determination of N in volumetric saturated reservoirs.
Case 3: Determination of N and m in gas cap drive reservoirs.
Case 4: Determination of N and W_e in water drive reservoirs.
Case 5: Determination of N, m, and W_e in combination drive reservoirs.
Case 6: Determination of average reservoir pressure $\bar{p}$.

4.4.1 Case 1: Volumetric Undersaturated Oil Reservoirs

The linear form of the MBE as expressed by Eq. (4.25) can be written as:

$$F = N[E_o + mE_g + E_{f,w}] + W_e \quad \textbf{(4.32)}$$

Assuming no water or gas injection, several terms in the above relationship may disappear when imposing the conditions associated with the assumed reservoir driving mechanism. For a volumetric and undersaturated reservoir, the conditions associated with driving mechanism are:

$W_e = 0$ since the reservoir is volumetric;
$m = 0$ since the reservoir is undersaturated;

$R_s = R_{si} = R_p$ since all produced gas is dissolved in the oil.

Applying the above conditions on Eq. (4.25) gives:

$$F = N(E_o + E_{f,w}) \tag{4.33}$$

or

$$N = \frac{F}{E_o + E_{f,w}} \tag{4.34}$$

with

$$F = N_p B_o + W_p B_w \tag{4.35}$$

$$E_o = B_o - B_{oi} \tag{4.36}$$

$$E_{f,w} = B_{oi} \left[\frac{c_w S_w + c_f}{1 - S_{wi}} \right] \Delta p \tag{4.37}$$

$$\Delta p = p_i - \bar{p}_r$$

where

N = initial oil-in-place, STB
p_i = initial reservoir pressure
$\bar{p}_r$ = volumetric average reservoir pressure

When a new field is discovered, one of the first tasks of the reservoir engineer is to determine if the reservoir can be classified as a volumetric reservoir, i.e., $W_e = 0$. The classical approach of addressing this problem is to assemble all the necessary data (i.e., production, pressure, and *PVT*) that is required to evaluate the right-hand side of Eq. (4.34). The term $F/(E_o + E_{f,w})$ for each pressure and time observation is plotted vs. cumulative production N_p or time, as shown in Figure 4.16. Dake (1994) suggested that such a plot can assume two various shapes:

(1) If all the calculated points of $F/(E_o + E_{f,w})$ lie on a horizontal straight line (see line A in Figure 4.16; it implies that the reservoir can be classified as a volumetric reservoir. This defines a purely depletion drive reservoir whose energy derives solely from the expansion of the rock, the connate water, and the oil. Furthermore, the ordinate value of the plateau determines the initial oil-in-place *N*.

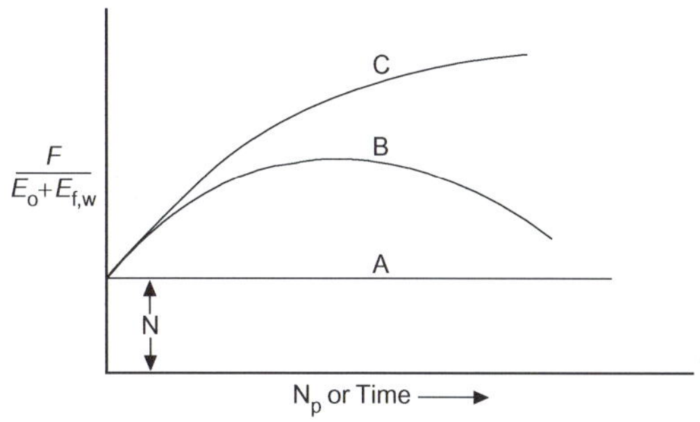

FIGURE 4.16 Classification of the reservoir.

(2) Alternately, if the calculated values of the term $F/(E_o + E_{f,w})$ rise, as illustrated by the curves B and C, it indicates that the reservoir has been energized by water influx, abnormal pore compaction, or a combination of these two. Curve B in Figure 4.16 might be for a strong water drive field in which the aquifer is displaying an infinite-acting behavior, whereas curve C represents an aquifer whose outer boundary had been felt, and the aquifer is depleting in unison with the reservoir itself. The downward trend in points on curve C as time progresses denotes the diminishing degree of energizing by the aquifer. Dake (1994) pointed out that in water drive reservoirs, the shape of the curve, i.e., $F/(E_o + E_{f,w})$ vs. time, is highly rate dependent. For instance, if the reservoir is producing at a higher rate than the water influx rate, the calculated values of $F/(E_o + E_{f,w})$ will dip downward, revealing a lack of energizing by the aquifer, whereas if the rate is decreased the reverse happens and the points are elevated.

Similarly Eq. (4.33) could be used to verify the characteristic of the reservoir driving mechanism and to determine the initial oil-in-place. A plot of the underground withdrawal *F* vs. the expansion term $(E_o + E_{f,w})$ should result in a straight line going through the origin with *N* being the slope. It should be noted that the origin is a "must" point; thus, one has a fixed point to guide the straight-line plot (as shown in Figure 4.17).

This interpretation technique is useful in that if the linear relationship is expected for the reservoir and yet the actual plot turns out to be nonlinear, then this deviation can itself be diagnostic in determining the actual drive mechanisms in the reservoir.

A linear plot of the underground withdrawal F vs. $(E_o + E_{f,w})$ indicates that the field is producing under volumetric performance, i.e., no water influx, and strictly by pressure depletion and fluid expansion. On the other hand, a nonlinear plot indicates that the reservoir should be characterized as a water drive reservoir.

Example 4.3

The Virginia Hills Beaverhill Lake Field is a volumetric undersaturated reservoir. Volumetric calculations indicate the reservoir contains 270.6 MMSTB of oil initially in place. The initial reservoir pressure is 3685 psi. The following additional data is available:

$S_{wi} = 24\%$, $B_w = 1.0$ bbl/STB
$c_w = 3.62 \times 10^{-6}\ \text{psi}^{-1}$, $p_b = 1500$ psi
$c_f = 4.95 \times 10^{-6}\ \text{psi}^{-1}$

The field production and *PVT* data is summarized below:

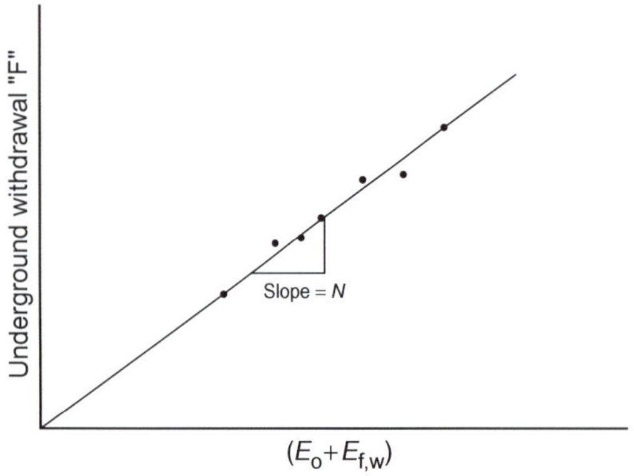

FIGURE 4.17 Underground withdrawal vs. $E_o + E_{f,w}$.

Volumetric Average Pressure	Number of Producing Wells	B_o (bbl/STB)	N_p (MSTB)	W_p (MSTB)
3685	1	1.3102	0	0
3680	2	1.3104	20.481	0
3676	2	1.3104	34.750	0
3667	3	1.3105	78.557	0
3664	4	1.3105	101.846	0
3640	19	1.3109	215.681	0
3605	25	1.3116	364.613	0
3567	36	1.3122	542.985	0.159
3515	48	1.3128	841.591	0.805
3448	59	1.3130	1273.53	2.579
3360	59	1.3150	1691.887	5.008
3275	61	1.3160	2127.077	6.500
3188	61	1.3170	2575.330	8.000

Calculate the initial oil-in-place by using the MBE and compare with the volumetric estimate of N.

Solution

Step 1. Calculate the initial water and rock expansion term $E_{f,w}$ from Eq. (4.37):

$$E_{f,w} = B_{oi}\left[\frac{c_w S_w + c_f}{1 - S_{wi}}\right]\Delta p$$

$$= 1.3102\left[\frac{3.62 \times 10^{-6}(0.24) + 4.95 \times 10^{-6}}{1 - 0.24}\right]\Delta p$$

$$= 10.0 \times 10^{-6}(3685 - \bar{p}_r)$$

Step 2. Construct the following table using Eqs. (4.35) and (4.36):

$$F = N_p B_o + W_p B_w E_o = B_o - B_{oi}$$

$$E_{f,w} = 10.0 \times 10^{-6}(3685 - \bar{p}_r)$$

$\bar{p}_r$ (psi)	F (Mbbl)	E_o (bbl/STB)	Δp	$E_{f,w}$	$E_o + E_{f,w}$
3685	—	—	0	0	—
3680	26.84	0.0002	5	50×10^{-6}	0.00025
3676	45.54	0.0002	9	90×10^{-6}	0.00029
3667	102.95	0.0003	18	180×10^{-6}	0.00048
3664	133.47	0.0003	21	210×10^{-6}	0.00051
3640	282.74	0.0007	45	450×10^{-6}	0.00115
3605	478.23	0.0014	80	800×10^{-6}	0.0022

3567	712.66	0.0020	118	1180×10^{-6}	0.00318
3515	1105.65	0.0026	170	1700×10^{-6}	0.0043
3448	1674.72	0.0028	237	2370×10^{-6}	0.00517
3360	2229.84	0.0048	325	3250×10^{-6}	0.00805
3275	2805.73	0.0058	410	4100×10^{-6}	0.0099
3188	3399.71	0.0068	497	4970×10^{-6}	0.0117

Step 3. Plot the underground withdrawal term F against the expansion term $(E_o + E_{f,w})$ on a Cartesian scale, as shown in Figure 4.18.

Step 4. Draw the best straight line through the points and determine the slope of the line and the volume of the active initial oil-in-place as:

$$N = 257 \text{ MMSTB}$$

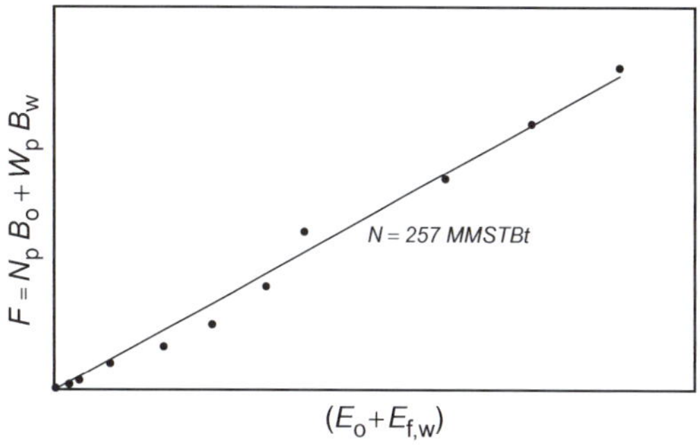

FIGURE 4.18 F vs. $E_o + E_{f,w}$ for example 4.3.

It should be noted that the value of the initial oil-in-place as determined from the MBE is referred to as the "effective" or "active" initial oil-in-place. This value is usually smaller than that of the volumetric estimate due to the oil being trapped in undrained fault compartments or low-permeability regions of the reservoir.

4.4.2 Case 2: Volumetric Saturated Oil Reservoirs

An oil reservoir that originally exists at *its bubble point pressure* is referred to as a "saturated oil reservoir." The main driving mechanism in this type of reservoir results from the liberation and expansion of the solution gas as the pressure drops below the bubble point pressure. The only unknown in a volumetric saturated oil reservoir is the initial oil-in-place N. Normally, the water and rock expansion term $E_{f,w}$ is negligible in comparison to the expansion of solution gas; however, it is recommended to include the term in the calculations. Eq. (4.32) can be simplified to give an identical form to that of Eq. (4.33), that is:

$$F = N(E_o + E_{f,w}) \quad \textbf{(4.38)}$$

However, the parameters F and E_o that constitute the above expression are given in an expanded form to reflect the reservoir condition as the pressure drops below the bubble point. The underground withdrawal F and the expansion term $(E_o + E_{f,w})$, see Eq. (4.38) are defined by:

F in terms of B_o	$F = N_p[B_o + (R_p - R_s)B_g] + W_p B_w$
or equivalently in terms of B_t	$F = N_p[B_t + (R_p - R_{si})B_g] + W_p B_w$
E_o in terms of B_o	$E_o = (B_o - B_{oi}) + (R_{si} - R_s)B_g$
or equivalently in terms of B_t	$E_o = B_t - B_{ti}$

and

$$E_{f,w} = B_{oi}\left[\frac{c_w S_w + c_f}{1 - S_{wi}}\right]\Delta p$$

Eq. (4.38) indicates that a plot of the underground withdrawal F, evaluated by using the actual reservoir production data, as a function of the fluid expansion term $(E_o + E_{f,w})$ should result in a straight line going through the origin with a slope of N.

The above interpretation technique is useful in that if a simple linear relationship such as Eq. (4.38) is expected for a reservoir and yet the actual plot turns out to be nonlinear, then this deviation can itself be diagnostic in determining the actual drive mechanisms in the reservoir. For instance, Eq. (4.38) may turn out to be nonlinear because there is an unsuspected water influx into the reservoir, helping to maintain the pressure.

Example 4.4

A volumetric undersaturated oil reservoir has a bubble point pressure of 4500 psi. The initial reservoir pressure is 7150 psia, and the volumetric calculations indicate the reservoir contains 650 MMSTB of oil initially in place. The field is a tight, naturally fractured, chalk reservoir and was developed without pressure support by water injection. The initial reservoir pressure is 3685 psi. The following additional data is available:[1]

$S_{wi} = 43\%$, $c_f = 3.3 \times 10^{-6}\ \text{psi}^{-1}$, $B_w = 1.0$ bbl/STB, $c_w = 3.00 \times 10^{-6}\ \text{psi}^{-1}$, $p_b = 1500$ psi

The field production and *PVT* data is summarized below:

p (psia)	Q_o (STB/day)	Q_g (MMscf/day)	B_o (bbl/STB)	R_s (scf/STB)	B_g (bbl/scf)	N_p (MMSTB)	R_p (scf/STB)
7150	–	–	1.743	1450	–	0	1450
6600	44,230	64.110	1.760	1450	–	8.072	1450
5800	79,326	115.616	1.796	1450	–	22.549	1455
4950	75,726	110.192	1.830	1450	–	36.369	1455
4500	–	–	1.850	1450	–	43.473	1447
4350	70,208	134.685	1.775	1323	0.000797	49.182	1576
4060	50,416	147.414	1.670	1143	0.000840	58.383	1788
3840	35,227	135.282	1.611	1037	0.000881	64.812	1992
3600	26,027	115.277	1.566	958	0.000916	69.562	2158
3480	27,452	151.167	1.523	882	0.000959	74.572	2383
3260	20,975	141.326	1.474	791	0.001015	78.400	2596
3100	15,753	125.107	1.440	734	0.001065	81.275	2785
2940	14,268	116.970	1.409	682	0.001121	83.879	2953
2800	13,819	111.792	1.382	637	0.001170	86.401	3103

Calculate the initial oil-in-place by using the MBE and compare with the volumetric estimate of *N*.

Solution

Step 1. For the undersaturated performance, the initial oil-in-place is described by Eq. (4.41) as:

$$N = \frac{F}{E_o + E_{f,w}}$$

where

$$F = N_p B_o$$

$$E_o = B_o - B_{oi}$$

$$E_{f,w} = B_{oi}\left[\frac{c_w S_w + c_f}{1 - S_{wi}}\right]\Delta p$$

$$= 1.743\left[\frac{3.00 \times 10^{-6}(0.43) + 3.30 \times 10^{-6}}{1 - 0.43}\right]\Delta p$$

$$= 8.05 \times 10^{-6}(7150 - \overline{p}_r)$$

Step 2. Calculate *N* using the undersaturated reservoir data:

$$F = N_p B_o$$

$$E_o = B_o - B_{oi} = B_o - 1.743$$

$$E_{f,w} = 8.05 \times 10^{-6}(7150 - \overline{p}_r)$$

$\overline{p}_r$ (psi)	F (MMbbl)	E_o (bbl/STB)	Δp (psi)	$E_{f,w}$ (bbl/STB)	$N = F/(E_o + E_{f,w})$ (MMSTB)
7150	–	–	0	0	–
6600	14.20672	0.0170	550	0.00772	574.7102
5800	40.49800	0.0530	1350	0.018949	562.8741
4950	66.55527	0.0870	2200	0.030879	564.6057
4500	80.42505	0.1070	2650	0.037195	557.752

The above calculations suggest that the initial oil-in-place as calculated from the undersaturated reservoir performance data is around 558 MMSTB, which is lower by

[1] Dake, L.P., 1994. *The Practice of Reservoir Engineering*. Elsevier, Amsterdam.

about 14% of the volumetric estimation of 650 MMSTB.

Step 3. Calculate N using the entire reservoir data:

$$F = N_p[B_o + (R_p - R_s)B_g]$$

$$E_o = (B_o - B_{oi}) + (R_{si} - R_s)B_g$$

$\bar{p}_r$ (psi)	F (MMbbl)	E_o (bbl/STB)	Δp (psi)	$E_{f,w}$ (bbl/STB)	$N = F/(E_o + E_{f,w})$ (MMSTB)
7150	–	–	0	0	–
6600	14.20672	0.0170	550	0.00772	574.7102
5800	40.49800	0.0530	1350	0.018949	562.8741
4950	66.55527	0.0870	2200	0.030879	564.6057
4500	80.42505	0.1070	2650	0.037195	557.752
4350	97.21516	0.133219	2800	0.09301	563.5015
4060	129.1315	0.184880	3090	0.043371	565.7429
3840	158.9420	0.231853	3310	0.046459	571.0827
3600	185.3966	0.273672	3550	0.048986	574.5924
3480	220.9165	0.324712	3670	0.051512	587.1939
3260	259.1963	0.399885	3890	0.054600	570.3076
3100	294.5662	0.459540	4050	0.056846	570.4382
2940	331.7239	0.526928	4210	0.059092	566.0629
2800	368.6921	0.590210	4350	0.061057	566.1154
Average					570.0000

It should be pointed out that as the reservoir pressures continues to decline below the bubble point and with increasing volume of the liberated gas, it reaches the time when the saturation of the liberated gas exceeds the critical gas saturation, and as a result, the gas will start to be produced in disproportionate quantities compared to the oil. At this stage of depletion, there is little that can be done to avert this situation during the primary production phase. As indicated earlier, the primary recovery from these types of reservoirs seldom exceeds 30%. However, under very favorable conditions, the oil and gas might separate with the gas moving structurally updip in the reservoir that might lead to preservation of the natural energy of the reservoir with a consequent improvement in overall oil recovery. Water injection is traditionally used by the oil industry to maintain the pressure above the bubble point pressure or alternatively to pressurize the reservoir to the bubble point pressure. In such type of reservoirs, as the reservoir pressure drops below the bubble point pressure, some volume of the liberated gas will remain in the reservoir as a free gas. This volume, as expressed in scf, is given by Eq. (4.30) as:

$$[\text{volume of the free gas in } scf] = NR_{si} - (N - N_p)R_s - N_pR_p$$

However, the total volume of the liberated gas at any depletion pressure is given by:

$$\begin{bmatrix}\text{Total volume of the}\\ \text{liberated gas in scf}\end{bmatrix} = NR_{si} - (N - N_p)R_s$$

Therefore, the fraction of the total solution gas that has been retained in the reservoir as a free gas α_g at any depletion stage is then given by:

$$\alpha_g = \frac{NR_{si} - (N - N_p)R_s - N_pR_p}{NR_{si} - (N - N_p)R_s} = 1 - \left[\frac{N_pR_p}{NR_{si} - (N - N_p)R_s}\right]$$

Alternatively, this can be expressed as a fraction of the total initial gas-in-solution by:

$$\alpha_{gi} = \frac{NR_{si} - (N - N_p)R_s - N_pR_p}{NR_{si}}$$

$$= 1 - \left[\frac{(N - N_p)R_s + N_pR_p}{NR_{si}}\right]$$

The calculation of the changes in the fluid saturations with declining reservoir pressure is an integral part of using the MBE. The remaining volume of each phase can be determined by calculating different phase saturation, recalling:

$$\text{Oil saturation } S_o = \frac{\text{Oil volume}}{\text{Pore volume}}$$

$$\text{Water saturation } S_w = \frac{\text{Water volume}}{\text{Pore volume}}$$

$$\text{Gas saturation } S_g = \frac{\text{Gas volume}}{\text{Pore volume}}$$

and

$$S_o + S_w + S_g = 1.0$$

If we consider a volumetric saturated oil reservoir that contains N stock-tank barrels of oil at the initial reservoir pressure p_i, i.e., p_b the initial oil saturation at the bubble point pressure is given by:

$$S_{oi} = 1 - S_{wi}$$

From the definition of oil saturation:

$$\frac{\text{Oil volume}}{\text{Pore volume}} = \frac{NB_{oi}}{\text{Pore volume}} = 1 - S_{wi}$$

or

$$\text{Pore volume} = \frac{NB_{oi}}{1 - S_{wi}}$$

If the reservoir has produced N_p stock-tank barrels of oil, the remaining oil volume is given by:

$$\text{Remaining oil volume} = (N - N_p)B_o$$

This indicates that for a volumetric-type oil reservoir, the oil saturation at any depletion state below the bubble point pressure can be represented by:

$$S_o = \frac{\text{Oil volume}}{\text{Pore volume}} = \frac{(N - N_p)B_o}{(NB_{oi}/(1 - S_{wi}))}$$

Rearranging:

$$S_o = (1 - S_{wi})\left(1 - \frac{N_p}{N}\right)\frac{B_o}{B_{oi}}$$

As the solution gas evolves from the oil with declining reservoir pressure, the gas saturation (assuming constant water saturation S_{wi}) is simply given by:

$$S_g = 1 - S_{wi} - S_o$$

or

$$S_g = 1 - S_{wi} - \left[(1 - S_{wi})\left(1 - \frac{N_p}{N}\right)\frac{B_o}{B_{oi}}\right]$$

Simplifying:

$$S_g = (1 - S_{wi})\left[1 - \left(1 - \frac{N_p}{N}\right)\frac{B_o}{B_{oi}}\right]$$

Another important function of the MBE is history matching the production–pressure data of individual wells. Once the reservoir pressure declines below the bubble point pressure, it is essential to perform the following tasks:

- Generate the pseudorelative permeability ratio k_{rg}/k_{ro} for the entire reservoir or for individual wells' drainage area.
- Assess the solution gas driving efficiency.
- Examine the field gas–oil ratio as compared to the laboratory solution gas solubility R_s to define the bubble point pressure and critical gas saturation.

The instantaneous gas–oil ratio (GOR), as discussed in detail in Chapter 5, is given by:

$$\text{GOR} = \frac{Q_g}{Q_o} = R_s + \left(\frac{k_{rg}}{k_{ro}}\right)\left(\frac{\mu_o B_o}{\mu_g B_g}\right)$$

This can be arranged to solve for the relative permeability ratio k_{rg}/k_{ro}, to give:

$$\left(\frac{k_{rg}}{k_{ro}}\right) = (\text{GOR} - R_s)\left(\frac{\mu_g B_g}{\mu_o B_o}\right)$$

One of the most practical applications of the MBE is its ability to generate the field relative permeability ratio as a function of gas saturation that can be used to adjust the laboratory core relative permeability data. *The main advantage of the field- or well-generated relative permeability ratio is that it incorporates some of the complexities of reservoir heterogeneity and degree of segregation of the oil and the evolved gas.*

It should be noted that the laboratory relative permeability data applies to an *unsegregated* reservoir, i.e., no change in fluid saturation with height. The laboratory relative permeability is most suitable for applications with the zero-dimensional tank model. For reservoirs with complete gravity segregation, it is possible to generate a pseudorelative permeability ratio k_{rg}/k_{ro}. A complete segregation means that the upper part of the reservoir contains gas and immobile oil, i.e., residual oil S_{or}, while the lower part contains oil and immobile gas that exists at its critical saturation S_{gc}. Vertical communication implies that as the gas evolves in the lower region, any gas with saturation above S_{gc} moves rapidly upward and leaves that region, while in the upper region any oil above S_{or} drains downward and moves into the lower region. On the basis of these assumptions, Poston (1987) proposed the following two relationships:

$$\frac{k_{rg}}{k_{ro}} = \frac{(S_g - S_{gc})(k_{rg})_{or}}{(S_o - S_{or})(k_{ro})_{gc}} \quad k_{ro} = \left[\frac{S_o - S_{or}(k_{rg})_{or}}{1 - S_w - S_{gc} - S_{or}}\right](k_{ro})_{gc}$$

where

$(k_{ro})_{gc}$ = relative permeability to oil at critical gas saturation
$(k_{go})_{or}$ = relative permeability to gas at residual oil saturation

If the reservoir is initially undersaturated, i.e., $p_i > p_b$, the reservoir pressure will continue to decline with production until it eventually reaches the bubble point pressure. It is recommended that the material calculations should be performed in two stages: first from p_i to p_b and second from p_b to different depletion pressures p. As the pressure declines from p_i to p_b, the following changes will occur as a result:

- Based on the water compressibility c_w, the connate water will expand with a resulting increase in the connate water saturation (provided that there is no water production).
- Based on the formation compressibility c_f, a reduction (compaction) in the entire reservoir pore volume will occur.

Therefore, there are several volumetric calculations that must be performed to reflect the reservoir condition at the bubble point pressure. These calculations are based on defining the following parameters:

- Initial oil-in-place N_i at the initial reservoir pressure p_i.
- Oil and water saturations ($S^{\backslash}_{oi}$ and $S^{\backslash}_{wi}$) at the initial reservoir pressure p_i.
- Cumulative oil produced at the bubble point pressure N_{pb}.
- Oil remaining at the bubble point pressure, i.e., *initial* oil at the bubble point:

$$N_b = N_i - N_{pb}$$

- Total pore volume at the bubble point pressure $(PV)_b$:

$$(PV)_b = \text{Remaining oil volume} + \text{Connate water volume} + \text{Connate water expansion} - \text{Reduction in PV due to compaction}$$

$$(PV)_b = (N_i - N_{pb})B_{ob} + \left[\frac{N_i B_{oi}}{1 - S^{\backslash}_{wi}}\right] S^{\backslash}_{wi} + \left[\frac{N_i B_{oi}}{1 - S_{wi}}\right](p_i - p_b)(-c_f + c_w S^{\backslash}_{wi})$$

Simplifying:

$$(PV)_b = (N_i - N_{pb})B_{ob} + \left[\frac{N_i B_{oi}}{1 - S^{\backslash}_{wi}}\right] \times \left[S^{\backslash}_{wi} + (p_i - p_b)(-c_f + c_w S^{\backslash}_{wi})\right]$$

- Initial oil and water saturations *at the bubble point pressure*, i.e., S_{oi} and S_{wi} are:

$$S_{oi} = \frac{(N_i - N_{pb})B_{ob}}{(PV)_b} = \frac{(N_i - N_{pb})B_{ob}}{(N_i - N_{pb})B_{ob} + [(N_i B_{oi})/(1 - S^{\backslash}_{wi})] \times [S^{\backslash}_{wi} + (p_i - p_b)(-c_f + c_w S^{\backslash}_{wi})]}$$

$$S_{wi} = \frac{[(N_i B_{oi})/(1 - S^{\backslash}_{wi})][S^{\backslash}_{wi} + (p_i - p_b)(-c_f + c_w S^{\backslash}_{wi})]}{(N_i - N_{pb})B_{ob} + [(N_i B_{oi})/(1 - S^{\backslash}_{wi})] \times [S^{\backslash}_{wi} + (p_i - p_b)(-c_f + c_w S^{\backslash}_{wi})]} = 1 - S_{oi}$$

- Oil saturation S_o at any pressure below p_b is given by:

$$S_o = \frac{(N_i - N_p)B_o}{(PV)_b} = \frac{(N_i - N_p)B_o}{(N_i - N_{pb})B_{ob} + [(N_i B_{oi})/(1 - S^{\backslash}_{wi})] \times [S^{\backslash}_{wi} + (p_i - p_b)(-c_f + c_w S^{\backslash}_{wi})]}$$

Gas saturation S_g at any pressure below p_b, assuming no water production, is given by:

$$S_g = 1 - S_o - S_{wi}$$

where

N_i = initial oil-in-place at p_i, i.e., $p_i > p_b$, STB
N_b = initial oil-in-place at the bubble point pressure, STB
N_{pb} = cumulative oil produced at the bubble point pressure, STB
S'_{oi} = oil saturation at p_i, $p_i > p_b$
S_{oi} = initial oil saturation at p_b
S'_{wi} = water saturation at p_i, $p_i > p_b$
S_{wi} = initial water saturation at p_b

It is very convenient also to qualitatively represent the fluid production graphically by employing the concept of the bubble map. The bubble map essentially illustrates the growing size of the drainage area of a production well. The drainage area of each well is represented by a circle with an oil bubble radius r_{ob} of:

$$r_{ob} = \sqrt{\frac{5.615 N_p}{\pi \phi h(((1 - S_{wi})/B_{oi}) - (S_o/B_o))}}$$

This expression is based on the assumption that the saturation is evenly distributed throughout a homogeneous drainage area where

r_{ob} = oil bubble radius, ft
N_p = well current cumulative oil production, bbl
S_o = current oil saturation

Similarly, the growing bubble of the reservoir free gas can be described graphically by calculating gas bubble radius r_{gb} as:

$$r_{gb} = \sqrt{\frac{5.615[NR_{si} - (N - N_p)R_s - N_p R_p]B_g}{\pi \phi h(1 - S_o - S_{wi})}}$$

where

r_{gb} = gas bubble radius, ft
N_p = well current cumulative oil production, bbl
B_g = current gas formation volume factor, bbl/scf
S_o = current oil saturation

Example 4.5

In addition to the data given in Example 4.4 for the chalk reservoir, the oil–gas viscosity ratios, a function of pressure, are included with the *PVT* data as shown below:

p (psia)	Q_o (STB/day)	Q_g (MMscf/day)	B_o (bbl/STB)	R_s (scf/STB)	B_g (bbl/scf)	μ_o/μ_g	N_p (MMSTB)	R_p (scf/STB)
7150	–	–	1.743	1450	–	–	0	1450
6600	44,230	64.110	1.760	1450	–	–	8.072	1450
5800	79,326	115.616	1.796	1450	–	–	22.549	1455
4950	75,726	110.192	1.830	1450	–	–	36.369	1455
4500	–	–	1.850	1450	–	5.60	43.473	1447
4350	70,208	134.685	1.775	1323	0.000797	6.02	49.182	1576
4060	50,416	147.414	1.670	1143	0.000840	7.24	58.383	1788
3840	35,227	135.282	1.611	1037	0.000881	8.17	64.812	1992
3600	26,027	115.277	1.566	958	0.000916	9.35	69.562	2158
3480	27,452	151.167	1.523	882	0.000959	9.95	74.572	2383
3260	20,975	141.326	1.474	791	0.001015	11.1	78.400	2596
3100	15,753	125.107	1.440	734	0.001065	11.9	81.275	2785
2940	14,268	116.970	1.409	682	0.001121	12.8	83.879	2953
2800	13,819	111.792	1.382	637	0.001170	13.5	86.401	3103

Using the given pressure–production history of the field, estimate the following:

- Percentage of the liberated solution gas retained in the reservoir as the pressure declines below the bubble point pressure. Express the retained gas volume as a percentage of the total gas liberated α_g and also of total initial gas-in-solution α_{gi}.

- Oil and gas saturations.
- Relative permeability ratio k_{rg}/k_{ro}.

Solution

Step 1. Tabulate the values of α_g and α_{gi} as calculated from:

$$\alpha_g = 1 - \left[\frac{N_p R_p}{NR_{si} - (N - N_p)R_s}\right] = 1 - \left[\frac{N_p R_p}{570(1450) - (570 - N_p)R_s}\right]$$

$$\alpha_{gi} = 1 - \left[\frac{(N - N_p)R_s + N_p R_p}{NR_{si}}\right] = 1 - \left[\frac{(570 - N_p)R_s + N_p R_p}{570(1450)}\right]$$

That is:

p (psia)	**R_s (scf/STB)**	**N_p (MMSTB)**	**R_p (scf/STB)**	**α_g (%)**	**α_{gi} (%)**
7150	1450	0	1450	0.00	0.00
6600	1450	8.072	1450	0.00	0.00
5800	1450	22.549	1455	0.00	0.00
4950	1450	36.369	1455	0.00	0.00
4500	1450	43.473	1447	0.00	0.00
4350	1323	49.182	1576	43.6	7.25
4060	1143	58.383	1788	56.8	16.6
3840	1037	64.812	1992	57.3	21.0
3600	958	69.562	2158	56.7	23.8
3480	882	74.572	2383	54.4	25.6
3260	791	78.400	2596	53.5	28.3
3100	734	81.275	2785	51.6	29.2
2940	682	83.879	2953	50.0	29.9
2800	637	86.401	3103	48.3	30.3

Step 2. Calculate the PV at the bubble point pressure from:

$$(PV)_b = (N_i - N_{pb})B_{ob} + \left[\frac{N_i B_{oi}}{1 - S^{\mathrm{I}}_{wi}}\right] \times \left[S^{\mathrm{I}}_{wi} + (p_i - p_b)(-c_f + c_w S^{\mathrm{I}}_{wi})\right]$$

$$= (570 - 43.473)1.85 + \left[\frac{570(1.743)}{1 - 0.43}\right] \times [0.43 + (7150 - 4500) \times (-3.3 \times 10^{-6} + 3.0 \times 10^{-6}(0.43))]$$

$$= 1.71 \times 10^9 \text{ bbl}$$

Step 3. Calculate the initial oil and water saturations at the bubble point pressure:

$$S_{oi} = \frac{(N_i - N_{pb})B_{ob}}{(PV)_b} = \frac{(570 - 43.473)10^6(1.85)}{1.71 \times 10^9} = 0.568$$

$$S_{wi} = 1 - S_{oi} = 0.432$$

Step 4. Calculate the oil and gas saturations as a function of pressure below p_b:

$$S_o = \frac{(N_i - N_p)B_o}{(PV)_b} = \frac{(570 - N_p)10^6 B_o}{1.71 \times 10^9}$$

Gas saturation S_g at any pressure below p_b is given by:

$$S_g = 1 - S_o - 0.432$$

p (psia)	**N_p (MMSTB)**	**S_o (%)**	**S_g (%)**
4500	43.473	56.8	0.00
4350	49.182	53.9	2.89
4060	58.383	49.8	6.98
3840	64.812	47.5	9.35
3600	69.562	45.7	11.1
3480	74.572	44.0	12.8
3260	78.400	42.3	14.6
3100	81.275	41.1	15.8
2940	83.879	40.0	16.9
2800	86.401	39.0	17.8

Step 5. Calculate the gas–oil ratio as function of pressure $p < p_b$:

$$\text{GOR} = \frac{Q_g}{Q_o}$$

p (psia)	**Q_o (STB/day)**	**Q_g (MMscf/day)**	**GOR = Q_g/Q_o (scf/STB)**
4500	–	–	1450
4350	70,208	134.685	1918
4060	50,416	147.414	2923
3840	35,227	135.282	3840
3600	26,027	115.277	4429
3480	27,452	151.167	5506

3260	20,975	141.326	6737
3100	15,753	125.107	7942
2940	14,268	116.970	8198
2800	13,819	111.792	8090

Step 6. Calculate the relative permeability ratio k_{rg}/k_{ro}:

$$\left(\frac{k_{rg}}{k_{ro}}\right) = (\text{GOR} - R_s)\left(\frac{\mu_g B_g}{\mu_o B_o}\right)$$

p (psi)	N_p (MMSTB)	S_o (%)	S_g (%)	R_s (scf/STB)	μ_o/μ_g	B_o (bbl/STB)	B_g (bbl/scf)	GOR = Q_g/Q_o (scf/STB)	k_{rg}/k_{ro}
4500	43.473	56.8	0.00	1450	5.60	1.850	—	1450	—
4350	49.182	53.9	2.89	1323	6.02	1.775	0.000797	1918	0.0444
4060	58.383	49.8	6.98	1143	7.24	1.670	0.000840	2923	0.1237
3840	64.812	47.5	9.35	1037	8.17	1.611	0.000881	3840	0.1877
3600	69.562	45.7	11.1	958	9.35	1.566	0.000916	4429	0.21715
3480	74.572	44.0	12.8	882	9.95	1.523	0.000959	5506	0.29266
3260	78.400	42.3	14.6	791	11.1	1.474	0.001015	6737	0.36982
3100	81.275	41.1	15.8	734	11.9	1.440	0.001065	7942	0.44744
2940	83.879	40.0	16.9	682	12.8	1.409	0.001121	8198	0.46807
2800	86.401	39.0	17.8	637	13.5	1.382	0.001170	8090	0.46585

If the laboratory relative permeability data is available, the following procedure is recommended for generating the field relative permeability data:

(1) Use as much past reservoir production and pressure history as possible to calculate relative permeability ratio k_{rg}/k_{ro} vs. S_o, as shown in Example 4.5.
(2) Plot the permeability ratio k_{rg}/k_{ro} vs. liquid saturation S_L, i.e., $S_L = S_o + S_{wc}$, on semilog paper.
(3) Plot the lab relative permeability data on the same graph prepared in step 2. Extend the field-calculated permeability data parallel to the lab data.
(4) Extrapolated field data from step 3 is considered the relative permeability characteristics of the reservoir and should be used in predicting the future reservoir performance.

It should be pointed out that it is a characteristic of most solution gas drive reservoirs that only a fraction of the oil-in-place is recoverable by primary depletion methods. However, the liberated solution gas can move much more freely than the oil through the reservoir. The displacement of the oil by the expanding liberated gas is essentially the main driving mechanism in these types of reservoirs. In general, it is possible to estimate the amount of gas that will be recovered during the primary depletion method, which can provide us with an estimate of an *end point*, i.e., maximum, on the oil recovery performance curve. A log–log plot of the cumulative gas (on the y axis) vs. cumulative oil (on the x axis) *provides the recovery trend of the hydrocarbon recovery*. The generated curve can be extrapolated to the total gas available, e.g., (NR_{si}), and to read the upper limit on oil recovery at abandonment.

Example 4.6

Using the data given in Example 4.5, estimate the oil recovery factor and cumulative oil production after producing 50% of the solution gas.

Solution

Step 1. Using in-place values from Example 4.5 and from the definition of the recovery factor, construct the following table:

Oil-in-place $N = 570$ MMSTB

Gas-in-solution $G = NR_{si} = 570 \times 1450 = 826.5$ MMMscf

Cumulative gas produced $G_p = N_p R_p$

Oil recovery factor $\text{RF} = \dfrac{N_p}{N}$

Gas recovery factor $\text{RF} = \dfrac{G_p}{G}$

Months	p (psia)	N_p (MMSTB)	R_p (scf/STB)	$G_p = N_p R_p$ (MMMscf)	Oil RF (%)	Gas RF (%)
0	7150	0	1450	0	0	0
6	6600	8.072	1450	11.70	1.416	1.411
12	5800	22.549	1455	32.80	4.956	3.956
18	4950	36.369	1455	52.92	6.385	6.380
21	4500	43.473	1447	62.91	7.627	7.585
24	4350	49.182	1576	77.51	8.528	9.346
30	4060	58.383	1788	104.39	10.242	12.587
36	3840	64.812	1992	129.11	11.371	15.567
42	3600	69.562	2158	150.11	12.204	18.100
48	3480	74.572	2383	177.71	13.083	21.427
54	3260	78.400	2596	203.53	13.754	24.540
60	3100	81.275	2785	226.35	14.259	27.292
66	2940	83.879	2953	247.69	14.716	29.866
72	2800	86.401	3103	268.10	15.158	32.327

Step 2. From the log–log plot of N_p vs. G_p and the Cartesian plot of oil recovery factor vs. gas recovery factor, as shown in Figures 4.19 and 4.20:

Oil recovery factor = 17%

Cumulative oil $N_p = 0.17 \times 570 = 96.9$ MMSTB

Cumulative gas $G_p = 0.50 \times 826.5 = 413.25$ MMMscf

4.4.3 Case 3: Gas Cap Drive Reservoirs

For a reservoir in which the expansion of the gas cap gas is the predominant driving mechanism, the effect of water and pore compressibilities as a contributing driving mechanism can be considered negligible as compared to that of the high compressibility of the gas. However, Havlena and Odeh (1963, 1964) acknowledged that whenever a gas cap is present or its size is to be determined, an exceptional degree of accuracy of pressure data is required. The specific problem with reservoir pressure is that the underlying oil zone in a gas cap drive reservoir exists initially near its bubble point pressure. Therefore, the flowing pressures are obviously below the bubble point pressure, which exacerbates the difficulty in conventional pressure buildup interpretation to determine average reservoir pressure.

Assuming that there is no natural water influx or it is negligible (i.e., $W_e = 0$), the Havlena and Odeh material balance can be expressed as:

$$F = N[E_o + mE_g] \tag{4.39}$$

in which the variables F, E_o, and E_g are given by:

$$F = N_p\left[B_o + (R_p - R_s)B_g\right] + W_p B_w$$
$$= N_p\left[B_t + (R_p - R_{si})B_g\right] + W_p B_w$$
$$E_o = (B_o - B_{oi}) + (R_{si} - R_s)B_g$$
$$= B_t - B_{ti}$$
$$E_g = B_{oi}\left(\frac{B_g}{B_{gi}} - 1\right)$$

The methodology in which Eq. (4.39) can be used depends on the number of unknowns in the equation. There are three possible unknowns in Eq. (4.39). These are:

(1) N is unknown, m is known;
(2) m is unknown, N is known;
(3) N and m are unknown.

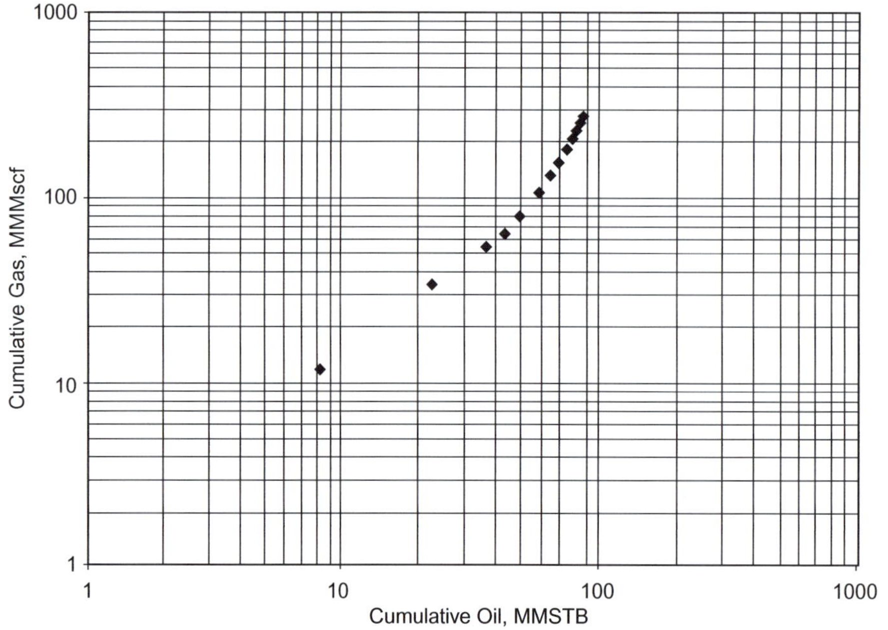

FIGURE 4.19 G_p vs. N_p, Example 4.6.

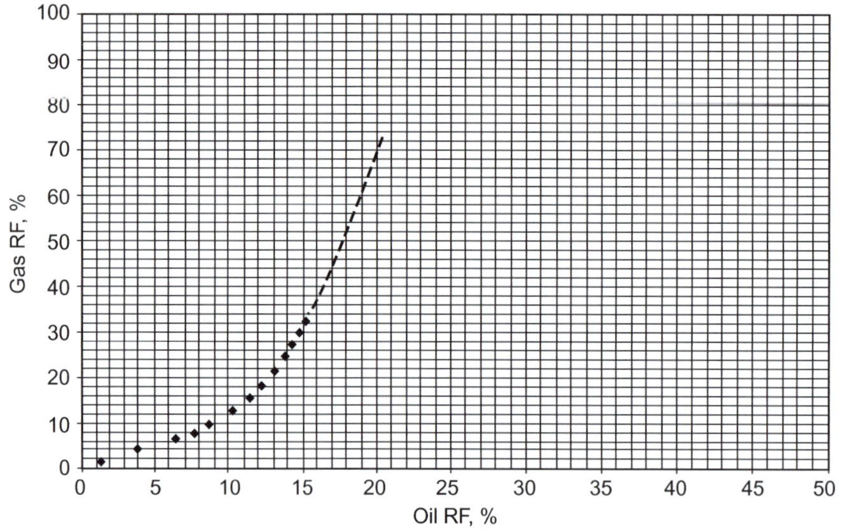

FIGURE 4.20 Gas recovery factor vs. oil recovery factor.

The practical use of Eq. (4.39) in determining the three possible unknowns is presented below.

Unknown N, known m: Eq. (4.39) indicates that a plot of F vs. $(E_o + mE_g)$ on a Cartesian scale would produce a straight line through the origin with a slope of N, as shown in Figure 4.21. In making the plot, the underground withdrawal F can be

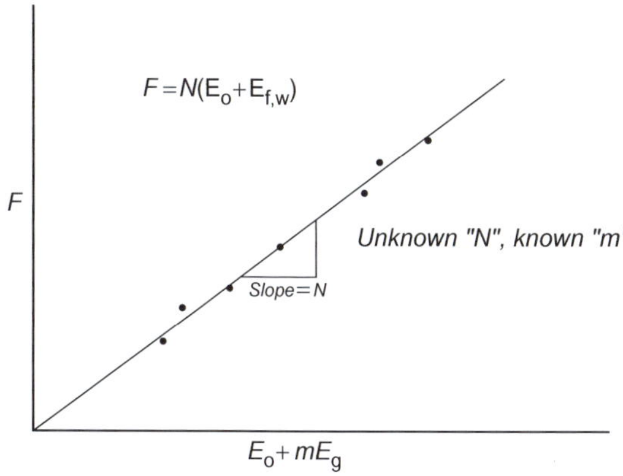

FIGURE 4.21 F vs. $E_o + mE_g$.

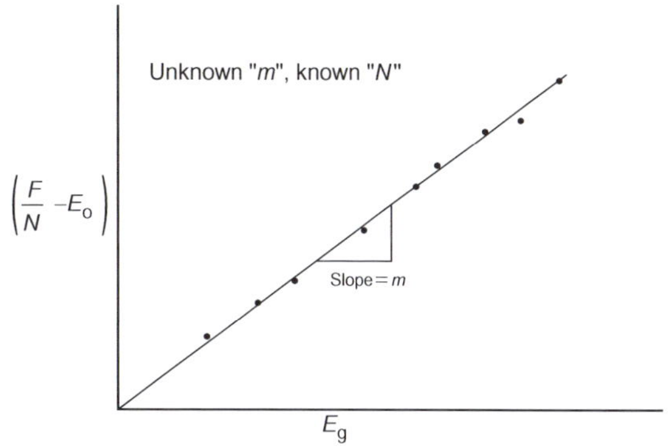

FIGURE 4.22 $(F/N - E_o)$ vs. E_g.

calculated at various times as a function of the production terms N_p and R_p.

Conclusion $N = \text{slope}$

Unknown m, known N: Eq. (4.39) can be rearranged as an equation of straight line, to give:

$$\left(\frac{F}{N} - E_o\right) = mE_g \tag{4.40}$$

This relationship shows that a plot of the term $(F/N - E_o)$ vs. E_g would produce a straight line with a slope of m. One advantage of this particular arrangement is that the straight line must pass through the origin that, therefore, acts as a control point. Figure 4.22 shows an illustration of such a plot.

Conclusion $m = \text{slope}$

Also Eq. (4.39) can be rearranged to solve for m, to give:

$$m = \frac{F - NE_o}{NE_g}$$

This relationship shows that a plot of the term $(F/N - E_o)$ vs. E_g would produce a straight line with a slope of m. One advantage of this particular arrangement is that the straight line must pass through the origin.

N and m are unknown: If there is uncertainty in both the values of N and m, Eq. (4.39) can be re-expressed as:

$$\frac{F}{E_o} = N + mN\left(\frac{E_g}{E_o}\right) \tag{4.41}$$

A plot of F/E_o vs. E_g/E_o should then be linear with intercept N and slope mN. This plot is illustrated in Figure 4.23.

Conclusions $N = \text{Intercept}$

$mN = \text{Slope}$

$$m = \frac{\text{Slope}}{\text{Intercept}} = \frac{\text{Slope}}{N}$$

Example 4.7

Reliable volumetric calculations on a well-developed gas cap drive reserve show the following results:

$N = 736$ MMSTB, $G = 320$ Bscf
$p_i = 2808$ psia, $B_{oi} = 1.39$ bbl/STB
$B_{gi} = 0.000919$ bbl/STB, $R_{si} = 755$ scf/STB

The production history in terms of parameter F and the *PVT* data are given below:

$\bar{p}$ (psi)	F (MMbbl)	B_t (bbl/STB)	B_g (bbl/scf)
2803	7.8928	1.3904	0.0009209
2802	7.8911	1.3905	0.0009213

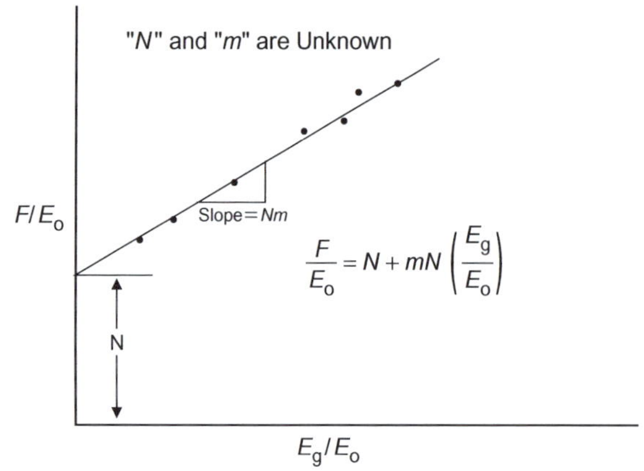

FIGURE 4.23 F/E_o vs. E_g/E_o.

2801	7.8894	1.3906	0.0009217
2800	7.8877	1.3907	0.0009220
2799	7.8860	1.3907	0.0009224
2798	7.8843	1.3908	0.0009228

Estimate the gas–oil volume ratio m and compare with the calculated value.

Solution

Step 1. Calculate the actual m from the results of the volumetric calculation:

$$m = \frac{GB_{gi}}{NB_{oi}} = \frac{(3200 \times 10^9)(0.000919)}{(736 \times 10^6)(1.390)} \approx 2.9$$

Step 2. Using the production data, calculate E_o, E_g, and m:

$$E_o = B_t - B_{ti}$$

$$E_g = B_{ti}\left(\frac{B_g}{B_{gi}} - 1\right)$$

$$m = \frac{F - NE_o}{NE_g}$$

$\bar{p}$ (psi)	F (MMbbl)	E_o (bbl/STB)	E_g (bbl/scf)	$m = (F - NE_o)/NE_g$
2803	7.8928	0.000442	0.002874	3.58
2802	7.8911	0.000511	0.003479	2.93
2801	7.8894	0.000581	0.004084	2.48
2800	7.8877	0.000650	0.004538	2.22
2799	7.8860	0.000721	0.005143	1.94
2798	7.8843	0.000791	0.005748	1.73

The above tabulated results appear to confirm the volumetric m value of 2.9; however, the results also show the sensitivity of the m value to the reported average reservoir pressure.

Example 4.8

The production history and the *PVT* data of a gas cap drive reservoir are given below:

Date	$\bar{p}$ (psi)	N_p (MSTB)	G_p (Mscf)	B_t (bbl/STB)	B_g (bbl/scf)
5/1/89	4415	–	–	1.6291	0.00077
1/1/91	3875	492.5	751.3	1.6839	0.00079
1/1/92	3315	1015.7	2409.6	1.7835	0.00087
1/1/93	2845	1322.5	3901.6	1.9110	0.00099

The initial gas solubility R_{si} is 975 scf/STB. Estimate the initial oil- and gas-in-place.

Solution

Step 1. Calculate the cumulative produced gas–oil ratio R_p:

$\bar{p}$	G_p (Mscf)	N_p (MSTB)	$R_p = G_p/N_p$ (scf/STB)
4415	–	–	–
3875	751.3	492.5	1525
3315	2409.6	1015.7	2372
2845	3901.6	1322.5	2950

Step 2. Calculate F, E_o, and E_g from:

$$F = N_p\left[B_t + (R_p - R_{si})B_g\right] + W_pB_w$$

$$E_o = B_t - B_{ti}$$

$$E_g = B_{ti}\left(\frac{B_g}{B_{gi}} - 1\right)$$

$\bar{p}$	F	E_o	E_g
3875	2.04×10^6	0.0548	0.0529
3315	8.77×10^6	0.1540	0.2220
2845	17.05×10^6	0.2820	0.4720

Step 3. Calculate F/E_o and E_g/E_o:

$\bar{p}$	F/E_o	E_g/E_o
3875	3.72×10^7	0.96
3315	5.69×10^7	0.44
2845	6.00×10^7	0.67

Step 4. Plot F/E_o vs. E_g/E_o as shown in Figure 4.24, to give:

$$\text{Intercept} = N = 9 \text{ MMSTB}$$

$$\text{Slope} = Nm = 3.1 \times 10^7$$

Step 5. Calculate m:

$$m = \frac{3.1 \times 10^7}{9 \times 10^6} = 3.44$$

Step 6. Calculate the initial gas cap gas volume G from the definition of m:

$$m = \frac{GB_{gi}}{NB_{oi}}$$

or

$$G = \frac{mNB_{oi}}{B_{gi}} = \frac{(3.44)(9 \times 10^6)(1.6291)}{0.00077}$$
$$= 66 \text{ MMMscf}$$

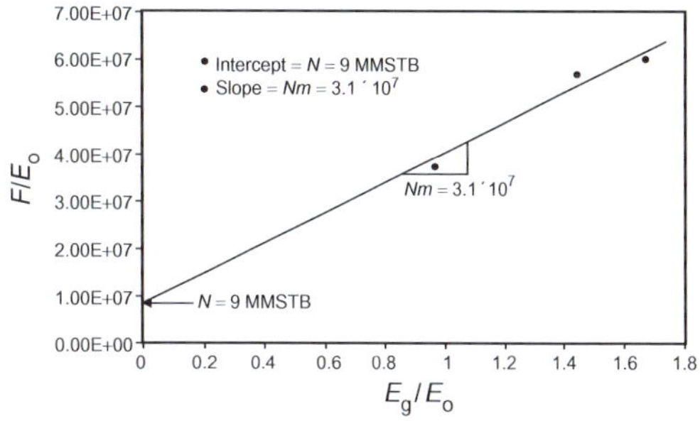

FIGURE 4.24 Calculation of m and N for Example 4.8.

4.4.4 Case 4: Water Drive Reservoirs

In a water drive reservoir, identifying the type of the aquifer and characterizing its properties are perhaps the most challenging tasks involved in conducting a reservoir engineering study. Yet, without an accurate description of the aquifer, future reservoir performance and management cannot be properly evaluated.

The full MBE can be expressed again as:

$$F = N(E_o + mE_g + E_{f,w}) + W_e$$

Dake (1978) pointed out that the term $E_{f,w}$ can frequently be neglected in water drive reservoirs. This is not only for the usual reason that the water and pore compressibilities are small, but also because water influx helps to maintain the reservoir pressure, and therefore, the Δp appearing in the $E_{f,w}$ term is reduced, or:

$$F = N(E_o + mE_g) + W_e \quad \textbf{(4.42)}$$

If, in addition, the reservoir has an initial gas cap, then Eq. (4.41) can be further reduced to:

$$F = NE_o + W_e \quad \textbf{(4.43)}$$

In attempting to use the above two equations to match the production and pressure history of a reservoir, the greatest uncertainty is always the determination of the water influx W_e. In fact, in order to calculate the water influx the engineer is confronted with what is inherently the greatest uncertainty in the whole subject of reservoir engineering. The reason is that the calculation of W_e requires a mathematical model that itself relies on the knowledge of aquifer properties. These, however, are seldom measured since wells are not deliberately drilled into the aquifer to obtain such information.

For a water drive reservoir with no gas cap, Eq. (4.43) can be rearranged and expressed as:

$$\frac{F}{E_o} = N + \frac{W_e}{E_o} \quad \textbf{(4.44)}$$

Several water influx models have been described in Chapter 2, including:

- the pot aquifer model;
- the Schilthuis steady-state method;
- the van Everdingen and Hurst model.

The use of these models in connection with Eq. (4.44) to simultaneously determine N and W_e is described below.

Pot Aquifer Model in the MBE. Assume that the water influx could be properly described by using the simple pot aquifer model as described by Eq. (2.5):

$$W_e = (c_w + c_f) W_i f (p_i - p)$$
$$f = \frac{(\text{Encroachment angle})^\circ}{360^\circ} = \frac{\theta}{360^\circ} \tag{4.45}$$
$$W_i = \left[\frac{\pi (r_a^2 - r_e^2) h \phi}{5.615}\right]$$

where

r_a = radius of the aquifer, ft
r_e = radius of the reservoir, ft
h = thickness of the aquifer, ft
ϕ = porosity of the aquifer
θ = encroachment angle
c_w = aquifer water compressibility, psi^{-1}
c_f = aquifer rock compressibility, psi^{-1}
W_i = initial volume of water in the aquifer, bbl

Since the ability to use Eq. (4.45) relies on the knowledge of the aquifer properties, i.e., c_w, c_f, h, r_a, and θ, these properties could be combined and treated as one unknown K in Eq. (4.45), or:

$$W_e = K \, \Delta p \tag{4.46}$$

where the water influx constant K represents the combined pot aquifer properties as:

$$K = (c_w + c_f) W_i f$$

Combining Eq. (4.46) with Eq. (4.44) gives:

$$\frac{F}{E_o} = N + K\left(\frac{\Delta p}{E_o}\right) \tag{4.47}$$

Eq. (4.47) indicates that a plot of the term F/E_o as a function of $\Delta p/E_o$ would yield a straight line with an intercept of N and a slope of K, as illustrated in Figure 4.25.

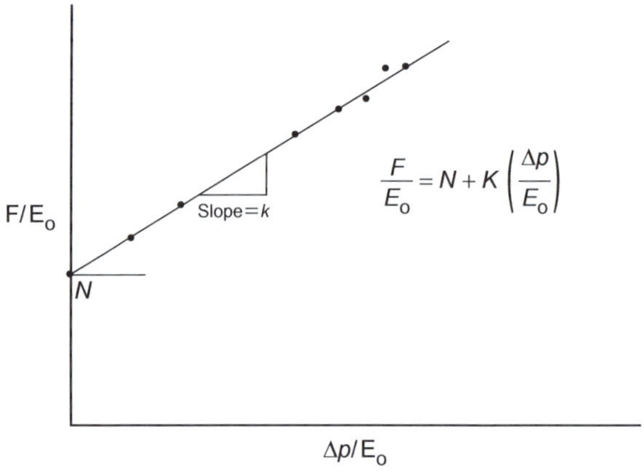

FIGURE 4.25 F/E_o vs. $\Delta p/E_o$.

If a gas gap with a known value of m exists, Eq. (4.42) can be expressed in the following linear form:

$$\frac{F}{E_o + mE_g} = N + K\left(\frac{\Delta p}{E_o + mE_g}\right)$$

This form indicates that a plot of the term $F/(E_o + mE_g)$ as a function of $\Delta p/(E_o + mE_g)$ would yield a straight line with an intercept of N and a slope of K.

The Steady-State Model in the MBE. The steady-state aquifer model as proposed by Schilthuis (1936) is given by:

$$W_e = C \int_0^t (p_i - p) dt \tag{4.48}$$

where

W_e = cumulative water influx, bbl
C = water influx constant, bbl/day/psi
t = time, days
p_i = initial reservoir pressure, psi
p = pressure at the oil–water contact at time t, psi

Combining Eq. (4.48) with Eq. (4.44) gives:

$$\frac{F}{E_o} = N + C\left(\frac{\int_0^t (p_i - p) dt}{E_o}\right) \tag{4.49}$$

Plotting F/E_o vs. $\int_0^t (p_i - p) dt / E_o$ results in a straight line with an intercept that represents the initial oil-in-place N and a slope that describes the water influx constant C as shown in Figure 4.26.

And for a known gas gap, Eq. (4.49) can be expressed in the following linear form:

$$\frac{F}{E_o + mE_g} = N + C\left(\frac{\int_0^t (p_i - p)dt}{E_o + mE_g}\right)$$

Plotting $F/(E_o + mE_g)$ vs. $\int_0^t (p_i - p)dt/(E_o + mE_g)$ results in a straight line with an intercept that represents the initial oil-in-place N and a slope that describes the water influx constant C.

The Unsteady-State Model in the MBE. The van Everdingen and Hurst unsteady-state model is given by:

$$W_e = B\Sigma\Delta pW_{eD} \quad (4.50)$$

with

$$B = 1.119\phi c_t r_e^2 hf$$

van Everdingen and Hurst presented the dimensionless water influx W_{eD} as a function of the dimensionless time t_D and dimensionless radius r_D that are given by:

$$t_D = 6.328 \times 10^{-3} \frac{kt}{\phi \mu_w c_t r_e^2}$$

$$r_D = \frac{r_a}{r_e}$$

$$c_t = c_w + c_f$$

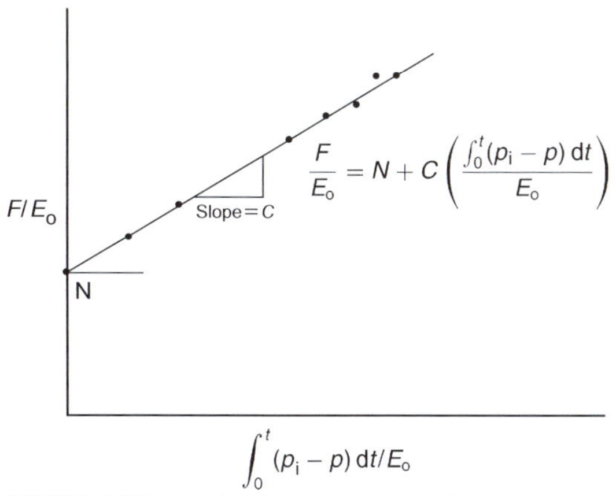

FIGURE 4.26 Graphical determination of N and c.

where

t = time, days
k = permeability of the *aquifer*, md
ϕ = porosity of the *aquifer*
μ_w = viscosity of water in the *aquifer*, cp
r_a = radius of the *aquifer*, ft
r_e = radius of the reservoir, ft
c_w = compressibility of the water, psi^{-1}

Combining Eq. (4.50) with Eq. (4.44) gives:

$$\frac{F}{E_o} = N + B\left(\frac{\sum \Delta pW_{eD}}{E_o}\right) \quad (4.51)$$

The proper methodology of solving the above linear relationship is summarized in the following steps:

Step 1. From the field past production and pressure history, calculate the underground withdrawal F and oil expansion E_o.

Step 2. Assume an aquifer configuration, i.e., linear or radial.

Step 3. Assume the aquifer radius r_a and calculate the dimensionless radius r_D.

Step 4. Plot F/E_o vs. $(\Sigma\Delta pW_{eD})/E_o$ on a Cartesian scale. If the assumed aquifer parameters are correct, the plot will be a straight line with N being the intercept and the water influx constant B being the slope. It should be noted that four other different plots might result. These are:

(1) complete random scatter of the individual points, which indicates that the calculation and/or the basic data are in error;
(2) a systematically upward-curved line, which suggests that the assumed aquifer radius (or dimensionless radius) is too small;
(3) a systematically downward-curved line, which indicates that the selected aquifer radius (or dimensionless radius) is too large;
(4) an S-shaped curve, which indicates that a better fit could be obtained if linear water influx is assumed.

Figure 4.27 shows a schematic illustration of the Havlena and Odeh methodology in determining the aquifer fitting parameters.

It should be noted that in many large fields, an infinite linear water drive satisfactorily describes the production-pressure behavior. For unit pressure drop, the cumulative water influx in an infinite linear case is simply proportional to $\sqrt{t}$ and does not require the estimation of the dimensionless time t_D. Thus, the van Everdingen and Hurst dimensionless water influx W_{eD} in Eq. (4.50) is replaced by the square root of time, to give:

$$W_w = B\sum\left[\Delta p_n\sqrt{(t-t_n)}\right]$$

Therefore, the linear form of the MBE can be expressed as:

$$\frac{F}{E_o} = N + B\left(\frac{\sum \Delta p_n\sqrt{t-t_n}}{E_o}\right)$$

Example 4.9

The material balance parameters, the underground withdrawal F, and the oil expansion E_o of a saturated oil reservoir (i.e., $m = 0$) are given below:

$\bar{p}$	F	E_o
3500	–	–
3488	2.04×10^6	0.0548
3162	8.77×10^6	0.1540
2782	17.05×10^6	0.2820

Assuming that the rock and water compressibilities are negligible, calculate the initial oil-in-place.

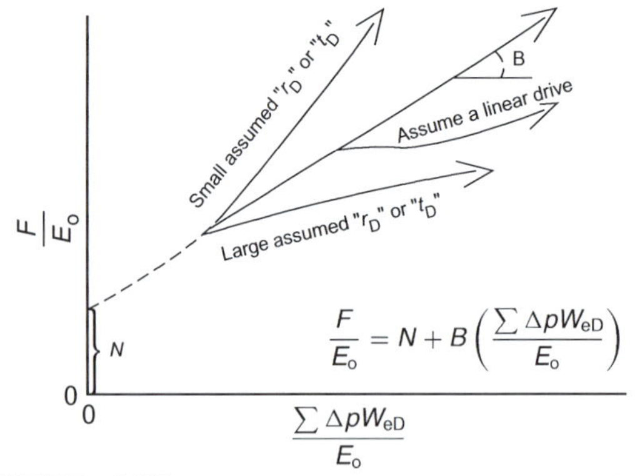

FIGURE 4.27 Havlena and odeh straight-line plot. *(After Havlena, D., Odeh, A.S., 1963. The material balance as an equation of a straight line: part 1. Trans. AIME 228, I-896).*

Solution

Step 1. The most important step in applying the MBE is to verify that no water influx exists. Assuming that the reservoir is volumetric, calculate the initial oil-in-place N by using every individual production data point in Eq. (4.38), or:

$$N = \frac{F}{E_o}$$

F	E_o	$N = F/E_o$
2.04×10^6	0.0548	37 MMSTB
8.77×10^6	0.1540	57 MMSTB
17.05×10^6	0.2820	60 MMSTB

Step 2. The above calculations show that the calculated values of the initial oil-in-place are increasing, as shown graphically in Figure 4.28, which indicates a water encroachment, i.e., water drive reservoir.

Step 3. For simplicity, select the pot aquifer model to represent the water encroachment calculations in the MBE as given by Eq. (4.47), or:

$$\frac{F}{E_o} = N + K\left(\frac{\Delta p}{E_o}\right)$$

Step 4. Calculate the terms F/E_o and $\Delta p/E_o$ of Eq. (4.47):

$\bar{p}$	Δp	F	E_o	F/E_o	$\Delta p/E_o$
3500	0	–	–	–	–
3488	12	2.04×10^6	0.0548	37.23×10^6	219.0
3162	338	8.77×10^6	0.1540	56.95×10^6	2194.8
2782	718	17.05×10^6	0.2820	60.46×10^6	2546

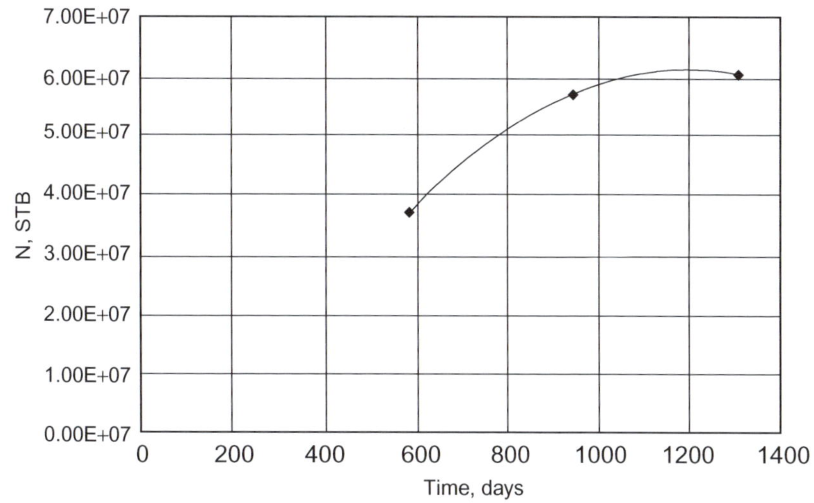

FIGURE 4.28 Indication of water influx.

Step 5. Plot F/E_o vs. $\Delta p/E_o$, as shown in Figure 4.29, and determine the intercept and the slope:

$$\text{Intercept} = N = 35 \text{ MMSTB}$$
$$\text{Slope} = K = 9983 \text{ MMSTB}$$

4.4.5 Case 5: Combination Drive Reservoirs

This relatively complicated case involves the determination of the following three unknowns:

(1) initial oil-in-place N;
(2) size of the gas cap m;
(3) water influx W_e.

The general MBE that includes the above three unknowns is given by Eq. (4.32) as:

$$F = N(E_o + mE_g) + W_e$$

where the variables constituting the above expression are defined by:

$$F = N_p\left[B_o + (R_p - R_s)B_g\right] + W_pB_w$$
$$= N_p\left[B_t + (R_p - R_{si})B_g\right] + W_pB_w$$
$$E_o = (B_o - B_{oi}) + (R_{si} - R_s)B_g$$
$$= B_t - B_{ti}$$
$$E_g = B_{oi}\left(\frac{B_g}{B_{gi}} - 1\right)$$

Havlena and Odeh differentiated Eq. (4.32) with respect to pressure and rearranged the resulting equation to eliminate m, to give:

$$\frac{FE_g^{\backprime} - F^{\backprime}E_g}{E_oE_g^{\backprime} - E_o^{\backprime}E_g} = N + \frac{W_eE_g^{\backprime} - W_e^{\backprime}E_g}{E_oE_g^{\backprime} - E_o^{\backprime}E_g} \quad (4.52)$$

in which the reversed primes denote derivatives with respect to pressure. That is:

$$E_g^{\backprime} = \frac{\partial E_g}{\partial p} = \left(\frac{B_{oi}}{B_{gi}}\right)\frac{\partial B_g}{\partial p} \approx \left(\frac{B_{oi}}{B_{gi}}\right)\frac{\Delta B_g}{\Delta p}$$
$$E_o^{\backprime} = \frac{\partial E_o}{\partial p} = \frac{\partial B_t}{\partial p} \approx \frac{\Delta B_t}{\Delta p}$$
$$F^{\backprime} = \frac{\partial F}{\partial p} \approx \frac{\Delta F}{\Delta p}$$
$$W_e^{\backprime} = \frac{\partial W_e}{\partial p} \approx \frac{\Delta W_e}{\Delta p}$$

A plot of the left-hand side of Eq. (4.52) vs. the second term on the right for a selected aquifer model should, if the choice is correct, provide a straight line with unit slope whose intercept on the ordinate gives the initial oil-in-place N. After having correctly determined N and W_e, Eq. 4.32 can be solved directly for m, to give:

$$m = \frac{F - NE_o - W_e}{NE_g}$$

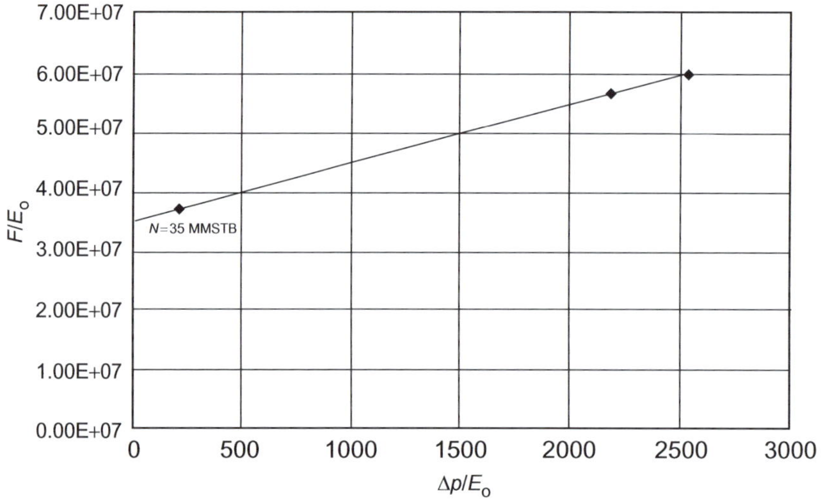

FIGURE 4.29 F/E_o vs. $\Delta p/E_o$.

Note that all the above derivatives can be evaluated numerically using one of the finite difference techniques; e.g., forward, backward, or central difference formula.

4.4.6 Case 6: Average Reservoir Pressure

To gain any understanding of the behavior of a reservoir with free gas, e.g., solution gas drive or gas cap drive, it is essential that every effort be made to determine reservoir pressures with accuracy. In the absence of reliable pressure data, the MBE can be used to estimate average reservoir pressure if accurate values of m and N are available from volumetric calculations. The general MBE is given by Eq. (4.39) as:

$$F = N[E_o + mE_g]$$

Solving Eq. (4.39) for the average pressure using the production history of the field involves the following graphical procedure:

Step 1. Select the time at which the average reservoir pressure is to be determined and obtain the corresponding production data, i.e., N_p, G_p, and R_p.

Step 2. Assume several average reservoir pressure values (i.e., $p = p_i - \Delta p$) and determine the *PVT* properties at each pressure.

Step 3. Calculate the left-hand side F of Eq. (4.39) at each of the assumed pressure. That is:

$$F = N_p[B_o + (R_p - R_s)B_g] + W_pB_w$$

Step 4. Using the same assumed average reservoir pressure values of step 2, calculate the right-hand side (RHS) of Eq. (4.39):

$$\text{RHS} = N[E_o + mE_g]$$

where

$$E_o = (B_o - B_{oi}) + (R_{si} - R_s)B_g$$

$$E_g = B_{oi}\left(\frac{B_g}{B_{gi}} - 1\right)$$

Step 5. Plot the left- and right-hand sides of the MBE, as calculated in steps 3 and 4, on Cartesian paper as a function of assumed average pressure. The point of intersection gives the average reservoir pressure that corresponds to the selected time of step 1. An illustration of the graph is shown in Figure 4.30.

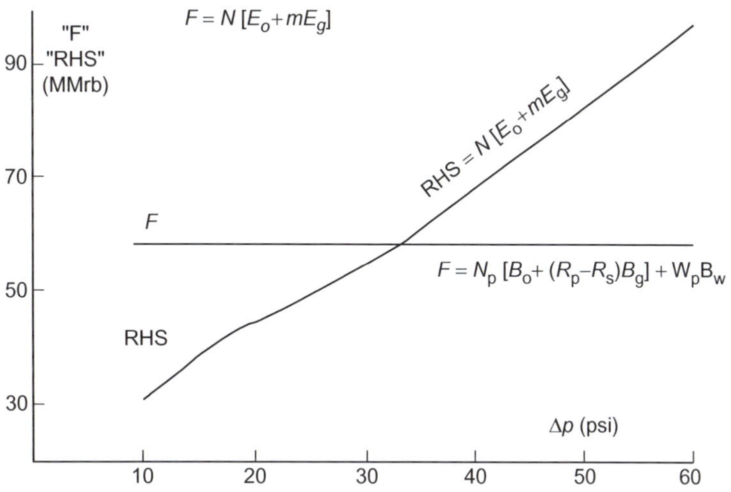

FIGURE 4.30 Solution of the material balance for the pressure.

Step 6. Repeat steps 1 through 5 to estimate reservoir pressure at each selected depletion time.

4.5 TRACY'S FORM OF THE MBE

Neglecting the formation and water compressibilities, the general MBE as expressed by Eq. (4.13) can be reduced to the following:

$$N = \frac{N_p B_o + (G_p - N_p R_s)B_g - (W_e - W_p B_w)}{(B_o - B_{oi}) + (R_{si} - R_s)B_g + mB_{oi}((B_g/B_{gi}) - 1)} \quad (4.53)$$

Tracy (1955) suggested that the above relationship can be rearranged into a more usable form as:

$$N = N_p \Phi_o + G_p \Phi_g + (W_p B_w - W_e)\Phi_w \quad (4.54)$$

where Φ_o, Φ_g, and Φ_w are considered *PVT*-related properties that are functions of pressure and defined by:

$$\Phi_o = \frac{B_o - R_s B_g}{\text{Den}} \quad (4.55)$$

$$\Phi_g = \frac{B_g}{\text{Den}} \quad (4.56)$$

$$\Phi_w = \frac{1}{\text{Den}} \quad (4.57)$$

with

$$\text{Den} = (B_o - B_{oi}) + (R_{si} - R_s)B_g + mB_{oi}\left[\frac{B_g}{B_{gi}} - 1\right] \quad (4.58)$$

where

Φ_o = oil *PVT* function
Φ_g = gas *PVT* function
Φ_w = water *PVT* function

Figure 4.31 shows a graphical presentation of the behavior of Tracy's *PVT* functions with the changing pressure.

Note that Φ_o is negative at low pressures and that all Φ functions are approaching infinity at bubble point pressure because the value of the denominator "Den" in Eqs. (4.55)–(4.57) approaches zero. Tracy's form is valid only for initial pressures equal to the bubble point pressure and cannot be used at pressures above the bubble point. Furthermore, shapes of the Φ function curves illustrate that small errors in pressure and/or production can cause large errors in calculated oil-in-place at pressures near

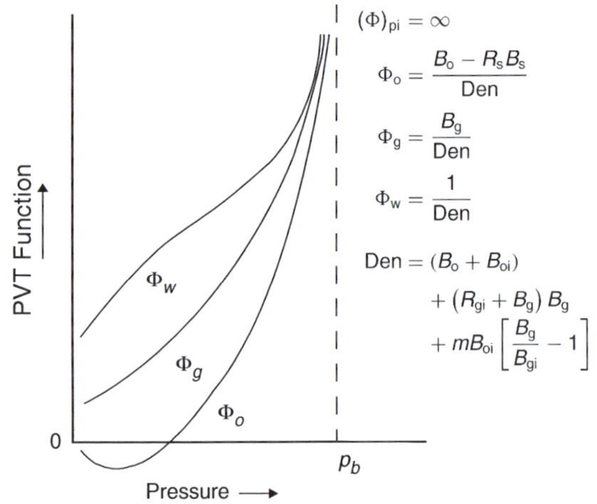

FIGURE 4.31 Tracy's *PVT* functions.

the bubble point. However, Steffensen (1987) pointed out that Tracy's equation uses the oil formation volume factor at the bubble point pressure B_{ob} for the initial B_{oi}, which causes all the *PVT* functions Φ to become infinity at the bubble point pressure. Steffensen suggested that Tracy's equation could be extended for applications above the bubble point pressure, i.e., for undersaturated oil reservoirs, by simply using the value of B_o at the initial reservoir pressure. He concluded that Tracy's methodology could predict reservoir performance for the entire pressure range from any initial pressure down to abandonment.

It should be pointed out that because the rock and water compressibilities are relatively unimportant below the bubble point pressure; they were not included in Tracy's material balance formulation. They can be included indirectly, however, by the use of pseudovalues of the oil formation volume factor at pressures below the initial pressure.

These pseudovalues B_o^* are given by:

$$B_o^* = B_o + B_{oi}\left(\frac{S_w c_w + c_f}{1 - S_w}\right)(p_i - p)$$

These pseudovalues include the additional pressure support of water and rock compressibilities in the material balance computations.

$$g_{gas} < \frac{dp}{dz} < g_{oil}$$

with

$$g_{gas} = \frac{\rho_g}{144} \quad g_{oil} = \frac{\rho_o}{144}$$

where

g_{oil} = oil gradient, psi/ft
ρ_o = oil density, lb/ft^3
g_{gas} = gas gradient, psi/ft
ρ_g = gas density, lb/ft^3
dp/dz = reservoir pressure gradient, psi/ft

The following example is given by Tracy (1955) to illustrate his proposed approach.

Example 4.10

The production history of a saturated oil reservoir is as follows:

$\bar{p}$ (psia)	N_p (MSTB)	G_p (MMscf)
1690	0	0
1600	398	38.6
1500	1570	155.8
1100	4470	803

The calculated values of the *PVT* functions are given below:

p	Φ_o	Φ_g
1600	36.60	0.4000
1500	14.30	0.1790
1100	2.10	0.0508

Calculate the oil-in-place N.

Solution

The calculations can be conveniently performed in the following tabulated form using:

$$N = N_p\Phi_o + G_p\Phi_g + 0$$

$\bar{p}$ (psia)	N_p (MSTB)	G_p (MMscf)	$(N_p\Phi_o)$	$(G_p\Phi_g)$	N (STB)
1600	398	38.6	14.52×10^6	15.42×10^6	29.74×10^6
1500	155.8	155.8	22.45×10^6	27.85×10^6	50.30×10^6
1100	803.0	803.0	9.39×10^6	40.79×10^6	50.18×10^6

The above results show that the original oil-in-place in this reservoir is approximately 50 MMSTB of oil. The calculation at 1600 psia is a good example of the sensitivity of such a calculation near the bubble point pressure. Since the last two values of the original oil-in-place agree so well, the first calculation is probably wrong.

4.6 PROBLEMS

(1) You have the following data on an oil reservoir:

	Oil	Aquifer
Geometry	Circular	Semicircular
Encroachment angle	–	180°
Radius, ft	4000	80,000
Flow regime	Semisteady state	Unsteady state
Porosity	–	0.20
Thickness, ft	–	30
Permeability, md	200	50
Viscosity, cp	1.2	0.36
Original pressure	3800	3800
Current pressure	3600	–
Original volume factor	1.300	1.04
Current volume factor	1.303	1.04
Bubble point pressure	3000	–

The field has been on production for 1120 days and has produced 800,000 STB of oil and 60,000 STB of water. Water and formation compressibilities are estimated to be 3×10^{-6} and 3×10^{-6} psi^{-1}, respectively.

Calculate the original oil-in-place.

(2) The following rock and fluid properties' data is available on the Nameless Field:

Reservoir area = 1000 acres, Porosity = 10%
Thickness = 20 ft, $T = 140°F$
$S_{wi} = 20\%$, $p_i = 4000$ psi
$p_b = 4000$ psi

The gas compressibility factor and relative permeability ratio are given by the following expressions:

$$Z = 0.8 - 0.00002(p - 4000)$$

$$\frac{k_{rg}}{k_{ro}} = 0.00127 \exp(17.269 S_g)$$

The production history of the field is given below:

	4000 psi	3500 psi	3000 psi
μ_o, cp	1.3	1.25	1.2
μ_g, cp	–	0.0125	0.0120
B_o, bbl/STB	1.4	1.35	1.30
R_s, scf/STB	–	–	450
GOR, scf/STB	600	–	1573

Subsurface information indicates that there is no aquifer and that there has been no water production.

Calculate:

(a) the remaining oil-in-place at 3000 psi;
(b) the cumulative gas produced at 3000 psi.

(3) The following *PVT* and production history data is available on an oil reservoir in West Texas.

Original *oil-in-place* = 10 MMSTB
Initial water saturation = 22%
Initial reservoir pressure = 2496 psia
Bubble point pressure = 2496 psi

Pressure (psi)	B_o (bbl/STB)	R_s (scf/STB)	B_g (bbl/scf)	μ_o (cp)	μ_g (cp)	GOR (scf/STB)
2496	1.325	650	0.000796	0.906	0.016	650
1498	1.250	486	0.001335	1.373	0.015	1360
1302	1.233	450	0.001616	1.437	0.014	2080

The cumulative gas–oil ratio at 1302 psi is recorded at 953 scf/STB. Calculate:

(a) the oil saturation at 1302 psia;
(b) the volume of the free gas in the reservoir at 1302 psia;
(c) the relative permeability ratio (k_g/k_o) at 1302 psia.

(4) The Nameless Field is an undersaturated oil reservoir. The crude oil system and the rock

type indicate that the reservoir is *highly compressible*. The available reservoir and production data is given below:

$S_{wi} = 0.25$,	$\phi = 20\%$
Area = 1000 acres,	$h = 70$ ft
$T = 150°F$,	Bubble point pressure = 3500 psia

	Original Conditions	Current Conditions
Pressure, psi	5000	4500
B_o, bbl/STB	1.905	1.920
R_s, scf/STB	700	700
N_p, MSTB	0	610.9

Calculate the cumulative oil production at 3900 psi. The *PVT* data shows that the oil formation volume factor is equal to 1.938 bbl/STB at 3900 psia.

(5) The following data[2] is available on a gas cap drive reservoir:

Pressure (psi)	N_p (MMSTB)	R_p (scf/STB)	B_o (RB/STB)	R_s (scf/STB)	B_g (RB/scf)
3330	–	–	1.2511	510	0.00087
3150	3.295	1050	1.2353	477	0.00092
3000	5.903	1060	1.2222	450	0.00096
2850	8.852	1160	1.2122	425	0.00101
2700	11.503	1235	1.2022	401	0.00107
2550	14.513	1265	1.1922	375	0.00113
2400	17.730	1300	1.1822	352	0.00120

Calculate the initial oil and free gas volumes.

(6) If 1 million STB of oil has been produced from the Calgary Reservoir at a cumulative produced GOR of 2700 scf/STB, causing the reservoir pressure to drop from the initial reservoir pressure of 400 psia to 2400 psia, what is the initial stock-tank oil-in-place?

[2]Dake, L., 1978. *Fundamentals of Reservoir Engineering*. Elsevier, Amsterdam.

(7) The following data is taken from an oil field that had no original gas cap and no water drive:

Oil pore volume of reservoir = 75 MM ft^3;
Solubility of gas in crude = 0.42 scf/STB/psi;
Initial bottom-hole pressure = 3500 psia;
Bottom-hole temperature = 140°F;
Bubble point pressure of the reservoir = 3000 psia;
Formation volume factor at 3500 psia = 1.333 bbl/STB;
Compressibility factor of the gas at 1000 psia and 140°F = 0.95;
Oil produced when pressure is 2000 psia = 1.0 MMSTB.
Net cumulative produced GOR = 2800 scf/STB

(a) Calculate the initial STB of oil in the reservoir.
(b) Calculate the initial scf of gas in the reservoir.
(c) Calculate the initial dissolved GOR of the reservoir.
(d) Calculate the scf of gas remaining in the reservoir at 2000 psia.
(e) Calculate the scf of free gas in the reservoir at 2000 psia.
(f) Calculate the gas volume factor of the escaped gas at 2000 psia at standard conditions of 14.7 psia and 60°F.
(g) Calculate the reservoir volume of the free gas at 2000 psia.
(h) Calculate the total reservoir GOR at 2000 psia.
(i) Calculate the dissolved GOR at 2000 psia.
(j) Calculate the liquid volume factor of the oil at 2000 psia.
(k) Calculate the total, or two-phase, oil volume factor of the oil and its initial complement of dissolved gas at 2000 psia.

(8) Production data, along with reservoir and fluid data, for an undersaturated reservoir follows. There was no measureable water produced, and it can be assumed that there

was no free gas flow in the reservoir. Determine the following:

(a) the saturations of oil, gas, and water at a reservoir pressure of 2258 psi;

(b) has water encroachment occurred and, if so, what is the volume?

Gas gravity = 0.78
Reservoir temperature = 160°F
Initial water saturation = 25%
Original oil-in-place = 180 MMSTB
Bubble point pressure = 2819 psia

The following expressions for B_o and R_{so} as functions of pressure were determined from laboratory data:

$$B_o = 1.00 + 0.00015p, \text{ bbl/STB}$$
$$R_{so} = 50 + 0.42p, \text{scf/STB}$$

Pressure (psia)	Cumulative Oil Produced (MMSTB)	Cumulative Gas Produced (MMscf)	Instantaneous GOR (scf/STB)
2819	0	0	1000
2742	4.38	4.380	1280
2639	10.16	10.360	1480
2506	20.09	21.295	2000
2403	27.02	30.260	2500
2258	34.29	41.150	3300

(9) The Wildcat Reservoir was discovered in 1970. The reservoir had an initial pressure of 3000 psia, and laboratory data indicated a bubble point pressure of 2500 psia. The connate water saturation was 22%. Calculate the fractional recovery N_p/N from initial conditions down to a pressure of 2300 psia. State any assumptions that you make relative to the calculations.

Porosity = 0.165
Formation compressibility = 2.5×10^{-6} psia^{-1}
Reservoir temperature = 150°F

Pressure (psia)	B_o (bbl/STB)	R_{so} (scf/STB)	Z	B_g (bbl/scf)	Viscosity ratio μ_o/μ_g
3000	1.315	650	0.745	0.000726	53.91
2500	1.325	650	0.680	0.000796	56.60
2300	1.311	618	0.663	0.000843	61.46

CHAPTER 5

Predicting Oil Reservoir Performance

Most reservoir engineering calculations involve the use of the material balance equation (MBE). Some of the most useful applications of the MBE require the concurrent use of fluid flow equations, e.g., Darcy's equation. Combining the two concepts would enable the engineer to predict the reservoir future production performance as a function of time. Without the fluid flow concepts, the MBE simply provides performance as a function of the average reservoir pressure. Prediction of the reservoir future performance is ordinarily performed in the following three phases:

Phase 1: The first phase involves the use of the MBE in a predictive mode to estimate cumulative hydrocarbon production and fractional oil recovery as a function of declining reservoir pressure and increasing gas–oil ratio (GOR). These results are incomplete, however, because they give no indication of the time that it will take to recover oil at any depletion stage. In addition, this stage of calculations is performed without considering:

- the actual number of wells;
- the location of wells;
- the production rate of individual wells;
- the time required to deplete the reservoir.

Phase 2: To determine recovery profile as a function of time, it is necessary to generate individual well performance profile with declining reservoir pressure. This phase documents different techniques that are designed to model the production performance of vertical and horizontal wells.

Phase 3: The third stage of prediction is the time–production phase. In these calculations, the reservoir and well performance data is correlated with time. It is necessary in this phase to account for the number of wells and the productivity of individual well.

5.1 PHASE 1. RESERVOIR PERFORMANCE PREDICTION METHODS

The MBE in its various mathematical forms as presented in Chapter 4 is designed to provide estimates of the initial oil-in-place N, size of the gas cap m, and water influx W_e. To use the MBE to predict the reservoir future performance, two additional relations are required:

the equation of producing (instantaneous) GOR;
the equation for relating saturations to cumulative oil production.

These auxiliary mathematical expressions are presented below.

5.1.1 Instantaneous GOR

The produced GOR at any particular time is the ratio of the standard cubic feet of total gas being produced at any time to the stock-tank barrels of oil being produced at the same instant—hence, the name instantaneous GOR. Eq. (1.53) describes the GOR mathematically by the following expression:

$$\text{GOR} = R_s + \left(\frac{k_{rg}}{k_{ro}}\right)\left(\frac{\mu_o B_o}{\mu_g B_g}\right) \tag{5.1}$$

where

GOR = instantaneous gas–oil ratio, scf/STB
R_s = gas solubility, scf/STB
k_{rg} = relative permeability to gas
k_{ro} = relative permeability to oil
B_o = oil formation volume factor, bbl/STB
B_g = gas formation volume factor, bbl/scf
μ_o = oil viscosity, cp
μ_g = gas viscosity, cp

The instantaneous GOR equation is of fundamental importance in reservoir analysis. The importance of Eq. (5.11) can appropriately be discussed in conjunction with Figures 5.1 and 5.2. Those illustrations show the history of the GOR of a hypothetical depletion drive reservoir that is typically characterized by the following points:

Point 1. When the reservoir pressure p is above the bubble point pressure p_b, there is no free gas in the formation, i.e., $k_{rg} = 0$, and therefore:

$$\text{GOR} = R_{si} = R_{sb} \tag{5.2}$$

The GOR remains constant at R_{si} until the pressure reaches the bubble point pressure at point 2.

Point 2. As the reservoir pressure declines below p_b, the gas begins to evolve from solution and its saturation increases. However, this free gas cannot flow until the gas saturation S_g reaches the critical gas saturation S_{gc} at point 3. From point 2 to point 3, the instantaneous GOR is described by a decreasing gas solubility, as:

$$\text{GOR} = R_s \tag{5.3}$$

Point 3. At this point, the free gas begins to flow with the oil and the values of GOR progressively increase with the declining reservoir pressure to point 4. During this pressure decline period, the GOR is described by Eq. (5.1), or:

$$\text{GOR} = R_s + \left(\frac{k_{rg}}{k_{ro}}\right)\left(\frac{\mu_o \beta_o}{\mu_g B_g}\right)$$

Point 4. At this point, the maximum GOR is reached due to the fact that the supply of gas has reached a maximum and marks the beginning of the blow-down period to point 5.

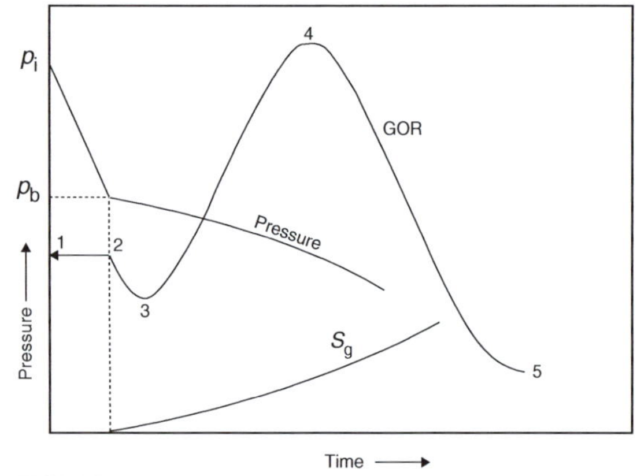

FIGURE 5.1 Characteristics of solution gas drive reservoirs.

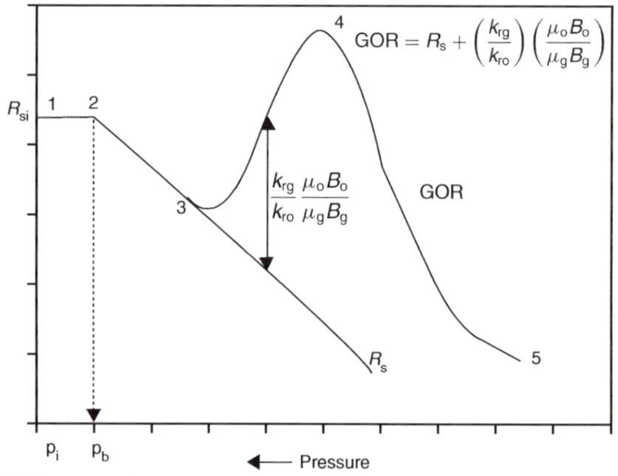

FIGURE 5.2 History of GOR and R_s for a solution gas drive reservoir.

Point 5. This point indicates that all the producible free gas has been produced and the GOR is essentially equal to the gas solubility and continues to decline following the R_s curve.

There are three types of GORs, all expressed in scf/STB, which must be clearly distinguished from each other. These are:

- instantaneous GOR (defined by Eq. (5.1));
- solution GOR, i.e., gas solubility R_s;
- cumulative GOR R_p.

The solution GOR is a *PVT* property of the crude oil system. It is commonly referred to as "gas solubility" and denoted by R_s. It measures the tendency of the gas to dissolve in or evolve from the oil with changing pressures. It should be pointed out that as long as the evolved gas remains immobile, i.e., gas saturation S_g is less than the critical gas saturation, the instantaneous GOR is equal to the gas solubility. That is:

$$\text{GOR} = R_s$$

The cumulative GOR R_p, as defined previously in the MBE, should be clearly distinguished from the producing (instantaneous) GOR. The cumulative GOR is defined as:

$$R_p = \frac{\text{Cumulative (total) gas produced}}{\text{Cumulative oil produced}}$$

or

$$R_p = \frac{G_p}{N_p} \tag{5.4}$$

where

R_p = cumulative GOR, scf/STB
G_p = cumulative gas produced, scf
N_p = cumulative oil produced, STB

The cumulative gas produced G_p is related to the instantaneous GOR and cumulative oil production by the expression:

$$G_p = \int_0^{N_p} (\text{GOR}) dN_p \tag{5.5}$$

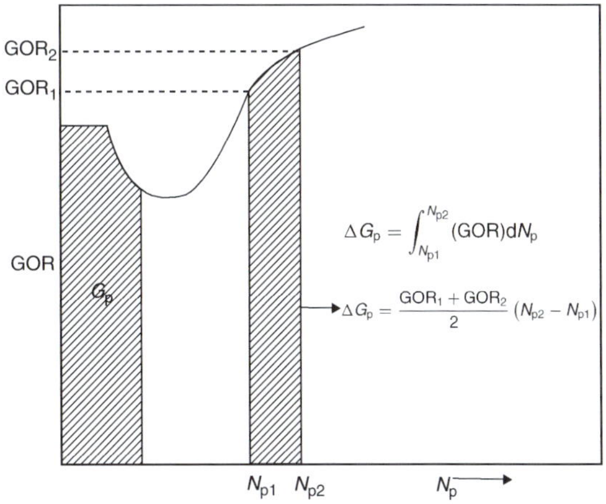

FIGURE 5.3 Relationship between GOR and G_p.

Eq. (5.5) simply indicates that the cumulative gas production at any time is essentially the area under the curve of the GOR vs. N_p relationship, as shown in Figure 5.3. The incremental cumulative gas produced, ΔG_p, between N_{p1} and N_{p2} is then given by:

$$\Delta G_p = \int_{N_{p1}}^{N_{p2}} (\text{GOR}) dN_p \tag{5.6}$$

This integral can be approximated by using the trapezoidal rule, to give:

$$\Delta G_p = \left[\frac{(\text{GOR})_1 + (\text{GOR})_2}{2}\right](N_{p2} - N_{p1})$$

or

$$\Delta G_p = (\text{GOR})_{\text{avg}} \, \Delta N_p$$

Eq. (5.5) can then be approximated as:

$$G_p = \sum_0 (\text{GOR})_{\text{avg}} \Delta N_p \tag{5.7}$$

Example 5.1

The following production data is available on a depletion drive reservoir:

p (psi)	GOR (scf/STB)	N_p (MMSTB)
	1340	0
2600	1340	1.380
2400	1340	2.260

	1340	3.445
1800	1936	7.240
1500	3584	12.029
1200	6230	15.321

The initial reservoir pressure is 2925 psia with a bubble point pressure of 2100 psia. Calculate cumulative gas produced G_p and cumulative GOR at each pressure.

Solution

Step 1. Construct the following table by applying Eqs. (5.4) and (5.7):

$$R_p = \frac{G_p}{N_p}$$

$$\Delta G_p = \left[\frac{(\text{GOR})_1 + (\text{GOR})_2}{2}\right](N_{p2} - N_{p1}) = (\text{GOR})_{\text{avg}}\,\Delta N_p$$

$$G_p = \sum_0 (\text{GOR})_{\text{avg}}\,\Delta N_p$$

P (psi)	GOR (scf/STB)	$(\text{GOR})_{\text{avg}}$ (scf/STB)	N_p (MMSTB)	ΔN_p (MMSTB)	ΔG_p (MMscf)	G_p (MMscf)	R_p (scf/STB)
2925	1340	1340	0	0	0	0	–
2600	1340	1340	1.380	1.380	1849	1849	1340
2400	1340	1340	2.260	0.880	1179	3028	1340
2100	1340	1340	3.445	1.185	1588	4616	1340
1800	1936	1638	7.240	3.795	6216	10,832	1496
1500	3584	2760	12.029	4.789	13,618	24,450	2033
1200	6230	4907	15.321	3.292	16,154	40,604	2650

It should be pointed out that the crude oil *PVT* properties used in the MBE are appropriate for moderate–low volatility "black oil" systems, which, when produced at the surface, are separated into oil and solution gas. These properties, as defined mathematically below, are designed to relate surface volumes to reservoir volumes and vice versa.

$$R_s = \frac{\text{Volume of solution gas dissolved in the oil at reservoir condition}}{\text{Volume of the oil at stock-tank conditions}}$$

$$B_o = \frac{\text{Volume of oil at reservoir condition}}{\text{Volume of the oil at stock-tank conditions}}$$

$$B_g = \frac{\text{Volume of the free gas at reservoir condition}}{\text{Volume of free gas at stock-tank conditions}}$$

Whitson and Brule (2000) point out that the above three properties constitute the classical (black oil) *PVT* data required for various type of applications of the MBE. However, in formulating the material balance equation, the following assumptions were made when using the black oil *PVT* data:

(1) Reservoir gas does not yield liquid when brought to the surface.
(2) Reservoir oil consists of two surface "components"; stock-tank oil and total surface separator gas.
(3) Properties of stock-tank oil in terms of its *API* gravity and surface gas do not change with depletion pressure.
(4) Surface gas released from the reservoir oil has the same properties as the reservoir gas.

This situation is more complex when dealing with volatile oils. This type of crude oil system is characterized by significant hydrocarbon liquid recovery from their produced reservoir gases. As the reservoir pressure drops below the bubble point pressure, the evolved solution gas liberated in the reservoir contains enough heavy components to yield appreciable condensate dropout at the separators that is combined with the stock-tank oil. This is in contrast to black oils for which little error is introduced by the assumption that there is negligible hydrocarbon liquid recovery from produced gas. Also, volatile oils evolve gas and develop free-gas saturation in the reservoir more rapidly than normal black oils as pressure declines below the bubble point. This causes relatively high GORs at the wellhead. Thus, performance predictions differ from those discussed for black oils mainly because of the need to account for liquid recovery from the produced gas. Conventional material balances with standard laboratory *PVT* (black oil) data *underestimate* oil recovery. The error increases for increasing oil volatility.

Consequently, depletion performance of volatile oil reservoirs below bubble point is strongly influenced by the rapid shrinkage of oil and by the large amounts of gas evolved. This results in relatively high gas saturation, high producing GORs, and low-to-moderate production of reservoir oil. The produced gas can yield a substantial volume of hydrocarbon liquids in the processing equipment. This liquid recovery at the surface can equal or exceed the volume of stock-tank oil produced from the reservoir liquid phase. Depletion-drive recoveries are often between 15% and 30% of the original oil-in-place.

For volatile oil reservoir primary-performance prediction methods, the key requirements are correct handling of the oil shrinkage, gas evolution, gas and oil flow in the reservoir, and liquids recovery at the surface. If

Q_o = black oil flow rate, STB/day
$Q_o^\backslash$ = total flow rate including condensate, STB/day
R_s = gas solubility, scf/STB
GOR = total measured gas–oil ratio, scf/STB
r_s = condensate yield, STB/scf

then

$$Q_o = Q_o^\backslash - (Q_o^\backslash \mathrm{GOR} - Q_o R_s) r_s$$

Solving for Q_o gives:

$$Q_o = Q_o^\backslash \left[\frac{1 - (r_s \mathrm{GOR})}{1 - (r_s R_s)}\right]$$

The above expression can be used to adjust the cumulative "black oil" production, N_p, to account for the condensate production. The black oil cumulative production is then calculated from:

$$N_p = \int_0^t Q_o \mathrm{d}t \approx \sum_0^t (\Delta Q_o \Delta t)$$

The cumulative total gas production "G_p" and the *adjusted* cumulative black oil production "N_p" is used in Eq. (5.4) to calculate the cumulative gas–oil ratio, i.e.:

$$R_p = \frac{G_p}{N_p}$$

See Whitson and Brule (2000).

5.1.2 The Reservoir Saturation Equations and Their Adjustments

The saturation of a fluid (gas, oil, or water) in the reservoir is defined as the volume of the fluid divided by the pore volume, or:

$$S_o = \frac{\text{Oil volume}}{\text{Pore volume}} \tag{5.8}$$

$$S_w = \frac{\text{Water volume}}{\text{Pore volume}} \tag{5.9}$$

$$S_g = \frac{\text{Gas volume}}{\text{Pore volume}} \tag{5.10}$$

$$S_o + S_w + S_g = 1.0 \tag{5.11}$$

Consider a volumetric oil reservoir with no gas cap that contains N stock-tank barrels of oil at the initial reservoir pressure p_i. Assuming no water influx gives:

$$S_{oi} = 1 - S_{wi}$$

where the subscript "i" indicates the initial reservoir condition. From the definition of oil saturation:

$$1 - S_{wi} = \frac{NB_{oi}}{\text{Pore volume}}$$

or

$$\text{Pore volume} = \frac{NB_{oi}}{1 - S_{wi}} \tag{5.12}$$

If the reservoir has produced N_p stock-tank barrels of oil, the remaining oil volume is given by:

$$\text{Remaining oil volume} = (N - N_p)B_o \tag{5.13}$$

Substituting Eqs. (5.13) and (5.12) into Eq. (5.8) gives:

$$S_o = \frac{\text{Remaining oil volume}}{\text{Pore volume}} = \frac{(N - N_p)B_o}{\dfrac{(NB_{oi})}{(1 - S_{wi})}} \tag{5.14}$$

or

$$S_o = (1 - S_{wi})\left(1 - \frac{N_p}{N}\right)\frac{B_o}{B_{oi}} \tag{5.15}$$

and therefore:

$$S_g = 1 - S_o - S_{wi} \tag{5.16}$$

Example 5.2

A volumetric solution gas drive reservoir has an initial water saturation of 20%. The initial oil formation volume factor is reported at 1.5 bbl/STB. When 10% of the initial oil was produced, the value of B_o decreased to 1.38. Calculate the oil saturation and the gas saturation.

Solution

From Eqs. (5.15) and (5.16):

$$S_o = (1 - S_{wi})\left(1 - \frac{N_p}{N}\right)\frac{B_o}{B_{oi}}$$

$$= (1 - 0.2)(1 - 0.1)\left(\frac{1.38}{1.50}\right) = 0.662$$

$$S_g = 1 - S_o - S_{wi}$$

$$= 1 - 0.662 - 0.20 = 0.138$$

It should be pointed out that the values of the relative permeability ratio k_{rg}/k_{ro} as a function of oil saturation can be generated by using the actual field production as expressed in terms of N_p, GOR, and *PVT* data. The recommended methodology involves the following steps:

Step 1. Given the actual field cumulative oil production N_p and the *PVT* data as a function of pressure, calculate the oil and gas saturations from Eqs. (5.15) and (5.16):

$$S_o = (1 - S_{wi})\left(1 - \frac{N_p}{N}\right)\frac{B_o}{B_{oi}}$$

$$S_g = 1 - S_o - S_{wi}$$

Step 2. Using the actual field instantaneous GORs, solve Eq. (5.1) for the relative permeability ratio, as:

$$\frac{k_{rg}}{k_{ro}} = (\text{GOR} - R_s)\left(\frac{\mu_g B_g}{\mu_o B_o}\right)$$

Step 3. The relative permeability ratio is traditionally expressed graphically by plotting k_{rg}/k_{ro} vs. S_o on semilog paper. This is obviously not the case in a gravity drainage reservoir and will result in the calculation of abnormally low oil saturation.

Note that Eq. (5.14) suggests that all the remaining oil saturation at any depletion stage is distributed uniformly throughout the reservoir. In dealing with gravity drainage reservoirs, water drive reservoirs, or gas cap drive reservoirs, adjustments must be made to the oil saturation as calculated by Eq. (5.14) to account for:

- migration of the evolved gas upstructure;
- trapped oil in the water-invaded region;
- trapped oil in the gas cap expansion zone;
- loss of oil saturation in the gas cap shrinkage zone.

Oil Saturation Adjustment in Gravity Drainage Reservoirs. In these types of reservoirs, the gravity effects result in much lower producing GORs than would be expected from reservoirs producing without the benefit of gravity drainage. This is due to the upstructure migration of the gas and consequent *higher oil saturation* in the vicinity of the completion intervals of the production wells that should be used when calculating the oil relative permeability k_{ro}. The following steps summarize the recommended procedure for adjusting Eq. (5.14) to reflect the migration of gas to the top of the structure:

Step 1. Calculate the volume of the evolved gas that will migrate to the top of the formation to form the *secondary gas cap* from the following relationship:

$$(\text{gas})_{\text{migrated}} = \left[NR_{si} - (N - N_p)R_s - N_pR_p\right]B_g - \left[\frac{NB_{oi}}{1 - S_{wi}} - (PV)_{SGC}\right]S_{gc}$$

where

$(PV)_{SGC}$ = pore volume of the secondary gas cap, bbl
S_{gc} = critical gas saturation
B_g = current gas formation volume factor, bbl/scf

Step 2. Recalculate the volume of the evolved gas that will form the secondary gas cap from the following relationship:

$$(\text{gas})_{\text{migrated}} = [1 - S_{wi} - S_{org}](PV)_{SGC}$$

where

$(PV)_{SGC}$ = pore volume of the secondary gas cap, bbl
S_{org} = residual oil saturation to gas displacement
S_{wi} = connate or initial water saturation

Step 3. Equating the two derived relationships and solving for secondary gas cap pore volume gives:

$$(PV)_{SGC} = \frac{\left[NR_{si} - (N - N_p)R_s - N_pR_p\right]B_g - \left[\dfrac{(NB_{oi})}{(1 - S_{wi})}\right]S_{gc}}{(1 - S_{wi} - S_{org} - S_{gc})}$$

Step 4. Adjust Eq. (5.14) to account for the migration of the evolved gas to the secondary gas cap, to give:

$$S_o = \frac{(N - N_p)B_o - (PV)_{SGC}S_{org}}{\left(\dfrac{(NB_{oi})}{(1 - S_{wi})}\right) - (PV)_{SGC}} \quad \textbf{(5.17)}$$

It should be noted that the oil recovery by gravity drainage involves two fundamental mechanisms:

(1) the formation of the secondary gas cap as presented by Eq. (5.17);
(2) the gravity drainage rate.

For an efficient gravity drive mechanism, the gas must flow upstructure while the oil flows downstructure, i.e., both fluids are moving in opposite directions; this is called the "counterflow" of oil and gas. Since both fluids are flowing, gas–oil relative permeability characteristics of the formation are very important. Since the gas saturation is not uniform throughout the oil column, the field calculated k_{rg}/k_{ro}, which is based on the material balance calculations, must be used. For the counterflow to occur, the actual reservoir pressure gradient must be between the static gradient of the oil and that of the gas. That is:

$$\rho_{gas} < \left(\frac{dp}{dz}\right) < \rho_{oil}$$

where

ρ_{oil} = oil gradient, psi/ft
ρ_{gas} = gas gradient, psi/ft
dp/dz = reservoir pressure gradient, psi/ft

Terwilliger et al. (1951) pointed out that oil recovery by gravity segregation is rate sensitive and that a rather sharp decrease in recovery would occur at production rates above the maximum rate of gravity drainage and, hence, production should not exceed this particular maximum rate. The maximum rate of gravity drainage is defined as the "rate at which complete counterflow exists" and expressed mathematically by the following expression:

$$q_o = \frac{7.83 \times 10^{-6} kk_{ro}A(\rho_o - \rho_g)\sin(\alpha)}{\mu_o}$$

where

q_o = oil production rate, bbl/day
ρ_o = oil density, lb/ft^3
ρ_g = gas density, lb/ft^3

A = cross-sectional area open to flow, ft^2
k = absolute permeability, md
α = dip angle

This calculated value of q_o represents the maximum oil rate that should not be exceeded without causing the gas to flow downward.

Oil saturation Adjustment due to Water Influx. The proposed oil saturation adjustment methodology is illustrated in Figure 5.4 and described by the following steps:

Step 1. Calculate the PV in the water-invaded region, as:

$$W_e - W_p B_w = (\text{PV})_{\text{water}}(1 - S_{wi} - S_{orw})$$

Solving for the PV of the water-invaded zone $(\text{PV})_{\text{water}}$ gives:

$$(\text{PV})_{\text{water}} = \frac{W_e - W_p B_w}{1 - S_{wi} - S_{orw}} \quad (5.18)$$

where
$(\text{PV})_{\text{water}}$ = pore volume in water-invaded zone, bbl
S_{orw} = residual oil saturated in the imbibition water–oil system

Step 2. Calculate the oil volume in the water-invaded zone, or:

$$\text{Volume of oil} = (\text{PV})_{\text{water}} S_{orw} \quad (5.19)$$

Step 3. Adjust Eq. (5.14) to account for the trapped oil by using Eqs. (5.18) and (5.19):

$$S_o = \frac{(N - N_p)B_o - \left[\dfrac{W_e - W_p B_w}{1 - S_{wi} - S_{orw}}\right] S_{orw}}{\left(\dfrac{NB_{oi}}{1 - S_{wi}}\right) - \left[\dfrac{W_e - W_p B_w}{1 - S_{wi} - S_{orw}}\right]} \quad (5.20)$$

Oil Saturation Adjustment due to Gas Cap Expansion. The oil saturation adjustment procedure is illustrated in Figure 5.5 and summarized below:

Step 1. Assuming no gas is produced from the gas cap, calculate the net expansion of the gas cap from:

$$\text{Expansion of the gas cap} = mNB_{oi}\left(\frac{B_g}{B_{gi}} - 1\right) \quad (5.21)$$

Step 2. Calculate the PV of the gas-invaded zone $(\text{PV})_{\text{gas}}$ by solving the following simple material balance:

$$mNB_{oi}\left(\frac{B_g}{B_{gi}} - 1\right) = (\text{PV})_{\text{gas}}(1 - S_{wi} - S_{org})$$

or

$$(\text{PV})_{\text{gas}} = \frac{mNB_{oi}((B_g/B_{gi}) - 1)}{1 - S_{wi} - S_{org}} \quad (5.22)$$

where
$(\text{PV})_{\text{gas}}$ = pore volume of the gas-invaded zone
S_{org} = residual oil saturation in gas–oil system

Step 3. Calculate the volume of oil in the gas-invaded zone.

$$\text{Oil volume} = (\text{PV})_{\text{gas}} S_{org} \quad (5.23)$$

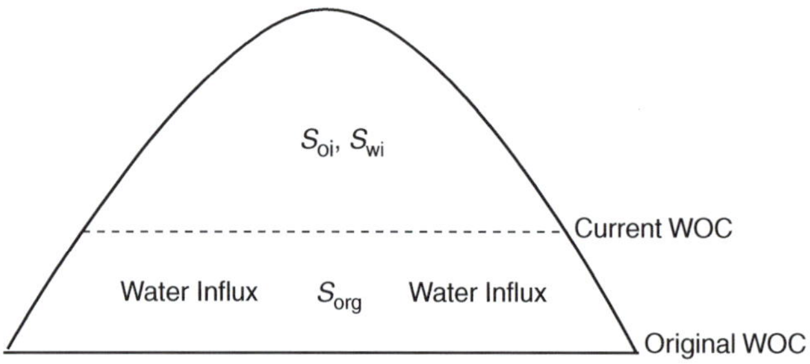

FIGURE 5.4 Oil saturation adjustment for water influx.

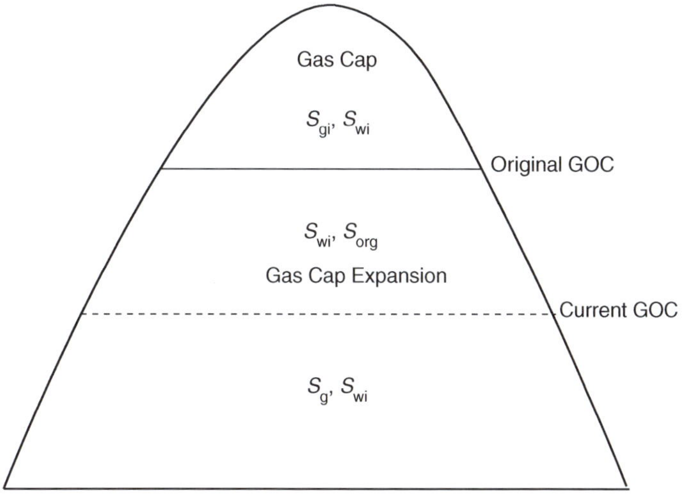

FIGURE 5.5 Oil saturation adjustment for gas cap expansion.

Step 4. Adjust Eq. (5.14) to account for the trapped oil in the gas expansion zone by using Eqs. (5.22) and (5.23), to give:

$$S_o = \frac{(N - N_p)B_o - \left[\dfrac{mNB_{oi}\left(\dfrac{B_g}{B_{gi}} - 1\right)}{1 - S_{wi} - S_{org}}\right] S_{org}}{\left(\dfrac{NB_{oi}}{1 - S_{wi}}\right) - \left[\dfrac{mNB_{oi}}{1 - S_{wi} - S_{org}}\right]\left(\dfrac{B_g}{B_{gi}} - 1\right)} \quad (5.24)$$

Oil Saturation Adjustment for Combination Drive. For a combination drive reservoir, i.e., water influx and gas cap, the oil saturation equation as given by Eq. (5.14) can be adjusted to account for both driving mechanisms, as:

$$S_o = \frac{(N - N_p)B_o - \left[\left(\dfrac{mNB_{oi}\left(\frac{B_g}{B_{gi}} - 1\right)S_{org}}{1 - S_{wi} - S_{org}}\right) + \left(\dfrac{(W_e - B_w W_p)S_{orw}}{1 - S_{wi} - S_{orw}}\right)\right]}{\left(\dfrac{NB_{oi}}{1 - S_{wi}}\right) - \left[\left(\dfrac{mNB_{oi}\left(\frac{B_g}{B_{gi}} - 1\right)}{1 - S_{wi} - S_{org}}\right) + \left(\dfrac{W_e - W_p B_w}{1 - S_{wi} - S_{orw}}\right)\right]} \quad (5.25)$$

Oil Saturation Adjustment for Shrinking Gas Cap. The control of the gas cap size is very often a reliable guide to the efficiency of reservoir operations. A shrinking gas cap will cause the loss of a substantial amount of oil, which might otherwise be recovered. Normally, there is little or no oil saturation in the gas cap, and if the oil migrates into the original gas zone there will necessarily be some residual oil saturation remaining in this portion of the gas cap at abandonment. As pointed out by Cole (1969), the magnitude of this loss may be quite large and depends on:

- the area of the gas–oil contact;
- the rate of gas cap shrinkage;
- the relative permeability characteristics;
- the vertical permeability.

A shrinking gas cap can be controlled by either shutting in wells that are producing large quantities of gas cap gas or returning some of the produced gas back to the gas cap portion of the reservoir. In many cases, the shrinkage cannot be completely eliminated by shutting in wells, as there is a practical limit to the number of wells that can be shut in. The amount of oil lost by the shrinking gas cap can be very well the engineer's most important economic justification for the installation of gas return facilities.

The difference between the original volume of the gas cap and the volume occupied by the gas cap at any subsequent time is a measure of the volume of oil that has migrated into the gas cap. If the size of the original gas cap is mNB_{oi}, then the expansion of the original free gas resulting from reducing the pressure from p_i to p is:

$$\text{Expansion of the original gas cap} = mNB_{oi}\left[\left(\frac{B_g}{B_{gi}}\right) - 1\right]$$

where

mNB_{oi} = original gas cap volume, bbl
B_g = gas formation volume factor, bbl/scf

If the gas cap is shrinking, then the volume of the produced gas must be larger than the gas cap expansion. All of the oil that moves into the gas cap will not be lost, as this oil will also be subject to the various driving mechanisms. Assuming no original oil saturation in the gas zone, the oil that will be lost is essentially the residual oil saturation remaining at abandonment. If the cumulative gas production from the gas cap is G_{pc} in scf, the volume of the gas cap shrinkage as expressed in barrels is equal to:

$$\text{Gas cap shrinkage} = G_{pc}B_g - mNB_{oi}\left[\left(\frac{B_g}{B_{gi}}\right) - 1\right]$$

From the volumetric equation:

$$G_{pc}B_g - mNB_{oi}\left[\left(\frac{B_g}{B_{gi}}\right) - 1\right] = 7758Ah\phi(1 - S_{wi} - S_{gr})$$

where

A = average cross-sectional area of the gas–oil contact, acres
h = average change in depth of the gas–oil contact, ft
S_{gr} = residual gas saturation in the shrinking zone

The volume of oil lost as a result of oil migration to the gas cap can also be calculated from the volumetric equation as follows:

$$\text{Oil lost} = \frac{7758Ah\phi S_{org}}{B_{oa}}$$

where

S_{org} = residual oil saturation in the gas cap shrinking zone
B_{oa} = oil formation volume factor at abandonment

Combining the above relationships and eliminating the term $7758Ah\phi$ gives the following expression for estimating the volume of oil in barrels lost in the gas cap:

$$\text{Oil lost} = \frac{\left[G_{pc}B_g - mNB_{oi}\left(\left(\frac{B_g}{B_{gi}}\right) - 1\right)\right]S_{org}}{(1 - S_{wi} - S_{gr})B_{oa}}$$

where

G_{pc} = cumulative gas production for the gas cap, scf
B_g = gas formation volume factor, bbl/scf

All the methodologies that have been developed to predict the future reservoir performance are essentially based on employing and combining the above relationships that include:

- the MBE;
- the saturation equations;
- the instantaneous GOR;
- the equation relating the cumulative GOR to the instantaneous GOR.

Using the above information, it is possible to predict the field primary recovery performance with declining reservoir pressure. There are three methodologies that are widely used in the petroleum industry to perform a reservoir study. These are:

(1) the Tracy method;
(2) the Muskat method;
(3) the Tarner method.

All three methods yield essentially the same results when small intervals of pressure or time are used. The methods can be used to predict the performance of a reservoir under any driving mechanism, including:

- solution gas drive;
- gas cap drive;
- water drive;
- combination drive.

The practical use of all the techniques is illustrated in predicting the primary recovery performance of a volumetric solution gas drive reservoir. Using the appropriate saturation equation, e.g., Eq. (5.20), for a water drive reservoir, any of the available reservoir prediction techniques could be applied to other reservoirs operating under different driving mechanisms.

The following two cases of the solution gas drive reservoir are considered:

(1) undersaturated oil reservoirs;
(2) saturated oil reservoirs.

5.1.3 Undersaturated Oil Reservoirs

When the reservoir pressure is above the bubble point pressure of the crude oil system, the reservoir is considered as undersaturated. The general material balance is expressed in Chapter 4 by Eq. (4.15):

$$N = \frac{N_p[B_o + (R_p - R_s)B_g] - (W_e - W_p B_w) - G_{inj}B_{ginj} - W_{inj}B_{wi}}{(B_o - B_{oi}) + (R_{si} - R_s)B_g + mB_{oi}\left[\left(\frac{B_g}{B_{gi}}\right) - 1\right] + B_{oi}(1+m)\left[\frac{(S_{wi}c_w + c_f)}{(1 - S_{wi})}\right]\Delta p}$$

For a volumetric undersaturated reservoir with no fluid injection, the following conditions are observed:

- $m = 0$;
- $W_e = 0$;
- $R_s = R_{si} = R_p$.

Imposing the above conditions on the MBE reduces the equation to the following simplified form:

$$N = \frac{N_p B_o}{(B_o - B_{oi}) + B_{oi}\left[\frac{(S_{wi}c_w + c_f)}{(1 - S_{wi})}\right]\Delta p} \tag{5.26}$$

with

$$\Delta p = p_i - p$$

where

p_i = initial reservoir pressure
p = current reservoir pressure

Hawkins (1955) introduced the oil compressibility c_o into the MBE to further simplify the equation. The oil compressed is defined as:

$$c_o = \frac{1}{B_{oi}}\frac{\partial B_o}{\partial p} \approx \frac{1}{B_{oi}}\frac{B_o - B_{oi}}{\Delta p}$$

Rearranging:

$$B_o - B_{oi} = c_o B_{oi}\,\Delta p$$

Combining the above expression with Eq. (5.26) gives:

$$N = \frac{N_p B_o}{c_o B_{oi}\,\Delta p + B_{oi}\left[\frac{S_{wi}c_w + c_f}{1 - S_{wi}}\right]\Delta p} \tag{5.27}$$

The *denominator* of the above equation can be regrouped as:

$$N = \frac{N_p B_o}{B_{oi}\left[c_o + \left(\frac{S_{wi}c_w}{1 - S_{wi}}\right) + \left(\frac{c_f}{1 - S_{wi}}\right)\right]\Delta p} \tag{5.28}$$

Since there are only two fluids in the reservoir, i.e., oil and water, then:

$$S_{oi} = 1 - S_{wi}$$

Rearranging Eq. (5.28) to include initial oil saturation gives:

$$N = \frac{N_p B_o}{B_{oi}\left[\frac{(S_{oi}c_o + S_{wi}c_w + c_f)}{(1 - S_{wi})}\right]\Delta p}$$

The term in the square brackets is called the effective compressibility and defined by Hawkins (1955) as:

$$c_e = \frac{S_{oi}c_o + S_{wi}c_w + c_f}{1 - S_{wi}} \tag{5.29}$$

Therefore, the MBE above the bubble point pressure becomes:

$$N = \frac{N_p B_o}{B_{oi}c_e\,\Delta p} \tag{5.30}$$

Eq. (5.30) can be expressed as the equation of a straight line by:

$$p = p_i - \left[\frac{1}{NB_{oi}c_e}\right]N_pB_o \qquad (5.31)$$

Figure 5.6 indicates that the reservoir pressure will decrease linearly with cumulative reservoir voidage N_pB_o.

Rearranging Eq. (5.31) and solving for the cumulative oil production N_p gives:

$$N_p = Nc_e\left(\frac{B_o}{B_{oi}}\right)\Delta p \qquad (5.32)$$

The calculation of future reservoir production, therefore, does not require a trial-and-error procedure, but can be obtained directly from the above expression.

Example 5.3

The following data is available on a volumetric undersaturated oil reservoir.

$p_i = 4000$ psi,	$c_o = 15\times10^{-6}$ psi^{-1}
$p_b = 3000$ psi,	$c_w = 3\times10^{-6}$ psi^{-1}
$N = 85$ MMSTB,	$S_{wi} = 30\%$
$c_f = 5\times10^{-6}$ psi^{-1},	$B_{oi} = 1.40$ bbl/STB

Estimate cumulative oil production when the reservoir pressure drops to 3500 psi. The oil formation volume factor at 3500 psi is 1.414 bbl/STB.

Solution

Step 1. Determine the effective compressibility from Eq. (5.29):

$$c_e = \frac{S_{oi}c_o + S_{wi}c_w + c_f}{1 - S_{wi}}$$

$$= \frac{(0.7)(15\times10^{-6}) + (0.3)(3\times10^{-6}) + 5\times10^{-6}}{1 - 0.3}$$

$$= 23.43\times10^{-6}\ \text{psi}^{-1}$$

Step 2. Estimate N_p from Eq. (5.32):

$$N_p = Nc_e\left(\frac{B_o}{B_{oi}}\right)\Delta p$$

$$= (85\times10^6)(23.43\times10^{-6})\left(\frac{1.411}{1.400}\right)(4000 - 3500)$$

$$= 985.18\ \text{MSTB}$$

5.1.4 Saturated Oil Reservoirs

If the reservoir originally exists at its bubble point pressure, the reservoir is referred to as a saturated oil reservoir. This is considered as the second type of solution gas drive reservoir. As the reservoir pressure declines below the bubble point, the gas begins to evolve from the solution. The general MBE may be simplified by assuming that the expansion of the gas is much greater than the expansion of rock and initial water and, therefore, can be neglected. For a

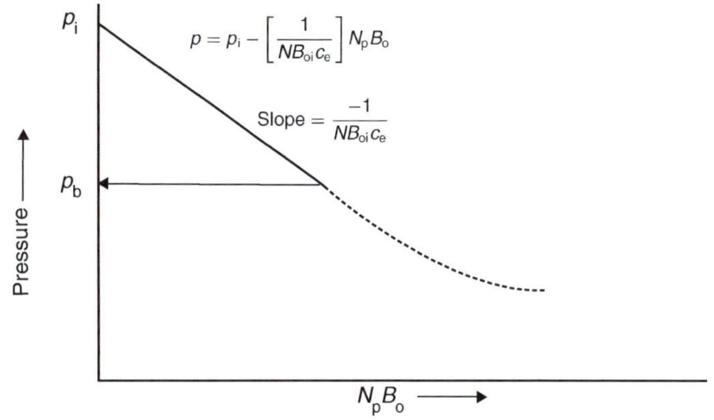

FIGURE 5.6 Pressure voidage relationship.

volumetric and saturated oil reservoir with no fluid injection, the MBE can be expressed by:

$$N = \frac{N_p B_o + (G_p - N_p R_s) B_g}{(B_o - B_{oi}) + (R_{si} - R_s) B_g} \quad (5.33)$$

This MBE contains two unknowns. These are:

(1) cumulative oil production N_p;
(2) cumulative gas production G_p.

The following reservoir and *PVT* data must be available in order to predict the primary recovery performance of a depletion drive reservoir in terms of N_p and G_p:

Initial oil-in-place *N*: Generally the volumetric estimate of oil-in-place is used in calculating the performance. However, where there is sufficient solution gas drive history, this estimate may be checked by calculating a material balance estimate.

Hydrocarbon *PVT* data: Since differential gas liberation is assumed to best represent the conditions in the reservoir, differential laboratory *PVT* data should be used in reservoir material balance. The flash *PVT* data is then used to convert from reservoir conditions to stock-tank conditions.

If laboratory data is not available, reasonable estimates may sometimes be obtained from published correlations. If differential data is not available, the flash data may be used instead; however, this may result in large errors for high-solubility crude oils.

Initial fluid saturations: Initial fluid saturations obtained from a laboratory analysis of core data are preferred; however, if these are not available, estimates in some cases either may be obtained from a well log analysis or may be obtained from other reservoirs in the same or similar formations.

Relative permeability data: Generally, laboratory-determined k_g/k_o and k_{ro} data is averaged to obtain a single representative set for the reservoir. If laboratory data is not available, estimates in some cases may be obtained from other reservoirs in the same or similar formations.

Where there is sufficient solution gas drive history for the reservoir, calculate k_{rg}/k_{ro} values vs. saturation from:

$$S_o = (1 - S_{wi}) \left(1 - \frac{N_p}{N}\right) \frac{B_o}{B_{oi}}$$

$$\frac{k_{rg}}{k_{ro}} = (\text{GOR} - R_s) \left(\frac{\mu_g B_g}{\mu_o B_o}\right)$$

The above results should be compared with the averaged laboratory relative permeability data. This may indicate a *needed adjustment* in the early data and possibly an adjustment in the overall data.

All the techniques that are used to predict the future performance of a reservoir are based on combining the appropriate MBE with the instantaneous GOR using the proper saturation equation. The calculations are repeated at a series of assumed reservoir pressure drops. These calculations are usually based on one stock-tank barrel of oil-in-place at the bubble point pressure, i.e., $N = 1$. This avoids dealing with large numbers in the calculation procedure and permits calculations to be made on the basis of the fractional recovery of initial oil-in-place.

As mentioned above, there are several widely used techniques that were specifically developed to predict the performance of solution gas drive reservoirs, including:

- the Tracy method;
- the Muskat technique;
- the Tarner method.

These methodologies are presented below.

Tracy Method. Tracy (1955) suggested that the general MBE can be rearranged and expressed in terms of three functions of *PVT* variables. Tracy's arrangement is given in Chapter 4 by Eq. (4.54) and is repeated here for convenience:

$$N = N_p \Phi_o + G_p \Phi_g + (W_p B_w - W_e) \Phi_w \quad (5.34)$$

where Φ_o, Φ_g, and Φ_w are considered *PVT*-related properties that are functions of pressure and defined by:

$$\Phi_o = \frac{B_o - R_s B_g}{\text{Den}}$$

$$\Phi_g = \frac{B_g}{\text{Den}}$$

$$\Phi_w = \frac{1}{\text{Den}}$$

with

$$\text{Den} = (B_o - B_{oi}) + (R_{si} - R_s)B_g + mB_{oi}\left[\frac{B_g}{B_{gi}} - 1\right] \quad (5.35)$$

For a solution gas drive reservoir, Eqs. (5.34) and (5.35) are reduced to the following expressions, respectively:

$$N = N_p\Phi_o + G_p\Phi_g \quad (5.36)$$

and

$$\text{Den} = (B_o - B_{oi}) + (R_{si} - R_s)B_g \quad (5.37)$$

Tracy's calculations are performed in a series of pressure drops that proceed from known reservoir conditions at the previous reservoir pressure p^* to the new, assumed, lower pressure p. The calculated results at the new reservoir pressure become "known" at the next assumed lower pressure.

In progressing from the conditions at any pressure p^* to the lower reservoir pressure p, consider the incremental oil and gas production as ΔN_p and ΔG_p, or:

$$N_p = N_p^* + \Delta N_p \quad (5.38)$$

$$G_p = G_p^* + \Delta G_p \quad (5.39)$$

where

N_p^*, G_p^* = "known" cumulative oil and gas production at previous pressure level p^*

N_p, G_p = "unknown" cumulative oil and gas at new pressure level p

Replacing N_p and G_p in Eq. (5.36) with those of Eqs. (5.38) and (5.39) gives

$$N = (N_p^* + \Delta N_p)\Phi_o + (G_p^* + \Delta G_p)\Phi_g \quad (5.40)$$

Defining the average instantaneous GOR between the two pressures p^* and p by:

$$(\text{GOR})_{avg} = \frac{\text{GOR}^* + \text{GOR}}{2} \quad (5.41)$$

the incremental cumulative gas production ΔG_p can be approximated by Eq. (5.6) as:

$$\Delta G_p = (\text{GOR})_{avg}\,\Delta N_p \quad (5.42)$$

Replacing ΔG_p in Eq. (5.40) with that of Eq. (5.41) gives:

$$N = [N_p^* + \Delta N_p]\Phi_o + [G_p^* + \Delta N_p(\text{GOR})_{avg}]\Phi_g \quad (5.43)$$

If Eq. (5.43) is expressed for $N = 1$, the cumulative oil production N_p and the cumulative gas production G_p become fractions of initial oil-in-place. Rearranging Eq. (5.43) gives:

$$\Delta N_p = \frac{1 - (N_p^*\Phi_o + G_p^*\Phi_g)}{\Phi_o + (\text{GOR})_{avg}\Phi_g} \quad (5.44)$$

Eq. (5.44) shows that there are essentially two unknowns. These are:

the incremental cumulative oil production ΔN_p;
the average gas–oil ratio $(\text{GOR})_{avg}$.

The methodology involved in solving Eq. (5.44) is basically an iterative technique with the objective of converging to the future GOR. In the calculations as described below, three GORs are included at any assumed depletion reservoir pressure. These are:

the *current* (known) gas–oil ratio GOR* at current (known) reservoir pressure p^*;
the *estimated* gas–oil ratio $(\text{GOR})_{est}$ at a selected new reservoir pressure p;
the *calculated* gas–oil ratio $(\text{GOR})_{cal}$ at the same selected new reservoir pressure p.

The specific steps of solving Eq. (5.44) are given below:

Step 1. Select a *new* average reservoir pressure p below the previous reservoir pressure p^*.

Step 2. Calculate the values of the *PVT* functions Φ_o and Φ_g at the selected new reservoir pressure p.

Step 3. *Estimate* the GOR designated as $(GOR)_{est}$ at the selected new reservoir pressure p.

Step 4. Calculate the average instantaneous GOR:

$$(GOR)_{avg} = \frac{GOR^* + (GOR)_{est}}{2}$$

where GOR* is a "known" GOR at previous pressure level p^*.

Step 5. Calculate the incremental cumulative oil production ΔN_p from Eq. (5.44), as:

$$\Delta N_p = \frac{1 - (N_p^* \Phi_o + G_p^* \Phi_g)}{\Phi_o + (GOR)_{avg} \Phi_g}$$

Step 6. Calculate cumulative oil production N_p:

$$N_p = N_p^* + \Delta N_p$$

Step 7. Calculate the oil and gas saturations at selected average reservoir pressure by using Eqs. (5.15) and (5.16), as:

$$S_o = (1 - S_{wi})\left(1 - \frac{N_p}{N}\right)\frac{B_o}{B_{oi}}$$

Since the calculations are based on $N = 1$, then:

$$S_o = (1 - S_{wi})(1 - N_p)\frac{B_o}{B_{oi}}$$

with gas saturation of:

$$S_g = 1 - S_o - S_{wi}$$

Step 8. Obtain the ratio k_{rg}/k_{ro} at S_L, i.e., at $(S_o + S_{wi})$, from the available laboratory or field relative permeability data.

Step 9. Using the relative permeability ratio k_{rg}/k_{ro}, *calculate* the instantaneous GOR from Eq. (5.1) and designate it as $(GOR)_{cal}$:

$$(GOR)_{cal} = R_s + \frac{k_{rg}}{k_{ro}}\left(\frac{\mu_o B_o}{\mu_g B_g}\right)$$

Step 10. Compare the estimated $(GOR)_{est}$ in step 3 with the calculated $(GOR)_{cal}$ in step 9. If the values are within the acceptable tolerance of:

$$0.999 \le \frac{(GOR)_{cal}}{(GOR)_{est}} \le 1.001$$

then proceed to the next step. If they are not within the tolerance, set the estimated $(GOR)_{est}$ equal to the calculated $(GOR)_{cal}$ and repeat the calculations from step 4. Steps 4 through 10 are repeated until convergence is achieved.

Step 11. Calculate the cumulative gas production:

$$G_p = G_p^* + \Delta N_p (GOR)_{avg}$$

Step 12. Since results of the calculations are based on 1 STB of oil initially in place, a final check on the accuracy of the prediction should be made on the MBE, or:

$$0.999 \le (N_p \Phi_o + G_p \Phi_g) \le 1.001$$

Step 13. Repeat from step 1 with a new pressure and setting:

$$p^* = p$$
$$GOR^* = GOR$$
$$G_p^* = G_p$$
$$N_p^* = N_p$$

As the calculation progresses, a plot of GOR vs. pressure should be maintained and extrapolated as an aid in estimating GOR at each new pressure.

Example 5.4

The following *PVT* data characterizes a solution gas drive reservoir. The relative permeability data is shown in Figure 5.7.

p (psi)	B_o (bbl/STB)	B_g (bbl/scf)	R_s (scf/STB)
4350	1.43	6.9 × 10	840
4150	1.420	7.1 × 10	820
3950	1.395	7.4×10^{-4}	770

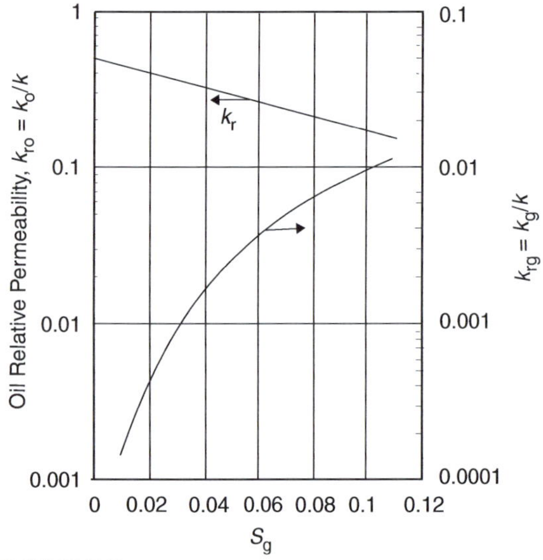

FIGURE 5.7 Relative permeability data for Example 5.4. *(After Economides, M., Hill, A., Economides, C., 1994. Petroleum Production Systems. Prentice Hall, Englewood Cliffs, NJ).*

3750	1.380	7.8×10^{-4}	730
3550	1.360	8.1×10^{-4}	680
3350	1.345	8.5×10^{-4}	640

The following additional data is available:

$N = 15$ MMSTB, $p^* = 4350$
$p_i = 4350$ psia, $\text{GOR}^* = 840$ scf/STB
$p_b = 4350$ psia, $G_p^* = 0$
$S_{wi} = 30\%$, $N_p^* = 0$
$N = 15$ MMSTB

Predict the cumulative oil and gas production to 3350 psi.

Solution

A sample of Tracy's calculation procedure is performed *at* 4150 psi.

Step 1. Calculate Tracy's *PVT* functions at 4150 psia. First calculate the term "Den" from Eq. (5.37):

$$\text{Den} = (B_o - B_{oi}) + (R_{si} - R_s)B_g$$
$$= (1.42 - 1.43) + (840 - 820)(7.1 \times 10^4)$$
$$= 0.0042$$

Then calculate Φ_o and Φ_g at 4150 psi:

$$\Phi_o = \frac{B_o - R_s B_g}{\text{Den}} = \frac{1.42 - (820)(7.1 \times 10^{-4})}{0.0042} = 199$$

$$\Phi_g = \frac{B_g}{\text{Den}} = \frac{7.1 \times 10^{-4}}{0.0042} = 0.17$$

Similarly, these *PVT* variables are calculated for all other pressures, to give:

p	Φ_o	Φ_g
4350	–	–
4150	199	00.17
3950	49	00.044
3750	22.6	00.022
3550	13.6	00.014
3350	90.42	00.010

Step 2. Estimate (assume) a value for the GOR at 4150 psi:

$$(\text{GOR})_{est} = 850 \text{ scf/STB}$$

Step 3. Calculate the average GOR:

$$(\text{GOR})_{avg} = \frac{\text{GOR}^* + (\text{GOR})_{est}}{2} = \frac{840 + 850}{2} = 845 \text{ scf/STB}$$

Step 4. Calculate the incremental cumulative oil production ΔN_p:

$$\Delta N_p = \frac{1 - (N_p^* \Phi_o + G_p^* \Phi_g)}{\Phi_o + (\text{GOR})_{avg}\Phi_g} = \frac{1 - 0}{199 + (845)(0.17)} = 0.00292 \text{ STB}$$

Step 5. Calculate the cumulative oil production N_p at 4150 psi:

$$N_p = N_p^* + \Delta N_p$$
$$= 0 + 0.00292 = 0.00292$$

Step 6. Calculate oil and gas saturations:

$$S_o = (1 - S_{wi})\left(1 - \frac{N_p}{N}\right)\frac{B_o}{B_{oi}}$$

$$= (1 - 0.3)(1 - 0.00292)\left(\frac{1.42}{1.43}\right) = 0.693$$

$$S_g = 1 - S_{wi} - S_o = 1 - 0.3 - 0.693 = 0.007$$

Step 7. Determine the relative permeability ratio k_{rg}/k_{ro} from Figure 5.7, to give:

$$\frac{k_{rg}}{k_{ro}} = 8 \times 10^{-5}$$

Step 8. Using $\mu_o = 1.7$ cp and $\mu_g = 0.023$ cp, calculate the instantaneous GOR:

$$(\text{GOR})_{cal} = R_s + \frac{k_{rg}}{k_{ro}}\left(\frac{\mu_o B_o}{\mu_g B_g}\right)$$

$$= 820 + (1.7 \times 10^4)\frac{(1.7)(1.42)}{(0.023)(7.1 \times 10^{-4})}$$

$$= 845 \text{ scf/STB}$$

which agrees with the assumed value of 850.

Step 9. Calculate cumulative gas production:

$$G_p = 0 + (0.00292)(850) = 2.48$$

Complete results of the method are shown below:

$\bar{p}$	ΔN_p	N_p	$(\text{GOR})_{avg}$	ΔG_p	G_p (scf/STB)	$N_p = 15 \times 10^6 N$ (STB)	$G_p = 15 \times 10^6 N$ (scf)
4350	–	–	–	–	–		
4150	0.00292	0.00292	845	2.48	2.48	0.0438×10^6	–
3950	0.00841	0.0110	880	7.23	9.71	0.165×10^6	37.2×10^6
3750	0.0120	0.0230	1000	12	21.71	0.180×10^6	145.65×10^6
3550	0.0126	0.0356	1280	16.1	37.81	0.534×10^6	325.65×10^6
3350	0.011	0.0460	1650	18.2	56.01	0.699×10^6	567.15×10^6

Muskat Method. Muskat (1945) expressed the MBE for a depletion drive reservoir in the following differential form:

$$\frac{dS_o}{dp} = \frac{\dfrac{(S_o B_g/B_o)(dR_s/dp) + (S_o/B_o)(k_{rg}/k_{ro})(\mu_o/\mu_g)(dB_o/dp) - ((1 - S_o - S_{wi})/B_g)(dB_g/dp)}{}}{1 + (\mu_o/\mu_g)(k_{rg}/k_{ro})} \quad (5.45)$$

with

$$\Delta S_o = S_o^* - S_o$$

$$\Delta p = p^* - p$$

where

S_o^*, p^* = oil saturation and average reservoir pressure at the beginning of the pressure step (known values)

S_o, p = oil saturation and average reservoir pressure at the end of the time step

R_s = gas solubility at pressure p, scf/STB

B_g = gas formation volume factor, bbl/scf

S_{wi} = initial water saturation

Craft and Hawkins (1991) suggested that the calculations can be greatly facilitated by computing and preparing in advance in graphical form the following pressure-dependent groups:

$$X(p) = \frac{B_g}{B_o}\frac{dR_s}{dp} \quad (5.46)$$

$$Y(p) = \frac{1}{B_o}\frac{\mu_o}{\mu_g}\frac{dB_o}{dp} \quad (5.47)$$

$$Z(p) = \frac{1}{B_g}\frac{dB_g}{dp} \quad (5.48)$$

Introducing the above pressure-dependent terms into Eq. (5.45) gives:

$$\left(\frac{\Delta S_o}{\Delta p}\right) = \frac{S_o X(p) + S_o(k_{rg}/k_{ro})Y(p) - (1 - S_o - S_{wi})Z(p)}{1 + (\mu_o/\mu_g)(k_{rg}/k_{ro})} \quad (5.49)$$

Given:

- initial oil-in-place N;
- current (known) pressure p^*;

- current cumulative oil production N_p^*;
- current cumulative gas production G_p^*;
- current GOR*;
- current oil saturation S_o^*;
- initial water saturation S_{wi}.

Eq. (5.49) can be solved to predict cumulative production and fluid saturation at a given pressure drop Δp, i.e., $(p^* - p)$, by employing the following steps:

Step 1. Prepare a plot of k_{rg}/k_{ro} vs. gas saturation.

Step 2. Plot R_s, B_o, and B_g vs. pressure and numerically determine the slope of the *PVT* properties (i.e., dB_o/dp, dR_s/dp, and $d(B_g)/dp$) at several pressures. Tabulate the generated values as a function of pressure.

Step 3. Calculate the pressure-dependent terms $X(p)$, $Y(p)$, and $Z(p)$ at each of the selected pressures in step 2. That is:

$$X(p) = \frac{B_g}{B_o}\frac{dR_s}{dp}$$

$$Y(p) = \frac{1}{B_o}\frac{\mu_o}{\mu_g}\frac{dB_o}{dp}$$

$$Z(p) = \frac{1}{B_g}\frac{dB_g}{dp}$$

Step 4. Plot the pressure-dependent terms $X(p)$, $Y(p)$, and $Z(p)$ as a function of pressure, as illustrated in Figure 5.8.

Step 5. Assume that the reservoir pressure has declined from initial (known) average reservoir pressure of p^* to a *selected* reservoir pressure p. Graphically determine the values of $X(p)$, $Y(p)$, and $Z(p)$ that correspond to the pressure p.

Step 6. Solve Eq. (5.49) for $(\Delta S_o/\Delta p)$ by using the current oil saturation S_o^* at the beginning of the pressure drop interval p^*:

$$\left(\frac{\Delta S_o}{\Delta p}\right) = \frac{S_o^* X(p^*) + S_o^*(k_{rg}/k_{ro})Y(p^*) - (1 - S_o^* - S_{wi})Z(p^*)}{1 + (\mu_o/\mu_g)(k_{rg}/k_{ro})}$$

Step 7. Determine the oil saturation S_o at the assumed (selected) average reservoir pressure p, from:

$$S_o = S_o^* - (p^* - p)\left(\frac{\Delta S_o}{\Delta p}\right) \tag{5.50}$$

Step 8. Using the *calculated oil saturation* S_o from step 7, the updated value of the relative permeability ratio k_{rg}/k_{ro} at S_o, and the *PVT* terms at the assumed pressure p, recalculate $(\Delta S_o/\Delta p)$ by applying Eq. (5.49):

$$\left(\frac{\Delta S_o}{\Delta p}\right) = \frac{S_o X(p) + S_o(k_{rg}/k_{ro})Y(p) - (1 - S_o - S_{wi})Z(p)}{1 + (\mu_o/\mu_g)(k_{rg}/k_{ro})}$$

Step 9. Calculate the average value for $(\Delta S_o/\Delta p)$ from the two values obtained in steps 6 and 8, or:

$$\left(\frac{\Delta S_o}{\Delta p}\right)_{avg} = \frac{1}{2}\left[\left(\frac{\Delta S_o}{\Delta p}\right)_{step\ 6} + \left(\frac{\Delta S_o}{\Delta p}\right)_{step\ 8}\right]$$

Step 10. Using $(\Delta S_o/\Delta p)_{avg}$, solve for the oil saturation S_o from:

$$S_o = S_o^* - (p^* - p)\left(\frac{\Delta S_o}{\Delta p}\right)_{avg} \tag{5.51}$$

Step 11. Calculate the gas saturation S_g and the GOR from:

$$S_g = 1 - S_{wi} - S_o$$

$$GOR = R_s + \frac{k_{rg}}{k_{ro}}\left(\frac{\mu_o B_o}{\mu_g B_g}\right)$$

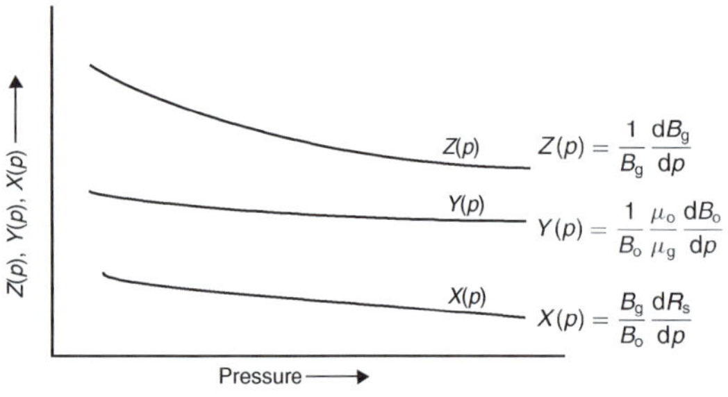

FIGURE 5.8 Pressure-dependent terms vs. *p*.

Step 12. Using the saturation equation, i.e., Eq. (5.15), solve for the cumulative oil production:

$$N_p = N\left[1 - \left(\frac{B_{oi}}{B_o}\right)\left(\frac{S_o}{1 - S_{wi}}\right)\right] \quad (5.52)$$

with an incremental cumulative oil production of:

$$\Delta N_p = N_p - N_p^*$$

Step 13. Calculate the incremental cumulative gas production by using Eqs. (5.40) and (5.41):

$$(GOR)_{avg} = \frac{GOR^* + GOR}{2}$$

$$\Delta G_p = (GOR)_{avg}\, \Delta N_p$$

with a total cumulative gas production of:

$$G_p = \sum \Delta G_p$$

Step 14. Repeat steps 5 through 13 for all pressure drops of interest and setting:

$$p^* = p$$

$$N_o^* = N_p$$

$$G_p^* = G_p$$

$$GOR^* = GOR$$

$$S_o^* = S_o$$

Example 5.5

A volumetric depletion drive reservoir exists at its bubble point pressure of 2500 psi. Detailed fluid property data is listed by Craft and his co-authors and given here for only two pressures[1] :

Fluid Property	$p^* = 2500$ psi	$p = 2300$ psi
B_o, bbl/STB	1.498	1.463
R_s, scf/STB	721	669
B_g, bbl/scf	0.001048	0.001155
μ_o, cp	0.488	0.539
μ_g, cp	0.0170	0.0166
$X(p)$	0.00018	0.00021
$Y(p)$	0.00328	0.00380
$Z(p)$	0.00045	0.00050

The following additional information is available:

$$N = 56 \text{ MMSTB}, \quad S_{wi} = 20\%,$$

Soi = 80%

S_g	k_{rg}/k_{ro}
0.10	0.010
0.20	0.065
0.30	0.200
0.50	2.000
0.55	3.000
0.57	5.000

Calculate the cumulative oil production for a pressure drop of 200 psi, i.e., at 2300 psi.

Solution

Step 1. Using the oil saturation at the beginning of the pressure interval, i.e., $S_o^* = 0.8$, calculate k_{rg}/k_{ro}, to give:

$$\frac{k_{rg}}{k_{ro}} = 0.0 \text{ (no free gas initially in place)}$$

Step 2. Evaluate $(\Delta S_o/\Delta p)$ by applying Eq. (5.49):

$$\left(\frac{\Delta S_o}{\Delta p}\right) = S_o^* X(p^*) + S_o^*(k_{rg}/k_{ro})$$

$$\frac{Y(p^*) - (1 - S_o^* - S_{wi})Z(p^*)}{1 + (\mu_o/\mu_g)(k_{rg}/k_{ro})}$$

$$= \frac{(0.8)(0.00018) + 0 - (1 - 0.8 - 0.2)(0.00045)}{1 + 0}$$

$$= 0.000146$$

Step 3. Estimate the oil saturation at $p = 2300$ psi from Eq. (5.51):

$$S_o = S_o^* - (p^* - p)\left(\frac{\Delta S_o}{\Delta p}\right)_{avg}$$

$$= 0.8 - 200(0.000146) = 0.7709$$

Step 4. Recalculate $(\Delta S_o/\Delta p)$ by using $S_o = 0.7709$, relative permeability ratio

[1]Craft, B.C., Hawkins, M., Terry, R., 1991. *Applied Petroleum Reservoir Engineering*, third ed. Prentice Hall.

k_{rg}/k_{ro} at S_o, and the *pressure-dependent PVT terms at* 2300 psi:

$$\left(\frac{\Delta S_o}{\Delta p}\right) = \frac{S_o X(p) + S_o(k_{rg}/k_{ro}) Y(p) - (1 - S_o - S_{wi})Z(p)}{1 + (\mu_o/\mu_g)(k_{rg}/k_{ro})}$$

$$= 0.7709(0.00021) + 0.7709(0.00001)0.0038 - \frac{(1 - 0.2 - 0.7709)0.0005}{1 + (0.539/0.0166)(0.00001)}$$

$$= 0.000173$$

Step 5. Calculate the average $(\Delta S_o/\Delta p)$:

$$\left(\frac{\Delta S_o}{\Delta p}\right)_{avg} = \frac{0.000146 + 0.000173}{2} = 0.000159$$

Step 6. Calculate the oil saturation at 2300 psi by applying Eq. (5.51):

$$S_o = S_o^* - (p^* - p)\left(\frac{\Delta S_o}{\Delta p}\right)_{avg}$$

$$= 0.8 - (2500 - 2300)(0.000159) = 0.7682$$

Step 7. Calculate the gas saturation:

$$S_g = 1 - 0.2 - 0.7682 = 0.0318$$

Step 8. Calculate cumulative oil production at 2300 psi by using Eq. (5.52):

$$N_p = N\left[1 - \left(\frac{B_{oi}}{B_o}\right)\left(\frac{S_o}{1 - S_{wi}}\right)\right]$$

$$= 56 \times 10^6\left[1 - \left(\frac{1.498}{1.463}\right)\left(\frac{0.7682}{1 - 0.2}\right)\right]$$

$$= 939,500 \text{ STB}$$

Step 9. Calculate k_{rg}/k_{ro} at 2300 psi, to give $k_{rg}/k_{ro} = 0.00001$.

Step 10. Calculate the instantaneous GOR at 2300 psi:

$$\text{GOR} = R_s + \frac{k_{rg}}{k_{ro}}\left(\frac{\mu_o B_o}{\mu_g B_g}\right)$$

$$= 669 + 0.00001\frac{(0.539)(1.463)}{(0.0166)(0.001155)}$$

$$= 670 \text{ scf/STB}$$

Step 11. Calculate the incremental cumulative gas production:

$$(\text{GOR})_{avg} = \frac{\text{GOR}^* + \text{GOR}}{2} = \frac{669 + 670}{2}$$

$$= 669.5 \text{ scf/STB}$$

$$\Delta G_p = (\text{GOR})_{avg}\,\Delta N_p$$

$$= 669.5(939500 - 0) = 629 \text{ MMscf}$$

It should be stressed that this method is based on the assumption of uniform oil saturation in the whole reservoir and that the solution will therefore break down when there is appreciable gas segregation in the formation. It is therefore applicable only when permeabilities are relatively low.

Tarner Method. Tarner (1944) suggested an iterative technique for predicting cumulative oil production N_p and cumulative gas production G_p as a function of reservoir pressure. The method is based on solving the MBE and the instantaneous GOR equation simultaneously for a given reservoir pressure drop from a known pressure p^* to an assumed (new) pressure p. It is accordingly assumed that the cumulative oil and gas production has increased from known values of N_p^* and G_p^* at reservoir pressure p^* to future values of N_p and G_p at the assumed pressure p. To simplify the description of the proposed iterative procedure, the stepwise calculation is illustrated for a volumetric saturated oil reservoir; however, *the method can be used to predict the volumetric behavior of reservoirs under different driving mechanisms.*

Step 1. Select (assume) a future reservoir pressure p below the initial (current) reservoir pressure p^* and obtain the necessary *PVT* data. Assume that the cumulative oil production has increased from N_p^* to N_p. Note that N_p^* and G_p^* are set equal to 0 at the initial reservoir pressure.

Step 2. Estimate or guess the cumulative oil production N_p at the selected (assumed) reservoir pressure p of step 1.

Step 3. Calculate the cumulative gas production G_p by rearranging the MBE, i.e., Eq. (5.33), to give:

$$G_p = N\left[(R_{si} - R_s) - \frac{B_{oi} - B_o}{B_g}\right] - N_p\left[\frac{B_o}{B_g} - R_s\right] \quad (5.53)$$

Equivalently, the above relationship can be expressed in terms of the two-phase (total) formation volume factor B_t, as:

$$G_p = \frac{N(B_t - B_{ti}) - N_p(B_t - R_{si}B_g)}{B_g} \quad (5.54)$$

where

B_{oi} = initial oil formation volume factor, bbl/STB
R_{si} = initial gas solubility, scf/STB
B_o = oil formation volume factor at the assumed reservoir pressure p, bbl/STB
B_g = gas formation volume factor at the assumed reservoir pressure p, bbl/scf
B_o = oil formation volume factor at the assumed reservoir pressure p, bbl/STB
B_t = two-phase formation volume factor at the assumed reservoir pressure p, bbl/STB
N = initial oil-in-place, STB

Step 4. Calculate the oil and gas saturations at the assumed cumulative oil production N_p and the selected reservoir pressure p by applying Eqs. (5.15) and (5.16) respectively, or:

$$S_o = (1 - S_{wi})\left[1 - \frac{N_p}{N}\right]\left(\frac{B_o}{B_{oi}}\right)$$

$$S_g = 1 - S_o - S_{wi}$$

and

$$S_L = S_o + S_{wi}$$

where

S_L = total liquid saturation
B_{oi} = initial oil formation volume factor at p_i, bbl/STB
B_o = oil formation volume factor at p, bbl/STB
S_g = gas saturation at the assumed reservoir pressure p
S_o = oil saturation at assumed reservoir pressure p

Step 5. Using the available relative permeability data, determine the relative permeability ratio k_{rg}/k_{ro} that corresponds to the calculated total liquid saturation S_L of step 4 and compute the instantaneous GOR at p from Eq. (5.1):

$$\text{GOR} = R_s + \left(\frac{k_{rg}}{k_{ro}}\right)\left[\frac{\mu_o B_o}{\mu_g B_g}\right] \quad (5.55)$$

It should be noted that all the *PVT* data in the expression *must be evaluated at the assumed reservoir pressure p*.

Step 6. Calculate again the cumulative gas production G_p at p by applying Eq. (5.7):

$$G_p = G_p^* + \left[\frac{\text{GOR}^* + \text{GOR}}{2}\right]\left[N_p - N_p^*\right] \quad (5.56)$$

in which GOR* represents the instantaneous GOR at p^*. Note that if p^* represents the initial reservoir pressure, then set GOR* = R_{si}.

Step 7. The calculations as performed in steps 3 and 6 give two estimates for cumulative gas produced G_p at the assumed (future) pressure p:

G_p as calculated from the MBE;
G_p as calculated from the GOR equation.

These two values of G_p are calculated from two independent methods, and therefore, if the cumulative gas production G_p as calculated from step 3 agrees with the value of step 6, the assumed value of N_p is correct and a new pressure may be selected and steps 1 through 6 are repeated. Otherwise, assume another value of N_p and repeat steps 2 through 6.

Step 8. In order to simplify this iterative process, three values of N_p can be assumed, which yield three different solutions of cumulative gas production for each of the equations (i.e., MBE and GOR equations). When the computed values of G_p are plotted vs. the assumed values of N_p, the resulting two curves (one representing the results of step 3 and the other representing the results of step 5) will intersect. This intersection indicates the cumulative oil and gas production that will satisfy both equations.

It should be pointed out that it may be more convenient to assume values of N_p as a fraction of the initial oil-in-place N. For instance, N_p could be assumed as $0.01N$, rather than as 10,000 STB. In this method, a true value of N is not required. Results of the calculations would be, therefore, in terms of STB of oil produced per STB of oil initially in place and scf of gas produced per STB of oil initially in place.

To illustrate the application of the Tarner method, Cole (1969) presented the following example.

Example 5.6

A saturated oil reservoir has a bubble point pressure of 2100 psi at 175°F. The initial reservoir pressure is 2400 psi. The following data summarizes the rock and fluid properties of the field:

Orginal oil-in-place = 10 MMSTB

Connate water saturation = 15%

Porosity = 12%

$c_w = 3.2 \times 10^{-6}\ \text{psi}^{-1}$

$c_f = 3.1 \times 10^{-6}\ \text{psi}^{-1}$

Basic *PVT* data is as follows:

p (psi)	B_o (bbl/STB)	B_t (bbl/STB)	R_s (scf/STB)	B_g (bbl/scf)	μ_o/μ_g
2400	1.464	1.464	1340	–	–
2100	1.480	1.480	1340	0.001283	34.1
1800	1.468	1.559	1280	0.001518	38.3
1500	1.440	1.792	1150	0.001853	42.4

Relative permeability ratio:

S_L (%)	k_{rg}/k_{ro}
96	0.018
91	0.063
75	0.850
65	3.350
55	10.200

Predict the cumulative oil and gas production at 2100, 1800, and 1500 psi.

Solution

The required calculations will be performed under the following two different driving mechanisms:

1. When the reservoir pressure declines from the initial reservoir pressure of 2500 psi to the bubble point pressure of 2100 psi, the reservoir is considered undersaturated, and therefore, the MBE can be used directly in cumulative production without restoring the iterative technique.
2. For reservoir pressures below the bubble point pressure, the reservoir is treated as a saturated oil reservoir and the Tarner method may be applied.

Oil Recovery Prediction from Initial Pressure to the Bubble Point Pressure:

Step 1. The MBE for an undersaturated reservoir is given by Eq. (4.33):

$$F = N(E_o + E_{f,w})$$

where

$$F = N_p B_o + W_p B_w$$

$$E_o = B_o - B_{oi}$$

$$E_{f,w} = B_{oi}\left[\frac{c_w S_w + c_f}{1 - S_{wi}}\right]\Delta p$$

$$\Delta p = p_i - \bar{p}_r$$

Since there is no water production, Eq. (4.33) can be solved for cumulative oil production, to give:

$$N_p = \frac{N[E_o + E_{f,w}]}{B_o} \tag{5.57}$$

Step 2. Calculate the two expansion factors E_o and $E_{f,w}$ for the pressure decline from the initial reservoir pressure of 2400 psi to the bubble point pressure of 2100 psi:

$$E_o = B_o - B_{oi}$$
$$= 1.480 - 1.464 = 0.016$$

$$E_{f,w} = B_{oi}\left[\frac{c_w S_w + c_f}{1 - S_{wi}}\right]\Delta p$$
$$= 1.464\left[\frac{(3.2\times10^{-6})(0.15)+(3.1\times10^{-6})}{1-0.15}\right]$$
$$\times(2400-2100) = 0.0018$$

Step 3. Calculate the cumulative oil and gas production when the reservoir pressure declines from 2400 to 2100 psi by applying Eq. (5.57), to give:

$$N_p = \frac{N[E_o + E_{f,w}]}{B_o}$$
$$= \frac{10\times10^6[0.016+0.0018]}{1.480} = 120,270 \text{ STB}$$

At or above the bubble point pressure, the producing GOR is equal to the gas solubility at the bubble point, and therefore, the cumulative gas production is given by:

$$G_p = N_p R_{si}$$
$$= (120,270)(1340) = 161 \text{ MMscf}$$

Step 4. Determine the remaining oil-in-place at 2100 psi:

$$\text{Remaining oil-in-place} = 10,000,000 - 120,270$$
$$= 9.880 \text{ MMSTB}$$

The remaining oil-in-place is considered as the initial oil-in-place during the reservoir performance below the saturation pressure. That is:

$$N = 9.880 \text{ MMSTB}$$
$$N_p = N_p^* = 0.0 \text{ STB}$$
$$G_p = G_p^* = 0.0 \text{ scf}$$
$$R_{si} = 1340 \text{ scf/STB}$$
$$B_{oi} = 1.489 \text{ bbl/STB}$$
$$B_{ti} = 1.489 \text{ bbl/STB}$$
$$B_{gi} = 0.001283 \text{ bbl/scf}$$

Oil Recovery Prediction below the Bubble Point Pressure. Oil recovery prediction at 1800 psi is performed with the following *PVT* properties:

$$B_o = 1.468 \text{ bbl/STB}$$
$$B_t = 1.559 \text{ bbl/STB}$$
$$B_g = 0.001518 \text{ bbl/scf}$$
$$R_s = 1280 \text{ scf/STB}$$

Step 1. Assume that 1% of the bubble point oil will be produced when the reservoir pressure drops 1800 psi. That is:

$$N_p = 0.01N$$

Calculate the corresponding cumulative gas G_p by applying Eq. (5.54):

$$G_p = \frac{N(B_t - B_{ti}) - N_p(B_t - R_{si}B_g)}{B_g}$$
$$= \frac{N(1.559-1.480)-(0.01N)[1.559-(1340)(0.001518)]}{0.001518}$$
$$= 55.17N$$

Step 2. Calculate the oil saturation, to give:

$$S_o = (1 - S_{wi})\left(1 - \frac{N_p}{N}\right)\frac{B_o}{B_{oi}}$$
$$= (1-0.15)\left(1 - \frac{0.01N}{N}\right)\frac{1.468}{1.480} = 0.835$$

Step 3. Determine the relative permeability ratio k_{rg}/k_{ro} from the tabulated data at total liquid saturation of S_L, to give:

$$S_L = S_o + S_{wi} = 0.835 + 0.15 = 0.985$$
$$\frac{k_{rg}}{k_{ro}} = 0.0100$$

Step 4. Calculate the instantaneous GOR at 1800 psi by applying Eq. (5.55), to give:

$$\text{GOR} = R_s + \left(\frac{k_{rg}}{k_{ro}}\right)\left[\frac{\mu_o B_o}{\mu_g B_g}\right]$$
$$= 1280 + 0.0100(38.3)\left(\frac{1.468}{0.001518}\right)$$
$$= 1650 \text{ scf/STB}$$

Step 5. Solve again for the cumulative gas production by using the average GOR and applying Eq. (5.56) to yield:

$$G_p = G_p^* + \left[\frac{GOR^* + GOR}{2}\right]\left[N_p - N_p^*\right]$$
$$= 0 + \frac{1340 + 1650}{2}(0.01N - 0) = 14.95N$$

Step 6. Since the cumulative gas production, as calculated by the two independent methods (steps 1 and 5), does not agree, the calculations must be repeated by assuming a different value for N_p and plotting results of the calculation. Repeated calculations converge at:

$$N_p = 0.0393N \text{ STB/STB of bubble point oil}$$
$$G_p = 64.34N \text{ scf/STB of bubble point oil}$$

or

$$N_p = 0.0393(9.88 \times 10^6) = 388,284 \text{ STB}$$
$$G_p = 64.34(9.88 \times 10^6) = 635.679 \text{ MMscf}$$

It should be pointed out that the cumulative production *above the bubble point pressure must be included when reporting the total cumulative oil and gas production*. The cumulative oil and gas production as the pressure declines from the initial pressure to the bubble point pressure is:

$$N_p = 120,270 \text{ STB}$$
$$G_p = 161 \text{ MMscf}$$

Therefore, the actual cumulative recovery at 1800 psi is:

$$N_p = 120,270 + 388,284 = 508,554 \text{ STB}$$
$$G_p = 161 + 635.679 = 799.679 \text{ MMscf}$$

The final results as summarized below show the cumulative gas and oil production as the pressure declines from the bubble point pressure:

Pressure	N_p	Actual N_p (STB)	G_p	Actual G_p (MMscf)
1800	0.0393N	508 554	64.34 N	799.679
1500	0.0889N	998 602	136.6 N	1510.608

It is apparent from the three predictive oil recovery methods, i.e., Tracy's, Muskat's, and Tarner's, that the relative permeability ratio k_{rg}/k_{ro} is the most important single factor governing the oil recovery. In cases where no detailed data is available concerning the physical characteristics of the reservoir rock in terms of k_{rg}/k_{ro} relationship, Wahl et al. (1958) presented an empirical expression for predicting the relative permeability ratio in sandstones:

$$\frac{k_{rg}}{k_{ro}} = \zeta(0.0435 + 0.4556\zeta)$$

with

$$\zeta = \frac{1 - S_{gc} - S_{wi} - S_o}{S_o - 0.25}$$

where

S_{gc} = critical gas saturation
S_{wi} = initial water saturation
S_o = oil saturation

Torcaso and Wyllie (1958) presented a similar correlation for sandstones in the following form:

$$\frac{k_{rg}}{k_{ro}} = \frac{(1 - S^*)^2[1 - (S^*)^2]}{(S^*)^4}$$

with

$$S^* = \frac{S_o}{1 - S_{wi}}$$

5.2 PHASE 2. OIL WELL PERFORMANCE

All the reservoir performance prediction techniques show the relationship of cumulative oil production N_p, cumulative gas production G_p, and instantaneous GOR as a function of the declining average reservoir pressure but do not relate the production to time. However, reservoir performance can be related to time by the use of relationships that are designed to predict the flow rate performance of the reservoirs' individual wells. Such flow rate relationships are traditionally expressed in terms of:

- the well productivity index;
- the well inflow performance relationship (IPR).

These relationships are presented below for vertical and horizontal wells.

5.2.1 Vertical Oil Well Performance

Productivity Index and IPR. A commonly used measure of the ability of the well to produce is the productivity index. Defined by the symbol J, the productivity index is the ratio of the total liquid flow rate to the pressure drawdown. For a water-free oil production, the productivity index is given by:

$$J = \frac{Q_o}{\bar{p}_r - p_{wf}} = \frac{Q_o}{\Delta p} \quad (5.58)$$

where

Q_o = oil flow rate, STB/day
J = productivity index, STB/day/psi
$\bar{p}_r$ = volumetric average drainage area pressure (static pressure)
p_{wf} = bottom-hole flowing pressure
Δp = drawdown, psi

The productivity index is generally measured during a production test on the well. The well is shut in until the static reservoir pressure is reached. The well is then allowed to produce at a constant flow rate of Q and a stabilized bottom-hole flow pressure of p_{wf}. Since a stabilized pressure at the surface does not necessarily indicate a stabilized p_{wf}, the bottom-hole flowing pressure should be recorded continuously from the time the well is allowed to flow. The productivity index is then calculated from Eq. (5.1).

It is important to note that the productivity index is a valid measure of the well productivity potential only if the well is flowing at pseudosteady-state conditions. Therefore, in order to accurately measure the productivity index of a well, it is essential that the well is allowed to flow at a constant flow rate for a sufficient amount of time to reach the pseudosteady state as illustrated in Figure 5.9. The figure indicates that during the transient flow period, the calculated values of the productivity index will vary depending upon the time at which the measurements of p_{wf} are made.

The productivity index can be numerically calculated by recognizing that J must be defined in terms of semisteady-state flow conditions. Recalling Eq. (1.148):

$$Q_o = \frac{0.00708 k_o h(\bar{p}_r - p_{wf})}{\mu_o B_o[\ln(r_e/r_w) - 0.75 + s]} \quad (5.59)$$

The above equation is combined with Eq. (5.58), to give:

$$J = \frac{0.00708 k_o h}{\mu_o B_o[\ln(r_e/r_w) - 0.75 + s]} \quad (5.60)$$

where

J = productivity index, STB/day/psi
k_o = effective permeability of the oil, md
s = skin factor
h = thickness, ft

The oil relative permeability concept can be conveniently introduced into Eq. (5.60), to give:

$$J = \frac{0.00708 hk}{[\ln(r_e/r_w) - 0.75 + s]}\left(\frac{k_{ro}}{\mu_o B_o}\right) \quad (5.61)$$

Since most of the well's life is spent in a flow regime that is approximating the pseudosteady state, the productivity index is a valuable

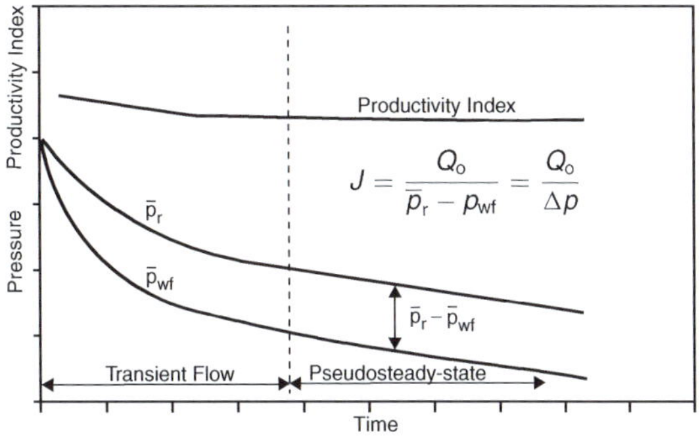

FIGURE 5.9 Productivity index during flow regimes.

methodology for predicting the future performance of wells. Further, by monitoring the productivity index during the life of a well, it is possible to determine if the well has become damaged due to completion, workover, production, injection operations, or mechanical problems. If a measured J has an unexpected decline, one of the indicated problems should be investigated. A comparison of productivity indexes of different wells in the same reservoir should also indicate that some of the wells might have experienced unusual difficulties or damage during completion. Since the productivity indexes may vary from well to well because of the variation in thickness of the reservoir, it is helpful to normalize the indexes by dividing each by the thickness of the well. This is defined as the specific productivity index J_s, or:

$$J_s = \frac{J}{h} = \frac{Q_o}{h(\bar{p}_r - p_{wf})} \quad (5.62)$$

Assuming that the well's productivity index is constant, Eq. (5.58) can be rewritten as:

$$Q_o = J(\bar{p}_r - p_{wf}) = J\,\Delta p \quad (5.63)$$

where

Δp = drawdown, psi
J = productivity index

Eq. (5.63) indicates that the relationship between Q_o and Δp is a straight line passing through the origin with a slope of J as shown in Figure 5.10.

Alternatively, Eq. (5.58) can be written as:

$$p_{wf} = \bar{p}_r - \left(\frac{1}{J}\right) Q_o \quad (5.64)$$

This expression shows that the plot of p_{wf} vs. Q_o is a straight line with a slope of $-1/J$ as shown schematically in Figure 5.11. This graphical representation of the relationship that exists between the oil flow rate and the bottom-hole flowing pressure is called the "inflow performance relationship" and referred to as IPR.

Several important features of the straight-line IPR can be seen in Figure 5.11:

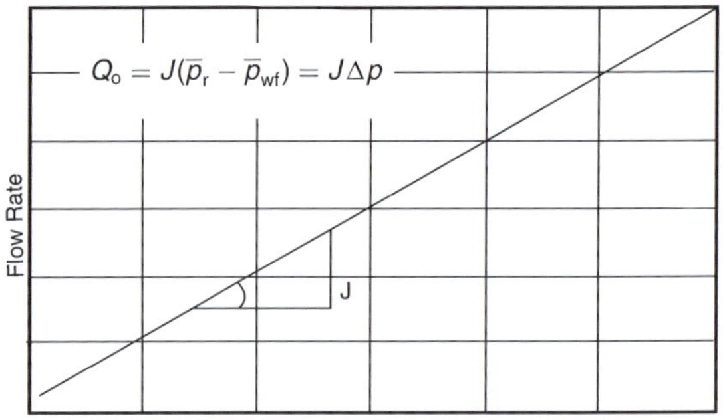

FIGURE 5.10 Q_o vs. Δp *relationship.*

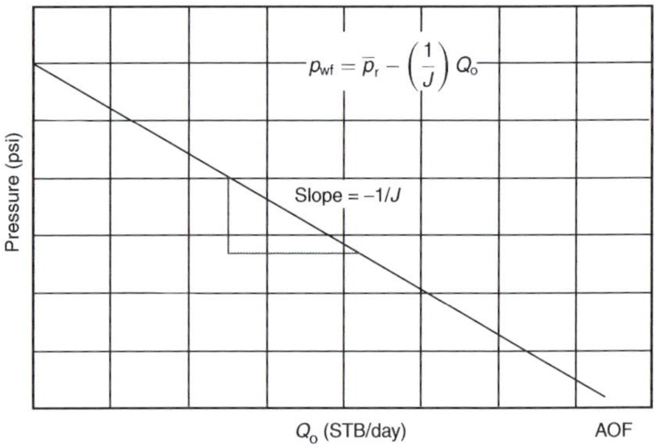

FIGURE 5.11 *IPR.*

- When p_{wf} equals the average reservoir pressure, the flow rate is zero due to the absence of any pressure drawdown.
- Maximum rate of flow occurs when p_{wf} is zero. This maximum rate is called "absolute open flow" and referred to as AOF. Although in practice this may not be a condition at which the well can produce, it is a useful definition that has widespread applications in the petroleum industry (e.g., comparing flow potential of different wells in the field). The AOF is then calculated by:

$$\text{AOF} = J\bar{p}_r$$

The slope of the straight line equals the reciprocal of the productivity index.

Example 5.7

A productivity test was conducted on a well. The test results indicate that the well is capable of producing at a stabilized flow rate of 110 STB/day and a bottom-hole flowing pressure of 900 psi. After shutting in the well for 24 hours, the bottom-hole pressure reached a static value of 1300 psi.

Calculate:

(a) the productivity index;
(b) the AOF;
(c) the oil flow rate at a bottom-hole flowing pressure of 600 psi;
(d) the wellbore flowing pressure required to produce 250 STB/day.

Solution

(a) Calculate J from Eq. (5.58):

$$J = \frac{Q_o}{\bar{p}_r - p_{wf}} = \frac{Q_o}{\Delta p}$$

$$= \frac{110}{1300 - 900} = 0.275 \text{ STB/psi}$$

(b) Determine the AOF from:

$$\text{AOF} = J(\bar{p}_r - 0)$$

$$= 0.275(1300 - 0) = 375.5 \text{ STB/day}$$

(c) Solve for the oil flow rate by applying Eq. (5.58):

$$Q_o = J(\bar{p}_r - p_{wf})$$

$$= 0.275(1300 - 600) = 192.5 \text{ STB/day}$$

(d) Solve for p_{wf} by using Eq. (5.64):

$$p_{wf} = \bar{p}_r - \left(\frac{1}{J}\right) Q_o$$

$$= 1300 - \left(\frac{1}{0.275}\right) 250 = 390.9 \text{ psi}$$

The previous discussion, as illustrated by the example, suggested that the inflow into a well is directly proportional to the pressure drawdown and the constant of proportionality is the productivity index. Muskat and Evinger (1942) and Vogel (1968) observed that when the pressure drops below the bubble point pressure, the IPR deviates from that of the simple straight-line relationship as shown in Figure 5.12. Recalling Eq. (5.61):

$$J = \frac{0.00708hk}{\ln(r_e/r_w) - 0.75 + s}\left(\frac{k_{ro}}{\mu_o B_o}\right)$$

Treating the term in the brackets as a constant c, the above equation can be written in the following form:

$$J = c\left(\frac{k_{ro}}{\mu_o B_o}\right) \tag{5.65}$$

with the coefficient c as defined by:

$$c = \frac{0.00708kh}{\ln(r_e/r_w) - 0.75 + s}$$

Eq. (5.65) reveals that the variables affecting the productivity index are essentially those that are pressure dependent, namely:

- oil viscosity μ_o;
- oil formation volume factor B_o;
- relative permeability to oil k_{ro}.

Figure 5.13 schematically illustrates the behavior of these variables as a function of pressure. Figure 5.14 shows the overall effect of changing the pressure on the term $k_{ro}/\mu_o B_o$. Above the bubble point pressure p_b, the relative oil permeability k_{ro} equals unity ($k_{ro} = 1$) and the term ($k_{ro}/\mu_o B_o$) is almost constant. As the

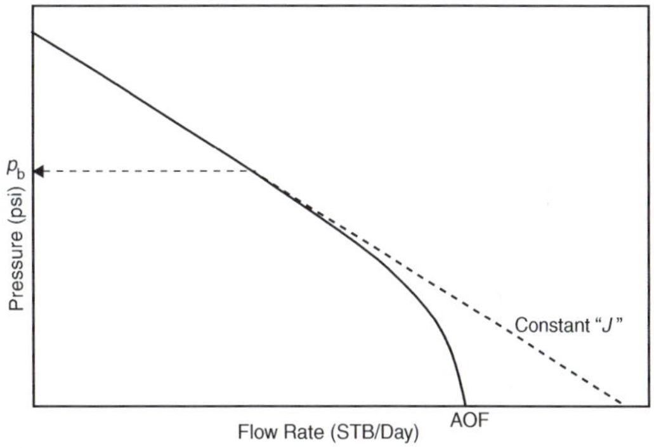

FIGURE 5.12 *IPR* below p_b.

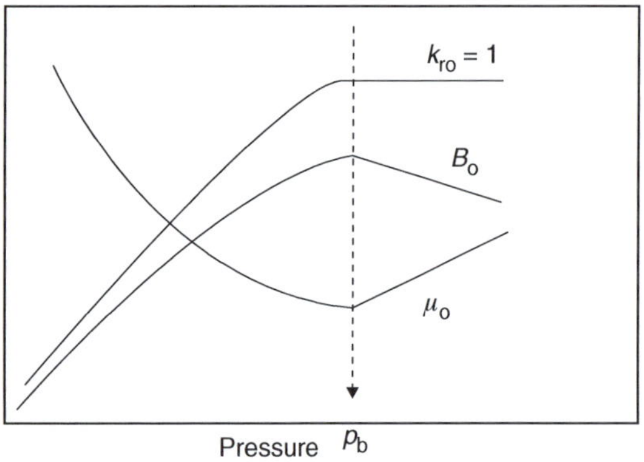

FIGURE 5.13 Effect of pressure on B_o, μ_o, and k_{ro}.

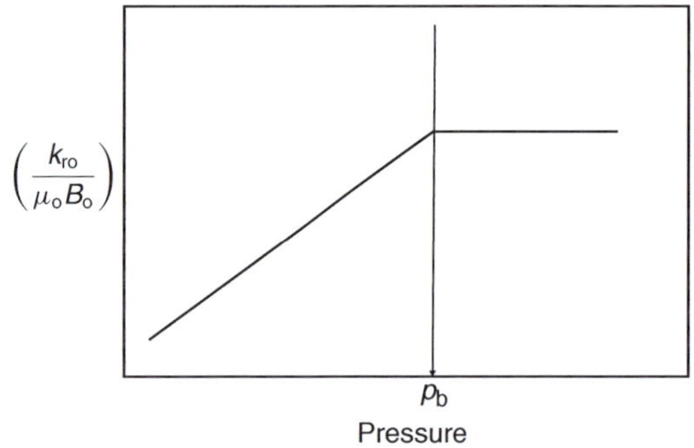

FIGURE 5.14 $k_{ro}/\mu_o B_o$ as a function of pressure.

pressure declines below p_b, the gas is released from the solution, which can cause a large decrease in both k_{ro} and $k_{ro}/\mu_o B_o$. Figure 5.15 shows qualitatively the effect of reservoir depletion on the IPR.

There are several empirical methods that are designed to predict the nonlinear behavior of the IPR for solution gas drive reservoirs. Most of these methods require at least one stabilized flow test in which Q_o and p_{wf} are measured. All the methods include the following two computational steps:

(1) Using the stabilized flow test data, construct the IPR curve at the current average reservoir pressure $\bar{p}_r$.
(2) Predict future IPRs as a function of average reservoir pressures.

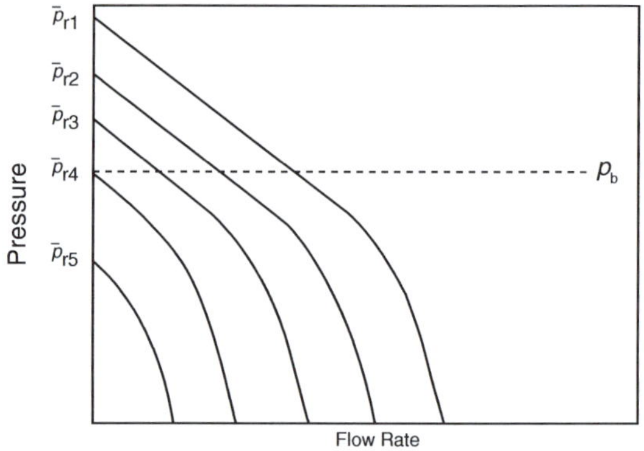

FIGURE 5.15 Effect of reservoir pressure on IPR.

The following empirical methods are designed to generate the current and future inflow performance relationships:

- the Vogel method;
- the Wiggins method;
- the Standing method;
- the Fetkovich method;
- the Klins and Clark method.

Vogel Method. Vogel (1968) used a computer model to generate IPRs for several hypothetical saturated oil reservoirs that are producing under a wide range of conditions. Vogel normalized the calculated IPRs and expressed the relationships in a dimensionless form. He normalized the IPRs by introducing the following dimensionless parameters:

$$\text{Dimensionless pressure} = \frac{p_{wf}}{\bar{p}_r}$$

$$\text{Dimensionless flow rate} = \frac{Q_o}{(Q_o)_{max}}$$

where $(Q_o)_{max}$ is the flow rate at zero wellbore pressure, i.e., the AOF.

Vogel plotted the dimensionless IPR curves for all the reservoir cases and arrived at the following relationship between the above dimensionless parameters:

$$\frac{Q_o}{(Q_o)_{max}} = 1 - 0.2\left(\frac{p_{wf}}{\bar{p}_r}\right) - 0.8\left(\frac{p_{wf}}{\bar{p}_r}\right)^2 \tag{5.66}$$

where

Q_o = oil rate at p_{wf}
$(Q_o)_{max}$ = maximum oil flow rate at zero wellbore pressure, i.e., the AOF
$\bar{p}_r$ = current average reservoir pressure, psig
p_{wf} = wellbore pressure, psig

Note that p_{wf} and $\bar{p}_r$ must be expressed in psig.

The Vogel method can be extended to account for water production by replacing the dimensionless rate with $Q_L/(Q_L)_{max}$ where $Q_l = Q_o + Q_w$. This has proved to be valid for wells producing at water cuts as high as 97%. The method requires the following data:

- current average reservoir pressure $\bar{p}_r$;
- bubble point pressure p_b;
- stabilized flow test data that includes Q_o at p_{wf}.

Vogel's methodology can be used to predict the IPR curve for the following two types of reservoirs:

(1) saturated oil reservoirs: $\bar{p}_r \le p_b$;
(2) undersaturated oil reservoirs: $\bar{p}_r > p_b$.

The Vertical Well IPR in Saturated Oil Reservoirs. When the reservoir pressure equals the bubble point pressure, the oil reservoir is referred to as a saturated oil reservoir. The computational procedure of applying the Vogel method in a saturated oil reservoir to generate the IPR curve for a well with a stabilized flow data point, i.e., a recorded Q_o value at p_{wf}, is summarized below:

Step 1. Using the stabilized flow data, i.e., Q_o and p_{wf}, calculate $(Q_o)_{max}$ from Eq. (5.66), or:

$$(Q_o)_{max} = \frac{Q_o}{1 - 0.2(p_{wf}/\bar{p}_r) - 0.8(p_{wf}/\bar{p}_r)^2}$$

Step 2. Construct the IPR curve by assuming various values for p_{wf} and calculating the corresponding Q_o by applying Eq. (5.66):

$$\frac{Q_o}{(Q_o)_{max}} = 1 - 0.2\left(\frac{p_{wf}}{\bar{p}_r}\right) - 0.8\left(\frac{p_{wf}}{\bar{p}_r}\right)^2$$

or

$$Q_o = (Q_o)_{max}\left[1 - 0.2\left(\frac{p_{wf}}{\bar{p}_r}\right) - 0.8\left(\frac{p_{wf}}{\bar{p}_r}\right)^2\right]$$

Example 5.8

A well is producing from a saturated reservoir with an average reservoir pressure of 2500 psig. Stabilized production test data indicates that the stabilized rate and wellbore pressure are 350 STB/day and 2000 psig, respectively. Calculate:

the oil flow rate at $p_{wf} = 1850$ psig;
the oil flow rate assuming constant J.

Construct the IPR by using the Vogel method and the constant productivity index approach.

Solution

(a) Step 1. Calculate $(Q_o)_{max}$:

$$(Q_o)_{max} = \frac{Q_o}{1 - 0.2(p_{wf}/\bar{p}_r) - 0.8(p_{wf}/\bar{p}_r)^2}$$
$$= \frac{350}{1 - 0.2(2000/2500) - 0.8(2000/2500)^2}$$
$$= 1067.1 \text{ STB/day}$$

Step 2. Calculate Q_o at $p_{wf} = 1850$ psig by using Vogel's equation:

$$Q_o = (Q_o)_{max}\left[1 - 0.2\left(\frac{p_{wf}}{\bar{p}_r}\right) - 0.8\left(\frac{p_{wf}}{\bar{p}_r}\right)^2\right]$$
$$= 1067.1\left[1 - 0.2\left(\frac{1850}{2500}\right) - 0.8\left(\frac{1850}{2500}\right)^2\right]$$
$$= 441.7 \text{ STB/day}$$

(b) Step 1. Apply Eq. (5.59) to determine J:

$$J = \frac{Q_o}{\bar{p}_r - p_{wf}}$$
$$= \frac{350}{2500 - 2000} = 0.7 \text{ STB/day/psi}$$

Step 2. Calculate Q_o:

$$Q_o = J(\bar{p}_r - p_{wf}) = 0.7(2500 - 1850)$$
$$= 455 \text{ STB/day}$$

(c) Assume several values for p_{wf} and calculate the corresponding Q_o:

p_{wf}	Vogel	$Q_o = J(\bar{p}_r - p_{wf})$
2500	0	0
2200	218.2	210

1500	631.7	700
1000	845.1	1050
500	990.3	1400
0	1067.1	1750

The Vertical Well IPR in Undersaturated Oil Reservoirs. Beggs (1991) pointed out that in applying the Vogel method for undersaturated reservoirs, there are two possible outcomes of the recorded stabilized flow test data that must be considered, as shown schematically in Figure 5.16:

(1) The recorded stabilized bottom-hole flowing pressure is greater than or equal to the bubble point pressure, i.e., $p_{wf} \geq p_b$.
(2) The recorded stabilized bottom-hole flowing pressure is less than the bubble point pressure, i.e., $p_{wf} < p_b$.

Case 1 $\mathbf{p_{wf} \geq p_b}$**.** Beggs outlined the following procedure for determining the IPR when the stabilized bottom-hole pressure is greater than or equal to the bubble point pressure (Figure 5.16):

Step 1. Using the stabilized test data point (Q_o and p_{wf}), calculate the productivity index J:

$$J = \frac{Q_o}{\overline{p}_r - p_{wf}}$$

Step 2. Calculate the oil flow rate at the bubble point pressure:

$$Q_{ob} = J(\overline{p}_r - p_b) \qquad (5.67)$$

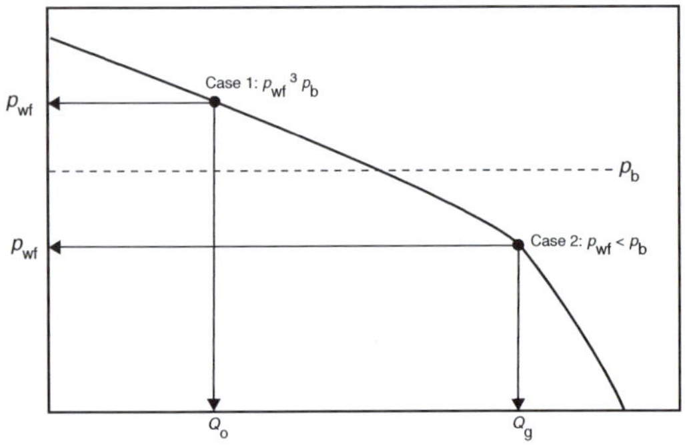

FIGURE 5.16 Stabilized flow test data.

where Q_{ob} is the oil flow rate at p_b.

Step 3. Generate the IPR values below the bubble point pressure by assuming different values of $p_{wf} < p_b$ and calculating the corresponding oil flow rates by applying the following relationship:

$$Q_o = Q_{ob} + \frac{Jp_b}{1.8}\left[1 - 0.2\left(\frac{p_{wf}}{p_b}\right) - 0.8\left(\frac{p_{wf}}{p_b}\right)^2\right] \qquad (5.68)$$

The maximum oil flow rate ($Q_{o\,max}$ or AOF) occurs when the bottom-hole flowing pressure is zero, i.e., $p_{wf} = 0$, which can be determined from the above expression as:

$$Q_{o\,max} = Q_{ob} + \frac{Jp_b}{1.8}$$

It should be pointed out that when $p_{wf} \geq p_b$, the IPR is linear and is described by:

$$Q_o = J(\overline{p}_r - p_{wf})$$

Example 5.9

An oil well is producing from an undersaturated reservoir that is characterized by a bubble point pressure of 2130 psig. The current average reservoir pressure is 3000 psig. Available flow test data shows that the well produced 250 STB/day at a stabilized p_{wf} of 2500 psig. Construct the IPR data.

Solution

The problem indicates that the flow test data was recorded above the bubble point pressure, $p_{wf} \geq p_b$, and therefore, the "Case 1" procedure for undersaturated reservoirs as outlined previously must be used:

Step 1. Calculate J using the flow test data:

$$J = \frac{Q_o}{\overline{p}_r - p_{wf}} = \frac{250}{3000 - 2500} = 0.5 \text{ STB/day/psi}$$

Step 2. Calculate the oil flow rate at the bubble point pressure by applying Eq. (5.67):

$$Q_{ob} = J(\overline{p}_r - p_b)$$
$$= 0.5(3000 - 2130) = 435 \text{ STB/day}$$

Step 3. Generate the IPR data by applying the constant J approach for all pressures above p_b and Eq. (5.68) for all pressures below p_b:

$$Q_o = Q_{ob} + \frac{Jp_b}{1.8}\left[1 - 0.2\left(\frac{p_{wf}}{p_b}\right) - 0.8\left(\frac{p_{wf}}{p_b}\right)^2\right]$$
$$= 435 + \frac{(0.5)(2130)}{1.8} \times \left[1 - 0.2\left(\frac{p_{wf}}{2130}\right) - 0.8\left(\frac{p_{wf}}{2130}\right)^2\right]$$

p_{wf}	Q_o
$p_i = 3000$	0
2800	100
2600	200
$p_b = 2130$	435
1500	709
1000	867
500	973
0	1027

Case 2 **$p_{wf} < p_b$.** When the recorded p_{wf} from the stabilized flow test is below the bubble point pressure, as shown in Figure 5.16, the following procedure for generating the IPR data is proposed:

Step 1. Using the stabilized well flow test data and combining Eq. (5.67) with Eq. (5.68), solve for the productivity index J, to give:

$$J = \frac{Q_o}{(\overline{p}_r - p_b) + (p_b/1.8)[1 - 0.2(p_{wf}/p_b) - 0.8(p_{wf}/p_b)^2]} \quad \textbf{(5.69)}$$

Step 2. Calculate Q_{ob} by using Eq. (5.67), or:

$$Q_{ob} = J(\overline{p}_r - p_b)$$

Step 3. Generate the IPR for $p_{wf} \geq p_b$ by assuming several values for p_{wf} above the bubble point pressure and calculating the corresponding Q_o from:

$$Q_o = J\overline{p}_r - p_{wf}$$

Step 4. Use Eq. (5.68) to calculate Q_o at various values of p_{wf} below p_b, or:

$$Q_o = Q_{ob} + \frac{Jp_b}{1.8}\left[1 - 0.2\left(\frac{p_{wf}}{p_b}\right) - 0.8\left(\frac{p_{wf}}{p_b}\right)^2\right]$$

Example 5.10

The well described in Example 5.8 was retested and the following results were obtained:

$$p_{wf} = 1700 \text{ psig}, \quad Q_o = 630.7 \text{ STB/day}$$

Generate the IPR data using the new test data.

Solution

Note that the stabilized p_{wf} is less than p_b.

Step 1. Solve for J by applying Eq. (5.69):

$$J = \frac{Q_o}{(\overline{p}_r - p_b) + (p_b/1.8)[1 - 0.2(p_{wf}/p_b) - 0.8(p_{wf}/p_b)^2]}$$
$$= \frac{630.7}{(3000 - 2130) + (2130/1.8)[1 - (1700/2130) - (1700/2130)^2]}$$
$$= 0.5 \text{ STB/day/psi}$$

Step 2. Determine Q_{ob}:

$$Q_{ob} = J(\overline{p}_r - p_b)$$
$$= 0.5(3000 - 2130) = 435 \text{ STB/day}$$

Step 3. Generate the IPR data by applying Eq. (5.63) when $p_{wf} > p_b$ and Eq. (5.68) when $p_{wf} < p_b$:

$$Q_o = J(\overline{p}_r - p_{wf}) = J\Delta p$$
$$= Q_{ob} + \frac{Jp_b}{1.8}\left[1 - 0.2\left(\frac{p_{wf}}{p_b}\right)^2 - 0.8\left(\frac{p_{wf}}{p_b}\right)^2\right]$$

p_{wf}	Equation	Q_o
3000	(5.63)	0
2800	(5.63)	100
2600	(5.63)	200
2130	(5.63)	435
1500	(5.68)	709
1000	(5.68)	867
500	(5.68)	973
0	(5.68)	1027

Quite often it is necessary to predict the well's inflow performance for future times as the reservoir pressure declines. Future well performance calculations require the development of a relationship that can be used to predict the future maximum oil flow rates.

There are several methods that are designed to address the problem of how the IPR might shift as the reservoir pressure declines. Some of these prediction methods require the application of the MBE to generate future oil saturation data as a function of reservoir pressure. In the absence of such data, there are two simple approximation methods that can be used in conjunction with the Vogel method to predict future IPRs.

First Approximation Method. This method provides a rough approximation of the future maximum oil flow rate $(Q_{o\,max})_f$ at the specified *future* average reservoir pressure $(\bar{p}_r)_f$. This future maximum flow rate $(Q_{o\,max})_f$ can be used in Vogel's equation to predict the future IPRs at $(\bar{p}_r)_f$. The following steps summarize the method:

Step 1. Calculate $(Q_{o\,max})_f$ at $(\bar{p}_r)_f$ from:

$$(Q_{o\,max})_f = (Q_{o\,max})_p \left[\frac{(\bar{p}_r)_f}{(\bar{p}_r)_p}\right]\left[0.2 + 0.8\frac{(\bar{p}_r)_f}{(\bar{p}_r)_p}\right] \quad \textbf{(5.70)}$$

where the subscripts "f" and "p" represent future and present conditions, respectively.

Step 2. Using the new calculated value of $(Q_{o\,max})$f, generate the IPR by using Eq. (5.66).

Second Approximation Method. A simple approximation for estimating future $(Q_{o\,max})_f$ at $(\bar{p}_r)_f$ was proposed by Fetkovich (1973). The relationship has the following mathematical form:

$$(Q_{o\,max})_f = (Q_{o\,max})_p \left[\frac{(\bar{p}_r)_f}{(\bar{p}_r)_p}\right]^{3.0}$$

where the subscripts "f" and "p" represent future and present conditions, respectively. The above equation is intended only to provide a rough estimation of future $(Q_{o\,max})$.

Example 5.11

Using the data given in Example 5.8, predict the IPR when the average reservoir pressure declines from 2500 to 2200 psig.

Solution

Example 5.8 shows the following information:

present average reservoir pressure $(\bar{p}_r)_p$ = 2500 psig;
present maximum oil rate $(Q_{o\,max})_p$ = 1067.1 STB/day

Step 1. Solve for $(Q_{o\,max})_f$ by applying Eq. (5.70):

$$(Q_{o\,max})_f = (Q_{o\,max})_p \left[\frac{(\bar{p}_r)_f}{(\bar{p}_r)_p}\right]\left[0.2 + 0.8\frac{(\bar{p}_r)_f}{(\bar{p}_r)_p}\right]$$
$$= (1067.1)\left(\frac{2200}{2500}\right)\left[0.2 + 0.8\frac{2200}{2500}\right]$$
$$= 849 \text{ STB/day}$$

Step 2. Generate the IPR data by applying Eq. (5.66):

$$Q_o = (Q_o)_{max}\left[1 - 0.2\left(\frac{p_{wf}}{\bar{p}_r}\right) - 0.8\left(\frac{p_{wf}}{\bar{p}_r}\right)^2\right]$$
$$= 849\left[1 - 0.2\left(\frac{p_{wf}}{2200}\right) - 0.8\left(\frac{p_{wf}}{2200}\right)^2\right]$$

p_{wf}	Q_o
2200	0
1800	255
1500	418
500	776
0	849

It should be pointed out that the main disadvantage of Vogel's methodology lies with its sensitivity to the match point, i.e., the stabilized flow test data point, used to generate the IPR curve for the well.

For a production well completed in a multilayered system, it is possible to allocate individual layer production by applying the following relationships:

$$(Q_o)_i = Q_{oT}\frac{[1-(\bar{S}_i f_{wT})](((k_o)_i(h)_i)/((\mu_o)_{Li}))}{\sum_{i=1}^{n\,\text{Layers}}[1-(\bar{S}_i f_{wT})](((k_o)_i(h)_i)/((\mu_o)_i))}$$

$$(Q_w)_i = Q_{wT}\frac{[(\bar{S}_i f_{wT})](((k_w)_i(h)_i)/((\mu_w)_i))}{\sum_{i=1}^{n\,\text{Layers}}[(\bar{S}_i f_{wT})](((k_w)_i(h)_i)/((\mu_w)_i))}$$

with

$$\overline{S}_i = \frac{(S_w)_i}{\sum_{i=1}^{n\,\text{Layers}} (S_w)_i}$$

where

$(Q_o)_i$ = allocated oil rate for layer i
$(Q_w)_i$ = allocated water rate for layer i
f_{wT} = total well water cut
$(k_o)_i$ = effective oil permeability for layer i
$(k_w)_i$ = effective water permeability for layer i
n Layers = number of layers

Wiggins Method. Wiggins (1993) used four sets of relative permeability and fluid property data as the basic input for a computer model to develop equations to predict inflow performance. The generated relationships are limited by the assumption that the reservoir initially exists at its bubble point pressure. Wiggins proposed generalized correlations that are suitable for predicting the IPR during the three-phase flow. His proposed expressions are similar to those of Vogel and are expressed as:

$$Q_o = (Q_o)_{max}\left[1 - 0.52\left(\frac{p_{wf}}{\overline{p}_r}\right) - 0.48\left(\frac{p_{wf}}{\overline{p}_r}\right)^2\right] \quad \textbf{(5.71)}$$

$$Q_w = (Q_w)_{max}\left[1 - 0.72\left(\frac{p_{wf}}{\overline{p}_r}\right) - 0.28\left(\frac{p_{wf}}{\overline{p}_r}\right)^2\right] \quad \textbf{(5.72)}$$

where

Q_w = water flow rate, STB/day
$(Q_w)_{max}$ = maximum water production rate at $p_{wf} = 0$, STB/day

As in the Vogel method, data from a stabilized flow test on the well must be available in order to determine $(Q_o)_{max}$ and $(Q_w)_{max}$.

Wiggins extended the application of the above relationships to predict future performance by providing expressions for estimating future maximum flow rates. He expressed future maximum rates as a function of:

- current (present) average pressure $(\overline{p}_r)_p$;
- future average pressure $(\overline{p}_r)_f$;
- current maximum oil flow rate $(Q_{o\,max})_p$;
- current maximum water flow rate $(Q_{w\,max})_p$.

Wiggins proposed the following relationships:

$$(Q_{o\,max})_f = (Q_{o\,max})_p\left[0.15\frac{(\overline{p}_r)_f}{(\overline{p}_r)_p} + 0.84\left(\frac{(\overline{p}_r)_f}{(\overline{p}_r)_p}\right)^2\right] \quad \textbf{(5.73)}$$

$$(Q_{w\,max})_f = (Q_{w\,max})_p\left[0.59\frac{(\overline{p}_r)_f}{(\overline{p}_r)_p} + 0.36\left(\frac{(\overline{p}_r)_f}{(\overline{p}_r)_p}\right)^2\right] \quad \textbf{(5.74)}$$

Example 5.12

The information given in Examples 5.8 and 5.11 is repeated here for convenience.

- current average pressure = 2500 psig;
- stabilized oil flow rate = 350 STB/day;
- stabilized wellbore pressure = 2000 psig.

Generate the current IPR data and predict the future IPR when the reservoir pressure declines from 2500 to 2000 psig by using the Wiggins method.

Solution

Step 1. Using the stabilized flow test data, calculate the current maximum oil flow rate by applying Eq. (5.71):

$$Q_o = (Q_o)_{max}\left[1 - 0.52\left(\frac{p_{wf}}{\overline{p}_r}\right) - 0.48\left(\frac{p_{wf}}{\overline{p}_r}\right)^2\right]$$

Solve for the present $(Q_o)_{max}$, to give:

$$(Q_{o\,max})_p = \frac{350}{1 - 0.52(2000/2500) - 0.48(2000/2500)^2} = 1264 \text{ STB/day}$$

Step 2. Generate the current IPR data by using the Wiggins method and compare the results with those of Vogel. Results of the two methods are shown graphically in Figure 5.17.

p_{wf}	Wiggins	Vogel
2500	0	0
2200	216	218
1500	651	632

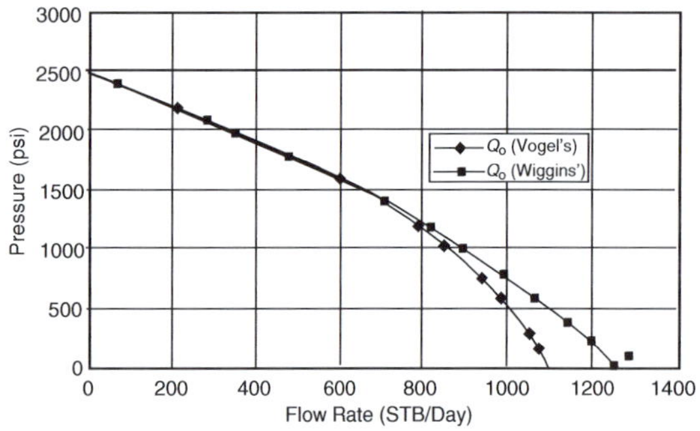

FIGURE 5.17 *IPR curves.*

1000	904	845
500	1108	990
0	1264	1067

Step 3. Calculate future maximum oil flow rate by using Eq. (5.73):

$$(Q_{o\,max})_f = (Q_{o\,max})_p \left[0.15\frac{(\overline{p}_r)_f}{(\overline{p}_r)_p} + 0.84\left(\frac{(\overline{p}_r)_f}{(\overline{p}_r)_p}\right)^2\right]$$

$$= 1264\left[0.15\left(\frac{2200}{2500}\right) + 0.84\left(\frac{2200}{2500}\right)^2\right]$$

$$= 989 \text{ STB/day}$$

Step 4. Generate future IPR data by using Eq. (5.71):

$$Q_o = (Q_o)_{max}\left[1 - 0.52\left(\frac{p_{wf}}{\overline{p}_r}\right) - 0.48\left(\frac{p_{wf}}{\overline{p}_r}\right)^2\right]$$

$$= 989\left[1 - 0.52\left(\frac{p_{wf}}{2200}\right) - 0.48\left(\frac{p_{wf}}{2200}\right)^2\right]$$

p_{wf}	Q_o
2200	0
1800	250
1500	418
500	848
0	989

Standing Method. Standing (1970) essentially extended the application of the Vogel method to predict the future IPR of a well as a function of reservoir pressure. He noted that Vogel's equation (Eq. (5.66)) can be rearranged as:

$$\frac{Q_o}{(Q_o)_{max}} = \left(1 - \frac{p_{wf}}{\overline{p}_r}\right)\left[1 + 0.8\left(\frac{p_{wf}}{\overline{p}_r}\right)\right] \quad (5.75)$$

Standing introduced the productivity index J as defined by Eq. (5.1) into Eq. (5.75) to yield:

$$J = \frac{(Q_o)_{max}}{\overline{p}_r}\left[1 + 0.8\left(\frac{p_{wf}}{\overline{p}_r}\right)\right] \quad (5.76)$$

Standing then defined a "zero drawdown" productivity index as:

$$J_p^* = 1.8\left[\frac{(Q_o)_{max}}{\overline{p}_r}\right] \quad (5.77)$$

where J_p^* is the current zero-drawdown productivity index. J_p^* is related to the productivity index J by:

$$\frac{J}{J_p^*} = \frac{1}{1.8}\left[1 + 0.8\left(\frac{p_{wf}}{\overline{p}_r}\right)\right] \quad (5.78)$$

Eq. (5.78) permits the calculation of J_p^* from a measured value of J. That is:

$$J_p^* = \frac{1.8J}{1 + 0.8(p_{wf}/\overline{p}_r)}$$

To arrive at the final expression for predicting the desired IPR expression, Standing combines Eq. (5.77) with Eq. (5.75) to eliminate $(Q_o)_{max}$, to give:

$$Q_o = \left[\frac{J_f^*(\overline{p}_r)_f}{1.8}\right]\left\{1 - 0.2\frac{p_{wf}}{(\overline{p}_r)_f} - 0.8\left[\frac{p_{wf}}{(\overline{p}_r)_f}\right]^2\right\} \quad (5.79)$$

where the subscript "f" refers to the future condition.

Standing suggested that J_f^* can be estimated from the present value of J_p^* by the following expression:

$$J_f^* = J_p^*\frac{(k_{ro}/\mu_o B_o)_f}{(k_{ro}/\mu_o B_o)_p} \quad (5.80)$$

where the subscript "p" refers to the present condition.

If the relative permeability data is not available, J_f^* can be roughly estimated from:

$$J_f^* = J_p^* \left[\frac{(\overline{p}_r)_f}{(\overline{p}_r)_p}\right]^2 \quad (5.81)$$

Standing's methodology for predicting a future IPR is summarized in the following steps:

Step 1. Using the current time condition and the available flow test data, calculate $(Q_o)_{max}$ from Eq. (5.75):

$$(Q_o)_{max} = \frac{Q_o}{(1-(p_{wf}/\overline{p}_r))[1+0.8(p_{wf}/\overline{p}_r)]}$$

Step 2. Calculate J^* at the present condition, i.e., J_p^* by using Eq. (5.77). Note that other combinations of Eqs. (5.75)–(5.78) can be used to estimate J_p^*:

$$J_p^* = 1.8\left[\frac{(Q_o)_{max}}{\overline{p}_r}\right]$$

or from:

$$J_p^* = \frac{1.8J}{1+0.8(p_{wf}/\overline{p}_r)}$$

Step 3. Using fluid property, saturation, and relative permeability data, calculate both $(k_{ro}/\mu_o B_o)_p$ and $(k_{ro}/\mu_o B_o)_f$

Step 4. Calculate J_f^* by using Eq. (5.80). Use Eq. (5.81) if the oil relative permeability data is not available:

$$J_f^* = J_p^* \frac{(k_{ro}/\mu_o B_o)_f}{(k_{ro}/\mu_o B_o)_p}$$

or

$$J_f^* = J_p^* \left[\frac{(\overline{p}_r)_f}{(\overline{p}_r)_p}\right]^2$$

Step 5. Generate the future IPR by applying Eq. (5.79):

$$Q_o = \left[\frac{J_f^*(\overline{p}_r)_f}{1.8}\right]\left\{1-20\frac{p_{wf}}{(\overline{p}_r)_f}-0.8\left[\frac{p_{wf}}{(\overline{p}_r)_f}\right]^2\right\}$$

Example 5.13

A well is producing from a saturated oil reservoir that exists at its saturation pressure of 4000 psig. The well is flowing at a stabilized rate of 600 STB/day and a p_{wf} of 3200 psig. Material balance calculations provide the following current and future predictions for oil saturation and *PVT* properties.

	Present	Future
$\overline{p}_r$	4000	3000
μ_o (cp)	2.40	2.20
B_o (bbl/STB)	1.20	1.15
k_{ro}	1.00	0.66

Generate the future IPR for the well at 3000 psig by using the Standing method.

Solution

Step 1. Calculate the current $(Q_o)_{max}$ from Eq. (5.75):

$$(Q_o)_{max} = \frac{Q_o}{(1-p_{wf}/\overline{p}_r)[1+0.8(p_{wf}/\overline{p}_r)]}$$
$$= \frac{600}{(1-(3200/4000))[1+0.8(3200/4000)]}$$
$$= 1829 \text{ STB/day}$$

Step 2. Calculate J_p^* by using Eq. (5.78):

$$J_p^* = 1.8\frac{(Q_o)_{max}}{\overline{p}_r}$$
$$= 1.8\left[\frac{1829}{4000}\right] = 0.823$$

Step 3. Calculate the following pressure function:

$$\left(\frac{k_{ro}}{\mu_o B_o}\right)_p = \frac{1}{(2.4)(1.20)} = 0.3472$$
$$\left(\frac{k_{ro}}{\mu_o B_o}\right)_f = \frac{0.66}{(2.2)(1.15)} = 0.2609$$

Step 4. Calculate J_f^* by applying Eq. (5.80):

$$J_f^* = J_P^* \frac{(k_{ro}/\mu_o B_o)_f}{(k_{ro}/\mu_o B_o)_p}$$
$$= 0.823\left(\frac{0.2609}{0.3472}\right) = 0.618$$

Step 5. Generate the IPR by using Eq. (5.79):

$$Q_o = \left[\frac{J_f^*(\bar{p}_r)_f}{1.8}\right]\left\{1 - 0.2\frac{p_{wf}}{(\bar{p}_r)_f} - 0.8\left[\frac{p_{wf}}{(\bar{p}_r)_f}\right]^2\right\}$$

$$= \left[\frac{(0.618)(3000)}{1.8}\right]\left\{1 - 0.2\frac{p_{wf}}{3000} - 0.8\left[\frac{p_{wf}}{3000}\right]^2\right\}$$

p_{wf}	Q_o (STB/day)
3000	0
2000	527
1500	721
1000	870
500	973
0	1030

It should be noted that one of the main disadvantages of Standing's methodology is that it requires reliable permeability information; in addition, it also requires material balance calculations to predict oil saturations at future average reservoir pressures.

Fetkovich Method. Muskat and Evinger (1942) attempted to account for the observed nonlinear flow behavior (i.e., IPR) of wells by calculating a theoretical productivity index from the pseudosteady-state flow equation. They expressed Darcy's equation as:

$$Q_o = \frac{0.00708kh}{[\ln(r_e/r_w) - 0.75 + s]}\int_{p_{wf}}^{\bar{p}_r} f(p)dp \quad \textbf{(5.82)}$$

where the pressure function $f(p)$ is defined by:

$$f(p) = \frac{k_{ro}}{\mu_o B_o} \quad \textbf{(5.83)}$$

where

k_{ro} = oil relative permeability
k = absolute permeability, md
B_o = oil formation volume factor
μ_o = oil viscosity, cp

Fetkovich (1973) suggested that the pressure function $f(p)$ can basically fall into one of the following two regions:

Region 1: *Undersaturated region*: The pressure function $f(p)$ falls into this region if $p>p_b$. Since oil relative permeability in this region equals unity (i.e., $k_{ro} = 1$), then:

$$f(p) = \left(\frac{1}{\mu_o B_o}\right)_p \quad \textbf{(5.84)}$$

Fetkovich observed that the variation in $f(p)$ is only slight and the pressure function is considered constant as shown in Figure 5.18.

Region 2: *Saturated region*: In the saturated region where $p < p_b$, Fetkovich showed that $k_{ro}/\mu_o B_o$ changes linearly with pressure and that the straight line passes through the origin. This linear plot is shown schematically in Figure 5.18 and can be expressed mathematically as:

$$f(p) = 0 + (\text{slope})p$$

or

$$f(p) = 0 + \left(\frac{1/(\mu_o B_o)}{p_b}\right)_{p_b} p$$

Simplifying:

$$f(p) = \left(\frac{1}{\mu_o B_o}\right)_{p_b}\left(\frac{p}{p_b}\right) \quad \textbf{(5.85)}$$

where μ_o and B_o are evaluated at the bubble point pressure. In the application of the straight-line pressure function, there are *three cases* that must be considered:

(1) $\bar{p}_r$ and $p_{wf} > p_b$;
(2) $\bar{p}_r$ and $p_{wf} < p_b$;
(3) $\bar{p}_r > p_b$ and $p_{wf} < p_b$.

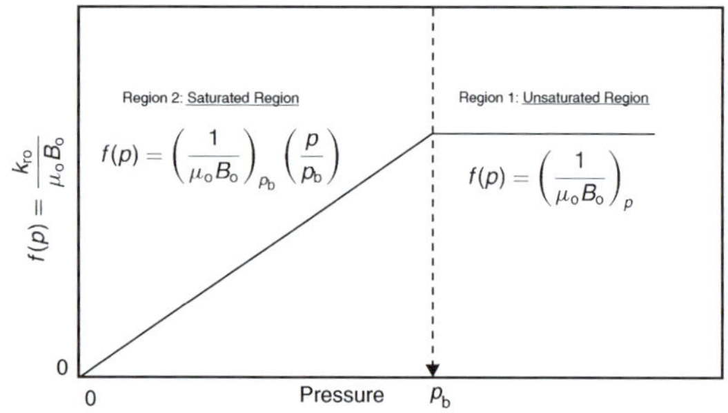

FIGURE 5.18 Pressure function concept.

These three cases are presented below.

***Case 1:* $\bar{p}_r$ and p_{wf} both are Greater Than p_b.** This is the case of a well producing from an undersaturated oil reservoir where both p_{wf} and $\bar{p}_r$ are greater than the bubble point pressure. The pressure function $f(p)$ in this case is described by Eq. (5.84). Substituting Eq. (5.84) into Eq. (5.82) gives:

$$Q_o = \frac{0.00708kh}{\ln(r_e/r_w) - 0.75 + s}\int_{p_{wf}}^{\bar{p}_r}\left(\frac{1}{\mu_o B_o}\right)dp$$

Since $(1/\mu_o B_o)$ is constant, then:

$$Q_o = \frac{0.00708kh}{\mu_o B_o[\ln(r_e/r_w) - 0.75 + s]}(\bar{p}_r - p_w) \tag{5.86}$$

and from the definition of the productivity index:

$$Q_o = J(\bar{p}_r - p_{wf}) \tag{5.87}$$

The productivity index is defined in terms of the reservoir parameters as:

$$J = \frac{0.00708kh}{\mu_o B_o[\ln(r_e/r_w) - 0.75 + s]} \tag{5.88}$$

where B_o and μ_o are evaluated at $(\bar{p}_r + p_{wf})/2$.

Example 5.14

A well is producing from an undersaturated oil reservoir that exists at an average reservoir pressure of 3000 psi. The bubble point pressure is recorded as 1500 psi at 150°F. The following additional data is available:

stabilized flow rate = 280 STB/day,
stabilized wellbore pressure = 2200 psi
$h = 20$ ft, $r_w = 0.3$ ft,
$r_e = 660$ ft, $s = -0.5$
$k = 65$ md, μ_o at 2600 psi = 2.4 cp,
B_o at 2600 psi = 1.4 bbl/STB.

Calculate the productivity index by using both the reservoir properties (i.e., Eq. (5.88)) and the flow test data (i.e., Eq. (5.58)):

Solution

From Eq. (5.87):

$$J = \frac{0.00708kh}{\mu_o B_o[\ln(r_e/r_w) - 0.75 + s]}$$
$$= \frac{0.00708(65)(20)}{(24)(1.4)[\ln(660/0.3) - 0.75 - 0.5]}$$
$$= 0.42 \text{ STB/day/psi}$$

From production data:

$$J = \frac{Q_o}{\bar{p}_r - p_{wf}} = \frac{Q_o}{\Delta p}$$
$$= \frac{200}{3000 - 2200} = 0.35 \text{ STB/day/psi}$$

Results show a reasonable match between the two approaches. However, it should be noted that several uncertainties exist in the values of the parameters used in Eq. (5.88) to determine the productivity index. For example, changes in the skin factor k or drainage area would change the calculated value of J.

***Case 2:* $\bar{p}_r$ *and* $p_{wf} < p_b$.** When the reservoir pressure $\bar{p}_r$ and bottom-hole flowing pressure p_{wf} both are below the bubble point pressure p_b, the pressure function $f(p)$ is represented by the straight-line relationship of Eq. (5.85). Combining Eq. (5.85) with Eq. (5.82) gives:

$$Q_o = \left[\frac{0.00708kh}{\ln(r_e/r_w) - 0.75 + s}\right]\int_{p_{wf}}^{\bar{p}_r}\frac{1}{(\mu_o B_o)_{p_b}}\left(\frac{p}{p_b}\right)dp$$

Since the term $[(1/\mu_o B_o)_{p_b}(1/p_b)]$ is constant, then:

$$Q_o = \left[\frac{0.00708kh}{\ln(r_e/r_w) - 0.75 + s}\right]\frac{1}{(\mu_o B_o)_{p_b}}\left(\frac{1}{p_b}\right)\int_{p_{wf}}^{\bar{p}r} p\,dp$$

Integrating:

$$Q_o = \frac{0.00708kh}{(\mu_o B_o)_{p_b}[\ln(r_e/r_w) - 0.75 + s]}\left(\frac{1}{2p_b}\right)(\bar{p}_r^2 - p_{wf}^2) \tag{5.89}$$

Introducing the productivity index, as defined by Eq. (5.81), into the above equation gives:

$$Q_o = J\left(\frac{1}{2p_b}\right)(\bar{p}_r^2 - p_{wf}^2) \tag{5.90}$$

The term $(J/2p_b)$ is commonly referred to as the performance coefficient C, or:

$$Q_o = C(\bar{p}_r^2 - p_{wf}^2) \quad (5.91)$$

To account for the possibility of nonDarcy flow (turbulent flow) in oil wells, Fetkovich introduced the exponent n in Eq. (5.91) to yield:

$$Q_o = C(\bar{p}_r^2 - p_{wf}^2)n \quad (5.92)$$

The value of n ranges from 1.0 for complete laminar flow to 0.5 for highly turbulent flow.

There are two unknowns in Eq. (5.92): the performance coefficient C and the exponent n. At least two tests are required to evaluate these two parameters, assuming $\bar{p}_r$ is known.

By taking the log of both sides of Eq. (5.92) and solving for $(\bar{p}_r^2 - p_{wf}^2)$, the expression can be written as:

$$\log(\bar{p}_r^2 - p_{wf}^2) = \frac{1}{n}\log Q_o - \frac{1}{n}\log C$$

A plot of $\bar{p}_r^2 - p_{wf}^2$ vs. q_o on a log–log scale will result in a straight line having a slope of $1/n$ and an intercept of C at $\bar{p}_r^2 - p_{wf}^2 = 1$. The value of C can also be calculated using any point on the linear plot once n has been determined, to give:

$$C = \frac{Q_o}{(\bar{p}_r^2 - p_{wf}^2)^n}$$

Once the values of C and n are determined from test data, Eq. (5.92) can be used to generate a complete IPR.

To construct the future IPR when the average reservoir pressure declines to $(\bar{p}_r)_f$, Fetkovich assumed that the performance coefficient C is a linear function of the average reservoir pressure, and therefore, the value of C can be adjusted as:

$$(C)_f = (C)_p \frac{(\bar{p}_r)_f}{(\bar{p}_r)_p} \quad (5.93)$$

where the subscripts "f" and "p" represent the future and present conditions.

Fetkovich assumed that the value of the exponent n would not change as the reservoir pressure declines. Beggs (1991) presented an excellent and comprehensive discussion of the different methodologies used in constructing the IPR curves for the oil and gas wells.

The following example was used by Beggs (1991) to illustrate the Fetkovich method for generating the current and future IPR.

Example 5.15

A four-point stabilized flow test was conducted on a well producing from a saturated reservoir that exists at an average pressure of 3600 psi.

Q_o (STB/day)	p_{wf} (psi)
263	3170
383	2890
497	2440
640	2150

Construct a complete IPR by using the Fetkovich method.

Construct the IPR when the reservoir pressure declines to 2000 psi.

Solution

(a) Step 1. Construct the following table:

Q_o (STB/day)	p_{wf} (psi)	$(\bar{p}_r^2 - p_{wf}^2) \times 10^{-6}$ (psi^2)
263	3170	2.911
383	2897	4.567
497	2440	7.006
640	2150	8.338

Step 2. Plot $(\bar{p}_r^2 - p_{wf}^2)$ vs. Q_o on log–log paper as shown in Figure 5.19 and determine the exponent n, or:

$$n = \frac{\log(750) - \log(105)}{\log(10^7) - \log(10^6)} = 0.854$$

Step 3. Solve for the performance coefficient C by selecting any point on the straight line, e.g., $(745, 10 \times 10^6)$, and solving for C from Eq. (5.92):

$$Q_o = C(\bar{p}_r^2 - p_{wf}^2)^n$$

$$745 = C(10 \times 10^6)^{0.854}$$

$$C = 0.00079$$

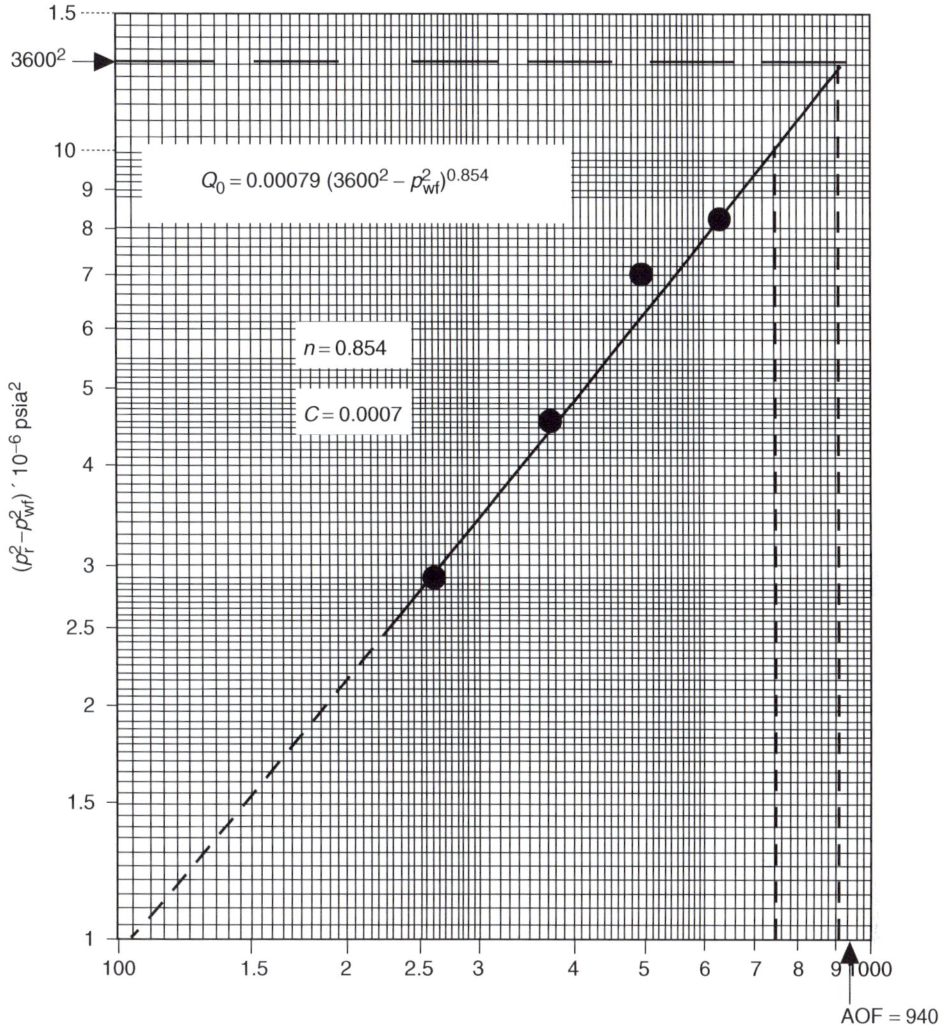

FIGURE 5.19 Flow–after-flow data for Example 5.15. (*After Beggs, D., 1991. Production Optimization Using Nodal Analysis. OGCI, Tulsa, OK*).

Step 4. Generate the IPR by assuming various values for p_{wf} and calculating the corresponding flow rate from Eq. (5.92):

$$Q_o = 0.00079(3600^2 - p_{wf}^2)^{0.854}$$

p_{wf}	Q_o (STB/day)
3600	0
3000	340
2500	503
2000	684
1500	796
1000	875
500	922
0	937

The IPR curve is shown in Figure 5.20. Note that the AOF, i.e., $(Q_o)_{max}$, is 937 STB/day.

(b) Step 1. Calculate future C by applying Eq. (5.94):

$$(C)_f = (C)_p \frac{(\overline{p}_r)_f}{(\overline{p}_r)_p}$$

$$= 0.00079\left(\frac{2000}{3600}\right) = 0.000439$$

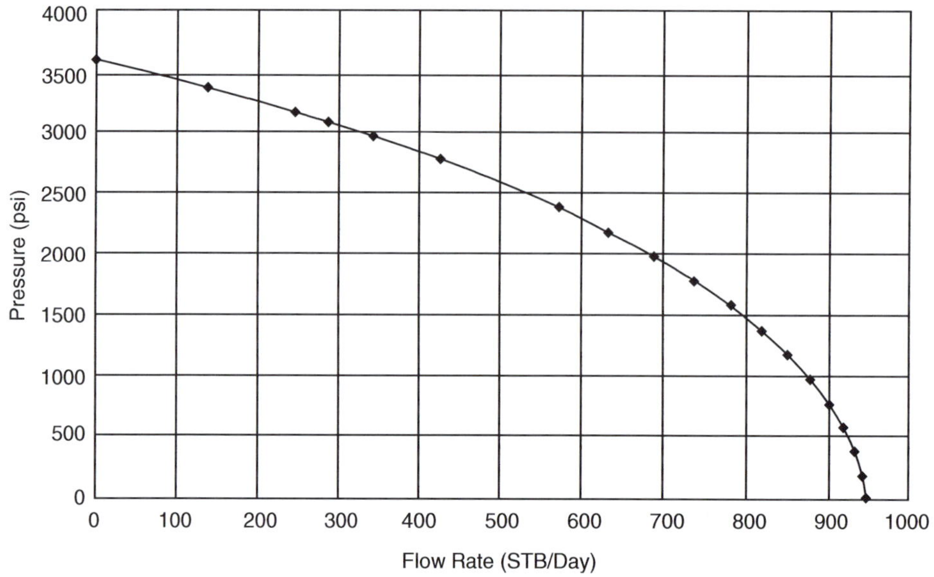

FIGURE 5.20 IPR using Fetkovich method.

Step 2. Construct the new IPR curve at 2000 psi by using the new calculated C and applying the inflow equation:

$$Q_o = 0.000439(2000^2 - p_{wf}^2)^{0.854}$$

p_{wf}	Q_o (STB/day)
2000	0
1500	94
1000	150
500	181
0	191

Both the present and future IPRs are plotted in Figure 5.21.

Klins and Clark (1993) developed empirical correlations that correlate the changes in Fetkovich's performance coefficient C and the flow exponent n with the decline in the reservoir pressure. The authors observed that the exponent n changes considerably with reservoir pressure. Klins and Clark concluded that the "future" values of $(n)_f$ and C at pressure $(\bar{p}_r)_f$ are related to the values of n and C at the bubble point pressure. Denoting C_b and n_b as the values of the performance coefficient and the flow exponent at the bubble point pressure p_b, Klins and Clark introduced the following dimensionless parameters:

- dimensionless performance coefficient $= C/C_b$;
- dimensionless flow exponent $= n/n_b$;
- dimensionless average reservoir pressure $= \bar{p}_r/p_b$.

The authors correlated C/C_b and n/n_b to the dimensionless pressure by the following two expressions:

$$\left(\frac{n}{n_b}\right) = 1 + 0.0577\left(1 - \frac{\bar{p}_r}{p_b}\right) - 0.2459\left(1 - \frac{\bar{p}_r}{p_b}\right)^2 + 0.503\left(1 - \frac{\bar{p}_r}{p_b}\right)^3 \quad \textbf{(5.94)}$$

and

$$\left(\frac{C}{C_b}\right) = 1 - 3.5718\left(1 - \frac{\bar{p}_r}{p_b}\right) + 4.7981\left(1 - \frac{\bar{p}_r}{p_b}\right)^2 - 2.3066\left(1 - \frac{\bar{p}_r}{p_b}\right)^3 \quad \textbf{(5.95)}$$

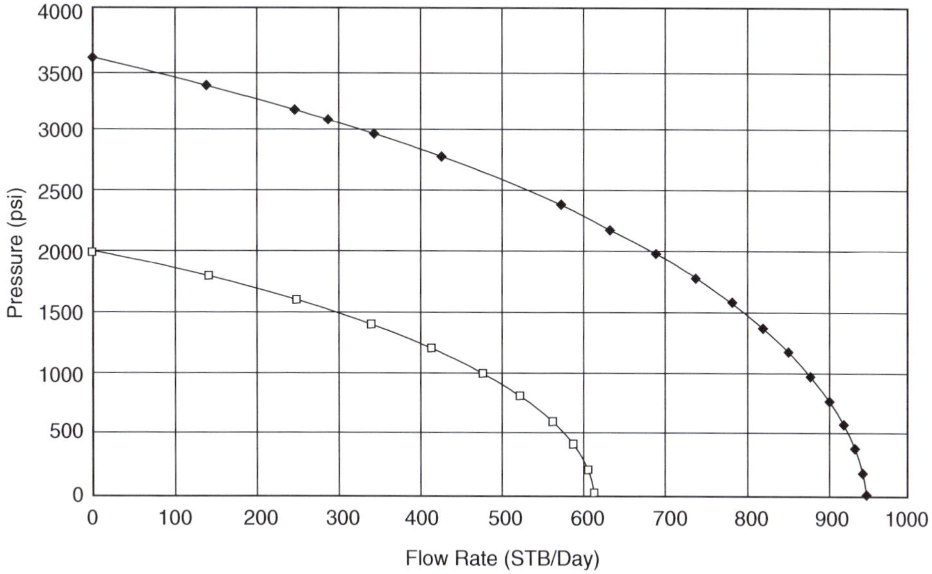

FIGURE 5.21 Future IPR at 2000 psi.

where

C_b = performance coefficient at the *bubble point pressure*

n_b = flow exponent at the *bubble point pressure*

The procedure of applying the above relationships in adjusting the coefficients C and n with changing average reservoir pressure is detailed below:

Step 1. Using the available flow test data in conjunction with Fetkovich's equation, i.e., Eq. (5.92), calculate the present (current) values of n and C at the present average pressure $\bar{p}_r$.

Step 2. Using the current values of $\bar{p}_r$, calculate the dimensionless values of n/n_b and C/C_b by applying Eqs. (5.94) and (5.95), respectively.

Step 3. Solve for the constants n_b and C_b from:

$$n_b = \frac{n}{n/n_b} \tag{5.96}$$

and

$$C_b = \frac{C}{(C/C_b)} \tag{5.97}$$

It should be pointed out that if the present reservoir pressure equals the bubble point pressure, the values of n and C as calculated in step 1 are essentially n_b and C_b.

Step 4. Assume future average reservoir pressure $(\bar{p}_r)_f$ and solve for the corresponding future dimensionless parameters n_f/n_b and C_f/C_b by applying Eqs. (5.94) and (5.95), respectively.

Step 5. Solve for future values of n_f and C_f from:

$$n_f = n_b\left(\frac{n}{n_b}\right)$$

$$C_f = C_b\left(\frac{C_f}{C_b}\right)$$

Step 6. Use n_f and C_f in Fetkovich's equation to generate the well's future IPR at the desired (future) average reservoir pressure $(\bar{p}_r)_f$. It should be noted that the maximum oil flow rate $(Q_o)_{max}$ at $(\bar{p}_r)_f$ is given by:

$$(Q_o)_{max} = C_f[(\bar{p}_r)^2]^{n_f} \quad (5.98)$$

Example 5.16

Using the data given in Example 5.15, generate the future IPR data when the reservoir pressure drops to 3200 psi.

Solution

Step 1. Since the reservoir exists at its bubble point pressure, $p_b = 3600$ psi, then:

$$n_b = 0.854 \quad \text{and} \quad C_b = 0.00079$$

Step 2. Calculate the future dimensionless parameters at 3200 psi by applying Eqs. (5.94) and (5.95):

$$\left(\frac{n}{n_b}\right) = 1 + 0.0577\left(1 - \frac{3200}{3600}\right) - 0.2459 \times \left(1 - \frac{3200}{3600}\right)^2 + 0.5030\left(1 - \frac{3200}{3600}\right)^6 = 1.0041$$

$$\left(\frac{C}{C_b}\right) = 1 - 3.5718\left(1 - \frac{3200}{3600}\right) + 4.7981 \times \left(1 - \frac{3200}{3600}\right)^2 - 2.3066\left(1 - \frac{3200}{3600}\right)^3 = 0.6592$$

Step 3. Solve for n_f and C_f:

$$n_f = n_b(1.0041) = (0.854)(1.0041) = 0.8575$$

$$C_f = C_b(0.6592) = (0.00079)(0.6592) = 0.00052$$

Therefore, the flow rate is then expressed as:

$$Q_o = C(p_r^{-2} - p_{wf}^2)^n = 0.00052(3200^2 - p_{wf}^2)^{0.8575}$$

The maximum oil flow rate, i.e., AOF, occurs at $p_{wf} = 0$, or:

$$(Q_o)_{max} = 0.00052(3200^2 - 0^2)^{0.8575} = 534 \text{ STB/day}$$

Step 4. Construct the following table by assuming several values for p_{wf}:

$$Q_o = 0.00052[3200^2 - (p_{wf})^2]^{0.8575} = 534 \text{ STB/day}$$

p_{wf}	Q_o
3200	0
2000	349
1500	431
5000	523
0	534

Figure 5.22 compares current and future IPRs as calculated in Examples 5.10 and 5.11.

Case 3: $\bar{p}_r > p_b$ and $p_{wf} < p_b$. Figure 5.23 shows a schematic illustration of Case 3 in which it is assumed that $p_{wf} < p_b$ and $p_r > p_b$. The integral in Eq. (5.82) can be expanded and written as:

$$Q_o = \frac{0.00708kh}{\ln(r_e/r_w) - 0.75 + s}\left[\int_{p_{wf}}^{p_b} f(p)dp + \int_{p_b}^{\bar{p}_r} f(p)dp\right]$$

Substituting Eqs. (5.84) and (5.85) into the above expression gives:

$$Q_o = \frac{0.00708kh}{\ln(r_e/r_w) - 0.75 + s} \times \left[\int_{p_{wf}}^{p_b}\left(\frac{1}{\mu_o B_o}\right)\left(\frac{p}{p_b}\right)dp + \int_{p_b}^{\bar{p}_r}\left(\frac{1}{\mu_o B_o}\right)dp\right]$$

where μ_o and B_o are evaluated at the bubble point pressure p_b. Rearranging the above expression gives:

$$Q_o = \frac{0.00708kh}{\mu_o B_o[\ln(r_e/r_w) - 0.75 + s]}\left[\frac{1}{p_b}\int_{p_{wf}}^{p_b} p\,dp + \int_{p_b}^{\bar{p}_b} dp\right]$$

Integrating and introducing the productivity index J into the above relationship gives:

$$Q_o = J\left[\frac{1}{2p_b}(p_b^2 - p_{wf}^2) + (\bar{p}_r - p_b)\right]$$

or

$$Q_o = J(\bar{p}_r - p_b) + \frac{J}{2p_b}(p_b^2 - p_{wf}^2) \quad (5.99)$$

Example 5.17

The following reservoir and flow test data is available on an oil well:

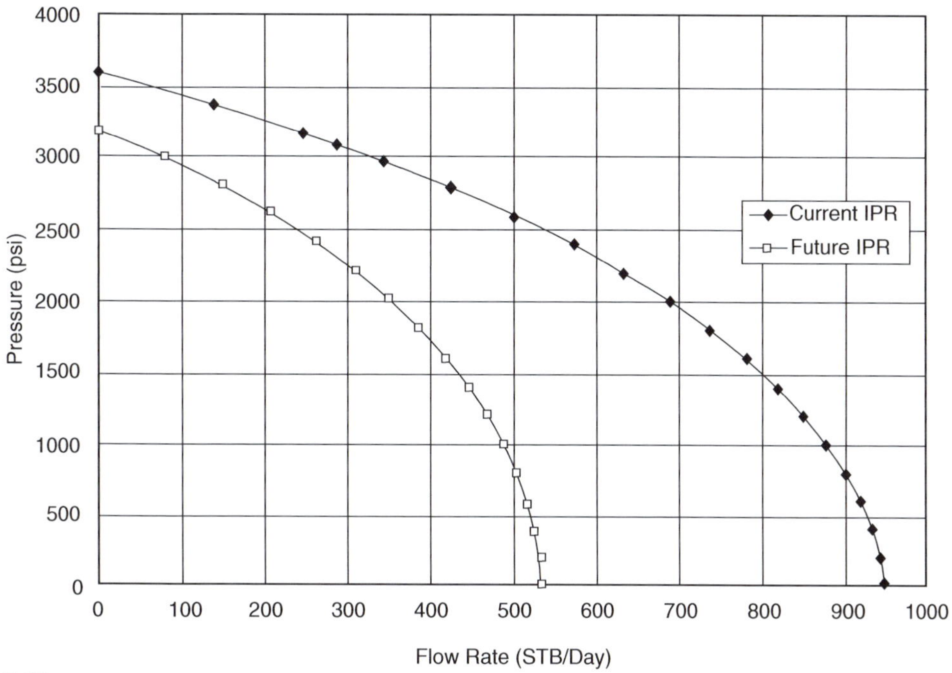

FIGURE 5.22 IPR.

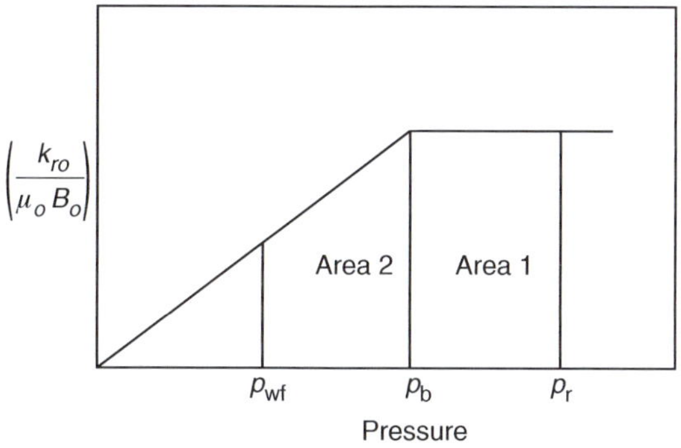

FIGURE 5.23 $(k_{ro}/\mu_o B_o)$ vs. Pressure for case 3.

pressure data: $\bar{p}_r = 4000$ psi, $p_b = 3200$ psi;
flow test data: $p_{wf} = 3600$ psi, $Q_o = 280$ STB/day.

Generate the IPR data of the well.

Solution

Step 1. Since $p_{wf} < p_b$, calculate the productivity index from Eq. (5.58):

$$J = \frac{Q_o}{\bar{p}_r - p_{wf}} = \frac{Q_o}{\Delta p}$$
$$= \frac{280}{4000 - 3600} = 0.7 \text{ STB/day/psi}$$

Step 2. Generate the IPR data by applying Eq. (5.87) when the assumed $p_{wf} > p_b$ and using Eq. (5.99) when $p_{wf} < p_b$. That is:

$$Q_o = J(\bar{p}_r - p_{wf})$$
$$= 0.7(4000 - p_{wf})$$

and

$$Q_o = J(\bar{p}_r - p_b) + \frac{J}{2p_b}(p_b^2 - p_{wf}^2)$$
$$= 0.7(4000 - 3200) + \frac{0.7}{2(3200)}\left[(3200)^2 - p_{wf}^2\right]$$

p_{wf}	Equation	Q_o
4000	(5.87)	0
3800	(5.87)	140
3600	(5.87)	280
3200	(5.87)	560
3000	(5.99)	696
2600	(5.99)	941
2200	(5.99)	1151
2000	(5.99)	1243

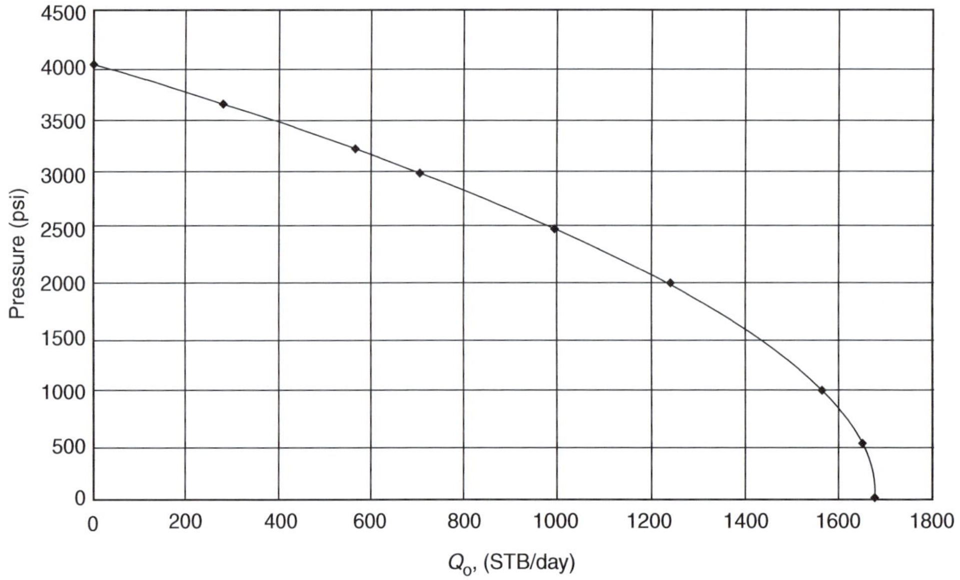

FIGURE 5.24 IPR using the Fetkovich method.

1000	(5.99)	1571
500	(5.99)	1653
0	(5.99)	1680

Results of the calculations are shown graphically in Figure 5.24.

It should be pointed out that the Fetkovich method has the advantage over Standing's methodology in that it does not require the tedious material balance calculations to predict oil saturations at future average reservoir pressures.

Klins and Clark Method. Klins and Clark (1993) proposed an inflow expression similar in form to that of Vogel's and can be used to estimate future IPR data. To improve the predictive capability of Vogel's equation, the authors introduced a new exponent d to Vogel's expression. The authors proposed the following relationships:

$$\frac{Q}{(Q_o)_{max}} = 1 - 0.295\left(\frac{p_{wf}}{\bar{p}_r}\right) - 0.705\left(\frac{p_{wf}}{\bar{p}_r}\right)^d \quad \textbf{(5.100)}$$

where

$$d = \left[0.28 + 0.72\left(\frac{\bar{p}_r}{p_b}\right)\right](1.24 + 0.001 p_b) \quad \textbf{(5.101)}$$

The computational steps of the Klins and Clark method are summarized below:

Step 1. Knowing the bubble point pressure and the current reservoir pressure, calculate the exponent d from Eq. (5.101).

Step 2. From the available stabilized flow data, i.e., Q_o at p_{wf}, solve Eq. (5.100) for $(Q_o)_{max}$. That is:

$$(Q_o)_{max} = \frac{Q_o}{1 - 0.295(p_{wf}/\bar{p}_r) - 0.705(p_{wf}/\bar{p}_r)^d}$$

Step 3. Construct the current IPR by assuming several values of p_{wf} in Eq. (5.100) and solving for Q_o.

5.2.2 Horizontal Oil Well Performance

Since 1980, horizontal wells began capturing an ever-increasing share of hydrocarbon production.

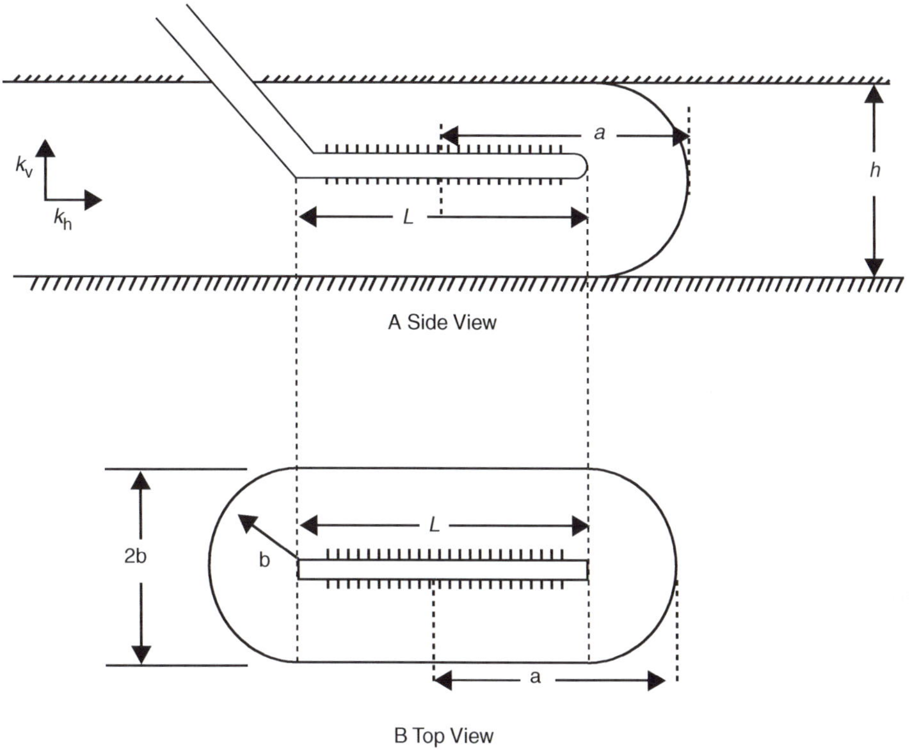

FIGURE 5.25 Horizontal well drainage area.

Horizontal wells offer the following advantages over vertical wells:

- The large volume of the reservoir can be drained by each horizontal well.
- Higher productions from thin pay zones.
- Horizontal wells minimize water and gas zoning problems.
- In high-permeability reservoirs, where near-wellbore gas velocities are high in vertical wells, horizontal wells can be used to reduce near-wellbore velocities and turbulence.
- In secondary and enhanced oil recovery applications, long horizontal injection wells provide higher injectivity rates.
- The length of the horizontal well can provide contact with multiple fractures and greatly improve productivity.

The actual production mechanism and reservoir flow regimes around the horizontal well are considered more complicated than those for the vertical well, especially if the horizontal section of the well is of a considerable length. Some combination of both linear and radial flow actually exists, and the well may behave in a manner similar to that of a well that has been extensively fractured. Sherrad et al. (1987) reported that the shape of measured IPRs for horizontal wells is similar to those predicted by the Vogel or Fetkovich methods. The authors pointed out that the productivity gain from drilling horizontal wells 1500 ft long is two to four times that from a vertical well.

A horizontal well can be looked upon as a number of vertical wells drilling next to each other and completed in a limited pay zone thickness. Figure 5.25 shows the drainage area of a horizontal well of length L in a reservoir with a pay zone thickness of h. Each end of the horizontal well would drain a half-circular area of radius b, with a rectangular drainage shape of the horizontal well.

Assuming that each end of the horizontal well is represented by a vertical well that drains an area of a semicircle with a radius of b, Joshi (1991) proposed the following two methods for calculating the drainage area of a horizontal well.

Method I. Joshi proposed that the drainage area is represented by two semicircles of radius b (equivalent to a radius of a vertical well r_{ev}) at each end and a rectangle, of dimensions $2b - L$, in the center. The drainage area of the horizontal well is then given by:

$$A = \frac{L(2b) + \pi b^2}{43,560} \tag{5.102}$$

where

A = drainage area, acres
L = length of the horizontal well, ft
b = half minor axis of an ellipse, ft

Method II. Joshi assumed that the horizontal well drainage area is an ellipse and given by:

$$A = \frac{\pi ab}{43,560} \tag{5.103}$$

with

$$a = \frac{L}{2} + b \tag{5.104}$$

where a is the half major axis of an ellipse.

Joshi noted that the two methods give different values for the drainage area A and suggested assigning the average value for the drainage of the horizontal well. Most of the production rate equations require the value of the drainage radius of the horizontal well, which is given by:

$$r_{eh} = \sqrt{\frac{43,560A}{\pi}}$$

where

r_{eh} = drainage radius of the horizontal well, ft
A = drainage area of the horizontal well, acres

Example 5.18

A 480-acre lease is to be developed by using 12 vertical wells. Assuming that each vertical well would effectively drain 40 acres, calculate the possible number of either 1000- or 2000-ft-long horizontal wells that will drain the lease effectively.

Solution

Step 1. Calculate the drainage radius of the vertical well:

$$r_{ev} = b = \sqrt{\frac{(40)(43,560)}{\pi}} = 745 \text{ ft}$$

Step 2. Calculate the drainage area of the 1000- and 2000-ft-long horizontal well using Joshi's two methods.

Method I. For the 1000-ft horizontal well and using Eq. (5.102):

$$A = \frac{L(2b) + \pi b^2}{43,560}$$
$$= \frac{(1000)(2 \times 745) + \pi(745)^2}{43,560} = 74 \text{ acres}$$

For the 2000-ft horizontal well:

$$A = \frac{L(2b) + \pi b^2}{43,560}$$
$$= \frac{(2000)(2 \times 745) + \pi(745)^2}{43,560} = 108 \text{ acres}$$

Method II. For the 1000-ft horizontal well and using Eq. (5.103):

$$a = \frac{L}{2} + b$$
$$= \frac{1000}{2} + 745 = 1245 \text{ ft}$$
$$A = \frac{\pi ab}{43,560}$$
$$= \frac{\pi(1245)(745)}{43,560} = 67 \text{ acres}$$

For the 2000-ft horizontal well:

$$a = \frac{2000}{2} + 745 = 1745 \text{ ft}$$
$$A = \frac{\pi(1745)(75)}{43,560} = 94 \text{ acres}$$

Step 3. *Averaging* the values from the two methods, the drainage area of the well is 1000 ft long:

$$A = \frac{74 + 67}{2} = 71 \text{ acres}$$

and the drainage area of 2000-ft-long well is:

$$A = \frac{108 + 94}{2} = 101 \text{ acres}$$

Step 4. Calculate the number of 1000-ft-long horizontal wells:

Total number of 1000-ft horizontal wells

$$= \frac{\text{Total area}}{\text{Draining area per well}}$$

$$= \frac{480}{71} = 7 \text{ wells}$$

Step 5. Calculate the number of 2000-ft-long horizontal wells:

Total number of 2000-ft horizontal wells

$$= \frac{\text{Total area}}{\text{Drainage area per well}}$$

$$= \frac{480}{101} = 5 \text{ wells}$$

From a practical standpoint, inflow performance calculations for horizontal wells are presented here under the following two flowing conditions:

(1) steady-state single-phase flow;
(2) pseudosteady-state two-phase flow.

The reference textbook by Joshi (1991) provides an excellent treatment of horizontal well technology, and it contains detailed documentation of recent methodologies of generating IPRs.

5.2.3 Horizontal Well Productivity under Steady-State Flow

The steady-state analytical solutions are the simplest form of horizontal well solutions. The steady-state solution requires that the pressure at any point in the reservoir does not change with time. The flow rate equation in a steady-state condition is represented by:

$$Q_{oh} = J_h(p_r - p_{wf}) = J_h \, \Delta p \tag{5.105}$$

where

Q_{oh} = horizontal well flow rate, STB/day
Δp = pressure drop from the drainage boundary to wellbore, psi
J_h = productivity index of the horizontal well, STB/day/psi

The productivity index of the horizontal well J_h can always be obtained by dividing the flow rate Q_{oh} by the pressure drop Δp, or:

$$J_h = \frac{Q_{oh}}{\Delta p}$$

There are several methods that are designed to predict the productivity index from the fluid and reservoir properties. Some of these methods include:

- the Borisov method;
- the Giger, Reiss, and Jourdan method;
- the Joshi method;
- the Benard and Dupuy method.

Borisov Method. Borisov (1984) proposed the following expression for predicting the productivity index of a horizontal well in an isotropic reservoir, i.e., $k_v = k_h$:

$$J_h = \frac{0.00708 h k_h}{\mu_o B_o\left[[\ln(4r_{eh}/L) + (h/L)\ln(h/2\pi r_w)]\right]} \tag{5.106}$$

where

h = thickness, ft
k_h = horizontal permeability, md
k_v = vertical permeability, md
L = length of the horizontal well, ft
r_{eh} = drainage radius of the horizontal well, ft
r_w = wellbore radius, ft
J_h = productivity index, STB/day/psi

Giger, Reiss, and Jourdan Method. For an isotropic reservoir where the vertical permeability k_v equals the horizontal permeability k_h, Giger et al. (1984) proposed the following expression for determining J_h:

$$J_h = \frac{0.00708Lk_h}{\mu_o B_o[(L/h)\ln(X) + \ln(h/2r_w)]} \quad (5.107)$$

where

$$X = \frac{1 + \sqrt{1 + [L/2r_{eh}]^2}}{L/(2r_{eh})} \quad (5.108)$$

To account for the reservoir anisotropy, the authors proposed the following relationships:

$$J_h = \frac{0.00708k_h}{\mu_o B_o[(1/h)\ln(X) + (\beta^2/L)\ln(h/2r_w)]} \quad (5.109)$$

with the parameter β as defined by:

$$\beta = \sqrt{\frac{k_h}{k_v}} \quad (5.110)$$

where

k_v = vertical permeability, md
L = length of the horizontal section, ft

Joshi Method. Joshi (1991) presented the following expression for estimating the productivity index of a horizontal well in isotropic reservoirs:

$$J_h = \frac{0.00708hk_h}{\mu_o B_o[\ln(R) + (h/L)\ln(h/2r_w)]} \quad (5.111)$$

with

$$R = \frac{a + \sqrt{a^2 - (L/2)^2}}{(L/2)} \quad (5.112)$$

and a is half the major axis of the drainage ellipse and given by:

$$a = \left(\frac{L}{2}\right)\left[0.5 + \sqrt{0.25 + \left(\frac{2r_{eh}}{L}\right)^4}\right]^{0.5} \quad (5.113)$$

Joshi accounted for the influence of the reservoir anisotropy by introducing the vertical permeability k_v into Eq. (5.111), to give:

$$J_h = \frac{0.00708kh_h}{\mu_o B_o[\ln(R) + (B^2h/L)\ln(h/2r_w)]} \quad (5.114)$$

where the parameters B and R are defined by Eqs. (5.110) and (5.112), respectively.

Renard and Dupuy Method. For an isotropic reservoir, Renard and Dupuy (1990) proposed the following expression:

$$J_h = \frac{0.00708kh_h}{\mu_o B_o[\cosh^{-1}(2a/L) + (h/L)\ln(h/2\pi r_w)]} \quad (5.115)$$

where a is half the major axis of the drainage ellipse and given by Eq. (5.113).

For anisotropic reservoirs, the authors proposed the following relationship:

$$J_h = \frac{0.00708hk_h}{\mu_o B_o[\cosh^{-1}(2a/L) + (\beta h/L)\ln(h/2\pi r'_w)]} \quad (5.116)$$

where

$$r_w^{\prime} = \frac{(1+\beta)r_w}{2\beta} \quad (5.117)$$

with the parameter β as defined by Eq. (5.110).

Example 5.19

A horizontal well 2000 ft long drains an estimated drainage area of 120 acres. The reservoir is characterized by an isotropic formation with the following properties:

$k_v = k_h = 100$ md, $h = 60$ ft,
$B_o = 1.2$ bbl/STB, $\mu_o = 0.9$ cp,
$p_e = 3000$ psi, $p_{wf} = 2500$ psi,
$r_w = 0.30$ ft

Assuming a steady-state flow, calculate the flow rate by using:

(a) the Borisov method;
(b) the Giger, Reiss, and Jourdan method;
(c) the Joshi's method;
(d) the Renard and Dupuy method.

Solution

(a) Borisov method:
Step 1. Calculate the drainage radius of the horizontal well:

$$r_{eh} = \sqrt{\frac{43,560A}{\pi}} = \sqrt{\frac{(43,560)(120)}{\pi}} = 1290 \text{ ft}$$

Step 2. Calculate J_h by using Eq. (5.106):

$$J_h = \frac{0.00708hk_h}{\mu_o B_o[\ln(4r_{eh}/L) + (h/L)\ln(h/2\pi r_w)]}$$
$$= \frac{(0.00708)(60)(100)}{(0.9)(1.2)[\ln\left(((4)(1290))/2000\right) + (60/2000)\ln(60/2\pi(0.3))]}$$
$$= 37.4 \text{ STB/day/psi}$$

Step 3. Calculate the flow rate by applying Eq. (5.105):

$$Q_{oh} = J_h \Delta p$$
$$= (37.4)(3000 - 2500) = 18,700 \text{ STB/day}$$

(b) Giger, Reiss, and Jourdan method:

Step 1. Calculate the parameter X from Eq. (5.108):

$$X = \frac{1 + \sqrt{1 + (L/2r_{eh})^2}}{L/(2r_{eh})}$$
$$= \frac{1 + \sqrt{1 + (2000/(2)(1290))^2}}{2000/[(2)(1290)]} = 2.105$$

Step 2. Solve for J_h by applying Eq. (5.107):

$$J_h = \frac{0.00708Lk_h}{\mu_o B_o[(L/h)\ln(X) + \ln(h/2r_w)]}$$
$$= \frac{(0.00708)(2000)(100)}{(0.9)(1.2)[(2000/60)\ln(2.105) + \ln(60/2(0.3))]}$$
$$= 44.57 \text{ STB/day}$$

Step 3. Calculate the flow rate:

$$Q_{oh} = 44.57(3000 - 2500) = 22,286 \text{ STB/day}$$

(c) Joshi method:

Step 1. Calculate the half major axis of the ellipse by using Eq. (5.113):

$$a = \left(\frac{L}{2}\right)\left[0.5 + \sqrt{0.25 + \left(\frac{2r_{eh}}{L}\right)^4}\right]^{0.5}$$
$$= \left(\frac{2000}{2}\right)\left[0.5 + \sqrt{0.25 + \left[\frac{2(1290)}{2000}\right]^2}\right]$$
$$= 1372 \text{ ft}$$

Step 2. Calculate the parameter R from Eq. (5.112):

$$R = \frac{a + \sqrt{a^2 - (L/2)^2}}{(L/2)}$$
$$= \frac{1372 + \sqrt{(1372)^2 - (2000/2)^2}}{(2000/2)} = 2.311$$

Step 3. Solve for J_h by applying Eq. (5.111):

$$J_h = \frac{0.00708hk_h}{\mu_o B_o[\ln(R) + (h/L)\ln(h/2r_w)]}$$
$$= \frac{0.00708(60)(100)}{(0.9)(1.2)\left[\ln(2.311) + (60/2000)\ln(60/((2)(0.3)))\right]}$$
$$= 40.3 \text{ STB/day/psi}$$

Step 4. Calculate the flow rate:

$$Q_{oh} = J_h \Delta p$$
$$= (40.3)(3000 - 2500) = 20,154 \text{ STB/day}$$

(d) Renard and Dupuy method:

Step 1. Calculate a from Eq. (5.113):

$$a = \left(\frac{L}{2}\right)\left[0.5 + \sqrt{0.25 + \left(\frac{2r_{eh}}{L}\right)^4}\right]^{0.5}$$
$$= \left(\frac{2000}{2}\right)\left[0.5 + \sqrt{0.25 + \left[\frac{2(1290)}{2000}\right]^2}\right]^{0.5}$$
$$= 1372 \text{ ft}$$

Step 2. Apply Eq. (5.115) to determine J_h:

$$J_h = \frac{0.00708hk_h}{\mu_o B_o[cosh^{-1}(2a/L) + (h/L)\ln(h/2\pi r_w)]}$$
$$= \frac{0.00708(60)(100)}{(0.9)(1.2)[\cosh^{-1}((2)(1327)/2000) + (60/2000)\text{Ln}(60/2\pi(0.3))]}$$
$$= 41.77 \text{ STB/day/psi}$$

Step 3. Calculate the flow rate:

$$Q_{oh} = 41.77(3000 - 2500) = 20,885 \text{ STB/day}$$

Example 5.20

Using the data in Example 5.19 and assuming an isotropic reservoir with $k_h = 100$ md and $k_v = 10$ md, calculate the flow rate by using:

the Giger, Reiss, and Jourdan method;
the Joshi method;
the Renard and Dupuy method.

Solution

(a) Giger, Reiss, and Jourdan method:

Step 1. Solve for the permeability ratio β by applying Eq. (5.110):

$$\beta = \sqrt{\frac{k_h}{k_v}} = \sqrt{\frac{100}{10}} = 3.162$$

Step 2. Calculate the parameter X as shown in Example 5.19, to give:

$$X = \frac{1 + \sqrt{1 + (L/2r_{eh})^2}}{L/(2r_{eh})} = 2.105$$

Step 3. Determine J_h by using Eq. (5.109):

$$J_h = \frac{0.00708k_h}{\mu_o B_o[(1/h)\ln(X) + (\beta^2/L)\ln(h/2r_w)]}$$

$$= \frac{0.00708(100)}{(0.9)(1.2)[(1/60)\ln(2.105) + (3.162^2/2000)\ln(60/(2)(0.3))]}$$

$$= 18.50 \text{ STB/day/psi}$$

Step 4. Calculate Q_{oh}:

$$Q_{oh} = (18.50)(300 - 2500) = 9252 \text{ STB/day}$$

(b) Joshi method:

Step 1. Calculate the permeability ratio β:

$$\beta = \sqrt{\frac{k_h}{k_v}} = 3.162$$

Step 2. Calculate the parameters a and R as given in Example 5.19:

$$a = 1372 \text{ ft}, \quad R = 2.311$$

Step 3. Calculate J_h by using Eq. (5.111):

$$J_h = \frac{0.00708hk_h}{\mu_o B_o[\ln(R) + (h/L)\ln(h/2r_w)]}$$

$$= \frac{0.00708(60)(100)}{(0.9)(1.2)[\ln(2.311) + ((3.162)^2(60)/2000)\ln(60/2(0.3))]}$$

$$= 17.73 \text{ STB/day/psi}$$

Step 4. Calculate the flow rate:

$$Q_{oh} = (17.73)(3000 - 2500) = 8863 \text{ STB/day}$$

(c) Renard and Dupuy method:

Step 1. Calculate $r_w^{\backslash}$ from Eq. (5.117):

$$r_w^{\backslash} = \frac{(1+\beta)r_w}{2\beta}$$

$$r_w^{\backslash} = \frac{(1+3.162)(0.3)}{(2)(3.162)} = 0.1974$$

Step 2. Apply Eq. (5.116):

$$J_h = \frac{0.00708(60)(100)}{(0.9)(1.2)\{\cosh^{-1}[(2)(1372)/2000] + [(3.162)^2(60)/2000]\ln(60/(2)\pi(0.1974))\}}$$

$$= 19.65 \text{ STB/day/psi}$$

Step 3. Calculate the flow rate:

$$Q_{oh} = 19.65(3000 - 2500) = 9825 \text{ STB/day}$$

5.2.4 Horizontal Well Productivity under Semisteady-State Flow

The complex flow regime existing around a horizontal wellbore probably precludes using a method as simple as that of Vogel to construct the IPR of a horizontal well in solution gas drive reservoirs. However, if at least two stabilized flow tests are available, the parameters J and n in Fetkovich's equation (i.e., Eq. (5.92)) could be determined and used to construct the IPR of the horizontal well. In this case, the values of J and n would account not only for the effects of turbulence and gas saturation around the wellbore, but also for the effects of the nonradial flow regime existing in the reservoir.

Bendakhlia and Aziz (1989) used a reservoir model to generate IPRs for a number of wells and found that a combination of Vogel's and Fetkovich's equations would fit the generated data if expressed as:

$$\frac{Q_{oh}}{(Q_{oh})_{max}} = \left[1 - V\left(\frac{p_{wf}}{\bar{p}_r}\right) - (1-V)\left(\frac{p_{wf}}{\bar{p}_r}\right)^2\right]^n \quad \textbf{(5.118)}$$

where

$(Q_{oh})_{max}$ = horizontal well maximum flow rate, STB/day
n = exponent in Fetkovich's equation
V = variable parameter

In order to apply the equation, at least three stabilized flow tests are required to evaluate the three unknowns $(Q_{oh})_{max}$, V, and n at any given average reservoir pressure $\bar{p}_r$. However, Bendakhlia and Aziz indicated that the parameters V and n are functions of the reservoir pressure or recovery factor, and thus, the use of Eq. (5.118) is not convenient in a predictive mode.

Cheng (1990) presented a form of Vogel's equation for horizontal wells that is based on the results from a numerical simulator. The proposed expression has the following form:

$$\frac{Q_{oh}}{(Q_{oh})_{max}} = 0.9885 + 0.2055\left(\frac{p_{wf}}{\bar{p}_r}\right) - 1.1818\left(\frac{p_w}{\bar{p}_r}\right)^2 \quad \textbf{(5.119)}$$

Petnanto and Economides (1998) developed a generalized IPR equation for a horizontal and multilateral well in a solution gas drive reservoir. The proposed expression has the following form:

$$\frac{Q_{oh}}{(Q_{oh})_{max}} = 1 - 0.25\left(\frac{p_{wf}}{\bar{p}_r}\right) - 0.75\left(\frac{p_{wf}}{\bar{p}_r}\right)^n \quad \textbf{(5.120)}$$

where

$$n = \left[-0.27 + 1.46\left(\frac{\bar{p}_r}{p_b}\right) - 0.96\left(\frac{\bar{p}_r}{p_b}\right)^2\right] \times (4 + 1.66 \times 10^{-3} p_b) \quad \textbf{(5.121)}$$

with

$$(Q_{oh})_{max} = \frac{J\bar{p}_r}{0.25 + 0.75n}$$

Example 5.21

A horizontal well 1000 ft long is drilled in a solution gas drive reservoir. The well is producing at a stabilized flow rate of 760 STB/day and a wellbore pressure of 1242 psi. The current average reservoir pressure is 2145 psi. Generate the IPR data of this horizontal well by using the Cheng method.

Solution

Step 1. Use the given stabilized flow data to calculate the maximum flow rate of the horizontal well:

$$\frac{Q_{oh}}{(Q_{oh})_{max}} = 1.0 + 0.2055\left(\frac{p_{wf}}{\bar{p}_r}\right) - 1.1818\left(\frac{p_w}{\bar{p}_r}\right)^2$$

$$\frac{760}{(Q_{oh})_{max}} = 1 + 0.2055\left(\frac{1242}{2145}\right) - 1.1818\left(\frac{1242}{2145}\right)$$

$$(Q_{oh})_{max} = 1052 \text{ STB/day}$$

Step 2. Generate the IPR data by applying Eq. (5.120):

$$Q_{oh} = (Q_{oh})_{max}\left[0.1 + 0.2055\left(\frac{p_{wf}}{\bar{p}_r}\right) - 1.1818\left(\frac{p_{wf}}{\bar{p}_r}\right)^2\right]$$

p_{wf}	$(Q_{oh})_{max}$
2145	0
1919	250
1580	536
1016	875
500	1034
0	1052

5.3 PHASE 3. RELATING RESERVOIR PERFORMANCE TO TIME

All reservoir performance techniques show the relationship of cumulative oil production and the instantaneous GOR as a function of average

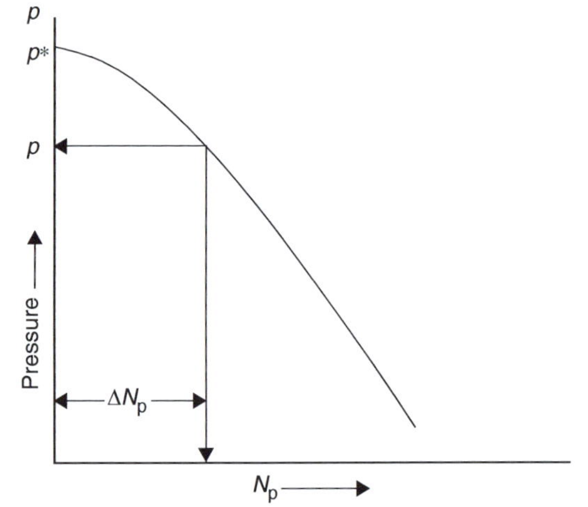

FIGURE 5.26 Cumulative production as a function of average reservoir pressure.

reservoir pressure. However, these techniques do not relate the cumulative oil production N_p and cumulative gas production G_p with time. Figure 5.26 shows a schematic illustration of the predicted cumulative oil production with declining average reservoir pressure.

The time required for production can be calculated by applying the concept of the IPR in conjunction with the MBE predictions. For example, Vogel (1968) expressed the well's IPR by Eq. (5.66) as:

$$Q_o = (Q_o)_{max}\left[1 - 0.2\left(\frac{p_{wf}}{\bar{p}_r}\right) - 0.8\left(\frac{p_{wf}}{\bar{p}_r}\right)^2\right]$$

The following methodology can be employed to correlate the predicted cumulative field production with time t:

Step 1. Plot the *predicted* cumulative oil production N_p as a function of average reservoir pressure p as shown in Figure 5.26.

Step 2. Assume that the current reservoir pressure is p^* with a current cumulative oil production of $(N_p)^*$ and total field flow rate of $(Q_o)_T^*$.

Step 3. Select a future average reservoir pressure p and determine the future cumulative oil production N_p from Figure 5.26.

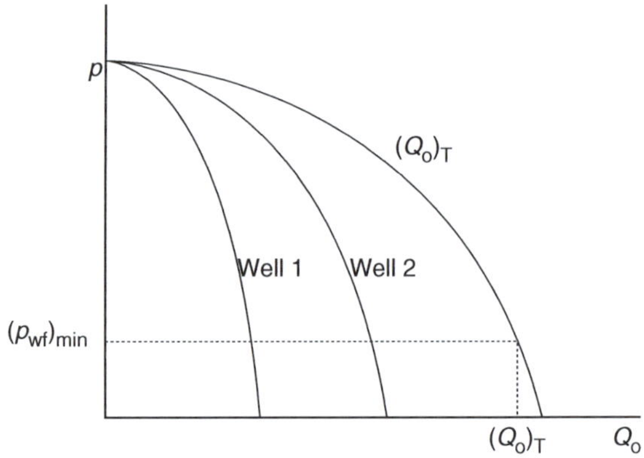

FIGURE 5.27 Overall field IPR at future average pressure.

Step 4. Using the selected future average reservoir pressure p, construct the IPR curve for *each well in the field* (as shown schematically in Figure 5.27 for two hypothetical wells). Establish the total field IPR by taking the summation of the flow rates of all wells at any time.

Step 5. Using the minimum bottom-hole flowing pressure $(p_{wf})_{min}$, determine the total field flow rate $(Q_o)_T$.

$$(Q_o)_T = \sum_{i=1}^{\#\,well} (Q_o)_i$$

Step 6. Calculate the average field production rate $(\overline{Q}_o)_T$:

$$(\overline{Q}_o)_T = \frac{(Q_o)_T + (Q_o)_T^*}{2}$$

Step 7. Calculate the time Δt required for the incremental oil production ΔN_p during the first pressure drop interval, i.e., from p^* to p, by:

$$\Delta t = \frac{N_p - N_p^*}{(\overline{Q}_o)_T} = \frac{\Delta N_p}{(\overline{Q}_o)_T}$$

Step 8. Repeat the above steps and calculate the total time t to reach an average reservoir pressure p, by:

$$t = \sum \Delta t$$

5.4 PROBLEMS

(1) An oil well is producing under steady-state flow conditions at 300 STB/day. The bottom-hole flowing pressure is recorded at 2500 psi. Given:

$h = 23$ ft, $k = 50$ md, $\mu_o = 2.3$ cp;
$B_o = 1.4$ bbl/STB, $r_e = 660$ ft, $s = 0.5$.

Calculate:

(a) the reservoir pressure;
(b) the AOF;
(c) the productivity index.

(2) A well is producing from a saturated oil reservoir with an average reservoir pressure of 3000 psig. Stabilized flow test data indicates that the well is capable of producing 400 STB/day at a bottom-hole flowing pressure of 2580 psig.

Calculate the remaining oil-in-place at 3000 psi.

(a) Oil flow rate at $p_{wf} = 1950$ psig.
(b) Construct the IPR curve at the current average pressure.
(c) Construct the IPR curve by assuming a constant J.
(d) Plot the IPR curve when the reservoir pressure is 2700 psig.

(3) An oil well is producing from an undersaturated reservoir that is characterized by a bubble point pressure of 2230 psig. The current average reservoir pressure is 3500 psig. Available flow test data shows that the well produced 350 STB/day at a stabilized p_{wf} of 2800 psig. Construct the IPR data, by using:

(a) Vogel's correlation;
(b) Wiggins method.
(c) Generate the IPR curve when the reservoir pressure declines to 2230 and 2000 psi.

(4) A well is producing from a saturated oil reservoir that exists at its saturation pressure of 4500 psig. The well is flowing at a stabilized rate of 800 STB/day and a p_{wf} of 3700 psig. Material balance calculations provide the following current and future predictions for oil saturation and *PVT* properties:

	Present	Future
$\bar{p}_r$	4500	3300
μ_o, cp	1.45	1.25
B_o, bbl/STB	1.23	1.18
k_{ro}	1.00	0.86

Generate the future IPR for the well at 3300 psig by using the Standing method.

(5) A four-point stabilized flow test was conducted on a well producing from a saturated reservoir that exists at an average pressure of 4320 psi.

Q_o (STB/day)	p_{wf} (psi)
342	3804
498	3468
646	2928
832	2580

Construct a complete IPR by using the Fetkovich method.
Construct the IPR when the reservoir pressure declines to 2500 psi.

(6) The following reservoir and flow test data is available on an oil well:

pressure data: $p_r = 3280$ psi, $p_b = 2624$ psi;
flow test data: $p_{wf} = 2952$ psi, $Q_o =$ STB/day.

Generate the IPR data of the well.

(7) A horizontal well 2500 ft long drains an estimated drainage area of 120 acres. The reservoir is characterized by an isotropic formation with the following properties:

$k_v = k_h = 60$ md, $h = 70$ ft
$B_o = 1.4$ bbl/STB, $\mu_o = 1.9$ cp
$p_e = 3900$ psi, $p_{wf} = 3250$ psi
$r_w = 0.30$ ft

Assuming a steady-state flow, calculate the flow rate by using:

the Borisov method;
the Giger, Reiss, and Jourdan method;

the Joshi method;
the Renard and Dupuy method.

(8) A horizontal well 2000 ft long is drilled in a solution gas drive reservoir. The well is producing at a stabilized flow rate of 900 STB/day and a wellbore pressure of 1000 psi. The current average reservoir pressure is 2000 psi. Generate the IPR data of this horizontal well by using the Cheng method.

(9) The following *PVT* data is for the Aneth Field in Utah:

Pressure (psia)	B_o (bbl/STB)	R_{so} (scf/STB)	B_g (bbl/SCF)	μ_o/μ_g
2200	1.383	727	–	–
1850	1.388	727	0.00130	35
1600	1.358	654	0.00150	39
1300	1.321	563	0.00182	47
1000	1.280	469	0.00250	56
700	1.241	374	0.00375	68
400	1.199	277	0.00691	85
100	1.139	143	0.02495	130
40	1.100	78	0.05430	420

The initial reservoir temperature was 133°F, the initial pressure was 220 psia, and the bubble point pressure was 1850 psia. There was no active water drive. From 1850 to 1300 psia, a total of 720 MMSTB of oil and 590.6 MMscf of gas were produced.

(a) How many reservoir barrels of oil were in place at 1850 psia?
(b) The average porosity was 10%, and connate water saturation was 28%. The field covered 50,000 acres. What is the average formation thickness in feet?

(10) An oil reservoir initially contains 4 MMSTB of oil at its bubble point pressure of 3150 psia with 600 scf/STB of gas in solution. When the average reservoir pressure has dropped to 2900 psia, the gas in solution is 550 scf/STB. B_{oi} was 1.34 bbl/STB, and B_o at a pressure of 2900 psia is 1.32 bbl/STB. Other data:

$R_p = 600$ scf/STB at 2900 psia, $S_{wi} = 0.25$;
$B_g = 0.0011$ bbl/SCF at 2900 psia.

volumetric reservoir no original gas cap

(a) How many STB of oil will be produced when the pressure has decreased to 2900 psia?
(b) Calculate the free gas saturation that exists at 2900 psia.

(11) The following data is obtained from laboratory core tests, production data, and logging information:

well spacing = 320 acres;
net pay thickness = 50 ft with the gas–oil contact 10 ft from the top;
porosity = 0.17;
initial water saturation = 0.26;
initial gas saturation = 0.15;
bubble point pressure = 3600 psia;
initial reservoir pressure = 3000 psia;
reservoir temperature = 120°F;
$B_{oi} = 1.26$ bbl/STB;
$B_o = 1.37$ bbl/STB at the bubble point pressure;
$B_o = 1.19$ bbl/STB at 2000 psia;
$N_p = 2.00$ MM/STB at 2000 psia;
$G_p = 2.4$ MMMSCF at 2000 psia;
gas compressibility factor, $Z = 1.0 - 0.0001p$;
solution, GOR $R_{so} = 0.2p$.

Calculate the amount of water that has influxed and the drive indexes at 2000 psia.

(12) The following production data is available on a depletion drive reservoir:

p (psi)	GOR (scf/STB)	N_p (MMSTB)
3276	1098.8	0
2912	1098.8	1.1316
2688	1098.8	1.8532
2352	1098.8	2.8249
2016	1587.52	5.9368
1680	2938.88	9.86378
1344	5108.6	12.5632

Calculate cumulative gas produced G_p and cumulative GOR at each pressure.

(13) A volumetric solution gas drive reservoir has an initial water saturation of 25%. The initial oil formation volume factor is

reported at 1.35 bbl/STB. When 8% of the initial oil was produced, the value of B_o decreased to 1.28. Calculate the oil saturation and the gas saturation.

(14) The following data is available on a volumetric undersaturated oil reservoir:

$p_i = 4400$ psi, $p_b = 3400$ psi;
$N = 120$ MMSTB, $c_f = 4 \times 10^{-6}$ psi^{-1};
$c_o = 12 \times 10^{-6}$ psi^{-1},
$c_w = 2 \times 10^{-6}$ psi^{-1};
$S_{wi} = 25\%$, $B_{oi} = 1.35$ bbl/STB.

Estimate cumulative oil production when the reservoir pressure drops to 4000 psi. The oil formation volume factor at 4000 psi is 1.38 bbl/STB.

CHAPTER
6

Introduction to Enhanced Oil Recovery

Primary oil recovery, secondary oil recovery, and tertiary (enhanced) oil recovery are terms that are traditionally used in describing hydrocarbons recovered according to the method of production or the time at which they are obtained.

Primary oil recovery describes the production of hydrocarbons under the natural driving mechanisms present in the reservoir without supplementary help from injected fluids such as gas or water. In most cases, the natural driving mechanism is a relatively inefficient process and results in a low overall oil recovery. The lack of sufficient natural drive in most reservoirs has led to the practice of supplementing the natural reservoir energy by introducing some form of artificial drive, the most basic method being the injection of gas or water.

Secondary oil recovery refers to the additional recovery resulting from the conventional methods of water injection and immiscible gas injection. Usually, the selected secondary recovery follows the primary recovery but may be conducted concurrently with the primary recovery. Water flooding is perhaps the most common method of secondary recovery. However, before undertaking a secondary recovery project it should be clearly proven that the natural recovery processes are insufficient; otherwise, there is a risk that the required substantial capital investment may be wasted.

Tertiary (enhanced) oil recovery is the additional recovery over and above what could be recovered by secondary recovery methods. Various methods of enhanced oil recovery (EOR) are essentially designed to recover oil, commonly described as residual oil, left in the reservoir after both primary and secondary recovery methods have been exploited to their respective economic limits. Figure 6.1 illustrates the concept of the three recovery categories.

6.1 MECHANISMS OF ENHANCED OIL RECOVERY

The terms enhanced oil recovery (EOR) and improved oil recovery (IOR) have been used loosely and interchangeably at times. IOR is a general term that implies improving oil recovery by any means (e.g., operational strategies, such as infill drilling, horizontal wells, and improving vertical and areal sweep). EOR is more specific in concept and it can be considered as a subset of IOR. EOR implies the process of enhancing oil recovery by reducing oil saturation below the residual oil saturation "S_{or}." The target of EOR varies considerably by different types of hydrocarbons. Figure 6.2 shows the fluid saturations and the target of EOR for typical light and heavy oil reservoirs and tar sand. For light oil reservoirs, EOR is usually

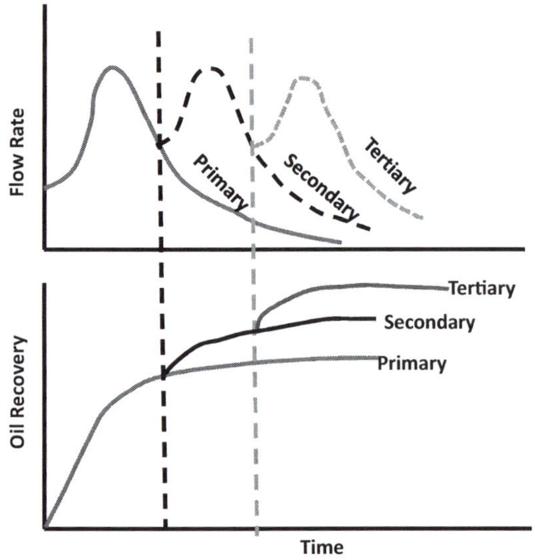

FIGURE 6.1 Oil recovery categories.

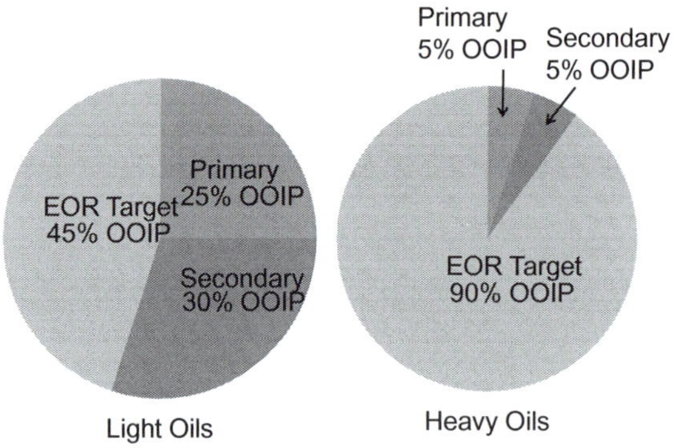

FIGURE 6.2 Target for different crude oil systems.

applicable after secondary recovery operations with an EOR target of approximately 45% original oil in place (OOIP). Heavy oils and tar sands respond poorly to primary and secondary recovery methods, and the bulk of the production from these types of reservoirs come from EOR methods.

The magnitude of the reduction and mobilization of residual oil saturation "S_{or}" by an EOR process is controlled by two major factors, these are:

- capillary number "N_c"; and
- mobility ratio "M."

The capillary number is defined as the ratio of viscous force to interfacial tension force, or

$$N_c = \frac{\text{Viscous force}}{\text{Interfacial tension force}} = \frac{v\mu}{\sigma}$$

Or equivalently as:

$$N_c = \left(\frac{k_o}{\phi\sigma}\right)\left(\frac{\Delta p}{L}\right) \tag{6.1}$$

where

μ = viscosity of the displacing fluid
σ = interfacial tension
v = Darcy velocity
ϕ = porosity
k_o = effective permeability of the displaced fluid, i.e., oil
$\Delta p/L$ = pressure gradient

Figure 6.3 is a schematic representation of the capillary number and the ratio of residual oil saturation (after conduction of an EOR process to residual oil saturation before the EOR process).

The illustration shows the reduction in the residual oil saturation with the increase in the capillary number. It is clear that the capillary number can be increased by:

- increasing the pressure gradient $\Delta p/L$;
- increasing the viscosity of the displacing fluid;
- decreasing the interfacial tension between the injection fluid and displaced fluid.

The reduction in the interfacial tension between the displacing and displaced fluid is perhaps the only practical option in reducing residual oil saturation by increasing capillary number. As shown in Figure 6.3, the capillary number has to exceed the critical capillary number to mobilize residual oil saturation. It should be noticed that by reducing the interfacial tension to zero, the capillary number becomes infinite, indicating complete "miscible displacement."

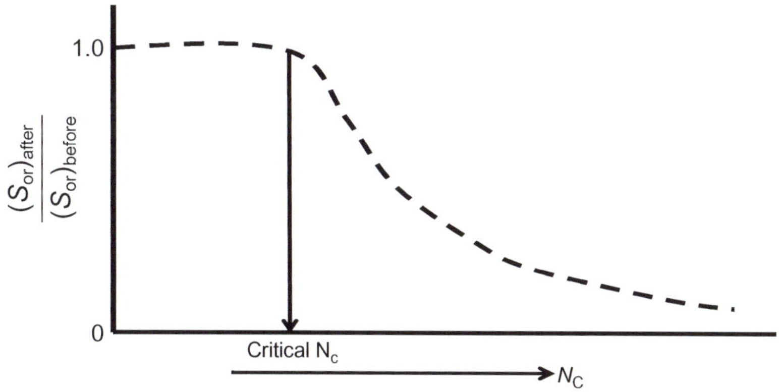

FIGURE 6.3 Effect of N_c on residual oil saturation.

Another important concept in understanding the displacing mechanism of an EOR process is the mobility ratio "M." The mobility ratio is defined as the ratio of the displacing fluid mobility to that of the displaced fluid, or:

$$M = \frac{\lambda_{\text{displacing}}}{\lambda_{\text{displaced}}} = \frac{(k/\mu)_{\text{displacing}}}{(k_o/\mu_o)_{\text{displaced}}}$$

where "k" is the effective permeability and "μ" is the viscosity. The mobility ratio influences the microscopic (pore-level) and macroscopic (areal and vertical sweep) displacement efficiencies. A value of $M > 1$ is considered unfavorable because it indicates that the displacing fluid flows more readily than the displaced fluid (oil). This unfavorable condition can cause channeling and bypassing of residual oil. Improvement in mobility ratio can be achieved by increasing the viscosity of the injection fluid, e.g., polymer flood.

6.2 ENHANCED OIL RECOVERY METHODS

All EOR methods that have been developed are designed to increase the capillary number as given by Eq. (6.1). In general, EOR technologies can be broadly grouped into the following four categories:

- thermal;
- chemical;
- miscible;
- others.

Each of the four categories contains an assortment of injection schemes and a different variety of injection fluids, as summarized below:

Thermal

- Steam injection
  - Cyclic steam stimulation
  - Steam flooding
  - Steam-assisted gravity drainage
- In situ combustion
  - Forward combustion
  - Reverse combustion
  - Wet combustion

Chemical Flood

- Polymer
- Surfactant slug
- Alkaline
- Micellar
- Alkaline-surfactant-polymer (ASP)

Miscible

- CO_2 injection
- Lean gas
- N_2
- Rich gas
- WAG flood

Others

- MEOR
- Foam

EOR Methods

As a first step in selecting and implementing an enhanced oil recovery method, a screening study should be conducted to identify the appropriate EOR technique and evaluate its applicability to the reservoir. Taber et al. (1997) proposed screening criteria for enhanced oil recovery methods that were developed by compiling numerous data from EOR projects around the world. Based on extensive analysis of the collected data, the authors listed the optimum reservoir and oil characteristics that are required for implementing a successful EOR project in a particular field, as shown in Table 6.1.

There is a vast amount of literature on the subject of EOR and its variations, including excellent reference textbooks by Smith (1966), Willhite (1986), van Poollen (1980), Lake (1989), Stalkup (1983), and Prats (1983), among others. Brief description and discussion of some of the listed EOR methods are presented next.

6.3 THERMAL PROCESSES

Primary and secondary recovery from reservoirs containing heavy, low-gravity crude oils is usually a very small fraction of the initial oil-in-place. It is not uncommon for a heavy oil of 13°API gravity to have a viscosity of 2000 cp at a reservoir temperature of 110 °F and have a viscosity of only 60 cp at 220 °F. This potential 33-fold reduction in oil viscosity corresponds to a 33-fold increase in oil production rate. The temperature in the reservoir can be raised by injecting a hot fluid or by generating thermal energy by burning a portion of the oil-in-place.

6.3.1 Cyclic Steam Stimulation

The cyclic steam stimulation (CSS) method, also known as "huff-and-puff" or "steam soak," consists of three stages:

- injection;
- soaking; and
- production.

In the initial stage, steam is injected into a well at a relatively high injection rate for approximately 1 month. At the end of the injection period, the well is shut in for a few days (approximately 5 days) to allow "steam soaking" to heat the oil in the area immediately around the wellbore. The well is then put on production until it reaches the economic flow rate and at this point, the entire cycle is repeated. The steam injection and soak may be repeated four to five times or until the response to stimulation diminishes to noneconomic level. In general, the process can be quite effective, especially in the first few cycles. Stimulating the well by the huff-and-puff process significantly improves oil rate by three means:

- Removing accumulated asphaltic and/or paraffinic deposits around the wellbore, resulting in an improvement of the permeability around the wellbore (i.e., favorable skin factor).
- Radically decreasing the oil viscosity, which in turn improves oil mobility and well productivity.
- Increasing the thermal expansion of the oil, which impacts the oil saturation and its relative permeability.

Many initial applications result in production increases considerably greater than those predicted by model studies. This is mainly due to well cleanup and permeability improvement around the wellbore. The improvement in the production rate associated with the decrease in oil viscosity and the removal of deposits can be approximated by applying the following simplified assumptions:

- Reservoir has been heated out to a radius "r_{hot}" to a uniform temperature.
- The heated oil viscosity out to radius "r_{hot}" is represented by $(\mu_o)_{hot}$ as compared with original oil viscosity of $(\mu_0)_{cold}$.
- Improved skin factor of S_{hot} as compared with original skin factor of S_{cold}.

The pressure drop between the drainage radius "r_e" and wellbore radius "r_w" can be expressed by:

$$(\Delta p)_{hot} = (p_e - p) + (p - p_{wf}) \tag{6.2}$$

TABLE 6.1 Summary of Screening Criteria for EOR Methods

Process	Crude Oil	Reservoir
N_2 and flue gas	• >35° API • <1.0 cp • High percentage of light hydrocarbons	• $S_o > 40\%$ • Formation: SS or carbonate with few fractures • Thickness: relatively thin unless formation is dipping • Permeability: not critical • Depth > 6000 ft • Temperature: not critical
Chemical	• >20° API • <35 cp • ASP: organic acid groups in the oi are need	• $S_o > 35\%$ • Formation: SS preferred • Thickness: not critical • Permeability > 10 md • Depth<9000 ft (function of temperature) • Temperature < 200° F
Polymer	• >15° API • <100 cp	• $S_o > 50\%$ • Formation: SS but can be used in carbonates • Thickness: not critical • Permeability > 10 md • Depth < 9000 ft • Temperature < 200° F
Miscible CO_2	• >22° API • <10 cp • High percentage of intermediate components (C_5–C_{12})	• So > 20% • Formation: SS or carbonate • Thickness: relatively thin unless dipping • Permeability: not critical • Depth: depends on the required minimum miscibility pressure "MMP"
First-contact miscible flood	• >23° API • <3 cp • High C_m	• $S_o > 30\%$ • Formation: SS or carbonate with min fractures • Thickness: relatively thin unless formation is dipping • Permeability: not critical • Depth > 4000 ft • Temperature: can have a significant effect on MMP
Steam flooding	• 10–25° API • <10,000 cp	• $S_o > 40\%$ • Formation: SS with high permeability • Thickness > 20 ft • Permeability > 200 md • Depth < 5000 ft • Temperature: not critical
In situ combustion	• 10–27 °API • <5000 cp	• $S_o > 50\%$ • Formation: SS with high porosity • Thickness > 10 ft • Permeability > 50 md • Depth < 12,000 ft • Temperature > 100 °F

where

p = pressure oil radius, r
p_e = average reservoir pressure

Applying Darcy's equation for pressure drop in the unheated and heated drainage areas, gives:

$$(\Delta p)_{cold} = \frac{(q_o)_{cold}(\mu_o)_{cold}\left[\ln(r_e/r_w) + S_{cold}\right]}{k_o h} \quad (6.3)$$

$$(\Delta p)_{cold} = \frac{(q_o)_{cold}(\mu_o)_{cold}[\ln(r_e/r_h) + S_{cold}]}{k_o h} + \frac{(q_o)_{hot}(\mu_o)_{hot}[\ln(r_{hot}/r_w) + S_{hot}]}{k_o h} \quad (6.4)$$

Assuming that the pressure drop across the radial system will be the same for the hot or cold reservoir case, Eqs (6.2)–(6.4) can be arranged to give:

$$(q_o)_{hot} = (q_o)_{cold}\left[\frac{(\mu_o)_{cold}[\ln(r_e/r_w) + S_{cold}]}{(\mu_o)_{hot}[\ln(r_{hot}/r_w) + S_{hot}] + (\mu_o)_{cold}[\ln(r_e/r_{hot}) + S_{cold}]}\right]$$

The above expression shows that the increase in the well productivity is attributed to the combined reduction in oil viscosity and skin factor.

After several applications of steam cycling process, the huff-and-puff application is converted to a steam flooding project.

6.3.2 Steam Flooding (Steam Drive)

Steam flooding is a pattern drive, a process similar to water flooding in that a suitable well pattern is chosen and steam is injected into a number of wells, while oil is produced from adjacent wells. Most steam floods are traditionally developed on ±5-acre spacing. The recovery performance from steam flooding depends highly on the selected flooding pattern, pattern size, and reservoir characteristics. The steam flood project typically proceeds through four phases of development:

- reservoir screening;
- pilot tests;
- fieldwide implementation; and
- reservoir management.

Most reservoirs that are subject to successful huff-and-puff operations are considered good candidates for steam flood. The process involves continuous injection of system to form a steam zone around the injector that continues to advance in the reservoir with injection. In typical steam drive projects, the injected fluid contains 80% steam and 20% water, i.e., steam quality of 80%. The majority of the steam drive field applications are typically conducted jointly with the huff-and-puff process, where the process is conducted on producing wells, particularly when the oil is too viscous to flow before the heat from the steam injection wells arrives.

As steam moves through the reservoir between the injector and producer, it creates several regions of different temperatures and oil saturations, as shown conceptually in Figure 6.4. The illustration identifies the following five regions, each with the associated temperature and oil saturation profiles:

- *Steam zone "region A"*

  As the steam enters the pay zone, it forms a steam-saturated zone around, with a temperature that is nearly equal to that of the injected steam. A typical temperature profile for the steam flood is shown by the upper curve of Figure 6.4. The profile shows the gradual transition from the steam temperature at the injection well to the reservoir temperature at the producing well. Due to the high temperature in region A, the oil saturation is reduced to its lowest saturation, as shown by lower saturation profile curve of Figure 6.4. This drastic reduction in the oil saturation is attributed to the following:
  - significant improvement in oil mobility by reducing viscosity;
  - steam distillation and vaporization of the lighter component in the crude oil. In the steam zone, the hydrocarbon recovery by

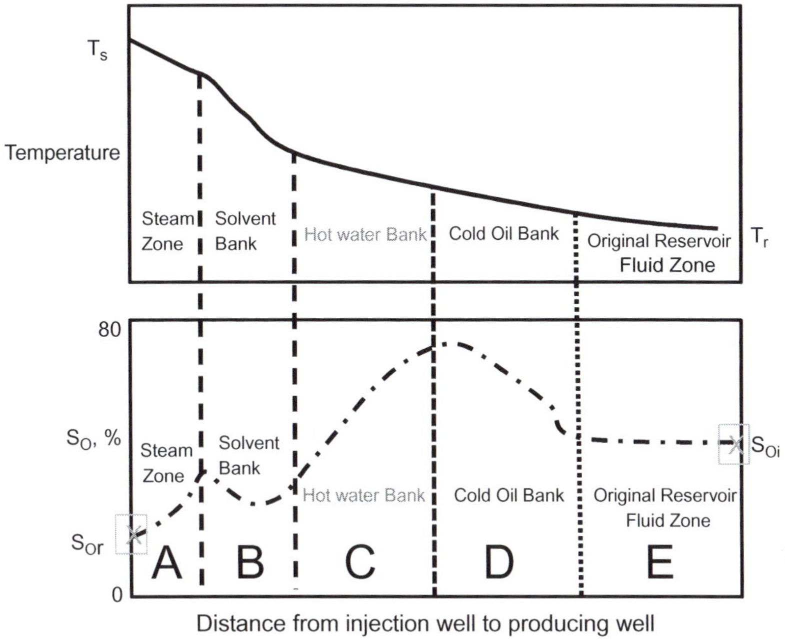

FIGURE 6.4 Temperature and oil saturation profiles.

steam is greater for lighter oils because they contain a greater fraction of steam-distillable components.

- *Hot condensate zone "regions B and C"*

  The hot condensate zone can be divided into a solvent bank "B" and hot water bank "C." As the steam zone moves away from the injector, its temperature drops due to the heat loss to the surrounding formation and at some distance from the injection well, the steam and some of the vaporized hydrocarbon vapor condense to form the hot condensate zone, i.e., regions B and C. The hot condensate zone can be described as a mix of solvent bank (condensed hydrocarbon fluid bank) and hot water bank. In this hot condensate zone, the solvent bank extracts additional oil from the formation to form a miscible hydrocarbon-slug drive that is miscible with the initial oil-in-place. This miscible displacement contributes significantly to the ultimate oil recovery process by steam injection.

- *Oil bank "region D"*

  As the mobilized oil is displaced by the advancing steam and hot water fronts, an oil bank with higher oil saturation than the initial saturation is formed in region D. The zone is characterized by a temperature profile ranging from the hot condensate zone temperature to that of the initial reservoir temperature.

- *Reservoir fluid zone "region E"*

  Region "E" essentially represents that portion of the reservoir that has not been affected or contacted by the steam. The region contains the reservoir fluid system that exists at the initial reservoir condition in terms of fluid saturations and original reservoir temperature.

Steam Recovery Mechanisms. Under steam injection, the crude oil is recovered under several combined recovery mechanisms, all with different degrees of contribution and importance. Essentially, there are five driving mechanisms that have been identified as the main driving forces:

- viscosity reduction;
- thermal expansion and swelling of the oil;
- steam distillation;
- solution gas drive;
- miscible displacement.

Figure 6.5 illustrates the contribution of each mechanism to the overall recovery by steam flooding of heavy oil. Each mechanism is discussed briefly below.

Viscosity Reduction. The decrease in oil viscosity with increasing temperature is perhaps the most important driving mechanism for recovering heavy oils. The net result of increasing temperature is the improvement in the mobility ratio "*M*," as defined previously by:

$$M = \left(\frac{k_w}{k_o}\right)\left(\frac{\mu_o}{\mu_w}\right)$$

With lower viscosity, the displacement and areal sweep efficiencies are improved considerably. As the oil is displaced from the high-temperature region to an area where the temperature may be considerably lower, the oil viscosity increases again and as a result, the rate of advance oil flow is reduced. Consequently, a large amount of oil accumulates to form an oil bank. This bank, often observed when steam flooding heavy oils, is responsible for high oil production rates and low water-oil ratios prior to heat breakthrough at the producing well.

Thermal Expansion. Thermal expansion is an important recovery mechanism in the hot condensate region with an oil recovery that depends highly on:

- initial oil saturation;
- type of the crude oil;
- temperature of the heated zone.

As the oil expands with increasing temperature, its saturation increases and it becomes more mobile. The amount of expansion depends on the composition of the oil. Because the ability of the oil to expand is greater for light oils than heavy oils, the thermal expansion is probably more effective in recovering light oils. In general, the recovery contribution from the thermal expansion ranges between 5% and 10%.

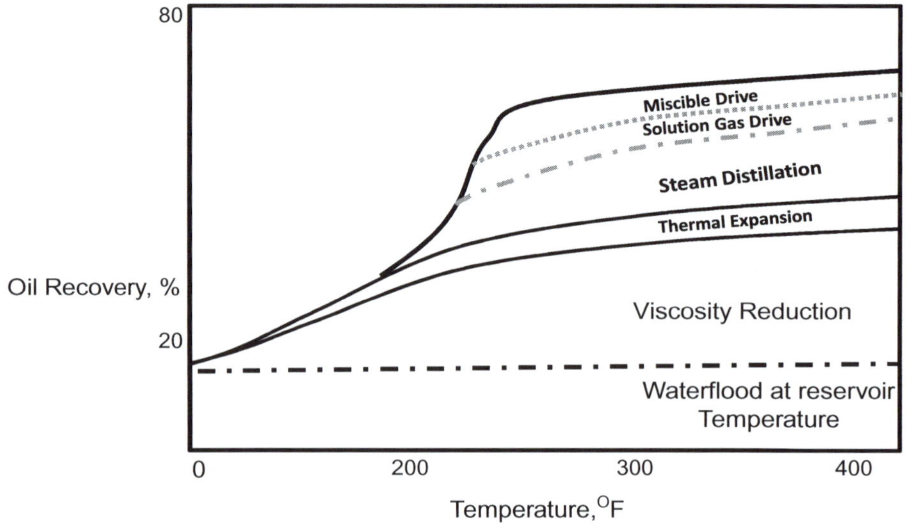

FIGURE 6.5 Contribution of steam flooding mechanisms to oil recovery.

Steam Distillation. Steam distillation is the main recovery mechanism in the steam zone. The distillation process involves the vaporization of the relatively light components in the crude oil to form a mixture of steam and condensable hydrocarbon vapors. Some of the hydrocarbon vapors will condense along with the steam and mix with the residual oil trapped by the advancing hot condensate region ahead of the steam zone. This mixing will create a solvent bank behind the hot condensation front. The distillation of the crude oil bypassed by the advancing hot condensation zone can result in very low ultimate residual oil saturations in the steam-swept zone. In principle, residual oil saturations can be essentially zero where the original crude oil has been mixed with large volumes of hydrocarbon condensate.

Solution Gas Drive. With increasing temperatures, solution gas is liberated from the oil. The liberated solution gas expands proportionally with the pressure decline gradient between the injector and the producer. This expansion in the gas phase provides additional driving forces that contribute to the oil recovery process. In addition, CO_2 can be generated during the injection process either from the high-temperature reaction with the formation containing CO_2 or from the oil that contains CO_2. If large quantities of CO_2 are liberated, it can contribute to an additional oil recovery due to its ability to reduce the viscosity of the oil as the liberated gas expands and contacts more of the original oil-in-place.

Miscible Displacement. In the hot condensate zone, the solvent bank generated by the steam zone extracts additional oil from the formation to form an "oil-phase miscible drive." Essentially, the steam zone "manufactures" a miscible-oil slug that can displace the oil it contacts with 100% displacement efficiency. The additional oil recovery due to this miscible displacement ranges between 3% and 5% of the original oil-in-place.

The main advantage of steam injection over other EOR methods is that steam flooding can be applied to a wide variety of reservoirs. However, there are two limiting factors that must be evaluated before considering steam flooding in a specific reservoir:

The total reservoir depth should be less than 5000 ft. This depth limitation is imposed by the critical pressure of steam (3202 psia).

The reservoir net pay should be greater than 25 ft. The limitation should be considered when evaluating a reservoir for steam flooding in order to reduce the heat loss to the base and cap rock of the pay.

In general, the following guideline summarizes the required screening criteria for steam injection:

oil viscosity < 3000 cp;
depth < 1500 ft;
API gravity < 30;
permeability > 300 md.

Thermal Properties of Fluids. The design of steam flood projects requires clear understanding of the physical and thermal properties of the steam, reservoir fluids, and solids. These properties are essential in performing the following two steam flooding calculations:

- estimating heat losses in order to properly compute the capacities of steam-generating equipment;
- evaluating the physical oil recovery displacement mechanisms by both steam and hot water.

Some of the needed thermal properties of fluids and solids are summarized next.

Thermal Properties of Steam and Liquids. When 1 lb of water at an initial temperature "T_i" is heated at a constant pressure "P_s," it will attain a maximum temperature "T_s," called the saturation temperature, before it is converted into steam. The amount of heat absorbed by the water "h_w" is called "*enthalpy*" or "*sensible heat*" and is given by the following relationship:

$$h_w = C_w(T_s - T_i) \tag{6.5}$$

where

C_w = specific heat of water, BTU/lb-°F
T_s = saturation temperature, °F
T_i = initial water temperature, °F
h_w = heat content of the saturated water (*enthalpy* of saturated water), BTU/lb

If 1 lb of saturated water at T_s is further heated at the same saturation pressure P_s, it will continue to absorb heat without a change in temperature until it is totally converted to steam. The amount of additional heat that is required to convert the water to steam (vapor) is called the enthalpy of vaporization or latent heat of steam "L_V," with total heat content h_s as given by:

$$h_s = h_w + L_V \tag{6.6}$$

where

h_s = steam heat content or enthalpy, BTU/lb
L_V = latent heat, BTU/lb

Further heating of the steam to a temperature T_{sup} above T_s, while maintaining the pressure at P_s, converts the steam from saturated to superheated steam. The heat content (enthalpy) h_{sup} of the superheated steam is given by:

$$h_{sup} = h_s + C_s(T_{sup} - T_s) \tag{6.7}$$

where

C_s = average specific heat of steam in the temperature range of T_s to T_{sup}

In the case of a wet steam with a steam quality of "X," the heat content (enthalpy) of the wet steam is given by:

$$h_s = h_w + XL_v \tag{6.8}$$

The volume of 1 lb (specific volume "V") of wet steam is given by:

$$V = (1 - X)\ V_w + X\ V_s \tag{6.9}$$

where

V_w = volume of 1 lb of saturated water
V_s = volume of dry steam

Standard steam tables are available that list steam properties; when not available, the following expressions can be used to obtain estimates of steam properties:

$$T_s = 115.1\ P_s^{0.225} \tag{6.10}$$

$$h_w = 91\ P_s^{0.2574} \tag{6.11}$$

$$L_v = 1318\ P_s^{-0.08774} \tag{6.12}$$

$$h_s = 1119\ P_s^{0.01267} \tag{6.13}$$

$$V_s = 363.9\ P_s^{-0.9588} \tag{6.14}$$

Another important property of steam is viscosity. For steam at pressure below 1500 psia, the following linear relationship expresses the steam viscosity μ_{steam} as a function of temperature:

$$\mu_{steam} = 0.0088 + 2.112 \times 10^{-5}(T - 492)$$

Example 6.1

Find the enthalpy of 80% quality steam at 1000 psia.

Solution

Step 1. Estimate the water sensible heat h_w by applying Eq. (6.11):

$$h_w = 91\ P_s^{0.2574} = 91(1000)^{0.2574} = 538.57\ \text{BTU/lb}$$

Step 2. Estimate the latent heat L_V by applying Eq. (6.12):

$$L_V = 1318\ P_s^{-0.08774} = 1318\ (1000)^{-0.08774} = 718.94\ \text{BTU/lb}$$

Step 3. Calculate the steam enthalpy h_s by applying Eq. (6.8):

$$h_s = h_w + XL_v = 538.57 + (0.8 \times 718.94) = 1113.71\ \text{BTU/lb}$$

The change in the temperature that is taking place in the reservoir during steam flooding has a substantial impact on the reservoir fluid properties. One of the most important fluid properties that are strongly temperature dependent is the oil viscosity. The importance of the viscosity is due to the fact that the flow of a given fluid in a porous media is inversely proportional to its viscosity. The considerable drop in heavy oil viscosities at elevated temperatures makes this parameter particularly significant in a steam flood.

Steam injection reduces the viscosity of both oil and water, while it increases that of gas. Ahmed (1989 and 2006) lists several correlations that can be used for estimating the crude oil viscosity when laboratory viscosity measurements are not available. However, these correlations are based on the assumption that the crude oil does not change in character either through cracking or distillation effects at high temperatures or through the development of suspended solids (e.g., asphaltene). Assuming heavy oil system with an oil gravity of °API, the viscosity of the oil at any temperature can be estimated from the following expression:

$$\mu_o = 220.15 \times 10^9 \left(\frac{5T}{9}\right)^{-3.556} [\log(\text{API})]^z$$

with

$$z = \left[12.5428 \times \log(5T/9)\right] - 45.7874$$

where the temperature "T" is expressed in °R.

The viscosity of the water at temperature "T" can be estimated from the following relationship:

$$\mu_w = \left(\frac{2.185}{0.04012(T - 460) + 0.0000051535(T - 460)^2 - 1}\right)$$

Another important property that is an integral part of oil recovery calculations by steam injection is the specific heat in BTU/lb-°F. In general, the specific heat "C" is defined by the ratio of the amount of heat required to raise the temperature of a unit mass of a substance by one unit of temperature to the amount of heat required to raise the temperature of a similar mass of a reference material, usually water, by the same amount. The specific heat of the saturated water can be approximated by the following expression:

$$C_w = 1.3287 - 0.000605\,T + 1.79(10^{-6})(T - 460)^2$$

where temperature T is in °R. For crude oil, specific heat can be approximated by the expression:

$$C_o = \left(\frac{0.022913 + 56.9666 \times 10^{-6}\,(T - 460)}{\sqrt{\rho_o}}\right)$$

where

C_w = specific heat of water, BTU/lb-°F
C_o = specific heat of oil, BTU/lb-°F
T = temperature, °R
ρ = density of oil in lb/ft^3

Heat Transfer During the Steam Injection Processes. Heat transfer is the movement of thermal energy from one position to another. Heat flow always occurs from a higher-temperature region to a lower-temperature region as described by the second law of thermodynamics. There are three distinct modes of heat transfer:

- Conduction—transfer of thermal energy between objects in direct physical contact.
- Convection—transfer of energy between an object (usually a fluid) and surroundings by fluid motion.
- Radiation—transfer of energy from or to a body by the emission of absorption of electromagnetic radiation. Radiation heat transfer in a reservoir during steam injection is not significant because there is insufficient void space for the electromagnetic radiation to propagate.

Conduction. Thermal conduction is responsible for heat losses to the overburden and underburden strata, as shown in Figure 6.6.

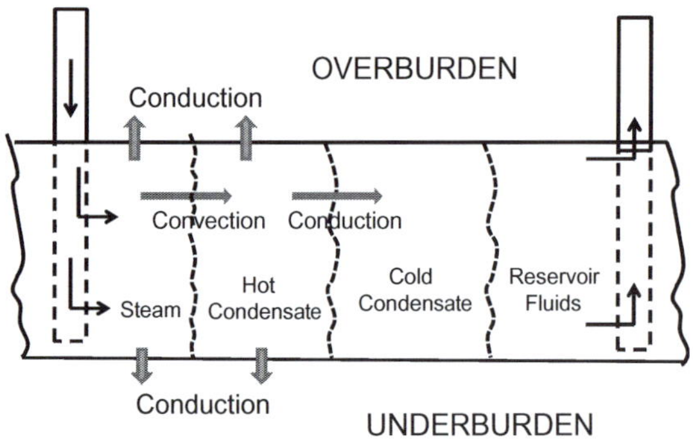

FIGURE 6.6 Steam flood heat transfer and losses.

Thermal conduction can also be important within the reservoir when fluid velocities are low. Conduction heat transfer is described by the following expression:

$$q_{\text{heat}} = \lambda A \frac{dT}{dx}$$

where

q_{heat} = rate of heat transfer in the x-direction, BTU/hour
λ = thermal conductivity, BTU/hour-ft-°F
A = area normal to the x-direction, ft^2
T = temperature, °F
x = length along the direction of heat transfer, ft

The thermal conductivity "λ" measures the ability of a solid to transmit 1 BTU of thermal energy in 1 hour through an area of 1 ft^2, if a temperature difference of 1°F is imposed across a thickness of 1 ft. Thermal conductivity of a porous rock increases with increase in:

- rock density;
- liquid saturation;
- pressure;
- saturating liquid thermal conductivity.

while it decreases with an increase in:

- temperature
- porosity

Listed below are selected values for thermal conductivities of various dry and saturated rocks.

Rock	Density, lb/ft^3	Specific Heat, BTU/lb-°F	Thermal Conductivity, BTU/hour-ft-°F	Thermal Diffusivity, ft^2/hour
(a) **Dry rocks**				
Sandstone	130	0.183	0.507	0.0213
Silty sand	119	0.202	0.400	0.0167
Siltstone	120	0.204	0.396	0.0162
Shale	145	0.194	0.603	0.0216
Limestone	137	0.202	0.983	0.0355
Sand (fine)	102	0.183	0.362	0.0194
Sand (coarse)	109	0.183	0.322	0.0161
(b) **Water-saturated**				
Sandstone	142	0.252	1.592	0.0445
Silty sand	132	0.288	1.500	0.0394
Siltstone	132	0.276	1.510	0.0414
Shale	149	0.213	0.975	0.0307
Limestone	149	0.266	2.050	0.0517
Sand (fine)	126	0.339	1.590	0.0372
Sand (coarse)	130	0.315	1.775	0.0433

Tikhomirov (1968) presented a relationship to estimate the thermal conductivity of water-saturated rocks as given by:

$$\lambda_R = \frac{6.36[\exp(0.6\rho_r + 0.6\,S_w)}{(0.556\,T + 255.3)^{0.55}} \tag{6.15}$$

where:

ρ_r = dry rock density, g/cm^3
S_w = water saturation, fraction
T = temperature, °F

In thermal recovery calculations, the value of λ_R is in the range of 1.0–1.4 BTU/hour-ft-°F.

The rate at which the thermal front propagates through the formation by conduction is governed by thermal diffusivity "D." The thermal diffusivity is defined as the ratio of thermal conductivity "λ_R" to the volumetric heat capacity of the rock and can be approximated by the following expression:

$$D = \frac{\lambda_R}{\rho_r\,C_r} \tag{6.16}$$

where

λ_R = rock thermal conductivity, BTU/hour-ft-°F
D = thermal diffusivity, ft^2/hour
ρ_r = rock density, lb/ft^3
C_r = specific heat of the rock, BTU/lb$_m$-°F

In most thermal recovery calculations, the value of "D" is about 0.04 ft^2/hour. The product $(\rho_r C_r)$ is the volumetric heat capacity of the rock, i.e., overburden.

The volumetric heat capacity of the saturated formation, "M" is given by:

$$M = \phi(S_o \rho_o C_o + S_w \rho_w C_w) + (1 - \phi)\rho_r C_r \qquad (6.17)$$

where

M = volumetric heat capacity of the formation, BTU/ft^3-°F
ρ = density at reservoir temperature, lb/ft^3
C = specific heat, BTU/lb-°F
S = saturation, fraction
ϕ = porosity, fraction

The subscripts o, w, and r refer to oil, water, and rock matrix, respectively. It should be pointed out that the heat capacity of the rock matrix, i.e., $(1 - \phi)\ \rho_r C_r$, accounts for about 75% of the total heat capacity, i.e., 75% of the heat injected into the formation is used for heating rock matrix.

Prediction of Steam Flood Performance

Thermal recovery processes such as cyclic and continuous steam injection involve heat and mass transport that can be described mathematically by a set of differential equations. These mathematical expressions can be solved by means of numerical simulation. The numerical simulation offers numerous advantages, including:

- ability to incorporate irregular injection patterns;
- taking into account the reservoir heterogeneity;
- imposing individual well constraints;
- changes in hydrocarbon base behavior;
- mechanism.

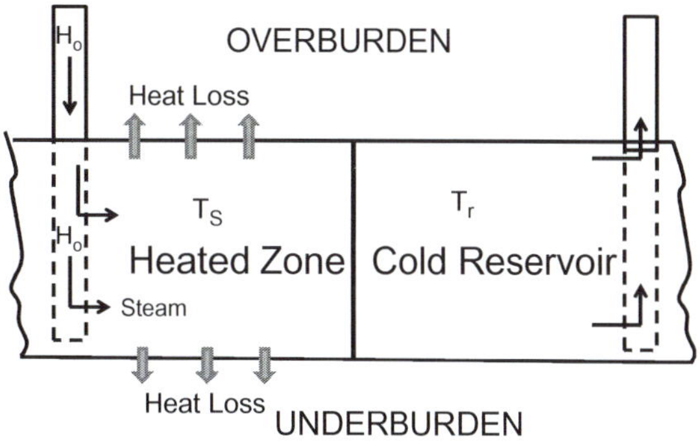

FIGURE 6.7 Marx and Langenheim conceptual model.

Numerical simulation, however, requires extensive reservoir data and lengthy calculations and specialized thermal software. Analytical models that yield acceptable results and serve as simple computational tools for practicing engineers are thus desirable.

The earliest model to be applied to steam flooding was that of Marx and Langenheim (1959) for predicting the growth of the steam zone in the reservoir during steam injection into a single well. Marx and Langenheim "M-L" applied the following assumptions in developing the model:

- Steam penetrates a single layer of uniform thickness.
- Heat losses are normal to the boundaries of the steam zone (i.e., at the cap rock and base rock) as shown in Figure 6.7.
- Temperature of the heated zone remains at steam temperature "T_s" and falls to reservoir temperature immediately outside the heated zone, as shown in Figure 6.8.

Marx and Langenheim heating model is based on the assumption that the injection of hot fluid into a well is at constant heat rate "H_o" and constant temperature "T_s," as shown conceptually in Figure 6.7. The authors proposed the following expression for calculating the heat injection rate:

$$H_o = \left(\frac{5.615\ \rho_w Q_{inj}}{24}\right)[h_s - C_w(T_r - 32)] \qquad (6.18)$$

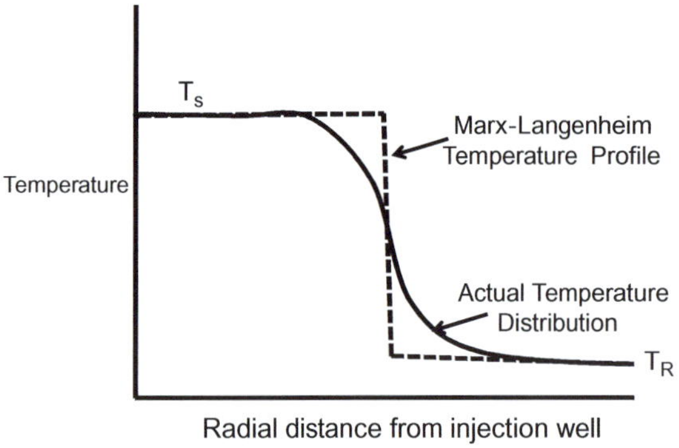

FIGURE 6.8 Marx and Langenheim temperature profile.

where

Q_{inj} = feed water rate into the steam generator, bbl/day
H_o = heat injection rate, BTU/hour
T_r = reservoir temperature, °F
ρ_w = water density, lb/ft^3
h_s = steam heat content or enthalpy as expressed by Eq. (6.8), BTU/lb
C_w = specific heat of water, BTU/lb-°F

Example 6.2

The feed cold-water rate into a generator is 1000 bbl/day. The outlet steam pressure, temperature, and quality are 1271 psia, 575 °F, and 0.73, respectively. Other pertinent reservoir and fluid properties are given below:

$C_w = 1.08$ BTU/lb-°F
$\rho_w = 62.4$ lb/ft^3
$T_r = 120$ °F

Calculate the heat injection rate in BTU/hour.

Solution

Step 1. Estimate the water sensible heat h_w by applying Eq. (6.11):

$$h_w = 91\ P_s^{0.2574} = 91(1273)^{0.2574} = 573.1 \text{ BTU/lb}$$

Step 2. Estimate the latent heat L_v by applying Eq. (6.12):

$$L_v = 1318\ P_s^{-0.08774} = 1318(1273)^{-0.08774}$$
$$= 861.64 \text{ BTU/lb}$$

Step 3. Calculate the steam enthalpy h_s by applying Eq. (6.8):

$$h_s = h_w + XL_v = 573.1 + (0.78 \times 861.64)$$
$$= 1245.2 \text{ BTU/lb}$$

Step 4. Calculate the heat injection rate H_o by applying Eq. (6.18):

$$H_o = \left(\frac{5.615\rho_w Q_{inj}}{24}\right)[h_s - C_w(T_r - 32)]$$
$$= \left(\frac{5.615(62.4)1000}{24}\right)[1245.2 - 1.08(120 - 32)]$$
$$= 16.79 \text{ MMBTU/hour}$$

Marx and Langenheim applied the heat balance on a single horizontal layer with uniform and constant properties to estimate the areal extent and propagation of the heated zone "$A_s(t)$" as a function of time "t," to give:

$$A_s(t) = \left[\frac{H_o MhD}{(4)\ 43,560(\lambda_R)^2 \Delta T}\right] G(t_D) \quad (6.19)$$

with the function $G(t_D)$ defined by:

$$G(t_D) = e^{t_D}\operatorname{erfc}(\sqrt{t_D}) + 2\sqrt{\frac{t_D}{\pi}} - 1 \quad (6.20)$$

The dimensionless time t_D is defined by the expression:

$$t_D = \left[\frac{4(\lambda_R)^2}{M^2 h^2 D}\right] t \quad (6.21)$$

where

$A(t)$ = cumulative heated area at time t, acres
$\operatorname{erfc}(x)$ = complementary error function
H_o = constant heat injection rate, BTU/hour
λ_R = thermal conductivity of the cap and base rock, BTU/hour-ft-°F
h = reservoir thickness, ft
t = time, hours

TABLE 6.2 Marx and Langenheim Auxiliary Functions

(1) t_D	(2) E_h	(3) G	(4) $e^{t_D}\text{erfc}\sqrt{t_D}$	(5) erfc t_D
0.0	1.0000	0	1.0000	1.0000
0.01	0.9290	0.0093	0.8965	0.9887
0.0144	0.9167	0.0132	0.8778	0.9837
0.0225	0.8959	0.0202	0.8509	0.9746
0.04	0.8765	0.0347	0.8090	0.9549
0.0625	0.8399	0.0524	0.7704	0.9295
0.09	0.8123	0.0731	0.7346	0.8987
0.16	0.7634	0.1221	0.6708	0.8210
0.25	0.7195	0.1799	0.6157	0.7237
0.36	0.6801	0.2488	0.5678	0.6107
0.49	0.6445	0.3158	0.5259	0.4883
0.64	0.6122	0.3918	0.4891	0.3654
0.81	0.5828	0.4721	0.4565	0.2520
1.00	0.5560	0.5560	0.4275	0.1573
1.44	0.5087	0.7326	0.3785	0.0417
2.25	0.4507	0.7783	0.3216	0.0015
4.00	0.3780	1.5122	0.2554	0.0000
6.25	0.3251	2.0318	0.2108	
9.00	0.2849	2.5641	0.1790	
16.00	0.2282	3.6505	0.1370	
25.00	0.1901	4.7526	0.1107	
36.00	0.1629	5.8630	0.0928	
49.00	0.1424	6.9784	0.0798	
64.00	0.1265	8.9070	0.0700	
81.00	0.1138	9.2177	0.0623	
100.00	0.1034	10.3399	0.0561	

ΔT = difference between steam zone temperature and reservoir temperature, i.e. $T_s - T_r$, °F

D = overburden and underburden thermal diffusivity as defined by Eq. (6.16), ft²/hour, i.e.:

$$D = \left(\frac{\lambda_R}{\rho_r C_r}\right)_{\text{Overburden}}$$

M = reservoir (formation) volumetric heat capacity, BTU/ft³-°F, as given by Eq. (6.17), i.e.:

$$M = \phi(S_o\rho_o C_o + S_w\rho_w C_w) + (1-\phi)\rho_r C_r$$

Values of "G," $e^{t_D}\text{erfc}\sqrt{t_D}$, and erfc t_D as a function of the dimensionless time are listed columns 3–5, respectively, in Table 6.2.

Effinger and Wasson (1969) proposed the following mathematical expression to approximate $e^{t_D}\text{erfc}\sqrt{t_D}$:

$$e^{t_D}\text{erfc}\sqrt{t_D} = 0.254829592y - 0.284496736y^2 + 1.42143741y^3 - 1.453152027y^4 + 1.061405429y^5 \quad (6.22)$$

with

$$y = \frac{1}{1 + 0.3275911\sqrt{t_D}}$$

Several important derivatives of Marx and Langenheim heating model include the following steam flood performance relationships:

(a) Oil flow rate, q_o

Assuming that all the movable oil is displaced in the heated area, the oil displacement flow rate is given by:

$$q_{od} = 4.275\left(\frac{H_o\phi\ (S_{oi} - S_{or})}{M\ \Delta T}\right)(e^{t_D}\operatorname{erfc}\sqrt{t_D}) \quad \textbf{(6.23)}$$

where

q_{od} = displaced oil rate in bbl/day
S_{oi} = initial oil saturation
S_{or} = residual oil saturation
H_o = heat injection rate, BTU/hour
ϕ = porosity

(b) Instantaneous steam-oil ratio "SOR" is given by the ratio of the steam injection rate (cold water equivalent) to that of the displaced oil rate; i.e.:

$$\text{SOR} = \frac{i_{steam}}{q_{od}}$$

(c) Total heat injected "H_{inj}":

$$H_{inj} = H_o t$$

where

H_{inj} = total heat injected, BTU
t = total injection time, hour

(d) Total rate of heat lost to the adjacent formations "H_{lost}"

The rate of heat lost in BTU/hour is given by the expression:

$$H_{lost} = H_o(1 - e^{t_D}\operatorname{erfc}\sqrt{t_D}) \quad \textbf{(6.24)}$$

(e) Total remaining heat in the reservoir "H_r"

The heat remaining in the heated zone in "BTU" is given by:

$$H_r = \left[\frac{H_o M^2 h^2 D}{4(\lambda_R)^2}\right] G(t_D) \quad \textbf{(6.25)}$$

(f) Reservoir heat efficiency "E_h"

Marx and Langenheim defined the reservoir thermal (heat) efficiency as the ratio of heat remaining in the reservoir to the total heat injected at time t, i.e.:

$$E_h = \frac{H_r}{H_o t} = \left[\frac{G(t_D)}{t_D}\right]$$

or equivalently as:

$$E_h = \frac{1}{t_D}\left[e^{t_D}\operatorname{erfc}(\sqrt{t_D}) + 2\sqrt{\frac{t_D}{\pi}} - 1\right] \quad \textbf{(6.26)}$$

Values of the reservoir heat efficiency E_h are conveniently listed in Table 6.2.

Example 6.3

Using the data and results from Example 6.2 with a constant heat injection rate of 16.78 MMBTU/hour, estimate the oil recovery performance using Marx and Langenheim method. The following additional data is available:

$\rho_o = 50.0$ lb/ft^3
$\rho_r = 167.0$ lb/ft^3
$\rho_w = 61.0$ lb/ft^3
$S_{oi} = 0.60$
$S_{or} = 0.10$
$S_w = 0.40$
$C_o = 0.50$ BTU/lb-°F
$C_w = 1.08$ BTU/lb-°F
$C_r = 0.21$ BTU/lb-°F
$T_r = 120$ °F
$T_s - 575$ °F
$\phi = 0.25$
$h = 40$ *ft*

Thermal diffusivity of base and rock "D" = 0.029 ft^2/hour BTU/lb-°F

Rock thermal conductivity $\lambda_R = 1.50$ BTU/hour-ft-°F

Solution:

Step 1. Calculate the reservoir volumetric heat capacity "M" from Eq. (6.17):

$$M = \phi(S_o\rho_o C_o + S_w\rho_w C_w) + (1-\phi)\rho_r C_r$$

$$M = 0.25(0.6\times50\times0.5 + 0.4\times61.0\times1.08) + (1-0.25)\times167\times0.21 = 36.64,\ \text{BTU/ft}^3-°F$$

Step 2. Perform the required recovery calculations in the following tabulated form:

Time, days	Time, hrs	Eq. (6.26) t_D	Eq. (6.21) E_h	Eq. (6.20) G	Eq. (6.22) e^{t}_{D} erfc $\sqrt{t_D}$	Eq. (6.19) $A_s(t)$, acres	Eq. (6.23) q_{od}	SOR
10	240	0.03468	0.88516	0.03069	0.82024	0.1305	472.5986	2.1160
30	720	0.10403	0.80642	0.08389	0.71937	0.3567	414.4810	2.4127
60	1440	0.20805	0.74212	0.15440	0.63889	0.6566	368.1123	2.7166
90	2160	0.31208	0.69912	0.21818	0.58682	0.9278	338.1078	2.9576
120	2880	0.41611	0.66639	0.27729	0.54825	1.1791	315.8879	3.1657
150	3600	0.52014	0.63987	0.33282	0.51773	1.4153	298.3007	3.3523
200	4800	0.69351	0.60442	0.41918	0.47799	1.7825	275.4050	3.6310
230	5520	0.79754	0.58678	0.46798	0.45867	1.9900	264.2736	3.7840
300	7200	1.04027	0.55265	0.57491	0.42220	2.4447	243.2585	4.1109
400	9600	1.38703	0.51515	0.71453	0.38349	3.0384	220.9583	4.5257
500	12000	1.73378	0.48593	0.84249	0.35435	3.5826	204.1655	4.8980
600	14400	2.08054	0.46211	0.96145	0.33127	4.0884	190.8666	5.2393
800	19200	2.77405	0.42494	1.17881	0.29644	5.0127	170.8029	5.8547
1000	24000	3.46757	0.39668	1.37551	0.27096	5.8492	156.1185	6.4054
1500	36000	5.20135	0.34721	1.80597	0.22843	7.6796	131.6138	7.5980
2000	48000	6.93513	0.31400	2.17766	0.20137	9.2602	116.0231	8.6190
3000	72000	10.40270	0.27040	2.81287	0.16767	11.9613	96.6090	10.3510
4000	96000	13.87027	0.24195	3.35594	0.14683	14.2706	84.5993	11.8204
5000	120000	17.33783	0.22138	3.83825	0.13232	16.3216	76.2408	13.1163
6000	144000	20.80540	0.20555	4.27657	0.12149	18.1855	69.9984	14.2860
7000	168000	24.27297	0.19285	4.68113	0.11301	19.9058	65.1110	15.3584
7300	175200	25.31324	0.18951	4.79699	0.11080	20.3985	63.8400	15.6642

Above results are presented graphically in Figure 6.9

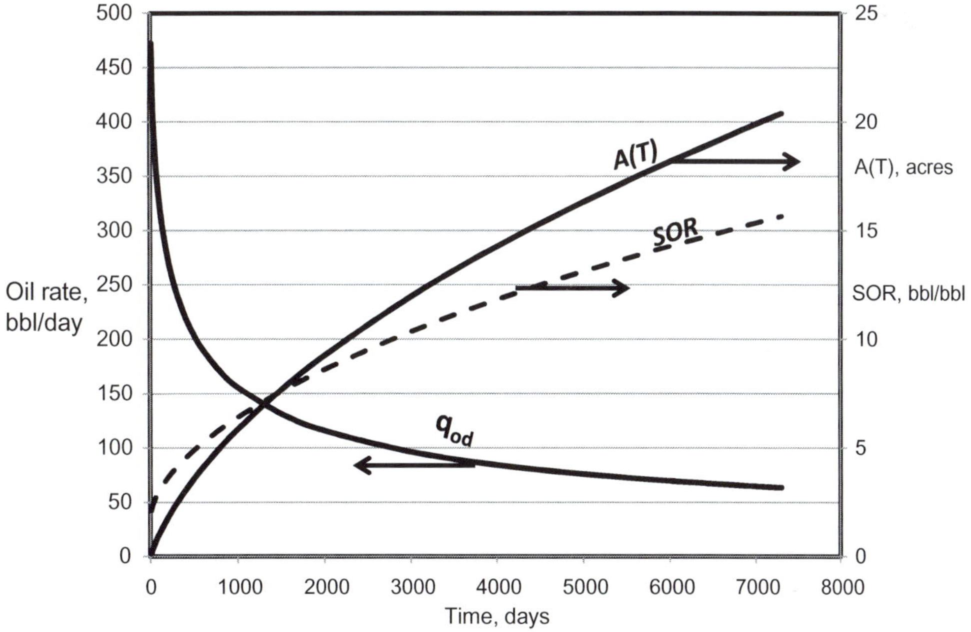

FIGURE 6.9 Marx and Langenheim performance profile.

Based on the observation that heated area measured in laboratory experiments tends to be lower than that predicted by Marx and Langenheim method, Mandl and Volek (1969) introduced the concept of the ***critical time*** "t_c" which identifies the time beyond which the zone downstream from the advancing hot condensation front is heated by a hot water moving through the condensation front. Prior to the critical time "t_c," all the heat in the reservoir is within the steam zone and performance results can be obtained by applying Marx and Langenheim method. The authors propose that when $t > t_c$, the heated area is given by:

$$A_s(t) = \frac{H_o MhD}{(4)\ 43{,}560(\lambda_R)^2 \Delta T} \left\{ G(\lambda_R) - \left[\sqrt{\frac{t_D - t_{cD}}{\pi}} \left(E_{hv} + \frac{[t_D - (t_{cD})^{-3}] e^{t_D} \operatorname{erfc}\sqrt{t_D}}{3} - \frac{t_D - t_{cD}}{3\sqrt{\pi t_D}} \right) \right] \right\}$$

The parameter "E_{hv}" represents the fraction of heat injected in latent form and is given by:

$$E_{hv} = \frac{1}{1 + (XL_v / C_w \Delta T)} \tag{6.27}$$

where

L_v = latent heat, BTU/lb
X = steam quality
ΔT = difference between steam zone temperature and reservoir temperature, i.e., $T_s - T_r$, °F
C_w = Average specific heat of water over the temperature range of ΔT, BTU/lb-°F

The following steps summarize the procedure for calculating the critical time t_c:

1. Calculate the fraction of heat injected in latent form from Eq. (6.27):

$$E_{hv} = \frac{1}{1 + (XL_v / C_w \Delta T)}$$

2. Calculate the critical complementary error function:

$$e^{t_{cD}} \operatorname{erfc}\sqrt{t_{cD}} = 1 - E_{hv} \tag{6.28}$$

3. Enter Table 6.2 with the value of $e^{t_{cD}} \operatorname{erfc}\sqrt{t_{cD}}$ and read the corresponding value of t_{cD}.
4. Calculate the critical time from Eq. (6.21), or:

$$t_c = \left[\frac{M^2 h^2 D}{4(\lambda_R)^2} \right] t_{cD}$$

Figure 6.10 shows a graph of the thermal efficiency of the steam zone "E_h" as a function of the dimensionless time "t_D" and "E_{hv}." The upper curve, i.e., $E_{hv} = 1.0$, follows Marx and Langenheim thermal efficiency as represented mathematically by Eq. (6.26) or the tabulated values listed in column 2 of Table 6.2. Marx and Langenheim heat efficiency is used as described previously in recovery calculations when the time "t" is less than the critical time "t_c." After the critical time, *the heat efficiency of the steam zone would follow the curve corresponding to* E_{hv} *as given by Eq. (6.28).*

Example 6.4

Using the data and results from Example 6.2, calculate the critical time

Solution

1. Calculate the fraction of heat injected in latent form from Eq. (6.27):

$$E_{hv} = \frac{1}{1 + (XL_V / C_w \Delta T)} = \frac{1}{1 + ((0.73)(861.64)/1.08(455))} = 0.439$$

2. Calculate the critical complementary error function:

$$e^{t_{cD}} \operatorname{erfc}\sqrt{t_{cD}} = 1 - E_{hv} = 1 - 0.439 = 0.561$$

3. Enter Table 6.2 with the value of $e^{t_{cD}} \operatorname{erfc}\sqrt{t_{cD}}$, i.e., 0.561, and read the corresponding value of t_{cD}, to give (by interpolation): $t_{cD} = 0.381$

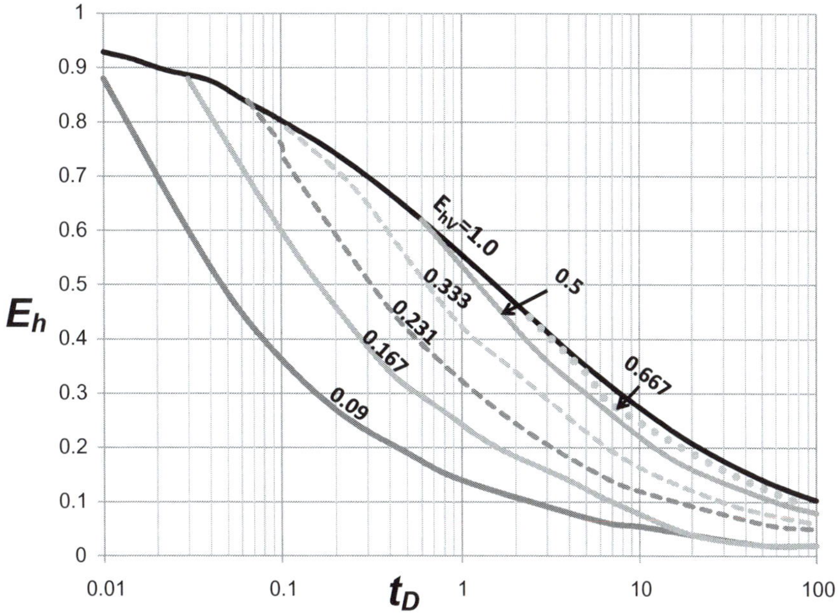

FIGURE 6.10 Fraction of heat injected remaining in the steam zone.

4. Calculate the critical time from Eq. (6.21), or:

$$t_c = \left[\frac{M^2 h^2 D}{4(\lambda_R)^2}\right] t_{cD} = \left[\frac{36.6^2(40^2)0.029}{4(1.5)^2}\right] 0.381$$
$$= 6906 \text{ hours} = 288 \text{ days}$$

The above example indicates that Marx and Langenheim heating model can be applied until the critical time of 288 days is reached with the values of Marx and Langenheim reservoir thermal (heat) efficiencies as calculated from Eq. (6.26) or read from the upper curve of Figure 6.10. After the critical time is reached, the thermal efficiency is read as a function of t_D tracing the appropriate E_{hv} curve.

It should be pointed out that the volume of the steam zone "V_s" is related to the fraction of the injected heat present in the steam zone, i.e., E_h, by:

$$V_s = \left[\frac{H_o t E_h}{43,560 M \Delta T}\right] t E_h \tag{6.29}$$

with the steam zone area as:

$$A_s(t) = \frac{V_s}{h} \tag{6.30}$$

where

V_s = volume of the steam zone as a function of time, acre-ft
H_o = heat injection rate, BTU/hour
t = time, hour
E_h = thermal efficiency of the steam zone at time "t"
$A_s(t)$ = areal extent of the steam zone at time t, acres
h = thickness, ft
M = the reservoir volumetric heat capacity "M" from Eq. (6.17), BTU/ft^3-°F

The cumulative oil produced "N_P" can then be calculated from the following relationship:

$$N_P = 7758\phi \frac{h_n}{h_t}(S_{oi} - S_{or})\ V_s$$

where

N_P = cumulative oil production, bbl
h_n = net thickness, ft
h_t = total thickness, ft

Field results indicate that the oil flow rate "q_{od}" as calculated from the Marx and Langenheim expressions generally overestimates steam flood performances as compared with observed field production data. To bring results into agreement with observed field data, the parameter "Γ" is included in steam flood performance equations, to give:

$$q_o = (q_{od})\Gamma \tag{6.31}$$

The parameter Γ is called capture efficiency and represents the fraction of oil displaced from the steam zone. The capture efficiency lies between 0.66 and 1. In some cases, a value of Γ greater than 1 suggests that more oil is produced than displaced from the steam zone. This is possible if significant gravity drainage forces influence and contribute to the production of the oil from outside the steam zone. The parameter Γ is usually set at 0.7. Jones (1981) and Chanadra and Damara (2007) suggest that the capture efficiency is a product of three dimensionless elements, each of which varies from a value of 0.0 to a value of 1. The capture efficiency is defined by:

$$\Gamma = A_{cD} V_{oD} V_{PD}$$

where

A_{cD} = dimensionless steam zone area
V_{oD} = volume of displaced oil produced, dimensionless
V_{PD} = fraction of pore volume filled with steam

These capturing parameters can be approximated by applying the following equations:

$$A_{cD} = \left[\frac{A_s(t)}{A[0.11 \ln (\mu_{oi}/100)]^{1/2}}\right]^2 \tag{6.32}$$

With the limit $0 \leq A_{cD} \leq 1.0$ and $A_{cD} = 1.0$ when $\mu_{oi} \leq 100$ cp. This dimensionless steam zone area parameter A_{cD} is designed to account for the dependence of early oil production rates on the pattern size and the domination of initial oil viscosity.

$$V_{pD} = \left[\frac{5.615(V_s)_{inj}}{43560 A\, h\phi\, S_g}\right]^2 \tag{6.33}$$

with the limit $0 \leq V_{pD} \leq 1.0$ and $V_{PD} = 1.0$ when $S_g = 0$. If significant amount of gas initially exists, the V_{PD} parameter describes the reservoir fill-up process that accounts for an additional response delay to steam injection due to the initial gas saturation.

$$V_{oD} = \sqrt{1 - \frac{7758\phi\, S_{oi} V_s}{N}} \tag{6.34}$$

with the limit $0 \leq V_{oD} \leq 1.0$. The V_{oD} parameter, as described by Eq. (6.34), indicates that the oil displacement/production is increasingly controlled by the oil remaining in the reservoir pattern at anytime during the steam injection process.
where

A = total reservoir (pilot) area, acres
N = initial oil-in-place, bbl
S_g = initial gas saturation
h_n = net thickness, ft
$(V_s)_{inj}$ = cumulative volume of the steam injected, bbl
V_s = volume of the steam zone as a function of time, acre-ft

Figure 6.11 is a graphical presentation of the three components comprising the capture efficiency.

6.3.3 Steam-Assisted Gravity Drainage

Steam-assisted gravity drainage (SAGD) is a thermal in situ heavy oil recovery process that was originally developed by Butler (1991). Butler proposed to use steam injection, coupled with horizontal well technology, to assist the movement of

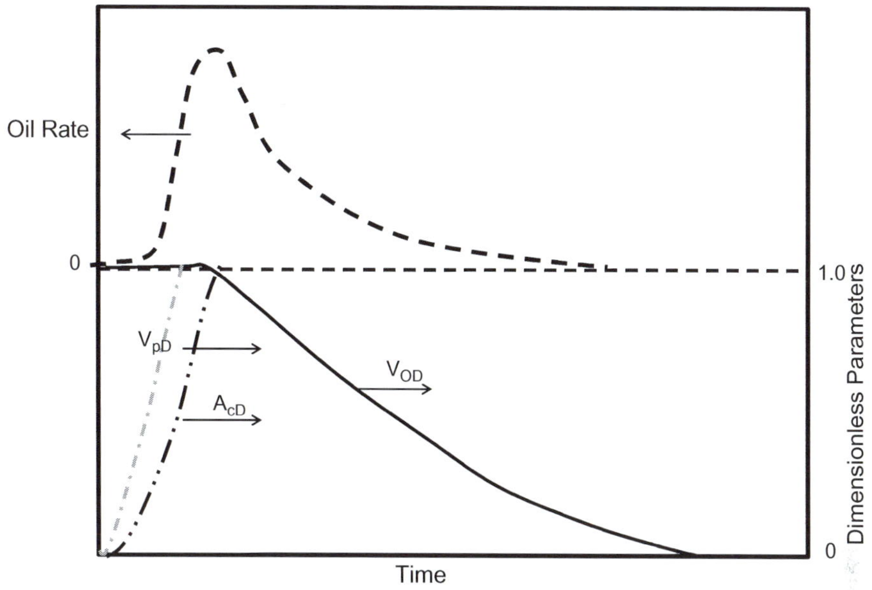

FIGURE 6.11 Graphical illustration of the capture parameters.

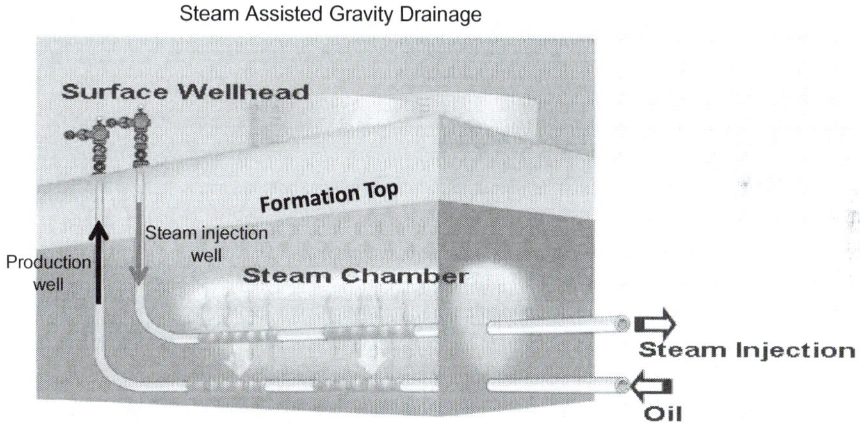

FIGURE 6.12 Schematic illustration of the SAGD concept.

oil to a production well by means of gravitational forces. The procedure utilizes a pair of parallel horizontal wells, one at the bottom of the formation and the other is placed about 10–30 ft above it. The wells are vertically aligned with each other with their length in the order of 3500 ft, as shown schematically in Figure 6.12. These wells are typically drilled in groups off central pads. The top well is the steam injector and the bottom well serves as the producer.

Initially, the cold heavy oil is essentially immobile. Therefore, an initial preheating stage is necessary to create a uniform thermohydraulic communication between well pair. In this start-up period, steam is injected in both wells to preheat the reservoir between the wells. This steam

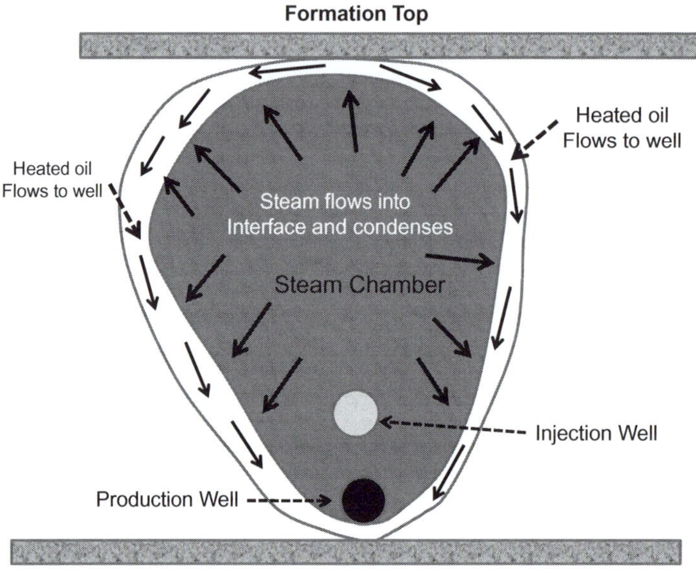

FIGURE 6.13 Schematic illustration of the SAGD mechanism.

circulation process in both the injector and the producer continues for approximately 2–4 months to enhance the oil mobility by reducing its viscosity. Once mobility has been established, steam is injected continuously into the upper well only. With the continuous injection of the steam, the steam rises to the top of the formation forming a "steam chamber" that grows vertically and horizontally. The injected steam will reach the chamber interface, heating the surrounding cold oil sand. The condensate and heated oil drain by gravity and flow towered the horizontal well near the base of the reservoir in countercurrent to the rising steam. It should be noted that since the flow path of oil and steam are separate, the displacement process is slow. However, the fingering problem that is traditionally associated with steam flooding is essentially eliminated, thereby improving the oil recovery efficiency by SAGD. The steam chamber expansion process and associated drainage flow are shown schematically in Figure 6.13.

Butler confirmed the SAGD concept by laboratory experiments and documented the following unique features of this thermal recovery process.

- Use of gravity as the primary driving force for moving oil.
- Large production rates obtainable with gravity using horizontal wells.
- Flow of the heated oil directly to the production well without having to displace uncontacted oil.
- Almost immediate oil production response (especially in a heavy oil reservoir).
- High recovery efficiency (up to 70%–75% in certain cases).
- Low sensitivity to reservoir heterogeneities.

6.3.4 In Situ Combustion

In situ combustion, or fire flooding, is a unique EOR process because a portion of the oil-in-place is oxidized and used as a fuel to generate heat. In the in situ combustion process, the crude oil in the reservoir is ignited and the fire is sustained by air injection. The process is initiated by continuous injection of air into a centrally located injection well. Ignition of the reservoir crude oil can either occur spontaneously after air has been injected over some length of time of it requires heating. Chemical

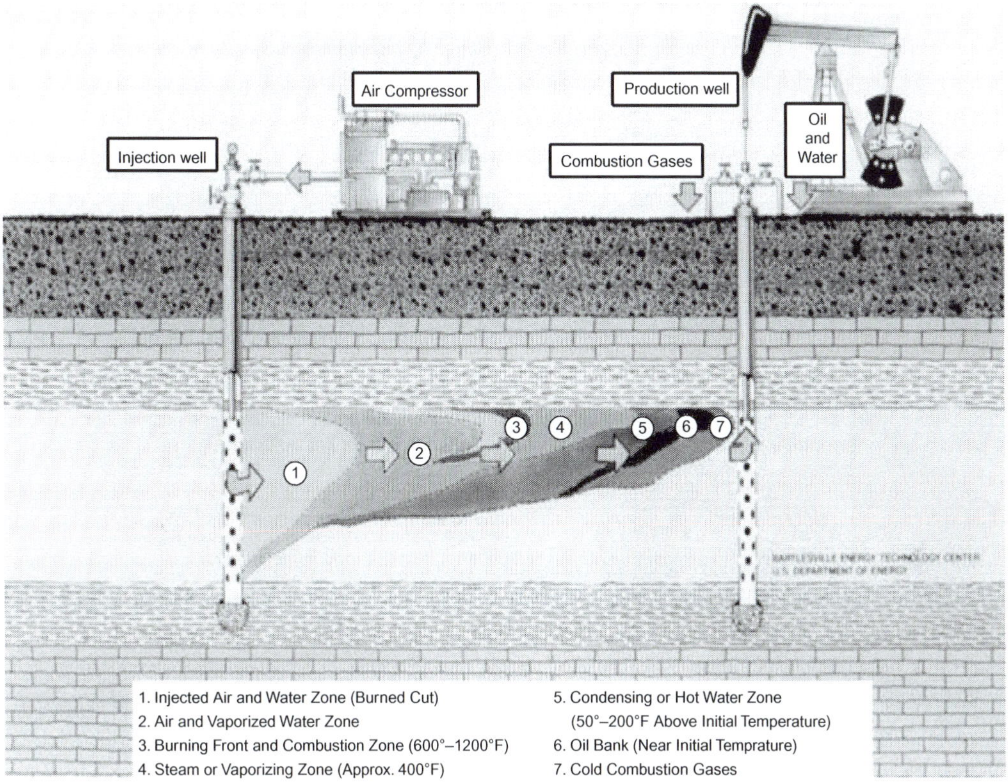

FIGURE 6.14 In situ combustion process. *(Courtesy Bartlesville Energy Technology Center, US Department of Energy, Washington, DC).*

reaction between oxygen in the injected air and the crude oil generates heat even without combustion. Depending on the crude composition, the speed of this oxidation process may be sufficient to develop temperatures that ignite the oil. If not, ignition can be initiated by:

- downhole electric heaters;
- preheating injection air; or
- preceding air injection with oxidizable chemicals.

Numerous laboratory experiments and field oil recovery data indicate generated heat and the vaporized hydrocarbon gases from the in situ combustion procedure will displace 100% of the oil from the reservoir that is contacted by the process.

There are three forms of in situ combustion processes:

- forward combustion;
- reversed combustion;
- wet combustion.

The above three processes are briefly discussed below.

Forward Combustion. The term "forward combustion" is used to signify the fact that the flame front is advancing in the same direction as the injected air. Figure 6.14 shows a schematic view of several distinct zones formed in an oil reservoir during the forward combustion process, while Figure 6.15 demonstrates the oil displacement mechanism and temperature profile associated with each of these zones.

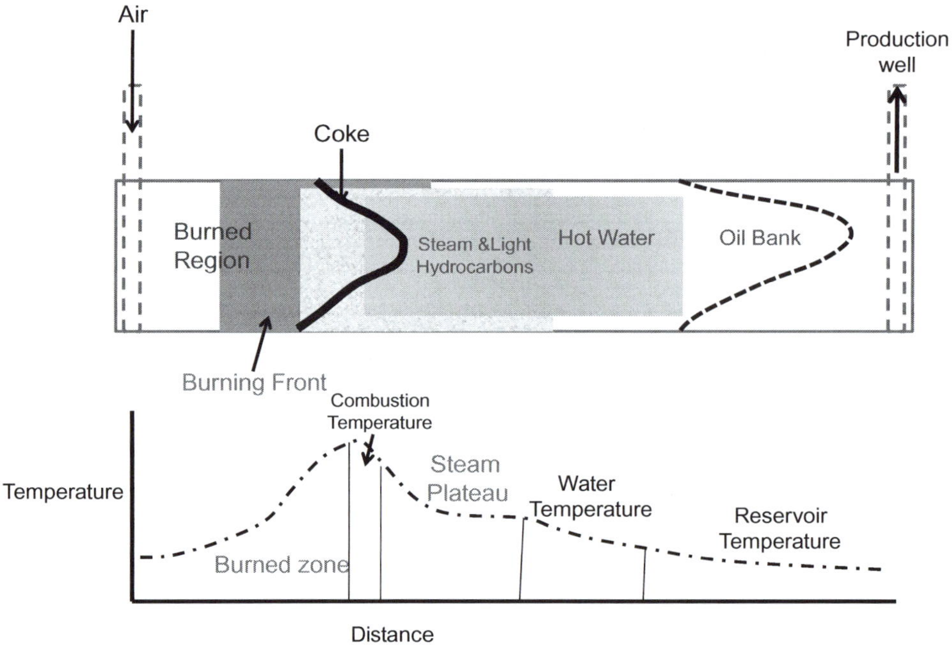

FIGURE 6.15 In situ combustion temperature zones.

As shown in Figures 6.14 and 6.15, there are seven zones that have been recognized during the forward combustion process, these are:

- *The burned zone:* The burned zone is the region that is already burned. This zone is filled with air and may contain a small amount of residual unburned organic materials; otherwise, it is essentially composed primarily of clean sand that is completely free of its oil or coke content. Because of the continuous air injection, the burned zone temperature increases from the injected air temperature at the injector to the temperature at the combustion leading edge.
- *Combustion front zone:* Ahead of the burned-out zone is combustion front region with a temperature variation ranging from 600°F to 1200°F. It is in this region that oxygen combines with fuel and high-temperature oxidation occurs.
- *The coke zone:* Immediately ahead of the combustion front zone is the coke region. The coke region represents the zone where carbonaceous material has been deposited as a result of thermal cracking of the crude oil. The coke residual fractions are composed of components with high molecular weight and boiling point temperatures. These fractions can represent up to 20% of the crude oil.
- *Vaporizing zone:* Ahead of the coke region is the vaporizing zone that consists of vaporized light hydrocarbons, combustion products, and steam. Temperatures across this zone vary from the high temperature of combustion to that necessary to vaporize the reservoir connate water.
- *Condensing zone:* Further downstream of the vaporizing region is the condensing zone, from which oil is displaced by several driving mechanisms. The condensed light hydrocarbons displace reservoir oil miscibly, condensed steam creates a hot water flood mechanism, and the combustion gases provide additional oil recovery by gas drive.

Temperatures in this zone are typically 50°F–200°F above initial reservoir temperature.

- *Oil bank zone:* The displaced oil accumulates in the next zone to form an oil bank. The temperature in the zone is essentially near the initial reservoir temperature with minor improvement in oil viscosity.
- *Undisturbed reservoir:* Further ahead of the oil bank lies the undisturbed part of the reservoir which has not been affected by the combustion process.

Reverse Combustion. The reserve combustion technique has been suggested for application in reservoirs that contain extremely viscous crude oil systems. The reverse combustion process is first started as a forward combustion process by injecting air in a well that will be converted later to a producer. After establishing ignition and burning out a short distance in the oil sand, the well is put on production and air injection is switched to another adjacent well. The air injection in the adjacent well displaces the oil toward the producing well passing through the heated zone while the combustion front travels in the opposite direction toward the air injection well. However, if the oil around the air injection well ignites spontaneously, the air (i.e., oxygen supply) is stopped and the process reverts to a forward combustion scheme.

Brigham and Castanier point out that the reverse combustion process has not been successful economically for the following two major reasons:

- Combustion started at the producer results in hot produced fluids that often contain unreacted oxygen. These conditions require special, high-cost tubular to protect against high temperatures and corrosion. More oxygen is required to propagate the front compared to forward combustion, thus increasing the major cost of operating an in situ combustion project.
- Unreacted, coke-like heavy ends will remain in the burned portion of the reservoir. At some time in the process, the coke will start to burn and the process will revert to forward combustion with considerable heat generation but little oil production. This has occurred even in carefully controlled laboratory experiments.

Wet Combustion. Heat utilization in the forward combustion process is very inefficient due to the fact that air has a poor heat-carrying capacity. Only about 20% of the generated heat during the forward combustion scheme is carried forward ahead of the combustion front where it is beneficial to oil recovery. The remaining heat is stored in the burned zone and is eventually lost to the cap and base rock of the pay zone.

Several variations of the in situ process have been proposed to utilize this lost heat. Water may be injected simultaneously or alternately with air, resulting in better heat distribution and reduced air requirements. In the burned zone, injected water is converted to superheated steam which flows through the flame and heats the reservoir ahead. This is called COFCAW (combination of forward combustion and water flood) process.

As the superheated system mixed with air reaches the combustion front, only the oxygen is utilized in the burning process. On crossing the combustion front, the superheated steam mixes with nitrogen from the air and flue gas consisting mainly of CO and CO_2. This mixture of gases displaces the oil in front of the combustion zone and condenses as soon as its temperature drops to about 400°F. The length of the steam zone is determined by the amount of heat recovered from the burned zone upstream.

Depending on the water/air ratio, wet combustion is classified as:

- incomplete when the water is converted into superheated steam and recovers only part of the heat from the burned zone;
- normal when all the heat from the burned zone is recovered;

- quenched or super wet when the front temperature declines as a result of the injected water.

When operated properly, water-assisted combustion reduces the amount of fuel needed, resulting in increased oil recovery and decreased air requirements to heat a given volume of reservoir. Up to 25% improvement in process efficiency can be achieved. Determination of the optimum water/air ratio is difficult because of reservoir heterogeneities and gravity override that can affect fluid movement and saturation distributions. Injecting too much water can result in an inefficient fire front, thus losing the benefits of the process.

In Situ Combustion Screening Guidelines
In general, in situ combustion process is a very complex process that combines the effect of several driving mechanisms, including:

- steam drive;
- hot and cold water flood;
- miscible and immiscible flood.

The process can achieve a high oil recovery in a wide variety of reservoirs, particularly if the reservoir criteria are:

- Reservoir thickness should be greater than 10 ft to avoid excessive heat losses to surrounding formation. However, very thick formations may present sweep efficiency problems because of gravity override.
- Permeability has to be large enough (greater than 100 md) to allow more flow of the viscous oil and achieving the desired air injectivity.
- Porosity and oil saturation have to be large enough to allow the economic success of the process.
- The reservoir depth should be large enough to confine the injected air in the reservoir. In general, there is no depth limit except that may affect the injection pressure.

Concluding Remarks. There are several disadvantages with the in situ combustion process including:

- formation of oil-water emulsions which cause pumping problems and reduce well productivity;
- production of low-pH (acidic) hot water rich in sulfate and iron that causes corrosion problems;
- increased sand production and cavings;
- formation of wax and asphaltene as a result of thermal cracking of the oil;
- liner and tubing failure due to excessive temperatures at the production wells.

The in situ combustion process has a tendency to sweep only the upper part of the oil zone; therefore, vertical sweep in very thick formations is likely to be poor. The burning front produces steam both by evaporating the interstitial water and by combustion reactions. The steam mobilizes and displaces much of the heavy oil ahead of the front, but when water condenses from the steam it settles below steam vapors and combustion gases, thus causing their flow to concentrate in the upper part of the oil zone.

Much of the heat generated by the in situ combustion is not utilized in heating the oil; rather, it heats the oil-bearing strata, interbedded shale and base and cap rock. Therefore, in situ combustion would be economically feasible when there is less rock material to be heated, i.e., when the porosity and oil saturations are high and the sand thickness is moderate.

6.4 CHEMICAL FLOOD

The overall recovery factory (efficiency) "RF" of any secondary or tertiary oil recovery method is the product of a combination of three individual efficiency factors as given by the following generalized expression:

$$RF = (E_A\,E_V)\,E_D$$

or

$$RF = (E_{Vol})\,E_D$$

In terms of cumulative oil production, the above can be written as:

$$N_P = N_S\, E_A\, E_V\, E_D$$

where

RF = overall recovery factor
N_S = initial oil-in-place at the start of secondary or tertiary flood, STB
N_P = cumulative oil produced, STB
E_D = displacement efficiency
E_A = areal sweep efficiency
E_V = vertical sweep efficiency
E_{Vol} = volumetric sweep efficiency

The areal sweep efficiency, E_A, is the fractional area of the pattern that is swept by the displacing fluid. The major factors determining areal sweep are:

- fluid mobilities;
- pattern type;
- areal heterogeneity;
- total volume of fluid injected.

The vertical sweep efficiency "E_V" is the fraction of the vertical section of the pay zone that is contacted by injected fluids. The vertical sweep efficiency is primarily a function of:

- vertical heterogeneity;
- degree of gravity segregation;
- fluid mobilities;
- total volume injection.

Note that the product of E_A and E_V is called the volumetric sweep efficiency "E_{vol}" and represents the overall fraction of the flood pattern that is contacted by the injected fluid.

The displacement efficiency E_D is the fraction of *movable oil*, i.e., $S_{oi} - S_{or}$, that has been displaced from the swept zone at any given time or pore volume injected. The displacement efficiency can only approach 100% if the residual oil saturation is reduced to zero. Chemical flooding has essentially two mail objectives:

- Increase the capillary number "N_C" to mobilize residual oil and improve the displacement efficiency. As shown previously in Figure 6.3, the residual oil saturation can be reduced by increasing the capillary number, as defined by Eq. (6.1) as:

$$N_c = \left(\frac{k_o}{\phi\sigma}\right)\left(\frac{\Delta p}{L}\right)$$

  The reduction in the interfacial tension between the displacing and displaced fluids is perhaps the only practical option in increasing the capillary number. The capillary number has to exceed the critical capillary number to mobilize residual oil saturation. It should be noticed that by reducing the interfacial tension to zero, the capillary number becomes infinite indicating 100% displacement efficiency, i.e., "*miscible displacement.*"
- Decrease the mobility ratio "M" for a better areal and vertical sweep efficiencies.

Chemical flooding is one of the enhanced oil recovery categories designed to increase oil recovery by improving sweep efficiencies, i.e., E_A, E_V, and E_D. Chemical oil recovery methods include:

- polymer;
- surfactant-polymer (variations are called micellar-polymer or microemulsion);
- alkaline (or caustic) flooding;
- Alkaline-surfactant-polymer (ASP) flood.

All the above methods involve mixing chemical with water prior to injection and, therefore, these methods require reservoir characteristics and conditions that are favorable to water injection. A brief discussion of each of the above flooding techniques is presented next.

6.4.1 Polymer Flood

Most enhanced oil recovery methods are directed to improve displacement efficiency by reducing residual oil saturation. Polymer flood, however, is designed to improve sweep efficiency by reducing the mobility ratio. The mobility ratio "M" is defined as the ratio of the displacing fluid mobility, $\lambda_{\text{displacing}}$, to that of the displaced fluid, $\lambda_{\text{displaced}}$. In traditional water flooding, the mobility ratio is defined mathematically as:

$$M = \frac{\lambda_{\text{displacing}}}{\lambda_{\text{displaced}}} = \frac{k_{rw}\,\mu_o}{k_{ro}\,\mu_w} \quad \textbf{(6.35)}$$

A mobility ratio of approximately 1, or less, is considered favorable, which indicates that the injected fluid cannot travel faster than the displaced fluid. For example, when $M = 10$, the ability of water to flow is 10 times greater than that of oil. Eq. (6.35) indicates that M can be made more favorable by any of the following:

- decrease oil viscosity, μ_o;
- increase the effective permeability to oil;
- decrease the effective permeability of water;
- increase the water viscosity, μ_w.

Little can be done to improve the flow characteristics of the oil in the reservoir, i.e., μ_o and k_{ro}, except by thermal recovery methods. However, a class of chemicals, i.e., polymers, which when added to water, even in low concentrations, increases its viscosity and decreases the effective permeability to water, resulting in a reduction in the mobility ratio.

The need to control or reduce the mobility of water led, therefore, to the development of polymer flooding or polymer-augmented water flooding. Polymer flooding is viewed as an improved water flooding technique since it does not ordinarily recover residual oil that has been trapped in pore spaces and isolated by water. However, polymer flooding can produce additional oil over that obtained from water flooding by improving the sweep efficiency and increasing the volume of reservoir that is contacted. Dilute aqueous solutions of water-soluble polymers have the ability to reduce the mobility of water in a reservoir, thereby improving the efficiency of the flood. Both partially hydrolyzed polyacrylamides (HPAM) and xanthan gum (XG) polymers reduce the mobility of water by:

- Increasing the viscosity of the injected aqueous phase
- Reducing the permeability of the formation to water. This reduction in the water permeability is fairly permanent, while the permeability to oil remains relatively unchanged. The reduction in permeability is measured in laboratory core flood and results are expressed in two permeability reduction factors:
  - Residual resistance factor "R_{rf}"
    Residual resistance factor is a laboratory-measured property that describes the reduction of water permeability after polymer flood. Polymer solutions continue to reduce the permeability of the aqueous phase even after the polymer solution has been displaced by brine. The ability of the polymer solution to reduce the permeability is measured in the laboratory and expressed in a property that is called the *residual resistance factor*. This permeability reduction factor is defined as ratio of the mobility of the injected brine before and after the injection of the polymer solution, i.e.:

$$R_{rf} = \frac{\lambda_w(\text{before polymer injection})}{\lambda_w(\text{after polymer injection})} \quad \textbf{(6.36)}$$

  - Resistance factor "R_f"
    The resistance factor "R_f" describes the reduction in water mobility and is defined as the ratio of the brine mobility to that of the polymer solution, with both mobilities measured under the same conditions, i.e.:

$$R_f = \frac{\lambda_w}{\lambda_p} = \frac{k_w/\mu_w}{k_p/\mu_p} \quad \textbf{(6.37)}$$

where

λ_w = mobility of brine
k_w = effective permeability to brine
μ_w = brine viscosity
λ_P = mobility of polymer solution
k_P = effective permeability to polymer solution
μ_P = polymer solution viscosity

This permeability reduction is essentially caused by the retention of polymer molecules in the reservoir rock. This is a combination of adsorption and entrapment, and it is not entirely reversible. Thus, most of the polymer (and the benefits it provides) remains in the reservoir long after polymer injection is stopped and the field is returned to water injection. Adsorption is the irreversible retention of polymer molecules on the rock surface. The amount of polymer adsorbed on the rock surface depends on the type and size of the polymer molecules, polymer concentration, and rock surface properties.

Two testing methods are used to measure the amount of adsorption, one is a static condition test and one is a dynamic test, which measures the adsorption in a core flood. Polymer molecules adsorb on the rock surface as a monolayer with the thickness equal to the diameter of the polymer molecules. Once the monolayer saturation level is reached, no more adsorption will occur.

Figure 6.16 shows the effect of mobility ratio improvement on the sweep efficiency for a five-spot water flood pattern. This graph correlates the areal sweep efficiency at various water cuts "f_w" as a function of the reciprocal of the mobility ratio "$1/M$." Table 6.3, as generated from Figure 6.16, compares areal sweep efficiency at various producing water cuts and mobility ratios. Both Figure 6.16 and Table 6.3 reveal the substantial impact of improving the mobility ratio on areal sweep efficiency. Table 6.3 shows that by reducing the mobility ratio from 10 to 1, twice the fractional areal is swept before water production begins.

Another important factor that adversely impacts efficiency is the viscous instabilities (*viscous fingering*) that are associated with displacement processes where the displaced fluid (crude oil) has a higher viscosity than the displacing

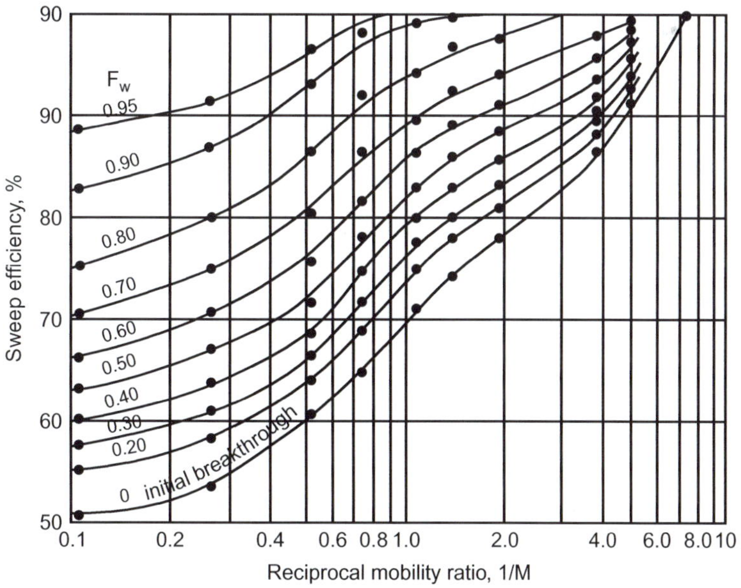

FIGURE 6.16 Areal sweep efficiency as a function of $1/M$ and f_w.

TABLE 6.3 Comparison of E_A for Varying M

M	$1/M$	E_A at Breakthrough	E_A at 95% Water Cut
10	0.1	0.35	0.83
2	0.5	0.58	0.97
1	1	0.69	0.98
0.5	2	0.79	1.00
0.25	4	0.90	1.00

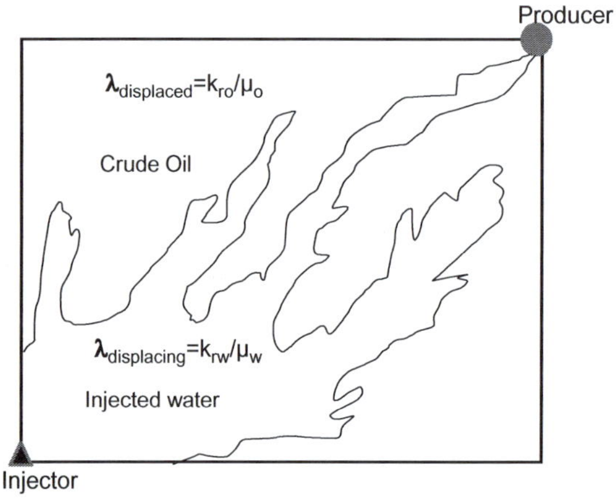

FIGURE 6.17 Viscous fingering in water flood.

fluid, e.g., water. The less viscous displacing fluid generally flows more easily than the more viscous displaced fluid, causing the occurrence of perturbations which finger through the crude oil system. Viscous fingering can have dramatic effect on the sweep efficiency of a displacement process. Unstable displacement processes due to viscous fingering are often associated with early breakthrough of the displacing fluid. Figure 6.17 illustrates viscous fingering in a quarter of a five-spot model. The stability of the displacing fluid front in terms of developing viscous fingering as a function of mobility ratio is conceptually illustrated in Figure 6.18. This illustration compares the stability of the leading edge of polymer flood that is recognized with a better mobility ratio than that of the traditional water flood. It clearly indicates the importance of improving the displacing phase mobility in order to achieve a better overall sweep efficiency that will be translated into a better oil recovery factor.

Polymer Properties. When a Newtonian fluid, e.g., water or oil, is subjected to a shearing force, it deforms or flows. There is a resistance to the flow which is defined as the ratio of the shearing force (shear stress) to the rate of flow (shear rate). For a Newtonian fluid, this ratio is constant and is defined as the apparent viscosity of the fluid. Mathematically, the viscosity is expressed as:

$$\mu_{app} = \frac{\text{shear stress}}{\text{shear rate}} = \frac{\tau}{v} \quad (6.38)$$

where

μ_{app} = apparent viscosity
τ = shear stress (pressure difference)
v = shear rate (flow velocity)

For the flow of the Newtonian fluids in porous media, the apparent viscosity can simply be expressed in terms of Darcy's equation as:

$$\mu_{app} = k\frac{\Delta p/L}{q/A} \quad (6.39)$$

where

μ_{app} = apparent viscosity
$\Delta p/L$ = pressure gradient
q/A = superficial fluid velocity
k = effective permeability

The apparent viscosity may also be expressed in terms of the resistance factor "R_f" as:

$\mu_{app} = \mu_w R_f$

The apparent viscosity as defined above includes the effect of permeability reduction due

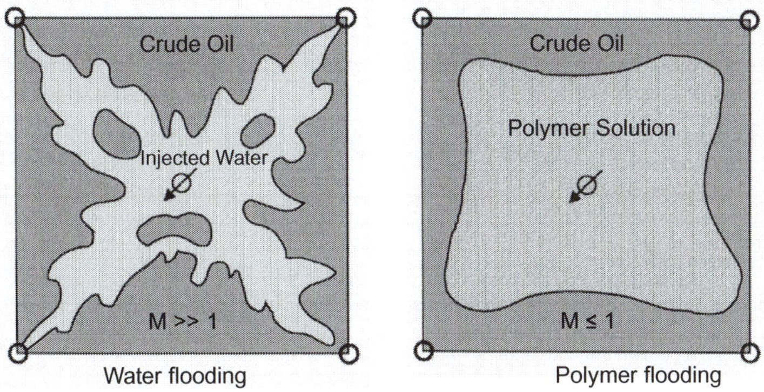

FIGURE 6.18 Comparison of viscous instability in water and polymer flood.

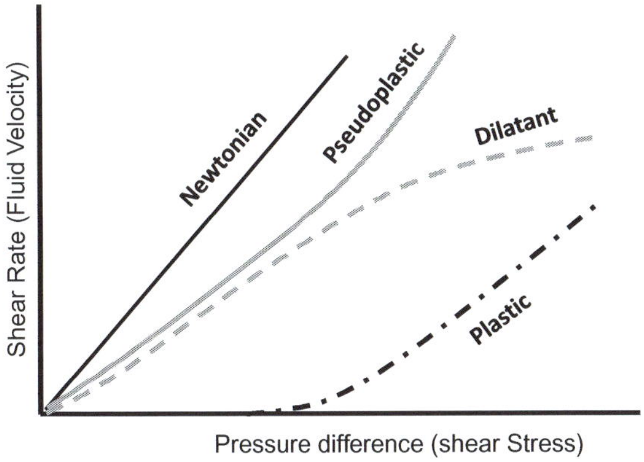

FIGURE 6.19 Newtonian and non-Newtonian flow.

to adsorption or plugging by the polymer solution.

Non-Newtonian fluids cannot be characterized by a single and constant viscosity value as calculated by Eq. (6.38) or Eq. (6.39) because the ratio of shear stress to shear rate is not a constant. As shown in Figure 6.19, the flow of these non-Newtonian fluids may follow one of the following complex fluid models:

- Plastic fluids: Drilling fluids exhibit the characteristics of the plastic fluids type. A pressure differential (shear stress) is required for these types of fluids to initiate the flow. As shown in Figure 6.19, the viscosity of the plastic fluid decreases with increasing flow rate.
- Dilatant fluids: This category of fluids is characterized by apparent viscosities that increase with increasing shear rate (flow velocity).
- Pseudoplastic fluids: Polymer solutions are generally classified as pseudoplastic fluids under most fluid injection and reservoir conditions. These types of fluids exhibit larger apparent viscosities when flowing at low velocities and lower apparent viscosity when flowing at high velocities.

The rheological behavior of the flow of polymer solution through porous media can be divided into four flow regions as shown in Figure 6.20. At low velocities, the apparent viscosity of the polymer solution will approach a maximum limiting value. For a larger range of velocities, the solution exhibits the flow characteristics of pseudoplastic and the viscosity decreases with increasing velocity. At a higher velocity, the viscosity of polymer solution approaches a minimum value that is equal to or greater than the solvent viscosity. At very high velocity, the viscosity increases with increasing velocity, with the polymer solution exhibiting the flow characteristics of the dilatant fluid. It should be noted that plotting of the resistance factor "R_f" as defined by Eq. (6.37) as a function of fluid velocity would produce a plot similar to that of a polymer viscosity curve with the exact four dividing flow regions. The viscosity

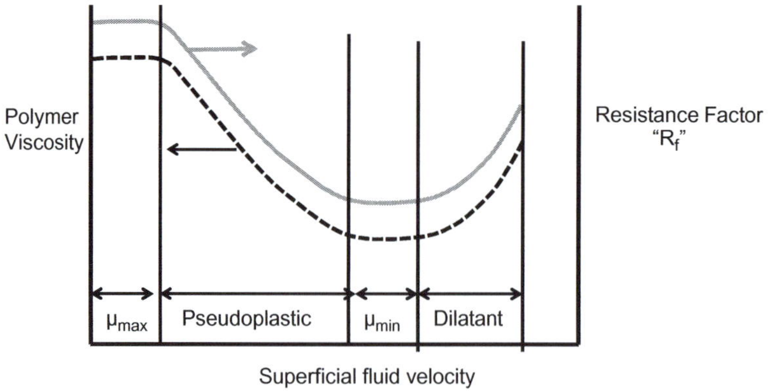

FIGURE 6.20 Rheological behavior of polymer solution in porous media.

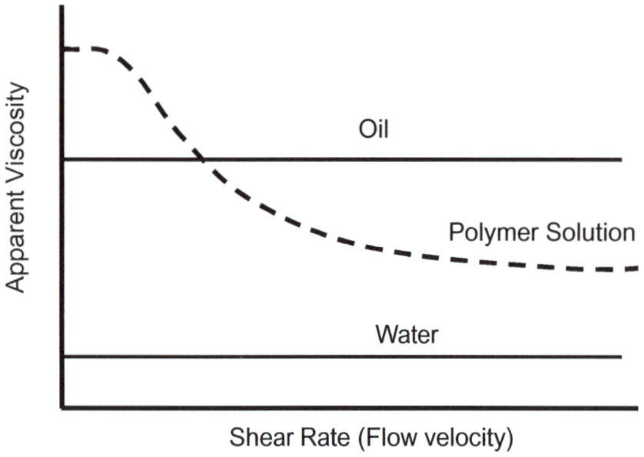

FIGURE 6.21 Apparent viscosity vs. flow velocity.

of the polymer solution can be estimated from the residual and resistance factors by applying the following relationship:

$$\mu_p = \mu_w \left(\frac{R_f}{R_{rf}}\right)$$

Despite the complex flow behavior of polymer solutions, the apparent viscosities of these fluids are significantly higher than the viscosity of water, even at high shear rates, as illustrated schematically in Figure 6.21.

The power law, as given below, is a simple mathematical expression that can be used to curve-fit a typical viscosity-shear rate experimental viscometer data:

$$\mu = K v^{n-1} \tag{6.40}$$

where

μ = viscosity
K = power-law coefficient
n = power-law exponent
v = superficial fluid viscosity

Based on the value of the exponent "n," the power law can have three different types of fluid:

Type of Fluid	n
Pseudoplastic	<1
Newtonian	1
Dilatant	>1

Gaitonde and Middleman (1967) modified the power-law relationship to model the flow of pseudoplastic behavior of the polymer solution through porous media. The modified power-law accounts for the porous media in terms of localized permeability and porosity over a wide range of flow rates. The modified expression is given by:

$$\mu_p = 0.017543\, K \left(\frac{9n+3}{n}\right) [150\, k_w \phi (1 - S_{orw})]^{\frac{1-n}{2}} (v)^{n-1} \tag{6.41}$$

where

μ_p = apparent viscosity of the polymer solution, cp
K = power-law coefficient from the viscometer experimental data, cp (second)$^{n-1}$

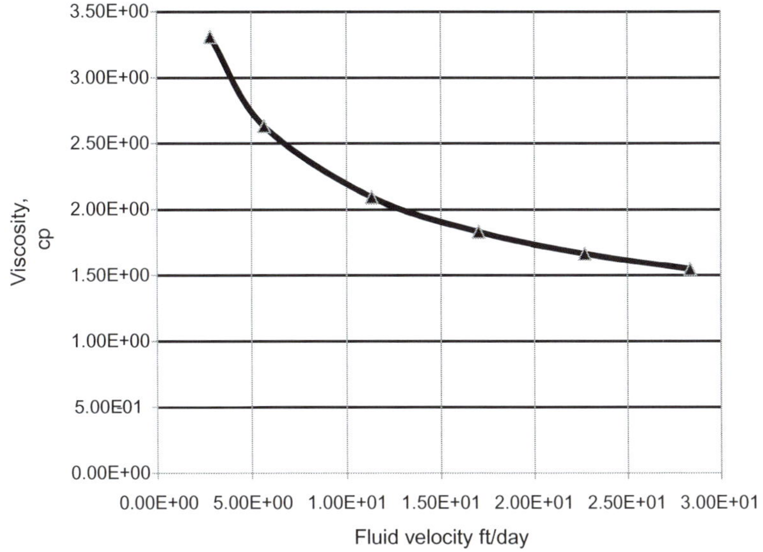

FIGURE 6.22 Polymer viscosity vs. flow velocity.

n = power-law exponent
ϕ = porosity, fraction
k_w = permeability of water, md
S_{orw} = residual oil saturation to water, fraction
v = superficial fluid velocity, ft/day

Example 6.4

The viscometric and core flood data for a polymer flood are given below:

Core data:

$k_w = 17$ md
$\phi = 0.188$
$S_{orw} = 0.32$

Polymer solution data:

Polymer concentration = 200 ppm
Viscosity of brine = 0.84 cp
$K = 7.6$ cp (second)$^{n-1}$
$n = 0.67$

Calculate and plot the polymer viscosity at the following superficial fluid velocity: 2.83, 5.67, 11.3, 17.0, 22.7, and 28.3 ft/day.

Solution

Step 1. Apply and simplify Eq. (6.41), to give:

$$\mu_p = 0.017543\, K\left(\frac{9n+3}{n}\right)\left[150\, k_w \phi (1 - S_{orw})\right]^{\frac{1-n}{2}} (v)^{n-1}$$

$$\mu_p = 0.017543\ (7.6)\left(\frac{9(0.67)+3}{0.67}\right)[150(17)(0.188)$$
$$(1-0.32)]^{\frac{1-0.67}{0.67}} (v)^{0.67-1} \mu_p = 4.668846\ (v)^{-0.33}$$

Step 2. Calculate the viscosity of the polymer solution at the required fluid velocity.

v, ft/day	μ_p, cp
2.83	3.31
5.67	2.63
11.3	2.10
17.0	1.83
22.7	1.67
28.3	1.55

Step 3. Plot the polymer rheological behavior as shown in Figure 6.22.

Polymer Displacement Mechanisms. Polymer flooding can yield a significant increase in oil recovery as compared to conventional water flooding techniques, as shown schematically in Figure 6.23. Polymer floods are conducted by injecting a slug of polymer solution (approximately 25%–50% of the reservoir pore volume) followed by chase water to drive the polymer slug and the developed oil bank toward the production wells. Because the chase water-polymer mobility ratio is unfavorable, the chase water tends to finger through the polymer slug and gradually dilutes the trailing edge of the slug. To minimize the effect of this unfavorable mobility ratio, traditionally a fresh water buffer zone contains polymer with a decreasing polymer concentration (a grading or taper) that separates the chase water from the polymer slug, as schematically shown in Figure 6.24. The grading of the buffer zone solution is designed in a way that the viscosity of leading edge of the buffer zone is equal to the viscosity of the polymer slug, while the viscosity of the trailing edge of the buffer solution is equal to the viscosity of the chase water.

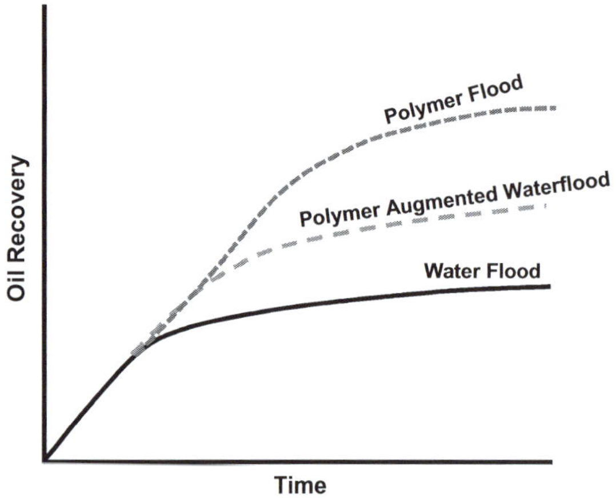

FIGURE 6.23 Oil recovery by polymer flood.

6.4.2 Surfactant Slug and Micellar Solution Flood

Oil recovery processes using surfactant are classified as:

- Surfactant-polymer (SP) slug
- Micellar-polymer (MP)
- Alkaline-surfactant-polymer (ASP)

One of the main ingredients of any of the above chemical flooding techniques is surfactant. Surfactants, or surface acting agents, are soaps or soap-like substances. They have the ability to change and reduce the interfacial tension properties of their solutions to a noticeable degree (even if they are present in minute amounts) to promote the mobilization and displacement of the remaining oil that it contacts.

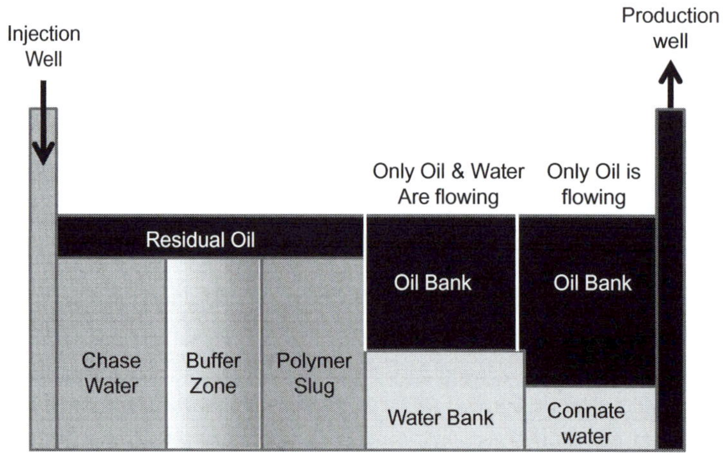

FIGURE 6.24 Polymer flooding as a secondary recovery process.

Surfactants are characterized by having an amphiphilic molecule. One end of this molecule is attracted to water (this is the hydrophilic end) and the other is attracted to oil (this is the olephilic end). It is this dual attraction nature of these surfactants that enables them to solubilize oil and water to form a miscible mixture. One type of surfactants that is commonly used in the industry is petroleum sulfonate. This chemical agent is produced from hydrocarbons ranging from LPG to the crude oil itself. The chemistry of the petroleum sulfonates is very complex and traditionally they have been described by their molecular weights that vary widely from 350 to 550.

The surfactant slug injection process consists of the following:

- **Preflush:** The objective of the preflush is to condition the reservoir by injecting a brine solution prior to the injection of the surfactant slug. The brine solution is designed to lower the salinity and hardness of the reservoir existing water phase so that mixing with surfactant will not cause the loss of the surfactant interfacial property. The preflush solution volume is typically ranging from 50% to 100% pore volume.
- **Surfactant slug:** The volume of the surfactant slug ranges between 5% and 15% pore volumes in field applications. Extensive laboratory studies show that the minimum slug size is 5% pore volume in order to achieve effective oil displacement and recovery.
- **Mobility buffer:** The surfactant slug is displaced by a mobility buffer solution with varying polymer concentrations between the slug and chase water. The mobility buffer solution separating chase water and the surfactant slug prevents rapid slug deterioration from the trailing edge of the surfactant slug. This process of injecting and designing a mobility buffer solution is an essential and integral process in all chemical flooding techniques in order to minimize the chemical slug size required for efficient and economical oil recovery.

Figures 6.25 and 6.26 show schematic illustrations of the chemical slug (surfactant, micellar, or ASP) injection mechanism when used as a *secondary recovery* or *tertiary recovery* process, respectively. Note that the chemical slug will miscibly displace and form a bank; the remaining oil and water with both phases are flowing simultaneously. When the chemical flood is used as a secondary recovery process, as shown in Figure 6.25, production wells will continue to produce at the preexisting decline rate until the breakthrough of the oil-water

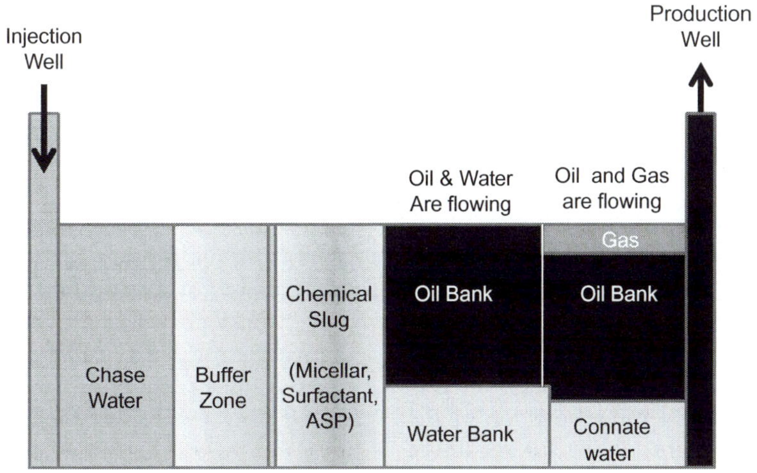

FIGURE 6.25 Chemical flood as a secondary recovery process.

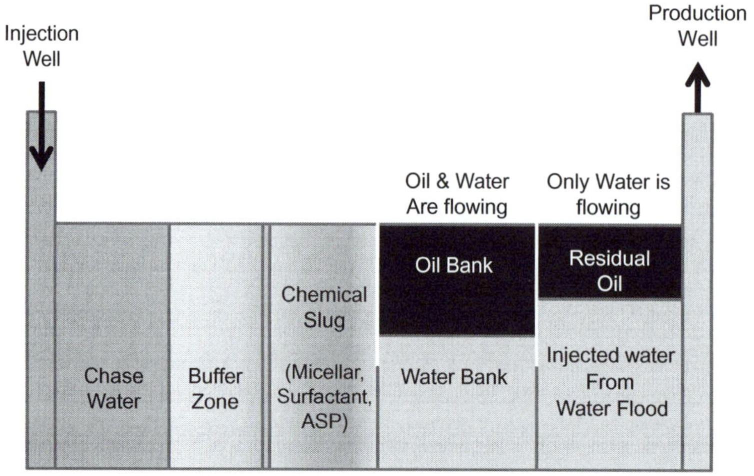

FIGURE 6.26 Chemical flood as an EOR process.

bank. An increase in the oil and water production rates signifies the field response to the chemical flood injection process. Figure 6.26 shows the chemical flood injection as a tertiary recovery process. It indicates that the prior water flood has displaced the oil to residual immobile oil saturation. Only water will be produced until the oil-water bank reaches the producing well. An economic parameter that must be considered during the process is the problem of handling the produced water.

Another variation of the surfactant slug process is labeled as *micellar* solutions, *microemulsions*, etc. The chemical slug essentially contains other chemicals added to it. The ***micellar solution*** is a type of surfactant solution slug that is composed of the following five main ingredients:

- surfactant;
- hydrocarbon;
- co-surfactant (alcohol);
- electrolyte;
- water.

Because one of the main problems in any chemical flood process is the adsorption of the chemical agent (surfactant) on the surface of the porous media, the co-surfactant (alcohol) is added to the solution slug to reduce adsorption of the surfactant to the reservoir rock. The electrolyte, such as sodium chloride or ammonium sulfate, is added to the micellar solution to adjust and control the changes in the viscosity of micellar solution as it contacts the reservoir water phase.

As in all types of chemical floods, often the composition of the reservoir water phase has an adverse effect on the injected chemical slug. Therefore, floods are traditionally started by first injecting a preflush bank of water which is compatible with the chemical slug solution and displaces the formation brine out of the reservoir.

The mobility is perhaps one of the most important controlling properties that must be considered when designing a chemical or miscible injection process. A properly designed chemical or miscible process must have a solution slug with mobility that is equal to or less than the mobility of the stabilized displaced fluid bank. Consider a set of hypothetical water-oil relative permeability curves as shown in the upper graph in Figure 6.27. Assuming constant oil and water viscosities, the total relative mobilities, i.e., $\lambda_w + \lambda_o$, are calculated from k_{rw} and k_{ro} curves as shown in the lower graph of

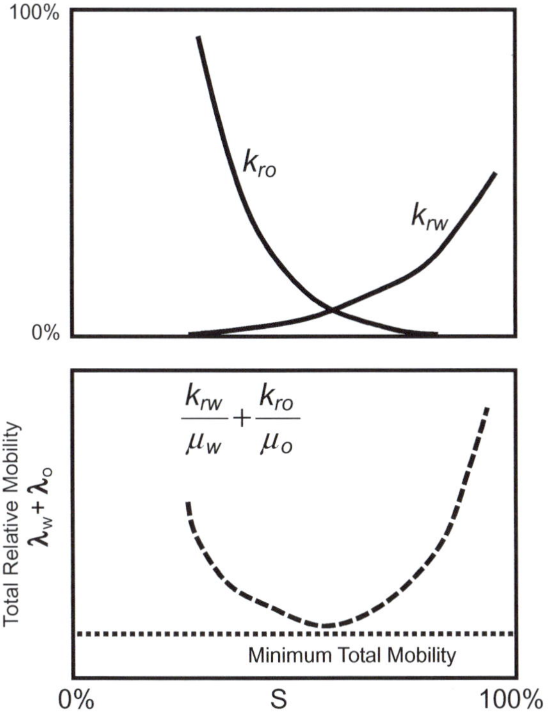

FIGURE 6.27 Total relative permeability vs. water saturation.

Figure 6.27, with the minimum total mobility as represented by a horizontal tangent to the resulting curve. The minimum total relative mobility is designated as the maximum required mobility of the chemical slug. The viscosity of the chemical slug is adjusted to the concentration of polymer in the slug to achieve the optimum solution mobility.

In addition, in order to ensure the stability of the entire displacing system, the mobility of the buffer solution must be equal to or less than the mobility of the micellar slug. However, due to the unfavorable mobility between the chase water drive and the mobility buffer solution, the drive water will penetrate and bypass the mobility buffer and the chemical slug. Therefore, the volume of mobility buffer must be large enough to protect the slug. For constant concentration of buffer solutions, the volume of the buffer solution usually will require 50%–100% of reservoir pore volume. However, using a polymer concentration grading of the mobility buffer solution will decrease the rate of chase water penetration and improve the economics of the process. Although different grading procedures can be used, a semilogarithmic relationship as shown in Figure 6.28 is a simple approach for designing the buffer zone.

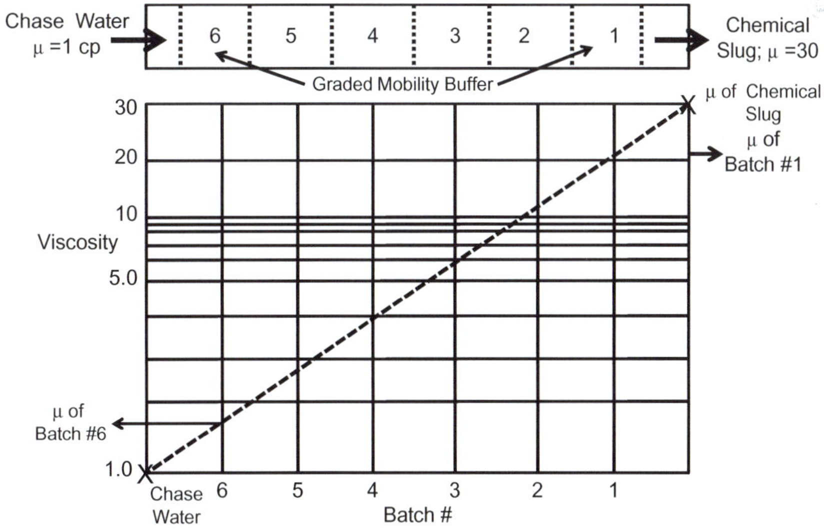

FIGURE 6.28 Grading the mobility buffer.

The application of the approach is illustrated through the following example.

Example 6.5

Given the following data, design a tapered buffer zone solution of 60% reservoir pore volume. Assume that six batches of equal volumes will be injected, i.e., each with 10% reservoir pore volume.

Viscosity of the chemical slug "μ_{slug}" = 30 cp
Viscosity of the chase water "μ_{Chase}" = 1.0 cp

Solution

- On a semilogarithmic paper, plot the viscosity of the chemical slug and chase water on the *y*-axis and connect with a straight line with the *x*-axis divided equally to six segments as shown in Figure 6.28.
- Read the required viscosity for each batch from the straight line to give.

Batch #	Viscosity
1	20
2	11
3	7
4	4.2
5	2.6
6	1.6

Application of Chemical Flooding. Chemical flooding is probably applicable to many reservoirs that have been successfully water flooded. In general, chemical flooding applications include the following:

- It is applicable to sandstone reservoirs, but is limited in use in carbonate reservoirs or where reservoir brines contain excessive calcium or magnesium ions. Adsorption of the surfactant is high in these type of reservoirs.
- The process is best applied to reservoirs with medium-gravity crude oils. Prospects with low-gravity crudes probably would not be economical. A low-gravity, high-viscosity crude oil would require increasing the viscosity of the micellar and polymer slugs for a favorable mobility ratio, resulting in higher costs.
- Chemical flooding process is technically applicable for secondary or tertiary recovery. If used for secondary recovery, it eliminates one set of operating costs; however, it still should be justified economically on the incremental oil it will recover above water flooding.

Advantages of Chemical Flooding

- Chemical solution followed by polymer buffer is an ideal displacing fluid. It proves the high unit displacement of miscible floods and high areal sweep efficiency. Mobility control, when using this process, allows optimization of areal sweep.
- The field operation is little different from a water flood except for the additional mixing and filtering equipment.
- In early surfactant flooding, the adsorption of the surfactant to the reservoir rock reduced the process to a water flood. Micellar solutions with proper co-surfactants and electrolytes limit the adsorption problem.

Disadvantages of Chemical Flooding

- The main disadvantage is the large amounts of high cost chemicals needed. Large expenditures must be made very early in the life of the project, most of it during the first year. The investment in chemicals is a function of the pore volume. The income is based on two parameters which are difficult to determine—the oil saturation in the reservoir and the amount of oil that the chemical flood will recover.
- When micellar flooding follows a depleted water flood, only water is produced for ½–2 years, depending on the residual oil saturation and pattern size. If the quality of the water prohibits its use in mixing the micellar and polymer solutions, the water must be disposed of during this mixing period.

6.4.3 ASP Flood

The alkaline-surfactant-polymer (ASP) technology uses similar mechanisms as in the micellar flood technique that is designed to mobilize the residual oil. The alkaline-surfactant-polymer technology is based on combining interfacial tension–reducing chemicals with mobility control chemical to improve the overall displacement efficiency and increase the incremental oil recovery. The technology relies on reducing the expensive surfactant concentration by 20–70 folds by adding the much lower-cost **alkali** as one of the main ingredients of the injected ASP slug. The alkali has the natural ability to generate in situ surfactants by interaction with the residual oil if the oil contains natural organic acids, most commonly the naphthenic acids. Therefore, lowering the slug cost significantly by reducing the amount of the required commercial surfactants. The additional benefits of using sodium carbonate (alkali) include the following:

- It reduces the adsorption of surfactant and polymer on the rock.
- It alters the wettability of the formation to become a "more water-wet" or to change the wettability from an oil-wet to a water-wet system.

The design of an ASP flooding process must achieve three main objectives:

- propagation of the chemicals in an active mode to contact and displace the residual oil with 100% displacement efficiency;
- complete volumetric coverage of the area of the interest by controlling the slug mobility through optimizing the polymer concentration in the solution;
- injection of enough chemicals and slug size to account for retention and slug breakdown by adsorption.

Achieving these objectives is significantly affected by the design and selection of the chemicals used in formulating the injected slug. The two most common alkaline agents used for ASP flooding are:

- soda ash (sodium carbonate [Na_2CO_3]) and
- caustic soda (sodium hydroxide [$NaOH$]).

As in all types of chemical flooding techniques, ASP flood proceeds in the four traditional distinct phases as shown in Figure 6.26:

1. *Preflush*: Often the composition of the brine in a reservoir has an adverse effect on the ASP solution. To correct this problem and separate the hard formation brine from the slug, floods are started by first injecting a preflush bank of water ahead of the slug. This preflush water, which is compatible with the ASP solution, flushes the formation brine out of the reservoir.
2. *ASP slug*: The slug size can range from 15% to 30% pore volume. The slug formulation is similar to that of the micellar slug except that much of the surfactant is replaced by the low-cost alkali, so slugs can be much larger but overall cost is lower. As the slug moves through the formation, it displaces 100% of the oil contacted in a miscible-type displacement. The areal sweep efficiency is controlled by the mobility ratio; that is, the mobility of the displacing fluid divided by the mobility of the displaced fluid. A predetermined amount of polymer is added to the ASP slug to adjust its mobility to approach, or be less than, the total mobility of the oil-water.
3. *Mobility buffer*: A displacing solution is required to displace the ASP slug through the reservoir. A favorable mobility between the displacing solution and the ASP slug is also desired. If water is used as the displacing fluid, an unfavorable mobility ratio might exist. This would result in reduced areal sweep efficiency and in water fingering through the ASP slug—diluting and dissipating the slug. To protect the slug, a mobility buffer of thickened water is injected immediately behind the slug. This thickened water is

a solution of water and polymer. The viscosity of the polymer bank is graded from high viscosity behind the ASP slug to a low value at the trailing edge of the polymer bank. This grading is accomplished by varying the polymer concentration in the solution. This graduated bank is less costly and achieves a more favorable mobility ratio between the chase water and the polymer bank. The minimum size of the polymer bank is in the range of 50% of pore volume.

4. *Chase water*: The mobility buffer is displaced by chase water until the economic limit of the project is reached.

Advantages of ASP Flood. ASP solution followed by polymer buffer is an ideal displacing fluid. It provides the high unit displacement of miscible floods and high areal sweep efficiency. Mobility control, when using this process, allows optimization of areal sweep. The field operation is little different from a water flood except for additional mixing and filtering equipment.

Disadvantages of ASP Flood. There are several disadvantages and limitations associated with the application of this technology as presented below:

- Large expenditures must be made very early in the life of the project, with most of it during the first year. The investment in chemicals is a function of the size of the slug. The income is based on two parameters that might be hard to determine: initial oil saturation at the start of the flood and the amount of oil that can be recovered.
- Another problem arises from the fact that when ASP is used in a depleted water flood reservoir, only water will be produced for ½–2 years depending on the pattern size. If the quality of this water prohibits its use in mixing the ASP slug, the water must be disposed of during this mixing period.
- The process is not well suited for carbonate reservoirs.
- Gypsum or anhydrite may precipitate in production wellbores.
- Degradation of chemicals may occur at high temperature.
- Chemicals used in ASP floods represent the most clearly identifiable group of potential hazards. These are of special concern because many will be left behind in the reservoirs after the recovery project is completed. If in the recovery process, chemicals escape to the environment in sufficient quantities, their presence can degrade water supply quality, among other hazards.

6.5 MISCIBLE GAS FLOOD

It is well known that oil and water do not mix. If these two fluids are poured into a bottle and allowed to settle, two distinct liquids are apparent, separated by a sharp interface. Oil and water are categorized as immiscible liquids. Similarly, oil and natural gas are also immiscible. The reduction of interfacial or surface tension between the displacing and displaced fluids is one of the major keys that contribute to the success of the injection process.

In contrast to the definition of immiscibility, two fluids are considered miscible when they can be mixed together in all proportions and resulting mixtures remain single phase. Gasoline and kerosene are examples of two liquids that are miscible. Because only one phase results from mixtures of miscible fluids, there are no interfaces and consequently no interfacial tension between fluids. It is apparent from Figure 6.3 and the definition of capillary number "N_c" as given by Eq. (6.1) that the interfacial tension "σ" between oil and displacing phase is eliminated completely (i.e., N_c becomes ∞ when $\sigma \approx 0$), residual oil saturation can be reduced to its lowest possible value with displacement efficiency E_D approaching 100%. This is essentially the objective of any form of miscible displacements (e.g., chemical, gas, etc.). Therefore, classifying a system as miscible or immiscible can have a substantial impact on ultimate oil recovery by fluid injection. The ultimate oil recovery factor "RF" for any forms of secondary or tertiary recovery processes is defined by:

$$RF = E_{VOL} E_D \quad (6.42)$$

with the volumetric sweep efficiency "E_{VOL}" defined by:

$E_{VOL} = E_D\, E_A$

where

RF = recovery factor
E_A = areal sweep efficiency
E_V = vertical sweep efficiency
E_{VOL} = volumetric sweep efficiency
E_D = displacement sweep efficiency

The oil recovery performance by the immiscible and miscible displacement processes indicated Eq. (6.42) shows that the oil recovery factor "RF" is limited and controlled primarily by the level achieved during the injection process by two factors:

a. The volumetric sweep efficiency "E_{VOL}": Depending largely on the mobility ratio and reservoir characteristics, this efficiency factor is usually less than 100% because of the following controlling parameters:
   - permeability stratification;
   - viscous fingering;
   - gravity segregation;
   - incomplete areal sweep and vertical and areal sweep efficiencies.
b. The displacement efficiency "E_D": The displacement efficiency is defined as the fraction of movable oil that has been displaced from a swept reservoir zone by a displacing fluid, i.e.:

$$E_D = \frac{S_{oi} - \overline{S}_o}{S_{oi}} \quad (6.43)$$

where

S_{oi} = initial oil saturation
$\overline{S}_o$ = remaining residual oil saturation in the swept area

Because high interfacial/surface tension presents between oil and conventional water or gas injection, high remaining oil saturation $\overline{S}_o$ always exists during this immiscible displacement process, and therefore E_D will never approach 100%. On the other hand, the miscible displacement process is designed to reduce the interfacial/surface tension to a significantly low value, resulting in a displacement efficiency approaching 100% with the substantial reduction in remaining (residual oil) saturation in swept areas of the reservoir.

6.5.1 Miscibility

For a miscible flood to be economically successful in a given reservoir, several conditions must be satisfied:

- An adequate volume of solvent must be available at a rate and cost that will allow favorable economics.
- The reservoir pressure required for miscibility between solvent and reservoir oil must be attainable.
- Incremental oil recovery must be sufficiently large and timely for project economics to compensate for the associated added costs.

There are two types of miscible displacements:

- **First-contact miscible (FCM) displacement:** Displacements in which the injection fluid and the in situ reservoir fluid form a single-phase mixture for all mixing proportions.
- **Multiple-contact miscible (MCM) displacement:** Processes in which the injected fluid and the reservoir oil are not miscible in the first contact but miscibility could develop after multiple contacts (***dynamic miscibility***). These processes are categorized into:
  - vaporizing lean gas drive, alternatively called "high-pressure lean gas injection";
  - condensing rich gas drive; and
  - combined vaporizing–condensing drive.

Some injection fluids, e.g., chemical floods and LPG, mix directly with the reservoir crude oil in all proportions and their mixtures remain single phase. This displacement process is classified as "***first-contact miscible displacement.***" Other injection fluids used for miscible flooding form two phases when mixed directly with the reservoir crude oil, i.e., they are not first-contact miscible. However, miscibility could develop after multiple contacts with reservoir crude oil and is termed "dynamic miscibility." This dynamic displacement mechanism during the process is described as the in situ manufacture of a miscible slug. Miscibility is achieved in this process by in situ mass transfer "***vaporizing***" or "***condensing***" of components resulting from repeated contacts of oil with the injection fluid. This process of developing miscibility is classified as "***multiple-contact miscible displacement.***" It should be pointed out that because mixtures in the reservoir miscible region remain as a single phase, the wettability of the rock and relative permeability lose their significance since there is no interface between fluids. However, the mobility ratio has a significant effect on the recovery efficiency simply because it is a strong function of the viscosity ratio of the miscible solution and the displaced oil.

In the remainder of this chapter, miscible injection fluids that achieve either first-contact or multiple-contact miscibility are called miscible "*solvents.*" Both types of miscible displacement are reviewed next.

First-Contact-Miscible Displacement. Liquefied petroleum gas (LPG) products such as ethane, propane, and butane are the common solvents that have been used for first-contact miscible (FCM) flooding, i.e., miscible with reservoir crude oils immediately on contact. The LPG products are miscible with oil only as long as they remain in the liquid state, i.e., when reservoir temperatures are below their critical temperatures and at pressures at or above their vapor pressures. The temperature isotherms for some of the commonly used LPG products are listed in Table 6.4.

For example, at 150°F, methane will remain in the gas phase regardless of the pressure; on the other hand, propane will remain in the liquid state at pressures ≥ 360 psia, while *n*-butane will remain in the liquid state at pressures ≥ 110 psia.

In practice, the LPG solvents are injected as a slug of liquid hydrocarbons of approximately 5% of reservoir pore volume and displaced by a less expensive chase gas such as lean or flue gas. However, the chase gas must be miscible with the hydrocarbon slug to prevent deterioration of the slug at the trailing edge. As noted in Table 6.4, pressures required to liquefy the LPG products and achieve first-contact miscibility are low; however, the required pressure to achieve miscibility between the chase gas and the LPG slug is much higher than the pressure required to liquefy these hydrocarbons. For

TABLE 6.4 Temperature-Pressure Relationship to Maintain Liquid State

Methane		Propane		*n*-Butane	
T, °F	P, psia	T, °F	P, psia	T, °F	P, psia
50	460	50	92	50	22
90	709	100	190	100	52
		150	360	150	110
		200	590	200	198
		206*	617	250	340
				300	530
				305*	550

*Critical temperature.

example, at a temperature 160°F, butane is miscible with methane as the chase gas at pressures greater than 1600 psia and with nitrogen (as a chase gas) at pressures greater than 3600 psia, even though the miscibility between butane and oil is attained at a pressure of approximately 125 psia. Therefore, one of the basic requirements for LPG slug injection is that the solvent slug must be miscible with both the reservoir oil and the drive gas. To improve the overall sweep efficiency by LPG process, the hydrocarbon slug is displaced by alternating the chase gas with water slug and eventually with continuous water injection, as shown in Figure 6.29.

Multiple-Contact Miscible Displacement. A valuable approach for representing the phase behavior of multicomponent hydrocarbon mixtures and their interaction with a displacing gas is the use of the pseudoternary diagram as shown in Figure 6.30. The components of the reservoir fluid are grouped into three pseudocomponents located on the corners of the ternary plot. One possible grouping that has been used frequently includes the following mixed components:

- **Component 1**: Represents a volatile pseudocomponent and is composed of methane, nitrogen, and carbon dioxide located on the uppermost corner of the triangle.
- **Component 2**: Represents a pseudocomponent that is composed of *intermediate hydrocarbon components* such as ethane through hexane. The component is located on the lower right corner of the plot. It should be pointed out that sometimes CO_2 is included with the intermediate components.
- **Component 3**: Is essentially the heptanes-plus fraction "C_{7+}" and is located on the lower left corner of the plot.

Each corner of the triangular plot represents 100% of a given pseudocomponent, progressive from 0% at the opposite side of each corner to 100% (usually with incremental step size of 10%). An area in the graph surrounded by the curve "ACB" is called the ***phase envelope*** and represents the phase behavior of mixtures with varying combinations of the three pseudocomponents. For example, point "Z" in Figure 6.31 represents a mixture that is composed of 50% of "$C_1 + N_2 + CO_2$," 20% of "C_2-C_6," and 30% of C_{7+}. At the pressure and temperature of the diagram, any system of the three components whose composition is inside the phase envelope, e.g., point Z, will form two phases, a saturated gas phase with a composition represented by point "Y" and a saturated liquid phase composition represented by point "X." The dashed line connecting points X and Y

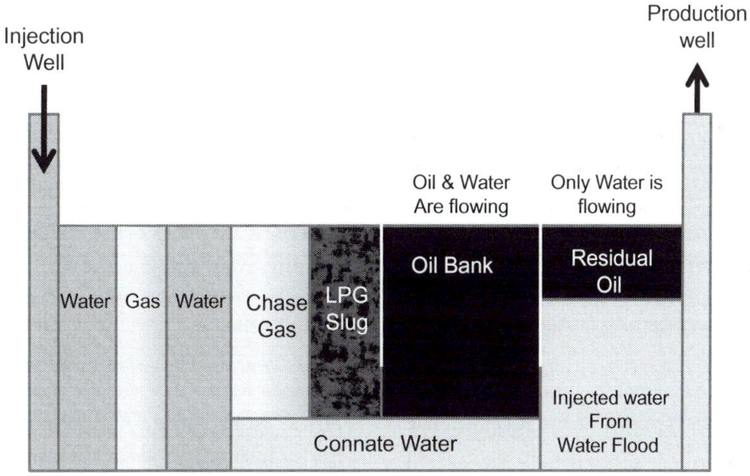

FIGURE 6.29 LPG flood as an EOR process.

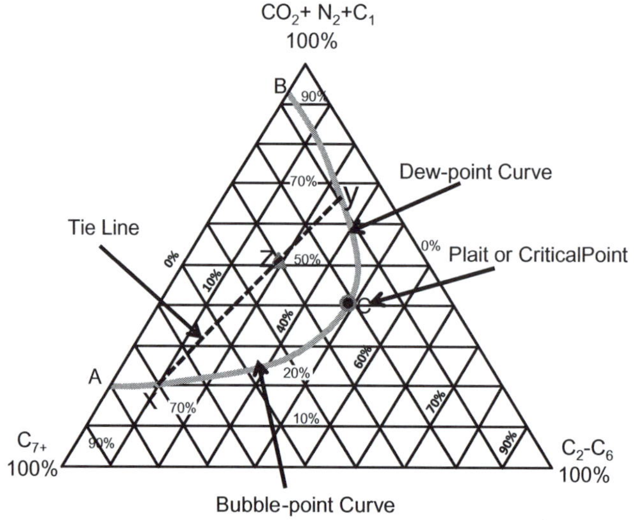

FIGURE 6.30 Ternary diagram.

and passing through point Z is called the ***tie line***. The segment "AC" of the curve is called the ***bubble-point curve*** and represents the composition of the saturated liquid, with the segment "BC" called the ***dewpoint curve*** representing the composition of the saturated gas. The dewpoint curve joins the bubble-point curve at the ***plait point*** (***critical point***) "C," which indicates that the compositions and properties of the equilibrium gas and liquid are identical.

There are two additional principles that must be recognized when representing phase behavior relations with the ternary diagram:

- In general, if the coordinates of the overall composition "z" of a hydrocarbon mixture place the mixture within the phase envelope, as shown in Figure 6.30, the mixture will form two phases (liquid and gas) under the prevailing pressure and temperature; however, when placed outside that phase envelope, the mixture will exist as a single phase.
- The ternary diagram shown schematically in (Figure 6.31 illustrates the concept and the basic requirement for achieving first-contact miscible (FCM) displacements and identifies the multi-contact miscible (MCM) region. The illustration shows two different mixtures of hydrocarbon gases and a crude oil system at a constant pressure and temperature. These three hydrocarbon systems are:
  - lean gas with a composition represent by A;
  - rich gas (LPG diluted with lean gas) represent by point D; and
  - original oil-in-place with a composition represented by point B.

The basic theory of miscible and immiscible displacement processes suggests that constructing a straight line between two points representing the compositions of the injected hydrocarbon system (point A or B) and that of the oil-in-place (point B) will identify the type of the occurring displacement process. If the constructing line crosses the phase envelope (i.e., the two-phase region), it indicates a multi-contact displacement process might occur between the injected gas and oil-in-place; if not, it indicates a first-contact miscible displacement process will be achieved. As shown in Figure 6.31, line AB crosses the two-phase region indicating that the injected gas of composition "A" is not a first-contact miscible with

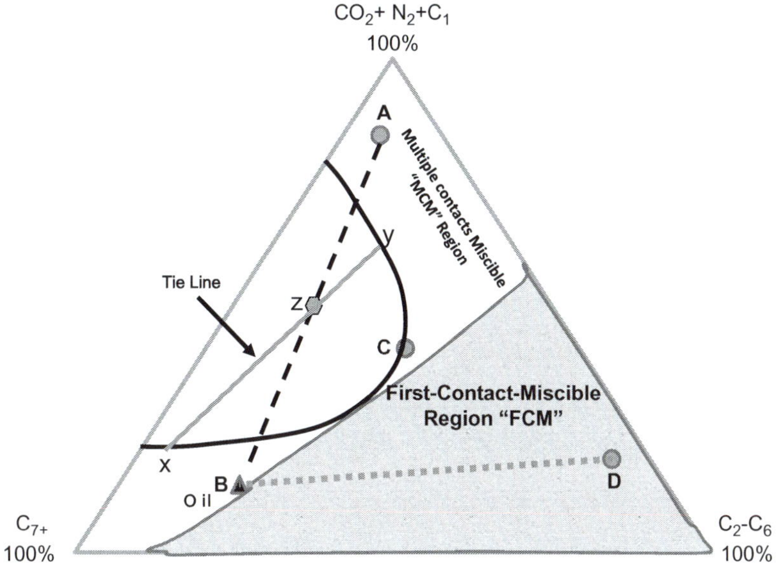

FIGURE 6.31 First and multiple-contact miscible regions.

crude oil of composition B. The combined overall mixture will form two phases with compositions represented by points x and y. The resulting tie line crosses the line "AB" at point "z," which represents the overall (combined) composition of the injected gas "A" and crude oil "B." Drawing the straight line "DB" between point "D" that represents the composition of the injected rich hydrocarbon (i.e., LPG + lean and nonhydrocarbon components), and point "B" shows that the line does not cross the phase envelope indicating that the injected fluid remains in the single-phase region and, therefore, will attain a first-contact miscibility with the crude oil system. It should be pointed out that any injected solvent slug, e.g., LPG, with a composition located in the shaded area of Figure 6.31, will achieve an FCM with the crude oil represented by point B.

Based on the above discussion, the differentiation between FCM and MCM is based on whether the straight line connecting the composition of the injected fluid with that of liquid will or will not cross the phase envelope. However, the size of the two-phase region

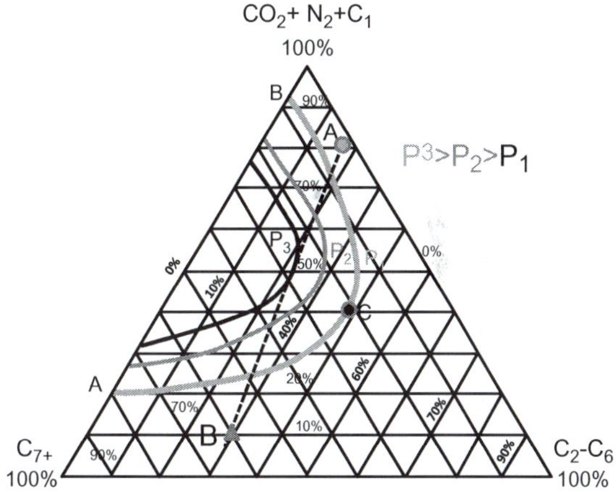

FIGURE 6.32 Effect of pressure on the size of the phase envelope.

(phase envelope) depends upon the pressure and temperature. For a constant temperature, the size of the two-phase region will shrink with increase in the pressure, as shown schematically in Figure 6.32. The illustration suggests that first-contact miscibility will be achieved at a pressure equal to P_3. This pressure is termed ***the minimum miscibility pressure*** "MMP."

CHAPTER

7

Economic Analysis

7.1 INTRODUCTION

This chapter covers the basics of oil and gas economic analysis, examines international petroleum fiscal regimes and discusses issues associated with reserve reporting. The oil and gas industry has invested billions of dollars in finding, discovering, developing, producing, transporting, and refining hydrocarbons for more than a century and has long been an enormous source of wealth creation. In countries such as the United States, where a great deal of the ownership of subsurface mineral rights is privately owned, individuals and corporations have been able to generate significant wealth through the extraction of oil and gas. In most countries, such mineral ownership resides with the state. Historically, major international oil and gas companies (IOCs) played a major role in taking the risks of exploring for oil and gas around the world providing risk capital, development capital, expertise, and personnel to many nations who were the owners of their resources. Such sovereign nations established national oil companies (NOCs) to both manage the relationships with IOCs and ultimately develop resources independently of the IOCs. The most technically advanced and financially capable of such NOCs now compete technically and financially on the global stage, to the point where there is now a recognized group of strongly capitalized international national oil companies (INOCs) competing with the IOCs for access to resources. Over the last few decades the "super majors" and recently even relatively small independent oil companies have found niche positions exploring for and developing oil and gas resources around the world. The playing ground has been leveled by the widespread access to technology often provided by universities, research organizations, and service companies. Service companies have also participated in developing and (less frequently) producing oil and gas fields around the world.

Reservoir engineering deals with all phases of the production of oil and gas. Most of this book deals with the physics associated with fluid flow in porous media, estimating future recoveries and enhancing both rates of recovery and ultimate recovery. Reservoir engineers *must* also fully comprehend the economics associated with oil and gas decisions. Lester C. Uren is credited with writing the earliest textbooks in petroleum engineering, and in the 1924 Preface of his book *Petroleum Production Engineering* he states:

> *"The engineer is both a technologist and an economist. In his professional work, his objectives require not only an application of science to the needs of industry but also achievement of these objectives within economic limits that will result in financial profit."*

In most cases, reservoir engineers serve as analysts who make recommendations individually or (more typically) as part of a team. The economic decisions they make must be comprehensible to decision makers, reflect value and risk correctly, and properly compare alternatives. While many economists have excellent skills in this area, it is the role of the integrated team to capture the best technical decisions and translate them into proper decisions. Integrated asset teams comprise reservoir and production engineers, geologists, geophysicists, and other specialists such as geomechanics and petrophysics experts.

In the next few examples, typical decisions that reservoir engineers routinely must evaluate are discussed. The way reservoir engineers evaluate cash flows, capital investments, and decisions under uncertainty (risk) is described here. Finally, a series of advanced topics including typical alternative schemes for shared risk and revenue is illustrated.

The following examples are used to illustrate typical questions a reservoir engineer may be called on to answer. We will revisit some of these examples to understand the calculation of economic parameters and decision-making criteria.

7.1.1 Tight Gas Optimal Spacing Example

In a tight gas reservoir, the ultimate theoretical *technical* recovery (neglecting reservoir heterogeneities, liquid loading, etc.) may not be a strong function of spacing. Figure 7.8 shows the results of a series of reservoir simulation runs in a tight gas reservoir with an average permeability of 0.008 mD and a thickness of 55 ft. The same hydraulic fracture lengths and the same absolute fracture conductivities are used in each case. These cases resulted in varying ratios of fracture length to drainage radius; however, this primarily affected the transient behavior and length of time before boundary effects were felt. The reservoir spacing for the cases ranged from 40 to 640 acres per well meaning that in one square mile there could be as few as one well or as many as 16 wells. In practice, these wells would be drilled over time, and the earliest wells would produce higher rates and ultimate recoveries than the later wells. In this simplified example, no maximum total production constraint is imposed on the entire field, i.e., it is assumed that there are no gathering systems, facilities, compression, or sales constraints. This example oversimplifies the problem for illustration purposes but can be solved more accurately by incorporating timing, heterogeneities, varying hydraulic fracturing results, well location issues, etc.

The gas rates for each case are plotted as a function of time; each case represents the combined production from all wells in the drainage area. The initial rate of the 40-acre case is approximately 16 times the initial rate of the 640-acre case because the early transient behavior of each well is essentially identical. The length of time required for each well to reach an estimated minimum rate of 20 Mcf/D varies from less than 9 years in the 40-acre case to more than 100 years in the 640-acre case. The simulations were not allowed to run in excess of 100 years. Ultimate recovery varies relatively little between these cases. The question the operator must ultimately answer is "How many wells should I drill?" (Figures 7.1 and 7.2)

The reader may have an instinctive belief that recovering 11.8 Bcf in 41 years (from four wells on 160-acre spacing) is significantly better than recovering a similar quantity in 87 years (from two wells on 320-acre spacing); does the improved value of accelerating gas recovery warrant the costs of two additional wells? It is more difficult to decide whether relatively modest acceleration, such as the difference in the 40- and 80-acre or 80- and 160-acre cases, is worth doubling or quadrupling the amount of cash required for capital investments. It is up to the integrated efforts of the reservoir engineer and economist (and many reservoir engineers function as both) to identify the optimal capital investment for this project and to rank such an investment with respect to alternative capital investments. The answer to this problem is a function of many factors including future product prices, limits on the surface constraints, regulatory constraints, the capital costs of the wells, operating expenses, completions

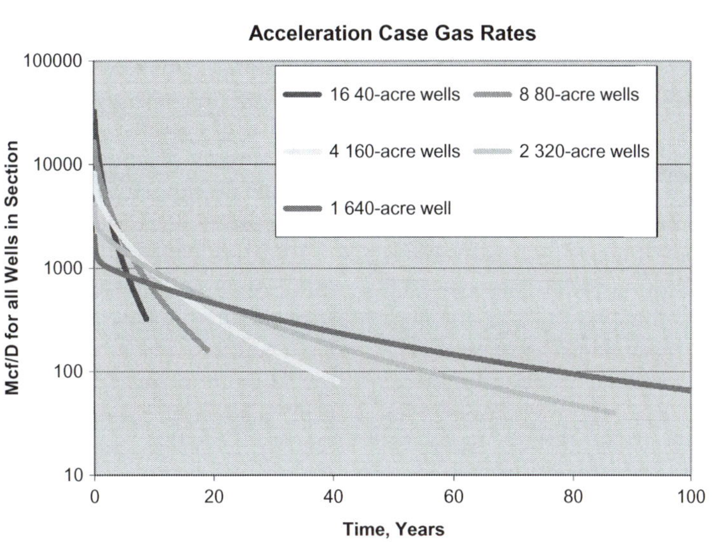

FIGURE 7.1 Comparison of rates as a function of time for different well spacings in homogeneous tight gas sand.

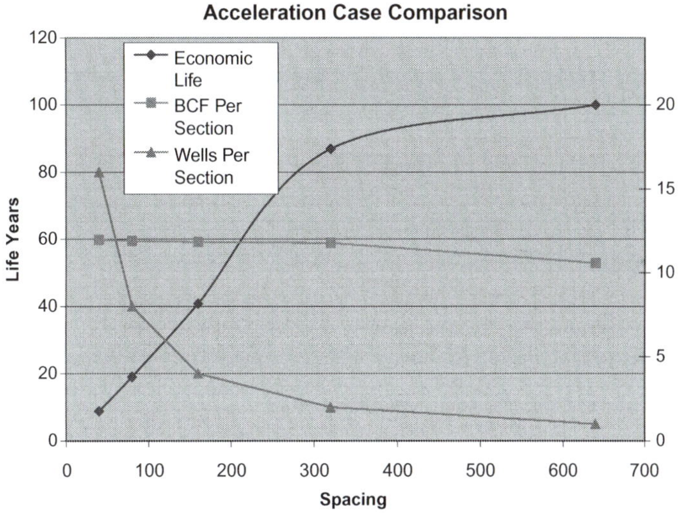

FIGURE 7.2 Economic life, ultimate gas recovery, and number of wells required as a function of well spacing.

including tubular and liquid handling issues, reservoir heterogeneities, etc. In practice, reservoir engineers and geologists often underestimate the level of complexity of reservoirs leading (in the case of most tight gas reservoirs) to an optimal well spacing that is tighter (higher well density) together than that indicated in a homogeneous reservoir model. Similarly, drilling additional wells in very similar geologic environments can lead to substantial efficiencies and optimization of drilling and completion practices. Higher well densities provide redundancy that may be beneficial if one or more producing wells fail mechanically and the cost to restore the damaged wells to production cannot be justified. Repairing or replacing wells late in a reservoir's life may also be technically more difficult as the low reservoir pressures present drilling and completion challenges. While these problems are commonplace, it is difficult to quantify which, if any, will occur.

7.1.2 Drill vs. Farmout Example

In this example, the reservoir engineer is called on to estimate the potentially recoverable resources in an exploration project that is located near a series of small discoveries. The engineer has a map generated (by geologists and geophysicists) from seismic, geological, and petrophysical data and has estimated a "most likely" gas-in-place based on the analysis of logs from the nearby discoveries, their actual gas–water contacts and spill points, anticipated pressures, etc. Well costs have been estimated, and a success case development plan involving the discovery well and one development well has been made. The company already owns leases for the prospect and has asked the engineer if the exploration prospect should be drilled. This prospect is considered an excellent analog to offset discoveries that have been drilled with very high success rates. There are varying estimates for the likelihood of a discovery ranging from 50% to 90%. In the case of an initial dry hole, it is anticipated that no further expenditures are likely, and the lease and prospect will be abandoned. Another operator has offered to "farm-in"[1] the acreage, assuming all exploration well costs in return for earning 75% of the block. After the other operator recovers 150% of their investments, the reservoir engineer's company share would rise to 40%. The reservoir engineer must now evaluate retaining all of the future ownership in the field while putting his company's capital at risk compared to retaining 25% (or ultimately more) of the block but without any risk.

What factors dictate the answer and determine the operator's decision? While the intrinsic economic attractiveness of the project and the "chance of success" are predominant, the capital situation of the company, its portfolio of investment opportunities, etc., are all important. In Figure 7.3, a net present value at 10%[2] (NPV10) is shown for the "drill" and "farmout" cases as a function of the probability of success. In this case "success" is a single case, *viz.* the discovery of the gas field precisely with the ultimate recovery and timing estimated pre-drill. The failure case is a single dry hole. In reality, reservoir engineers evaluate numerous alternative cases including a continuum of potential cases, and advanced techniques described later are used (Figure 7.3).

Which one is preferable, drilling the well or farming out? It is clear that at low chances of success, the farmout case is always superior, while as the probability of success approaches unity, the drill case becomes increasingly better. But what about a case with 60% probability of success (40% probability of a dry hole)? The NPV10 is slightly greater than farming out. However, there is a reasonably large amount of risk being taken for a small incremental benefit.

7.1.3 Value of Advanced Technology

A final example illustrates the tie between uncertain reservoir conditions and decisions related to the development of the reservoir. In this case, the operator has only one available surface location to develop two distinct fault blocks and will drill a multilateral well. A low-cost option is proposed by the drilling department in which the well is completed in a fashion that does not allow selective shutoff of either lateral. The reservoir engineer has identified strong water drives in most of the fault blocks but has had limited success in predicting when water production will start and has

---

[1] A "farm-in" or "farmout" is when one party (the farmor) holding the rights to drill on a lease assigns some fraction of these rights to another party (the farmee). These agreements can include assigning obligations, cash, warrants, information, technology, etc., in addition to the drilling and production rights. Terms range from simple to bewilderingly complex.

[2] For the purpose of this example, a 10% discount rate has been used. The appropriate discount rate may be different and is a function of the systematic risk associated with the investment opportunity

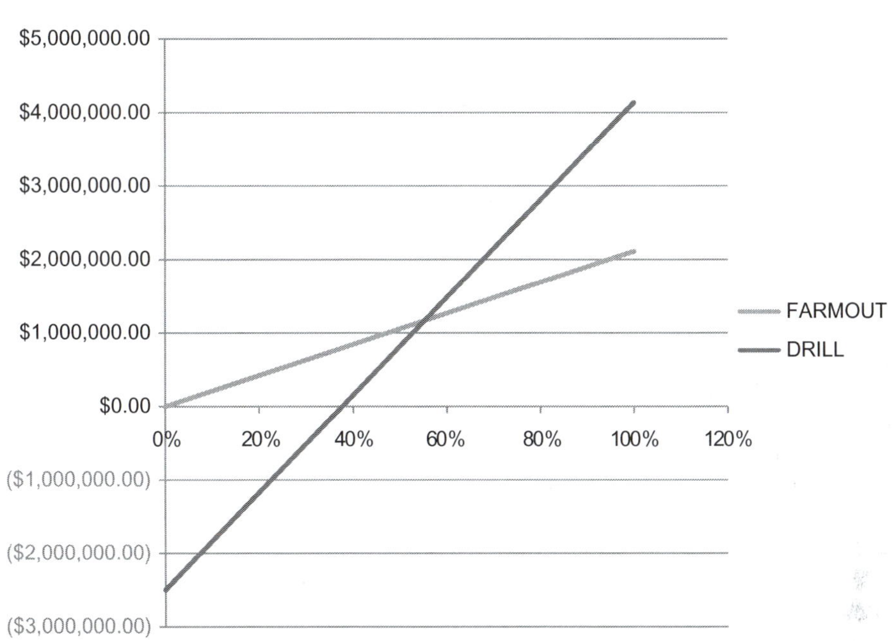

FIGURE 7.3 NPV of drill and farmout cases vs. chance of dry hole.

recommended a more costly, intelligent well completion coupled with inflow control devices (ICDs) that allows the operator to selectively choke either lateral from the surface. Because of the significant incremental cost, management needs to know how much extra oil will be produced and how much operating costs will be reduced based on lower water-handling. In practice, advanced simulation techniques, such as experimental design, can be used to evaluate a wide range of potential outcomes and estimate an "expected" case result. In this example, only the "no water for first 10 years" case yields a large negative NPV10 for the intelligent completion case, and the operator can see the advantage of the intelligent well completion with ICDs.

7.2 EVALUATION CRITERIA AND CASH FLOW ANALYSIS

7.2.1 Payout

In this section, many of the common measures of cash flow are defined and illustrated. Perhaps the most common economic measure and one of the simplest is "payout." Also known as payback time, it is defined as the length of time required to recover invested capital and expenses. In other words, it is the length of time to reach a cumulative cash flow of zero. Payout makes the most sense for economic projects in which there is an initial negative cash flow followed by a period of (in aggregate) positive cash flows. Payout comes in several versions, including discounted payout.

In an example (Table 7.1) we will revisit several times during this and following chapters, an initial investment is followed by a series of monthly cash flow dues to oil production. This example case can be found on the website of the Society of Petroleum Evaluation Engineers (SPEE)[3] along with a number of guidelines that are relevant to reserve reporting, oil and gas property evaluation, and economic appraisals. Some of the common assumptions in the case include (see Table 7.1):

[3]SPEE "...was organized exclusively for educational purposes and to promote the profession of petroleum evaluation engineering, to foster the spirit of scientific research among its members, and to disseminate facts pertaining to petroleum evaluation engineering among its members and the public." http://www.spee.org/index.html

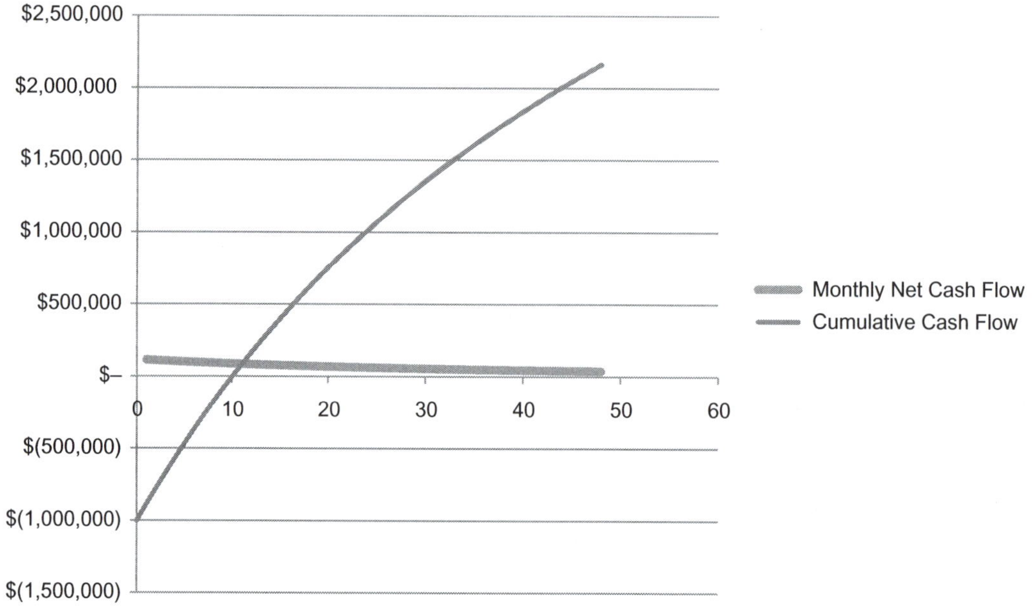

TABLE 7-1

Initial Oil Rate	200 BOPD
Investment	$1,000,000
Decline rate	0.3 $year^{-1}$
Hyperbolic exponent	0.5
Constant oil price	25 $/bbl
Net revenue interest	85%
Working interest	100%
Operating costs	2000 $/month
Severance taxes	10%
As of date	1/1/2000
Evaluation date	1/1/2000

From these assumptions, we can calculate monthly cash flows. These monthly cash flows can be accumulated as in the above figure.

In this example, a visual inspection of the cash flows shows that the payout is in approximately 10 months. A close examination of the monthly cash flows shows that the cumulative cash flow in 10 months is a negative $6401, while at the end of month 11 it is positive $78,717. Common practice for economic evaluators is to interpolate (linearly) which would result in a payout of just less than 10.1 months or 0.84 years. Payout can be expressed in any common time units. It is a common practice in many software programs to linearly interpolate payout at the finest level of timing used in the evaluation. If that is annual, the linear interpolation assumption could introduce a noticeable error. Payout measures are inherently imprecise because the actual cash flow timing is almost never precisely correct. While a well may produce daily, operators rarely get paid daily for that production. Many analysts (including the published SPEE example) ignore the variation in days/month and leap year. Others take these variations into account. Those who do account for this should not use fixed expenses as $/month. Payout should not be reported to an excess number of significant figures.

After-tax calculations of payout are commonplace as are "discounted" payouts. Unless otherwise noted by the analyst, "payout" is assumed to be a before-tax undiscounted measure. Company procedures and practices will dictate consistent approaches to calculating each of these economic parameters.

Payout seems simple; however, there are a host of circumstances that lead to inconsistencies in its calculation. These include:

- questions about when to start the clock;
- situations with multiple future investments including multiple times when the cumulative cash flow reaches zero;
- situations where the cumulative cash flow is always positive;
- incremental evaluations.

The initial investment in the prior example was assumed to be at "time zero," and the SPEE example does not specify the investment to be a drilling cost. It is possible that a well or a field producing 500 BOPD could be purchased on a given date and then it could be generating cash for the account of the owner the very same day. In the case of drilling a well, a certain amount of time is required to drill, complete, equip, and begin producing the oil. In more complex situations, an operator might have the following series of expenditures:

- Obtain offshore studies and "spec" seismic.
- Invest months of geological and geophysical time to analyze and identify potential blocks in an upcoming offshore bid round.
- Bid on several blocks, winning at least one.
- Conduct a 3-D seismic study on the newly acquired block(s) and conduct further Geological and Geophysical (G&G) studies along with engineering studies for drilling, completing, and producing any potential discoveries.
- Drill one or more exploration wells.
- Drill appraisal wells as needed.
- Test appraisal wells as needed.
- Conduct further studies and sanction a development project.
- Build and install an offshore platform with corresponding facilities. Complete the drilling and equipping the necessary wells.
- Construct oil and/or gas pipelines or alternative methods to transport products to market. This may include onshore facilities, such as crude stabilization or even electric power generation, to monetize natural gas in areas without ready markets.
- Start production.

It is entirely possible that the time period for this project could run for in excess of 10 years prior to any positive cash flows being generated. For preparing economics of a bid for a lease sale, it is typical to use as 'time zero" the date the bid will be submitted, and prior expenditures are usually ignored. The logic in this approach is that the prior expenses are part of more general 'exploration costs,' such as employing internal expertise. However, once the first exploration well is to be drilled, it is not unusual to "reset the clock" and recalculate payout at the time the exploration well is to be drilled. This is often repeated at the major capital investment steps. The logic is that prior expenditures are "sunk costs" and that the current decision is being made on the basis of the immediate decisions to be made. In such cases, after-tax calculations must correctly incorporate the tax implications of legitimate alternatives such as abandoning a block and writing off prior expenditures.[4]

A general rule is that the appropriate evaluation time is at the beginning of "substantial" expenditures. In the offshore case, this could have been the lease bonus or the exploration well. In the case of construction of a facility such as a gas processing plant, it would generally not be until construction actually commenced and would depend on the terms of the contract stating when payments are due. In all cases, the analyst should clearly state when payout starts and the assumptions involved.

In many cases, a moderate-sized project is commenced to fully evaluate a technology or to generate early cash flow while additional opportunities to develop more of the field mature. One example is a thermal recovery project in which a few patterns are steamflooded under existing permits and full-scale development

[4]The handling of both the financial and tax calculations vary by country and often by the specific circumstances of the operator.

might not be permitted (or possible) for a few years due to regulations, available fresh water, processing facilities, etc. If the steamflood project is being evaluated, the first project might very well pay out prior to expenditures taking place for the larger project. Payout can be stated for the small project, and an incremental approach can be used to show the larger project payout, resetting the clock to time zero when the large expenditures commence.

Payout is rarely the determining or sole economic criterion in complex or large projects. It is best suited for simple projects such as a workover in which the investments are relatively small and the characteristics of the resultant cash flows are familiar to decision makers. Payout's major weakness is that it gives decision makers no idea how much money the project actually makes. Other weaknesses include only a simplistic piece of information about timing.

7.2.2 Number of Times Investment Returned

NTIR or number of times investment returned is similar to the profit-to-investment ratio. The PI ratio is often called ROI; however, this is discouraged because of inevitable confusion with DCFROI. It is a simple, undiscounted measure of the total net cash flows *other than the "investment."* In the prior example, the cumulative undiscounted cash flow ultimately produces $4,914,952 in net cash flow or $5,914,952 in positive cash flow following from a $1,000,000 investment (negative cash flow). The NTIR in this case is 5.9. The profit-to-investment ratio is defined as the cumulative profit divided by the investment and would be 4.9. In a project that just recovers its investment, the NTIR is 1.0, while the PI ratio is 0.0.

NTIR is a poor choice when comparing projects with different profiles. In the case of the tight gas case illustrated previously, the NTIR for the 640-acre case is going to be many, many times higher than that of the infill cases. NTIR can become very large when investments are small. It is sometimes the case that other investments are made over time. Some analysts include these in cash flows, while others include them in the initial investments. NTIR is most useful when comparing projects that have similar cash flow profiles and timing. A comparison of a large number of economic evaluations of Austin Chalk horizontal wells showed that NTIR ranked projects nearly the same as other more sophisticated techniques. NTIR may also be useful when there are high decline rates and low escalation rates leading to the majority of cash flows occurring early in the life of the project. NTIR is particularly misleading in very long-life projects with escalated prices.

While the authors do not care much for either NTIR or PI ratio as evaluation criteria, the main thing to remember is to only use *one* of them along with other measures of economic value. Decision makers can get used to either, but using both will inevitably lead to more confusion than these criteria are worth.

7.2.3 Discounting of Future Cash Flows, Time Value of Money

Money has a time value. This is self-evident. Would you prefer to have $1000 paid to you annually for the next 30 years or to receive $30,000 today? While there may be certain circumstances where the delayed payments are preferable,[5] it is generally obvious that a dollar

[5]One example might be when there are large taxes due on a large amount of money received, but those taxes would be much less over time. Imagine the case of $100,000 paid annually or $3,000,000 paid today. A highly progressive tax scheme might make the former preferable. For the purposes of this discussion, taxes have been neglected. In another example, a spendthrift might easily waste the larger sum paid today due to a lack of discipline (imagine a lottery winner). If the amounts are large, the individual decision maker to receive the funds might not value the sum of the future payments much more than a single payment, i.e., the recipient has a nonlinear utility function. For the purposes of this discussion, we will generally assume that the decision makers have a unit slope, linear utility function over the range of the decisions, and are financially disciplined. Exceptions to these assumptions will be examined in Section 7.7.3.

received today is worth more than a dollar received in the future, assuming that the current and future dollars have the same tax impacts. The concepts of discounting and compounding are closely linked. Most people readily understand the concept of compounding and that if a sum of money (say \$100) is put into a bank account that pays 5.0% interest, then in one year's time the future value (FV) of the \$100 would be \$105. At the end of 2 years, the FV would be \$110.25. There are, of course, variations on compounding, and not all 5.0% interest rates are created equally. The general relationship for compounding is:

$$\mathrm{FV} = \mathrm{PV}(1 + i)^n$$

The value i is called the *compounding* rate and is used to *compound* present values (PVs) to determine future values. The compounding rate is often referred to as the interest rate. Interest rates are intrinsically associated with lending, and discussions of interest rates are often confused with the specific financial instruments, monetary policy, inflation issues, etc. Some of the economic indicators (GRR) we will discuss require compounding, and it is more appropriate to use the compounding rate than the term *interest rate* for such indicators.

Discount rate has a specific meaning in the realm of oil and gas evaluation, which is different than other common meanings.[6] Discount rates are the rates used to convert future values to present values. Evaluations often use more than one discount rate, and the calculation of PVs as a function of discount rate is common. Most oil companies (or banks, regulators, investors, and others who review evaluations and recommendations) have one or two specific discount rates that are considered. There is also a particularly important discount rate used by corporations known as the weighted average cost of capital (WACC).

[6]Other meanings deal with the interest rate charged to banks for short-term borrowing directly from the federal reserve, fees charged for accepting credit cards, etc.

The present value of a series of discrete cash flows over n time periods at a given discount rate i is calculated as:

$$\mathrm{PV} = \sum_{t=0}^{n} \frac{\mathrm{FV}(t)}{(1 + i)^t}$$

The time periods can be any convenient ones but *should be consistently applied and clearly stated.* Annual discounting is common for long-lived projects, particularly for those with slowly changing cash flows over time. Monthly discounting may be as (or more) commonplace than is annual discounting. The following section discusses the specific discount approaches and their relative merits.

Is there a "correct" discounting method? While some may not agree, there is in fact a "best" method and that is a discounting method that *most nearly approximates the actual cash flows.* If payments are made annually on a given anniversary date (as in a typical lottery payment), then annual discounting (specifically annual end-of-period discounting) exactly models the cash flow and would be the most accurate way to value that cash flow stream.

Because of accounting practices along with production reporting practices, many analysts argue in favor of monthly discounting. In evaluations of properties with rapidly declining (or increasing) cash flows over a short time period near the beginning of the evaluation, monthly discounting will more accurately rank projects than will annual discounting. But if monthly discounting is better than annual discounting, why not weekly, daily, or by the microsecond? Cash flows received from oil and gas operations might well be as infrequent as monthly but in fact are nearly continuous.

In the case of continuous discounting of future cash flows:

$$\mathrm{PV} = \int_0^T \mathrm{FV}(t)\, e^{-\lambda t}\, dt$$

Where $\lambda = \ln(1 + i)$. Continuous discounting methods applied to discrete cash flows have a simple relationship, and the continuous discounting approach is simply a variation on the discount rate used. The technique is most helpful in cases where the cash flows can be described analytically as continuous functions, and its use is relatively uncommon in oil and gas evaluations.

Because monthly discounting is so widely used, several characteristics of evaluation approaches should be discussed, including:

- How to handle varying days per month and per year?
- How to reconcile monthly interest rates with annual discounting approaches?
- What monthly interest rate to use if the monthly interest rate method is to be used?

Some analysts use the actual days per month corresponding with the specific calendar dates of the evaluation including leap year.[7] Others use equally sized months and account for leap year by using years of length 365.25 days. A month would then have 365.25/12 days or 30.42 days. For the purpose of ranking and evaluating oil and gas investment decisions, these methods are generally identical. In the calendar-correct approach, the variations in days per month will result in variations in estimated volumes per month that would look a bit odd if portrayed graphically with equal spacing for months. Graphical displays should reflect elapsed days or account for monthly spacing properly. Similarly, cost estimates based on so many $/month per well or per facility would lead to slightly odd cash flow estimates, *particularly as the project reaches the economic limit.* These can all be resolved and neither method is radically better than others. The method used must be communicated clearly.

In this chapter, some examples (such as the SPEE examples) use 365.25 days per year and equally sized months. Reservoir simulation output, such as the tight gas example, will tend to use the actual calendar days. In the latter case, an operating expense of X $/month could be treated as $(X/30.42) \times$ actual days in each month. Either method may be sufficiently accurate. Consistency and clearly stated assumptions are important.

Discounting Monthly Cash Flows. One method of discounting monthly cash flows is to use the number of years (a noninteger) corresponding to the months and use the annual discount factor.[8] Another alternative is to use a monthly discount rate and to use the integer months for the calculation. In the latter case, there are two common methods to convert the annual discount rate to a monthly discount rate. In the former, the "APR" or "annual percentage rate" familiar from home loans and credit cards is used. The other is the "effective monthly interest rate." This approach results in the compounded monthly interest rate being equal to the annual interest rate. A simple example of these approaches follows. The annual discount rate used is 10% with 12 monthly cash flows of $1000 to be discounted using the end-of-period approach. Method A uses the noninteger years approach, and the first month's discounted cash flow would be:

$$\text{PV} = \frac{\$1000}{(1+0.1)^{1/12}} = \$992.09$$

In Method B (the APR approach), the monthly interest rate would be 0.1/12 or 0.8333%. In Method C (the effective monthly interest rate approach), the monthly interest rate is calculated from:

$$i_{\text{monthly}} = \frac{\ln(1+i)}{12}$$

or

$$= \frac{\ln(1.1)}{12}$$

or

$$= 0.7943\%$$

[7]A few go so far as to reflect the 23- and 25-hour days corresponding with changes for daylight savings time, a detail that most analysts ignore in property evaluations.

[8]See the SPE-recommended practices for more detail.

TABLE 7-2

Annual discount rate			**10%**		
Monthly discount rate APR			**0.8333%**		
Monthly discount rate effective			**0.7943%**		
Month	**Years**	**Cash Flow**	**Method A (Monthly, Noninteger)**	**Method B (Monthly, APR)**	**Method C (Monthly, Effective)**
1	0.083	\$1000.00	\$992.09	\$991.74	\$992.12
2	0.167	\$1000.00	\$984.24	\$983.54	\$984.30
3	0.250	\$1000.00	\$976.45	\$975.41	\$976.55
4	0.333	\$1000.00	\$968.73	\$967.35	\$968.85
5	0.417	\$1000.00	\$961.07	\$959.36	\$961.22
6	0.500	\$1000.00	\$953.46	\$951.43	\$953.64
7	0.583	\$1000.00	\$945.92	\$943.56	\$946.13
8	0.667	\$1000.00	\$938.44	\$935.77	\$938.67
9	0.750	\$1000.00	\$931.01	\$928.03	\$931.28
10	0.833	\$1000.00	\$923.65	\$920.36	\$923.94
11	0.917	\$1000.00	\$916.34	\$912.76	\$916.66
12	1.000	\$1000.00	\$909.09	\$905.21	\$909.43
		\$12,000.00	\$11,400.49	\$11,374.51	\$11,402.78

In these last two methods, the first month's cash flow would be calculated as:

$$PV = \frac{\$1000}{(1 + 0.08333)^1} = \$991.74 \quad \text{APR approach}$$

$$PV = \frac{\$1000}{(1 + 0.07943)^1}$$
$$= \$992.12 \quad \text{Effective monthly rate approach}$$

The APR method is more intuitive, while the effective monthly rate approach is closer to the annual approach, particularly when mid-period discounting is used. See Table 7.2.

At higher discount rates, the methods result in greater differences.

7.2.4 Period Discounting

Assuming that all cash flows occur at the end of the time period (end-of-period or EOP) is the most conservative approach for *positive* cash flows and the most optimistic one for negative cash flows. Annual end-of-period (ANEP) is still widely used; however, mid-period (MP) discounting (assuming that all cash flows occur in the middle of the time period) has become increasingly common. While mathematically possible, beginning-of-period discounting is unusual and is not recommended except in special cases.

In the previous discussion of discounting approaches, the noninteger annual rate method and two monthly rate approaches were used that illustrated monthly EOP approaches. To do the exercise with MP discounting, the years in case A would be lowered by $1/(2 \times 12)$ or 0.042. The integer months would be changed from 1, 2, 3, . . ., 12 to 0.5, 1.5, 2.5, . . ., 11.5.

7.3 PRICE ESCALATION AND CONSTANT PRICE CASES

Changing product prices for oil and gas are a fact that is now ingrained in the thought processes of evaluators. Price forecasting can be based on many approaches ranging from fundamental supply and demand predictions through trend analysis. No method has a stellar track record over a significant time period. One approach to deal with such uncertainty is the so-called "constant price" approach. This approach is in fact a forecast, and in many cases it may be as good as any other. Economic evaluations, which include the impact of inflation on the various costs and revenues associated with a project or business opportunity, are more common than are constant price cases.

Most large corporations use an official corporate forecast for consistency among economic evaluations, typically with a maximum price or a fixed number of years of escalations. Some companies include both an escalated price case and a fixed price case to illustrate the impact of price escalation. In constant price cases, escalators for operating costs are also usually set to zero with the only increases in operating costs due to (for example) increased water production. Constant prices do generally reflect contractual changes in prices or costs. Many constant price cases are evaluated at WACC or corporate discount rate minus the inflation rate as the "constant price" discount rate. Some evaluations, such as SEC reserve evaluation, require the use of constant price (flat price) cases. Constant (flat) price cases can be compared to cases showing price and cost inflation to illustrate the value being created intrinsically vs. that created by assumptions on pricing.

7.3.1 SPEE Guidelines for Escalations

The SPEE guidelines for reporting price escalations can be found on their website; however, the focus of these guidelines deals with publishing evaluations. The important factor is a consistent application of escalation approaches and clear communication about the assumptions. Some excerpts of the relevant SPEE recommendations include:

> *"In keeping with general practice, the application of escalation factors should be assumed to start with the second time period. The application of escalation factors should be based on the size of the smallest time period being evaluated. The most commonly encountered time period sizes are monthly and annual, although quarterly or semi-annual time periods may be encountered."*

> *"If monthly cash flows are used, escalation should take place in a 'stair step' fashion on a monthly basis. Thus if prices are assumed to increase at 6% per year, the monthly increase would be based on an effective annual rate of 6% per year with prices increasing every month."*

> *"If annual cash flows are used, escalation should take place in a 'stair step' fashion on an annual basis. Thus, if prices are assumed to increase 6% per year, the price is held constant at the escalated rate for the entire year, then increased 6% for the following year[1]."*

7.4 PRESENT VALUE

Present Value (PV) is particularly useful in ranking comparably sized projects with similar investment requirements. *Net* present value refers to total of all cash flows to the party being evaluated. It can be a pre-tax or after-tax number. The terms PV and NPV are often used interchangeably with some companies using NPV for after-tax calculations. A specific discount rate is also associated with NPV and that rate (expressed as an annual percentage) is often appended as in NPV10 to mean the net present value at a 10% discount rate. It is also important to state clearly the "as of" date of any present value calculation.

In the following example, a series of cash flows is to be compared using the evaluation tools we have discussed so far. Each project has a "time zero" investment *that is not discounted.* A common error among spreadsheet users is to discount all cash flows using (in the case of Microsoft Excel) functions like NPV, which will discount the first cash flow by one time period.[9]

[9]So if the "time zero" cash flow of negative -1000 appears in A1 and annual positive cash flows of 300 each appear in A2 through A6, the correct EOP discount method is not `=NPV(0.1,A1:A6)` or $124.76, it is `=NPV(0.1,A2:A6)+A1` or $137.24. ANMP results in a higher NPV10 of $192.74. This is because the positive cash flows (of $1500) are discounted one half year less (roughly speaking) using ANMP than ANEP.

Time zero cash flows must be added (or subtracted) from the discounted cash flows in a spreadsheet approach. These functions often employ EOP discounting. To use another method, the actual discount factors or equations would need to be included. See Table 7.3 below.

TABLE 7-3

Year	Annual Cash Flows Project A	Project B	Project C	Project D
0	−1000	−1000	−1000	−5000
1	500	100	600	3000
2	500	100	500	2500
3	500	1400	400	20,001
Undiscounted total	500	600	500	25,001
NPV15 ANEP	$141.61	$83.09	$162.82	$814.09
Payout (years)	2	2.5	1.8	1.8
NTIR	1.5	2.57	1.5	1.5

In this case, NPV15 ranks Projects A, B, and C fairly as does simple payout. NTIR suggests B is the best project, but few people would select B over A or C. Project C is superior to A or B at all discount rates above 7%. At lower discount rates (e.g., 5%), Project B ranks ahead of Project C for both NTIR and NPV.

NPV has a number of significant advantages as an economic indicator. These include an indicator of the total wealth being indicated. Unlike DCFROI (discussed in Section 7.4.3) it is not a trial-and-error solution. Even complex cash flows yield unique solutions without the possibility of multiple solutions. Now it is easy to use in conjunction with both probabilistic analyses and with conventional risk analysis. It can be used to compare alternatives when all cash flows are positive or all cash flows are negative. It fairly evaluates lease vs. purchase decisions and forms the basis of evaluation of oil and gas properties.

NPV's greatest drawback is that it does not indicate the rate of cash generation and investment efficiency. Project D in the prior example seems superior to all other methods at all discount rates based solely on NPV. Project D is, of course, just five times Project C. Its greater NPV is due solely to the larger investment. NPV alone cannot rank such projects fairly. Economic parameters that incorporate investment efficiency are required.

In calculating NPV, the appropriate source of the cash flows is net cash flow or NCF. It is inappropriate to calculate NPV based on financial book profit (net income) or cash flow from operations. Net income is not the equivalent of NCF, and CFOPS does not incorporate the relevant cash outflows for investments in fixed or working capital.

In the SPEE example (Table 7.1), a series of assumptions were provided that can be used to illustrate the differences between various discounting approaches. The cumulative undiscounted cash flow is $4,914,952. While there is no end point specified in the table, the economic limit can be either calculated in advance or identified based on the monthly calculated cash flows. To calculate it in advance with complexly varying escalations and changing net interests may not be practical. However, in this case, we can calculate it as:

$$\text{Economic limit} = \frac{\text{Monthly opex}}{(\text{Oil price} \times (\text{NRI}) \times (1 - \text{SEV}))}$$
$$= \frac{2000}{(25 \times 0.85 \times (1 - 0.1))} = 104.58 \text{ BOPM}$$

For an average month, this results in 3.43 BOPD. The time to abandonment can then be calculated from the hyperbolic decline equation:

$$t = \frac{(q_i/q_o)^n - 1}{nD_i}$$

Substituting these values results in 37.17 years. While it is *often* possible to perform this calculation in advance, in practice many analysts prefer to make monthly or annual forecasts and simply cutoff production when cash flows become negative. Analysts should be cautious about making such determinations and have realistic understandings of the actual nature of operating expenses. It is not obvious from a lease P&L (profit and loss) statement which expenses are fixed and which are variable, which of the variable expenses are phase dependent, and what the correct forecast is. It is not the historical expenses that are critical in an evaluation, it is the future ones. In many cases, a single well having a negative cash flow does not cause operators to immediately shut-in or abandon the well. The monthly operating costs might not decrease by the estimated amounts as (for example) the salary of a pumper watching 30 wells does not decrease if one is shut-in or abandoned. In some cases, the operating costs vary by volumes of produced fluids, by groups of shared facilities, offshore platforms, gathering systems, etc.

It is not the historical expenses that are critical in an evaluation, it is the future ones. In many cases, a single well having a negative cash flow does not cause operators to immediately shut-in or abandon the well.

In the SPEE case, we calculate NPV for both monthly and annual forecasts of the production and calculate both NPV and DCFROI. NPV10 is calculated using each of the methodologies discussed. These various approaches make significant impacts on DCFROI as well as NPV. The SPEE-recommended evaluation practice is discussed in the following section.

7.4.1 SPEE-Recommended Evaluation Practice

Cash flows calculated on a monthly basis should be discounted no earlier than the end of the month. When monthly compounding is used, annual interest rates should be converted to effective monthly interest rates through the equation:

$$i_m = (1 + i_y)^{1/12} - 1$$

The methodology used for discounting should be discussed in either the cover letter or the body of the reserve report in such a manner that the user of the report can easily understand the assumptions used. Suggested language for the discussion would be: "The cash flows in this report were determined on a monthly basis and discounted using an interest rate of $X\%$ per annum compounded annually. Cash flows for a month were assumed to occur at the end of the month in which the hydrocarbon was produced."

Cash flows calculated on an annual basis should be discounted using mid-period discounting. The cover letter or body of a reserve report incorporating annual cash flows should discuss the methodology used in a manner that leaves the user of the report with a clear understanding of the issue. Suggested language for the discussion would be: "The cash flows in this report were determined on an annual basis and discounted using an interest rate of $X\%$ per annum compounded annually. Cash flows resulting from production for a period were assumed to occur at the middle of the period in which the hydrocarbon was produced."

Regardless of whether the cash flows from production are modeled monthly or annually, lump-sum cash flows, such as a lease bonus, property purchase, or major investment that will occur at a given date, should be modeled at the date of anticipated occurrence.

7.4.2 Discounted Payout

Payout fans will recognize that payout can also be determined using discounted cash flows. A discounted payout basically answers the question: "When will I have a positive NPV at a given discount rate?" It suffers from all of the weaknesses of payout other than the fact that payout is on undiscounted cash flows. It has an additional drawback in that it is less familiar. Payout is simple; discounted payout requires a specific discount rate and specific discounting approach just as does NPV. Discounted payout is not widely used.

7.4.3 Discounted Cash Flow Return on Investment

Discounted cash flow return on investment (DCFROI) also referred to as discounted cash flow rate-of-return (DCFROR) or internal rate-of-return (IRR), is defined as the discount rate at which the NPV is zero. It is very attractive to

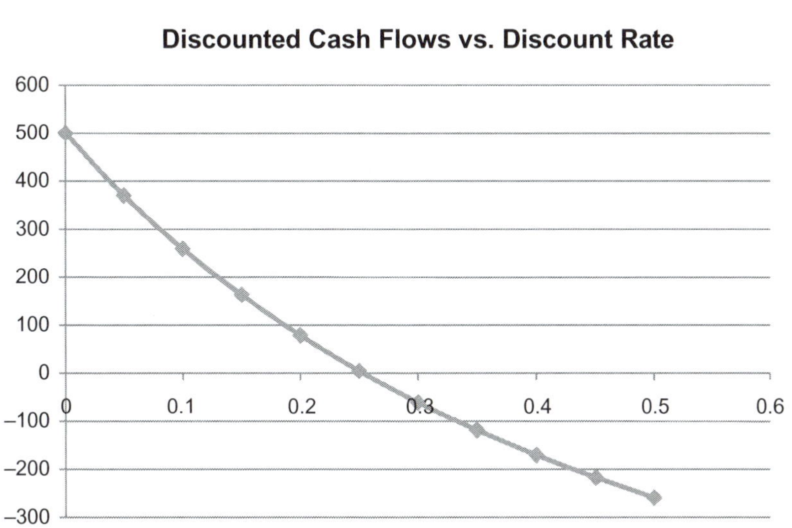

FIGURE 7.4 Illustration of calculating DCFROR.

decision makers because of its instinctive similarity to the interest rate at a bank. Oil and gas property cash flows are not the same as putting money in a bank and receiving small interest payments that are then reinvested over time. DCFROI has several other drawbacks; however, it is one of the most popular investment evaluation tools, particularly for investment efficiency. Let us return to Project C in the prior example and continue to use ANEP discounting but vary the discount rate from 0% to 50% (Figure 7.4).

Values above approximately 25% for the discount rate result in negative values for NPV. Values less than 25% have a positive NPV. Using trial-and-error, the discount rate that sets the NPV to 0 is 25.35%.

Project D has the identical DCFROI to Project C because all cash flows are just five times larger. Project A has a DCFROI of 23.38%, and Project B has the lowest DCFROI at 18.37%. The "best" project will, in fact, depend on the company's applicable discount rate, which for this purpose we will take to be the cost of capital.

Now examine the cash flows from Project C. If you were able to invest money in a bank at 25.35%, would you expect your cash flows to look anything like those in Project C? The decision maker who uses DCFROR to rank projects will need to understand its strengths and weaknesses fully.

There are certain types of cash flow schedules in which more than one discount rate results in a zero value for NPV. Examples of such cases include rate acceleration incremental evaluations and cases requiring large investments sometime during the life of the project. A nonunique solution may occur whenever there is more than one change of sign in the cumulative cash flow. In other words, if the cumulative cash flow starts out negative, turns positive, and turns negative again (perhaps turning positive again later), a nonunique solution or a multiple rates-of-return solution is possible. The incremental cash flow from evaluating an acceleration project is typical of such a project. Because some companies utilize DCFROI in comparing alternatives, reporting only one of the values where the discount rate yields a zero NPV would be misleading.

The SPEE conducted a study based on comparison of various economic evaluation softwares at the SPEE Petroleum Economics Software Symposium 2000 held on March 2, 2000 in Houston, TX. Some software products reported only one value; others printed both. Some issued cautions. The SPEE "Recommended Evaluation Practice #9 – Reporting

Multiple Rates of Return" recommends the following evaluation practice:

> *"In cases where multiple rates of return exist, the reported economic summary should alert the user of the report that multiple rates of return exist (in lieu of printing a single rate of return). In these cases the summary output should also refer the reader to the present value profile data. A suggested presentation for such an alert on the summary output might be as follows:*
>
> *IRR: Multiple rates of return may exist, see present value profile plot."*

DCFROI has other problems as a tool for ranking projects. It cannot be calculated in the following situations:

(a) When cash flows are all negative (dry hole cases, leasing costs, projects that fail to generate any positive cash flows, etc.).
(b) When cash flows are all positive (farmout of a lease, which becomes a producing property, situations without investments involved that generate production, etc.).
(c) When cash flows are inadequate to achieve simple payout. If the cash flows do not recover the investment cost, they fail to generate a positive DCFROI (a well that fails to recover its drilling and completion expenses prior to abandonment, etc.).

Another complaint that many people express about DCFROI is what is purported to be the inherent assumption that all cash flows are reinvested at the same discount rate as the DCFROI rate. This is based on the fact that all cash flows are discounted at the same rate. In this argument, DCFROI is overly optimistic for high rates-of-return because when the oil company invests a certain amount of money into a project with a high DCFROI the cash flows that are returned to the company are reinvested at an arguably lower rate-of-return. This confusion comes about primarily because many people view DCFROI as somehow being equivalent to a bank interest rate. If an oil company does Project C in our prior example, it *is not* the same thing as investing $1000 at 25.35% interest. The company in Project C invested $1000 and returned $1500 in 3 years. Had they invested $1000 in a bank at 25.35% they would have received $1969 at the end of 3 years. Of course the company did not leave the money it generated during the first 3 years in a sock in their office desk. But the return they achieved on reinvested funds is irrelevant unless you are trying to estimate future wealth of the corporation rather than the marginal contribution of the project.

This argument forms much of the basis for the use of return on discounted cash outlays (RODCO). On the other side of this argument is the common sense analogy to interest rates. If the bank loans someone $1000 at 10% interest rate, the return of interest and principal discounted at 10% will result in a zero NPV as long as the discounting procedures match. It does not matter what the bank or the borrower does with other investments; the specific transaction has a 10% DCFROI. The key issue to remember is that DCFROI is not the equivalent of a bank interest rate and should generally be used to compare similar projects for investment efficiency.

Very high values of DCFROI can be obtained. When values are calculated above 100% per year, it is recommended to simply report 100%+ instead of the high values. DCFROI is a more realistic measure of financial attractiveness than NTIR or payout, primarily because it incorporates the time value of money. It is an excellent measure of capital efficiency to compare alternative projects with comparable life spans and cash flow patterns. It cannot be calculated for certain types of projects and can lead to multiple solutions in others.

7.4.4 Net Present Value and Discounted Cash Flow Return on Investment

When investment opportunities have a DCFROI greater than the cost of capital and the NPV is

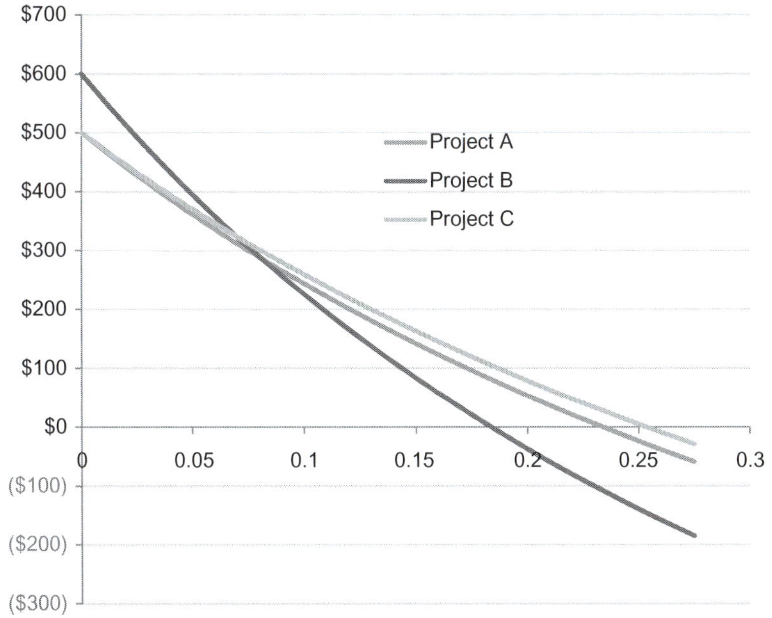

FIGURE 7.5 NPV vs. discount rate for three projects.

positive, both of these measures yield the same accept or reject decisions if there are no restriction on the number of acceptable projects. In cases of capital allocation, the ranking of projects may not be identical for these two measures.

Consider the cash flows for Projects A, B, and C. The following figure shows the NPV as a function of discount rate for each project (Figure 7.5).

Project C has the highest value of discount rate at which the NPV equals zero and thus the highest DCFROI. If the corporate discount rate is 10% or 15%, Project C also generates the highest NPV (and PVR since the investment values are identical). If the corporate discount factor was less than 6.9%, Project B actually leads to the highest NPV and PVR.

Caution in Evaluating Infill Gas Programs

There have been many infill gas programs in fields producing at low gas rates in which the total field production failed to increase as much as the production from the infill wells. Some analysts have suggested that lower production from infill wells might have reflected significant reservoir interference and (thus) less incremental recovery from the infill program than anticipated. While there is "no arguing with the sales meter," it is often the case that the infill wells producing at somewhat higher flowing pressures raised the surface gathering pressure enough to back off existing wells. Some existing wells may also have been producing just above liquid loading rates. In such cases, there was real interference, but it was 'surface' interference which can technically (if not economically) be remediated by the optimization of gathering systems, compression, and other measures to remediate liquid loading in gas wells.

7.5 RATE ACCELERATION INVESTMENTS

An acceleration investment may be defined as a supplementary investment made for the purpose of increasing the rate at which the income is received from a project already in place. Typically in the minerals industry, this has involved the investment in some process to speed up the production of reserves for which a basic recovery capability is already available. The drilling of infill wells to accelerate the depletion of a reservoir being adequately drained by existing wells at a slower rate provides a good example. In reality, infill wells often lead to increased recoveries due to previously unseen reservoir heterogeneities, the ability to lower total reservoir pressure due to smaller drainage area, improved volumetric sweep efficiency, etc.

The basic approach to evaluating a rate acceleration project is simply a comparison of alternatives. The net present value of the rate acceleration case must be calculated and compared to the net present value if the base or unaccelerated case (i.e., continuing existing operations). If the NPV of the rate acceleration case exceeds the NPV of the base case, then the project is a candidate for consideration. DCFROI is generally not attractive for evaluating such projects for multiple reasons. In our initial tight gas spacing case, we compared a tight gas case on well spacings ranging from 40 to 640 acres in a homogeneous reservoir. We had several unrealistic assumptions including that all of the wells were being drilled simultaneously and put on production at the same time, that all hydraulic fractures were of identical length and conductivity, and (most importantly) that permeability was isotropic and homogeneous. We also assumed that there were no rate limitations due to surface facilities or contract constraints.

TABLE 7-4

Assumptions	
Well cost (prior), $	2,950,000
Well cost (optimized), $	1,300,000
Operating costs, $/well/month	2000
Royalty, %	20
Production taxes, %	7.5
Initial gas price, $/Mcf	4.50
Escalation rate, %	4.0
Maximum years for escalation	20
Discount method	MMP
Discount rate (annual %)	10

If we make a few other simplifying assumptions, we can calculate the NPV10 for each case (see Table 7.4).

With these assumptions we can calculate monthly gas prices, operating costs, production, cash flows, and all other criteria needed for the evaluation. In the above table, the most glaring assumption is the drilling cost. In the discovery well, numerous cores, advanced logs, testing, and so forth raised drilling costs substantially. The lower cost is a "target" cost if many wells are drilled based on a combination of improved drilling performance, optimized casing designs, improved hydraulic fracturing designs, etc. *Major reductions in well construction costs are almost always possible when many wells are drilled and suitable engineering analysis is applied.*

The following table shows the NPV as a function of discount rate for the above cases based on the target well cost. It is clear that at typical corporate discount rates of 10–15% the 80- and 160-acre cases result in the highest values of NPV. Because all of these cases have high values of DCFROI, this criterion does little to measure the capital efficiency. The PVRs for each case favor the larger spacings as shown in the following table:

TABLE 7-5

Well spacing, acres	40	80	160	320	640
NPV ratio	0.6	1.9	3.9	6.2	8.5

While the 640-acre case generates the most "bang for the buck," what does that mean? In this case, it means that we would like to have as many sections as possible to develop. But for a given section, how do we determine if the 320-acre decision (two wells per section) is preferable to the 640-acre decision? It clearly generates greater NPV10; so how good a decision is it? (Figure 7.6)

The simple approach is to subtract the 640-acre cash flows from the 320-acre cash flows to get an incremental cash flow analysis. This can also be done for the 160-acre vs. 320-acre and 80-acre vs. 160-acre case as given in Figure 7.7.

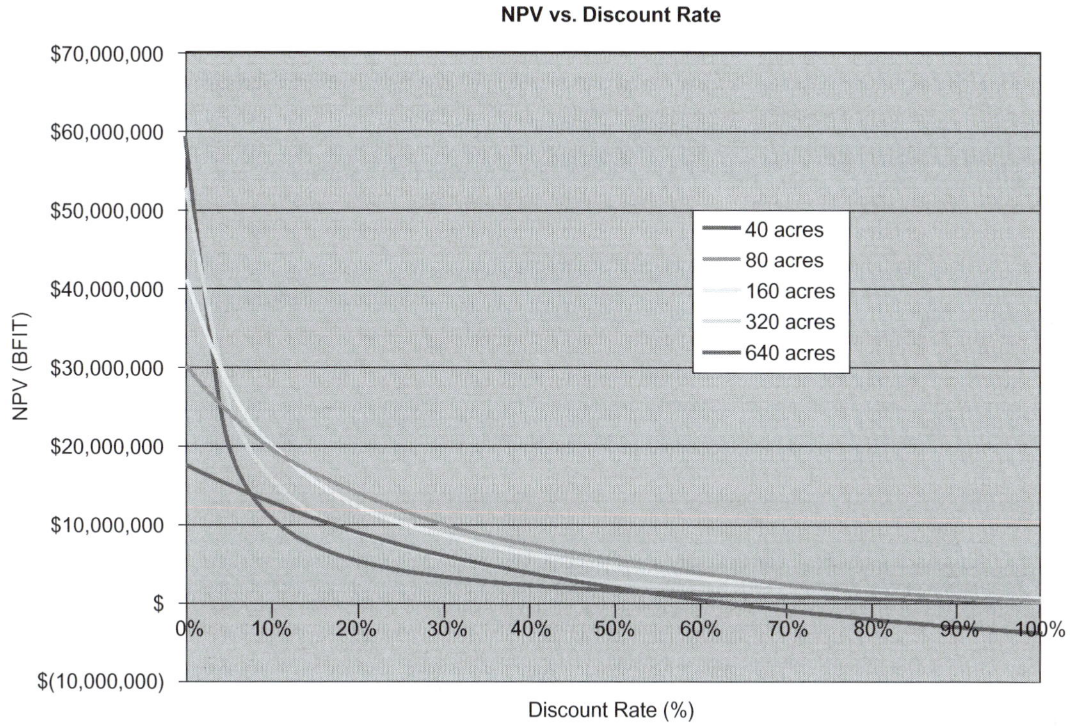

FIGURE 7.6 NPV vs. discount rate for tight gas cases.

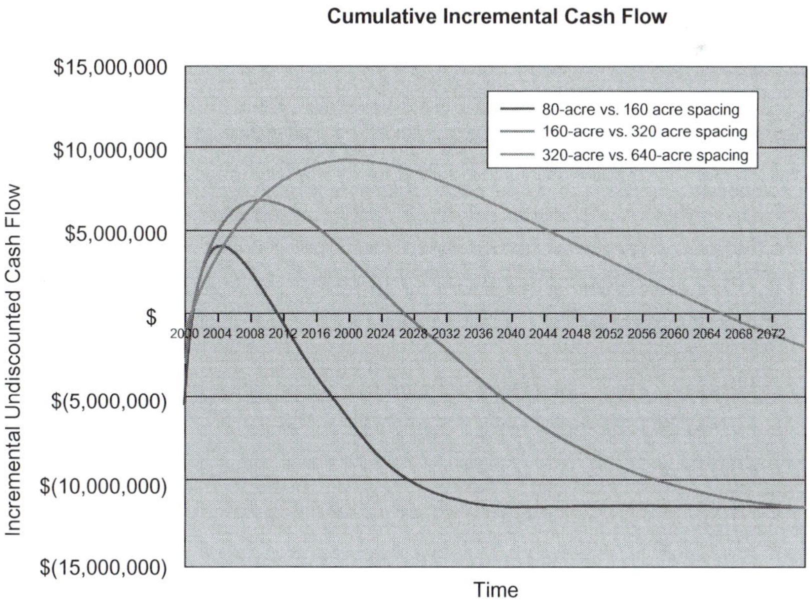

FIGURE 7.7 Incremental cash flow for infill wells.

How do we appropriately decide which is best? The incremental NPV10 and NPV10 ratios are as follows:

TABLE 7-6

Comparison	320–640	160–320	80–160
Incremental NPV10	$5,251,031	$3,871,971	$(230,042)
Incremental NPV10 ratio	$4.0	$1.5	$0.0

While the 320-acre case on its own had a NPV10 ratio of more than 6.0, it still has a high degree of capital efficiency incrementally over the 640-acre case. The 160-acre case generates a significant amount of additional NPV10; however, it comes at the cost of two additional wells and has only modest incremental capital efficiency. Even in sensitivities where the 80-acre case would generate greater NPV10 than the 160-acre case, the 80-acre case will have relatively low capital efficiency.

Sensitivities in such cases often provide useful insights. The next series of figures shows how varying the well cost assumptions and the initial gas price assumptions affects the decision. As expected, higher well costs favor wider spacings, while higher gas prices favor tighter spacings. In reality, higher gas prices may be correlated to higher well costs, as market competition for available rigs and hydraulic fracturing units raises components of well costs (Figures 7.8 and 7.9).

7.5.1 Present Value Ratio (PVR)

While NPV fails to deliver a measure of capital efficiency, the Present Value Ratio index calculates a measure of investment efficiency that is very useful in ranking projects with significant capital investment. It is the ratio of the discounted (after-tax) net cash generated by a project to the discounted pre-tax cash outlays (or investment). Discounting for both measures is at the corporate discount rate. Note that the

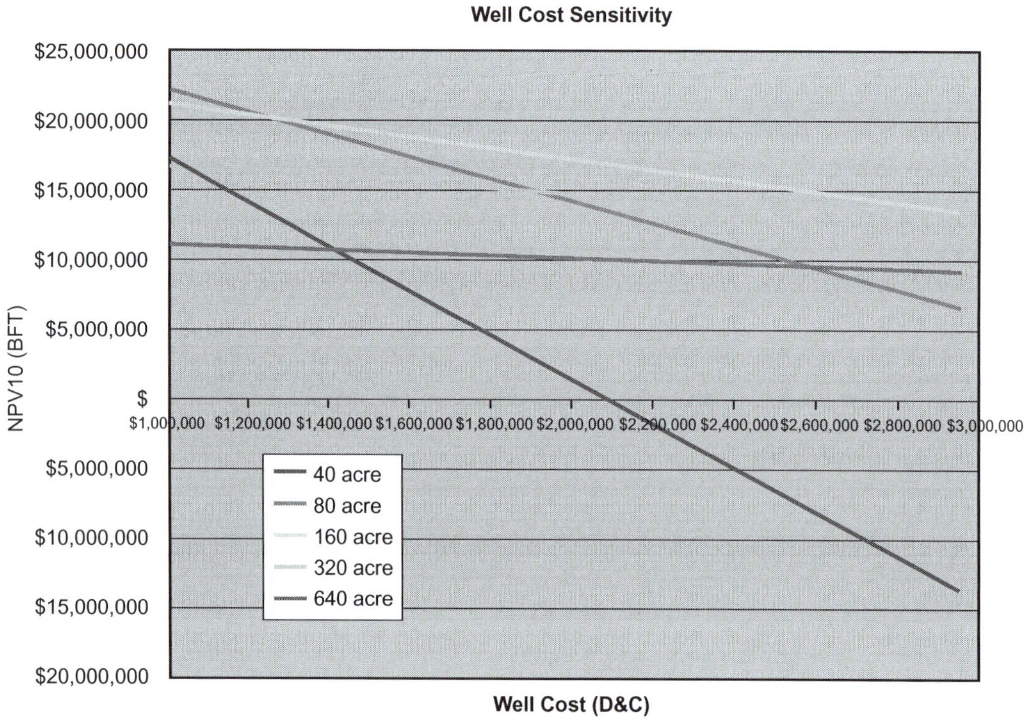

FIGURE 7.8 Well cost sensitivity.

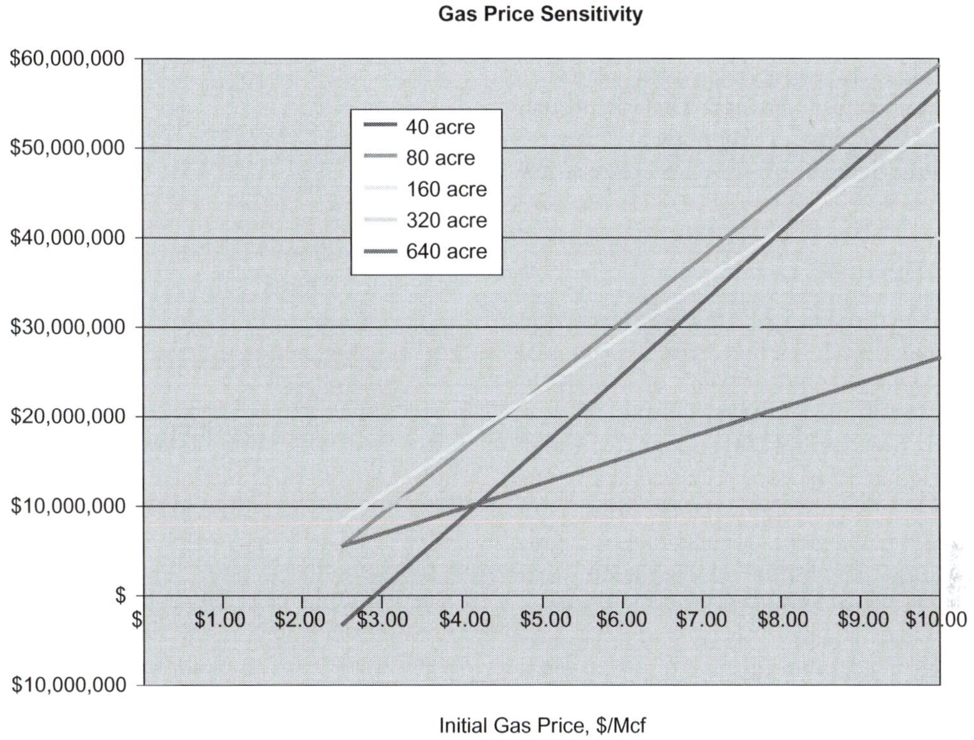

FIGURE 7.9 Gas price sensitivity for tight gas spacing.

numerator is not revenue, but net cash generated. Operating expenses would be subtracted from the revenue along with taxes, royalties, etc., and not discounted back as part of the investment. Some companies use a version of PVR that is one plus this definition and is analogous to a discounted version of NTIR. A project with a PVR of one is equivalent to a project with an after-tax (ATAX) PV equal to zero and a DCFROI equal to the discount rate.

PVR has many of the advantages of NPV in that there is no confusion about corporate reinvestment rates, no multiple solutions, etc. In the examples with Projects A–D, the PVR always ranks the projects in a way that generates the greatest NPV "bang for the buck."

7.5.2 Growth Rate-of-Return (GRR)

PVR has the weakness that it does not have the same intrinsic feel of an interest rate as does DCFROI. GRR is a measure that translates cash flows into an interest rate-like measure that will always rank projects the same way as PVR.

To calculate GRR, all *positive* cash flows are compounded forward at the corporate discount rate to some time horizon, say t years in the future. Cash flows past that date are discounted back to the point t. This calculates the total equivalent amount of cash generated (say B) at time t assuming all cash flows are reinvested at the corporate discount rate. The negative net cash flows (excluding operating costs, taxes, etc.) are discounted back to time zero to get an equivalent time zero investment I. If we were to put these I dollars in the bank, and they grew to B at time t, the interest rate required would be the GRR. For ANEP compounding, the equation is:

$$\mathrm{GRR} = \left(\frac{B}{I}\right)^{1/t} - 1$$

GRR requires some "getting used to" by management as it pushes high DCFROI projects downward and low (compared to the corporate discount rate) projects upward. There is no theoretically correct answer for how and what t should be, and few people have an intrinsic feel for GRR.

7.5.3 Perpetuity

A perpetuity is a series of cash payments that continues indefinitely. While there are no real perpetuities, the theoretical value of a perpetuity can be useful in approximating the value of certain cash flow streams, including real estate and the terminal value of a going concern. The valuation of a perpetuity assumes either constant periodic payments at regular time intervals infinitely into the future or payments that increase or decrease with a given growth rate g. The value of the perpetuity is finite because payments received in the distant future are discounted to negligible present values. The theoretical value of a perpetuity is:

$$PV = \frac{P}{r}$$

where

PV = present value of the perpetuity
P = the periodic payment
r = discount rate or interest rate

If the payments grow at rate g, the above equation becomes, for $r>g$:

$$PV = \frac{P}{(r-g)}$$

It is obvious that perpetuity approaches are not useful as the growth rate g approaches or exceeds the discount rate. If $r=g$, the PV becomes infinite as it is equivalent to an infinite series of effectively undiscounted cash flows.

Typical evaluations of a potential merger or acquisition develop forecast of cash flows for 5–10 years out and then to add a "perpetuity" value to account for the remaining life. If forecasts of net cash flow are made for 10 years, then the perpetuity value would be calculated based on the 10th year's cash flow.

$$PV(acq) = PV_1 + PV_{12} + \cdots + PV_{10} + \left(\frac{P_{10}}{(r-g)}\right) \text{discount factor for year 10}$$

Perpetuity concepts yield potentially unrealistic values as the anticipated growth rate (g) approaches the cost of capital. Thus terminal values based on this approach should be compared with those based on other approaches to valuations of going concerns.

7.6 WEIGHTED AVERAGE COST OF CAPITAL

7.6.1 Conceptual Framework

Several concepts of financial theory are used to determine the weighted average cost of capital (WACC). This cost is the weighted return an investor requires from all sources of funding for a corporation, including both debt and equity. This approach makes the most sense for publicly traded firms, and underlying assumptions include the following concepts:

(1) The first concept of these assumptions is that the value of a firm at any point in time is equal to the market value of its debt and equity capital. That is, the whole is equal to the sum of its parts. The value of any part of a corporation's capital structure, whether it be a bond, commercial paper, common stock, or any other component of the capital structure, is determined by the current financial parameters, such as the prime interest rate, level of economic activity, etc.
(2) The second concept of financial theory used to determine cost of capital is that the cost of capital is the after-tax weighted average marginal cost of debt and equity capital. Weighted CC means that the weights are the percentage of debt or equity capital in the total corporate structure. The marginal cost of debt capital is the interest rate to be paid on the next dollar of borrowed funds; the marginal cost of equity is the return that the next shareholder expects from the purchase of a corporation's stock. Marginal costs (and not historical costs) of debt or equity are used because these determine the

market value of the total debt or equity components of the capital structure.

In an algebraic form, the after-tax weighted average cost of capital (WACC) is:

$$\text{WACC} = \frac{D}{V} r_d (1 - \text{MTR}) + \frac{E}{V} r_e$$

where

D = the market value of all interest-bearing liabilities in the capital structure
E = the market value of all equity securities
V = the value of the firm or $D + E$; hence $1.0 = D/V + E/V$
r_d = the marginal cost of debt
MTR = the marginal tax rate
r_e = the marginal cost of equity

While the marginal cost of debt is fairly easily obtained from financial markets for public trading, the marginal coast of equity is more elusive.

(3) This brings us to the third financial concept used to determine the weighted average cost of capital; the capital asset pricing model (CAPM).

Through CAPM, the marginal cost of equity can be calculated. The model uses historical and readily available current information in the calculation. (There are other models for calculating the cost of equity.)

Algebraically, CAPM is:

$$r_e = r_f + \beta \times (r_m - r_f)$$

where

r_e = the marginal cost of equity
r_f = the risk-free interest rate, usually defined as the interest rate on 13-week T-Bills[10]
r_m = the expected return on the market current return on some stock index, such as the S&P 500 or the DJIA
β = a coefficient that measures the tendency of a security's return to move in parallel with the overall markets' return

A value of β equal to 1.0 means that the security's return precisely mirrors the market movement. A β of 0.0 would be completely uncorrelated. A value of β greater than 1 would suggest higher volatility than the market (up and down). In principle, a negative β possibly means that a stock tends to rise when the market falls and vice versa. $\beta \times (r_m - r_f)$ is the market's evaluation of the variability in the expected return or, in other words, the risk premium for that security.

7.6.2 Value of a Corporation

An estimate of the market value of a corporation can be made by determining the market value of its capital elements. Pertinent information can be obtained from various filings and market information. This value *is not* the same as what the company might sell for to a potential buyer! It is essentially what it is selling for today.

7.6.3 Market Value of Debt

The various components of a company's interest-bearing liabilities have varying types of instruments, maturity dates, and coupon rates. Market values for listed securities may be obtained from public sources, while the market values for other debt instruments can be estimated from direct bid/asked information or yield information for similar securities.

Examples of securities include equipment obligations, debentures, notes, refunding mortgage bonds, commercial paper, pollution control bonds, convertible debentures, sinking fund debentures, private placement notes, current portion of long-term debt (LTD), and a bewildering array of other options. The book and market values of these securities are used to estimate the market value of debt.

To determine the marginal cost of debt, the assumption is typically made that the next

[10]Recent events associated with the downgrade of US sovereign debt may result in an alternative definition for r_F.

dollar of borrowed capital would come from the mix of debt instruments currently in the debt capital structure. The costs of such debt are typically weighted in current proportions with advice from corporate financial expertise as to future debt funding plans.

The CAPM is one method of calculating the marginal cost of equity. While others are typically used in establishing a value of WACC for a corporation, CAPM is sufficient to illustrate the concept. We have the following formula for the calculation:

$$r_e = r_f + \beta \times (r_m - r_f)$$

T-Bills typically can be used to establish the value of (r_f). Assuming that the historical relationship of ($r_m - r_f$) remains fixed during rapidly changing financial conditions, all that is needed to calculate the marginal cost of equity is β.

Several values for β were available. These include public sources and regression correlation coefficients of stock price and market prices over time frames ranging from a month to several years.

7.6.4 Market Value of Equity

The market value of equity can be estimated by multiplying the number of outstanding common shares by the price per share. In many cases, a company's stock may be selling at a substantial premium (or discount) over book value, while total debt may be worth a discount (or less frequently a premium) from book value. The effect of valuing each component of capital structure on a market value basis is to lower (or raise) the market debt to capitalization ratio when compared to its value on a book value basis. For some companies, there is less debt due to inflation and more equity due to expectations of financial performance.

7.6.5 Value of the Firm

Taking the value of the firm as the sum of the market value of its debt and equity, we can estimate the market and book values of the firm. Typically there is a large disparity, and most firms trade well above book value.

7.7 RISK ANALYSIS

Nearly all business decisions are made under conditions of uncertainty. Decision making under uncertainty implies that adequate information for ensuring the right decision is lacking, and two or more outcomes are possible as a result of the decision. Petroleum exploration is a classic example of decision making under uncertainty. The following discussion of risk analysis will be phrased in terms of petroleum exploration, although they may be applied to manufacturing, marketing, and service company decisions. It is recommended that exploration wells and programs be evaluated using expected value economics that account for the probabilities of realizing various outcomes.

Risk analysis provides a more thorough and comprehensive approach to evaluate and compare the degree of risk and uncertainty in a project than the methods previously discussed. The intended result is to provide the decision maker with more insight into the potential profitability and the likelihood of achieving various levels of profitability than do traditional methods of investment analysis.

The application of Monte Carlo simulations is one of the most important tools that reservoir engineers can master in the field of modern risk analysis.

Conventional methods of analysis usually involve only cash flow and rate-of-return considerations. The added benefit of risk analysis to the decision maker's process is the quantitative review of risk and uncertainty and how these factors can be incorporated into the process of developing and implementing investment strategies. Risk and uncertainty cannot be eliminated from business decision making by such analysts, or by any other method of investment review. The advantage of decision analysis is its use as a tool to evaluate, quantify, and understand risk so that management can devise and implement strategies that will allow the company to minimize its exposure to risk.

Decision analysis is a multidisciplinary science. It involves aspects of many different disciplines, including probability and statistics, economics, engineering, geology, finance, etc. Certain statistical methods of decision analysis provide excellent ways to evaluate the sensitivity of various factors in a risk-based economic analysis.

Several petroleum industry methods for handling risk are briefly described below:

- *Arbitrary decision minimums*—In some instances, risk is treated by raising the minimum DCFROI to accept the project. For example, a normal hurdle rate of 15% could be arbitrarily specified at 30% for projects with higher level risks or uncertainty. Such a procedure reflects the need to have the return commensurate with the degree of risk. Although directionally correct, this method does not explicitly consider the varying levels of risk between competing investments. This is not generally recommended.
- *Allowable dry holes*—Some express relative degree of risk in exploratory drilling projects by a parameter defined as the allowable dry holes or dry hole capacity. In this approach, the analyst computes an estimated NPV that would result from a prospect, if successful. This NPV is then divided by the cost of an exploratory dry hole. The result is a multiple of how many times the present value cash flow from a discovery exceeds the dry hole costs. This approach does not yield any information about the probability of discovery but gives management an insight as to how many exploratory wells it could afford to drill based on the value created by one discovery. As such, it provides a relative indicator of the affordable risk of competing exploration areas. This approach can be useful in certain cases but it has several limitations. It does not explicitly account for estimated probability of discovery, nor does it provide a specific "go/no go" decision criterion. Further, it does not tell us how much greater than some specified number the allowable dry holes' multiple or success capacity must be in order to achieve an objective level of profitability.
- *Simulation techniques*—The concept of simulation allows the analyst the option of describing risks or uncertainty in the form of distributions of possible values of the uncertain parameters. These distributions are combined by a computer (Monte Carlo simulation) to yield the distribution of the possible levels of profitability that could be expected from an investment opportunity. The application of Monte Carlo simulations is clearly one of the most important tools that reservoir engineers can master in the field of modern risk analysis. Additional discussions of Monte Carlo simulations are provided in Section 7.7.4.
- *Expected value economics*—This cornerstone of decision analysis is the expected value concept, a method of combining probability estimates with quantitative estimates, which results in a risk-adjusted decision basis. The concept is not meant to be a substitute for manager's judgment but rather a tool to allow evaluation and comparison of the possible outcomes of different investment alternatives.

The decision to drill an exploration well can result in a dry hole, discovery of a giant field, or something in between. Each outcome has some likelihood of occurring, yet no outcome is certain to occur. Many take the view because of the inherent subjectivity involved in assigning probability estimates; expected value analysis has little to offer. There is no doubt that assigning probabilities to the possible outcomes of a drilling prospect is difficult. Sometimes it is not even feasible to define all the possible outcomes. The benefit of expected value analysis, however, is confirmed by its application to repeated trials. If a firm consistently strives to maximize the expected value of many projects over the long run, it can be shown that the firm will do better utilizing risk-weighted value economics.

7.7.1 Adjusted Discount Rates

Many types of risk analysis approaches are used and if the management of a company is successful with one approach it may be hard to justify changing it. One example of an approach that is not recommended, yet still relatively commonplace, is the adjustment of discount rates to account for risk. A manager may wish to see infill wells and workovers evaluated at an NPV10, while exploration wells would be evaluated on NPV25. Intermediate levels of risk might be evaluated with more discount rates. In principle, if the reservoir engineers and analysts conducting the evaluations use consistent approaches for estimation, the adjusted discount rates might be correct. The authors are generally skeptical that this is the case. If we revisit the "drill vs. farmout" case presented earlier, it can be left to the reader to evaluate the cases at 25% and decide when it is preferable to drill or farmout. The "risk the discount rate" approach often incorrectly assesses risk in when comparing cases. Rather, it is strongly recommended that either the "risk" be overtly applied to cash flows or the Monte Carlo simulation approaches be used for risk analysis.

7.7.2 Sensitivity Analysis

A common approach to risk and the handling of unknowns includes sensitivity analysis in which relevant input parameters are modified and the impact of these changed assumptions are displayed as a function of the changed parameter(s). Previous figures such as those that showed different cash flows or economic results for different well spacings are typical. In many cases, one parameter may be modified that actually is not independent of the other parameters. For example, in the following tornado chart, the initial rate is varied as is the net thickness. There can be many reasons that wells with the same thickness have different initial rates; the varying rates may be a function of permeability, viscosity, skin, etc. This particular type of chart is often used to quickly show the most important factors driving variations (in this case for NPV10) in economic value (Figure 7.10).

Simply varying one variable may have complex results that are not always handled consistently in such sensitivities. If the engineer introduces a net-to-gross sensitivity, the hydrocarbons in place are obviously changed. What

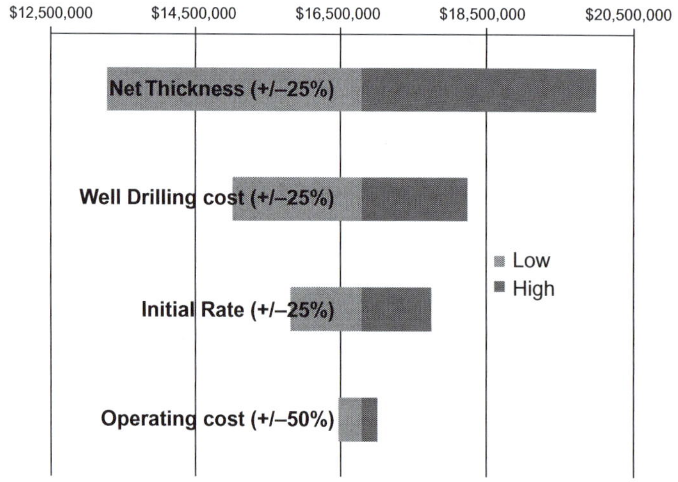

FIGURE 7.10 Example tornado chart for net present value.

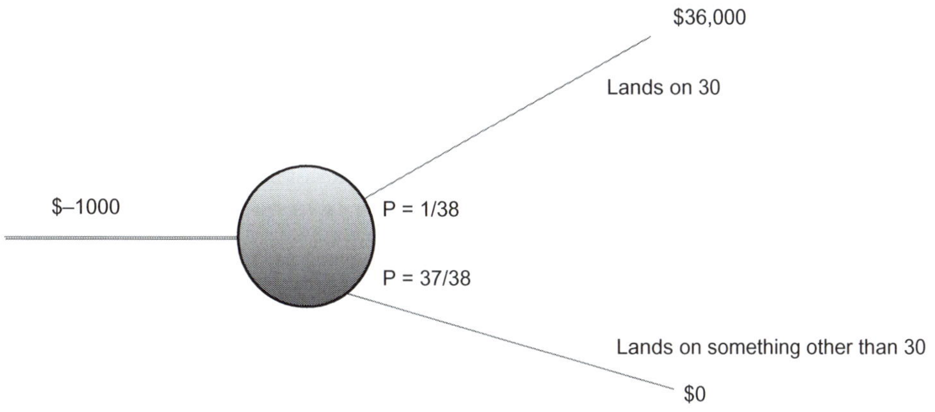

FIGURE 7.11 Probability node.

about the initial rates? Interaction of the aquifer? Fluid handling and artificial lift issues? The simple sensitivity introduces numerous assumptions that need to be handled in a consistent and easily understood manner. This is particularly important when conducting sensitivity analyses of reservoir simulation forecasts for reservoirs with significant production histories. Consider a case in which a very good history match is believed to have been obtained. In such a case, there might still be questions about many of the parameters. If the engineer simply changed (for example) the residual oil saturation as a sensitivity case without reconstructing the history match, unrealistic variations in future recoveries are likely. Had the engineer used a different value for S_{or} and "rematched" the cases, it is likely that the forecast results would show *less* variation from the history-matched case than would a forecast that simply changed the value for S_{or}, exaggerating the sensitivity of the reservoir simulation to errors in assumptions.

7.7.3 Decision Trees and Utility Theory

Decision trees are a useful way to describe alternative scenarios and select the decision that maximizes the NPV or whatever the decision maker is trying to optimize. In a subsequent section, we will see that "utils" can express the relative desirability of various outcomes. In decision theory, the most desired outcome is based on the goals and preferences of the decision maker. The reservoir engineer can use decision trees to describe complex scenarios with multiple decisions and multiple probabilities. This discussion can be considered only a brief introduction. In constructing a decision tree, we use rectangles to represent decision nodes and circles to represent probability nodes. Two or more decisions can be associated with each decision node, and multiple nodes can be associated with a probability node. A probability node representing betting \$1000 on number "30" at a roulette wheel[11] in Las Vegas is shown in Figure 7.11

The single bet on number 30 can easily be evaluated as to its expected value as follows:

$$\mathrm{EV} = -1000\$ + \left(\frac{1}{38}\right) \times \$36,000 + \left(\frac{37}{38}\right) \times 0 = -\$52.63$$

[11]American-style roulette wheels have 38 equally probable outcomes. A bet placed on a single number pays \$35 for each dollar bet if successful. There are many alternative bets such as betting on black or red which would pay \$1 for each dollar bet. In both success cases you keep your bet. As there are two green numbers, 18 black and 18 red, the EMV of betting \$1000 on red or black is $-1000 + (18/38) \times 2000 = -\$52.63$. European style roulette wheels have only one green slot and 37 possible outcomes with slightly different payoffs that are nonetheless EMV negative for participants.

In other words, a single bet of $1000 on number 30 (or any other number) has a negative expected value of $52.63. Similar analyses will show negative expectations for each of the gambling games explaining the fabulous hotels and inexpensive "all you can eat" buffets in Las Vegas. But is it crazy to play roulette or make other decisions selecting lower expected values than other alternatives? No, the decider may have a different use for $35,000 than $1000. Maybe he owes a debt that is immediately due and has a major negative result if he is unable to generate $35,000 right away. This particular preference for risk is actually unusual; most people have less utility for expected outcomes that have large negative impacts. This analysis does not mean that *every player will lose money playing roulette*. It is a relatively straightforward exercise to model a roulette wheel with various strategies in which a significant fraction of the players win.[12] It is the aggregate EMV of all players over the long run that is negative.

Suppose someone gives you the chance to play a game in which a fair coin is flipped. In the case of heads,[13] you receive $2 and for tails you get nothing. You will no doubt be happy to play this game as it has an expected monetary value (EMV) of $1. How much would you be willing to sell your ticket for? It is unlikely anyone will pay you much more than $1, and if you sell it for much less you are "giving away" EMV. Now consider another game. In this game you have to buy a ticket. In this game a heads pays $3 and a tail pays $1. How much would you be willing to pay for this ticket? The EMV of this game is $2, and if you pay any less than that you are (on an expected value basis) gaining money. Would you pay more than $1? If you paid $1, the second game becomes equivalent to the first with the net result of a head being $3 − 1 = $2 and the result of a tail would be $1 − 1 = $0. Is there a difference in how much you are willing to sell your ticket for in the first game and what you are willing to pay for it in the second game? Decision makers often make decisions on other than an expected value basis based on how much investment exposure is necessary.

Let us consider another set of decisions. In the first option, you pay €1000 by investing in a very small percentage (0.1%) of a drilling well that you anticipate has a 50% chance of success (or a coin flip for heads if you prefer). In the case of a discovery you win a series of cash flows with an NPV of €4000, while a dry hole pays nothing. The EMV is $0.5 \times €4000 - 1000 = 1000$. Are you interested in this investment? If you believe these numbers and have €1000 to invest, it is an obvious decision to participate in the project. Now let us look at the 100% working interest position. In this case, you need to invest €1,000,000 and have a 50% chance of €4,000,000. Assume that your net worth is just enough that you could come up with the money by mortgaging your house, cashing in your retirement, and borrowing all of the money that you can; it is unlikely that you would accept such an investment opportunity. A single investment or a series of investments that has the potential to bankrupt an investor is known as "gambler's ruin." Your utility for a positive €1,000,000 is considerably less than 1000 times greater than it is for €1000. By analyzing your responses to a series of similarly constructed alternatives, an individual with game theory expertise could construct your "indifference curve." Your personal utility and indifference curves and those of the decision maker are not as important as are the utility functions of the corporation. For our purposes, we will assume that the corporation has a unit slope linear utility function and makes its decisions entirely on EMV. Exceptions to this would only occur for massive investments.

In the drill vs. farmout example, we had a decision tree, see Fig. 7.12.

---

[12]Consider the trivial case of 38 players each betting $1000 on each of the 38 spots and play one time. One person will walk away with $36,000 (his bet plus his $35,000 in winnings), while 37 players lose their bets. The house makes $2000.

[13]Coin collectors refer to the obverse and reverse of a coin rather than "heads" or "tails," but we will use the more common convention.

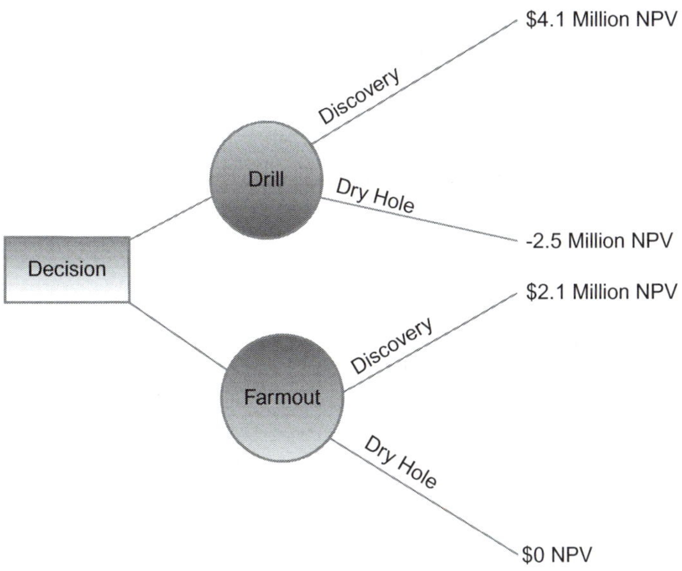

FIGURE 7-12 Decision tree.

There were only two decisions: drill and farmout. The probability nodes were only dry hole or discovery. The analysis of a decision tree proceeds from right to left as the EMV is calculated for each probability node. The expected value of each probability node is replaced with its expected value, and the highest EMV decision node is selected. There can be multiple probabilities at each probability, and the probability node can be replaced by Monte Carlo simulations. In fact, the entire decision tree can be replaced by Monte Carlo simulations with a distribution of decisions being made and the corresponding variability in results conveyed to decision makers.

7.7.4 Monte Carlo Simulations

The concept of using simulations to estimate variability and risk has evolved in acceptance and popularity over the author's careers. An understanding of the concepts involved is so fundamental for reservoir engineers that we have chosen to illustrate these concepts very simply. A number of excellent references are given for further study. Commercial software is available that can be used to assist in performing the calculations; these tools can generally be adapted to complex problems relatively easily. The most difficult and most important part of the process is developing credible, realistic distributions *incorporating as much data as possible*. Practicing Monte Carlo simulations while ignoring available data (or failing to make the effort to obtain and use such data) is not just sloppy but potentially costly and misleading.

Monte Carlo simulations refer to the computational methods that rely on random (or nearly random) sampling of distributions of independent variables to repeatedly solve complex equations. A repeated series of these calculations results in a distribution of answers that reflect the range of solutions possible if the input distributions accurately reflect the variability of the independent variables. If one or more of the independent variables are actually correlated (positively or negatively) to another variable, modified techniques must be employed to handle such partial dependencies.

Let us consider a simple example in which two 6-sided dies are rolled simultaneously, *and the numbers are added together*. An integer ranging 1 through 6 appears on each face of each roll of each die and this occurs randomly and independently of each other. It is obvious that the integer results 2–12 are the only possible rolls. We know that a "2" occurs only when each die rolls a "1." The probability of a 2 is then $(1/6) \times (1/6) = 0.027778$. If we rolled these dice 300 times, then (on average) we would expect the following results, see Table 7.7 below:

TABLE 7-7

Roll	Probability	Occurrences in 300 rolls
2	0.02777778	8.33
3	0.05555556	16.67
4	0.08333333	25
5	0.11111111	33.33
6	0.13888889	41.67
7	0.16666667	50
8	0.13888889	41.67
9	0.11111111	33.33
10	0.08333333	25
11	0.05555556	16.67
12	0.02777778	8.33

Obviously, there can only be an integer number of occurrences of a specific roll. Although we know this situation analytically, let us use Monte Carlo simulations to do the same exercise. While this can be done many ways, entering the `EXCEL` function `=int(rand()*6+1)` will result in a pseudo-random integer that closely approximates the roll of one die. Doing this in two columns and adding the results is very straightforward `EXCEL` coding but will generate new distributions of answers with every recalculation. Thus, if the reader repeats this experiment, the answers obtained should be approximately similar but will not precisely reproduce the results that follow. In this case, we "rolled the dice" 300 times for two independent variables and added them together. The input distribution for each variable was a probability density function with the probability (1/6) associated with each outcome (1 through 6). The output was the simple sum of the two sampled distributions. Figure 7.13 shows the results of three trials of 300 rolls, each along with the theoretical distribution. While the three trials cluster about the theoretical answers, some anomalies are noticed. In the first 300 rolls there were significantly more rolls of "9" than "8," which is (a little) surprising (see Fig. 7.13).

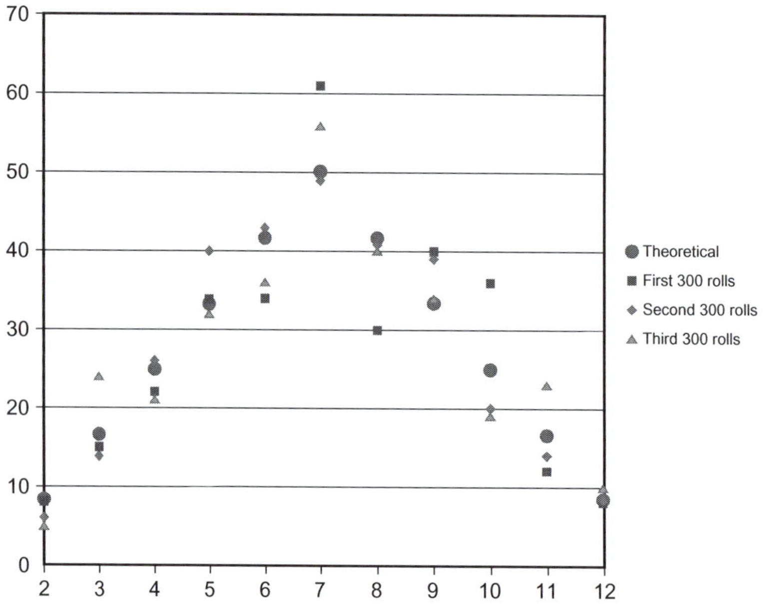

FIGURE 7-13 Results of three trials of 300 rolls

The solution to this problem is simply to do more rolls. If instead of 300 rolls we chose to do 5000 rolls, the results show very good agreement with the theoretical solution (see Fig. 7.14).

This agreement would further improve if we did 10,000 or more simulations. The mean, mode, and median of the resulting distribution is 7, and any reader familiar with the game of "craps" and the house odds in gambling establishments can show that the entire value of each bet results in a negative EMV for the participants. This does not mean that all participants lose money any more than does our prior illustration with the roulette wheel. But for repeated play, the EMV is negative, independent of strategy.

This exercise has been very simple but extending it slightly can illustrate another important point. What if instead of *adding* the two dice we *multiply* them together? We know that the output will vary between integers 1 and 36 and will not contain all of the integers in between. The probability of 1 and 36 will remain 0.027778 as in the extreme values. The resulting analytic solution and a 5000 roll simulation results are shown in Figure 7.15.

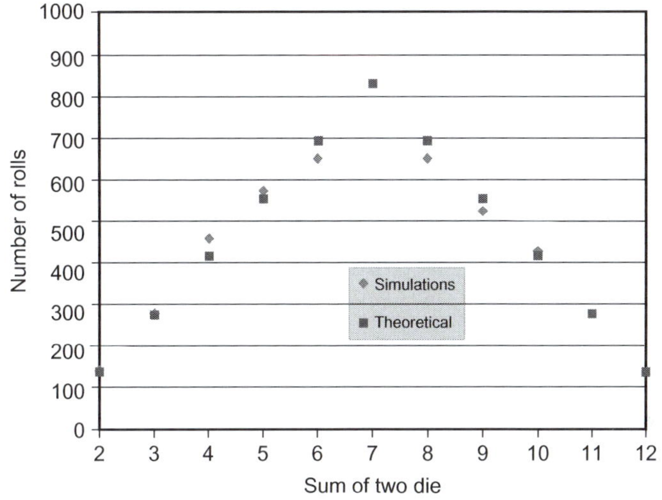

FIGURE 7-14 Results of 5000 rolls

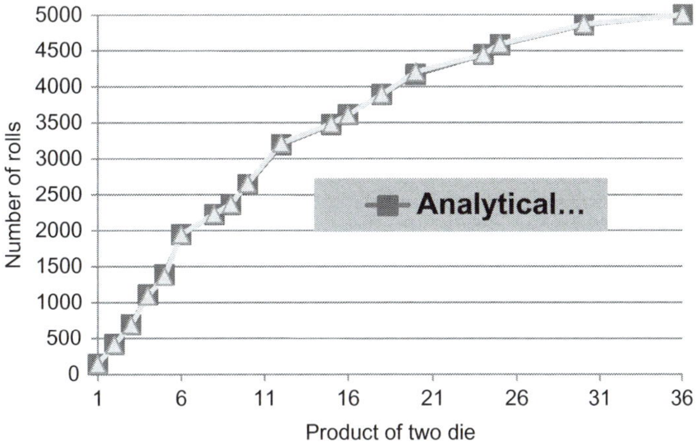

FIGURE 7-15 Analytical solution and a 5000 roll simulation results

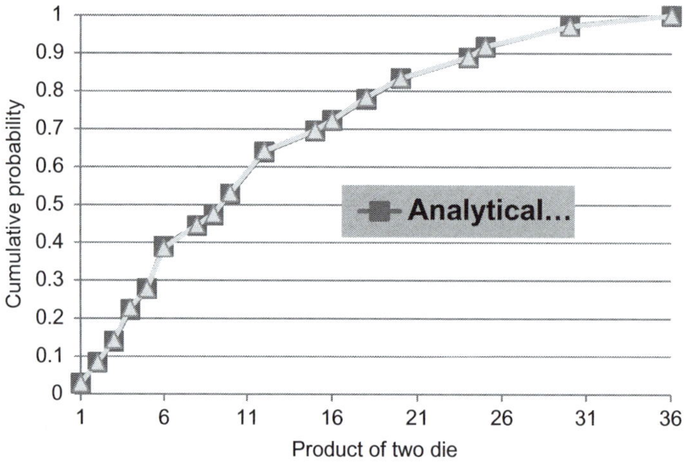

FIGURE 7-16 Cumulative probability

In this case, we have used the *cumulative density function* rather than the probability density function. It is also common to display the above graph not in number of realizations, but in the cumulative probability that it is less than or equal to a certain value as shown in Figure 7.16

This graph has the advantage that the ordinate can be understood as probabilities and that we can easily note that the median value of this exercise is between 9 and 10, the mean is approximately 12.25, and the distribution is bimodal. This characteristic in which the mean is larger than the median is common in Monte Carlo simulations of hydrocarbon recoveries and is also noted in distributions of reserve sizes. While some authors like to invoke mathematical explanations, it is clear enough that many things are not normally distributed, but are distributed in a way that is skewed. Income, wealth, and the height of adults are distributed in this way; the mean exceeds the median. The mean can be heavily influenced by a relatively small number of large positive values.

In the oil and gas cases, Monte Carlo simulations can be used to estimate hydrocarbons in place, recoverable hydrocarbons, number of wells required and future rates, capital and operating expenses, and future net cash flows. They form part of the process in developing bid strategies and can be used in almost any decision under uncertain conditions. In the simplest case, we can calculate the OIP in a potential prospect. Oil-in-place (N) is calculated (in oilfield units) as:

$$N = \frac{7758 \varnothing A h S_o}{B_o}$$

Examining the variables used to calculate the oil-in-place, the area and the net thickness for a potential prospect are obviously the average over the field or drainage area, and the product of the two could be replaced with an estimate of the bulk volume of net pay. The oil saturation and porosity must also be averages, but in many cases these are not independent as lower porosities tend to be correlated with lower oil saturations. How do we get the appropriate distributions for these variables? This turns out to be nontrivial and simply guessing a minimum, maximum, and most likely value, and using a triangular distribution may or may not be better than a single value estimate. Best practice is the careful analysis of porosity data and the distribution of declustered data.

Declustering is absolutely essential due to sample bias. Offset and trend production may be disproportionate from wells in the most productive fields. These data tend to skew distributions to the most heavily sampled data (this example may be the highest porosity data). As this chapter is not meant to be a primer on Monte Carlo simulations, the details of the methodologies employed in developing probability distributions, their relative merits, means of handling partial dependencies, etc., are not included here. Similarly, there are excellent commercial software tools available to handle such evaluations.

7.8 TYPES OF INTERNATIONAL PETROLEUM FISCAL REGIMES

7.8.1 Background

In most countries, including the United States on federal, state, and bureau lands, the rights to explore for and produce hydrocarbons and other minerals belong to the state or occasionally its sovereigns. Historically, IOCs provided several things that independent nations determined to be worth giving producers significant rights and share of the profits from exploration and production. These included:

(1) a willingness to take large risks and expose significant capital searching for hydrocarbons;
(2) technical expertise in exploration and production including technology not available to the country;
(3) massive capital required to develop large fields and a willingness to invest those funds years in advance of revenues;
(4) highly trained and experienced people capable of managing such major projects;
(5) access to refineries and distribution systems to refine, upgrade, and market oil and gas produced.

However, it became apparent that simply turning over rights to an IOC in return for just cash (and in some cases, a minor share of the cash being generated) did nothing for the host nation. Their staff remained inexperienced and with the oil or gas being exported no industry was being created locally. While the share of profitability began to be changed in the 1950s, issues of control, involvement of citizens in more than just low-level roles and development of local industry and infrastructure did not really develop until the 1960s. At this point, sovereign nations usually established one or more NOCs with the goal of addressing all these issues and changed the way that IOCs were allowed to operate in their country.

7.8.2 Generic Contract Styles

There are a host of alternative contractual bases for IOCs to explore for and/or produce the hydrocarbons in a sovereign country. These are generally described in this section, and the specifics of a few countries' current or recent systems are compared. The laws and details of these agreements vary on a frequent basis; so, any such summary is likely to be out of date at any time. While there are many general types of agreements, the basic differences can be summarized by their approaches in the four following areas:

- *Ownership*: Are the hydrocarbons owned by the oil company in the ground or at the wellhead or elsewhere, or are they owned by the state throughout?
- *Payment*: Is payment made by companies receiving hydrocarbons/by lifting hydrocarbons they own, or in lieu of payment for cost and profit recovery?
- *Profit drivers*: Is the contract structured such that the oil companies are fully exposed to price risk, or are their returns fundamentally driven by payments based on the amount of money invested?
- *Operational freedom*: How do all contractual and administrative terms combine to affect the degrees of freedom with which companies can operate and vary their investment decisions within the country?

It should also be noted that there is no one best approach. None of the specific approaches discussed is necessarily more or less generous than the others as the specific levels of payments and handling of risk can and do vary greatly from country to country and contract to contract.

Although this section is entitled "Types of International Petroleum *Fiscal* Regimes," it might also be entitled simply "Types of International Petroleum Regimes"; the point being that it is not always easy to separate the fiscal terms from the legal and contractual structure under which they exist.

Typically, there are three "headline" styles of petroleum regime: concessions, Production Sharing Contracts (PSCs), and service contracts. A discussion of the general features of each of these follows, but typically under a concession arrangement the fiscal components are handled separately from the award of rights to explore and produce, while under PSAs and service contracts the fiscal structure is usually tightly interwoven with the underlying contracts specifying each party's rights.

However, as with any generalization, care must be taken as it is possible to construct any of the headline regime styles to look and act very much like another; in particular, the financial returns from each may be very similar notwithstanding more obvious differences. Indeed, when countries look to update or modify their petroleum contractual or fiscal regime, they are always "benchmarking" it against that of other countries, and aspects are "borrowed" from one to another regardless of the headline contract style involved.

7.8.3 Concessions

While many major fields were developed under the concession model, this is generally a historical artifact. In the first two thirds of the twentieth century, it was not uncommon for sovereign nations to grant large concessions to operators. This was in a time of relatively low and nearly constant oil prices and a time when access to refineries and transportation and distribution systems were nearly as important as exploration and production expertise in generating value from an oil and gas field.

Concessions were large grants of acreage rights, occasionally for an entire nation's onshore or offshore rights. They had long duration, sometimes as much as 50–99 years. The recipient of the concession had complete oil and gas rights in the concessions including all management decisions. The host nation was typically paid a flat royalty per barrel or percentage of revenue. In many cases, the taxes imposed on the IOC by its own home government were higher per barrel of oil than the host government received from that same barrel. Eventually, the inequities of such contracts resulted in either the renegotiation of the terms and conditions or the replacement of the IOC. In some cases, the host government unilaterally abrogated a concession or "nationalized" the assets previously belonging to the IOC.

7.8.4 Joint Ventures

Typical joint ventures for development share the risks and benefits from oil and gas development. The NOC partner may receive a relatively large initial payment for the execution of the JV, and the contractor group partners may carry 100% of exploration costs and potentially all costs "to the tanks" for first oil. Subsequent capital and operating costs are shared in the proportions of the JV ownership. Management decisions for the field and staffing of the JV are also shared with the host government, typically via the NOC as the JV partner. There is nonetheless a clear separation between the government as a taxing and licensing authority and the government-owned IOC JV partner. Some portion of the exploration and development "carried costs" are typically reimbursed by the NOC partner to the contractor group in either cash or oil. Ownership of the government share of the oil is independent of the contractor group ownership. The contractor group is typically entitled only to book reserves for their share of the JV's gross reserves less any government royalty and potentially the reimbursable costs if they are repaid from crude oil.

7.8.5 Tax/Royalty Schemes

Tax/royalty schemes grew out of concession systems. The concept of tax and royalty schemes is easy to describe in that the government owners of the minerals lease tracts for exploration and development either directly to an IOC contractor group through negotiations or through some sort of competitive bidding. An initial cost plus acreage rental payments plus fixed or variable royalties is a typical scheme. The government taxing authorities tax the contractor group members based on their profitability from the block.

The US Outer Continental Shelf (OCS) mineral leases represent a tax/royalty scheme. While most OCS leases contain a competitive bid and fixed royalty payments, tax/royalty schemes can include work commitments, variable royalties, net profit interests, etc.

A number of countries with tax/royalty regimes include, in addition to corporation tax, various forms of "rent" or taxes to capture a greater share of the economic benefit arising from operations, whether these result are simply from highly profitable fields or from windfalls, such as high petroleum prices. Examples include the UK's Petroleum Revenue Tax (PRT), Norway's Supplemental Petroleum Tax (SPT), Brazil's Special Participation (SP), Australia's Petroleum Resource Rent Tax (PRRT) and Alaska's Production Tax (known as ACES). In the cases of UK, Norway, and much of offshore Australia, no royalty at all is now levied, and the countries rely on "rent" and income taxes for virtually all of their share of profits.

Leases granted under a tax/royalty style arrangement are quite different from the old-style concession agreements, even though the term "concession" may still be used (as is permit or license). While details vary from one jurisdiction to another, they all contain significant term provisions, usually involving relinquishment of some part of the acreage at various stages so that only the immediate producing area remains held for a long time (typically the life of production). In some jurisdictions, minimum work obligations will also apply to different holding periods.

Operators are able to book their "net" reserves that are 100% of the gross reserves less royalty.

7.8.6 Production Sharing Contracts

With increasing world demand and global distribution more commonplace, the advantages IOCs bring can be summarized as people, technology, management, and capital.

The first production sharing contracts (PSCs) were signed in 1967 with Indonesia. The two parties to the PSC are the NOC and the IOC, referred to as the "Contractor." Unlike tax/royalty systems, PSCs (in some countries also known as Production Sharing Agreements, PSA) generally transfer title to the produced hydrocarbons at the export point (compared to at the wellhead in tax/royalty systems where the resource in the ground is owned by the state). PSCs typically differ from *service* contracts in that reimbursement to the IOC is in kind and the parties to the PSC own the rights to their share of the oil.

In general, PSCs divide gross production into what are frequently referred to as cost oil (oil or gas applied to reimburse costs; for simplicity here both are referred to as "oil") and profit oil (that in excess of cost oil), with the contractor receiving its compensation from cost oil and a share of the remaining profit oil.

As is indicated by their name, PSC is a contract that includes provisions covering the way that matters will function operationally and financially including the following:

- Descriptions of the acreage conveyed in the PSC and the term (duration) of the agreement.
- A lengthy set of definitions to such terms as "arm's length sales" and "community and social programs."

TABLE 7-8 Example Profit Oil Split

Production Range (BOPD)	BOPD	Government Share (%)	Contractor Share (%)
—	25,000	68	32
25,000	50,000	71	29
50,000	75,000	73	28
75,000	100,000	77	23
100,000	Higher	80	20

- A schedule for relinquishment of the acreage. Over time, certain percentages of the acreage must be returned to the host government. These often correspond with terms of work commitments.
- Work obligations or minimum expenditures as a function of time (typically referred to as the *work commitment*). Participation with the host government. While all PSCs have the host government, typically through the NOC, as a partner some also allow the host government to participate as *a working interest* partner. In all cases, the contractor group carries all of the exploration and appraisal costs *to a certain* point, typically the approval of the field development plan; where the state is also a working interest partner it may pick up its pro rata share of costs thereafter.
- Definition of discovery and of commerciality. Many PSCs require the contractor group to declare a commercial discovery and submit for government approval a field development plan upon commerciality. A number of things are typically triggered at each of these events. The obligations for a noncommercial discovery are also described (typically surrendering the acreage associated with the noncommercial discovery no later than the end of the exploration period).
- Cost oil and profit oil splits. This fairly unique aspect of PSCs describes how the contractor group is compensated for its expenditures.
  - A certain percentage of oil production is considered "cost oil" from which recoverable costs can be recouped by the contractor group. Most PSCs allow a wide range of costs to be recovered including personnel costs for studies; modern PSCs may limit recoverable costs in a variety of ways. Generally "operating" costs are 100% recoverable immediately from available cost oil, although the amount of production dedicated to cost recovery may be limited, to ensure that the host government always receives a share of what is being produced. By the time production commences, the contractor group will typically have spent a significant amount of capital in exploration costs, drilling, completion, and equipment costs. These costs may have to be depreciated over time for cost recovery; in some cases, an uplift is allowed in recognition of the delay (time value of money) incurred in cost recovery. In general, though this acts very much like depreciation for corporate tax purposes.
  - Cost oil in excess of what is needed for cost recovery is referred to as *ullage*. It may be handled in many different ways including having a specified split between the government and the contractor group (e.g., 90/10), going entirely to the host government or (more typically) being added to and split as part of the profit oil. When a PSC has a high portion of the oil allocated to cost recovery, the *ullage* may be a very significant portion of total revenue.
  - Profit oil is the portion of hydrocarbons reserved to compensate the contractor group for taking risks and succeeding. In many cases, the percentage of the profit oil going to the contractor group is based on production rates (by quarter or by month). Alternatively, it may be a fixed split between the host government and the contractor group.

- As an alternative to modifying the profit oil split by production, it may be split according to a profitability concept. In some cases, this may be rate-of-return; in others (and probably more common), it may be what is known as an R Factor. The specific definition of R Factor may vary between contracts, but in general, it is the ratio of cumulative revenue to cumulative costs: an R Factor of 1.0 is achieved when the operation pays out (on a cash basis, ignoring the time value of money).
- Taxes. Most PSCs specify that the contractor group is subject to all local taxes. In some cases, the PSC states that all taxes (and specifically corporate income taxes and any other production related taxes) are to be paid for by the host government entity and not the contractor group. If any petroleum income taxes or other taxes are deemed payable by the contractor group, they are typically spelled out in the PSC.
- Rights and obligations. Most (if not all) PSCs grant the right to freely export contractor's share of petroleum and retain abroad proceeds from the sale of hydrocarbons. However, some PSCs restrict the amounts of local and foreign currencies to be used or expatriated. Many waive or limit import duties.
- Some PSCs may also have an obligation to sell a portion of hydrocarbons to the local market, which typically means at a price lower than what is attainable in export markets. The impact of this may be factored into the profit oil equation.
- Bonuses, royalties, and other payments. Most PSCs require specific bonus payments initially and at certain time or production hurdles. While some PSCs have no royalty provisions, many have a basic royalty in order to ensure a certain level of cash flow to the host government. Most require some form of investment in scholarship or educational programs annually, typically increasing once production is achieved. While host governments prefer to have the contractor group simply write a check and allow ministerial control over such funds, oil companies need to exercise caution in this area. It is generally preferable to take an active role in education and scholarships to avoid any potential for corruption, although some governments see this as interfering in local affairs. Payments for social programs in the country similar to scholarship and educational programs are also typical. Acreage rentals for exploratory areas are also typical as are development and production rentals. In general, bonus payments and royalties *are not recoverable* from cost oil.
- Most PSCs have extensive *local content* provisions requiring hiring and training of nationals, as well as commitments to utilize certain national companies and partnerships, for a significant portion of the total expenditures.
- PSCs will also contain extensive legal discussions around dispute resolution, termination of the agreement, governing law, marketing, force majeure, etc.
- A key area is also contract stability and dispute resolution. Contract stability is something that seeks to protect the IOC from a government changing contract terms at a later date. While very important to IOCs in emerging markets, these are controversial and can be very difficult to write. Nations such as those in North America and North Sea countries would not accept the limitation on sovereign rights, for example. Increasing dispute resolution includes arbitration clauses. Even if the contract is written under the law of the host country, both parties may agree to disputes being heard by one of the international arbitration bureaux in a neutral location.

7.8.7 Ring Fencing

Ring fencing is simply the level at which each fiscal or administrative component is to be calculated or administered. The level can be as low as the field/deployment area or up to the entire contract area, in some cases the entire region or country. While PSC fiscal components of cost oil, profit oil, royalties, taxes, and bonuses can be ring-fenced, the concept applies to other fiscal arrangements. It is common that more than one commercial field is discovered within one PSC area. Ring fencing allows the costs and production to be pooled along with certain exploration expenses. This removes some of the risk for the contractor group and encourages them to make additional expenditures. It can also act as a fiscal incentive to further activity (e.g., if the host nation is taking 80% of profit oil, drilling a new exploration well means that the state is also paying 80% of the exploration costs); for this reason, some countries are reluctant to allow such ring fencing as it may significantly reduce their revenue in the short term.

7.8.8 Reserve Treatment and Issues for PSCs

The key point related to the booking of reserves for PSCs is that the terms of the PSCs both in letter and in spirit must be read carefully to accurately reflect what reserves can be booked by the contractor group. Typically reserves are booked by what is known as the entitlement method. This looks to the volume of barrels that an IOC can lift as a result of its financial entitlements under the PSC (cost oil plus profit oil). The dollar amount, calculated on a pre-tax basis, is equivalent to a certain barrel volume at prevailing world prices. This is the "entitlement" that is booked.

An unusual behavior of PSCs is the impact of product prices on the reported reserves by the contractor group. In tax/royalty schemes, a decrease in product prices lowers the economic limit and decreases the reserves. This has an impact only in later life situations and is more significant for wells and fields with shallow decline rates. However, marginal, undeveloped reserves may not be economic, such as an expansion of a steamflood or additional undrilled well locations.

In PSCs, the lower hydrocarbon prices may actually *increase* the reserves an operator may book. While the value of profit oil reduces with reducing prices, cost oil remains constant (reflecting unrecovered costs, unrelated to price), and therefore, more barrels are required to pay for the same cash entitlement. Assuming the development of the PUDs becomes marginal, the contractor group may be able to nonetheless drill them (with host government approval and modifications of contract terms) while getting (more barrels of) cost oil and corresponding profit oil to pay for the activity that would not have been undertaken in a tax/royalty scheme under comparable circumstances. Essentially, the host government subsidizes such development and may wish to do so to reach production rate, recovery, and even employment objectives.

7.8.9 Service Contracts

Service contracts differ from PSCs primarily based on the fact that reimbursement from service is typically in cash and the contractor group has no rights to the produced hydrocarbons. Service contracts can be pure service contracts in which there is (for example) a flat fee per barrel or a risk service contract in which the fee is tied to production or other measures of performance. Pure service contracts may or may not be operated in conjunction with purchase agreements for the produced crude oil; where such "back to back" lifting agreements exist and where the service contract has at-risk components, the service contract may in fact be very similar to a PSC. In any event many of the concepts and clauses of a PSC may also be found in a risk service contract.

Hybrid agreements may incorporate any aspects of the various agreements; the major limitations to such agreements are governmental regulations and laws specifying how terms and provisions are to be applied. Reserve booking by the contractor group for service contracts may be more difficult than in other international fiscal regimes, but not always impossible and will typically follow the entitlement method described in Section 7.8.8.

7.8.10 Issues with PSC and Service Contracts

PSC and service contract terms often leave production levels in the hands of the contractor group. When there is no ring fencing, the contractor group may be in a position in which it must limit investments in one discovery that would recover only its costs very slowly due to high operating expenses or declining production rates. Decisions toward the end of the agreement can be held hostage to extension discussions.

When successful fields are discovered and a contract has generous terms for cost recovery, the operator has relatively little incentive to minimize costs and may test complex and expensive technologies in fields where they may or may not be applicable. On the other hand, generous cost recovery terms encourage experimentation and capital risk-taking, which may well benefit the host government.

When the host government participates as a working interest partner in development, the host government (or its NOC) will have to actually write checks for large amounts. In some cases, the host government is very slow in making such payments and may owe large amounts of cash to the contractor group. This changes and complicates the dynamics between the host government, the NOC, and the contractor group with respect to capital decisions, operating practices, and contract extensions.

7.9 COUNTRY EXAMPLES

The following five examples provide overviews of petroleum regimes in five countries. These are not to be considered definitive interpretations but as overviews designed to show the complexities and variations in realistic example cases. They are then compared at a summary level. Regimes are continuously changing and the engineer is strongly counseled to obtain current advice on contractual details and practicalities of implementing and evaluating any oil and gas property to obtain up-to-date and accurate interpretations.

7.9.1 Brazil (Excluding Pre-Salt)

Outside of the recently discovered pre-salt play, Brazil operates a concession/royalty/tax system, the legislation for which was only passed in 1997. Prior to that the system reflected a monopoly held by the national oil company, Petrobras. While Petrobras remains Brazil's NOC, apart from being endowed with a commanding acreage position from its legacy status, its rights are identical to all other companies. Within the pre-salt play area, which is subject to a PSC regime, Petrobras also enjoys special treatment.

Licenses are awarded on the basis of competitive auction to international oil companies.[14] All unlicensed acreage is held by a newly created government agency, the Agência Nacional do Petróleo (ANP), which is separate from the Ministry of Mines. In its first round (in 1999), the ANP awarded licenses on the basis of an open auction, with bids weighted 85% to cash and 15% to commitments to spend exploration and development monies with Brazilian suppliers. The detail of award criteria has changed with subsequent rounds, but it still represents an open competitive bidding environment. Essentially, the cash bonuses offered by companies reflect an offer by the companies to pay away some of the fiscal rent that they can imagine themselves earning, when risks and costs are taken into account.

The 1998 Petroleum Law established the basic principles under which licenses would be held and awarded, although the licenses themselves are held under a model Concession Agreement that was drawn up following adoption of the

[14]As of the time of writing there has been a deferral of licensing following discovery of the pre-salt play, although there have been suggestions that this will resume during 2011.

Petroleum Law. This Agreement is fundamentally the same for all players (Petrobras and the licensees from the first round), save some clauses that reflect terms specific to the round/award process. Petrobras is free to bid in competition or in consortium with other oil companies for new licenses, but must pay its way at all stages of the exploration and exploitation process.

Under the 1998 Petroleum Law companies are required to pay a royalty, special petroleum tax known as "Special Participation," rentals, and bonus to acquire the license. A subsequent Presidential Decree established the calculation of Special Participation, which aims to be a form of rent tax, with a sliding scale taxation rate based on the productivity and profitability of individual fields. Royalty rates are set as legal minimum and maximum of 5% and 10%, respectively, although the ANP has the right to vary these as it deems appropriate and necessary. Rentals are established according to a schedule, and the signature bonus is bid.

Within the Concession Agreement the rights and obligations of the oil companies are established, including minimum levels of expenditure during each license period. These commitments are established prior to bidding on a round. Contract duration is up to 9 years for exploration in three phases, and a further 27 years for each development. Companies have the right to export or sell domestically all their production. Administration of the Concession Agreement, and collection of royalty and special participation is undertaken by the ANP. Issues of general taxation and employment are the responsibility of other government departments.

7.9.2 Indonesia

Indonesia is the originator of the classic Production Sharing Contract or PSC. Under this contract foreign oil companies fund all aspects of exploration and exploitation on behalf of Pertamina, the national oil company. In return, companies are entitled to lift hydrocarbons according to a "production sharing" formula.

The formula has changed slightly over the years and is currently different for oil and gas, and traditional and frontier areas. However, classically it is known as an "85:15" split, by which it is intended that after allowing sufficient liftings for recovery of the oil companies' (Contractors') costs, Pertamina (representing the State) will receive 85% of the hydrocarbons and the contractor 15%.

The formula is slightly more complex than this. Indonesia does not have a royalty called such, but it has a concept called "First Tranche Petroleum," by which the first 20% of production is shared between Pertamina and the oil companies according to the 85:15 (or other applicable) split. Thereafter, the distribution looks to the recovery of operating and investment costs, the latter by reference to an amortization schedule, although an uplift or "investment credit" of up to 27% is allowed. Some regimes also have explicit limitations as to the amount of production that can be allocated to cost recovery. Indonesia effectively limits this to 80% by application of First Tranche Petroleum and by amortization rather than immediate recovery of costs. After computation of cost recovery, all remaining production is shared according to the production split.

Indonesia imposes a further level of take, through a mechanism it calls Domestic Obligation. This requires companies to provide a proportion of their production to the domestic market at below world market prices. Although not explicitly styled as "government take," it effectively acts like as one.

Corporation tax does apply to oil companies in Indonesia, but it is deemed to be paid on behalf of the companies by Pertamina (this mechanism applies in a number of countries). In practice, there is an official corporation tax rate, and the 85:15 split is adjusted such that after application of the tax rate the split remains 85:15. The net result is the same, but it allows companies much greater flexibility in arranging taxation affairs with their home jurisdiction, and is a very important component of the structure.

Contract awards in Indonesia have typically been by a combined process of competition and negotiation. Indonesia has not typically held "rounds" in the way other countries have, but it has opened areas to the industry and then negotiated directly with the interested companies. Contract duration has varied over the years, but at present it is 3-year exploration (with the possibility of a second 3-year term) and 20-year development.

7.9.3 United Kingdom

The United Kingdom offers a concession-style agreement, and the general style is applicable to other countries as diverse as Argentina, Australia, Norway, and the United States. UK licenses are for the most part awarded on the basis of competitive work program bids in licensing rounds. However, there is an element of discretion in award that takes account of the companies' overall performance previously in the country. It is also possible for companies to apply directly for open areas that adjoin existing acreage and where they can identify extensions to discoveries. The final approval for award lies with the responsible Minister or Secretary, though as a practical matter all recommendations are made by the civil servants in the ministry (currently the Department of Energy and Climate Change).

Companies undertake exploration obligations at the time of license award, but beyond that they are free to decide on the level of activity they wish to undertake. Field developments must be approved in advance, but license terms for the residual part (after initial relinquishments) of blocks are relatively long at 30–40 years, and companies generally have opportunity to defer or advance the timing of development activity as they see fit.

Terms applicable vary according to the vintage of the license granted, and in the case of field developments, the date on which development approval was granted. The detailed terms have changed many times over the last 30 years. Presently a license covers a single block in mature areas, though it may cover several blocks in frontier areas. A "block" covers 10 minutes of latitude by 12° of longitude, except where it is the award of a previously relinquished part of a block when it may have an irregular shape.

Early field developments were subject to royalty, at 12.5% of the wellhead value of the crude. However, for all fields approved for development after 1982, royalty has been abolished, and it was abolished in 2002 for all fields. Petroleum Revenue Tax (PRT, a form of rent tax) also applies at a 50% rate to all fields approved prior to 1993, but is not applicable to fields approved after that. It is ring-fenced on a field-by-field basis.

Corporation tax (CT) applies to all companies operating in the United Kingdom, and there are special rules that apply to the depreciation of certain assets. Thus, there are fields in the United Kingdom that have PRT and CT or CT only. Although the United Kingdom went through a phase of lowering taxes such that in the late 1990s new fields paid only a 30% CT rate, this was increased to 40% from 2003 and 50% from 2006, although this was accompanied by allowing all capital costs to be immediately depreciated. Thus, companies effectively pay no tax to the government until all costs are recovered. In addition, companies that explore and have no production (or liability to CT) may uplift their exploration by a small amount for several years when they do get to deduct them.

Ownership of all assets lies with the company making the investment, and title to the hydrocarbons passes to the company when produced. There is no national oil company in the United Kingdom (though there was for a while in late 1970s/early 1980s). The overall system is controlled by a number of government ministries or independent agencies, such as the Department of Energy and Climate Change (licensing and general approvals), Department of the Environment, Health and Rural Affairs, Customs and Revenue, and the independent Health and Safety Executive.

7.9.4 Iraq Service Contracts

In 2008, Iraq embarked on the first of what, to date, has been three rounds of service contract awards to the international petroleum industry.

Each of the contracts has been slightly different, but all have followed the same general principle of service contract structure; namely, that the IOCs (treated in consortium as a Contractor) fund all activity in return for a recovery of their costs plus a fixed fee per barrel of production. The fee per barrel was set pursuant to competitive bidding between companies. Companies can receive their compensation either in cash or by lifting oil to the value of the fees to which they are entitled.

The first round of service contracts involved the large legacy fields of Iraq, each producing anywhere from around 200,000 to 1,000,000 BOPD. The contract required that IOCs enter into a contract whereby, in conjunction with the incumbent Iraqi Ministry of Oil affiliate then operating the field, they would invest and increase oil production over the following 20 years.

The Contractor fees (cost recovery and per barrel payment) comes out of a percentage of "incremental" production, defined as actual production in excess of a contractually defined baseline, itself defined as a percentage annual decline over starting production. That not recoverable in one accounting period could be carried forward to the next.

No payments for cost recovery could be started until the Contractors had increased production by a modest threshold amount (it accrued up to that point), and the per barrel fee only applied once this threshold had been exceeded.

Although defining contractual baselines can be a difficult and contentious aspect of such contracts, the impact of this was mitigated here as IOCs knew in advance of contract award what that baseline was to be.

Contracts were awarded through a competitive bidding process whereby consortia bid a combination of the production plateau they believed they could achieve and the per barrel fee they would accept. Points were awarded for each of the two bidding parameters based on the relative bids of others, and the consortium with the most points was declared the winner of the contract.

However, a further stage in the process required that the fee per barrel does not exceed a maximum previously set (but not disclosed) by the Ministry of Oil. In the first round all per-barrel fee bids exceeded the Ministry maximum. The winning bidder then had the opportunity to accept the Ministry maximum in lieu of that included in their bid. Four consortia accepted the Ministry-set fee as a result. Although the maximum fee varied by contract area, it was typically in the order of $2 per barrel.

The consequences of this are that the fiscal terms for the contract are set only in part by the state. In significant part, they are also set by the industry through the competitive tender process. The contract also contains a provision that penalizes companies that do not make their plateau production bid an incentive, both to plan and maintain that level operationally and not to bid too high in the auction.

The second and third rounds of service contract offerings had detailed differences, but involved the same general fee structure and bidding process. Most of the fields on offer in these two rounds had very limited or no production; so, all future production was available for paying fees. Further, following the "signals" of the level of fees that the Ministry would accept, many of the bids were at or below the maximum set by the Ministry, although with these undeveloped fields the maximum fee was slightly higher, up to $6 per barrel in some cases.

All these contracts are (at the time of writing) in an early stage of execution, and it remains to be seen as to what issues may emerge. In addition to the fiscal terms as described above, contractors also had a minimum expenditure obligation over the initial 3-year period of the contract. The plateau bid has to be achieved within 6 years of contract award, and then maintained for a contractually defined period (ranging from 6 to 13 years). While plans for future development of the field are generated by the Contractor, they must still be approved by the local Iraqi operating company, with ongoing governance being carried out through a management committee (joint Contractor and Iraqi) and an annual work program and budget approval process.

7.9.5 Summary

At the beginning of this section, we discussed the relative components of the various schemes, including ownership, the manner in which entitlement to hydrocarbons is given or calculated, the manner in which payment is made, and operational freedom.

No one single parameter defines a contract style. There is a tendency to observe that tax/royalty contracts are the favored contract styles in OECD countries, that they take less than PSCs or service contracts, and that they tend to offer greater operational freedom. However, there is, in practice, no limitation on what may be contained in any one contract style, and, as noted previously, it is quite possible to take features from one contract and move them to another.

Production sharing has elements in it that are different to tax/royalty, but from a purely fiscal perspective it operates very similarly. A tax/royalty regime with a high production tax may leave the IOC no more in profit than a PSC; it just gets to that end point somewhat differently (see Fig. 7.17).

As important as looking at the overall level of government take is when the take occurs. Regimes may have very similar levels of government take overall, but timing can cause returns to be very different. Conversely, very different levels of government take can yield the very same return.

The illustrations below show four different fiscal structures, where the government variously takes different amounts of bonus, royalty, production tax, and income tax or operates in a PSC or a service contract mode. All yield the same rate-of-return to the IOC; however, but because the timing of the overall government take is quite different.

Figure 7.18 shows the profit split on the same field for four different levels of government take, all yielding a 20% rate-of-return to the IOC. The second chart explains why, showing the

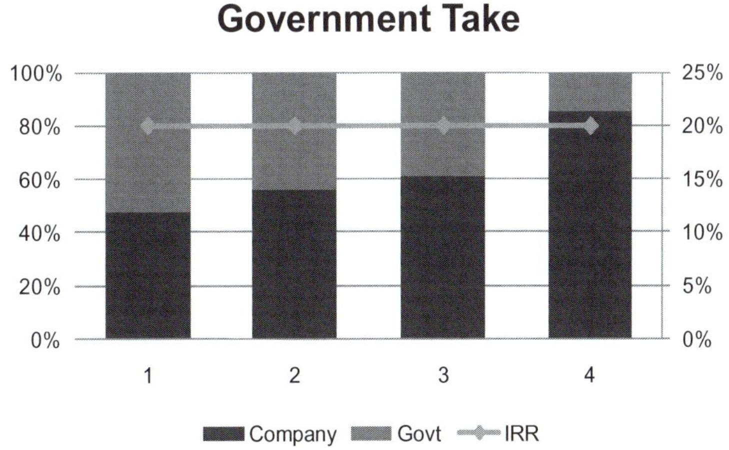

FIGURE 7.17 A tax/royalty regime with a high production tax.

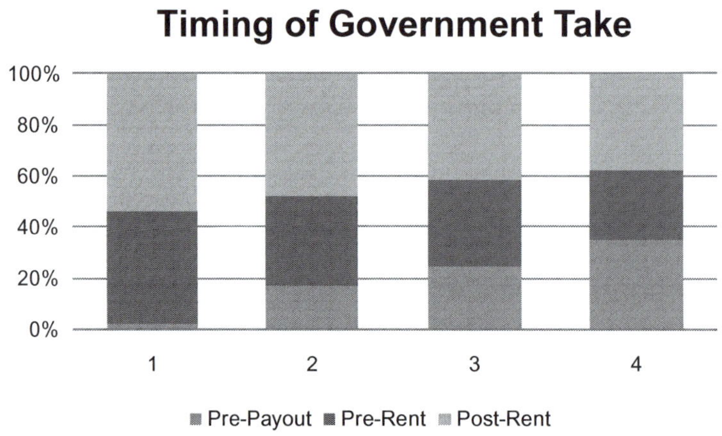

FIGURE 7.18 Profit split on the same field for four different levels of government take.

timing of that government take, the profit was divided into three phases: that taken by the government prior to the investor recovering its investment (pre-payout), that recovered post-payout, but prior to the IOC receiving an appropriate return (pre-rent), and that received after it has received that return (post-rent).

For IOCs, the most attractive structure is in line with the first example; most of the government's take occurs after they have recovered an appropriate return. On the other hand, governments need to balance gaining the maximum return they can with the timing of that return, as early as possible but not too early, or they drive down the IOCs' return to the point of uncompetitiveness.

There is no exact science in designing fiscal structure. It is a judgment made on a number of factors including nonfiscal ones, such as geological potential, the cost structure of exploration and production in the area in question, and the overall business and contractual attractiveness. Establishing a competitive contract requires comparing the contract to those on offer elsewhere in countries likely to be competing for the same investment dollars; even then such competitiveness is unlikely to stay static over time, but move with oil or gas price, changing geological potential or cost structure and competition from elsewhere.

7.10 GENERAL RESERVE BOOKING ISSUES

No textbook on reservoir engineering would be complete without discussing the estimation of recoverable hydrocarbons and of reserve booking. Reserve booking is governed by multiple entities throughout the world, and while many countries use similar guidelines, the specific details may change regularly. Second, the specific technologies used to estimate resources and reserves are addressed only indirectly in this book; these technologies are specific to the fields and reservoirs being evaluated.

7.10.1 Petroleum Resources

The term "petroleum resources" refers to the remaining recoverable hydrocarbons within the Earth. Petroleum itself is defined as naturally occurring hydrocarbon in any phase. Nonhydrocarbons often associated with petroleum include carbon dioxide, hydrogen sulfide, nitrogen, sulfur, and helium. The nonhydrocarbon diluents may have commercial value. As the quantities of petroleum cannot be known with precision, we invariably deal with estimates of these volumes. The relevant estimation process also deals with estimating the volumes of the hydrocarbons that can be technically and commercially recovered and marketed as well as the timing and valuation of those volumes. Even if the reader is not interested in the details of reserve evaluation, booking, and reporting, they should remember that the requirements to be "reserves" are (broadly) hydrocarbons that are:

- discovered;
- recoverable;
- remaining (as of a specified date);
- commercial (based on a specific project).

The hydrocarbons that do not qualify as reserves should be referred to as "resources." Terms to avoid include:

- remaining reserves (all reserves have an "as of" date and are remaining at that date);
- recoverable reserves (there is no such thing as reserves that are not recoverable);
- commercial reserves (if noncommercial they are not reserves);
- remaining recoverable reserves (doubly redundant);
- certified reserves (reserve certification has a role, but it does not change the reserve status).

All of these are simply "reserves." Similarly, the following terminology does not apply to reserves and should be avoided:

- initial or ultimate reserves (Estimated Ultimate Recovery (EUR) is accepted terminology);
- technical reserves (sometimes used to refer to noncommercial resources that could be physically recovered. These are resources, not reserves);
- geological reserves or "in-place" reserves (often confused with initial oil- or gas-in-place)
- prospective/undiscovered/speculative reserves. (These undiscovered volumes are prospective resources.)

7.10.2 Guidelines for Resource Estimation and Accounting

Numerous organizations have set forth standards and guidelines for these processes. The reader will certainly want to be familiar with the Petroleum Resources Management System[15] (PRMS, discussed in Section 7.10.3) as sponsored by the combined efforts of the Society of Petroleum Engineers (SPE), the American Association of Petroleum Geologists (AAPG), the World Petroleum Council (WPC), and the Society of Petroleum Evaluation Engineers (SPEE). Securities' regulators (such as the Securities and Exchange Commission in the United States) in various countries may have specific definitions that are *broadly similar but* different in material aspects from the definitions and approaches within the PRMS. For example, the SEC requires disclosure of proved reserves by all companies that are publicly traded in the United States and only beginning January 1, 2010 allowed for reporting of probable and possible reserves at reporting entity's discretion. The Australian Stock Exchange (ASX) requires either proved or proved plus probable.

Specific governmental definitions also exist outside of security-related guidelines. Finally, individual companies may track resource or reserve estimates in somewhat different ways internally. The reserve estimator will have to have a solid working knowledge of those regulations and definitions relevant to their efforts and will have to stay abreast of changes in those rules.

Who Books Reserves?. Reserve and resource estimation is done both within oil and gas companies and externally. The internal staff may be supplemented by contractors. The responsible

[15]http://www.spe.org/industry/reserves/docs/Petroleum_Resources_Management_System_2007.pdf

teams often report through engineering management but are increasingly part of the company's finance organizations. Internal processes vary widely. In many cases, operating units (geographically defined business units) prepare the reserve estimates, which are then reviewed by others. Some companies use internal groups to calculate companywide reserves, and others combine the approaches.

External consultants can be used in a variety of roles. Some companies use external firms to actually prepare all reserve estimates; some consultants review or audit some fraction of the company's prepared estimates; and some consultants audit the processes and procedures used as opposed to the actual numbers. External consultants may also serve as advisors to or as participants in a company's internal reserves auditing committee.

Section "References and Resources" on the SPEE website[16] contains a series of notes on "best practices" including reserve estimation and reporting. A series of SEC documents related to proved reserve disclosures can be found on the SPEE website[17] and the SEC website. Canadian regulation information as well as other relevant reserve estimation-related material can also be found on the website.

7.10.3 Resource Classification Framework

Figure 7.19 illustrates the SPE/WPC/AAPG/SPEE resources' classification system associated with the Petroleum Resource Management System (PRMS). Much of the following discussion is from PRMS which can be found on the respective society's website. The system includes these resource classes: production, reserves, contingent resources, prospective resources, and unrecoverable petroleum. The "range of uncertainty" is meant to describe varying levels of uncertainty as to the actual quantities (potentially) recoverable from a *specific project*. The "chance of commerciality" is meant to convey the likelihood (increasing from bottom to top) that the project will be developed and will reach commercial producing status.

On the left-hand side of the figure, the total petroleum initially in place (PIIP) can be divided into that which has been discovered and that which is as yet undiscovered. The discovered PIIP is either sub-commercial or commercial depending on the current economic, technical, regulatory, and other factors.

Undiscovered PIIP that could be technically recoverable (once discovered!) is known as "prospective resources." These have a relatively low chance of becoming commercial (on any given project) as their actual identity has not yet been confirmed. There are specific guidelines to describe discovery status that relate to the significance of evidence from exploratory and/or delineation wells and measurements obtained from those wells.

Discovered PIIP that is technically recoverable, yet sub-commercial, is referred to as contingent resources. A natural gas field located far from existing markets or infrastructure is often referred to as "stranded gas" and is but one example of a contingent resource. Another contingent resource would be the potential-increased recovery due to an EOR process that cannot yet be economically justified. Sub-commercial discovered PIIP that cannot be technically recovered based on existing or developing technologies (e.g., residual oil and tar mats) is not a contingent resource.

Discovered PIIP that is commercial and has not yet been produced as of a given date is referred to as reserves. Indications of commerciality include:

- evidence to support a reasonable timetable for development;
- a reasonable assessment of the future economics of such development projects meeting defined investment and operating criteria;

[16]http://www.spee.org/ReferencesResources/index.html
[17]http://www.spee.org/ReferencesResources/SECGuidelines.html

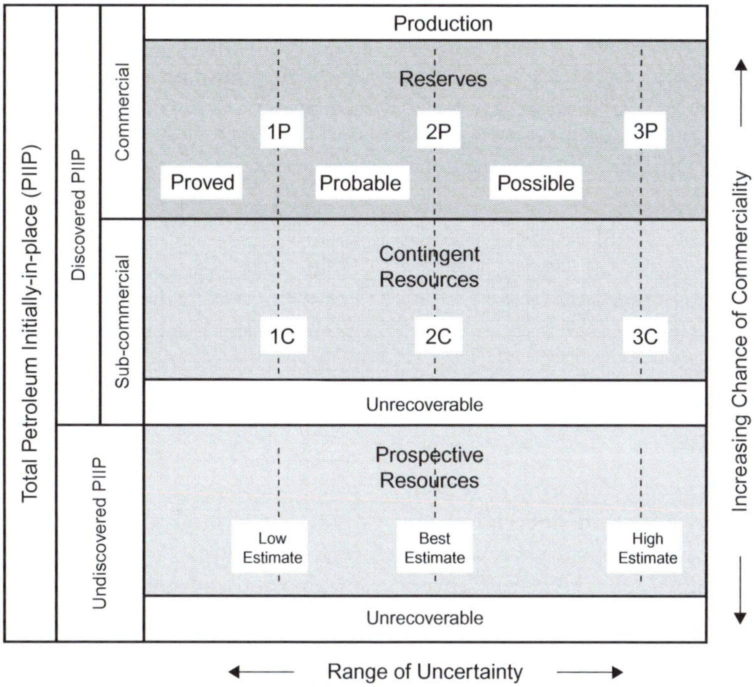

FIGURE 7.19 Resources classification framework. *(Source: PRMS Figure 1.1).*

- a reasonable expectation that there will be a market for all or at least the expected sales quantities of production required to justify development;
- evidence that the necessary production and transportation facilities are available or can be made available;
- evidence that legal, contractual, environmental, and other social and economic concerns will allow for the actual implementation of the recovery project being evaluated.

Note that just as in the term "reasonable certainty," there is a considerable judgment factor in many of these indicators of commerciality. Classification as reserves implies a high degree of confidence that the project can be commercially producible. Actual evidence of this is often necessary, and the requirements for such evidence may vary.

The discovered PIIP that is commercially recoverable plus the cumulative production is the Estimated Ultimate Recovery (EUR). This is also an estimate, and while it refers to the volumes of commercially and technically recoverable petroleum prior to the start of production, the value of this estimate changes as reserve estimates change. The variability in uncertainty of reserves, contingent resources, and prospective resources is often more difficult to describe as there is a more subjective component in describing uncertainty.

7.10.4 A Note on Risk and Uncertainty

There is often a great deal of confusion regarding the terms risk and uncertainty. Some view risk as "the probability that something bad is going to happen." Many view economic risk as "variability in return." For the discussions in reserve reporting, we mean the following:

- Risk is the probability of a discrete event occurring, e.g.:
  - o The risk that drilling a well results in a discovery (or conversely a dry hole).

  - The risk that a PSC will be extended.
- Uncertainty is the full range of outcomes if that event occurs.
  - The uncertainty in the discovery case is the entire range (often expressed probabilistically) of the recovery from the discovery.
  - The variability in future recovery associated with the additional years (perhaps with less attractive financial terms.).

7.10.5 Project-Based Resource Evaluations

While the resource classification system addresses the commerciality and uncertainty of various resource estimates, it is the net resources that are often of the most importance to a given evaluation. Figure 7.20 is useful for understanding the *net* recoverable resources and illustrates the project concept.

The reservoir contains the quantities of petroleum and its characteristics that generally define the potential recovery from the reservoir under a given set of actions. The property refers to a lease or license area and dictates specific fiscal terms that govern ownership of produced minerals. A given project refers to the well(s), reservoir(s), and economic actions (such as drilling wells, waterflooding, etc.) along with the corresponding produced volumes, product prices received, operating and capital expenses, etc. The project is the link between the petroleum accumulation and the decision-making process.

7.10.6 Project Maturity Sub-Classes and Reserve Classes

Field development projects may be divided according to sub-classes that reflect further details in the level of maturity of the resource as it increases in the likelihood of commerciality (see Fig. 7.21).

An exploration concept may mature from basin studies and large play areas as more geological and geophysical data is acquired and interpreted and as more exploration wells are drilled. Similarly, contingent resources, such as the stranded gas example, may move from "development not viable" for the most remote discoveries with no realistic hope of reaching markets to "on hold" when pipelines are in the region, and a few other small discoveries suggest the potential to extend pipeline access to the otherwise stranded gas. As the economic case is made and regulatory and partner approvals are sought, the project may move to a "development pending" status waiting for final commitments that might include final approvals, additional delineation of resource flow capacity, higher quality cost estimates, etc. As the project is deemed commercial, the reserve status may increase from "justified for development" meaning that it is

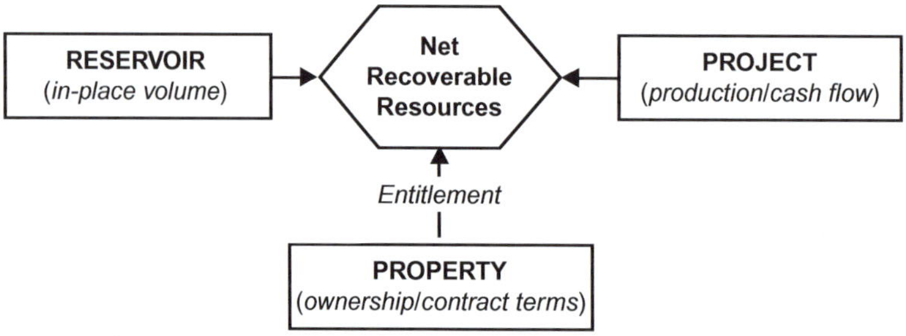

FIGURE 7.20 Resources evaluation data sources. *(Source: PRMS Figure 1.2).*

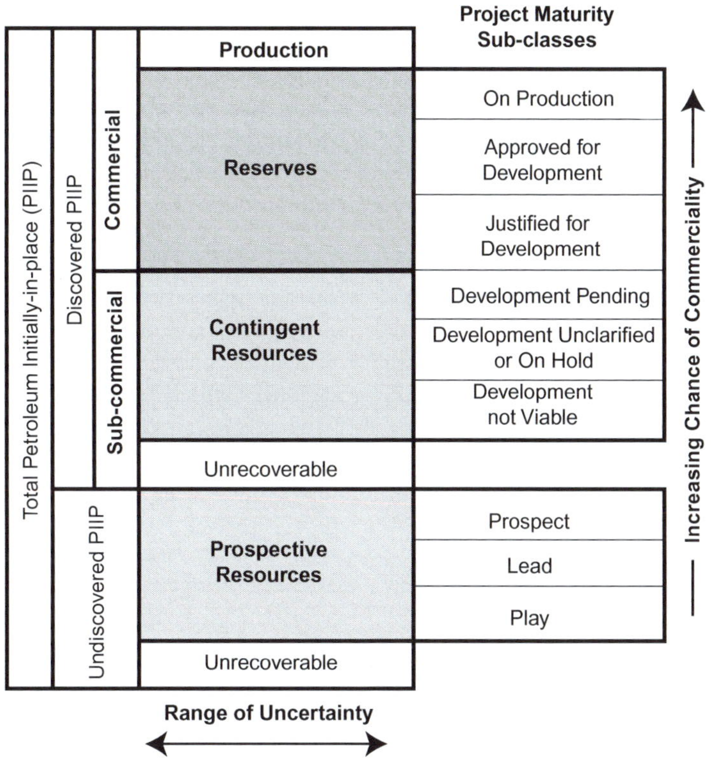

FIGURE 7.21 Sub-classes based on project maturity. *(Source: PRMS Figure 2.1)*

ready for final approvals, approved, executed, and ultimately "on production."

Note that none of these project maturity sub-classes address the uncertainty in resource or reserve volumes. There can still be proved, probable, or possible reserves in several different reserve sub-classes.

Developed reserves are expected quantities to be recovered from existing wells and facilities. Developed producing reserves are expected to be recovered from completion intervals that are open and producing at the time of the estimate. Developed nonproducing reserves include shut-in and behind-pipe reserves. Undeveloped reserves are quantities expected to be recovered through future investments. Reserves in undeveloped categories must make reasonable progress toward being developed if they are still to be considered reserves. Historically, Proved Undeveloped Reserves (PUDs) have received a great deal of attention because they are proved reserves and should have the same relatively low variability in uncertainty as other proved reserves that are developed and producing. Recent SEC rule changes limit how long a particular PUD can be on the books prior to being developed. Both probable and possible reserves can have the same range of variability in their level of development.

7.10.7 Resource and Reserve Uncertainty

The horizontal axis in the resource classification diagram (Figure 7.21) relates to uncertainty in the volumes of recoverable petroleum. This uncertainty may arise due to the potential variability in petroleum in place, the recovery factor applicable to the reservoir, and the efficacy of recovery and development approaches. The variability in uncertainty can be represented by either deterministic or probabilistic approaches. PRMS suggests that the appropriate cumulative probability thresholds for low, best, and high cases are 90%, 50%, and 10%, respectively, for the probabilistic approach. That is, the low case should have a 90% probability that the future recovery will equal *or exceed* the low case estimate. For reserves, these estimates are referred to as 1P, 2P, and 3P, respectively, as cumulative values. The incremental terms are proved, probable, and possible. Thus, proved plus probable reserves are the 2P values, adding the possible reserves is the 3P value.

Similarly, the low, best, and high estimates applied to the contingent resources are the 1C, 2C, and 3C cumulative cases, respectively, and the terms for the prospective resources are low, best, and high. No incremental definitions are specified for the contingent and prospective resources' classifications.

PRMS goes into considerable detail on each of these prior topics and further addresses incremental projects, compression, infill drilling, EOR, and unconventional resources. It also addresses economic considerations and reserve reporting practices along with general discussions of various approaches to reserve estimation. Finally, the document provides the following tables, which summarize the various classifications for recoverable resources (Tables 7.1–7.3).

7.11 HISTORICAL SEC RESERVE REGULATIONS

For publicly traded companies in the United States, proved reserves must be reported annually according to the guidelines established by the SEC. SEC proved reserve disclosures are governed by Regulation S-X, Reg. § 210.4-10, as explained and discussed by the SEC staff in various special SEC releases, including Staff Accounting Bulletins. The SEC promulgated Rule 4-10(a) in 1978. These regulations were ultimately revised in 2009 with an effective date of January 1, 2010. This narrative will illustrate some historical issues and summarize the current status of the US regulations.

Rule 4-10(a) set forth certain accounting and reporting standards for companies engaged in oil and gas production including definitions of proved oil and gas reserves. Prior to the modernization changes as of January 1, 2010, Rule 4-10(a) defined "proved oil and gas reserves" as "the estimated quantities of crude oil, natural gas, and natural gas liquids which geological and engineering data demonstrate with reasonable certainty to be recoverable in future years from known reservoirs under existing economic and operating conditions, i.e., prices and costs as of the date the estimate is made." Rule 4-10 (a) excludes from this definition quantities, "the recovery of which is subject to reasonable doubt because of uncertainty as to geology, reservoir characteristics, or economic factors."

The SEC engineering staff released interpretations of Rule 4-10(a) on June 30, 2000 and March 31, 2001. These interpretations were a significant transformation[18] in how proved reserves were to be estimated and reported pursuant to Rule 4-10(a). The Staff Accounting Bulletins redefined and modified interpretations of the regulations. The SEC Staff guidance acknowledged that there were confusion and inconsistency in the oil and gas industry regarding the precise application of Rule 4-10(a)'s "reasonable certainty" standard.

In addition to the lack of clarity of the "reasonable certainty" standard, numerous changes in technology (such as 3-D seismic

[18]SPE 96382, "The Material Issues Involved in the Discussion Surrounding the Virtues of the SPE/WPC Definition of Proved Oil and Gas Reserves vs. the Corresponding Definition Set Out by the United States Securities and Exchange Commission (SEC), Cline, W. B., Rhodes, B.C., and Hattingh, S.K., presented at the SPE Annual Technical Conference and Exhibition held in Dallas, TX 9–12 October 2005.

TABLE 7.11 Recoverable Resources Classes and Sub-Classes*

Class/Subclass	Definition	Guidelines
Reserves	Reserves are those quantities of petroleum anticipated to be commercially recoverable by application of development projects to known accumulations from a given date forward under defined conditions.	Reserves must satisfy four criteria: they must be discovered, recoverable, commercial, and remaining based on the development project (s) applied. Reserves are further subdivided in accordance with the level of certainty associated with the estimates and may be subclassified based on project maturity and/or characterized by their development and production status.
		To be included in the reserves' class, a project must be sufficiently defined to establish its commercial viability. There must be a reasonable expectation that all required internal and external approvals will be forthcoming, and there is evidence of firm intention to proceed with development within a reasonable time frame.
		A reasonable time frame for the initiation of development depends on the specific circumstances and varies according to the scope of the project. While 5 years is recommended as a benchmark, a longer time frame could be applied where, for example, development of economic projects is deferred at the option of the producer for, among other things, market-related reasons, or to meet contractual or strategic objectives. In all cases, the justification for classification as reserves should be clearly documented.
		To be included in the reserves' class, there must be a high confidence in the commercial producibility of the reservoir as supported by actual production or formation tests. In certain cases, reserves may be assigned on the basis of well logs and/or core analysis that indicate that the subject reservoir is hydrocarbon-bearing and is analogous to reservoirs in the same area that are producing or have demonstrated the ability to produce on formation tests.
On production	The development project is currently producing and selling petroleum to market.	The key criterion is that the project is receiving income from sales, rather than the approved development project necessarily being complete. This is the point at which the project "chance of commerciality" can be said to be 100%.
		The project "decision gate" is the decision to initiate commercial production from the project.

**Source: PRMS Table 1.*

(Continued)

TABLE 7.11 (Continued)

Class/Subclass	Definition	Guidelines
Approved for development	All necessary approvals have been obtained, capital funds have been committed, and implementation of the development project is under way.	At this point, it must be certain that the development project is going ahead. The project must not be subject to any contingencies such as outstanding regulatory approvals or sales contracts. Forecast capital expenditures should be included in the reporting entity's current or following year's approved budget.
		The project "decision gate" is the decision to start investing capital in the construction of production facilities and/or drilling development wells.
Justified for development	Implementation of the development project is justified on the basis of reasonable forecast commercial conditions at the time of reporting, and there are reasonable expectations that all necessary approvals/ contracts will be obtained.	In order to move to this level of project maturity, and hence have reserves associated with it, the development project must be commercially viable at the time of reporting, based on the reporting entity's assumptions of future prices, costs, etc. (forecast case), and the specific circumstances of the project. Evidence of a firm intention to proceed with development within a reasonable timeframe will be sufficient to demonstrate commercially. There should be a development plan in sufficient detail to support the assessment of commerciality and a reasonable expectation that any regulatory approvals or sales contracts required prior to project implementation will be forthcoming. Other than such approvals/contracts, there should be no known contingencies that could preclude the development from proceeding within a reasonable timeframe (see Reserves class).
		The project "decision gate" is the decision by the reporting entity and its partners, if any, that the project has reached a level of technical and commercial maturity sufficient to justify proceeding with development at that point in time.
Contingent resources	Those quantities of petroleum estimated, as of a given date, to be potentially recoverable from known accumulations by application of development projects, but which are not currently considered to be commercially recoverable due to one or more contingencies.	Contingent resources may include, for example, projects for which there are currently no viable markets, or where commercial recovery is dependent on technology under development, or where evaluation of the accumulation is insufficient to clearly assess commerciality. Contingent resources are further categorized in accordance with the level of certainty associated with the estimates and may be subclassified based on project maturity and/or characterized by their economic status.

(*Continued*)

TABLE 7.11 (Continued)

Class/Subclass	Definition	Guidelines
Development pending	A discovered accumulation where project activities are ongoing to justify commercial development in the foreseeable future.	The project is seen to have reasonable potential for eventual commercial development to the extent that further data acquisition (e.g. drilling, seismic data) and/or evaluations are currently ongoing with a view to confirming that the project is commercially viable and providing the basis for selection of an appropriate development plan. The critical contingencies have been identified and are reasonably expected to be resolved within a reasonable timeframe. Note that disappointing appraisal/evaluation results could lead to a re-classification of the project to "on hold" or "not viable" status.
		The project "decision gate" is the decision to undertake further data acquisition and/or studies designed to move the project to a level of technical and commercial maturity at which a decision can be made to proceed with development and production.
Development unclarified or on hold	A discovered accumulation where project activities are on hold and/or where justification as a commercial development may be subject to significant delay.	The project is seen to have potential for eventual commercial development, but further appraisal/evaluation activities are on hold pending the removal of significant contingencies external to the project, or substantial further appraisal/evaluation activities are required to clarify the potential for eventual commercial development. Development may be subject to a significant time delay. Note that a change in circumstances, such that there is no longer a reasonable expectation that a critical contingency can be removed in the foreseeable future, for example, could lead to a reclassification of the project to "not viable" status.
		The project "decision gate" is the decision to either proceed with additional evaluates designed to clarify the potential for eventual commercial development or to temporarily suspend or delay further activities pending resolution of external contingencies.
Development not viable	A discovered accumulation for which there are no current plans to develop or to acquire additional data at the time due to limited production potential.	The project is not seen to have potential for eventual commercial development at the time of reporting, but the theoretically recoverable quantities are recorded so that the potential opportunity will be recognized in the event of

(Continued)

TABLE 7.11 (Continued)

Class/Subclass	Definition	Guidelines
		a major change in technology or commercial conditions.
		The project "decision gate" is the decision not to undertake any further data acquisition or studies on the project for the foreseeable future.
Prospective resources	Those quantities of petroleum which are estimated, as of a given date, to be potentially recoverable from undiscovered accumulations.	Potential accumulations are evaluated according to their chance of discovery and, assuming a discovery, the estimated quantities that would be recoverable under defined development projects. It is recognized that the development programs will be of significantly less detail and depend more heavily on analog developments in the earlier phases of exploration.
Prospect	A project associated with a potential accumulation that is sufficiently well defined to represent a viable drilling target.	Project activities are focused on assessing the chance of discovery and, assuming discovery, the range of potential recoverable quantities under a commercial development program.
Lead	A project associated with a potential accumulation that is currently poorly defined and requires more data acquisition and/or evaluation in order to be classified as a prospect.	Project activities are focused on acquiring additional data and/or undertaking further evaluation designed to confirm whether or not the lead can be matured into a prospect. Such evaluation includes the assessment of the chance of discovery and, assuming discovery, the range of potential recovery under feasible development scenarios.
Play	A project associated with a prospective trend of potential prospects, but which requires more data acquisition and/or evaluation in order to define specific leads or prospects.	Project activities are focused on acquiring additional data and/or undertaking further evaluation designed to define specific leads or prospects for more detailed analysis of their chance of discovery and, assuming discovery, the range of potential recovery under hypothetical development scenarios.

TABLE 7.12 Reserves Status Definitions and Guidelines*

Status	Definition	Guidelines
Developed reserves	Developed reserves are expected quantities to be recovered from existing wells and facilities.	Reserves are considered developed only after the necessary equipment has been installed, or when the costs to do so are relatively minor compared to the cost of a well. Where required facilities become unavailable; it may be necessary to reclassify developed reserves as undeveloped. Developed reserves may be further subclassified as producing or non-producing.
Developed producing reserves	Developed producing reserves are expected to be recovered from completion intervals that are open and producing at the time of the estimate.	Improved recovery reserves are considered producing only after the improved recovery project is in operation.
Developed nonproducing reserves	Developed non-producing reserves include shut-in and behind-pipe reserves.	Shut-in reserves are expected to be recovered from (1) completion intervals that are open at the time of the estimate but that have not yet started producing, (2) wells that were shut-in for market conditions or pipeline connections, or (3) wells not capable of production for mechanical reasons. Behind-pipe reserves are expected to be recovered from zones in existing wells that will require additional completion work or future recompletion prior to start of production.
		In all cases, production can be initiated or restored with relatively low expenditure compared to the cost of drilling a new well.
Undeveloped reserves	Undeveloped reserves are quantities expected to be recovered through future investments.	(1) From new wells on undrilled acreage in known accumulations, (2) from deepening existing wells to a different (but known) reservoir, (3) from infill wells that will increase recovery, or (4) where a relatively large expenditure (e.g., when compared to the cost of drilling a new well) is required to (a) recomplete an existing well or (b) install production or transportation facilities for primary or improved recovery projects.

** Source: PRMS Table 2.*

TABLE 7.13 Reserves Category Definitions and Guidelines*

Category	Definition	Guidelines
Proved reserves	Proved reserves are those quantities of petroleum, which, by analysis of geoscience and engineering data, can be estimated with reasonable certainty to be commercially recoverable, from a given date forward, from known reservoirs and under defined economic conditions, operating methods, and government regulations.	If deterministic methods are used, the term reasonable certainty is intended to express a high degree of confidence that the quantities will be recovered. If probabilistic methods are used, there should be at least a 90% probability that the quantities actually recovered will equal or exceed the estimate.
		The area of the reservoir considered as proved includes (1) the area delineated by drilling and defined by fluid contacts, if any, and (2) adjacent undrilled portions of the reservoir that can reasonably be judged as continuous with it and commercially productive on the basis of available geoscience and engineering data.
		In the absence of data on fluid contacts, proved quantities in a reservoir are limited by the lowest known hydrocarbon (LKH) as seen in a well penetration unless otherwise indicated by definitive geoscience, engineering, or performance data. Such definitive information may include pressure gradient analysis and seismic indicators. Seismic data alone may not be sufficient to define fluid contacts for proved reserves (see "2001 Supplemental Guidelines," Chapter 8).
		Reserves in undeveloped locations may be classified as proved provided that:
		• The locations are in undrilled areas of the reservoir that can be judged with reasonable certainty to be commercially productive. • Interpretations of available geoscience and engineering data indicate with reasonable certainty that the objective formation is laterally continuous with drilled proved locations.
		For proved reserves, the recovery efficiency applied to these reservoirs should be defined based on a range of possibilities supported by analogs and sound engineering judgment considering the characteristics of the proved area and the applied development program.
Probable reserves	Probable reserves are those additional reserves that analysis of geoscience and engineering data indicates are less likely to be recovered than proved reserves but more certain to be recovered than possible reserves.	It is equally likely that actual remaining quantities recovered will be greater than or less than the sum of the estimated proved plus probable reserves (2P). In this context, when probabilistic methods are used, there should be at least a 50% probability that the actual quantities recovered will equal or exceed the 2P estimate.

Source: PRMS Table 3.

(*Continued*)

TABLE 7.13 (Continued)

Category	Definition	Guidelines
		Probable reserves may be assigned to areas of a reservoir adjacent to proved where data control or interpretations of available data are less certain. The interpreted reservoir continuity may not meet the reasonable certainty criteria.
		Probable estimates also include incremental recoveries associated with project recovery efficiencies beyond that assumed for proved.
Possible Reserves	Possible Reserves are those additional reserves which analysis of geoscience and engineering data indicate are less likely to be recoverable than Probable Reserves.	The total quantities ultimately recovered from the project have a low probability to exceed the sum of Proved plus Probable plus Possible (3P), which is equivalent to the high estimate scenario. When probabilistic methods are used, there should be at least a 10% probability that the actual quantities recovered will equal or exceed the 3P estimate.
		Possible Reserves may be assigned to areas of a reservoir adjacent to Probable where data control and interpretations of available data are progressively less certain. Frequently this may be in areas where geoscience and engineering data are unable to clearly define the area and vertical reservoir limits of commercial production from the reservoir by a defined project.
		Possible estimates also include incremental quantities associated with project recovery efficiencies beyond that assumed for Probable.
Probable and possible reserves	(See above for separate criteria for probable reserves and possible reserves.)	The 2P and 3P estimates may be based on reasonable alternative technical and commercial interpretations within the reservoir and/or subject project that are clearly documented, including comparisons to results in successful similar projects.
		In conventional accumulations, probable and/or possible reserves may be assigned where geoscience and engineering data identify directly adjacent portions of a reservoir within the same accumulation that may be separated from proved areas by minor faulting or other geological discontinuities and have not been penetrated by a wellbore but are interpreted to be in communication with the known (proved) reservoir. Probable or possible reserves may be assigned to areas that are structurally higher than the proved area. Possible (and in some cases, probable) reserves may be assigned to areas that are structurally lower than the adjacent proved or 2P area.
		Caution should be exercised in assigning reserves to adjacent reservoirs isolated by major, potentially sealing, faults until this reservoir is penetrated and evaluated as commercially productive. Justification for assigning

(Continued)

TABLE 7.13 (Continued)

Category	Definition	Guidelines
		reserves in such cases should be clearly documented. Reserves should not be assigned to areas that are clearly separated from a known accumulation by non-productive reservoir (i.e., absence of reservoir, structurally low reservoir, or negative test results); such areas may contain prospective resources.
		In conventional accumulations, where drilling has defined a highest known oil (HKO) elevation and there exists the potential for an associated gas cap, proved oil reserves should only be assigned in the structurally higher portions of the reservoir if there is reasonable certainty that such portions are initially above bubble point pressure based on documented engineering analyses. Reservoir portions that do not meet this certainty may be assigned as probable and possible oil and/or gas based on reservoir fluid properties and pressure gradient interpretations.

interpretation, reservoir simulation, and computer technology improvements) were contributing factors to confusion and inconsistencies prior to the SEC Staff guidance. The industry changed dramatically throughout the 1980s and 1990s, and the use of these new technologies *significantly improved "reasonable certainty"* as perceived by technology leaders. Examples of these technology changes follow.

Three-dimensional (3-D) seismic has been described as the most exciting development in Earth Sciences over the past century,[19] and its influence on oil and gas exploration and development cannot be understated. Advances in computational techniques allowed this powerful technology to advance rapidly until by the mid- to late-1990s it was routinely used and provided tremendous improvements in visualization of oil and gas reservoirs.

Advances in drilling technology, field development technology, and well intervention technology pushed the boundary of deepwater oil and gas activities even deeper. While offshore oil and gas activities have occurred for more than 50 years, deepwater drilling commenced only in the 1970s and deepwater production in the 1990s. A host of new technologies for exploiting such discoveries evolved over this time period, including tension leg platforms, compliant towers, floating production and storage and offloading vessels (FPSO), spars, etc.

Reservoir simulation and related technologies got benefited from the computer revolution and many other advances. In 1978 when 4-10(a) was published, the state-of-the-art in reservoir simulation technology[20] could only routinely run reservoir simulation models with on the order of 10,000 grid cells. No realistic geological or geostatistical models had been developed. Also unavailable were detailed quantitative geological modeling, structural and fault modeling,

[19] *3D seismic technology: the geological 'Hubble'*, Joe Cartwright and Mads Huuse, Basin Research (2005) 17, 1–20, doi: 10.1111/j.1365-2117.2005.00252.x

[20] SPE 38441, "Reservoir Simulation: Past, Present and Future," J.W. Watts, *SPE Computer Applications*, Volume 9, Number 6, December 1997, pp. 171–176.

and other advanced reservoir characterization tools. Today, the ability to generate geological models with hundreds of millions of grid cells, conduct rapid streamline models to select and upscale selected models, and routinely conduct simulation studies with tens if not hundreds of millions of grid cells is possible.

In 1978, essentially all wells were either vertical wells or "deviated" wells drilled from a central platform. Horizontal well technology revolutionized reservoir access. Contributing technologies that evolved during the late 1980s through the turn of the century included medium- and short-radius drilling tools, MWD directional equipment, advanced downhole mud motors, and a host of completion tools.

New reservoir monitoring technologies including surface and subsurface tilt meters, downhole microseismic monitoring, 4-D seismic, gravity meters, and others meant that operators could "see" the movement of fluids in ways not previously possible.

The SEC Staff guidance provided the first significant interpretation by the SEC of Rule 4-10 (a) since its issuance in 1978. The guidance provided additional information for the industry to consider their decisions regarding how to estimate proved reserves. Several key areas were given significant attention. This additional information caused a sea change in the way the industry interpreted Rule 4-10(a)'s "reasonable certainty" requirement for the estimation of proved reserves.

Rule 4-10(a)'s definition of proved oil and gas reserves requires "reasonable certainty" that oil and gas will be recovered "in future years . . . under existing economic and operating conditions, i.e., prices and costs as of the date the estimate is made." In the SEC Staff guidance, the phrase "existing economic and operating conditions" was given additional detail as follows (emphasis added): "Existing economic and operating conditions are the product prices, operating costs, *production methods*, recovery techniques, *transportation and marketing arrangements, ownership and/or entitlement terms and regulatory requirements* that are extant on the effective date of the estimate."

This interpretation adds to the "existing economic and operating conditions" referenced in Rule 4-10(a), "i.e., prices and costs." The "existing conditions" language of Rule 4-10(a) previously had been interpreted to include the normal conditions associated with oil and gas discoveries, i.e., that transportation and marketing arrangements were often developed in parallel with delineation and development planning.

Rule 4-10(a) states: "Reservoirs are considered proved if economic producibility is supported by either actual production or conclusive formation test." At the time Rule 4-10(a) was promulgated, the term "formation test" was widely used to describe a test in which samples were retrieved via a wireline-conveyed tool.[21] Common industry practice allowed booking proved reserves for formations with such conclusive formation tests when petrophysical analyses of the well logs and experience in the area resulted in "reasonable certainty." In many wells, multiple zones are present that are typically produced consecutively, from deepest to shallowest. The various, as yet, nonproducing zones were routinely booked as proved undeveloped (or perhaps proved developed nonproducing—behind pipe) based on petrophysical analysis and formation tests, such as the wireline-conveyed formation tester.

Yet, the SEC Staff guidance changed the landscape on this issue as well, stating:

> "*Proved reserves may be attributed to a prospective zone if a conclusive formation test has been performed or if there is production from the zone at economic rates. It is clear to the SEC staff that wireline recovery of small volumes (e.g. 100 cc) or*

[21]The typical use of this term is to refer to a device to obtain fluid and pressure samples such as Schlumberger's MDT tool or Baker Hughes' RCI tool. Pressure samples alone could also be obtained. This tool can be used to estimate fluid contacts and hydraulic connectivity.

> *production of a few hundred barrels per day in remote locations is not necessarily conclusive. Analyses of openhole well logs which imply that an interval is productive are not sufficient for attribution of proved reserves.*"

The additional (and often unnecessary) steps of conducting full blown completions and conventional production flow tests were not initially accepted by many engineers. Industry personnel cited the massive additional costs (particularly for deepwater wells), the additional environmental risks, and the fact that the only reason to conduct the tests would be to book proved reserves. The wasted costs and technical and environmental risks of this interpretation ultimately led the SEC staff in 2004 to waive the production flow test requirement for the deepwater Gulf of Mexico when openhole logs, seismic surveys, cores, and wireline tests supported the proved reserve estimate. The staff specifically did not waive this requirement for deepwater wells elsewhere or for any other oil and gas fields. From a technical point of view, this seems wholly arbitrary as the physical indications from openhole logs, seismic surveys, cores, and wireline tests hold identical relevance offshore Nigeria, in the shallow Gulf of Mexico, and onshore Louisiana as they do in the deep Gulf of Mexico. Ultimately, the SEC's 2009 revisions allowed the use of "reliable technology" to achieve the conclusion of producibility.

Similarly, in the SEC accounting staff's ASR 257 issued in 1978, certain proved reserves clearly and explicitly did not require a conventional production flow test:

> "*In certain instances, proved reserves may be assigned to reservoirs on the basis of a combination of electrical and other type logs and core analyses which indicate the reservoirs are analogous to similar reservoirs in the same field which are producing or have demonstrated the ability to produce on a formation test.*"

The SEC Staff guidance, by contrast, requires "overwhelming" support for the use of log and core analysis in lieu of a conventional production flow test, and even states it would be a "rare event" for this to be satisfactory in an exploratory situation. This aspect of the SEC Staff guidance imposed stricter requirements for "reasonable certainty" than the industry had previously understood Rule 4-10(a) to require.

7.11.1 Proved Area Definition ("Offsets")

The prior Rule 4-10(a) provided that the proved area of a reservoir includes "the portion delineated by drilling" and "the immediately adjoining portions not yet drilled, but which can be reasonably judged as economically productive on the basis of available geological and engineering data." Industry practice prior to the SEC Staff guidance allowed for the use of seismic and pressure data, as well as geological knowledge, to inform reasonable judgment of economical productivity. This allowed for technical justification of drainage areas for proved reserves.

The SEC Staff guidance, however, "emphasize[d]" that "proved reserves cannot be claimed more than one offset location away from a productive well if there are no other wells in the reservoir, even though seismic data may exist" and added that seismic data could not be the sole indicator of proved reserves "beyond the *legal* ... drainage areas of wells that were drilled (emphasis added)." This interpretation of the SEC staff required that no more than one legal (as defined by local regulatory spacing rules) offset be booked as proved undeveloped regardless of how low the technical risks and how otherwise certain the geological and engineering prospects of producibility are.

Such an interpretation was internally contradictory. Geological certainty of continuity is based primarily on the technical understanding of the depositional and diagenetic history of the oil- and gas-bearing formations. Regulatory spacing rules often dictate larger spacing (more distance between wells) early in the life of fields and tighter spacing (less distance) later in the life of the reservoirs for reasons that have little to do with geological uncertainty. Thus, as more information is known about the reservoir performance, this guidance suggests that less area qualifies as proved reserves. Notably, the SEC Staff guidance states elsewhere that "[t]he concept of reasonable certainty implies that, as more technical data becomes available, a positive, or upward, revision is much more likely than a negative, or downward, revision."

Additionally, the legal spacing rules suggested by the SEC staff's interpretation make sense only in the onshore United States and Canada where such rules are prevalent, and do little but confuse the analysis elsewhere. Where such legal spacing units are not defined, the SEC guidance suggests that "technically justified drainage areas" might never be acceptable.

7.11.2 2009 SEC Changes

On December 12, 2007 the SEC issued a "Concept Release" for public comment on potential changes to the reserves' booking policies, procedures, and standards. On June 26, 2008 the SEC announced that it had proposed revised oil and gas company reporting requirements. It is important to note that the SEC did not adopt PRMS even though they made numerous changes in that direction. The SEC has stated that it had revised its proposals so that the final definitions are more consistent with terms and definitions in the PRMS to improve compliance and understanding of the new rules. The detailed rules are available on the SEC website.[22]

The summary of areas of changes is as follows:

- Technologies may now be used to estimate proved reserves if those technologies have been demonstrated empirically to lead to reliable conclusions about reserves' volumes. The technologies or computational techniques applied must be field-tested and demonstrated to provide reasonably certain results, with consistency and repeatability in the formation being evaluated or in an analogous formation. A general discussion of the technologies included is to be disclosed for new reserve bookings or material additions to existing reserves, but the individual properties need not be identified. It will not be surprising if future Staff comments make changes in the "reliable technology" category as they did with respect to testing requirements for deepwater wells. The onus is on the oil companies to demonstrate that any technology applied is "reliable" (as defined), and the SEC has stated that they will not provide a list of technologies that they would consider as meeting this requirement.
- Companies have the option to disclose their probable and possible reserves to investors. Prior rules limited disclosure to only proved reserves.
- Resources such as oil sands may now be classified as oil and gas reserves. Previously, these resources were considered to be mining reserves. Companies will be able to report "saleable hydrocarbons, in the solid, liquid, or gaseous state, from oil sands, shale, coal beds, or other nonrenewable natural resources which are intended to be upgraded into synthetic oil or gas, and activities undertaken with a view to such extraction." Companies are prohibited from including "coal and oil shale that is not intended to be converted into oil and gas as oil and gas reserves." Volumes of synthetic oil and gas are to be identified separately.
    - For oil, gas, or gas liquids, natural or synthetic, delivered to a main pipeline, a common carrier, a refinery, or a marine terminal, the volumes of upgraded product are to be reported.
    - For natural resources that are intended to be upgraded into synthetic oil or gas by a third party, the "raw" product as delivered is to be reported.

[22]http://www.sec.gov/rules/final/2008/33-8995.pdf

- Companies are required to report the independence and qualifications of a preparer or auditer, based on current Society of Petroleum Engineers criteria.
- Companies that rely on a third party to prepare reserves estimates or conduct a reserves audit have additional disclosure obligations.
- Companies reporting oil and gas reserves must now use an average price based upon the prior 12-month period rather than a year-end single-day price, to maximize the comparability of reserve estimates among companies and mitigate the distortion of the estimates that arises when using a single pricing date. The average price to be used will be based on the unweighted arithmetic average of the price on the first day of the month during the prior 12-month period. As in the prior rules, prices governed by specific contractual arrangements (excluding escalations based on future conditions) override the 12-month average approach. Note that (as contrasted to SEC rules) PRMS forecast cases are based on "the entity's reasonable forecast of future conditions, including costs and prices, which will exist during the life of the project (forecast case)." PRMS also suggests that evaluators may examine constant price cases.

On December 29, 2008, the SEC announced that it had unanimously approved revisions designed to "modernize" its reporting requirements. Although the SEC stated that these regulations had not been modified in more than 25 years, the staff guidance and other bulletins had in fact made numerous changes. The effective date of the new rules is for annual reports (10-K and 20-F) effective December 31, 2009 or later filed on or after January 1, 2010.

There are a host of additional changes in the new regulations, and the individual(s) responsible for reserve booking will need to be familiar with the details. A few of the most relevant areas that differ from PRMS include:

- Economic limits in the new rules may not include income from the sale of nonhydrocarbon products. These are allowable under PRMS, and the exclusion under SEC 2009 rules will generally lead to understatements of reserves reported.
- Reserve definitions under PRMS require the "firm intention" to proceed with the development based on a series of criteria that includes:
  - evidence that *legal, contractual, environmental, and other social and economic concerns* will allow for the actual implementation of the recovery project being evaluated (SEC 2009 only specifies the legal right to produce);
  - evidence that the necessary production and transportation facilities are available or can be made available.
  - a reasonable expectation that there will be a market for all or at least the expected sales quantities of production required to justify development.
  - evidence to support a *reasonable timetable* for development. (SEC 2009 specifies a 5-year timetable. This may be reasonable for some projects but, while arbitrary, is widely accepted.)
  - a reasonable assessment of the future economics of such development projects meeting defined investment and operating criteria.
  - While PRMS requires firm intention, SEC 2009 specifies a "reasonable expectation" that is not explicitly defined; however, in the FAQs issued in October 2010, the SEC indicated that they expect a "final investment decision" to support the "adoption of a development plan."
- The definitions of proved and probable for SEC 2009 differ subtly from PRMS; in some cases, these will limit proved and probable bookings more than does PRMS.
- Additional reporting details in SEC 2009 are required in several areas:
  - Material changes occur in proved undeveloped reserves including proved undeveloped reserves converted into proved developed reserves.

  - The amount and timing of investments and progress made to convert proved undeveloped reserves to proved developed reserves including, but not limited to, capital expenditures.
  - Reasons why material amounts of proved undeveloped reserves in individual fields or countries remain undeveloped for 5 years or more after disclosure as proved undeveloped reserves.
- Estimates of resources other than reserves are not to be disclosed except when required by foreign or state laws or to parties interested in acquisition of the reporting company's securities.
- Reserve-reporting disclosures need to identify the countries with more than 15% of the company's reserves (on a proved oil-equivalent basis) with some exceptions related to foreign restrictions on those disclosures.
- The SEC 2009 rules also allow disclosures of reserves based on other price and cost criteria, including management's price forecasts. It is hoped that companies will use this option to provide PRMS-like disclosures.
- Companies will also need to disclose and describe the internal controls the registrant uses in its reserves estimation effort, along with the qualifications of the technical person primarily responsible for overseeing the preparation of the reserves estimates. If the registrant represents that a third party conducted a reserves audit, the qualifications of the technical person primarily responsible for overseeing such reserves audit must be disclosed.

CHAPTER 8

Financial Analysis

Petroleum engineers do not practice their skills in a vacuum and must have a suitable level of familiarity with other disciplines such as petrophysics, geology, and geophysics to be successful. Ultimately, the oil and gas business is about money, and petroleum engineers are likely to interact with financial professionals, particularly if they are promoted to any significant leadership positions during their careers. This section is intended to convey some of the typical financial issues usually encountered by reservoir engineers but is by no means all inclusive. The various rules and practices also vary by country and to a lesser extent within an oil company. Almost all Mergers and Acquisitions (M&A) activity involving oil and gas properties requires reservoir engineers. We will also cover some of the M&A issues beyond the standard evaluation approaches.

8.1 FIXED CAPITAL INVESTMENTS

When conducting economic evaluations, it is necessary to distinguish the purposes of cash outlays. Cash outlays for current period expenses, fixed capital, and working capital have different tax consequences, and hence, a different effect on the after-tax cash flows of the investment opportunity. There is also a difference in how these various cash outlays are treated in calculating economic indices.

Cash outlays for fixed assets are known as fixed capital investments. Fixed assets have an estimated useful life in excess of 1 year in the operation of a business and are not intended for sale. A compressor purchased to be part of a production facility is fixed capital since it is relatively permanent over the course of its useful life, used in the operation of the business, and is not intended for sale. A compressor manufactured or purchased by a company whose business is the selling of compressors is an item of working capital because, while is a relatively permanent item (when placed into service), the intent of the business operation is the resale of the property. The compressor is an element of working capital, specifically finished goods inventory. Ownership of land is a special case. While it is permanent and may not be for sale, it can be treated as an investment if intended for future resale. Undeveloped land held for the purpose of producing goods or providing services is a fixed asset.

Fixed assets may be either tangible or intangible. An asset is generally tangible if it has physical substance (e.g., well-logging tools). An asset is intangible (e.g., a patent on well-logging technology) if its value resides not in any physical properties but in the rights that its possession confers upon its owner.

Most fixed assets have a limited useful life. The cost of such an asset (less any salvage value) is generally charged off (for financial reporting and often for tax purposes) gradually against any revenues derived during the period of its useful life. The words commonly used to describe such a systematic assignment of fixed asset costs to expense are depreciation, depletion, and amortization (often referred to as DD&A).

Fixed assets are classified with respect to their nature and/or projected use. Tangible assets, such as a plant or production facility, are generally subject to depreciation. This includes process units (both on-site and off-site), equipment, furniture, buildings, and their fixtures. While land is a tangible fixed asset, it is not subject to depreciation. Depletion is the "write-off" process for exhausting assets such as minerals (including oil and natural gas). Depletion may exist as either percentage (statutory) or cost depletion. While many oil companies no longer receive the statutory depletion allowance, such treatment remains more common for other minerals and in other countries.

Certain exploration and production expenditures, such as lease bonuses maybe depleted based upon a units-of-production (UOP) basis. In the case of oil and gas properties, this would be based on the hydrocarbons produced as a fraction of proved reserves. A low estimate of produced reserves would result in inappropriately rapid depletion on a UOP basis and vice versa. Intangible assets are normally subject to amortization. These include patents, copyrights, and leasehold improvements. Goodwill and trademarks are two examples of intangible assets that are not subject to amortization for tax purposes.

8.1.1 Cost Basis

Fixed assets are usually valued on a cost basis. Generally, the cost of an asset is equal to the cash consideration parted with when acquiring an asset. As applied to fixed capital, acquisition costs include all expenditures made in acquiring the asset and putting it in place and in condition for use as intended in the operating activities of the business. Thus, the cost of a pumping unit includes freight, installation costs, allocated overhead (when applicable), and any other relevant costs in addition to invoice prices for the unit. A distinction must be made between plant hardware acquisition costs and land costs because expenditures for tangible fixed assets are subject to depreciation, whereas expenditures for land are not.

8.1.2 Cash Flow Consequences

Expenditures for fixed assets affect after-tax cash flow in two ways. First, the expenditure to acquire the asset reduces NCF by the full amount of the expenditure in the calendar year in which the expenditure is made. The second effect of expenditures for fixed assets applies to income tax computation. The IRS and many foreign governments allow a deduction for a portion of the fixed capital expenditure in any one period. The cost of the asset is recovered over time as allowed by government regulations.

Detailed tax calculations for the United States, much less internationally, are beyond the scope of this text. However, knowledge of the specific tax issues applicable to an oil and gas evaluation is essential to the analyst. In many cases, economic software tools have mind-boggling arrays of potential tax options. While these can easily be identified for one specific type of routine evaluation, the analyst is wise to seek competent tax and financial assistance for larger and out of the ordinary evaluations.

In most cases, the only tax consideration required by the analyst *is the marginal impact on taxes*. As a result, many corporate taxation complexities can be overlooked. When conducting an evaluation of an entire company (particularly for many M&A activities), it is important to correctly integrate all significant tax calculations consistently.

Accounting standards and/or government regulations require different fixed capital accounting policies for tax and financial reporting purposes. An example of this is start-up costs which are part of a fixed asset's acquisition cost. Precise start-up costs are not normally determined until well after start up has occurred, but as a guideline for tax planning purposes, start-up costs should be amortized over the minimum allowable period, which starts with the commencement of the new activity. For financial reporting purposes, start-up costs are typically capitalized and depreciated over the life of the fixed asset. By writing off the start-up costs for tax reporting purposes as rapidly as possible, tax deductions are accelerated. Financial reporting guidelines dictate timing for depreciation that is often different than for cash taxes.

8.1.3 Maintenance Expense

Routine maintenance expenditures generally are those done on a periodic basis. They are charged to operating expenses in the period that they occur. Although major maintenance expenditures, such as turnarounds, are not an everyday occurrence, for planning purposes, they can be assumed to be occurring continuously. Money reserved[1] for turnarounds is actually accrued expenses (noncash charges) in any given period. The actual cash flow does not occur until the turnaround is undertaken—say every 18 or 24 months. However, incorporation of the precise cash flow pattern into the economic evaluation requires clairvoyance. Many companies make the simplifying assumption that turnaround expenses are continuous for economic evaluation purposes.

8.1.4 Additions of Fixed Capital

Maintenance actions, which significantly enhance the operation of an investment (extend its life or increase its capacity), may become a fixed capital investment and have their own depreciation schedule. If the need for such expenditures is known, these should be reflected in the economic evaluation as fixed capital expenditures. Spare parts are generally thought of as elements of working capital until they are placed in use. If they meet the criteria of being relatively permanent in nature (typically by having a useful life in excess of a year), they become a fixed asset when placed in service, and their cost is recovered via depreciation.

[1]In finance and accounting, the term "reserves" is often used to refer to money set aside or accounted for that is associated with a specific project or entity. Thus, "reserves" for maintenance is money accounted for to cover future maintenance and has nothing whatsoever to do with oil and gas reserves.

8.1.5 Working Capital

Working capital can be defined for economic evaluation purposes as current assets minus current liabilities. Noncash current assets include raw material inventory, finished product inventory, accounts receivable, and spare parts inventory. Current liabilities include accounts payable. The levels of inventory and the number of days of accounts receivable and payable should be based upon expected operating situations and industry standard practices. When working capital is relatively small (as in the drilling and production of wells), it can be ignored for economic evaluations.

8.2 FINANCIAL REPORTING

Although some of the following examples are based on US examples, every country has its own accounting and tax rules. In many cases, they may be identical or nearly identical to US or UK rules but this should not be assumed. Nonetheless, the concepts from US rules serve to illustrate the types of considerations used in most other nations.

8.2.1 Generally Accepted Accounting Principles (GAAP)

In the United States, financial reporting for stockholders, lenders, and the public is governed by "GAAP" (Generally Accepted Accounting Principles), promulgated by the accounting profession and the Securities and Exchange Commission (SEC). The corporate Federal tax return is regulated by the Internal Revenue Service (IRS). Most other countries have analogous regulatory and advisory bodies;

GAAP is used as an example and the illustrations in this chapter may not reflect recent changes in either IRS or GAAP rules. Frequently, the earnings and taxes reported for the tax return will differ from those reflected on the financial (book) statements. This disparity results from certain tax regulations that vary from GAAP.

GAAP statements are meant to reflect operating results according to a consistent set of guidelines. Tax rules are designed to generate revenue for the government, and tax payers may legitimately defer or minimize tax payments in several ways to maximize cash profitability. When we refer to a company's sales, profits, and inventory, we are generally referring to the financial reporting numbers as opposed to either the tax or the cash numbers. It is the tax (cash) numbers that we will generally be using for calculating economic indicators.

The balance sheet summarizes the financial position of company at a specific point in time and is organized by assets, liabilities, and equity. The income statement (often "P&L statement") presents financial results for a specified time period and is composed of revenues, less expenses (or charges) to determine net income. While reservoir engineers need not be fully competent in finance and accounting, the ability to read a balance statement and an income statement is valuable.

8.2.2 Net Income

Revenue is defined as the gross income returned by an investment or produced by a given source. Net income is what remains of the revenue after the deduction of all expenses, outlays or losses, and taxes. A basic accounting principle used in determining net income is "matching." Stated simply, it involves the matching of expenses within the same period as the related revenues. If there is no basis to associate revenues with an expenditure, the resulting expense should be recognized when incurred.

8.2.3 Timing Differences

As a result of matching, expenses in any one time period may not necessarily equal the cash spent. This is due to the fact that often cash spent in one period may benefit several periods. In those instances, the cash expenditure is charged to expense over the years benefited. Ultimately, all cash spent will be recognized as expense. The following example illustrates this principle.

The expansion of a gas plant would require cash to be spent during the construction period. This would be reflected as cash flow during the period the money is spent as a use of funds. However, the gas plant will produce revenues over several years. These construction costs should be matched with the related revenues. To facilitate this matching, the construction costs are capitalized as an asset on the balance sheet as the cash is spent. These costs will then be expensed as depreciation, depletion. and amortization (data) expense over the estimated economic life of the asset, called the life of the asset.

For economic evaluation purposes, the investment is treated as a cash expenditure when it occurred. If that was over more than a few months, the analyst will have to make a decision as to treating the entire cost as a one-time investment or spread it over some time. For tax purposes, the incremental (cash) taxes estimated to be due should be included in the evaluation.

It is also common for a timing difference to occur between revenues and cash receipts. The basis upon which a sale is recorded is the transfer of title, not the receipt of cash. An example is the sale of a product to a customer who is extended 45 days of credit to pay for the product. The seller would record the sale when title transferred—such as when the product is loaded on the customer's barge—not when the cash payment is received.

An additional financial report is the cash flow statement. Cash flow is defined as the net cash received and disbursed by a company during a given time period. It is not the same as net income. Cash flow information is necessary to present net cash outflow or receipt. The classic presentation of cash flow involves adding to net income any noncash expenses or changes, such as depreciations, and subtracting cash outlays not currently expensed, such as a capital asset purchase. The cash flow statement is for management purposes only and does not satisfy GAAP. GAAP requires a funds statement, which is somewhat different.

8.2.4 Depreciation, Depletion, and Amortization (DD&A)

Earnings used for a company's tax return will frequently differ from those reported for the financial (book) statements. This disparity is often the result of "timing differences" caused by differing tax regulations and GAAP principles. DD&A expense may also be used to illustrate tax timing differences. For financial reporting purposes, depreciation expense reflects the expensing of cash outlays of earlier periods and is thus a current expense without a current cash payment.

Tax rules often allow more depreciation expense earlier in the life of an asset than is recognized for financial reporting. Over the life of the asset, each method will (usually) yield the same total depreciation but in different amounts individual periods. The example in Table 8.1 illustrates timing differences.

As shown in the table, total depreciation expense over the life of the asset is the same under both methods. In the early years of the asset, more depreciation is recorded for tax purposes than for financial reporting purposes, thereby reducing taxable income, which in turn reduces cash required to pay taxes in those early years. More (cash) tax will be paid in later years; ultimately, the same tax will be paid under each method assuming tax rates remain unchanged. However, since less cash is paid for tax expense in the early years, the company has the benefit of the cash during the early years.

For illustration purposes let us assume that this project generates $500,000 in the first year with a 20% cash margin. The book and cash taxes for the first year are then:

TABLE 8.1

Cost	$50,000
Salvage value	0
Useful life	4 years
Tax rate	40%

Year	Depreciation		Excess Tax
	Financial Reporting Straight Line	Tax Reporting Accelerated	Depreciation over Financial
1	$12,500	20,000	$7500
2	$12,500	15,000	$2500
3	$12,500	10,000	$(2500)
4	$12,500	5000	$(7500)
Total	$50,000	$50,000	–

Note: Numbers in parentheses represents negative numbers.

TABLE 8.2

	Book	Tax	Difference
Revenue	$500,000	$500,000	–
Cash expenses	$(400,000)	$(400,000)	–
Depreciation	$(12,500)	$(20,000)	$(7500)
Operation income	$87,500	$80,000	$(7500)
Federal taxes	$35,000	$32,000	$(3000)
Net income	$52,500	$48,000	$(4500)

Note: Numbers in parentheses represents negative numbers.

8.2.5 Deferred Tax

The reader may be disappointed to know that deferred taxes do not change when cash taxes are due! The term "deferred" refers to the difference between the two separate tax calculations for financial (book) reporting and tax reporting purposes. The deferred tax in this example results from using accelerated depreciation for tax purposes and the straight line method for financial reporting. For tax reporting purposes, the use of the accelerated depreciation approach gives rise to $3000 less payable to the IRS (the current tax) in year 1. The $35,000 must be recorded as taxes for financial reporting; however, only $32,000 was paid in cash during year 1. The difference is recorded on financial books as deferred taxes. Deferred taxes can be either positive or negative.

In the example cases, there is no ultimate difference in taxes paid; however, several situations exist that may give rise to permanent differences in deferred taxes. Examples that may give rise to permanent differences include:

(1) tax exempt interest (municipal, state) income;
(2) changes in the tax code;
(3) amortization of goodwill;
(4) certain fines or penalties;
(5) dividends received;
(6) excess percentage depletion over cost depletion.

If a project or asset is terminated prior to the end of its estimated life, its remaining value is normally expensed, for both tax and accounting purposes. In addition, if a project or asset continues functioning subsequent to the end of its estimated life, no further depreciation will be expensed for either accounting or tax purposes since the asset's full cost has already been expensed.

8.2.6 Cash Flow Generation

To arrive at a current year cash generation figure, book income is adjusted for the noncash expenses by adding back the depreciation and deferred taxes.

This brief discussion of deferred taxes is simplified because any one project or asset could involve several differences between financial reporting and tax reporting. A few examples of timing differences include:

Financial Reporting	Tax
Using straight-line depreciation method.	Using an accelerated depreciation method.
Capitalizing intangible drilling and development costs (IDCs) when incurred and expensed.	Expensing 70% of IDC when incurred and amortizing the remaining 30% over 60 months.
Expensing geological and geophysical costs when incurred.	Capitalizing geological and geophysical costs, then amortizing such costs over production.
Capitalizing interest expense as part of a project construction cost, to be depreciated over the life of the asset.	Expensing interest expense when incurred.

8.3 MERGERS AND ACQUISITIONS

Acquisitions and mergers have become very common around the world with potential acquirers often sought at significant premiums over the preannouncement market prices of their common stocks. Many of the more complex tax, legal, and accounting questions occur in acquisition evaluations. Reservoir engineers play a key role in acquisitions of E&P assets and need to be broadly aware of the major issues in such activities. Many of these items may not be considered reservoir engineering or even economics; however, the reservoir engineer not only contributes to, but often leads such M&A teams. While this section primarily discusses exploration and production assets, these comments generally apply to other types of business opportunities as well.

An acquisition study involves three basic segments: (1) the search for candidates, (b) business and financial appraisal including due diligence, and (3) negotiations. The search for candidates can involve both inside personnel and outside use of finders, investment houses, and banks. After potential candidates have been screened, the acquiring firm conducts a business and financial appraisal of the candidate firm. These appraisals include the target firm's assets, its markets, and potential to the buyer. This is essential to determine the value of the company. The third segment, negotiating the acquisition from the time of initial acquaintance to finalizing the transaction, is the most critical part of the process. Analysts who do not have extensive experience in these highly specialized areas should obtain professional expertise as early as possible.

8.4 OVERVIEW OF E&P ACQUISITION ENVIRONMENT

Exploration and producing property acquisitions are completed in various forms. These range in scope from entire ongoing concerns to purchases of individual property. Solicitation methods vary widely, ranging from public solicitation of bids from the business community to private negotiations with a single firm. The level of competition is often closely associated with the solicitation method used by the divesting company. Unsolicited and often unfriendly tender offers to purchase companies have also been used to acquire firms uninterested in being acquired.

There are circumstances in which firms decide to divest all or part of their exploration and production assets. Some reasons for such divestitures include:

- decisions to concentrate efforts in nonpetroleum industries;
- decisions to concentrate on other segments of the petroleum industry;
- decisions to eliminate operations/interest in specific geographic areas;
- inadequate capital base to properly develop exploratory discoveries;
- inadequate technical resources to properly exploit reserve potential;
- requirements for cash to reduce excessive debt;
- market conditions in which demand for properties allows sellers high price premiums ("sell at the top");
- special tax position enhanced attractiveness of divestitures;
- estate planning requirements for private companies;
- opportunities for two companies to create substantial synergies if merged, either through cost reductions, improved performance, and/or decreased competition.

Many factors contribute to the attractiveness of acquired exploration and production assets. Acquisition provides a means for quickly reducing deficiencies in production, reserves, and leasehold position. The long lead time often required for maturation of the exploration project programs can be circumvented through acquisitions. Acquisitions of an entire company can result in additions of personnel and expertise. Increased control over raw materials required by a firm's downstream operations may also be a strong reason for acquiring producing properties. Acquisitions can often provide immediate net income growth plus opportunities for both short- and long-term growth. A firm which possesses significant capital and/or technical strengths can acquire development, operations improvement, and enhanced recovery opportunities from companies that are unable to fully exploit them. Increased raw material control can strengthen expansion opportunities in gas processing, pipelines, and other downstream opportunities. Acquisitions are a means of attaining large blocks of acreage and of expanding into strategically desirable locations.

Competition for exploration/production assets has been extremely fierce. Many firms have the necessary resources and interests to seriously compete for producing properties. Principal competition for large acquisitions will include major oil companies and large independents. However, even small independents and non-oil firms have made substantial purchases of E&P assets. These companies determine that it is sometimes more economical to buy reserves than to explore for them.

8.4.1 Tax Consequences

An acquisition can either be at the stockholder level (the acquisition of stock) or at the corporate level (the acquisition of assets). When the acquiring corporation talks about acquiring a company, it can acquire the stock of the selling corporation by negotiating directly with its stockholders. In a sense, the selling corporation itself is not involved at all. In this case, the selling corporation becomes a subsidiary of the buying corporation. On the other hand, the buying corporation can talk to the selling corporation (not directly to the selling corporation stockholders as above) and buy the assets of the selling corporation. Under these circumstances, the selling corporation continues to exist except that now its assets no longer consist of inventory, machines, reserves, etc., but of cash, notes, stock, or whatever the buying corporation paid for the assets of the selling corporation.

Frequently, an acquisition will be referred to as taxable or tax free. This is really referring to the immediate tax effects to the seller of the company. A tax-free or nontaxable transaction is actually a tax-deferred transaction to the seller. In a simplified case, if the shareholders of a selling company sell their shares for cash for a value higher than they paid for them, the difference in that price will be taxed (potentially at capital gains rates depending on governing tax laws) immediately, with some exceptions. On the other hand, the selling shareholders might be able to trade their stock to the buying corporation and not incur any immediate taxes. These gains may in fact be deferred until the shareholders of the selling corporation who now own some of the stock of the buying corporation dispose of that stock. These laws vary by country and over time, so current, competent tax counsel is important.

If a seller seeks a tax-free transaction, the acquiring company must generally issue equity securities for a substantial part of the purchase price. The acquiring company may either acquire the seller's stock or its net assets as long as equity securities are used as consideration for the acquisition. If the transaction is tax free to the seller, then the buyer succeeds to the seller's tax basis, which will likely be a low basis. In a taxable transaction, the acquirer may either purchase the stock or the net assets of the seller. With the use of a taxable transaction, the acquirer must allocate the purchase price to all assets on the basis of fair market value; any portion of the purchase price that cannot be assigned to specific assets is considered goodwill for tax purposes. No tax deductions can be made for goodwill. It is very important that amounts be assigned to specific assets making up the total purchase price in order that there will be no substantial difficulty with the taxing authorities regarding deductibility of future tax write-offs. This issue varies by country, so local advice is recommended.

For an acquisition of producing properties, the portion of the purchase price that is not assigned to specific tangible assets can generally be assigned to the reserves of the acquired company, which then will have a tax basis equal to the remainder of the purchase price. This purchase price can typically be written off on a units-of-production basis. The reserves portion of an acquisition will amount to the majority of the value of an oil company. In an acquisition study, future reserves may be estimated, which require any of the following: additional development, exploratory drilling, water flooding, or enhanced recovery.

8.4.2 Accounting for Acquisitions

The accounting for acquisitions is no less complex than the tax and legal considerations. This discussion is a brief overview and reflects US accounting issues; these vary by country and are subject to change. Prior to 2001, either pooling or purchase accounting may have been used for mergers and acquisitions. since then only purchase accounting is permitted.

8.4.3 Pooling Accounting

Although no longer applicable in the United States, it is appropriate to have some understanding of pooling accounting. When using pooling accounting, assets are carried forward at the previous owner's carrying value. Purchase accounting required "fair market value". This precludes charges to earnings for depreciation where fixed assets are carried in accounts at amounts substantially lower than the present fair market value.

8.4.4 Purchase Accounting

Under purchase accounting rules, the cost of an acquisition is determined by the fair value of consideration given or the value of the assets received, whichever is more clearly evident. A portion of the cost is then allocated to each individual asset based upon the fair value of the particular assets and liabilities, i.e., present values. Intangible assets, which can be identified and named, also have fair value ascribed to them. *The excess of cost over fair value is assigned to goodwill.* Goodwill is evaluated at the end of each subsequent year for potential diminution in value.

8.4.5 Due Diligence in Acquisitions

Due diligence refers to the analysis of supporting information and is an important part of the M&A process. It is typically done after deal terms have been discussed and agreed upon; however, some aspects may be completed prior to reaching definitive deal terms. In the case of an acquisition, due diligences is primarily the responsibility of the buyer. In mergers or acquisitions in which the acquiring company uses its own stock for the purchase, both parties will perform due diligence. Due diligence is a coordinated set of activities and analyses conducted by a team, which is sponsored by the buyer to closely examine the acquired company's assets, financial records, contractual obligations, and the processes and systems of its various functional areas. Due diligence is usually conducted by accountants and attorneys, both in-house and outside of the company, and by representatives from the various functional areas of the acquiring company. Outside experts or consultants may be retained for one or more of the functional areas depending on the circumstances of each acquisition. The due diligence team is constituted at the time the formal due diligence process begins. The team consists of one or more representatives of the following areas, as applicable:

- accounting;
- corporate development;
- customer contracts;
- engineering;
- ethics and compliance;
- geology, geophysics
- HR;
- HSE
- investor relations;
- intellectual property (IP);
- information technology (IT);
- legal;
- operations;
- real estate;
- reserves;
- risk management;
- security;
- supply chain;
- tax;
- treasury.

Due diligence team members are expected to make a comprehensive review of the target with respect to their area of responsibility. Among other things, team members are expected to do the following to the extent applicable to their area of expertise:

- Visit the acquired company's principal facilities.
- Meet and interview management.
- Contact customers, supplier and business partners, subject to confidentiality considerations.

- Review contracts, internal processes and policies, and other documents included in the due diligence document requests.
- Catalog all documents received/reviewed.
- Make follow-up requests of the target as deemed necessary.
- Prepare a formal due diligence report for their area of responsibility.

The lead persons of the various due diligence sub-teams will be expected to write a report and may be asked to present an oral report to the executive sponsor and selected members of the due diligence team at a final due diligence meeting.

8.4.6 Valuation

Some conceptual comments are offered to help in the determination of the value of a firm to be acquired or merged. The first of these comments is that cash flow is the important criterion to be used in determining the financial attractiveness of an acquisition or merger candidate. As in other forms of economic evaluation, say a new business, the net present value or DCFROI indices are the most relevant. The incremental cash flow associated with the acquired or merged company is the basis for the economic evaluation process.

The nature of the incremental cash flow from an acquisition may, however, be different from the normal incremental cash flows analyzed by the reservoir engineer. The incremental cash flow associated with an acquisition or merger is not just the candidate's cash flow without the acquisition or merger. Synergy between the acquiring and selling firms could result in an incremental cash flow that is not available to either as stand-alone companies. Some financial theorists claim that synergy is the only justification for an acquisition or merger. This synergistic cash flow can arise for several reasons. Cost savings are a common occurrence, or perhaps the combined organization is of sufficient size to capture opportunities that are prohibitively large for each company as stand-alone organizations (critical mass). Regardless of their origin, analysts conducting the economic evaluation should include such synergistic cash flows in their study.

A second comment regarding valuation techniques is that the candidate's cost of capital (discount rate), and not the acquirer's, should be used in determining the value of the acquisition candidate. This is done to obtain a valuation that is consistent with the financial securities market (an alternate source of funds). A final comment about the valuation of an acquisition or merger candidate is that the acquiring firms can only acquire the equity of the candidate—the banks and other creditors already own the debt!

Conceptually, the value of a firm can be thought of as the present value of the candidate's net cash flows, with synergy if any. Or:

$$V = \sum_{i=0}^{N} \frac{\mathrm{NCF}_i}{(1 + \mathrm{COC}_{\mathrm{candidate}})^i}$$

where NCF_i is the forecast net cash flow in period i and $\mathrm{COC}_{\mathrm{candidate}}$ is the cost of capital (discount rate) of the company to be acquired.

By definition, the value of a firm is equal to the market value of its debt and equity, or:

$$V = D_{\mathrm{m}} + E_{\mathrm{m}}$$

Therefore, to determine the maximum price an acquirer would be willing to pay (based upon the accuracy of assumptions), the market value of the debt is subtracted from the value of the firm, or:

$$V_{\max} = \sum_{i=0}^{N} \frac{\mathrm{NCF}_i}{(1 + \mathrm{COC}_{\mathrm{candidate}})^i} - D_{\mathrm{m}}$$

CHAPTER

9

Professionalism and Ethics

9.1 WHAT IS A PROFESSION?

A profession is a career or occupation based on specialized education and training. The purpose of a profession is to provide disinterested counsel or service for a defined compensation independent of other business gain. Examples of professions include accounting, surveying, medicine, dentistry, actuarial science, law, architecture, and engineering. Professions share several characteristics including:

- being a full-time occupation;
- having a specialized course of study;
- being governed by local and national associations;
- having codes of professional conduct; and
- having state or other governmental licensing regulations.

The existence of state regulatory bodies governing the practice of a profession (and deciding who can be admitted into a profession) limits access to that profession. This bestows a limited monopoly on the practice of that profession. If, e.g., the state or other governmental entity requires a medical doctor to approve prescriptions for certain medicines, the required training, testing, and licensing of medical doctors provides a kind of limited monopoly. Similarly, requiring a licensed professional engineer to certify a certain type of document or to certify design aspects of buildings restricts those who can practice certain aspects of the profession.

Can you imagine a situation in which a company would employ medical doctors or attorneys who were not educated and licensed to governing standards? They would be limited to only doing certain internal activities that did not affect the public and would not enjoy the privileges of the profession. We would assume that would be unusual. But in the case of petroleum engineering, it is in fact the norm! Most states do not require the licensing of engineers who are employees of a company that does not offer to perform engineering services to the public. By contrast, almost all petroleum engineers in Canada seek to become licensed. Some people refer to licensed professional engineers as "registered" engineers; the term licensed more correctly conveys the concept. The authors strongly encourage professional licensure of practicing engineers.

9.2 ETHICS

In this section, we address engineering ethics while trying to focus on the specific items most critical for petroleum reservoir engineers. The Society of Petroleum Engineers (SPE)[1] is the largest professional organization that represents

[1]The mission of the SPE is "...to collect, disseminate, and exchange technical knowledge concerning the exploration, development and production of oil and gas resources, and related technologies for the public benefit; and to provide opportunities for professionals to enhance their technical and professional competence."

petroleum engineers including more reservoir engineers than any other organization. Engineering ethics deals with the standards of professional conduct for engineers with respect to the engineer's responsibility to the public, to his employer and clients, and to the profession of engineering. The SPE Guide for Professional Conduct summarizes these obligations.

9.2.1 Guide for Professional Conduct

Preamble.. *Engineers recognize that the practice of engineering has a vital influence on the quality of life for all people. Engineers should exhibit high standards of competency, honesty, integrity, and impartiality; be fair and equitable; and accept a personal responsibility for adherence to applicable laws, the protection of the environment, and safeguarding the public welfare in their professional actions and behavior. These principles govern professional conduct in serving the interests of the public, clients, employers, colleagues, and the profession.*

The Fundamental Principle.. *The engineer as a professional is dedicated to improving competence, service, fairness, and the exercise of well-founded judgment in the ethical practice of engineering for all who use engineering services with fundamental concern for protecting the environment and safeguarding the health, safety, and well-being of the public in the pursuit of this practice.*

Canons of Professional Conduct

- *Engineers offer services in the areas of their competence and experience, affording full disclosure of their qualifications.*
- *Engineers consider the consequences of their work and societal issues pertinent to it and seek to extend public understanding of those relationships.*
- *Engineers are honest, truthful, ethical, and fair in presenting information and in making public statements, which reflect on professional matters and their professional role.*
- *Engineers engage in professional relationships without bias because of race, religion, gender, age, ethnic or national origin, attire, or disability.*
- *Engineers act in professional matters for each employer or client as faithful agents or trustees disclosing nothing of a proprietary or confidential nature concerning the business affairs or technical processes of any present or former client or employer without the necessary consent.*
- *Engineers disclose to affected parties any known or potential conflicts of interest or other circumstances, which might influence, or appear to influence, judgment or impair the fairness or quality of their performance.*
- *Engineers are responsible for enhancing their professional competence throughout their careers and for encouraging similar actions by their colleagues.*
- *Engineers accept responsibility for their actions; seek and acknowledge criticism of their work; offer honest and constructive criticism of the work of others; properly credit the contributions of others; and do not accept credit for work not their own.*
- *Engineers, perceiving a consequence of their professional duties to adversely affect the present or future public health and safety, shall formally advise their employers or clients, and, if warranted, consider further disclosure.*
- *Engineers seek to adopt technical and economical measures to minimize environmental impact.*
- *Engineers participate with other professionals in multidiscipline teams to create synergy and to add value to their work product.*
- *Engineers act in accordance with all applicable laws and the canons of ethics as applicable to the practice of engineering as stated in the laws and regulations governing the practice of engineering in their country, territory, or state, and lend support to others who strive to do likewise.*

—Approved by the Board of Directors, September 26, 2004

The SPEE has also published[2] an extensive documentation on Ethics including a discussion of special issues for expert witnesses. State licensing boards often provide both guidelines and training in practical ethics applications.

Common concerns for engineers in ethics issues in the practice of engineering or in expert witness situations are all addressed in the SPE Canons and the SPEE Principles. The engineer never tries to practice outside his areas of expertise when offering services "to the public" or to clients. Full and accurate disclosure of qualifications and experience are essential. The engineer's résumé should be kept up to date, not for the purpose of seeking employment elsewhere, but to accurately summarize his experience and capabilities. An engineer working for a large oil company may well be asked to work outside his areas of expertise and certainly may do so as long as his employer knows that he is learning as he goes and he should not be in responsible charge of such projects until he has gained greater experience. The consulting engineer must not offer engineering services except when he is fully qualified to do so.

The engineer must be fastidious in avoiding even the appearance of a conflict of interest. Such conflicts can arise subtly and things that may appear not to be a conflict to the engineer may appear so to his client. It is important to address any potential conflict as early as possible. Are you involved in making a decision on a vendor but have a relative or close friend as an employee of that vendor? Do you own any shares (usually outside of a mutual fund) in any company that you have the potential to do business with? Have you been the beneficiary of any significant entertainment or other thing of value from someone your company may do business with? In one case, an oil company employee (A) was recommended for a significant SPE award by another oil company engineer (B) who subsequently was involved in negotiations with (A). If you were A's supervisor, would you want to know that it was B who recommended A?

[2]On their website at: http://www.spee.org/images/PDFs/ReferencesResources/SPEE%20Discussion%20and%20Guidance%20on%20ethics.pdf

9.3 THE ENGINEER AS AN EXPERT WITNESS

There is an old joke about experts. The definition of an expert is broken down into two parts, "ex" meaning a "has been" and "spurt" meaning a "drip under pressure." In fact, "expert" refers to having specialized skills and expertise that enable others to rely on the expert's efforts and his opinions. Engineers and specifically reservoir engineers are called on as experts under a number of circumstances to provide expertise, recommendations, and testimony. These might include recommendations before a regulatory body, as testimony in litigation, in arbitration or mediation, or before government agencies promulgating regulations or investigating. In each case, engineering testimony may ultimately sway major decisions whose financial impact on his or her employer or client could dramatically exceed the value of purely technical recommendations. No matter how intelligent, well educated, and published an engineer might be, it is his or her credibility and communication skills that are most valuable in the role of expert.

The vast bulk of reservoir engineering expert testimony is in litigation matters. Far more lawsuits are filed than there are arbitration cases or even contested regulatory hearings. In lawsuits, there may be one or more plaintiffs and one or more defendants. Each side may hire a host of experts in a wide variety of areas. The reservoir engineer as expert is often asked to opine about his interpretation of facts, views of whether or not certain actions or failure to perform certain actions rose to certain (often legally defined) standards. He may be asked to estimate reserves (whether proved, probable, possible, or some other definition) and the value of a certain property. He is often asked to estimate what the quantity or value of recoverable hydrocarbons were or are under certain leases and what the impact of various actions might have been. He can be asked to hypothesize within his area of expertise and estimate values under alternative scenarios. One of the authors was asked to opine as to what the level of understanding of reservoir heterogeneity and its impact on recovery were in the early 1930s!

> *The most important part about serving as an expert witness is to tell the truth. The way a reservoir engineer understands an oil and gas system may mean that explaining what the truth is to nonexperts may require analogies or simplifications. But the essential thing is to never stray from the truth.*

None of these opinions are being offered at random. Each tends to be a part of a strategy by one side to prevail over the other (or at least minimize damages). There is something *very important* for engineering experts to recognize. Your job as an expert is not to be an advocate for your side! Your job is to use your expertise to present clear and compelling testimony that the conclusions you have reached are correct. The most important part about serving as an expert witness is to tell the truth. The way a reservoir engineer understands an oil and gas system may mean that explaining what the truth is to nonexperts may require analogies or simplifications. But the essential thing is to never stray from the truth.

Your job may well include pointing out factual and technical errors made by "the other side." There are definite risks in doing this. The expert runs the risk of losing his own credibility if he is perceived to be an advocate. Few things diminish an expert's perception of independence than repeated or *ad hominem* attacks on the work of other experts, no matter how well deserved.

When consultants are hired as independent experts, they are often retained by the attorneys representing the client. Remember that *you* don't "represent" the client. The attorneys represent the client and they will not be under oath during your testimony. Even if you are testifying on behalf of your employer as an expert, you must be able to withstand questioning by opposing counsel as to your credibility and your conclusions. It is not uncommon for counsel to suggest that since you owe your livelihood to your employer, you would be willing to say or do anything to help them avoid a highly negative outcome in the litigation matter and the resulting "stain" on your career. You will need to be able to assert your nonnegotiable commitment to the truth in a convincing manner.

There are several occasions in which a reservoir engineer might be called on to testify (usually by deposition but also at trial) in a capacity other than as an expert. Fact witnesses testify as to actual facts that they have knowledge of and do not render opinions. Opposing counsel may well try to solicit expert opinions from a fact witness who is not obligated to provide them. Company employees are often called on to answer questions (interrogatories) on behalf of their employer and to testify about certain records.

9.3.1 Credibility and Credentials

Your credentials are essential to establishing credibility. Without proper credentials, you may not be allowed to testify as an expert. You will want to have a résumé that is current and emphasizes your education, experience, special qualifications, publications, professional society memberships and activities, awards, etc. The expert witness résumé is different from the résumé an engineer might use to seek a new job. The sole goal of the expert witness résumé is to convince the reader that the expert is very well qualified to render an opinion on the matters at hand.

The expert witness résumé will have absolutely no typographical errors and avoids aggrandizement. In particular, the expert should avoid touting prior success in litigation matters. "I have testified in ten lawsuits and my clients have won more than $200 million in those suits"—While this might seem to be a good thing (and the attorneys who hire experts might well be impressed), the opposing counsel may skewer this expert who is clearly an advocate and willing to say or do anything to keep his record intact for his side!

Activity in appropriate professional societies or analogous organizations is important. Recognition by peers, peer-reviewed papers, participation as a leader or organizer in workshops and forums all improve the expert witness résumé. These activities do much more than that. They make it possible for the expert to interact with other professionals who are likely to be experts in the field. They may provide referrals of consulting work or be useful sources of information. Unlike nonexperts, expert witnesses are expected to be able to analyze and evaluate the expertise of others and can rely on "hearsay" evidence. Engineering experts who solely testify in litigation matters would be wise to have a conventional consulting practice of some type to avoid being labeled a "professional expert" and the negative perception that conveys.

Because the credibility of an expert witness is of absolute importance, a few ethical issues for engineers are worth revisiting. The engineer as expert witness never testifies beyond his area of expertise. It is hard to overstate the importance of this fact. Some engineers may be tempted to do "too much," stretching their real expertise for a variety of reasons.

Everything the expert witness does can ultimately affect his credibility. Expect that anything you say or write or post in a blog, forum, Facebook entry, Youtube video, past testimony or reports, letter to the editor, or any other venue will be found by opposing counsel and potentially become a source of embarrassment, contradiction, or other issue in credibility. The same can be said of any unusual hobbies, affiliations, with extreme or even mildly controversial organizations, etc. Marketing efforts that make the expert appear to be an advocate or tout the expert's success can also diminish credibility.

9.3.2 Compensation and Payments

Expert witnesses are typically compensated at hourly or daily rates which must be disclosed. If the expert's rate is low compared to other experts (even if it is far more than jurors might earn), the jury may discount the testimony of the bargain expert. High rates for experts who are well prepared and professional do not generally diminish the jury's opinions of the expert and may, in fact, enhance it. There might be a negative perception by the jury if the expert's rate is vastly greater than all other experts. Regardless, the client paying the bills will want the lowest reasonable rates possible.

Compensation for expert testimony should be limited to "time and materials." There should not be compensation "at risk," depending on the outcome of the case. Plaintiff's attorneys routinely accept cases in which all or most of their compensation rests on a successful outcome in the matter. The expert hired by the plaintiff's attorneys should not be compensated in this way as it has the perception (if not the reality) of coloring his testimony.

Further, it is critical that the expert insist on being paid in a timely manner. If the expert has a great deal of money owed to him, particularly by a plaintiff or a smaller company, opposing counsel can make a great deal of it.

> Q: Now Mr. Engineer, you stated that so far you have invoiced Plaintiff Corp. a little over $210,000, correct?
> A: Yes.
> Q: Is that a pretty significant amount of money for you?
> A: Well, yes.
> Q: and if Plaintiff Corp. doesn't prevail in this lawsuit, you might not get paid, correct?
> A: I am confident that I will get paid.
> Q: Good for you. But if Plaintiff Corp. wins, there won't be any problem getting paid now, will there?
> A: Like I said, I am confident that I will get paid.
> Q: But it is clearly in your own personal financial interest to testify to anything you can to make sure you get this "significant amount of money" paid to you, isn't it?

The engineering expert who works for a large consulting firm is less sensitive to this exposure than the expert working for a small firm with fewer clients. One approach for an engineering expert is to have a "replenishing retainer" in a reasonable amount. The anticipated monthly expenditures are often considered reasonable amounts. Alternatively, it is appropriate to insist that the firm or company hiring the expert pay all invoices in a very timely manner, usually within 30 days.

9.3.3 The Expert Report

Most experts will be required to generate expert reports. These are unlike conventional technical reports.[3] The typical expert report references the litigation and provides (among other things):

- the expert's qualifications;
- information and materials considered in forming opinions;
- compensation;
- background facts;
- summary of opinions;
- basis for the expert opinion (which can and should be as long as is needed to support the opinions);
- a signature, often with the Licensed Engineering stamp or seal;
- the expert witness résumé;
- a list of other cases in which the expert has provided testimony at trial or by deposition (if required);
- tables and figures as required supporting the opinions.

The expert report should provide each of the opinions the expert intends to testify about; in practice, closely related opinions formed after completion of the report may be able to be given at trial. The expert report will be provided to opposing counsel who will ask their experts to either formally rebut this report or help the attorney pick it apart during deposition and/or trial. The expert will be cross-examined and may be examined at length over the details of the report.

The expert must be thorough in his analysis of the facts in forming an opinion. Where possible he will have visited the field or examined the wells or tools in question, carefully documenting his observations. He insists on being able to take the time required to formulate an opinion that is defensible and compelling. The expert is wise to avoid accepting cases with low budgets or deadlines that do not enable him to complete his work. Accepting such cases will necessarily result in opinions that are not as thorough as the expert would normally do. Such opinions will not be sufficiently compelling or defensible.

The expert is careful to state conclusions clearly and to back them up with the necessary facts. It is not sufficient that you are a leading expert in your field to reach a conclusion. The facts and assumptions supporting that conclusion must be clearly articulated and able to stand up to cross-examination. This may be difficult for highly technical individuals without much experience as expert witnesses.

The expert engineer is not an attorney and is not expected to understand every nuance of the law. However, it is vital that the engineer understand enough of the relevant law to make sure that his conclusions rise to the appropriate legal standards of sufficiency. If opposing counsel asks a question using legal jargon with which you are unfamiliar, it might be appropriate to ask for a definition, or at least preface an answer with your "layman's understanding" of the law. It is vitally important that the engineer avoid using legal expressions and/or technical jargon in his report *that he does not fully understand*. In one example, an expert was questioned over some legal phrases he had used in his conclusions. He could not explain them. He had gotten the advice to use those terms from the attorney who had in fact written some of the sentences. "What other parts of your report

[3]Sadly, the art of writing technical reports is fading at many oil companies where E&P has apparently come to mean "e-mail and PowerPoint."

did your attorney write for you?" was the embarrassing follow-up question.

When asked about an invention, the attorney cross-examined an unprepared witness who was only offered to describe how a given technology was used. Questions he was asked included phrases such as "undue experimentation" and "obviousness." The witness didn't recognize these terms as having specific legal meanings in patent law and answered them (ignoring his own company's counsel's objections) in a way that was inconsistent with the legal definitions.

You may have to produce any drafts that you make of your expert report. Do not destroy drafts that you do make. However, it is the author's practice not to make any drafts. In the process of developing the report, the same document is appended, edited, and updated as the opinions are being formed. Copies of drafts are not provided and not printed out. On the occasions prior to the report being finalized, the attorneys retaining the expert can read the current version of the report on the expert's computer, beamed up on the wall in their office or remotely via modern desktop-viewing software (e.g., Webex, GotomyPC, etc.) and make any comments. You can also anticipate being asked to disclose any suggested revisions, additions, or clarifications made by the attorneys. It is easy to explain that you added a few more sentences to explain a certain part of the opinion or reorganized the flow of the document at their suggestion. Accepting wholesale editing, revisions of the substance (particularly elimination of alternative theories and so forth) is unacceptable.

9.3.4 Depositions

Most lawsuits do not go to trial. A great number of them do proceed to the "deposition" stage. A deposition is sworn testimony taken in advance of a trial. In general, the expert will only be examined by opposing counsel at a deposition with "his side" reserving questions for trial. Experts can expect depositions to have a court reporter present and (increasingly) a videographer who will record their testimony. Deposition testimony can be replayed (if videotaped) or read back at trial. If the expert says at trial, "I am not sure that anyone can estimate that accurately," he may be faced with seeing himself on television saying, "I have estimated that number and believe it to be within the range of 40 to 45%." Deposition testimony is always printed out and the expert has a certain amount of time to correct errors in the transcript. Do not waive your rights to review, correct, and sign the deposition transcript.

At one time many attorneys had the philosophy that "you can't win in a deposition, you can only lose." Increasingly, as legal sophistication has grown more complex, deposition testimony has become a larger part of motions and other processes after deposition and before trials. Since most cases never proceed to trial, "making the case" at deposition becomes increasingly vital. Many cases are won or lost at the deposition stage. Regardless of whether at trial or deposition, the expert listens carefully to the question, takes as much time as necessary, and responds thoughtfully. While some questions may require a prompt and forceful answer, they are infrequent.

The expert should remember that in almost all of the cases, the attorneys with whom he is working *do not represent him*. The expert is not usually represented by counsel. The attorney who hires the expert represents the client and will generally provide some assistance during the deposition. The expert's conversations with counsel may or may not be privileged (as "work product") and the expert may have to testify (if asked) about any conversations with counsel or others even during breaks.

Do not be embarrassed to ask for a break at any time during a deposition that you need to stretch your legs or go to the bathroom. There are many subtle and not so subtle means that deposing counsel may use to wear down, agitate, or confuse experts. Do not volunteer extraneous information or try to assist counsel in formulating questions. Do not "agree" to limit your answers in any way, but particularly do not agree to "yes", "no," or "I don't know." Do not hold the deposition in your office and bring with you only those documents you are required to bring. You will typically have had to produce certain documents and it is essential that you do produce those. They can be produced as you normally keep them and it is generally not necessary that you produce reports that are organized in a specific way if that is not how you retain documents in the normal course of business.

You may have to produce each copy of a report that has any differences at all. If you have a copy of opposing expert's reports with your notes written on them, you may have to produce that copy. Keep this in mind before scrawling "What an idiot!" on a particularly poor conclusion.

9.3.5 Direct Examination

Although most cases do not go to trial, the expert witness prepares for any litigation, arbitration, or similar as if he will have to be cross-examined in front of a jury, panel of experts, or judge(s) as appropriate. While some lawsuits (and this depends on the specific country and legal system involved), particularly intellectual property (IP) cases are heard by a judge, many civil cases are still decided by a jury. While the whole range of jurisprudence cannot be discussed here, some parts are particularly relevant to an expert witness. The majority of reservoir engineering expert witness roles will (hopefully!) be in civil or regulatory cases, rather than criminal cases. In such cases, the plaintiff (or in the case of a regulatory hearing the applicant) will present their case through direct examination of a series of witnesses including experts.

Do not treat this as a battle of wits ... You are not trying to "win" but rather to convey to the judge or jury that your conclusions are sound and can be relied on when deciding the facts.

A direct examination consists of counsel asking a series of questions to the witness who responds to them. Counsel should not essentially testify and ask the witness, "That's correct, now isn't it?" In some courts, counsel can ask very broad questions and the witness will be allowed to expound at length on his theories, assumptions, methodologies, and conclusions. Examples of direct examination questions that allow the expert to explain things to the jury include:

Q: Dr. Jones, will you please explain to the jury what hydraulic fracturing is, how it works and why you believe that XYZ Corporation acted imprudently?
A: Let me start with answering that question about hydraulic fracturing. Natural gas in the Cotton Valley sand reservoir that is the subject of this dispute occurs in very small pores in sandstone rock located about 10,000 feet below the earth's surface. I have a sample of that rock here. You can see...
Q: Ms. Ingleson, do you believe that the Upper and Lower Slippery Rock formations were improperly commingled by the defendants?
A: Yes I do.
Q: Please explain.

Most courts allow "demonstrative exhibits" ranging from blown up copies of logs and maps to 3-D physical models to elaborate computer animations. The expert will make sure that any demonstratives he uses are clear, readily understandable for the information intended to convey, and working flawlessly. It is the experience of the authors that jurors tire of elaborate 3-D animations if used excessively. Properly operating physical exhibits that make a very specific point can be extremely convincing to jurors whose expertise does not include reservoir simulation, pressure transient analysis, or other technologies that may be used.

The direct examination should offer no surprises to the expert. It is reasonable that counsel runs through the anticipated questions and responses in advance; however, the expert should not sound overly rehearsed. The expert should have his file at his fingertips and be able to back up and illustrate his assumptions and conclusions.

9.3.6 Cross-Examination

Following direct testimony, the opposing counsel has the right to ask questions of the expert. The rules and approaches vary in such examinations. Leading the witness is allowed. If an expert does not answer the actual question asked (a very common tactic), the cross-examining counsel may repeatedly ask the same question "...and I will keep on asking it until you answer the question I am asking, not the one your attorney wants you to answer." Note that during cross-examination, counsel may seem friendly, hostile, bored, or confused. This is likely to be theatrical rather than real in the case of testimony in front of a jury. Cross-examination is in many ways the easiest part of the job for the well-prepared expert. Do not treat this as a battle of wits or as a chess match. You are not trying to "win" but rather to convey to the judge or jury that your conclusions are sound and can be relied on when deciding the facts. The expert should remember the following:

- Listen *very carefully* to the question and answer the question asked. You might be surprised how often counsel will just make a statement and not have actually asked a question. Even more often there are multiple questions and you will want to make sure which question(s) you are answering.
- Do not volunteer extra information in your answers unless you are certain that such information is necessary to communicate your conclusions.
- Do not be forced into a "yes," "no," or "I don't know" response. If a yes or no would be misleading to the jury feel free to say that you need to give a complete answer otherwise it could be misleading.
- Don't try to "help" counsel (appearing to be) struggling to formulate a question.
- Wait a little before answering to allow your client's counsel to register any objections. There may be something you need to hear in those objections and some of the objections should be raised for various legal reasons.
- In jury cases, listen to the questioner but answer speaking to the jury as a group. Your credibility with them is essential.
- Avoid the temptation to answer sarcastically or condescendingly. This is almost always perceived badly by juries and can come back and bite you.
- Get to the point quickly and do not try to be too cute by not answering the question.
- Simplify the complex. The engineering expert needs to be able to teach and explain complex concepts in a way that can be understood by jurors with little or no knowledge of the specific issues and varied educational levels.
- Answer the entire question. If the cross-examining attorney interrupts you before completing your answer you have every right to say "I haven't finished my answer."
- If you have to criticize the work or conclusions of opposing experts, do so clearly and explain what errors or faulty assumptions they made. Avoid criticizing them, their reputations, or their credentials unless you have solid factual evidence that is responsive to a question.
  - Q: Now Dr. Smith, you have stated that our expert Mr. Jones made numerous factual errors in his conclusions about the amount of hydrocarbons our client claims that the client you represent, Massive Oil Company, stole from us by drilling across the lease line. Are you and your lawyers trying to say that Mr. Jones is incapable of correctly calculating these numbers or are you calling him a liar? Or is this maybe just two experts with a different point of view and the truth is probably somewhere in the middle?

- (This question is based on a real question in a real case. What are the strengths and weaknesses of this expert's response?)
- A: Before I can answer your questions I need to clarify a few things so that my answer will not be misleading to the jury. You referred to "my lawyers" and the "client I represent." I was hired by the DC&H law firm to render an opinion as to what, if any drainage occurred in this matter and how much, if any damages arose from such drainage. They are not my lawyers and I do not represent the company. As you know from my scientific analysis and conclusions I do not believe that there was any theft of hydrocarbons and I have demonstrated that MOC did not drill across any lease lines. As to your question about Mr. Jones' capabilities, I note that he is not a member of the SPE or SPEE and is not a licensed petroleum engineer. He is a facilities engineer by background and experience. His CV doesn't show me any past jobs where he was in responsible charge of reserve calculations or economic evaluations. While I conducted an integrated 3-D reservoir simulation based on a carefully constructed static model, Mr. Jones used a method of ratios of areas on a map that failed to account for a variety of geological features and is theoretically incorrect. So whether or not Mr. Jones is capable of performing the correct calculations has yet to be seen, but I do not believe that he has the experience to evaluate my work or perform similar work. As to your second question, I do not believe that the differences in conclusions between Mr. Jones and I can fairly be represented as a difference of opinion between two experts. My conclusions are based on solid science and under cross-examination of Mr. Jones it became clear that his conclusions are based on theoretically and practically deficient approximations.
- If permitted, there may be a time when you can get up and draw something or derive something that might be responsive to a question being asked. Once when asked (for whatever reason) what the porosity of a stack of equally sized spheres would be, one of the authors responded "I do not recall the number, but I could derive it." The opposing counsel gleefully agreed to allow this and provided a large pad visible to the jurors and markers. After successfully deriving this simple equation, the deflation in opposing counsel's attitude was sufficiently visible to the jury that the expert's conclusions were all accepted by the jury.
- Even if you can't draw or derive, the most effective experts will find reasonable ways to get up out of the witness box. Coupled with effective visual aids and demonstrative examples, the effect will be memorable in establishing the expert's credibility and helping jurors to remember the expert's testimony. All demonstratives must be flawless in their execution and clear in the message they are designed to convey.
- Generally you should take the advice of your counsel but remember, you are the expert and your actions are independent. One expert spent the morning at trial in brilliant direct testimony. On returning to the trial for cross-examination (and having been reminded that he was still under oath), he had his suit buttoned and unbuttoned it just before sitting down. At that point he smiled at the jury and made very brief eye contact with them. The cross-examination was as follows:
  - Q: Dr. Roentgen, I noticed that you unbuttoned your suit coat just before you sat down and that you smiled at the jury just after you sat down—is that correct?
  - A: Yes sir.
  - Q: Did Mr. Dewey or one of the other attorneys on your side tell you to do that?
  - A: uh, yes. . .sort of, well yes.
  - Q: No further questions.

- Numbered lists are excellent ways to answer questions. They show the jury that your conclusion is well backed up by facts and that it is easy to understand.
  - Q: Now Dr. Muskat, you can't actually tell whether or not Cornershoot Oil and Gas actually crossed the lease line on this horizontal well, can you?
  - A: Actually I can and there are five reasons, *viz.*
    - First, the original directional survey clearly shows that the well path never crossed the lease line,
    - Second, repeated case hole surveys have been conducted and while they could not physically go the entire length of the well path, they covered more than half of the horizontal lateral distance and closely match the original directional survey,
    - Third, in order for the ....

9.3.7 Intellectual Property

While intellectual property (IP) refers to a variety of things including copyrights, trademarks, etc., for most reservoir engineers as expert witnesses, IP is synonymous with patents. A patent is granted by a government to one or more inventors as a way to encourage and reward innovation. It is the *exclusive* legal right to use whatever is protected by the patent (whether it is a design for a packer or a process for optimizing oil recovery) for a period of time. The owner of the patent can use it exclusively, license it to others, or even simply prevent others from using it. The inventor does have to disclose the patent in appropriate detail that sufficiently talented readers could reproduce the patented tools or processes without "undue experimentation."

Patent law has a whole set of nomenclature unto itself. There are ordinary words that have specific legal meanings such as "obviousness" or "experimentation." Less common phrases such as "one of ordinary skill in the art" suggest specific legal definitions and the expert may be called on to assist in a "Markman hearing" or similar activity in order to establish what claim and patent language really means. The meanings of specific words is often critical in determining validity of a patent or whether or not infringement occurred.

9.3.8 Junk Science

Engineers as experts must utilize well-established and accepted scientific and engineering principles in reaching their expert conclusions. "Junk science" should be rejected by the Judge and not permitted before jurors. Examples that might suggest unacceptable testimony include:

- testimony about a methodology that is not based on peer-reviewed and published methods;
- testimony about a methodology whose error rate cannot be estimated;
- testimony about a methodology that can't be tested according to accepted scientific principles;
- testimony about a methodology that is not generally accepted by the scientific community.

A methodology that has only been used in other lawsuits would be particularly suspect. While various legal issues in the United States are at play, the challenging of an expert's testimony under this general header is often called a "Daubert[4] challenge" or a "Frye test," the former becoming more commonplace. Sophisticated courts globally are adopting similar rules to preclude speculative technology and pseudoscience. The judge is usually the sole determinant of the success of a Daubert challenge. An expert engineer who loses such a challenge may not be able to testify again as an expert witness as opposing counsel will invariably bring up how the expert's testimony was "thrown out for being junk science in such and such a case." While there are

[4]Daubert is pronounced "daw" like the first syllable in daughter and "burt" as in Bert and Ernie, the two reservoir engineers of Sesame Street fame. Don't assume a French pronunciation in this case.

few cases in which an expert will want to hire his own counsel, a serious Daubert challenge may well be one of them.

9.4 FCPA CONSIDERATIONS

No discussion about reservoir engineering and ethics would be complete without mention of the United States Foreign Corrupt Practices Act, commonly referred to as the "FCPA." Enacted in 1977, the FCPA has become a significant regulatory enforcement tool of the US Department of Justice, particularly in the last 10 years. Much of the enforcement focus has been aimed at the oil and gas industry, particularly since oil and gas exploration and production in countries known for corruption has increased. Most countries that do significant business globally have similar statutes or can be expected to have them.

The statute is an attempt by the United States to eliminate corruption in the international business arena by punishing US companies and citizens who attempt to act corruptly in business transactions outside the United States. The FCPA makes it a criminal offense for a US publicly traded company, its officers, directors, employees, representatives or stockholders, and US citizens to bribe (or offer to bribe) a foreign public official in order to obtain or retain business or obtain an improper business advantage. The FCPA also contains provisions requiring publicly traded companies to maintain accurate books and records and a system of internal controls to ensure that the assets of the company are utilized in accordance with management's direction.

It is important to note that no value has to actually change hands in order to violate the antibribery provision of the FCPA. An offer, scheme, or promise to pay or give something of value (even in the future) may constitute a violation of the antibribery section of the FCPA. Moreover, bribery can take many forms, including the payment of money or anything else of value, such as "in kind" items or services. Bribes, kickbacks, giving, or promising to give, anything of value in an attempt to influence the action or inaction of a foreign government official can lead to a violation of the FCPA. This prohibition extends to payments made through consultants, agents, or any other representative when the person or company that benefits from the payment knows, or has reason to believe, that some part of the payment will be used to bribe or otherwise influence a foreign public official. An improper act by an agent will generally be interpreted as an improper act by the company employing the agent.

The definition of a foreign public official under the FCPA is broad and includes any officer or employee of a foreign government or any department agency or instrumentality of the foreign government. This includes employees of state-owned companies and members of royal families, who may lack "official" authority but maintain ownership or managerial interests in government industries or government-controlled companies as well as foreign political parties. It also includes customs, immigration, and transportation officials in foreign countries.

The FCPA is not an obscure statute that is rarely enforced. The US Department of Justice continues ramping up enforcement proceedings against US companies and citizens, *particularly in the petroleum industry*. Corporate employees, managers, and officers may be imputed with sufficient knowledge for a violation of the antibribery section of the FCPA if they deliberately insulate themselves through willful blindness, deliberate ignorance of, or conscious disregard of suspicious actions by the company or its personnel. This can lead to a criminal prosecution of the corporate entity and the individuals associated with the improper actions.

The magnitude of FCPA penalties has risen dramatically in recent years. In late 2008, Siemens AG pleaded guilty to a series of FCPA violations and agreed to pay at least $800 million in fines and disgorgement of prior profits. Halliburton agreed in early 2009 to pay $559 million to the US government to settle charges that one of its former units bribed Nigerian officials associated with the construction of a gas plant. Numerous companies will pay significant fines related to FCPA violations associated with the "oil for food" program in Iraq. A significant number of individuals have been and are being prosecuted as well, with many serving jail time.

Corruption of public officials robs nations of jobs and prosperity and is morally and ethically repugnant. The reservoir engineer working in the global environment may see instances of corruption that range from the very modest to criminal acts and must be prepared and knowledgeable enough to behave with the utmost integrity. As statutes and situations change, the engineer is wise to update himself regularly on relevant statutes dealing with the countries in which he is employed and working.

9.5 ETHICS GONE AWRY, ETHICAL DILEMMAS

What goes wrong and why? While overt unethical behavior that rises to the level of fraud occurs, there are a wide range of other ethical lapses. The following two cases are inspired by actual cases with the circumstances changed to protect the guilty; all the names are fictitious. The first case has enough issues in it that it is provided as a starting place for discussions. A few additional cases are inspired by cases copyrighted by the National Society of Professional Engineers.[5] While the original cases may or may not be inspired by actual cases, the three presented here are wholly fictitious.

9.5.1 The Case of the Unintended Consequences of Success Bonuses

Larry Transient and Chuck Strikeslip worked together as engineering and geological consultants, respectively, for many years at a major oil company. When it was acquired and they were asked to move in order to stay employed in the new company, they "took the package" and each started consulting firms. While they have separate companies, they share an office and routinely work together on projects. They often let one or the other companies do the invoicing and both have business cards for their colleague's company.

They both bill consulting by the hour, but Chuck spends his unbilled time generating prospects and has successfully sold several of them for combinations of cash and overrides. An investor (Bob Bigbelt) has engaged both of them to help him "get in the oil business" and has committed to pay a minimum number of consulting hours each month at a small discount from their "going rates." Bob has also orally committed to what could be potentially large success bonuses based on the investor obtaining certain levels of reserves (both proved and probable). Larry and Chuck don't have much experience in the kind of prospects Bob wants them to look at, but they have both "seen a lot" in their careers.

After killing a lot of deals, Chuck spots one that has some really complex-looking seismic but (at first review) appears to have an excellent-looking structure with a fault and three-way structural closure below an idle oilfield in a Latin American country. The nearest wells are almost 50 miles away but are at approximately the same depth and produce pretty well. Unfortunately, they don't have access to the seismic on the nearby fields and the nearest deep sonic logs raised more questions than answers. Chuck initially considers this as no

[5]The copyright notice associated with the NSPE cases states "This opinion is based on data submitted to the Board of Ethical Review and does not necessarily represent all the pertinent facts when applied to a specific case. This opinion is for educational purposes only and should not be construed as expressing any opinion on the ethics of specific individuals. This opinion may be reprinted without further permission, provided that this statement is included before or after the text of the case." The cases are found at http://wadsworth.com/philosophy_d/templates/student_resources/0534605796_harris/cases/Cases.htm#Cases%20exclusively%20on%20the%20site and were originally titled "Whose Witness?" "Gift Giving," and "Forced Ranking." The website has numerous example cases of ethical controversies involving engineering.

more than a good lead. Nonetheless, Bob is ecstatic. His investors seem to be willing to "roll the dice" and ask Larry to do a "success case" evaluation which (to no one's surprise) looks *very* attractive. Bob is working on making the deal happen and his investors are pushing for results. Larry observes that Chuck's maps look better with every revision. Chuck now refers to a "drillable prospect." Larry has never questioned Chuck's objectivity before, but now he is a little concerned.

The deal looks like they will have to buy the old field and Bob wants a rejuvenation plan for the field. According to the records, some of the wells were producing as much as 35–50 BOPD when the field was abandoned. When Larry suggests a site visit, Bob declines as he has hired Paulo Producer (an "operations guy") in country to check everything out. Paulo will also share the success bonus. While Bob was supposed to get "all the records" of the producing field, what he got were some structure maps, illegible logs in a many time photocopied cross-section, some decline curves, and a series of well tests dated long after production was supposed to have ended. There was no clear indication of what workovers had been done or when. Bob's instructions were to come up with a rejuvenation plan including data gathering but to make sure and present a "success case" for the rejuvenation plan.

Larry's success case looks fantastic. He and Chuck can now calculate that they could each potentially earn more on this project than they had managed to accumulate in their retirement plans. Larry and Chuck are now supposed to fly to New York to meet with the investors before the group finalizes an offer for the properties. This shocks Larry. Larry has essentially done the evaluation using cost numbers he referred to as "placeholders" and were essentially semieducated guesses. Paulo hasn't answered any questions about the conditions of the field or provided any more data. His comfort level with Chuck's maps is (for the first time in his career) not very high. Larry realizes that he would never even have dreamed of recommending a deal like this to his prior employer.

Then he learns that Chuck's wife is ill and that Chuck's mediocre insurance is not going to pay nearly all of the costs he is about to incur. Larry's retirement plan is off 35% from its peak. But whenever he brings up the risks and uncertainties to Bob, he gets the same response. "Great Larry, I am glad you have identified these. We are going to take the right steps to mitigate these. Just make sure you lead off with the success case because these guys want to first see the 'size of the prize'."

With everyone wanting to make the deal happen, it hardly seems fair that Larry has to stop it. The plane to New York leaves tomorrow. What does Larry do?

What went wrong here? A lot. First, Larry and Chuck have significant conflicts of interest. Even if all of the investors knew about them (which it turns out they did not), such conflicts can cloud judgment and color thinking. Larry and Chuck were both making recommendations outside their real areas of expertise. Neither had "helped set someone up in the oil business" before and the type of prospects and operations were new to them. As a result, Larry underestimated costs by large margins. The conditions on the ground resulted in multiyear delays before any efforts could be made to restore production. Workovers on several wells resulted in high water cuts and uneconomic wells. Paulo never gave any negative information to Larry or Chuck, but a trip to the field would have shown Larry the rusted heater treaters, leaky water tanks, and obvious signs of salt water leakage over the years.

Were Larry and Chuck using best practices in the evaluations? Of course not. Placeholders as cost estimates? Larry knew that real oil companies would have to spend the time and money to justify projects. They were instead using what their client *asked them to use*. This is a significant ethical breach. The engineer's duty is to use best practices, not just "do what they are asked." Were Larry and Chuck objective? The question answers itself.

Larry and Chuck got on the plane and they made their presentations. Larry made a strong point that there were major risks in the project and they could not be certain of being able to successfully rejuvenate the field and that exploration was intrinsically risky. One investor asked Larry how many times he had been involved in rejuvenating fields and he honestly answered "more than a dozen." The next question was "how many of them were unsuccessful?" to which he honestly answered "none of them." He wanted to point out the differences but the investors only asked about the success cases. They asked how much "running room" these opportunities had and Bob produced a map neither Larry or Chuck had ever seen before showing five more fields with "rejuvenation potential."

Larry and Chuck got down payments of more than $200,000 each on their success bonuses when the deal closed. The total success fee would be tied to results but Bob reminded them that "It should have two commas in it!" They got little further consulting work until something went awry and Bob occasionally paid them for their time. He made it clear that he thought they should be working "like he was" to capture the potential of the project and earn the rest of the success bonuses. A year later when the project had become a money sinkhole, the first workover produced over 98% water from a zone that had no record of injection or production and was updip to most of the estimated incremental oil. Larry was dispatched to the field and uncovered most of the bad news. Paulo had basically been dishonest from day one, holding back records with potentially damaging news. Paulo was fired and Bob hired a less optimistic operations guy. An excellent study was unearthed just after Paulo was fired showing very small opportunities for rejuvenation. Even these turned out to be optimistic. Bob was fired by the investor group when Chuck finished remapping the deep prospect. It had disappeared. What should have been done differently and when?

9.5.2 How Much of Your Expertise Belongs to Your Employer?

Employees are to act in "professional matters for each employer or client as faithful agents or trustees disclosing nothing of a proprietary or confidential nature concerning the business affairs or technical processes of any present or former client or employer without the necessary consent." Exactly what knowledge of business affairs or technical processes belong to the employer and what is the knowledge and experience of the engineer? Consider the following illustrations.

Acquisition Expertise Walks out the Door?

- Engineer Lisa heads the M&A group for Buy-um-up Oil & Gas. As such, she has studied essentially all of the small independent oil and gas operators in North America. She knows their assets, production, and potential. Her bosses rely on her analysis to negotiate deals and have given her increasingly responsible roles. She has built an amazing database but no one else but Lisa can really navigate it. Her contact list includes all of the top managers of the large independents.

  At the SPEE Conference, Lisa met the President of BigAcqPetro. They are very interested in hiring her and Lisa is ready to make the move which allows her to relocate to a very desirable location and make a lot more money. BAP has made it clear that they want her to do "essentially the same things" she is doing at BOG but on "a larger scale." Her boss at BOG went ballistic when she tendered her resignation as they were in the middle of an acquisition project. Lisa points out that they are always in the middle of such a project and that the projects take months to complete. Her boss takes some time and comes back with the following list of things Lisa cannot do at her new employer and he wants her to agree to them in writing. These include:

  - She cannot take her database or any information from the database with her.

- She has to stay at BOG long enough to train someone to be able to use the database (no one has been identified yet).
- She cannot take her contacts list with her.
- She cannot work with BAP on acquiring any companies BOG has considered for acquisition.
- She cannot use BOG's "approach" to acquiring fields for BAP.

Which (if any) of these requirements is Lisa ethically obligated to do? Should she sign an agreement? What should BAP do?.

Top Employee Starts Consulting Firm

- Technology Bob worked for a major oil company in the R&D department where he helped develop an intelligent well system that was subsequently licensed to a service company, IntelliWellGroup. He joined IWG and built the consulting arm of the service company that was instrumental in helping them sell more tools. The software he developed allows operators to optimize how intelligent wells are developed. IWG management allowed him a lot of time to write papers and attend conferences where he became well known in the intelligent well community. Bob has decided to start an independent consulting company along with two of his employees, effectively gutting IWG's consulting arm.
- Bob knows he cannot take any trade secrets with him including the software he developed at IWG. He does think he can recreate similar software and maybe even better software that can handle IWG's as well as competitor's tools. Bob knows he learned this capability on IWG's dime and wonders about his ethical obligations.
- Bob is a little concerned about hiring his coworkers. He had planned not to do so but to hire some people from other companies. However, when he hinted about his plans to the guys working for him at IWG, they were enthusiastic about joining him. Bob realizes that IWG will not only lose some consulting revenue but may lose market share.
- Bob knows IWG's strategies for getting the most money from oil companies and expects he will be hired to help get "better deals" from IWG and its competitors.
- In the past, Bob's team has always recommended IWG. Now he plans on being agnostic as to vendor and being able to recommend any vendor without bias. He realizes that his past association with IWG may mean that he will have to bend over backwards to give other vendors a fair chance in his recommendations.
- Bob plans on contacting the other industry experts he has met at SPE ATWs, forums, and other meetings to market his consulting services.

What ethical concerns should Bob consider? Which of these are likely to lead to ethical breaches? What limits to Bob's plans would you suggest?

9.5.3 Whose Witness Anyway?

Maria Hotshot is the acknowledged world expert in shale gas wells from an engineering perspective. She has designed and evaluated every aspect of horizontal wells and hydraulic fracturing and has spent the last decade of her 30 years in the industry working on shale gas projects around the world. While she is highly sought after as a consultant, it is her expertise and demeanor as an expert witness that has enabled her to testify at several high-profile lawsuits. After one case, her testimony was so effective that the opposing (losing) side immediately asked to retain her on another matter. After lunch, Maria returns to her office to find telephone messages and e-mails. The first are from a local plaintiff's attorney. His e-mail text was as follows:

> Dr. Hotshot,
> My name is Mark Cheatham and I represent Shale Guys, Inc (SGI). We have a great deal of acreage in the Marcellus and

had willing investors to help us drill that acreage. Instead we farmed out some of the acreage to Big Oil & Gas Company as they represented they had great expertise in developing this acreage. They were obligated to drill several wells to earn certain interests but they really messed up. They didn't do a geomechanical study and drilled the well improperly. They also messed up the frac jobs. Now they refuse to drill the other wells and our investors have soured on the deal because of the awful job BOGCO did. SGI is going to have to drill wells to maintain their valuable acreage position. Please call me immediately to discuss this further as we are sure we'd like to hire you in this case. We expect you can show what they did wrong and calculate how much damages we have suffered as a result of their failure to act as a prudent operator.

Sincerely,
Mark Cheatham
Sr. Counsel
DC&H Law Firm

The other message was from a large firm Maria knew well and had been deposed by. It simply said that they represented a large shale gas producer in a case against another company and wanted to discuss the possibility of retaining her and to call.

Questions:

1. Whose expert witness would you prefer to be? Why?
2. Which call will you return first? Why?
3. In deciding which, if either, caller to favor as expert witness, what kinds of questions will you ask yourself? What ethical questions, if any, will you ask?

It is important that the expert not allow attorneys to explain their theories of the case in too much detail prior to being retained. In some cases, attorneys deliberately expose potential experts to information that might taint their abilities to serve for the other side. Others might retain multiple experts only to eliminate them as potential experts for the other side. This is particularly risky in areas where a few experts are particularly well known and few alternatives exist.

Finally, it is not advisable that the expert solely represents plaintiffs or solely represents defendants. Experts who do will be characterized as "hired guns" that only can see cases from one point of view.

9.5.4 Forced Rankings[6]

Part One.. Jim Peters leaned back in his office chair and sighed with relief. Supervisor of the specialized petrophysics group, he had just finished writing the last of the annual performance appraisals on his 12-person team for BOGCO. Nearing the end of his first year as supervisor, this was Jim's first experience in appraising employees. Nevertheless, he felt he had done his appraisals well. He had held a thorough performance review discussion with each individual, going over progress toward specific annual objectives established early in the year. These discussions were open, frank, and, Jim believed, of value to him and to each employee.

BOGCO's appraisal forms required giving each employee a ranking of: High Achiever, Excellent, Satisfactory, Marginal, or Deficient. Each ranking requires a supporting, written justification. Jim ranked 8 of his 12 people at either High Achiever or Excellent. He ranked only one as Marginal, and he ranked the other three as Satisfactory. A ranking of Deficient was interpreted as a signal the employee was going to be terminated.

Jim delivered his appraisals to his immediate supervisor, Jason "Mac" McDougal, manager of Advanced Reservoir Characterization at BOGCO. Mac had to review Jim's appraisals along with those of his other sections, approve them, and submit them to the Human Resource

[6]This example is based closely on the NSPE examples: http://wadsworth.com/philosophy_d/templates/student_resources/0534605796_harris/cases/Cases/case60.htm

Director for BOGCO's E&P group. Jim assumed Mac would quickly and easily approve his appraisals.

Much to Jim's surprise, Mac stormed into Jim's office a few days later, threw the appraisals on his desk and exclaimed: "Jim, these appraisals just won't do! You're overrating your people! You know I have to force-rank everyone in ARC and turn that ranking in with all the appraisals. It looks to me like you've tried to assure that all your people will be placed high in the forced ranking. I want these appraisals rewritten and your ratings adjusted to something that more closely approximates a 'normal distribution'—you ought not to have more than a couple of High Achievers and probably a couple of Marginals or Deficients. I want the revised appraisals back on my desk by the end of the working day tomorrow! Understand?"

Jim felt frustrated, disillusioned, and disappointed by this turn of events. It seemed to him the appraisal system was being manipulated to produce an expected result and was not truly reflecting the performance of people. He also felt pressed for time, since he had the next day fully committed to other projects. What options do you think Jim has? Which do you think he should select? Explain.

Part Two.. Jim Peters worked late into the night to meet Mac McDougal's deadline. As he approached the end of his task, he grew careless and changed two Excellents to Satisfactory without changing his comments on their performance.

Jim submitted his revised appraisals to Mac and went back to the daily routine of supervising the Petrophysics Section. Mac appeared satisfied with the revised appraisals and submitted them (along with his forced ranking) into the normal chain of approval. Some weeks later, the appraisals were returned to Jim, who then scheduled individual appointments with each of his people to inform them of their appraisal ratings and discuss plans for subsequent improvement.

Jim's individual meetings went reasonably well until he met with Pete Evans. (Pete's appraisal was one that had the changed rating without revised comments.) Pete listened to Jim as they reviewed the appraisal and finally burst out, "Jim, your comments seem to sound like I'm an excellent performer but you only rated me 'satisfactory'!" Jim had been afraid of such an observation but hadn't carefully thought out a response. He simply blurted out, "I had to reduce most of the ratings I gave to conform to the distribution management expects!"

Pete stormed out of Jim's office muttering: "I thought my appraisal was supposed to motivate me to improve! It sure as heck didn't!" Discuss Jim's handling of his reappraisal task. What might he have done differently that would have had better results? What would you suggest that he do now? What changes, if any, do you think XYZ should make in its appraisal system? What ethical issues are raised?

9.5.5 Gifts and Entertainment

Everyone agrees that we should not steal. Everyone agrees that we should not accept bribes. But not every case is black or white. Consider the following in the context of "not stealing":

1. breaking into a store and taking $3000 in merchandise;
2. "borrowing" a friend's car and failing to return it;
3. taking a bicycle that someone had forgotten to lock;
4. developing a computer program on company time for your company, and then patenting a considerably improved version of the program under your own name;
5. borrowing a book from a friend, keeping it by mistake for a long time and then failing to return it (you discover the book after your friend has moved overseas, and you decide to keep it);
6. using some ideas you developed at Company A for a very different petrophysical application at Company B;
7. using some management techniques at Company B that were developed at Company A;

8. picking up a quarter that you saw someone drop on the street;
9. failing to return a sheet of paper (or paper clip) you borrowed;
10. picking up a quarter that someone (you don't know who) has dropped on the street.

Essentially, no one views 1–3 as anything but theft (stealing). Similarly, the last two seem unlikely to be considered theft. Probably most people would consider example 4 a type of theft. Example 5 is something many of us might have done. We might say that the action is justified, because the expense and trouble to us of returning the book is probably greater than the value of the book to our friend. This might be especially true if we knew the book was old and out of date. We would probably resist the use of the word "theft" to describe our action. Examples 6 and 7 might be considered less clear examples of theft than example 8, except for the potentially large amount of money involved in these two examples.

One of the considerations that makes the determination of what is and is not theft so difficult is that there is no single criterion that can be used to decide the issue. The most obvious such criterion is the monetary value of the property in question. But this criterion will not always work. Snatching a dollar bill from an old lady is more clearly an example of theft than using an idea one developed at Company A for a very different application at Company B, even though the latter example involves vastly greater sums of money than the first. A variety of considerations are relevant, monetary value being only one.

Similar consideration can be applied to bribery. We all agree that accepting bribes is a violation of professional ethics, but we may not always find it easy to determine what is and is not a bribe. Certainly, not all examples of accepting gifts and amenities qualify as accepting bribes, just as not all cases of taking another's property should be considered theft. Determining when a rule against taking bribes is being violated requires common sense, discrimination, and powers of moral deliberation. These kinds of abilities should be a part of one's professional training. The following examples are designed to serve as discussion points for evaluating a "gray" issue.[7]

9.5.6 The Bribery Coast

Case C-X.. Tom had been one of the most successful leaders of drilling, reservoir, and completion teams in the company. No one was surprised when he was promoted and named the asset manager of a large team designed to drill hundreds of deep horizontal wells in a sour tight gas play. He had full P&L authority and supervised the drilling, G&G, reservoir, completion and production engineering, and field operations leaders. Although his HR, legal, and financial leaders technically reported to their functional leads, they all considered themselves to be part of his team in no small part because of Tom's inspirational style of leadership.

Tom's prior experience proved invaluable as the team exceeded expectations time and again. As he examined drilling performance, he identified three new technologies in bits, directional drilling, and completions that were particularly applicable. All three were provided by the same vendor. After a few test wells, the performance of these technologies was proven and even led to improved HSE performance.

After spending quite a bit of money on the new tools, the executive sales representative (Jim) of the service company supplying the technology introduced himself to Tom and invited him on a very nice fishing trip to South America. While Tom had no direct purchasing responsibilities and had only wanted the technologies to improve performance, he did love fishing. Tom's company policies do not specifically prohibit such gifts and he knows people at his level had accepted similar entertainment

[7]This example is modeled loosely on the NSPE example in: http://wadsworth.com/philosophy_d/templates/student_resources/0534605796_harris/cases/Cases/case72.htm

before. He also knows that his new supervisor had accepted far more generous entertainment in the past. Should he accept the fishing trip?

The first thing to notice about this case is that it is *not* a paradigm case of bribery. In fact, it is not a case of bribery at all. We might define bribery as remuneration for the performance of an act that is inconsistent with the work contract or the nature of the work one has been hired to perform. Tom did not act contrary to his obligations to his employer, and in fact he acted in accordance with his obligations. Furthermore, the gift was offered after Tom's recommendations and without any prior knowledge and expectation of the gift.

Case C-1.. The following might qualify as a paradigm case of a bribe. We shall call it C-1.

Tom was promoted to asset manager and went to lunch with his old friend Jim. Jim started taking Tom and his wife to fairly lavish dinners and offered him a number of amenities if he would recommend the bits and directional work from his company. The technology was nearly comparable but more expensive than what they had been using. Tom was willing to try them based on Jim's recommendations. Jim then invited Tom on a very nice fishing trip to South America.[8]

Even though the original case (C-X) is not a bribe, it does involve accepting a large gift that has some analogies with a bribe. In order to see this, consider the following case, which has very few, if any, analogies with a true bribe. We shall call it C-10.

Case C-10.. Tom was promoted to the new asset manager position and realized right away that existing technologies and processes used by the company would not allow him to reach the company's aggressive production and profitability goals. Tom's prior experience proved invaluable as the team exceeded expectations time and again. As he examined the drilling performance, he identified three new technologies in bits, directional drilling, and completions that were particularly applicable. All three were provided by the same vendor. After a few test wells, the performance of these technologies was proven and even led to improved HSE performance.

After a large contract had been let to the vendor, the salesman came by and introduced himself, giving Tom a flash drive with the company's logo on it worth less than twenty dollars.

It is obvious that there is a continuum of cases here, all the way from C-1 which is a clear case of bribery to C-10 which is clearly not a case of bribery. Now the question is: What place in the continuum do we assign to C-X? Should we call it C-2, indicating that it is very close to C-1 so that it should probably be considered morally wrong? Or should it be labeled C-9, in which case it probably should not be considered morally wrong? Or should we give it some number in between? Finally, in C-X, should Tom take the fishing trip?

In order to begin thinking about this question, consider some of the characteristics of C-1.

1. While Tom may not have had direct authority for specifying the bits and directional work, his reputation allowed him to do so.
2. The salesman approached Tom and made the offer before the work was specified or purchased.
3. The work performed was not clearly superior to less costly alternatives.
4. There was a causal relationship between the offer of the amenities and Tom's decision. In other words, Tom requested that Jim's services be used as a direct result of Jim's offer.
5. Even though C-1 involves bribery, the company will probably benefit from an ongoing cordial relationship with suppliers, which Tom specified. For example, obtaining service will probably be easier. (We shall assume this to be the case.)

[8]If the "nice fishing trip" appears to be too valuable to be justified by many of the participants, it can be replaced by "golf at an exclusive country club" or "tickets to the Kentucky Derby." The important thing is that the thing being offered is more valuable than routine entertainment but not completely over the top.

6. Tom rarely accepts amenities from suppliers with whom he does not do business. (We shall assume this.)
7. Knowledge of the gift may influence others to buy from Jim, even if Jim's product is not the best.
8. The gift was for a substantial amount of money.
9. Even if there had been no actual corruption, there was certainly the appearance of corruption. For example, consider IBM's test: "If you read about it in your local newspaper, would you wonder whether the gift just might have something to do with a business relationship?" By this test, there was the appearance of corruption. (In this case, of course, the appearance was not misleading.)

C-10, we shall assume, shares only characteristic #5 with C-1, but C-X shares characteristics 5 and 7–9 with C-1. We shall assume that in C-X, Tom often accepts amenities from suppliers after a deal is completed, even if the supplier has not won the sale.

How would you evaluate Tom's action in C-X? Do you think it crosses the line between morally acceptable and morally unacceptable conduct? We shall give several arguments on both sides and then leave it to you to come to a final decision. But before giving these arguments, one observation may be helpful.

There are going to be cases that are unclear in terms of their moral permissibility. Although there are clear cases at either end of the spectrum and even some near the middle that can be decided with reasonable conviction, there are some cases that are so ambiguous that they must be decided arbitrarily. This is true in the law, and it is also true in morality.

To use a common analogy, there is dusk or twilight as well as daytime and night. If there is any specific point in time where night becomes day or day becomes night, it must be set arbitrarily. This does not mean that there is no difference between day and night, however. It does not even mean that some of the areas of transition are unclear. Most of us would probably say sunrise is more like day than night. Late dusk is probably more like night than day. But what about late evening? Better still, precisely when does day become night? We could set a time. We could say that it is when we have to turn on our car lights when driving, for example. But for other purposes, a different time might be more appropriate. So the moment of transition is not only arbitrary, but it may be arbitrarily set at different times for different purposes.

Now there are a number of cases which we could probably agree on in terms of their location in the continuum. For example, if the offer of the trip had been made before Tom made his recommendation, we would probably agree that Tom should not have accepted the offer. We might want to call this C-2. Let us suppose that Tom still recommended Jim's services because he genuinely believed they were the best, so that Jim's offer of a trip to South America was not the cause (or at least not a necessary precondition) of Tom's recommendation. In this case, Jim's offer clearly was a bribe, but Tom did not make his decision because of the bribe. Nevertheless, Tom probably should not accept the offer.

Now let's consider some of the arguments, beginning with arguments that C-X should be considered morally impermissible.

Arguments For and Against Tom's Action

Arguments Against Tom's Action of Accepting the Invitation

1. The tendency in Western morality—and probably morality throughout the world—is to increase restrictions on bribery. This implies that the restrictions on actions closely related to bribery would also be increased. Most large oil companies have well-defined rules in gift giving and gift receiving.
2. The size of the gift is morally troubling.
3. Knowledge of the gift could influence others to buy from Jim, even if Jim's products are not the most appropriate for them. This might operate as a kind of bribe-ahead-of-time for other people in Tom's business unit, even if Tom had no idea he would be offered

a trip. They might say, "If we buy from Jim, we can expect a nice gift."

4. Knowledge that Tom accepts after-the-fact gifts may give him a reputation as being someone who can be "bought."
5. In morality, one of the important questions to ask is whether you would be willing for others to do the same thing you did. If every salesman offered gifts to people who bought—or recommended the purchase of—his products, and every purchaser accepted the gifts, the practice would of course become universal. Our first reaction is to say this would neutralize the influence of the gifts. You could expect a bribe from somebody, no matter whose product you recommended. Thus, Tom might have been offered a nice trip to South America by whatever salesman made the sale. But this begins to look like extortion if not bribery: a salesman has to offer something to even have his product considered. Furthermore, smaller companies might not be able to offer the lavish gifts and so might not have their products considered. This would harm the competitive process. Furthermore, the gifts would probably tend to get larger and larger, as each salesman tried to top the other one. Thus, the general acceptance of the practice would have undesirable consequences.

Arguments in Favor of Tom's Action of Accepting the Invitation

1. We have already pointed out that Tom's action cannot be an example of accepting a bribe in the true sense of the term. In order to be a true bribe taker, Tom would have had to make his decision because of Jim's offer. Since the trip was offered after Tom's decision and Tom did not know about the trip ahead of time, the trip could not be a bribe in the true sense.
2. Tom's company may stand to benefit from the personal relationship between Tom and Jim. It may make it easier to get replacement improved service and to get other types of service from Jim's company.
3. Business life should have its "perks." Business and professional life involves a lot of hard work. Fishing trips and similar amenities add spice to life that is important in terms of job satisfaction and productivity. In fact, accepting such gifts was apparently the norm at Tom's company at his level and higher.
4. Accepting the kinds of gifts that Tom took advantage of is quite common in Tom's industry. It adds very little to the cost of the product. Any industry large enough to provide these complex and costly products and service in the first place would be able to afford such gifts without financial strain.
5. It is true that, in taking the moral point of view, we must assume that everybody has a right to do what we do. But if every salesman offered trips and every person in Tom's position accepted them, no harm would result. Things would equalize. There might be a kind of "extortion" here, but this is just a word. You have to ask what harm is done.

Note that each side has legitimate reasons for the decision made. Many shades of gray could be introduced in this case such that there will always be a marginally acceptable case for any reader. While we can agree that even the perception of bribery is to be avoided, there is always a softer version of the story that makes the decisions borderline appropriate.

The reader is encouraged to think responsibly about this issue; it is at least similar to issues they will almost certainly encounter as soon as they begin their professional careers. These cases are based loosely on an NSPE copyrighted case, which provides considerable additional content, references, and discussions on gift giving and bribery.[9]

[9]http://wadsworth.com/philosophy_d/templates/student_resources/0534605796_harris/cases/Cases/case72.htm

REFERENCES

Agarwal, R.G., 1980. A new method to account for producing time effects when drawdown type curves are used to analyze pressure buildup and other test data. SPE Paper 9289, presented at SPE–AIME 55th Annual Technical Conference, Dallas, TX, September 21–24, 1980.

Agarwal, R.G., Al-Hussainy, R., Ramey Jr., H.J., 1970. An investigation of wellbore storage and skin effect in unsteady liquid flow: I. Analytical treatment. SPE J. 10 (3), 279–290.

Agarwal, R.G., Carter, R.D., Pollock, C.B., 1979. Evaluation and performance prediction of low-permeability gas wells stimulated by massive hydraulic fracturing. J. Pet. Technol. 31 (3), 362–372, also in SPE Reprint Series No. 9.

Al-Ghamdi, A. Issaka, M., 2001. SPE Paper 71589, presented at the SPE Annual Conference, New Orleans, LA, September 30–October 3, 2001.

Al-Hussainy, R., Ramey Jr., H.J., Crawford, P.B., 1966. The flow of real gases through porous media. Trans. AIME 237, 624.

Allard, D.R., Chen, S.M., 1988. Calculation of water influx for bottomwater drive reservoirs. SPE Reservoir Eval. Eng. 3 (2), 369–379.

Anash, J., Blasingame, T.A., Knowles, R.S., 2000. A semianalytic (p/Z) rate-time relation for the analysis and prediction of gas well performance. SPE Reservoir Eval. Eng. 3, 525–533.

Ancell, K., Lamberts, S., Johnson, F., 1980. Analysis of the coalbed degasification process. SPE/DPE Paper 8971, presented at Unconventional Gas Recovery Symposium, Pittsburgh, PA, May 18–12, 1980.

Arps, J., 1945. Analysis of decline curve. Trans. AIME 160, 228–231.

Beggs, D., 1991. Production Optimization Using Nodal Analysis. OGCI, Tulsa, OK.

Begland, T., Whitehead, W., 1989. Depletion performance of volumetric high-pressured gas reservoirs. SPE Reservoir Eval. Eng. 4 (3), 279–282.

Bendakhlia, H., Aziz, K., 1989. IPR for solution-gas drive horizontal wells. SPE Paper 19823, presented at the 64th SPE Annual Meeting, San Antonio, TX, October 8–11, 1989.

Borisov, Ju. P., 1984. Oil Production Using Horizontal and Multiple Deviation Wells (J. Strauss, Trans. and S.D. Joshi, Ed.). Phillips Petroleum Co., Bartlesville, OK (the R&D Library Translation).

Bossie-Codreanu, D., 1989. A simple buildup analysis method to determine well drainage area and drawdown pressure for a stabilized well. SPE Form. Eval. 4 (3), 418–420.

Bourdet, D., 1985. SPE Paper 13628, presented at the SPE Regional Meeting, Bakersfield, CA, March 27–29, 1985.

Bourdet, D., Gringarten, A.C., 1980. Determination of fissure volume and block size in fractured reservoirs by type-curve analysis. SPE Paper 9293, presented at the Annual Technical Conference and Exhibition, Dallas, TX, September 21–24, 1980.

Bourdet, D., Alagoa, A., Ayoub, J.A., Pirard, Y.M., 1984. New type curves aid analysis of fissured zone well tests. World Oil, April, 111–124.

Bourdet, D., Whittle, T.M., Douglas, A.A., Pirard, Y.M., 1983. A new set of type curves simplifies well test analysis. World Oil, May, 95–106.

Bourgoyne, A., 1990. Shale water as a pressure support mechanism. J. Pet. Sci. 3, 305.

Brigham, W., Castanier, L. In-situ Combustion, Petroleum Engineering Handbook, vol. V. SPE, Dallas, TX, p. 1367 (Chapter 16).

Butler, R.M., 1991. Thermal Recovery of Oil and Bitumen. Prentice Hall, Englewood Cliffs, NJ.

Carson, D., Katz, D., 1942. Natural gas hydrates. Trans. AIME 146, 150–159.

Carter, R., 1985. Type curves for finite radial and linear gas-flow systems. SPE J. 25 (5), 719–728.

Carter, R., Tracy, G., 1960. An improved method for calculations of water influx. Trans. AIME 152.

Chatas, A.T., 1953. A practical treatment of nonsteady-state flow problems in reservoir systems. Pet. Eng. B-44–B-56, August.

Chaudhry, A., 2003. Gas Well Testing Handbook. Gulf Publishing, Houston, TX.

Cheng, AM., 1990. IPR for solution gas-drive horizontal wells. SPE Paper 20720, presented at the 65th SPE Annual Meeting, New Orleans, LA, September 23–26, 1990.

Cinco-Ley, H., Samaniego, F., 1981. Transient pressure analysis for finite conductivity fracture case versus damage fracture case. SPE Paper 10179.

Clark, N., 1969. Elements of Petroleum Reservoirs. Society of Petroleum Engineers, Dallas, TX.

Closmann, P.J., Seba, R.D., 1983. Laboratory tests on heavy oil recovery by steam injection. SPE J. 23 (3), 417–426.

Coats, K., 1962. A mathematical model for water movement about bottom-water-drive reservoirs. SPE J. 2 (1), 44–52.

Cole, F.W., 1969. Reservoir Engineering Manual. Gulf Publishing, Houston, TX.

Craft, B., Hawkins, M., 1959. Applied Petroleum Reservoir Engineering. Prentice Hall, Englewood Cliffs, NJ.

Craft, B.C., Hawkins, M. (Revised by Terry, R.E.) 1991. Applied Petroleum Reservoir Engineering, second ed. Prentice Hall, Englewood Cliffs, NJ.

Culham, W.E., 1974. Pressure buildup equations for spherical flow regime problems. SPE J. 14 (6), 545–555.

Cullender, M., Smith, R., 1956. Practical solution of gas flow equations for wells and pipelines. Trans. AIME 207, 281–287.

Dake, L., 1978. Fundamentals of Reservoir Engineering. Elsevier, Amsterdam.

Dake, L.P., 1994. The Practice of Reservoir Engineering. Elsevier, Amsterdam.

Dietz, D.N., 1965. Determination of average reservoir pressure from buildup surveys. J. Pet. Technol. 17 (8), 955–959.

Donohue, D., Erkekin, T., 1982. Gas Well Testing, Theory and Practice. International Human Resources Development Corporation, Boston, TX.

Duggan, J.O., 1972. The Anderson 'L' – an abnormally pressured gas reservoir in South Texas. J. Pet. Technol. 24 (2), 132–138.

Earlougher, Robert C., Jr., 1977. Advances in Well Test Analysis, Monograph, vol. 5. Society of Petroleum Engineers of AIME, Dallas, TX.

Economides, M., Hill, A., Economides, C., 1994. Petroleum Production Systems. Prentice Hall, Englewood Cliffs, NJ.

Economides, C., 1988. Use of the pressure derivative for diagnosing pressure-transient behavior. J. Pet. Technol. 40 (10), 1280–1282.

Edwardson, M., et al., 1962. Calculation of formation temperature disturbances caused by mud circulation. J. Pet. Technol 14 (4), 416–425.

Theory and Practice of the Testing of Gas Wells. 1975. third ed. Energy Resources Conservation Board, Calgary.

Fanchi, J., 1985. Analytical representation of the van Everdingen-Hurst influence functions. SPE J. 25 (3), 405–425.

Fetkovich, E.J., Fetkovich, M.J., Fetkovich, M.D., 1996. Useful concepts for decline curve forecasting, reserve estimation, and analysis. SPE Reservoir Eval. Eng. 11 (1), 13–22.

Fetkovich, M., Reese, D., Whitson, C., 1998. Application of a general material balance for high-pressure gas reservoirs. SPE J. 3 (1), 3–13.

Fetkovich, M.J., 1971. A simplified approach to water influx calculations – finite aquifer systems. J. Pet. Technol. 23 (7), 814–828.

Fetkovich, M.J., 1973. The isochronal testing of oil wells. SPE Paper 4529, presented at the SPE Annual Meeting, Las Vegas, NV, September 30–October 3, 1973.

Fetkovich, M.J. 1980. Decline curve analysis using type curves. SPE 4629, SPE J., June.

Fetkovich, M.J., Vienot, M.E., Bradley, M.D., Kiesow, U. G., 1987. Decline curve analysis using type curves – case histories. SPE Form. Eval. 2 (4), 637–656.

Gaitonde, N.Y., Middleman, S., 1967. Flow of visco-elastic fluids through porous media. Ind. Eng. Chem. Fund. 6, 145–147.

Gentry, R.W., 1972. Decline curve analysis. J. Pet. Technol. 24 (1), 38–41.

Giger, F.M., Reiss, L.H., Jourdan, A.P., 1984. The reservoir engineering aspect of horizontal drilling. SPE Paper 13024, presented at the 59th SPE Annual Technical Conference and Exhibition, Houston, TX, September 16–19, 1984.

Godbole, S., Kamath, V., Economides, C., 1988. Natural gas hydrates in the Alaskan Arctic. SPE Form. Eval. 3 (1), 263–266.

Golan, M., Whitson, C., 1986. Well Performance. second ed. Prentice Hall, Englewood Cliffs, NJ.

Gomaa, E.E., 1980. Correlations for predicting oil recovery by steamflood. J. Pet. Technol. 32 (2), 325–332. 10.2118/6169-PA, SPE-6169-PA.

Gray, K., 1965. Approximating well-to-fault distance from pressure build-up tests. J. Pet. Technol. 17 (7), 761–767.

Gringarten, A., 1984. Interpretations of tests in fissured and multilayered reservoirs with double-porosity behavior. J. Pet. Technol. 36 (4), 549–554.

Gringarten, A., 1987. Type curve analysis. J. Pet. Technol. 39 (1), 11–13.

Gringarten, A.C., Bourdet, D.P., Landel, P.A., Kniazeff, V.J., 1979. Comparison between different skin and wellbore storage type-curves for early time transient analysis. SPE Paper 8205, presented at SPE–AIME 54th Annual Technical Conference, Las Vegas, NV, September 23–25, 1979.

Gringarten, A.C., Ramey Jr., H.J., Raghavan, R., 1974. Unsteady-state pressure distributions created by a well with a single infinite-conductivity vertical fracture. SPE J. 14 (4), 347–360.

Gringarten, A.C., Ramey Jr., H.J., Raghavan, R., 1975. Applied pressure analysis for fractured wells. J. Pet. Technol. 27 (7), 887–892.

Gunawan Gan R., Blasingame, T.A., 2001. A semianalytic (p/Z) technique for the analysis of reservoir performance from abnormally pressured gas reservoirs. SPE Paper 71514, presented at SPE Annual Technical Conference and Exhibition, New Orleans, LA, September 30–October 3, 2001.

Hagoort, J., Hoogstra, ROB, 1999. Numerical solution of the material balance equations of compartmented gas reservoirs. SPE Reservoir Eval. Eng. 2 (4), 385–392.

Hammerlindl, D.J., 1971. Predicting gas reserves in abnormally pressure reservoirs. Paper SPE 3479, presented at the 46th Annual Fall Meeting of SPE–AIME. New Orleans, LA, October, 1971.

Harville, D., Hawkins, M., 1969. Rock compressibility in geopressured gas reservoirs. J. Pet. Technol. 21 (12), 1528–1532.

Havlena, D., Odeh, A.S., 1963. The material balance as an equation of a straight line: part 1. Trans. AIME 228, I–896.

Havlena, D., Odeh, A.S., 1964. The material balance as an equation of a straight line: part 2. Trans. AIME 231, I–815.

Hawkins, M., 1955. Material balances in expansion type reservoirs above bubble-point. SPE Transactions Reprint Series No. 3, pp. 36–40.

Hawkins, M., 1956. A note on the skin factor. Trans. AIME 207, 356–357.

Holder, G., Anger, C., 1982. A thermodynamic evaluation of thermal recovery of gas from hydrates in the earth. J. Pet. Technol. 34 (5), 1127–1132.

Holder, G., et al., 1987. Effect of gas composition and geothermal properties on the thickness of gas hydrate zones. J. Pet. Technol. 39 (9), 1142–1147.

Holditch, S. et al., 1988. Enhanced recovery of coalbed methane through hydraulic fracturing. SPE Paper 18250, presented at the SPE Annual Meeting, Houston, TX, October 2–5, 1988.

Hong, K.C., 1994. Steamflood Reservoir Management. PennWell Books, Tulsa, OK.

Horn, R., 1995. Modern Test Analysis. Petroway, Palo Alto, CA.

Horner, D.R., 1951. Pressure build-up in wells. Proceedings of the Third World Petroleum Congress, The Hague, Sec II, 503–523. Also Pressure Analysis Methods, Reprint Series, No. 9. Society of Petroleum Engineers of AIME, Dallas, TX, pp. 25–43.

Hughes, B., Logan, T., 1990. How to design a coalbed methane well. Pet. Eng. Int. 5 (62), 16–23.

Hurst, W., 1943. Water influx into a reservoir. Trans. AIME 151, 57.

Ikoku, C., 1984. Natural Gas Reservoir Engineering. John Wiley & Sons, New York, NY.

Jones, J., 1981. Steam drive model for hand-held programmable calculators. J. Pet. Technol. 33 (9), 1583–1598. 10.2118/8882-PA, SPE-8882-PA.

Jones, S.C., 1987. Using the inertial coefficient, b, to characterize heterogeneity in reservoir rock. SPE Paper 16949, presented at the SPE Conference, Dallas, TX, September 27–30, 1987.

Joshi, S., 1991. Horizontal Well Technology. PennWell, Tulsa, OK.

Kamal, M., 1983. Interference and pulse testing – a review. J. Pet. Technol. 2257–2270, December.

Kamal, M., Bigham, W.E., 1975. Pulse testing response for unequal pulse and shut-in periods. SPE J. 15 (5), 399–410.

Kamal, M., Freyder, D., Murray, M., 1995. Use of transient testing in reservoir management. J. Pet. Technol. 47 (11), 992–999.

Kartoatmodjo, F., Schmidt, Z., 1994. Large data bank improves crude physical property correlation. Oil Gas J. 4, 51–55.

Katz, D., 1971. Depths to which frozen gas fields may be expected. J. Pet. Technol. 24 (5), 557–558.

Kazemi, H., 1969. Pressure transient analysis of naturally fractured reservoirs with uniform fracture distribution. SPE J. 9 (4), 451–462.

Kazemi, H., 1974. Determining average reservoir pressure from pressure buildup tests. SPE. J. 14 (1), 55–62.

Kazemi, H., Seth, M., 1969. Effect of anisotropy on pressure transient analysis of wells with restricted flow entry. J. Pet. Technol. 21 (5), 639–647.

King, G., 1992. Material balance techniques for coal seam and Devonian shale gas reservoirs with limited water influx. SPE Reservoir Eval. Eng. 8 (1), 67–75.

King, G., Ertekin, T., Schwerer, F., 1986. Numerical simulation of the transient behavior of coal seam wells. SPE Form. Eval. 1 (2), 165–183.

Klins, M., Clark, L., 1993. An improved method to predict future IPR curves. SPE Reservoir Eval. Eng. 8 (4), 243–248.

Lake, L.W., 1989. Enhanced Oil Recovery. Prentice Hall, Englewood Cliffs, NJ.

Langmuir, I., 1918. The constitution and fundamental properties of solids and liquids. J. Am. Chem. Soc. 38, 2221–2295.

Lee, J., 1982. Well Testing. Society of Petroleum Engineers of AIME, Dallas, TX.

Lee, J., Wattenbarger, R., 1996. Gas Reservoir Engineering, 5. Society of Petroleum Engineers, Dallas, TX, SPE Textbook Series.

Lefkovits, H., Hazebroek, P., Allen, E., Matthews, C., 1961. A study of the behavior of bounded reservoirs. SPE. J. 1 (1), 43–58.

Levine, J. (1991). The impact of oil formed during coalification on generating natural gas in coalbed reservoirs. The Coalbed Methane Symposium, The University of Alabama, Tuscaloosa, AL, May 13–16, 1991.

Makogon, Y., 1981. Hydrates of Natural Gas. PennWell, Tulsa, OK.

Mandl, G., Volek, C.W., 1967. Heat and mass transport in steam-drive process. SPE Paper 1896, presented at the Fall Meeting of the Society of Petroleum Engineers of AIME, New Orleans, LA, October 1–4, 1967. doi:10.2118/1896-MS.

Marx, J.W., Langenheim, R.H., 1959. Reservoir heating by hot fluid injection. Trans. AIME 216, 312–314.

Mattar, L., Anderson, D., 2003. A systematic and comprehensive methodology for advanced analysis of production data. SPE Paper 84472, presented at the SPE Conference, Denver, CO, October 5–8, 2003.

Matthews, C.S., Russell, D.G., 1967. Pressure Buildup and Flow Tests in Wells, Monograph, vol. 1. Society of Petroleum Engineers of AIME, Dallas, TX.

Matthews, C.S., Brons, F., Hazebroek, P., 1954. A method for determination of average pressure in a bounded reservoir. Trans. AIME 201, 182–191, also in SPE Reprint Series, No. 9.

Mavor, M., Nelson, C. (1997). Coalbed reservoirs gasin-place analysis. Gas Research Institute Report GRI 97/0263, Chicago, IL.

Mavor, M., Close, J., McBane, R., 1990. Formation evaluation of coalbed methane wells. Pet. Soc. CIM, CIM/SPE Paper 90-101.

McLennan, J., Schafer, P., 1995. A guide to coalbed gas content determination. Gas Research Institute Report GRI 94/0396, Chicago, IL.

McLeod, N., Coulter, A., 1969. The simulation treatment of pressure record. J. Pet. Technol. 951–960, August.

Merrill, L.S., Kazemi, H., Cogarty, W.B., 1974. Pressure falloff analysis in reservoirs with fluid banks. J. Pet. Technol. 26 (7), 809–818.

Meunier, D., Wittmann, M.J., Stewart, G., 1985. Interpretation of pressure buildup test using in-situ measurement of afterflow. J. Pet. Technol. 37 (1), 143–152.

Miller, M.A., Leung, W.K., 1985. A simple gravity override model of steam drive. Paper SPE 14241, presented at the SPE Annual Technical Conference and Exhibition, Las Vegas, NV, September 22–25, 1985. doi:10.2118/14241-MS.

Muller, S., 1947. Permafrost. J.W. Edwards, Ann Arbor, MI.

Muskat, M., 1945. The production histories of oil producing gas-drive reservoirs. J. Appl. Phys. 16, 167.

Muskat, M., Evinger, H.H., 1942. Calculations of theoretical productivity factor. Trans. AIME 146, 126–139.

Myhill, N.A., Stegemeier, G.L., 1978. Steam-drive correlation and predication. J. Pet. Technol. 30 (2), 173–182. 10.2118/5572-PA, SPE-5572-PA.

Najurieta, H.L., 1980. A theory for pressure transient analysis in naturally fractured reservoirs. J. Pet. Technol. 32, 1241–1250.

Neavel, R., et al., 1986. Interrelationship between coal compositional parameters. Fuel 65, 312–320.

Nelson, C., 1989. Chemistry of Coal Weathering. Elsevier Science, New York, NY.

Nelson, R., 1999. Effects of coalbed reservoir property analysis methods on gas-in-place estimates. SPE Paper 57443, presented at SPE Regional Meeting, Charleston, WV, October 21–22, 1999.

Neuman, C.H., 1974. A mathematical model of the steam drive process applications. SPE Paper 4757, presented at the SPE Improved Oil Recovery Symposium, Tulsa, April 22–24, 1974. doi: 10.2118/4757-MS.

Ostergaard, K. et al., 2000. Effects of reservoir fluid production on gas hydrate phase boundaries. SPE Paper 50689, presented at the SPE European Petroleum Conference, The Hague, The Netherlands, October 20–22, 1998.

Palacio, C., Blasingame, T., 1993. Decline-curve analysis using type-curves analysis of gas well production data. SPE Paper 25909, presented at the SPE Rocky Mountain Regional Meeting, Denver, CO, April 26–28, 1993.

Papadopulos, I., 1965. Unsteady flow to a well in an infinite aquifer. Int. Assoc. Sci. Hydrol. I 21–31.

Payne, David A., 1996. Material balance calculations in tight gas reservoirs: the pitfalls of p/Z plots and a more accurate technique. SPE Reservoir Eval. Eng. 11 (4), 260–267.

Perrine, R., 1956. Analysis of pressure buildup curves. Drilling and Production Practice API 482–509.

Petnanto, A., Economides, M. (1998). Inflow performance relationships for horizontal wells. SPE Paper 50659, presented at the SPE European Conference held in The Hague, The Netherlands, October 20–22, 1998.

Pinson, A., 1972. Conveniences in analysing two-rate flow tests. J. Pet. Techol. 24 (9), 1139–1143.

Pletcher, J., 2000. Improvements to reservoir material balance methods. SPE Paper 62882, SPE Annual Technical Conference, Dallas, TX, October 1–4, 2000.

Poston, S. (1987). The simultaneous determination of formation compressibility and gas in place. Paper presented at the 1987 Production Operation Symposium, Oklahoma City, OK.

Poston, S., Berg, R., 1997. Overpressured Gas Reservoirs. Society of Petroleum Engineers, Richardson, TX.

Pratikno, H., Rushing, J., Blasingame, T.A., 2003. Decline curve analysis using type curves – fractured wells. SPE Paper 84287, SPE Annual Technical Conference, Denver, CO, October 5–8, 2003.

Prats, I.N., 1983. Thermal Recovery, Monograph 7. SPE, Dallas, TX.

Pratt, T., Mavor, M., Debruyn, R., 1999. Coal gas resources and production potential in the Powder River Basin. Paper SPE 55599, presented at the Rocky Mountain Meeting, Gillette, WY, May 15–18, 1999.

Ramey Jr., H.J., 1975. Interference analysis for anisotropic formations. J. Pet. Technol 27 (10), 1290–1298.

Ramey, H., Cobb, W., 1971. A general pressure buildup theory for a well located in a closed drainage area. J. Pet. Technol. 23 (12), 1493–1505.

Rawlins, E.L., Schellhardt, M.A., 1936. Back-pressure Data on Natural Gas Wells and Their Application to Production Practices. US Bureau of Mines, Monograph 7.

Remner, D., et al., 1986. A parametric stuffy of the effects of coal seam properties on gas drainage efficiency. SPE Reservoir Eval. Eng. 1 (6), 633–646.

Renard, G.I., Dupuy, J.M., 1990. Influence of formation damage on the flow efficiency of horizontal wells. SPE Paper 19414, presented at the Formation Damage Control Symposium, Lafayette, LA, February 22–23, 1990.

Roach, R.H., 1981. Analyzing geopressured reservoirs – a material balance technique. SPE Paper 9968, Society of Petroleum Engineers of AIME, Dallas, TX, December, 1981.

Russell, D., Truitt, N., 1964. Transient pressure behaviour in vertically fractured reservoirs. J. Pet. Technol. 16 (10), 1159–1170.

Sabet, M., 1991. Well Test Analysis. Gulf Publishing, Dallas, TX.

Saidikowski, R., 1979. SPE Paper 8204, presented at the SPE Annual Conference, Las Vegas, NV, September 23–25, 1979.

Schilthuis, R., 1936. Active oil and reservoir energy. Trans. AIME 118, 37.

Seidle, J., 1999. A modified p/Z method for coal wells. SPE Paper 55605, presented at the Rocky Mountain Meeting, Gillette, WY, May 15–18, 1999.

Seidle, J., Arrl, A., 1990. Use of the conventional reservoir model for coalbed methane simulation. CIM/SPE Paper No. 90-118.

Sherrad, D., Brice, B., MacDonald, D., 1987. Application of horizontal wells in Prudhoe Bay. J. Pet. Technol. 39 (11), 1417–1421.

Slider, H.C., 1976. Practical Petroleum Reservoir Engineering Methods. Petroleum Publishing, Tulsa, OK.

Sloan, D., 1984. Phase equilibria of natural gas hydrates. Paper presented at the Gas Producers Association Annual Meeting, New Orleans, LA, March 19–21, 1984.

Sloan, E., 2000. Hydrate Engineering. Society of Petroleum Engineers, Richardson, TX.

Smith, C.R., 1966. Mechanics of Secondary Oil Recovery. Robert E. Krieger Publishing, Huntington, NY.

Smith, J., Cobb, W., 1979. Pressure buildup tests in bounded reservoirs. J. Pet. Technol. August.

Somerton, D., et al., 1975. Effects of stress on permeability of coal. Int. J. Rock Mech. Min. Sci. Geomech. Abstr. 12, 129–145.

Stalkup Jr., F.I., 1983. Miscible Displacement, Monograph 8. Dallas, TX, SPE.

Standing, M.B., 1970. Inflow performance relationships for damaged wells producing by solution-gas drive. J. Pet. Technol. 22 (11), 1399–1400.

Steffensen, R., 1987. Solution-gas-drive reservoirs. Petroleum Engineering Handbook. Society of Petroleum Engineers, Dallas, TX, Chapter 37.

Stegemeier, G., Matthews, C., 1958. A study of anomalous pressure buildup behavior. Trans. AIME 213, 44–50.

Strobel, C., Gulati, M., Ramey Jr., H.J., 1976. Reservoir limit tests in a naturally fractured reservoir. J. Pet. Technol. 28 (9), 1097–1106.

Taber, J.J, Martin, F, Seight, R., 1997. EOR screening criteria revisited. SPE Reservoir Eval. Eng. 12 (3), 199–203.

Tarner, J., 1944. How different size gas caps and pressure maintenance affect ultimate recovery. Oil Weekly, June 12, 32–36.

Terwilliger, P., et al., 1951. An experimental and theoretical investigation of gravity drainage performance. Trans. AIME 192, 285–296.

Tiab, D., Kumar, A., 1981. Application of the pD function to interference tests. J. Pet. Technol. 1465–1470, August.

Tracy, G., 1955. Simplified form of the MBE. Trans. AIME 204, 243–246.

Unsworth, J., Fowler, C., Junes, L., 1989. Moisture in coal. Fuel 68, 18–26.

van Everdingen, A.F., Hurst, W., 1949. The application of the Laplace transformation to flow problems in reservoirs. Trans. AIME 186, 305–324.

van Poollen, H.K., 1980. Enhanced Oil Recovery. PennWell Publishing, Tulsa, OK.

Vogel, J.V., 1968. Inflow performance relationships for solution-gas drive wells. J. Pet. Technol. 20 (1), 86–92.

Walsh, J., 1981. Effect of pore pressure on fracture permeability. Int. J. Rock Mech. Min. Sci. Geomech. Abstr. 18, 429–435.

Warren, J.E., Root, P.J., 1963. The behavior of naturally fractured reservoirs. SPE J. 3 (3), 245–255.

Wattenbarger, R.A., Ramey Jr., H.J., 1968. Gas well testing with turbulence damage and wellbore storage. J. Pet. Technol. 20 (8), 877–887.

West, S., Cochrane, P., 1994. Reserve determination using type curve matching and extended material balance methods in The Medicine Hat Shallow Gas Field. SPE Paper 28609, presented at the 69th Annual Technical Conference, New Orleans, LA, September 25–28, 1994.

Whitson, C., Brule, M., 2000. Phase Behavior. Society of Petroleum Engineers, Richardson, TX.

Wick, D. et al., 1986. Effective production strategies for coalbed methane in the Warrior Basin. SPE Paper 15234, presented at the SPE Regional Meeting, Louisville, KY, May 18–21, 1986.

Wiggins, M.L., 1993. Generalized inflow performance relationships for three-phase flow. Paper SPE 25458, presented at the SPE Production Operations Symposium, Oklahoma City, OK, March 21–23, 1993.

Willhite, G.P., 1986. Water Flooding. SPE, Dallas, TX.

Willman, B.T., Valleroy, V.V., Runberg, G.W., Cornelius, A.J., Powers, L.W., 1961. Laboratory studies of oil recovery by steam injection. J. Pet. Technol. 13 (7), 681–690.

Yeh, N., Agarwal, R., 1989. Pressure transient analysis of injection wells. SPE Paper 19775, presented at the SPE Annual Conference, San Antonio, TX, October 8–11, 1989.

Zuber, M. et al. (1987). The use of simulation to determine coalbed methane reservoir properties. Paper SPE 16420, presented at the Reservoir Symposium, Denver, CO, May 18–19, 1987.

INDEX